EXPERIMENTAL PSYCHOLOGY WITH ADVANCED EXPERIMENTS

VOLUME - I

M. RAJAMANICKAM
M.A., M.Litt., Ph.D.
Formerly Dean, Faculty of Education and Professor
and Head of the Department of Psychology,
Annamalai University

CONCEPT PUBLISHING COMPANY, NEW DELHI-110059

Cataloging in Publication Data—DK
Courtesy: D.K. Agencies (P) Ltd. <docinfo@dkagencies.com>

Rajamanickam, M. (Muniappa), 1925-
Experimental psychology with advanced experiments / M. Rajamanickam.
v. cm.
Includes bibliographical references.
Includes index.
ISBN 8180691101 (set)

1. Psychology, Experimental. I. Title.

DDC 150.724 21

ISBN 81-8069-113-6 (Set) - P.B.
ISBN 81-8069-114-4 (Vol. 1) - P.B.
ISBN 81-8069-115-2 (Vol. 2) - P.B.

First Published 2005

Published and Printed by
Ashok Kumar Mittal
Concept Publishing Company
A/15-16, Commercial Block, Mohan Garden,
NEW DELHI-110059 (India)

Phone : 25351460, 25351794
Fax : 091-11-25357103
E-mail : publishing@conceptpub.com

PREFACE

Experimental Psychology had its origin in Fechner's Psychophysical Theory and Psychophysical Methods in the third quarter of nineteenth century. Gustav Theodor Fechner (1801-1887), a German Physicist and Philosopher, was very much interested in studying the relationship between mind and body and the problems involved in it. He was working on it for more than ten years and at last was able to evolve a theory. Even before this time when Fechner was concentrating on mind-body problems, there were also other scientists working on the relationship between the body-soul or spirit (mind). Fechner has perceived this problem from different angle with his experience and knowledge in Physics and Philosophy.

The problem, he specifically studied was the amount of physical energy that could have corresponding effect on mental energy. From this idea he was able to formulate a hypothesis relating physical energy and mental energy. What amount of physical energy that could cause changes in the mental energy? He was able to succeed on this attempt and also could evolve a formula which was called Fechner's Law. Later on he has modified it.

There were also many contemporary scientists during Fechner's time, like Weber, Wundt and others, but they were not working in the same line as Fechner did. Psychologists at this time strongly believed that experiments conducted according to the Psychophysical and Scaling methods yielded reliable and accurate in.ormation. Psychophysical methods keep one actively involved in the laboratory works to watch the new results. Any experiment conducted in the laboratory situation provides exciting experience whether it is based on Psychophysical or Scaling methods or an expeiment in any other field, yields new and accurate information.

Experimental Psychology is as much important as any other subject in a course of study. Since experimental results provide new information to every area of the subject experiments should be considered as the foundation for the area of any subject or discipline.

The discipline of Psychology was very much nourished by rich laboratory works in European countries during the last quarter of nineteenth and first quarter of twentieth centuries.

Indian Psychologists in their efforts to build up the discipline of psychology are very much anxious about the performance of psychological experiments in psychological laboratories. They expect that the experiments should go according to the correct procedures in the same manner as experiments conducted in Physics laboratories, Chemistry laboratories and other biological laboratories with all seriousness. Unless proper seriousness is shown in the laboratory situation in the performance both by the teacher and the student, psychological experiments could never fulfil the requirements. Indian psychological laboratories have come up very slowly. They are still with very few instruments and many universities never thought of having a Department of Psychology with a laboratory. Consciously or unconsciously, an idea is set up in the minds of the teachers who handle the practical classes and the students who take up practicals that *it is a leisure time activity. This attitude is prevailing in many Indian Universities*. This is a very dangerous attitude which should be thoroughly avoided.

This can be done only by giving more importance to practical classes with all seriousness.

In this book, '*Experimental Psychology with Advanced Experiments*' the experiments included and discussed are from various areas and these experiments can be conducted with different types of tools and instruments. The experimenter may find not only interest in it but also some kind of inspiration to conduct similar experiments more and more. The experimenter may feel that he/she has discovered something new and also may feel happy that he/she has made some new contribution to the field of Psychology and got the reward for what he/she has perspired. Therefore, it is immanent that any student of Psychology who conducts experiments with all seriousness and sincerity according to the procedure can certainly discover something new and may feel very proud of it.

There are twenty chapters with sixty experiments in these two volumes, which are covering the areas of sensory experiences, physiology emotions, and mainly the cognitive areas like perception, learning and remembering, thinking, language comprehension, problem solving, nature of intelligence and its measurement, creativity and its measurement, measurement of human aptitudes and special abilities, motivation and achievement, assessment of personality types and traits, social behaviour and environment and behaviour.

In certain areas where suitable test materials were not available I, myself had designed, developed and validated the tests and included in this book with examples and interpretation of the results. Most of the experiments included in this book are such that could reveal the skills of the participant and the efficiency of the experimenter. Therefore, I am quite confident that the experimenter may feel ease and grace with this book in all his experimental works. The Psychophysical Experiments are not included in this book. All the Psychophysical Experiments and the Scaling experiments are included in my another book, '*Modern Psychophysical and Scaling Methods and Experimentation*'.

This book is written with utmost care and untiresome efforts after analysing the syllabuses of Experimental Psychology of many Indian Universities with the view of making it as a comprehensive guide, and as an alternative to the existing study materials available for Experimental Psychology. Therefore, I strongly believe that this book would certainly meet the requirements of students of Psychology at all levels which may serve as a textbook as a manual, as a self-guide and as a reference work to the teacher who handles the Experimental Psychology classes.

I am extremely grateful to Mr. Ashok Kumar Mittal of Concept Publishing Company for having envisaged so much interest as to bring out this voluminous book, *Experimental Psychology with Advanced Experiments* in two volumes in an excellent manner. Mr. Mittal, an outstanding Indian publisher, a man of fertile imagination and lofty thoughts with creative spirits, has already published quite a large number of books for the benefits of Indian intellectuals and for the colleges and universities. He deserves all admiration for the wonderful services rendered for the cause of education of this Great Nation by way of publishing valuable scientific and creative works with international standard.

Chennai-40
January, 2004

M. Rajamanickam

CONTENTS

Volume - I

LIST OF EXPERIMENTS

Volume - I

LIST OF FIGURES

Volume - I

1

TEACHING OF PSYCHOLOGY

Experimental Psychology is taught in colleges and universities to various heterogeneous groups and the method of teaching is the conventional lecture method. The lecture method of teaching is the oldest and widely adapted method. The ancient Greeks were using the dialectical method, a logical examination of testing one's kowledge as a method of teaching. Socrates and his followers Plato and Aristotle were using this method. In ancient India, the *Gurukula* method was widely prevalent, by which the student lived with his teacher till the completion of his education and the teacher dictated everything he knew whenever he found time.

The notes taking method was adapted in the medieval period. Whatever the teacher dictated was taken by the student by writing in palm leaves and the same was preserved for generations. But in modern times various techniques were introduced in teaching a subject. The discussion method, seminar method, workshop method, student centred teaching method, problem solving method, using the mass media of communication like the newspaper or periodicals and publishing lessons in them, the radio programme method, the television programme method, the computer programme method are all used as methods of teaching in place of the direct teaching method. Apart from these indirect methods of teaching, reading books directly and learning, correspondence course method, the open university method are some of the methods of teaching adapted by many student-learners in these days without having any direct contact with the teacher in the classroom situation. Psychology is also learned through all these methods.

But what is the best method of teaching Psychology? With regard to teaching methods, there are diverse views among the scholars of the present time. But the best method of teaching depends upon the teacher who teaches the subject. It is true that some teachers put the least possible efforts in their work and leave everything to the students' efforts. This is not the case with all the teachers. There are teachers who put all their best in teaching and make it the best and stimulate the students for better understanding and thinking. They even change the existing attitudes about the lecture method of teaching.

Investigations were conducted to find out the best method of teaching by comparing the effects of various methods. But a substantial conclusion was not evolved on this issue. In teaching science subjects, the lecture method of teaching is followed by demonstrations of the subject matter and practicals are conducted on every matter learned by the students. The practical or the experiment is conducted to get the subject matter more familiarized, so that the student may remember it forever. Therefore, experiments are introduced as supplement to lecture method. The experiments are also intended to reinforce the subject matter taught by the lecture method. The teacher, while explaining an idea through some demonstration, makes the demonstration as a practical method of teaching which is supposed to produce better effect in student's learning. This type of teaching stimulates the student further for some new scientific inventions and discoveries. Thus the demonstration of an idea through the instruments and apparatuses becomes a practical method or experimental method of teaching.

Teaching of Psychology in these days is in no way different from the teaching of other physical and biological sciences. Physics studies the physical events and matter and events may occur in the physical world naturally or may be caused purposely as desired. The nature and characteristics of these events in the physical world become the subject matter for Physics. Similarly, in the study of Psychology the nature and characteristics of the events that cause perceptible behaviours in an organism become the subject matters of Psychology. Some of the events are studied with instruments and tools and others by way of observation. After the establishment of the psychological laboratory all sensory responses (events) are subjected to cause and effect and psychologists were able to demonstrate certain ideas or the subject matter through the instruments and tools in the laboratory experiments. The student in a psychological laboratory observes the occurrence of some new event or response in a person which may stimulate his thought process and when he personally performs and repeats the same practical in his own way, he learns better.

Similarly, in a social situation many things are occurring in human behaviour. The individual's reaction in a group or social situation or the behaviour of a group may be studied by observation and by using some designed tests and conducting investigations. This is also an experiment in a social situation. The teacher may demonstrate these psychological phenomena and the student may comprehend them in an effective manner.

Thus the method of teaching Psychology has become an experimental method in Psychology. The experimental method of Psychology keeps the experimenter actively involved in the problem solving task of the Psychological process. Since enormous amount of work has been carried out in Psychology using the experimental method, the experimental method itself has become a field of Psychology and it has been designated as Experimental Psychology. More emphasis is put on both laboratory experiments and on field experiments. Therefore, Experimental Psychology is both a method of Psychology and a field of Psychology.

Boring, the great American psychologist, as early as 1924 has identified a branch of Psychology which he designated as Experimental Psychology and stressed that the goals of Experimental Psychology are to examine the human behaviour empirically and to formulate hypotheses and verify them. Boring also set the topics of Experimental Psychology. According to his specification the components that qualify Experimental Psychology are sensation, perception, feelings and emotion, learning, memory, attention, action and thinking. Therefore, Experimental Psychology is greatly interested in the empirical study of various aspects of human behaviour by subjecting them to experimentation for fulfilling the educational objectives. The student is expected to comprehend fully the objectives, the methods, procedure, collection of data and interpreting them by applying statistical treatment to the data in such a way that he should be able to explain them by himself to another person.

Thus the laboratory experience or study is a means of getting acquaintance with the methodology and acquiring the skills to interpret the results analytically and also to integrate several diverse materials for the purpose. Therefore, the experimentally based method of teaching Psychology supposed to provide the student with various approaches to deal with the problem. The student by these means becomes not only well learned, but also technically fully conversant and competent to meet any problem and also gets ready to offer critical views and to formulate creative thinking.

Though the lecture method of teaching Psychology has been variously criticised it cannot be totally rejected or substituted by any other method of teaching. It depends upon the competency of the teacher who uses the lecture method and the method by itself is not defective.

In teaching Psychology the lecture method should always be supplemented with demonstration of the relevant experiment. There may be some intriguing issues which require someone to explain and to simplify certain complicated matters. The teacher is the person immediately available for the student to help him in such cases and the teacher in the next class may integrate the matter in his lectures and clear the doubts. The teacher through his years of experience and study might have known many new things which the student is not aware of them or it may take many years for him to learn so much information. The teacher, who has dedicated himself for the teaching profession may wish to give the student all the best he has with him, would certainly incorporate all of them on some relevant context in his lectures and the student no doubt would be very much benefited. Whatever new and effective methods of teaching are invented, the lecture method would continue and survive forever and no new method of teaching can substitute it. Psychologically, there are many things that a student learns in the classroom lecture method by having direct contact with the teacher and with other learning materials, which the distance education student or the student who learns indirectly any discipline totally misses them.

Experimental Psychology is the fundamental field of study for all the fields of Psychology since it forms the basis for the empirical study, dealing with experiments, formulating theories and attributing characteristics to Psychology for qualifying it as a science. Experiments provide the experimenter with new data for describing and interpreting human behaviour. Exponents in the field of Experimental Psychology have put much of their efforts to equate Psychology with some of the physical and biological sciences through the experimental evidences. It is often criticised that human reaction to a stimulus is subjective in nature and that it cannot be studied objectively. But experimental psychologists have demonstrated that human behaviour can be studied objectively. The role of the early and modern psychophysicists in this respect are to be greatly admired. The contributions made by the psychophysicists, like Stevens and others have solved several problems in the fields of vision (light), sound and other sensory areas and in reaction time.

Experimental Psychology had the strong impact of Watson's behaviourism and the S-R theory for about fifty (1910-1960) years. Though the impact was in varying forms, the functions of psychological enquiry was continued and survived undisturbed for a long time The impact of the after effect of the World War II suddenly highlighted new trend as to how humans operate the machines, a question that has become the cause of a new approach in the study of psychological processes. The advancement that took place in engineering field brought out more improved models of processes and the same were utilized for describing the fundamental nature of human behaviour which resulted in the emergence of a new field of psychology called the Cognitive Psychology. There is a need to understand human behaviour at action or at the application level. This trend was gaining more importance in Experimental Psychology.

Generally, cognition refers to the various aspects of acquisition of knowledge. It is more concerned with knowing and perceiving the psychological processes than with movements and responses. Cognitive psychologists are of the view that the human mind is not a passive receptor of stimuli from the outside world. The human mind is actively processing every information it receives and transforming it into some kind of new forms. Therefore, in general, Cognitive Psychology may be defined as the branch of Psychology that studies all aspects of mental processes that transform the sensory inputs in various forms of knowledge. It is the study of the psychological processes that contribute to the acquisition of knowledge. The psychological components or the subject matters of Cognitive Psychology are, the sensation, perception, learning, memory, feeling and emotion, thinking, imagery, intelligence, problem solving, language comprehension, information processing, the organization of information and decision-making. Some of the components stated here are in addition to what Professor Boring has said in his Experimental Psychology as the subject matter. Most of the components of Cognitive Psychology are dealt with in this book.

Learning is a cognitive activity, but learning and teaching always go together. Teaching is an art of imparting new knowledge in some specific area to those who are devoid of it and to those who are in need of it. Teaching is an effective process of communicating some information, especially of an abstract quality. It is giving instruction or lessons in some specific subject. Teaching need not be confining to only classroom situation, it may be also referring to imparting of some highly intellectual or technical subject matter which is done normally with visual aids and experimental demonstrations. Therefore, teaching of Psychology should be always accompanied by some visual aids and experimental demonstrations, so that the information communicated may be be easily registered in the memory of the learner.

EXPERIMENTAL PSYCHOLOGY

Introduction

Experimental Psychology, now, has grown and developed into a broad field of Psychology by the relentless and consistent efforts made by the modern psychologists during the course of last one hundred years. It has its beginning in Fechner's Psychophysics in 1860. His attempts to study sensations, judgment of weights, brightness, loudness and other magnitudes were like physical sciences. Since he was not very much successful as he expected to study the sensory reaction in the same manner as physicists study matter, he changed his approach. He felt that there is relationship between physical events and mental functions. He tried to establish it by experiments and quantification. It is felt that Experimental Psychology must study the conscious experiences and this beginning was found only in Psychophysics. Psychophysics is an area of application of psychophysical methods.

Fechner has observed that every physical event should have a corresponding psychological event. In other words every physical event is having corresponding psychological response. A sensation is depending directly upon a physical stimulus that is aroused. For example, loudness depends upon the intensity of the auditory stimulus. This is verifiable, reproducible and reliable. This is a simple experience. Similarly other sensations like light, weight are also depending upon the intensity or strength of the physical stimulus aroused in the situation. Therefore, the magnitudes are based upon the intensities of the physical stimuli. This appears to be the natural beginning to start with Experimental Psychology. Till the beginning of twentieth century psychophysics consistently engaged in this venture. Therefore, psychophysics and psychophysical methods were used for the purpose of formulating laws and principles relating to sensory experiences to the relevant stimulus properties. The important characteristics of the application of psychophysical methods are to make the judgment simple and to detect the presence or absence of sensation. Fechner was able to evolve three metods for studying various sensations and judgments. They were the methods of limits, method of constant stimuli or constant method and method of average error.

Following these, several different psychophysical methods were developed by various modern psychophysicists. These methods were based upon the criticisms levelled against Fechner's methods. Woodworth (1947) one of the outstanding American psychologists has stated that Fechner's psychophysical interpretation of his results are found to be not acceptable to psychologists, but his methods are still called psychophysical. Many of the psychologists have used them and many have contributed to further development and improvement of his methods.

Whatever type of criticism made against Fechner's psychophysics and psychophysical methods his theoretical aspects of psychophysics become the basis for experimental approach for studying and understanding the sensory experiences and judgment. Though Fechner's critics like Stevens (1975) and others have severely criticised Fechner's psychophysics and psychophysical methods and provided substitutes, still Fechner's methods are alive and used in psychological experiments.

Psychophysical Tradition

Many of the early scholars in the field of Psychology were German scientists. For example, Weber (1795-1898), Fechner (1801-1887), Helmholtz (1821-1894), Woundt (1832-1920), G.E. Muller, Ebbinghaus (1850-1909) were all German scientists. Then came the American and British psychologists, William James (1842-1910), Cattell, (1860-1944), Watson (1878-1958), Francis Galton (1822-1911) and others.

A few years earlier to Fechner, Weber in Germany has conducted experiments on various sensations. E.H. Weber was professor of Anatomy at the University of Leipzig. The sensations he studied were visual, auditory, tactual and muscular. He used some of the principles of psychophysics in studying the relationship between stimulus and response or judgment. But he has not given any name to his methods. It was only Fechner who came out with a name 'Psychophysics'. Weber has already proposed one of the important problems of Psychophysics, the just noticeable difference (JND). He stated that just noticeable increment in a stimulus is proportional to the stimulus. Having this as the basis, Fechner developed his famous psychophysical theory using the logarithmic law. Accordingly Fechner has stated that the strength of the sensory process is proportional to the logarithmic of the stimulus and then Fechner (1889) defined Psychophysics as "an exact science of the fundamental relations of dependency between body and mind".

The Weber's law $\frac{\Delta S}{S} = K$

where Δ stands for increment in the stimulus,
S stands for stimulus value, and
K stands for a constant and it is also called Weber's ratio.

According to Weber's fraction the visual perception of brightness is approximately 1/100. This means that the light of 100 candle power must be increased by one candle power for perceiving the increased brightness. A brightness of 200 candle power must be increased by 2 candle power. But Weber's fraction varies with different sense modalities. Fechner, later on elaborated Weber's principle into a general law. This is called Weber-Fechner law. It is expressed as $S = K \log R$. Where, S stands for sensation (the perception of real change), R stands for the stimulus, and K stands for a constant multiplier.

Psychophysics studies the relationship between the given physical stimulus and the corresponding effect upon the organism, that is the response and also the effects of the stimulus changes and the response changes or perceptual changes. In the study of these relationships three important questions are generally asked.

(1) What is the least or minimum amount of stimulus value required to produce a response in an organism? Or what is minimum amount of stimulus value that will produce a perception?
(2) How strong or intense should be the stimulus? What sort of stimulus is necessary?
(3) How it should be manipulated?

To find out answer for these questions the minimum or the least amount of stimulus which can be effective has to be determined. Therefore, this question is concerned with absolute limen or absolute threshold. It is also stated as the determination of stimulus limen. That is the stimulus presented to a person is sufficient enough to arouse response in him fifty per cent of the times.

The other question is, what is the minimum or least stimulus difference between two stimuli in the same sensory modalities that is capable of producing an awareness in the organism as to state that there are two different sense impressions. How much brightness should a light have in order to state that it is brighter than the other light close to it? This is to determine just noticeable difference in the stimulus. This is also called determination of differential limen.

The third problem is, the judgment of relationship between stimuli. That is, under what conditions the stimuli are to be judged as equal? To what extent the respondents are committing errors in making judgments, while judging one stimulus as 'greater' or 'smaller' or 'equal'? These are the problems associated with psychophysical experiments.

There are three psychophysical methods which are based upon the problems. They are (1) Method of minimal changes or the method of just noticeable differences, (2) Method of constant stimuli, and (3) Method of Average Error.

The study of human behaviour was first started with the study of sensations experimentally by the early German scientists. These studies were conducted in a piecemeal fashion till a systematic psychological laboratory was set up by Wilhelm Wundt in 1879 in University of Leipzig. Soon after this several psychological laboratories started and experiments were conducted on different kinds of sensations. Each one started specializing in certain area. Helmholtz (1867), a German physicist concentrated on vision and the famous Young-Helmholtz theory of colour vision has emerged. The American psychologist Cattell went to Germany to see Wundt's psychological laboratory and had association with Wundt. Then he went back to America started his own laboratory and also invented several instruments. He made excellent contribution to the field of reaction times, sensation, and association. Ebbinghaus, had concentrated on learning and memory. His experiments using non-sense syllables for memorising and his curve of forgetting were main contributions. Ebbinghaus was a pioneer in the study of memory and his famous work "*On memory*" was published in 1885.

Study of Mental Abilities

The British psychologist Francis Galton has concentrated on measurement of mental abilities and also introduced the Questionnaire method in Experimental Psychology. He devised several tests and also invented many intruments in his laboratory. At the same time in Germany Oswald Kulpe (1862-1955) following Wundt's principles, set up his own experimental laboratory and concentrated on the study of memory, thinking and judgment.

Having started pscyhological laboratories following Wundt, experimental psychologists in Europe and America adopted different approaches. They have started studying human behaviour under controlled conditions and in experimental conditions. They have introduced animals in their laboratories for studying their behaviour to compare it with human behaviour. Since Galton was interested in the study of individual differences, Individual Psychology and Experimental Psychology were growing side by side. The Individual Psychology later is named as Differential Psychology, studying individual difference. Experimental Psychology with its findings and results nurtured general psychology. The subject matters for topics like nervous system, sensations (seeing, hearing, tasting, smelling and touching), attention, perception, learning, memory, forgetting, thinking and problem solving, feeling and emotion and motivation are all derived from Experimental Psychology.

Experimental Psychology and Factor Analysis

The individual psychology mainly was depending on the measurement of mental abilities and statistical treatment and analysis. Galton used also correlation methods to associate the variables. He wanted to determine how the stature of the son depends upon the stature of the father. His famous work, *Inquiries in the Human Faculty* which was brought out in 1883 was the result of his study of individual differences. Later on Spearman published a book in 1927, which had its root in Galton's work. Spearman in his correlation analysis was able to see that a positive correlation between variables was due to the presence of a general ability common to the variables. He named this aspect as '*G*' factor which he called general intelligence. He has also brought out the '*S*' factor, which he stated as specific factor, the special ability which is specific for individuals. This is called Spearman's famous Two-factor Theory which was published in 1904.

This work of Spearman had resulted in starting factor analysis in Experimental Psychology. Following this extensive works were carried out to enable us to state the matter more accurately. There has been large number of research volumes of books monographs and papers published on human abilities during the first and second quarters of twentieth century. In most of the works the technique employed was known as factor analysis. This was used to trace out the underlying the mental factors. Factor analysis starts with a table of correlations between scores of different variables. It needs much larger table with large number of test scores of variables. Factor analysis is a systematic procedure by which the meaning of a test is examined by correlating it with other variables. It is important that the researcher should give a large number of tests to the same person or group of people. The analysis may determine how many abilities could be measured reliably. It may also provide the information about the individual characteristics and what the given tests measure. Factor analysis is a mathematical technique which reveals what the tests have in common or what common elements the tests measure.

In England under the leadership of Cyril Burt and in United States under the leadership of Thurstone the factor analysis studies were started. Burt after years of study brought out the work *The Factors of the Mind* in the year 1941. In America Thurstone, first brought his book *Vectors of Mind* in 1935 and later on his monograph, *The Primary Mental Abilities* in 1938. Following this Holzinger and Harman published their work *Factor Analysis : A Synthesis of Factorial Methods*. Following these works Thurstone has published a serious works on factor analysis, *Factorial Studies on Intelligence* in 1941, *A factorial study of perception* – as a psychometric monogaph in 1944, *Factor analysis of body type* in 1946, and lastly *Multiple Factor Analysis* in 1947.

After these there were several authors brought out books, monographs and papers on factor analysis making some revision and improvements in the previous works. Some of them to be mentioned are Fruchter, Guilford (1948), Thompson (1948), Cattell (1952) and Gorsuch (1974).

The factors obtained from factor analysis have been interpreted in various ways. Apart from Spearman's *G* factor (the general intelligence) and the specific to factors S_1, S_2, S_3 (special abilities) which constitute the total ability *A*, $(G+S_1+S_2+S_3=A)$, it is also interpreted that between the general factor and the specific factors there are abilities which are less than the general factor and less than the specific factors. These factors are now called group factors. Present day scholars state that there are three forms in factor analysis. They are common factor variance, specific variance and error variance. The common factor variance is the agreement between two or more variables. The greater the variance the greater the agreement among them. The specific variance is not having any relationship with other variables. In every variable there may be some degree of specific variance, which is not related to common factor. Thirdly, there is some identifiable variance which has nothing to do with common factor or specific factor variances. This may be due to the faking answers given by the subjects or may be due to incorrect or contradictory answers.

Most of those who applied factor analysis in their studies accept the group factor theory suggested by Spearman and his co-workers. The group factor is something different from the *G* factor and *S* factors. Later on it was also felt that there can be more than one common factor in the performances of certain tests. This phenomenon is stated as multiple factor. The multiple factor theory permits the division of obtained score into true and error components. The true variance is further divided into common factor variances and specific variances. Generally the common factor variance is contributed by all the tests. The specific variance is contributed by two parallel forms of the tests. The error variance also is contributed by all the tests.

The proportion of common factor vairance is stated as communality of the test and it is indicated as h^2. The proportion of specific variance is S^2 and by adding this to the common factor variance we get the true variance. The specific variance and the error variance may be combined and it is called uniqueness. It is determined by detecting h^2 from $1(1-h^2)$. Thurstone's book on Multiple Factor Analysis was published in 1947. He provided in this book enormous amount of material on factor analysis.

A number of factor analysis methods have been brought out by scholars in this field. But all the methods start with the correlation matrix. The important methods are the principle components method proposed by Hotelling (1933), the principle axes method proposed by Kelley (1935), the summation method given by Burt (1941) and the centroid method proposed by Thurstone (1947). The centroid method of factor analysis proposed by Thurstone is widely applied in researches.

Factors are extracted from the correlation matrix. The readers can refer to any one of the books mentioned in this section. It is always safe to receive guidance from experienced scholars in this area. Otherwise it may look very laborious work to extract a factor if the tests are many. It is also possible to use factor analysis with few tests. But the results may not be reliable. If the researcher has understood the method it would be very interesting to extract the factors in the factor analysis. In the application of factor analysis we may not be knowing how many factors are present in the correlation variances. But the assumption is that there are as many common factors as the rank of the correlation matrix or the number of tests. But it will not be possible to extract so many factors. There is a stage at which we must stop the extraction of factors. After extracting every factor the values in the correlation matrix is going on reducing and some values may even vanish. Extraction of factors may be stopped when the matrix values become too weak. Thurstone (1940) and Fruchter (1954) have suggested a formula to be applied as to the sufficiency of factor extraction. But where to stop factor extraction is or when to cease extracting centroid factors is still a baffling issue. Thurstone (1958) says that a satisfactory criterion along these lines has not yet been developed.

The factor analysis begins with correlation matrix. Correlation matrix are prepared according to the technique we choose for the factor analysis. There are *Q* technique, *R* technique and *P* technique. In the matrix there are as many columns as the tests we use for the study and also as many rows as the persons or occasions. In the *Q* technique the correlations are between rows. This means that correlations are made between the scores of the persons. In the *R* technique the correlations are made between columns. That is, a person's score in one test is correlated with the score of another test. Every test score of the individual is correlated with the score of every other tests. There can be any number of tests. The *P* technique is intended to analyse the characteristics of a single person from the population. The occasions are used in the place of tests. There can be several occasions. It may be daily, weekly or monthly testing on the same individual. For example, the individual might have been tested everyday for 25 days. This is like obtaining scores from 25 tests and these sets of scores are correlated to elicit the person's characteristics.

Rotation

Rotation is another step in treating the data in factor analysis. Whether to rotate the extracted factors for analysis or analyse the factors as they are? There is difference of opinion with regard to rotation of the extracted factors. Burt and other British psychologists state that there is no necessity to rotate the factors. Already the data had undergone several step of statistical treatments. What is obtained is the final result and, therefore, there is no need to rotate the factors. But the American psychologists following Thurstone's principle state that rotation should be made to make the result psychologically meaningful. In the rotation of factors there is no change in the communalities *(R^2)*. The communalities for both unrotated factors and rotated factors are the same. But there are some changes in the values of rotated factors.

There are two types of rotations. One is the graphic Orthogonal Rotations and the other is Oblique Rotations. Both methods of rotations are to be made only after extracting sufficient number of factors from the correlation matrix. But if the communalities (k^2) of the unrotated factors and the rotated factors should not have much difference. In many of the studies it was found that both are having almost the same value. Any difference in the communalities is attributed to the rotation method. The researcher, therefore, can verify his results by comparing the two communalities. Factor analysis is a highly sophisticated statistical treatment to the experimental data and also the results are very much reliable. Factor analysis primarily was started in the field of mental ability testing. But now it is applied in almost all the areas as it has become essential to know the common factors operating in the functions of a group or in several tests. Experimental Psychology in the second and third quarters of this century is very much benefited by factor analysis. Further, it is accepted that a more scientific and highly mathematical principles can be applied in Experimental Psychology and reliable and meaningful results can be obtained which may be useful in solving several individual and group problems.

Diversified Thinking

Modern authors on Experimental Psychology mainly deal with what is current trend in the university curriculum and also what the researchers' interests are in their work. As result of this modern book on Experimental Psychology deals with topics relating to non-laboratory experimentation. This is not a healthy sign for a young science like psychology. This approach will not fulfil the requirements of experimental psychology. The development of certain new areas like the humanistic psychology created a sense of apathy and a feeling of wearisome towards rigorous experimentation and a sense of likeness for philosophical interpretation. This kind of thinking has been aroused by the scholars who have been indulged in the study of personality characteristics of the people. Still philosophical interpretation can be given to the traits or characteristics derived by the experimental methods. But a total discard of experimental method to the study of personality or other behaviour patterns is a non-scientific approach which may be harmful to Experimental Pscyhology in the long run. Abraham Maslow (1974) did not say that Experimental Psychology studies only mentally unhealthy people. He only stressed the study of normal people who are mentally healthy.

The contributions made by modern experimental psychologists like Woodworth (1947), Stevens (1959), Helson (1958, 1964), Galandar (1962), Candland (1968) to the development of Experimental Psychology have provided enormous amount of source materials to all fields of Psychology. Every experiment conducted in the laboratory situation provides some new information and, therefore, the laboratory experimentation at all levels is the scientific approach to the study of human behaviour and scientific methods are applied reliably. McGuigan (1978) states that we would agree with the statement made in some circles that psychologists do not formally go through the steps of scientific methods strictly in conducting their researches.

EXPERIMENTS AND EXPERIMENTAL PSYCHOLOGY

Experimental Psychology had its beginning in the study of sensory process in the laboratory situation. Experiments were conducted first with sensations. An organism reacts to a stimulus if it is effective. A presented stimulus or an accidentally occurred event if it is not effective, then the organism will not react to it. In this case there is no difference between human beings and animals. But in the beginning psychological experiments were conducted only on human beings.

What is an Experiment?

An experiment is a planned investigation to provide evidence for or against a hypothesis. It is an act of conducting an investigation or test. It is an attempt at something new or different. It is an effort to evolve something original. An experiment may be planned and conducted based upon some natural event or a hypothetic assumption. Whatever it is something new is evolved from an experiment. Basically an experiment is an event created by the experimenter to test his natural curiosity. A curiosity may be aroused naturally on the basis of some past experience gained by the experimenter in some situation. Whatever it is, an event is caused or occurred in a situation which is planned with a view of observing the consequence of the event. But an event caused by an experimenter is different. The experimenter is causing the events in a controlled situation in a laboratory and he is waiting curiously the consequence of it. The experimenter may change the pattern of causing the events with the view of observing the subsequent consequences. He may be manipulating the materials used in his experimentation in varied ways. These are common characteristics of an experiment. These are applicable to any physical sciences and also Psychology.

Experimental Psychology

Definitions

Experimental Psychology is defined in various ways. The early authors defined Experimental Psychology in terms of its content. But the present trend is to define Experimental Psychology in terms of scientific methodology. The psychological problems should be studied systematically giving more importance to the methods by which informations are collected. Therefore, Experimental Psychology may be defined as the specialized branch of Psychology in which the scientific methods are used to collect information about the problem studied. The experimenter can also determine the variable that is having control over the psychological reactions.

Sheridan (1979) states that 'Experimental Pscyhology is simply that branch of Psychology in which experimental methods are used'. According to him it is a comprehensive discpline and the experimental pschologists can raise and answer questions about hatred, memory, meditation, vision, mutual attraction, decision making and any other psychological subject matter. His definition is just simple. "Whenever experimental method is applied to psychological events, it is an instance of Experimental Psychology". But some of the experimental psychologists are not satisfied with this type of definition since Experimental Psychology serves as a source of information for all fields of Psychology. This definition seems to encompass all the fields of Psychology into its fold. The argument is that the journal of

Experimental Psychology is not dealing with its subject matter so widely as to include papers relating to other experimental areas. It seems to confine itself only on general experimental psychology and does not include fields like Experimental Social Psychology, Experimental Physiological Psychology, Experimental Clinical and Counselling Psychology and other applied areas. Therefore, the scope of Experimental Psychology seems to be limited to certain psychological phenomenon. Sheridan states that such things are accidental and Experimental Psychology like most scientific discplines started first with simplest potential topics, but later on included all the fields and now the full range of Experimental Psychology is made known. In every field of Psychology there are complex psychological problems and experimental psychologists carry out researches on all such complex problems and, therefore, there cannot be any special claims for clinical and social psychology to exclude from this fold. Therefore, Experimental Psychology includes the widest possible range of areas wherever experiments are conducted applying scientific methods.

In the study of Experimental Psychology the techniques of experimental methods are applied for collecting the data. In the experimental techniques, the external conditions are controlled and the sources of errors are eliminated. The effects of independent variables upon the dependent variables are measured. An experiment has been defined scientifically by the physicist Bridgman (1927). The approach is called operationism or operationalism. His definition of experiment is defined in terms of words and concepts. The words and concepts are involved in describing operations. For example, let us operationally define a material body, a pencil of six inches long. How to define a pencil operationally. The suitable way is first to formulate operations. For defining operationally, the first thing is to measure it with a scale and it is found that it is six inches long. Secondly, the operational question is, what is it made of. It is made of or composed of wood and lead. Thirdly, our operational question is, how it looks like. It looks like a cylinder. By defining the pencil, a six inches long, made of wood and lead and cylindrical shape, we differentiate it from all other material items in the world. We may provide more specifications for the operational definition like texture, colour, diameter and the use of it etc. In this way we make an operational difinition of an item of pencil.

But Candland (1968) states that operational definition has certain objections also. Every characteristic we mentioned here about the item may be questioned. What is meant by six inches long? What is wood? What is lead? What is meant by cylindrical in shape? We will have to provide operational definitions for these items also, if we want to answer these questions. But this type of problem should not interfere in our attempt to provide the use of operational definition. Therefore, we must be precise in providing our operational definition. In Psychology we apply operational definition for the concepts like learning, memory, intelligence, motivation and so on. There may be disagreement with the experimenter. Critics may question everything and the experimenter has to clear all the doubts raised by them. Operational definition has some advantages if systematically applied. The experimenter comes to know very well, what he is doing when he uses a concept and also reveals to others what it is. Secondly, the experimenter may find out that he eliminates the unnecessary concepts which are interfering in the concepts he is dealing with. Lastly, experimenter is made to understand the implications of his experimental design clearly. We cannot be always sticking to the operational definition. The problem of regression, no doubt may interfere and it may result in providing only very little information. Therefore, we may sometimes resort to the conventional definitions. But when we are designing experiments and in developing theories the principles of operationalism furnish the Experimental Psychology with a scheme of clear thinking. We also formulate clear hpyotheses by using the terms of operational definition. We pay special care in defining the independent and the dependent variables in terms of operational definition. For example, this occurs in the case of defining the variables in an experiment in which the independent variable of bright light outside may distract the attention of the students in the classroom situation.

Experimental Psychology as a Method

Experimental Psychology applies scientific methods to study its subject matter. We state that Psychology is a science and, therefore, it must possess all the characteristics of a science. Psychology possesses all the characteristics of a science. But that does not mean that we can treat the mind in the same manner as scientists deal with matter. Certain instruments used by physicists can be used in the psychological laboratory. Physicists measure the matter with scales and weigh its weight in balances. Such measurements are not possible in psychological laboratories. Though psychologists also use the same type of instruments the methods and procedures are different. Psychologists apply the instruments on human beings for eliciting their nature of reactions. The procedures and purpose vary in psychological experiments. But the psychologists apply the rules of scientific methods to their subject matter. By which they discover the existing relationships that control behaviour. The behaviour may be simple or complex. Whatever it is, psychologists try to find out what factors determine the actions in the organisms.

Variables

The common characteristic of scientific method is solving problems and more rigidly following the procedure and more diligently controlling the variables. A problem exists and the experimenter is interested to find out a solution to it. For which he states a tentative solution or answer to the problem which is stated as hypothesis. It is also important to see the antecedent of the behaviour. Under what condition it occurs. He tries to find out the relationship that occurs among the variables. The general principle of cause and effect is applied here. The cause is a variable and the effect is a variable depending upon the first variable. That which one causes an event or action or behaviour is an independent variable and the effect of it is called dependent variable.

Variable is defined in various ways. A variable is a characteristic or attribute that can vary values in a test. Or that which makes changes in an event or action or behaviour is a variable. For example, the way in which the students are responding to the college bell, or the speed with which the railway passengers are reacting to the railway guard's whistle on the platform are the types of variables in which psychology is interested. There are two types of variables. One is independent variable and the other is dependent variable. These two are always related. If there is no relationship betwen the independent and dependent variables there is no subject matter for the psychologist to study.

An organism reacts or responds to something in a natural situation. For example, the passengers on the platform are hurriedly running to get into the train has the antecedent cause – the guard's whistle. The whistle is the independent variable and the passengers' reaction is the dependent variable. But the same independent variable will not arouse any response in passengers who are waiting for another train or in a person who is buying a book in the book stall. Therefore, there must be relationship betwen the variables. Psychologists in the laboratory are conducting experiment in a control situation and manipulate the independent variables to arouse response in the subject. Only when the subject reacts there is relationship and response becomes the dependent variable. The responses in the subject may be made to change by presenting the independent variable in varying degrees. In all such experiments what the psychologist wanted is to discover a principle relating the independent and dependent variables. Thus the variables for the experimental psychologists are stimuli and responses, the events in the laboratory or environment and the individual's responses to the events. The relationship between the independent and dependent variables can be formulated as an empirical law provided if the relationship confirms the result of the experiment as desired by the experimenter.

Control of Experimental Variables

In order to elicit the effectiveness of the experimental variables it is important to have the experimentation controlled. That is, we are introducing controls in the experimental situation. The assumption is, whether the observed events in the experiment are really due to the procedure adopted or some other factors are resoponsible for the events. To test this the experimenter should take into account two situations, the experimental situation and the control situation. Two groups of people may be brought into the experimental situation, one group may be stated as experimental group and the other group may be stated as control group. Suppose, an experimenter wanted to know whether the memory pills really improve the memory power of the students. To test this hypothesis the experimenter gives memory pills regularly for certain period to the experimental group, whereas the control group is kept away from the supply of the memory pills. After a few days both the groups were given the same memory test and the scores of these two groups of students are compared. If sufficient number of subjects were used in the experiment and if there were no bias in testing a realiable conclusion about the effects of memory pills on this test of the behaviour of the groups could be derived from the comparison of the test scores of the two groups. Any difference in the scores between the experimental and the control groups must be due to the introduction of the experimental treatment. The difference is caused by the effect of the memory pills.

Other Variables

The purpose of introducing control conditions in an experiment is to avoid certain intervening variables that may have some effect upon the subjects. If the other variables are not controlled we do not know whether the response is due to the effect of independent variable or other intervening variables. For example, a variable known as confounded variable is an uncontrolled factor. It may have some effect as the independent variable. Then the experimenter may not be in a position to state whether the response is due to independent variable or the confounded variable. Some authors state the relationship between the independent variable and dependent variable is the same as the relationship between predictive variable and predicted variable. Because the independent or the predictive variable is also making control over the response or values of the predicted variable.

OBJECTIVES OF EXPERIMENTAL PSYCHOLOGY

(1) The primary objective of experimental psychology is only the study of behaviour of the organism systematically. This is also one of the objectives of Psychology in general. But Experimental Psychology has certain specific objectives. The first objective is to formulate hypothesis and to measure the psychological events and verify the hypothesis. (2) The other objective is to predict the conditions under which the specific psychological events would occur. (3) The objective is also to predict the behaviour of the population from the result obtained from the sample. To what extent such result will occur? There might have been the interference of extraneous variables in the result however carefully the experiment was planned. Therefore, the result may be erroneous as these variables are uncontrolled. Therefore, the experimenter has to test whether the changes in the dependent variable is due to independent variable only and not due to uncontrolled or extraneous variables. For this we must seek the assistance of statistical techniques. By this technique the experimenter can estimate the probability of changes in the dependent variable caused by the uncontrolled variables. If the probability is found to be very low like 5 per cent or one per cent he can conclude that the change in the dependent variable is due to the independent variable. By this means the experimenter ascertains that the findings of the experiment are reliable. (4) The fourth objective is that if one is desired to produce the same psychological events with the same technique he will be able to produce it. (5) The objectives of science are the Measurement, Analysis and Synthesis of the subject matter. Experimental Psychology also fulfils these objectives in its functions. The rsults obtained in the experiment by measurement are analysed. The results are explained and interpreted and the underlying principles or laws in the process are evolved, justified and established. It is also possible to verify the law by repeating the same experiment or put it in solving some problems in practical situations. By synthesis, things which are of similar type or synonymous are brought together and made an integrated whole thereby giving a convincing explanation.

The experiments presented in this book may be conducted with the same objectives. They are from almost all the important areas in accordance with our definition of experimental psychology. Certain areas might be given more emphasis as they have more feasibility for experimentation in the psychological laboratories.

FORMULATION OF PROBLEMS

The experimenter or researcher is generally interested in some scientific enquiries. Because he has already acquired certain amount of knowledge by reviewing the scientific literature. These scientific literature are mainly the scientific journals. In the case of psychological problems the experimenter mainly reviews throughly, the journals of psychological abstracts, reviews, comments and criticisms. Having collected certain amount of knowledge in this way he might have understood something about the problem. This literature may provide him information about several things not only for the formulation of problems, hypotheses, experimental designs, characteristics of variables and many things. The experimenter or the researcher on collecting valuable information and acquiring knowledge wanted to start some scientific enquiry.

While reviewing the literature the experimenter might have observed some phenomena and he wanted to visualize them for his purpose. He may find himelf at a difficult position. He is not in a position to answer certain questions and issues. He finds that the knowledge he has acquired is not sufficient to answer such questions or explain the issues. Therefore, a problem exists. That is the experimenter is not able to answer certain questions that arise in his attempt of making scientific enquiry. The experimenter should formulate a problem in such a way that an answer he finds to it should be useful to the humanity. Some may evolve simple problems and may attempt to answer them. But the outcome of them are of no use to anyone. Therefore, one may try hard or perspire and formulate important problem. The solution or answer to the problem should be long ranged and have much fruitful consequences.

Experimenters and researchers should have fertile and rich creative imagination. In formulating important and much useful problem the experimenter may also face some risks. The problem does not end with the formulation alone. It also follows the ways and means of finding out correct answer or solution. How a person becomes conscious of a problem? While reviewing the literature the experimenter goes through many experiments or researches conducted in the past and the problems answered or solved. He could have seen the problems there similar to one he has and can select any of them to formulate a problem area in which he is interested. There are various areas which are classified and published under different heads, in the journals of psychological abstracts.

While reviewing the abstracts the experimenter might have noticed that out of several hundred published articles some may be having some gaps. The results may not be providing us satisfactory answer. Further, some of the results of the investigation may not be agreeable to us. Further there may be also facts that the investigator not properly presented. *The results are not meaningfully explained.* The experimenter having observed these types of problems in the results of the past researches, he can make a note of all these and formulate a problem in which he is interested and wanted to solve it or find an answer to it.

A problem is anything, matter or person etc., that is difficult to deal with, solve or overcome.

For example, a problem child, a puzzle, question set for solution, all these are problems. A problem that is posed to us may be solvable or answerable or unsolvable or unanswerable. This means that the knowledge that we have acquired is not sufficient or adequate to solve it or to find out answer for it. Therefore, we must try to solve the problems by exploring the possibilities by some investigations or experiments. Towards such, we must formulate a problem which should not be too lengthy and ambiguous. A problem is that which baffles us and causes cofusion without providing any scope for an answer. We will have to search for an answer. Even while searching for an answer we do not get it suddenly, we take time and slowly process for it mentally. The solution for the problem may also need a comprehensive way of thinking and planning. Our search for the solution or answer may result in an exploring activity or an experiment. For example : (1) Is it true that students coming from rural areas are having low intelligent quotient? (2) In the school final examination more percentages of girls get pass than the boys, why? (3) There are more suicides among women in India, why? (4) In spite of punishments to law-breakers the crime rates increase in our country, why? (5) It is said that more youths are engaged in criminal activities, why?

Problems like these cannot be answered readily. Investigations or experiments or surveys should be conducted by formulating hypotheses to get correct asnwers for the problems.

EXPERIMENTER AND HIS HYPOTHESES

Hypothesis is the basic and most fundamental aspect of an experiment in Psychology. Therefore, in formulating hypotheses, the experimenter should formulate suitable hypothesis. He should have gained sufficient knowledge in certain specific area. Therefore, the experimenter first of all should review as far as possible all the available pyschological literature. Literature always provide all the required information to the experimenter, investigator or to the researcher. The literature, here, we are referring to, are mainly the journals of psychological Abstracts, Reviews and Summaries and other journals and books dealing with earlier research materials. The experimenter or researcher should not expect everything he wants from these literatures. They may provide information relating to the topics of the studies made earlier, the problems. Hypotheses, samples, sample sizes instruments used and the procedure of data collection. They may have also the important results briefly summarised.

While reviewing the psychological literature the experimenter or researcher can observe the works carried out earlier similar to his works or theme. From these he can select the relevant topic or title with problem, the independent and dependent variables quite suitable to the study. The experimenter should not choose the same topic or problem that he observed in the literature. He can evolve the idea about the problems and hypotheses that were not attempted earlier. He can choose more than one and finally select one in which he is very much interested and which he can carry out without facing much problem in conducting experiment or investigation. The psychological literature mainly the abstract, reviews comments and criticisms should not be taken as they are available. It is for the experimenter to evolve something new, not attempted earlier for his experiment or research.

The objectives of the experimenter or researcher or investigator, should be explorative, or descriptive or experimental. Whatever type of objective the experimenter or researcher should see that what has planned to study should be original and not attempted by others earlier. Therefore, the experimenter or researcher should formulate the hypotheses of his study that is planned and work out the details for it, such as the independent and dependent variable, the relevant population, the size of the population, the instruments to be used for collecting data and so on.

An experimenter or the researcher may be very much interested in one area or topic and review such literature thoroughly. In such cases he may evolve several new ideas for several new hypotheses, other than those found in the literature. Therefore, by this means the experimenter may go ahead of formulating some new hypotheses and follow some of them. But unless the validity of some experimental result or an established theory is highly suspected there is no meaning in selecting an experiment by evolving an idea from the literature and formulating hypothesis.

A hypothesis may be pertaining to anything. There is no limitations or restrictions about what can be hypothesized. A hypothesis need not be true. Many of the hypotheses which were true, were not known to the experimenters. Therefore, a hypothesis is a tentative statement about some phenomena that the experimenter or researcher wanted to verify and accept it or reject it. A hypothesis also may be evolved from some ideal sources. We may find that some people wanted to do research or experiment on their own ideas. Sheridan (1979) states that there is nothing in picking up a research problem suggested by someone else. We can observe that many people often have good ideas for research, but they have not got any ambition or desire or capacity to put into action. Prople by themselves do not provide us any research ideas, but we sometimes get research ideas in some indirect ways. During the course of our conversation with others we discuss many issues and problems and arguing about many things. Ideas come to us from rich fertile sources. There is nothing wrong in making use of such new ideas and to formulate new hypotheses.

There are excellent suggestions we find sometimes in everyday life with rich source of ideas that were not brought into experimental situations. Beliefs and opinions or even faiths may be subjected to experimental or research verification. Many psychological experiments and researches we find were from such sources and many experimental hypotheses also were from such sources.

Many theories in every area or topic that we come across were the accepted truth after experimental studies. The theories were tested and accepted as valuable fact or truth. Though the theories are guiding research of the present generation they were once assumptions and tested with hypotheses. They were repeatedly tested and the same results were obtained. Therefore, they are valid and true. Scientists often state that experiments are conducted to test the hypothesis and, therefore, hypotheses are derived by reviewing the acceptable theories.

Types of Hypotheses

Hypotheses are of various types. Any hypothesis can be fit into an investigation. It is generally stated that a hypothesis is a tentative statement about something to be tested. A hypothesis is a suggested explanation for a group of phenomena or facts, either accepted as a basis for further verification or accepted as likely to be true. It is an assumption used in an argument, a supposition. Logically a hypothesis is an unproved theory. A hypothesis is a test of basic reasoning. A hypothesis may be also stated as a testable statement that is evolved as a possible or latent solution to a problem.

A hypothesis may be stated simply in terms of, "if and then". For example, "if I want to catch the train at 8 O'clock in the morning to go to Mumbai, then I must be there in the railway station not later than 7.30 a.m. itself." "If one wants to study medicine by joining a medical college, then he must be proficient in biological sciences."

The other way of stating a hypothesis is : Low income groups in India have more children than the high income groups. There are more suicide incidents in poor families than in rich families. The crime rates are increasing in urban areas than in rural areas. Children from broken homes become delinquents. People from very low income areas often migrate to industrial areas with the view of getting jobs.

Thus hypotheses may be pertaining to anything. There is no limitation about which is to be hypothesised. It need not be always true. The important types of hypotheses we shall discuss here. The classification of hypotheses is mainly based upon the conveniences of the researchers. A hypothesis is a statement relating two variables of which the first one is independent variable and the second one is dependent variable. If '*A*' exists, then '*B*' also exists. McGuigan (1978) states that if certain conditions hold, then certain other conditions should also hold. In the "if and then" relationship, according to him, let '*A*' stand for the first set of conditions then '*B*' for the second set of condition. In this '*A*' is referred to the antecedent condition of the hypothesis and '*B*' is referred to consequent condition of the hypothesis. Therefore a hypothesis proposes a relationship between two variables. According to Black and Champion (1976) the primary hypothesis are three. They are research hypotheses, null hypotheses and statistical hypotheses.

The Research Hypotheses

These hypotheses are also called working hypotheses. They are derived from the established theories mainly about some phonomena of social type. The researcher believes that his research hypotheses are true. It is also believed that since the hypotheses are tentative statement and they are derived from the theories, the theories also may be regarded as tentative supposition about thngs. Only when they were tested and accepted as true it is true, and free from any objections. A hypothesis is subjected to testing whether it is acceptable or unacceptable. The hypothesis : Students who enter the plus two class in rural areas are stated to be of average intelligence. This can be tested by administering an intelligence test and if their I.Q.s. were assessed between 90-95, then the hypothesis is said to be confirmed. But if the IQ of these students is found to be much higher, (say 100-115), or much lower, (say 80-85) then the hypothesis is not supported or disconfirmed.

The Null Hypotheses

The Null Hypotheses are also stated in the form of statements about something to be tested. But such hypotheses are always stated in the reverse form to the research hypotheses. The null hypothesis refutes or disagrees with what is clearly stated in research hypothesis. In other words we may state that the independent variable is not having any effect upon the dependent variable. The same example may be stated for null hypothesis. The students who enter the plus two class in rural areas are not of average in intelligence. But according to the test results the null hypothesis is not correct. The null hypothesis refutes the results of the research hypothesis or regarded as not true. Therefore, we may conclude that if it is not true, then the intelligence of the boys who enter in the plus two class in rural areas are on the average. Therefore, the statement that the boys who enter the plus two class in rural areas are of average in intelligence is accepted or regarded as confirmed. The experimenter or the researcher faces two opposite situations, one being that the plus two boys' intelligence is on the average and the other not on the average. The experimenter finds two statements one for confirmation of the hypothesis and the other for refuting the hypothesis. These two cannot go together. Therefore, he must take the result in such a way that the denial of the null hypothesis is for confirming the research hypothesis, or it may be stated, that the null hypothesis confirms an alternative (research) hypothesis. But the null hypothesis has certain limitations.

The null hypothesis states that there is no difference between the mean scores of the urban students and the rural students in their learning ability. The experimenter conducts an experiment and wanted to test and determine whether the mean scores of the two groups of urban and rural students differ in their learning abilities. The hypothesis stated is that "there is no difference between the mean scores of urban and the rural students in their learning abilities. The null hypothesis, therefore, stated in the form of symbolic statement $\mu_1 - \mu_2 = 0$, μ the Greek letter which means mu. μ_1 is the population mean for group 1 and μ_2 is the population mean for group 2. If the difference between the population means is $\overline{x}_1 - \overline{x}_2$ is very small, then there is no difference between the two group means ($\mu_1 - \mu_2 = 0$). The hypothesis is confirmed. But if the mean difference between the two groups is greater ($x_1 - x_2$), then the hypothesis is refuted ($\mu_1 - \mu_2 = 0$). The null hypothesis was first used by Ronald A. Fisher (1953).

Statistical Hypothesis

A third hypothesis, we shall discuss here is the statistical hypothesis. According to Winters (1967), it is about the statistical population and this population may refer to people or animals, birds or things. The experimenter's observations on people or things are converted into numerical values. Conclusions are made on the basis of the numerical quantities. The hypothesis formulated is that students who come from high income families are more intelligent than the students who come from low income families. The first group is designated as group *A*, and second group as group *B*. *A* test was conducted to assess the Intelligence quotieant (IQ) by administering a suitable Intelligence test. The scores of each group were obtained and the average IQ was worked out to see whether group *A* has scored high or group *B*. If the high income group is really more intelligent than the low income group, then the average IQ of the group *A* should be considered higher than the group *B*. The IQ difference between the groups should be statistically significant either at one per cent ($P > .01$) level or at five per cent level ($P > .05$). But if the difference between the means is not statistically significant than the hypothesis is not confirmed. It means that there is no difference between the two groups (*A* and *B*) in intelligence.

Other Hypotheses

There are also other hypotheses which the experimenters and the researchers use in their studies. An experiment is conducted for testing an empirical relationship between variables. This relationship is stated as hypothesis. Therefore, a hypothesis is empirical which means that it directly refers to some data that were obtained from observations. When the hypothesis is subjected to empirical test to measure the relationship between the independent variables and the dependent variables then we call it as empirical hypothesis. Empirical means derived from experiments or related to observations. An experience is based upon practical experience rather than scientific proof. It is derived from experience. Generally, stating about the empirical hypothesis, in all the experiments, the experimenters conduct or perform practicals in laboratories and the empirical hypotheses are tested. In statistical hypothesis also what the experimenter or researcher tests, is the empirical hypothesis. In Experimental Psychology empirical is considered to be an important concept.

One hypothesis that is spoken by the experimenters is the universal hypothesis. Since the word universal refers to the whole, for all kinds that which present everywhere. Therefore the universal hypothesis clearly states the degree of generality. The universal hypothesis asserts the relationship between variables or among all the variables. The relationship is stated for all the times and in all the places. It is present everywhere. It is stated with 'If and then'. If *a* is true *b* also is true. Some psychologists quote the maze learning as an example for universal hypothesis.

If *a* is for the chimpanzees in one cage to reach the banana at a height with two sticks and two boxes without mistakes and *b* is for the other chimpanzees, in the other cage, will learn to reach the banana without mistakes. Here we are speaking about all chimpanzees in all the places are trained to reach the banana. This universal hypothesis in psychological experiments is very much limited. We may state that practically it is not applied in any type of experiments or researches.

The existential hypothesis is also similar to the universal hypothesis, is not very much in application. It is a statement that asserts that a relationship is held for at least in one case. It was Hull (1952) who postulated this hypothesis in the context of his learning experiments. He states that this hypothesis may be very much useful in psychological experiments. Psychologists stress sometimes that a given phenomenon occurs in an experiment but it is not known how often it occurs. The existential hypothesis may be applicable only in certain specific experimental situation.

FORMULATION OF PROBLEMS AND HYPOTHESES

In the preceding section we have discussed in detail about the formulation of experimental problems and hypotheses. We have stated that the experimenter or researcher should first of all go through the psychological abstracts for selecting the relevant problems and hypotheses. This may be only of some help to the experimenter or to the researcher. Experienced scholars in experiments and in research rarely seek experimental or research information purely from the psychological abstracts. But for the beginners it may be deceptive. The journals of Psychological Abstracts or journals of Educational Abstracts or any other abstracts do not provide all the essential information required by the experimenters and researches. They need very much the problems, hypotheses, the selection of population, size of sample, instruments, procedure of data collection, the specific statistical treatment to the data and the important results. But psychological Abstracts provide information like the name of the author of the article published, the title of the paper, sample and a few results from the paper. These informations are very inadequate for an experimenter or researcher who is very anxious to conduct an experiment or research.

The abstracts published in the psychological Abstracts fail to provide the essential information required for the experimenter or the researcher. The reason is that abstracts are not prepared by experts. The editor of the journal of psychological Abstracts is having a team of persons to assist him and whom the editor distributes the articles published in various journals. Generally, they do not go through the whole article and collect essential information for publication. They abstract the articles according to their own idea and there is no standard procedure in abstraction of the articles. As result of it the essential informations are missed in the most of the abstracts. Further, they abstract the articles from only some journals and thoroughly omit all the articles from some journals. Because of this, articles from certain specific areas in which the experimenter or researcher is interested do not find place in the journals of abstracts.

Therefore, it is better that the experimenter or the researcher should directly refer to the articles published in the journals. He/She can collect the issues of current psychological journals and go through the articles thoroughly once. Back issues up to 5 years also may be collected and go through the articles and select the problems and hypotheses which to some extent similar to his/her planned experiment or research work. By this method there is every possibility to collect very valuable problems or hypotheses which he/she can design experiments or surveys.

The experimenter or researcher can also contact research experts and can take advices. The teacher and other members of the faculty and also those who are having deep knowledge in the area in which the experimenter or the researcher is interested may provide valuable information. One of the best source of information is the textbooks and the books in the respective areas. The library catalogue may provide information about the titles of the books in the specific area. The bibliographies in the books and references of the articles, journals of psychological reviews, psychological bulletin, all may serve as sources of information for the experimenter or the researcher. In all these sources rich and reliable experimental designs may be available.

EXPERIMENTAL DESIGN

The experimenter has a problem and he has to find out an answer or a solution for it. The answer or solution would not come to him unless he takes steps to deal with it. It is by conducting an experiment he can find out a solution to his problem. Therefore, the experimenter has to design an experiment or more than one experiment. Experimental design involves various properties like the independent variables, dependent variables, the population, sample, instruments or test materials and so on. He has to mainpulate the independent variable which is relevant to his experiment and control other variables. The experimenter has to assess the effects of the independent variable upon the dependent variable to find out solution for his problem. In other words, the experimenter assesses the effect of the controlled or manipulated independent variable upon the subjects in the form of reactions or effects. Therefore, experimental design may be defined as the method or process of assigning tasks or treatments to subjects.

There are many experimental designs and they are classified on the basis of the properties such as the number of independent variables, number of dependent variables, the process of assigning task to the experimental subjects and so on.

The important purpose of experimental design is to find out the relationship between the independent and the dependent variables. If we know the nature of relationship between these variables, we can predict the future events in a sample of population. The function of the independent variable may provide us with the knowledge for further plan. It is not enough if we predict the future event. But we must control it if it is undesirable event. For example, adding too much of sugar in the regular food is the cause of diabetes. The immediate action is to control it. First of all we must control the outcome or dependent variable, the effect. We must find out which specific independent variable or variables can be put into action to control the effect of the diabetes. The next step is the manipulation of the controlling of independent variable or variables. Thus only in the experimental method we can acquire adequate knowledge about the controlling variable and what the controlling variable is. But it is very difficult task to identify the confusing variables in the experimental situation and to eliminate them. Experimental design helps us to tell others what we are doing in the experimental situation. Controlling of the variables is not missed even by oversight. One of the important characteristics of experimental design is that there are standard statistical treatments applied to the data. According to standard experimental design we come to know how to assign tasks to subjects. By this means we can obtain clear conclusion whether the independent variable controls and predicts the level of dependent variable in the experiment.

Experimental design includes assigning the independent variables to the subjects. Suppose, for example, in a laboratory or experimental situation we have 25 graduate students who have willingly came forward to serve as participants. The experiment is of manipulating some instruments. Now the experimenter should decide and assign what the subjects should do with the variable or variables in order to bring out the expected effects or responses in the dependent variable or variables.

Contemporary scientific experimental study is characterised by some kind of study plan.

This study plan conventionally named as the experimental design. Generally experimental designs are the plans that specify how data should be collected and analysed. There is no one experimental design universally applicable for all experimental or research purposes at any particular period. Every kind of experimental design functions to facilitate the experimenter to conduct the experiment in different ways and at different levels of complex state. Therefore, selecting the appropriate experimental design is depending upon the type of problem the experimenter wishes to answer or solve.

In contemporary experimental field there are several experimental designs used. But, however, one type of experimental design as an alternative may be best in a particular case. The decision to select the best depends upon the strengths, weaknesses and the functions of each experimental design and the experimenter's interest and ideal selection of it.

Functions of Experimental Design

The important functions of experimental or research designs are depending upon the complexity accompanied with the experimenter's decisions. Black and Champion (1976) state that there are three experimental functions. But there are several functions. The most important function of the experimental design is that it provides the experimenter the 'blueprint' for studying the problem. The blueprint function of the experimental design is the plan similar to the builders have like the drawings, sketches etc. In the case of an experimenter or researcher, it refers to the description of the population he studies; the type of samples, the sample size, the tools used for data collection, the description of them, the procedure of data collection etc., and how the tools or instruments are administered, how the collected materials are analysed, the type of statistical treatment to the data and so on. Secondly, the experimenter or researcher should know the limitation of his study. He should ascertain the boundaries of his study and control his study area within that limit. Thirdly, the experimental design provides the experimenter some expertness in finding out solution to his problem or question. Further, the experimental design may also provide some estimation or cost of his study.

The Objectives of Experimental Design

The experimenter who studies a problem should have an objective. Selltiz *et al.* (1959) state that the experimental design is closely connected with the experimenter's objectives. They state that the objectives of the experimental or research designs are exploratory, descriptive and experimental in nature. In accordance with these types of designs they also state that the important purposes of research are to become familiar with the phenomena or the problem or getting clear understanding, to explain things, matters, to decide relationship between variables and to test the hypotheses formulated, already.

It is observed that most of the *experimental designs* are based upon exploratory objectives. According to this the investigator has no idea or knowledge about the problem what he studies. The experimenter may not be having any familiarity with the target group or other people and they do not provide him any scope for conducting any investigation. Therefore the exploratory investigation has some important outcomes. The investigator by chance has the opportunity to discover something new which may be very important factor and he may make it later on more improved and valuable one. The discovery sometimes may occur even incidentally or accidentally which the investigator can make it sophisticated later on. Exploring is examining, searching or trying to find out suitable answer or solution to a question or problem. It may be also manipulating of some variables with a view of finding out the solution.

The European navigators from fifteenth century to eighteenth century have seriously organized expeditions in high seas with an object of exploring new lands. Therefore, exploration is discovering something new which did not exist earlier. This may take place even in a laboratory situation and the experimenter's objective is to find out something new that is, an answer or solution to his question or problem.

The descriptive objective studies provide enormous amount of information. By describing the characteristics of a sample or population the experimenter is able to conduct the experiments in an improved manner. Unlike exploratory objectives, the descriptive objectives are more specific. The study is directed to more particular aspect. Selltiz *et al.* (1959) state that descriptive design can bring out the relationship between variables accurately and may set up scope for more detail investigation in due course.

The experimental objectives mainly deals with controls of variables. The experimenter can observe the effects of variables upon other variables in the controlled situation. It may be holding one variable constant and others freely. In a laboratory situation, the experimenter, in a sound cage experiment, controls the subject's eye-sight by blindfolding and making the subject to attend only to the sound he produces. There may be also other sound variables outside, which are controlled. This experiment is conducted under controlled condition.

The experimental objective design mainly is to conduct the experiment in a controlled condition, by controlling other independent variables and to observe the effects of the specific controlled variable. By this means the effects emitted in the dependent variable are accurately assessed. But many experimental designs have the characteristics of all the three objectives.

Classification of Experimental Designs

Experimenters classify the experimental designs in a variety of ways. This classification of experimental designs depend upon the variables the experimenters think suitable for their purpose. It also depends upon the information the design communicates easily in the experimental situation. The classification of designs also depends upon the manner in which the subjects assigned treatments, whether it is done randomly or according to some plan. The number of independent factors also is considered in the classification, whether there is only one independent variable or several variables and how the variables are manipulated.

Single Randomised Group Design

When it is not possible to have a two-group design the experimenter can go with a single, the experimental group. A group of subjects are randomly selected from a population who have no knowledge about computerization. They were examined with some oral and written questions. Their responses were recorded. The experimental group was given some lessons about computerization in television for a period of one month. At the end of one month they were again tested with the same type of questions and the responses were recorded systematically. The group's response before the television programme and after the television programme were recorded and compared. The difference between the pre-television programme and after the television programme were statistically treated and compared. The hypothesis is that after attending the television programme the group performed better in the test. This was tested and the hypothesis was confirmed since the subject's performance was very much better.

Randomised Two-Group Design

One important experimental design is complete randomised two-group design. The subjects are drawn randomly from a population or group. The two groups may be of equal in number. The Random Number Table may be used and a number may be assigned to each subject. One group may be designated as experimental group and the other as control group. Both groups were given the same test (independent variable) and the results were obtained. Later on the experimental group was given some training or lesson regarding the same test, but the other group (control) was kept away. After the training both groups were given the same test and the results were obtained and compared. It was found that the experimental group had performed the test much better than the control group. This shows that the training or lesson the experimental group received exerted some influence in the performance of experimental group while the control group which was devoid of the training had not done better in the test.

Randomised Three-Group Design

In designing a three-group experiment the experimenter chooses ten subjects for each group from a student population. He wanted to test whether these three groups differ in their understanding of a lesson. All the 30 subjects were selected from Table of Random Numbers and each one was assigned a number and also assigned to a group among *A*, *B* and *C*. Group *A* is made to attend a regular lecture class for a lesson on computer. Group *B* is made to listen to a television programme in which the same lesson was shown. Group *C* is made to study the same lesson at home. They were not told that they will have a test. After two days the same students of all the three groups were collected and they were asked to take up a test in the lesson they have learned. Thus the three independent variables such as direct lecture, television programme and home study were used in this experiment. The answers were assessed and marks were assigned and compared to find out which group preformed better. This was to test to what extent the independent variable exerted influence in the dependent variable.

In the same manner a randomised four group design and a radomised five group design also can be made from the same population. But care should be taken to vary the independent variables. The sample size also should be the same as far as a possible. Let us design a randomised four-group design.

Randomised Four-Group Design

The population is university students studying in the second year Psychology B.A. Degree course. There were 20 students, ten boys and ten girls. They were randomly divided into four groups, five in each group as *A, B, C* and *D* by using the Table of Random Number. But the boys and the girls were put in separate groups. They were to attend some psychology practical class. Group *A* attended the practical demonstration class directly. Group *B* listened to the practical demonstration class in television. Group *C* attended practical class only by lecture, but without demonstration. Group *D* is made to know the practical by studying the written material. The students of different groups were not allowed to mingle with one another and to have any discussion about the practical. They were not allowed to go out. After two hours break, one student from each group was asked to go to the laboratory and perform the practical and the performance was assessed according to the procedure by four lecturers. When everything was over the students were allowed to go away. Then the responses were obtained and converted them into numerical values and a statistical treatment was given to the data. The scores of four groups were compared with each other to see which group performed better and which group badly. A rank order was given to the four groups. The hypothesis was that if a group is allowed to observe the demonstration of the practical directly then the group will perform the practical much better than any other group. This hypothesis was tested in this experiment and the hypothesis was accepted as the group which witnessed the practical class directly performed the experiment much better than the other three groups and it was also found that every group differed from every other group. The independent variable, directly witnessing the practical demonstration, had significant effect upon the variable, the practical performance. It was because of the different independent variables for the four groups the performance also varied. Therefore, the experimental hypothesis was confirmed and found to be true.

Factorial Design

Factorial design also is a type of experimental design. In the experimental designs we discussed earlier one independent variable that was used in the experiments. It may be varied in two ways. If the independent variable is varied in two ways or in more ways it may be called as factorial design. It is also possible to study more than one independent variable in a single experiment. A complete factorial design is one in which all the combinations of the selected values of the independent variable are used. Therefore, we may state that factorial designs are the methods of multifactor designs. In the laboratory situation in the experimental psychology practical class the students may sometimes come across the multifactor experiments. Several independent variables may be used in a single experiment.

For example, in the two-point threshold experiment, the experimenter fixes the two points at different distances and thus makes each one an independent variable. He is manipulating the instrument Aesthesiometer to give pressure on the participant's skin surface. The Aesthesiometer has two pin points and the participant has to receive two touches or pricks and has to state whether he/she has received two touches or one touch from the instrument for every time out of 100 times. If the experimenter has given pressures or pricks in six places, the responses for them are from six different stimulation variables. The distances he made are 4 mm, 6 mm, 8 mm, 12 mm, 14 mm, 16 mm. The distance between 2 mm will not cause any difference since the distance is very short. Therefore, the experimenter changes 2 mm as the minimum distance after few trials on the participant. After giving pressure with 2 mm, he increased the distance in the Aesthesiometer by 2 mm. The participant is blind folded and he has to state when he received the prick or the pressure on his forearm skin surface. The 6 stimuli or the independent variables are presented not in the regular order, but in a jumbled order each one presenting several times. Thus this experiment is multifactor design. The experimenter works out the number of times each variable is responded with 'Two' or 'One'. In this experiment each variable is presented 100 times. Here there are 6 independent variables varied in these ways.

In a factorial design experiment there are four groups. Three groups are given learning task. These groups may be having equal number of subjects or varied number of subjects. A lesson is given for learning. Two groups were asked to study the lesson by whole learning method and the other group by part learning method. The third group is asked to learn by unspaced learning method and the fourth group by spaced learning method. Next day a test was conducted for all the four groups to see which group has learned the matter better. This design is 2x2 factorial design. There are four independent variables in this test, the four methods of learning. The result is assessed for each group. Marks are assigned to each individual of the groups. There were 6 students in each group. By statistical analysis we can find out which group is superior in learning and which is the least and where the other two groups stand and also the level of difference between groups and among the groups. The same experiment can be designed by adding two more independent variables such as male and female subjects and the design may be stated as 2x2x2 factorial design. Two more independent variables may be added and the group may be divided on basis of age, that is students under 14 years and above 14 years of age. This design is stated as 2x2x2x2 1factorial design. By dividing the groups on the basis of age, the number of variables may be increased. In the statistical analysis we can find out the group which is superior in learning and also where the other groups stand. The difference among the groups and between groups in learning may be assessed. On the basis of these results we can search for reason why the particular group is superior and the other groups are falling back. We can take any one of the independent variable, for example, one specific method of learning, or the gender, or the age and explore the reason for the better or poor performance of the groups.

2

TREATMENT OF THE EXPERIMENTAL RESULTS : Statistical Methods

INTRODUCTION

Psychology studies its subject matter experimentally just like all other sciences. But its subject matter is human being, the behaviour of man. Human behaviour cannot be studied in the same manner as other scientists study their subject matter. Psychology has its own methods and procedures. There are two approaches in psychological experimentation. One is guided by rules and principles and the other is governed by the activities of a person as chosen by the experimenter. Thus one type of experimentation is system controlled and the other type is person controlled. Whatever type of pscyhological experimentation it is, the main purpose is to find out some answers to some questions raised in the situation or to find out solution to some problems. Whether one applies system centred approach or person centred approach it is intended to find out some answer or to discover solution to the problem. Sometimes it will not be possible to find out an answer to a question or to discover a solution to a problem. The experimenter will have to work hard to evolve a conclusion. Because certain questions or problems by nature cannot be answered by psychological experimentation. The experimenter needs certain amount of reasoning. The experimenter expects correct conclusion to the experimental results, for which the logical reasoning is sought. The results are treated with logical sequence for arriving at the satisfying correct conclusion. If the experimenter believes that the conclusion is correct by some means than the psychological experimentation becomes system controlled. But the experimental result should be more reasonable and should have practical significance and should be repeatable. The essential fact about the experimental conclusion is that the validity should be established in such a way as to generalize the results from the specific facts. But however rigorous reasoning process is made with regard to the validity of the experimental result, it is subjected to certain amount of uncertainty. It is because that any result obtained by inference, however rigorously argued, the element of uncertainty is always accompanying it. The experimental result should not be accompanying it. Therefore, if the experimental results are subjected to the element of uncertainty then the experimenter should resort to the statistical analysis.

THE PURPOSES OF STATISTICS IN PSYCHOLOGICAL EXPERIMENTATION

Modern Psychology is very much enriched by its association with several other sciences like physiology, biology, neurology, animal psychology and statistics. Statistics has played an important role in the development of Psychology. From the beginning of twentieth century (1900-2000) to the end of it the role of statistics in the study Psychology and in psychological research was very tremendous. Psychological measurements become impossible without the application of statistics. The accuracy of psychological experimentation mainly depends upon the statistical analysis. Modern literature on Psychology presents enormous amount of new and interesting information by applying statistical methods and analyses.

In these days the students of Psychology is supposed to possess a considerable amount of knowledge in statistics to understand the lessons in Psychology books. Every modern Psychology book is written with statistical interpretation to explain the experimental results. In this type of interpretation and explanation there is certain amount of accuracy of knowledge and clarity. It becomes easy to remember the subject matter.

Further, the application of statistics to the psychological experimentation facilitates and speeds up the research works. Scholars engaged in research work find the statistical knowledge is very much useful, in designing their experiments. Designing a psychological experiment itself is the application of statistics to experimentation. The treatment of the experimental results can be explicitly done if only an experiment is designed according to the statistical techniques.

In addition, psychological experimentation should be planned in scientific methods. It involves persuing from known facts to unknown consequences. What was observed as a fact must be tested and repeated. Thus in this way the research is going from one particular known fact to other unknown processes and generalizes the conclusions. In all these the statistical knowledge serves as a guiding principle. Therefore, a sound knowledge in Statistics is absolutely necessary.

PSYCHOLOGICAL MEASUREMENT

The subject matter of Psychology is human behaviour. Human behaviour is assessed by suitable psychological tests. What a psychological test measures is the individual's behaviour or activities in a particular situation. An individual's particular activity is measured by a psychological test and compared with other individuals. When a test is administered to an individual or to a group of individuals the result we obtain is called score. The score is always in the numerical form. Suppose, for example a person is asked to perform a task of answering certain questions. There may be right answers and wrong answers. Marks are assigned to right responses and these marks are scores. If the test is administered to a group of persons then there may be more number of scores. Each individual has a score. If the group consisted of some 60 persons, then there will be 60 scores. Here is an illustrative example.

A test was conducted in a classroom, choosing 60 students randomly. There were 30 boys and 30 girls. The test consisted of 100 one-word answer questions and for every correct answer one mark was assigned and thus a student who answers all 100 questions correctly gets 100 marks. There were wrong answers also. When the answers of all the 60 students were checked and marks were given, it is likely, that the scores may vary from 0 to 100. If all answers were right then the score was 100 and if all were wrong then the score was 0. In this case the scores ranged from 32 to 92, 32 being the lowest and 92 being the highest. The scores of all the students fell between 32 and 92. This means that every individual's score varied from every other student. Thus the scores of 60 students in the one-word answer test are given in Table 2.1.

Table 2.1 : Showing the scores of 60 students in the test of one-word answer

Girls - 30					Boys - 30				
62	86	58	67	64	57	35	70	61	59
47	80	63	92	87	60	59	47	52	77
69	61	71	60	59	46	81	62	50	40
61	53	75	68	72	64	48	52	67	60
68	73	32	64	57	82	54	41	66	74
54	76	73	69	79	67	37	73	78	58

The table scores range from 92 to 32, the highest being 92 and 32 being the lowest. Thus the range is very wide (92–32 = 60). A range of 60 scores unit should include all the scores. It is not possible. Therefore, it is required to condense the range with frequency distribution by grouping the individual scores into intervals. Grouping the scores into class intervals simplifies the frequency distribution without sacrificing the accuracy. In this example the range of scores is 60. It is better to start with the lowest scores and to have the multiples of them for the size of class intervals. It is convenient to have the class size with odd numbers like, 3, 5, 7, 9, so that the mid-point of the class will be whole number without any fraction. If the interval between two scores is 3, then start with 6, 9, 12 and so on. If the interval is 5 then, start with 10, 15 and so on. In this example to have the class interval, we can divide the range 60 by suitable size. Three is too small and 7 is too large, but 5 is suitable and, therefore, the range 60 is divided by 5 (60 ÷ 5 = 12). The highest score is 92 but assume it as 95 and the lowest 32 as 30 and we get 13 class intervals. We cannot have 12 class intervals because there are 2 scores above 90 and 2 scores below 34. The class intervals are worked out on this basis and presented in Table 2.2.

Here all the 60 scores are brought under 13 groups. In other words all the numbers come under 13 class intervals, having five scores in each class interval. Thus the class intervals are 30-34, 35-39, 40-44, 45-49, 50-54, 55-59, 60-64, 65-69, 70-74, 75-79, 80-84, 85-89, 90-94. The lowest intervals is 30-34. In order to have continuity we must have the lowest class interval to start with 29.5 which ends with 34.5 and the second interval starts with 34.5 and ends with 39.5.

Table 2.2 : Showing the Frequency Distribution for 60 scores

Class Interval	Mid-points	Scores	Tallis	Frequency	Deviation
90-94	92	92	I	1	+6
85-89	87	86, 87	II	2	+5
80-84	82	80, 82, 81	III	3	+4
75-79	77	76, 75, 79, 78, 77	卌	5	+3
70-74	72	73, 71, 73, 72, 70, 73, 74	卌 II	7	+2
65-69	67	69, 68, 67, 68, 69, 67, 66, 67	卌 III	8	+1
60-64	62	62, 61, 61, 63, 60, 64, 64, 60, 64, 62, 61, 60	卌 卌 II	12	0
55-59	57	58, 59, 57, 57, 59, 59, 58	卌 II	7	-1
50-54	52	54, 53, 54, 52, 52, 50	卌 I	6	-2
45-49	47	47, 46, 48, 47	IIII	4	-3
40-44	42	41, 40	II	2	-4
35-39	37	35, 37	II	2	-5
30-34	32	32	I	1	-6

GEOMETRICAL PRESENTATION OF THE DATA

The experimental data may be treated and presented geometrically and arithmetically. Let us first present them geometrically. Normally in experimental works the Cartesian coordinates are used by placing the values (scores) of the independent variables on the horizontal axis (*X*) and the values (scores) of dependent variable on the vertical axis (*Y*). The frequency distribution is very thin at the ends and bulky at the centre. The mid-points are 92, 87, 82, 77, 72, 67, 62, 57, 52, 47, 42, 37 and 32. This means that the scores are distributed evenly and symmetrically around the centre. In making use of the class intervals we must note down the exact point of class intervals. For example, in the case of class interval 30-34 the score of 30 is really from 29.5 to 30.5. In the case of interval with scores 35, it starts from 34.5 and ends with 35.5 and in the case of 39 it is from 38.5 to 39.5. This is the general principle to start with .5 step below the ends.

Table 2.3 : Showing the Frequency Distribution and other Scores of the one-word answer Test

Class Interval	Mid point of Class Interval	Frequency	Smoothed F Scores	P of F $\frac{f}{N}$	Cumulative Frequency Scores	Cumu-lative Propor-tion	Upper Limit of the Intervals
(1)	(2)	(3)	(4)	(5)	(6)	(7)	(8)
90-94	92	1	1.3	.017	60	1.000	94
85-89	87	2	2.0	.033	59	.983	89
80-84	82	3	3.3	.050	57	.950	84

Contd. ...

Table 2.3 *Contd...*

(1)	(2)	(3)	(4)	(5)	(6)	(7)	(8)
75-79	77	5	5.0	.083	54	.900	79
70-74	72	7	6.7	.117	49	.817	74
65-69	67	8	9.0	.133	42	.700	69
60-64	62	12	9.0	.200	34	.567	64
55-59	57	7	8.3	.117	22	.367	59
50-54	52	6	5.7	.100	15	.250	54
45-49	47	4	4.0	.067	9	.150	49
40-44	42	2	2.8	.033	5	.083	44
35-39	37	2	1.7	.033	3	.050	39
30-34	32	1	1.3	.017	1	.017	34
		60	60.0	1.000			

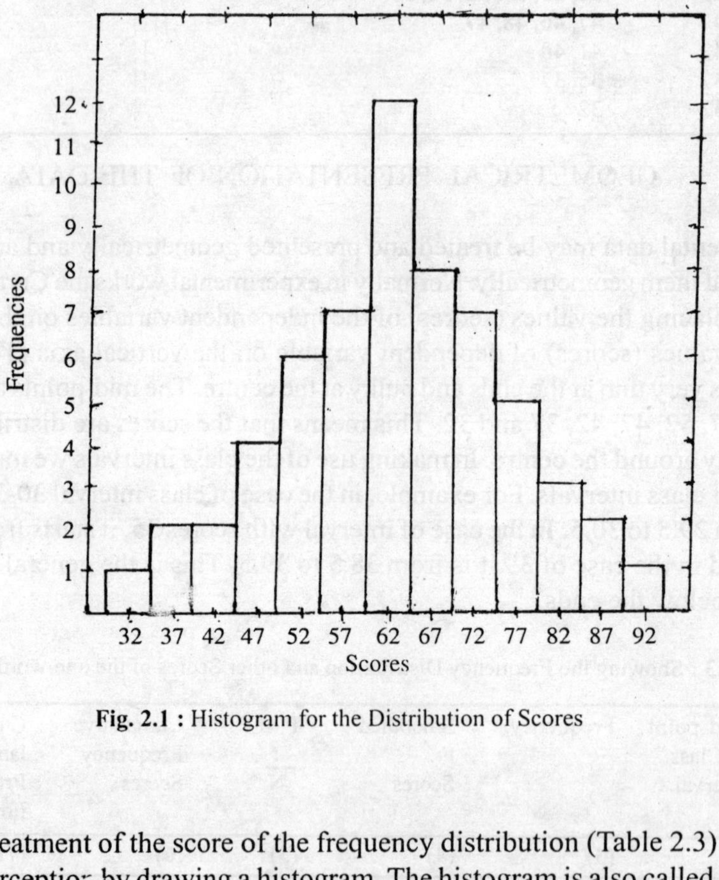

Fig. 2.1 : Histogram for the Distribution of Scores

Histogram

A geometrical treatment of the score of the frequency distribution (Table 2.3) can be given to have visual perception by drawing a histogram. The histogram is also called column chart or stemplot. This is done by placing the grouped class intervals on the horizontal axis (*X*) and the frequencies on the vertical axis (*Y*). The *X*-axis is called abscissa drawn horizontally and the *Y*-axis is called ordinate drawn vertically. These two values, the abscissa (*X*) and the ordinate (Y) are called Cartesian coordinates. On the horizontal axis the mid-points 92, 87, 82, etc., of the class intervals should be noted down. The mid-points as the single score may be taken as the representatives of the scores included in every class interval. Every mid-point has the corresponding frequencies. The mid-points and the frequencies are coordinated. For example in Table 2.2, 1 and 92 are coordinates, 2 and 87 are coordinates etc. They are from the columns of 2 and 5 of Table 2.2. Now, the plottings are made for both variables. The frequencies are plotted against the corresponding mid-points. By joining the plottings of X-axis and Y-axis a rectangle is formed. For every mid-point and a frequency a rectangle is formed. Every score in the distribution represents in the form of a small rectangle. The histogram is a geometrical illustration to show that the scores are heavily collected at the centre and become thinner and thinner at the two ends (Fig. 2.1). The histogram is a series of vertical bars drawn to the frequencies with a common boundary line between two nearby frequencies. The height of the each bar is the frequency and each bar occupies certain area and the sum of the small bars of the histogram is equivalent to total number of cases (Fig. 2.2). If we divide the histogram by N then there will be small rectangles at the rate of one for each score.

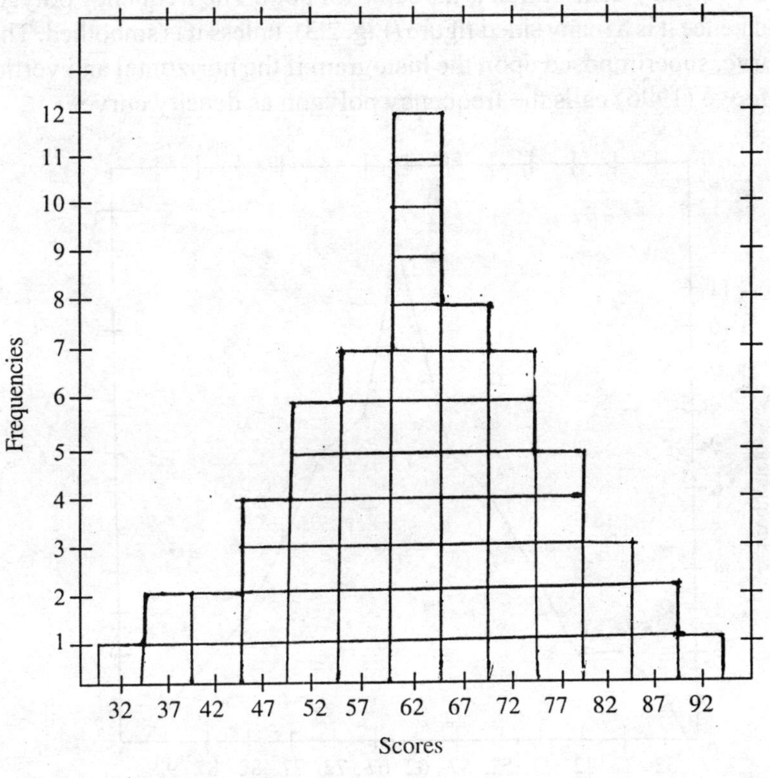

Fig. 2.2 : Histogram for the Distribution of Scores given in Table 2.3 with frequencies and the corresponding areas

The small column of the histogram is the expression of the proportion to the whole area of the histogram. This may be obtained by dividing each frequency of the class interval by the total scores, 60. The proportion for each frequency is presented in Table 2.3 column 5. The assumption is that the sum of the cumulative proportion of the small rectangles of the histogram should be 1.00 (Table 2.3, column 7).

Frequency Polygon

The frequency polygon is a graph or figure drawn by plotting the frequency distribution. The frequencies are plotted above the mid-points of the intervals along the horizontal axis (*X*). In drawing the frequency polygon the same frequency distribution scores (given in Table 2.3) on the horizontal and vertical scales are used. In the frequency polygon the frequencies are marked by dots directly above the mid-point of the class intervals. These dots are joined by straight line. On joining all the dots marked above the mid-points the same line may be extended by one class interval below and one class interval above. By this means the polygon is completed by touching base line on the horizontal axis. Now the frequency polygon is complete and it can be observed that the area of the frequency polygon is the same as the histogram if the frequency distribution is the same for both. The frequency polygon is having many facets and hence it is a many sided figure (Fig. 2.3), unless it is smoothed. The frequency polygon is a curve, superimposed upon the histogram if the horizontal and vertical lines are disregarded. Moove (1996) calls the frequency polygon as density curve.

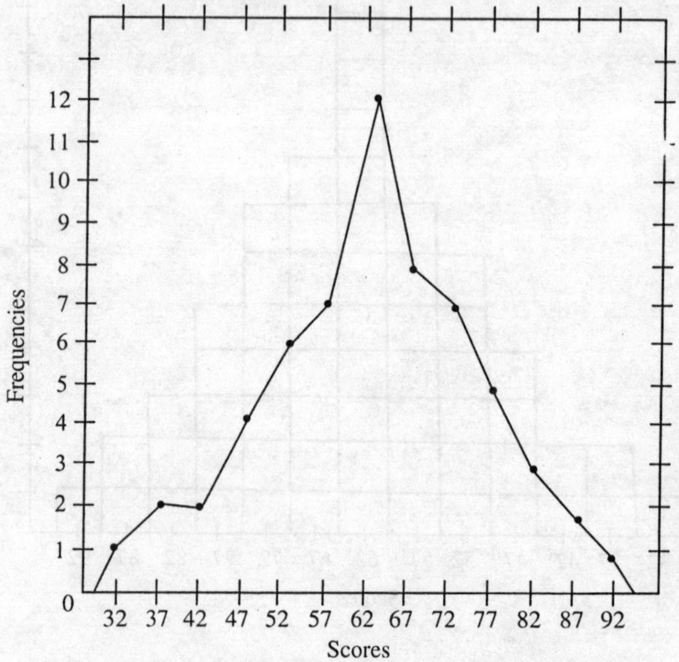

Fig. 2.3 : Frequency Polygon for the Distribution of Scores

Smoothed Frequency (Polygon) Curve

The smoothed frequency polygon is a curve drawn to the distribution of scores (*X*) and the frequencies (*Y*). The irregularities found in the figure of histogram and the frequency polygon are regularised. A regular shape is given to the frequency polygon by the process of smoothing the curve. This is done by taking the average of 3 class intervals. The frequencies of three corresponding class intervals are summed up and divided by three. The frequency value of the respective class interval, one frequency value above of it and one frequency value below of it are summed up and average of these three values (frequencies) is marked at the centre (middle) interval for smoothing.

This may be explained with an example from the data given in Table 2.3. The smoothed values for the interval as shown above are worked out. The frequency value for the interval 35-39 is two and for the interval 30-34 it is one, and frequency value for 41-45 it is two. By summing up these frequencies we get 1+2+2 = 5, which is divided by 3 and we get 5 ÷ 3 = 1.67. Similarly the values are worked out for all the class intervals and presented in Table 2.3 (column 4). Thus the smoothed frequency curve is a curve which is superimposed upon the frequency polygon if we disregard the facets in the polygon. The area of both is the same (Fig. 2.4).

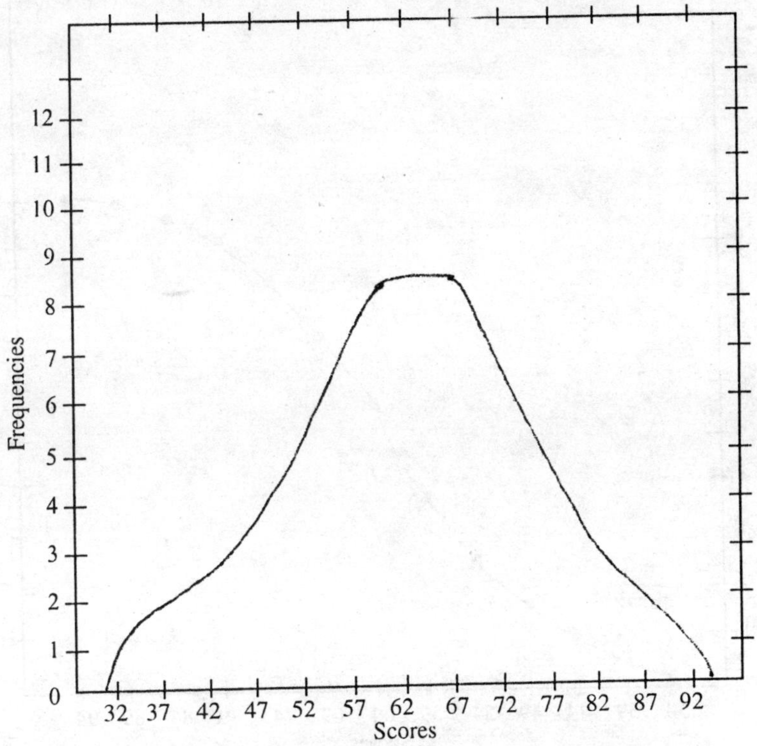

Fig. 2.4 : Smoothed Frequency Polygon to the Scores given in Table 2.3

Ogive Curve or Cumulative Proportion Graph

The frequency distribution data can be used to draw another curve called Ogive Curve or Cumulative Proportion Graph. It is also called Percentage Graph. The cumulative frequencies we get by adding successive frequencies from the lowest class interval to the highest class interval. For example, the cumulative frequency for the interval 50-54 is 15 (Table 2.3 column 6). This value is obtained by adding the frequencies falling below the class interval of 50-54 (1+2+2+4+6). The cumulative frequency for the highest class interval is 60. (Table 2.3, column 6). For obtaining the cumulative proportions the cumulative frequencies should be divided by total number of scores, the *N*. There is another procedure to obtain this value. Since the *N* is constant the value of 1/*N* could be estimated and multiplied by respective cumulative frequency.

For example, for the class interval 50-54, the cumulative frequency is 15. This can be divided by the *N* to obtain the cumulative proportion 15/60 = .25. According to the other procedure, dividing 1 by 60 and multiplying by 15. In this way we have (1/60)15=.25 the same value. We can work out for all the cumulative frequencies and present them in Table 2.3 column 7.

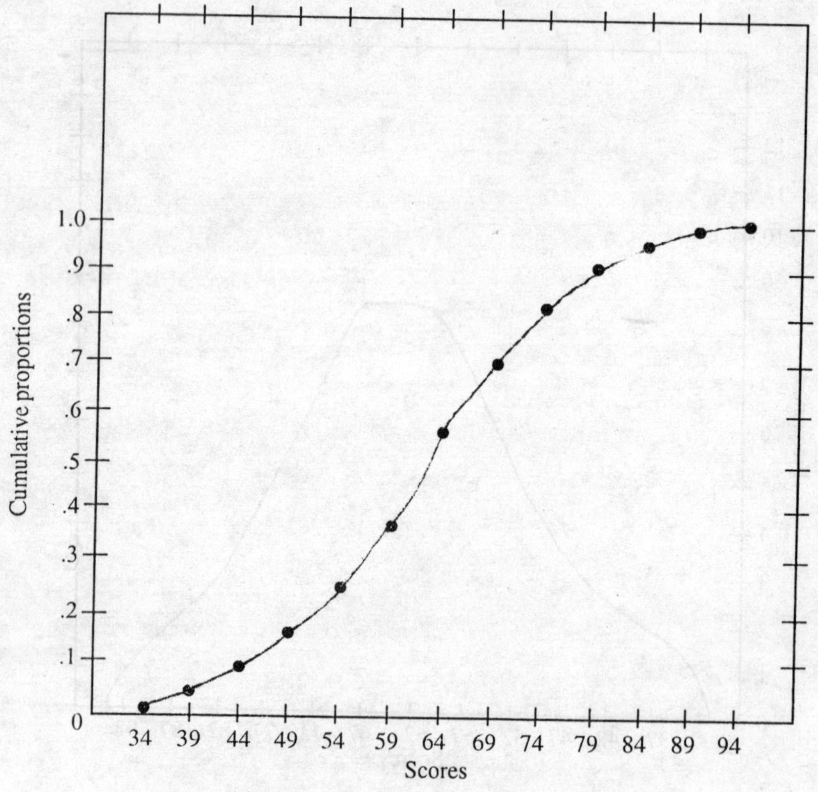

Fig. 2.5 : Ogive Curve or Cumulative Proportion Graph

The cumulative proportion graph can be drawn for the scores and to the cumulative proportions. The cumulative curve is also called Ogive Curve. We need not use the mid-points of the class intervals in plotting the points for the cumulative proportion curve, as we have used in other figures. The upper values of the class intervals are used for plotting this curve (Table 2.3 column 8). Thus cumulative proportion graph is drawn to the upper limit of the class intervals and to the cumulative proportions. The curve is almost like the learning curve of the increasing return (Fig. 2.5).

Normal Probability Curve

The experimenter or a researcher after conducting the experiments tries to test his experimental data by applying some statistical method. He works out the class intervals for the scores and then the frequency distribution. The best representation of the frequency distribution, one expects, is the normal distribution. This means that the scores should be evenly or symmetrically distributed. When a curve is drawn to the distribution, it should be generally normal. There may be few frequencies very low and few very high at the ends. This means that some students scored very high, some students scored very low in the test and most of them fall in the middle. A normal curve is obtained from the frequency distribution of the experimental data if the frequency distribution is very much symmetrical. The curve is generally called Normal Probability Curve. It is also called Bell Shaped Curve because of its shape. It is also called Gaussian Curve (Fig. 2.6).

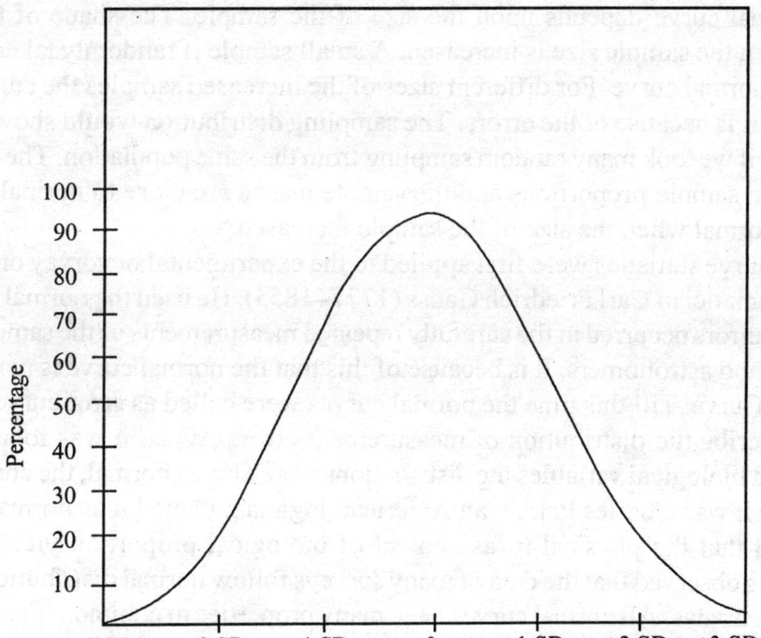

Fig. 2.6 : Normal Probability Curve

The Normal Probability Curve has high peak at the centre. Garrett (1975) states that the normal curve is widely used in psychological experiments and measurements. According to him many human traits like height, weight, learning habits, intelligence, mechanical skills and personality characteristics almost have normal distribution in large groups. We often check many distributions taking into account the normal curve as the standard or model and agree or accept the distribution. Generally all the biological and psychological traits in large groups are normally distributed.

The normal curve is used as a standard in developing and validating mental and educational tests. If the tests do not yield the normal curve, then the test developer has to revise his tests. The normal curve is very much valuable when inferences are made from few cases for assessing the characteristics of a large population. The normal curve is also defined as the systematic bell shaped curve representing the normal distribution and showing the probability associated with the value of each random variable.

Sheridan (1972) states that in Gaussian distribution, in the sets of observation, it often occurs that when there are many random factors, each one is having small influence on the observations. As a result of it some influences are falling in positive and some in negative directions. This means that in the well controlled experiments, systematic errors are taken out. It is for this reason that mathematicians follow the Gaussian distribution.

The normal probability curve can be drawn to the scores obtained in the experiment. The important data required are the frequency distribution of scores and the data worked out in Table 2.3. From these data we will have to work out the mean, the standard deviation (SD) and the percentages of the frequencies (Fig. 2.6).

But the normal curve depends upon the size of the sample. The shape of the curve varies as and when the sample size is increased. A small sample if randomly taken will not have the correct normal curve. For different sizes of the increased samples the curves yield different shapes. It is because of the errors. The sampling distribution would show how the curve would vary if we took many random sampling from the same population. The sampling distribution of the sample proportions and the sample means are close to normal and also would be more normal when the size of the sample increased.

The normal curve statistics were first applied to the experimental or survey or research data by the mathematician Carl Friedrich Gauss (1777–1855). He used the normal curves to explain the small errors occurred in the carefully repeated measurements of the same quantity by the surveyors and astronomers. It is because of this that the normal curve is named after him as Gaussian Curve. Till this time the normal curves were called as error curves as they were used to describe the distribution of measurement errors. When it was found that in psychological and biological variables the distributions were almost normal, the curves were called as normal. It was Charles Peirce, an American logician, named it as normal curve.

It is believed that the physical measurement of biological proportion yields normal distribution. It was observed that the data of many surveys follow normal distribution but not the data of all the surveys. All normal curves have many properties in common. For example, in normal distribution about 68 per cent of the observations or scores fall within one standard deviation (SD) of the mean. In all normal distributions about 95 per cent of the observations or scores fall within two standard deviations of the mean. In all normal distributions about 99.7 per cent of the observations or scores fall within three standard deviations of the mean. But no distribution can be exactly normal. All are only approximation (Fig. 2.7).

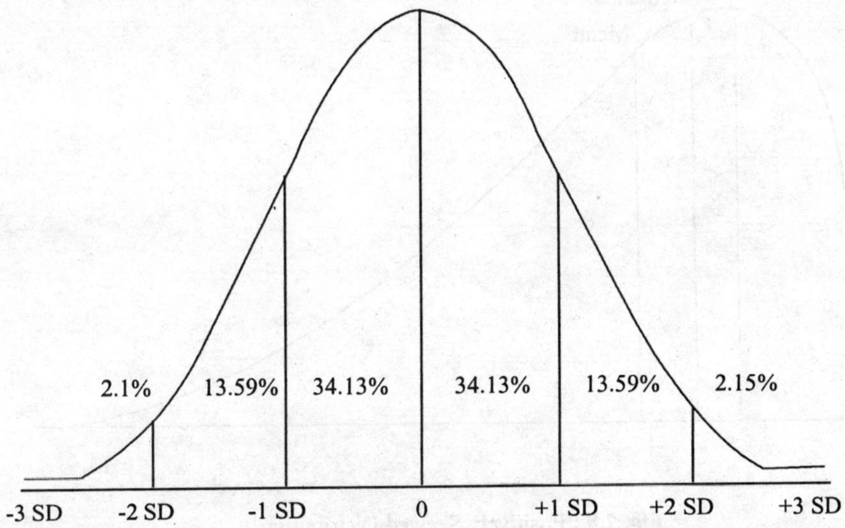

Fig. 2.7 : Proportions under the Normal Curve at various distances from the Mean

Skewness

The normal probability curve, though it is considered as the best representative of frequency distribution, sometimes it is deviating from the normal level and we do not get the symmetrical distribution. This occurs when the frequency distributions are off the centre or imbalanced on the sides. Such distribution is skewed. A distribution is skewed in the direction of its skinny tail.

Generally, it is observed that large number of scores fall at the higher (high score) end or lower (low score) end or on either side of the distribution. In such cases the curve may be lopsided and needs to be skewed. This curve is called skewed curve and the distribution is called skewed distribution. In a distribution more number of cases may fall at the end of lower scores or on the left and when it is skewed it is called positively skewed (Fig. 2.8). This means that a large number of individuals in a test have obtained low scores. Sometimes more scores in the distribution may fall at the higher (high scores) end on the right end and when it is skewed it is called negatively skewed or negative skew (Fig. 2.9). This means that a large number of individuals have obtained high scores.

Skewness is a statistical method used to normalize asymmetrical curve. It is a frequency polygon which is asymmetrical or non-symmetrical or a distribution off the centre as compared with normal curve. The scores of the heterogeneous group is likely to have the peak of a distribution at the lower (scores) end (on the left) or at the higher (scores) end (on the right).

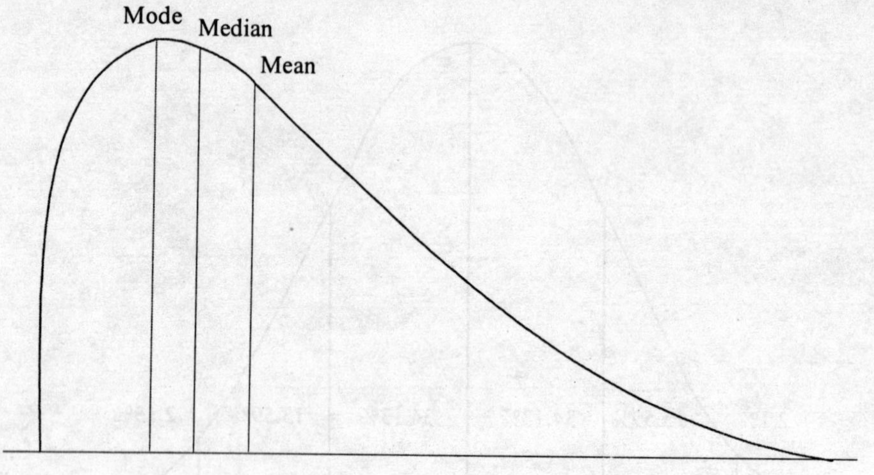

Fig. 2.8 : Positively Skewed Distribution

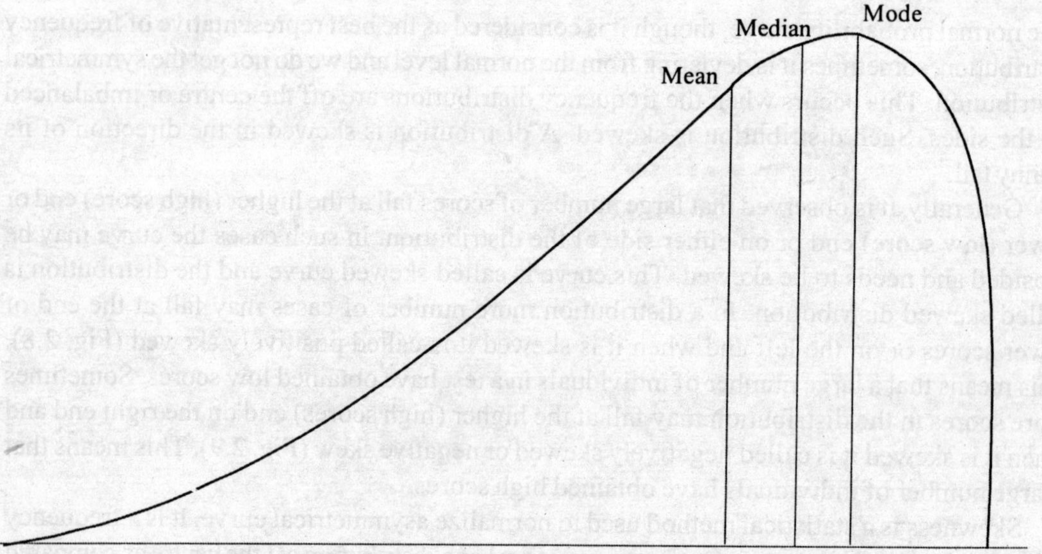

Fig. 2.9 : Negatively Skewed Distribution

This means that the scores of most of the subjects may be low or high. In the case low scores the curve is positively skewed and in the case of high scores it is negatively skewed (Figs. 2.8 and 2.9).

The sampling distributions in small samples may deviate from normality in various ways. Gulford and Fruchter (1987) state that these deviations occur in various statistics like the standard deviations, correlation coefficients and in proportions. In all these treatments also a skewness is applied. Therefore, wherever the sample distributions depart from the normality skewness statistics is applied to bring the distribution means to the normal.

Kurtosis

Kurtosis is another statistical method applied when the sampling distributions depart from the normality. Kurtosis is more drastic than the skewness in sampling distribution. The departing from the normality is taking place in three forms in sampling distribution. The middle part is very steep in departing in Kurtosis. Kurtosis is a statistical method which is sharpening of a peak as a curve of density function or distribution, especially in comparison with that of a normal density or distribution with the same variance. Kurtosis is a deviation from the normal distribution in which a sharp peak is observed and thus Kurtosis is also called as *peakedness* or flatness in the frequency distribution compared to the normal distribution. Kurtosis in a distribution occurs when it is more peaked than the normal curve. That means that the frequency distribution in Kurtosis goes higher than the normal distribution. Therefore, the peak of the curve is higher than the peak of the normal curve. This may be called *leptokurtic* (Fig 2.10). This type of Kurtosis may occur when a very large number of cases

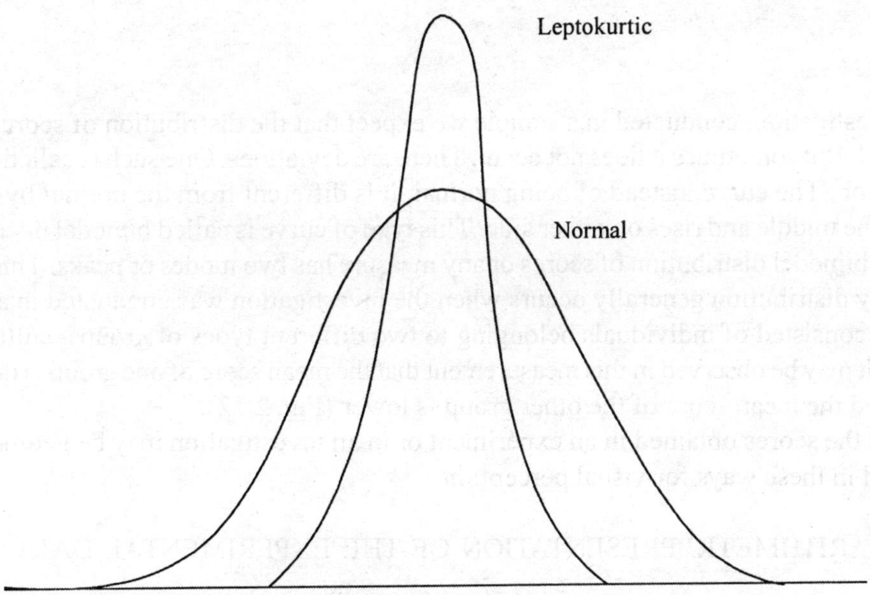

Fig. 2.10 : Kurtosis : Leptokurtosis

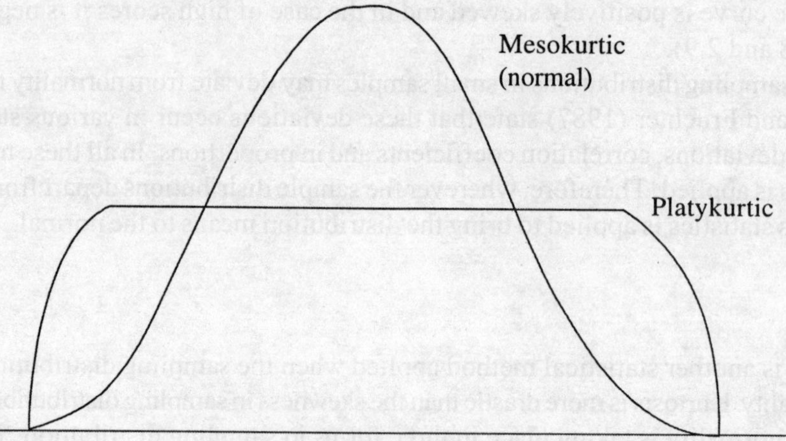

Fig. 2.11 : Kurtosis Distribution : Platykurtic and Mesokurtic

fall close to the mean and when only few cases fall above or below this point. Another type of occurrence is also possible. That is, the frequency distribution may yield a curve which is flat, flatter than the normal curve. This curve deviates more from the normal curve, which may be called *Platykurtic.* In this case a large number of cases fall above the mean and below the mean. Since the curve is flat the number of cases falling near the mean also are almost the same, or less. But if the distribution is normal the curve is called *Mesokurtic.* In mesokurtic there is no such deviation of cases in the frequency distribution (Fig. 2.11).

Bimodel

In an investigation, conducted in a sample we expect that the distribution of scores should be normal. But sometimes it does not occur. There are deviations. One such is called *bimodel* distribution. The curve instead of being normal, it is different from the normal by taking a steep in the middle and rises on either side. This type of curve is called bimodel distribution. Thus the bimodel distribution of scores or any measure has two modes or peaks. This type of frequency distribution generally occurs when the investigation was conducted in a sample which is consisted of individuals belonging to two different types of groups, culturally or socially. It may be observed in this measurement that the mean score of one group is decidedly higher and the mean score of the other group is lower (Fig. 2.12).

Thus the scores obtained in an experiment or in an investigation may be geometrically presented in these ways for visual perception.

ARITHMETIC PRESENTATION OF THE EXPERIMENTAL DATA

In the preceding section we have presented the experimental data geometrically. The data

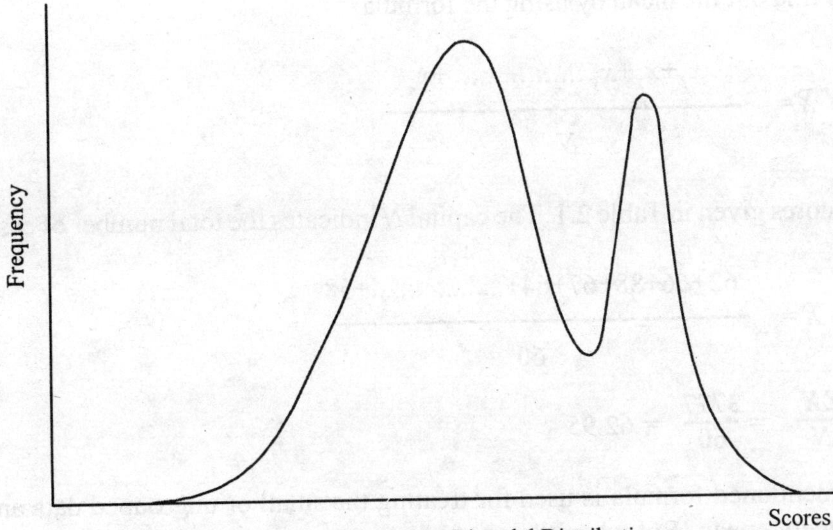

Fig. 2.12 : Bimodel Distribution

were illustrated with figures of various types. The same data we shall present here arithmetically. The important or the basic statistics are the means, medians and the modes. These three measures generally refer to the central values of the data. In other words they are called the central tendencies.

Mean

The mean is the average of the data or central value. The mean is a set of observations or scores. It is the arithmetic average and is the sum of observations or scores divided by the number of observations or cases. The mean is the most common measure of the central values. Means are the common measures. The mean is easily obtained by dividing the number of observations or cases if the number of cases are small. The mean is always indicated by the notation : $\bar{x}$. If the number of observations or scores are, x_1, x_2, x_3, x_4... etc., then their mean is :

$$\bar{X} = \frac{x_1 + x_2 + x_3 \ldots\ldots\ldots\ldots\ldots x_n}{N}$$

Using the subscript notation $\bar{X}$ is common in statistics and it is pronounced as X-bar ($\bar{X}$). The bar over the X indicates the mean of all the observations or scores.

This formula is also stated as $\frac{\Sigma X}{N} = M$

Σ indicates the total of all Xs.
N the total number of cases or observations.

Now, let us find out the mean by using the formula

$$\bar{X} = \frac{x_1 + x_2 + x_3 .... + x_n}{N}$$

for the data or scores given in Table 2.1. The capital N indicates the total number of cases or observations.

$$\bar{X} = \frac{62+66+88+67+64+..................+58}{60}$$

$$\frac{\Sigma X}{N} = \frac{3777}{60} = 62.95$$

The above mentioned formula is used for treating the small or ungrouped data and to obtain the arithmetic mean. But when the observations or scores are in large number we must group the data for finding out means. There is another formula to find out the arithmetic mean for the grouped data :

The formula is $\bar{X} = \frac{\Sigma f x_c}{N}$

Σ = indicates the sum
X = each score in the measurement
N = number of cases or scores
f = the frequency distribution, number of scores in an interval
c = the class interval in the grouping (mid-point).

Table 2.4 : Showing the mean of the grouped data

Class Interval	Mid-point of Class Interval	Frequency	fx_c
90–94	92	1	92
85–89	87	2	174
80–84	82	3	246
75–79	77	5	385
70–74	72	7	504
65–69	67	8	536
60–64	62	12	744
55–59	57	7	399
50–54	52	6	312
45–49	47	4	188
40–44	42	2	84
35–39	37	2	74
30–34	32	1	32
		N = 60	$\Sigma fx_c = 3770$

$$\text{Mean} = \frac{\Sigma f x_c}{N} = \frac{3770}{60} = 62.83$$

The data used here are the same as found in Table 2.3 and we treated them to find the mean from the grouped data. There may be a very small difference between the means obtained by the ungrouped method and grouped method. For the ungrouped data the mean is 62.95 and for the grouped data it is 62.83. This very negligible difference is due to the mid-points. There may be some scores above the mid-point and some below the mid-point and some may fall on the mid-point. This small difference may be occurring only in some cases and not in all cases.

The procedure of working out the class interval and the frequency distribution was already described in the preceding section. In finding out the arithmetic mean from the grouped data we are taking into consideration only three characteristics of the Table 2.4. The mid-point, the frequency, and the fx_c, multiplied sum of the mid-point by the frequencies. Each mid-point is multiplied by the corresponding frequencies and we get the sum fx_c for each class interval. Then the fx_c values are summed up and we got Σfx_c value, 3770 and this is divided by the number of cases or $N(60)$. Thus the obtained mean is $\frac{\Sigma fx_c}{N} = \frac{3770}{60} = 62.83$ which is almost the same as the ungrouped mean 62.95.

Median of the Measurement

The mean is not the only way of determining the centre of a distribution. There is another way of indicating the middle value in a distribution. Statistically it is called median. The median is one of the central tendencies in the measurement. It is the mid-point of the measurement. It is the mid-point when the scores are arranged in increasing order. In this arrangement we can find that half of the scores are smaller and other half of the scores are larger and there is a mid-point. This is the median of the distribution. This median is generally called as M in short. It is the middle score in a set of scores when arranged in an increasing order. Generally the median is fifty per cent of point in a frequency distribution. This means that half of the scores are above this mid-point and half of the scores below this mid-point.

In determining the median of the measurement we will have to arrange the scores in an increasing order. When the number of scores or the N is in odd numbers then there is always one number in the centre of the arranged list. This is the median, M. But if the N is in even numbers there is no one middle point. Instead there is a middle pair. In such cases the median is the average or mean of the middle pair, it is a point half way between the two even scores. The formula to find out the median in a distribution of scores is $(n+1)/2$.

Here n is the number of cases or scores. This formula can be applied when the scores are arranged in increasing order and when the scores are in even numbers. When there are large number of cases or scores then we can use the class interval and frequency distributions.

Table 2.5 : Computation of Median

Class Interval	Frequencies		Cumulative Frequencies
90-94	1		60
85-89	2		59
80-84	3		57
75-79	5		54
70-74	7		49
65-69	8	−26	42
60-64	12		34
55-59	7		22
50-54	6		15
45-49	4		9
40-44	2		5
35-39	2		3
30-34	1	−22	1
	N = 60		

26 cases above the class interval 60-64 where the median falls.

Median $64.5 - \frac{4}{12} \times 5 = 62.83$

22 cases below the class interval 60-64 where the median falls.

Median $59.5 + \frac{8}{12} \times 5 = 62.83$

In this experimental study there are 60 cases and the median is supposed to have 50 per cent of cases above the mid-point and 50 per cent of cases below the mid-point. Since the number of cases are 60, even numbered we cannot have the median exactly at the mid-point. Therefore, to locate the median we will have to work out through the class interval and frequencies. It is observed that 26 cases fall above the median and 22 cases below the median and the median falls in the class interval of 60 – 64 with 12 frequencies. The real values of 60-64 are 59.5 and 64.5. In the cases of upper measures we have only 26 and we need 4 more cases from the class interval 60-64 (59.5 – 64.5). The frequency is 12, and 4 cases we will have to take from 12. If we take the upper value, 64.5 we will have to deduct 4 cases from 12. Thus we have $64.5 - \frac{4}{12} \times 5 = 62.83$**. The product of** $64.5 - \frac{4}{12}$ **should be multiplied by 5 the value of class interval.**

In the case of lower measure we have only 22 cases and we need 30. Therefore, 8 cases should be taken from the class interval 60 – 64 (59.5 – 64.5) with the frequency of 12. Thus we get the median for the lower measuring $59.5 + \frac{8}{12} \times 5 = 62.83$**. Since the class interval has 5 cases the products should be multiplied by 5. In both upper and lower measures we have the median as 62.83. These two should be the same. Otherwise there is something wrong in the computation of the median. Here we have both the same value. Therefore, the**

median is correct.

The data sometimes may be in few numbers and the median for the same has to be ascertained. The principle for this is the same as for the grouped data. If there are 15 observations the median falls at the centre at 8 since the scores or observations are in odd number. But the observations or scores are even numbers, then the formula $(n+1)/2$ is applied. Before applying this formula the scores should be arranged in the increasing order from smallest to the largest.

For example there are 5 observations : 4, 7, 8, 5, 6 the first step is to arrange then in increasing order as : 4, 5, 6, 7, 8. The number of observations here is odd, having one number at the centre 6. The number six is having 2 observations on either side. Therefore, the median is 6.

But if the observations or the n is in even numbers, then there is no one middle observation. There are two numbers in the middle. In such case the formula $(n+1)/2$ should be used. For example : 4, 3, 9, 8, 6, 7, 5, 2.

These numbers should be arranged in an increasing order.

2, 3, 4, 5, 6, 7, 8, 9

In this list there are two middle numbers 5 and 6, having 3 numbers on either side. Therefore 5 and 6 should be added and divided by 2.

$$\frac{5+6}{2} = \frac{11}{2} = 5.5 \text{ M}$$

The median falls at 5.5 = between 5 and 6.

When we apply the formula $(n+1)/2$ to the first example :

There are 5 observations $(5+1)/2 = 6/2 = 3$. This means that the median falls on the 3rd position and therefore the median is 6. In the case of the second example when we apply the formula $(n+1)/2 = (8 + 1)/2 = 9/2 = 4.5$. This means that the median falls at 4.5th number. Fourth number is 5, .5 is between the numbers 5 and 6. *4.5 is the position on which the median falls* and itself is not median. Therefore the median is 5.5.

Mode

The mode is a statistic which is often estimated in a frequency distribution. In a frequency distribution it is the most often occurring score. It is any value that is occurring most often among the observations. Therefore, a mode may be defined as the point at which the maximum frequencies fall in a distribution of measurement. Though the mode is different from the mean and median, it is somewhat closer to them. The mean and the median may be one in a measurement, but sometimes there may be two modes. In a frequency distribution it is the highest point with more number of cases or scores.

In determining the mode in an ungrouped data it is easy if the scores are few or the sample is very small. We can pick out some and decide the mode if that particular number has occurred more number of times. But if the scores are many or the sample is large, then it becomes necessary to group the data. In grouping and in frequency distribution, the class interval which is having the highest frequency becomes the mode. If we draw a frequency polygon to the data the peak-point is the mode. The mode may fall quite close to the mean and to the median. Sometimes all the three may coincide in a distribution of the data. The mode may fall on the left or on the right of the mean. But the median always falls between the mean and the mode (Fig. 2.13). The estimation of mode can be illustrated by making use of the data give in Table 2.6.

Table 2.6 : Showing the determination mode

Class Interval	Frequencies	Midpoint
90-94	1	92
85-89	2	87
80-84	3	82
75-79	5	77
70-74	7	72
65-69	8	67
60-64	12	62
55-59	7	57
50-54	6	52
45-49	4	47
40-44	2	42
35-39	2	37
30-34	1	32
	N = 60	

The highest frequency is 12 which falls in the class interval of 60-64. The mid-point of the class interval 60-64 is 62 and this is the mode in this distribution. If a graph is drawn to this data, the curve is frequency polygon. The mode is the highest point, 12 in the curve falling in the class interval of 60-64.

Mean, Median and Mode

The mean, median and the mode are three central values generally determined in the measurements. In our illustration we find that the grouped arithmetic mean is 62.83, the median 62.83 and the mode is 62.00. These three central values do not differ very much, from each other. The values are almost the same (62.83, 62.83, 62.00, Fig. 2.13). The ungrouped arithmetic mean is 62.95.

Among these three central values the mean is more important than the other two values. In most of the measurements experimenters stop with mean and they do not go far median and mode. Generally the mean is used for determining the deviation values, specifically the standard deviation (SD) and the standard error of the mean and also for several other statistics.

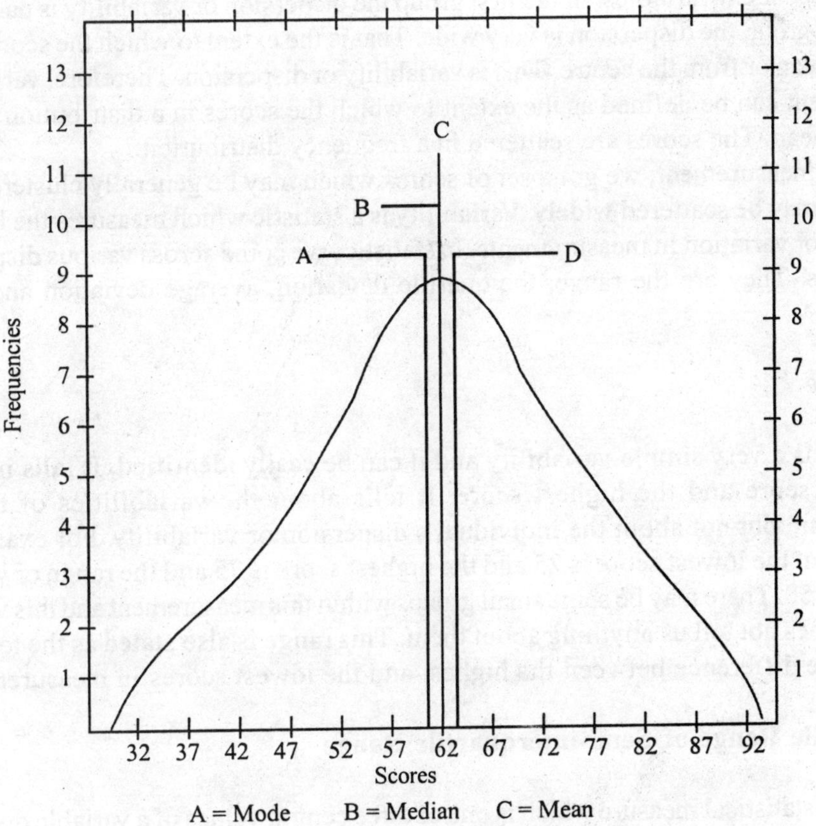

Fig. 2.13 : The Place of Three Central Tendencies, A Mode 62.00, B Median 62.83, C Mean 62.83, D Ungrouped Arithmetic Mean 62.95

MEASURES OF VARIABILITIES OR SPREAD OF SCORES

An experimenter or a researcher after completing his/her work gets the data in bulk. The data as they are have no meaning unless they are treated with suitable statistical methods. The data or scores may range from the lowest to the highest. There may be very wide range between the lowest to the highest. If the scores are few there is no problem. For example, there are two distributions *A* and *B*. *A* has the scores as 7, 12, 17, 18, 21 and *B* has the scores as 2, 6, 7, 20, 27. These two distributions have the same mean and also same median. But they are varying in several ways. In the first distribution the numbers are somewhat closer. But in the case of second distribution the numbers are scattered. There is more distance between the numbers than the numbers of the first distribution. The distribution *A* has less variability or dispersion than the second. The differences between the scores of the first distribution and the second distribution are due to the influence of independent variables in the measurements. Therefore, the variabilities or dispersions in the measurements are caused by the independent variables. In the first group the dispersion or variability is narrow, but in the second group the dispersion is very wide. That is the extent to which the scores disperse from the mean or from the centre. This is variability or dispersion. Therefore, variability in a measurement can be defined as the extent to which the scores in a distribution spread out from the mean. The scores are scattered in a frequency distribution.

In any measurement, we got a set of scores which may be generally clustered at some points and may be scattered widely. Variability is a statistic which measures the level of the dispersion or variation in measurements. In statistics we come across various dispersions or variabilities. They are the range, the quartile deviation, average deviation and standard deviation.

The Range

The Range is a very simple variability and it can be easily identified. It falls in between the lowest score and the highest score. It tells about the variabilities of the whole measurement, but not about the individual's dispersion or variability. For example, in a measurement the lowest score is 25 and the highest score is 75 and the range or variability is (75–25) = 50. There may be some small groups within this measurement and this variability, the range does not tell us anything about them. This range is also stated as the total range. Range is the difference between the highest and the lowest scores in measurement.

The Quartile Range of Semi-interquartile Range

Quartile is a statistical measure which is one of three central values of a variable dividing the distribution into four groups with equal frequencies. The qualities are always indicated by the letter Q. The semi-interquartile range is half of the range betwen 25th percentile and 75th percentile. The quartile range or deviation is the middle position falling at the 50th percentile between 25th and 75th percentiles. It is just like the median which divides the frequency distribution into two equal halves. The quartile deviation divides the frequency distribution into four equal parts. The quartiles stand for points in the dispersion or variability scale. The first deviation point is Q_1, the second deviation point is Q_2 and the third is Q_3. Thus there are four divisions.Quartile deviation is obtained by applying the formula, Q or $Q_2 = \frac{Q_3 - Q_1}{2}$

Since Q_2 is obtained by this formula by dividing the range between Q_3 and Q_1 into two, it is called interquartile range and the statistic which measures the variabilities is called semi-interquartile ranges the Q (Q_2). Let us illustrate this with an example using the data given in Table 2.3. If the scores are few we can arrange them in an increasing order from the lowest to highest and divide them into four equal parts. But if the range is very wider with many number of scores we must group them and work out the frequency distribution to find out the quartile ranges.

Table 2.7 : Showing the determination of the Quartile Range in a Distribution

Class Interval	Frequencies Distribution	
90-94	1	
85-89	2	
80-84	3	
75-79	5	
70-74	7	Q_3 lies in the class interval $69.5 + \frac{4}{7} \times 5$
		$= 69.5 + 2.86 = 72.36$
65-69	8	
60-64	12	
55-59	7	
50-54	6	Q_1 lies in this class interval $49.5 + \frac{6}{6} \times 5 = 54.5$
45-49	4	
40-44	2	
35-39	2	
30-34	1	
	N = 60	

Dividing the N by 4 we get $60 \div 4 = 15$. In the case of Q_1 we will have to find out where one-fourth of the cases, 15 fall. This is worked out by using the values of class interval and the frequencies. By counting the cases upward from the bottom, 15 cases fall at the class interval 50-54. According to the formula, $49.5 + \frac{6}{6} \times 5 = 54.5$, we get the first quartile, $Q_1 = 54.5$. In the case of Q_3 let us count 15 cases from top to downward. Fifteen cases fall in the class interval 70-74 with a frequency of 7. From the top to downward we get 11 cases and we need 4 more cases. We will have to get it from the frequency of 7 of the class interval 70-74. According to the formula, we have the value $69.5 + \frac{4}{7} \times 5 = 72.36$, Q_3 is 72.36. In this way we have worked out and obtained Q_1 and Q_3. Therefore $Q(Q_2)$ is

$$\frac{Q_3 - Q_1}{2} = \frac{72.36 - 54.50}{2} = \frac{17.86}{2} = 8.93$$ and the semi-interquartile range is 8.93.

Average Deviation

In every measurement we get the scores of the individuals and we work out the mean. The individual's scores may be either above the mean or below the mean and some may fall in the mean. If the individual's score is above the mean it is deviating from the mean positively and if the individual's Score is below the mean it is negatively deviating from the mean. Every score has a deviation positively or negatively. If we sum up all the deviations and divide it by the number of cases we get zero. The plus deviations cancel out the minus deviations. Let us illustrate this with an example. We have some individual scores as : 5, 7, 6, 4, 8, 9, 3. The formula to find out the average deviation (AD) is $\Sigma x / N$, where Σ stands for the sum, x is for the individual score and N is for number of cases.

$\frac{5+7+6+4+8+9+3}{7}=\frac{42}{7}=6.$ The mean of this distribution is 6. Now we can subtract the mean from each score to find out the differences. The differences we have : –1, +1, 0, – 2, +2, +3, –3. Let us now sum up all the deviations and divide it by 7.

$$(-1)+(+1)+0+(-2)+(+2)+(+3)+(-3)=\frac{-6+6}{7}=\frac{0}{7}=0.$$

The average deviation in this distribution is 0. The positive deviation +6 cancels out the negative deviation –6. Therefore the deviation is 0. This is in the case of ungrouped data.

The average deviation for grouped data also can be determined. The formula is $AD=\frac{\Sigma fx}{N}$, where Σ is the sum of scores, *f* is the frequency distribution and *N* is the number of cases.

Standard Deviation

Standard Deviation (*SD*) is a measure of variability in a frequency distribution. This is most frequently used measure in experiments and in researches. The standard deviation is considered to be one of the important and reliable measures in statisics. This is the indication of the degree of dispersion in a measurement and is often indicated by sign σ or *S*. The standard deviation would also indicate whether the sample chosen from a population is reliable or not. If several samples are taken from the same population the *SD* may be very small. If the population is heterogeneous the *SD* ma vary widely. Then there may be greater dispersion. The *SD* would also determine the mean, whether the mean is widely spread and also would indicate the range of scores. Standard deviation is always indicated by the square root of the variance as S^2. Let us illustrate it with an example in an ungrouped data.

The given data are : 245, 284, 270, 320, 310, 335, 350. The number of cases are 7. Let us sum up and divide it by 7 to get the mean.

$$\frac{245+284+270+320+310+335+350}{7}=\frac{2114}{7}=302$$

The mean of this distribution is 302.

The deviations are spread from the mean both positively and negatively.

Table 2.8 : Showing the Standard Deviation for an ungrouped data

Observations	Deviations from the Mean $x - \bar{x}$	Standard Deviations $(x - \bar{x})^2$
245	$245–302 = (-57)^2$	= 3249
284	$284–302 = (-18)^2$	= 324
270	$270–302 = (-32)^2$	= 1024
320	$320–302 = (+18)^2$	= 324
310	$310–302 = (+8)^2$	= 64
335	$335–302 = (+33)^2$	= 1089
350	$350–302 = (+48)^2$	= 2304
$\bar{X} = 302$	0.0	SUM = 8378

The formula to determine the S is = $\sqrt{\frac{\Sigma(\bar{x} - x)^2}{N}}$. The sum is divided by N–1. Therefore,

$$S^2 = \sqrt{\frac{\Sigma(X - \bar{X})^2}{(7-1)}} = \sqrt{\frac{\Sigma(X - \bar{X})^2}{6}} = \sqrt{\frac{8378}{6}}$$

$S^2 = \sqrt{\frac{8378}{6}} = \sqrt{1393}$. The standard deviation is the square root of the variance. Therefore

$S = \sqrt{1393} = 37.33$

In this ungrouped data the mean is : 302

The standard deviation S or σ is = 37.33

The properties of standard deviation are the mean $\bar{X}$ and the S which measure the spread of the scores from the mean. If S is 0 then there is no spread or deviation. This may occur rarely when all the scores are the same. Therefore, when the *SD* is 0 then there is no spread or deviation. It may be stated in other words as $S>0$. The S may become very large when the scores deviate or spread very widely from the mean.

Table 2.9 : Computation of standard deviation in a grouped data

Class Interval	Frequency (f)	X	fx	fx^2
90 – 94	1	+6	6	36
85 – 89	2	+5	10	50
80 – 84	3	+4	12	48
75 – 79	5	+3	15	45
70 – 74	7	+2	14	28
65 – 69	8	+1	8	8
60 – 64	12	0	0	0
55 – 59	7	– 1	– 7	7
50 – 54	6	– 2	– 12	24
45 – 49	4	– 3	– 12	36
40 – 44	2	– 4	– 8	32
35 – 39	2	– 5	– 10	50
30 – 34	1	– 6	– 6	36
	$N = 60$		$\Sigma fx = +10$	$\Sigma fx^2 = 400$

The standard deviation is strongly influenced by the extreme scores. This is true also with the mean. A few scores on the extremes can make the *S* very large and it may also increase the error in the mean.

Let us illustrate the standard deviation or spread or the dispersion in a grouped data. The scores as given in Table 2.3 are used here for working out the standard deviation (SD).

Let us now work out the standard deviation for a grouped data. The formula is : *SD*

$$\text{or} \quad \sigma = h\sqrt{\frac{\Sigma f(x)^2}{N} - \frac{(\Sigma fx)^2}{N}} -$$

Where h = number of cases in each interval

f = Frequency

x = deviation scores

N = Number of cases

Σ = the sum.

Let us apply this formula to the data.

$$\text{S.D.} = 5\sqrt{\frac{(400)}{60} - \frac{(10)^2}{60}} = 5\sqrt{6.67 - 1.67} = 5\sqrt{5.00} = 5 \times 2.24 = 11.20$$

The standard deviation of this distribution is = 11.20.

Standard deviation is a numerical measure of spread for a class of symmetrical distribution, the normal distribution. Numerical value of the centre and the spread provide specific facts about a distribution. It indicates the degree of dispersion or variability from the mean numerically. Standard deviation is involved in numerous statistical measurements. It is the average of the dispersion of scores from the mean. Guilford and Fruchter (1987) state that standard deviation may differ in small groups, which are less than 30 cases. But in large groups drawn from a population the standard deviations do not differ very much from each other.

VARIABLE, VARIANCE, VARIABILITY, DEVIATION AND DISPERSION

In psychological and educational studies while using the statistical methods we often come across the concepts, variable, variance, variability, deviation and dispersion. These concepts need certain explanation to avoid some confusion among the experimenters or the researchers.

Variable is commonly used in most of the psychological measurements. *Variable* means anything that can change the result and assume different numerical values. In other words, any response or behaviour which can take different degrees, or defects of various amounts caused by an independent variable in performance (dependent variable) which may be investigated.

Variance is the act of varying the quality, state or degrees of being divergent or discrepancy. It is an instance of diverging. Statistically it is a measure of dispersion, obtained by taking the mean of the squared deviation of the observed values from their mean in a frequency distribution.

Variability means thing that varies, anything changeable. In a measurement all the scores are not the same. Every score varies from the other score. The change or difference may take place due to several reasons and the changes also may be of different types. Therefore, variability generally refers to the qualities inherent in the scores for change or deviate.

Deviation is a general concept. In statistics it refers to the scores in a measurement going away from the mean. It may take place on either side of the mean depending upon the range of the scores and the extremities of the scores.

Dispersion generally means to break up and scatter away. The scores around the mean are clustered. But most of the scores break up and scatter away from the mean. This is statistically stated as dispersion. Standard deviation is the sum of the square of all the dispersion scores divided by the total number of cases and root of it.

The *SD* is always associated with the mean and from the mean the *SD* is derived. The standard deviation has mathematical relationship with the normal curve. The relationship between the normal curve and the standard deviation is indicated by the proportion of cases falling at various levels. The mathematical relationship indicates that in any normal curve about 68.2 per cent of cases fall between +1 and –1 standard deviation from the mean. It is observed that about 95 per cent of cases fall between +2 and –2 standard deviations from the mean and almost all the cases, about 99.8 per cent of cases fall between +3 and –3 standard deviations from the mean. For example, if an individual's score is one standard deviation above the mean. In our illustrations given above the mean score is 62.83 (Table 2.4) and the +SD is 11.20 (Table 2.9) respectively. Now we take the scores of two individuals, as 86 and 54 and interpret them. The mean score of the group is 62.83 and the SD is 11.20. The first student with his score 86 is 1 SD above the mean 62.83+11.20 = 74.03 (Table 2.9). The second student with the score 54 is 1 SD below $(\overline{X} = 1SD)$ the mean score 62.83 – 11.20 = 51.63

The standard deviation is an essential characteristics in a distribution and the individual's scores are often stated as in SD units, as one SD above or one SD below the mean score.

In statistical analysis the individuals are often reported in terms of SD units. Apart from this, the individual's standard scores are also worked out for various purposes. Suppose, for example, a student has scored 92. The group mean is 62.83 and the SD is 11.20. His position is 92.0 – 62.83 = 29.17. He has 29.17 points above the group mean. With 29.17 points, the student goes above the position which is equal to $\frac{29.17}{11.20} = 2.60$. This means that the student's score is 2.63 standard deviation above the mean. In the case of another student whose score is 54, his position is 62.93 – 54.00 = 8.83. His score is one *SD* below the mean. With 8.83 points, the student goes below the mean which is equal to $\frac{8.83}{11.20} = 0.788$. This means that the student's score is 0.788 standard deviation below the mean of the group. This is how the standard score is used to interpret the standard deviation in a distribution.

CORRELATION METHODS

Experimenters and researchers use various statistical methods to treat their data obtained from their experiments and studies. The application of the type of statistical method depends upon their problems, hypotheses and experimental designs. So far they use one independent variable in their studies and observe the reactions or effects there is no problem in their choice of statistical methods. We display the data of a single variable by histogram and summarise them numerically or by the mean and standard deviation. But if we have data of two variables we prepare graphs and work out numerical values that describe the relationship between the variables. The effective way of displaying the relation between two variables of numerical values is the scatter diagram. A scatter diagram shows the relationship between two variables measured on the same individuals when both variables have ordinal or interval scales. The values of one variable appear on the horizontal axis and the values of the other variable appear on the vertical axis in a scatter diagram. Each individual data appears as a point in the diagram.

The points at the scatter diagram show the direction, form and strength of the relationship between two variables. In scatter diagram what the experimenter expects is the straight line relationship between the two variables. A straight line relation between the two variables is strong for which the points should lie very close to a straight line. If the points are widely scattered then the straight line is wrecked and the relation between the variables may be zero or negative. A scatter diagram would display the level of association between two variables. The numerical measures can help us to describe the level of association. The formula used in scatter diagram to compute the correlation coefficient is

$$r_{XY} = \frac{\frac{(\Sigma fx'y')}{N} - \frac{(\Sigma fy')}{N}}{\sqrt{\left[\frac{\Sigma fx^2}{N} - \frac{(\Sigma fx)^2}{N}\right]\left[\frac{\Sigma fy^2}{N} - \frac{(\Sigma fy)^2}{N}\right]}}$$

The level of relationship or association between the measures of two variables is statistically called as coefficient of correlation. Psychologists are very much interested in finding out the level of association between the variables, they handle in their studies. It has become very important to know the relationship for predicting certain events unknown from the known. Thus there is one statistic involving both the variables and provides us a single value. This statistic is called correlation coefficient. This statistical method has provided with very wider scope for determining the psychological values. The relationship between variables is a matter of degree and the statistical index of this relationship is called coefficient of correlation. This is computed using the measurements of both the variables and it is generally indicated by a symbol *r*. This symbol stands for the Pearson product moment coefficient of correlation. This value provides us the extent to which two variables are strongly or linearly related. It is a very valuable statistic that is applied in a variety of ways.

Correlation coefficient is an index which expresses the degree of relationship between two sets of measures. It is always indicated in decimal fraction. Generally, a correlation is a measure of the extent or degree to which two variables are related.

The general meaning of correlation is that it is a mutual or reciprocal relationship. It is the degree of quantitative association between two variables or attributes. Correlation is positive or direct when variables move in the same direction. It is negative or adverse when they move in opposite direction.

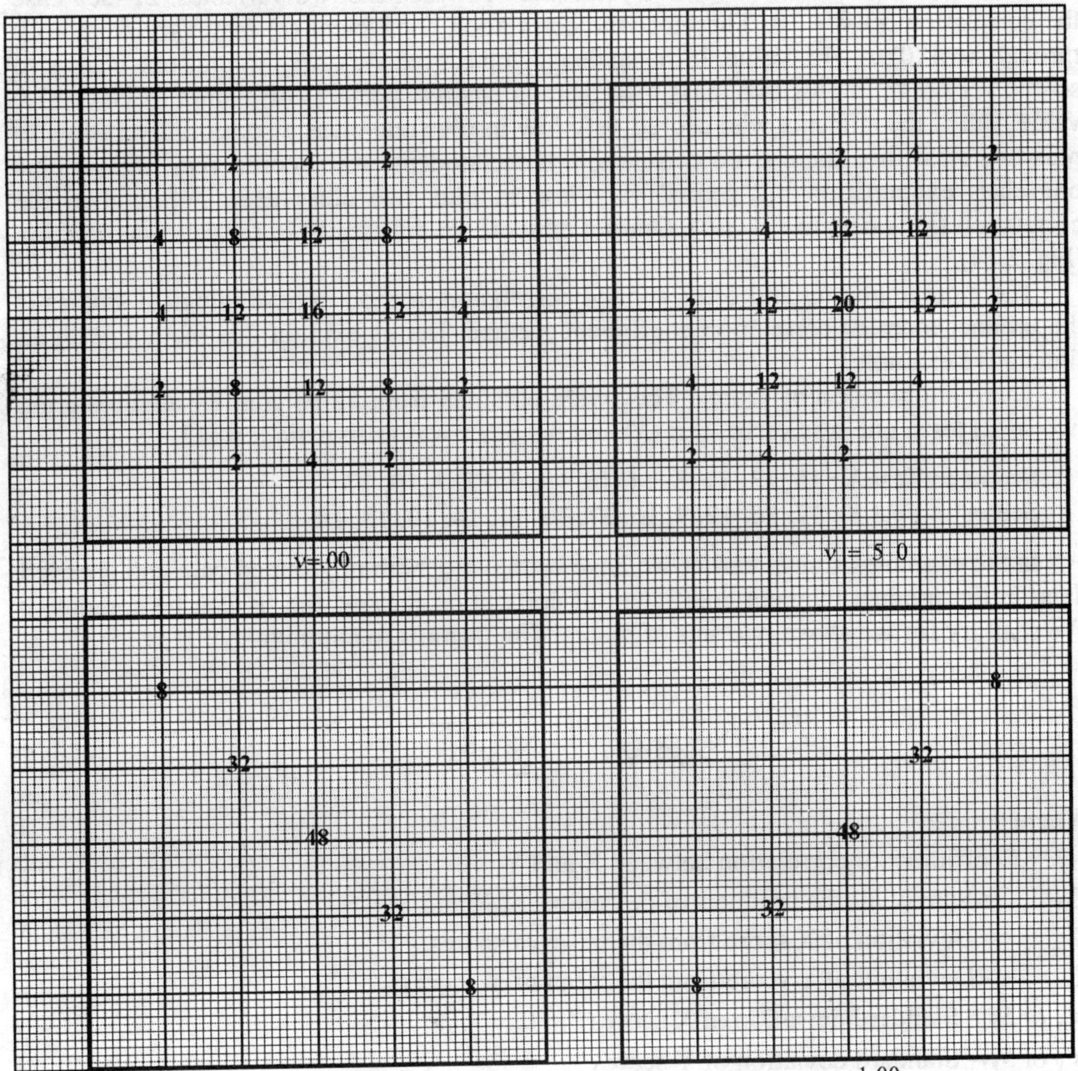

v = −1.00 v = +1.00

Fig. 2.14 : Graphical Presentation of Correlation

The scores of any two variables obtained from the same population or sample are generally used in computing the correlation coefficient. For example the average marks of all the subjects of a group of students, obtained in plus one class of the final examination and the average marks of all the subjects obtained by the same students in an examination when they were in plus two class were taken and the '*r*' was computed. The average marks of plus one class is one variable and the average marks of plus two is another variable. What type of association these two variables are having? It is tested in the correlation method. If the correlation coefficient is perfect and positive the value is +1.0. But if there is no perfect correlation then the value of '*r*' is very low or zero (0). But there is also a chance for a negative correlation between the two variables. In such case the correlation (*r*) is –1.0. Therefore the correlation coefficient (*r*) may vary from +1.0 through zero (0) to –1.0 value. If the students who scored very high in plus one class also scored very high in plus two class then there is positive perfect correlation to yield +1.0 value. But if the students who scored very low in plus one class also scored very low in plus two class then there is no correlation. If the students who scored very high in plus one class examination but scored very low in plus two class examination then the correlation (*r*) is negative, (–1.0). One set of score go drastically in the reverse direction (Fig. 2.14).

The most important and often applied correlation method in psychological studies is the product moment correlation coefficient method. Let us explain this method with an illustrative example.

There are two sets of scores. They were obtained from 12 persons by administering two small tests, one was verbal test (*X*) and the other was arithmetic test (*Y*).

Scores in Test *X* : 15, 13, 11, 8, 14, 16, 19, 9, 12, 7, 10, 17
Scores in Test *Y* : 17, 15, 12, 9, 16, 17, 20, 11, 14, 10, 13, 18

In the examples shown earlier we have been using both ungrouped data and grouped data. In the product moment coefficient of correlation method the '*r*' can be computed. We shall try for the above two sets of scores. The formula to compute the '*r*' is :

$$r_{XY} = \frac{\Sigma XY}{N\sigma_X\sigma_Y} \text{ (or) } \frac{\Sigma XY}{NS_X S_Y}$$

Where

r_{XY} = Coefficient of correlation between X and Y.

Σ_{XY} = Sum of all the products of deviations of X and Y.

X^I = Deviation of X score from the mean of scores.

Y^I = Deviation of Y score from the mean of Y scores.

σ_X or S_Y = Standard deviation of X scores

σ_X or S_Y = Standard deviation of Y scores

N = Total number of scores.

In computing the '*r*' we must first of all arrange the two sets of scores in columns, making sure of *X* and *Y* scores are together in pairs.

Table 2.10 : Showing the computation of product moment coefficient of correlation of two sets of scores obtained from 12 persons

S. No.	*X*	*Y*	*X*'	*Y*'	*X*²	*Y*²	*X'Y'*
	1	2	3	4	5	6	7
1	15	17	2.50	2.00	6.25	4.00	5.00
2	13	15	0.50	0.00	0.25	0.00	0.00
3	11	14	-1.50	-1.00	0.25	1.00	1.50
4	8	9	-4.50	-6.00	20.25	36.00	27.00
5	14	16	1.50	1.00	2.25	1.00	1.50
6	16	19	3.50	4.00	12.25	16.00	14.00
7	18	22	5.50	7.00	30.25	49.00	38.50
8	9	11	-3.50	-4.00	12.25	16.00	14.00
9	12	14	0.50	-1.00	0.25	1.00	0.50
10	7	10	-5.50	-5.00	30.25	25.00	27.50
11	10	13	-2.50	-2.00	6.25	4.00	5.00
12	17	20	4.50	5.00	20.25	25.00	22.50
Sum	150	180	0.00	0.00	141.00	178.00	157.00
					Σ X²	Σ Y²	XY
	Mean 12.50	15.00					

In columns 1 and 2 we have the raw scores of *X* and *Y* variables. The mean for both variables were worked out and given at the bottom. The raw scores of *X* variable are deducted from the mean and presented in column 3, and the raw scores of *Y* variable are deducted from the mean and presented in column 4. Columns 5 and 6 are the squares of the variables *X'* and *Y'*. Column 7 is the multiplied sum of the deviations given in columns 3 and 4.

The SD is worked out for both the variables *X* and *Y* by using the formula given here. The r_{XY} is computed by using the formula

$$r_{XY} = \frac{\Sigma XY}{NS_xS_y}$$

$$S_X = \sqrt{\frac{\Sigma x^2}{N}} = \sqrt{\frac{141}{12}} = \sqrt{11.75} = 3.42$$

$$S_Y = \sqrt{\frac{\Sigma y^2}{N}} = \sqrt{\frac{178}{12}} = \sqrt{14.83} = 3.85$$

$$r_{XY} = \frac{\Sigma xy}{NS_xS_y} = \frac{157}{12(3.42)(3.85)} = \frac{157}{12(13.17)} = \frac{157}{158.0} = 0.99$$

This method of product moment correlation usually is applied when the number of cases are few. This method requires standard deviation of both the variables. There is another alternative method of product moment coefficient of correlation. This method does not require the standard deviation of the variables. We can compute the correlation coefficient with only three elements. This is also applied to small samples.

The formula is : $r_{XY} = \frac{\Sigma xy}{\sqrt{(\Sigma x^2)(\Sigma y^2)}}$

This is applied to the above data.

$$r_{XY} = \frac{157}{\sqrt{(141)(178)}} = \frac{157}{158.4} = 0.99$$

We get the same value r as above 0.99. In this method the computation work is somewhat reduced. The Pearson product moment correlation method has another alternative method. This method can be used on small groups and also on somewhat larger groups. This does not require to work out the standard deviations for the scores. The correlation coefficient is computed from original scores. The formula is :

$$r_{XY} = \frac{N\Sigma XY - (\Sigma X)(\Sigma Y)}{\sqrt{\left[N\Sigma X^2 - (\Sigma X)^2\right]\left[(N\Sigma Y)^2 - (\Sigma Y)^2\right]}}$$

where

X = Original score of X variable.

Y = Original score of Y variable.

Σ = Sum of the score of the respective variable

N = Total number of cases or scores.

It would be more clear if we illustrate this method with an example. We have the data obtained from 15 students who have willingly come forward to take up the tests. One is general knowledge test and the other is arithmetic tests. The students who took the test were between 15 and 17 years of age. They were all belonging to the same cultural group. Two teachers conducted the tests and obtained the scores. The scores are given here. The maximum marks in the tests were 50.

Marks in General Knowledge test (X) : 45, 48, 39, 38, 39, 43, 46, 38, 36, 39, 37, 44, 36, 44, 25.

Marks in Arithmetic test (Y) : 38, 43, 36, 29, 33, 39, 48, 34, 30, 33, 42, 45, 40, 48, 31

Table 2.11 : Computing of correlation coefficient by Pearson's Product Moment Correlation method

S. No.	X	Y	X^2	Y^2	XY
1	2	3	4	5	6
1	45	38	2025	1444	1710
2	48	43	2304	1849	2064
3	39	36	1521	1296	1404
4	38	29	1444	841	1102
5	39	33	1521	1089	1287
6	43	39	1849	1521	1877
7	46	48	2116	2304	2208
8	38	34	1444	1156	1292
9	36	30	1296	900	1080
10	39	33	1521	1089	1287
11	37	42	1369	1764	1554
12	44	45	1936	2025	1980
13	36	40	1296	1600	1440
14	44	48	1934	2304	2112
15	25	31	1225	961	1085
Sum	597	569	24801	22143	23279
	ΣX	ΣY	ΣX^2	ΣY^2	ΣXY

First of all the scores of *X* and *Y* variables were arranged in columns 2 and 3 in pairs against the subjects who obtained the scores. Then the *X* and *Y* scores were squared and entered in columns 4 and 5. Thirdly, the *X* and *Y* variable scores were multiplied and presented in column 6. The values in columns were summed up and entered at the bottom. Now, we have five measures $\Sigma X, \Sigma Y, \Sigma X^2, \Sigma Y^2, \Sigma XY$.

Applying the formula to the data for obtaining the correlation coefficient

$$r_{XY} = \frac{N\Sigma XY - (\Sigma X)(\Sigma Y)}{\sqrt{[N\Sigma X^2 - (\Sigma X)^2][(N\Sigma Y^2 - (\Sigma Y)^2]}}$$

$$r_{XY} = \frac{15 \times 23279 - (597)(569)}{\sqrt{[15 \times 24801 - (597)^2][15 \times 22143 - (569)^2]}}$$

$$r_{XY} = \frac{349185 - 339693}{\sqrt{[372015 - 356409][332145 - 323761]}}$$

$$r_{XY} = \frac{9492}{\sqrt{[15606][8384]}} = \frac{9492}{\sqrt{130840704}}$$

$$r_{XY} = \frac{9492}{11438} = 0.829 = 0.83$$

The correlation coefficient between the two variables is 0.83. It is highly positive correlation. When the scores of *X* variable increased, the scores in *Y* variable also were increased. But to what extent this correlation is significant ? We shall discuss this in the later section.

Scatter Diagram

We have already referred to the scatter diagram correlation method. We have shown only the simple method of plotting the scores in a diagram and also stated the nature of positive correlation and negative correlation. But using a scatter diagram, plotting the *X* and *Y* variables scores and computing the correlation coefficient is a laborious task. Therefore, it is not shown here. The experimenter or researcher who is interested in it can refer to the authors, J.P. Guilford and B. Fruchter (1987), Fundamental Statistics in Psychology and Education or Thorndike and E. Hagan (1961) Measurement in Psychology and Education or M. Rajamanickam (2001) Statistical Method in Psychological and Educational Research.

Spearman's Rank Order Correlation Method

In the preceding section we have been discussing about the Pearson product moment coefficient of correlation method. This has been very widely applied in experiments, researches and in classroom works. Pearson product movement correlation method is applied in psychological and educational measurements mostly. When there was necessity to compare measurement of two variables, it is applied to the measurements of small samples and also large samples. But in certain measurements product moment correlation method is not suitable. There are some other methods of correlation, applied in certain specific cases. One of the important correlation methods very often applied is the Spearman's (1904) rank different correlation method.

Spearman's rank difference method of correlation is applied mostly in qualitative assessments. The variables used in the measurements are mostly qualitative in nature. This method disregards the values or scores in numerical form. The method converts them into ranks. This is very convenient method for small samples below 30 in number. Generally in our everyday life we often evaluate things and persons only qualitatively, on the basis of their quality or performance or efficiency. While comparing between two or more articles we decide by stating, "this is best", this is second and so on. In assessing the performance of persons or workers in industry or in any work, we decide, "it is first, it is second" and so on. In international matches and games we see often that ranks are assigned to players. Particularly in cricket, and tennis, ranks are assigned to top ten players every year. In schools and colleges students are assessed on the basis of the marks obtained in examinations. Thus the rank order is applied in most of the life aspects.

Spearman's rank order method is very much useful for these purposes. Pearsons are placed in ranks on the basis of skills, talents and efficiencies. In Spearman's rank order method the data is presented in ordinal values as 1st, 2nd, 3rd and so on. Then these ordinal values are converted into ordinal numbers as 1, 2, 3 etc. It is also called rank difference correlation method. Let us illustrate this method with an example.

Two sets of scores are required for this. But the scores are not used in computing the correlation. Ranks are given to them by arranging them from highest (top) to lowest (bottom). Then ranks are given to each score on the basis of its place among the scores. Let us use the same data as given in Table 2.11 for this purpose. The data were obtained from 15 students.

Test X: 45, 48, 39, 38, 39, 43, 46, 38, 36, 39, 37, 44, 36, 45, 35

Test Y: 38, 43, 36, 29, 33, 39, 48, 34, 30, 33, 42, 45, 40, 48, 31

In assigning ranks to the scores there may be two or three scores of the same values. This is one problem the experimenter may find in this method. But it is very simple and can be easily solved. Among the two with the same value which one should be first and which one should be the second. In the case of 3 scores having the same value which one should be first and which one should be second and which one should be third.

At this stage we must assume that one of them should be first and the other second. Similarly in the case of three scores of the same value we must assume that anyone of the three should be first and the rest as second and third. But who is to be first or which score. In the case of three who is to be first or last among the three? This is solved by giving the ranks serially and dividing the sum ranks by 2. In the case of 3 giving ranks serially and dividing the sum of them by three.

Table 2.12 : Spearman's Rank Difference Correlation Coefficient between two sets of scores (N=15)

S. No.	Score in Test X	Score in Test Y	R (x)	R (y)	Diff. (D)	D^2
1	2	3	4	5	6	7
1	45	38	3.5	8.5	5.0	25.00
2	48	43	1	4	3.0	9.00
3	39	36	8	10	2.0	4.00
4	38	29	10.5	15	4.5	20.25
5	39	33	8	12	4.0	16.00
6	43	39	6	7	1.0	1.00
7	46	48	2	1.5	0.5	0.25
8	38	34	10.5	11	0.5	0.25
9	36	30	13.5	14	0.5	0.25
10	39	38	8	8.5	0.5	0.25
11	37	42	12	5	7.0	49.00
12	44	45	5	3	2.0	4.00
13	36	40	13.5	6	7.5	56.25
14	45	48	3.5	1.5	2.0	4.00
15	35	31	15	13	2.0	2.00
						191.50

In the computation of Spearman rank difference correlation procedure it is necessary to convert the scores into ranks. It is illustrated in Table 2.12. In the conversion of scores into ranks we find that in four places in the first test the same scores repeated, 45, 38, 36 and 39. There are two 45, two 38, two 36, scores and the score 39 is repeated in three places. The scores of two 45 get the rank of 3 and 4. These two ranks are summed up as 7 and divided by two and we get the ranks 3.5 and 3.5. In the case of two 38 scores they get the ranks 10 and 11 and the total 21, is divided by two and we get the ranks 10.5 and 10.5. The other two scores 36 appear in two places and get the ranks of 13 and 14 and the total 27, is divided by two and the ranks are 13.5 and 13.5. But the other score 39 appears in three places and the ranks assigned to them are 7, 8 and 9. The total of these three is 24 and it is divided by 3 and we get the ranks as 8, 8, 8. The same procedure is followed in the case of the scores of test Y. This is how the scores are converted into ranks for both the tests. The difference between the ranks of X and Y tests were worked out (D) and the same were squared, and summed up ΣD^2, and entered at the bottom. The sum of D^2 in our illustration is 191.5.

The formula to compute the coefficient of correlation according to Spearman is :

$$r_{XY} = 1 - \frac{6\Sigma D^2}{N(N^2-1)} = \frac{6(191.5)}{15(15^2-1)}$$

$$r_{XY} = 1 - \frac{1149.0}{15(225-1)} = \frac{1149.0}{15(224)}$$

$$= 1 - \frac{1149.0}{3360} = 0.343$$

$$\rho = r_{XY} = 1 - 0.343 = 0.657 = 0.66$$

According to Spearman's rank difference correlation coefficient method the two variables *X* and *Y* associate at 0.66 level. It is a positive correlation. Spearman's correlation coefficient is stated as ρ. There is not much difference between Pearson's product moment correlation coefficient (0.69) and the Spearman's rank difference correlation coefficient (0.66). But Spearman's correlation coefficient is little less than Pearson's correlation coefficient. There is no formula to test the level of confidence for the Spearman's correlation.

There are also other correlation methods applied in psychological and educational studies. Apart from the two correlation methods we have discussed above, there are Kendall's rank correlation coefficient, contingency coefficient of correlation. Phi-coefficient or four fold point correlation, biserial correlation, point-biserial correlation and tetrachoric correlation.

ANALYSIS OF VARIANCE AND F RATIO

Analysis of variance is a statistical method often applied in these days for the experimental data by the experimenters and researchers. But experimenters invariably were using the '*t*' test in their treatments to the data when their sample is small and when there were only two groups. In finding out the difference between two groups in their experiments they were testing only a single mean difference. If there were more than two groups or more and if the samples were large then the '*t*' test method cannot be used. The analysis of variance is the method applicable when there were more number of groups and more number of mean differences. It is also possible to compare more group mean in '*t*' test. For example if there were 3 groups, we can compare group 1 with group 2, group 1 with group 3, and group 2 with group 3. But when there were five groups or more it is not possible to use '*t*' test or the critical ratio. Because the member of comparisons become $\frac{5(5-1)}{2}=\frac{20}{2}=10$. There will be 10 comparisons for five groups.

When there were 9 groups in the investigation then the number of comparisons between groups would be $\frac{9(9-1)}{2}=\frac{9\times 8}{2}=\frac{72}{2}=36$ comparisons. To work out 36 comparisons between groups in this context is very unscientific. Therefore, the analysis of variance is preferred. Whether the groups involved in the measurement differ from each other or whether there is agreement between groups is tested in applying analysis of variance. In short analysis of variance is a multi-group analysis. It mainly deals with the mean scores of the groups and also the variances.

Analysis of variance is a statistical method of analyzing the total variance involved in an experimental data into components. It is by computing the F-ratio between the components included in the experimental data, various empirical problems or questions can be solved or answered. It determines whether three or more groups included in an experiment significantly differ or not. Generally, in analysis of variance the total sum of squares of the sample is analysed into groups of two, or three or four or any number of groups. Analysis of variance, as it is, is a single composite test which can be used to test simultaneously the whole sample distribution of a measurement and explain the characteristics of the sample. The data might have been obtained from 3 or 4 or 5 samples. Therefore, the analysis of variance tries to ascertain whether the variances from various samples are identical or different or heterogeneous. Analysis of variance also tries to test several samples having the same characteristics and concludes whether there is any difference among or between the means of those groups and also determines whether the means of the groups differ from the population mean.

The simplest application of analysis of variance is to test two means obtained from two groups whether these two groups significantly differ or identical. But it involves more computational labour. It is better to apply the '*t*' test in the case of two means or groups. But the '*t*' test can be conveniently used for two or three means or groups and not more than that. Therefore, in the case of more than three means or groups analysis of variance is safe and also reliable.

In analysis of variance there are two important measures of elements. They are one, sum of squares (based upon the variation) within the groups. There may be three, or four or five or any number of group means. The other is the sum of squares based upon the variation among or between group means. The sum of squares of these two values is the estimation of the total population variance. Here between groups refers to the number of groups included in the measurement and within groups refers to the number of cases or participants included in all the groups or the total numbe of subjects of the study. The formula used to determine the degrees of freedom for between groups is $r - 1$, and for within groups is $N - 1$. (r is for the number of groups and N is for total number of cases or participants).

The total sum of squares is analysed into two values, such as the squares between groups or among groups. When there are two means we state it as between groups. When there are three and more groups we state it as among groups. The second value is stated as sum of sequence within the groups. If the mean difference among several groups are large, then the sum of squares among the groups is also large. But, if the mean difference among the groups are small then the sum of squares among the groups is also small. If the individuals were treated alike the groups differ considerably. Then the sum of squares within the group is large. But if the individual group variances are large, then the sum of the squares within the group also is large. The larger the sum of squares among the groups and the smaller the sum of squares (occurred) within groups, if occurred in the measurement, then there is more likelihood that the means of the groups differ significantly. It is expected that the sum of squares or variances between or among groups is always larger than the sum of squares or variances within groups. The two estimates, among the groups and within groups may vary within the sample limits as every individual score varies from every other individual. This results in the mean score differences within groups and also among the groups. With these estimates the null hypothesis is tested. The 'F' ratio can be computed by dividing the larger variance by the smaller variance (within groups). Let us illustrate this with an experimental data.

A small problem solving test was conducted among 40 higher secondary students studying in a semi-urban school. They were randomly chosen for this test. Two teachers were assigned the task of conducting the test and collect data. There were four groups of students, ten boys and ten girls from urban areas, ten boys and ten girls from rural areas.

The null hypothesis was : There is no difference among the mean scores of the urban male and female students and rural male and female students in the problem solving test.

The students were of the same age group drawn from varied cultural areas. Analysis of variance was applied to test this hypothesis. The means of any number of groups can be tested with this statistics and it also tests the homogeneity of the sample.

Table 2.13 : Score obtained by four groups in a Problem Solving Test

	Group A		Group B		Group C		Group D	
S. No.	X	X^2	X	X^2	X	X^2	X	X^2
1	12	144	14	196	23	529	23	529
2	12	144	14	196	18	324	24	576
3	9	81	15	225	18	324	22	484
4	12	144	16	256	18	324	22	484
5	15	225	15	225	18	324	24	576
6	13	169	14	196	18	324	22	484
7	12	144	14	196	18	324	19	361
8	12	144	15	225	18	324	22	484
9	13	169	14	196	19	366	23	529
10	15	225	18	324	20	400	24	576
11	14	196	17	289	23	529	23	529
12	15	225	18	324	18	324	22	484
13	14	196	17	289	16	256	23	529
14	15	225	18	324	19	361	24	576
15	17	289	17	289	22	484	23	529
Σx	**200**		**236**		**286**		**340**	
Σx^2		**2720**		**3750**		**5512**		**7730**
N	**15**		**15**		**15**		**15**	
$\bar{x}$	**13.3**		**15.7**		**19.06**		**22.67**	

Let us now apply the equation first to the data for obtaining the total sum of squares

$$\Sigma x_1^2 + \Sigma x_2^2 + \Sigma x_3^2 + \Sigma x_4^2 - \frac{(\Sigma x_1 + \Sigma x_2 + \Sigma x_3 + \Sigma x_4)^2}{N}$$

$$2720+3750+5512+7730-\frac{(200+236+286+340)^2}{60}$$

$$19712-\frac{(1062)^2}{60}=19712-\frac{1127844}{60}=19712-18797.4=914.6$$

The second equation is to compute the sum of squares among/between groups. It is :

$$SS = \frac{(\Sigma x_1)^2}{n_1} + \frac{(\Sigma x_2)^2}{n_2} + \frac{(\Sigma x_3)^2}{n_3} + \frac{(\Sigma x_4)^2}{n_4} - \frac{(\Sigma x_1 + \Sigma x_2 + \Sigma x_3 + \Sigma x_4)^2}{N}$$

$$= \frac{(200)^2}{15} + \frac{(236)^2}{15} + \frac{(286)^2}{15} + \frac{(340)^2}{15} - \frac{(200+236+286+340)^2}{60}$$

$$= \frac{40000}{15} + \frac{55696}{15} + \frac{81796}{15} + \frac{115600}{15} - \frac{(1062)^2}{60}$$

$$= \frac{40000}{15} + \frac{55696}{15} + \frac{81796}{15} + \frac{115600}{15} - \frac{1127844}{60}$$

$$= 2666.6 + 3713.07 + 5453.07 + 7706.7 - 18797.4$$

$$= 19539.44 - 18797.4 = 742.04$$

In the third stage, sum of squares within the groups is computed. It is by subtracting the SS among groups from the total sum of squares : 914.6 – 742.04 = 172.56

Table 2.14 : Summary of the sums of squares of the analysis of variance for four groups

Source of Variation	Sum of Squares	Degrees of Freedom	Mean Squares	F-Ratio	Level of Significance
Among Groups	742.04	3	247.35	80.33	.001
Within Groups	172.56	56	3.08		
Total SS	914.60	59			

The next step is to find out the degrees of freedom for three sources of variations.

The formula to find out the degrees of freedom for the total sum of squares is : $N-1$

The formula to find out the degrees of freedom for the sum of squares among groups is : $r-1$

The formula to find out the degrees of freedom for the sum of squares within groups is : $N-r$

Where N stands for the total participants, r for number of groups. Thus we have the degrees of freedom for the total SS = 60 – 1= 59

Degrees of freedom for among/between groups = 4 – 1 = 3

Degrees of freedom within groups = 60 – 4 = 56

Therefore, the degrees of freedom among the groups and within groups is = 56 + 3 = 59

The F-ratio is obtained by dividing the mean squares among groups 247.35 by the mean squares of within groups 3.08. It is $247.35 \div 3 = 80.31$. The F-ratio 80.31 is significant or not? (The experimenter or the researcher may refer to the F-ratio Table in any psychological statistics book). This can be tested by referring to the F-ratio Table, whether it is significant at 5 per cent (0.05) level or at one per cent (0.01) level or at 0.001 level. We can find out the Table figure for 3-59 degrees of freedom. The number 3 is the numerator goes from left to right in the F-ratio Table, and 59 is the denominator goes from top to bottom in the Table. Against the degrees of freedom 3-59, we can see that there are three F-ratios, 2.76, 4.13 and 6.60. These table figures indicate the levels of significance. Since there is no degrees of freedom for 59 we can take it as 60. Hence 3.60 degrees of freedom in the column indicates 2.36 as 5 per cent significant. Since our F-ratio is 80.31 which is much greater than the Table value (6.60). We can conclude that the F-ratio is significant at 0.001 level.

This means that the four groups included in this experiment differ among them at 0.001 level. Every group varies significantly from every other group in the performance.

Having known by the application of analysis of variance that the four groups included in this experiment differ significantly, we must know to what extent these four groups differ. If the sample for all the four groups are drawn randomly from the same population there would not be so much differences among the groups. The analysis of variance is also a test of the sample. It is also possible to state whether the samples of the four groups were drawn randomly from a homogeneous population or heterogeneous population. In this case, it was from a heterogeneous population. The groups were randomly drawn from different sources, mainly varied environmental and socio-economic statuses. It is because that the mean variances among the groups are greater while the mean variances within groups are smaller. The individual differences within the groups also are smaller. The dispersion of scores within the groups is very narrow. Since the differences among the groups are highly significant the null hypotheses is rejected.

Computation of Critical Ratio or '*t*' Test

But the extensity of difference between groups are not known. The analysis of variance cannot tell us this. Therefore, we must go for some other statistical technique to find out the level of differences between the groups. The '*t*' test or the critical ratio are the two techniques suitable for this purpose. Since there are 4 groups in this experiment and each group is different from the other groups, we will have to compare the means between any two groups. Then, we have to make $\frac{N(N-1)}{2}=\frac{4(4-1)}{2}=\frac{4\times3}{2}=\frac{12}{2}=6$ comparisons.

The '*t*' requires the values of M_1 and M_2 the respective SD_1 and SD_2 standard error of the mean, standard error of the mean difference (SEMD) and also the *N*. Therefore, we will have to compute the SDs for all the four groups from the original measurement and also the SEMs. The formula for SD is

$$\text{SD or } \sigma \text{ or } S=\frac{1}{N}\sqrt{N\Sigma X^2-(\Sigma X)^2}$$

In this method the standard deviation is computed without the knowledge of deviations from the original and ungrouped data.

Where N = Number of cases

X = Original score from the measurement

Σ = Summing up

The same scores given in Table 2.13 can be used here for working out the means and standard deviations.

Table 2.15 : Showing the scores of four groups of students obtained in a Test

S. No.	Group A		Group B		Group C		Group D	
	X	X^2	X	X^2	X	X^2	X	X^2
1	12	144	14	196	23	529	23	529
2	12	144	14	196	18	324	24	576
3	9	81	15	225	18	324	22	484
4	12	144	16	256	18	324	22	484
5	15	225	15	225	18	324	24	576
6	13	169	14	196	18	324	22	484
7	12	144	14	196	18	324	19	361
8	12	144	15	225	18	324	22	484
9	13	169	14	196	19	361	23	529
10	15	225	18	324	20	400	24	576
11	14	196	17	289	23	529	23	529
12	15	225	18	324	18	324	22	484
13	14	196	17	289	16	256	23	529
14	15	225	18	324	19	361	24	576
15	17	289	17	289	22	484	23	529
Σ	200	2720	236	3750	286	5512	340	7730
N	15		15		15		15	
$\overline{X}$	13.3		15.7		19.06		22.67	

$$SD_1 = \frac{1}{15}\sqrt{15(2720)-(200)^2} = \frac{1}{15}\sqrt{40800-40000} = \frac{1}{15}\sqrt{800} = \frac{28.25}{15} = 1.89$$

$$SD_2 = \frac{1}{15}\sqrt{15(3750)-(286)^2} = \frac{1}{15}\sqrt{56250-55696} = \frac{1}{15}\sqrt{554} = \frac{23.54}{15} = 1.57$$

$$SD_3 = \frac{1}{15}\sqrt{15(5512)-(286)^2} = \frac{1}{15}\sqrt{82680-81796} = \frac{1}{15}\sqrt{884} = \frac{29.73}{15} = 1.98$$

$$SD_4 = \frac{1}{15}\sqrt{15(7730)-(340)^2} = \frac{1}{15}\sqrt{115950-115600} = \frac{1}{15}\sqrt{350} = \frac{18.71}{15} = 1.25$$

Two important values required for our purpose are the standard errors of the means and the standard error of the mean differences. The standard error of the mean can be estimated from the known statistics. The formula is :

$$SEM = \frac{SD}{\sqrt{N-1}}$$

The standard deviation we have already estimated from the original measurements for four groups. The N is the same for all four groups. Applying the formula to the data we have the standard error of the mean for all four groups.

$$\text{Group A : SEM} = \frac{SD}{\sqrt{N-1}} = \frac{1.89}{\sqrt{15-1}} = \frac{1.89}{\sqrt{14}} = \frac{1.89}{3.74} = 0.505$$

$$\text{Group B : SEM} = \frac{SD}{\sqrt{N-1}} = \frac{1.57}{\sqrt{15-1}} = \frac{1.57}{\sqrt{14}} = \frac{1.57}{3.74} = 0.42$$

$$\text{Group C : SEM} = \frac{SD}{\sqrt{N-1}} = \frac{1.98}{\sqrt{15-1}} = \frac{1.98}{\sqrt{14}} = \frac{1.98}{3.74} = 0.53$$

$$\text{Group D : SEM} = \frac{SD}{\sqrt{N-1}} = \frac{1.25}{\sqrt{15-1}} = \frac{1.25}{\sqrt{14}} = \frac{1.25}{3.74} = 0.33$$

We shall estimate now the standard error of difference between means. The formula is

$$\text{SEM}_{\text{Diff}} = \sqrt{\text{SEM}_1^2 + \text{SEM}_2^2}$$

Where SEM_1 = Standard Error of the first mean
SEM_2 = Standard Error of the second mean

Applying the formula to the data we can work out six SEM_{Diff} as six comparisons are to be made.

$$\text{Groups A \& B : SEM}_{\text{Diff}} = \sqrt{.505^2 + .42^2} = \sqrt{.255 + .176} = \sqrt{.431} = .656$$

$$\text{Groups A \& C : SEM}_{\text{Diff}} = \sqrt{.505^2 + .53^2} = \sqrt{.255 + .281} = \sqrt{.536} = .732$$

$$\text{Groups A \& D : SEM}_{\text{Diff}} = \sqrt{.505^2 + .33^2} = \sqrt{.255 + .109} = \sqrt{.364} = .603$$

$$\text{Groups B \& C : SEM}_{\text{Diff}} = \sqrt{.42^2 + .53^2} = \sqrt{.176 + .281} = \sqrt{.457} = .676$$

$$\text{Groups B \& D : SEM}_{\text{Diff}} = \sqrt{.42^2 + .33^2} = \sqrt{.176 + .109} = \sqrt{.285} = .534$$

$$\text{Groups C \& D : SEM}_{\text{Diff}} = \sqrt{.53^2 + .32^2} = \sqrt{.281 + .109} = \sqrt{.390} = .624$$

Having obtained the essential values for estimating the '*t*' ratio or CR, now we shall apply the formula for the same.

$$t \text{ or CR} = \frac{M_1 - M_2}{\text{SEM}_{\text{Diff}}}$$

There are six comparisons between the means of the groups. They are groups A and B, A and C, A and D, B and C, B and D and C and D.

$$\text{'}t\text{' for A and B} = \frac{M_1 - M_2}{\text{SEM}_{\text{Diff}}} = \frac{15.7 - 13.3}{.656} = \frac{2.4}{.656} = 3.66$$

$$\text{'}t\text{' for A and C} = \frac{M_1 - M_2}{\text{SEM}_{\text{Diff}}} = \frac{19.6 - 13.3}{.732} = \frac{6.3}{.732} = 8.61$$

$$\text{'}t\text{' for A and D} = \frac{M_1 - M_2}{\text{SEM}_{\text{Diff}}} = \frac{22.67 - 13.3}{.603} = \frac{9.37}{.603} = 15.53$$

$$\text{'}t\text{' for B and C} = \frac{M_1 - M_2}{\text{SEM}_{\text{Diff}}} = \frac{19.06-15.7}{.673} = \frac{3.36}{.673} = 4.99$$

$$\text{'}t\text{' for B and D} = \frac{M_1 - M_2}{\text{SEM}_{\text{Diff}}} = \frac{22.67-15.7}{.534} = \frac{7.17}{.554} = 11.31$$

$$\text{'}t\text{' for C and D} = \frac{M_1 - M_2}{\text{SEM}_{\text{Diff}}} = \frac{22.67-19.06}{.624} = \frac{3.68}{.624} = 5.89$$

Where M_1 is the mean of the first group.
M_2 is the mean of the second group.

The means of M_1 and M_2 are not fixed. Any one of means becomes M_1 and any of means becomes M_2, depending upon the size. The greater mean is M_1 and the smaller mean is M_2. We can observe that the '*t*' values seem to be considerably higher, higher than the Table values 1.96 (5 per cent significant) and 2.58 (one per cent significant).

Let us present the results in a Table for clear observation and understanding.

Table 2.16 : Showing scores and the '*t*' values of the four groups comparing the mean scores between any two groups

Groups	*N*	Mean	S.D.	SEM	$M_1 - M_2$	SEM_{Diff}	'*t*' ratio	Level of Significance
A	15	13.3	1.89	.505	AB:2.14	0.656	3.66	.01
B	15	15.7	1.57	.42	AC:6.30	0.732	8.61	.001
C	15	19.06	1.98	.53	AD:9.37	0.603	15.53	.001
					BC:3.36	0.673	4.99	.001
D	15	22.67	1.25	.33	BD:7.17	0.534	11.31	.001
					CD:3.68	0.624	5.89	.001

There is an alternative method for estimating the '*t*' ratio or CR.

$$\text{It is : } t = \frac{M_1 - M_2}{\sqrt{\frac{SD_1^2}{N_1} + \frac{SD_2^2}{N_2}}}$$

This method does not require the standard error of the mean difference. But this method is somewhat cumbersome and also the '*t*' value may vary from the first method to a small extent.

The analysis of variance indicated that all four groups are heterogeneous groups. The F-ratio was highly significant. On the basis of this it is tried to see to what extent the difference between groups occur. It is tested with the '*t*' test and the results are presented in Table 2.16. It is observed that only between groups *A* and *B* the difference is at one per cent level ($p < .01$). In all other cases the differences between the groups are highly significant at .001 level. This means that even if the investigation is conducted 1000 times among these groups the difference may exist for 999 times. Therefore, the null hypothesis is rejected.

Chi Square

Experimental data in the fields of psychology and education are treated with various types of statistical methods for obtaining accurate and reliable results. All those methods we have discussed in the preceding sections were intended to serve some limited purpose. But sometimes we need a statistic which can deal with wide areas which can serve several purposes. The statistic that can be useful for our purpose to provide more information involving more than one set of data that can be tested and the level of significance can be ascertained is the Chi Square. The Chi Square statistic can be applied to any simple data. Guilford and Fruchter (1987) called the Chi Square as the versatile statistics. Because one of the important properties of this statistics is that it can combine several statistics or values in the same test. Further, in Chi Square statistic a hypothesis can be tested with more than one set of experimental data.

One of the general features of the Chi Square statistic is that it does not apply the data as they come. It mainly deals with frequencies. If the data are not in the form of frequencies the Chi Square statistic converts them into frequencies for its purpose. The Chi Square is always stated in the form of χ^2. But the Chi Square can be explained by using the standard score *Z*. Therefore, the formula stated is :

$$\chi^2 = Z^2 = \frac{(X-u)^2}{\sigma^2} \ or \ \frac{(X-\overline{X})^2}{\text{SD}}$$

Where X = Score obtained in a measurement

u or $\overline{X}$ = Mean Score

σ^2 or SD = Standard Deviation

The formula indicates the mathematical relationship of χ^2 to Z^2 with one degrees of freedom.

The Chi Square, the *Z* and the '*t*' ratio have certain common characteristics. The Chi Square is a sampling statistic, similarly the *Z* has sampling distribution. But there is some difference between Chi Square and the *Z* as sampling distribution. The *Z* is normal, whereas the Chi Square is not normal and it is directly related to Z^2. Hence the sampling distribution of Chi Square is not normal. The relation of Chi Square to the *Z* is only one degree of freedom, whereas in the case of Chi Square there are more hypotheses to be tested and hence there are more degrees of freedom in Chi Square.

But with regard to *Z* and the '*t*' ratio, both are treated in the same manner in sampling distribution. The *Z* score is a score obtained by determining the deviation of a score from the mean and dividing it by the standard deviation of the measurement, whereas the '*t*' or critical ratio is obtained from the mean differences and divided by the standard error of the mean differences

$$t \text{ or } c = \frac{(M_1 - M_2)}{\text{SEM}_{\text{Diff}}}$$

The '*Z*' is often equated with critical ratio. But statisticians state that the critical ratio is applied in the case of larger sampling while '*t*' is applied to smaller sampling below 30 in number. But researchers invariably apply the '*t*' ratio irrespective of the sampling size and also equate it with *Z*. But the Chi Square is equated with only Z^2 and not with *Z*.

Generally Chi Square is a statistic. It is a test derived from the distribution to compare the goodness of fit of theoretical and observed frequency distributions. Chi Square distribution is a *continuous single parameter distribution* derived as a special case of the gamma distribution and used especially to measure the *goodness of fit* and to test hypotheses and to obtain confidence of intervals for the variance of a normally distributed variables. *Gamma distribution* is also a statistic which is a *continuous two-parameter distribution* from which the Chi Square and the experimental distributions are derived. The Chi Square is mainly associated with the goodness of fit. Goodness of fit is a statistic, the extent of which the observed sample values of a variable approximate to the values derived from a theoretical density or frequency.

Let us explain the Chi Square statistic with an illustration. An investigation was conducted in a group of 60 Engineering College students. The problem investigated was : "Air pollution in big cities was caused by motor transports by emitting heavy smokes of various types". The responses were collected from 60 students and the Chi Square treatment was given. There were both affirmative responses and negative responses. The hypothesis was that there is no difference between the two types of responses. This means that 50 per cent of the subjects would accept it and 50 per cent of the subjects would reject it. The frequencies are equal 50-50 per cent or 60 × .5 = 30. The expected frequency is 30. Thus the mean of the population distribution is 30. There were 40 students agreed the idea and 20 students disagreed the idea. The Chi Square statistic states that the mean is the expected frequency (*fe*), and the obtained frequency is treated as *fo*. We will have to test whether the obtained frequency *fo* significantly differ from the expected frequency *fe*, 30.

The formula to test the hypothesis is $Z = \dfrac{fo - fe}{\text{NPq}}$

Where fo = obtained frequency

fe = expected frequency

N = number of cases

P = 0.5 proportion

q = 1-0.5

$$Z = \frac{40-20}{\sqrt{60 \times 0.5 \times 0.5}} = \frac{20}{\sqrt{15}} = \frac{20}{3.87} = 5.17$$

Now we can refer to the '*t*' ratio Table to find out the level of significance. The Table value 2.58 is significant at one per cent level and 3.29 is significant at .001 level. But the obtained *Z* value is 5.17 which is much higher than the two Table values. Therefore, the obtained *Z* value is significant at .001 level. In this measurement the mean is 30 and the variance is 3.87. This is obtained by the formula :

$$\text{NPq} = \sqrt{60 \times .5 \times .5} = 3.87$$

The important finding in this measurement is that the obtained value of frequency is much higher than the expected value or frequency. In this example illustrated the *Z* and the Chi Square are the same, since the degree of freedom is one. If $X^2 = Z^2$, then for determining the Z we will have to work out the root of Chi Square.

If the degrees of freedom is more than one, then we cannot equate Chi Square with Z value. The Chi Square test can be applied to test a hypothesis by combining the test scores which have more than two responses or more than two tests. We can explain this with an example of a test with more than two responses.

A test was conducted among 96 post-graduate students in a college and there were both men and women students in equal strength. The problem investigated was with regard to the use of computer in institutions and in marketing places. The use of computer in these days is very essential as it simplifies and speeds up every work. They were asked to respond with 'Agree', 'Disagree' and 'No idea'. The hypothesis was, there is no gender difference in the frequencies of all the three responses taken together. The formula applied for estimating the Chi Square is :

$$\chi^2 = \frac{\Sigma(fo - fe)^2}{fe}$$

The response frequencies were analysed and presented in a tabular form. There were response frequencies in all the three categories from both genders.

Table 2.17 : Showing the obtained Frequencies and Expected Frequencies for all the three Responses of Male and Female Students

Response	*fe*			*fo*		*fo – fe*		$(fo-fe)^2$		$\frac{(fo-fe)^2}{fe}$		
	(1)			(2)		(3)		(4)		(5)		
	M	W	Both	M	W	M	W	M	W	M	W	Both
Agree	16	16	32	20	12	+4	– 4	16	16	1.00	1	2.00
Disagree	16	16	32	8	24	– 8	+8	64	64	4.00	4	8.00
No idea	16	16	32	20	12	+4	– 4	16	16	1.00	1	2.00
Sum	48	48	96	48	48	0	0			6.00	6	12.00 χ^2

M = Men Students, W = Women Students

Table 2.17 illustrates the distribution of frequencies for both men and women students and for the three response categories. The expected frequency (*fe*) in column 1 is 48 for men and 48 for women. The expected frequency in the three response categories is 16 for agree, 16 for disagree and 16 for no idea for both men and women.

The obtained frequencies in all the three response categories vary. For men the obtained frequency for 'Agree' 20, 'Disagree' 8, and for 'No idea' 20. These obtained frequencies vary in the case of women. It is 12 for 'Agree', 24 for 'Disagree' and 12 for 'No idea'. The frequencies for *fo–fe* are obtained by deducting the expected from the obtained in all three response categories for both men and women and entered in column 3. There are both plus values and minus values in all the three response categories and the sum value at the bottom is zero (0). The plus value cancel out the minus values.

The differences of *fo–fe* are squared and entered in column 4 as $(fo–fe)^2$. These values in column 4, are divided by the respective expected values for all the three response categories for both men and women and entered in column 5 as $\frac{(fo - fe)^2}{fe}$. The summing up was made for both men and women and for the three response categories. The sum of the sums is the Chi Square and it is (6+6) 12.

Determination of the degrees of freedom for Chi Square is very simple. It requires the number of rows and the number of columns. Here the rows are the response categories, and the columns are the expected and obtained values. Therefore, the formula for degrees of freedom : $(r-1)(c-1) = (3-1)(2-1) = 2 \times 1 = 2$. Therefore, for 2 degrees of freedom the Chi Square is 12. Which is significant at one per cent level. Therefore, the null hypothesis that there is no gender difference in the problem of using computer in institutions and in marketing places is rejected. But there is difference between men and women regarding the agreement and disagreement. More frequency for men (20) agree and more frequency of women disagree (24). The Chi Square is high (8.0) for both men and women for the response of disagree. It is also found that more men were not able to decide the problem.

DETERMINATION OF ERRORS IN MEASUREMENTS

We have discussed in the proceeding section various aspects of inferential statistics. In psychological studies the application of inferential statistic is very essential, without which there can be no psychological or educational measurements. The inferential statistic is also called as the population statistic. By applying the inferential statistic we can draw conclusions and generalize the characteristics of a population. It is by testing some sample from the population, we can infer the characteristics. There are various sampling techniques. We believe that a small sample may represent a large population. Therefore, a sample may be defined as the small representation of cases of a large population. There are two broad types of sampling techniques. One is the probability of sampling technique and the other is non-probability of sampling technique. All the sampling techniques however carefully planned are not absolutely free from the occurrence of the error in the measurement. In the study of human behaviour it is hypothetical that all test scores and measurements are subject to the occurrence of errors. These errors are occurring in a variety of ways from different sources. Some of the sources are the sampling plan, the procedure of collecting the sample, the instruments used for collecting the data, the limitation of it, the procedure of measurement, and the non-representation groups in the population, all may contribute to the occurrence of error variance. These sources of error may affect the set scores, the means and the standard deviations, the correlation coefficient, the reliability and the validity of measures. In estimating the errors in measurement authors have different opinions.

But if the sample should be free from errors some of the important characteristics like the randomness, probability, representativeness and adequacy should be emphasized and followed. However carefully all these procedures are followed some kind of dispersion is taking place in sampling and this makes the occurrence of errors and thus the samples become biased. If the sampling is inadequate and the procedures are not correctly or unconsciously followed then the confidence and the reliability of the results are affected. This is generally called sampling error. The sampling error and the error in measurements are different aspects. The sampling error is mainly affecting the reliability of the measurement. According to Gullikson (1950), the accuracy of the test scores depends upon the reliability of measurement and the accuracy of validity depends upon the error of estimation. Thus the sample results help the experimenter or the investigator to generalize the characteristic of the population.

An investigator, for example, wanted to test the mental ability of the urban and rural boys and girls, studying in 10th Standard in one of the States in India, specifically in Tamil Nadu. He cannot do this all over Tamil Nadu as the population may go in lakhs. Therefore, he has chosen one district at random among 25 districts and 20 schools out of 100 schools, as the representative of the total student population of the State. It will not be possible to conduct a survey among all the 10th Standard students of both boys and girls in all the schools throughout the State. But the investigator should fully ascertain that the sample he has chosen is the correct representative of all the urban and rural schools and the whole State. In this way the investigator is making his sample as random, representative and stratified. Since the investigator is fully convinced with his sample that it would identify the characteristics fully the student population of his study he has chosen one district out of many, 20 schools out of 100 and 10th Standard specifically of both boys and girls. This is supposed to be free from sampling error.

However carefully the sampling was planned in a population, there are certain variables interfering in the process of investigation such as choosing certain specific area, the school, including some unknown persons, not including certain areas or schools or students may contribute to the sampling error. Statisticians have devised methods of eliminating such sampling errors or biased errors. They are of the view that the probability of simple random sampling and controlled or stratified random sampling methods may help us to eliminate the errors in sampling.

Standard Error of the Mean

The other important error to be estimated is the standard error of the mean. This is normally estimated in small or larger samplings. This is estimated from deviation scores. On obtaining the mean, whether it is for ungrouped data or grouped data, the standard deviation is estimated. Any score deviation from the mean either positively or negatively is a deviation. The deviation for such score is squared and summed up and divided by the *N* and the root of which is taken as the standard deviation. The formula is : SD = $\sqrt{\frac{(x-\bar{x})^2}{N}}$. For grouped data the formula is different. SD = $h\sqrt{\frac{\Sigma f(x)^2}{N} - \frac{(\Sigma fx)^2}{N}}$

Where *h* is the class interval in grouped data.

Standard deviation can be estimated from the original data if the group is very small. The formula is :

$$\text{SD} = \frac{1}{N}\sqrt{\Sigma x^2 - (\Sigma x^2)}$$

Where x = original score

N = Number of cases

Σ = Summing up.

The values required for estimating the standard error of the mean are the standard deviation and the (*N*) number of cases of the sample.

The formula is : $\text{SEM} = \dfrac{SD}{\sqrt{N-1}}$

The mean of the measurement is the mean of the sample. It should be noted that the number of cases of the sample determines the size of the standard error. If the sample is smaller, then the standard error of the mean is greater. Similarly the greater the sample size the smaller the standard error of the mean. The sample mean is more likely to deviate from the true mean in a smaller sampling and also the dispersion would be very wider. Therefore, we may state that if the representative sample is larger then there is chance for greater identity of characteristics with the population from which the sample is drawn. Therefore, for any reliable conclusion of any study we should have the sample larger, representative and randomly drawn from the population, strictly according to the sampling procedures. It is also observed that the standard deviation of a random sample drawn from a larger population is smaller than the standard deviation of the sample drawn from larger population. The reason is that there may be extreme dispersions in the larger population which may be reduced or missing in the random sampling.

The standard error of the mean can be also estimated from the sum of squares of deviations of the sample. The formula is :

$$\text{SEM} = \sqrt{\frac{X^2}{N(N-1)}}$$

Where, SEM = Standard error of the mean

X^2 = Sum of the squared deviations

N = Number of sample size

Generally the standard error of the mean is the Standard Error of the sample. Therefore, the standard errors may be defined as the estimate of the sampling error in an investigation.

Errors of Measurement

Psychological and educational measurements applying the statistical methods are incomplete without making any reference to the measurement of errors. But it is pity to see that some experimenters and researchers often fail to estimate the error variance and discuss their results relating to the errors. They must discuss the nature of errors in their measurements.

Whether the error variance is smaller or greater it should be presented in the Table and discussed. Sheridan (1979) states that there is no way of avoiding the errors in psychological measurements. It is true that we cannot also totally eliminate errors in psychological and educational measurements. But we can as far as possible reduce the errors in most of the measurements. The main sources of error are the sampling procedure and the instruments used for data collection. Experts in the field of measurements often warn that the instruments or test materials should be previously tested before using them in any measurement. It is because that every instrument has certain limitations. The error may be with instrument or with the person who makes use of it, a small error in a correct measurement, if neglected by the person thinking that it is only a minor thing, may be accumulated into a bigger source of error in the process of scoring and collecting the individuals scores.

Whatever the manner the errors in measurement occur there are two types of errors often mentioned in psychological measurements. They are one, the systematic error and the other unsystematic error or random error and it is also called the chance error. These errors can never be eliminated from the measurements because psychological measurements are not made with yardsticks or with weighing machines. It is the measurement of behaviour as revealed by other human beings and assessed by psychological instruments and scales which are not calibrated like linear scales or weighing machines. Psychological a scales and instruments are designed and developed by humans. They are only hypothetical. Therefore, there is every chance for error by using these instruments and scales in the measurement of human behaviour.

It is because of these the measurements made with these instruments or scales make the results systematically as over estimation or under estimation. In the case of over estimation of the true measurement the error is consistently positive with (+) plus sign and in the case of under estimation of the true measurement the error is consistently negative with (–) minus sign.

Sometimes the overestimation and underestimation of the true measurement may be occurring with equal frequencies. Both the positive errors and negative errors may be occurring equally and the positive and negative errors are summed up and equated. If these two errors are found to be equal, then the positive error cancels out the negative error and the sum of errors is (0) zero. This we call as unsystematic or random error. These unsystematic or random errors in any measurement cannot be absolutely eliminated. In Psychophysical methods these types of errors are estimated. The formula to find out the unsystematic or random error is :

$$e = x - x_t$$

Where e = random or unsystematic error

x = observed score or measurement

x_t = the true score or measurement

There is another error called circumstantial or ambient error which may occur in a measurement due to circumstances like change of situation, atmospheric pressure like heat, temperature and sound.

According to Candland (1968) standard errors and sampling errors are different things. He states that standard errors of measurement refer to inconsistent data obtained when the same instrument on the same subjects used for the second time but sampling errors refer to the inconsistent results if the same instrument applied to different samples of subjects. But Ebel (1966) states that standard error of measurement is estimated from the standard deviation of the errors of measurement related to the test scores in set. Therefore, standard error of measurement is estimated by multiplying the standard deviation of the score by the square root of one minus the reliability coefficient.

The formula is : $Se = S_t \sqrt{1 - r_{tt}}$

Generally standard error refers to the error variance in a measurement.

3

SENSORY EXPERIENCE

INTRODUCTION

The sense organs are said to be the gateways of our knowledge. The sense organs receive stimuli of various types and convert them into experience and make us aware of what has happened outside. We know that the world in which we live is full of energies and these energies attack the organisms in various ways. Physicists say that we are constantly assailed by cosmic rays, light rays, sound waves and heat waves. Therefore, our bodily parts come into contact with this information. If we want to survive in this world we must understand the information these forces brings to us. Man is always very much interested in formulating laws about the physical events he is observing. He is also very anxious what goes on around him in the world. The European scientists like Helmholtz (1821-1894) (1867) and the physiological psychologists believed that the first step to understand the behaviour is to understand how we come to know the external world. They said that it is only by perceiving through the sense organs. According to them all knowledge and all behaviour come to us because of what is perceived of the external world by seeing, hearing, tasting, smelling and touching.

If we go deep into the sensory psychology we find the names of many scientists, associated with the subject, Physics. For example, in the field of vision we come across the names of Young, Maxwell, Newton and Ernest Mach. Thomas Young is known to psychologists as the inventor of the theory of colour vision. Maxwell published many papers on colour vision. Newton also wrote extensively in the area of colour vision and Ernest Mach is known for his classical work, 'The Analysis of Sensation', had made many contributions on the study of vision. Therefore, the contributions made by scientists from various areas enriched the sensory psychology and we are able to interpret and explain the functions of the sensory apparatus.

Since our sense organs are specialized in a particular area they receive the respective information from outside in the form of stimuli and convert them into sensations. These sensations become the basis for all our reactions and behaviour in the outside world. The sun sends energy to earth in the form of light wave and heat wave and these become stimuli to our eye and skin. The conversion of this stimulus into sensation, into experience and ultimately into knowledge is an elaborate process involving four important stages. They are sensory stage, attentive stage, perceptual stage and conceptual stage. Therefore, in the knowledge getting process sensation is only the first stage. If a sensation is to be meaningful it should be attended to and then understood. A sensation is a kind of awareness to sensory stimulation. It is aroused in an organism by a stimulus. A stimulus may be internal or external. The sensation or the experience we get is the act of the organism.

An organism responds to many stimuli. The organs of response are of three types. They are receptors, connectors and effectors. The sense organs are called receptors, muscles and glands are called the effectors and the brain and the nervous system are called the connectors. The three-fold division receptors, connectors and effectors in human behaviours are quite similar to roughly the action of the telephone exchange. The stimuli are like the incoming calls sent to the central nervous system (the brain) like the telephone exchange and sent out through the muscles and glands like the outgoing lines to the place where it should go. Sensation is the first response of the brain to the stimulus, which strikes on the organism.

According to popular belief man is said to possess five senses. But recent physiological research has made it clear that man has eight sensory receptors. Hence man has eight senses. They are vision, audition, gustation, olfaction, and tactual sensation. The other three are internal such as organic, kinesthesis and vestibular. Further, each sense organ is specialized in one particular type of sensations. The ear cannot see nor the eyes could hear. It should be noted that even when a sense organ is capable of receiving an appropriate stimulus it could respond effectively only according to certain law. For example, there is no sense of hearing to both very feeble sound and very loud noise. Similarly there is no sense of seeing to very dim lighting situation and very bright light.

Our response to the world is based upon what the senses report to us. We do not make indiscriminate responses to mass of stimuli or do not respond to everything at once. We make more specific integrated response that has relevance for our purpose. Our senses function according to four laws, says Woodworth (1952). It is selective and turns only to specific stimulus. It is sensitive only to a particular kind of stimulus. Its responses vary to different intensities of the stimulus, and lastly, it responds to different qualities of the stimulus. Thus by responding to the relevance of our conduct and leaving the rest of the things to serve as background we not only responding to the stimulus, but also come into contact with the environment and the process we call as perceiving.

Table 3.1 : Sense organs, the stimuli and the corresponding sensations

S. No.	Sense organs.	Stimuli	Sensations
1	The eyes	Ether vibration	Visual (Seeing)
2	The Ears	Air vibration	Auditory (hearing)
3	The Tongue	Chemical substances dissolved in saliva	Gustatory (tasting)
4	The Nose	Chemical substances in gaseous form	Olfactory (smelling)
5	The Skin	Changes in temperature or objects coming into contact with the skin	Tactual or Cutaneous (touching)
6	Stomach and back of the throat	Churning of the walls of the stomach and certain chemical changes in the back of the throat	Organic (internal)
7	Muscle spindles and tendons in the joints	Pressure on the muscle, spindles and the tendons near the joints	Kinesthetic (muscle)
8	The semi-circular canals in the ear	Disturbances in the fluid of the semi-circular canals	Bodily equilibrium (balance)

The psychologists are interested in all these eight sensations. How these sensations are caused? How the various forms of physical energy such as cosmic rays, light, waves, heat waves and sound waves affect the receptor cells within the sense organs? How the cells convert these energies into information and send them to the central nervous system. Therefore, it is important to know something about the mechanism that takes place in various processes of seeing, hearing, tasting, smelling and touching. The mechanism of sensation is more complex. If a sensation is to occur in human organism, stimulus from the external world should come into contact with a sense-organ. This contact will result in the release of some kind of nervous energy which is transmitted by the sensory nerves to an appropriate region in the cortical area of the brain. When this has happened there will be an awareness in the organism of experiencing a sensation.

PSYCHOPHYSICS AND STIMULUS

Absolute Thresholds

In any sensation there is a relationship between the physical event, the stimulus and the corresponding psychological event, the response. A study of this relationship between the physical events, the stimuli such as light, sound, touch and the corresponding sensation is known as psychophysics. An equation is drawn between the physical stimulus and the response to the stimulus in psychophysics. It is mainly the determination of thresholds. Threshold is the amount of stimulus value that is required to arouse a response. Threshold is the minimum amount of stimulus value that is adequate to elicit a response or sensation. Psychophysicists speak of two kinds of threshold, one is absolute threshold (AL) and the other is differential threshold (DL).

The absolute threshold is also stated as of two kinds, one is the stimulus threshold or lower threshold (RL), which is just noticeable magnitude and sometimes not noticed. The other is the absolute threshold (AL) which is also stated as the upper threshold or terminal threshold (TL). The terminal threshold is a point statistically determined. The stimulus at this point is too high that it is perceived more than fifty per cent of the times.

Vision : The eye is the sensory apparatus for the sensation of vision. Light is the stimulus to which the eye responds. Candland (1968) states that though most of the stimuli perceived by the eye are produced by light, light is not the only stimulus that can make the eye to respond. Bell (1948) and Muller (1938) have produced evidence that visual sensation occurs whether stimulation is light or pressure. A bright light or bright colour according to them is seen even after closing the eyes as after images.

The parts of the eye and the mechanical structure and function of the eye may be had from some text book of Psychology. The retina which is the third layer in the eye is the most important part in the visual sensation. This retina receives stimulus from the outside world. In the retina area there are cells which are sensitive to light and they are closely packed together in the fovea region. The rods are sensitive to normal day light vision and the cones are sensitive to colours.

Visual Sensation

Light is the stimulus for the eye and this stimulus enters eye from any point and can vary in three ways such as in wavelength, in energy and in the amount of mixture. The energy or the intensity of the stimulus is similar to brilliance of the visual sensation. The brilliance of visual sensation may vary from dark to light. The wavelength of the stimulus is similar to hue of the visual sensation. The hue of the visual sensation may vary from red through yellow to violet (seven colours red, orange, yellow, green, blue, indigo and violet). In the light stimulus there is a mixture of wavelength which is similar to saturation of the colour sensation. A pure colour, is saturated and a dull colour is unsaturated. A saturated colour need not be very bright. These are the general relationship between physical light stimulus and visual sensation. But the relationship is much more complex. Brilliance is not only depending upon the intensity of the stimulus, but also upon the wavelength. The wavelengths vary from colour to colour. For example, the wavelength red is about 700 millimicrons which is the longest and the wavelength of violet is about 400 millimicrons which is the shortest.

The colour zone in the visual sensitivity can be measured and colour blindness also can be detected by administering suitable tests. The widely used tests for mapping the colour zone and colour blindness are the Ishihara test and perimeter.

Auditory Sensation (Hearing)

The ear is the specialized sensory apparatus of auditory sensation. Sound wave is the stimulus that could arose response in an organism if it possesses sound hearing sensation. The process of hearing begins when the sound waves attack the ear drums and cause it to vibrate. The human ear consisted of a complex mechanism. It is divided into three parts as outer ear, middle ear and inner ear. The outer ear catches the sound waves and carry them to the middle ear. The middle ear consists of the ear drum or tympanic membrane. The sound waves push this membrane to vibrate. The vibration is transmitted to the ossicles. Ossicles consists of a chain of three little bones in the middle ear such as malleus (hammer), incus (anvil) and the stapes (stirrup). These three bones are known by their shapes as hammer, anvil and stirrup. The middle ear is connected to the throat by means of the eustachian tube which supplies air to the middle ear constantly. The inner ear spreads right back into the skull. It is made up of three parts like the semi-circular canals, the vestibule and the cochlea. The semi-circular canals are responsible for the sense of balance or equilibrium of the body. The vestibule is a receiving and transmitting centre. The cochlea is the real organ of hearing and it is the main part in the inner ear. It is having the shape of shell or snail. The basilar membrane is another important part which is running parallel similar to the structure of cochlea. It consists of thousands of hair like fibres which vibrate like the piano strings. These fibres are cells which receive the sound vibrations and also stimulate adjacent fibres of the auditory nerve and the nerve impulses reach the brain.

The inner ear is a watertight compartment and it is filled with a saline liquid known as perilymph. Sound waves gathered by the outer ear pass along the auditory canal. They strike the tympanic membrane and push it and it vibrates. Then the three bones of the middle ear are moved and the stirrup closes the oval window. Now the air waves pass into the auditory area of the brain through cochlea and we hear the sound.

Sound is physically a wave motion and it travels at a speed of 1100 feet per second. The tones or noises are of three qualities, such as pitch, loudness and timbre. A tone is described as high, medium or low pitch. The pitch of a tone depends upon the vibration of the sounding matter. A slow vibration produces a low tone and a rapid vibration produces a high tone. It has been estimated that upper limit of man's tone is 25,000 vibrations per second. By means of an audiometer the extent and degree of hearing loss can be measured.

The loudness or intensity of a tone depends upon width of the vibration of the sounding matter or wave amplitude. Loudness is measured on the decibel scale.

Timbre is the quality of the tone. The difference between the same note on a violin and on a flute is one of timbre. Timbre depends upon the frequency and the intensities of the tone.

Studies on sound show that sounding two tones together may result in masking. The louder sound may mask the softer one. Experiments show that low tones mask higher tones to a greater extent than the higher tones mask low tones. It was also observed that complex tones mask pure tones.

Perception of Sound

It is easy to judge whether a tone is from a nearby place or a distant place by its loudness. Loud tone is likely to be quite close and a faint tone should be from far away. Sound can be located whether it is coming from right or left or from above the head. This phenomenon is studied in the laboratories by the use of an instrument called sound cage. The participant is blind folded and clicking sounds are produced at different points of the sound cage. The perception of sound depends upon the auditory cues. Sounds are best located if they are from left or right than in the front or back because the sounds directly hit the ears.

The Skin Senses

A touch is made up of several different sensations. These are called cutaneous sensations. These are connected with touch or tactual sensations. But in reality skin has four senses each with a separate quality and different sense organs. The four sensations are (a) cold, (b) warmth, (c) pressure (touch) and (d) pain. The specific sense organs associated with these sensations are called spots. They are located below the outer surface of the skin. Any specific area of the skin contains all four spots in different proportions. This can be located with suitable stimulus. But the various areas of the body differ in sensitivity to cold, heat, touch and pain. Experiments can be conducted on skin senses by selecting a part of the body, for example, the forearm. The area can be mapped for skin receptors. The presence or absence of the receptor cells for pressure and pain can be determined by applying a sharp needle to the skin. Every time the needle touching the skin the participant reported whether he/she felt the pressure or pain or not. When all the pressure and pain spots are located the experimenter can map the area from cold spots or warm spots. This can be performed by a cooled or heated rod to a specific temperature. The participant should report whether he is experiencing cold or hot. This is how the spots can be experimentally located on the skin surface and the existence of four skin receptors for cold, warm, touch and pain can be established.

The Chemical Senses

Olfaction and Gustation are called chemical senses. These are activated by chemical substances. Unlike other areas there are not many studies on these two areas. The information available on these two are limited. It is very difficult to conduct experiment on these two senses.

Olfaction

The sense of smell is known as olfactory sensation. We know very little about this sensation. The organ of smell is the olfactory membrane which is situated in the upper nasal cavity. According to experimental evidences there are six elementary sensations. They are (1) spicy (pepper) or cloves, (2) fragrant (flowery or rose), (3) ethereal (fruity orange or lemon), (4) resinous (turpentine), (5) putrid (fowl - rotten eggs), (6) burned (scorched as in tarry substance).

Different parts of the nasal membrane are connected with different types of smells. But most of the smells we perceive in everyday life are combinations of the six smells in different degrees. Man's sense of smell is very poor compared to that of animals and birds. The sense of smell sometimes serves as a caution against bad air and food.

There are methods of measuring olfactory thresholds. An instrument called olfactometer is used for the measurement of olfactory sensations. This instrument is a modified form of the instrument proposed by Zwaademakeer, an early research worker in Olfaction. Henning (1916) classified smells into six categories. But a more recent classification is made by Crocker and Henderson. According to them there are four basic qualities. An apparatus has been designed by Magnen (1942-43) using an air whistle that sounds as the subject inhales. Geldard (1953) states that there are thousands of specific smells and there are difficulties to classify them into certain categories. He made nine categories of the odours and states that still hundreds of odours could not be brought under this categories. Candland (1968) states that there is no standard method for naming the olfactory sensations.

Gustation

The sense of taste is known as gustatory sensation. We experience numerous tastes. Every food is having a unique taste. It is not only the tongue, but also the smell and touch sensations make us to think of taste. There are four elementary tastes like the sweet, sour, bitter and salt. The other tastes are only combination of these four tastes. A skilful cook can produce hundreds of tastes from these four tastes.

The real sense organs of taste are the taste buds which lie underneath the surface of the tongue. Hence the taste buds of the tongue are the receptors of the sense of taste. The tongue is full of countless pits and the taste buds are actually in these pits. The taste buds are branches of the sense cells. Each cell is having a tip that runs into the sides of the pit where it is exposed to substances. We can taste liquids quickly. But solid food must dissolve well in saliva to have the taste sensation. The sensory cells thus made to function and these arouse the sensory nerves and sensory nerves carry the impulses to the brain. Gustation is a very complicated sensory system. There are three cranial nerves carry the sensation of taste to the brain.

It is observed that taste receptors change with age. One study reported by Wenger *et al.* (1956) produce evidences for this. Temperature seems to modify the taste considerably. We realize sweetness on the tip of the tongue, sourness on the sides, bitterness on the back and saltness both on the tip and sides of the tongue.

Experiments on gustation are very few and very little information are available. One of the earliest studies was made by Carl Pfaffmann (1941) revealed that individual fibres are specialized in the way they respond. Some taste fibres respond only to acids or to sour stimuli. Other fibres respond to acid, salt and so on. According to his study no fibres found to be sensitive to stimulation by sugar.

Kinesthesis

Kinesthesis means movement. It is the sense that provides us information about the position and movement of our bodily parts. This enables us to know the difference in the weights of several objects by lifting them with our hands. This sensation is different from skin sensation. Judging the weight and objects is purely a matter of pressure. We lift an object and say how heavy it is. But for this we will have to move our hand up and down several times. Therefore, we have to suppose that we have a special equipment to find out movement and weight over and above the help offered by the sensation of touch. This special equipment may be called kinesthetic sense. This sensation originates in the receptors in the muscle spindles, tendons and joints in general. These receptors transmit information about whether the arm is stretched or relaxed, the fist is taut or loose, the stomach muscles are tight or slack and so on. Actually in the absence of the kinesthetic sensation it will not be possible for a man to say where he is going and what he is doing. Kinesthesis indirectly provides us information regarding our position. One cue may indicate the arm is fixed and the head is bent. We perceive the stretching and relaxing of the muscles and the joints.

The stimuli that arouse responses in our senses are emanating from external world and also for internal world. On the basis of this Sherrington (1906) divided senses into two types as extroceptive and introceptives. The extroceptives are external to the body and the introceptives are for within our body. Senses such as vision, audition, olfaction, gustation, and cutaneous senses are relating to the stimuli from external to the body, other senses are relating to the private stimuli, perceived within the body by the individual.

Vestibular

This is responsible for sensitivity to the balance, position and movement. The receptors are located in the non-auditory part of the labyrinth of the inner ears on each side of the head. There are three tube-like structures called semi-circular canals. One of the canals is oriented in the horizontal plane, one in the vertical plane and third in the sagital plane. The position of these canals will be determined by the position of the head to the surface of the earth. The vestibule and the semi-circular canals are more important because they provide sensory information for the maintenance of posture, equilibrium and steadiness of movement. Meuller (1969) states that this can be seen mainly in birds and fish. They rely on it to get information which way is up and in which direction they are going while flying in the air and swimming in water. The cat while falling from the tree correctly lands on its feet.

The vestibule and the semi-circular canals in the inner ear are responsible for the sensation of equilibrium or balance. There are receptors in the vestibules and semi-circular canals. They are filled with fluid and at the base of each of these canals there is bulb like part called ampulla. According to Easton (1978) within the ampulla is a small patch of sensory hair cells and each one of these is fixed in a gelatinous cap (cupula) that extends above the hair cells to the wall of the ampulla. The cupula is formed as a complete boundary across the ampulla. The sides of the cupula are free except at their attachment to the hair cells, but closely connected with the sides of the ampulla. When the fluid moves in the semi-circular canal the cupula is pushed by the pressure of the moving liquid. The hair like structures are stimulated by the liquid movement caused by rotation and they serve as sensory apparatus. When constant speed is maintained in rotation the cupula is displaced with constant acceleration. At zero acceleration the cupula is returning to its original position within 30 seconds. When the head is at rest the nerve fibres from the semi-circular canals send steady stream of impulses to the motor area of the brain that controls the movements of the body.

When a person sites at a merry-go-round for few minutes by the swift rotation the semi-circular canals are affected and the person complains of giddiness. When it is stopped, within few seconds he restores the normal position. Suppose, if the semi-circular canals are affected or injured then a person automatically loses his sense of balance and proper motor adjustments of the body become impossible.

Organic Sensations

Sensations from internal organs are called organic sensations. It includes a variety of sensations from the internal organs such as hunger, thirst, nausea, heart burn, suffocation and the vogue bodily sensations that is caused by emotional and tones. Hunger is a sensation aroused by the contraction of the stomach walls. The low water position in the human organism and consequent dryness of the throat and mouth give rise to the sensation of thirst. There may be special organs to inform us about disturbances in the respiratory and circulatory systems, but we have very little experimental evidences.

EXPERIMENT 1 : DETERMINATION OF TWO-POINT THRESHOLD

INTRODUCTION

The relation between variations in physical stimulation and the reported experience is one of the oldest problem in psychological studies. It is with regard to the nature and meaning of subjective experience. The problem is ascertaining the lawful relationship between the measurable characteristics of the stimulus on one hand and the reportable attributes of sensory experience on the other hand. In studying the relation between the characteristics of the stimulus and the attributes of experience certain experimental questions are raised.

What is the minimum amount of stimulation required for the detection of a stimulus? What type of stimulus is required? How much intensity is required in order that a subject may reliably distinguish between its presence or absence ? One important point is that the minimum amount of stimulation required will vary with the conditions of the testing situation. For example a dim light is required in a dark room than in a bright room. A feeble voice is needed in a sound controlled room than in a noisy room. So also with very sensory experience like hearing or seeing, smelling, tasting and touching.

The detection of minimal stimulus differences in every sensory area was the problem for psychophysicist for a long time. The minimal difference quantitatively or qualitatively needed between two stimuli as to facilitate the subject to reliably cognize the difference. This minimum value of the difference will vary from person to person, from situation to situation. The experimental problems of psychophysicist are not limited to the determination of stimulus and stimuli differences that are detectable. It is also important to judge the stimuli required well above the minimum for discrimination. In what situation or condition two stimuli judged as equal? How accurately a subject may respond to a stimuli when presented with twice as intense as another stimuli?

DETERMINATION OF THRESHOLD

A weak stimulus may fail to arouse a response in an organism. But a strong or intense one never fails to produce a response. There is a line separating these two kinds of stimuli. But they can never be sharply identified. Because, the transition from one to the other is gradual and continuous. In the experimental situation there cannot be anyone stimulus value which represents the minimum necessity for a response all the times. Therefore, it is generally stated as the absolute threshold which yields a response fifty per cent of the time. The concept of a threshold has been arousing the interest of psychologist and physiologist for a long time. A threshold is that point at which a subject perceives and reports the experience of some phenomenon such as discrete on-off sequence flashes of light or sound or a pin-prick. The on-off interval of flashing of light or the presence or absence of the pin-prick reported by the subject is defined as the threshold value for the particular phenomenon. Therefore, a threshold is defined by Woodworth (1954) as the boundary which is separating the stimuli that elicit one response from the stimuli that elicit a different response.

The threshold is defined always in terms of the physical values. The observer should be able to respond fifty per cent of the times. In an experiment, for example dealing with the recognition of differences in the weight of two objects, a constant stimulus weight was placed in one hand of the subject and a steadily increasing weight placed in the other hand. The point here is that, where or at which place the subject has reported that the stimuli were of different weights should be considered as the threshold. Here we are concerned with determination of threshold. According to Candland (1968) determination of threshold means the detection of changes in the stimulus which elicit different responses. Epstein and Shontz (1971) state that threshold refers to the minimum stimulus value at which the response will occur. The absolute threshold is the minimum value at which the presence of stimulus is detected.

The threshold are of two kinds. One is absolute threshold (AL) and the other is differential threshold (DL). In the absolute threshold, we can observe two kinds of threshold. The first is the stimulus threshold or lower threshold (RL) and the second is the terminal threshold or upper threshold (TL).

The terminal threshold refers to the point at which the stimulus is perceived sometimes and not perceived sometimes. For example in the Mueller-Lyer Illusion board an individual may equate 19 mm as equivalent to 20 mm and sometimes 21 mm as equivalent to 22 mm. In the case of this individual's judgment 19 is the lower threshold and 21 is the upper threshold, whereas the threshold is at 22 mm.

The differential threshold (DL) is the stimulus difference which is judged fifty per cent of the time. The presence of the change is felt by the subject fifty per cent of the time. This is the least (minimum) amount of change required in the physical stimulus for the observer to report 50 per cent of the time that there is a change. In the case of weight the difference in weight is just noticeable and it is termed as JND.

Psychophysicists have tried many ways of establishing threshold levels for cutaneous and other sensitivities. It has been noticed with all sensations that if the stimulus value is below certain level of intensity, though it is above the zero point a response in the sense organs will not be stimulated. There cannot be any sensation corresponding to the intensities of stimuli below certain point, that is below the threshold. It is generally called by the psychophysicists as 'absolute threshold'.

The absolute threshold is the minimum intensity of the stimulus that is necessary to arouse a response in an organism in any sensory apparatus. The threshold varies from species to species in animal world. In the case of human beings the threshold is very low. But there are individual differences. What is threshold for one person is not threshold for another person not only in the case of cutaneous sensation, but also in all other sensations like, seeing, hearing, smelling, tasting etc.

Here we are interested in the study of characteristics of experiences which could enable us to distinguish among the skin senses. The skin sense is called cutaneous sense.

Candland calls the cutaneous sensibility as the forgotten sense. He states that man is very much concerned over the deficiencies of other senses, but he seldom cares the cutaneous sensibility. Whether the day is hot or cold, high humidity or low humidity is realized by us only through our cutaneous senses. Much of the world information are provided by our senses of skin. The human skin is a very big data processing instrument. The human skin is richly supplied with nerve endings and receptor cells. Postman and Egan (1949) state that it is not grouped into a single sense organ as in the case of eye or ear. The skin senses do not contribute sensory systems which can be independently defined. The skin senses must be defined and distinguished from each other in terms of varieties or dimensions of response to stimulation of the skin. On the basis of these response variations four skin senses have been distinguished. They are pain sense, pressure sense, warmth sense and cold sense. All the four senses can be traced by mapping with a stiff hair or sharp needle or pin or some suitable instrument with one or two needles. The accuracy of spotting these sense areas depends upon the intensity of sensitivity spots. The sensitivity spots are closely packed together in certain parts of our body. Particularly in the finger tip, lips are highly sensitive to pressures. Therefore, absolute sensitivity tends to increase with the density of the spots. The threshold for pressure depends upon the rate application for the force. If the force is rapidly applied the threshold value is small and if it is slowly applied it is larger.

If the area of the contact with the stimulus is very small (hair stimuli) the absolute threshold is independent of the extent of contact area. After obtaining the absolute threshold with a given hair if we use bigger hair with double the diameter we need a double force for eliciting a response. The two-point threshold experiment is based upon the pressure exerted on the skin and determining the sensitivity of the subject. A compass needle or aesthesiometer is used. The two points of the compass or the aesthesiometer with certain minimum amount of distance between them be adjusted and pressed on some part of the skin, particularly on the forearm and the subject has to report two discrete pressure sensations. If these two points are brought closer then there may be only one sensation. Since it may vary depending upon the density of the spots on the skin surface the average values are always preferred from a normal human being. Clinical studies show that persons suffering from organic psychoses reveal different type of responses.

Several studies have been conducted to identify the various cutaneous sensations. Weddell (1941, 1947) tried to isolate the various sensations of pain in this experiments. He assumed two different qualities of pain sensation, the pricking sensation and the burning sensation. The pricking pain sensation was of short duration which makes immediate withdrawal respone by the subject. In this case the subject was not able to identify the stimulus and further sensory information he did not provide. Instead, the burning pain sensation was of longer duration and it had a deeper sensation. The stimuli used in these experiments were chemical, electrical and mechanical types.

The tactile sensitivity of 24 men who had undergone above the elbow amputations of one arm was studied by Haber (1955, 1958). He tested each subject for sensitivity to light touch using the tips of a set of hairs by the classical psychophysical method, two-point threshold. He also tested the ability to locate a single stimulated point accurately on the skin. Measures were taken near the trunk of the amputated arm and away from the trunk. Results of this study revealed that the part away from the trunk had more sensitivity than the part near the trunk of the amputated arm. Haber (1956 and Simmal 1958, 1959, 1961, 1962, 1963) have interviewed subjects about their phantom experiences. Haber found remarkable similarity of their reports of phantom limb sensations. Simmal observed in his studies that when limbs are lost early in life phantom sensations did not occur whereas in the case of sudden loss it was felt.

One of the methods the psychophysicists often use for establishing threshold levels for cutaneous sensation is the method of limit. The stimuli according to this method are presented both in ascending and descending magnitudes of difference so as to establish a range of values which might determine the mean value to represent the threshold.

EXPERIMENT

Problem

The purpose of this experiment is to determine the two-point threshold value for cutaneous sensations in a person.

Method

Participant

Choose a person who has no knowledge about this experiment and the apparatus. Describe his/her characteristics. Check whether he/she has any skin deformities. Collect other information like, sex, age, qualification, occupation etc.

Materials

An Aesthesiometer, blindfold

Description of the Materials : The aesthesiometer is just like the compass having two needles. There is a screw with which the experimenter can increase or decrease the width between the two needles. There is scale in mm/cm for adjusting the distance between the two sharp points. State the instruments' manufacturers, brand, model etc. Also state where it is available. Draw the diagram of the instrument.

Procedure

The experiment is conducted in two series, ascending series and descending series. The distance between the two needles should be increased or decreased every time with 5 mm interval. There should be ten ascending series and ten descending series with the same interval. This should be prearranged by the experimenter and entered in a table before starting the experiment. When this is ready, the subject should be blind-folded and be seated in front of the apparatus comfortably. Then the experimenter should ask the participant to sit erect and stretch his forearm (left or right). The forearm should not have any support underneath. The experimenter should choose a hairless part in the forearm and should give the following instructions :

> "You will be receiving needle touches on your forearm. You will have to tell me whether you have received any touch. You may be receiving one touch or two touches. If it is one touch, say, one, if it is two, say two."

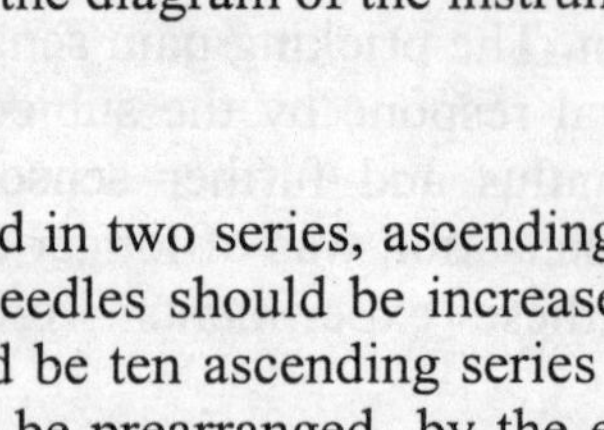

Fig. 3.1 : Aesthesiometer : Two-Point Threshold (Experiment 1)

For the ascending series the experimenter should begin with the aesthesiometer set at 0 mm. After the subject has responded with one or two five times, widen the two pins by 5 mm and touch the subject's skin. The responses should be recorded on the record sheet or table. For every point at the aesthesiometer five trials should be taken. The experimenter should go on increasing 5 mm between the needles for every successive point. When the experimenter receive the response as two from the subject for the first time in all the five trials he should record it and should not stop. He should continue the procedure at some more points by increasing the interval between the needles by 5 mm When the subject responded as two at three consecutive steps, stop the ascending series.

For the descending series the needles in the aesthesiometer should be set at widest possible point. At every point five responses should be recorded. The experimenter should go on decreasing the interval by 5 mm in the aesthesiometer and continue the trials until the subject responded as one at three consecutive points. For example, the subject might have responded as one at 10 mm the experimenter should not stop at this point. He should reduce the interval to 5 mm and get the response as one. Again he should reduce the interval by 5 mm, that is by making 0 mm and get the response. All these responses should be recorded in the appropriate spaces on the record sheet.

Results

In the record sheet there will be five trial entries for each point in the aesthesiometer. In the ascending series for the 0 mm there will be five entries, for 5 mm five entries, for 10 mm five entries, like this. Similarly in the descending series if the widest point is 40 mm, then there should be five entries for 40 mm, or if the widest point is 35 mm then there should be five entries for 35 mm, five entries for 30 mm, 5 entries for 25 mm. This should go on till it is completed with 0 mm.

Table 3.2 : Showing individual data with stimulus value and responses in the two-point threshold experiment

S. No.	Ascending Series							Descending Series						
	Stimulus Value mm	Response 1 or 2 1	2	3	4	5	Total 1or 2	Stimulus Value mm	Response 1 or 2 1	2	3	4	5	Total 1 or 2
1	00 mm							40 mm						
2	05 mm							35 mm						
3	10 mm							30 mm						
4	15 mm							25 mm						
5	20 mm							20 mm						
6	25 mm							15 mm						
7	30 mm							10 mm						
8	35 mm							05 mm						
9	40 mm							00 mm						
	Threshold Value Mean (mm)							Threshold Value Mean (mm)						

One mean value can be obtained for the threshold (in mm) from the ascending series and another for the threshold from the descending series. Check whether there is any difference in the threshold of the ascending series and descending series. Find out the mean threshold value for the subject. Give the individual data in Table 3.2 showing the stimulus values and the responses for both the ascending and descending series. Collect the group data of the threshold for both the series and give it in Table 3.3.

Discussion

Discuss the individual's data. How the subject was responding to each series. Was he responding to each trial instantly or with hesitation or with short interval? Is there any difference between the ascending series and descending series in the threshold value? What is the threshold value? Compare it with the group mean threshold. Where he stands in the group? Whose threshold value is the smallest and whose value is the widest? Why? Can you determine the reason? Compare male and female subjects.

Table 3.3 : Showing the group data with the threshold values for ascending and descending series

S. No.	Name	Threshold value in mm	
		Ascending series	Descending series
1.	K.R.		
2.	R.V.		
3.	M.S.		
4.	P.K.		
5.	M.R.		
6.	K.V.		
7.	B.V.		
8.	P.C.		
9.	T.R.		
10.	C.S.		
	Total		
	Mean		

Conclusions

State the threshold of your subject. What the threshold of the group? Is there any individual difference in the group? Whose threshold is the lowest and whose is the widest? Is there sex difference in the threshold value?

EXPERIMENT 2 : ASSESSING KINESTHETIC SENSITIVITY

INTRODUCTION

Objects occupy space and we locate them only in space. Our perception of objects is taking place through our senses. We see, hear, touch, smell and taste objects in our perceptual process. Objects occupy a definite position outside and we perceive them for our purpose correctly. We are able to locate the object in space. We also locate the parts of our own body in relation to the space outside and other objects. The way in which the living organism moves and behaves quite aptly in a space is of three-dimensional and to locate objects accurately is a great wonder. Perception of space is a highly complicated process and the operation is taking place in the organism by integration. That is, the perceiver must be engaged in a regulated and integrative activity in the space.

Human beings for their spatial adjustment mainly depend upon their eyes. All other perception, judgment and thinking are based upon their visual world only. In spite of the vital importance of the visual perception, human beings also make use of other important sensory system as the source of knowledge. They provide the perceiver with information about the location of the objects and events in space. We see events in space, but we also hear them from where they are coming. Is it from right or left, from above or below, from front or back. Further we feel them as they touch our hand, arm or other parts of our body.

KINESTHETIC SENSES

When we need to locate objects in a particular part of our body we need the aid of certain sensory systems which are particularly adopted to the function. The first system that plays important role in this function is our kinesthetic system. 'Kinesthesis' means sense of movement. Meuller (1969) states that to deal with the kinesthetic and vestibular senses is one of the most difficult task in sensory psychology. The reason might be that the psychophysical information in these areas are more limited than the other sensory areas. But this system is one that provides us with data that are necessary for maintaining the most delicate characteristics of our normal and coordinated behaviour patterns. The word derived from Greek word *'Kincin'* which means something resulting from motion. Our muscles, tendons and joints are firmly surrounded by or equipped with receptors which respond to movements of the parts. The proprioceptive impulses coming from the muscles are responsible for the maintenance of posture. Kinesthetic system indicates us about the movements of our bodies in space. The vestibular system is aiding this system. The semi-circular canals and the vestibule of the inner ear contain receptors which are highly sensitive to movements.

The vestibular system, semi-circular canals and the kinesthetic sense are discussed always together. Because they are all inter-related. Any damage in the semi-circular canals, damages tone of the three canals on one side of the head inhibits voluntary flight in bird. This has been demonstrated by Flourens (1830) one hundred and seventy years ago. Even in the case human beings any damage to these system cause lack of information from them.

Any damage to the semi-circular canals in just one side of the head results in causing extreme dizziness, nausea and vomiting. There is change in heart rate, blood pressure and perspiration. Any person with this damage will not be in a position to maintain normal erect posture of the body.

It is equally difficult to get more information about kinesthetic sense if it is damaged like the vestibular sense. There are not many studies involving these sensory systems. Experiments on these senses are extremely difficult to see how the environmental factors are influencing the functions of these senses. Further it is not easy to detect the changes in the bodily positions without the tactile system. The lifted weights experiments in psychophysics for determining the threshold becomes the classic in the study of kinesthetic senses (Woodworth and Schlosberg, 1961).

The Physiological Functions of Kinesthesis

There are also two other systems apart from the tactile that are important, involved in this kind of discriminations of kinesthetic cues. The first one is called the muscle sense and the second is called joint sense. The muscle sense is very often involved and provide information necessary for fine skilled movements. In many of the reflex movements this muscle sense is very active. For example, in the stretch reflex and knee jerk.

The stretch reflex is the simplest form of reflex movement. The most commonly occurring stretch reflex is the knee jerk. The physiological mechanism involved in this is that when a muscle is stretched it tends to contract reflexly. When the muscle fibres stretched the sensory equipment in and around muscle fibres are stimulated. Then in the sensory nerves impuleses are generated to serve the muscle tissue and the nerve fibres pass the information to the spinal cord and stimulate activity in motor nerve fibres leading to the muscle fibres. This aspect has been extensively studied by Pattern (1960). The stretch receptors are the sense cells involved in this type of reflex action. If the muscle is stretched the next occurring contraction of the muscle will decrease the stimulation of these stretch receptors. But this contraction increases the activity of the sense cells located in tendon receptors. When there is tension these sense cells are responding to it.

Studies on Kinesthesis

The determination of threshold for kinesthetic sensitivity involve several problems. They are related to the internal position of the receptors. According to Wenger and Jones (1956) the Golgi Corpuscles (organs) are carrying on a threshold about 100 times more than that of spiral endings. The tendons contain Golgi organs or receptors. The Golgi tendon organ is another receptors mechanism stimulated by stretch. The receptor is located in the tendon of the muscle. This is stimulated by stretch from an external pull on the tendon or the pull that caused by the muscle contraction. Candland (1968) states that there is sufficient reason to assume that the muscles are more sensitive to movement than the tendons. According to Leukel (1968) the tendon organ has higher threshold than the receptors in the muscle spindle.

The Golgi corpuscles are widely distributed in the subcutaneous layers and in the muscle and joints and found throughout the body. They have both kinesthetic and cutaneous functions. Laidlaw and Hamilton (1937) quote from Goldscheider that of the nine joints Goldscheider tested the shoulder was the most sensitive part of the body, the wrist was the second and the ankle was the last. The main joint of the big toe and hip were found to be as sensitive as the shoulder. The most common test is to use fingers for kinesthetic discriminations. In one of the studies of kinesthetic discrimination Dietze (1961) used rods, asked the subjects to hold it between the thumb and the forefinger and asked them to judge the length of the rods. He used constant method with standards of 10, 30 and 50 mm. Dietze obtained fine discrimination in kinesthetic finger-span discrimination.

In another experiment of the same type, Stevens and Stones (1959) measured the apparent thickness of the stimuli held between thumb and the middle finger using magnitude judgment. The results were found in terms of Weber's fraction. The subjective increase in size was in agreement with the increase in the width of the stimulus. In another experiment Oberlin (1936) assessed the DL for kinesthesis by using lifting light weights. Oberlin hypothesis was that whatever differences exist in judgment are due to kinesthetic sensitivity. But he found that the Weber's fraction to be between 1/6 and 1/8 with standards ranging from 3 to 600 gm. Further he observed that the shoulder was the most sensitive part and the wrist the least sensitive part of the body.

In the studies of kinesthetic sensitivity it was observed that kinesthetic illusion similar to visual vertical–horizontal illusions occur. Reid (1954) has reported that in one of his experiments he blindfolded the subjects to manipulate a stylus in one direction and than asked them to make the same movement on the right angles. Reid found that the horizontal movements are underestimated and the vertical movement are overestimated.

In our everyday life we will have to make adjustment to depth and other visual situations. This is done by motor response and motor responses always require kinesthetic function. There are not many studies on kinesthetic deficiency. It was viewed that pathological states in kinesthetic sensitivity is very rare. Kinesthetic deficiency takes place if the disease is sufficient enough to damage kinesthesis. Paralysis or some other serious disorders may cause disorder in the kinesthetic sensitivity. In the disorder called tabes dorsalis the neurons in the spinal cord dealing with kinesthetic sensation are destroyed. The individual with this disease is able to walk only by looking at his feet. Since he has no kinesthetic cues he has to walk only by looking at his feet. He moves his steps by looking at his feet. This is an important characteristics of kinesthesis. In this way we know that the kinesthetic sensitivity is mainly concerned with balance of the body, bodily posture and movements.

Therefore, there are several major sensory systems participate in the adjustments which human beings make to locate of objects in the environment. The main one is the visual and the other two are auditory and somesthetic systems. These three systems play important role precisely and successfully in our spatial orientation. There is intersensory cooperation continuously taking place as we move about in the environment.

In all kinds of muscular activities which are involving skills, kinesthetic sensitivity plays its role. In various standing postures, movements and positions of the body kinesthetic sensitivity is involved. We are able to judge the weight of some material. In moving a matter forward or backward kinesthetic sensitivity comes into action. When we stretch one arm to catch something we accurately put our hand on it. We don't stretch too long or too short and then adjust our arm, but we correctly put our hand on the object. In all these activities the role of kinesthetic sensitivity is very important.

EXPERIMENT

Problem

Assess the accuracy with which your subject judges space with his kinesthetic sensitivity.

Method

Participant

Choose a person who has no knowledge about this experiment and collect his personal information like, age, sex, qualification and any observable deformity.

Materials

A quadrant graduated wooden or card board of 1-1/2′ or 2′ square with angles marked from 0° to 90°, (Fig 3.2) and Blindfold.

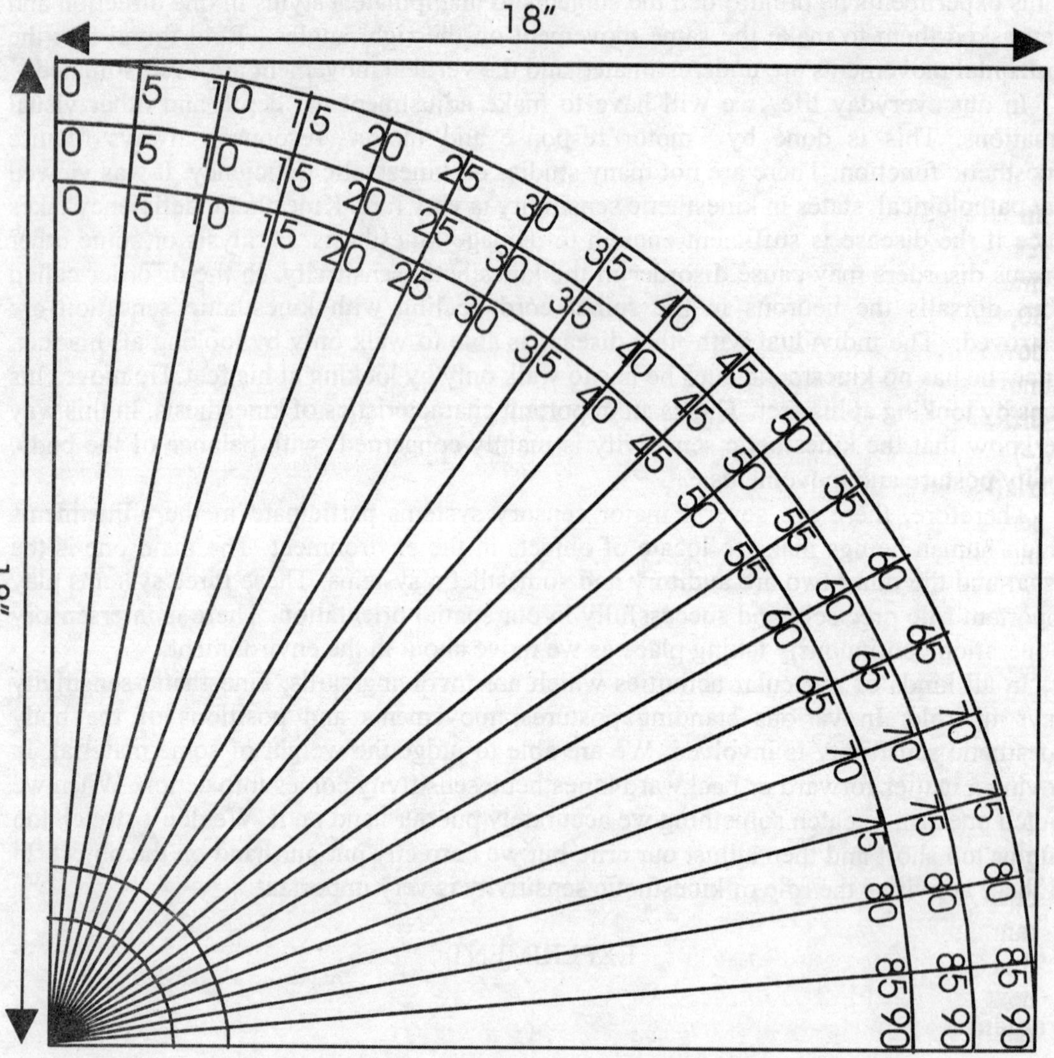

Fig. 3.2 : Quadrant Graduated Board

Procedure

The quadrant graduated board with angles marked (Fig 3.2) from 0° to 90° is placed on a table of flat surface. The participant is made to sit comfortably in front of it. The participant has to place his elbow at the origin of the arc and his middle finger tip should be showing the marking of the degree. He is asked to study the arc on the board and move his forearm from 0° to 90° point in a clockwise manner along with the markings and back to 0°. After this he is blindfolded and the following instructions are given to the participant keeping his middle finger on the zero point. There are two series and this is the first series which is called passive movement.

> "I will lift your forearm, move to the right and place it at a participant and lift it and bring back to the original position and place it. Then you will have to move your forearm yourself to the same place and put your middle finger in the same positions and indicate it".

The experiment should predetermine the position at random order the positions with an interval of 5° in between each position on the board from the 0° point to 90° in his record sheet. Each is to be used two times. In this way if each 5° position is used two times, there are 18 from 5 to 90 (5, 10, 15, 20, 25, 30, 35, 40, 45, 50, 55, 60, 65, 70, 75, 80, 85, 90) positions and for two times it becomes 36 trials. The experimenter should note down the errors committed by the subject in a tabular column. There may be positive movement, moving beyond the position for which a plus (+) sign is given to the difference. There may be negative movement moving to below position, for which minus (–) sign is given to the difference. The experimenter should note down like this for all 36 trials. After completing this first series the experimenter should arrange for the second series which is called active movement. The participant is blindfolded and the following instructions are given to him.

> "In all the previous trials, first I lifted and placed your middle finger to a position and brought it back to zero position and you moved your finger to indicate the position on your own. Now I shall call a particular position and you will have to move your forearm and place your middle finger in the particular position and bring it back to the original position and wait for the next trial."

Then the experimenter should ask the participant to start by saying, "ready go to 20°". The participant then moves his middle finger to the position of 20° and bring it back to zero degree. In the same manner the experimenter should call each position at random two times to make the trials 36 as before. The experimenter should see carefully that the participant brings his arm back to zero position every time before he starts the next trial. Just like in the case of the first series the experimenter should note down the readings in the same manner for the second series also. When all the trials are over the participant can remove the blindfold and he may be asked to give his introspective report and allow him/her to go away.

Results

The difference between the original position where the experimenter placed the participant's finger and the place where the participant himself placed his finger is to be noted down by the experimenter for every trial and recorded in Table 3.4 for both the series.

Table 3.4 : Showing the scores of the individual for both the series

S. No.	Horizontal passive series		Horizontal active series	
	Stimulus in Degree	Response Error (+ or –)	Stimulus in Degree	Response Error (+ or –)
1				
2				
3				
4				
5				
6				
7				
8				
9				
10				
11				
12				
13				
14				
15				
16				
17				
18				
19				
20				
21				
22				
23				
24				
25				
26				
27				
28				
29				
30				
31				
32				
33				
34				
35				
36				
Total				
Mean				

There may be positive error for which a plus (+) sign is given and there may be negative error for which a minus (–) sign is given. The first series is called *passive movement* and the second is called *active movement.* Since the movement is made horizontally the series are called *horizontal passive series* and *horizontal active series.* Find out the total error for both the series and work out the mean scores and record them in Table 3.4. State how many positive errors your participant has committed in the passive and how many in active series. How many negative error he/she has committed in passive series and in active series? How many trials in both series are correct?

Collect the group data for both series from other experimenters and enter in Table 3.5

Table 3.5 : Showing the mean scores (Errors) of the group on both the series

S. No.	Name	Horizontal Passive Series	Horizontal Active Series
1.			
2.			
3.			
4.			
5.			
6.			
7.			
8.			
9.			
10.			
	MEAN		

Work out the mean for both passive and active movements. Compare the means and state which has more positive errors and which has more negative errors. State the highest and the lowest positive errors and also the highest negative and lowest errors. How many of them in the group are more close to the group mean and how many more away from the group mean?

State whether your participant was reducing the errors from trial to trial or increasing the errors. If so in what direction, state it.

Discussion

Discuss the data of the passive and active series and state whether your subject is having more positive scores or negative scores. State what the positive scores indicate and what the negative scores indicate. Describe then the nature of kinesthetic sensitivity. Discuss the group data. How many of them are above the group mean in the passive series and in active series? Are the plus errors more or minus errors more in both series? What it indicates? Discuss it. As a whole, is the group overestimating the positions or underestimating the position? How many of them made correct movements without error? State their backgrounds. Who holds the best kinesthetic sensitivity and who holds very bad kinesthetic sensitivity? State the reason.

Conclusions

1. What is the nature of the judgment of your subjects?
2. State whether your subject overestimated or underestimated the positions.
3. Are there any correct estimates among the 36 trials?
4. What is the estimate of the group? Is it over or under?
5. Where your subject stands in the group? In which rank?
6. How many subjects made correct movements without error?

4

PHYSIOLOGY OF EMOTIONS

INTRODUCTION

Emotion is a subject matter that is often referred to by the general public and the psychologists in different situations. In the common sense world the concept of emotion and feeling are used interchangeably and any feeling is referred to some emotional shade in our everyday activities. It is true that most of our experiences are filled with a feeling-tone which may be pleasant or unpleasant to some extent. The outcomes of emotional experiences are called feelings. We often describe feelings as emotions. But it is very difficult to differentiate these phenomena clearly and with experimental evidences. In the earlier Psychology books emotion is compared with motivation. But modern Psychology books do not say anything about activation instead of emotion. The evolutionary thought process in various areas of psychology that is in progress often replace concepts and phrases from the experimental point of view. Still some modern psychologists equate stress with emotion.

Emotion as Motive

Authors of textbooks in Psychology do not agree among themselves with regard to the definition of emotion. Some of the early psychologists treat emotions as motives since emotions show all the characteristics of motivated behaviour. According to them motive means an inner drive or impulse, which causes the organism to act in a specific way. It is causing motion or movement. Since emotion also is having similar meaning 'move' or 'act' they say that emotion is a kind of motive. Candland (1968) says that motivation has endured many problems that had also afflicted emotion. Many early theorists have regarded emotional states as motivational. Generally, emotion is a motivated state followed by increased awareness in man. It is recognized and identified by the behaviour of the organism. There will be either approach or withdrawal behaviour. Leukal (1968) says that much of this behaviour is autonomic activity and there are widespread activities in the central nervous system. For example, see the emotional states of anger, fear or joy. All these involve high degree of motivation. The emotional state of anger involves approach and aggressive behaviour. Fear involves avoidance or withdrawal behaviour and joy involves peaceful and friendly behaviour.

These emotional states can be observed in the behaviour of the organism. In all these emotional states internal physiological changes such as increase of heart rate, rapid breathing, stomach disorders and many things are taking place. These changes are the indications of the increased autonomic activity and increased function of the central nervous system.

Emotion as Conscious State

Philosophers and early psychologists have been dealing with emotion as a *conscious state* of the organism. In perceiving the world different kinds of awareness are taking place and thereby there are different conscious states. Every conscious state is an emotion causing state. Different conscious state causes different kinds of emotional behaviour. A conscious state is an event occurring in an organism as the result of perceiving the event in the outside world. Therefore, a conscious state is enough to cause an emotion in an organism and for the emotional behaviour. According to William James emotion is a conscious state, which causes emotional behaviour. But all the emotional states are not conscious states. Many of the emotional states are unconscious in the sense that they are not expressed verbally nor overtly. For example, persons who are aggressive and hostile do not express their emotional states outwardly. The persons who have become angry or hostile may even deny their emotional states. Therefore, an emotional state may be having enough conscious state of awareness or may not be having it and also a conscious state may not be necessary for a particular kind of emotional behaviour.

But defining emotion as conscious state is not supported by overt behaviour in all cases. For example, the aggressive behaviour of anger, the avoiding or withdrawal behaviour of flight or the pleasant or friendly behaviour of joy is ignored in many cases. The physiological changes and measurement may indicate the presence of strong emotion. But it is not covertly or overtly expressed. A person who has to learn to control his emotion or a poker faced person can avoid emotional behaviour either in action or in his expressionless face. All the internal physiological responses are there. But no one can observe any change in him outwardly. Persons who are surcharged with strong racial and communal prejudices often act with poker faces in civilized societies.

Bertrand Russell (1951) says that the theory of emotion has been radically changed after the discovery of the part played by the endocrine glands. Cannon's theory and works have been known widely. Russell states that certain secretions from the glands into the blood are essential physiological conditions of the emotions. Physiological changes correlate highly with these adrenal secretions in emotional states. The active emotions are fear and anger. They demand certain kind of behaviour towards an object. But when this behaviour is not called for, these emotions are not felt. The emotions are not overtly expressed. Other emotions like melancholy do not demand an object like anger and fear. Any disorder in liver causes the emotion of melancholy. The emotions that do not demand an object of stimulation are those emotions that do not call for any action. For example, the adrenal glands are stimulated and secrete hormones on seeing a frightful object, a tiger, or on seeing a person who enraged you. Hormones are secreted in an organism on the very presence of a cognitive element. The full emotion of fear or anger arises because the emotion is connected itself to the object which has made the glands to secrete the hormones. This type of behaviour occurs in natural setting. In an artificial setting hormones may be made to secrete, but the cognitive element, which causes fear or anger is not present and, therefore, the emotion fails to arise.

Emotion as Activation

Modern psychologists speak emotion in terms of activation and arousal. Activation refers to general body tension or arousal. It enables the body to move and act. It is very difficult to know the cause and effect. Because it deals with complex set of internal physiological changes. The pattern and the degree of activations are primarily responsible for the feelings and emotions that play very great roles in our everyday life. Candland (1968) states that emotion has its origin in Latin word *movere* which means 'to move out'. In the course of development emotion was given the meaning of 'physical agitation'. Activation theorists considered emotion as a change in psychological and physiological activation of the organism. They treated the physiological changes as the dependant variables.

According to Moskowitz and Orgal (1969) activation is referred to the states of general body tension or arousal in a complex organism, which is accompanied by various types of motivation. Most of human behaviour is followed by activation, because it enables the body to act. Activation takes place whenever physiological factors press for a motivated response. If the internal need is stronger there is greater degree of activation. The more desire one has for a job the greater the degree of tension a person develops. Therefore, different forms of motivation may produce different patterns and degrees of activation. For example, a painful stimulus may produce a pattern of activation called fear. A frustrating incidence or situation may produce another pattern of activation called anger. A sex-related motive may produce a pattern of activation called love. Therefore, the primary basic biological drives or motives make the organism to move and act to satisfy the needs and maintain the constant states for survival.

The activation theorists state that emotion is an act of adjustment. When an organism is in intense emotional state, for example, rage or fear the organism gets itself physiologically ready to meet with the external situation. Bills (1927) states that the secretion of more hormonic of epinephrine in the endocrine glands prepares the organism for more efficient function either for a fight or for a flight. This type of activation, Bills says, prepares the organism for more efficient functioning, it takes place also in verbal memorization and simple mathematical works. These, the activation theorists call as second type of dynamic processes, the energy mobilizing processes. The emotions of rage, fear, love, hate, joy and jealousy are viewed by them as energy mobilizing factors. Activation and arousal functions not only release physical energy for efficient behaviour of the organism but also produce the feelings, which are called emotions.

The strong emotions like the pain, rage, fear according to Cannon (1913) are all followed by a common pattern of physiological changes that prepare the organism for efficient behaviour, like fight or flight. Cannon's views at present are included in the commonly accepted idea of emotion as activation and arousal.

Emotion as Stress

Very recently a team of psychologists brought out an exciting theoretical idea, which is explained by a concept and is quoted by Epstein and Shonz (1971) as the revolutionary new concept. The concept is 'the stress'. Stress is used as a substitute for emotion like anxiety, conflict, emotional distress, frustration and ego-threat.

The concept stress has been in application for long time in the field of psychology, particularly in Clinical Psychology, which mainly dealt with abnormalities of personality, drug addiction, neuroses, psychoses and feeblemindedness of individual cases. Discussing stress in general or experimental psychology and applying it on normal behaviour of human beings is a new approach. Stress generally is defined as any stimulus or situation that disturbs the dynamic equilibrium of the systems of the organism. According to Coleman (1974), stress is an adjustive demand that requires coping behaviour on the part of the individual. Coleman states that the important stresses may derive from severe physical handicap that prevents the individual's activities. The reaction to stress is emotion. The emotional states like anger, fear and anxiety are reactions to stress and these reactions may vary greatly from grief to extreme emotional state like mobilizing energy in the organism to meet the situations. Thus stress is employed to substitute emotions and also activation.

Appley and Turnbull (1967) state that stress has been used to describe the stimuli that are new, intense rapidly changing, sudden, unexpected, persistent, conflict producing, boring, deficient frightening and absent. According to them it has been applied to almost all types of emotional responses. This has been measured by secretion of the hormones of endocrine system, respiratory rates, brain waves, skin resistance, speech disturbance, repetitive action, fatigue, rate of error in performance. Appley and Turnbull state that it is unfortunate, no basic principle has been formulated with all these. Moreover individuals under stress do not exhibit in any common way of behaviour. Under stress situation some respond with increased heart rate and no skin response. Some show decreased efficiency and accuracy. These responses are not responses of abnormal behaviour. These responses are found in any normal human being under stressful situation. Therefore, stress need not be totally associated with maladjustment illness.

For example, anxiety has been classified as a stress emotion. Psychologists have classified emotions into two kinds, one as positively toned emotions and the other as stress emotions. The positively toned emotions are love, affection, joy exhilaration and the stress emotion are anger, fear, anxiety, jealousy and grief. The later emotions are the responses to stressful situations or stimuli. A stress may be very simple like missing a bus or a severe stress like missing the whole property. The type of reaction may vary in each case. The first may cause simple frustration, but the second may cause severe frustration, may put the man into depression and grief. The stress activates and arouses the emotions and these emotions depend upon the severity of stress. Therefore, in this way, according to Brady *et al.* (1958) stress causes excessive hydrochloric acid secretion (HCl) through reticular-hypothalamic activation. In this, reticular system is having more influence upon endocrine and visceral regulation and in this way plays important role in causing physiological changes in emotions and activation states.

Physiological Approach

The physiological approach to the study of human behaviour had its origin when Charles Darwin (1809-82) spoke about posture and bodily expression in man and animals in the work (1872) *Expression of the Emotion in Man and Animals* and the works of William James (1890), Carl Large (1894-1900), Cannon (1871-1945) and Bard and Mount Castle

(1947). The physiological approach did not stop with the study of expression in emotion in man. This approach is also extended to the study of personality characteristics.

At the same time the Russian physiologists were studying the interrelations between excitatory and inhibitory processes in cerebral cortex. Pavlov was working on dogs. Baglioni and Vkhtomsky were working on the nature of reflex act on frogs, Durmishyan on dogs and cats and also on the effects of spinal transaction in man. Pavlov had demonstrated experimental neurosis in dogs. When Pavlov increased the vibration in his experimental procedure to produce saliva in dogs the extreme excitatory made the dogs to suffer extreme disorganization of behaviour. The inhibitory processes have failed to go with rapid vibrations. As a result of this the dog became extremely excited, straining too much and also struggling to escape from the painful situation and also bitten the experimenter. This showed the signs of anxiety attack. Pavlov (1928) states that this form of experimentally produced neurotic behaviour in dogs by extreme excitatory procedure is quite similar to the human neurosis of psychasthenia.

Eysenck, who developed interest to make his study of personality more of biological, was working on Jungian tradition and on the tradition of classical Greek physicians. Pavlov was able to demonstrate and identify sanguine temperament in extreme excitatory neurotic animal and afterwards choleric temperament experimentally. Many years later Eysenck analyzed Pavlov's experimental neurosis and categorized it as melancholic. Eysenck (1957) states that there are certain inconsistencies in Pavlov's description of his four types of nerves. But, however, Pavlov's contribution to personality on the basis of excitatory and inhibitory processes becomes the basis for the observable personality types and development of neurotic behaviour. Eysenck wished to apply the conditioning data of Pavlov accurately in his study of personality characteristics and classification.

All these physiological studies in some way have become the source information for Eysenck (1967) to revise his theory of personality. He postulated relationship between his personality dimensions (introversion-extroversion) and the ascending raticular activating system (ARAS) of the brain. Eysenck has found the physiological causes of excitation and inhibition in the ARAS and based his introversion and extroversion on this.

Eysenck observed that his revised personality dimension consists of mainly the differences in *emotional activation.* According to him, the *neurotics are highly emotional* compared to the normals. He states that the visceral brain is thought to function as the mediators of this type of behaviour. Every emotional behaviour and emotional awareness or activation is followed by the physical events and they are mediated by autonomic nervous system. Eysenck developed his dimensions of personality biologically by making use of the knowledge of structure of the *hypothalamus and limbic system relating to emotional states.* It is to be pointed out that the hypothalamus and other visceral brain units put them into action through the autonomic nervous system. The neural effects range from activation of glands and muscles to heart rate. If there is low threshold of visceral brain activation in an organism, the organism displays high emotional activation which is susceptible to neurotic disorders.

In extreme emotional activation states, *i.e.* extreme fear, intense rage, the normal functioning between the arousal of ascending raticular activating system (ARAS) of the cortex and the visceral brain's emotional activation of the autonomic nervous system break down. The individual is emotionally active and the personality dimensions lose their independence (Eysenck, 1957). This is to show how Eysenck has made use of the findings of physiological experiments in his dimensions of personality in which the role of emotion is involved.

Emotion and the Autonomic Nervous System

The earlier studies in the physiology of emotion were made out of the curiosity to know whether there are any important internal bodily parts involved in the emotional life of man. Most of the researches on the physiology of emotional responses were based upon the physiology of rage. In the extreme emotional state of anger there is vigorous activity in the visceral motor system and the autonomic nervous system (ANS) and it is defined as the visceral motor system. The ANS consists of two divisions, one is called sympathetic nervous system and other is called parasympathetic nervous system. Many writers in the field of psychology and physiology have identified the autonomic nervous system with emotional states.

In the emotional state what happens to the visceral organs of the body? We all recognize many bodily changes when we are aroused. There is increased heart rate, breathing rate, increase of blood pressure, sweating, trembling, tingling of the skin, tension in the muscles, digestion is affected, circulation of the blood becomes disrupted and blood is sent to the parts of the body which demand urgently. These bodily changes to a large extent are the products of the function of the autonomic nervous system. This nervous system is so named because it is not under voluntary control.

The autonomic nervous system consists of bunches of nerve cells and nerve fibres connected with central nervous system, the brain and the spinal cord. The autonomic nerves run to the heart, blood vessels, lungs, stomach, intestine and other visceral parts. These nerves also run to several glands, muscles of the hairs, and to the iris of the eye. These nerves run to smooth muscles and to glands. The autonomous nerves stem from the cells of the brain stem, spinal cord and these are parts of the general nervous system. In a sense it is dominant over the central nervous system as the activity of the ANS may coerce the activities of the central nervous system.

The autonomic nervous system consists of three divisions such as the *upper, middle* and *lower*. The upper division comes out from the brain stem and the lower division which comes out from lower part of the spinal cord plays an important part in stimulating the glands of the stomach for more gastric juice, the duct glands, unstriped muscular of the vital organs, and slows down the heart-beat among other effects. Therefore, the upper division is active in digestive activities. The upper and lower divisions of the autonomic nervous system constitute the *parasympathetic nervous system.* This system is mainly concerned with vegitative activities such as nutrition, oxidation, elimination, growth, self-repair, sex and with urgent needs of the organism in normal situations. It also plays an important role in dealing with emergency function.

The middle division of the autonomic nervous system is called the *sympathetic system* consists of sympathetic nerves, which come out from the spinal cord. This lies in between the cranial and sacral system, at the level of the chest. Nerve fibres are leading to those organs which are active in emergency responses of the body.The sympathetic system increases heart rate, increases blood pressure, checks the stomach activity, secretes more adrenin.

The acts of sympathetic and parasympathetic systems are antagonistic in the nature of their functions. For example, if the sympathetic system increases heart rate, the parasympathetic reduces the heart rate. It is only for maintaining the balance the shift from one division to other takes place. One increases the excited or emotional state and the other reduces the emotional state. It is possible because that every visceral organ of the body receives one nerve fibre from sympathetic system and one nerve fibre from parasympathetic system (Fig.4.1). The autonomic nervous system is not directly under the control of the central nervous system. It is independent to some extent.

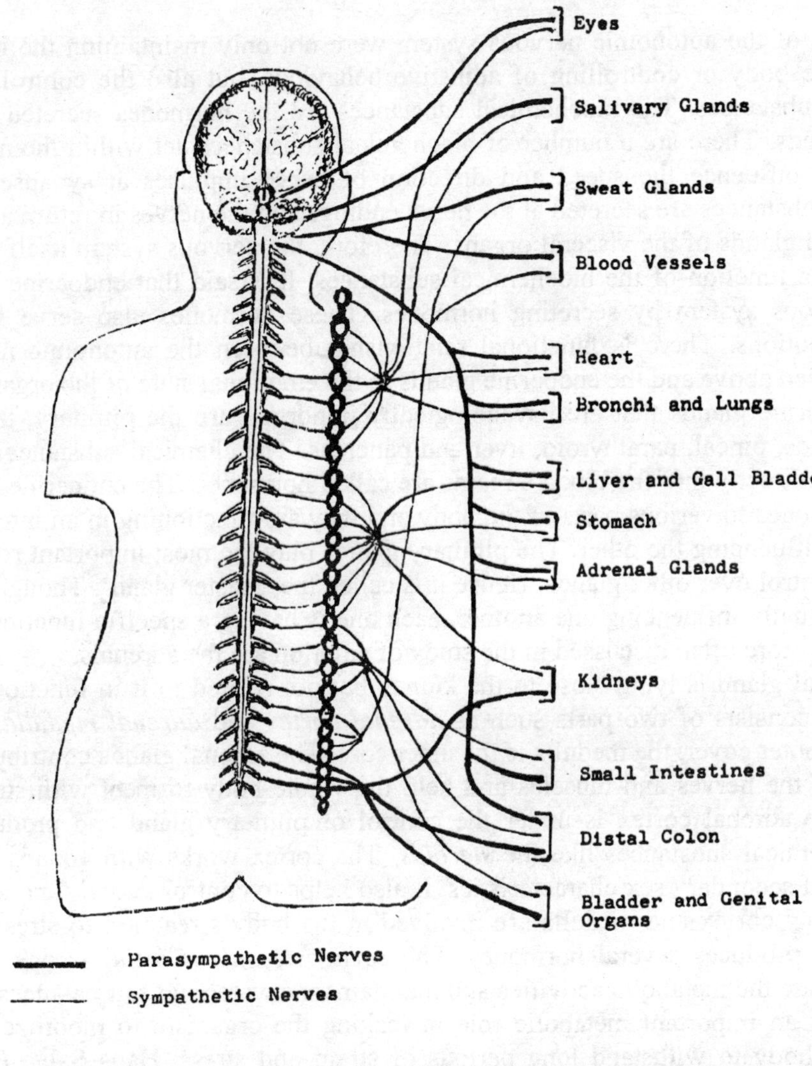

Fig. 4.1 : Autonomic Nervous System and Major Pathways of the Sympathetic and Parasympathetic Nerves

Sternback (1966) states that the sympathetic nervous system is having a mass action, but the para-sympathetic nervous system is having selective action – it selects its target organ for specificity. For example, when an organism is aroused emotionally, the reaction tends to spread all over the body. It also involves the cardiovascular system, the digestive system, the lungs, and endocrine glands and not any one specific organ. It should be noted here that the reactions of autonomic nervous system such as heart rate, blood pressure, and other changes associating the emotional states of the organism and the internal balance or the homeostatis of the body are also maintained by the autonomic nervous system.

Emotion and the Endocrine Glands

The functions of the autonomic nervous system were not only maintaining the internal balance of the body or controlling of adjustive behaviour, but also the controlling of biochemical substances. The biochemical substances are the hormones secreted by the endocrine glands. There are a number of biochemical substances act within the nervous system. They influence the speed and direction of nerve impulses at synapses. The biochemical substances are secreted at the nerve endings and the nerves in return activate the organs and glands of the visceral organs. Therefore, the nervous system itself cannot act without the function of the biochemical substances. It is said that endocrine glands bath the nervous system by secreting hormones. These hormones also serve for the arousal of emotions. There is functional relationship between the autonomic nervous system discussed above and the endocrine glands in the emotional state of the organism.

The endocrine glands that are psychologically important are the pituitary, thyroid, adrenals, gonads, pineal, parathyroid, liver and pancreas. The chemical substances these glands secrete directly into the blood streams are called hormones. The endocrine glands circulate hormones to various parts of the body and they are functioning in an integrated manner, one influencing the other. The pituitary glands play the most important role and it is having control over other glands. Hence it is called the 'master gland'. Though these glands are mutually influencing one another, each one is having a specific function. The glands that are more often discussed in the study of emotion are the adrenals.

The adrenal gland is lying close to the kidney but not related to it in function. The adrenal gland consists of two parts such as *adrenal cortex* and *adrenal medulla.* The adrenal is the outer cover, the medulla is the inner core. The adrenal glands contribute the functioning of the nerves and muscles and help the whole body to meet with stressful situations. The adrenal cortex is under the control of pituitary gland and produces a number of chemical substances like the *steroids.* The cortex works with gonads in the development of secondary sex characteristics. It also helps to control sexual functioning. Both the adrenal cortex and medulla are involved in the body's reaction to stress. The adrenal cortex produces several hormones. The *corticosteroids (Cortin)* is one which serves to regulate the metabolic activities and maintains the water and salt balances. This hormone plays an important metabolic role in making the organism to mobilize more energy in the body to withstand long periods of strain and stress. Hans Salya (1956) observed this in his work on physiological stress. Following the results of Salya's work many psychophysiologists assessed the stress reaction by measuring the amount of corticosteroid hydrocortisone and other hormones secreted into the blood or excreted in the urine.

The bio-chemical substance secreted by the other part of adrenal gland, the medulla is called *adrenin.* It is also called *adrenaline.* This is a very powerful chemical substance, which is having much influence in the emotional life of an organism, particularly in the emotion of fear and anger. Lazarus (1976) states that according to recent discovery the hormone secreted by adrenal medulla is called *noradreline*. The substances are known as *'catcholamines'* — a chemical composition. These substances have overlapping effect and they are associated with different emotional states. Ax (1953) states that *adrenaline* is believed to be the most important in the emotional state of *fear* and *noradrenaline* is found to be associated with the emotion of *anger*. Levi (1965) and Frankenhauser (1975) state that in their studies there is no clear distinction in the functions of these two hormones. There is overlapping between the effects of autonomic nervous systems and the hormones secreted by medulla. The two systems seem to function together in emotional states.

There is some variations in the nomenclature of the biochemical substances secreted by the adrenal medulla. Apart from the names stated above, as adrenaline and noradrenaline the names of *epinephrine* and *norepinephrine* also used by some authors (Leukal, 1964). Epinephrine is a hormone secreted by adrenal medulla whose effects are quite similar to sympathetic nervous system. This hormone arouses the visceral organs. It increases heart rate, increases the blood pressure and increases the beating rate. The epinephrine secretes more hormones in the emotional reactions of fear. It is also a transmitter at the nerve endings of the sympathetic nervous system.

The norepinephrine secreted by medulla, has the same effect as sympathetic nervous system, raises the blood pressure by squeezing the blood vessels leading to the skin and visceral organs. Norepinephrine releases more hormones in the emotional state of anger. This makes the parasympathetic stimulation to lower the blood pressure more in fear than in anger. It is because the parasympathetic system slows the heart and actively opposes the effects of epinephrine. From the effects it is clear that the *adrenaline* and *noradrenaline* are the same as the *epinephrine* and *norepinephrine* substances.

We have discussed the relationship between emotions and bodily changes involving the autonomic nervous system and the adrenal glands. Based upon this relation John Mason (1970) carried out research work in the field of psychoendocrinology. According to him the internal environment is regulated by the biochemical substances and the autonomic nervous systems, the operation of the visceral organs such as heart and digestive system are regulated in proper order, and a condition called homeostasis is maintained for preserving the life and health. But when this order is dislocated or gone out of order the consequence is illness. Mason states that the stress emotions are not functioning in the territory of one discipline. It involves many and each one should specialize in some parts or the whole.

Emotion and Hypothalamus

In the study of the physiology of emotion one of the important parts that is often associated with it is the hypothalamus part of the brain. The hypothalamus is a small area of the brain located in the midbrain, below the thalamus. It contains a collection of cell aggregates or nuclei. It was studied very early by Goltz (1892) and recognized as important in the emotional life of an organism.

The hypothalamus contains important control centres for hunger, thirst, satiation, sleep and wakefulness. It exerts a general control over the sympathetic and parasympathetic activities of the autonomic nervous system. Therefore, it is prime important in the regulation of motivated behaviour including sexual arousal and emotions. Cannon (1927) thought that the primary centre of emotion must be thalamic. He also assumed the relationship between thalamus and the cortex. The functions of thalamic are inhibited by the cortex in normal state. But in emotional state the inhibition in cortex ceases to exist. This has given the strong idea that the thalamus is the primary portion of the arousal of emotion. Therefore, the hypothesis Cannon (1931) formulated was whether there are any particular parts which are responsible for the arousal of emotions, in the idea that an organism in possession of such parts in the brain should show the emotional reactions, whereas an organism deprived of the parts should not. Cannon and his coworker Bard have conducted a series of experiments on dogs by slicing out the brain backward. The dog was showing emotional reactions till the portion in between the cerebral hemispheres and the midbrain was not sliced out. When the diencephalon, the posterior part of the forebrain (from which the thalamus and the hypothalamus develop), was removed then the dog did not display any emotional reaction. This result led Cannon to the idea to ascertain that the hypothalamus portion is the definite emotional centre of the brain.

There were serious objections to Cannon's theory of emotion. Masserman (1943) has verified the result evolved by Cannon and stated that to locate emotion in the hypothalamus is a difficult task. He states that stimulating hypothalamus is a mechanical act. It is induced and the response is diffused, stimulus bound and has no strong emotional indication. There were also criticism against Masserman's statement by Grossman (1967) and others. Masserman's electrical stimulation of the hypothalamus did not interfere with normal ongoing behaviour of the animal. Therefore, Grossman states that Masserman has failed to prove a conditioned emotional response to the stimuli in the animal. The hypothalamus is receiving continuous attention from the physiologists and psychologists who are interested in the motivational and emotional processes.

Emotion and the Limbic System

If emotion is to be understood through the activity of central nervous system, a system having inter-relationship with emotional centres becomes necessary. This system, should include all brain centres and any stimulation or removal of any part should alter emotional behaviour. This is a very difficult task to formulate a system. But, however, some investigations have been made to bring them under a system, which is called limbic system.The management of emotional behaviour of an organism is the utmost important for the survival of the organism. The regulation of emotional reactions in relation to the environment is one of the functions that the brain must perform. This is the important function of limbic system.

The limbic system is a ring of interconnecting pathways and centres that include the septum, cingulate gyrus, hippocampus, enterocrinin area, amygdala and anterior thalamus (Fig. 4.2). Papez named it like this because it looks like a ring, round in the area of cerebral hemisphere. Most of these pathways are associated with smell input. But Papez (1964) says that they are involved in emotional experience. This system has both input and output excitations. It receives input from rhinencephalon mainly concerned with smell sensory areas of neocortex, the BSRF and the hypothalamus. The limbic system sends our excitations to most of the areas of neocortex, the BSRF and the hypothalamus. Papez states that this type of input and output forms circuit and the circuit provides a neural function for emotional behaviour. In emotional states the hypothalamus controls the autonomic nervous system and two of the important systems are found in limbic system.

The limbic system, which includes all these pathways, is regarded as the seat of emotions according to some investigators who made studies on this in first and second quarter of this century. Most of the parts of the limbic system lie in the cortex area. Some of it includes sub-cortical structures below the outer shell, having neural connection with hypothalamus. Limbic means border, hence limbic system is a borderline system. Papez's (1964) contributions attempt to delineate the specific cortical mechanisms which take part in the regulation of emotion. Herrick (1933) said that the limbic system may influence the internal apparatus of general bodily attitude and emotional tone. Papez was influenced by Bard's experiments in the demonstration of the importance of hypothalamic mechanisms in the emotional behaviour of anger. Magoun (1963) states that since certain particular neural structures of the limbic system involved in certain emotions has not been clarified, it is now believed that the entire limbic system may be involved in the regulation of emotional states.

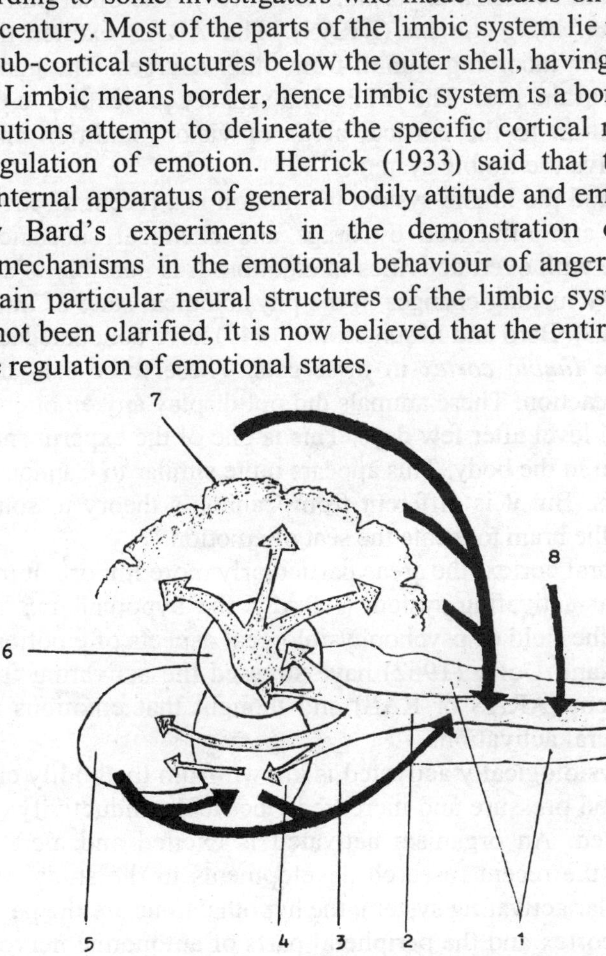

Fig. 4.2 : Visceral Brain and Limbic System

(1) Brain Stem. (2) Ascending Reticular Activating System (ARAS). (3) Medial Forebrain Bundle. (4) Lateral Hypothlamus. (5) Rhinencephalon. (6) Thalamus. (7) Neocortex. (8) All Sensory Systems.

Papez felt that the expression of emotion depends upon the *integrative action of the hypothalamus*. He states that the cortex of the *cingular gyrus* is the region which receives the emotional experience as the result of impulses sent from the hypothalamus region. The cingular gyrus radiates the emotional process to other regions in the cerebral cortex. This adds the shades of emotion in other areas.

The Ascending Reticular Activating System (ARAS) and Emotion

The role of *ascending reticular activating system* (ARAS) and the *diffuse thalamic projection system* (DTPS) in emotion has an activation pattern. The ARAS is a system of many fibres in the brain stem, which activates the whole brain, specifically the cortex. The ARAS originates in the central parts of medulla, midbrain and diencephalon. The ARAS is a part of *brain stem reticular formation* (BSRF). The other part of BSRF is the *descending reticular activating system* (DRAS). The ARAS excites the cells of the cerebral cortex and lowers incoming stimuli from other sources. The important function of ARAS is to keep the brain alert. The DTPS is a sensory projection to the cortex. This carries sensory information to the cortical areas of vision, audition and somasthesis. These two systems involve the limbic system.

The role of ARAS and the limbic system in emotion is a recent discovery. The limbic system and the ARAS are influenced by drugs and chemical substances and exhibit changes, which correlate with overt activities in emotion. It is possible to alter the level of arousal in an organism by making changes in the physiological state of limbic system and in reticular system (ARAS). Bard and Mountcastle (1947) have conducted an experiment in which they removed the *limbic cortex* in three cats. These three cats showed lowered threshold for the anger reaction. These animals did not display any emotional reactions till they return to the normal level after few days. This is one of the experiments conducted to locate the seat of emotion in the body. This appears quite similar to Cannon's experimental result conducted on dogs. But it is different from Cannon's theory to some extent. This involves a wider area of the brain to locate the seat of emotion.

Apart from the cerebral cortex, the areas particularly more important in the arousal of emotion are the reticular activating system (ARAS), the hypothalamus and the limbic system. Researchers, in the field of psychophysiological aspects of emotion like Lindsley (1951), Malmo (1959) and Duffy (1962) have stressed the activating functions of the reticular activating system (ARAS or RAS) and thought that emotions are essentially heightened states of general activation.

A person who is physiologically activated is known from the bodily changes such as increased heart rate, blood pressure and increased electrical conductivity of the skin that he is emotionally aroused. An organism activated is excited and alert behaviourally. Therefore, according to the recent research developments to the study of emotions the limbic system, the reticular activating system, the hypothalamus are the parts of the brain, apart from the cerebral cortex and the peripheral parts of autonomic nervous system and the endocrine system are the important parts involved in the emotional states and experience of an organism. The discussions we made so far are the recent and latest developments taken place in the study of physiology of emotion.

Theories of Emotion

From the foregoing discussions we can understand that right from the beginning when psychologists and physiologists started studying emotions they have been trying to locate it in some parts of the body. This resulted in evolving theories of emotion, beginning with the American psychologist William James and the Danish physiologist, Lange, independently of each other put forth a theory called James-Lange theory which is in other words called the peripheral theory. They held the view that emotions are sensational states. Emotional states often result from sensations, which are arising in the body in general.

The James-Lange theory considers that when a person sees an emotion-producing stimulus, a mass of physiological changes readily takes place in his body. The changes are rapid heart beat, dryness in the throat, erection of hairs all over the body and many other changes. These physiological changes stimulate the nerve impulses, which reach the brain communicating sensory impressions and showing the nature of the inner physiological changes. Emotion, according to James-Lange, simply consists of these masses of sensory impressions, which are stimulated by the physiological sudden change. Therefore, it means that physiological sudden change takes place first and then follows the experience we call emotion.

The physiological sudden change which takes place on seeing the fear or anger-causing stimulus is the basis of the emotional state. This can be demonstrated in physiological experiments. Therefore, for James and Lange emotion is an organic state. It finds its origin in the body. Sherrington (190[illegible]) experimentally verified this theory on dogs. A nerve-cutting operation was made and the dog was deprived of all sensations. This loss of sensations did not affect the dog's emotional behaviour. Therefore, sensations have nothing to do with emotions. American psychologist Cannon (1878-1945) continued Sherrington's (1927) experiment by cutting the sympathetic nerve in a cat and the cat continued to show the emotion of anger. These experimental evidences showed that the organic state of an organism is not the essential part in the emotional state. If the physiological changes are the evidences for the emotional state, these changes like the increased heart rate, increased blood pressure, increased breathing rate are also taking place when the body is put into physical exercises.

These experimental results are intended to show that the emotional state must be elsewhere and not in the periphery. This hypothesis has been studied by Cannon. Cannon's strong point was that when the nerve impulses were cut off the emotional behaviour of the animal remains unchanged. Therefore, the seat of emotion must be in the brain. Similar view was expressed by Bard (1904). Therefore, according to Cannon-Bard theory there are certain centres in the brain, which are responsible for the emotional responses of the organism. Cannon believed that thalamus must be the seat of emotion.

Cannon (1927, 1928) conducted experiment on cats and cut the cerebrum and cerebral cortex of the animal and made it a decerebrated and decorticated animal. The animal showed anger for any stimulus, though not well coordinated. The anger terminated as soon as the stimulus was removed. He and his colleague found the importance of the sub-cortical areas such as thalamus and hypothalamus on the emotional state. However on removing the thalamus the animal did not show the rage reactions. The hypothalamus was found to be integrating the various autonomic and somatic reactions that are vital in emotional states. Therefore, the hypothalamus is an important neural way station in the process of emotion and it is also stated as the *switchboard* of the autonomic nervous system. Later on it was found that the integration made at the hypothalamus level was not complete and a system that could integrate the normal emotional state is sought.

Recently it was felt that there must be a system within the central nervous system, which seemed to be important in the emotional behaviour of the organism.

A group of psychologists who were not satisfied with the existing theories of emotion continued further with all the accumulated knowledge to probe into the matter and was able to come out with some solution and with a new type of interpretation for emotion. They evolved a theory called activation theory and viewed emotion as an extreme arousal and it is an act of adjustment. For them activation is a general body tension, a state of general excitation. Activation occurs whenever physiological factors activate a motivated response. The stronger the internal need the greater the degree of activation. The strong intensity aspect of motivation is revealed in the form of feeling and emotion.

An organism must be alert or aroused to deal with the external stimulation. A part of the brainstem which is intimately involved in the state of activation and *quiescence*, is the *reticular activating system*. It_is the part of the brainstem. Mass of gray matter of brainstem made up of short branching golgi type II cells. The BSRF arouses the brain activity and stimulates or inhibits extensor motoneurones. This system is of two parts one is *ascending reticular activating system* (ARAS) which arouses activity in the whole of brain, specially the cortex and the other is *descending reticular activating system* (DRAS) which is an outflow for the BSRF, increases the muscle tone and increases the sensory feedback and maintains the state of wakefulness. In the state of alert the BSRF activates the cortex via ARAS.

The reticular activating system and the limbic system are involved in activating the emotional states of an organism. The limbic system is a ring of interconnecting pathways and centres that includes parts like septum, cingulate gyrus anterior thalamus and several other parts. It has two functions, through singular cortex the organism can be tamed extremely and an injury on septum may cause the animal highly excitable. Kluver and Bucy (1939) have demonstrated this in monkeys. They removed the end portion of temporal lobes and certain areas of the limbic system. The animal failed to recognize the common objects visually and became hyper sexuality. The lack of eating behaviour was due to the lesion in the limbic system. This means that the more correct place appears to be the hypothalamus.

Therefore, according to the activation theory the limbic system, the brainstem, reticular formation (ARAS and DRAS) and the hypothalamus apart from the cerebral cortex, are intimately connected with emotional behaviour of the organism.

EXPERIMENT 3 : MEASUREMENT OF PHYSIOLOGICAL CHANGES IN EMOTION

INTRODUCTION

Psychologists try to study human responses more specifically to make their studies accurate and precise. There are certain behavioural patterns, which are very difficult for them to study, as they desired. Emotion is one such phenomenon which is not brought under the control of more specific study and classification. Emotion may be defined as a startled pattern or stirred up state of the organism manifesting in various forms like anger, fear, excitement, thrill, joy etc. When an organism is in a highly aroused state or highly depressed state expressing certain responses physically or verbally we say that the organism is in emotional states. Attempts have been made for many years to relate such emotional states with specific situation or with specific bodily change. Such attempts did not yield any fruitful results. But it is observed that the normal function of the organism is much disturbed under extreme stressful state.

Present day authors state that it is activation. It is a state of general body tension or arousal accompanying types of motivation in complex organs, which is called activation. Activation is making the body to move and act (Moskowitz and Orgal, 1969). Activation occurs whenever the physiological factors initiate motivational response. Woodworth (1952) states that emotion is moved stirred up state of the individual, it is a stirred up state of feeling. He says that the heart and stomach and other internal organs of the body are very much disturbed in emotional state. Candland (1968) says that there is no one universally accepted definition of emotion. He considers that emotion is a deviation from the normal state of the organism. Murphy (1964) states that emotion involves everything that is going on in an organism. It involves memory, thinking, imagination and perception of one's environment.

There are physiological studies on emotion made by William James, Carl Lange, Walter B., Cannon, Phillip Bard, Stanley Schachter Christina Maslach and others. They were able to demonstrate that in emergency situation there were syndromes of specific reactions. These reactions are mainly mediated by autonomic nervous system and the hormone of adrenal glands. These reactions are involved with the respiratory system, the circulatory system and the digestive system. Even though these reactions are really taking place under extreme situations there are no experimental evidences to relate the specific stressful situations with the specific bodily symptoms. There are highly similar bodily reactions in excitement, anger and fear, which serve to mobilize the bodily energy of the organism to meet with the stressful situations. These reactions are having several forms. Some of the observable changes are increase of pulse rate, increasing tension of the skeletal muscles, increase of blood pressure, increasing the blood clotting when the body is wounded and reducing the activities of the digestive system. But it is very difficult to say which of these changes specifically is taking place in any particular stressful situation. But we will have to differentiate the emotional states and explain the subject matter.

The different emotional states expressed by the body in the stressful situations would serve only as the partial information for our purpose of explaining the emotional states. The other part of the knowledge of our judgment of emotional states is received from the stimulus situation in which the organism is put up. Another source of information we receive to explain the emotional states is the culture in which the organism is born and brought up. What sort of reaction the culture expects from the organism in a particular stressful situation? We know this, because we might have witnessed on several occasions.

We might have faced severe insult and the reaction to this ordinarily expected is the emotion of anger or rage. The insult resulting in anger is further aided in our judgment as to the overt behaviour of the person. In a given situation to overt behaviour of an organism to anger may be flight or attack. The habit of response of an individual may be anything. But it is helping us to explain and interpret the emotional behaviour with special reference to anger. There are various terms used to differentiate emotions which are lying deep in our responses to stressful situations and we have learned to employ various suitable responses. For example, in the case of insult a man may prone to utter words of violent attack and we describe that the man is angry. The continuous use of these general descriptions of anger may be getting more and more information by adding to it from different instances constitute a well differentiated explanation of emotion. This may help anyone to explain what an emotion of anger is. Therefore, what ordinarily called an emotion of anger, involves various aspects. They are the stimulus situations, which ordinarily should be stressful or provoking, the verbal response of the organism to the situation, a series of internal physiological changes, the overt behaviour (*i.e.,* flight or attack) of the organism in the situation and there are culturally and socially determined responses of the organism. The emotional reactions may vary culture to culture. In some culture it is to be controlled and there should be least expression and in some cultures full expression is given. Therefore, it will not be possible to relate or to correlate, the emotional responses to specific situations generally. All the informations collected with regard to any emotion should be synchronized for correct description.

Candland (1968) discusses four important methods of measuring emotion. They are (1) Physiological measures, (2) Projective testing, (3) Ablations and lesions and (4) Posture.

The physiological measures involve recording the physiological changes that take place in the emotional states. The emotion is the independent variable and the dependent variables are measurement of blood pressure, recording heat rate, muscle tension and physiological responses. The apparatus and instruments used for this purpose are atoximeter, automatograph, electroencephalograph, phethysmograph and the GSR, pneumograph, sphygmomanometer and sphygmograph.

Results of the studies of body sway have been published in various literatures that ataximeter measures the degree of body Sway of a person in experimental situations. Hull's (1973) study brought out some interesting results. The automatograph is an instrument, which records the amount of movement caused upon the plate of the instrument upon which the subject is placing his/her palm or fingers. It is observed that shaking is occurring in the palm or fingers due to the production of emotion by the microswitches placed under the plate.

The electroencephalograph is not normally used in the study of emotional responses. But, still it is believed that it records changes in emotion. Lindsley (1951) states that there is interference of emotional responses in the EEG recordings. The phethysmograph is one of the apparatuses commonly used to measure emotional responses by recording the contraction and dilation of blood vessels. Woodworth and Schlosberg (1954) have measured the changes in electrical resistance of the skin of the subjects by using the galvanometer.

Changes in the respiratory system are measured by using the pneumograph. This apparatus is most commonly used in psychological laboratories. This instrument helps to measure the rate and depth of respiration.

The sphygmomanometer is an instrument normally used by the physicians in hospitals. But it is very much useful in psychological experiment in the study of emotional responses of measuring blood pressure. Altman and Dittmer (1964) measured the blood pressure at different emotional states in various species. The sphygmograph is an instrument normally used for recording the rate of pulse beat and amplitude.

There are several projective and non-projective tests to measure the emotional responses of individuals. The widely used tests are the Thematic Apperception Test and the Rorschach Ink Blot Test, and Taylor Manifest Anxiety Scale. Duffy (1962) used Taylor's scale in his study and correlated emotionality with classical conditioning.

In the method of ablation and lesions, the brain centres are stimulated using them as independent variables and the reaction to this is studied as dependent variable for which several tasks are assigned. The training is given mainly on escape and avoidance. This is based upon the findings that certain areas of central nervous system are correlated with emotional response. Bard and Mountcastle (1947) studied this by removing the limbic cortex in three cats and observed that they had shown lower threshold for the emotional reaction of anger.

The bodily parts of the animals and human beings express the emotional reactions in various ways. Certain emotions are expressed in faces. Certain by way one stands. This can be observed in dogs, cats and other animals. In human beings emotional expression can be seen in the eyes, mouth, looks and the standing posture or movements of the bodily parts like showing the fist and so on.

Measuring bodily changes in emotion : Therefore, the bodily changes should be first investigated. Since the nature of the problem is such, that the investigation should be conducted in the physiological or psychological laboratories. Some of the bodily expressions of the emotional reactions like respiratory changes, increased blood pressure, increased heart rate, pulse rate and skin resistance are often measured in psychological studies of emotions.

The respiratory changes are important bodily changes that take place in emotional states. In the respiratory changes there are two important phases of respiratory cycle observed. They are the inhalation and exhalation of air. Inspiration refers to rapidly in taking of air and immediately expiration takes place more gradually. A breathing cycle is complete with one inhalation and one exhalation of air. In quantitative studies the relative amplitude and the relative duration of the inhalation and exhalation have been used. The respiratory cycle is quantified by the ratio of duration of inspiration to the duration of expiration. It is *I/E* and is called the inspiration/expiration ratio. Woodworth (1938) has discussed this in detail.

EXPERIMENT

Problem

To measure the changes in respiration evoked by stimulus to produce emotional reactions.

Method

Subject

Choose a person who has no knowledge about this experiment. Collect his personal information, like, age, sex, qualifications, height, weight, whether wear spectacles, etc.

Materials

Polygraph with all accessories or a pneumograph, a tambour, connecting tubes, blind-fold, kymograph set, recording parts, smoke paper, etc. and an improved model of Polygraph.

Description of the Apparatus : The pneumograph instrument consists of simply curled rubber tube to be fastened around the chest of the subject. Through an airtight joint this tube let to a small chamber of air. It is a small metal cup covered with rubber diaphragm. A recording lever, on the movement of the diaphragm makes record on the smoke paper of the revolving drum. The rise and fall of the

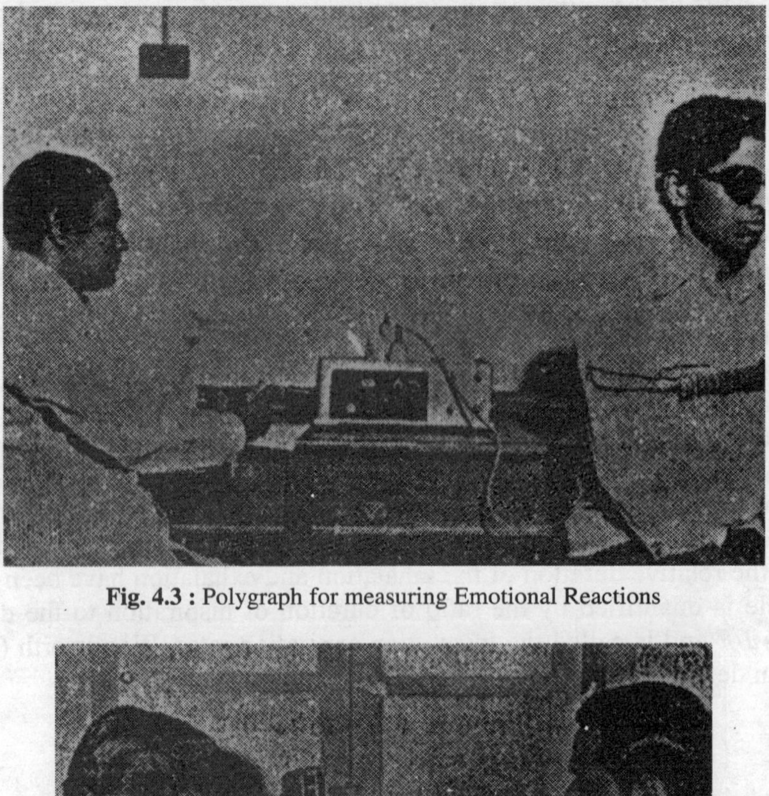

Fig. 4.3 : Polygraph for measuring Emotional Reactions

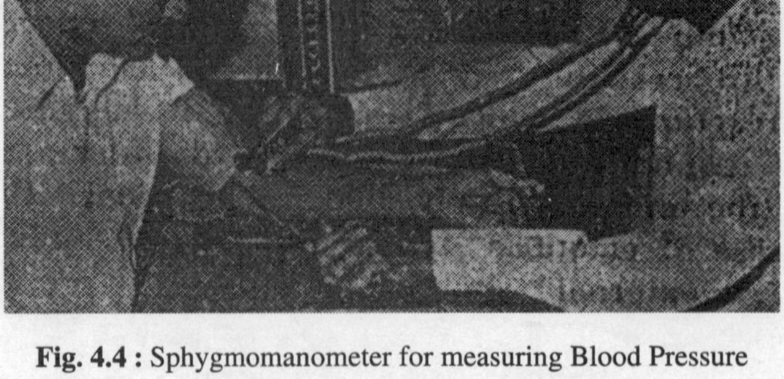

Fig. 4.4 : Sphygmomanometer for measuring Blood Pressure

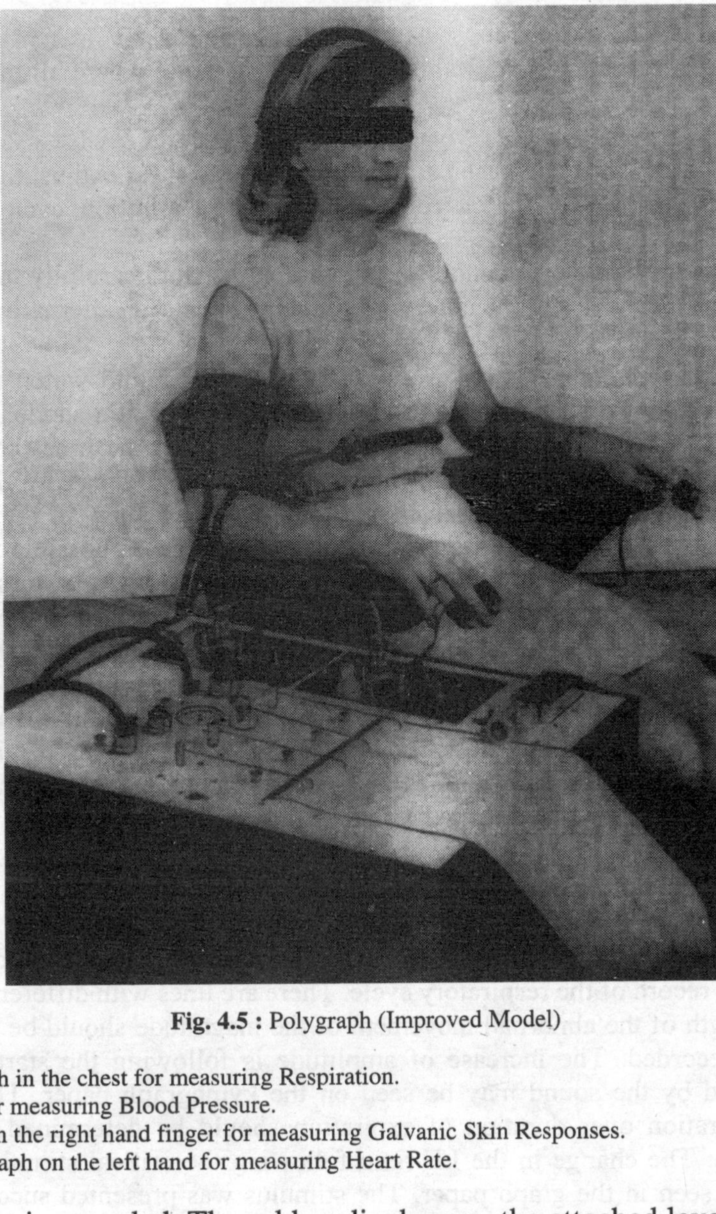

Fig. 4.5 : Polygraph (Improved Model)

1. Pneumograph in the chest for measuring Respiration.
2. Arm Cuff for measuring Blood Pressure.
3. Electrodes on the right hand finger for measuring Galvanic Skin Responses.
4. Plethysmograph on the left hand for measuring Heart Rate.

subject's chest is recorded. The rubber diaphragm, the attached lever and chamber of air (metal cup) constitute tambour. When there is change in the chest movement the lever is making a move and makes a marking in the drum. The I/E ratio may vary in different situations. The I/E ratio may be altered by the tension producing stimuli. A loud noise raised suddenly and unexpectedly, a threat and all these produce change in the breathing cycle and can be recorded. The I/E ratio indicate tensions and expectations. The I/E ratio is also used for lie detection. Other materials like electric shocking device, loud sound producing instrument and blind-fold may be also used.

Procedure

The subject should be seated comfortably in front of the table where all the devices for the experiment are made.

1. The pneumograph is to be fastened around the subject's chest. Adjust the parts of the apparatus in such a way that a regular inspiration-expiration cycle is recorded properly by the writing lever.
2. The electrodes of the Psychogalnometer should be fastened carefully to one hand of the subject, the electrodes to the other for producing shock (an alternative method).

When all the arrangements are over the experimenter should watch the subject's breathing record until a regular rhythm is established. Then all of a sudden without any information to the subject a loud sound be produced, at the same time the key for the signal maker should be depressed. By this means it is possible to locate on the record the exact time at which the stimulus was given and to gauge the latency of the response. In order to get correct result it is better to raise the sound as close as possible to the end of an expiration period. A remarkable increase in the amplitude of the inspiration is more likely to appear. Repeat this procedure several times for recording. The degree of unexpectedness will be higher in the beginning and will be reduced after that. Postman and Egan (1949) suggest that the loud noise is a reliable stimulus and a variety of instruments may be used for this purpose. The most important aspect in this is the sudden onset of the stimulus and high intensity of sound. To compensate the subject's habituation the sound should be given at the regular intervals. The magnitude of the respiratory response will decline during the course of the experiment.

Results

The recording made by the writing lever on the kymograph paper should be studied carefully. It is the record of the respiratory cycle. There are lines with different magnitude (height). The length of the abnormal movement of the magnitude should be measured in millimetre and recorded. The increase of amplitude is following the startle or shock stimulus produced by the sound may be seen on the kymograph paper. The I/E ratio, duration of inspiration over duration of expiration should be determined and plotted against each time. The change in the I/E ratio following the critical stimulus (the loud sound) would be seen in the graph paper. The stimulus was presented successively for which the subject learns to adopt to the stimulus situation and there will be decrease in the effectiveness — that the amplitude may be reducing gradually. Measure the length of the function of time recorded in the kymograph paper during the normal course and give it in Table 4.1, and the same during the stimulus situation in Table 4.2. Find out the difference between the two.

Collect the group data of the same from other experimenters and put it in Table 4.3, and find out the mean. Draw a curve showing the height of the amplitude and the function time.

Table 4.1: Showing Ratio of the Respiratory Cycle with the function of time during normal course

No. of cycle	Inspiration length of function of time (in mm)	Expiration length of function of time (in mm)	Ratio I/E
1			
2			
3			
4			
5			

Table 4.2 : Showing the Ratio of the Respiratory Cycle with the function of time stimulus (sound) situation

No. of cycle	Inspiration length of function of time (in mm)	Expiration length of function of time (in mm)	Ratio I/E
1			
2			
3			
4			
5			
6			
7			
8			
9			
10			

Table 4.3 : Showing the I/E Ratio for the group both in normal course and in stimulus situation

S. No.	Name	I/E ratio in normal course	I/E ratio in stimulus situation	Difference
1	S.V.			
2	P.T.			
3	R.M.			
4	K.R.			
5	P.K.			
6	A.L.			
7	G.G.			
8	S.M.			
9	S.T.			
10	B.S.			
Mean				

Discussion

Discuss the result in the light of the data given in Tables 4.1 and 4.2. State your subject's I/E ratio during the normal course and during the experimental situation. State whether there is any difference between the two situations. Find out whether your subject has shown any alarming sign in this respiratory cycle.

Discuss the group data in terms of the mean scores. The I/E ratio may be varying from person to person. Whose I/E ratio is too high and whose is too low? Discuss the cause for this.

Discuss the curve drawn to the height of the amplitude and to the function of the time of the respiratory cycle for your subject for both the normal course and stimulus situation.

Conclusions

State I/E ratio of the respiratory cycle of the subject during the normal course and during the stimulus situation. Was there any difference? State whose I/E is high in the group and whose is low? Where your subject stands?

EXPERIMENT 4 : MEASURING STATE TRAIT ANGER

Man, just like all other creatures, is motivated for action. Since man is a complex living organism, often goes into the states of general body tension, which is said to be the state of arousal. In other words it may be called activation. Activation is the concept newly introduced in the study of human behaviour to indicate the response to some stimulus. Activation makes the body to move and act. This is a very difficult matter for study since it deals with a complex set of internal and physiological changes, which are very difficult to observe and assess. Activation may function with different patterns and degrees and which may have deep and complicated effects on human behaviour. The patterns and degrees of activation are mainly responsible for the feelings and emotions that play greater role in the everyday life of man. It is observed that when the physiological factors initiate the organism for motivated response activation takes place. If the internal need is stronger than the degree of activation is greater. In other words if the chosen stimulus is more powerful the response or the activation is also stronger.

Activation is converted into activity. The organism is made to move and act towards the stimulus — the goal. If the need is very much for the organism or if the inner deficiency is very severe the organism is motivated to act very vigorously. Activation may be an act of movement or even running. The speed of action is in direct proportion to the internal deficiency of the organism. Animals act more vigorously in times of the feeling of hunger. The more severely the animal feels hungry the greater the speed it takes in its running.

This is true in the case of human beings also. Man becomes more nervous and tensional when he faces competitive situations. A man may be very much interested and highly motivated for some lucrative job. When he faces the interview he becomes more nervous and tensional whether he would get it or not. Sportsmen, athletes and those who participate in competitive games often put in the state of activation and become tensional. The failure of which may lead to frustration or depression.

The organism may be aroused or activated in different forms depending upon the various forms of motivation. For example, painful experiences are caused by certain stimuli, which may produce the pattern or activation of fear. Another pattern of activation is the anger. When the motivated activity is thwarted by some obstacles the organism gets frustrated and may become aggressive and the resulting pattern of activation is anger. At this stage the man who is subjected to this situation may even assault or attack the object of the obstacle. Another pattern of activation is love. It is a sex-oriented motivation, which may become stronger if the loved object is having greater attraction. Grief or sorrow is also a pattern of activation, aroused in a man when he hears a sad news or great disappointment. All the emotional states, fear, anger, love and grief are classified into two categories as pleasant emotions and unpleasant emotions. For example, fear, anger, grief are unpleasant emotions, and love, affection and joy are pleasant emotions.

All the emotions that are activated are having physiological bases. When a man experience an intense emotion such as anger, or fear or love he is aware of a number of bodily changes in him. There is rapid heart beat, heavy breathing, increased muscle tension, perspiration, dryness in the throat, gooseflesh, erection of hairs and so on. Every emotion arouses certain physiological changes in its intense state. All the physiological changes that occur during intense emotional state are activated by the sympathetic system of the autonomous nervous system. This system makes the body alert to face any emergency situation. For example in the case of anger, the person is activated to attack the object of his anger. In the case of fear he takes precautionary measures to escape from the danger. Therefore, the physiological changes prepare the body for the action. During this time the important changes aroused by the sympathetic nervous system are : (1) increasing blood pressure and heart rate, (2) rapid respiration, (3) dilation of the eyes, (4) decrease of electrical resistance of the skin, (5) increase of blood sugar to provide more energy to the body, (6) clotting of blood more quickly, (7) decrease or stoppage of blood in the gastro-intestinal tract and diverting blood from stomach and intestines and sending it to the brain and skeletal muscles, (8) causing gooseflesh, and (9) erection of hairs on the skin. As soon as the action is over by the sympathetic system, emotional state becomes calm and the parasympathetic system comes into action and the body returns to its normal condition. If any intense emotional reaction persistently continued it may cause some bodily ailment. The famous German philosopher Emmanuel Kant (1724-1804) who lived in eighteenth century has stated that emotion is the disease of the mind.

The intense emotional states followed by the physiological changes are measured by an instrument called polygraph. Since polygraph is able to identify the level of emotional states it is used as an instrument in certain criminal investigations. This is commonly known as the "lie detector". Though the validity of the polygraph is not so much trusted in crime detections and not accepted as the valuable evidence in the courts, it serves as a tool for research studies for measuring emotional levels in stressful situations. The physiological changes caused by drugs are also identified by this instrument. But whatever the polygraph records, is not referring to any specific emotion. The readings taken from the polygraph are only general. We cannot say whether the particular readings are referring to the emotion of anger or fear or grief or love. Cannon (1927) states that the physiological changes that occur with the emotion of anger also occur with the emotion of fear. The adrenal glands secrete the hormones in both emotional states of fear and anger. Therefore, the physiological changes indicate that an intense emotion is aroused in the organism and it does not refer to any specific emotion. But the study of the specific emotion of anger has become more important in these days as anger is often associated with some serious consequences.

Early authors on books of Psychology have given more importance to feelings and emotions and extensively dealt with them. They defined emotion as a stirred up state of the organism. It is said that emotion adds spices to life. There is both pleasure and pain. But modern or present day authors do not pay so much attention to emotions. They combine emotions either with motivation or with activation. As result of it studies specifically on the emotions are becoming rare. But we cannot neglect it. Anger or rage is an emotional response to interference or frustration. No human being is free from this emotional response of anger or fear. You get angry with your friend when he fails to understand your argument. The child gets angry if you interfere with his toys. The mild animal cat gets angry if you go near her kitten. The hen gets angry even on her mistress if she touches the chicken. Even Saint Vishwamitra and Saint Paul were not free from getting this emotion of anger.

The emotion of anger is expressed in various ways when a person's self esteem is injured. For example, if you snatch a toy from a child's hand he at once shrieks, shouts, jumps, kicks, bites, rolls on the ground and so on. Anger is one of the emotions that mobilize the whole energy of the body for action. The instinct theorists have stated that anger is the outcome of the instinct of pugnacity. It is an innate hereditary impulse to fight and destroy the enemy. Sigmund Freud (1856-1939) the founder of psychoanalysis says that anger is a form of death instinct. The instinct theory fails to convince us with its explanation that anger is also an instinct like hunger and thirst. The frustration-aggression hypothesis explains that the emotion of anger is the consequence of frustration to some goal seeking activity. When you are hungry and food was not served you get angry. Your anger mobilizes more energy in your body to attack and remove the obstacles on the way of your goal. This anger is the consequence of frustration. In your everyday life you might have experienced several times such anger. Though anger is an unpleasant emotion in some cases it is a potential aid in dealing with problems. In winning a game or in facing some difficult problem you are aggressively attacking the opponents.

There is no inherited facial pattern of anger. It is very difficult to distinguish one emotional response from another. Because emotional responses vary in intensity and human beings conceal emotional expressions. It is true that we express our emotion of anger in some way in our everyday life. But that is not the universal way of response. The emotion of anger involves threats to bodily equilibrium. It is observed in several studies that chronic high blood pressure is associated with persistent anger. Spielberger (1984) states that the rage we expend on others seems to lash back and strike us in the heart muscle. Experts who dealt with coronary heart diseases (CHD) have taken long time to realize that apoplexy (paralysis), cardiovascular attacks, myocardial and endocardial infractions and such internal disasters seem to happen due to frequent burst of intense emotion of anger. A man's 'intense urge to kill' another kills himself. This intense urge is the violent emotion of anger.

Heart research is a subject matter belonging to medical field. But in what context psychologists are interested in the study of heart diseases? Going back in the history of heart ailment and the consequences to the period of Second World War may provide many interesting instances and statistics that the involvement of psychologists in this matter is also important. In has been estimated that at the end of World War II an American male in his prime service meets with heart attack thrice as much as Denish, the Swedish and five times as much as the Japanese. In United States the death rate of heart attack in 1900 was 20 per cent and it was 55 per cent in 1950s. The main causes were found to be physiological such as high blood pressure, high cholesterol, smoking and difficulty in sugar metabolism. These findings were widely published as risk factors and people were warned to be free from these factors to get rid of heart attacks. It was believed that nutrition and exercise can prevent heart attacks in healthy people.

People resorted to dieting and jogging. Recent scientific researches thrown some more light on this problem. Control of feelings and emotion was found to be as important as exercise and control of diet.

Two Sanfrancisco cardiologists have made extensive study on heart attacks. They have classified people. into two categories, the Type A and Type B. The A type people, even before heart attack had been uncommonly ambitious, striving, competitive, hard driving and hurried. The Type B in contrast to Type A were relaxed. The type theory was confirmed after conducting another study after few years that Type A people had heart attacks twice as much as Type B people. Several studies following this found out three major components in Type A person's behaviour who had heart attacks. They were (1) a competitive hard driving way of life, (2) a sense of time urgency and speed, impatience and obsession, (3) excessive involvement in his job. Spielberger (1984) states that Rosenman and others further continued the research on Type A behaviour by using factor analysis with the view of eliminating the characteristics that do not contribute to heart attacks. Out of 44 characteristics of Type A that were included in the factor analysis Rosenman and his coworkers have found only seven proved to be related to coronary heart diseases (CHD). These were: (1) Potential hostility, (2) Getting angry more than once a week, (3) Directing the anger outward, (4) Getting irritation at having to wait in lines, (5) Competitiveness in peer group games, (6) Explosive voice, and (7) Vigorous replies to interview questions.

Spielberger had observed in this study that a single emotion was running through these seven deadly characteristics, hostility, irritation, impatient, explosive and other symptoms. The underlying strongest emotion in these deadly characteristics was anger, which was present in Type A people. Spielberger states that the strongest emotion of anger is also present in Type B people who suffer from heart attacks. This means that the findings of these studies made it clear that anger is the major component of Type A people and anger and hostility are the important factors contributing to heart troubles.

Therefore, the phenomenon of anger has become important subject matter for the psychologists to examine and to see whether the function of anger as an independent factor contributes to coronary heart diseases along with high blood pressure, heart rate, cholesterol and smoking.

EXPERIMENT

Problem

Assess the state trait anger of a person by using suitable test materials and state whether he/she is hot-headed or safe person.

Method

Participant

Choose a person who has no knowledge about this test. Collect personal information such as age, sex, married or single, qualification, occupation, interests, family size, order of his/her birth, income of the family, etc.

Materials

State Trait Anger Scale (STAS), Manual, Test, and scoring key.

Description of the Materials : The State Trait Anger Scale was designed and developed by Charles D. Spielberger in collaboration with Perry London. In 1960s a group of cardiologists were engaged in an extensive research work on heart attacks. After several repetitive studies it was found that some seven deadly characteristics were mainly contributing to heart attacks and Spielberger had observed that for all the seven characteristics the emotion of anger was the strongest characteristics, which was contributing to heart attacks. Therefore, he felt that the level of anger should be measured in persons. As there was no test material to measure the level of anger Spielberger who had spent two decades on the measurement of anxiety and its effects on health had designed the State Trait Anger Scale. This scale at present is used all over the world.

The scale consisted of ten statements, all referring to the emotion of anger. In each statement an idea about anger is expressed. There are four response categories such as almost never, sometimes, often, almost always (Appendix-I) . The participant is asked to state the response suitable to him in each statement. One mark is given to the first response, almost never, 2 marks for sometimes, 3 marks too often and 4 to almost always (Appendix-II). The total marks constitute the individual's score in this scale.

Modification

The State Trait Anger Scale designed originally for the American population and, therefore, there are some slang and informal wording in these ten statements. These are not clearly understandable to the Indian population. Therefore, certain changes in the words were made without damaging the meaning of the words. The modified form was administered among 250 Indian University students and it was found that they were able to answer them without any difficulty and they have expressed their feelings. The modified version is given in the appendices.

Procedure

The participant is made to sit comfortably in front of a table and the following directions were given:

> "There are a number of statements in this sheet. These statements are what people use to describe about themselves. There are 4 responses given under each statement such as, almost never, sometimes, often, and almost always. Read each statement carefully and underline the response with which your feeling agrees. Answer all the statements in the same manner. There is no right or wrong answer. Do not spend too much time on each statement, but give the answer which seems to describe what you generally feel."

As soon as the participant had responded to all the statements, the experimenter should collect the test sheet and score according to the procedure.

Results

Each response the participant has underlined is assigned with the numerical value as given in the manual and the total marks for all 10 statements is the individual score. The highest possible score is 40 and the lowest possible score is 10. Indicate what the individual score is. Collect the group data and present them in Table 4.4. Find out how many are above the mean and how many are below the mean.

Table 4.4 : Showing the scores of the group on State Trait Anger Scale

S. No.	Respondents	Total score	Type		Percentile Score	Remarks
			Hot Headed	Safe		
1	R.V.					
2	M.S.					
3	R.K.					
4	P.V.					
5	S.V.					
6	B.V.					
7	K.R.					
8	A.R.					
9	B.C.					
10	P.K.					
	Mean					

Discussion

Discuss the individual's score in relation to the norms. Whether he scored above or below the percentile rank given by the author of the scale. A person who scores below 13 is in the safe area. He is free from the consequences of anger of heart diseases. He will not respond to any provoking situations. But a score of 23 and above puts the person in hot headed. He gets angry very easily and provoked by situations.

Relate other personal factors for low scoring and high scoring. What type of personal factors facilitate low score and what type of high score. Compare your participant with the group. Where he/she stands? How many of them in the group get angry easily or hot headed. Find reasons for this if any and discuss.

Conclusions

1. State whether you are participant in hot headed or in safe zone.
2. How many in the group are hot headed and safe persons.
3. State the dominating factor contributing to hot head.

APPENDIX-I

STATE TRAIT ANGER SCALE (MODIFIED VERSION)

There are 10 statements given here below. Under each statement there are 4 responses. Read each statement carefully and underline any one of the responses with which your feeling agrees. Do it for all 10 statements.

1. I am quick tempered.
   Almost never, sometimes, often, almost always.
2. I have a fiery (hot spirited fire like) temper.
   Almost never, sometimes, often, almost always.
3. I am a hot-headed (rash tempered) person.
   Almost never, sometimes, often, almost always.
4. I get angry when I am made to go slow by other's mistake.
   Almost never, sometimes, often, almost always.
5. I feel annoyed (irritated) when I am not given recognition for doing good work.
   Almost never, sometimes, often, almost always.
6. I fly off (loose temper) the handle.
   Almost never, sometimes, often, almost always.
7. When I get mad (angry) I say nasty things.
   Almost never, sometimes, often, almost always.
8. It makes me furious (very angry) when I am criticized; in front of others.
   Almost never, sometimes, often, almost always.
9. When I get frustrated, I feel like hitting someone.
   Almost never, sometimes, often, almost always.
10. I feel infuriated (get very angry) when I do a good job and get a poor evaluation.
    Almost never, sometimes, often, almost always.

APPENDIX-II

KEY

Response	Almost never	Sometimes	Often	Almost always
Marks for Statements 1 to 10	1	2	3	4

APPENDIX-III

Percentile Ranks	Trait Anger	
	Female	Male
95	28	28
85	23	23
75	21	21
50	18	17
25	15	14
15	14	13
5	12	11

5

PERCEPTION

INTRODUCTION

Human behaviour constitutes various kinds of activities. These activities are classified as motor activities like walking, running, speaking, cognitive activities like learning, remembering, thinking, perceiving, and emotional activities like laughing, weeping feeling sad or happy. Perception is classified as one of the cognitive activities. In other words it may be also stated as the knowledge getting activity. Bertrand Russell (1872-1970) says that perception is a process of discovering of some kind of knowledge by an observer. Perception cannot be dealt with as it appears to the perceiver, but as tested by an outside observer (1952).

One may think that a man can perceive anything that he notices through his sense organs. Perception is not the activity of the sense organs alone. Though the sense organs are necessary, a person cannot perceive by sight the objects, which are not attended to and not in his field of vision. For example, a man may look straight at a thing without perceiving it. This type of instance might have occurred in the life of every human being. Therefore, we cannot say that a man is perceiving objects by observing through his sense organs alone. The observer can say that a man is perceiving something if only the man responds quite appropriately to the event. Suppose, for example, you are asking a person, "Please lend me your pen for a few seconds" and if he lends it thereupon, it is highly reasonable to say that he perceived what you said. A person and yourself were sitting on a bench in a park. After a few minutes he got up and walked away leaving his bag on the bench. Then you clap your hands to call him, as you do not know his name. Thereupon if he turns and looks at you, then it may be said that the person has perceived your clapping. Like these in many cases we can obtain practical certainty as to some of the things that other people are perceiving what is happening around them.

Perception is one of the oldest subjects on which man has been pondering for a long time for correct knowledge and has conducted quite a number of research studies. Therefore, it has a long history of theoretical and factual information. Various schools of experimental research have studied different perceptual problems. This has resulted in bringing about tremendous amount of information specifically in the area of perception and in general increased the vitality of Psychology on par with biological and physical sciences. Though perception is considered to be one of the topics of psychology, the study of perception started long before the emergence of the scientific study of

Psychology. The old problem on which the study of perception started was, "How man comes to know this world?" This was a highly philosophical problem.

WHAT IS PERCEPTION ?

Perception generally means attaching meaning to an object observed by our senses in the outside world. It need not be only seeing an object, but also hearing, smelling, tasting and touching something. In attaching meaning to an object our past experience or knowledge plays an important role. Physiological psychologists stress the role of the Central Nervous System (CNS) in the process of perception. According to Leukel (1968) in human beings the functions of central nervous system is frequently altered by learning. Sensory impressions interact with memory traces of the past experience to form perception. Therefore, the perception of a stimulus or stimulus pattern or an organism's reaction to the stimulus or stimulus pattern is based upon the past experience with the stimulus or the stimulus pattern of the same type. The organism reacts only to similar type of stimulus and also to similar type of stimulus from other sources. An organism does not react appropriately to an unknown or unfamiliar stimulus.

Perception is a complex process, which is a synthesis of two or more stimuli. Man is surrounded by multitude of stimuli or forces. He either converts them into some kind of knowledge for his use or avoids them if they are dangerous for him. A person knows that there is an object outside of him at a given distance in a particular direction with certain qualities, which are important for him. There may be varieties of stimuli which come in complex form and stimulate him. However complex they are, he was able to organize them in someway automatically and he was able to describe them. This process of organization has taken place in him to give the correct meaning to the object. This is what takes place in the process of perception. Perception is the process of integration and interpretation of what the senses received. The response of the organism to a stimulus is specific and serves some purpose on some particular occasion. Thus the response is selective, purposeful and relevant to our needs. Therefore, we select and organize those stimuli, which are needed for our purpose and leave the rest at the background of our perceptual field. Moskowitz and Orgel (1969) are of the view that perception is a wide range of responses to a stimulus or set of stimuli, which integrates the information beyond that what stimulus has provided. The information and meanings obtained from stimuli may be also added to this when necessary. Garrett (1975) states that perception is an organizing process by which we come to know objects in their appropriate identity. Perception is not operating like an adding machine. Impressions are not cumulative. The mind integrates and interprets whatever it receives at present with the past and makes them meaningful.

Perception is defined in various ways. Wickens and Meyer (1961) state that perception is a process, which gives significance to the stimuli making the perceiver to react to the event in an adoptive manner. Gardner Murphy (1964) says that perception is way of establishing an agreement with the environment playing certain distinct part as required by the perceiver. He emphasizes four different aspects in perception such as similarity, quality, past experience and set. According to him all these four aspects are functioning in perception in some way. It is true that all these four aspects are present in any perceptual process. If anyone of them is lacking perception is not complete and perfect. Perception, according to Stagner and Karwoski (1952), is the process of obtaining knowledge of the external (world) objects and events by means of senses. Man makes use of his sensations meaningful and arrives at certain conclusions about the objects and events of the world. If the object and events he perceived are real and well established he does not doubt his sensory experiences and the knowledge.

The physical objects and events that he perceives at present may be new and unfamiliar and also not well established. Further, they are far away from him. In such cases he cannot put too much of trust on his senses and sensory experience. Here, his knowledge is subjected to verification as chances of errors may be possible in it.

In our perceptual process our past learning and experience play important roles by providing the meaning and the use of the objects and events. Let us think for a moment that an uneducated person from a remote part of the country where there were no electricity, motor cars, buses, trains, industries, big buildings, sea and other modern features etc., and was suddenly brought into a big city like Madras. Under such absolutely new situation what he would look at? He would be in a state of buzzing. He is lost in the midst of innumerable number of colours, contours, brightness and shapes, and the rapid movement of crowds of people everywhere, continuous and non-stop movements of various types of transports. The person is visually and auditorily stimulated by many things and everything is new for him. He is not able to identify any of the objects. Even though he could make use of them and also know the language that everyone speaks there he is not in a state of answering any of the questions relating to the objects, persons, places and events. He must see, use, handle, and investigate the objects and learn to answer any question related to the objects and events. It is easy for such persons and children to name the colours, which are attractive. But in the case of size, shape, distance and motion they must learn and acquire knowledge. But it may take longer time for them to identify the shapes. William James (1842-1910) says that a child's world is a big booming, buzzing and confusion (1890). Perception occurs only when the stimulus or a part of the stimulus is used as a cue to predict the presence of the total stimulus situation. In other words, it is like predicting the presence of the object, for example the flowers at a distance, or like a skilful cook preparing a delicious dish for a dinner. Both can be perceived from the smell without seeing the objects.

We are making severe perceptual reaction to things and events in our everyday life. When the dark clouds are collected in the sky and the cool breeze blows we at once understand that there is going to be rains and take precautionary measures. You are waiting for a particular train on the platform and the railway bell is ringing, then you make all the preparations to get into the railway compartment thinking that the train is arriving at a distance. In all these incidents we are combining several aspects in our perceptual process. Perception reduces the chances of our failure to achieve something and simplifies life process and makes our adjustment more easy and effective. Therefore, perception may be also defined as the process or way of knowing the objects and events in the external world through our senses and making them a meaningful whole with the help of our past learning and experience. Woodworth (1952) says that the individual is dealing with the environment constantly. He is acting on the objects in the environment effectively and by this means he comes to know those objects to some extent, a part of it and not completely. The individual's relation with the environment is active-give-and-take relation. Woodworth explains his definition of perception through a formula: W-S-O_W-R-W. Here, the W stands for the world or environment, the S stands for stimulus which reaches O, the organism, and O is responding to it by his muscular movements, that is his reaction to the stimulus and his reaction produces effects in the world or environment. The O_W is the set of the individual or organism for the environment. Thus, according to Woodworth, perception is the process of getting to know the environment (objects and objective facts) by the use of the senses.

In the process of perception we only speak about what the perceiver (the human being) is doing in his interaction with the environment. The individual's nature of interaction depends upon his sensitivity. It is not with the same type of sensitivity that individuals interact with the environment. Everyone may have different sensitivity in his interaction with the objects or objective facts. Russell (1952) says that perception is a species of wider sensitivity. A rosy plant with beautiful roses may stimulate with different perceptual meaning in persons who observe it with different sensitivities. For example, a poet, a painter, a botanist, a gardener, a young woman and a young man sees the same object in the environment with different sensitivity. The meaning each one gives to the roses and rosy plant varies. The poet takes it to his world of poetic imagination, the painter plans how to make it a classical painting, the botanist wants to investigate the specific species it belongs to, the gardener wishes as how to multiply it and make it marketable, the young woman wishes to have it to decorate her hair, the young man may imagine to please his girl friend with it. It is possible that an object of this type of observation may stimulate persons with different sensitivity. But whatever the sensitivity each person has the perceived object has some meaning. They all come to know that it is a rose plant, having many beautiful roses. Russell says that sensitivity is not restricted to living objects. A material object also may be sensitive to certain stimulus when it is present. A material object behaves in a noticeably different way when the stimulus is absent. Russell quotes many material objects as to show how they are reacting to the presence and absence of the stimulus, like the sensitivity of the photographic plate to light, the thermometer which is sensitive to temperature, the barometer which is sensitive to atmospheric pressure and so on. In all these cases every material instrument react or perceives the stimulus to which everyone is sensitive. But, is that all in perception? In perception there is something more in human perception than what we find in the material instruments. We may say it is consciousness, which can integrate and interpret the aspects of the stimulus. The material instrument is sensitive to only one aspect, whereas man is sensitive or reacting to various aspects of the stimulus.

The material instruments are sensitive to only one aspect of the stimulus. Man may be sensitive to many aspects of the stimulus as he is endowed with many senses. But there are individual differences. All individuals are not always sensitive to all aspects of the objective and events in the outside world. As a result of this what one person perceives in an object is not perceived by another person in the same object or he may perceive a different meaning in it. Garrett (1975) says that the variations in perceptual process depends upon age, sex, intelligence and background of the experience.

SENSATION AND PERCEPTION

In the preceding section while discussing about perception we have been using the concepts senses and sensations. This naturally infers that there is relationship between sensation and perception. In the perceptual process what is occurring first is sensation. We see with our eyes the objects and events in the outer world. Therefore, sensation is the primary response of the sense organ and perception is the secondary response, which provides meaning to the object. This distinction is only for a theoretical consideration. Practically or experimentally we cannot separate these two processes, sensation and perception. Strictly speaking the relation between sensation and perception is not simple. Sensation is a response to a specific aspect of the stimulus or stimulus pattern such as visual response to light, auditory response to sound and olfactory response to smell. Perception, on the other hand, is a response to the total environmental situations. Perception relates our behaviour to a complex pattern of stimuli. Sensation refers to the function of sense organs with their nerves and nerve centres, whereas perception refers to the objects of the world known through the sense organs.

Sensation occurs only when an object or stimulus is present in front of us, whereas perception may occur even in the absence of the object or stimulus by mentioning the name of it verbally. Boring (1942) has made the distinction between sensation and perception. According to him sensation is referred to the action of the receptor when stimulated by a particular object while perception is referred to the meaning given to the sensation.

ATTENTION AND PERCEPTION

Both attention and perception are taking place in the organism. Attention takes place first and then perception occurs. Therefore, it may be stated that attention precedes perception. Attention is active and also quick and rapid. It may shift rapidly and it exists relatively for a short duration. An object may appear only for a short duration and our sense organs are very attentive to the movements of the objects. In sports ground we may be witnessing the races and we may be very attentive to see who comes first or second. We do not turn our attention even for few seconds away from the tracks. At the end we give a complete note of the track events, which we may state as perception. Attention may be stated as the part of perception since it exists only for a short duration. Perception is a mediating process of the antecedent attentive processes. Attention is also selective in a situation. In the midst of many noises we may be attentive to a particular voice, neglecting all others and when we describe it we mediate what we heard in that particular voice.

Attention may be captured and held by the exciting factors in the outside world and also strongly motivated from within the organism. There may be several attention getting characteristics in the stimulus. They are mainly intensity, contrast, change, size, novelty, repetition and duration. But these are also contributing to the process of perception as antecedent causes. It is very difficult to bring out clear cut distinction between attention and perception. Sensation, attention and perception all have many common characteristics.

PERCEPTION OF FORM OR ORGANIZATION OF PERCEPTION

The perception of form is discussed by some authors under the heading of organization of perception. The reason is that perception of form involves various principles of perceptual organization. Man, though living in a complex environmental situation, is able to react to certain stimuli and give organization and meaning in his perceptual process. There may be various kinds of stimuli. They may be having different shapes, sizes, forms and colours. They may be square, round, curved or angular. Some of the stimuli may sound with very high or low intensities and the sound may be continuous or intermittent. Many stimuli may be reaching our senses simultaneously. All these types of stimuli are responded by us in an appropriate manner. There is no confusion or buzzing in our sensory process. In spite of innumerable number of stimuli and of stimulations we are able to make out meaning and things emerge with organized pattern. This process of organized pattern of objects becomes the basis for our appropriate reaction to the situation and to give form and meaning to perception.

In the study of perception the Gestalt psychologists (Kohler, 1887-1968) have made extensive work and made many contributions (1929). Gestalt is a German word, which means form, organized whole or configuration. They have determined how sensory impressions are organized. They argue that objects are perceived as complete, unitary shapes and not as elementary components adding together. According to them the whole is more than the sum of its parts. In other words, a perception is more than the sum of a number of sensations. They have established that there is a tendency in human beings to follow certain basic principle to organize the sensations.

Figure-Ground Relationship

In our everyday experiences we perceive objects against some background. The buildings stand out against the ground, the hill stands out against the sky. The man stands out against the wall. This figure-ground perception is the fundamental principle in the organization of sensations. It is the most primitive process in perception. The separation of figure from the ground is basic in every perceptual process. Morris (1973) states that this perceptual process is a development in the organism independent of practice or training on the part of the individual.

There is also another fundamental principle in figure-ground perception. That is, if we look at a stimulus pattern or the object, the figure-ground relationship emerges first before we perceive all other aspects of the stimulus pattern. This has been experimentally tested by using the instrument tachistopscope. Hundreds of examples can be given from our daily life experience for the figure-ground relationship. Figure-ground perception, according to Moskowitz and Orgel (1969), depends upon certain physical qualities of the stimulus and in many situations, figure-ground perception is partly determined by factors such as eye movements, sensory adaptation, learning and motivation. In visual perception a figure is often differentiated from its background by its qualities such as size, shape and colour. When an object approaches closer and closer the background goes on receding to a distance. The famous ambiguous figure used by Munn (1962) is a good example to illustrate figure-ground perception (Fig. 5.1). If you see the vase in the figure the side view of the two faces disappear and if you see the faces the vase disappears. When one is perceived the other becomes the ground. Familiar objects or figures are more easily separated from their background than unfamiliar objects or figures.

Fig. 5.1 : Figure-Ground

It can be observed that all our perceptual processes are involved in organizing the parts of the object into some pattern familiar to us and giving a definite form to it and also we name it. What has finally taken place out of the perceptual process is that a structured form comes into being. This, in short, can be called perception of form. Perception of form is not a sudden discovery on the part of the individual. It is determined by several factors with which we were familiar in the past. For example, in the perception of form the factors already present there in the object are the size, shape, colour, the visibility of its contours and its brightness. The object may be in a completely structured form or may be in separate parts, sometimes it may be in a disorganized form or it may be in the midst of some other objects. In whatever manner an object is seen outside, the form of the object is perceived. For example, in a room there are several objects lying in various forms. A person is brought in and he was asked, "Do you see a chair here ?" The person, then looking around, says, "Yes, there is a chair, behind the table". In another room a chair is separated into its parts and lying there. The person was asked, "What do you see there?" He says, "I see a chair in parts lying there". In these cases the person's past experience or learning and familiarity with the object help him to perceive the object as chair. The chair and the chair in parts stand out from the background.

Further, in the perception of form, there is a tendency in the individual to follow certain basic principles to organize the various parts of the field into most useful arrangement.

In the perceptual process there is a tendency to see things as a whole. Separate parts or figures are organized into a pattern or structure. There appears to be a built-in tendency in the nervous system to simplify the organization of perception. When we receive the sensations of an object in an incomplete form we try to complete it and see it as a whole. This is also called perception of form. In the perceptual process the incompleteness of the object is overlooked and the figure or the object is perceived as a completed one. The

Gestalt psychologists call this closing the gap, a tendency to see the objects in complete form (Fig. 5.2).

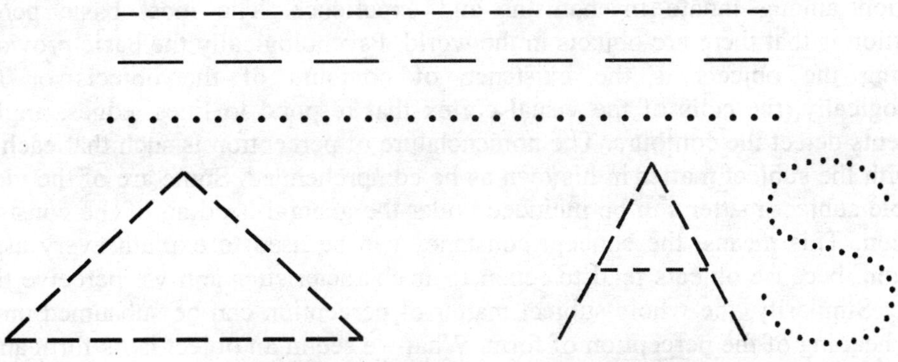

Fig 5.2 : Completing the Form and closing the gap.

Continuity

There is a tendency to see objects in a continuous pattern. Even if one is missing we see them all that they should go with other elements in such a way as to have a *continuity*. For example, in the figure we see the line as continued and the curve as continued.

Proximity

When objects are put close to each other we perceive them as one. This we called *proximity*, which is a tendency to group the objects on the basis of relative distance. Certain individual items may vary to some extent, still we group them in a single pattern. Proximity may be spatial or temporal. When a cracker is fired in a dark place we see the bright light and hear the loud noise. We see these two as parts of the same element.

Similarity

Another important factor in the organizational process is *similarity*. Objects that are having similar characteristics have greater chance to be grouped. Similarity means that objects may resemble each other in several respects such as size, shape, colour, intensity, weight, odour and other physical attributes.

One of the important psychological issues that has been debated for long time is the human form perception. Whether form perception is innate or acquired through experience? Some studies state, it is innate and some state that it is acquired. Recent studies on this show that there is no correct answer for this. According to Fauz (1961) form perception is present even in new born organism if it is exposed to light. He states that young chicken were able to perceive forms, small circles, spheres and also three-dimensional spheres. Similar studies were also made on human infants and it was found that one week infants were able to perceive forms and also showed preference among the forms. They preferred forms like human faces to disorderly pattern.

In Riesen's (1950) study of form perception among the infant chimpanzees which were kept in dark for seven months it was observed that they required training for weeks to develop normal form perception. Therefore, it appears that development of form perception like the development of other aspects of perception is the result of complex interaction among innate mechanisms and experience. The most basic perceptual explanation is that there are objects in the world. Psychologically the basic provision for perceiving the objects is the existence of contours of the objects or figures. Physiologically, the cells of the visual cortex that respond to lines, edges, angles and movements detect the contours. The nomenclature of perception is such that each author deals with the subject matter in his own as he comprehended. Some are of the view that the whole subject matter can be included under the general heading of the constancy of perception. This means, the concept constancy can be used to explain every aspect of perception, because objects tend to retain their characteristics and we perceive them as they are. Similarity, the whole subject matter of perception can be subsumed under the general heading of the perception of form. What we see in an object is its form and form includes the whole object.

PERCEPTION OF SPACE, DEPTH OR DISTANCE

The phenomenon of perception is both philosophical and psychological. Let us not go into the philosophical intricacies. The psychological comprehension has yielded sufficient information to understand what can be subsumed under the heading of perception. In the preceding section we have discussed the perception of form. As we stated earlier the most basic perceptual explanation is that there are objects in the world. The cells of our visual cortex detect the contours of the objects outside. Outside means the space. Objects exist in the space. How do we perceive the objects in the space? Every object occupies space. The world in which we live is spatial in character. This spatial character involves two more phenomena, apart from space, called distance and depth. Therefore, while discussing perception we involve also these three concepts and state as perception of space, perception of distance and depth perception. Some authors discuss the whole subject matter under one heading either under perception of space or perception of distance or depth perception. This may cause confusion to a student who studies it under the heading of perception of space when he comes across perception of distance and depth perception. Most of the textbook authors do not clarify this matter to dispel confusion or doubt from the minds of the students. Practically perception of space, perception of depth and perception of distance all refer to the same thing. Having known something about the processes of form perception and perception of objects, the next question is, how do we know where the objects are, at what distance and at what angle?

John Locke (1632-1704), the British philosopher, had stated in his famous work that perception has to take into account three factors as essential (1690). They are, the mind, the object, and the appearance of the object to the mind (or idea). The idea of the object we perceive is in the mind. In perception the idea of the object we see is in the mind only when the object as well as the mind is present. For example, in imagination and memory we have ideas of the absent objects. Locke's theory of perceptual knowledge is called as the theory of representative perception. It is not direct. The idea represents the object to the mind and mediate between mind and material world. We experience directly only our own ideas and we experience material objects indirectly through these ideas.

The world is mainly spatial in character and it is a ***three-dimensional***. We cannot imagine or think of a world without three dimensions. When we look around, we see objects in three dimensions, at specific distances with all their characteristics like, size, shape, colour and contours. The three dimensions of the object are ***height, width*** and ***depth***. We localize objects up and down (height), to right and left (width) and also near and far (depth or distance). Any solid object is seen having the depth. Thus there is a spatial framework and we see the real world only in this manner. The frame is extended to up and down and to left and right. In the words of Postman and Egan (1949) it is anchored vertically and horizontally. There is high degree of stability in these two directions and they provide total field situation for localization of objects.

For example, we perceive an object, a room, which is a three-dimensional one. The room has a floor and a roof, and walls on the sides – left and right. There is also space between the side walls and the room has a length. The vertical directions up and down, horizontal direction left and right and the room is extended to certain distance. Thus we see the rooms as three-dimensional. At the end of the room a man is standing and he is 16 feet away from the entrance of the room. This 16 feet distance is stated as space, depth and distance. A solid object, for example, a cubical wooden block has three dimensions, height, width and depth. In order to understand how we are able to perceive the world it is important to know how we observe the elementary sensations.

This process can be also explained by showing the distinction between ***distal stimulus*** and the ***proximal stimulus*** fields. There are energies in the physical world. Particularly the actual object from the outer world that would excite the sense organs is referred to as distal stimulus. Hochberg's good example is the door bell. The door bell is the distal stimulus. This is the distal stimulus object. A man presses the door bell. The pressing hits a light, sounds the bell, and gives sensation to the finger. The light waves of the light, the sound waves of the bell and the touch sensation at the finger are all called the proximal stimulation.

When these energies strike the sense organs the receptors excite the retina area of the eye, they bring about the patterns of stimulation which are referred to as ***proximal stimuli***. The patterns of light energies reflected by the object, the book, that we are seeing are ***distal stimuli*** and the light waves in the form of image of the book that strike the retina area are the ***proximal stimuli***.

Any proximal stimulation is an elementary physical event from the object. But how any pattern of proximal stimulation at the sensory apparatus, the eye, can be analysed with simple elementary physical events? Hochberg (1969) states that any distal object can be duplicated by providing suitable proximal stimulation at each sense organ. We can analyse any complex stimulus distribution into simpler physical events for each sense. For example, a proximal distribution is produced at the observer's eye by a singer. This can be also produced quite similar to that by a television picture. The pattern of light energy (proximal stimulus) which reaches the eye can be dealt with a system which may break it down into different intensities at different points in space. Thus we may have a low intensity (dark patch), high intensity (a high patch) and several intensities in between these two. Other methods of analyses can also be made for light and also for sound. This can be also demonstrated in the case of sizes. The same retinal images can be produced by any number of objects of different sizes. The British philosopher Bishop George

Berkeley (1685-1753) first used this principle in his studies of perception. He also wrote about the primary cues and the secondary cues in the perception of distance.

But our sensory systems are not directly in contact with the worldly objects and their properties. Because of this psychologists call them as distal stimuli or objects. They stimulate only our nervous system indirectly by reflecting light energy or sound energy, which may reach our sense organs sometimes and sometimes not. The ***energy patterns*** that reach and affect our sense organs are called ***proximal stimuli***. Strictly speaking we can know only about the distal stimuli or the distal physical world – or the world of objects, space and motion. It is through these proximal stimulus distribution which are acting on our senses that we come to know the world. We will not observe anything if there is any interference in this proximal pattern of stimulation. But if proper proximal stimulation is provided the object can be observed even if it is really absent in the situation.

In perception several major questions can be raised. We observe clear ***difference between distal and proximal stimulus fields***. The first and most important is, how can we perceive the three-dimensional world (distal stimulus) with two-dimensional pattern of our retinal stimulation (proximal stimulus)? The objects or scenes, which we observe outside are not flat like the photographs we see in books and papers. The objects have depth. Man's biological make up is such that the retina of his ***eyes has only two dimensions, height and width and not depth***.

Normally, a person has to face certain situation constantly, in which he has to judge the distance between himself and the objects. While we are walking through the crowded streets we are able to avoid dashing against other people, motor cars, or other moving vehicles. It is because, the perception of distance helps us. If we want to pick up something from the floor we judge the distance and stretch our arm quite appropriately. Trains running fast stop at railway station platforms without any confusion. Aeroplanes from the sky land in airport runways without any deliction, buses stop at bus stops correctly. In all these instances persons involved in them make judgments of distances accurately. Everyday we may be making some such judgments regarding the depth of the objects in space. How deep the well is? How much the barrel can hold? We can explain the process as involved in all these.

How do we perceive space and distance and how do we convert the two-dimensional pattern of retinal stimulation into a three-dimensional pattern and perceive the world? Our ability to perceive the visual depth and distance is due to the fact that we have two eyes and retina of the two eyes are separated by a distance of about 2.5 inches.

While focusing our both eyes on the same object each eye receives a slightly different view of the object. This difference is called ***retinal disparity***. But though we are not aware of it, the right eye sees more of the right side objects and the left eye more of the left side objects. But when the brain and nervous system combine these two scenes (the retinal disparity) as a single view the objects receives a solidity, and becomes three-dimensional. In this way the retinal disparity provides a cue for the perception of depth and distance. When we look at two objects at different distance the images of the objects do not fall on the retina one behind the other. Because, the retina is flat, the image of the far away object can be fallen on the left or right, below or above of the image of another object. Our visual images have only two dimensions. Therefore, a three-dimensional world is a puzzle for us. How it occurs? The two-dimensional images that fall on the retina are converted into three-dimensional perceptions. The two-dimensions provide stimulus information (cues) to evoke a third dimension. This third dimension is depth perception or perception of distance.

The information (cues) provided by the two dimensions fall under two sets. One is related to the (1) organism and the other is related to (2) environment.

The Organismic Cues

The organismic cues are also called primary cues. They are binocular cues, accommodation, retinai disparity and convergence. The organismic or primary cues to perceive distance depends upon the functions of the visual system. These are with reference to the eyes, cues provided by the eyes to perceive the depth and distance in space.

Binocular Cues

This is referring to the cues provided for perceiving the depth and distance by our two eyes. In the perceptual process the two eyes may have certain advantages. Two eyes may provide wider visual area and we can perceive depth and distance more accurately than with one eye.

The Environmental Cues

The environmental cues are monocular cues, superposition, perspective, light and shadow, movement and texture gradient. These are also called secondary cues to distance. The environmental cues depend upon certain physical characteristics of object in the world.

Monocular cues

The information received from the objects by using one eye are called monocular cues. We can make accurate judgments about distance and depth with two eyes. But the monocular cues also often help us to judge distance and depth successfully. We cannot judge certain things very accurately with our two eyes. For example, in shooting, the aim is made accurately only with one eye. The watch maker uses one eye only and also in certain highly technical jobs monocular view alone is used.

Superposition

The superposition sometimes occurs in our perception when we see two objects simultaneously. Some authors (Candland, 1968) state that this is interposition. If two objects are on the same line of our vision one is superimposed on the other and it partially blocks the distant one. The first one is nearer to us and the second is farther away. We could also see the difference between the two objects, that the nearer object is continuous and unbroken and the distant object appears broken by the nearer object.

Perspective

It is also stated as *linear perspective,* a technical word used in painting and drawing for the estimation of distance and depth. Artists can indicate the distant objects and nearby objects in their paintings. The nearer objects are painted as larger and the distant objects as smaller. This is how artist evoke depth perception in the perceivers. Two lines that run parallel to each other may appear to come together at a point on the horizon, is also a cue to depth perception. These converging lines are providing information important to the aeroplane pilots. The pilot could understand the length of the runway, which he is going to take. These cues to judge the distance and depth are called linear perspectives.

Aerial Perspective

There is also aerial perspective which serves as cue in the perception of distance and depth perception. Objects and buildings farther away appear hazier and less clean, whereas nearby objects and buildings look clean and brighter. This we may call aerial perspective.

Light and Shadow

These two are important cues in depth perception. When shadow appears on the part of the object it is farther away. A two dimensional drawing may have a three dimensional look if lights and shadows are distributed appropriately. By this method a geometrical figure may give an appearance of a solid matter with three dimensions. A circle drawn on a paper may look like a flat disc. But by giving appropriate shadows on the edges it may look like a round ball. The shadow gives a different look to an object.

Movement

The speed and direction with which an organism moves in the space serve as cues to perceive distance. While you are travelling in a fast moving train or bus you might have often observed through the windows that objects like the trees and the telegraphic posts close to the railway lines and road side are appearing running fast in the opposite direction and the distant objects are slowly moving with you. In reality the objects are not moving, but only the person is moving. Thus movements serve as cues to perceive distance. Candland (1968) says that the process involved in parallax is important for understanding the visual objects. It has special reference to depth perception. Parallax refers to changes in the scene or objects in the outside world as a result of observer's movement. If one is in motion then motion parallax takes place.

Texture Gradient

Shadows influence the texture of a place or object. Shadows produce a relief effect on objects. Most surfaces are having a texture. The texture of the object provides certain cues for the perception of distance. The nearer objects appear rough, irregular and full of dust while the distant objects look smooth, regular and clean. For example, roads, lawns, house, mountains and valleys have these characteristics.

Accommodation

This is a monocular cue for depth perception. Accommodation in perceptual process specifically refers to the change of the shape of the eye lens. There is a change in its curvature. This function in the lens is needed to focus different objects on the retina area. If the object is quite nearer, the lens bulges and becomes round and if the object is far away, the lens becomes flat. The muscular system of the eye also contributes to the observation of visual distance. The kinesthetic receptors in these muscles get tension when the eye is viewing nearby object, but for far away object it is not. The tension in eye muscle is the cue, which indicates that the object is quite close.

Retinal Disparity

The two eyes receive always slightly different views of objects. The difference between the images of the two eyes is called retinal disparity. The right eye could collect more information from right side and left from left side. These two informations are different to some extent, but we are not aware of it. This can be demonstrated experimentally. Hold your pointing finger three inches away from your nose. Close your left eye and see the finger with right eye and then close your right eye and see the finger with left eye. Repeat it many times. What could you see? You might have seen your finger moving fast from left to right and from right to left. When you see with your left eye the finger look slightly on the right and when you see with your right eye the finger is seen slightly on the left side. It is not moving really. This is because of the retinal disparity. But when you see the finger with both eyes the two different images merge into one and you see the finger is not moving. This is called *stereoscopic vision.* The two images of the two eyes are given a three-dimensional effect in depth perception. The two eyes may have overlapping perceptual fields or slightly different pictures. The binocular perception removes these differences.

Convergence

Convergence means bringing together at a point. Sometimes we are bringing together the two eyes to look at an object. When we look at objects that are farther away from us the two lines of our vision of both eyes are parallel. But if the objects are moving towards us at a close distance our eyes tend to converge. They slightly turn inward towards each other. The kinesthetic sensations control this movement. Thus convergence is a *binocular cue* that provides additional information for the perception of distance and depth. Berkeley as early as 1709 has pointed out that convergence becomes a cue for the perception of distance through learning, in his classical work, *An Essay Towards a New Theory of Vision*. According to him convergence is a sensation and a child links it to extent his reach to an object in the bed. The child also uses it as a memory till he uses convergence alone to the distance of a toy. The two eyes do not converge independently and create two separate images on the two retinas.

PERCEPTION OF MOTION

In the preceding section we have discussed perception of form and perception of distance and depth in space. In these, the subject matter is concerned with static objects and not with moving objects. We keep ourselves constantly moving place to place. The space in which we move is filled with variety of objects moving most of the times. People walk on the street, some fast while others slowly, cars run fast making sounds along the highway, trains run on the rails, birds fly in the sky, aeroplanes fly in the air, ships move in the sea and so on. In all these persons, organisms or objects keep them in motion. The motion or movement of persons or objects in space is a part of perception of objects in space. The perception of motion is determined by certain factors. Postman and Egan (1949) say that the conditions that determine the perception of stationary objects also seem to be operating in the perception of motion. Therefore, it becomes necessary for us to know the various types of motions that could be perceived and the factors that determine perception of motion in space.

One of the important characteristics of our eye is that it is able to perceive motion and provides us with information regarding the destination and the movement of people and vehicles towards us. We are also able to see the speed of the oncoming motor cars, so that any untowards can be avoided. We see the motion pictures and television scenes without any difficulty and could make out the correct meaning of them. The perception of motion is a basic human problem, but we rarely give importance for the study of it. The opposite of motion is stability and we study both. But the ability to see and detect the movement of the object is more important than seeing and detecting the stationary objects. Our eyes are in constant movement. As the eyes are analyzing the world, images of objects move backward and forward over the retinas of the eyes. In this process the rods and cones are stimulated in the retinal areas. Thus all the proximal stimuli are moving and shifting while the eyes are watching the objects in motion. When an object moves across the field of vision, the rods and cones are stimulated in quick succession and motion is perceived. At the same time the distant objects like the buildings, trees, lamp posts appear motionless (stationary). But when the eye moves the images of the stationary objects also stimulate the rods and cones in succession. Thus a complex process, perceiving moving and static objects, is taking place and it takes place automatically, because of the system of compensation. That is, the nervous system coordinates the movements of the eyes and head and changes in the retinal pattern also take place. The rods and cones are stimulated and there is movement of the retinal image, which is in agreement with the movement of the eye and the object is perceived as stable. But if the eye and head movement is abnormal then the compensation is affected and stability disappears.

For example, if you sit in a revolving chair and rotate it for few minutes you will see the world is going round and not stable. But still we perceive the world as stable. This ability to compensate the illusion caused by the eye movements is learned, when we were infants. The perception of motion is relative. We state that an object is moving because another is stable in the same field. This means that a moving object is producing a moving image in the retinal area in contrast to the stable retinal image in the stationary field. The perception of real motion depends upon the relationship between objects within the visual field. The accurate perception of what is moving and what is stationary depends greatly upon the frame of reference in which it is perceived.

The only information we have for accurate perception of motion is the visual field. In the absence of frame of reference illusory perception of motion occurs. One assumption we often make in the absence of frame of references is that large objects are stationary and small objects are moving. For example, a person sees a small spot of light on a large dark screen background and large screen is slowly moved while the spot of light is stationary. The person sees that the spot of light is moving. An illusion of movement called autokinesthetic effect has taken place. This is induced movement of the object. Another illusion of motion that is often experienced is the moon-cloud perception. On a cloudy evening when the moon is visible with a thick cover of moving clouds, the moon appears to be fastly gliding in and out of the clouds in the opposite direction. In spite of our knowledge that it is the clouds that are in fast motion rather than the stationary moon we assume that the moon is in motion. Wallach (1959) states that the clouds are in the field of vision and they become the frames of reference. In reality the frames of reference are stationary and hence we are inclined to see the clouds as stationary and the moon appears to be moving.

In the perception of motion, the motion receptors, which are located in the semi-circular canals of the inner ear are involved. These receptors are stimulated whenever the body is involved in motion. It is this stimulation that provides us the feeling that we are in motion. Any irregular or violent motion may produce motion sickness or a vomiting sensation (nausea). The motion receptors feedback the sensory system. This feedback system contributes to the perception of illusions of motion, which is a tendency to retain the (illusory) motion even after the motion is ceased. Because of feedback we are conditioned to also other forms of stimulation. Hence the train even after it has come to complete halt the illusion of motion is continuing. Further we are also not sure whether it is our train is moving or the train on the next track is moving.

There is another kind of illusion of motion or apparent motion called ***stroboscopic motion***. This forms the basis mainly for films. This illusion of can motion be produced by presenting separate stimuli in quick succession. This illusion of motion is adopted mostly in motion pictures, advertisements and televisions. In all these, separate stills of pictures, each one slightly different from the preceding ones is used. They are presented so rapidly that the pictures blend with one another and maintain a sequence of the events.

The simple form of stroboscopic motion known as the ***phiphenomenon***, an illusion studied extensively and demonstrated in the laboratory situations. The flicker-fusion apparatus experiment is a good example to demonstrate the phiphenomenon. This is a very sophisticated instrument for this purpose. All other instruments are crude. Light is flashed in a fraction of a second and the observer will have to judge whether it is flickering or fusion by seeing through the viewing chamber.

The instrument is designed in such a way that the light is made to flash only for a fraction of a second. It appears to the observer as if the light is burning continuously without any break. But actually there is break between two flashes. The time interval between flashes is the crucial factor in this illusion.

In motion picture and television only stills of different pictures are presented in quick succession to produce the illusion of motion. The gap between one still picture and another still picture is not observed and we see as if there is a continuous motion of the pictures and we do not see several still pictures appearing on the scene one by one. How this illusion in caused? Several studies have been made on this phenomenon. Gibson (1954) says that the temporal-spatial relationships produce the illusion of motion. According to him there is a small real difference between real movement and the illusion of motion in the temporal patterns of retinal stimulation.*

We see in certain situations that objects standing still are seen as moving. Max Wertheimer (1912) has demonstrated this. For example, there are two lights *A* and *B* at a close distance of six inches. These lights are flashed one after the other. The time interval separating the two flashed is about 1/10 of a minute (6 seconds). In this condition we do not see the first light and then the second light. But one light is moving from the field of vision *A* and *B*. This becomes the basis for electrician to produce electrical signboards in which the arrow is moving. The distance between the two lights and the time interval between the two flashes are the crucial factors in this illusion. Within certain limits if the distance between the two positions is longer the time interval also should be optimum between the two flashes to produce the illusion of motion.

There are quite number of such demonstrations on phiphenomenon. Max Wertheimer (1912) and Korte (1965) have formulated the laws of perception of motion and phiphenomenon. Korte's laws of movement state that the preception of motion is determined by three variables such as intensity of illumination, distance between the stimuli and the time interval between the onset and termination of the lights. Therefore, the relationships among the three variables the light, the distance and the time were worked out and which is known as Korte's laws. Fernberger (1934) has produced three-dimensional apparent motion. Cohen (1958) has demonstrated the influence of illumination in the perception of autokinetic effect.

PERCEPTUAL CONSTANCY

Perception is the process of organization of our sensory experiences by which we become aware of the worldly objects, their qualities and relationships and attach meaning and stability to them and it is this perception that mediates human behaviour. One of the Gestalt psychologists, Kurt Koffka (1886–1941), (1950) states that the basic psychological problem in perception is concentrating on the world of sight. But the fundamental fact regarding the perceptual process is object achievement. In other words, it is the reorganization of an object as enduring and necessary. If an object has been

---

* For further discussion of the function of Flicker-Fusion apparatus, *see* M. Rajamanickam, *Modern Psychophysical and Scaling Methods and Experiments.*

perceptually organized as stable and permanent it is stated as constant. This persists in spite of differences. Therefore, the tendency to see an object with its permanent characteristics, under any varied altered conditions of light, distance and position is called object constancy. Similarly we have the tendencies to see the objects with the same brightness, colour, shape and size.

But, how does the individual is able to interpret the constantly changing variables of proximal stimuli in such a way that these same stimuli accurately give back the constancies of the objects of the world? Let us discuss how this process is taking place with special reference to brightness, size and shape.

The organism makes use of certain characteristics in the process of complete perception. These characteristics are both internal and external which we have already discussed in the perception of space. These characteristics are generally called cues. In any visual perception the primary cue is the visual field or frame of reference. It is the complete view of the situation. In perceptual constancies the organism makes the comparison steadily between the specific object with the total frame of reference or situation.

Brightness Constancy or Light Constancy

In perceptual constancies our past experience plays greater role in identifying the quality of the object. Any new impression on our sensory apparatus is judged in terms of our past experience. For example a piece of coal is perceived as black. It is observed as black always. Whatever type of light is focused on it, it is seen as black. The coal in a bright lighting situation may reflect good deal of more light but still it is seen as black. And also a piece of chalk is seen as white and it is always seen as white. Whatever type of light is flashed on it and even if it is taken to dark room, it is seen as white. It is because of our life time experience in dealing with these objects as black and white. Therefore, we perceive them with their original characteristics. This we call as brightness constancy. When some bright light falls on an object in a situation we can see that there are two levels of light intensity present in the situation. The first level is the intensity of light falling on the object – it is called *illumination*. This is also referred to as the incident of light. The object, which received the illumination, reflects certain amount of the incident of light. The brightness of an object depends upon how much light is reflected from the surface of the object into our eyes. The percentage of the total light that an object reflects or the ratio of the reflected light to the incident of total light is called the ***albedo of the object***. If the absolute intensity of illumination is increased the albedo remains constant. For example, if the albedo of a white sheet of paper is 90 per cent in sunlight the albedo is still 90 per cent when taken indoor where there is low illumination. Therefore, even if the absolute amount of brightness is altered the albedo of the object is not changed. The albedo is a physical property on the surface of the object and independent of the absolute brightness of light, which falls on the object. Hence, it is the albedo that accounts for the brightness constancy.

Hochberg (1969) states that it is not brightness constancy as interpreted by some of the authors, but it is lightness constancy. According to him brightness and lightness are not referring to the same thing. Lightness is a continuum running from white surface to black surface, whereas brightness refers to the apparent amount of light coming from the object. A piece of coal in the sunlight and a piece of coal in the shade are of equal lightness, but the coal in the sunlight will appear much brighter. The explanation given by Hochberg is that when light energy falls upon the surface of an object some of the light energy is reflected and some of the light energy is absorbed if the object is transparent. If the object reflects most of the light energy it is seen as white, instead if an object is not reflecting any light energy it appears black. The percentage of the light the surface of the object reflects is its albedo. Even though the amount of light from the object reaching the eye varies widely the physical characteristic (albedo) is observed accurately.

Colour Constancy

Our perception and judgment about colour of an object is based upon the fixed physical property of the object. In judging the colour of an object an organism takes into account the amount of light reflected by the object. Therefore, illumination is a factor in judging the difference between two colours. Colour constancy is sometimes misunderstood with brightness constancy. The reason is that if there is any change in the illumination it may alter the apparent colour of the object. But the colour is not altered. For example, a green coat appears to be green and continues to look green wherever you carry it, whether it is to a bright sunlight area, or to somewhere indoors or to a dark room. Even if you place it in the midst of other colours, under a yellow light the coat looks green. The quality of greenness is more fixed. It may be light tint, or deep shade you see it always green. Therefore, there is a tendency to see the familiar objects in the same colours, irrespective of altered illumination on the object. This is called colour constancy.

Since we are familiar with objects and their colours we see them in their colours under whatever variety of lighting conditions they are placed. Two 'ifferent coloured objects under the same coloured light can be perceived with their original colours provided there are sufficient contrasting appearance and shadows. A black dog is seen by the owner as black whether he looks at it in bright sunlight, or in a dull illuminated room, or in the light under a yellow room light. He is depending on his memory about the colour of his dog. The memory is one of the contributing factors of colour constancy. Unless the object is a colour absorbing one our memory can be trusted in perceiving the same colour. If blue colour is projected on yellow wall it results in grey. This is because of mixing one colour with another colour. However we are able to identify the colours of the familiar objects, this effect will persist. But, whatever result takes place in perceiving the colour, the organism takes into account only the total amount of light in responding to the amount of light reflected by the object – the albedo. That is, in judging the colour of an object, Postman and Egan (1949) state that perception neglects the factor of illumination at least in part. Therefore, the judged difference between the two colours of the objects is based upon the difference between the albedos of the two objects and not based upon the absolute light. Since it is only perceiving the colour, the illumination of the object is discounted and any changes in the illumination do not have any effect in our perception of colour.

Size Constancy

In our visual field we may observe tremendous amount of changes in the size of the objects. The changes of size in our normal visual perception are not noticed. Whatever changes in the size of the objects take place the observed size of the objects remains constant. Distant objects look smaller, while nearby objects look larger. Both the objects may be of the same size. Let us see a moving object, the motor car on a highway. When it is close to us it is larger with its normal size. As and when it moves more and more away from us it becomes smaller and smaller and ultimately it even loses its real shape and looks like an indistinguishable object. This means that distant object looks smaller and nearby object looks larger. Actually there is no change in the object. The change is taking place in our eyes. The perception of size is closely related to the perception of distance. The law of visual angle connects these two. The law is that an object must look smaller and smaller as it recedes from the eye and the decrease in size is proportional to the distance. Holway and Boring (1941) have designed an experiment to demonstrate the size constancy in which the retinal image of the object remained constant in size irrespective of the object's distance from him. There are rows of trees on the road. A tree at 300 feet away looks smaller than a tree at 20 feet away. Why the trees of the same size look with different sizes? It is because of the size of the retinal image in our eyes varies inversely with the distance of an object observed. As and when the object recedes from us the retinal image is going on decreasing in size. If the object is nearer the image in the retina is larger and we see the object larger. When the object is far away the retinal image is too smaller and we see the object as smaller. But in our perceptual process there is a tendency to compensate the difference in the size of the retinal image. This tendency is to see the size of the object to remain constant by making changes in the size of the proximal stimulus is called size constancy.

The size constancy also includes certain related tendencies to see the larger objects as larger and smaller objects as smaller with the same proximal stimulus sizes or retinal images. It is because we know the true physical sizes of the objects and also we are familiar with size cues. For example, for familiar objects such as people, animal, houses, automobile learning is important to judge the size. When we recognize an object we judge it with an apparent size irrespective of its distance. We have cues to judge distances. But if the cues for distance are eliminated then the size constancy breaks down. The shape of the objects appear relatively constant irrespective of the positioning space within certain limits. It has been estimated that the real size of the object can be correctly judged upto a distance of half of a kilometre.

Shape Constancy

We observe objects often from different points or angles. The retinal images also vary accordingly. From whatever point or angle we look at an object we perceive it with the same shape. It is the point or angle from which we see an object causes change in the shape of an object. Actually the shape of the object is not changing. Changes take place in the retinal images. From whatever the angle the object is projected the object is perceived in the same shape. Hence there is a tendency in our perceptual process to observe objects with the same shape from whatever the angle it is presented to our view. There are changes in the retinal images as the shape of the object changes at every angle. The angles distort the retinal images. Generally the shape of an object is observed in relation to its adjustment within the total field.

We could notice several instances of shape constancies. The shape of an object may give us different shape while the object is moving either from left to right or right to left or receding farther away from us or approaching towards us from far away. At every position the object may appear with different shape. But we see it with its original shape and characteristics. For example, a door is seen with rectangular shape when it is seen in normal position. When it is opened, the door is perceived with the same rectangular structure in shape, even though the shape of the door as projected on the retina goes through series of twisting out of the shape. The rectangular shape of the door becomes a trapezoid. The edge of the door towards us looks wider than the edge hinged with the frame. Slowly the trapezoid becomes thinner and thinner and all that is projected on the retina is vertical line, which is the thickness of the door. These changes we can observe and distinguish. But we do not accept it, our experience is that the door moves forward and backward on its hinges and is not changed. We see the door with its original shape. This tendency to see the objects with their original shape irrespective of their angle positions, disregarding the apparent change of its shape is called shape constancy.

A rupee coin lying on the table may project its image on the retina as an ellipse or an oval shape and we see the coin on the table appearing like an ellipse. But the completeness of the round outline of the rupee coin is maintained and we see it always as round.

Thus objects are perceived by us as they are. The order and stability of the perceptual world greatly depends upon perceptual constancies. It is because of this that we perceive objects as existing on the same size, shape and colour whatever the image of the object's size and shape fall on the retina area.

PERCEPTION AND LEARNING

In the preceding section the phenomenon of perception has been discussed under various topics like perception of form or organization, perception of distance and depth, perception of movement and perceptual constancies. The problem facing us now is, how to explain what has happened? This is with regard to the individual who perceives the objects in the outside world. How correctly the person perceives the world? How it becomes possible for him? Has he got any innate tendencies to perceive and judge objects correctly? Or has he learned to perceive and interpret the objects in the visual world? This is a problem of nature-nurture and there has been an endless discussion of this issue among the philosophers of seventeenth and eighteenth centuries. A group of idealistic philosophers like Rane Descartes (1594–1650) a French philosopher and mathematician and also founder of modern philosophy, and Emmanuel Kant (1724–1804), a German philosopher, who were also called as nativists, claimed that ideas exist innately in the mind.

But against this idea, the triad of British philosophers, John Locke (1632–1704), George Berkeley (1685–1753) and David Hume (1711–1776) who were called empiricists approached the problem with more empirical and scientific outlook. All these three approached philosophy exclusively from the side of psychology and they were concerned to study man rather the universe. Locke (1690) in his famous work *An Essay Concerning Human Understanding* states that we come to know the material objects only through their sensed appearances, which could vary through what is called our ideas of them. Three factors are necessary if perception should take place, such as the mind, the object, and the appearance of the object to the mind. Real knowledge according to Locke is the agreement of our ideas with things in the physical world and he says that knowledge is generated by experience. Thus he combines both the idealism and the empiricism. He is a dualist. Berkeley states that when we have an idea of anything (in perception) it is a specific object that is present in our mind. It is not the full house we perceive, but a part of it, a square (one side). We cannot have abstract ideas. Our ideas are sensational in character. But Hume (1740) though an empiricist was also a rationalist. His sceptic philosophy was explained in his famous work, *A Treatise on Human Nature*. He has stated that human knowledge is that which can be perceived by the senses.

The early psychologists like Hering and Helmholtz were also holding opposed views on this. Most contemporary psychologists like Max Wertheimer, one of the founders of Gestalt psychology and others are of the view that in perception there is integration of these two views. Gestalt psychologists believe in insight and human beings and certain animals learn by insight and also solve the problems by insight. Has this insight any relationship with innate ability or idea as conceived by the nativists? The Gestalts did not explain this phenomenon. We cannot completely reject the idealistic view that it is the innate idea that is projected upon the object. The argument is that we are not aware of a computer fifty years ago. The idea existed in the mind of a scientist and he projected this idea on material in the form of an instrument and it came into existence. This was how every scientific inventions came into being. The computer as a material object was not in existence previously. It existed only in the mind in the form of an idea.

Some equate insight with intuition. Then it becomes a philosophical phenomenon. Henri Bergson (1859–1941), the French philosopher states that there is a kind of knowledge, which is absolute, immediate, direct and non-symbolical – which is intuition or intuitive knowledge. According to him intuition gives us the object in its completeness. Science bases its theories on metaphysical intuition and its concepts are influenced by that intuition and are also constructed on its basis. Bergson states that the great classical theories of science were aroused from an original intuition and even Newton and Galileo have built up their theories on an insight which came to them in an immediate and illuminating flash. Intuition is made possible if we already have fund of experience and observation. Science would be most beneficial if it collected a great mass of observations and then strive to achieve intuitions or insights. Bergson states that though intuition is difficult to achieve, itself is a simple and not mysterious, but a faculty, which we all at some time have exercised it in an illuminating moment.

To the objection raised against Bergson's view of intuition as the foundation of scientific truths, Bergson seemed to have indicated a situation where Newton has a sudden flash of intuition in which the reality of gravitation was disclosed to him and his analyses and theory were only supplementary processes. Now such a thing as insight in science is undisputed. We may call it intuition if we wish. It is a spontaneous occurrence of a fertile idea to a scientific man.

The passage in the preceding section denotes, how intuition and insight have been treated synonymously in knowledge getting process. The psychological meaning of insight or insightful learning in knowledge getting process was not thoroughly disregarded in the philosophical interpretations. In the correct perception of objects, their size and distance the psychological explanation is that man makes use of signs and symbols and he learns the meaning of the same from experience. Woodworth (1952) states that one of the main tasks of the psychologist in the study of perception is to discover the signs that are often used by everyone in perceiving the objects. Psychologists often speak of cues instead of signs and symbols. The multitude of signs and symbols that human beings use in their life have to be only learned.

Cues, or signs help us in learning to discriminate objects perceived. Garrett (1975) considers that perception and learning are parts of the same process. A young child is not able to discriminate between similar objects. There is only the process of generalization. For a child anything that is used for sitting is a stool. He learns slowly that there are also chairs and benches used for sitting. Sensory experiences first lead to generalization and then discrimination takes place. This process occurs both in animals and human beings. Discrimination is a selective response which takes place as a result of learning and maturity. Discrimination of certain cues needs specialized knowledge and persons not adequately trained will not pay any attention to them and they will not make use of the information.

Visual Deprivation and Perception

The study of perception is also associated with deprivation of sensations. Persons who are deprived of senses cannot learn to perceive anything in this world. But a man would not have lost all his senses. A person who has lost his vision may learn to perceive certain objects by making use of other senses like touch, hearing, smell or taste. Studies have been made specially with reference to visual deprivation. With regard to visual deprivation and perception, the often quoted illustrative example has been a letter received by John Locke from his friend about a born blind man. The letter was that a born blind man, when grown into an adult was taught to distinguish between a cube and a sphere of the same metal by his sense of touch. The man has learned which is cube and which is sphere. Suppose if the cube and the sphere were placed on the table and the man all of a sudden got his eyesight, could he distinguish and tell which is globe and which is cube? Locke told his friends that he could not distinguish them. This reply of John Locke was from an empiricist's point of view. He can say which is globe and which is cube only by examining the objects by his touch sensation.

In one of the experiments conducted by Senden (1960) on persons blind from birth and got their vision by operation when they were adults it was observed that the patients got confused first of the visual stimuli. But the person was able to distinguish the figure-ground relationship and analyse them and understand the moving figures with his eyes. These abilities appeared to be innate. But he cannot identify objects by sight alone those which were familiar to him by sense of touch such as keys, knives, and faces. He cannot distinguish a triangle from a square without tracing the outline with his finger and counting the number of corners. Even after learning to identify objects his sense of generalization was poor.

The results of these studies suggest that when blind adults got their vision suddenly their perceptions develop gradually from primitive visual experience of distinguishing the figure-ground relationships and colour to more detailed perception by practice. It takes longer time for them to distinguish shapes. But these evidences are not sufficient to conclude that these are due to innate visual abilities.

An English patient who lost his vision of both eyes at the age of ten months, had his vision restored surgically at the age of fifty two was tested by Gregory and Wallace (1963) with several visual tests after one month of the operation. He described his first visual experience that he had learned a voice coming to him from the front and to one side. He said that when he turned to the direction of the sound he saw a dim indistinct shape and said he realised that it must be a face. He said that he did not know that voice, come from the faces. He judged the lines in Muller-Lyer illusion having the same length. He showed no sign of depth perception. He could not make out anything from the photographs. He misjudged depth. He thought that it is easy to step down from a second story window to the ground a distance of forty feet. The patient has learned to read a watch by feeling the position of the hands. He learned to draw a picture of a bus and he has improved the quality of his drawings.

The patient at first was very much delighted for having got the vision and was enjoying the sight on the road, the moving cars and the trucks through the window of the building from the fourth story. The incidents gave him a mixed feelings of joys and sorrows. He was very active making simple things in the shed, riding bicycle with his friend who was guiding him by putting his hand on his shoulder. He could cross the streets in heavy London traffic. But after getting his vision he was very much frightened to cross the streets even with the help of two people. The world for him was a drab and full of defective objects. Gregory (1977) states that people after many years of blindness become psychologically depressed on recovery of sight. They often realize what they have missed in their life, more than their visual experience they were denied of opportunities during their blindness to do things. The patient, whom Gregory tested, instead of putting on lights, he sat in darkness and gradually given up active living and died after three years. These studies on people who were visually deprived indicate that learning is important for perception. Further, other evidences indicate that some aspects of perception mainly depends upon experience.

Perception in Infants

The new born infant, as Gardner Murphy (1964) said, has no shape or clear perception of anything from all the evidence we have. In the words of William James (1890) for the new born infant the world is a huge bellow, a blooming, buzzing and confusion. For the new born child every sensory system should function soon after the birth and by making use of these sensory functions the child should learn more complex perceptual process by developing more perceptual abilities. The child misses many things in this world if he is deprived of any of the senses or failed to develop a particular faculty by learning and experience right from childhood period. Experiments have been conducted by Bornstein, Kessen and Weiskolf (1976) on four-month old babies by the method of habituation whether the babies see anything new. In their test of visual sensitivity of these babies, they observed that the babies were able to distinguish the psychological primary colours, red, green, blue and yellow. Wertheimer (1961) has observed the auditory response in his own new born child that his daughter was able to locate the sounds ten minutes after her birth. Fantz (1961) studied visual acuity in infants. He observed that visual acuity in early infancy period was very poor. According to him babies do not adapt to focus on objects at varied distances until they are about one month old.

The perception of depth of infants was tested by Gibson and Walk (1960) in infants aged 6 to 14 months by placing them on the centre of a visual cliff. The mothers called the child from cliff side and from shallow side. All the infants crawled on the shallow side, but refused to crawl on the deep side. This shows that at this age the infants have depth perception. But depth perception cannot be tested on infants until they are old enough to crawl.

Recognition and Identification

In our perceptual process we sometimes make quick perceptual judgment and sometimes we hesitate or wait in making judgment if we are not fully sure of the cues, we have received in our perceptual process. In the perceptual process we are learning to abstract the meaningful characteristics of a person, object or event and if we find them suitable we also apply them to different situations. In this process of perception and association the important factors operating are recognition and identification. Moskowitz and Orgal (1969) state that recognition and identification is a process of conceptual learning, the learning quite similar to the concept formation. We perceive people and objects, name them, classify and categorize them.

For example, while walking on the road, among many men, women, and animals some one catches your attention and you recognize him as your friend. This recognition is not by the colour of the shirt he was wearing or the two wheeler on which he was riding or the pattern of moving. It is by the experience you built with him. This experience is based upon learning. The several instances of meeting him, the conversations with him, and also sharing in his interests – all provide you more cues to recognize and identify him. Sometimes the information we receive may be limited and recognition and identification of the person or objects may be very difficult. We will have to integrate the available information with our past experience and perceive the person or the objects. Here we are trying to build up complete perception with incomplete information. This is the ability human beings are developing by learning and also learn to reintegrate the incomplete cues to make it complete and meaningful. This ability to reintegrate incomplete information to build up complete perception is not always successful.

The information you get sometimes may have certain similarity or resemblance to some well known friend or object and you might have recognized him or the object wrongly. Sometimes you might have even tapped the person on the back or shoulder with more certainty and on knowing that you made false perception of the person you are put into a greater awkward position. This is because of our past experience and inadequate information that we commit the mistake of wrongly recognizing persons and objects. This may be also stated as false perception or illusion. We are coming to this point later on.

FACTORS OF PERCEPTION

Most of our sensory inputs are visual in nature and our visual perception varies from individual to individual. But perception is more than the messages provided by our senses and each one perceives something new. Therefore, perceptual responses are used as dependent variables in psychological studies. This means that these factors as independent variables that determine perceptual responses. It is because of this that perceptual responses are used for studying various aspects of human beings. The projective tests are used to assess the personality traits of an individual. Individual perceptions of these test materials vary considerably from one individual to another. Further the perceptual responses of the individuals are used to uncover certain information relating to their learning, experience, interest, social attitudes, prejudices, motives, values and other important determinants of perception. Therefore, there are several factors or conditions that determine our perceptual process. These factors facilitate our perception sometimes and make it accurate and some make it inaccurate and lead to false perception. The important facts that are operating in our perceptual responses are set or motivation, expectations, cognitive style and culture and society.

Set or Motivation

Perception is influenced by set or motivation. The set or motive may be a goal seeking activity. Therefore, Murphy (1964) says that since perceiving is influenced by the goals which one seeks, the set is within the individual and motivation is an important kind of set. Perception is greatly influenced by the individual's needs and desires. When the individual is in need of something he will perceive the same. His perceptual process is influenced by needs and motives. This process may be stated as perceptual set. The individual is set to perceive what he needs or desires. If the perceptual set is strong it may increase some aspects of behaviour in the organism and decrease other aspects of behaviour in order to achieve the goal.

For example, when a person is feeling hungry he is more likely to see food-related objects. Experiments have been conducted to determine the influence of physiological needs or motives on perception. An ambiguous figure was perceived by him as related to food. The person thinks and talks about food. McClelland, *et al.* (1948) have conducted an experiment to study the motivational process. In their experiment persons who had not eaten food for several hours were made to see blurred pictures. These persons were starved for varying hours ranging from one hour to sixteen hours. Person who starved for sixteen hours perceived images of food in the pictures more often than the persons who starved for less number of hours. The physiological motive for food made some persons to perceive images of food in the picture. The role of set or motive is to produce a tendency in starving person to perceive images of food objects in passing clouds, also a tendency in the frightened persons to perceive thieves in the shades and in darkness. The thief or the culprit who is escaping from the police has a tendency to perceive every tree on the road as policeman.

Everyone of us has a perceptual set or motive different from other persons and hence there is individual differences in the perception relating to the individual motives. In one of the experiments conducted by Levine *et al.* (1942) eighty different objects were presented one by one to a group of subjects who were made to sit behind a ground-glass screen. These subjects went there without food for varying lengths of time. Black and white drawings of various objects were made to appear in indistinct form on the screen. The subjects who had been deprived of food for different intervals of time were asked to identify the objects they saw on the screen. As the food deprivation increased upto six hours the subjects have developed a tendency more and more to interpret these indistinct drawings as food objects. They were almost like the explorers or the caravan travellers who used to have visions of rich fields and cool springs at the time of hunger and thirst. The food deprivation as a motive created a tendency in these subjects to perceive food in ambiguous figures. This tendency to perceive, remember, forget, think and imagine by the influence of motive is called ***autism***.

The perceptual set or motive may change from time to time. We never perceive anything without motive or remain indifferent. We are happy at a new opportunity and miserable when it is lost. The intensity of motive mostly determine the gist of our life. In the study conducted by Franklin *et al.* (1948) the army men were kept on a semi-starvation diet for months as a part of an assignment in the World War II. It was observed that they began to think of food, dream of food and also obsessed with problem of food. In all their conversations food was the principal topic. As the hunger stress continued more dreams about food were reported. Every psychological activity was identified by the matter of food. In our everyday life we might have had some such experiences in which motivation playing important role in perception.

Expectations

Human perceptions are influenced by motivational factors. The role of sets or motives in various physiological states make us to perceive what we need. This has been discussed in the preceding section. Human beings have a tendency to look ahead for certain things when they are in need of them or expect certain things out of their learning experience or advance knowledge. Edward C. Tolman (1886-1959) verified this hypothesis in his experiments on rats. Tolman's (1932,1951) view varied little from Hull's view (1943, 1951). Hull has stated that the drive is developed in the organism due to lack of homeostasis and it resulted in increasing the motivational level of the organism. The hungry rat becomes restless and its exploratory activity is increased. Thus there is relationship between hunger and increased activity. This means according to Hull the hungry rat searched for food. Tolman states that the hungry rat expects food and when food is found the expectation is confirmed. To Tolman this confirmation of expectations is the basic aspect of motivation. The organism wants food, it expects food and gets it and on getting it the organism confirms its expectation.

According to Tolman the organism does not respond randomly to motivation, but expects confirmation and, therefore, it searches to confirm the expectations by perceiving its food. Since the rat has known its food and it is not a new thing for it, the rat expects or searches to perceive its food. Thus Tolman's expectancy theory explains the role of expectation in perception. Having Tolman's expectancy theory into confidence experiments have been conducted to test the role of expectation in our perceptions. Siipola (1940) tested in his experiment how individuals respond to certain words when they were provided with the names of certain categories. In this study, the experimenter used two groups of subjects. He told the first group of subjects that they would be presented with words related to *animals* and they should report what they saw in the screen. Thus the experimenter has created a sense of expectation in the minds of the subjects and they were expecting to see the words of animals in the screen. But at the time of presenting the words, instead of words related to animals some combination of letters (non-sense syllables) that did not give any meaning were shown on the screen. The combination letters wharl was perceived by the subjects as whale, dack was perceived as duck and sael as seal.

The second group was told that they would be shown on the screen some words referring *boats*. The same letter combinations, wharl, dack, sael were shown to them. The subjects of the second group perceived and reported the non-sense syllable wharl as wharf (platform where ships are loaded and unloaded), dack as deck (floor of a ship) and sael as sail. Because the subjects were guided to expect something related to animals and boats they perceived words related to animals and boats even in non-sense syllables. Thus expectations serve as the determining factors in our perceptions. We are expecting to perceive certain things in the outside world and we perceived them even though they were not the real objects.

Culture and Society

The culture in which one is born and brought up makes him to adapt the way of that culture and the individual's perceptions are very much influenced by that culture. An individual cannot act anything away from his culture. For example, the language one speaks, the style of dressing, the cooking and food habits, receiving the guests and treating them in all these individual perceives the situation through his cultural norms. The individual may use various kinds of cultural cues in his perceptions.

Our perceptual process is always guided by our cultural norms. We receive suggestions from outside in the form of words and action indirectly. We accept what others do in our society and follow the same. Therefore, the culture and society in which we live determine many of our perceptions and we develop a tendency to perceive things as other people around us perceive them. Sherif (1936), an American psychologist has conducted an experiment to measure social effects on individual's perception. It was a dark room experiment to study the individual's autokinetic effect (self-moving effect). A point of light was fixed on the wall of a dark room and individuals were asked to watch. It was suggested to them that the point of light is moving in small steps and everyone should measure the distance in inches and report the distance the light moved. The experimenter has put all the subjects together and asked them to measure the moving distance and report. Now the judgment of the individual persons varied. This means that the individuals have modified their perceptions and measurement on the basis of the suggestions received from others or in conformity with other opinions. Sheriff has demonstrated how delicately the apparent movements of the light had social stimulation upon the individual subjects. The result of this experiment suggest that a person tends to perceive what others perceive when he is a member of a group. This experiment has been repeated by a number of experimenters and no one has failed under this condition to get the autokinetic effect from the subjects.

When the individual subject is alone in the experimental situation and judged the moving distance, even if it is for hundred trials the average estimation is stable. But when he is brought into a group situation with others his estimation receives the influence of his group members and his judgment figure represents the group average. When we have the tendency to perceive an object as others in our group perceive it, we are inclined to this group effect. It was observed that when Sherif's subjects were tested alone, they were clinging to the group tendencies even after some time. Bovard (1948), on verifying the results of Sherif's experiments, has observed that there was a greater individual difference in the responses of the subjects to the group pressure.

Therefore, there is a universal tendency to perceive things as the members of our group perceive them. Murphy calls this as ***'society shared autism'***. Parents or society transmit the cultural values and social norms to the children. Similarly old members of the society transmit cultural values and social norms to the young members and to the new members of the society and these values and norms determine the perceptions.

ERRORS IN PERCEPTION

Perception appears to represent or reflect the characteristics of the object from the environment with wonderful accuracy. We also trust that our perceptual experience provides us with high degree of accuracy of the external world. The view, that the retinal image is sincerely reflecting the physical stimulus and, therefore, our perception is an accurate imitation or copy of this retinal image, was prevailing for a long time. Hence the accuracy of our perception was explained. But the optical illusions we experience posed them as a problem and was a challenge to this view. Because, there are experiences that have failed to reflect the physical object and its image accurately on the retina area. If this would have taken place correctly illusion would not have occurred. Therefore, it was viewed that optical illusions are perplexing exceptions to the general law of perception. But this simple view about illusions was not accepted by the modern thinkers.

According to Postman and Egan (1949) visual perceptions is not the exact reproduction of retinal image. The retinal image is the first stage in the perceptual process and it is only the raw material presented to the retina area in the form of an image of the stimulus. The visual perception is organized only after this stage in which several factors are involved. Perception of an object mainly depends upon (1) the retinal image of the distal object, (2) the nature of proximal stimulation which reaches the sense organ, (3) the stimulation of the total visual receptor area, (4) the functioning of sense organ to convert the proximal stimulation into nerve impulses, (5) the sensory nerves transmit impulses to the brain, the outer layer of the cerebral cortex and (6) stimulating the region which causes the visual experience or perception. In these processes there is no problem for an illusion. Then how it occurs? Boring (1942) states that perception depends upon subjective factors and upon the stimulus. If the stimulus is not accurately reflected in our perception due to subjective factors every perception is an illusion.

In a sense Boring is correct. We have discussed in the preceding section the factors that determine our perception such as learning, past experience, recognition and identification, set or motive, expectations, culture and society. In the perceptual process the factors mostly seem to be involved are learning, past experience, motive and expectations. These are subjective factors and we apply these variables in our perceptions. Any one of the factors or two or three factors in combination may determine our perceptual process within us. Whatever accurately the distal object is presented in the total visual receptor area the perceptual process that takes place after this is the process of organization in which the personal factors are involved to attach meaning for the object. We may attach meaning to the object or stimulus with our learning or past experience, or motive or expectations. In this organizational process anyone of the factors or two may dominate and determine the meaning of the object. It is at this stage we may be committing a mistake of interpreting the cues differently. This mistaken interpretations of the cues or objects or stimuli with our subjective experience or knowledge may be called as illusions.

Objects in the outside world may have various forms, shapes, sizes, styles and appearances. Our visual sensory apparatus present them to us as they are with all their characteristics and we apply our past experience and knowledge and judge the objects what they are. If it is correct, then there is no problem. But if it is not then our judgment becomes wrong. If the object is not the same, the real object, then our perceptual judgment is wrong. This we call illusion. Woodworth (1952) states that an illusion is an error of perception. According to him, we are getting a false impression of the objects or facts that are presented to our senses. The perceiver falls into a trap and attributes false meaning to the object from the signs he received. It is the ambiguity of the object that makes one to commit the error in perception. Or it may also be stated that, because the individual has failed to carefully perceive the objects that lie in his visual field or because of his insensitivity to the objects outside the world error in perception occurs. It is not the object or the ambiguity of the object that is misleading the person, it is the person who fails to perceive it correctly and attach the meaning to it. Therefore, it is purely subjective and the object has nothing to do with the perception of the individual. Objects may look bright, or dull, clear or ambiguous it is the perceiver who should make use of his sensory apparatus accurately and attach a correct meaning to the object.

In all our perceptions, our past experience and knowledge and expectations play important role. For example, at twilight while walking on the road we may be suddenly alarmed of seeing a snake at a distance and when we go nearby and examine it is not a snake, but a piece of rope. Similarly a shell may appear like a silver coin. The shell and silver coin have certain similarities. The rope and the snake have certain similarities. The perceiver sees the shell and the rope with his past experience with silver coin and snake. Further the place may not be bright area as to see the objects clearly and hence his vision may not be very acute. This kind of error in perception is general, because many people may have the same type of error in perception. Caravan travellers often see oases on the desert areas. All these are errors in perception. Therefore, what we call illusions are mostly subjective, a failure to observe the objects correct. Garrett (1975) says that illusions are confused or misinterpreted perception. He calls this equivocal perceptions. According to Kendler (1962) an illusion is a representation of a discrepancy between physical and psychological measurement. Woodworth says that some illusions are due to external, physical causes and some are due to the causes within the organism. If it is within the organism, it may be due to the defects in sensory apparatus of the organism or may be due to error in perception. The study of illusions has certain values. We come to know that perceptual experiences are mostly determined by subjective factors and also by the stimulus properties.

ILLUSION

Illusions may be caused in every perceptual constancy. The object fails to maintain its constancy and we perceive the object differently. The rope appearing like a snake and the shell appearing like a silver coin are good examples for this. Similarly the size of the object is not maintained and it appears smaller at far off places. From certain angle we do not see the object with its original shape. For example the opening door is having the shape of a trapezoid. Murphy (1964) says that illusions are not actually unusual happenings of nature, but they are simple instances. The ordinary constancies that were maintaining well most of the time deceive us in certain special cases. Our contact with the environment most of the times was correct. We depend upon our senses through our past experience and most of the objects were given the correct meaning. The illusions seem to be the instances that we sometimes fail to predict the correct meanings of the objects with our past experience in a new situation. Therefore, it is generally easy for anyone in a new situation to get misguided.

Psychologists have investigated the phenomenon of illusion very extensively as it has been perplexing them for long time and the studies are going still in psychological laboratories. When constancies fail in our perceptual process illusions occur. The study of illusions reveal how the subjective factors play important role in our perceptual process. Though the objects we perceive do not change when we perceive them the properties of the objects sometimes misguide the perceiver and, therefore, the physical properties of the objects also determine our perceptual experience. Postman and Egan (1949) state that there are two general kinds of illusion. This has been often observed in the study of perceptions. They are illusions of extent and illusions of direction.

Illusions of Extent

This type of illusion is a common occurrence in everyday life. We often see in every instance that one of the two identical sizes or objects appears to be larger than the other. When two physical objects or stimuli of equal in size are judged as different in size or length illusion occurs. Generally the objects or stimuli in these situations are always parts of some complex design or figure. Some of illusions of extent are the Muller-Lyer illusion, the moon illusion, the Ponzo illusion, the height-width illusion, the Jastrow illusion and the Sanders Parallelogram.

The most widely known illusion for a long time is the Muller-Lyer illusion. In the Muller-Lyer illusion there are two horizontal lines, one with feather heads on either side and one with arrow heads on either side. The two lines are equal in length (Fig. 5.3). But people perceive the feather head line as longer than the arrow head line. The slanting position and the size of the arrows contribute to the illusion. The study of illusions are revealed in many fascinating forms. Every author has a form of his own illustration.

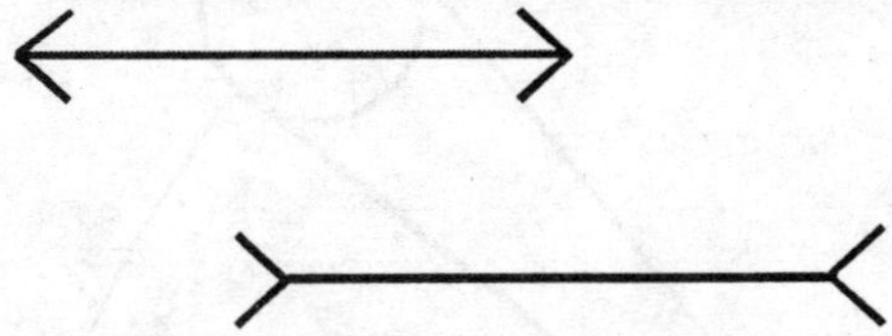

Fig. 5.3 : Muller-Lyer Illusion

The moon illusion is another often quoted one. The moon appears much larger near horizon than it looks in the sky though the retinal image is the same in both cases. The moon illusion is always associated with Ptolemy (2nd C.A.D.) a Greek astronomer and mathematician lived in the second century A.D. According to him, an object seen through space that is filled with objects appear to be farther away than an object seen through empty space. The horizon moon is seen through objects-filled space and there are many secondary ones mostly like the texture gradients, linear and aerial perspective to indicate the distance and, therefore, it is seen as larger, while the zenith moon is seen through the empty space and there are cues to indicate the distance and it is seen straight up. Further the imaged arched roof of the visual sky, the orbit of the zenith moon is misjudged. Therefore, the zenith moon appears to be nearer and smaller than the horizon moon (Fig. 5.4). Many factors like the cues to perceive objects, distance, size, past experience and judgment are involved in the moon illusion. The moon illusion occurs due to the angle of the vision at which we perceive the moon. It is because of this, the horizon moon becomes smaller and smaller to the size of the zenith moon and is moving from far off distance to closer and closer to the place of the zenith moon. The principle is not only applicable to moon but it can also be applied to the sun. The sun at the horizon looks larger and at for off. At the zenith it looks smaller and nearer. Therefore, the change in size and distance of the moon and the sun depends upon the angle at which we perceive them.

The third important illusion often discussed is the Ponzo illusion. Ponzo illusion is also called the rail-road track illusion. The main principle involved in Ponzo illusion is that two vertical lines with equal distance between them are perceived as coming close or converging at a distance. The railroad track is a good example for this. The two lines converging at a distance provide a powerful cue to distance, linear perspective. We know from our experience that the railroad lines at distance are of the same size and have the same distance between lines as the lines nearest to us. But it appears to us that the two railway lines are coming closer and closer and meeting at a distance even though they never meet. Since the back, the horizontal line projects the same retinal image as the front line the back line is perceived to be longer and at more distant place by the firmly established habit (Fig. 5.5).

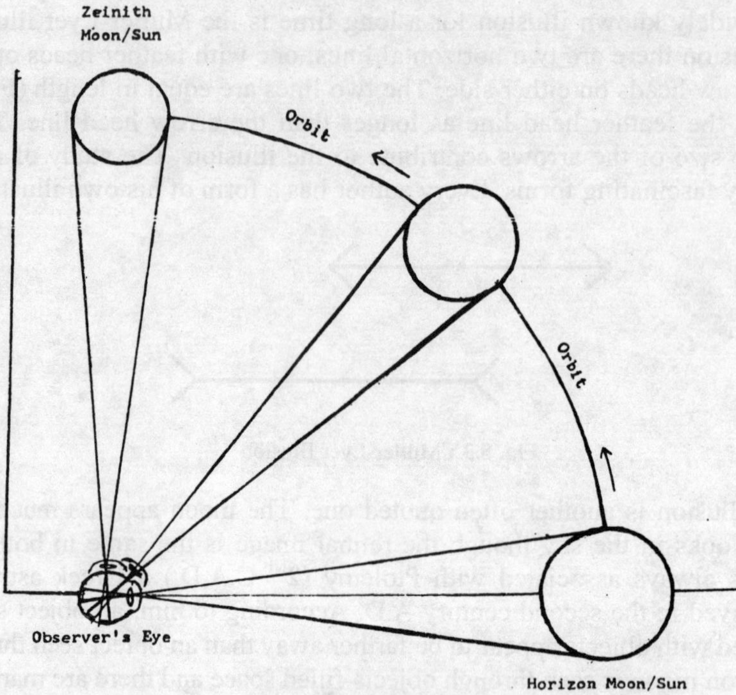

Fig. 5.4 : Moon/Sun Illusion.

Fig. 5.5 : Rail-Road Track or Ponzo Illusion

The height-width illusion can be illustrated with one horizontal line and a vertical line. The vertical line meets the horizontal line at the centre. Both the lines are of the same length. But the vertical line looks longer than the horizontal line. The reason for this kind of illusion is that the eye movement is easier to make horizontally than to make vertically to the same length. Therefore, the vertical line is perceived taller than the horizontal line. This kind of illusions generally arise also in the height and width of buildings (Fig.5.6).

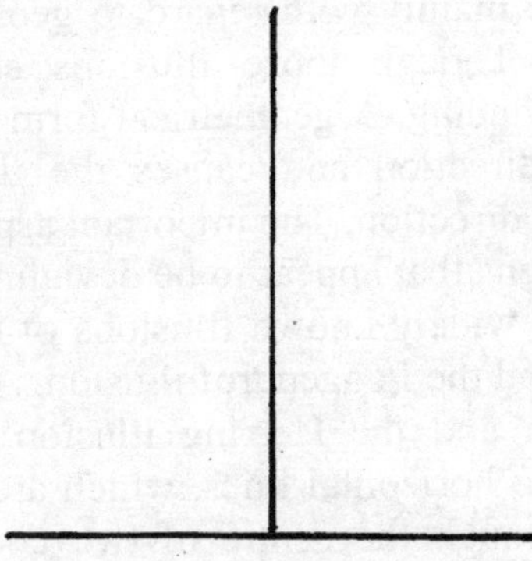

Fig. 5.6 : Height-Width Illusion

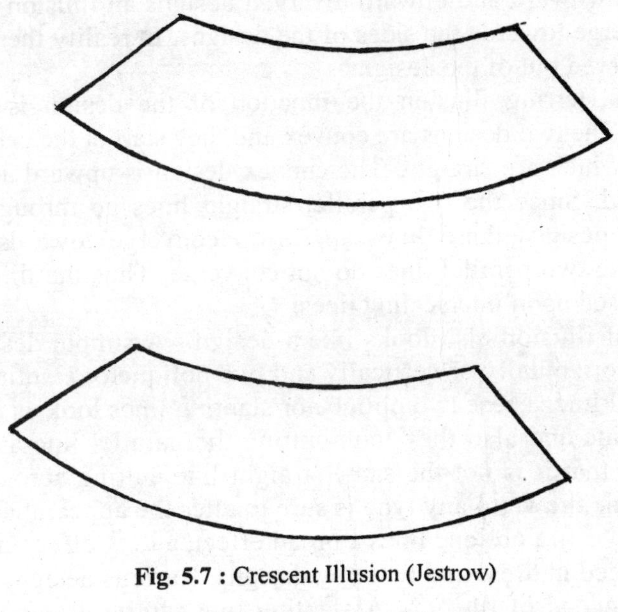

Fig. 5.7 : Crescent Illusion (Jestrow)

The Jestrow illusion consists of two crescents of the same size. But they do not look of the same size to an observer. They are presented one below and one above. The narrower part of one crescent is placed against the wider part of the other crescent. The observer's eye moves immediately from the narrow-part of the lower crescent to the wider part of the upper crescent. This creates an impression that the upper crescent is bigger than the lower crescent (Fig. 5.7). The Senders Parallelogram is also used to illustrate the illusion of extent. In this the two figures are drawn to different sizes. One looks as rectangular and the other as square. But the two diagonals are equal in length though they look differently.

Illusions of Directions

The illusions of directions are mainly with regard to geometrical forms. They are also called geometrical illusions. Though those illusions studied for a long time the explanations are not fully convincing. A geometrical form has apparent directions, but it deviates from the expected direction and causes the illusion. There are not many illustrations for the illusions of direction. The important aspect in the illusion of direction is the geometrical lines or designs that appear to be deviating from the normal course and causing the illusion. The most widely known illusions of direction are the Wundt lines illusion, the Herring illusion, and the Poggendrof illusion.

The Wundt Lines illusion and the Herring illusion have certain similarities. In Wundtian illusion there are two horizontal lines, which are straight and parallel to each other. These two lines run through the centre of two concave designs. Two concave designs of the same size meeting at a place between the two parallel lines. The upper portion of the design is diverged downward and the lower portion is diverged upward and the two meet at equal distance from the starting points. Since the two horizontal lines pass through the downward and upward diverged designs an illusion arises that the two lines appear to diverge towards the sides of the designs. In reality these lines are straight and parallel if observed out of the designs.

In the case of Herring illusion the function of the design is quite opposite to Wundtian illusion. The two designs are convex and they start at the centre in between the two parallel lines, which are straight. The convex design is upward and another convex design is downward. Since the two parallel straight lines go through the upward and downward convex designs these lines appear to converge towards the sides of the designs. Actually the two parallel lines do not converge. Thus the illusion arises. These two illusions are based upon intersecting lines.

The Poggendrof illusion also looks like a design. A simple design is drawn with two parallel lines horizontally or vertically and two oblique or slanting lines are drawn one on each parallel line. These two oblique or slanting lines look as if they are parts of the same straight line and also they look cutting the parallel line at wrong point and give an impression that it is not the same straight line cutting across the two parallel lines. An oblique line drawn to any type is sure to alter the appearance of an object and cause an illusion. By extra oblique lines a broad effect, a long effect and also a lopsided effect can be produced in dress and building designs. Various designs may be drawn on the principle of Poggendrof illusion. A slanting line can be drawn as if it is passing through two parallel lines or a design or two giving the look as if it cuts at wrong points. Similarly a square can be provided having apparent bending on all four sides (Fig. 5.8).

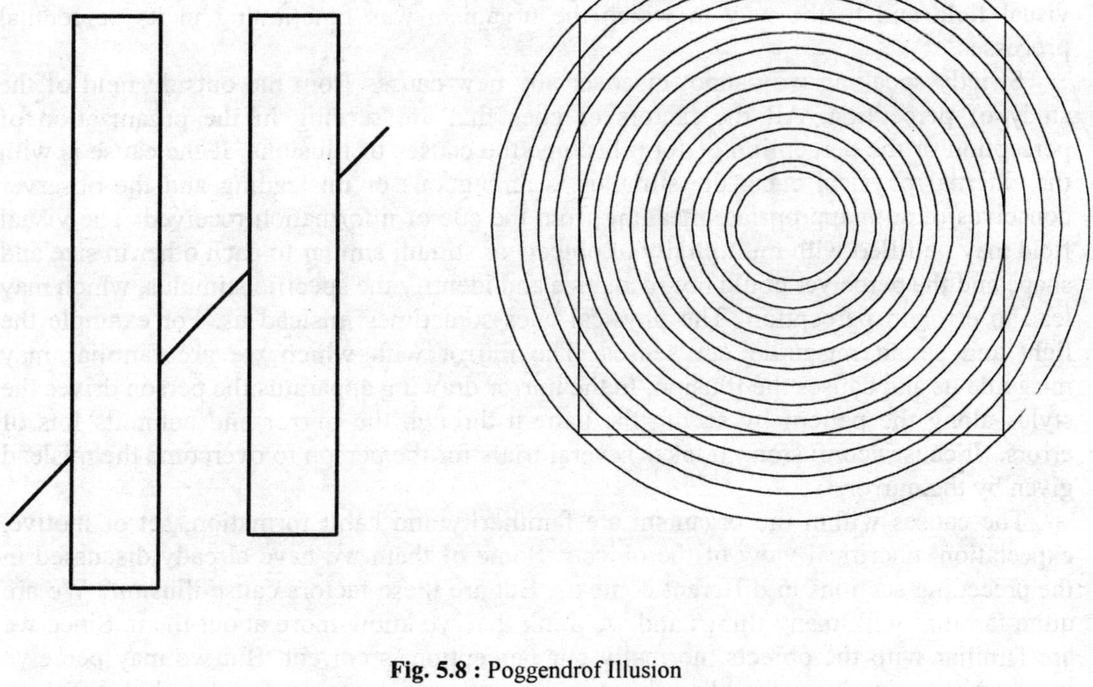

Fig. 5.8 : Poggendrof Illusion

Other Types of Illusions

Illusions arise when we fail to maintain constancies in our perceptual experiences. Hence, there are also illusions of object, movement, shape, size, colour and so on. We sometimes perceive movements in objects when actually there are no physical movements. On the motion picture screen we see only still photographs. But we see movement of persons and objects in a continuous manner. A series of discrete still pictures are presented with so short intervals that we combine them in our perceptual process and see the figures as moving. This movement is only apparent and not real. Therefore, this movement is an illusion. This apparent movement is named as the phi-phenomenon by the German psychologist Max Wertheimer (1912) one of the founders of the Gestalt Psychology.

Causes of Illusion

The causes of illusion are rooted in our perceptual processes. We know that every object or stimulus is consisting of several contours, which shut in all round and make it a full figure or a pattern. A pattern is a composition or organization of separate parts. With our learning or past experience we tend to perceive in objects in the environment a pattern or structure of form for which we attribute characteristics and names. Most of the time we are successful in this endeavour, unless the object or stimulus is absolutely new in the situation. But sometimes, we fail to cognize or assess the correct meaning of the object. The causes may be many and may be attributed to both the external physical cues, the visual field and to the way in which the organism was functioning in its perceptual processes.

Strictly speaking we cannot discover any new causes from the outside field of the study of perception. All the factors or cues that are serving in the organization of perception or the perception of form become the causes of illusions. If the cause is with the external physical cues, the stimulus is ambiguous or misleading and the observer conceives of an inappropriate meaning from the cue of information received. The visual field may be filled with multiplicity of objects or stimuli similar to each other in size and shape and the perceiver could not recognize and identify the specific stimulus, which may lead to error in perception. The physical cues sometimes mislead us. For example the light and sound misguide our senses. The mirror with which we are familiar may misguide us and causes the illusion. In the mirror drawing apparatus the person drives the stylus along the pattern by seeing the pattern through the mirror and commits lots of errors. It causes confusion. It takes several trials for the person to overcome the mislead given by the mirror.

The causes within the organism are familiarity and habit formation, set or motive, expectation, uncritical view of the objects. Some of them we have already discussed in the preceding sections in different contexts. But are these factors cause illusion? We are quite familiar with many things and we think that we know more about them. Since we are familiar with the objects, normally our perception is correct. But we may perceive incorrectly a stimulus resembling the stimulus with which we are familiar, but different. This type of illusion commonly occur in proof reading. Woodworth (1952) calls this proof reader's illusion. This type of illusion also occurs in reading typed materials. This type of errors sometimes escape even from the professional proof readers. However carefully the proof reading is repeated but errors sometimes escape the notice and books are printed with some errors. This is because of illusion caused by similar letters and words. Here the familiarity causes to overlook the similarities in letters.

The readers also commit the same type of errors. We have developed the habit of reading every printed material rightly. If there is a printing error in a word we overlook the error because of the habit we have developed to read everything correctly. For example, we read the letter composition wharl as whale, and the nonsense syllable, dack as duck. In one of the experiments conducted by Siipole, his subjects, college students read the nonsense syllables wharl as wharf and dack as deck, sael as sail. This type of illusion is caused due to the habit we have developed to read everything correctly.

The set we made already to perceive some object may misguide us in certain situation if an object is similar to what we set to perceive appear. The object which has certain resemblance to the object we want to see is often mistaken by us. Since we set up to see it and anxiously expecting for it, anything that resembles the object is perceived by us as the correct object. But it is not the correct object. Thus the set plays role in causing an illusion. For example, the caravan travellers develop a motive or set to see water. At a distance something appears like water and they are disappointed when they went nearby and it was only a mirage. The set was the cause of this type of illusion. Persons starved for days without food often see image of food in pictures. The thief who is going out after a night's burglary, out of fear is mistaking every lamp post for a policeman. When we make ourselves ready to perceive certain fact we accept even some minor or inadequate stimulus as the evidence of the fact. The example, a mother who is busy in her kitchen is mistaking any howling noise for her baby's cry, out of anxiety already she set for the care of the baby.

Expectation as the cause of illusion is not much different from the set or motive. We are expecting to perceive something very anxiously and any small cue or information makes us to accept and we attribute meaning to it. But subsequently we come to know that is different thing and not what we expected. Tolman's expectancy theory is explained elsewhere here with special reference to perceptual process. The hungry rat is expecting food and when the event occurred the expectation is confirmed. Instead of the real event if some new consequence is occurred then the expectation is not confirmed, it is disconfirmed. This decreases the probabilistic expectation. Just like the rat expects food on some inadequate cues and gets disappointed on some new event other than food, human beings also build up their expectation. They get misguided on some small cues when they were expecting something they wanted. Experiments have been conducted to assess how far expectations cause illusions. Siipola (1940) used two groups of subjects in his experiment. He informed his first experimental subjects that he is going to present words related to animals and asked them to reproduce those words. Thus he created a sense of expectation in the minds of the subjects and they were expecting the words of the animals on the screen. Siipola instead of presenting the words of animals, presented nonsense syllables with letters similar to the names of some animals. When he presented the composition of letters wharl, dack and sael the subjects reproduced them as whale, duck and seal. Similarly expectation was created in the minds of the second group of subjects by telling them that they would be presented with words related to boats. But Siipola presented the same nonsense syllables, wharl, dack and sael. But the subjects reproduced these words actually related to boats as wharf, deck and sail. These illusions were caused by expectations. Siipola's experiment is also a method of suggestion. He has suggested to his subjects that they are going to see the names of animals and words related to boat. The subjects acted uncritically on this suggestion and perceived the words wrongly. This suggestion may also cause error on perception.

We perceive certain objects as they were presented to us and accept them without going into the complexities or the stimuli. We perceive certain stimuli uncritically and in an unanalysed manner. The important cue is concealed and we are carried away by visual outward appearances. The real cue for identifying the object can be traced only by close observation and by going into the details. Since the exploration of the fact involves difficult task many do not take the risk and tend to avoid it. Most of the illusions, which the psychologists speak of, belong to this category. Most of the comparative judgment of similar objects involves some kind of intricacies. For example, the Muller-Lyer illusion forms, the staircase illusion, Pongo illusion, the Herring illusion, the Necker cube illusion, the Wundtian lines illusion and the Sanders parallelogram, all of this type. They need a critical and analytical approach to trace out the conceded facts. Illusions of these types, which need critical and analytical exploration, have certain practical importance.

The architects, planners and dress designers generally must understand the complexities of the objects in which the real facts are concealed in some manner to increase perfection in their works.

Hallucination

While we discuss the problem of illusion we have stated that suggestions also may cause error in perception. Any uncritical act on suggestion may make the person to ignore the correct cues in perception. In deep hypnotic state it is possible to produce gross illusions and hallucinations. Though these two are different mental states both have certain similarities with regard to perception. Both are in this way are interactions of the outer and inner determinants of perception.

A real perception is one that agrees with the physical stimulus in the environment. An error in perception is one that is determined by the individual's memory. An illusion is also mistaking one thing for another. An illusion, a false impression of the objects that are presented to the senses. But hallucination is a form of perception in which a person sees and hears something but not in reality. There is nothing around him and no one perceives the same thing anywhere. This is stated as an abnormal symptom. The person sees visions and hears voices. This is an abnormal hallucinatory experience. This differs from common hallucinatory experience. People are having this kind of experience in their life in some way or other. But it varies in degrees from person to person. Therefore, a hallucination may be defined as a sensory experience in the absence of the corresponding external stimulus. It is a misinterpretation of memory or imaginary experience perceived as true. Hallucinations are some kind of peculiar misperceptions and do not represent objective reality. Therefore, illusions and hallucinations are certain physiological and psychological states of organism. It is a way of responding to familiar sensations and experiences by the individual in the absence of any stimuli.

EXPERIMENT 5 : PERCEPTION OF SPACE IN DEPTH

INTRODUCTION

Every object occupies some space in the outer environment. Whether it is small or big, moving one or static, solid or liquid finds a place in the environment. We can locate an object if only it had occupied a space and we perceived it. Objects cannot exist in a vacuum. Therefore, we can locate objects in space for granted only in a spatial world and we could not imagine anything other than this. If we begin to think about how living organisms learn and manage to behave adequately in space and move about successfully without any confusion in three-dimensional space and locate objects correctly is astonishing and mysterious for us. There is also precision in all the movements the organisms make. Space is a boundlessly expanded phenomenon within which things are contained. To perceive such a space accurately is highly complicated and anyone single sensory apparatus may not adequately help us in this task of perception. Therefore, the operation of the integrative activity of our senses alone could perform the task successfully. The perceiver should be accomplished with the sensitivity of all his sensory apparatus.

In the perception of space the eyes are playing important role. Boring (1942) says that the functional properties of the optical process make sure of the stability of our visual world and the accuracy of the spatial adjustments. Objects fixed in space appear so and also remain so and also the spatial relations between and among objects show similar stability. One important stabilising property of the optical process is the retinal image. Woodworth (1954) says that the formation of the retinal image is a long chain of events necessary for the perception of an object. If the retinal image did not represent certain features of the physical object a stable visual world is not possible.

When the physical object is perceived by an organism, the eye as an optical instrument forms a retinal image and maintains the spatial relations. Whatever the part of an object outside, is also represented in the retinal image. There will be no change in retina. But the physical object in the retina area falls upside down. It is inverse. The inversion of this image does not present any process in the analysis of perception.

If an object recedes from the eye the size of the retinal image decreases in accordance with the law of the visual angle. The law is that the linear size of the optical image is inversely proportional to the distance of the object. If we double the distance of an object from the eye we have half the height and width of the retinal image. But it does not mean that we see everything in accordance with the law. For example, a man 30 feet away does not appear half as tall as he does at 15 feet.

The visual system according to Bartly (1941) is having the resolving power. This is one of the important mechanisms that is taking place in the visual system or in the eye. When the eye is focused upon two objects in the environment two retinal images are formed and the visual system resolves to a greater degree. Only when we see two objects we could perceive the relationship between them. Therefore, the visual perception of space according to Hilgard *et al.* (1975) is based upon the spatial relations among the objects. The visual system provide basic conditions for the visual perception of space. Our visual perceptual world is always *tridimensional*. Objects are seen not only on the (1) *left and right* or (2) *up and down* but also seen (3) *near and far away.* We also see solid objects as solid. That is, the *thickness of the object or the depth* is also seen. Therefore, there is a spatial framework which may be stated as three dimensions. In the perception of space the *depth of the object is called the third dimension*. In the spatial framework an important property that fixes up the main sides, the *vertical and horizontal* is, this dimension or the depth which locates the objects. That is, *how far away the objects we located from the eye* is determined by the third dimension which we may also state as the *perception of distance* or *perception of space in depth.*

The important point in the perception of depth is, how the two-dimensional retinal images lead to the perception of space in depth. The two dimensional images provide certain clues and indications to infer a tridimensional space in the environment. Any perceived object will cast its image on the retinal area of the two eyes separately and differently. But what happens is, the person sees a single solid object in three dimensions. There is a process of resolution which takes place in the brain. The two retinal images are fused in the brain into one image.

An object is seen with one eye is called monocular vision. Clues for third dimension or for distance are limited. Certain clues from two eyes, which is called binocular vision, ought to be added to it. We know that the retinal image is larger for the nearer object and smaller for distance object. Objects may be bigger or smaller and bigger objects overlap smaller objects if they are located in the same place. Moskowitz and Orgal (1969) state that the contours of overlapping objects provide clues for distance. The object whose contour is continuous at the point of overlapping is seen as closer. Further, the decrease in size and separation of objects provide clues to the perception of distance.

The clues for depth perception also come from the shadows of the objects. The shadow cast by a tall tree may be different from that of a short tree. The relative movement of the objects also provide clues for the distance. That is, an object moving fast and rapidly is seen as nearer that one moving slowly. The brightness or the colour of the objects may serve as clue in the perception of distant objects. Nearby objects appear more brightly than the distant objects (Kendler, 1963).

EXPERIMENT

Problem

Study the perception of space (depth) of a person. Demonstrate how the two-dimensional retinal images lead to the perception space in depth.

Method

Participant

Choose a person who has no knowledge about this experiment. Collect his personal information like, age, sex, education, whether wears spectacles or not.

Materials

Howard Dolman Depth perception apparatus, metre scale chin rest, blind-fold.

Description of the Apparatus

The Howard-Doman apparatus is an improved model of the old Depth perception apparatus. The parts of the apparatus fixed on a wooden board. The wooden board is of 4 ½ feet long and 1 feet wide. At the centre one metre scale is fixed. There are two iron rods on either side of the scale. They are painted with white colour. One of the rods can be moved forward and backward by a string. The string is connected to the rod and two ends of the string are brought outside the apparatus to some extent for dealing with them by the subject. At the back end of the apparatus a black screen is there. In the front there is a wooden plank standing with silt at the centre through which the rods can be seen by the subject. The chin rest is kept little away from the apparatus for keeping the head of the subject at a fixed position.

Fig. 5.9 : Howard-Dolman Depth Perception Apparatus (Open Model)

Fig. 5.10 : Depth Perception Apparatus (Closed Model)

Procedure

The experimenter should set the apparatus correctly and check whether everything is in order. Then he should make the subject to sit comfortably in front of the apparatus. The apparatus also should be placed at a higher level so that the subject's eyes are on same level of the slit of the front wooden board. If the room is not sufficiently illuminated the electric bulb arranged in the apparatus should be put on to provide proper illumination. The experimenter should make the subject to sit 2½ metres away from the front part of the apparatus and he should be given with the two ends of the strings in his right and left hands. By pulling one string the iron rod will move forward, and by pulling the other string the rod will move backward. By this means the iron rod may be adjusted as required.

The experiment is conducted in three stages. (1) The first is the binocular vision with and without head movement while making the judgment. (2) The monocular vision is made with right and left eyes separately without head movement. (3) Again the monocular vision is made with frequent head movement. The head movement of the subject is controlled by using the chin rest. Now the subject's task is to make the two rods in alignment for each other by adjustment. Then the experimenter should give the following instructions to the subject.

> "Please sit erect. Look at the slit in the wooden board. You are able to see two white rods standing vertically. There is a black screen behind. Hold the two strings one with your right hand and one with your left hand. By pulling the right, the rod moves forward while for the left the rod moves backward. Your task is to make the two rods on par with each other by making the adjustment with the help of the two strings. Now, look the two vertical rods are in different places. When I say ready go you can start the work."
> "Ready, go"

The experimenter first begins with the binocular vision without head movement by fixing a chin rest to the head of the subject. When the subject adjusted the moving rod to the level of the fixed rod the experimenter notes down the difference in mm from the metre scale. Then moves the rod and sets in another place. Then again the subject adjusts by moving forward and backward and sets the rod on par with the fixed rod. There should be 15 trials without head movement, alternatively arranging for forward and backward movements of the rod. In the alternative procedure the chin rest may be removed and the subject can make movement of his head in making the adjustment. The subject should carry out the task in the same manner and the experimenter should place the rod in the different places after every trial for forward and backward movements. With head movement another 15 trials should be conducted.

As soon as the first stage is over the experimenter should arrange for the second stage of the experiment. This is monocular vision with right and left eyes without head movement. While one eye is used the other eye is blind folded. The same procedure is followed. First for the right eye and then for the left eye without head movement 15 trials will be taken. The experimenter should keep the chin of the subject on a chin rest and also keep the rod in different places.

The third stage of the experimenter is monocular vision with head movement for right and left eye separately. When all the three stages of the experiment are over the experimenter should get the introspective report from the subject and allow him to go away. On the whole the experimenter should take (60+60+60) 180 trials to complete this experiment.

Results

The data of the binocular vision with and without head movement of the subject are to be recorded in Table 5.1. The mean score for the individual data should be worked out. The data of monocular vision for the right and left eyes without head movement should be recorded in Table 5.2. The mean score of the 15 trials should be worked out.

The data of the monocular vision for the right and left eyes with head movement should be recorded in Table 5.3. The mean score for all the 15 trials should be worked out. Collect the group data of the depth perception experiment for all the three stages of the experiment conducted with head movement and without head movement and record them in Table 5.4. Analyse the results giving the data.

Table 5.1 : Showing the individual data of the *binocular* vision with and without head movements

Trial No.	Without movement		Without movement	
	Forward in mm	Backward in mm	Forward in mm	Backward in mm
1				
2				
3				
4				
5				
6				
7				
8				
9				
10				
11				
12				
13				
14				
15				
Mean				

Table 5.2 : Showing the individual data of the *monocular* vision with right and left eye without head movement

Trial No.	Right eye		Left eye	
	Forward in mm	Backward in mm	Forward in mm	Backward in mm
1				
2				
3				
4				
5				
6				
7				
8				
9				
10				
11				
12				
13				
14				
15				
Mean				

Table 5.3 : Showing the individual data of the *monocular* vision with right and left eye with head movement

Trial No.	Right eye		Left eye	
	Forward in mm	Backward in mm	Forward in mm	Backward in mm
1				
2				
3				
4				
5				
6				
7				
8				
9				
10				
11				
12				
13				
14				
15				
Mean				

Table 5.4 : Showing the group data (mean scores) of the Depth perception experiment with *binocular* and *monocular* vision

Visual condition	Trial No.	Without movement		With movement	
		Forward in mm	Backward in mm	Forward in mm	Backward in mm
Binocular vision	15				
Monocular vision					
Left eye	15				
Right eye	15				

With regard to the collection of group data the experimenter should work out the mean scores of the group for the binocular vision with movements and without movements separately and for the monocular vision of the right and left eyes without movement and with movement separately and should give them in Table 5.4. The mean scores for the forward movements of the rod and the backward movements of the rod also should be given separately in Table 5.4.

Discussion

Discuss the data of the experimental results as given in Tables 5.1, 5.2 and 5.3 for the individual. The error caused in the judgment of the moving variable by the subject is given in mm. Discuss the data obtained from the trials without movement and compare them with the data obtained from trials with movement. Whether the head movement of the subject reduced or increased the extent of error? Also discuss whether the subject commits more errors in forwarding movement of the rod or in backward movement of the rod. Find out reason for the errors caused in all 4 variables (with movement, without movement, forward and backward).

Discuss the results given in Table 5.2 comparing the right eye with left eye. Which monocular vision commits more errors? Find out whether the forwarding movement of the rod causes more errors, or the backwarding movement. Find out the reason and discuss the outcome.

Examine the data as given in Table 5.3 for the right and left eye for the trials with head movement. With head movement which eye judgment causes more errors? Why? State the cause. Also whether the forwarding movement of the rod caused more errors or the backward movement of the rods caused more errors. State the reason. On the whole state whether the binocular vision's judgment is more accurate or the monocular vision's judgment.

Discuss the group data in general and state how many of them with binocular vision commit less error and how many of them with monocular vision. Which eye, whether the right or left was found to be advantageous? State whether the errors are less in the backward movement or in the forward movement.

Compare the performance of your subject with that of the group and where he/she stand in the correct judgment of the moving variable.

Conclusions

State whether your subject is more of binocular vision or monocular vision. In what stage he/she judges the object accurately. State whether the right eye is more accurate or the left eye. Whether the subject is accurate in forward movements or backward movements. State whether with head movement vision is accurate or without head movement.

What is the group's general performance? Have they committed more errors with binocular vision or with monocular vision? In what condition (with head movement and without head movement) they are accurate in the judgment?

EXPERIMENT 6 : PERCEPTION OF SIZE CONSTANCY

WHAT IS PERCEPTION? AN INTRODUCTION

We acquire enormous amount of experience and knowledge out of our contact we have with the world around us. Our contact with the world takes place through our senses only. We will have to depend upon our senses to move about and get things done mostly on our sense of sight. Many things in the world we come to know only by seeing and understanding though other senses are also involved. But seeing alone is not sufficient for us to know everything about the world. The eyes may be seeing something, but nothing might have taken place. Many times we all might have experienced this phenomenon. If one sees something outside and have relevant response, then the seeing is having some meaning. The sense of sight must produce some meaningful effect in the organism. The organism must attach some meaning to the incidence of sight. This we call perception. The concept, perception, is having very wider meaning.

Perception has been classified as one of the cognitive activities. All cognitive activities are knowledge getting activities. The most basic process behind perception is the interpretation of sensory information obtained through seeing, hearing, tasting, smelling and touching. So we recognize that there is something out there. Another important aspect of perception is the circumstances under which the perception occurs.

In the perceptual process the sensory information are interpreted. Some meaning is attached to the sensory information. By this means, we could say, what is that we are seeing. Therefore, perception has reference to persons, places, objects, events and relationships. Perception is the process of integration and interpretation of what the senses receive. It is the response to the world through our senses simultaneously. The response we make is specific and it serves some purpose on some particular occasion. Therefore, the response we made in the perceptual process is selective, purposive and relevant to our needs. We select and respond to objects outside and organize only those things which are needed for our purpose and leave the other things in the background of our perceptual field. According to Moskowitz and Orgal (1969) perception is a wide range of response to a stimulus or set of stimuli which integrates the information beyond that what the stimulus contains. This means that the meanings of other stimuli already obtained from our past experiences are also added to this. Kimble *et al.* (1985) define perception in simple manner. According to them, perception is the interpretation of sensory information. Stagner and Karwoski (1952) state that perception is the process of obtaining knowledge of external objects and events by means of senses.

While perceiving something we attach some meaning to the object and interpret what it is and how it is relevant to the purpose. In this we are also adding something from our past experience to it. We may be also attributing to the object something more than what we see in the object at the particular time. The quality of the object that we have known on an earlier occasion is also added to it and we perceive the object with its full meaning. For example, we see a piano in a store. When we see it we not only get its physical appearance, but also its sound. How it would sound when it is played. We can also recollect a performance in which a famous pianist played it with all his skills. We may even go beyond this and say various types of piano performances. All these put together may be called perception. According to Gestalt psychologists perception is more than the sum of a number of sensations.

FACTORS DETERMINING PERCEPTION

In the perceptual process the sense of sight plays an important role. Though the stimuli outside attract us for perception, we have seen that it is selective and purposive. Whether we are purposively attending to an object or not there are certain factors that are determining our perception. They are classified as objective factors and subjective factors. We may be also selectively attending to an object. But sometimes the direction of attention is not completely under our control. There may be some stimuli in the environment very strong, for example loud noises, bright lights, strong odours and the like. These forces may distract our focus of perception.

The important objective factors are size, intensity, motion or change, **novelty**, repetitions, duration, and contrast.

The stimuli which are larger in size is more likely to be perceived than the smaller stimuli in the environment. A loud noise is clearly perceived than a feeble voice. A moving object is perceived clearly than a static one. An object in an unfamiliar place is perceived better than an object in a familiar place. An object presented intermittently is perceived at once than one presented permanently. An object is perceived clearly if it is presented for a minimum duration of time than an object disappears in a fraction of a minute. An object having contrasting characteristics in a situation is better perceived than one which is of similar type in the environment.

The subjective factors are motivation, previous experience and perceptual set. Motivation is a force that impels one for achieving his goal. A man who is very angry is motivated to see and talk about food. Our past experience facilitates us for accurate perception. Set is related to previous experience. Set is a readiness of an organism to pay interest to perceive certain objects. All these processes are also applicable for determining attention.

Organization of Perception or Perception of Form

In addition to the factors that we have discussed above there are also various other factors in the environment as well as in the organism give organization and meaning to perception. In many instances these organizations of factors of perception become crucial in determining the perceptual world. These organizing factors are shapes, forms, sizes and colours of the objects. Objects may be square, curved, round or angular. We are able to perceive objects with organized pattern though there are innumerable numbers of stimuli and welter of stimulations. The Gestalt psychologists were very much interested in the study of perception. They say that there is a tendency in our organism to follow certain basic principles to organize the sensations for perceptions. Organization of perception may be also stated as the perception of form. Form perception involves organizing separate figures or parts into a pattern or structure and may be called as form. In human beings there is a built-in tendency in the nervous system which simplifies the perceptual process.

Figure-ground

In our perception the first process is to isolate the object from the background. Moskowitz and Orgel (1969) state that it is the most primitive process in perception. The separation of figure or object from the ground is basic in every perceptual process. This perceptual process is a development in an organism independent of training or practice. It is found in human beings. Figure-ground perception in part depends upon certain specific physical characteristic of the object.

When we look at a stimulus pattern the figure-ground relationship emerges first before all other aspects are perceived. In visual perception an object is often differentiated from the background by its quality such as shape, size, colour and the like.

An object always stands out of its background. A book with red wrapper on the table stands out as a distinct figure against blue sunglassed table top.

Closure

In our perceptual process there is always a tendency to perceive objects as a whole. We may be also receiving sensations in an incomplete form. But we try to complete it and see it as full. The incompleteness is overlooked. The concept closure refers to the tendency to fill up the gap in the stimulus presented in the perceptual field. This is the principle evolved by the Gestalt psychologists.

Continuity

This is a tendency to bring together those elements that should go with other elements in such a way that we have a continuity of a line, curve, circle, to a movement in the same direction. We try to bring out a maximum possible regularity in the situation. But this principle of continuity sometimes may lead to inaccurate perception.

Proximity

Proximity means nearness. Perceptual organization occurs when objects are put close to each other, we do not perceive them as separate objects. Therefore, proximity is a tendency to group objects or figures on the basis of relative distance. The individual items may vary to some extent. Still we bring together and group them to a single pattern. Proximity may be spatial or temporal.

Similarity

This is the tendency to group figures of the same type together and perceive them as belonging to the same group. Generally objects that are more similar to each other will have greater chance to be grouped. Similarity means objects or persons have the same type of quality, one resembles others in several respects. There may be physical attributes, size, shape, colour, intensity and the like.

Perception of Distance or Depth

We know that the world in which we live is spatial in character. The world is three-dimensional with space. Everything we see, hear, smell, touch and handle are found in space. Every object occupy a definite space. It is a very difficult matter to explain how the human beings and other organisms move about and locate objects without any confusion in the three-dimensional space. The perception of space is highly complicated. What is going on the space is an automatic integrative activity. Automatic means we do not pay any special attention in our everyday activities. We are performing everything accurately. We judge the objects in front of us correctly; move on the road without causing any untoward incidents, we correctly judge the ups and downs and act accordingly. We also guide people to various directions. All these are taking place in the spaces around us. We are able to adjust ourselves correctly in land and also in water and air. For all these spatial movements we mainly trust upon our eyes.

When we look around us we see objects in three dimensions. The three dimensions are height, width and depth or distance. The room in which we sit and work is a three-dimensional. The room has a floor, the walls on the sides and the space in between the side walls. An object also has three dimensions like, length, width and thickness or the height. In perceiving the space, the third dimension, the important functioning is taking place in the visual system. In the construction of the spatial world the visual system provides us with the information about the location of the objects, events and movements in space. Along with our visual system there are also other sensory systems like the hearing and touching which are also helping us to locate the objects. From the sound we can understand whether it is from left or right, front or back, up or down. But how do we perceive space and distance? It is by translating the two-dimensional pattern retinal stimulation into a three-dimensional view, we perceive the space and distance.

Our ability to perceive depth or distance in space has something to do with our two eyes, which are separated, by a small distance. When we focus our two eyes on the same object each eye receives the information little differently from other eye. The retina of the eye is two-dimensional surface. A man with one eye can have all the experiences of a man who has two eyes. But a man with two eyes has certain advantage over the person with one eye by having a wider visual area and he can see more at one glance and has the benefit of *stereoscopic vision.* In stereoscopic vision the two eyes cooperatively provide the experience solidity and the distance. In the stereoscopic vision two flat figures, one in each eye, are presented before the eyes. These two figures are combined and provide the experience of depth. This we call three dimension. This is different from the flat figure received by a single eye. Objects at various distance stimulate the two retinas. The degree of stimulation depends upon the distance of the objects from the eyes. Since the two eyes are separated by a small distance the two eyes do not get exactly the same view. The figure in the left eye is slightly different from that of the right eye. This is called *retinal disparity.* The retinal disparity is a binocular vision. The three-dimensional experience of the single flat figures separately perceived in each eye is different from the stereoscopic experience in which both the figures are integrated. The stereoscopic function takes place when the slightly different figures of each eye are combined. Therefore, three-dimensional experience can be obtained in a stereoscopic vision and also in a monocular vision.

Another important point is that when we look at two objects at two different distances the image of the objects do not fall on the retina one behind the other based upon the distance in the environment. Since the retina is flat the image of the far away object may fall on the left or on the right, or below or above the image of the another nearby object. Why it is like that? The reason is that the human visual images are having only two dimensions. Postman and Egan (1959) state that the framework of the eye is anchored by two main directions, the vertical and the horizontal. The two dimensional images that fall on the retina area are converted into three-dimensional perceptions. The two dimensions provide stimulus information for evoking a third dimension. This is perception of depth or distance or space. The information falls under two sets, one is related to the environment and the other related to the organism.

The environment cues are monocular ones. They are interposition, perspective: linear perspective and aerial perspective, light and shadows, texture gradiant and movement. The organismic cues are binocular cues, accommodation, retinal disparity and convergence.

1. *The environmental cues* are primary distance cues and they are also *monocular cues.* Each eye gets the cues separately. The environmental cues depend upon physical aspects of the objects. The retinal disparity is a monocular one. Monocular cue depends upon our past experience. It helps us to judge distances accurately. The artists present the depth on two-dimensional picture by using certain principles. The *interposition,* as a cue, makes the nearest object as continuous and unbroken while those far away broken by the nearby objects. *Perspective* is a technique used by artist to show nearby objects larger and distant objects as smaller. In the case of linear perspective the edges of the roads, or railway lines which run parallel appear to meet at a distance. In the case of *aerial perspective,* distant objects, buildings, and hills appear hazier and less clear whereas nearby buildings look clear and bright. *Light and shadow* differentiate the distance of the objects. Objects having shadows are far away. The *Texture gradiant* of the object serves as cue to our perception of distance. Nearby objects look rough, irregular and full of dust whereas the distant objects look smooth, regular and finer. *Movement* indicates the distance of the object. Nearby vehicles look moving faster while the far away vehicles look moving slowly.

2. *The organismic cues* are secondary distance cues and they are also called ***binocular cues***, provided by the eyes. Accommodation in perceptual process refers to the change of the shape of the lens of the eyes. The lens bulges for nearby objects and flattens for far away objects. The nearby objects may also cause tension in the eyes. The retinal disparity is the difference in the perception between the two eyes. The two eyes receive slightly different images of the same object in normal binocular vision. The right eye receives more information from the right side and the left eye receives more information from left side. The nervous system combines these two visions as a single view and a three-dimensional view is provided and the object has solidity. The convergence is a non-visual cue. Beyond a distance of thirty feet the lines of the sight of the two eyes looking at an object are parallel. But to look at very close object the two eyes must turn towards each other. Muscles attached to the eyeballs turn them inward and they are in a state of tension to look at a nearby objects and the tension varies to look at far away objects.

In the studies made by Gibson and Walk (1960) it was observed that there was perception of depth or distance in new born animal and in new born human infants even though they had no chance to learn to perceive depth or distance. Therefore, Moskowitz and Orgal (1969) state that the central nervous system provides a primitive physiological cue to the perception of depth or distance.

Perception of Constancies

In the preceding sections we have discussed various aspects of perception. The important aspect is perception of distance or depth. Perception of distance is closely related to the main problem of this experiment, the perception of size and size constancy. The visual angle always links these two together. The world in which we live and see is by and large a stable one. We also see that objects retain their shape, size and colour. The characteristics of objects appear to remain approximately constant though the corresponding proximal stimulation changes. Proximal refers to the *space between the object from which the stimulation is emitted and the place from where the organism receives the stimulation.* These phenomena are general, invariably called object constancies. When we state that an object is constant, it implies that it is not changing as far its real character is concerned.

By the mechanism of constancies the individual gets the correct information about the world, as it is necessary to his effective actions and survival. This means that the individual recognizes the utility of the constancies. If the constancies of the objects in the world are not maintained there would be lots of confusion. For example, it is a matter of our common experience that a table looks rectangular and we retain the idea that the concerned table is always rectangular. The position of the table may be changed and may be placed in different styles. When it is kept at a distance it may look like a horizontal blank with four legs. But still we see it as a table. Even if it is dismantled, we may say that a table of rectangular size is broken. In whatever manner the shape of the table is distorted we see it as a table. Men, animals, trees, houses — all retain their known sizes, shapes, and proportions when perceived them from different distances and also under different conditions of light. Hochberg (1964) says that there is something in fact that does remain constant in the retinal image in each instance of constancies.

Our perception of objects, and events takes place within our perceptual field of space and time. In a room where we sit and work, we see quite number of objects of different types. Our perception is generally directed more towards objects or things rather than their qualities or sensory features. But we are also aware of the qualities of the objects like the colour, shape and size. Hilgard *et al.* (1975) state that our perceptual experiences are not separated from us. They build up a world of cognizable things. The objects perceived continue to exist in our memory and when we meet the same object we recognize it. Even in the absence of the object we think of it and remember its qualities. A well known or a familiar object is stable and permanent irrespective of its brightness, colour, shape, size and the distance at which it appeared. Therefore, there is a tendency in us to see the colour of a familiar object irrespective of its appearance in different shades of lighting conditions. This is called ***colour constancy***. We have also a tendency to see objects with the same shape as we perceived them and we do not change the shape of the objects irrespective of the angle from which we perceive.The perception of a table explained above is a good example for shape constancy. In our perceptual field we see that tremendous changes in the size of objects take place. In the normal perception we do not notice the change of size of the objects. But the observed size of an object tends to remain constant. This tendency to perceive an object with the same size in spite of changes in distance is called ***size constancy***. Therefore, there is a tendency in human perception to see objects as they are in colour, shape and size in spite of some changes in the environmental situations.

Perception of Size Constancy

In the case of size constancy the important factor that is likely to alter the size of an object is the distance. By altering the distance an object is made to look smaller and also larger. But human beings tend to see with its original size regardless of its proximal stimulus changes. Postman and Egan (1949) state that the size of an object and the distance are connected by the law of visual angle. The law is that an object becomes smaller and smaller as it moves away from the eye longer and longer in distance. Size constancy also includes the related tendency to see large objects as large and smaller objects as smaller even when their proximal stimulus sizes (distances) are the same.

The physical size of an object always remains constant. But the retinal image of the object is going on decreasing with the distance. The size of the retinal image is larger for the object very close to the eye. The size of the retinal image for the far away object is very small. But the retinal image may reduce in size only to certain distance. The law of the visual angle does not function beyond certain stage. The size of the retinal image no more reduces for very long distant object. Upto certain distance the size of the retinal image reduces rapidly proportional to the distance, after that the reduction takes place very slowly. Beyond certain distance no reduction takes place in proportion to the distance. According to Postman and Egan (1949) the cessation of the diminution of the size in the perceived object in proportion to the distance is called *size constancy.*

If we observe the size of the retinal image in the eye of a person we can see that the size of the retinal image varies inversely with the distance from which an object is seen (Fig. 5.11). When the object is near the size of the retinal image is large. When the object is far away the size of the retinal image in the eye is smaller. The perceived size of the object neither varies with the visual image nor changes with distance.

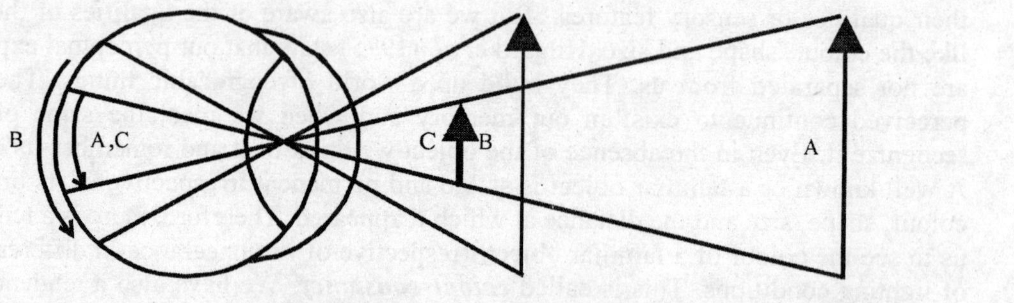

Fig. 5.11 : Retinal Image and Distance

Perception of the size of an object depends upon the information we receive about the distance. The more information we receive about the distance the more the actual size is perceived. When the cues of the distance are eliminated the perceived size cannot be judged correctly unless the object is familiar. If the objects are familiar with us then we can judge the appropriate size of it even if the distance cues are eliminated. This means that size constancy is developed largely out of experience. According to Bower (1966) size constancy develops with age for more longer distances.

Experiments can be designed to investigate the perceived size as a function of binocular and monocular visions under varied conditions at various distances. The aim, therefore, may be to obtain the functional relationship between perceived size of an object under different conditions and distances. The hypothesis that the perceived size of an object is inversely related to the distance may be verified.

EXPERIMENT

Problem

Determine how accurately a person perceives the size of the standard stimulus at different distances under several viewing conditions.

or

Determine the functional relationship between perceived size of an object under varied conditions and distances.

Method

Participant

Choose a person who has no knowledge about this experiment and collect his personal information like, sex, age, qualification, occupation or parent's occupation whether he/she wears eye glass, and deformities, if any.

Materials

Materials for the experiment consists of (1) 9 round black card boards. Each round card board varies in size from the other by one centimetre in diameter. The sizes of the round cards are 4 cm, 5 cm, 6 cm, 7 cm, 8 cm, 9 cm, 10 cm, 11 cm and 12 cm in diameter. These round cards are variable stimuli. (2) The size of the standard round black cardboard is having 8 centimetres diameter. All the variable round black cardboards and the standard black cardboard are mounted on large white cardboards of 30 cm × 30 cm size to have visible background for the stimulus. (3) One metre scale. (4) One blind-fold. (5) Black coated tunnel view tube. (Model-materials — Appendix-I).

Procedure

The room where this experiment is conducted should be a long enough and well illuminated. The area should be free from any materials or objects. The participant should be comfortably seated and the standard stimulus should be placed two metres away in front of him on the right or left at 45°. and clamped with a stand at a height on Bar with the eyesight of the participant. It should be arranged in such a way that the participant should turn his head 45° on the left or right to fix his eyes on the standard stimulus (Appendix–II).

Then the variable stimulus should be clamped with stand at a height on bar with the eyes of the participant and it should be placed directly in front of the participant at a distance of two metres. The experiment is conducted at four varied distances of two metres, four metres, six metres and at eight metres and also with two conditions as *approaching condition* and *receding condition* and each condition with *ascending series* and *descending series*. Therefore, the experimenter should prepare everything in advances. Since the variable stimuli will have to be placed at four different distances like two metres, four metres, six metres and eight metres the experimenter should measure the distances with a metre scale and mark the four places where the variable stimulus should be placed in succession. A slanting line may be drawn at 45° from the participant to place the standard stimulus at a two metre distance.

The idea of keeping the standard stimulus constantly at the nearest point and varying the distances of variable stimulus is that it is the test of perceptual size constancy. The size of an object seen at close distance may appear smaller at a longer distance. It is to see whether the participant is able to judge this correctly or carried away by the illusion caused by the distance. In addition the participant should judge the larger as larger and the smaller as smaller even if the distances are varied. Therefore, it is to test the ability of the participant how far the size of the perceived object is maintained correctly at varied distances and conditions. If the standard stimulus is kept at close proximity at every distance with the variable stimulus the purpose of the experiment is defeated. Then it is not the test of the ability of size constancy. It becomes different experiment.

The experimenter should prepare a Table for making entries of the participant's judgments. Then the following instructions should be given to him: The experimenter first begins with approaching condition and secondly with receding condition.

> "Now you are seeing one round black pattern on a white board at your right at two metres away which is called standard stimulus. At your front at a distance two metres another black round pattern on the white board is kept. This is called variable stimulus. I will be going on changing the variable stimulus and also the distance. Your task is to say whether the variable stimulus in the front is equal to the standard stimulus. If equal say 'yes'. If not, say whether it is larger than or smaller than the standard stimulus. As soon as the task is over at one position, I will move the variable stimuli to the next positions, step by step. After finishing the task at the last position, I will move the variable stimuli forward to the front position and step by step approach to the nearest position. Then again I will move backward step by step and on reaching the last position and move forward to the nearest position. Your task at all places is to say whether it is equal. If not say whether the variable stimulus is larger or smaller. There are three conditions of viewing the stimulus. First, you will see with both eyes. Then, secondly, you will have to see it with one eye and judge. Thirdly, you will have to see with one eye through the funnel shaped viewing tube. Now let us start."

I. Approaching Condition : The experimenter first starts the *binocular vision unimpeded*. It is approaching condition in ascending series. The standard stimulus is kept at a distance of two metres 45° on the right of the participant and the variable stimulus is directly in front of the participant at an eight metre distance. The experimenter first presents the smallest stimulus 4 cm diameter card and asks the participant : "Now, make your judgment whether these two are equal". The participant states, "not equal, the variable is smaller". If the participant says that the variable stimulus is *equal* to the standard stimulus the experimenter should write the letter '*E*' in his record sheet. If he says that the variable stimulus is *smaller* the experimenter should write the letter '*S*' and if he states *larger* the experimenter should write the letter '*L*' for the judgments. The letter '*E*' stands for equal, '*S*' stands for smaller and '*L*' stands for larger. Then the experimenter presents the next larger stimulus, 5 cm diameter card and the participant gives the statements. In the same manner he should present all the nine cards one by one and the participant's statement should be recorded by him in his Table.5.5. When all the nine stimuli are presented in ascending series the experimenter begins the descending series in the same position and presents the largest size card (12 cm diameter) first and goes on decreasing the size one by one. On presenting all the nine stimuli in descending series he moves the variable stimulus forward to the next position to six metre distance but the standard stimulus will be in the same position at two metre distance at 45° on the right. Then he starts the ascending series with the 4 cm diameter stimulus card and goes on presenting all the stimuli one by one and enters the statements of the participant in his Table. Then starts the descending series by presenting the largest size card first and goes on decreasing the size of the cards.

Then the experimenter moves the variable stimulus to the third position, 4 metre distance, carries out the same task by presenting all the nine variable stimulus cards one by one to the participant with ascending and descending series and records his statement in the Table. On presenting all the nine stimuli he moves the variable stimulus to the last position to two metre distance. The same performance the experimenter repeats with the nine variable stimulus cards one by one with ascending and descending series and gets the statements of the participant recorded in his Table. The approaching condition with ascending and descending series is now completed.

Now the experimenter begins the *receding condition with ascending and descending series.* Since it is the receding condition, he begins at the nearest position, at two metre distance. He presents all the nine variable stimuli one by one and completing them he moves the variable stimulus backward to the next position to 4 metre distance. But he is not moving the standard stimulus. It is at the same position, at two metre distance at 45° slanting position. He presents all the nine stimuli in ascending and descending series and the statements of the participant are recorded. He carried out this till he reaches the last position 8 metre distance. Thus he completes the receding condition with ascending and descending series. The experimenter now has completed one approaching and one receding condition with ascending and descending series in binocular view.

II. Receding Condition : The next viewing condition is *monocular vision.* In this viewing condition one of the eyes of the participant is covered by the blind fold. The experimenter blind folds the participant's left eye and the participant can see only with his right eye. The variable stimulus is kept directly in front of the participant at eight metre distance and the standard stimulus kept at 45° two metre away on the right of the participant. In the monocular viewing condition it is now approaching condition with ascending and descending series. The same procedure is followed at all 4 positions. On reaching the last position two metre distance and on recording the statement of the participant the experimenter begins the receding condition with ascending and descending series and carries out the same procedure by moving the variable stimulus from two metre to 8 metre position.

III. Viewing Condition : The third viewing condition is *monocular vision through tunnel tube.* One eye is blind-folded in this viewing condition as in condition II and the participant should see the stimulus only through the tunnel hole. The experimenter once again arranges that the variable stimulus in front of the participant at eight metres away for approaching condition. The experimenter blind-folds the left eye of the participant and hands over the tunnel tube to the participant and directs him to carry out the same task. The experimenter begins the approaching condition with ascending and descending series and repeats the same in the receding condition and records all the statements of the participant in his Table. The important aspect of this experiment is that the experimenter presents the variable stimuli first in increasing order and then in decreasing order in the same position. He repeats the same thing at all four positions, i.e. at 8 metres, 6 metres, 4 metres and at 2 metres distances. He carries out the same both in approaching and receding conditions for all the three viewing conditions. The experimenter should prepare three separate tables of similar type of three viewing conditions. As soon as all the three steps of performances are over the participant may be allowed to go away.

Results

The experimenter has indicated with the letter '*E*' if the variable stimulus is judged as equal, with the letter, '*L*' if the variable stimulus is judged as larger and with the letter '*S*' if the variable stimulus is judged as smaller. First, he makes entries for the approaching condition for all the four distances and then he makes entries for the receding condition for all four distances. This procedure is followed for all the three viewing conditions.

The average is worked out for ascending and descending series separately for each distance and then the average is worked for approaching and receding conditions separately. Then the overall average for each distance separately should be worked out. The overall average for each distance is given in the last row. This is the average size of the variable stimulus judged by the participant as equal to the standard stimulus at each distance under various conditions and series. This may be stated as the size constancy (Table 5.5).

The entries in Table 5.6 are the judgments made by the participant on the nine variable stimuli in comparison with the standard stimulus at each distance under approaching and receding conditions when the stimuli are presented in ascending order and in descending order. The data are presented under various groups. Now the average for each column should be worked out which becomes the average sizes of the variable stimuli judged as equal to the standard stimulus by the participant at each of the distance under various conditions of observation. First, we get the average for each series and then the average for approaching and receding conditions separately and then the overall average for each distance. So we get three types of averages. The overall average for each distance is the size the participant observed. This observation is made by the participant with one eye.

Model Table

Table 5.5 : Showing the judgments of the participant in the unimpeded ***Binocular vision*** at 4 distances under various conditions of observation (Size of the standard stimulus = 8 cm diameter). (*Note* : First, enter approaching ascending and descending judgments for all 4 distances and then enter receding-ascending and descending judgments for all 4 distances)

S. No	Sizes of the Stimuli in cm Diam.	2 metres				4 metres				6 metres				8 metres			
		Approaching		Receding		Approaching		Receding		Approaching		Receding		Approaching		Receding	
		AS	DS	AS	DS	AS	DS	AS	DS	AS	DS	AS	DS	AS	DS	AS	DS
1	4 cm	S	S	S	S	S	S	S	S	S	S	S	S	S	S	S	S
2	5 cm	S	S	S	S	S	S	S	S	S	S	S	S	S	S	S	S
3	6 cm	S	S	S	S	S	S	S	S	S	S	S	S	S	S	S	S
4	7 cm	S	S	E	S	E	S	S	S	E	E	E	E	S	S	S	S
5	8 cm	E	E	E	E	E	E	E	E	E	E	E	L	S	S	S	E
6	9 cm	L	L	E	E	L	L	E	E	L	L	L	L	E	E	E	E
7	10 cm	L	L	L	L	L	L	L	L	L	L	L	L	L	L	E	E
8	11 cm	L	L	L	L	L	L	L	L	L	L	L	L	L	L	L	L
9	12 cm	L	L	L	L	L	L	L	L	L	L	L	L	L	L	L	L
Average		8	8	8	8.5	7.5	8.5	8.5	8.5	7.5	7.5	7.5	7.0	9.0	9.0	9.5	9.0
Average		8		8.25		8		8.5		7.5		7.25		9.0		9.25	
Overall Average		8.11				8.25				7.38				9.13			

Table 5.6 : Showing the judgments of the participant in the ***Monocular vision*** at 4 distances under various conditions of observation (*Note* : First, enter in the Table Approaching ascending and descending judgments for all 4 distances and then enter receding-ascending and descending judgments for all 4 distances). (Size of the Standard stimulus = 8 cm diameter)

S. No	Sizes of the Stimuli in cm Diam.	2 metres				4 metres				6 metres				8 metres			
		Approaching		Receding		Approaching		Receding		Approaching		Receding		Approaching		Receding	
		AS	DS	AS	DS	AS	DS	AS	DS	AS	DS	AS	DS	AS	DS	AS	DS
1	4 cm																
2	5 cm																
3	6 cm																
4	7 cm																
5	8 cm																
6	9 cm																
7	10 cm																
8	11 cm																
9	12 cm																
Average																	
Average																	
Overall Average																	

Table 5.7 : Showing the judgments of the participant in the ***Monocular vision through tunnel tube*** at 4 distances under various conditions of observation. The left eye of the participant is blind folded. (*Note* : First, make entries of Approaching-asendency and descendency judgments in the table for all 4 distances and then enter receding-ascendency and descendency judgments for 4 distances) (Size of the Standard stimulus = 8 cm diameter)

S. No	Size of the Stimuli in cm Diam.	2 metres				4 metres				6 metres				8 metres			
		Approaching		Receding		Approaching		Receding		Approaching		Receding		Approaching		Receding	
		AS	DS	AS	DS	AS	DS	AS	DS	AS	DS	AS	DS	AS	DS·	AS	DS
1	4 cm																
2	5 cm																
3	6 cm																
4	7 cm																
5	8 cm																
6	9 cm																
7	10 cm																
8	11 cm																
9	12 cm																
Average																	
Average																	
Overall Average																	

In Table 5.7 the entries are the judgments made by the participant with the monocular vision through tunnel tube on the nine variable stimuli in comparison with the standard stimulus at 4 distances under approaching and receding conditions when the stimuli are presented in ascending order and descending order. The data are presented under four groups. In this experimental condition participant's vision is hindered in two ways. He is viewing with one eye and also through the tunnel tube. Therefore, there is more chance for the participant to make incorrect judgments of the variable stimuli in comparison with the standard stimulus.

Three types of averages are worked out for the data in columns. One average is for the ascendancy and descendency, the second average is for the approaching and receding conditions and the third is overall average for the distance.

A comparative analysis of the judgments of the variable stimuli in comparison with the standard stimulus in all the three viewing conditions at four distances is made. The relevant data are presented in Table 5.8. In Table 5.8 the participant's average judgments in all the three views reveal his perceptual size constancy. Whether this person over-estimates or underestimates the variable stimuli in comparison with the standard stimulus is revealed.

The experimenter should collect the group data from other experimenters and present them in Table 5.9. The data include the average of the three viewing conditions and the average judgments made at all four distances. Work out the mean for every individual for all 4 distances and also the mean for each view at each of the 4 distances.

Table 5.8 : Showing the average judgments of the various stimuli in comparison with the standard stimulus made by the participant in three viewing conditions at 4 distances.

S.No.	Distance in metres	Viewing conditions			Average
		Binocular vision unimpeded	Monocular vision	Monocular vision Tunnel view	
1	2 m				
2	4 m				
3	6 m				
4	8 m				
Average					

Table 5.9 : Showing the average judgments of the group of the variable stimuli in comparison with the standard stimulus in three viewing conditions of the Binocular, Monocular, Monocular Tunnel view and at four distances, 2 m, 4 m, 6 m, 8 m

S. No	Name	Distance												Average
		2 metres			4 metres			6 metres			8 metres			
		Binocular	Monocular	Monocular Tunnel	Binocular	Monocular	Monocular Tunnel	Binocular	Monocular	Monocular Tunnel	Binocular	Monocular	Monocular Tunnel	
1	P.V.													
2	K.R.													
3	M.S.													
4	R.V.													
5	R.M.													
6	S.V.													
7	T.N.													
8	K.M.													
9	V.R.													
10	M.V.													
11	S.R.													
12	R.A.													

Discussion

State what is constancy and explain the place of constancy and the size of constancy in human perceptual process. State and discuss whether the judgments of ascending series is more closer to the standard stimulus or the judgment of descending series when all the series put together in bonocular vision. State the reason, if any. Similarly, if you find any difference between approaching condition and receding conditions state the reason. Do the judgments vary in accuracy when the distances are increased in the experimental situation. If there is anything, state the reason for the same. Postman and Egan (1949) state that the problems of perception of distance and the perception of size are closely related. Do you find that this hypothesis is accepted in your experiment? Explain it.

In judging the variable stimuli in comparison with the standard stimulus do you find any difference between binocular vision and monocular vision and also with the monocular vision through tunnel view? Which viewing condition facilitates for accurate judgment? Discuss the general performance of your participant in this experiment and state in which viewing condition he/she has performed well and where his/her performance is poor? State the reason.

Compare your participant's performance with the members of the group and state where he/she stands? Whether your participant's perception of size constancy is positively oriented or negatively oriented. How many of them in the group positively judge the variable stimuli and how many of them negatively judge? State the reason for this. State and discuss whether distance is having any influence in the judgment. Do you find that binocular vision is having any advantage over the monocular vision and monocular tunnel view? How many of them in the group had the advantage of the binocular vision? State the reason. How many of them in the group made correct judgment of the standard stimulus at 8 metres distance. Discuss this phenomenon in comparison with the judgment made by others. State the reason, if any.

Do you find any individual difference in this size constancy experiment? What personal factors contributed to perceptual size constancy. Do you find any difference in the perceptual constancy due to age, sex, qualification or some personal deformities in this experiment. Workout this in your data and discuss.

Conclusions

1. What is the overall average of perceptual size constancy of your participant in three viewing conditions?
2. What is the size constancy in your participant's judgment at each of the distances?
3. State whether distance is a determining factor of perceptual size constancy.
4. State the range of perceived size constancy of the members oı the group.
5. How many of them in the group made correct judgment of the standard stimulus?
6. Find out and state whether age, sex, qualification have any effect in the perceptual size constancy.

APPENDIX - I

Fig. 5.12 : Standard Stimulus (8 cm in diameter)

Fig. 5.13 : Variable Stimulus (4 cm in diameter)

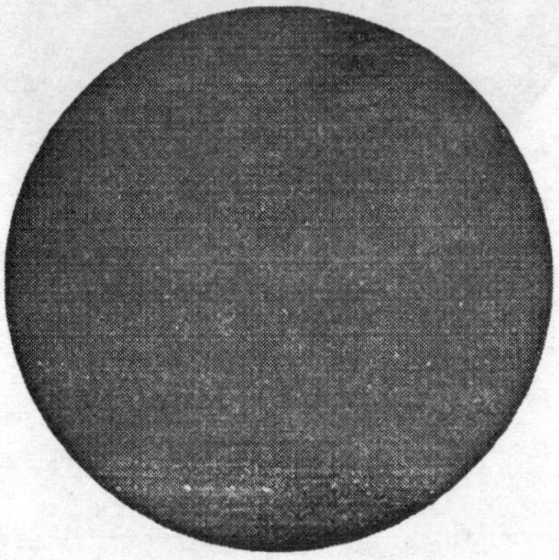

Fig. 5.14 : Variable Stimulus (5 cm in diameter)

Fig. 5.15 : Variable Stimulus (6 cm in diameter)

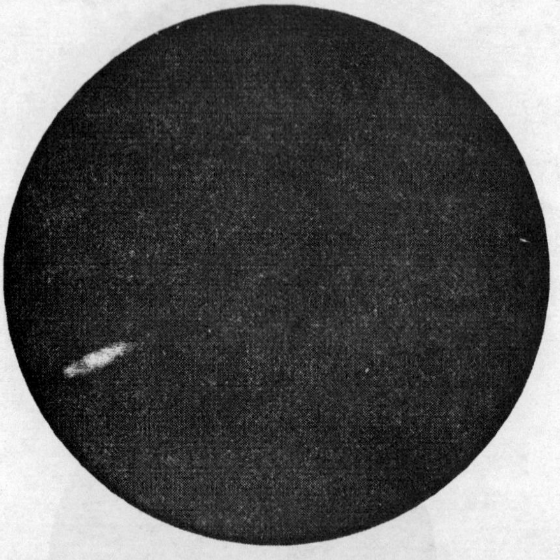

Fig. 5.16 : Variable Stimulus (7 cm in diameter)

Fig. 5.17 : Variable Stimulus (8 cm in diameter)

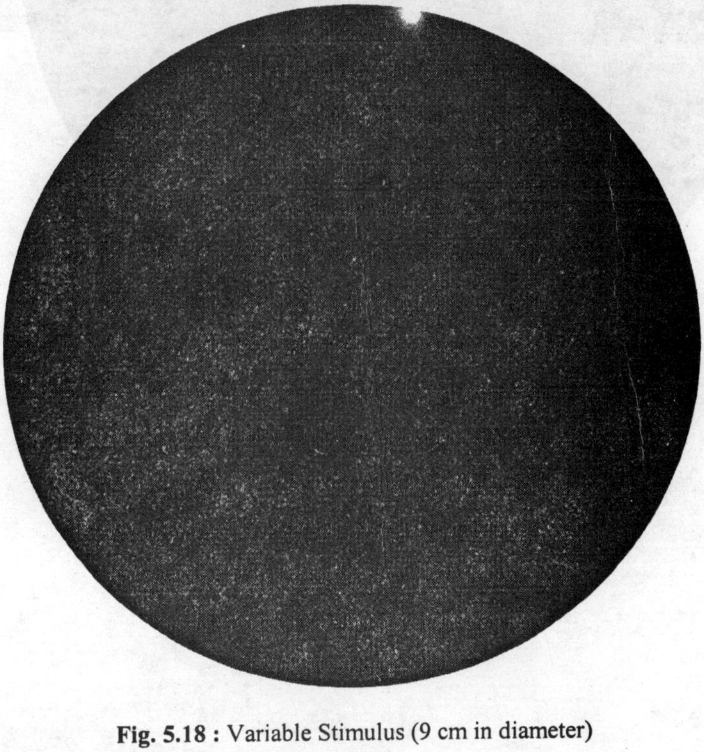

Fig. 5.18 : Variable Stimulus (9 cm in diameter)

Fig. 5.19 : Variable Stimulus (10 cm in diameter)

Fig. 5.20 : Variable Stimulus (11 cm in diameter)

Fig. 5.21 : Variable Stimulus (12 cm in diameter)

APPENDIX-II

PERCEPTION

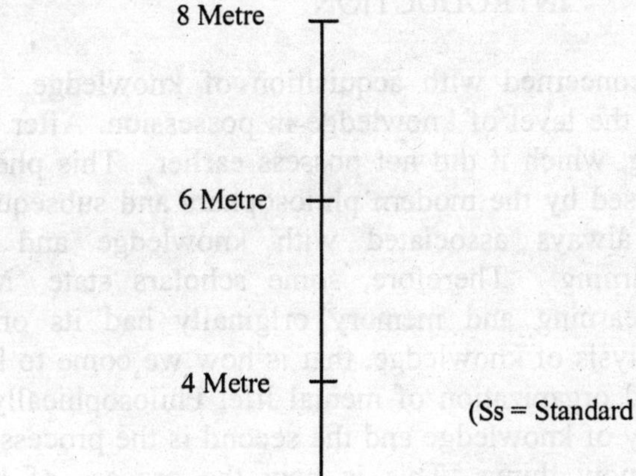

Fig. 5.22 : Physical Plan of the Experiment 6 : Size Constancy

6

LEARNING, TRANSFER OF TRAINING, REMEMBERING AND FORGETTING

INTRODUCTION

Learning, we all know, is concerned with acquisition of knowledge. Acquisition basically refers to changes in the level of knowledge in possession. After learning, the organism possesses something, which it did not possess earlier. This phenomenon of learning has been long discussed by the modern philosophers and subsequently by the psychologists. Learning is always associated with knowledge and memory is synonymously used with learning. Therefore, some scholars state 'Memory and knowledge'. The study of learning and memory originally had its origin in two philosophical sources, the analysis of knowledge, that is how we come to know things, and the analysis of nature and organization of mental life. Philosophically the first is called epistemology, the theory of knowledge and the second is the processes of mental phenomenon. We learn to know things. This is how the process of learning and knowledge are related. To learn is to gain knowledge. The early question raised was, how our knowledge arises and what is the relationship between knowledge and organization of the mind. Two major opposing philosophies have dealt with this question in detail. One is Empiricism and the other is Rationalism. They have been fighting constantly with one another for centuries.

The philosophy of empiricism advocates that all knowledge derived from only experience. It gives more emphasis to sensory experience. All our concepts and ideas are derived from the impressions of our senses. The concepts and ideas are direct copies of sensory impressions (simple ideas) or they are combination of several simple ideas (complex ideas). The sensory impression of an object may be broken into a set of sensory qualities. The qualities of the object from which sensory impressions are received may be colour, smell, taste, size, texture, distance and so on. These sensory qualities are associated in the mind. For example the idea of an object, a chair is a complex one. It can be reduced to more simple ideas by association. The knowledge acquired about the chair can be expressed to other relevant ideas — complex ideas in association, by stating that chairs are furniture, made out of wood, used for sitting, and so on. This sort of knowledge may be acquired verbally by listening to someone telling all such facts.

In the explanation given about the empiricism there are four important characteristics such as sensationalism, reductionism, associationism and mechanism. All knowledge is derived from sensory experience (sensationalism), all complex ideas stored in a basic stock of single ideas may be reduced to qualities of the object (reductionism), the ideas connected through action of association by continuous experience and the mind is like a machine built out of simple elements (mechanism).

Empiricism advocates two basic ideas for learning. One is the simple ideas which are internal representation which may be also called memory images. These ideas originate in the sensory impressions stored in the memory. The other is the complex ideas formed by connecting together the simple ideas that are experienced continuously in memory. The memory records the events that occurred first with the second by associations that event *A* was followed by the event *B*. This is how the sensory impressions and copying effect in memory are functioning. The British philosophers, Hobbes, Locke, Hume, James Mill and John Mill have elaborated this idea into a philosophy of Empiricism.

Rationalism is of the view that knowledge is acquired by reason without regard to experience. Reason alone is the only valid basis for knowledge, belief and not spiritual revelation, or intuition, or sense data. The rationalists have entirely different conclusion about the role of sense data or sensory impressions in the acquisition of knowledge or understanding the reality. The empiricists state that our ideas are passive copies of sensory impressions, whereas for the rationalist sensory impressions are unstructured, blurred, diffused and chaos that only provide raw material. The raw data can be interpreted only according to some innate perceptual assumption. Immanuel Kant (1724-1804), the German philosopher who lived in eighteenth century states in his famous work called *Critique of Pure Reason* (1781) that the element, which corresponds to sensation can be called the matter. The element with its various factors arranged in certain ways relative to one another has its form. Without this form there is no order for sensations and form itself cannot be sensation. The matter is presented to entirely later on, secondly the form is the first to appear in the mind. Therefore, it must be considered first apart from sensation. According to Kant material is known empirically prior to the experience.

The rationalists state that the object or matter is already in existence in our mind and we project the idea of the physical object outside into an order of particular qualities (that the qualities the object possesses). The material will always maintain its ideal identity. Whatever changes take place in the material the identity is reserved. For example, in the case of the example chair, we project the idea of chair and we see a chair. We know the characteristics of the chair even before we see it, that it has four legs, rests for our back and hands and how it is made of. Even before seeing the object we come to know what it is from the idea. Therefore, reason is the prime source of knowledge and not experience.

The rationalistic philosophy was developed by a group of philosophers of the modern period and they have explained how we acquire knowledge. Some of the important philosophers are, Descartes (1596-1650), Leibniz (1646-1716), Immanuel Kant and Berkeley (1685-1753). They advocated that only mind is real and objects outside are only ideas. These philosophers are also called idealists.

The two opposing philosophies, the empiricism and rationalism explained above with reference to the process of acquisition of knowledge serve as the valuable background in understanding modern psychological concepts like experience, knowledge, memory, learning and ideas. Many of the learning theories are classified on the basis of these philosophies. Associationism is one of the important components of empiricism and all the behaviouristic theories of learning are based upon associationalism. For example, therefore, Pavlov, Guthrie, Hull, Skinner and the functionalists Titchner, John Dewey (1859-1952), Galton (1821-1917) and Cattall (1860-1944) all have their source in empiricism. John Dewey, an American pragmatic philosopher, formulated an instrumentation theory of learning through experience. The Gestalt psychologists, Tolman and modern cognitive psychologists are all classified as rationalists in learning.

The organism does not react in the same manner all the times. There are changes in its reaction to situations. That means something has taken place in the organism to make changes in its state. This something may be stated as knowledge. How this has come to the organism. From the source discussed above we may state that some sensory impressions might have got into the organism and organism has acquired some knowledge which makes it to change its state. Or, it may be due to the reasoning out the events that occur outside, the organism might have acquired knowledge. The nature of knowledge the organism has acquired is so complex that it is very difficult to explain. Knowledge comes to us in hundred different ways and hundred different things we come to know. If every person narrates his personal experience of his life events there can be no reason for us to doubt or deny his statements. He speaks out his knowledge from memory. Hilgard and Bower (1975) state that what we speak of memory trace is the internal representation of the specific information stored at a time. The word memory or remembering is commonly used in a number of different senses. This has to be distinguished. According to Bertrand Russell (1872-1970), an English philosopher and mathematician, who have simplified many philosophical principles, has also stated how we come to know by our knowledge and what degree of reliance we can place upon its various parts. He also discusses about perception, memory, the way in which animals and infants learn and the part played by words. Regarding memory, he states that more empirically, there is a broad sense, in which the word memory applies to the power of repeating any habitual act previously learned, and a narrow sense, in which the word memory applies only to recollection of past events. People often speak of remembering or memory only in this broad sense.

This broad sense of remembering is applied to animals and infants also. The dog remembers his master and also his name when called. Darwin said that plants have memory. Samuel Butler (1612-1680) an English poet and atirist, had attributed the instinctive behaviour to memory of ancestral experience. He used the word memory in its widest sense. Bergson (1859-1941), the French philosopher, who sought to bridge the gap between metaphysics and science has brought out one main work in memory and matter. He states that true memory is restricted to the recollection of a past occurrence. The event remembered has occurred only once. This cannot be called as habit in behaviourists sense. According to behaviourism all memory consists of retaining of a habit. Since memory is absorbed in the study of habit, it does not deserve any special study. Watson (1878-1958), the founder of the school of behaviourism, states that he does not use the term memory in his behaviourism, as it has no place in an objective psychology. Watson says that if memory is used in the sense of recollection of past events, it may be called as verbal habit.

Suppose, for example, long ago Mr. *X.* sees a man Mr. *Y* on the street. Both of them formed mutual verbal and manual or physical habits towards each other and part with.

After few months of absence they see each other again somewhere. At this time they call out not only the old verbal habits, but also many types of bodily and visceral types of responses. Watson says what we mean memory is nothing but the fact that when we meet a stimulus again after an absence of few months we react with the old habits. We speak the old words and show the old visceral and emotional behaviour pattern that we have learned to do in the presence of the stimulus when we met for the first time. This theory seems to be agreeable in many ways. Watson says, memory is not a 'mystical faculty'. We are supposed to remember always everything that should remember if suitable stimulus is presented. Remembering is a specific causative act and such an act is observable externally. This is how Watson was able to explain and establish his theory. Russell says that it is quite possible by the behaviourist methods that much memory can be adequately explained as habit, but there seems to be great difficulties in the view that memory consists of entirely of habits.

Just like, experience, knowledge and memory, philosophers of different periods have extensively discussed about learning. Russell (1952) says that learning denotes modification of behaviour as the result of experience. He considers that the understanding of behaviourist psychology is vital for proper understanding of philosophy for an objective study of knowledge and inference. The objective study is one in which the observer is different from the observed. Both need not be the same person. If both are same then it becomes subjective study. The animal learning has been studied very much in recent years. The experiments on animals need great patience and observation. Certain results have been obtained as desired by the experimenters. The Russians, the Americans and the Germans studied animals. The results of these studies revealed the rational characteristics of the experimenter. The problem is, that the person who naturally sets up an experiment to an animal depends upon his own philosophy. What the Russians expect from the dogs, what the Americans expect from the cats and what the Germans expect from the monkeys. These expectations probably remain cause for the differences in the results. Though each one gets one result from his experiment and different from the other they are not altogether rejected. They can be compared provided an experimenter could provide complete survey of the whole field.

Animal learning started scientifically beginning with Thorndike's (1911) Animal Intelligence. He conducted experiments on cats with mazes. He formulated two laws of learning, the laws of effect and the laws of exercise. According to the law of effect, the animal makes several responses in the same situation and it is more firmly connected with the situation which is followed by satisfaction or the response which provides the animal with food. Such responses are strengthened and other responses, which are not satisfying, are weakened. According to the law of exercise, the animal is strengthening its connection with practice and weakening its connection when practice is discontinued. Strengthening means that the response is made when the situation recurs.

The rise of behaviourism, the importance given to conditioned response, and the attacks by the Gestalt psychologists and other criticisms have made Thorndike to revise his laws. The law of effect was revised and the law of exercise was practically given up in place of a law of learning. The kernel of Thorndike's laws is that an animal tends to repeat the responses which has brought it pleasure (food), and animal tends to repeat the response what it has done often before.

In the law of effect Thorndike speaks of the animal's satisfaction and discomfort, which are subjective in nature. How can one verify whether the animal is satisfied or in the state of discomfort. Therefore, his results are not objectively verified and it is not objective psychology. But we can only infer since the animal tends to behave so as to make the events to recur. This is what Thorndike means 'give satisfaction' to the animal as it gets the cheese. The animal tends to repeat the acts as it desires by which finds satisfaction. In this way the objectivity in Thorndike's law can be explained.

In animal learning every student of Psychology knows pretty well Kohler's (1913) experiments with chimpanzees. Kohler put the banana in different difficult situations and he observed how the chimpanzees were able to reach the banana and manoeuvred the situation, which Kohler called 'insight'. Kohler states that the banana is objective. The chimpanzee in the experimental situation acts freely on its own accord to achieve its goal. It is not by accident nor by trial and error the chimpanzee behaves in solving the problem. The animal's behaviour definitely seems to be intelligent. The animal takes into account the whole situation in dealing with the problem. Kohler states that genuine problem solving activities do not improve by practice or repetition. The solution comes perfectly on the first occasion. If efforts go wrong by repetition the organism loses the excitement or satisfaction of its success. Thus the accounts given by Kohler about his chimpanzees' behaviour are totally different from that of the rats in mazes and cats in puzzle boxes. It appears that the conclusions drawn for the American experiments are as desired by the experimenters. All the experiments seem to have one type of problem and drawn one type of conclusion and they are made applicable to all problems of animal learning.

Therefore, there appears to be two broad ways of learning, one is by experience which includes what we have discussed earlier, association and other variables, the other is learning by insight. Learning by experience can be done by most of the vertebrates, that is most of the organisms, which are having backbone or spinal column. But learning by insight is taking place in chimpanzees and in most highly developed apes. Learning by insights so far is not tested among cats and rats. Therefore, any animal lower than anthropoid is not capable of insight learning. In insight learning certain amount of intelligence is involved and it is classified a: the rationalist end. Since Kohler's chimpanzees and the anthropoid apes are capable of insight learning are we to assume that they are also capable of reasoning? There must be certain amount of reasoning in their behaviour; otherwise learning by insight is not possible.

There are several criticisms against Kohler's theory of insight learning and the behaviour of his chimpanzees. The new acts in chimpanzees are composed of a number of parts and each of which might have occurred on some former occasion. Series of the acts resulted together in obtaining the banana. Thorndike thinks that the solution came by pure chance and the baffled animal's acts are random acts. The acts are combined by chance. Each act performed earlier might have been successful for some other purpose. But now they are combined for the first time ending with success in the right order. Kohler says that it is not correct to state that the solution is chance success. The chimpanzee's acts are directed to the objective. Every stage is clearly perceived in the behaviour of the chimpanzee. It behaves in the same manner what a human being would do in similar situation. The behaviour appears to be a complete attempt to the solution. It is seen that no part of the behaviour appears to be arranging the parts accidentally for producing the successful effort. The solution does not appear to be a disordered one and the consequences of blind impulse on the part of the chimpanzees. What was going on was continuous smooth action. Therefore, the statement that the solutions are the results of mere chance is not acceptable argument.

HUMAN LEARNING

Human learning takes place in two ways, one is by experience and the other is by insight. The experience is based upon the theory of empiricism and insight is based upon the theory of rationalism. All other theories are brought under these two phenomena. We have discussed at length the problem of learning in animals or non-verbal numan. The experimenters in the study of animal learning will have to observe only some kind of overt behaviour of the animals taking into account only a specific learning of the animal. Therefore, in the case of animal learning an operational definition is not only a difficult business for them but also not applicable. But in the case of human learning a working definition of learning becomes essential and it has to be applied at all levels. A working definition of learning according to Hilgard and Bower (1975) is that learning refers to the changes in subject's behaviour in a given situation and the change is brought out by his repeated practices in that situation and this behaviour change on any account should not be attributable to the native response tendencies, maturation or any of the states or influence of fatigue or drugs.

The concept native response tendencies refer to innate behaviour. They are also called unlearned activities such as the reflex and pupillary movements. It is also stated as instinctive behaviour. Birds and animals act on instincts. The instinctive behaviour pattern is innate, which means that it forms its origin in the nervous system of the members of the species. It is stated as independent of experience of learning. According to Beach (1955) and Hinde (1966) the expressions of instinctive behaviour pattern depend very much upon varied experience and learning. The innate constitution of the animal forces some slender selective tendencies to develop in certain ways. But the tendencies can be influenced by extreme environmental conditions.

Growth and maturation serve as important modifiers of behaviour in competition with learning. If behaviour matures through different stages of the individual without any intervention of practice then behaviour is very much influenced by maturation and by learning. But, however, many of the individual activities develop through the combination of both learning and maturation. For example, the language development in the child is a good example to illustrate this process. The child is able to speak words only when there was adequate development in its body and nervous system.

Fatigue is both a bodily and mental state that often interferes with learning. In motor activities and sensory activities there is often loss of energy and efficiency. The work becomes slow and also erroneous. The organism may hesitate to continue the work. Production is reduced. If the response needs greater effort than fatigue effect is faster. When there is work decrement it is better to give some rest. This rest should be to the extent of recovering from the fatigue effect.

Generally, learning is indicated by the performance of the organism. The phenomena of learning *versus* performance are like disposition and actualization. What a person knows about something and what he actually does. An individual might have learned something but he may lack in performance. Performance may be reduced if there is a failure of retention of learned material. Students who write examinations face this sort of critical situations. They are not able to reproduce the learned material in writing the examinations. The process of retention or remembering is not strong. Motivation is supposed to be the psychological factor that is responsible for converting knowledge into action.

It is also possible that performance under the influence of drugs, alcohol and illness may be reduced. The individual may fail to perform the learned act when he is in such a state. It is not that he has not learned the act originally or forgotten what he has learned. When the individual restored his normal state he may return to the normal level of performance without any training. But it is also observed that if the original learned behaviour level could not be recovered immediately after the cessation of the drug or alcohol effects then it means that the drug or alcohol had some kind of effect on the memory, either by blocking or destroying the memory traces or preventing its expression.

As far as possible the definition of learning given above included all the factors that may operate in the process of learning. The interpretations and illustrations are also seemed to be quite agreeable. Difference of opinion or controversies are possible if one gives preference to a particular theory. This attitude leads one to pay more attention upon one kind of learning situation and ignoring of others. But every learning theory tries to answer the question or controversies raised by psychologists and non-psychologists. For example, behaviourism tried to explain every aspect of learning and also human behaviour with its stimulus response theory. Some of the later behaviourists, who were not followers of Watson in the strict sense, have dealt with every aspect of human behaviour though their major experiments were on animals. Watson himself has demonstrated many of the components of his theory on human beings. Thorndike has modified his theory of learning according to the later development that took place during his time and on the basis of criticism from various sources. The Gestalt psychologists have explained every aspect of learning on the basis of insight learning. Different theories can co-exist in learning.

Man interacts with his environment for several purposes. All of them are not psychologically important. Some of his reactions may be for fulfilling his basic needs like food, water, oxygen, shelter and so on. Some of his reactions to the environment may be for fulfilling his desires, motives, interests, aspirations, goals. He may easily achieve certain things and find satisfaction and in case of other things he has to work hard or struggle continuously to have his desire fulfilled or achieving the goal. He also finds out ways and means of avoiding painful events or dangerous situations. These processes are psychologically important and can be brought under the heading of learning. Learning is a continuous process in the life of an organism. Particularly in the case of man, he learns right from his birth till his death. During this period the effects of learning are manifested in his behaviour in various ways.

Generally there is change in his behaviour from time to time due to learning. In his effort to find out a solution to a difficult problem that he faces in his life situation he may be making several attempts. Every time he finds something new, acquiring a new skill to face the problem situation. Therefore, learning is a continuous process in the life of an organism and this learning leaves permanent change in the behaviour of the organisms. If learning does not make a permanent change or if the organism does not retain the previous experiences then learning has not taken place. Gardner Murphy (1964) says that the term learning covers every modification of behaviours to meet the environmental requirements from simple habits to acquiring complex skills and the most difficult personal philosophies of life.

The learning process takes place in an organism out of the problem of needs or wants and pressures and tensions from within and from outside. The need may be expressed in the form of tension or restlessness and imbalance. The restless movements lead the organism to the goal and the organism is relieved of its tension and finds satisfaction. In this process of learning, wrong or erroneous efforts are eliminated and right efforts are retained. The organism makes use of this skill in all its future efforts. This is what takes place in learning process.

Learning is defined in various ways. It is a kind of mental process that leaves permanent changes in the life of an individual. It is also defined as the modification of behaviour. It is the process by which the organism is able to respond adequately in a given situation as the result of past experience. Woodworth (1952) views learning as a form of exercise and also as a process of development. According to him exercise strengthens the activity that is exercised. But exercise by itself does not add anything new to the individual's knowledge. Therefore, the process of acquiring new knowledge is the process of learning.

Learning is not confined to school or classroom learning. It is much more than school learning. A sweet song heard several times here and there has been learned and get into our memory without any intention. All skills, all habits and all acquaintances with people and things are learned. A new learned act may be considered of some skill in which two or more acts may be combined. Ordinarily only one act is performed separately. For example the typist learns to type with both hands at once. Sometimes the skill learned to perform in part of the body may be transferred to another part of the body. There are several kinds of learning in human beings such as place learning, tool learning and things learning. All these learnings yield cumulative effect in the practice of the organism. All learning involves simple types of learning. In addition to the simple types there are other complex type of learning. The complex type of learning is also stated as higher units of skilful activity. To meet some complex demands or problem solving situation different types of skill is acquired.

Human beings learn to operate the instruments such as typewriter, piano, violin, telegraphy, aeroplane and so on. These are complex combination of skill and involve muscular movements and sequences. Another complex activity in which great skills are involved is the language spoken and written and also in singing. Language involves ability to acquire symbols and skills to organize them into meaningful units. Once the individual learns this he is capable of meeting the demands of complex situations. In speech or singing the mouth, tongue, teeth and the larynx and other speech apparatus are skilfully operated. For example in the case of typing man does not acquire the skill all of a sudden. The person who learns typewriting makes first the space learning. He understands the arrangement of letters in the keyboard and knows to do the job of typing which finger to use for each letter. On practice the learner may establish more familiarity between the letters and his fingers. This may facilitate the learner to adopt the blind system or touch system by which he can place his every finger of the letter he wanted without even seeing the keyboard. Thus he can build up speed in typing and increase his efficiency and skill for letter habit to word habit.

Any learning involving skilful activity needs more practice. Otherwise it will slowly diminish and ultimately forgotten. Therefore, practice is an essential component in learning skilful activities. In the case of typewriting also the person should continue the task. If there was a break in the practice or performance for a period of one or two years he loses touch and he cannot perform the task in the same speed as he was performing previously. Therefore, practice makes perfection in any learning of complex type or skilful activities. Perfection brings qualities in the nature of work, increases quantity of work or production, increases accuracy and speed in the work. The work performed would be clean with neat execution without mistake. All these characteristics are applicable to the typist to the telegraphists who sends telegraphic messages and also all works involving complex skills.

Practice, however, increase efficiency, quality or quantity in work, there is a limit for the maximum perfection in performance. What is maximum perfection in person is not the same in the case of another person. It is because there is individual difference. Individuals differ physiologically and mentally. These differences limit the maximum perfection in the individuals. There is limit imposed upon the individuals by the characteristics of their sensory apparatuses, nervous system and muscular system. The individual's reaction in handling the instruments, his coordinating skills all may have influence upon the practice and perfection. For example, in the case of typing the typist should coordinate his eyes and hands, fingers, and muscles. The tailor should coordinate his eyes, hands, legs, and muscles. In throwing a dart at the target, in aiming the bullet at the target in a firing range, and in aiming an arrow — in all these complex skills there is a coordinating efforts. More than one bodily parts are involved in these. Several factors are involved in these apart from practice.

Woodworth (1952) observes that all learning after certain level finds a limit, called plateau. According to him there are two main causes for this. One is poor motivation and the other is poor method of work. The individual's motivation is not strong enough to make further progress in the work and the method used is also not efficient enough to push the person forward for better performance. Gardner Murphy (1964) states that the resulting occurrence of flat spaces or plateaus is often to be lack of interest or effort on the part of the individuals.

TRANSFER OF TRAINING

One of the purposes of education is to prepare the young child to become a successful citizen. Hence what one learns in school must be effective and useful for the individual in a variety of life situations. If what is learned in one situation is applicable in another situation in part or fully then it is said to be transfer of training. Actually transfer of training is the result of formal discipline according to some system. This is called faculty psychology. According to faculty psychology mind is divided into faculties or compartments like observation and learning, memory, reasoning and judgment. According to this system any of these faculties can be strengthened by use of exercise. It is like the strengthening the muscles through exercise. For example, the games and sports activities or any physical work regularly performed becomes exercise to the muscles and if they are effective the muscles get strengthened. This is also applicable to intellectual activities or to the faculty of the mind which is dealing with through processes.

According to the British educational system, till recently, the study of Latin, Greek and mathematics was considered to be very essential for giving training to the mind. It is believed that they have special value. The special value is that they are very difficult to learn and in learning them one needs a systematic and logical arrangement of the facts to be learned and application of the same in learning them. In place of Latin and Greek now-a-days mathematics and memorizing passages are believed to be subjects for general memory trainings. Still, in place of mathematics, poetry or some parts of Shakespeare's plays are prescribed for memorizing which are supposed to be building up the young students' memory power and also training their logical thought processes.

But modern Psychology does not accept this system of interpretation of the mind and training. Mind is not consisting of number of faculties. Since modern psychologists depend upon experiments and research findings they are of the view that learning can be improved by training when there is similarity of learning cues or stimuli and similarity of responses. For example, if one learns Latin, he can also learn French through the process of stimulus generalization and discrimination. A motor car driver can easily learn to drive heavy transport like a bus or lorry. There may be some minor changes which he can understand easily and apply his previous experiences suitably.

Now-a-days everybody, whether educated or uneducated speaks that the present educational system is not useful for life. Mainly the politicians even some educationalists often speak in the same fashion. It is absolutely incorrect to speak like that. Where it is not useful? Is it not useful really?

In our everyday life we could see that transfer of the previous knowledge is taking place when the same material or the same method is put into use in two situations. Certain specific skill of knowledge is needed in the new situation. What we have learned in schools and colleges are transferred appropriately into business and in professional careers of our life. For example, the knowledge acquired in the study of mathematics, physics is transferred to study engineering, and the knowledge acquired in the study of biology and chemistry is transferred to the study of medicine. Similarly the study of commerce and accountancy is transferred for business. The study of arithmetic, language and grammar are not without any use in our life. The knowledge or arithmetic, language and grammar is very much useful in every vocation. So also the study of civics, culture, history and geography and other school subjects helps everyone to become more intelligent and successful citizen in this competitive world. All these are the outcomes of the present educational system. Without studying mathematics and physics how can one become an engineer? Without the study of biology and chemistry one can never think of a physician. Without language and grammar how can you communicate your thought either in speaking or writing without mistakes. If you do not know the history of your nation you are unfit to be called as the citizen of this nation. Therefore, the knowledge of every subject studied in schools and colleges is transferred appropriately to meet some new situations.

The skill acquire in one type of activity is useful in some other new activity. Every activity in our everyday life is not repetition or recapitulation of an earlier learning. But we perform certain new activities even though they were not learned earlier. How this occurs? It is by transferring the skill learned earlier to the present new activity appropriately. Whether it is simple or complex learning it involves transfer. For example, the typist can learn to play piano, much quicker than a non-typist by transferring his skills. A skilful singer of one language can sing a song of another language without much effort. It is also observed that students who are proficient in Latin and Greek learns English very quickly and easily. In all these transferring the skill appropriately is involved. This transfer of skill takes place right from childhood period. Therefore, transfer, according to Murphy, means the effect of training in one function on performance in another function in another part of the body. That is, an activity performed and practised in one part of the body is made to perform the same activity in another part of the body is called transfer of training or learning. Because there is one learning, which is solely learned for a specific purpose or situation.

We can observe that in our every new learning certain amount of old elements may be present. Murphy (1952) states that the experiments conducted by Thorndike and Woodworth and many others who followed them confirm their results indicating that identical elements are found in both old and new materials learned which become more essential for transferring the old to new learning. Therefore, the new is not really new. The new is effective because the old is present in it. Murphy says that in spite of all these in learning there is always a process of fresh integration, reorganization and insight. This interpretation can be given to all learnings.

The insight learning advocated by the Gestalt psychologists cannot be stated that it is fully a fresh start in learning. The chimpanzee's attempt to reach the banana at a height by inserting one stick into the hole of another stick and making it longer is not a new invention. Similar incidents, putting smaller into the larger hole is a common everyday event. The chimpanzee's such old experience might have appeared in the experimental situation also. Therefore, the chimpanzee's activity of learning is not solely new to the situation. Some of the elements in old learning is combined to this new situation in solving the problems which the Gestalts called insightful learning. It is a kind of integration or reorganization. This may take place not only in the life events of human beings, but also in the life of animals. In human abstract thinking this process is taking place, associating the old elements with new and giving new synthesis. Therefore, the transfer process in learning in the everyday life of human beings or in animals is of the same type. The elements of the old learning in combination with the elements of new learning in some fashion provide a new synthesis or integration.

This principle can be observed in the application in our everyday life affairs not only in learning subjects in school and colleges, or learning new skills in vocational situation or learning to solve some difficult problem but also in our contacts with new situations in our habits in social situations and in our development of conduct, character and other personality traits.

According to William James (1908) the process of learning is given back in the change of the learner's behaviour and once learning has taken place the individual is not the same person, he is now a different person, though the details of learning may be forgotten totally. Whether learning has occurred or not can be evaluated only in the change of behaviour of the individual. The learning may be formal education or informal education. If learning has really occurred it is having the effect very widely. It may be through generalization or transfer of training. Formal educational systems are having their basic purpose of giving instructions in skill and knowledge. These instructions are intended to bring about changes in the life of an individual. Therefore, the important outcome of formal education is to make accurate generalization from the formal classroom to the various aspects of the individual's life situations where the skill and knowledge can be applied at the time of learning and in future. Therefore, transfer of training is an instance of accurate generalization in the learner's life and it is aimed as an end product of formal education or instruction in schools and colleges.

In the preceding section we did come across two important concepts, accurate *generalization* and *transfer of training.* Though those two concepts appear to have certain similarities, they are different in some sense. In learning and transfer of training it becomes necessary to differentiate these two concepts.

Generalization

This is a process occurring in human learning and behaviour. Man is often interested in applying the learned material in his real life situations to solve problems. Therefore, it is revealed in his behaviour in some form eiher perfectly or imperfectly. Generalization is in an intellectual process used for evolving laws anu principles from the learning one has made and acquired knowledge. Since it is evolving from learning it takes place right from childhood and is going on extending its application as and when the individual learns more and more.

We generalize several things in our life situations. But a single event or matter is not sufficient for generalization. This may lead to dangerous ends. Sometimes generalization is not accurate. The information that we have received on some issue may not be sufficient to draw a conclusion either it is good or bad. On the basis of partial truth sometimes over-generalization is made. In psychological researches we use statistical treatment to the data and conclude certain results. We say one per cent significant or five per cent significant. Such results we take into account as reliable or reliably significant and we generalize the result of the sample to the whole population. In this type of generalization, though the result is reliable, it is important to check up whether the sample is drawn by the investigator according to the principle using the probability sampling plans. The positive feature is that the collected samples are considered as representative and, therefore, they are generalizable to the population from which they are drawn. The investigator should be able to state how accurate his sample and its characteristics are representative of the parent population and its characteristics. But instead if the samples are drawn according to some procedure, or non-probability sampling plans the results do not form the basis for estimating the sample characteristics for generalization. By following the second procedure the investigators are committing grave mistakes with their dubious results. In such cases generalizations is dangerous. Therefore, unless adequate evidences are acquired generalizations should not be made. Studies have been conducted by Bayler (1936), Bedell and Nelson (1954), Chaplin

(1953), Yarrow (1948), Maier (1933), Schroder and Rottar (1952), Stroud (1954), Kingsley and Garry (1957) and Judd (1908) on generalization process.

Results of these studies revealed that in some cases there were under-generalization. The individuals did not effectively use the results of their learning in their life situations. Certain bright students were able to discover the generalization skills independently of teaching, whereas certain students were not capable of generalization and they needed more direct help from the teachers. The level of generalization depends upon the intellectual levels of the learners and not depending upon what is taught. Bedell and Nelson have observed in their studies that specific type of learning was more beneficial to specific outcomes (vocabulary) than to more generalized process (comprehension). And also if there is comprehension then there is positive effect in teaching word-attack skills. The students must comprehend the elements in the learning material as common. They must understand the similarities between the materials learned earlier and later. Thus generalization process occurs.

In the studies conducted by Schroder and Rotter it was observed that students make better performance when the problem is not easily solvable. They presented to a group of students problems of varied types, which did not find a common solution. This lead to greater flexibility. In these cases the students did not undergeneralize nor over-generalize. The generalization process was used by most efficient approaches directly.

The problem of generalization in the life of human beings is not a must. But generalization takes place. The more we direct our efforts to generalization the more accurately it takes place. Many of our efforts of deal with our everyday problems are made without much conscious efforts to generalize. Since we find certain similarities in our accumulated learned materials and experiences generalization takes place automatically without much effort. Stroud states that the more extensive reading and study the more accurate generalization it takes place. Students with such broad background sources are in better positive position in drawing valid generalization than one uses only one source of information. Edwards and Scannell (1975) have observed that ge‑erally highly educated persons generalize more efficiently and adequately than the persons who are less educated. This is because they specifically encounter the subjects and transfer directly. It becomes easy for them as they have encountered many ideas. In this way they have more extensity of background experience for generalization. In this respect the old saying is right that the more you learn the more you know your ignorance. By learning more and more one may become wise and not ignorant. Therefore, one develops his knowledge for better generalization in spite of forgetting many specific things.

Generalization can be developed by teaching. This has been endorsed by Kingsley and Garry in their studies. Two elements were found to be influencing the development of generalization. The first element is locating essential features. The second is isolating them from the irrelevant features in specific cases. Generalization occurs when these two elements are employed in a behavioural situation.

Transfer of Training

This is a specific reaction of the learner to some problem situations. This is not the same situation in which the individual has learned originally. All the subjects studied in schools and colleges are devised for application in different situations. It is expecting that the individual should be able to apply the knowledge of the subjects quite appropriately in life situations. Every aspect of the knowledge acquired through learning has to be transferred in some form. For example, the arithmetic learned in school is transferred in various ways in our life situations. The writing is used to express our feelings and thoughts. In the case of writers all those written by them become permanent and preserved for long time.

Every learning should have good return by transfer. Suppose, for example, if the school and college learnings are not transferred appropriately to other areas of life situations everything would appear new for a person and he has to learn everything fresh and this may make him very unsuccessful in his life. If everything is to be learned fresh without any transfer from the previous learning man has to be going on learning throughout his life and this life span would not be sufficient. Therefore, the formal education is the representation of the attempt to teach things which the student has to learn intrinsically. Stroud (1956) states that if insistence upon this type of affair is made, then the end result is claiming for intellectual freedom. However the spread of formal learnings in the life of an individual and the effect of the same determine whether change at the intellectual level has taken place or not. Therefore, there cannot be any difference of opinion among the psychologists who have specialized in the field of learning whether transfer of training is a must or not. But what is needed is, how to improve or make the transfer of training to the maximum possible extent. This can be done only by the method of teaching. The students should be made to know the importance of the particular subject matter. Then transfer of the subject mater to relevant situation is possible. Since this is not easy and time consuming, directly teaching for the occurrence of transfer is preferred. As the public and some educationalists think transfer of training is not taking place automatically. There should be conscious efforts in transfer of training in dealing with complex and difficult real life situations.

Some of the psychologists who have specialized in the field of learning presume that there are some factors determine the transfer of training. They thought that transfer process is a complex one and the factors involved in it should be identified in understanding the problem. The complexity and the factors involved in the transfer of training were revealed by the study made by Archer (1930). The important aim of Archer's study was to determine the *effect of the study of words upon derived forms* and the *effect of the study of derived forms upon base words.* The second important aim was to assess the *effect of the study of base words and derived forms upon similar and different forms.* He also determine the influence of grade, sex, intelligence, nationality and word difficulty.

There was three experimental conditions, such as base forms, derived forms and similar forms in Archer's experiment. He also added a fourth condition called different forms. The base forms are selected group of words, derived forms are by adding 's', 'ed', 'ing', to the base words. Similar forms are also derived forms of words which come from the base word but not part of the words of derived group. The different forms are the words which differ from the base word or derived or similar forms.

This can be illustrated with an example. Let us use the words 'produce' and 'enumerate'. Here 'produce' and enumerate' are the base words as shown in Table 6.1.

Table 6.1 : Showing the base word and derived words

Experimental term	Word group	Word group
Base form	Produce	Enumerate
Derived form	produces	Enumerates
Derived form	produced	Enumerated
Derived form	producing	Enumerating
Derived form	production	Enumeration
Similar form	reproduce	Reenumerate
Different form	driving	Counting

The base words can be selected with any number of syllabus or having prefixes. But they should be viable for a adding 's', 'ed', 'ing' and 'tion' to the base words. In the illustration given here the base words are 'produce' and 'enumerate'. By studying these words the effects should form derived forms. From the base word 'produce' the derived forms are four such as 'produces', 'produced' 'producing' and 'production'. The similar form for the base word is 'reproduce'. This is not the part of the derived forms. It is also a kind of similar derivation from the base word. But the 'different form' differs from the base word. It is also different from derived form and similar form. Archer also tried to measure the effect of studying derived forms upon the base words.

Having observed the transfer effect, Archer also measured the effect of studying of base words and derived forms upon 'similar' and 'different' forms by using more difficult words than the words used in the first experiment. He used the derived forms by adding 's', 'ed' and 'ing' to the base words. He observed the transfer ability in children. That is, the ability to pronounce the derived forms after studying the base words. This ability was revealed even when the children did not know the spelling of the words and without actually studying the derived forms, following the principle. Transfer of training occurred directly on words. This sort of transfer can be extended to other functions. This may be called positive transfer.

Let us quote one example to illustrate this phenomenon. The base word, for example, is 'Germinate'. The children do not have the ability to give the correct spelling of this word. But when they are asked to give the derived forms without studying them they are able to give the derived forms such as, 'germinates', 'germinated', 'germinating' and 'germination' by adding 's', 'ed', 'ing' and 'tion' correctly following the previously learned models of derived forms.

The training given on one function often improves that function and transfer is positive. But when it is applied to another function there is loss of ability and there is no improvement in training or transfer. Then negative transfer is taking place. Generally words ending with 'e' when 'ing' is added the last letter 'e' has to be dropped and 'ing' should be added. That is, 'generate' should be spelled as 'generating'. Instead the children often spelled as 'generateing' without dropping the letter 'e'. They carried out the models correctly, but the derived form is not correct. This is called negative transfer. The learner should know the rules under which this is to be done. Archer states that though there may be this kind of negative transfer children learn to generalize.

Most of our learning are readily transferable to some of the life situations. It is taking place not in the original in form, but with some modification to suit the situation. The effect of learning one task may on the learning of another task is called transfer of learning or training. If the previous learning effect is successfully applied in the second act it is called positive transfer. That is, learning of one task facilitates the learning of another task. But transfer of training is not always positive. If transfer is based upon the component of generalization process, the material should be carefully mastered and crystallized in concrete situations. Otherwise it may lead to confusion and negative transfer may occur. Or when there is some interference negative transfer may take place. Quite a number of negative transfers are occurring in our everyday life and we also correct them. When one learning inhibits another learning negative transfer occurs. Similarity between learned materials also may lead to negative transfer. There is also a possibility of another transfer called zero transfer. The training one receives in one activity is not having any influence in another activity. Since experiments are conducted to measure the amount of positive or negative transfer, much is not said about zero transfer in learning experiments. Further we cannot conclude so rigidly that a particular learning or training has no effect.

Positive Transfer

Positive transfer effect occurs if the one or the previous learning facilitates for gaining a new skill or for solving of a new problem in the life of an organism. The individual is placed in a new situation and he performs significantly in a better manner without having any previous training in the particular task. A person playing many instruments has very small difficulty in learning to play a new instrument than a person who has not learned to play any instrument. The principles of learning to play instruments has been already acquired by this person and what is required here is to generalize and transfer the skill to the present task. In the case of this person whose previous learning is instrumental in the new task he can reduce the amount of time and number of errors involved in learning the new task. This we call as positive transfers. We expect that positive transfer should occur in education in the formal school education and in vocational education or training.

The infant begins to learn within few months after his birth. But he is unable to transfer all of them. It is limited to few needs. As and when he gains experience he slowly generalize events and reacts more actively and accurately and transfer in his life takes place. Most of the child's transfers are positive. Once positive transfer occurs it becomes a guideline for further activity or further successful transfer.

The positive transfer may occur from simple motor activity to higher cognitive activity. The positive transfer has an influence upon the mental functions, which incorporate the past experience to further actions. A positive transfer of a training is said to be a short-cut way for future activity. Certain values are socialized in every society. Attitudes, habits, prejudice, emotional reactions are generalized and the individual's actions are predetermined and they influence his future behaviour. In such cases there is no further learning. The individual's reaction is the direct positive transfer. In the socialized processes like habits, attitudes and prejudices generalization and positive transfer appear to be the same. For example, in the case of an attitude formed earlier towards some values is predetermined and generalized and transfer is positive and direct. But in certain specific situations, it is necessary to control transfer if the stimuli are having similarities. If a stimulus arouses a response, other stimuli of similar nature also evoke the same response. The teacher can devise a method to control the transfer that belongs to some other stimulus having similar characteristics.

There are several studies on positive transfer, conducted by Munn (1932), Webb (1917), Bruce (1933), Yum (1931), Bunch (1936), Frutchey (1937), Muehl (1960, 1961), Pimsleur and Bonkowski (1961) and Gladis (1960) using various types of study materials. In some of these studies the experimenters used teaching methods. Munn used eye-hand coordination method. The instrument used was cup and ball. He compared the performance of the experimental group with that of the control group and observed that the experimental group was found to be significantly more successful than the control group. This shows that positive transfer has resulted as the consequence of practice. Webb studied sensory motor transfer in maze tracing. He used six different mazes and observed the effects of learning in one maze upon learning on another maze. He used both human beings and animals in his experiment. There was positive transfer effect upon learning but in varying degrees. Webb states that the degree of positive transfer depends upon the degree of positive and negative elements existing in mazes. If the mazes had more similarity than the positive transfer effect was greater.

Evidences are provided by Pimsleur and Bonkowski for positive transfer in teaching foreign languages. They used pairs of non-sense syllables and colours. The materials were both visual and auditory modalities. The subjects were a group of college students. Half of the group of students learned the material first by visual presentation and then relearned by auditory presentation. The other group learned first orally and secondly, by visual. There was positive transfer in both ways in both groups. But it was observed that the students who had oral presentation first and secondly, visual presentation did better than the other group. Gladis has studied the effects of positive transfer among the students of different grades. He found that there was significant differences in the positive transfer among the students of different grades. Therefore, positive transfer of training is not occurring in equal degrees among the students of all grades. Probably the higher grade students might have made greater degree of positive transfer than the lower grade students. The higher grade and age might have facilitated them to acquire more knowledge and this accumulated knowledge might have helped to greater degree of positive transfer.

Negative Transfer

We expect that every learning should have the transfer effect and particularly the positive transfer effect. But all learnings do not have transfer effect and all transfers are not always positive. When some interference occurs in our learning negative transfer takes place. We can quote numerous examples of negative transfer in our everyday life. Negative transfer is the reverse of positive transfer. In learning some new task the previously learned material may interfere and hamper the learning of new task. This we call as the negative transfer. In other words, if the past experience interfered with the learning of a new skill or solution of a new problem, the learning of a new skill is not taking place. This process is called negative learning. The individual in a new situation performs very poorly than he would perform without previous training. This means that something in him prevents his better performance. Therefore, negative transfer is the reflection of reduction in learning one task as a result of some previous learning.

In learning, generally, a stimulus is attached to a response. But if a new response is attached to the old stimulus then the effect is negative transfer. Some animal, say, a horse was introduced to a small child by saying, "this is a horse. It has four legs" a picture of a horse or a real horse may be shown to the child. The child identifies the four legs with the horse. Later on a donkey was introduced to the child. Then a process of generalization goes on in the child's mind and the child may state that the donkey also is a horse as it has four legs. The child's generalization and transfer is that all four legged animals are horses. The child does not differentiate the donkey from the horse from other characteristics. The previous response is applied to the new stimulus. This is how negative transfer is taking place in our learning and transfer of training. But with continued practice this kind of negative transfer effect may be changed over to positive transfer. Kliene (1921) state that negative transfer is only temporary and can be made better by suitable methods designed by the teacher.

Negative transfer is quite common among children and persons beginning to learn. Studies and experiments have been conducted by Siipola (1940), Siipola and Israel (1933), Archer (1930), Harder (1937), Jones (1961), Gladis (1960), Kersh (1962), Yonge (1966) and others using different types of material. Siipola used code learning and his subjects had learned in the first task. The subjects did not often completely reproduce the responses that were learned first. There was difficulty for the subjects in the response generalization in contrast to the stimulus generalization which caused them to commit the errors. Siipola and Israel in another experiment have observed that such errors were considerably reduced by repeated practice and the new stimulus can arouse the old response. The classroom teaching may be much benefited by the findings on negative transfer.

Adult persons with accumulated knowledge are believed to be making positive transfer successfully. But this assumption was not correct. In the study made by Jones negative transfer was found among the adults in cognitive areas than spelling. In the study made by Gladis there was positive transfer among the students of 5th and 7th grade and negative transfer among the students of 3rd grade. The development factors were operating in the student's learning. Gladis has observed that negative transfer was found to be more often in young children than the older children. Yonge has called the negative transfer as the functional fixedness. According to him, "functional fixedness" is an instance in which previous experience has a negative effect on problem solving behaviour. Yonge's subjects, 115 female undergraduates involved in a task which provided them a number of varied experiences with the object. They have acquired several degrees of experience with the object. This was related to success on a criterion task to the structure of experiences. It was determined whether the "functional fixedness" was decreased due to increased variety of experiences. It was found that the increased variety of experiences did not reduce functional fixedness. It was also observed by him that in the occurrence of positive or negative transfer the perception of the total task was more important than the degree of practice with the task.

Zero Transfer

Zero transfer in learning means that the individual's present performance in the new situation is neither facilitated nor hampered by the past learning. There is neither positive transfer nor negative transfer from one situation to another situation. Though it is spoken in learning and transfer it is very difficult to demonstrate zero transfer experimentally. Edwards and Scannell (1968) state that it is better to state it as indeterminate transfer than to call it as zero transfer. The problem is that we are not able to determine the level of transfer experimentally. Probably the method used to study this effect may be inaccurate or the instrument used to measure the effect may be erroneous. But, however studies have been conducted for measuring the zero effect in learning. In the studies made by McGeoch (1928), Schultz (1960), Hudgins (1960) and others certain conclusions were drawn to this effect.

The ability of children to report narration and interrogation was tested by McGeoch. Practice was given to the children of different age on the assumption that practice effects may improve their ability to report more efficiently. The experimenter was able to see only a small insignificant improvement. There was lack of positive transfer. There was small improvement with increase of age. Therefore, the presence of zero transfer is not sure. Schultz in his study compared the average grade-point of first semester college students who have not done algebra in their high school with that of students who have done algebra for one year in high school. There was very small insignificant differences in achievement between the two groups of students. Schultz states that one year study of algebra in high school did not improve their achievement in college. Therefore, the previous learning of algebra did not show any positive or negative transfer significantly. There was only zero transfer. Zero transfer may occur in many situations.

LEARNING TO LEARN

Psychologists have recently evolved a method called learning to learn in transfer of training. Persons were given to learn successive lists of verbal materials for a period of days. After the period they were given another list of verbal materials for learning. It was found that they were able to increase the speed of the subsequent list of materials even though the lists were not of similar types. There was positive transfer in their learning. The persons seem to have learned a technique or a method to the task that could aid their performance in the later tasks of the same type. This has been demonstrated by Davenport *et al.* (1920) in their experiments with monkeys.

This assumption has been investigated among young children and found that it involved a number of factors. One factor Davenport and others observed was learning to relax in the experimental situation. The second factor was to ignore distracting forces and other irrelevant stimuli. The third factor is learning to identify the relevant cues in the situation.

Learning Principles of Transfer

Transfer of training can be successfully made by learning the principles that underlie in the previous learning. If one knows correctly the principle involved in his learning in the old situation the same can be appropriately applied to the new situation. The principle of his previous learning should be logically analysed for applying the same to the new situation. For example, there is some similarity between the typewriter and the piano. In both there is a keyboard. In typing the typist uses his fingers without seeing the keyboard and the letters. Similarly the pianist use his fingers on the keyboards without seeing the keys. If the typist is given the task of learning to play piano he can learn it quickly by using the principles he has learned in typewriting. He can apply the principles he has learned in typewriting correspondingly to learn piano. He is also applying the process of transfer of the skills from the typewriting task to learn to play piano. Quite a number of inventions and discoveries are based upon this principle. In many of them we can see the principle of logical sequence is applied appropriately by reasoning.

Persons engaged in archery and shooting have learned to apply certain principles in shooting the arrow from the bow. Similarly shooting with rifles at targets on land and sea require the application of certain principles. They must learn the theory of refraction. According to this theory the waves of light, sound, heat and air while passing from one medium to another tends to bend which causes displacement of objects at a distance. This is taking place both on land and sea. Those who have learned this principle may act appropriately in aiming at the target according to the refraction effect in air and sea. Those who are not aware of this principle may commit the mistake of not giving any allowance for the displacement of object. As a result of this their aims become erroneous. Hendrickson and Schrolder (1941) conducted one experiment on two groups of boys in shooting. They were asked to shoot with rifles at a target under water. The experimental group, before taking the shooting practice have studied theory of refraction of light and they understood the apparent displacement of objects under water. The control group did not receive such explanation. In the experiment the boys of the experimental group have learned to hit the target under water in the same number of trials as the boys of the control group did. The learning of the refraction theory did not have any influences in the task of the experimental group. The experimenters have made some change in the situation after the boys had attained proficiency in hitting the target. The depth of the target was put at a deeper level. There was positive transfer from the first task to the second in both the groups. But the experimental group exhibited much greater amount of transfer than the control group. This shows that the knowledge of the refraction theory facilitated them to master the new task in few trials than the control group.

CONCLUSIONS

Therefore, the process of transfer is very widely spread in all learnings. It is occurring in some form in our everyday life, consciously or unconsciously. In elderly people as they accumulate more and more knowledge and experience in various ways in various walks of life transfer of one skill from the previous experience to the present task becomes easy and also sometimes automatic. In our educational system we expect that the formal school education should provide the skill to the students to transfer their experience and knowledge appropriately to their life situations later on to solve difficult problems.

We have discussed here the various components involved in transfer of previous learning to new situations. The transfer effect of the previous learning mainly rests upon the extensity of the process generalization the individual could make in his performance. Since generalization may proceed easily in any direction there may be both positive transfer and negative transfer. If there is no generalization process in learning there is no transfer which we may call zero transfer, though experimental evidences are lacking to support zero transfer. Teaching methods in schools should be so designed as to facilitate the transfer effects to prevail in all learnings by including in the minds of the student the process of generalization.

FORGETTING

We learn everyday many new things and we do not remember all of them. When we wanted to make use of them in certain situations we are successful sometimes and we fail many times. Where those things have gone? We say that they are forgotten. We have not really forgotten all of them, but we fail to recall or bring back the material at the particular moment. Remembering and forgetting are the processes of the memory. Memory consists of remembering what has been learned previously. Therefore, memory according to Woodworth (1952) consists of learning, retention and remembering of what has been learned earlier. This learning need not be the formal learning like the learning of a poetry or a passage from an article or an arithmetical problem. It can be even seeing the faces of some persons, learning the names of them or witnessing an event or a scene from a play. All these go into our memory and whenever we are in need of them we bring back and make use of them or narrate. In this process if we are successful our memory is said to be strong and if we fail to reproduce them our memory is said to be poor. When we fail to reproduce or recognize the matter or faces, or names we say that it is forgotten. Therefore, forgetting in other words may be stated as the failure of memory.

Forgetting is a universal human experience that every man forgets something at sometimes. Garrett (1975) says that it is a convenient excuse for man to say that he has forgotten it. It is often stated that memory for names is always poor but good for faces. Woodworth (1952) says that in the memory process retention is inactive and remembering is active. Even in remembering there are two functions operating. One is recalling and the other is recognizing.

This remembering or recalling and recognizing depends upon the interval time between learning and the act of remembering. Woodworth explains this through the formula $L - I - R$, where L stands for learning, I stands for interval and R stands for remembering. The shorter the interval the greater the retrieval and the longer the interval the smaller the retrieval. Therefore, it is the time that seems to determine the amount of retrieval and the amount of forgetting of the learned materials.

What is Forgetting ?

Forgetting is used for both good and bad purposes. A man can escape from certain critical situation by saying that he has forgotten it. It is involved in learning and it is opposite of remembering. Forgetting is a failure to recall the learned materials. Forgetting is also stated as the failure of retention or loss of memory. There are two aspects in memory. One is short-term memory (STM) and the other is long-term memory (LTM). When the learned material is stored in the memory only for short period to the recall, it is called short-term memory, while the learned material is stored in the memory relatively for a longer period prior to the recall, it is called long-term memory. Forgetting actually means the difference between what has been learned originally and what has been remembered. This may be stated by a formula $F = L - R$, where F stands for forgetting, L for the amount of material originally learned and R for the amount retained.

We try to restore what has been forgotten by frequently reviewing it. However, we can control our forgetting much more effectively if we know more about the principles of forgetting. But the problem we are facing in our everyday life is that what we wanted to remember is not remembered and what we wanted to forget is not forgotten. Men have painful feelings and experiences and suffer mentally when they could not forget the most unhappy events of their life. They even pray to God to bless them with the energy and strength to forget the wretched things they have experienced in their life. Everyone of us at one time might have had this kind of wish. The English poet, Keats (1795-1821) had many unhappy events in his life and he had suffered a lot by remembering all those things. He prayed to God to bestow on him an insensible state to be free from the painful memories. He was very much jealous of the trees as they were not having such painful memories and they are fortunate. The English romanic poet, Byron (1788-1824) also might have had such unpleasant memories in his life and he was struggling to forget them. He expressed his feelings through one of the characters of his epic *Childe Harold's Pilgrimage*. If we feel sorry for forgetting something valuable we also feel sorry for not forgetting more sorrowful and painful experience of our life.

Why Do We Forget ?

There are several reasons for our forgetting the learned materials. The experiments conducted to measure the amount of materials remembered or retained are also the experiments to measure the amount of materials forgotten. When we forget more materials that we have learned ? Time is one of the important factors that determines the amount of materials retained or forgotten. The more the time passes the more materials we forget. Woodworth's retention formula, $L - I - R$, is a good explanation for this. As the interval time or duration increases the amount of material retained decreases and forgetting increases. One answer given by Underwood (1964) for this question, why do we forget, is that, that all forgetting occurs basically from interference between one material with another that man is putting in his memory system.

But why the interval time decreases the amount of materials retained and increases the level of forgetting? The reason is that we are not keeping quiet after learning the material. We keep ourselves in active participation of several activities and also learn many new things. Many events of our life may disturb our memory traces and damage retention process, when retention process is damaged it contributes to forgetting. Further, as time lapses the neural traces of our previous experience grow fainter and weaker. Because of the weakness of neural traces the ability to recall the learned material is lost and the recall fails. The process of forgetting is going on progressing. As the time passes on the past events become old and the recent day events become new. Therefore, the events that took place recently make the neural traces stronger for the new events. As a result of it recent and new events are remembered and past and old events are forgotten.

Experimental evidences show that the rate of forgetting depends upon how we learn the material. If a material is learned more thoroughly and adequately there is less chance for forgetting due to time. Some of us can recite even now the passages learned thoroughly in the days of our childhood.

First of all we forget the learned material if we do not understand the meaning of the material properly or if the passage is not correctly understood. Secondly, we forget the material if we do not recite and review them sufficiently after learning. Thirdly, forgetting may take place if the material is not learned according to certain principles. For example, material learned by spaced method is not forgotten as compared to unspaced method. Similarly, the process of forgetting is slower for the materials learned by whole method than by the method of part learning. Another important point in the rate of forgetting is what the person does after learning the material.

Do we forget everything ?

The simple answer to this question is that we are not completely forgetting everything. We know from our experience that we remember or retain many things but we might have failed to recall all those things we wanted them on a particular occasion. Sometimes we might have met one of our old friends after a very long time. We might have recognized him from the face, but we might have failed to recall his name. It is not forgotten. It is stored in our memory, but there is a failure of recalling his name. Students who are preparing for examinations are more often able to recognize the answer to some particular question, but at the same time they are not able to recall the answer in the examination hall. This shows that even though the answer is not recalled it is retained in the memory and not forgotten. If suitable cue is provided it is very easy to recall it.

We know that it is true that in some situations events and experiences are not recalled as if a total forgetting has taken place. Psycholanalysts say that these events and experiences can be recalled in exceptional circumstances such as dreams, in the state of somnambulism (sleep walking), in the delirious condition of a fever, in the state of automatic writing, in the state of crystal gazing, in the hypnotic states and in psycholanalysis.

Further, we forget things that we do not want to remember or retain. Therefore, forgetting is selective. Garrett (1975) says that whatever explanation can be given for forgetting, there is a good evidence that forgetting often may be selective and purposeful. He says that this is a common experience among the general population. In normal people's life there are highly emotional episodes. They do not continue this forever. They tend to lose or forget their emotional episode with time. It is a natural way of protecting the individual from disorganization or the destruction of the self. This sort of selective forgetting serves the purpose of survival of individuals. Therefore, there is a tendency in man to forget certain events that can cause him emotional upsets and revengeful feelings. It is a kind of wilful forgetting. In such cases there is always lack of reinforcement. Woodworth (1952) says that forgetting is not wholly disadvantageous. The tendency to forget should not be regarded as weak memory or weakness of the biological organism. In fact forgetting is going back to the normal state of the memory, a retroactive process.

TYPES OF FORGETTING

Forgetting is a common human experience. Most of our forgettings are taking place as we do not have any involvement in the material we have learned. We neither always guarantee that we will certainly remember nor we wanted to forget that material. As a consequence of it the originally learned material fades away from our memory. Forgetting is a process of the function of the memory and it may be classified into several categories. The theories discussed in the earlier textbooks are decay *versus* interference, repression *versus* selective learning. The decay or disuse is also called atrophy. Another type of classification is normal forgetting *versus* abnormal forgetting or pathological forgetting. The classifications made from the recent studies are passive forgetting, active forgetting and not remembering. Let us discuss a few types of forgetting.

Passive Forgetting

The passive forgetting is also called normal forgetting. Most of our forgettings are probably passive forgettings. Something we learn today and then we keep quite for a long time without making use of them and we also do not make any rehearsal. We are also engaged in various kinds of other activities. As a result of this the material fades away from our memory and has been forgotten.

The main cause of passive forgetting is disuse. Instead, if we keep ourselves in continuous rehearsal forgetting would not occur. But Osgood (1953) and few others reject this hypothesis saying that disuse or non-rehearsal is the cause of forgetting. Ekstrand (1967) had conducted an experiment to test the disuse hypothesis. According to him the hypothesis is that after learning a material, if the retention is tested in two groups after a period of interval and of disuse, the amount of material forgotten should be the same for both groups. Two groups of subjects were asked to learn 12 pairs of words. The first group learned the list of words immediately prior to retiring for a night sleep in the laboratory. The second group learned the list of words in the morning prior to starting their normal activities of the day. Both the groups were asked not to rehearse the material after learning was over. After eight hours of their learning both groups were tested. The first group, which slept that night was able to recall 89 per cent of the material and forgot only 11 per cent. The second group, which continued their normal activities was able to recall 77 per cent and forgot 23 per cent of the material. The subjects who spent the interval time in sleep forgot only half of the material forgotten by the subjects who were awake. This shows that the effect of sleep on forgetting is little when compared to the waking state. Therefore, disuse hypothesis seems to be not correct.

Interference

The results of the experiment described above also suggest that something other than disuse must be the cause of forgetting. Forgetting may be due to some interference. It is the interference of activities. Why the subjects who slept for eight hours were able to reproduce more amount of material than the subjects who were in waking state? The reason is that during sleep there is less interference. Therefore, the interference theory seems to be having a better reason to explain the cause of forgetting. There are two kinds of interference, one is retroactive interference and the other is proactive interference. They are also called retroactive inhibition and proactive inhibition.

The interference theory states that what a person does during the interval period is mainly responsible for the forgetting. If the time between learning the material and recalling the same goes wider and wider the amount of material to be recalled is also going on reducing. During this period the person might have been involved in various kinds of learning activities and these recent learnings may be having more easy access for recalling than the learning taken place earlier. Further, different memory traces may be overlapping over the other and disturbing each other.

TYPES OF INTERFERENCES

Interferences are of two types. One is retroactive inhibition and the other is proactive inhibition. In the retroactive inhibition new learnings interfere with old learnings. In the proactive inhibition old learnings interfere with new learnings. These two types of interferences can be explained through some experimental designs.

Retroactive Inhibition

This can be diagrammatically illustrated by designing an experiment as given below :

Experimental Plan for demonstrating Retroactive Inhibition

Group	Phase 1	Phase 2	Phase 3
Experimental Group	Learn List –A	Learn List – B	Recall List – A
Control Group	Learn List – A	Rest or Unrelated Activity	Recall list – A

The subjects of both the experiment and the control groups were first asked to learn a list of non-sense syllables from the list A. Then the experimental group was again asked to learn a list of non-sense syllables from a second list B. The control group was not given any task and they were taking rest. After a period of interval the subjects of both the groups were asked to recall the syllables from list A. It was found that the control group which has not learned the syllables from list B was able to recall the non-sense syllables from the list A significantly much more better than the experimental group. The reason for the experimental group for falling behind the control group in recalling the learned material from the list A has been the interference of the material they have learned from the list B. The non-sense syllables of the list B interfered with the syllables of the list A. This interference caused the experimental group to fall behind the control group in recalling the syllables from list A. The new learning (list B) inhibited the recalling of the old learning (list A). But in the case of the control group there was no such interference. Therefore, the subjects were able to recall more syllables. Melton and Irwin (1940) have demonstrated this with several re-learning trials.

Proactive Inhibition

The proactive inhibition is another type of interference in which old learnings interfere with the recalling of new learnings. This can be also diagrammatically illustrated with an experimental design.

Experimental Plan For demonstrating Proactive Inhibition

Group	Phase 1	Phase 2	Phase 3
Experimental Group	Learn List A	Learn List B	Recall List B
Control Group	Rest or Unrelated Activity	Learn List B	Recall List B

The experiment designed to test the proactive inhibition also had two groups of subjects, one experimental group and the other control group. The experimental group was given the task of learning the syllables in list A, while the control group was at rest. After a period of interval both the groups were asked to learn the syllables from the list B. After few hours of interval the subjects of both the groups were asked to recall the material from the list B. It was found that the control group was able to recall more amount of materials than the experimental group. This was because of the interference of the material the experimental group has learned earlier from list A. This process is called proactive inhibition. The material learned previously interferes with the present learning. If there was no interference of the previous learning the subjects of both the groups would have recalled the same amount of material. Since this has not taken place there must be some cause for this and it is the interference of the material learned earlier.

Active Forgetting

Many of our forgettings are due to our active involvement in many things. We exercise all our abilities in something in which we are interested. This makes us actively to forget what has been learned earlier. There are two types of active forgetting. One type of active forgetting is caused as a result of deliberate suppression. The learned material is voluntarily put out of mind, allowing ourselves to learn new information. After some time when return to the material learned earlier we find that it is no longer there for recalling.

The second type of active forgetting is the motivated forgetting. A person is often motivated by the need to express the memories that are likely to cause anxieties. This forgetting takes place due to unconscious repression. Sigmund Freud (1904) in his work, *Psychopathology of Everyday Life* explains this type of forgetting. According to Freud we often fail to recall things that we know perfectly well. As a result of it we are sometimes get into some awkward and embarrassing position. There is a reason for this, but we do not want to express or disclose the motive behind this.

Freud in his work had described several instances in which men and women forget the most important events of their lives and all of them had the causes rooted in the unconscious mind. Everyone of us might have had similar instances in our lives what Freud had stated. Murphy (1964) says that men often forget wifes' birthdays, names of close friends and telephone number of rival business associates. These types of forgettings are occurring due to some kind of motivation in our unconscious mind and it is blocking us from recall. We deliberately push those things to the unconscious level as we do not want to remember them. There is a kind of anxiety or worry if we remember them. Therefore, repression acts chiefly as a determining factor of what is to be forgotten. It acts on a selective principle of forgetting and saves the individual from the outcomes of anxieties.

Not Remembering

There are instances in the everyday life that people are escaping from certain embarrassments by saying, "Sorry I don't remember it", or by saying , "Oh, I totally forgot it." By saying, "don't remember" the person has totally removed the matter from his memory. By saying, "I forgot", the person expresses that he has the matter in his memory, but has failed to react to the situation. In the first statement there is a denial of the matter. In the second, there is a failure to recall. What is involved in not remembering is the difficulty in retrieval. At the moment there is no suitable cue or key by which it can gain access to the material stored earlier. The word 'tip of the tongue' (TOT) is in familiar usage, which is the expression of a failure to recall a word or name which one has known or with which he is familiar, Brown and McNeill (1966) have conducted experiments on the phenomenon of tip of the tongue (TOT) and found that a person knows the word or the matter in spite of his failure to recall it. He was able to recall it without further additional training. The person has the knowledge of it and if it is referring to an object he could even describe about it. Therefore it is the tip of the tongue (TOT).

EXPERIMENT 7 : STUDY OF LEARNING BY INSIGHT – BOLT HEAD MAZE

INTRODUCTION

Organisms often make modifications in their responses and behaviour through learning. The modifications are taking place continuously in the life of the organisms. Right from birth the organism acquires new responses and adopts the old ones to the changing situations. Learning involves retention. If there is no retention then learning has no value. Learning is a universal human experience taking place in every area of behaviour like the motor activities, verbal skills, attitudes and interests and problem solving. Learning is a complex activity. It is studied analytically under laboratory conditions.

Psychologists generally refer learning to measurable changes in behaviour as a result of practice. An individual who has learnt a passage from a book can repeat the passage without any error. The amount of material which he can reproduce without mistake is the measure of his retention. Retention may be defined as the continuation of the measurable modifications in behaviour that has been acquired through training or practice. In learning the individual fixes the materials in his memory. Good fixation is good registration. Fixation, registration and retention are sometimes used synonymously. Most of the learning activities are taking place gradually and they take certain amount of time to reach a level. Learning modifies the responses acquired by the organisms. There are different kinds of learning. They are classical conditioning, trial and error learning and insightful learning.

Insightful learning involves mainly in problem solving. In problem solving the person must discover and fixate the correct response and also eliminate the erroneous responses. Therefore, the problem is solved with certain amount of trial and error. There is certain amount of trial and error activity in solving the problem by insightful learning.

Insight means the ability to see and understand clearly the inner nature of things. It is synonymous to intuition. It refers to an instance of understanding a problem. The way to solve the problem is understood all of a sudden while manipulating shrewdly.

Generally an insight is stated as the sudden strike of an idea to solve a problem while an organism is contemplating over it.

According to Garrett (1961) insight is the sudden perception of relationships, which till this time was not comprehended. The sudden strike of an idea may not be the cent per cent correct procedure to solve the problem. The organism may also make improvement on it to be very much successful. The important process involved in insightful learning is to understood the relationship between the relevant aspects of the problem solving situation. The sudden grasping of the whole background in relation to the problem is an important process in insightful learning. All of a sudden the man may say, "Ah! I got the idea." Insightful learning is not taking place without any purpose. The organism wants to solve a problem in a situation in which he is in a fix. He set up his mind firmly on it. According to Candland (1968) insight is sudden solution to a problem. Murphy (1964) says that it is the realization of the meaning for the use of an object or situation.

Insightful learning may be also stated as the function of reformulation of the idea in a problem situation. Even though it appears to be taking place in the apparent absence of any overt behaviour there are instances to show that insightful learning was taking place with the help of certain aids. Apart from this, learning by insightful represents the culmination of the use of reasoning, imagery and other cognitive processes in the achievement of the solution to the problem. The classical examples often quoted are the experiments conducted by the German psychologists Kohlar, Kaffka and Max Werthimer on chimpanzee. In some of the experiments conducted by Kohlar (1925) the chimpanzee subjects were found to be solving a variety of difficult problems. The chimpanzee, Sultan, was able to learn by insight in such a way that it could get the incentive, the banana. When a single pole could not reach the banana, the chimpanzee did not keep quite. It worked out how to reach the banana. It was able to observe that one pole could be inserted into another and could be made longer to reach the banana and it succeeded in getting it. This was how the animal solved the problem. This we call by insightful learning. Woodworth (1952) says that when the chimpanzee passed from a state of helplessness to an assured mastery of the problem — the dramatic moment may be called a moment of insight. Insight is sometimes foresight and sometimes hindsight. Woodworth says, when the animal ran with a joint stick to the front of the cage to reach for the banana it showed foresight. This is the indication of foreseeing the success. The animal was able to see that the way is clear to its goal. Foresight is observing the way to the goal before trying it, whereas hindsight is seeing the way of lead whether it is good or bad. If the whole situation is clear without any hindrance then there is good chance to achieve the goal — this is foresight. But if the situation is difficult and needs ways and means to overcome the obstacles then it is called hindsight.

Postman and Egan (1949) say that an organism learns by doing, but the organism learns best by performing the task by insight. According to them insight means that the organism becomes fully aware of the principles that are guiding for a successful performance. These experiments still remain as classical in psychological studies.

In animal learning the most important instrument used for experimentation is the maze. In testing the intelligence of the white rats the T-shaped maze are used. While testing human beings who are illiterates or those who are unable to follow verbal instructions due to language problems different types of maze and puzzle boxes are used. There are tests like the puzzle boxes, step mazes, temporal mazes, stylus mazes, and finger mazes which are quite commonly used in psychological laboratories in testing human beings. There are also printed mazes and the subject will have to trace the path way with a pencil. In all maze learning it is believed that the insight plays an important role. The subject with insight learns to run the maze easily and quickly with minimum errors and in a very short time. Solving problems by insightful learning is based upon three important factors. They are, (a) insightful learning depends upon the arrangement of the situation, (b) if the solution is made once it can be repeated, (c) the achievement of a correct solution with insightful learning may be applied to other new situations.

EXPERIMENT

Problem : Study the role of insightful learning of a person and the influence of trial and error learning in it.

Method

Participant

Choose a person who has no knowledge about this experiment and describe him giving the details like his age, sex, educational qualifications, whether wearing eye glass etc.

Material

Two step maze/bolt head maze, metal stylus, battery box, electric counter, stop clock, piece of wires. Draw the diagram of the step maze.

Description of the Materials : The two step maze or bolt head maze is having two patterns, one is systematic and the other is unsystematic. It is a wooden box model square in size having bolt heads on the top in rows and columns. At one corner of the maze board there is a screw. The maze board nay be first connected to one of the screws of the battery box with an electric wire. Another screw of the battery box may be connected to one of the screws of the electric counter with an electric wire. Another screw of the electric counter may be connected to the metal stylus. If the metal stylus touches the bolt head of the maze a number in the counter should run. If this takes place then the circuit is complete and the instrument is ready for action. The experimenter should test this more than once. The bolt head maze is having one pattern systematically arranged with the bolts. The other pattern is unsystematically arrange with the bolts.

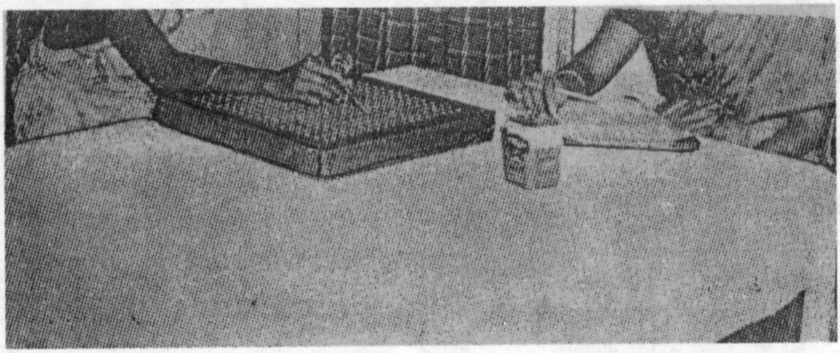

Fig. 6.1 : Learning by Insight : Steps Bolt Head / Step Maze : Experiment : 7

Procedure

Make the apparatus ready with required electric connections and check whether the circuit is complete and works. Make the participant to sit comfortably in front of the instrument. Then give the following instructions:

> "Look, there is a 'starting' on the board with a latter 'S' and there is the goal at the top with a letter 'G'. There are a number of bolt heads arranged with equal distance and with a particular pattern. There is one correct path to reach the goal from the starting point. If you go on the correct path there will be no error. If you go on a wrong path a number in the counter would run with a sound. In such a case you will have to come back to the original position and try through another path to reach the goal. Go from one screw to the next, step by step without skipping. You must move to the next screw to the right or left, above or below. Do not move diagonally. In the case of both the patterns you will have to follow the same procedure. When I say, ready go, you will have to start from the starting point '*S*'."

When the participant starts moving, the experimenter should start the stop clock to run and note down the time and error for each trial. The experimenter should proceed till the participant reaches the goal without any error in two successive trials in the systematic pattern. Then the participant should be asked to go to the unsystematic pattern and the task should be carried out till the participant completes two successive trials without error. When the task is over the experimenter should get the introspective report from the participant.

Results

1. Give the individual scores (times and error) obtained from the records of the systematic and unsystematic patterns. Find out the total time and error and work out the mean. State whether the time and error is reducing or increasing from trial to trial. State in how many trials the participant has taken to reach the goal without error in both the patterns. State whether the participant followed trial and error method or did the task by insightful learning method, give the data in Table 6.2.
2. Collect the group data and work out the mean and compare your participant's task on both the patterns and give the data in Table 6.3.
3. Draw a curve for both the patterns plotting the time and error.
4. Analyse the results in the light of the theory discussed in the introduction.
5. Draw the curve for the group data plotting the mean scores of time and errors.

Table 6.2 : Showing the individual data of the role of insightful learning experiment

S. No.	Trial No.	Systematic pattern		Unsystematic pattern	
		Time	Error	Time	Error
1	1				
2	2				
3	3				
4	4				
5	5				
6	6				
7	7				
8	8				
9	9				
10	10				
	Total				
	Mean				

Table 6.3 : Showing the group data of the role of insightful learning experiment

S. No.	Name	Systematic pattern		Unsystematic pattern	
		Time	Error	Time	Error
1	B.V.				
2	M.S.				
3	A.R.				
4	P.K.				
5	K.R.				
6	R.K.				
7	B.K.				
8	K.S.				
9	V.S.				
10	M.K.				
	Total				
	Mean				

Discussion

Discuss the individual's performance on both the patterns. How was his performance in the beginning? How he was making improvement in the task? Has he tried to solve it by insightful learning or by trial and error? Was there any stage in which he tried to reveal his insight? Discuss the results of the experiment in relation to the introspective report of the participant. Generally, was the learning quick or slow? Was there any nervousness in the participant? If so, at what stage he was able to overcome?

Discuss the group mean in general and in relation to your participant in specific. How many of them above group mean and below group mean in both the scores, time and error. Compare your subject with the group mean. Discuss whether your findings agree with the result of previous studies in the same type of experiments. Discuss the difference in scores on the basis of other individual variations. Is there any difference between the sexes. Discuss the curve and what shape the curve displays. What it indicates?

Conclusions

As a whole what was the prominent prevalence in your experiment? Whether insightful learning or trial and error. What the participant has revealed in his introspective report. Did the experiment proved his thinking factually? Where your participant stands in the group data? How many of them below the mean and how many of them above the mean? What is the mean time for a trial? What is the mean error for a trial in the group data?

EXPERIMENT 8 : STUDY OF HABIT INTERFERENCE : PLAYING CARDS

INTRODUCTION

Human life involves continuance of certain works throughout the life, performing in the same manner of making certain small or big changes in them. When some works continued in the same manner for some days or months a kind of blind performance of the work is taking place. That is the work or task is performed without much effort or careful attention. Since the same task was repeated several times a kind of familiarity in it is taking place. The movement of hands or legs in the task is taking place automatically without any error or mistake. This phenomenon is called habit in our everyday life. Therefore, we may state that a habit is an automatic response in a given task in a familiar situation. The term generally refers to the tendency to evoke a response in a situation.

The important characteristic involved in the habit formation is repetition. A task is repeated several times without any change, results in the formation of a habit. Therefore, habits are usually acquired through repeated training and learning. Habit formation is mainly taking place in motor activities. Therefore, habits are usually referred to motor responses. When any approach is made towards something without much thinking or effort as the consequences of habit, it is called an attitude. Attitudes are formed towards persons, places, objects and works due to habits since habits involve repetitive training and learning.

Many of the everyday works performed by human beings in their work situations involve habit formation and attitude development. Performance becomes simple and easy if habit once is formed in the work. In the habit formed work interference may have little effect and efficiency may be increased. For example, the typist is quite familiar with the clattering noise of the typewriter. The typewriter's noise may be disturbance for a boy who is sitting nearby and studying his lessons. But the noise is not a disturbance to typist. The absence of the typewriter's noise may be a loss of something in the work situation for the typist. He wants that noise, because it increases his efficiency. This is the habit he has developed in his work situation. Many of the works in the work situation are of the same type.

Woodworth (1952) associates habit with attention and intention. According to him it is a permanent factor. The individual has learned to pay attention to certain things and learned to disregard certain things. He says that the car driver has the habit of listening to the sounds of his motor car, since the sound can tell him whether there is any defect or runs freely. Children develop habits of attention and inattention through the influence of older people who used to indicate them that are worthy and not worthy.

Habits, according to Murphy (1964) are acts regularly or customarily repeated. Gerrett says that habit is sequence of activities learned well and mostly automatic in function. According to him habit is the name of a behaviour automatic and often repeated. Man is said to be a creature of habit. Man's everyday life is fully based upon his habits. Right from the time he gets up from bed till he goes to sleep he lives only on habits. He has dressing habits, eating habits, habit of going to work place, habit of going to particular place to buy a particular brand of article and many other everyday activities are taking place automatically without thinking.

Since habit is a learned activity it is also possible to unlearn. Garrett says that the word habit carries an unpleasant meaning for many people. It also implies bad habits or good habits, pleasant habits or unpleasant habits. Though bad or unpleasant habits are to be eliminated some stick to them not knowing the harmful effects. For example, the habit of smoking too much, going motion picture often, sitting before the T.V. all the times, drinking alcohol and so on.

In the view of Hall (1970), it is the learning element that is often called as habit. It may be thought of as an association between stimulus and a response. According to him habit and drive are abstract concepts. The function of the same can never be directly observed.

Learning theorists like, Guthrie (1952) while defining learning he involves the habits. According to him learning involves the formation habits. Habit is having an association between a stimulus and a response. Habits are acquired through contacts. Habits require only one representation of a stimulus for an appropriate response to be followed. A habit is always strong and can never be broken. Lefrancois (1979) says that habits are replaced and not broken. For example, if the habit of smoking is broken the person develops nail biting habit.

In the study of animal learning it is observed that animals also acquire habits. In Skinner's rat experiments the rat is trained to press the bar to get its food. The rat by operating the lever several times learns and sets up a habit that the pressing of the lever would fetch its food and, therefore, whenever, it is hungry it presses the lever in the Skinner box. Similarly Pavlov's dog learned to salivate for food at the appropriate time. The food is served everyday to the dog at 12 noon. After few days the dog learns that food is coming to him at 12 noon and begins to salivate as dogs normally salivate on seeing their food. Since the dog has learned to slivate at 12 noon even if the food is not served at 12 noon the dog never fails to salivate. It salivates at 12 noon even when there was no food. It is because that the dog has set up a habit for salivation. Therefore, habit formation is the consequence of repetition and learning of a particular task.

It may be also stated that habits are acquired readiness responses of an organism. The response is already present in it. But it has to be activated to manifest as behaviour. In the case of animals drives activate the behaviour of habit. There should be a need. In the case of bar pressing rat and salivating dog it is the food deprivation and hunger drive that makes the rat to press the lever and the dog to salivate. The human beings also respond in the same manner. Therefore, habit formation is a universal learning experience found in all living organisms. Allport (1924) says that habit formation is a kind of conditioning. Many of the human activities are performed by blind conditioning only. This we call as habit.

Once a habit is set and the organism is functioning according to it, there may be also conflict interfering in it. Another response similar to the first one may interfere. This causes conflict on the situation. The second or new response interferes in the task as some small change is made in the work situation. Something similar to the habit response is interfering in it. This is called habit interference. That is, something interferes in the habit and here we are to study the extensity of the interference in the habit already formed.

EXPERIMENT

Problem

Study the effect of change in the stimulus situation on the habit already set in your subject.

Method

Participant

Choose a person who has no knowledge of this experiment and collect his personal information such as his age, sex, qualification, occupation etc.

Materials

A card or wooden board of 2′ × 2′ size in which four divisions made. In each division the figures of spade, diamond, lovers and heart are painted in appropriate colours of red and black. A pack of playing cards of 60 in number, 15 in each category and a stop clock.

Procedure

The board should be placed on the table in front of the subject with spades at the left hand top position. The subject should stand comfortably to perform the task. The experimenter should remove the 60 cards from the pack, and hand over them to the subject. He should give the following instructions to the subject.

> "Now you are going to perform a task. There are 60 playing cards in your hand. First of all shuffle them very well. Then remove the top card, turn and see as to which group it belongs to and then deal with the card by putting it in the appropriate cell on the board. If you happened to put it in a wrong cell you can correct it by putting it in the correct cell. Then you can go to the next card and do the same thing. Once you have gone to the next card and dealt with it you should not try to correct the previous card that was already dealt with in a wrong cell. Complete all the 60 cards in the same manner. When one trial is over collect all the cards again and shuffle them very well and get ready for the second trial. Repeat the task in the same manner ten times. Do it as fast as you can. When I say 'ready go' you must start placing the cards in the appropriate cells."

The Experimenter can make the Test Sheet by two Square feet (2′×2′)

Fig. 6.2 : Habit Interference (Experiment 8)

"Now, ready go".

As soon as the subject starts the task the experimenter should start the stop clock and record the time for each trial and the number of errors in each trial.

When the subject completes ten trials the experimenter should turn the Board 90° to the right, so that the figure 'clover' comes to the top left hand position. When this adjustment is over the experimenter should ask the subject to shuffle the cards well and deal with the cards as before. After turning the Board by 90° to the right the subject performs only one trial, i.e. 11th trial and then stops the task. The experimenter records the error committed by the subject and the time taken for the task separately for the 11th trial. Then the experimenter should get the introspective report from the subject and the performance about the 11th trial and allows him to go away.

Results

The experimenter should record the errors committed by the subject in each trial and also the time taken for each trial in a Tabular column 1 for the 10 trials and work out the mean. Similarly the errors and the time for the 11th Trial should be recorded separately in Table 6.5.

Record the time taken and the number of errors committed for the 10th trial and the 11th trial separately and find out the difference between these two trials in Table 6.6. It is expected that in the 11th trial the time taken by the subject and the number of errors should be more than the previous trials. Collect the group data with mean time and mean errors of the Habit interference experiment and enter them in Table 6.7, and work out the group mean for time and error.

Analyse the results of the individual data. State the mean score of the ten trials and also the score of the 11th trial. State also the difference between the 11th trial and the 10th trial. Check whether the time and errors for the 11th trial are more than the 10th trial as expected in this experiment. State the mean score of the group and compare it with your subject's mean score.

Discussion

Discuss the results as given in Table 6.4, 6.5, and 6.6 for the individual. Analyse whether the time and errors are increasing or decreasing after every trial. If there was any change state the reason. Give a comparative analyses of the 11th trial and 10th trial and if there is any difference in time and errors between them find out the cause of it and discuss. Discuss the group data with reference to your subject in time and errors. Discuss how many of them stand above the group mean and how many below the group mean. Who takes shortest time, who takes longest time, who commits more errors and who commits less errors. Discuss everything in the light of the notes given in the introduction and the nature of performance of your subject.

Table 6.4 : Showing the individual data of the first ten trials of the Habit Interference experiment

Trial No.	Time taken in seconds	No. of errors
1		
2		
3		
4		
5		
6		
7		
8		
9		
10		
Total		
Mean		

Table 6.5 : Showing the individual data of the 11th trial

Trial No.	Time taken in seconds	No. of errors
11		

Table 6.6 : Showing the difference between the 10th trial and the 11th trial of the performance of the individual

Trial No.	Time taken in seconds	No. of errors
11th		
10th		
Difference		

Table 6.7 : Showing the group data of the habit interference experiment

S.No.	Name	Mean time for 10 trials	Time for 11th trial	Difference	Errors for 10th trial	Errors for 11th trial	Difference
1	S.V.						
2	P.K.						
3	B.S.						
4	R.M.						
5	A.L.						
6	M.S.						
7	K.R.						
8	T.P.						
9	R.B.						
10	T.V.						
	Mean						

Conclusions

State whether your subject has taken more time and committed more errors in comparison with other subjects. State whether there was any change in the performance of your subject between the original position and the changed position in the experimental situation.

EXPERIMENT 9 : TAPPING TEST – LEARNING

INTRODUCTION

Most of the physical works involve using our hands. It may be left hand or right hand or using both the hands. Some works require only left hand and some require only right hand. For example while reading we use left hand to hold the book and while writing we use right hand. Some people may be using left hand in place of right hand even in writing work. Such people are rare. For example, the American President, George Bush writes with his left hand only. But he shakes hands with his right hand only. Some cricket and tennis players are left handed. This sometimes looks for us an unusual. The reason for using left hand in most of the works by some people may be due to practice right from childhood period. It is also a kind of learning. In some it becomes a habit. Whichever hand one uses in his personal works, it is left to his conveniences. But if one is an employee working in an organization the work may be required to perform only with his right hand or left hand or engaging both the hands equally. Men working with sewing machines in dress manufacturing industries cannot be using left hands in place of right hands and *vice versa*. In such cases the efficiency of left handed people may not be of the same level as of the right handed people. There are many industrial works in which handedness increases efficiency of the work and production.

Therefore, some organizations are interested in the handedness of the persons to be recruited for some jobs. There is no doubt that the preferred hand may move fastly, accurately and preciously compared to the non-preferred hand. Since many people use both the hands equally in all kinds of works it looks as if there is no difference between the left hand and the right hand. But it is possible to test this and find out which hand functions more efficiently. Psychological tests have been devised for various purposes right from testing sensory reactions to aptitudes of various types. Generally, it is observed that handedness is also a kind of skill one has developed in some task by learning or developing a habit by doing the task repeatedly (Metfessel, 1932).

In psychological laboratories there are tests and instruments like Whipple's steadiness test, Eye-Hand Co-ordination test, Tracing test, Mirror drawing test, reaction-time apparatus, O'Connor Finger Dexterity test and so on. Every test is designed to study some aspect of human performance. Similarly a test has been devised to assess the handedness of a person. It is called Whipple's tapping board.

EXPERIMENT

Aim

The aim of this experiment is to study the hand tapping speed of a person and to determine whether he/she is right-handed or left-handed.

Method

Sample

Choose a person who has no knowledge about this experiment and describe his or her personal characteristics such as age, sex, qualification, technical experience, if any, socio-economic status etc.

Materials

Whipple's tapping board, tapping stylus, electric counter, stop clock, battery box, 6 or 9 volt batteries and wires.

Description of the Materials

The Whipple's tapping board is a wooden board of 55 cm by 10 cm in size. It is of 3 cm thickness, can be conveniently placed on a work table. It has two metal plates 10 cm square at each end. The two metal plates are connected to a screw at the centre at the back of the board by copper wires. The two metal plates are also connected to the battery box. The tapping system is also connected to the electric counter. The metal stylus is connected to one end of the board on a screw. When the stylus touches the metal plate on either end a number in the counter runs. If this takes place the circuit is complete. The experimenter, first should test this, whether everything is correct.

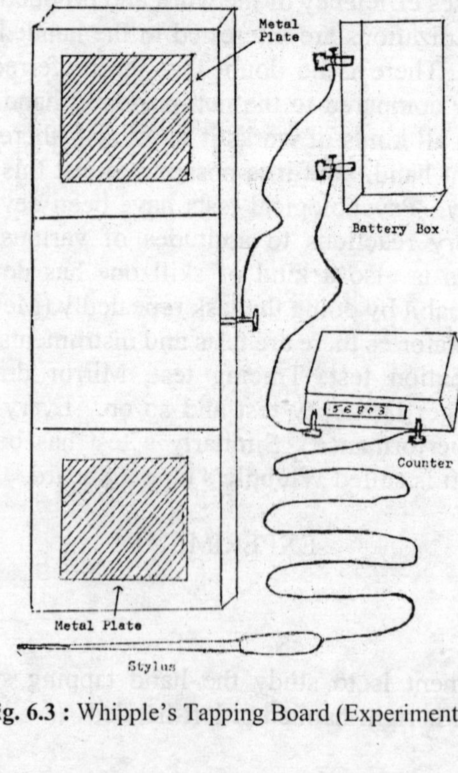

Fig. 6.3 : Whipple's Tapping Board (Experiment : 9)

Procedure

The person should be seated comfortably in a well illuminated room in front of a work table. The tapping test board should be placed in front of him within his reaching distance. Then the following instructions should be given.

> "In front of you there is a wooden board having metal sheet fixed at the either end (The experimenter should point out this to the respondent). This is a metal stylus and you can hold it with right hand. Your task is to tap the two metal plates, up and down alternatively as fast as you can for 75 seconds. After that you will have to do the same thing with your left hand. (The experimenter can demonstrate the task once with his right hand and left hand). When I say 'ready go', you must start it and when I say stop, at once you must stop the tapping. There are going to be five trials."

Precautions

The experimenter should keep watching the stop clock that the task is performed only for 75 seconds. After performing the task with right hand a short interval of half a minute should be given to the respondent. Then he should start with his left hand. After every trial a rest break of half a minute is essential. The respondent should not experience any discomfort nor there should be any distraction to his attention. The experimenter should see that the counter runs for each tapping made by the respondent. The experimenter should say after every 5 seconds, "now the tapping number is ..." and the reading in the counter at that point is to be taken as closely as possible and recorded in his note book. The readings are cumulative and the total must be taken at the end. The number of taps for every 5 seconds is to be taken by subtracting from the other. This is to be done for every 5 seconds for both the hands separately.

Results

The experimenter should work out the number of tappings for right hand and left hand and should give it in Table 6.8. All the five trials should be shown. The general tapping speed also may be given in the table. The mean tapping speed for right and left and for both the hands should be shown in the table. It should be stated whether there was increase of speed trial to trial or not. Collect the group data and present them in Table 6.9.

Table 6.8 : Showing the tapping speed of the respondent with right and left hands

Trial No.	Right hand	Left hand	General speed Both the hands
1	70	60	65
2	75	62	69
3	76	63	70
4	78	65	72
5	80	70	75
Total	379	320	699
Mean	76.0	64.0	70.0

Table 6.9 : Showing the group data of the tapping test

S. No.	Name	Mean tapping speed		General tapping speed – Mean
		Right hand	Left hand	
1	R.S.	76	64	70
2	T.K.	78	72	75
3	S.K.	80	74	77
4	C.K.	82	78	80
5	R.G.	84	80	82
	Mean	80	74	77

Discussion

The hypothesis of this experiment is that in tapping task the right hand differs in speed from that of the left hand. This is verified in this experiment. Observe which hand tapping is higher? That hand is supposed to exhibit more efficiency in the respondent. If that hand is right he is right handed person, if it is left hand he is left handed person. If the person is left handed his picking habits can further be tested by asking him to pick up an object from the ground or on the table etc. The participant's general tapping speed can be compared with the group and other members of the group. A graph may be drawn to show the progress the participants have made in the task. Discuss the results using the data given in Tables 6.8 and 6.9.

Conclusions

State the speed of your participant. Is it increasing or decreasing? Is he right handed or left handed? Has he learned to progress in the task trial after trial? Where he stands in the group in tapping speed? Is he above mean or below mean?

AN ALTERNATIVE EXPERIMENT

In the Whipple's Tapping test, it is also possible to test the handedness by using both right and left hands with two sets of Tapping Boards simultaneously. The same procedure can be followed and data also may be collected in the same manner. The results of this procedure can be discussed with the single handed tapping and compared. What is the efficiency of the respondent in the simultaneous tapping? Discuss it.

EXPERIMENT 10 : STUDY OF BILATERAL TRANSFER – MIRROR TRACING

INTRODUCTION

Generally, if learning is having any effect it must have some permanent changes in the behaviour of an organism. The effects of learning are also having some cumulative effects in the human individuals. These learning effects permanently modify the behaviour. When an organism faces new tasks and new problems its behaviour in the situation may be seriously influenced by the results of past learning and also by the cumulative experience. It is through such cumulative experience of learning that intellectual development and growth, progressive improvement of skill and creative thinking are taking place. It is true that sometimes past experience hinders the acquisition of new skills. The effect of past learning on the new situation is designated as transfer of training. Transfer of training is considered to be one of the prevalent characteristics in the behaviour of the organism. It is the transfer of training that asserts the continuity and the perceivable development of habits of increasing complexity. The effects of past learning on new learning situation is classified into three categories. According to Edward and Scannell (1975) there are three types of transfer. They are positive transfer, negative transfer and no transfer or zero transfer.

The positive transfer occurs if there is positive acquisition of a new skill or a new problem is solved. The learner in a new situation performs significantly better without the benefit of the past learning. The negative transfer occurs if the past learning slows down or reduces the acquisition of new skill in the solution of a new problem. The learner in a new situation performs poorly compared to the previous learning. The zero transfer refers to no improvement in learning. The present learning is neither aided nor hindered by the past learning.

What is transferred in transfer of training is the identical elements. The important point here is that since sign and the amount of transfer effects vary with specific materials and conditions transferring the amount of transfer should be proportional to the number of elements common to the activities between which transfer occurs. A study has been conducted by Bruce (1933) using pairs of non-sense syllables for learnings. The non-sense syllables were related with lists of different syllables. Bruce observed that in case of stimulus generalization there was high positive transfer. Yum (1931) used meaningful materials in pairs instead of non-sense syllables to study the transfer effect. There was highest transfer with same words as stimulus in Yum's study. The greatest advantage of positive transfer was observed by Frutchy (1937) in more generalized behaviour of human beings. Woodworth (1952) is of the view that the amount of transfer is proportional to the identities and similarities between the two tasks performed.

According to Postman and Egan (1949) an activity if practised with a particular part of the body generally makes easy the performance of the same activity with another part of the body. This is the effect of training the organism has received earlier. They call this as cross-education. The positive transfer effects are between the same parts of the body, that is from hand to hand, eye to eye, and from foot to foot. When the cross education is from one part of the body to its counterpart it is called *bilateral transfer*. Munn (1932) has investigated sensory motor transfer with college students. Transfer can also be from non-symmetrical part of the body. It may be from hand to foot and foot to hand. Bilateral transfer has been demonstrated in a variety of situations, particularly with perceptual motor activities. One of the most common experiments in this study is the sensory motor transfer of training conducted by Webb (1917) in maze tracing. One of the patterns very often used in most of the psychological laboratories is the mirror tracing. The subject's task is to trace a pattern, usually the star pattern. The star pattern is not having the direct view. It is reflected in the mirror. The subject is to aim at the target by tracing from the starting point. This is very difficult at the beginning for many subjects.

It may take quite long time as the task is to go on the reverse in the spatial direction as seen in the mirror.

The task is quite suitable for the study of bilateral transfer as it can be performed with the same two bodily parts — hands or feet. Another feature of this test is that subjects usually begin with a very slow start, with very little skill and improve the performance considerably after some practices. Candland (1968) says that mirror tracing experiment is particularly suitable for studying the effects of transfer and it is mostly used for this purpose. Bilateral transfer may be examined by making the subject to use the other hand after he has performed the task with his preferred hand.

The conventional method in psychological experiments is to have an experimental group and a control group. Instead of a group the experimenter can have one experimental subject and one control subject. The experimental subject is given a pre-test with his non-referred hand (*e.g.* left hand). Then a training with some trials is given with the preferred hand (*e.g.* right hand). Then the post-test with the non-preferred (left hand) is given. Whereas the control subject receives only pre-test and after sometimes post-test with the non-preferred hand without any practice. There is every chance for the experimental subject to show some gain in the post-test from the pre-test. *The gain is the indication of positive bilateral transfer*. In the case of the control group this gain may be very little or absent. This bilateral transfer or cross education may be performed with many activities besides the mirror tracing. Some of perceptual motor tasks are tracing a maze pattern, ball tossing, tapping, pursuit of rotating target and so on.

The conclusions of cross education need not be generalized too much. Because transfer from one situation to another does not depend upon the physical identity alone. When there is transfer from one eye to the other different pathways of stimulation are involved. Both eyes may not be having the same visual acuity. Similarly when a skill is transferred from one limb to another different bodily muscles are involved. But with all these the bilateral transfer is most common reliable type of transfer in cross education experiments.

EXPERIMENT

Problem

Measure the transfer effect from practising a task with one hand to the performance of the same task with the other hand.

Method

Participants

Choose two persons who have no knowledge about the experiment and the instrument. One may be experimental subject and the other may be control subject. Note down their age, education, whether wearing spectacles and other particulars.

Materials

The mirror tracing apparatus board with star pattern, stop clock, electric counter, battery box, stylus and wires.

Description of the Material

The mirror tracing apparatus is a wooden board of rectangular in shape. At the centre there is a star pattern. There is a mirror in the back and a screen above the pattern. The pattern is a groove with 2 cm deep into which a stylus can be made to run. The groove is having metal sides and it is connected to a screw on the board. One end of the wire is connected to the battery box and the other end to the board. The one end of another wire is connected between the battery box and the electric counter. A third wire is connected between the electric counter and the stylus. Now if the stylus touches the metal groove numbers in the counter will run. If this takes place the apparatus is ready for operation. The experimenter should check the working condition of the instrument. It should be so adjusted that the subject does not see the star pattern directly, but only in the mirror.

Fig. 6.4 : Mirror Tracing Apparatus : Bilateral Transfer

Fig. 6.5 : Pattern of Mirror Tracing : Experiment 10

Procedure

The subject is to be seated in front of the apparatus comfortably. The screen prevents him from seeing the star pattern directly. He can see it only in the mirror. The stylus, which is connected to the counter and battery box, is given to the person and he is to hold it with his non-preferred hand. The following directions are given to the participant.

"There is a star pattern, seen in the mirror. You will have to trace the star pattern into the groove with the stylus. The stylus should be moved on the pattern without touching the sides. If the stylus touches the sides of the groove an error is made and it is recorded in the counter. Therefore, try to drive the stylus in the groove without touching the sides. You will have to start from this place (The experimenter points out the place in the pattern) and go round the pattern and reach this place (The experimenter shows the place). When I say ready go, you will have to start."

As soon as the experimenter said, 'ready go', he also starts the stop clock to run and also observes the recording of the error in the counter. When the subject completes one trial with his non-preferred (left hand) hand the experimenter stops the stop clock and notes down the time taken to complete the trial and also the number of errors recorded in the counter. Then the control group subject is asked to trace the pattern in the same manner with his non-preferred hand. Then the experimenter asks the experimental subject to hold the stylus with his preferred hand (right hand) and to carry out the same task in the same pattern as he did with his left hand (Table 6.10). The experimenter starts the stop clock as soon as the subject began to trace the pattern with his right hand. The subject carries out 20 trials and time and error for each trial were noted down. Then, again the subject is asked to trace the pattern in the same manner with his non-preferred (left hand) one trial. The stop clock is made to run and the error and time are noted down. Again the control group subject was also asked to perform the task with his non-preferred hand. He was not given any trial with his preferred hand (Table 6.11).

Results

The scores in this experiment are the number of errors recorded in the counter and the amount of time taken by the subject for each trial to trace the pattern. Therefore, each trial with non-preferred and preferred hand is scored in terms of (1) the number of errors and (2) the time required for completing the pattern (Tables 6.10 and 6.11).

A learning curve should be drawn to show the course of practice with the preferred hand by plotting the number of errors and time scores against the trial number. This is the learning for the preferred hand. The amount of transfer is to be determined for both the experimental and control subjects. In the case of the experimental subject the scores of pre-training and post-training should be taken into account for determining the amount of transfer. In the case of control subject the scores of the first trial and the last trial with his non-preferred hand are taken into account. There is no training score with the preferred hand for the control subject.

Table 6.10 : Showing the scores of non-preferred hand by the experimental subject

Trial	No. of Errors	Time in seconds
First Trial		
Last Trial		
Gain or Loss After Training		

Deduct the scores of the first trial from the last trial to find out the gain after the training.

Table 6.11 : Showing the scores of the preferred hand by the experimental subject

Trials	Errors in number	Time in seconds
1		
2		
3		
4		
5		
6		
7		
8		
9		
10		
11		
12		
13		
14		
15		
16		
17		
18		
19		
20		
Mean Score		

Table 6.12 : Showing the scores of the non-preferred hand of the control subject

Trial	No. of Errors	Time taken in seconds
First Trial		
Second Trial		
Gain or loss		

Table 6.13 : Showing the group data of the non-preferred hand of the experimental (group) subjects

S. No.	Subject	First Trial		Last Trial		Gain or Loss	
		Error	Time	Error	Time	Error	Time
1	P.K.						
2	S.S.						
3	A.L.						
4	K.R.						
5	M.S.						
6	B.V.						
7	G.G.						
8	T.V.						
9	E.M.						
10	R.V.						
	Mean						

Table 6.14 : Showing the mean scores of the preferred hand of the experimental group for 20 trials

S. No.	Subjects	Errors in number	Time in seconds
1			
2			
3			
4			
5			
6			
7			
8			
9			
10			

Table 6.15 : Showing the scores of the non-preferred hand of the control group

S. No.	Subjects	Errors in numbers		Time in seconds		Gain or loss	
		Ist Trial	IInd Trial	Ist Trial	IInd Trial	Error	Time
1							
2							
3							
4							
5							
6							
7							
8							
9							
10							

Discussion

First draw the learning curve showing the number of trials and the errors and time taken for each performance of the experimental group with the preferred hand. Discuss the scores after every trial whether there is an improvement on learning or no learning.

Discuss the experimental subject's scores obtained from the performance with non-preferred hand. The difference in the scores (error and time) between the first trial and the last trial is the amount of gain made by the training if it is plus. There may be also minus score which mean loss or no gain.

Discuss the scores obtained from the performance of the preferred hand whether the subject was making any improvement or not after every trial. Discuss the scores taking into account the mean score of the 20 trials.

Examine the control subject's scores of the performance with non-preferred hand and state whether there was any learning or not.

Discuss the experimental group's scores obtained from the non-preferred hand taking into account the mean scores of the group.

Examine the mean scores of the performance obtained from the preferred hand of the experimental group for the 20 trials taking into account the mean scores for the whole group.

Discuss the mean scores of the non-preferred hand of the control group.

Conclusions

State whether your experimental subject has made any gain after the training with preferred hand. Where he stands in the group mean? Who has taken least amount and time and minimum number of mistakes?

EXPERIMENT 11 : TRANSFER OF TRAINING – CUP AND BALL.

INTRODUCTION

Learning is reflected in the behaviour of the organism in some form. It is by way of showing some change in its behaviour that the learning is reflected. The change in the behaviour is the result of learning and acquisition of some knowledge. The effects of learning are going on adding day by day and become cumulative in time and quantity. When an individual learns a list of words or a passage his experience is more or less modified permanently. When an individual learns a material or faces new problems his behaviour is affected by the effects of the previous learning. Sometimes past learning (experience) may hinder the new learning instead of aiding it. The effect of past learning on new learning is said to be transfer of training.

Transfer of training is one of the most important psychological problems of learning. It is an important pervasive characteristics of behaviour. Because it is transfer of training that ascertains the continuity of our knowledge and development of habits, the development of consistent personality traits, organization and integration of knowledge — all depend upon transfer of training.

The idea behind this concept is that whatever you learn must be useful to you later on. Transfer of training may be defined as the application of the training received previously in one situation to other forms in another situation. The new learning may be applied sometimes more rapidly in one situation and sometimes slowly in another situation. Transfer of training received greater importance in the field of education, engineering and technology. Educationalists in the early days believed that the study of mathematics led to the development of acquiring mastery over Greek and Latin languages because Greek and Latin produced attractive syntax and good English vocabulary. They thought that it produces positive transfer in education. But researches conducted on this in the first quarter of this century have proved that this belief is wrong.

All learning, it is believed, is transferable to new situations. But the amount of transfer often may vary from person to person and situation to situation. The effects of past learning on new learning may take place in three ways. They are positive transfer, negative transfer and zero transfer.

Positive transfer occurs when the organism's performance is benefited by the previous learning in a new situation. The individual when put in a new situation performs significantly better than he would do without the benefit of the past learning. Negative transfer occurs when an organism's performance in new situation is hindered by previous learning. Generally negative transfer effect is inferred if past experience is slowed down the new learning or interferes with the solution of a new problem. When an individual is placed in a new situation the learner performs very poorly than he would do without previous learning. There are not many experimental evidences for zero transfer of training. Zero transfer means that no transfer effects, from one situation to another situation can be detected. Generally zero transfer effect means that the performance in the new situation is neither benefited nor hindered by the previous learning.

The important rule of transfer of training is the degree of similarity of stimuli and responses. The phenomenon of stimulus generalization may be applied in transfer of training. Therefore, according to Edwards (1975) if the similarity is greater the generalization also is greater.

Another important rule of transfer of training is the occurrence of negative transfer when the desired response is different from that of one learned previously in that situation. The more dissimilar the new response is to the previous response the greater the negative transfer occurs. In human behaviour transfer of skill is important. The problem of transfer is not only changing simple responses or skills but also it is something more than this. What kind of skills can be transferred best to a new job in an industry? If an employee possesses a particular skill which is not needed to his new job, will this skill be transferred positively or negatively in a training programme: This kind of problem depends upon the attitudes and expectations of the learner.

The emotional reactions which are conditioned in one situation form the foundation of prejudice in another situation. Racial prejudices developed against the out group find their sources in the hatred sometimes turned against a member of his own group. The psychological process of scape-goat functions in the same manner. Political leaders often make use of rage or fear responses conditioned previously to activate people against the people in power.

We can design experiments to study the transfer of training. Some kind of training is given to an experimental group and the performance of this group is compared with a control group which did not receive any training or experience. If the performance of the experimental group is superior in similar task in comparison with the control group it is evident that the experimental group had the benefit of the training. There is positive transfer of training. But if the performance of the experimental group is inferior to that of the control group there is negative transfer. We cannot bring all cases of transfer of training under either stimulus similarity or response similarity. They both interact sometimes to produce transfer.

One of the theories of transfer of training is the theory of identical elements. In every transfer of training the sign and the amount of transfer effect differ with specific materials and conditions of training. Therefore according to this theory the amount of transfer is proportional to the number of elements in common between the materials transfer takes place.

If we want to design an experiment along the lines of this theory we must analyse learning one *(L_1)* and learning two *(L_2)*. The two are the learning or activity *from which* there is transfer and the learning or activity *to which* there is transfer. The activities should be broken down in stimulus and response elements. They should be broken down in such a way that there should be common elements in both. The transfer may be more efficient if the number of elements are common in learning one *(L_1)* and learning two *(L_2)*. But complete identity of stimuli and responses between L_1 and L_2 is not possible. It is better that the identical elements should be simple. It depends upon the requirements of specific experimental problem. Behaviour is a continuous process which involves more of muscular movements and certain specific verbal units. The identical elements the experimenter uses should be highly flexible. In some cases there may be limited acts and in some cases there may be specific movements. In other cases an element may be broadly conceived a general mode of attack. According to Postman and Egan (1949) the term element refers to minimal units. Woodworth (1938) speaks them of as components.

The study of transfer of training experimentally always involve the analysis of activities into units. It may be units or components, according to the needs of a given experiment. Different levels, like specific and complex may also prevail in the analysis. In the experimental situation we may break down a broad unit into more specific and limit one if it may serve for our purpose. But it will not be possible for us to say correctly in a transfer experiment what is transferred and which unit is transferred into which unit.

With regard to the experiment on transfer of training the experimenter's task is to identify the variables that function in the transfer of training and to explain. The variables that function in the transfer of training are classified in various ways. The classification depends upon the stimulus materials, type of experimental procedure and type of activity.

In learning to perform a task or acquiring some skill there are certain general modes of attack in solving a problem successfully. This is done through transfer that such general modes of attack are learned. In the measurement of sensitivity psychophysicists were able to find out that experienced observers could make reliable judgments with small thresholds in sensory systems than those without experience. It is because that experienced observers have learned how to utilize signs or cues with greater efficiency. They have also learned to eliminate irrelevant signs or cues and also they were able to ignore distractions. These profitable effects of the past experience on sensory systems are due to the transfer of observational components. Renshow (1945) in his experiments with tachistoscope was able to observe that persons who were trained recognize the stimulus items presented in a tachistoscope at a very rapid speed than those who were not trained.

Psychologists who are specialized in learning and memorizing state that men become more skilful in memorizing by taking tests on memorizing. Learning more lists of non-sense syllables facilitates one to learn new lists more quickly. Improvement was found to be fast in the first few lists. Later on memorizing was slow and improvement was reduced. In memorizing non-sense syllables the learner aopts some specific skills and attitudes. He utilizes rhythm and grouping methods and finds out the types of associations useful and also identifies the factors that hinders the learning process and eliminates them. Learning non-sense syllables is not new to him and, therefore, deals with the materials with more confidence.

In many of the activities, continued practice provides the skill to perform perceptual motor task efficiently. In the mirror tracing or in the finger maze we could observe certain awkward movements in the beginning. After few trials with error movements there is increase of speed and accuracy in tracing the mirror and the maze because of the continued practice. This process we can observe in almost all perceptual motor tasks, eye-hand coordination tasks and similar activities. Webb (1917) has studied the problem of sensory motor transfer in a study on maze tracing.

The effects of transfer of training may be more specific in some cases and less specific in other cases. Transfer of learned materials from one situation to another may be made in various ways. In certain learning we must know how to observe the particular or specific event. In making judgments in sensory systems we have seen that trained persons were able to make correct judgments than those who were not trained. For example, if the stimulus materials are of different weights the observer out of experience may make the judgment correctly and carry out the same experience in judging other types of materials. Similarly in the case of judging the length of lines. In all such cases usually there may be positive transfer. But the skills that are transferred do not make much improvement as much as the practice provides in the direct practice upon the specific activity or problem.

SIMILARITY OF STIMULI

In generalization one of the aspects is to find out the similarity in stimuli or identical characteristics in stimuli of both past learning and present learning. We have also seen that similarity makes transfer effect easy. But similarity of stimulus elements in two or more situations cannot be said that it would always provide positive transfer effect. If transfer is to occur the similarity of the stimuli in two or more situations should function as equivalent for the individual. The individual should observe that stimuli are functioning in the same manner. Identical characteristics in stimuli can be discovered only after much training and careful analysis. In two stimulus positions S_1 and S_2 if the behaviour of the organism is not altered the stimuli S_1 and S_2 are considered as similar or equivalent. If the stimulus S_1 position is progressively altered in its characteristics like size, light and weight till there is a significant change in the response to the stimulus then similarity in stimuli ceases to exist. Stimulus positions equivalent means perceiving a relationship common to two positions. Experiments can be designed to perceive the relationships common to two different situations and can be shown how transfer is depending upon it. There may be experiments in which physical identity of the stimuli is different and other relationships may be common.

Transfer effects may be successful on perceiving common features in stimuli placed in different situations. But it is not always the case. Transfer effects may take place in different situations even if the organism fails to perceive the common characteristics in stimuli or fails to perceive the two stimuli in two situations are different.

It has been found that the study of principles of successful transfer as compared with rote learning in the field of memory had superior transfer effects.

Training one part of the body in a particular activity enhances the performance of the same activity in another part of the body. This type of transfer effect is called cross-education. Generally positive transfer effects can be obtained from the counterparts of the body. This is usually from one eye to another eye, one hand to another hand and one foot to another foot. This type of transfer is called bilateral transfer. The transfer can also be from hand to foot and *vice-versa*. Several experiments have been designed to demonstrate cross-education and bilateral transfer. The experiments are mainly with perceptual motor activities or sensory motor learning. One important experiment in the study of cross-education is the bilateral transfer in cup and ball. This involves eye-hand coordination. This experiment was designed by Munn (1932). Munn in his experimental study was able to find out that his experimental subjects have discovered the types of movements which led to success while receiving the training with left hand. Then they applied the same to the final trials with the left hand. It means that some mental meditation process set up effective principles to use in the motor activity.

TRANSFER OF TRAINING — CUP AND BALL EXPERIMENT

Problem

Determine the effects of training with the preferred hand upon the non-preferred hand in a person and state whether there is any transfer effects in the performance and also eye-hand coordination.

Method

Participant

Choose a person who has no knowledge about this experiment. Collect personal information such as age, sex, qualification and other characteristics whether the person is wearing eye glass, his handedness, previous training in motor activities etc. It is better to have an experimental subject and one control subject. The experimenter can serve as the control subject. It can be one person in each group.

Apparatus

Wooden cup and wooden ball, one half metre scale. The ball is tied to the handle of the cup and suspended by a string of 18 inches long. The cup is at the top and there is a handle below. The cup is approximately 3′′ diameter and 3′′ deep. The size of the ball is such that it can go into the cup comfortably.

Fig. 6.6 : Cup and Ball : Transfer of Training : Experiment 11

Procedure

The participant is seated comfortably on an armless chair having sufficient space to move his right and left hands freely in all directions. Before starting the experiment the experimenter should prepare the scoring sheet as required for this experiment for both the control group and for the experimental group. When everything is ready, the experimenter should start the experiment. He must measure the length of the string between the cup and the ball with the help of a half-metre scale and verify whether it is 18 inches long. Otherwise it should be adjusted. When he is satisfied with it, he should give the following instructions.

"Hold this cup at the handle below between your thumb and fingers. A round wooden ball is tied at the handle and is suspended by a string of 18 inches long. Your task is to flip the ball into the air and to make ball to fall into the cup. If the ball falls into the cup you get 3 points, if the ball bounces out 2 points, if the ball hits the mouth of the cup 1 point and if the ball misses the cup 0 point. You would perform this task first with your left (non-preferred) hand. You can have 5 trials and then proceed on the task."

The experimenter at this stage can demonstrate one or two trials, so that the participant can understand the mode of handling the material. After the participant has performed 5 trials the experimenter must be ready to record the points in the scoring sheet.

> "You have completed 5 trials. When I say, ready go, you must carry out the task with 25 trials in the same manner as I have shown you. Ready, go."

The experimenter should note down the points for each trial in the scoring sheet. As soon as 25 trials are over the experimenter should ask the participant to stop it. Now the experimenter as the control subject should perform 25 trials with his left (non-preferred) hand the participant may be asked to make entry of the points for each trial.

The next step is to carry out the trials with the (preferred) right hand. The participant should be asked to start the trials in the same manner as he did previously. The following instructions should be given to the participant.

> "Now you will have to perform the task with your right (preferred) hand 250 trials. At the end of every 25 trials I will say stop and you must stop and take one minute rest and then continue the task with a break of one minute for every 25 trials till you complete 250 trials. Ready go."

The experimenter should be ready with prepared scoring sheet with columns and rows. He should note down the points for every trial. The points should be separately entered for every 25 trials. When all the 250 trials are over the experimenter should stop the performance.

Then the experimenter as a control subject should perform another 25 trials with his left (non-preferred) hand and the participant may be asked to make entries of the points. On completing 25 trials with his left hand the experimenter should ask the participant, the experimental subject, to perform another 25 trials with his left (non-preferred) hand. He should give the cup and ball in the hands of the participant with the following instructions:

> "Now your task is to carry out another 25 trials with your left (non-preferred) hand in the same manner as you did previously. Ready, go."

The experimenter should note down the points for each trial in the column of the table prepared for this. When the participant completed all 25 trials with his left hand for the second time the experimenter can collect the material from his hands. Then he may ask him whether he has followed any specific principle in performing the task with his left hand in the second time after the 250 trials. The experimenter may note down this statement and include it in the discussion of the results. Then he can allow the participant to go away.

Results

Work out the experimental subject's scores from the scoring sheet and enter in Table 6.16. In the first column put the experimental subject's score with left hand for 25 trials. On the right there are 10 columns for entering the scores of every 25 trials. Lastly in the 12th column enter the scores of the last 25 trials with left hand of the experimental subject. Now find out the total score for every 25 trials.

Find out the scores of the control subject and for the experimental subject from the scoring sheet for the first 25 trials and for the second 25 trials and enter them in Table 6.17. Find out the total score for both conditions before and after.

Collect the group data for the experimental group and enter them in Table 6.18. Find out the total score and also the mean score for every 25 trials. Find out the difference between before and after training.

Table 6.16 : Showing the data of the Experimental subject in the cup and ball experiment in 3 conditions

Trials	Left hand before training	Trials with Right hand										Left hand after training
		25	50	75	100	125	150	175	200	225	250	
1.												
2.												
3.												
4.												
5.												
6.												
7.												
8.												
9.												
10.												
11.												
12.												
13.												
14.												
15.												
16.												
17.												
18.												
19.												
20.												
21.												
22.												
23.												
24.												
25.												
Total												
Mean												

Table 6.17 : Showing the data of the control and experimental subjects before and after experience in the cup and ball experiment

Trials	Control subject		Experimental subject	
	Left hand before training	Right hand after training	Before experience	After experience
1.				
2.				
3.				
4.				
5.				
6.				
7.				
8.				
9.				
10.				
11.				
12.				
13.				
14.				
15.				
16.				
17.				
18.				
19.				
20.				
21.				
22.				
23.				
24.				
25.				
Total				
Mean				

Table 6.18 : Showing the group data of the experimental subjects in the Cup and Ball Experiment

S. No.	Name	Left hand before training	Trials with Right hand										Left hand after training	Difference
			25	50	75	100	125	150	175	200	225	250		
1.	P.S.													
2.	R.T.													
3.	G.G.													
4.	S.B.													
5.	T.S.													
6.	G.S.													
7.	V.S.													
8.	S.R.													
9.	S.L.													
10.	P.K.													
11.	K.R.													
12.	B.V.													
13.	R.S.													
14.	M.S.													
15.	M.R.													
Total														
Mean														

Collect the group data for the control group for both the conditions and enter them in Table 6.19. Work out the total and the mean score for the whole group. Find out the difference and see whether there is any gain in the second 25 trials or loss. Draw a graph showing the trials on the X-axis and the scores on the Y-axis for the experimental subject and for the experimental group in the same graph.

The assumption is that the training given to part of the body may be transferred to the other symmetrical part of the body. This assumption is tested in this experiment. The training given to the right hand should increase the efficiency by the process of transfer. The non-preferred hand's performance before training and after training will determine this. The experimental data should be statistically treated and to be given in Table 6.16 to 6.19. The data in these Tables may illustrate the results.

Discussion

The data in Table 6.16 may provide a complete picture of the experimental subject's performance in three conditions. In the first column the performance made by the left (non-preferred) hand in number of points for each trial should be noted down for 25 trials. The next ten columns should show the points for 250 trials with right (preferred) hand. The last column should show the points for the trial made by left (non-preferred) hand.

Table 6.19 : Showing the group data of the control group in the Cup and Ball experiments in 2 conditions

S. No.	Name	Left hand Ist 25 trials	Left hand 2nd 25 trials	Difference	Gain or Loss
1.	S.S.				
2.	S.M.				
3.	R.K.				
4.	P.M.				
5.	S.R.				
6.	V.K.				
7.	R.M.				
8.	B.A.				
9.	C.D.				
10.	M.P.				
11.	V.S.				
12.	R.M.				
13.	S.T.				
14.	G.T.				
15.	K.V.				
Total					
Mean					

The total and mean scores for these three conditions may show whether there was any improvement in the performance made by the left hand for the second time. That is the difference between the scores in the last column and in the first column. If the scores in the last column are higher than the first column, then improvement has taken place and the training taken by the subject with his right hand is transferred to the left hand. This shows the positive transfer.

Then compare the scores of the control subject between the first condition and the second condition. The control subject performs trials in both conditions with his left hand. See whether there is any improvement and if so, whether it is positive or negative. Compare this score with the experimental subject.

Find out whether there is any improvement in the performance of the experimental subject after every 25 trials. Whether the points after every 25 trials are increasing or decreasing. Find out the reason for this. State who has scored the highest and who has scored the lowest. Compare your experimental subject's data with the group data. What is mean score for the group after every 25 trial and what is mean score for your experimental subject. Whether he is above average or below average after every 25 trials and on the whole. How many of them scored above the average and below the average. Among the experimental subjects what is the rank occupied by him. Find out what principle your subject has followed to improve his performance. What principle other subjects have followed.

Work out the mean scores for the control group for the first 25 trials and for the second 25 trials. Compare the group mean of the control group with that of the experimental group of the performance made by left (non-preferred) hand. Find out whether there is any difference between these two groups at any level. State the reason for this. Explain the nature of the graph drawn for the individual data and for the group data. Whether the curves are of the same type or varies. Why? State.

Conclusions

1. State whether there was any transfer effect in the performance of the experimental subject and also among the subjects of the experimental group.
2. State whether the transfer effect is positive or negative. If so state the reason.
3. State whether experimental subject has followed any specific principle in the task. If so, state, what it was.
4. State how many of the experimental subjects exhibited positive transfer effect in this task.
5. State whether there was any general principle prescribed among the experimental subjects in this task.
6. State whether there was any difference between the experimental subjects and the control subjects in the second 25 trials with left hand.

EXPERIMENT 12 : PAIRED-ASSOCIATE LEARNING
(Immediate Memory Span for Numbers)

Learning is a continuous process in the life of a human being. He learns everyday something for his good or bad. This does not mean that learning leads one to bad things. It is upto him to make use of the learned materials. Harriman (1947) states that learning is a broad term referring to a modification of behaviour as a result of experience. According to Kimble (1961) learning refers to a more or less permanent change in behaviour which occurs as a result of practice. Underwood (1949) defines that learning is the acquisition of new responses or the enhanced execution of old ones. Learning is thus defined in various ways. But how to know whether one has learned something or not? It will not be possible for us to know whether one has learned or not unless it is tested. Therefore, Candland (1968) says that learning is an inferred process and one cannot observe it directly. If we say that the rat has learned to press the bar more often the measure of its learning is its performance with the bar. This means, one will have to measure the performances and judge that learning has taken place or not. The student says that he has learned everything required for his examination. The evidence of his learning is known only from his performance in the examination. Therefore, the effect of learning is inferred or known by the learner's performance.

All human learning is verbal learning and it provides valuable information. The method of performance is more important in human learning. There are various methods of human learning and the performance is emphasized in the study of human learning. In order to measure the amount of acquisition we must measure the amount of materials retained. Right from the last quarter of nineteenth century onwards considerable attention was given on the question of the laws of association. It was felt by the empiricist philosophers that contiguity of experienced events to be the necessary conditions for association formation. The empiricists have stated that the degree of association or the amount of memory may vary with the vividness of the experience. They emphasized the closeness of time. The degree of association or the amount of memory depends upon the duration of time. This view of the empiricists has generated much experimental research on learning and memory. Therefore, it was associationism that led to the experimental investigation on learning and memory.

The first experiment on human memory was conducted by the German scientist Hermann Ebbinghaus in 1885. In his studies Ebbinghaus wanted to measure, the way in which memory operates, in detail. He had first small unrelated responses and from them later on obtained data which produced general laws which can determine the development of human memory and learning. Thorndike (1898) at the same time concentrated on instrumental learning. Fechner (1860), Thorndike and Ebbinghaus all were interested in inventions. Ebbinghaus wanted to differentiate the process of learning through experiments. For which he invented the non-sense syllables. Following this he developed several methods for the study of memory. Ebbinghaus has conducted his experiments on memory on a limited subjects in order to avoid the criticism of the results. Therefore, he conducted memory experiments on himself. Then he has to develop methods of control and these methods are widely used today everywhere bearing his name.

The method of measuring acquisition is differentiated from the method of measuring retention. But these two methods are inter-related. There are several methods of measuring retention and in every method the stimulus is presented as the material. It is true that meaningful materials are easily memorized than non-sense syllables. Materials which are interesting to us are more quickly learned than the materials which are not interesting to us. Psychologists tried to discover the principle that operates in this process. They call this learning as verbal learning. Any learning that is dominated by language behaviour is called verbal learning. There are several simple forms of verbal learning. Among these, three forms have been receiving greater attention of the experimenters. They are (1) paired-associate learning, (2) serial learning and (3) free learning.

Method of Paired-Associate Learning

An apparatus called memory drum can be used to study all the three forms of learning if we want to present the material visually. We shall deal with serial learning and free learning in the subsequent sections. In the method of paired-associate learning the subject is presented with pairs of words or items or numbers. His task is to learn which two words or items or numbers go together. This method is very widely used in experiments on verbal learning. It is to study the development of associations between items. The person is asked to learn either by hearing or by seeing the items presented to him in pairs. This method is used in varied ways. Therefore, it is better to explain one standard procedure applied in this method. In paired-associate learning stimuli are presented in pairs. But the order of the pairs is not very important. The subject is only required to associate one of the pairs with the other word or number. Normally a number of pairs are learned at the same time, that is a stimulus is presented, the person responds and feedback also is given. Paired-associate learning also involves stimulus discrimination learning. Paired-associate learning is useful for our everyday experiences. For example, to remember of the telephone numbers, car numbers, in learning foreign language, in recognizing pair of names of the persons of the same family, in all these cases the paired-associate learning is very much applied and is useful. D'Amato (1970) says that the paired-association procedure provides greatest opportunity for analysis of verbal learning in general.

Experiments in laboratory situations have been conducted on verbal learning and this resulted in emerging a most useful theory of forgetting called the interference theory (Hilgard and Bower, 1975). Of all the theories, the interference theory is currently more dominating over other theories in experiments on forgetting. The reason must be that it is an association theory. The basic idea in this is that there is functional connection, an associative bond between two or more elements. The elements may be ideas, words, numbers, names of persons or situation and stimuli and responses. For example, the use of letters *A, B, C, D*, etc. may be used to represent the elements or ideas or items and the symbols *A-B, C-D* represent the association of bonds between *A* and *B* and *C* and *D*. This association of bonds might have been already established. But the strength of the association of bond may vary due to the amount of practice. Experiments have been designed to test this theory. The paired-associate learning method of experiment is a best illustration for the interference theory of forgetting. According to this method, the subject is made to learn a list of pairs of words or names or numbers or notations like, *A-B, C-D, E-F, G-H* and so on and then tested after a limited amount of time the level of retention of the learned material or the amount of material forgotten. Bower has conducted an experiment applying one element model in paired-associate learning on 30 college students using Greek letters. Several others (Kintsch and Morris, 1964; Bernbach, 1965 and Olson, 1969) also tried this one element model in paired-associate learning method. They have found one thing in common that there were two alternative responses, like, either seen the particular stimulus — the non-sense syllable or not seen.

From the studies made by these scholars it is understood that paired-associate learning can be a complex process, because it involves variables such as stimulus discrimination, response learning, associative processes, skilfully forming judgments and so on. Therefore, in order to avoid or to minimize the intervention of complicating factors, it is better to have simplified experimental procedure in paired-associate learning method. Some of the methods are employed for the purpose of separating the particular variable in learning. The amount of material the subject learns and the amount of material he has retained can be measured by a suitable method. This is assessing the immediate memory span of the subject by the method of paired-associate learning.

Immediate Memory or Short-Term Memory

The problem of retention in learning and forgetting has been investigated for long time. The question is, whether the difference in recalling the learned material is greater when it is measured immediately after learning or after some time? Therefore, the immediate memory has been experimentally studied for many years for ascertaining the level of memory span. Several tests have been developed to measure memory span and memory span tests have been combined with intelligence tests to see whether memory span is related to intelligence and also to see whether immediate memory span is a factor that facilitates one's ability for learning. Since intelligence tests are to be completed relatively in a given time the immediate memory span is also determined in these tests.

The immediate memory is also called Short-Term Memory (STM). In the experiments on immediate memory or short-term memory the subject has to recall the information within few seconds after he has briefly learned the matter. If the information is very short involving one or two words or non-sense syllables then the subject's immediate memory may be perfect. It is also possible for him to recall the information perfectly within an interval of 30 to 60 seconds. But during this period of interval there should not be any distractions in the subject's attention. If there is any distraction in the experimental situation or if he is given any other task during the interval period then his attention for the first learned material loses its strength and the ability to recall goes down suddenly within few seconds. This is a common occurrence in this type of testings. For example on hearing a telephone number from other end if we do not carefully note down at once and if you had short conversation with somebody and then go for the number we forget it. This kind of short memory experiences we have with street numbers, house numbers and also names of new persons. But we remember and recall our own phone number any time, whatever the distractions are there and also we know everything relating to it, because it is stored in the long-term memory (LTM).

Experiments on short-term memory have been conducted by Peterson and Peterson (1959) using non-sense trigram, a three-letter combination (*e.g.* CHQ). They did not call this as non-sense syllable, but non-sense trigram as the principle of combining the non-sense syllable is not applied by using the 5 vowels A, E, I, O, U. The subject in their experiment was presented with a non-sense trigram CHQ for a short time and then he was engaged in some mental arithmetic work, successfully subtracting 3 from a random number and then he was asked to recall the trigram. The arithmetic work prevented the subject from rehearing the trigram. The trigram was correctly recalled after every interval of 3 seconds upto 16 seconds in spite of the successive subtracting mental work. But the percentage of recall after 18 seconds was practically zero. The reason might have been that when the rehearsal was prevented or when distracting material was presented the short-term memory becomes weak and get decayed rapidly and exist only for a short time.

This phenomenon has aroused the curiosity of many researchers to continue their work on this line. Bower (1967), Atkinson and Shiffrin (1971), Peterson (1963), Rumelhart (1967), Falkenberg (1972) have made remarkable contributions on verbal learning and on short-term memory. Atkinson and Shiffrin (1965, 1968) have made most extensive and systematic study of short-term memory. According to them short-term memory is an active store house, but with a limited capacity that can only hold or register a few items or information at any one time. The capacity is limited to the amount of information in each item and also limited by the complexity of the information. The short-term memory is supposed to be a place in which active rehearsal of the verbal material or 'going over' is taking place. This rehearsal or 'going over' (thorough inspection) takes time. This time is very short. The time may be sufficient to go over or rehearse only a few items or words may be, two to four pairs items or non-sense trigram. Since next item, pair of words will be presented soon it has to deal with it.

If distracting tasks, like arithmetic works or any other form of distracting tasks are presented more attention is needed as it reduces the rehearsal of the material to be retained. Therefore, the person has to pay special attention to a particular item or material as and when presented. Otherwise the item or material is lost from the short-term memory or replaced by subsequent appearance of the new item or material. If the material is of greater in difficulty more attention is paid for rehearsal. As result of this the material to be remembered is lost or disappear from the STM.

A number of studies have been made and theories formulated to distinguish between the STM and long-term memory (LTM). One important contribution made by Atkinson and Shiffrin (1971) is to be worth mentioned here. From the results of their studies they were able to postulate two storage mechanisms, one for STM and one for LTM. There is also interaction between these two memories. Learning is required for both STM and LTM for storing the material. But the STM is considered as quickly decaying system. Whereas the LTM is considered as essentially a permanent store house. The incoming information into the STM is not attended to with sufficient rehearsal and hence it begins to disappear or fade away. But certain selected information may be paid more attention and rehearsal may be made and prevented from decay for a short period of time. But more attention and rehearsal is possible if only the items are few. When the items are too many there will not be anytime for rehearsal and an item not rehearsed rapidly fades away.

Investigations on the functions of memory and on learning processes are going on to find out explanations for some pressing issues. These are specially based upon the views expressed on chemical changes that take place in the central nervous system. Basically there are two kinds of views about the neural basis for the retention of learning. One is the *dynamic view* and the other is *structural view*. According to the dynamic view it is presumed that an experience sets up a continuing electrical activity in the appropriate neural circuits. These active neural circuits which are persistent, bring in relation with the existing of our memory. When this active trace or circuit process stops, the memory is lost or decayed. But the structural view which is opposed to it states that learning consists in some structural, physical or biochemical change in the nervous system. This chemical change will remain even after the original neural circuits have returned.

But the problem is, how a chemical change can be so rapid as to cause rapid forgetting that occurs in short-term memory and also adequately stable to allow retention in long-term memory.

Muller and Pilzecker (1900) have tried to explain these. According to them the neural activity is responsible for the physical change of storing the incoming message, the experience each stimulus leaves its mark upon the neural system. The physical change exists only for a short time after the experience and then the mark decays shortly. The neural activity is preserved by repeated associations and, it becomes more fixed and makes the retention stranger. There is a process called progressive fixation and also called consolidation. But if the existing neural activity is disturbed rapidly by some intervention then the physical change becomes a weaker one and the retention also becomes poor. Hebb (1949, 1961) has stated the same hypothesis, putting more emphasis on ***neuro-physiological explanation***. He assumed a transient trace and a permanent trace. The first is quite similar to short-term memory and the second is like the long-term memory. Some state this as stimulus trace theory (Candland, 1958).

The hypothesis, that events once registered in memory are not lost, has been counteracted by various views. *The decay* theory stressed that there is actual loss or erosion of memories. The *motivated notion of forgetting* states that though information is stored permanently in memory certain emotional disturbances are making the retrieval inaccessible. According to the interference view it is presumed that interference of certain materials actually destroy the memory traces or prevent the retrieval process. Thus there are two types of forgetting suggested by Tulving and Madigan (1970). They are the loss of information and inability to retrieve the information. These two are called *trace-dependent forgetting* (loss of information) and *cue-dependent forgetting* (inability to retrieve information). The trace-dependent forgetting is actual decay of the memory trace. It seems that it is quite similar to short-term memory. In the cue-dependent forgetting information is stored in memory and it is similar to long-term memory. But with all these we do not find a conclusive answer for the problem.

In paired-associate learning there are three important variables. One is the amount of information or stimuli presented and the second is the rate at which they are to be presented and the third is the rate at which the recall is to be made. The more the amount of materials presented the less amount recalled and the faster the rate of presentation the lower the learning and smaller the amount of recall. The shorter the recall time the fewer materials recalled. But the reserve result also is reported. Conrad and Hille (1958) state that the faster the presentation better the recall.

Therefore, paired-associate learning may be stated as the learning of stimulus-response pairs as in the case of learning a foreign language vocabulary or some telephone number. When the first member of pair, the stimulus, is presented the subject's task is to provide the second member, the response. The paired-associate learning is used to test the memory span of an individual. Immediate memory has been studied for many years in the past and the human memory span was measured. The memory span for an individual is the number of items that he is able to recall and reproduce immediately after few seconds of learning the material. From the memory span experiments we come to know that all normal adult human beings have the capacity to recall and reproduce on the average seven or plus or minus two. Sir William Hamilton an Irish mathematician and physicist in 1859, has stated that if a handful of six or seven marbles are thrown on the floor it is possible to see the number correctly at a single look without counting them. When more are thrown the number cannot be perceived instantly. Ebbinghaus first started his experiments on memory only on himself and found that it was seven. Miller's work verified this. Miller (1956) states that the immediate reproductive memory is limited in terms of number of 'chunks' of elements in preference to the number of single elements.

In this way a list of 7 two syllable words is remembered in the same manner as a list of one-syllable words or 7 single letter is remembered. The reason is that all these are treated as single units. The inference is that the immediate recall is limited to 7. In most of the studies the memory span was consistently 7. If the person wants to learn 14 unrelated words he has to organize the lower order unit into 7. On the basis of the experimental results Miller has designed a study in 1956 with a title "The magical number seven plus or minus two. Some limits on our capacity for processing information". Some experimental psychologists call this as the measurement or retention (Postman and Egan, 1949).

Experiments for measuring immediate memory span of a person by using the method of paired-associate have been conducted in various ways. In measuring immediate memory span the feedback system is not followed. Investigators in the study of memory span use various procedures and materials. Since there is no strict and fast rules like the psychophysical methods investigators take liberty to measure the immediate memory span in their own ways. For example, one procedure is, using a list of numbers in pair and the numbers may range from 3 digits to 10 digits. In every digit there can be two pairs of numbers. There can be 10 pairs of number in the list. Some investigators use single list of numbers, non-sense syllables or words and not in pairs.

Another investigator may use a list of non-sense or meaningless syllables, or trigram (3 letters). The list may be consisted of 10 such pairs of trigram or 10 trigrams and 10 numbers. The trigram is a 3 letter word composed to have a single speech sound which has no meaning. But some use two meaningful words as stimuli or one meaningful word and one number as a pair. In all these there should be ten items. Some investigators suggest 12 items in the list (Postman and Egan, 1949). According to Runquist (1966) and D'Amato (1970) the stimulus is shown first and after two seconds the response is shown. They associate numbers and words. A pair of item consists of a number (stimulus) and a word (response) while presenting the number alone the participant should learn to anticipate and say the correct response.

These items or stimuli may be presented orally or visually. In the case of auditory stimuli the experimenter will have to read out every pair one by one at the rate of 2 seconds between pair and the subject has to repeat within two seconds. In the case of visual stimuli the material in pairs can be presented in the apparatus called memory drum.

EXPERIMENT

Problem

To measure the immediate memory span of an individual for numbers, using stimuli orally/visually.

Method

Participant

Choose a person who has no knowledge about this experiment and describe his personal characteristics like, age, sex, educational qualification, present occupation, has any physical defects or wears spectacles and so on.

Materials

(1) A list of numbers in pairs, ranging form 3 digits to 10 digits. In each digit there should be two sets of pairs. For example, in 3 digits there should be 478-539 and 694-782, one set after the other. The first number is the stimulus and the second number is the response. (2) Stop clock. (For materials for conducting the experiment see Appendix 1). Memory drum also can be used for this experiment.

Procedure

The experimenter should decide whether he/she is going to present the material orally or visually. If a memory drum is available he can make use of it for visual presentation. If it is not available he can present the stimuli orally. It is to be noted that in the study of immediate memory span (STM) the material should be exposed only once.

(i) Oral Presentation of the Stimuli : In Paired-Associate Learning

The participant should be seated comfortably in a well illuminated room. There should not be any auditory disturbance or visual movements in the room. The experimenter should prepare the scoring sheet and keep ready everything. The list of pairs of numbers should not be seen by the participant. It is better to have a small wooden screen between the experimenter and the participant. Then the experimenter should give the following instructions to the participant:

"Please listen carefully. Now I am going to give you a pair of numbers. You will have to repeat the same by speaking out audibly. Then I will call the first number of the pair and you will have to speak out the second number of the pair. In every pair of numbers, I will be increasing the digits. Therefore, in every pair after repeating the two numbers you will have to give me the second number when I called the first number."

Table 6.20: Showing the participant's immediate memory span in the paired-associate learning method (The same Table can be used as scoring sheet). The Table is given as a model.

S. No.	No. of Digits	Pair of numbers presented first	Stimulus number	Response made by subject	Result Right or Wrong
1	3	276 – 396 438 – 547	276	396	Right
2	4	5734 – 6893 5385 – 7946	5734 6385	6893	Right
3	5	82643 – 95321 67953 –76842	82643 67953	95321	Right
4	6	706253 – 635478 953142 – 726384	706253 953142	635478	Right
5	7	4680579 – 8649753 3579468 – 9750864	4680579 3579468	8529753 9750864	Wrong Right
6	8	69398105 – 19753208 81427689 – 20835719	69398105 81427639	19753208	Right
7	9	673954108 – 745693268 819035976 – 326349357	672954108 819035976	746583268 325249547	Wrong Wrong
8	10	6326345972 – 2764385910 7541236809 – 8674679531	6326345972 7541236809	2754375910 8647579531	Wrong Wrong
The largest digit Number to which correct response made – 8					

Note. : Two pairs of numbers are given in every cell. But only one pair the first pair can be used for experiment. In case if the participants fails to give correct response for the first pair the second pair of the same digits can be used. The first number is the stimulus and the second number is the response. We are using eight items upto 10 numbers for the experiment on the idea that the memory span may not go more than 10.

If the participant fails to give the second number of a pair the experimenter should give a second pair of numbers of the same digits. If he successfully presents the second number then the experimenter can go to the next digit. But when the participant fails to reproduce the second number in both the pairs of the same digit then the experimenter should not stop the experiment. He must go till the end. Sometimes the participant may succeed with longer number where he has failed with shorter ones. The experimenter should ascertain that the participant has correctly understood the task. The experimenter can give a 'ready' signal and start speaking out the first pair of numbers from the list, *i.e.*

276-396, etc. After completing all the numbers of the first according to the procedure the experimenter should get an introspective report and allow him to go away.

Results

The responses of the participant should be presented in Table 6.20. The number of correct reproductions of the second number is the span of memory of the participant. It should be the number of digits just preceding the one at which the participant has failed to reproduce the second number in both the cases.

State the memory span of your participant. The memory span is the number of digits the participant has correctly repeated. The group data should be collecteu from other experimenters and should be presented in Table 6.21.

Table 6.21: Showing the group data of the immediate memory span in the paired-associate learning test for numbers

S. No.	Name	Immediate memory span for number	Remarks, if any
1	S.R.		
2	B.V.		
3	M.S.		
4	T.S.		
5	K.R.		
6	P.V.		
7	K.S.		
8	R.S.		
9	S.T.		
10	V.R.		
11	R.M.		
12	B.T.		
	Mean		

Discussion

Discuss the individual's memory span in relation to the group members. Did you observe that your participant was struggling in reproducing any of the numbers. State the digits in which he/she had any difficulty in repeating the number. Was your participant paying careful attention to the experiment or he/she has taken it casually. Was your participant's memory span low, or on the average or high. If lower than the group mean what was the reason? Did you find out any cause for this from his personal characteristics. How many in the group above the group mean and how many below the group mean. If there was any reason state it in remark column and discuss it. What was the highest memory span and lowest memory span in this experiment? Who were they? State the reason, if any. What is the rank of your participant in the memory span. Compare your results with the results of earlier studies. State the variables such as age, sex, qualification, occupation in which the group member vary in their memory span.

Conclusions

1. State your participant's memory span according to the method of paired-associate learning. 2. State the group mean of the memory span. State the variables on which individuals differ.

(ii) Visual Presentation of the Same Stimuli in Paired-Associate Learning

If memory drum is available the same numbers may be displayed in the window of the drum. The numbers may be in typed or printed form. First the pair of numbers may be arranged in the same order as shown for the oral type. Just below the pair only the stimulus, *i.e.* the first number should be placed. In the window in one exposure the pair of numbers should be displayed and in the next exposure only the stimulus number

Table 6.22 : Showing the procedure for presenting the pairs of numbers and the stimulus number for the visual presentation in Memory Drum for measuring the memory span (A model for preparing the Table and for entering the results)

S. No.	No. of digits	Pair of numbers presented	Response Number	Result right or wrong
1	3 S	276 – 396 276 ?	396	Right
2	4 S	5734 – 6893 5734 ?	6893	Right
3	5 S	82643 – 95321 82643 ?	95321	Right
4	6 S	706253 – 635478 706253 ?	635478	Right
5	7 S	4680579 – 8649753 4680579 ?	8649753	Right
6	8 S	69398105 – 19753208 69398105 ?	19793208	Wrong
7	9 S	672954108 – 745693268 672954108 ?	945683268	Wrong
8	10 S	6326345972 – 2764385910 6326345972 ?	2794376810	Wrong
		The longest digit number to which correct response - 7		

Note : In oral presentation we had two pairs of numbers for each digit. But there, for visual presentation we have one pair of numbers for each digit. This arrangement is from experimental point of view. The 'S' in second column is referring to stimulus number.

should be displayed. This type of arrangement should be made for all the pairs of numbers. The rotation of the drum should be adjusted in such a way that every exposure should be for two seconds. Thus the pair numbers are exposed for two seconds and immediately the stimulus number is presented for two seconds and at once the participant should give the other number either orally or by writing in a sheet of paper quickly before the next exposure takes place. The experimenter should switch on the drum saying the participant "ready go". This procedure should be continued till all the eight pairs of numbers are presented to the participant. Woodworth says that sometimes participant may fail to produce shorter number, but may succeed with longer numbers. Therefore, whether he is right or wrong all eight pairs of numbers should be presented.

Discussion

Since the visual presentation of the stimuli in paired-associate learning method for measuring the immediate memory span is an alternative procedure for conducting the experiment and collecting data, the other steps are not explained here. The treatment of the data, including collecting the group data, and discussion should be the same as shown for the oral presentation procedure shown above.

(iii) List of Single Number presenting Orally (Auditory)

But some investigators use list of single numbers for measuring immediate memory span. The paired-associate learning method is more difficult than using the single numbers for measuring immediate memory span. Therefore, immediate memory span can be also measured by presenting single numbers with the digits ranging from 3 to 10 instead of using pairs of numbers. There should be 3 sets of numbers. The numbers may be presented one by one to the participant clearly and distinctly and he may be asked to reproduce orally. When he fails to reproduce the correct numbers the experimenter should not stop and he should go till the end. Sometimes the participant may fail to reproduce shorter number, but he may succeed with longer number (Woodworth, 1947). The participant may succeed with 8 digit but fail with 7th digit. Therefore, an average of the 3 sets can be obtained. A credit of 1/3 for each correct reproduction or the average of the 3 sets may be taken as the immediate memory span. The numbers may be presented with two seconds interval.

(iv) List of Single Number presenting in Memory Drum (Visually)

The same 3 sets of single numbers used for oral presentation may be visually presented in a memory drum for measuring immediate memory span. The numbers from 3 digits to 10 digits may be typed or printed with equal interval space and affixed in the drum. The drum may be adjusted in such a way that it exposes every number for two seconds. After giving caution 'ready go' the experimenter should switch on the drum. The experiment can be stopped only when all the numbers are presented visually. The participant may sometimes produce longer number but fail to reproduce shorter number. The immediate memory span is the true average of all correct reproduction in all 3 sets of numbers.

Table 6.23 may be used as scoring sheet or as the Table for collecting data in presenting single number orally or visually for measuring immediate memory span of a person.

Results

Table 6.23 : Showing the procedure of presenting lists of single number orally or visually for measuring the immediate memory span (A model for oral and visual presentation – Single number separate Tables should be prepared for auditory and visual responses). The experimenter can use the following lists of numbers for conducting his experiment

S. No.	Digit No.	Stimulus No.	Response	Result
1	3	276	276	Right
2	4	5734	5734	Right
3	5	82643	82643	Right
4	6	706253		
5	7	4680579		
6	8	69398105		
7	9	672954108		
8	10	6326345972		
The longest digit correct reproduced is :				
1	3	438	438	Right
2	4	6385	6385	Right
3	5	67953	67953	Right
4	6	953142	953162	Wrong
5	7	3577468		
6	8	81427639		
7	9	819035976		
8	10	7541236809		
The longest digit correctly reproduced is :				
1	3	641		
2	4	2730		
3	5	85943		
4	6	706394		
5	7	1358796		
6	8	43287491		
7	9	640265129		
8	10	4794386215		
The longest digit correctly reproduced is :				

Reproduced longest digit in the first set : x_1

Reproduced longest digit in the second set : x_2

Reproduced longest digit in the third set : x_3

$$\text{Mean : } \frac{x_1 + x_2 + x_3}{N} = \frac{x_1 + x_2 + x_3}{3} = \frac{\Sigma x}{N}$$

The mean of the three sets is the immediate memory span of the participant. The same procedure can be applied to both auditory and visual methods for measuring the immediate memory span.

The group data may be collected from other experimenters and the mean immediate memory span may be worked out and the results may be discussed.

Discussion

If there is no variation in the performance of the participant in all 3 sets the results might show that the participant reproduced in all 3 sets upto 7 digits and nothing beyond that point. Sometimes variation may occur. For example, in the first set the participant was able to reproduce 7 digits, in the second set he reproduced again 7, in the third set he has failed to reproduce the 8 digit, but succeeded with the 9 digit. In such case the true mean span should be worked out. It is found that up to 6 digits there is no variation in all three sets. Variation takes place at 7. Therefore, the basic value of the span is 6. In two sets the digits is 7 and in one it is 9. Therefore, the immediate memory span is 6 + $\frac{1+1+1}{3} = 6 + \frac{3}{3} = 7$.

State whether there is any variation in the response of the participant in the 3 sets of numbers? Has the participant struggled in any set, if so at which digit and what could be the reason for it. At how many places the participant committed mistakes in smaller digit numbers. Find out the variables which had influence in the performance of the participant and discuss them.

State your participant's immediate memory span in the single number presentation all the 3 sets put together. What is the group mean? Whose memory span is the highest and whose is the lowest. How many above the group mean and how many below the group mean. Where your participant stands in the group (rank). Compare your results with that of the earlier studies. If there is any difference state the reason for the same.

Conclusions

1. What is the immediate memory span of your participant?
2. What is mean immediate memory span of the group.
3. What are the factors that caused the variation in the memory span?

APPENDIX – 1

MATERIAL FOR THE METHOD OF PAIRED-ASSOCIATE LEARNING — EXPERIMENT USING NUMBERS

The numbers should be presented in pair and it is better to have the pair in a box as shown below. The stimulus number is shown separately below the Box.

S. No.	Digits	Paired Numbers	Stimulus Number
1	3	[276 – 396]	[276]
		438 – 547	438
2	4	[5734 – 6893]	[5734]
		6385 – 7946	6385
3	5	[82643 – 95321]	[82643]
		67953 – 76842	67953
4	6	[706253 – 635478]	[706253]
		953142 – 726384	953142
5	7	[4680579 – 8649753]	[4680579]
		3579468 – 9750864	3579468
6	8	[69398105 – 19753208]	[69398105]
		81427639 – 20835719	81427639
7	9	[672954108 – 745693268]	[672954108]
		819035976 – 326349357	819035976
8	10	[6326245972 – 2764385910]	[6326345972]
		7541236809 – 8674679531	7541236809

Note : The paired numbers *within the Box* should be presented first. Then the stimulus number in the Box the experimenter should call for. Other paired numbers are alternative numbers. If the participant has failed to give the correct response number for the Stimulus in Box then the alternative paired numbers may be presented and the corresponding stimulus number may be called for.

EXPERIMENT 13 : PAIRED ASSOCIATE LEARNING

(Immediate Memory Span)

(Non-sense Syllables)

Immediate memory span is measured by applying paired-associate method for non-sense syllables. Ordinarily the numbers of different digits were used for measuring immediate memory span. It is found that the non-sense syllables may also cause same the amount of difficulty as the numbers for the participants in reproducing them. Even meaningful words which are unrelated may be used to measure immediate memory span. Kimble, Garmezy and Zigler (1984) state that one important aspect of short-term memory is that it deals with the limited amount of material before it fades away. Further the material should be exposed only once.

We can remember an average of about seven numbers, or letters or simple words with *just one presentation.* Miller (1956) has stated that memory span is dependable upto seven and plus or minus two. Therefore, it is believed that the span of short-term memory is the same for letters and words. The words must be learned as words and not collection of letters. If we remember words as letters we would remember only few of them as in the case of numbers. Even in the case of using letters, if they are non-sense syllables the effect is the same as numbers. Ebbinghaus (1850-1909) has realised that meaningful materials were easily learned and remembered. He wanted to control this problem and, therefore, eliminated meaning in words. Therefore, he devised three-letter consonant vowel sequences which he called non-sense syllables like ZIV, WEH, LUQ etc. These non-sense syllables had rapidly fading effect in the memory. The more time passes the more materials are forgotten.

The human memory system has a remarkable thing. It is the accuracy with which people can judge or predict the learning rate of various verbal units. Underwood (1966) was able to see in one of his studies the correlations as high as .92 between estimated and actual rates of learning three-letter verbal units. Miller (1956) states that the human memory system has limited capacity for holding materials. The limit is seven items for adults. But by combining words into 'chunks' human memory span can be expanded.

The limited capacity of the human memory system has been clearly recognized by experimenters. They determined the length of the list of verbal units to be used in verbal learning experiments. Postman and Riley (1957) state that like pairs of two digit numbers and two trigrams or two non-sense syllables are learned at about the same rate. But unlike pairs, like combining number-trigram or trigram-number or number — non-sense syllables vary in rate of learning. Postman and Riley have observed that subjects preferred to learn like pairs first though they were not easier to learn than the unlike pairs.

Therefore, it is better to use like pairs in immediate memory span experiments and it is better to use the paired-associate materials either both as numbers or both as non-sense syllables or both as meaningful words. In this experiment there can be ten pairs of non-sense syllables or three-letter trigrams as stimuli and responses. The first trigram serves as stimulus and the second as response. These ten pairs may be in printed or typed form. There should be sufficient space between the pair of trigram (non-sense syllable). Then leaving sufficient amount of space below the first stimulus, the second pair of trigram should be put. In the same manner all the ten pairs of non-sense syllables should be arranged.

These ten pairs of non-sense syllables (trigrams) may be presented to the participant in a memory drum. The memory drum should be so adjusted that it should make a move after every two seconds displaying in the window a pair of non-sense syllable. In the second move after two seconds the second pair of non-sense syllables should appear in the window. This procedure should be carried out till all the ten pairs of non-sense syllables are exposed for two seconds.

When the first pair of the non-sense syllable appeared in the window the participant should carefully watch and learn to associate both. Then immediately after two seconds the second pair of non-sense syllables will appear for two seconds. In this way each pair of non-sense syllable is exposed to the participant only once for two seconds and he has to learn and associate the stimulus syllable with the response syllable. After everything is over the experimenter should stop for six seconds and then start presenting the stimulus syllable alone and the participant's task is to call the particular response syllable with which the stimulus was presented.

Experimenters use various procedures to measure the immediate memory span for non-sense syllable. What is described here is one standard procedure. The other procedure is to present all the pairs of non-sense syllable once, each pair for two seconds. In the pair the stimulus is a number and the response is a word. The stimulus will appear alone and will disappear. The response will appear next. The subject has to associate the stimulus and the response as pair even though both were not seen as pair in the apperture. This means that the subject should anticipate the response (Atkinson, 1972).

A third procedure also experimenters use. All the ten pairs of non-sense syllables are presented to the participant once, one by one. Then the stimulus syllable is presented one by one and the participant is asked to reproduce the other syllable of the pair. But the stimuli are presented not in the same order as the pairs of syllables were presented. The original order of the ten stimuli is changed into different order.

EXPERIMENT — NON-SENSE SYLLABLES

Problem

To measure the immediate memory span for an individual using the method of paired-associate learning for non-sense syllables.

Method

Participant

Choose a person who has no knowledge about this experiment and describe his/her personal characteristics like, age, sex, qualification, occupation, whether he/she has any handicap, wearing eye glasses and so on.

Materials

(1) A list of pairs of non-sense syllables not less than ten. In a pair there should be two non-sense syllables, the first is the stimulus and the second is the response. Just above the first pair there should be one *X* mark which should be the indication that pairs are going to appear below. (2) *Memory drum* with all accessories. (3) Stop clock. (The list of 10 pairs of non-sense syllables are given in the appendix.)

Fig. 6.7 : Memory Drum Apparatus Experiments 12, 13, 14, 15, 16 and 17. The Memory Drum Apparatus is for measuring Memory Span of Various types.

Procedure

The participant should be seated in front of the memory drum comfortably in a well illuminated room. There should not be any disturbances or distractions visually or auditorily. The experimenter should have made all preparations for conducting the experiment. The pairs of non-sense syllable should be in printed or typed form with sufficient interval between any two pairs. The list should be pasted in the rotating drum having full visibility of the two syllables in the window. The drum should be adjusted in such a way that a pair of non-sense syllables should be exposed only for two seconds.

After having made this preparation the experimenter should give the following instruction to the participant :

> "Please look at this small window of the drum (showing it). You will see first one *'X'* mark appearing, which is the indication that pairs are going to appear below. Then you will see two non-sense syllables together in a line with a hyphen between them. They are put in a rectangular box with the view of showing them specifically. The first word is the stimulus and the second is the response. So I will present them in pairs only for a period of two seconds. I will show you ten such pairs. You will have to observe each pair carefully and learn to associate them. After presenting all ten pairs of these non-sense syllables I will stop it for few seconds. Then I will show you only the first syllable of the pair. Then your task is to produce the corresponding response syllable of this particular stimulus. As soon as you see the

stimulus or the first syllable of the pair you at once call the response syllable within two seconds before the appearance of the second stimulus. You might have understood everything now. If you have any doubt you can ask me."

"When I say 'ready, go', you must watch the *'X'* mark and then the pairs of non-sense syllables in this window carefully and register in your memory. 'Ready go'."

The experimenter at once should start the memory drum to move step by step for two seconds exposing a pair of syllables at every step. When all the ten pairs are presented the experimenter should stop for a period of six seconds and then start presenting the stimulus of each pair alone. The subject at once should call out the response of the particular stimulus. The experimenter with a list of pairs in hand should check up whether it is right or wrong. If wrong what response he/she has called for that stimulus. After presenting all the ten stimulus syllables alone the experimenter should collect the introspective report from the subject and allow him to go away. If the experimenter asked the participant to write down the responses in a sheet of paper, then he should collect the paper at once from the participant.

Results

The experimenter, then scrutinizes the responses given by the participant and finds out with how many stimulus syllables he/she has correctly associated the response syllables and present them in Table 6.24.

State how many stimuli the participant has associated incorrectly with the responses. Whether there was any change in the responses or in the letters of the non-sense syllables. The number of correct stimulus-response associations may be taken as the immediate memory span. This result may be presented in Table 6.24. To know that sort of responses the participant has given for each stimulus the experimenter should enter those responses against the response column 4.

Table 6.24 : Showing the responses made by the participant and the immediate memory span in the method of paired-associate learning experiment for non-sense syllables

Sl. No.	Paired-associates and the stimuli	Stimulus syllable	Response syllable by participant	Result
1	2	3	4	5
1	XEW – MVX	XEW		
2	CAJ – KEZ	CAJ		
3	VUQ – ZUY	VUQ		
4	DIJ-1 – FAP	DIJ		
5	WOX – BOZ	WOX		
6	KEW – ZIV	KEW		
7	HOX – GAF	HOX		
8	QUT – FOV	QUT		
9	LEH – QAM	LEH		
10	JIS – XAD	JIS		
Number of correct associations – Memory Span :				

Collect the group data from other exprimenters and present them in Table 6.25. Work out the mean immediate memory span for the group for non-sense syllables in the method of paired-associate learning.

Table 6.25 : Showing the group data of the experiment of immediate memory span by the method of paired-associate learning for non-sense syllables

S. No.	Name	Immediate memory span	Remarks
1	P.K.		
2	S.R.		
3	M.S.		
4	K.R.		
5	P.V.		
6	T.V.		
7	A.R.		
8	P.C.		
9	P.L.		
10	A.L.		
	Mean		

Discussion

Discuss the individual's performance in the paired-associate learning. Whether he is benefited or not? What is his/her immediate memory span? How was the group performance? What is the group mean? Whose memory span in this type of test is the highest and whose is the lowest? Where your participant stands? Was there any change of response of one stimulus to another? Instead of giving the response syllable has anybody given the stimulus as the response. Did you observe any factor interfering the performance? Whether any of the personal characteristics had any influence in enhancing the memory span. Discuss your results with the results of earlier studies.

Conclusions

1. State your participant's immediate memory span by the method of paired-associate learning in non-sense syllables.
2. What is the group's immediate memory span?
3. How the individuals differ in their memory span.
4. Is there any sex difference?

APPENDIX – I

Non-sense Syllables : Material for the Method of Paired-Associates Learning Experiment

Note : After preparing the materials for the experiment it is better to have the pairs of non-sense syllables in a Box as shown here. Before pasting the list in the memory drum the pairs should be in a box. The distance between the pairs should be such that in one move the first pair should be exposed and in another move the second pair should be exposed.

X

1. SEW – MUX
2. CAJ – KEZ
3. VOQ – ZUY
4. DIJ – FAP
5. WOS – BOZ
6. KEW – ZIV
7. HOX – GHF
8. QUT – FOV
9. LEH – QAM
10. JIS – XAD

Stimulus Syllables

The two stimulus syllables should be arrnaged in the same manner as the pairs, one by one with equal distance and at one move only one stimulus syllable should be exposed to which the subject should produce the corresponding response syllable.

1. SEW
2. CAJ
3. VOQ
4. DIJ
5. WOS
6. KEW
7. HOX
8. QUT
9. LEH
10. JIS

EXPERIMENT 14 : PAIRED-ASSOCIATE LEARNING (Immediate Memory Span) (Meaningful Words)

The method of paired associate learning has been used in various ways. The important aspect of this method is that materials should be presented in pairs, one of the pairs as stimulus and the other as response. These two are given only one exposure for two seconds to the subject for learning and then the subject is tested whether he/she has rightly learned or not. This is for studying the short-term memory (STM) or to measure one's immediate memory span. This is the essence of this method. But investigators administer the stimulus in varied ways.

While studying the immediate memory span of a person, either for numbers or for non-sense syllables or for meaningful words he/she must be given a chance to learn the material first. Since it is paired-associate learning method the subject should learn the material first in paired position. The materials should be presented or exposed to him in pairs for a standard time of two seconds. Therefore, materials should be always presented in pairs. If this is not taking place in the experimental procedure or if any investigator deviates from this logical conclusion it will not be called as paired-associate learning method. Therefore, the material should be presented in pairs first to the person for learning.

Secondly, testing takes place only after learning. The whole test materials, probably eight items, are presented to the subject once. Then we test, what the person has learned when the materials were presented for a standard time of two seconds. The testing procedure is by presenting the first item of the pair, the stimulus and we expect the subject to reproduce the response word or the other item of the pair within the short period of two seconds. The subject's response may be right or wrong. But we are testing his memory immediately. Therefore, after presenting all the pairs of items (numbers, or non-sense syllables or meaningful words) each for two seconds, the experimenter is presenting the stimulus alone and the subject has to reproduce the other member or the response word. This procedure is continued till all the stimuli are presented and the response words are obtained. Then the experimenter should check and find out the number of correct responses out of the pairs of items presented to the subject. This must be the correct procedure of paired-associate learning method.

While discussing the tasks in learning experiments, D'Amato (1970) states that because the term 'paired-associates' implies that each varbal stimulus item is the only signal for producing the corresponding verbal response the presentation of this stimulus item is the only basis on which the subject can give the appropriate response. We are not dealing with only one pair of items, so that the subject may not be paying sufficient attention. We are dealing with quite a number of pairs of items, from eight to eighteen and this is also informed to the subject and, therefore, he *has to be more attentive.* In paired associate learning procedure there is no chance for the process of anticipation on the part of the subject. Anticipation is one of the components of serial learning.

But, since there is no hard and fast rule laid down with regard to the experimental procedures for the study of verbal learning there are variations in the ways of presenting paired-associates. In paired-associate learning method some investigators like Runquist (1966) used dissimilar stimuli.

In Runquist's study for paired associate learning, numbers are used as stimuli and words are used as responses. The subject is asked to associate words and numbers. The list consisted of only eight pairs of items. In his procedure the number (the stimulus) will always appear in the window alone. Then the word is exposed for two seconds and the subject is asked to associate or relate the word with the number. When the response word is exposed the stimulus number will not be there. As the procedure goes the stimulus number and the response word were never exposed together. Then how should we call this procedure as paired-associate learning? There is no rationale in the procedure followed by Runquist and the procedure to be called as paired-associates. But the subject is asked to see the number and word as pair even though they were not presented together as pair.

Another point to be seriously thought about is that according to this procedure all the eight pairs of items are presented to the subject once and then the stimulus number is presented alone. The subject should anticipate the response word on the other side. After presenting all the eight pairs once, a second trial is taking place. In between the first trial and the second trial how much time is permitted as interval. The standard interval is said to be six seconds. The subject is tested in the second trial. Therefore, the time interval should be mentioned.

Lockhead (1962) has made a small change in the Runquist procedure. He first presented all the pairs of items once and then presenting stimulus one by one in a randomized order. He has not made mention about the time. In this procedure sometimes the subject is given the stimulus-response pair and sometimes only the stimulus. It is supposed to be the delayed knowledge of results and this impaired the learning. The paired-associate learning method has been applied still in various ways by Cook and Spitzer (1960), Hawker (1964), Sidowski (1968), Jung (1964), Postman and Goggin (1966), Kendler (1965), Gibson (1940) and others.

The variations in paired-associate learning method mainly depend upon the hypothesis each investigator has formulated for his study. But the study of short-term memory with the view of measuring the immediate memory span by the paired-associate learning method for numbers, non-sense syllables and meaningful words can be also designed still differently. The chance for presenting the stimulus as a response may also be eliminated. The meaningful words used in pairs should be thoroughly unrelated and should have equal difficulty and should be objectives.

According to Hall (1970) paired-associate learning is thought of as two step task. In the first step the subject must learn what are the various response words. Then he must learn to which stimulus the particular response word belongs to. Therefore, the first stage is response learning and the second stage involves forming an association with the stimulus. In the study of verbal learning several variables are related to learning. One of the important variables is the meaningfulness. Here meaningfulness refers to the number of other units associated with a particular word. For example, for many people more words may be associated with 'crow' than with 'cube'. Therefore, crow is more meaningful. In paired-associate learning meaningfulness of the response words is strongly related to the rate of learning. This means that if the response words are highly meaningful than the response learning stage becomes much easier. The meaningfulness of the stimulus words also has a positive effect upon the learning rate.

EXPERIMENT

Problem

To measure the immediate memory span for an individual using the method of paired-associate learning for meaningful words.

Method

Participant

Choose a person who has no knowledge about this experiment. Describe his/her personal characteristics such as age, sex, qualification, occupation, whether he/she has any handicap or wear eye glasses and so on.

Materials

(1) Memory Drum with all accessories. (2) A list of ten pairs of meaningful words. The pairs of words should be unrelated adjectives, and of increasing difficulty. (3) Stop clock (The list is given in the Appendix).

Procedure

The experimenter should make all arrangements to conduct this experiment in a memory drum. The experimenter should design, ten pairs of meaningful words. The two words in a pair should not be related in any form. The response word may be more meaningful. The stimulus words should not have any relation with the response words and also the response words in pairs should not be related in any form. These pairs of words should be in printed or typed form with sufficient amount of interval between them. Such a list should be pasted in the rotating drum with full visibility for the two words in the window. The two words should appear parallel in a Box with a small dash (–) between them. The timing should be so adjusted that the drum should move one step after every two seconds. Then the participant should be made to sit comfortably in front of the memory drum in a well illuminated room where there should not be any disturbances or distractions visually or auditorily. The experimenter, then, should give the following instructions to the participant:

> "Please look at this small window (pointing out the window in the memory drum apparatus). Two words may be appearing, one against the other. There will be a hyphen between the two words and they are within a rectangular box. They will be exposed to you only for two seconds. There will be ten such pairs of words. As soon as you see the pair of words for two seconds you must learn them and register in your memory. The first word is called stimulus word and the second word is called response word. Every

pair is presented only for two seconds. After presenting all the ten pairs of words to you in the windows, I will stop for few seconds (six). Then the stimulus or the first word of the pair will appear in the window alone. Your task is to tell me at once the second or response word with which this particular stimulus word was paired and presented to you. You must reproduce the response word within two seconds.

"When I say ready go, you must observe the pair of words and learn them. Now, "ready go". At once the experimenter should start the memory drum to rotate exposing each pair for two seconds."

As soon as all the ten pair of words are presented according to the procedure the experimenter should start presenting the stimulus word alone and ask the participant to call out the response word for the particular stimulus word. After presenting all ten stimulus words and getting the response words from the participant, the experimenter should get the introspective report from the participant and allow him/her to go away.

Results

Then the experimenter should scrutinize the responses and find out how many responses he/she has given correctly.

Table 6.26 : Showing the responses made by the participant and the immediate memory span in the method of paired-associate learning for meaningful words. The order of the stimulus words are given here after conducting the experiment (See Appendix)

Sl. No.	Paired associates and stimulus words	Response word	Remarks
1	2	3	4
1	Minor – Few S. Minor		
2	Brief – Hideous S. Brief		
3	Wicket – Quite S. Wicket		
4	Kindly – Guilty S. Kindly		
5	Modest – Satirical S. Modest		
6	Envious – Gabby S. Envious		
7	Furious – Perishable S. Furious		
8	Quadruple – Metrical S. Quadruple		
9	Rebellious – Theoretical S. Rebellious		
10	Vituperative – Generous S. Vituperative		
	Number of Correct Associations		

Note : S : Stimuli words. Write the reponse word given by the participant against the stimulus word in column 3.

The results should be presented in Table 6.26. State whether there was any change of response for any stimulus and state how many words the participant associated correctly with the stimuli. Did you observe whether the changed order of the stimulus words had any effect in the participant response? In Table 6.26 enter the response words the participant has given for each stimulus word. This will help the experimenter to analyse the answers. What sort of response words he/she is associating with what sort of stimulus words.

Collect the group data from other experimenters and present them inTable 6.27. Work out the mean immediate memory span for the group for meaningful words in the method of paired-associate learning.

Table 6.27 : Showing the group data of the experiment of immediate memory span by the method of paired-associate learning for meaningful words

S. No.	Name	Immediate memory span	Remarks
1	P.K.		
2	S.R.		
3	M.S.		
4	K.R.		
5	P.V.		
6	T.V.		
7	P.C.		
8	A.R.		
9	A.L.		
10	V.R.		
	Mean		

Discussion

State the participant's immediate memory span according to the method of paired associate learning for meaningful words. State whether you have noticed any specific characteristics of your participant while performing the task. What is the mean memory span for the group. Who has achieved the highest and who has the lowest. State their personal characteristics. Where your participant stand in the group? Are the results in agreement with the results of previous studies or differ. If so, state the reason. Are the members of the group differ in their memory span on the basis of their age, sex, qualification or due to any other factors. What sort of effect the order of change had in the subject's response when they were tested.

Conclusions

1. What is the memory span of your participant in paired-associate learning method for meaningful words?
2. What is the group mean of memory span?
3. Whose memory span is highest and whose is the lowest?
4. Is there any variation in the memory span due to sex or age?

APPENDIX-1

(i) **Meaningful Words :** Material for paired-associate learning method. At the time of presenting the pairs in the memory drum put them in Boxes as shown below :

1. Minor – Few
2. Brief – Hideous
3. Wicket – Quite
4. Kindly – Guilty
5. Modest – Satirical
6. Envious – Gabby
7. Furious – Perishable
8. Quadruple – Metrical
9. Rebellious – Theoretical
10. Vituperative – Generous

(ii) The stimulus words in the following order should be presented immediately after the above paired word are presented to the participant. The order should not be the same as given above for administering the test. Therefore, the following order may be used.

1. (5) Modest
2. (9) Rebellious
3. (2) Brief
4. (3) Wicket
5. (7) Furious
6. (1) Minor
7. (10) Vituperative
8. (4) Kindly
9. (6) Envious
10. (8) Quadruple

EXPERIMENT 15 : SERIAL LEARNING : NON-SENSE SYLLABLES

Learning in the common sense world goes on several names. Here we are using the term learning in the meaning of memorising, because in any memory experiment a person is asked to learn a given material and reproduce it. He at once puts himself actively in memorising the passage. He will learn something if the material is unfamiliar and new. Since unfamiliar materials are not wholly found, meaningless words or trigrams or non-sense syllables and numbers are used. In memory tests meaningful and familiar materials are also used if lengthy materials are to be learned or memorized. The basic idea in such tests is that the subject should make some errors in the reproduction of the material. If the passage is lengthy this aim is achieved. However, the results obtained from different individuals with meaningful material would not be having equal influence in everybody.

But the non-sense syllables or trigrams such as DUX, TUK, WIZ, YEG, KIQ, XAN may have equal difficulty to all the subjects as they are not familiar to them. Ebbinghaus believed that everyone will start with a blank mind in learning the non-sense syllables. He has designed 2300 non-sense syllables for his experimental purpose. Meaningful materials are easily memorized and recalled them the non-sense syllables. The other factors in memorizing meaningful materials are rhythm and rhyme. This is the reason we more easily memorise poetry than prose passages. Meaningful prose passage is more quickly memorised than meaningless words. The materials which are interesting to us are more quickly memorized than the materials which are not interesting to us. Psychologists tried to discover the principles that are operating in these processes. Any learning in which language is involved is called verbal learning. We learn everything only through language. Therefore, any learning that is determined by language behaviour is called verbal learning. There are three simple forms of learning widely used in experiments. They are (1) paired-associate learning, (2) serial learning, and (3) free learning or rote memorization.

We have discussed in detail about the paired associate learning and also described few experimental procedures. Here we shall deal with serial learning. The method of serial learning differs from paired-associate learning method in certain respects. Whether we use numbers or non-sense syllables or meaningful words they are presented in a chain like manner. In serial learning the items are held constant in presentation from trial to trial and the participant should call the items in the same order. Then only it is correct.

In the method of serial learning the subject is presented with a series of items. Each item is presented individually for short period of two seconds. The subject is asked to remember the order of the stimuli. The stimuli may be numbers, or non-sense syllabus or meaningful words. The same apparatus, the memory drum can be used for presenting the stimuli. A familiar example of serial learning is memorizing a poem, or a musical note or a multiplication table. A person who learns a musical composition, learns it in a sequence of 1, 2, 3, 4, 5, 6, 7, 8, 9, 10 and so on. He learns it serially. We can observe that in the method of serial learning there is also the elements of paired-associate learning. While learning the material according to serial learning the person is learning to pair 1 and 2, 2 and 3, 3 and 4 and so on. The important difference between the paired-associate learning and serial learning is that in the former stimuli are presented in pairs whereas in the latter stimuli are presented individually.

The individual at the beginning of the experiment might have had some acquaintance or might not have had with any of the individual items. He is having acquaintance with numbers and meaningful words everyday. But not with non-sense syllables. These may be new to him. They are also new to everybody. But whatever his experience with numbers or meaningful words or non-sense syllables his task here is to connect the words with each other. The individual may memorise anything like a list of non-sense syllables, a piece of poetry, or a prose passage or it may be a list of numbers. All these are examples of serial learning.

In memorising poetry and prose the whole series is characterised by meaning and there is also associations between members of the materials. But in the case of non-sense syallables each component part has no meaning, but the whole series has something — it is formation of association in memorizing between the members of the materials of the non-sense syllables. Therefore, the study of verbal learning is more concerned with conditions that bring association between members of the series. This process of association is taking place mainly due to practice or relearning. Investigators have designed different methods of presenting the stimuli for serial learning experiments.

In the method of serial learning a full list of items is presented to the participant one by one at a regular rate of two seconds per item for learning. The full series of material is presented once and then the participant is asked to state the item before it appears in the window. The use of non-sense syllables will have equal difficult effect to all subjects. Therefore, a series of non-sense syllables may be constructed for this serial learning. The symbol *X* is given at the top of the list. The *X* symbol serves as a cue for the first syllable of the list.

The memory drum apparatus is used for this experiment. The list of non-sense syllables, the number of syllables may be from 8 to 18, in printed or typed form with equal interval, should be pasted in the drum in serial order. Young (1968) and Glanzer and Peters (1962) suggest eight words. Morris (1973) states that there should be twelve non-sense syllables. Atkinson (1957) used eighteen unrelated words. Therefore, there is no fixed number of items that can be used in this experiment. It is for the experimenters to decide as to how many words should be used in his experiment. The drum will move at the rate of two seconds for every item. Then every item is exposed for two seconds. The *X* symbol will appear first in the window and after two seconds the first item will appear and the subject has to learn and memorize it before a second item appears in the window. After two seconds a second item appears and the subject has to learn it and expect for a third item. In the same manner all the 12 non-sense syllables should be presented once to the subject. It is supposed at this stage that the subject has learned all 12 items as they were all exposed to him each item for a period of two seconds.

Now the participant should be instructed to watch the window carefully. As soon as he sees the *X* symbol he should say what item is going to appear next. He is expected to say using the *X* symbol as a cue to the first item. The subject may be asked to write down or speak out the item quickly before it appears in the window. The subject has given the correct item or not, he can verify it when the correct item appears next. Now the first item is a cue to the second item and he has to state the second item and write it down before it appears in the window and he can verify whether he is correct or not. In the same manner all the items of the list should be presented once. After presenting all the items once there should be short rest period for six seconds before starting the test trial. The participant has no basis for making any correct response in the first trial because he/she has not seen the item previously. In the subsequent trials he can make more and more correct responses. Every time when the item is presented it is both a relearning and a test of retention. This procedure in this manner is repeated till the subject could reproduce the whole list of non-sense syllables correctly in the same order in one or two successive trials. Some suggest two errorless series of reproduction. The only the subject is said to have learned the list fully. This method is also called the method of anticipation as the subject anticipates next item.

EXPERIMENT : SERIAL LEARNING : NON-SENSE SYLLABLES

Problem

To study the effect of serial learning in serial anticipation method in a person for non-sense syllables.

METHOD

Participant

Choose a person who has no knowledge about this experiment. Collect the personal information such as, age, sex, qualification, occupation, whether he/she has any handicap or wear eye glasses.

Materials

(1) A list three-letter non-sense syllables. The list can have twelve syllables. In the list just above the first syllable one *'X'* mark should be made as the indication that syllables are going to appear below. (2) A memory drum with standard exposing device of the verbal materials and accessories. (3) Stop clock.

A strip of paper in which the twelve non-sense syllables are typed or printed with equal spacing between words and should be pasted around the drum. The drum should move in constant speed and expose each syllable for a short period of two seconds through the window. The experimenter can provide the participant sheets of paper and pencil if he/she has to write down the responses (non-sense syllables).

Materials for Alternative Experiments

(i) This experiment may be also conducted *with meaningful words.* A list should have meaningful unrelated words. The number of words may be 8 or 12 or 18. The words should have uniform difficulty level. It may be also in the form of increasing difficulty. (See Appendix-2).

(ii) The serial learning may be studied by the anticipation method *using also numbers.* The numbers should have equal difficulty level. There should be a cue number and it should be presented first before the serial learning items appear in the window of the memory drum (See Appendix-3).

Procedure

The experimenter should make preparation for this experiment with all requirements. Then he should make the participant to be seated comfortably in front of the memory drum in a well illuminated room. There should not be any disturbances or distractions visually or auditorily. Then he should provide the participant with a sheet of paper and pencil and give him the following instructions :

"This is an experiment on learning. A series of non-sense syllables will be shown to you through this window (showing the window). The non-sense syllable is a three-letter word, has no meaning. First of all you must look at these syllables carefully. I will present to you all the syllables once only for a short period of two seconds. First one *'X'* symbol will appear in the window. It is the indication that the syllables are going to follow one by one. Therefore you must anticipate the syllable in two seconds. When you see the symbol you must speak out/write down the first syllable and when the first syllable appears you must speak out/write down the second syllable. When the second syllable appears you must speak out/write down the third syllable. In the same manner all the items of the list will be presented to you. Please remember, your task is to speak out/write down each syllable before it appears in the window. We will have to continue this procedure till you speak out/write down all the syllables in the same trial."

Then the experimenter should verify whether he/she has understood the procedure correctly. He can also allow the participant to ask any questions, if he has any doubts. Having fully satisfied with the instructions and the participant's readiness the experimenter can start the memory drum to move at the standard rate of two seconds interval. At the time of starting the experimenter can give the caution : "Ready go".

The experimenter should also decide the criterion whether he should have all twelve syllables correctly reproduced or can be reduced to 7 or 8 or 9 or 10 and to have one errorless trial or two or three. He also must decide whether he wants to have the participant to write down the responses or speak out them. If writing he must provide the subject with sheet of paper and pencil.

After having presented all twelve syllables once according to the procedure, the experimenter should watch the first trial and tell him: "Now the first trial begins. You must carry out the task as you have been instructed". As and when the participant speaks out/writes down the syllable the experimenter should keep a list of syllables and verify whether it is correct or what syllable the participant reproduces. It is better to have two successive errorless trials. At some stage if the participant was not able to reproduce all the twelve syllables in two errorless successive trials the experimenter can reduce the number of syllables to 7 or 8 and on two successive errorless trials he can stop the experiment. The experimenter can collect the sheet of paper in which the participant has written down the syllables. He can get the introspective report from the participant and allow him to go away.

Alternative experiments

The same procedure may be followed for conducting experiment for meaningful words and also for numbers in all respects.

Results

The experimenter, should scrutinize the scoring sheet and present the results in Table 6.28. He should show what sort of wrong syllables the participant has given for each syllable. In the table the correct anticipation or reproduction may be indicated by a check mark (√) and an incorrect response by a cross mark (×). It is also better to state the three letter syllable, so that we may know the type of syllables he is getting confused. How many syllables he/she has reported in the first trial? In the next five trials are the syllables increasing or decreasing? What is the total number of errors in all the trials? How many trials the participant takes to reproduce all the syllables without error. Where he/she commits more errors and what are the letters, the consonants involved in them. Draw a curve for the individual data using the number of trials and number of words for each trial and number of errors in each trial.

Table 6.28 : Showing the individual's response in the serial anticipation method of learning experiment for the non-sense syllables

S. No.	Non-sense syllables	Trial numbers												Trials to complete the task	Total errors
		1	2	3	4	5	6	7	8	9	10	11	12		
1	FHE														
2	QAM														
3	CIJ														
4	DUY														
5	HOX														
6	GAH														
7	JIS														
8	XAD														
9	WUF														
10	KEJ														
11	WOX														
12	QUV														
Number of words in each trial															
Number of errors in each trial															

Table 6.29 : Showing the group data in the method of serial learning experiment for non-sense syllables

S. No.	Name	Number of trials to complete the task	Number of errors on the whole	Percentage of errors
1	R.S.			
2	B.V.			
3	M.S.			
4	K.R.			
5	P.K.			
6	A.L.			
7	R.V.			
8	K.B.			
9	M.R.			
10	T.V.			
11	G.G.			
12	T.A.			
	Mean			

Collect the group data from other experimenters and present them in Table 6.29. Present the trials each person has taken to complete the task without error in Table 6.29. Also state which syllable has taken more trials from the participants.

Draw a graph using the non-sense syllables on the X-axis and the number of trials on the Y-axis.

Discussion

Discuss the number of trials your participant has taken to reproduce all the non-sense syllables without error. Was he/she benefited by the feedback system of the experiment. At which stage he/she committed more mistakes, in the beginning, or in the middle or at the end. Did you observe him/her struggling with any specific type of syllables.

What is the group mean trials? Who has taken largest number of trials and who has taken lowest number of trials? Is there anybody who could not complete the task successfully? State the reason for the same. What is the mean error of the group for the whole syllables. In which syllable the mean errors are higher and in which they are lower. Who has committed more errors, how many and who has committed the least, how many? State the reason if any, what are the personal characteristics of the subjects that might have caused more errors and more trials. Is there sex variation. Discuss the graph. What type of shape the curve takes place? Why. Explain.

Conclusions

1. After how many trials your participant has succeeded the errorless anticipation.
2. What is the group mean trials for completing the task?
3. Who has taken more trials and who has taken less trials?
4. Who has committed more errors and who has committed the least errors, how many?
5. Is there sex variation or variation due to age, qualification.

APPENDIX – 1

Non-sense Syllables to be used as material in the Serial Learning experiment in the anticipation method.

X
FEH
QAM
CIJ
DUY
HOX
GAH
JIS
XAD
WUF
KEJ
WOX
QUV

The experimenter can reduce the number of non-sense syllables if he wishes, according to situations or according to the capacity of the participants. But the minimum must be *eight* items and maximum *eighteen.*

APPENDIX – 2

Meaningful unrelated words to be used as material in the serial learning experiment in anticipation method. These words are adjectives. In the top of the words there should be a signal indication *'X'* to show that words are following below. Among these words the experimenter can select any number of words required for his experiment.

X

1.	Lean	21.	Nervous
2.	Major	22.	Parted
3.	Optical	23	Sagacious
4.	Passive	24.	Utter
5.	Reticent	25.	Negative
6.	Throaty	26.	Thumping
7.	Uncertain	27.	Wakeful
8.	Vacant	28.	Zealous
9.	Yester	29.	Lactic
10.	Lively	30.	Degenerate
11.	Juvenile	31.	Scholastic
12.	Horrendous	32.	Vaulting
13.	Fusible	33.	Bumbling
14.	Documentary	34.	Involuntary
15.	Commensurate	35.	Grumpy
16.	Bounden	36.	Exorbitant
17.	Mural	37.	Contemptible
18.	Quick	38.	Marshalling
19.	Rakish	39.	Circumstantial
20.	Anthropoid	40.	Pre-requisite

APPENDIX-3

Numbers to be used as material in Serial Learning Experiments in anticipation method. Since it is serial learning all the numbers or words should be reproduced by the subject. This is *not immediate memory span test*. Therefore, the numbers can be with any number of digits ranging 5 to 7 or of the same digits. But the number of items may be 8 to 12. Trials can be repeated till the subject reports correctly all the number included in the list. The experimenter may select the number of items required for his purpose. First a signal *'X'* should be displayed to indicate that numbers are following :

X

1. 6 2 6 7 8 7
2. 8 2 8 2 3 8
3. 7 3 7 3 4 6
4. 5 9 6 4 7 6
5. 4 3 8 5 4 6
6. 3 4 6 7 8 7
7. 3 6 3 5 7 9
8. 9 4 5 7 3 4
9. 2 2 2 6 4 5
10. 1 7 6 9 8 4
11. 7 2 4 6 4 8 7
12. 7 4 7 6 7 8 6
13. 6 7 8 5 4 3 2
14. 2 6 5 3 4 2 5
15. 9 7 5 4 7 7 6
16. 8 4 8 6 3 5 3
17. 7 3 7 6 6 9 3
18. 2 9 6 3 6 4 7
19. 6 4 6 7 8 8 5
20. 5 2 5 6 8 7 9

EXPERIMENT 16 : SERIAL LEARNING : MEANINGFUL WORDS

Research studies that were carried from 1900 on verbal learning have produced innumerable amount of information about various aspects of learning. The information were dealing with the conditions that contribute easy learning, learning methods, conditions that affect learning and cause forgetting. In the study of learning psychologists have used simple verbal materials and numbers. They were mainly list of meaningless words, non-sense syllables, or meaningful words, numbers, and numbers and words or words and numbers. Psychologists sometimes preferred for their experiments on learning verbal units with which the subject is fully unfamiliar. They found that numbers and non-sense syllables had the same effect in degrees of difficulty.

The three types of verbal learning that were receiving much greater attention were paired-associate learning, serial learning and rote memorization or free learning. In the preceding sections we have dealt with paired associate learning for the measurement of immediate memory span for numbers, non-sense syllables and for meaningful words and serial learning with non-sense syllables. Here we shall deal with serial learning with meaningful words.

Serial learning means, learning a list of words which must be learned in the same order in which they are presented. Several variables have been found associated with verbal learning. One of the important variables is the meaningfulness of the material to be learned. It is defined in terms of the number of other units that are associated with a particular word. For example people may associate many words with 'education' than with the word 'epoch'. Therefore, words like 'education' are considered to be more meaningful. More meaningful words have several advantages in verbal learning. Hilgard and Bower (1975) have stated that verbal learning studies were carried out largely by a group of psychologists called functionalists during the years of 1900 to 1930. The central idea they had in mind was to perform detailed experimental analysis of important learning skills. Therefore, they wanted to analyse closely any given task like serial learning.

They thought that the task of serial learning may be dissected into a number of components and be experimentally analysed. The researchers who made experimental analysis about learning have thought that it is wise to avoid the theories that are associated with controversies in animal learning. The researchers who were engaged in human verbal learning thought that the empirical analysis of specific learning task may reveal truth if studies were made patiently (McGeogh and Irion, 1952).

General associationism is the background for verbal learning research. Ebbinghaus has started his work on memory only with this associationism. The basic idea he had was that remembering could be reconstructed through associations among ideas. These associations are recorded in memory by frequent occurrence of two ideas. The frequent occurrence or contact is caused either by time or space or the external events that could arouse similar ideas in memory. By whatever means the association occurs the associating experience of ideas *X* and *Y* together was taken for granted to establish an association between them. It may take place internally. The association between *X* and *Y* may vary in strength. Some elements may be strongly attached easily with *X* and not with *Y*. Similarly some elements may have strong attachment with *Y* and not with *X*. The stronger the *XY* association the more it may win in competitions with alternative responses.

Serial learning is also one form of serial memorization. Memorizing poetry, memorizing some lines of dialogue of a play, songs, musical notes all may be called serial learning. In memory experiments, in laboratory a list of words is memorized from the beginning to the end. In this form of learning each word in the list is serving also as a stimulus to the next word to follow. In serial learning the position of the item in the list is one of the most powerful variable that determine the rate of learning.

The first few items in serial learning are learned more rapidly and the last few items also are learned more rapidly. But the items in the middle and also beyond the middle are learned least rapidly or slowly. Sometimes incorrect responses also given. This process of occurrence is known as the *serial position effect*. A single word can be used as a serial list of letters. Jenson (1962) has used a single word as a serial list of letters in his experiment. The subjects, instead of words as responses were asked to use the letters. The participant should spell a word correctly and the letters of the word should be learned in a particular sequence. Jenson has shown the serial position effect of letters within words. According to his study there were few spelling errors in the beginning letters of the words. There were fewer spelling errors at the end letters of the word. But the spelling errors were greatest for letters near the middle of the words.

The serial position effect which has taken place in a single word may also take place in the list of words used in serial learning and also in paired-associate learning. In the serial learning the participants may commit few errors in the first few words and still more errors in the end of the list. But in the middle they may commit more errors particularly near the middle of the list. This phenomenon can be tested in laboratory experiments conducted with a list of words in paired-associate learning experiment and in serial learning experiment. Glanzer and Peters (1962) have verified this phenomenon and found what was true with a single word as a serial list of letters also was found to be true with a list of words.

The method of serial learning may be also called serial anticipation method as the participant has to anticipate the next word appearing in the apperture of the memory drum. In serial learning method the participant is required first to learn a list of words when they are exposed to him. These words should be unrelated to each other. The number of words may be .twelve to eighteen. The words are presented to the participant one at a time in the window of the memory drum for only two seconds. When the first word is presented the participant should learn it and expect what should be the next word appearing in the window. After two seconds the second word appears and the participant should again learn it. In the same manner all eighteen words are presented once in the window and the participant should learn them all.

After presenting all the words once there should be a short rest period of six seconds. Then the experimenter should start the trial and ask the participant to watch at the window of the drum a cue or signal that the first word is going to appear and as soon as he sees the cue or signal he should tell the word before it appears in the window. On seeing the first word he must anticipate and tell the next word using the first word as the stimulus before it appears in the window. Then using the second word as a stimulus he should tell the next word before it appears in the window. In the same manner the participant should make correct anticipation for all the eighteen words.

Then the experimenter should check up the correct response words. If all are correct he can make another correct trial for all the words. If they are not correct the experimenter should carry out the trial till the participant make correct responses for two successive trials. The participant has no chance of making all correct responses as he/she has not had any familiarity with the words or seen the words earlier. He/she has just seen each word only once in the window of the memory drum. In subsequent trials he/she might have tried to make more correct responses. As and when the number of trials increase the number of errors also might have reduced and the number of correct responses increased. The experimenter can watch the difference among the beginning trials and the subsequent trials and the ending trials. A curve can be drawn to the number of trials and to the number of correct responses the participant has made.

EXPERIMENT

Problem

To study the effect of serial learning by serial anticipation method for meaningful words.

Method

Participant

Choose a person who has no knowledge about this experiment. Describe his/her personal characteristics such as age, sex, qualification, occupation, experience, whether he/she has any handicap, wear eye glasses and so on.

Materials

(1) Memory drum with all accessories. (2) A list of 18 meaningful and unrelated words. (3) Stop clock.

Procedure

The experimenter should make ready to conduct this experiment with all the materials required. He/she must also check first whether they are in order. He/she must have the appropriate scoring sheet for making entries of the responses. The experimenter should adjust the memory drum that it moves one step for every two seconds. Then he should make the participant to sit comfortably in a chair in a well illuminated room. There should not be any visual or auditory disturbances or distractions. Having first satisfied with the experimental conditions the experimenter should give the following instructions to the participant :

> "Please look at this window of the memory drum (pointing out the place). I will first present one signal or cue in this window for two seconds. You must anticipate a word that is going to appear in the window. As soon as the signal disappears the word will appear only for two seconds. You must use this word as a stimulus and anticipate the next word. As soon as the first word disappears a second word would appear for only two seconds. In the same manner eighteen words one by one will appear only for two seconds. Your task is to learn every word and remember it. When all the eighteen words are presented to you, after a rest period of six seconds, I will start presenting the materials in the window once again. The signal will appear first in the window. As soon as it appeared your task is to call the first word at once and when it appeared you must use it as the stimulus and at once call the second word before the second appears in the window. If the same word what you called appeared in the window you are right, if some other word appeared you are wrong. You have committed an error. That is you did not call the correct response word. If you have delayed in calling the response word and if that word has

appeared in the window then also you have committed an error. Therefore, you must call the response word before the appearance of the next word."

"You must carry out this task till all the eighteen words are correctly reproduced without any error in two successive trials. Now you must start your task. Now, ready go."

The experimenter at once start the memory drum to make every movement at the rate of two seconds and also start the stop clock. He must maintain the time also. The experimenter should make a note of every call of the response made by the participant. When all the eighteen words are presented once he must stop the task for six seconds and then start the trial. He must continue the trials till the participant makes correct responses of all eighteen words for two successive trials. As soon as it takes place he should obtain the introspective report from the participant and allow him/her to go away. The experimenter, then check and scrutinize the errors and the number of trials.

Results

The experimenter should work out the number of errors in each trial and also the numoer of trials the participant has taken to make two successive correct responses. These results should be presented in Table 6.30. In the scoring sheet the correct response may be indicated by a

Table 6.30 : Showing the individual's response in the serial learning experiments by the method of anticipation for meaningful words

S. No.	Meaningful words	Trials : Words recalled															No. of trials for correct response	No. of errors to SL position
		1	2	3	4	5	6	7	8	9	10	11	12	13	14	15		
1	Sordid																	
2	Tender																	
3	Listless																	
4	Pedestrian																	
5	Vital																	
6	Quiet																	
7	Ashore																	
8	Embattled																	
9	Instant																	
10	Negligible																	
11	Worse																	
12	Haughty																	
13	Yeasty																	
14	Determinate																	
15	Melodious																	
16	Clear																	
17	Judicial																	
18	Occult																	
No. of words recalled																		
No. of words not recalled																		

Note : The experimenter, if he wishes, can reduce the number of meaningful words to 12 or 10, depending upon the experimental situation. Find out the number of words recalled in the first trial, in the first 5 trials, in the last 5 trials and in the middle trials.

check mark (√) and an incorrect response by a cross mark (×). It is also better to note down the incorrect response word the participant has made for a stimulus. This will facilitate the experimenter to know at which place the participant finds difficult to make correct responses. Draw a curve for serial position of the word and number of errors for the individual data.

The group data may be collected from other experimenters and the results may be presented in Table 6.31. The mean errors committed by the group and the mean trials for the group also may be worked out.

Table 6.31: Showing the group data of the serial learning experiment by the method of anticipation for meaningful words

S. No.	Name	Number of trials for correct response	Number of correct response in trials				Number of errors for Serial position	Number of errors in all trials
			Item 1	First 5 Items	Last 5 Items	Middle Items		
1	P.K.							
2	K.R.							
3	M.S.							
4	R.T.							
5	S.L.							
6	M.R.							
7	S.N.							
8	R.V.							
9	K.V.							
10	V.R.							
11	C.P.							
12	S.L.							
	Mean							

Note : ***Serial position effect*** **means that the position of each word may have different effect in recalling. Words in the first few positions and words in the last few positions may have some advantage in recalling. But words in the middle may cause trouble in recalling and they may cause errors. This is called serial position effect.**

Work out the mean number of trials for the whole group, mean errors for all the trials and also mean number of errors for serial position of the items and for the first trial, first 5 trials, last 5 trials and middle trials. Draw a curve for the group data using number of trials for correct response and number of errors in all trials.

Discussion

State how many trials your participant has taken to give correct response of the words. State the number of errors in each trial and also the number of errors for serial position of each item or word. At which position he/she committed more errors, *i.e.* in the beginning words or in the end words or in the middle words. Has your participant given any words other than the words used in the experiment? If so state what sort of words he has produced. Has the participant used the stimulus to call the response word accurately? Has he/she followed the directions in the task? Was the anticipation method helpful for him/her for serial learning? What is the shape of the curve? Why it is like that? Explain.

What is the mean number of trials for the group? Who has taken less number of trials and who has taken more number of trials? Has anybody failed to complete the task? Where your participant stands in the group? What is his/her rank in the group? For the whole group which word has taken more number of trials to evolve correct response? State the reason. Who has recalled more words in the first trial, how many, who recalled the lowest number of words, how many. Whether there is any difference in the successive trials, first 5, last 5, and middle trials.

Are there any influencing variables in this experiment in serial learning? What personal factor do you find having influence in serial learning? Is there sex difference? Compare your results with the results of earlier studies given in the introd'ctory section. What is the shape of the curve for the group data.

Conclusions

1. How many trials your participant has taken to complete the task?
2. What is the highest number of trials a subject has taken and what is the lowest number of trials.
3. How many of them above the group mean in the number of trials.

APPENDIX – 1

Meaningful Words for Serial Learning Experiment by the Method of Anticipation

After giving suitable instructions the signal or the cue should be presented first to indicate that words are following. The first trial is provided for learning. Then the subsequent trials should be accounted for the experiment. The memory drum should be used as shown in the procedure. The number of words may be reduced according to the experimental situation.

Caution : *'X'*

1. Sordid
2. Tender
3. Listless
4. Pedestrian
5. Vital
6. Quiet
7. Ashore
8. Embattled
9. Instant
10. Negligible
11. Worse
12. Haughty
13. Yeasty
14. Determinate
15. Melodious
16. Clear
17. Judicial
18. Occult

EXPERIMENT 17 : FREE LEARNING : MEANINGFUL WORDS

Human beings are capable of exhibiting larger amount of learned responses than rats, cats and monkeys. Human beings present more problems in learning than animals. The main reason is that most of the human learning is verbal learning. Verbal learning involves language and the human being should reveal everything he has learned verbally through language. Animals do not explain anything verbally though they make some sound. Whatever thorough knowledge we have about animal behaviour or learning it will not help us to explain human behaviour or learning. But a correct understanding of the principles of human learning may help us to some extent to conduct experiments on animals and to understand animal learning. Because the determining factors of human learning are so complicated than the determining factors of animal learning. As a result of this experimenters in human learning have evolved a variety of experimental tasks involving associative learning. A model has been given by Feigenbaum and Simon (1961) which is called *Elementary Perceiver and Memorizer*. This is stated in the abridged form as *EPAM*. This model of associative learning has aroused great interest among the psychologists.

This experimental task included paired-associate learning, rote learning, serial learning, recognition learning, immediate or short-term memory tasks and long-term memory task or learning to read text and the like. According to this model the material or stimuli may be presented through some instrument in any form either visually or verbally. Later on, Simon and Feigenbaum (1964) have revised their model which mainly dealt with learning a list of non-sense syllable pairs. According to this model when the subject is presented with a stimulus his task is to anticipate and produce the correct response. Later on experimenters have applied this model to various other variables also.

In verbal learning, it has been observed that several factors were found to be operating and each one determines the learner's ability in responding successfully to the stimuli. This has been tested by all the methods shown above. But psychologists who study human verbal learning do not show the same amount of interest to all the forms of learning. Candland (1968) says that among the variety of methods of stimulus presentation two have received greatest attention of psychologists. They are paired-associate learning and serial learning. Hall (1969) and others state three types of verbal learning, like paired-associate learning, serial learning and free learning or rote memorization. Some who have worked extensively in the field of learning, like D'Amato (1970) state that there are four main verbal learning tasks such as paired-associate learning, serial learning, free learning (free recall) and verbal discrimination learning.

We have discussed in the preceding sections the first two major types of learning. Here we shall deal with the free learning or free recall. In verbal learning the simplest verbal learning task is the free learning.

In free learning method the material or stimulus is generally a list of meaningful words or a passsage or a piece of poetry. The stimulus may be presented visually or verbally to the participant. The words or items may be ten to twenty. The memory drum instrument that is used in other learning experiments can be used in the free learning experiment also. The words may be printed or typed with equal spacing to suit the memory drum window. The memory drum should be adjusted in such a way that it makes one move for every two seconds exposing one word in the window. The list of words should be fastened in the drum. In the list in front of the first word there should be one cue or a signal like this *'X'* to indicate that test words are going to appear in the window. The participant should be instructed that he should watch in the window the signal and subsequently the appearance of the words and he should learn each word and when all the 20 words are exposed in the window there should be an interval time of six seconds. Then the subject should be asked to call or write the words he/she saw in the window one by one as it appears in his memory. And also he should be told that he need not present the words in the same order. He can produce them in whatever the order feasible to him. But he must produce all the words. If the participant fails to reproduce all the words in the first trial, trials may be repeated till he/she reports the whole list of words in some order once or twice.

Authors like Hall (1969) suggest to allow the participant to learn the material in whatever manner he/she wishes. According to Hall the list of words may be given to the participant and he/she may be asked to learn the list in any order. The time may be allowed at the rate of two seconds per word. When the subject has learned all the words once he/she may be asked to report as many words as possible in whatever order suitable to him/her. The free learning method was suggested by Underwood (1964).

Though many experimentalists have been using a single presentation of the words for recalling them, others have made changes in the performance. They preferred successive presentation or multi-trial presentation for free recall. In this procedure the order of presentation of the words to the participant may differ from trial to trial. But the participant is not restricted in the order of recalling the words for each trial. That is, the recalling of words need not be in the same order since the order of the presentation of the words from trial to trial are changed. Then in each trial the order of the report also the participant can change. Therefore, in free learning experiments there can be presentation of a single trial and also multi-trial or successive presentation for recall. This is to be decided by the experimenter.

In one of the multi-trial recall procedures Tulving (1966) has presented his subjects 22 meaningful words, mostly nouns. Each word is a common word having high frequency of words. These were selected from the teacher's word book of Thorndike and Lorge (1944). In Tulving's study there were twelve successive trials and the words were presented at the rate of one per second. Then the participants were allowed sixty seconds for recalling the words orally. The participants on the first trial have recalled an average of 9.25 words correctly. After each successive trial their progress was improved and at the end of twelfth trial the participants have recalled on an average of more than eighteen words out of 22. Tulving has observed that the participants have shown much interest in recalling the learned words. It was also found that they reorganize the words for recall in an orderly fashion which seemed to be related to the improvement they made in recall trial to trial in a multi-trial recall procedure.

From the analysis of free learning experimental studies it appears that there should be some relationship between the two process theory of memory and free learning. In the studies made by Deese (1972), Murdock (1974), Postman (1969) and Glanzer (1972) it was observed that the probability of recall varies with serial position of an item in a list. In a typical free learning experiment a list of unrelated words was presented to a participant one at a time at the rate of the standard time. After presenting all the words the participant tried to recall as many words as possible in any order he wished. From the result it was found that the probability of recalling each word was due to its function of its place in the list. The curve took a 'U' shape for the experimental results. The beginning words and the ending words in the list were recalled more rapidly than the middle or the near middle words.

This means that the increased recall for the first or beginning few words in the list was due to the *primacy effect*. The recall for the last few words was still higher than the beginning words in the list. This effect was attributed to *recency effect*. The recency effect was due to retrieval from short-term memory (STM). But if there is any other mental task between learning and recalling the material this retrieval effect in the short-term memory is lost. In other words the recency effect is eliminated when there was some mental work. If there was recall the retrieval was from long-term memory (LTM) only.

Experiments on free learning apart from the procedures explained here are conducted in varied ways. One such procedure is to present more than one list of words to the participant for recalling. More than one list means there can be three or four lists of unrelated words presented in one session. Craik (1970) in his experimental studies has presented many lists of words in one session in a free recall experiment to his participant. After presenting the words of all the lists at the end of the session the participant was asked to recall all the words presented. The assumption is that the recalling should be from long-term memory only. The result was that the recency effect of the serial position of the words appears to be the retrieval from both short-term memory and long-term memory.

EXPERIMENT (FREE LEARNING)

Problem

To assess the effect of free learning in a person for a single trial and for multi-trials for meaningful words.

Method

Participant

Choose a person who has no knowledge about this experiment. Describe his/her personal characteristics like, age, sex, qualification, occupation, experience and so on.

Materials

(1) Memory drum. (2) A list of meaningful unrelated words from 12 to 22. All the words should be nouns. (3) Stop clock. (For list of words see Appendix).

Procedure

The experimenter should make arrangements to conduct this experiment. A list of meaningful unrelated words should be arranged in printed or typed form with sufficient interval spacing between words. The memory drum should be adjusted in such a way that the drum moves one step after every two seconds. The list of words should be fastened in the drum and should have full visibility of the words in the small window of the drum. The experimenter should check up with the stop clock before starting the experiment whether the movement of the drum coincides with timing of the stop clock. Having satisfied with all this, the experimenter should make the participant to be seated in a well illuminated room before the memory drum. There should not be any disturbances or distractions either visually or auditorily. Then the following instructions should be given to the participant :

> "Please look at this window of the memory drum (showing point). This is an experiment on learning. I will present a word for two seconds and it is seen in this window. You must at once learn it and keep it in your memory. After two seconds another word would appear. Again you must learn it. In the same manner I shall present eighteen words, each exposing only for two seconds. As soon as it is over your task is to report as many words as possible either by speaking out or by writing them without taking much time. It need not be in the same order. You can report them in any order. When 1 say ready, go, the drum will move and you must start learning all the words one by one. When all the 18 words are presented to you I will stop the drum for few seconds. Then I will ask you to call out the words or if you wish you can write them on a piece of paper. Now you might have understood everything. As soon as I say ready, go, you must start your task. Now, ready go."

The experimenter at once should start the memory drum to move step by step exposing one word for every two seconds. He/she should watch the behaviour of the participant. If he is going to report verbally the experimenter should prepare himself to check the words with a list whether it is correct or not. If he is going to write it then he should check that he is not taking more time. When all the words are presented then the experimenter should stop the moving the memory drum. After six seconds he should ask the participant: "Now report those words you have seen in the window".

The experimenter should check up the number of words the participant has reported and note down them. He should also see the order he followed. There might not have been any chance for the participant to report all the eighteen words correctly as he has not seen them previously. Then the experimenter should start a second trial, asking the participant to learn the words once again and report them in the same manner in whatever order he/she can do it. Now the experimenter should give a second instructions to the participant: "In the first trial you have not reported all the words you have learned. Therefore, now we shall go on with more number of trials till you report all the words without error once. Now we start the second trial, ready go."

Then the experimenter should start the second trial, third trial successively as many trials as the participant takes to complete the task. He follows the multi-trial recall procedure. The participant on reporting all the words once successfully without error once the experimenter can stop the task and obtained the introspective report and allow him/her to go away. The correct responses may be marked with a check mark (√) and incorrect responses with a cross mark (×) in the scoring Table.

Results

The experimenter should, then, scrutinize the responses given by the participant for every trial and enter them in Table 6.32. How many trials he/she has taken to complete the task? Whether there is any difference between the first trial and subsequent trials? Has he/she followed any order of the words? Has he/she reported anytime words other than presented?

Find out the response for the first trial for each word and present them in Table 6.33. Find out also the number of correct responses for each word in the first 5 trials, last 5 trials and the middle trials and the number of errors on the whole for each word and put them in Table 6.33.

Table 6.32 : Showing the individual responses in a single trial and multi-trial free learning experiment for meaningful words

S. No.	Meaningful words	Trials															Number of trials for correct response
		1	2	3	4	5	6	7	8	9	10	11	12	13	14	15	
1	Diagram																
2	Mansion																
3	Pesticide																
4	Theatre																
5	Harvest																
6	Whale																
7	Rumour																
8	Plough																
9	Furniture																
10	Bicycle																
11	Soliloquy																
12	Apricot																
13	Jungle																
14	Valence																
15	Zenith																
16	Kettle																
17	Yeast																
18	Fertilizer																
No. of words for each trial																	

Table 6.33 : Showing the responses of the individual in Free Learning. Experiment for meaningful words in a single trial and multi-trial recalling procedures

S. No.	Meaningful words	Response in First Trial Correct/Not	How many correct responses in First Five Trials	How many correct responses in last Five Trials	How many correct responses in Middle Trials	Number of errors for each word on the whole
1	Diagram					
2	Mansion					
3	Pesticide					
4	Theatre					
5	Harvest					
6	Whale					
7	Rumour					
8	Plough					
9	Furniture					
10	Bicycle					
11	Soliloquy					
12	Apricot					
13	Jungle					
14	Valence					
15	Zenith					
16	Kettle					
17	Yeast					
18	Fertilizer					
	Total					
	Mean					

Find out the total for each category and also the mean. Compare the correct responses in first trial with the first 5 trials, last 5 trials and the middle trials and also the errors on the whole for each word for the individual.

The experimenter, from the responses in Table 6.32 may work out the number of trials the participant has taken for correctly reporting all the 18 words. He can also work out the total errors in each trial, so that he can see whether there was any progress in the successive trials. He can also work out the rate of reduction of the errors in the subsequent trials. The experimenter can collect the group data from other experimenters and present them in Table 6.34.

Table 6.34 : Showing the group data in a single trial and multi-trial free learning experiment for meaningful words

S. No.	Name	No. of words in Ist trial	No. of trials for correct response	Total No. of errors	No. of words in the Ist 5 trials	No. of words in the last 5 trials	No. of words in the middle trials
1	A.R.						
2	S.V.						
3	P.K.						
4	K.R.						
5	M.S.						
6	B.V.						
7	T.P.						
8	R.S.						
9	V.M.						
10	T.D.						
11	G.G.						
12	K.L.						
	Mean						

Work out the mean scores for the group in every category. Check whether there was any successive improvement after every stage.

Discussion

Discuss the individual's performance in the free learning experiment. Whether in free learning the free recalling procedure has facilitated the individual in recalling more words in each trial. How many words the individual has reported after the first trial out of 18 words. Whether there was any increase of words in the successive trials. After the first five trials how many words on the average the individual has reported correctly. What is the average number of words the individual has reported in the middle trials? What is the average number of words in the last five trials the individual has reported? The number of middle trials depends upon the total number trials the individual has taken on the whole to complete the task correctly. Had the serial position of the words caused any trouble to the participant. If so state the words.

What is the mean trials for the group? State also the mean error of the group? Who has taken highest number of trials and how many? Who has completed in lowest number of trials and how many? Where your participant stands in the group? What is his/her rank? Who has reported highest number of words in the first trial and how many trials he/she has taken to complete the task correctly? Who has reported the lowest number of words in the first trial and how many trials he/she has taken to complete task correctly? Compare these two persons throughout, *i.e.* in the first five trials, in the last five trials and in the middle trials and see whether there was any increase or decrease in their responses in the successive trials.

In the group who has committed the highest number of errors and how many? Who has committed the lowest number of errors and how many? Compare these two persons in all successive trials whether the high and low difference goes beginning with first trial till the end. What were the factors that seemed to have interfered with the participants free learning? Was there any variation in the performance due to sex, or other personal deformities. Compare your results with the results of previous studies shown in the introduction. Find out whether there was any primacy effect and recency effect in the responses of the subjects.

Conclusions

1. How many trials your participant has taken to complete the task correctly? How many errors he/she has committed?
2. What was the highest number of words reported in the first trial?
3. Who has taken highest trials and how many and who has taken lowest trials and how many?
4. What is the mean trial and mean error for the group?
5. Was there any improvement in successive trials generally for the subjects?

APPENDIX – 1

List of unrelated meaningful words to be used as material in Free Learning Experiments in a single trial and multi-trial recalling procedures. Here, there are 30 such words are given. The experimenter can use any number of words 12 to 22 in his experiment depending upon the experimental situation. The required number of words may be selected and typed with sufficient spacing between words to suit the step by step movement of the memory drum. If memory drum is not used the same list may be given to the participant and ask him to learn at the rate of two seconds per word. The experimenter must use stop clock in this case.

List of unrelated meaningful words (Nouns)

1. Harvest
2. Apricot
3. Tollgate
4. Soliloquy
5. Rumour
6. Bicycle
7. Pesticide
8. Mansion
9. Furniture
10. Groove
11. Plough
12. Theatre
13. Fertilizer
14. Stone
15. Nickel
16. Jungle
17. Kettle
18. Whale
19. Diagram
20. Zenith
21. Valence
22. Curtain
23. Yeast
24. Elevation
25. Gallantry
26. Orchard
27. Tarpaulin
28. Unitarian
29. Innovation
30. Laxity

EXPERIMENT 18 : THE EFFECT OF MNEMONIC STRATEGIES ON MEMORY

INTRODUCTION

Human beings are very much interested in remembering what they have learned and to retrieve whenever they wanted to make use of the learned materials. There are several methods of learning for this purpose. One of the methods formulated is the paired-associate learning. In this method an explicit stimulus is provided for each response term. A number of pairs are learned at the same time by the subject by using anticipation method. The stimulus is presented and the subject responds. The paired-associate learning has become very popular in human learning. The paired-associate learning has been divided into three components by the researchers in verbal learning. The first is, learning to discriminate among the stimuli; the second, learning the responses as units; and thirdly, learning to associate the correct response unit to each stimulus. McGuire (1961) has illustrated these processes in his experiments.

In paired-associate learning a kind of association formation is taking place. It is hooking up the discriminated stimuli in the whole response unit. This process has been used very much in research works. In the association formation the first concern is the time course of formation of an association over successive practices. Whether there is any increment in the practices or all are learned in one trial or none is learned. Hull in his experiment on verbal learning observed that there was gradual increment in learning. Guthire and others have expressed the other view, all-or-none view. The other issue in association formation is the role of mediators. It is like constructing a bridge between the stimulus and response. The mediator is referring to a bit of knowledge or already existing association that the subject has come with a help to learn the pair of words. The subject might say that he used certain aid. For example, to learn the pair HOX-COCK, GAH-GOAT. The non-sense syllable HOX is remembered as COCK and the non-sense syllable GAH is remembered as GOAT. These mediators indicate a person's knowledge to find out a link between the stimulus and the response. The use of mediators is called mnemonic device.

WHAT IS MNEMONIC DEVICE?

Literally mnemonic means aid. That which is used to aid one's memory. Mnemonics is also meant the act of practice of improving or aiding memory. It is a system of rules to aid the memory. Psychologists who were studying learning and memory have devised ways and means of learning difficult materials and to recall them whenever one wanted to make use of the materials. The student of psychology is quite familiar with paired association method in learning. Ebbinghaus used non-sense syllabus in his experiments to eliminate the possibility of the meaning of the words having any influence in learning the non-sense syllables. How to remember non-sense syllables? It is by transforming them into familiar words that non-sense syllables could be learned. It is also observed that this type of measure of word-making correlated highly with the recalling ability of non-sense syllables in verbal learning tasks. This method suggests that the learning persons convert non-sense syllable into a word and uses process of transformation and store this code in memory. This is decoded at the time of recall. In this way the persons who learn non-sense syllables store the codes in memory and with the help of them recall the materials. In the case of HOX-COCK and GAH-GOAT, persons may even recall COC rather than HOX, GOA rather than GAH without the last letters K and T of the words.

These mediators occur in learning and the learners may use them in whatever manner convenient to them. Particularly in learning new and unfamiliar materials the learners can easily assimilate them into something familiar by using mediators. For example, a person while learning a pair of non-sense syllable DUY-GOAT may use two partly familiar connections, associating D with tiger and can state it as tiger-GOAT. Tiger stands for DUY. By mentioning tiger the non-sense syllable DUY is remembered. In this the old association is chained with the new one. This procedure is more efficient in learning than learning new association directly. Similarly in learning the pair XAD-COCK the learner may use two partly familiar connections X-fox and can state fox-COCK. The fox stands for the non-sense syllable XAD. When the word fox is mentioned the person at once remembers the non-sense syllable XAD. In these examples the non-sense syllables DUY and XAD can be easily recalled by this method of association. Tiger and fox are used as aiding words for recalling the non-sense syllables DUY and XAD.

In learning new pairs of meaningful words also the same method can be adapted, but little differently. The pair of words can be made into a meaningful sentence. The two concepts can be linked into a possible interaction. The tiger is hunting the goat, the fox runs after the cock. This type of mediators improve paired associate learning. This type of action oriented linkings may serve as better mediators in learning than simply linking by conjunctions. The learner may also have visual images of the interaction. This may greatly facilitate the learner to store the materials in memory. Bobrow and Bower (1969) state that when the learners generate their own sentences or their own images the benefit is greater than using the experimenters sentences. Mnemonic devices are the techniques for deliberately storing the material to be learned into a form suitably associated through familiar concepts. It may be done by sentence generating or generating visual imagery.

The mnemonic techniques have been found quite effective in improving the memory. Bower (1970) and Cermek (1975) have experimentally verified the effects of these techniques in the laboratory situations. The development of cognitive theory and cognitive psychology made several innovations in the field of learning and memory. The Gestalt psychologists state that organization of the learning material determine the strength of memory and recall of the learned material.

The mnemonic devices are found to be good methods in helping us to remember materials quickly and easily. It is a method of organizing materials in a convenient manner. A person, in mnemonic device, consciously puts certain system or order in the material he learns. He can use meanings, rhythms and groupings as means of organising the material. It is said that the ancient Greek and Roman orators were using the mnemonic devices to remember long speeches as they have to deliver them without written notes. A simple mnemonic device is making sentences or words out of the material to be recalled.

Mnemonic devices are the techniques by which material is made easier to remember. It is a system of improving memory. The technique involves a set of symbols that can substitute the material to be remembered. For example, a number can be used as a substitute for a word. The learning material is organized in such a way that it helps us to remember the material easily and quickly. The learner when uses the mnemonic device he purposely puts some order in the material to be learned. He may use any systematic order. He may use meanings or groupings or rhythm as means.

The simplest mnemonic devices are rhythms and jingles to remember the data and facts. The colours of the spectrum are remembered if we know the word VIBGYOR. The word stands for colours like Violet, Indigo, Blue, Green, Yellow, Orange and Red. Some of the psychological tests are remembered in the same manner. For example, tests like MMPI, FACT, EPPI, DAT, TAT all remembered by some mnemonic devices.

The organizational process has been revealed in the free recall of the learned materials by Mandler (1967, 1968) and others in test situations. Studies on free recall have stated repeatedly that the persons who learn new materials systematically group words together during recall, even though these words were presented to them in a haphazard way. The learners grouped most often on the basis of the similarity of meaning of the words. But simply exposing a person to a list of words without proper instruction to learn does not promote good free recall, whereas allowing the person simply to sort out the words into categories on the basis of the similarities of meaning of the words results in good incidental recall (Mandler, 1967). All these results show that learning is greatly enhanced by organizing the materials to be learned. Hence there is growing interest in organization of materials in learning and memorising. Therefore, it has become necessary for the psychologists to make use of the mnemonic devices and to analyse the effect of them.

Mnemonic devices are the techniques that are enabling a person to enrich upon the material to be learned. The devices are only helping the person to remember the material strongly. The techniques also help the person to classify and organize the material. There are several mnemonic devices formulated on the basis of the type of materials to be recalled or on the basis of requirements. Bower (1970) has deviced one mnemonic device which has been analysed extensively. It is the 'pegword' method. This method can be used in learning lists of items such as list of shopping articles, historical events, points in a speech to be delivered, legal procedure codes to be quoted in a court and so on.

In using the pegword method the person first learns a list of rhymic pegwords. They are paired with concrete images. They can be 10 or 15 rhymic pegwords. Then the person while learning a list of new items he uses the pegwords upon which he can hook the items of the list successively. As the first word in the list occurs the person is to call to his mind the first pegword and form a vivid mental image of a scene having interaction between the two objects, the pegword and the list word. The image may be odd and fantastic as the person wished at the time of the event. The same procedure is followed for each of the list of words, imaginatively pairing. The person when he wants to receive the list of words in his memory, is needed only to go through the list of pegwords which were already learned well and he tries to recall the imaginary scene created upon the pegword image. When the imaginary scene is remembered he can identify the desired word in the list. This type of recalling the learned words is found to be very effective compared to other normal recalling methods. The mnemonic device functions very efficiently, because there is a systematic retrieval scheme. The person himself has associated the pegwords with the list words and created a visual imagery between these two. Let us see a few examples of pegword and mnemonic image scene after Bower's model (see Table 6.35).

Table 6.35: Showing Four Items of shopping list and the Four Pegword images

Pegword	Listword	Mnemonic Image
One – Pin	Paper	Pin fastening together papers
Two – dairy	Cheese	Cheese is bought in dairy
Three – grease	Wheel	Grease applied to wheel
Four – ring	Jewellery	Jeweller sells ring

When the first list word 'paper' is presented, the learner is to call to his mind the first pegword 'pin' and form vivid mental image of the interaction of between the two objects. The mnemonic image the person has made in his mind is that 'pin fastening together papers', Therefore when the word 'paper' appeared at once in his vivid mental image the pegword 'pin' also appears. This is how the stored words are recalled by the learner.

Pegword method is one of the techniques of mnemonic device for learning a series of items. When you deliver a speech a series of points may be required to be quoted, or a lawyer in the court may be required to cite the laws relating to his case he is arguing, or you may be having a shopping list in your memory when you go for shopping or the student who writes the final examination may be required to reproduce the related points in sequence — these are some of the examples in which the pegword method can be applied. What is required in the pegword method, is first of all the person has to memorise a list of ten or more pegwords using some system, particularly the rhymic system, associating the number rhymically with a word. In the examples the rhymic association is shown 1- pin, 2- dairy, 3- grease, 4- ring. These words are then used as pegwords on which the images are hanging and stand for recalling the successive list words. The tactic is to relate or associate the image of the first pegword with a concrete image of the first list word. The second image of the pegword with the second listword and so on. As the listwords are presented the person should take *five seconds* for each item for creating the image between the two words. After the presentation is over the learner may recall and recite to himself the list of pegwords and verify what did he visualized with each one, *i.e.* pin, dairy, grease and ring.

Mnemonic devices are guiding methods to improve one's learning and recall. They can be used to solve memory problems in practical situations. The mnemonic methods vary with the nature of materials to be learned. Apart from the method suggested above for the materials stated, in learning telephone numbers the method is different. The telephone numbers may range from 6 to 8 in our country. These numbers may be broken into small units of 2 or 4 digits. Then each digit may be converted into a consonant sound. That is, all o's are p's, all 1's are n's, all 2's are r's, all 3's are s's, all 4's are t's and so on. Then vowels whatever is suitable can be added to each unit to make each one a meaningful word.

There is also one technique called number peg method. This is a very simple method. Every number is attached to a picture similar to the shape of the number. For example, 1 is a pillar, 2 is a cock, 3 is a jug and so on. Medical students use mnemonic techniques for memorising large number of anatomical structures. In learning the names of the craniel nerves that are afferent and efferent connections to the brain the medical students use the mnemonic techniques.

The recent trend in research in the field of learning and memory is mental imagery. The pegword method is an outcome of the mnemonic imagery. In our everyday life imagery as a mnemonic device, is the association of physical characteristics of individuals in learning their names. In memory tests, respondents are asked to use their visual images when memorising list of words. The words that are memorised are imagined as visual pictures or images. This method makes memorization easier and makes the recall reliable.

The pegword method of mnemonic device is a suitable method to be recommended to the college and university students who are to remember large amount of information and to those who must remember large volume of things in some particular order. The academic materials can be shaped in a pegword form. It will be easy for the students to learn more materials if the abstract ideas are converted into concrete images. This may improve the students' memories.

Psychologists have made use of mnemonic devices in their experiments among college students in teaching and learning foreign language vocabularies. Atkinson (1975) and Raugh and Atkinson (1979) in their studies made educationally significant demonstration in making use of mnemonic devices in teaching Spanish words with English equivalent words. Bower and Hilgard (1986) state that the mnemonic technique has made dramatic success in laboratory work in teaching Russian and Spanish vocabulary to the experimental college classes at the Stanford University in 1970s. Tulving and Madigam (1970) called this type of mnemonic strategy as elaborative coding. Because the respondent produces additional information about the pegword and listword by which the materials can be recalled easily. The mnemonic technique may be designed in whatever suitable manner according to the learning material.

EXPERIMENT

Aim

The aim of the present study is to analyse the effect of mnemonic devices on memorisation of some information and to compare recall in respondents who have used mnemonic method to memorise the information in study.

Method

Participants

Obtain a minimum of four persons (if possible 16 persons) for this experiment. Out of four two will serve as experimental subject and the other two will serve as control subject (if 16, eight experimental and eight control).

Apparatus

List of pairs of words to be memorised, record sheet, stop watch. (For pairs of words see Appendix, List l, 2 and 3).

Description of the Apparatus: The test material consists of two lists of 10 pairs of words. The third list consists of ten pegwords. The experimenter himself can prepare the 3 lists of words. The pairs of words may be on the same line as stated in the introduction or whatever suitable to the experimenter. *The list of words are given in Appendix.*

Procedure

The test may be conducted in a group situation or with one or two persons. The two groups, the control group and the experimental group should be tested separately in this test.

Control Group

The participants of control group may be first allowed to enter the room be seated comfortably. Provide with a sheet of paper and pencil to each of them. The experimenter should hold the first list of pair words and give the participants the following instructions:

"I will read out to you now a list of pair of words, two words at a time slowly. You are to listen carefully. When I finish reading all the pairs of words I expect you to remember the words I have read together. For example if the pair of words butter-market or television- magazine, are presented to you, I will read out only the first word butter and you will have to write down market on your sheet of paper and when I say television you write down magazine on your sheet of paper. There will not be any repetition. I will read out only once."

The experimenter now should read out the pairs of words from the first list at the rate of one pair for every 10 seconds. The experimenter should look at the stop clock and maintain the time precisely in the presentation of words. After having presented all the word pairs the experimenter may allow 30 seconds to the participants to get ready to listen to the words in the second list. He must maintain the time accurately keeping a watch at the stop watch. Then the experimenter should give the following instructions to the participants :

"Listen ! Now I will read the first word of each of the pair words presented to you earlier. I just read out the first word, *but not in the same order*. When I say the first word you should write down the word that was paired with it, that is the second word. Now ready. Make it a point that you should write only the second word."

If the participants ask any doubts or questions the experimenter can clarify them. If there were no doubts or questions then the experimenter should read out the stimulus word of each pair from the list 2. Between each stimulus word the experimenter should keep 10 seconds interval. After presenting all the 10 stimulus words and when the participants have written the words the experimenter should collect the sheet of papers and allow them to go away.

Mnemonic Strategy or Experimental Group

Allow the experimental group now to enter the room and let them be seated comfortably. Read out the following instructions to them:

> "Now, I will present to you a list of 10 words. You should rhymically associate the word with the number. You should make concrete images for all 10 words. For example, the first word is gun, you should associate a concrete image between the two, like one–gun. Imagine, for what purpose the gun is used. It is for shooting birds or for hunting. The word *gun* is used here as an aid *to hook* another *associative word shoot or hunt*. You must create a concrete image in this way for every aid. There should be a kind of interaction between the aid word and the list word which I am going to present later on. Therefore, now your business is to learn the pegwords and rhymically associate with the number and later on with list word. You can take 10 seconds for each word. Now ready."

The experimenter, then presents the third list with 10 words to the participants. These words may be printed or typed in separate sheets as given in Appendix. They may be asked to create concrete image for each word associating the number as explained in the example. The experimenter should maintain the time accurately at the rate of 10 seconds for each word. On the whole after 100 seconds the experimenter should take back the list of words given to the participants and give the following instructions :

> "You have gone through the aid or pegwords, one to ten in the list and might have formed some concrete image on each word combining the number. Now I shall present to you pairs of words. The first words of the pairs you have already learned. Now you will hear the first word and its pair word together. Listen carefully, what words are pair words for the words you have learned. I will read out only once. You should not write anything now in your sheet of paper. Now ready."

Then the experimenter should read aloud each pair of words from the list I. While reading, the experimenter should keep 10 seconds between pairs. He should take 100 seconds for presenting all 10 pairs of words. After completing this task the experimenter should give the following instructions to them.

> "You have learned both the pegwords or the first word and their pair words. Now I will read out the pegwords or the first word *not in the same order*, but in a jumbled order. You should write down only the other pair words for each pair in your sheet of paper. You should also write down against the response word the connecting mental image between two objects. What sort of connection the two objects can have. Now ready."

The experimenter should read loudly *only the peg (stimulus) words* from the list 2. He should maintain 10 seconds in between one word and the other. After 100 seconds, when the participants have written all the words the experimenter should collect the sheets of paper from them and allow them to go away.

Results

The experimenter should first score the results of the control group by counting the number of correct pairs by comparing with the pairs as given in list l. One point should be given to every correct response. The maximum score is 10. The individual's score is the total number of correct response words. This should be presented in Table 6.36.

Table 6.36 : Showing the scores of the control group (2 subjects) in Mnemonics and Memory Test. The order of words as given in Table of the Appendix.

S. No.	Participant 1			Participant 2		
	Pegwords (Stimulus)	List words (Response)	Correct Response	Pegwords (Stimulus)	List word Response	Correct Response
1						
2						
3						
4						
5						
6						
7						
8						
9						
10						
Total number of correct responses				Total number of correct responses		

Table 6.37 : Showing the scores of the mnemonic Experimental group (2 subjects) in the Mnemonics and Memory Test. The order of words are given in list 2 of Appendix.

S. No.	Respondent-1			Respondent-2		
	Pegwords (Stimulus)	List words Response	Mnemonic image with correct response	Pegwords (Stimulus)	List words Response	Mnemonic image with correct response
1						
2						
3						
4						
5						
6						
7						
8						
9						
10						
Total number of correct responses				Total number of correct responses		

The results of the mnemonic (Experimental) group should be scored as did for the control group. Since the order of the words is changed the results should be compared with list 2 of the Appendix. For every correct response one point is given. The maximum score is 10. The data are presented in Table 6.37.

Find out the scores of the control group and the mnemonic group and present them in Table 6.38. The number of correct responses for each person and work out the mean score. Present the data on bar graphs.

Table 6.38 : Showing the scores of the control and mnemonic groups in the Mnemonics and Memory Test (group data assuming eight persons in each group)

S. No.	Name	Control group Number Correct (Score)	Name	Mnemonic Group Number Correct (Score)
1.	B.V.		S.R.	
2.	K.R.		T.V.	
3.	P.V.		V.R.	
4.	M.S.		S.V.	
5.	T.S.		M.S.	
6.	K.P.		N.K.	
7.	P.K.		R.K.	
8.	M.R.		V.K.	
	Total			
	Mean			

State the highest and the lowest scores of the control group. Who scored high and who scored low. State the highest and the lowest score of the persons in the Mnemonic group. State the mean score for the control and the Mnemonic groups. Whose mean score is high?

Discussion

Discuss the result of your participant(s) of the control group. How many word pairs he has recalled correctly. Where he stands in the group. Has he got reliable memory in free recalling of response words.

Discuss the result of your participant(s) of the Mnemonic group. How many word pairs he has recalled correctly. Was he benefited by the mnemonic aid. Has he used the pegwords to hook the list words. Has he formed any mental image of some vivid scene of interaction between the (pegwords and list words) two objects. In comparison how many of his mnemonic images are in line with list mnemonic images.

Analyse the group data comparing the highest score with the lowest score in using mnemonic strategy. Do you find anything interesting from the results obtained in the mnemonic memory test.

APPENDIX – 1

List 1 : List of word pairs to be read loudly to the control group first and secondly the mnemonic group

S. No.	Pairs of words		
1	hen	-	jackal
2	dairy	-	cheese
3	grease	-	wheel
4	ring	-	jewellery
5	hive	-	honey
6	egg	-	chicken
7	lemon	-	market
8	kite	-	boy
9	lion	-	circus
10	demon	-	evil

List 2 : List of stimulus words in jumbled order to be read loudly to control group first and secondly to the mnemonic group

S. No.	Peg (Stimulus) words	Response words (not to be read)
1	lemon	market
2	grease	wheel
3	ring	jewellery
4	hen	jackal
5	hive	honey
6	demon	evil
7	lion	circus
8	egg	chicken
9	kite	boy
10	dairy	cheese

List 3 : List of pegwords to be presented first to mnemonic group for learning

S. No.	Pegwords		
1	One	-	hen
2	Two	-	dairy
3	Three	-	grease
4	Four	-	ring
5	Five	-	hive
6	Six	-	egg
7	Seven	-	lemon
8	Eight	-	kite
9	Nine	-	lion
10	Ten	-	demon

Notes to the Experimenter

1. It is better to conduct this experiment on a small group of 8 experimental subjects and 8 control subjects. If this is not feasible a minimum of two subjects can be collected for each. Otherwise there should be at least one control subject and one experimental subject to conduct this experiment.
2. The experimenter first of all should thoroughly go through the introduction part and should know correctly how to conduct this experiment.

7

THINKING, IMAGINATION AND LANGUAGE COMPREHENSION

INTRODUCTION

Man is a social animal and also thinking animal. He expresses his thinking behaviour in various ways. To express his thinking meaningfully man needs a medium called language. Human beings express all their thinking behaviours through some language with which they are acquainted or learned to speak and write. Thinking is a most complex type of human behaviour. No two persons think alike. For example on seeing an object, the tree on the road two men think in two different ways. One may think that it provides shade or fruits, another may think that it can be used for making furniture. In the same way human being think in varied ways. Thinking is considered to be the highest form of mental activity. We may be thinking many things in a day, but nobody can say what each one of us was thinking the whole day. Unless it is revealed in some form we cannot say what a person was thinking all the times.

THINKING AND RELATED PROCESSES

Thinking has been classified as one of the cognitive processes or activities. Thinking, though it is stated as the cognitive activity, is an inexplicable phenomenon. Psychologists have studied extensively about thinking. What is going on inside the body while a person is thinking? Are there any physiological changes in some parts of the body? Actually the common sense view is that we think with our mind. It is primarily an intellectual function. We generally locate the mind in the brain or the central nervous system. Garrett (1975) states that the mental arithmetic is often thought of different from ordinary arithmetic. In mental arithmetic the work is done with head, whereas in ordinary arithmetic the figure work and calculations are done in paper. The first activity is mental, it is private and is going on inside, whereas the second is physical, outside, and also observable. Mental arithmetic involves counting within and it is talking to oneself. While the mental arithmetic is going on, the student may be counting his fingers, looking high, frowning, moving the lips and so on. This may draw the muscles in various parts of the body in the hands, eyes, mouth, neck and also in the throat. Thinking activity in this way is sure to involve more muscles. Thus we think with our fingers, hands and with our brains and we learn with our muscles and nervous systems. Sometimes our thinking is also expressed in our bodily features. When there was no language, our thinking was expressed in signs and through some symbols. The dumb person thinks through his fingers, hands, head, lips, eyes and other bodily parts. The dumb person learns to think in that fashion. Prior to the invention of language all human beings were thinking in this fashion only.

One of the eminent American psychologists, John B. Watson (1878-1948) the founder of behaviourism states that thinking is silent speaking. Watson (1924) explained the intangible processes like thinking and feeling in terms of movements of the muscles of the gut. He states that speaking involves movements of muscles of the throat. According to him thought is only sub-vocal speech. Therefore, thinking must involve movement of the throat muscles.

Jean Piaget (1896-1980), one of the foremost developmental psychologists of this century has made many contributions on various aspects of cognitive development. Piaget's (1926,1928, 1932, 1952) approach to the study of children was from biological orientation. There are four stages of development. The first is a sensory-motor stage, the second pre-operational stage which includes pre-conceptual and intuitive, the third is concrete operations stage and the fourth is formal operations stage and the thinking process is evolved only in the last stage. According to him children do not acquire operational thinking until they are seven years old. Till then their conceptual behaviour is full of contradictions and errors of logic. The thinking is dominated by perception. But after the age of four the children's thinking becomes somewhat more logical and more perceptual than reason. Piaget states that the period between 4 to 7 years is an intuitive period of thought. At this stage the child is more egocentric in problem solving. The child at the age of seven to eleven years of age is in the stage of *concrete operational thinking*. The child applies logic to classes and to members. His thinking becomes relatively decentralised. The thinking is not ego-centre nor based upon perception. The child deals only with real or with what is capable of imagining. The final stage of development of thinking is labelled as *formal operations*. If the problem is not a concrete, but a verbal type then it becomes difficult for the child to solve it.

In an experiment children were tested with coloured disc. They were asked to combine each colour with every other colour in all possible ways. They were not able to achieve a complete and systematic solution. Even earlier they have not done it systematically and also did not exhaust all possibilities. But they were able to believe it in *formal operations*. In this way a distinction between formal operation and concrete thinking was made. Piaget refers to this thinking involving the solution of this type or similar problems as combinational thinking. The development of formal operational stage in the school child is a significant one, because on the earlier stages the child is not understanding many concepts or understands it completely.

A child can understand concepts such as proportion in objects like size or height only in the second-order of formal operations. This is beyond the comprehension of a child at the concrete operational level. A second order of operation of thinking is required when the problem is beyond the product of first order of thinking.

Therefore, according to Piaget there are four stages of development to reach the formal thinking stage. The concrete operations stage is the third stage, which is lasting from the age of *seven* or *eight* to *eleven* or *twelve* years. This is largely characterized by the child's ability to deal with concrete problems and objects. The fourth is the formal operations stage, which is lasting from the age of eleven or twelve to fourteen or fifteen years. This stage is characterized by the child's increasing ability to apply logical thought processes.

Thinking, according to Morse and Wingo (1970), is the deliberate application of reflection to the solution of a problem. Why a man is thinking about some matter now, when ordinarily he does not do it at this particular time. The answer is that he is thinking because he has faced with a problem, which he must solve. Ordinary method of applying habit is not sufficient to solve the problem. Therefore, thought takes place in him and thinking takes place in facing problem. Normally we sit and think only when we face something serious or a problem specifically related to us.

Young children are taught and trained to think. The primary purpose of education is to train for thinking and acting accordingly. Teachers often admonish students to think clearly. Sometimes we may remark a person that he is not having a clear thinking. It means that he is oscillating between the decision whether 'to do' or 'not to do'. This, we may call unclear thinking. Sometimes we may even modify our thinking or improve it. According to Garrett (1975) thinking is an implicit or inner behaviour which is not readily observable. Thinking is carried out by slender muscular movements and by mental images which are representing various kinds of symbols. Symbols stand as substitutes for objects, persons, places, experiences, activities, words, numbers, arithmetical signs, badges, flags, songs and slogans and so on. We think through images and symbols.

Mental Images

Everyone of us may have many mental images of some type. It is very difficult to define mental image correctly. Generally images are mental pictures. We may form images of objects, persons and some experiences. Therefore, an image is a personal experience of an object, person, or event or scene. One may have even an image of a musical performance or a musical note. Thus images may be visual, auditory, olfactory, or gustatory or even sensation of touch. We may think of a dinner in which some delicious dishes were served. The taste of the food often may make us to think of the incidence. Similarly, a beautiful picture or scenery may make us often to think of it and we may enjoy it. These are all images. In this way every sensory experience may provide images for our thinking.

Human beings acquire knowledge of the world through the process of sensation and perception. Man acquires knowledge through the process of the formation of images. The formation of image is not possible without keen sensitivity to objects, places, persons or events. We see often objects, places and persons and witness events and hear voices. After few minutes, or hours or days we wish to know them. What we had was only sensory experiences and now we wish to recollect those sensory experiences by forming a picture of those objects in our mind in the absence of those objects. This, we call as the mental image. Mental image is a process of visualization of our past experience in the form of an object. The real object is not there. An image may be also defined as an artificial representation of the external form of any object. In a way it is a kind of thinking.

Moskowitz and Orgel (1969) state that in the history of Psychology at one point ideas are equated with images which are sensory-like experiences which can almost play at our will in the absence of the appropriate stimulus. All of us might have had sometimes some vivid visual, auditory, kinesthetic and other experiences which we may call forth or which may appear even without our will and all of us might have utilized such images on certain occasions for our symbolic thinking.

All images are private events and, therefore, it will not be possible to make objective investigation. Textbook writers give different interpretation for images. Moskowitz and Orgel (1969) state that symbols are ideas and the symbolic process is referred to as an ideational process and ideas are equated with images. They say that in the introspective observational studies it is convincingly stated that images are not present in the normal intellectual activity for symbolic interpretations. Therefore, they are of the view that something more than just images are required which may be stated as symbolic process. In such cases we are depending upon introspective reports of the subjects which are not trustworthy. The same images are stated as symbolic process and ideational processes. Because all these are ultimately based upon only our sensory experiences.

The concept, images are invariably used in the psychological studies, particularly in the study of memory and visual experiences. The concepts used to substitute images with *symbolic process* or *ideational process* appear to be incongruous. Images may be classified as memory images, after images and eidetic images.

Memory Images

When we see some objects or persons or events outside we do not completely discard them from our memory. We know that memory is the function that stores what we have learned or experienced on an earlier occasion. It is the storage of the effects of our experience. Therefore, we have the ability to trace the object or event and restate our past experience in the form of an image after some time. This is a mental picture of the object perceived earlier. This image may be clear and sometime not clear. They may be vague impressions. Some persons may have vivid, sharp and definite picture of the object or event like photographs. Because of this people's images may differ significantly. Two persons may give two different versions of the same object or event they have witnessed some time back. This depends upon what kind of image each one had of the object or event, depends upon several factors like past experience, motivation, and learning and age. Each one may use different style of language to describe the image he has formed of the object. The verbal description of the image one formed is not the same as that of the other. Therefore, the two descriptions vary from each other and, therefore, the two images may vary from the object. All the mental images are the memory images.

Whatever we perceive in the outside world persists for certain amount of time. The objects that we saw earlier make impressions on the sense organs and the impressions do not fade away from the brain. These impressions fade away gradually and disappear. It is just like the impressions take a few seconds to stimulate the brain centre, the fading of the impressions also takes few seconds. The effects continued in the sense organ or in the brain centre. The image of the object remains for few seconds in our senses, which is sometimes called after-image. Even after the disappearance of the object we may be having the effects of the object perceived or the person with whom we were having conversation. The person's voice may be continued to hear and his face continued to appear before our eyes. Gardner Murphy (1964), calls it the 'mind's eye'. Such after-effects are called *memory images.* The impressions of the voices and faces we sometimes call back after they were completely disappeared and we call these reappearance of the objects as memory images. It is arousing the impressions from the brain centre, which were registered when the objects were perceived originally. The sensory impressions, which existed for few seconds, are called after-images. This takes place without any effort on the part of the person. The memory is the second stage, which is the re-arousal of the sensory impressions. This is how the after-image or the sensory impressions may be differentiated from the memory images. Francis Galton (1822-1911) was the first to conduct experiments on the images. The information Galton (1883) collected were about the frequencies and the vividness of imagery. His study mainly dealt with memory images and not after-images. Memory images depend upon voluntary recall. In Galton's study the subjects were asked to recall the appearance of 'this morning breakfast table' for collecting data about the pictures of the 'mind's eye'. He has designed a scale to measure the degree of vividness of the image. He has given zero mark for no image and five for an image, which had, equal clarity to the original object. The results had a very wide range. It was a great surprise to Galton that some had no imagery at all while others had vivid images as hallucinations. But in Galton's interpretation of the data both the memory images and after-images or sensory images were involved. Therefore, a clear-cut distinction between memory images and after-images becomes very difficult.

After-Images

After-image is mainly associated with sensory impressions. This is a kind of image formed of an object in a person he saw for few seconds and when the object is disappeared. The sensory impressions continue to exist even after the object ceased to exist. The cones and rods in the retina area of our eyes retain the image of the object for few seconds. According, to Garret (1975) after-image is the prolongation of *sensory experience* after the object or stimulus has ceased to affect the sense organs. Murphy (1964) says that after-image is the *sensation persisting* after the stimulus has ceased to exist. He states that the visual after-images are positive if they are of the same as the original stimulus and negative if the colour of the object differs.

When our eyes are exposed to some bright light for a few minutes and if we turn our eyes away from the bright light we experience a short dark interval, the resulting sensation is the persistence of the stimulus, the bright light with less brightness. This is called positive after-image. This image is existing only for a few seconds. The positive after-image is a reproduction of the original sensation after the disappearance of the stimulus. Both Garrett and Murphy have stated that it is the persistence of the sensory experience or sensation of the object or stimulus for a short period. Therefore, an after-image may be defined as the sense experience that occurs even after the stimulus had ceased to exist. Woodworth (1951) says that a better name for after-image is *after-sensation.*

The after-images may also be stated as *sensory images*. Murphy (1964) calls this as *visual images*. Visual images may apply to only seeing. Therefore, after-images may be called as sensory images as it applied to also other sensations like, hearing, smelling and tasting. Several psychologists like Francis Galton (1883), Perky (1910), Titchener (1909), Theodor Reik (1949), Ellson (1942), have conducted experiments on after-images. Morris (1973) states that persons who had recurrent after-images report of ghosts and apparitions. Some of the experimenters reported that *visual imagery was very vivid* and most common. Other images, like auditory, tactile, kinesthetic, gustatory and olfactory were in this order, approximately. It was also found that those who were high in one of imagery also found high in other types.

Variations in imagery have been observed in people. In one of the Binet's (1857-1911) study it was found that *musicians* who dealt with rhythms had more *vivid auditory images* that the psychologists had. Children had more vivid imagery than adults. Concrete objects produce better vivid imagery than abstract ideas. It has been observed in the studies of *thinking and problems* solving that images are likely to occur more if there is any blockage in the flow of thought. It was also observed that *training, interest* and *motivation* seem to have some influence upon after-images. Starving subjects often reported thoughts of food and their after images for food were more vivid than normal human beings. Clinical studies show that men who have strong sex drives have more visions of women and images of sexual embraces. The after-images are modified on various ways depending upon situation and the conditions of the whole organism.

Eidetic Images

Some persons, particularly, children have better images of all types than the adults. But human beings on the whole have capacity for developing images, which may be quite similar to *photographs*. This type of images are called *eidetic images*. Persons who are having such experiences are called eidetics. Woodworth (1952) states that the eidetic images are called the primary memory images.

What has been said about images refer to recalling of the stimulus or object ceased to exist. The eidetic images are different from those images. An eidetic imagery has an unusual clarity and vividness, in character. A person with eidetic imagery is said to have a 'photographic memory'. Therefore, an eidetic imagery may be defined as unusually sharp and detailed visual representation of something that has been seen earlier. This something may be a picture, a sculpture, a scene or a page of text perceived visually. The eidetic imagery enables a person to describe features of an image in minute details. It has been observed that children between ten and fourteen years of age seem to show eidetic imagery frequently. Woodworth (1952), states that children under fourteen years of age, fifty per cent of them, if they were examined with complex objects or pictures by allowing them to observe for half a minute and then closed their eyes, can see the objects as if they are still before them. They can answer any question about the objects. Woodworth states that this type of image is not strictly photographic, but it is *plastic moulded imagery* and may be modified. Eidetic imagery is very common in the later childhood period and it fades away during adolescent period.

In the case of an object or person, in the eidetic imagery it may grow larger, may become regular or may change. In the case of colours, more bright colours may change to different shade of colours. Psychologists have shown much interest in the study of eidetic imagery. It is closely related to both after-images and to the actual percepts. The word percept refers to the object to be perceived. Meenes and Morton (1936) have studied the eidetics and non-eidetics jointly and independently. They have compared the after-images of two groups, one eidetics and the other non-eidetics. They have observed that the after-images of the eidetics were found to be superior to that of the non-eidetics in several respects. The images that were aroused suddenly remained for longer duration, more positive and with more details in the case of eidetics, whereas in the case of the non-edetics they were not so. After a week's time they were again tested and the eidetics were found to be again superior to the non-eidetics in their ability in reproducing their images with details.

In one of the studies conducted by Haber (1969) out of 500 school children 20 were found to be with eidetic imagery. He made every child look at a picture part by part for 30 seconds. Then he removed the picture. The child was asked to look at the blank frame of the picture and report what he saw in the picture. The children required 3 to 5 seconds to look at the frame of the picture closely to reproduce an image of the picture even though it was familiar to them. The images faded away after they have described the picture. They could not retain the images for long time to report. They eased the images by looking away from the blank frame of the picture. There was individual differences in the quality of eidetic imagery.

Haber has observed in his study that one of the girls, served as a subject was able to move and reverse the images and could reproduce them after several weeks of the study. Three children were able to produce the eidetic images as three-dimensional objects. Some children were able to form a new combination of the picture by superimposing the eidetic image of one picture into another. One interesting result, Haber observed in his study has been that the eidetic children were found to be in no way better than the non-eidetic children in memory tests. This shows that there may be several factors that determine the eidetic images of the individuals.

Associated Imagery

Apart from these three imagery there is also another type of imagery called associated imagery. It is possible to have imagery of every sensory experience. But every image may be associated with every other image with the same modelity and also with image of other sense organs. It is easy to associate the visual and the auditory images of an object, for example, the musical instrument. Similarly, it is easy to associate visual, gustatory and olfactory images, for example a delicious dish prepared by a skilful cook for a dinner colourfully combing it with palatable taste and pleasant smell.

ASSOCIATED THINKING : IMAGINATION, DAYDREAMS AND DREAMS

In the preceding section we have discussed in detail about the images and the kinds of images. We must know that imagination is not possible without the formation of mental images and the formation of mental images is not possible without keen sensitivity to objects, places, persons, and events. Therefore, an image is a collection or reconstruction of a sensory experience, which is often visual. An image may be also stated as an artificial imitation or representation of the external form of any object already seen. Fantasies or reverie or day-dreaming, dreaming and imagination are also described as thinking of some type. These are called associative thinking. These types of thinking are unlimited in their scope and they are not directed to solve any problem. Fantasy, reverie, day-dreaming and imagination — all refer to the same type of thinking.

Imagination

Imagination may be defined in various ways since it is based upon the formation of mental images. It is manipulation of images in whatever manner we wanted. Imagination may be defined as an act of forming a mental concept of what is not actually present to the senses. The important process that takes place in imagination is the rearrangement of the materials or images we have experienced in the past in relation to the present. This arrangement is based upon recency, frequency and vividness of our experiences. A recently perceived attractive object can be combined with similar object perceived in the past.

In our imagination we combine several of our past and present experiences in a new or in a peculiar fashion. We wish to see something out of the old objects. We do not care whether such combinations are possible in the world or not. But we wish to see it. It is interesting to hear the story of the mermaid in a poem written by an English poet. A mermaid is an imagined sea creature with head of a beautiful woman and body of a fish. A mermaid is an imagined creature in which half of a woman's body and half of a fish's body are combined. This idea often arises in our minds, because our poets and novelists often present our reading materials comparing woman's eyes to fishes. Adolescent boys and girls form fantastic imaginations. Children are highly imaginative. But children's imaginations may lack organization. They often find companions in their imaginations. Such imaginations express the needs of the children. When their needs are not fulfilled they often imagine companions, food and comforts. But the main causes of imagination in children and adults are the inner world of needs, worries and pre-occupations with outer world of sensory stimulation. These are all aimless thinking.

Imagination and Memory

These two appear to be the same. But they are different. Imagination is a way of thinking ourselves, presenting that has not existed previously and it can never exist, whereas memory is a way of presenting to ourselves something that has been learned or experienced earlier without any change or modification. Therefore, imagination is different from memory. But the degree of changes or rearrangements made in the original materials may vary in imagination from memory. The elements of imagination are old, but the combination of the materials presented is completely new. In memory we reproduce the materials learned earlier without any change. If you make any change or make modifications other than the original, it becomes imagination.

The materials of the memory are true. They are presented to you as they existed and you reproduce the same. Imagination is private, subjective and not real, whereas memory is objective and real. It is very difficult to explain to a blind man how an elephant would look like. You know the story of four blind men and an elephant. How fantastically they have imagined about an elephant? Therefore, if there is no visual perception it is possible to have fantastic imaginations from the verbal statements one has received. Gardner Murphy (1964) states that when the elements of imagination are combined in new ways by the application of the laws of association, the same laws are also applicable to memory. But in the case of imagination it is differently dealt with. Therefore, study of memory differs from imagination.

Study of Imagination

The process of imagination has been tried to study scientifically by the psychologists. The role of drives has been used in this study. Persons subjected to prolonged hunger often imagine of rich dinner. Persons put to more physical discomforts often imagine bodily pleasures. Our imaginations sometimes are utterly free and uncontrolled. Our thoughts may sometimes run riotously and fantastically. Nothing can prevent their free flowing. All our day dreams or fantasies are free imaginations. Though our ideas run very wildly in our imagination there is a procedure in it. A set is operating in it. Generally the roles of drives are used to study imaginations. The sensory drives, emotional or visceral drives and also activity drives play important role in our imaginations. Our sensory drives for seeing and hearing also guide our imaginations. You might have developed strong desire to see and enjoy a beautiful garden or building or to hear and enjoy some sweet music. When there is no chance for physically enjoying such things you sit and imagine about one you have enjoyed in the past.

Similarly there are drives that would make you to aspire for power and prestige and when it is not possible in actual life you make great achievements in your imagination. The sex urge is a strong drive that would carry everyone into the world of imagination where he finds his desires are fulfilled. You may be very much interested in some game to play and if you are deprived of it you indulge in imagining about all your skills and play the game and find the pleasure.

The process of imagination can be studied at close range by the method of chain association test. According to this method the experimenter presents a single word to the person and asks him to provide a chain of words associated with the word. By this means the each word the person has provided earlier is used by him as a clue for the next word. For example, a person may draw a chain of words like this: For the word, Student, the chain of words is – Student – school – classroom, teacher – headmaster – lessons – books – notebooks – examination – results – pass – fail – holidays and so on. For the word Boy – girl – love – marriage – father – mother – family – prayer – temple – God – religion and so on. The person may stop for a moment at some word, not knowing how to continue and himself may not be knowing that he has stopped because there is mental blockage. Psychologists call this as *interference* and they are interested in this. This interference reveals the preoccupation of the person.

In another type of study a child may be asked to tell a story of a bird or animal or human being. A picture of the same may be also shown to him to make out a story from it. In this type of study, we can find out the relationships between his imagination and events of his life. Such stories reveal the subject's desires, worries and his preoccupations, which he hesitates to reveal in ordinary face to face situations. The story events may reveal even the materials he has stored in his unconscious mind. Morgan and Murray (1935) have developed a test called Thematic Apperception Test (TAT). This test serves several purposes in psychological studies. It consists of a series of standardized pictures in which there will be one or more human figures in different poses. The participant is shown the pictures one by one and he is asked to make out a story for each picture. The person can make out a story out of his free imagination, which may reveal his interests ambitions and worries.

Daydreams

Daydreaming is a kind of imagination, which takes place without any particular objective in mind. Garrett (1975) says it is relatively aimless thinking. A student, sitting in the classroom, not listening to the teacher's lecture, is involved in some kind of mental activity which has no relevance to the present situation. He is engaged in daydreaming. He is not completely unaware of the situation, but he is pre-occupied with something. He may be thinking of the game he lost on the previous day and the mistakes he committed for the defeat. Therefore, there is going on some kind of mental activity which has nothing to do with the present situation. Woodworth (1952) calls this as *ideational activity*. Ideation is a mental activity having no ideas. In this activity there is neither muscular movement nor sensory stimulus present. The person lives in an inner world of his own, which is free from the real world. This kind of mental activity is called day dreaming. It is a kind of imagination in waking state. Daydreams, reveries and fantasies all generally refer to the same type of mental process.

Fantasy and Daydreams

Daydreams and fantasies are sometimes treated synonymously. Fantasy is the general term used to indicate free imagination. Fantasy in a highly developed form is called daydream. It is the result of forming representations of things not actually present. *Fantasy* is an imagination unrestricted by reality. It refers to a series of pleasing mental images, usually serving to fulfil a need, which was not gratified in real life. Fantasy is a creation of imagination in an unusual or amusing way. It is a mental activity forming such images. *Daydream* is a reverie. It is pleasant dreamlike state indulged in while awake. Daydream is an inactive reverie. A pleasant wish or scheme that is unlikely to be fulfilled. All the poetic materials are the results of daydreaming. In daydreams you exclude the external stimuli for the moment.

A student at the time of sitting in a classroom and listening to the teacher, his mind, starts wandering somewhere else on hearing some familiar song broadcasted from a nearby place. Or while he is listening to a story, told by his teacher he might have gone immediately into a daydream, that is associated with an event of a story told by his grandfather or he may associate it with story of a motion picture that he has seen a few days ago. For a moment he forgets what the real world is demanding of him and do something else. He relaxes his mind and he allows his ideas to wander freely. In this process one idea drags another idea and one memory recalls another and a chain-like process is taking place.

In this type of process of free association if you review your thoughts and experiences in a sequential manner long forgotten life events are also recalled. Here your thoughts are not controlled and directed towards any specific purpose. You are manipulating your ideas in whatever manner you wanted. Walls, iron bars, seas and oceans cannot prevent them. You can fly in the air, live under the sea, build castle in the air and live there.

Children and Daydreams

Children often resort to daydreaming and adolescents more frequently go into daydreaming. Generally, when children are prevented from eating sweet-meat or playing with toys and weapons, they have them in plenty in their daydreams. Sex starved young men and women gratify their desires in their daydreams. A person achieves in his daydreams all those things that he cannot achieve in his real life. There is continuity of thought in daydreams and the thoughts are going on working out various themes in which the person plays the important role. In this kind of free imagination a man's long cherished hopes and desires are fulfilled. He sees himself always engaged in heroic activities and make great success. He defeats his enemies, wins the heart of the most beautiful girls, get the first prize of ten lakhs of rupees in the lottery, passes in the examination in the first class with a first rank, exhibits unchallenging ability of some kind. In the daydream he plays always the role of conquering hero.

But, sometimes a person's daydream may lead to unresolved conflicts and he imagines all sorts of disappointments and worries and thus he plays the role of a suffering hero. He becomes unpopular among his friends and not respected by them, loses his good fortunes in the race or lottery, meets with heavy loss in business and his girl friend disappoints him after giving him hopes. All these materials of the daydreams come from our everyday experiences. Daydreams usually serve the purpose of wish fulfilment. There is always a plot in a daydream. It has a continuity. Man can escape from difficulties and responsibilities and from everyday boredom of his real life for few minutes in his daydream. It is also a kind of indoor game and he finds some kind of pleasure in it. But if the daydreaming goes to the extreme level it leads to several types of maladjustment. The person may be often playing the roles of both the conquering and suffering heroes and lose all contacts with the real world. Finally he may develop delusions and persecutions which are supposed to be abnormal symptoms.

Dreams

Let us move from daydreams to dreams. Daydreams we have while we are in waking state. But dreams we have while we are sleeping. It need not be only the night time. Even during day time if we sleep we may have dreams. All of us, without any exception might have had dreams while sleeping. The brain activity at this time is at a very low level. We will not have any contact with the environment. Whatever objects, persons, places we see in the dream may not be clear. We can never perceive a person's face clearly in dreams. It will be very crude or bizarre type. Dreams were considered as mysterious matter inaccessible for study until the early twentieth century when Sigmund Freud (1856-1939) had published his famous work, *The Interpretation of Dreams* in 1900. Dreams are regarded as a kind of personal experience man gets during his sleep.

Dream is a kind of imagination that goes on when one is asleep. Sleep is a state contrasted with the waking state. In the state of sleep there is a great deal of reduction of awareness and mental activity. A person may choose either to sleep or to be in waking states. But there are borderline stages between sleeping and waking states. Dreams are considered as the private property of the individual. For a long time it is considered that dreams cannot be observed objectively and assessed. But experimental psychologists have studied the dream experiences. According to them dreams may be defined as the appearance of *rapid eye movement of a sleeping person.*

Experimental Studies of Dreams

In the studies made by Rechtschaffer (1967) the subject regularly reported dreams when awakened from REM (Rapid Eye Movement) sleep. The REM were recorded in the Electro-Oculogram (EOG). The dream reports were less frequent when awakened from Non-Rapid Eye Movement (NREM) sleep. The duration of eye movements and the duration of dreams reported were found to be consistent. The direction of the eye recorded and the actions described in the dream reports were found to be closely associated. Vigorous eye movements were found to be the indication of the active dreams and less vigorous activity of the eye was the indication of passive dreams. Therefore, it was observed in the experiment that between the dream activities and the movements of the eyes there were consistent and reasonable relationships. During dream activities the eyes were found to be moving either passively or vigorously depending upon the nature of the dreams. This study also rejects the notion that the dream reports obtained from REM sleep are memories of previous dreams, and the subject makes out story on the spot.

In the dream state the activities of the brain are reduced to a low level. There is no contact with the environment. But some stimuli may be having some kind of influence upon the organism in a most unnatural manner. For example, a feeble tapping of some metal may be taken for a thunder storm. The sound of an alarm timepiece may be taken for a musical instrument and so on. Dreams do not have plots or sequence of events. But sometimes a dream broken in the middle, when the sleeper was disturbed, may continue if the sleeper continued his sleep at once. The dreamer may find the events of the dreams are real at the moment and he may wish to continue if it is a pleasant dream.

Several experimental studies have been made on dreams. The results show that dream is a kind of thought process — thinking often characterized by vivid sensory images mainly visual and auditory. Dream is associated with brain wave that is accompanied by rapid vertical and horizontal eye movements. Berger and Oswald (1962) state that there is significant correlation between eye movement and dream content. The eyes are moving when the sleeper is having a dream. In the waking state our eyes are moving in relation to objects. Similarly our eyes are moving when we watch an event, or object or person in the dream. Dement and Wolpert (1958) have studied various aspects of dreams experimentally. According to Dement (1965) there are four periods of dreaming in seven hours sleeping period. The first period lasts for ten minutes and begins within an hour after the onset of sleep. The second period may be lasting for an hour with an interval of about hundred minutes. Generally the third and fourth periods may be longer than the first and second periods. According to recent researches, as stated by Hartmann (1967) monkeys are capable of visual dreams. In one study made by Hartmann *et al.* (1968) it was observed that elephant also dreams while sleeping. Animals like dogs and cats also have dreams of simple type. Foulkes (1962), Foulkes and Vogel (1965), Verdone and Wheaton (1963), Cartwright and Monroe (1968) and many others have studied experimentally mental activities of various types during sleeping and dreaming state.

Theorists like Calvin Hall (1953, 1966) maintains that dreams are basically expressive in nature. According to Hall, a person describes what is in his mind and he works out solutions to his problems imaginatively in his dreams. In the dream, the person uses symbolic images. This is found to be more efficient than the images of real objects. A real object may appear in a symbolic form in the dream. In the studies made by Hall and Van de Castle (1966) it was observed that dreams do not directly reveal their meanings. The dream contents are always symbolic. The dream events cannot be brought under ordinary rules of logic. They state that people forget their dreams mostly. Some say that they never dream. It is not correct. They forget totally everything and claim that they did not dream. But it is very difficult to remember correctly the dream events and describe them. It is because of this that Sigmund Freud started studying his own dreams. He thought that he could more trust his own dreams than to depend upon the reports of others, which are not trustworthy for this purpose.

Studies on dreams were also made at the close of the nineteenth century. The recorded eye movements are the indications of the dream activities of a sleeping person is a claim made by the experimental psychologists. But this idea was already suggested by Ladd as early as 1892 itself before the modern instruments were invented to study the dreams. During the same time Calkins (1893), Weed and Hallam (1896) have also studied dreams and stated that the external stimuli had influence in dreams and it is possible to have the influence of external and internal stimulations in dreams.

But every normal person must have dreams every night during his sleep. A normal person in his twenties spends about twenty per cent of his night in dreaming. It is an important psycho-physiological function and any suppression of dreams may cause serious mental disturbances. Persons who suffer from mental illness like acute paranoid, psychosis spend about fifty-nine per cent of their total sleeping time in dreaming. Dream is a kind of hallucination–if it is abnormal. According to Freud dreams are the guardians of our sleep. Therefore, everyone should have dream in his sleep.

Causes of Dreams

There are three important causes of our dreams. They are: (1) wish fulfilment, (2) mastery impulse, and (3) physiological discomforts.

1. Everyone may be dreaming everyday while sleeping. They can themselves verify whether their dreams have any relationship with their desires they had on earlier occasions. The dreams may not always reveal their desires in the original form. But in disguised form. Most of our dreams are wish fulfilment. Children often dream of things that are desired by them. Starving people often dream of eating dinners. Sex craved youths often dream of fulfilling their sex desires. Explorers of Arctic regions dream of warm weather and green fields. The man who feels very thirsty dream of drinking gallons of water. The desires and physiological needs that are not fulfilled in the real life are fulfilled in the dream life.
2. Psychologists attribute several causes for our dreams. Man's impulses also are found to be one of the causes of dreams. It is also a kind of desire, but ıt is acting in different manner. The strong impulses are for dominance over others become active in dream. The servant who is teased by his master everyday slowly develops strong impulses to get rid of the teasing. It is possible if only he becomes a master, which is not possible in his real life. Therefore, his strong impulses becomes active in his dreams and he teases his master one day by saying 'you shut up'. The servant's impulses find outlet in this way in the dream.
3. The physiological discomforts one is experiencing at the time of sleeping also becomes one of the causes of dreams. Horton's (1919) experiment is a good example to illustrate this point. The man who sleeps by putting a heavy blanket over his feet dreams of getting into the hands of his enemies as his feet were tied together and he was not able to run and escape. Horton's (1919) experiment is a good example to illustrate this point.

The effect of heavy blanket is, pressing the sleeping man's feet. This is reflected in the dreaming man's image in two ways. In one dream image he plays the role of a flying Mercury. He is entirely free, but one of his legs is fixed to the pedestal. Since his leg is fixed he could not fly. In the other dream image the dreaming man is running behind a street tram car, but he could not run because his movement is prevented by a heavy outer garment whose heavy ends rolled round his feet and he could not run. The man had two dream images in which his legs are held up and he could not freely move. These two dream images are caused by the physical discomfort experienced by the man by the heavy blanket on his feet.

THEORIES OF DREAM

Though there were some mentions made about dreams in the earlier studies it was only after Sigmund Freud's study of dreams that men began to think of it in a psychological way. Freud brought his theory in his work, *The Interpretation of Dreams* in 1900. His dream theory is the most influential one for more than half a century. According to Freud man's unconscious impulses are mainly responsible for the dreams. Man's desires, mostly sexual, are not gratified in the real life as it is against his cultural norms. Therefore he suppresses them into the unconscious level. These suppressed desires that lay in the unconscious level are called *latent content.* During sleep since the mind is not active the desires from the unconscious mind try to come out in a *disguised form.* The latent content when come out in disguised form is called *manifest content.* This manifest content is the dream. This is the expression of the man's desire in a disguised form, Freud calls it as *sublimated form.* All those not acceptable in real life are converted into socially acceptable form. The dream content needs interpretation to find out the actual content or the desire of the person. Freud developed the technique of interpreting the dream contents and was able to tell what the man actually desired — the real meaning of that latent content.

In Freud's theory of dreams there are three important processes of function which may be called the mechanisms. They are condensation, displacement, and symbolization. Condensation is combining several ideas into few forms. This means, a single dream word may have several meanings in the dream. In displacement one object or matter stands for another. That is, a motive that is not socially acceptable is presented in an acceptable form. Symbolization refers to the representation of some ideas or events in some other forms. The dreams are the symbolic expressions. The real content is represented in some other form. It needs interpretation of the dream content to find out the real desire or wish of the dreamer. Freud says that most of them are sexual desires. A man who had fallen in love with his neighbour's wife, had dreams in which he was driving his neighbour's car. The neighbour's car is the symbolic representation of his neighbour's wife. This is how every dream content was a symbolic representation of the real object. If the raw materials appear in the original form in the dream, the dreamer's sleep get disturbed and the dreamer sometimes get frightened and awakened. This means that sometimes all the raw materials do not appear in the disguised form. Some appear in the original form. Therefore, the dream mechanism sometimes fails. As a result of this Freud (1933) has revised his dream interpretation in his later works.

Some of Freud's followers like Carl Jung (1875-1961), Alfred Adler (1870-1937) and Erikson (1902-) had difference of opinion with Freud with regard to the dream contents and had developed their own schools. Erikson (1954) had stated that the dream contents are much more than the disguised wish fulfilment. Hall (1953) has stated that the dream contents sometimes appear in the original forms.

THINKING AND RELATED PROCESSES

We have been discussing about associative thinking in the preceding section. Any treatise on psychology of thinking is not complete without involving all those processes related to thinking or part and parcel of thinking. Therefore, it is better to bring out everything that becomes basic function and the structure of thinking. It is an accepted phenomenon that we think through images and symbols and express our thinking through some symbolic process. In this way our images–*mental images, imagination, daydreams, dreams form bases of our thinking.* Thinking involved mainly other psychological processes like perception and learning. Though man is distinctively a thinking animal, he has to learn to think. Gardner Murphy (1954) says that one is not a born thinker. According to him man has to learn to think just he has to learn to perceive. He rejects the popular view that perceiving, learning and thinking are basically different aspects. He says that perceiving and thinking are closely related forms of learning to deal with the environment.

It is always a problem for a man where to begin thinking, since thinking extends over a very wide varieties of activities, ranging from a simple activity such as where to hang the calendar on the wall to inventing some new medicine to cure the AIDS or solving some difficult mathematical problems. The argument is always there before beginning any such activities. The process of the development of thinking is taking place only in the formal operation stage. According to Piaget (1926, 1928, 1932, 1952) who studied elaborately about the developmental stages of the child states that there are four stages of development to reach the formal thinking stage. The process of formal thinking begins at the age of seven or eight and goes through eleven or twelve. The concrete operational thinking is taking place in this stage and the child learns to apply logic to the classes and numbers. According to Piaget in the fourth stage, formal operational stage is the final stage of development for thinking at the age of eleven or twelve to fourteen and fifteen. He deals with real and also capable of imagining. He deals with concrete problem and finds it difficult if it is of verbal type. Piaget says that the process of development of thinking goes with the development of intelligence and maturity. Therefore, in human being both thinking and imagining take place simultaneously, almost at the same age. The age range may vary culture to culture, but the four stages which we have discussed earlier may be taken as authentic statements. All the earlier thinking and imagination of the child are bizarre and aimless types.

The stage of concrete and formal thinking and imagining, according to Piaget's study, take place in a systematic manner at the age of eleven or twelve when the child acquired sufficient mental and physical development. At this stage the child is capable of perceiving and learning many things, including the application of logic, language and arithmetics which are supposed to be difficult areas for the child. Therefore, the child perceives, learns, and learns to think and imagine and also daydreams and dreams. Analysing all these processes critically the order appears to take place naturally in human organism. It is perceiving, learning, thinking, imagining, daydreaming and dreaming. Perceiving and learning facilitate thinking and imagining.

There is something in common between imagination and daydreams. Both imagination and daydreams are acts of forming mental concepts, which are not actually present to the senses. Both are taking place in waking state and may be dissolved at any moment. But these two also differ in some respects. Imagination is seeing something new from the old objects and events, whereas in daydreams the concerned person allows his mind to roam about anywhere aimlessly while engaged in some kind of work. He finds pleasure in some similar external object or even while dealing with some work. He may recall both pleasant and painful experiences of his own life and try to continue in the manner in which he is interested. Therefore, he lives in an inner world of his own which frees him from the real world. Imaginations sometimes, may serve some purpose of solving an important problem or the contents may serve as materials for some creative works. Writers, poets, artists often go into imagination for creating something new whereas in daydreams this is not taking place but sometimes it is very difficult to give a clear cut distinction between these two phenomena. Similarly it is also very difficult to differentiate thinking and imagining.

Creative Products and Dreams

When we go into the personal experiences of some famous writers, novelists, poets, artists and scientists interesting notes are available regarding the way in which they got the source materials for their writings. The English writer and novelist, Robert Louis Stevenson (1850-1894) who lived only for a short time of forty-four years has produced enormous amount of materials. His famous novels, *Treasure Island* and *Kidnapped* reveal that he was a person who could imagine and write all sorts of things real and fantastic, exciting and horrifying poetic and prosaic, adventurous and exploratory themes. He revealed his astonishing inventions in his stories and verses with which he could enchant all sorts of people. He thought that his dreams were best sources of his writing materials and what he was imagining during his working states were also drawn from his dream processes. It appears that he was able to actualize his dreams and make them best stories. He was able to recall his dream contents and rearrange them into a story plots. Stevenson's dream contents and the imaginations of the waking states have served as story materials for Stevenson and he was able to rearrange them, as he desired. This is not a fictitious statement about Stevenson. These were the real events of Stevenson's life. Many writers and poets have drawn their story and poetic materials from their dreams. How are we to explain this process? We cannot reject this information on the ground that they are only dreams and nothing to do with thinking.

Many eminent English writers like Charles Dickens (1812-70) madc use of his dream materials in his novels. Charles Lamb (1775-1834) an English essayist and critic had several unpleasant events in his life. He often dreamt and the dream materials facilitated his thinking to write his autobiographical essays under the heading of *Essays of Elia* (1823). Therefore from the life events of may eminent writers and poets we come to know that their dreams were serving as sources for their writings. The dream materials as they appear in the dreams do not serve as plots for the writer. The dream materials may not be having any sequences and may not be in agreement with reality. They may be highly fantastic, amazing and also based upon impossibilities. A student who is a sportsman had narrated one of his dreams. He said that he had wonderful dreams yesterday night in sleep that he was jumping from a mountain top into the sea nearby and fought with a huge fish and was riding on its back to the shore. This might have been his wish, but this is not possible. One can imagine such incidents and the same may appear in his dream. But such dreams may provide good plot to the storywriter. This type of rough and impossible dreams materials may be rearranged in such a way that adventurous stories can be written by a writer. The writer makes a rough story first on this type of dream events and thinks over it how to relate the materials to the realities and make out a good novel. To write a story or novel, or a poem the writer needs a good plot and when it is provided by a dream he finds delight and elaborates and develops it by thinking in all possible ways to make the plot into an enchanting story or play or poem. Thus dream materials and imaginations serve as rough short composition of thought processes for formal elaborate thinking. Many immediately forget the dream materials and also what they have imagined at one time. But some have got the ability to recall most of the materials. Some eminent writers and poets have been doing this and through this they were able to establish their names as famous poets and writers.

Experimental psychologists who studied Rapid Eye Movement (REM) and dreams through various instruments like Electroculogram (EOG) and Electroencophalogram (EEG) did not attempt to record to see what goes on during imagination and thinking. Informations are available only on the dream processes and we do not know whether the same type of recording can be made also for imagination and thinking. Watson (1924) states that thinking involves the movement of the muscles of the gut. Watson asserts that speaking involves movement of the muscles of the throat. Since thought is only subvocal speech, thinking must involve movements of the throat muscles. Garrett (1975) states that thinking is carried on by slight muscular movements. If thinking is taking place by the movements of throat muscles and thinking is subvocal speech imagining also must be taking place in the same manner. According to Watson, speech is louder thinking and thinking is silent speaking. Similarly imagining also must be an activity of silent thinking and it is performed by the movements of throat muscles and slight muscular movements. Therefore, thinking, imagining, daydreaming and dreaming are all aspects of thinking. They are all inter-related and very difficult to isolate one from the other. The outcomes of all these processes are almost the same.

TOOLS OF THINKING

Man always is involved in various kinds of activities. Man is a doing animal. Whether the activity is simple or complex depends upon his previous thinking. Since man is a thinking animal, he thinks and performs activities. He has planned something earlier and now he is executing the same. Therefore, planning is a form of thinking. Thirdly, man is a talking animal. He speaks out what he has thought little earlier or few seconds ago. Man's thinking resulted in inventing the language through which he communicates his thoughts to other. We know that talking also is a kind of thinking. Man directs his behaviour by his thinking. In the process of thinking there are three important aspects. They are *symbols, concepts* and *language.* These three are called tools of thinking. Woodworth (1952) states that memory might be also called a tool of thinking. It is a better tool chest as we can draw our previous experience or acquired knowledge.

This may be happening in your everyday life. Your friend might have visited you and you were just looking at something and doing nothing. Your friend might have asked you, what were you doing about? Your answer normally might have been "I was thinking about my teacher." What does this thinking involve? Let us analyse this. First of all at the time of thinking about your teacher you might have had the image of your teacher–the physical appearance; and then particularly his face, eyes and voice. Sometimes your image might have been on his clean and neat dress, the manner in which he moves and calls the names, writes on the board and so on. All these are images of your teacher and you make a symbolic representation of all his features. Why the freatures are symbolic representations? Because you are not seeing them at this time. You have already formed an image of his features and, therefore, while mentioning the name of the teacher all of the features associated with him are representing as symbols.

Secondly, you fix certain concepts that you wanted to use when you think of him. When you want to place him in certain categories or classes, you connect him with your previous experiences with other teachers or persons in your college. He is tall like Arun, he speaks in a shrill voice, walks fast, very loyal to the institution, and honest and sincere in his duties, very punctual like the principal. Here, he is categorised. He is brought into some type by which persons may be identified and classified. We are using certain concepts referring to some categories or classes. Your teacher also is identified with one of the categories.

Thirdly, you might by using certain words from the language you know. With the help of language you identify him in your thinking and reveal what you thought about him. You think in terms of your language and the words you use. Therefore, your thinking when expressed through your language is having certain limitation. Woodworth (1952) states that we are testifying the presence of our thinking, 'the silent speech' inwardly whether our expression of language is correctly agreeing with our thinking.

Symbols

Symbols and Meaning

Man has tremendous amount of capacity to learn and also has ability to make use of what has been learned. More than this he has unlimited capacity to expand his learning ability by the process of symbolic process or symbolization. Man's ability to learn and to remember is simplified by the symbolic process. Without symbols we would be limited in our ability in communicating our thoughts and solving our problems. A symbol is an internal representation of a stimulus or an aspect of stimulus, or a response or response aspects. This stimulus or response or the aspects of stimulus or response is not present at the moment. Therefore, a symbol stand for any stimulus or response in its absence.

The response might have already occurred in an event or situation or is about to occur in certain situation. What has been called symbols are often been stated as also ideas and what we call symbolic process may be also referred to as an ideational process in the history of Psychology. We may be having sometimes vivid visual auditory, olfactory, kinesthetic and other sensory experiences, which may appear in our imagination (mental images) even without calling for them. We might have utilized such images on certain occasions for symbolic process or thinking. We have dealt with images and imagination in detail elsewhere.

We might have made a symbolic statement that there was hurricane yesterday. In this case a response (hurricane) has occurred. We know the situation that has caused the hurricane yesterday. It was the depression of weather that has caused the hurricane. The depression of weather was the stimulus and the hurricane was the response. Having in our mind the event that has already occurred we can make a statement in this way: "A hurricane is going to take place today as there is heavy depression of weather." This statement is symbolic statement. What has occurred yesterday by means of depression of weather and the hurricane stands as symbol in our mind. We have already formed these two stimulus and response symbols. We know what is the symbol for depression and what is the symbol for hurricane.

Moreover, a symbol may also indicate the quality of a stimulus or response, or situation or event. When that quality is not present in the stimulus or situation we cannot identify the object. In such cases we can make a symbolic statement in this way: "The way of approach to solve the problem is not correct." Thus a symbol provides correct cues to approach a problem. Therefore, a symbolic process may be defined as the utilization of symbols in making decisions, solving problems or adapting to the situations. We use symbols for objects, persons, places and to ideas. Symbols help us to talk about New Delhi, the Parliament House, the Red Fort, while we are sitting at Madras. We speak about the President of India, while we are sitting in the classroom. We also decide what we should buy before going for shopping, predict how the student is going to solve the particular problem.

Symbols help us to make our communication being easy and successful. Human activities in the everyday life would be much affected without symbols. It would not be possible for us to discriminate one thing from the other. Since every object is given a name there is no confusion in handling any object or material. We may be requiring many things or material to start with an office. We are able to list out everything, what we require, in a paper and collect the materials. At the time of listing out the materials we do not see any of the materials, but we write down only the names which are the symbolic representations for the objects. Since every matter is symbolically representing our plans and activities are going without any confusion and also by using the symbols things move fastly and our communication takes place correctly and successfully. Every object is referred to by a symbol.

The word 'World Map', for example, is a large printed sheet in different colours. The object is called a World Map. The World Map is not the symbol for the world. We think of the real world which consists of lands, mountain., seas, animals, human beings, birds, plantations and many things. By the use of the world map we see where the particular country is located, or where the city is, or the sea is and so on by the use of words — the language.

Sometimes we find it difficult to think of a problem. It is because of our thoughts may be going on wandering away from the problem as there are certain stimuli distracting us from the present situation. Woodworth (1952) says that this type of difficulties may be avoided to some extent by the use of symbols which are present in front of us for the objects which are not there or absent. The use of symbols may help us to overcome the disturbances or distractions caused by the various stimuli in our thought process.

Symbolization in Children

The process of thinking has been viewed by Jerome Bruner (1864,1968) as human evolutionary and inventive behaviour. According to him the child progresses through three stages in representing the world. *First,* the world is represented in terms of activities of looking, seeing, smelling, hearing, tasting, manipulating, walking around and so on. This stage is called *enactive representation.* Children represent the world in terms of activities. *The second* stage of representation system involves images-pictures that look like the real object. This stage is called *iconic representation. The third* stage is very sophisticated of human representational systems which is called *symbolic representation.* The iconic representation refers to images. It stands for a real resemblance to the objects they represent. But symbolic representation is purely arbitrary. This is the final stage in the development of a child's representation of the world in terms of arbitrary symbols. According to him symbolic representation includes the use of *language terms* as well as theoretical or hypothetical systems. According to Bruner any pattern for behaviour may be stated as symbolic model, if it is not a person. For example, books, written verbal instructions, motion picture, television are all symbolic models. The human person is not directly involved in these.

According to Piaget (1954) symbolization begins to take place in the life of the child during the sensory motor period. The sensory motor period is from birth to two years. There is no language used at this stage. But symbolization takes place. Symbolization is the internal representation of objects and events. This is through imitation. Thinking is an internalized activity. When the children represent to themselves a real activity internalization begins to take place. The child first becomes aware of the activities relating the objects and events that are present before him/her in the internalization process. In the second stage the child does not perceive any object or event before him/her. In the absence of the object or event the child can imitate. This is internal imitation, which is a symbolic representation of aspects of the environment. This is also called as *deferred imitation* by Piaget. Deferred imitation means the ability to imitate people or events when they were not present. It is also the beginning of language. Finally, the words are going to replace the concrete actions or images. From this simple relationship between activity and environment the processes like imitation, symbolic process and language are evolved. Hunt (1961) states that the more new things a child has seen and heard, he is interested in more new things to see and he has to cope with more variations in reality.

In most of the symbols we use language. Since man is using language for his communication of ideas or thought he uses (verbal) language words to represent objects and events, his thinking goes on mostly in terms of language. A symbol expressed through language or a word conveys the correct meaning. The symbol provides information about some object or event and makes us to evolve more meaning from it by which either we accept and act appropriately or reject.

The use of symbol is not restricted only to the known language of words. There are also other symbolic languages, such language of logic, mathematical and statistical symbols and astronomical symbols. There is also separate language system in telegraphic communications, radio and electrical communications. In all these they use symbolic words and phrases. The shorthand, a system of rapid handwriting employing simple strokes and symbols to represent words or phrases. In defence services and in police forces they use code words, a system of using symbolic words to send secret information in the communication. The real word of the language is not used. For every real word, they use a symbolic word, the meaning of which is not understood by others. On receiving the code message in symbolic words they convert them into correct meaning.

Apart from using language words as symbols for the real words there are also other symbols in the form of signs, sign boards. In road traffic we see some signs on the boards which stand for some real meaning. The sign of a cross stands for the symbolic meaning 'road closed', therefore do not proceed further, the figure of a school boy on the board is the symbolic expression that it is 'school area', drive slowly. There is also a figure of horn with a cross on it on roads. This is the symbolic caution for not making any horn in that area, probably the hospital area. We often see on the electrical posts a figure with a skull and two bones. This is a warning, symbolically that there is danger if you go nearby. In the case of such signs and boards you need not stand there for few seconds and read the instructions. The figure and signs stand for the cautions and warnings. On seeing these symbols people understand the real meaning of them and react to them appropriately. In all these, language is not used, but symbols are used to convey the meaning. We might have also seen the red flags on the road, which the symbolic expression of danger if you proceed further beyond that place. Mostly, where there are firing ranges there may be red flags at certain distances away from that place.

The military and police personnel wear different types of dress. The dress symbolically indicates that they are military and police personnel. They also wear in their dresses certain sign in the form of stars, crowns, and swords. These signs stand symbolically for expressing their ranks. Thus symbols are used in various ways to communicate the correct meaning and information. There are no disputes and difference of opinions with regard to the using of signs and symbols for certain specific purpose.

What we call symbols are referring to only ideas and the symbolic process may be referred to the ideational process. Ideas are often considered as synonymous with images. Images refer to sensory experiences, which cannot be experienced in the absence of an appropriate external stimulations. We have sometimes vivid visual, auditory, kinesthetic experiences in our imagination. We utilize such images for our symbolic thinking occasionally. It is often stated that the prefrontal area of the cerebral cortex (area of the frontal lobe) is mainly responsible for the formation and utilization of symbolic processes. However, the efficient use of symbols increases as a function of the cerebral cortex in the development of the individual. In all human symbolic activities language is greatly involved. In certain situations you may use language, but in certain situations you may use certain visual cues — a visual image of the place you wanted to visit. In either case you use symbols in solving your problem. Man has invented language only upon symbols and symbolic processes. The symbols and symbolic processes in turn influence the development of language.

Meaning : Semantic Differential

Since human beings are using language in all their activities the meaning of the words they use sometimes receive greater importance. A word in a language is having some meaning. It may *suggest an idea* which is called *connotative*. A word may *indicate its meaning* explicitly or *overtly* — which is called *denotative*. If all symbols are referred to only specific objects or events there is no difficulty in using the words. For example, chair, bench, school, etc. or walk, run etc. These are denotative. They are specific, something which all of us know. But in the case of many words connotative meanings go with denotative meanings or connotative meanings are attributed to denotative meanings.

In our usage of language words we evaluate certain words or we prefer to use a word to indicate certain meaning. The connotative meaning is emotionally surcharged. It may vary from person to person and from culture to culture. What is accepted in one group is not accepted in another group. There may be difference of opinion and misunderstanding because of the different connotatives of the words among different groups of people and culture. The important point in this phenomenon is that the meanings of certain words are going on changing which may give different reactions in the people who are using those words. For example, the word 'Chairman' was used to address a person who was presiding a meeting till very recently whether it was presided by a man or woman. The meaning of the word 'Chairman' has irked a section of the population —particularly the woman population. Therefore, it was replaced with the word 'Chairperson', the meaning of which is applicable to both men and women.

In our culture, for example, 'give me your pen' does not mean anything. But in European culture it has different meaning. The correct word for them is 'Lend me your pen.' Similarly the word 'vice' and 'nice' have both good and bad meanings. Therefore when using such words as symbols for indicating an object or event or quality, the meanings of such words become problems. Observing the confusions caused by different meanings of the words Osgood, Suci and Tennenbaum (1957) have studied these aspects and carried out factor analysis of the effective aspects of large number of words in order to measure variations in their contents and the information communicated by them. Based upon this study Charles Osgood (1962) and his associates have formulated a method called 'Semantic Differential' for indicating the meaning of the words more precisely.

In determining the connotations of words by the method of semantic differential, a person is asked to rate a word in between two adjective bipolar pairs of words. The two bipolar positions are divided into seven parts indicating varying the degree of acceptance. The connotating meaning is expressed in terms of three basic dimensions such as *evaluative dimension* (good or bad, refined or rough), *potency dimension* (long or short, heavy or light) and *active dimension* (busy or rest, fast or slow). The person's rating may fall near one of the contrasting pair of adjectives. The average of the placements made by a group for each concept on a Semantic Differential Scale may be taken as the connotations of the group for a word. Osgood (1967) studied this with .arious cultural groups and the results were found to be quite convincing.

CONCEPTS AND CONCEPT FORMATION

Concept

Concept refers to the formation of some abstract idea or thought in the mind. It is a general abstract idea of something formed mentally, combining all specific parts and characteristic features. Fundamentally concept refer to an idea. It also refers to an idea representing the meaning of universal term. Therefore, we can also state that the meaning of the word is a concept. A concept is an abstract idea or abstraction formed by generalizing from many experiences with particular things and events. It is general in its application and does not refer to only one particular member of a class. It also includes the essential qualities of a class or species. Concepts are mainly referring to categories for classifying specific objects, persons, animals, and events on the basis of certain common elements or qualities.

A group of objects or persons of the same type having the same qualities may be categorised under one head and a name may be assigned to them. This we call as a concept. This concept is symbolically standing for a group of objects or persons. Therefore, a concept also is a symbol. Gardner Murphy (1964) states that a concept is a symbol which stands for a specific quality possessed in common by a number of stimuli or objects. According to Moskowitz and Orgel (1969) concept is a response or a reaction made to some quality or attribute of a situation instead of to the whole situation and the same can be generalised or attributed to a new and different situation having the same quality or attribute. According to them concept can be developed by repeated experience in human learning situation.

There are objects of different types in the world in which we live in. All these objects possess certain common qualities. It may be in size, shape, colour or in fineness. The common qualities possessed by some objects make them different from objects. When a person learns to respond to certain common qualities specific to a group of objects and learns to ignore the objects possessing other, a different type of qualities, he is said to have learned a concept. For example, *animal*, it is a concept, a general concept. The description is that all those creatures which are having four legs and a tail are called animals. *Dog* is a concept, specifically referring to a particular kind of animal. House, automobile, and chair are all simple concepts. Belief, faith are abstract concepts. Shorter, longer, smoother are concepts based upon common relations.

The child slowly learns to comprehend persons, objects and events. It is very difficult to say at what age the child makes transfer of one situation to another and group the common qualities the two situations have. The child not only finds common quality in two things and makes transfer, but also has to find names for them.

The adult persons throughout their life experiences may be trying to find out real fact or information in a number of different events or situations by which they can be quite confident that the rule that is applicable to one event or situation may be also applicable to all other events or situations. By this means they are trying to find out some general principle common to all. This we may call generalization. In all our psychological studies we are only trying to find out the common variable or factor that is operating in the whole group and classify people on the basis of this generalization.

Concept Formation

Concept formation takes place from the childhood period throughout the life experiences of human persons. The nature of concept formation varies according to age and experience. When a symbol stands for a class or a particular category of objects or events with certain common properties we call it a concept. Concept formation may be also stated as the process of discovering some characteristics that is common to series of objects. It mainly refers to the description of the common elements. It differentiates the quality of one object from another. For example, the child gradually learns to comprehend persons, objects and events. He sees a pencil and understands that it is a pencil and it is used for the purpose of writing. He could differentiate a pencil from an ordinary piece of stick or fountain pen. Secondly, when he sees the pencil, he sees it as red in colour and he differentiates it from a green pencil or from other colours. In this context the child has learned the concept of red colour or redness and also has learned to differentiate the properties or qualities of the same type of objects.

The child thus learns that pencil may be red, green, blue or yellow in colour. Pencils are made of wood and lead. This is another common characteristic the child uses to differentiate pencils of different colours, from fountain pens, ballpoint pens, sketch pens and other writing materials. Thus we see that children learn colour concepts and concepts of concrete objects more easily. When a child develops the concept of a chair he/she deals with it as a furniture. All those objects that have a seat, four legs, a back and two arms are called chairs. Whenever, a child sees such an object he calls it a chair. If he/she hears the voice of some one saying 'chair' he/she forms the physical picture or the image of the chair and is able to describe it. Other furnitures may also be having a seat and four legs. They are not called chairs for the child. In this way when correct generalizations of the objects with special reference to certain common characteristics are developed they are called concepts formation. The process of concept formation is also quite similar to the process of discrimination of learning. Concepts are formed on the basis of common properties and we allow the name of the concept to cover all the objects that have the same attributes or properties.

Experiments conducted by Kendler and Kendler (1961) show that children who have not acquired relevant verbal chains or stimulus-response associations to represent the stimuli cannot form concepts. The previous learning in the form of language is an important internal condition for the formation of concepts. According to Gagne (1970), there are four external conditions that make easy the concept formation. (1) To elicit relevant stimulus-response association from the child all objects should be presented at the same time. To teach the child odd or strange two objects should be identical and one must be odd or different from the other two. (2) In this, varieties of objects are presented, but a reward is placed only with the odd object. (3) Additional examples may be added to verify or test the child's understanding of the concept. (4) The same can be repeated to reinforce for the formation of the concept.

There are two important psychological processes operating in the concept formation. They are generalization and discrimination. In all the concept formation processes the child or an adult person uses both generalization and discrimination.

Generalization

The term generalization appears to be simple. But it is very difficult to define, that can be accepted by all. We can illustrate it through a simple definition. Generalization in concept formation may be defined as the process of detection by the person of a principle or characteristic common to a class of objects, events or problems. It is widely inclusive of relationship among the object or events. Therefore, all the principles or laws of science, the rules of grammar, the mathematical theorems, the principles of arts, in fact, a great part of all these that are taught and learnt are generalizations. Because they mainly express relationships which is based upon generalizations. For example, in geometry, it is often stated that the square of the hypotenuse is equal to the sum of the squares of the other two sides, is generalization about a right angle triangle.

Generalizations need not be always verbal. It can be also expressed one's emotional level. A person may state that 'the whole world is hostile, all the ways for happy living are closed, all are conspiring against me'. Here words have little use, the person's feeling level is generalized. Generalization can be formed through experience. We can make generalization by abstracting and reorganizing the elements of experience. Our generalizations are mainly based upon our perceptions and feelings. People who have strong conviction in tradition generalize certain everyday events. For example, while going out on some business, the appearance of a cat, firewood cart, is considered as a sign of bad omen. The racial, religious, national and caste prejudices are generalized attitudes. Beliefs and attitudes of this type are based upon deep-rooted false generalizations that have been accepted uncritically.

Generalization in the beginning is very rough or crude. A little child on seeing a horse for the first time might call it a cow or a donkey if he has already learned something of the concepts–cow and donkey. The new animal is related to the old in some manner. But when he comes to know that it is not the same animal that he saw earlier, he is ready to learn the new name and differentiates it from the others. The child has generalized by reacting to similarities, that is, all animals have four legs and one tail. The more nearly the same or alike the new stimulus to the original there is more likelihood of substituting it for the original. This principle is called generalization.

Discrimination

Discrimination is a process complementary to generalization. The main difference between these two processes is that in generalization the reaction is to *similarities,* whereas in discrimination the reaction is to *differences*. Discriminations involve acquiring the ability to differentiate among similar stimuli in order to respond correctly to those stimuli.

The child mostly learns the concepts by seeing and by asking questions. Sometimes we teach concepts to a child. For example, a cow is introduced to the child. The cow can be assigned to the general *concept as animal* or mammalian. Cow is an animal, it has four legs, a tail just like other animals and also it has two horns. Secondly, the concept, 'cow' helps us to differentiate it from other animals, because it is specifically stated as 'cow' among the general species, the animals. From the personal experience, reading and from some other sources the child may understand that most of the cows are harmless animals. Therefore, the reaction to a cow on seeing is not with any fear or alarm that human beings normally react to an elephant or tiger on seeing it.

Thirdly, the concept cow may be used for *abstracting* without making mention to any specific animals. In this way the cow has been discriminated from other animals that have certain common features with the cow. This is called abstraction. Abstraction is the process of isolating certain characteristics of an object of the same class of objects and disregarding all other characteristics of the same object at the moment. In the case of the cow we are isolating it from other animals that it is soft and quiet and it supplies milk. This cow is discriminated from also other cow-like animals, for example, a bull, or the cow not supplying milk.

Lefrancois (1979) cites a common task to illustrate the process of discrimination. It is a motor discrimination. The task is selecting the appropriate key from a number of similar, but different keys. It is also possible to find out the suitable key by the process of trial-and-error. But here what required is that the individual has to learn to discriminate among a number of keys. One factor that should be present for discrimination is the presence of the related individual chains (stimulus-response bonds). A chain is a series of stimulus response bonds, in the bonds *each response in succession serves as stimulus for the following response.* Gagne (1970) used this term in the context of learning. A chain may be also stated as simple stimulus-response associations. When there are more than three stimuli present in the situation the process is called multiple discriminations. Several external conditions are also considered to be important in the discrimination process.

Among several stimuli, each one, that is to be discriminated should be presented for the purpose of evolving the stimulus-response association (the chain) suitable for it. Discrimination is a complex process. It is very much prevalent in school learning. It involves to make different type of responses to printed letter, numbers, words, phrases and figures of geometric type. It also involves learning to discriminate between classes of things and also to identify similar objects. Generally, to discriminate is to respond to differences among the stimuli and learning a concept to respond to similarities.

According to Gagne (1970) we must make use of concept to get mastery over a principle. For example, the concept of dog, the man is having a dog, and his relationship with the dog, must be known comfortable without any trouble or pain. The little child is required to make multiple discrimination and also appropriate discrimination. The dog is different from small animals like the cat in several ways. First of all the child must differentiate the dogs from non-dogs. For this he must make different verbal responses that describe the differences. Then he must associate the words to things. In this there must be verbal associations. These verbal associations are based on combinations or simpler stimulus-response associations. For example, this is a dog, he is different from other animals like the cat, donkey and horse. He has a name, as 'Tiger'. Mr. X is the master, when he calls it by the name 'Tiger' the dog fondly runs by his side wagging his tail, expressing his affection. He watches the house. He is very alert. Thus the child makes discriminations and uses different concept which he has learned earlier.

CONCEPT ACQUISITION IN LANGUAGE AND MATHEMATICS

Piaget's study of the four developmental stages throws much light upon the development of thought process. In the act of thinking emphasis has been made on language and mathematics. Therefore, without discussing these phenomena any attempt to explain thinking is incomplete. Piaget (1965) has published his studies 'The language and thought of the child' after completing his experiment on two six years old boys in morning classes at school. Whatever the boys said was recorded in complete detail. The data were analysed and classified. The language of the two subjects was classified into two categories, as egocentric language and socialized language. These two together is called spontaneous language.

In the egocentric language there is no reference to objects or to any one who is listening to his talk. The child speaks only about himself. Three elements were observed in the child's egocentric talk. Firstly, there was repetition with sense and non-sense language. Secondly, it was monologue, it was like the child's louder thinking. Thirdly, there was collective monologue talk of another person's physical presence. He spoke loudly in the presence of others.

In the socialized language there are five functions. The first is adapted information. The listener's view is admitted. The second is criticism. In this there is importance for affective aspects. Other's view are criticized and the self is ascertained. Thirdly, there are commands, threats and requests. There is need for interaction between one child and the other. Fourthly, there is raising of many questions and fifthly, the answers are expected from others, other children. This is how language development in children is taking place. Piaget states that the adult thinks socially even when he is alone, but the child under seven years old thinks ego-centrically even in the presence of others.

There are behavioural outcomes within the context of language development. Mathematics, for example, creates interest in two ways. One is the context, that is to comprehend the contents of mathematics and the second is the desire to utilize it. Piaget (1953) states that what the child learns about the mathematical concept is gained spontaneously and independently of teaching by somebody. At various developmental stages the child acquires more concepts of his own. Piaget says that the use of mathematics is a natural development.

According to Piaget for the development of *number concept* the child should understand the fundamental law of the sources of quantity. This is not based upon numerical principle, but it is a *logical concept.* Children very easily put equal quantity of material of the same type, whether it is number or in mass. At the age of seven children evolve this spontaneously and thus learn *number concepts.* Piaget says that even in the case of geometry no teaching is required to find out the spatial relationships. Piaget and his associates, to their surprise, found that children develop geometrical order in the reverse order of normal course of discovery. They found that there were more children from topological geometry to projective geometry. That is, they draw small circle inside the large circle. The approach made by children is close to logical construction of science. The concept acquisition, according to Piaget is mainly related to age and mental development.

From the studies made by Piaget, it appears that thinking or thought process is mainly based upon physical and mental developments. In other words, it is based upon intelligence and maturity. The experiments conducted by Piaget and his associates on children to find out how they acquire concepts in language and mathematics exemplify this view. Piaget (1961) says that thinking is a dynamic act and it is understanding and understanding is the development of transformations. According to him the dynamics of thought are dealing with transformation, which are revealed in the acts of various types. For example, dismantling a Television box is to understand how it functions.

Characteristics of Concepts

Concepts are formed through learning and experience and the concept formation is essential for an individual to react in an appropriate manner and also to a number of varied situations, responding to events and stimuli. In all these there are certain common characteristics and the individual is responding to them by the process of generalization and discrimination. The nature of experience the individual has acquired or learned seems to identify two important characteristics in the concepts. They are breadth and clarity.

The Breadth

The breadth of the concept refers to what it contains. It means that it has a very wider range. The organism has acquired a variety of experience from which the breadth has developed. For example, animal is broader than house, man is broader than son, river is broader than a canal, colour is broader than green. This is how concepts are learned through variety of experiences.

The Clarity

A concept has a characteristic called clarity. It refers to the clear quality the object possesses. Clarity refers to the fine edge of the concept's boundaries. This quality of the concept helps us to easily and accurately differentiate the concepts. Further, we appropriately use the concept without any mistake. We do not misuse the concept because of this sharpness of the ends. For example, a sword is a clear concept than pardon, and an eyesore is sharper than headache, an object at five-metre distance is sharper than object at one-kilometre distance.

Man has greater advantage over animal in his ability to form concepts. He has another greater advantage of possessing a mechanism for the expression and remembering of a greater number and variety of concepts that is not possessed by any other species. Because of this mechanism adults, through years of experience and learning complex skills, have greater level of concept formation than children. This mechanism is language. Children take some years to learn language and then learn concepts.

TYPES OF CONCEPTS

Teaching and learning are concerned with concept formation in an endless manner. We can say when concept formation takes place in human beings. But we cannot say when or at what age of a man concept formation ceases to develop. Because now and then new concepts of complex types arise and man begins to learn all the related concepts. For example, man's invention of the computer puts man constantly in learning new concepts. He has to learn them.

How for a child succeeded in the development of concept formation depends upon whether he has a clear concept of animals like, the cat, dog, horse, donkey and cow, or inanimate objects, like, chair, table and bench, material like pencil and pen, number concepts like, four, five and six, colour concepts like red, green, blue and yellow, relative concepts like, large, small, long and short, equal and so on. He must separate concept gloomy from the related concept of cheerful. He must also see the difference between abstract concepts like justice, mercy, kindness and faith.

In making the child to master certain particular concept we must choose some circumstances. What important concepts are needed in certain situations? The situation may facilitate the child's ability to comprehend and learn the concepts. But even by this method, some concepts may be easier to learn than other concepts. The teacher should have some control over the situations and the concepts at every time as to which concepts the child should learn now and which when the child becomes older. In learning different types of concepts, age and experience play important role. In deciding the difficulty of a concept in a particular situation, it is also important to see whether the child is ready to comprehend such concepts. In this context we must see whether the external conditions that could facilitate concept formation.

In concept attainment it is true that we use relevant quality or characteristics to detect or identify the particular class or category. But it is also important what *rule* or *principle* is exercised in this task must be ascertained. If we comprehend the rules or principles that apply to them we can identify different types of concepts. It means that in concept attainment there are different types of concepts.

In the concept attainment it is important to decide whether a concept is conjunctive type or disjunctive type or simple type or rational type. Thus there are three types of concepts. But authors discuss the types of concepts according to their conveniences. The types have been classified on the basis of the qualities attributed to the concepts. Let us discuss them in an order convenient to us.

Simple Concept

Some of the concepts are very simple. The concept, for example, red has only one attribute. It is redness. Red is the quality found in many things. For example, fire is red, blood is red, the danger signal is red. In some big cities there are red light area stating that it is a dangerous zone.

Conjunctive Concepts

Conjunctive concepts are those concepts that have two or more attributes or qualities in them. Conjunctive concept refers to the characteristics possessed by persons, animals, birds, objects and events in common which can bring together under one class or category. For example, a conjunctive concept such as 'cow' supposed to possess all the attributes. First of all it is an animal, polite animal, having all the physical characteristics like four legs, one tail, two horns and gives birth to a calf and yields milk. It is a mammalian. The cow has clear cut distinction from other animals and no animal can duplicate it. Any animal which is lacking any of these attributes is not called cow. Similarly the concept of dog may be classified as conjunctive concept. Just like the cow, the dog has clear cut attributes to be called as 'dog'. Though there may be several varieties of cows and dogs they all have the same characteristics.

Disjunctive Concepts

Disjunctive concepts are those concepts that have one of these features or the other feature or all the features put together may bring into a category or may qualify the concepts to be called as disjunctive concepts. For example, in cricket game a man who is batting is said to be 'out'. The umpire decides how the batsman is out? It can be wicket out, or catch out, or run out, or stump out. Either one of these attributes would justify the umpire in classifying this event as an 'out'. Similarly, you may call a person by using the concept as your 'uncle'. The attribute here is not clear cut just like father and son. The person may be your father's brother, or your mother's brother, or your aunt's husband, or your mother's sister's husband. Anyone of these relationships qualifies him to be called as uncle. Out of several attributes, one of them is correct. There are more complicated type of concepts for which there are other rules for defining them.

Rational Concepts

The concepts which require certain rational bases for classification may be called as rational concepts. The rational basis are inferred or logically derived. The attributes are generally comparative in nature. For example, there are three lights illuminating a place. The first one is bright, the second is brighter and the third is the brightest. The concept brighter is attributed to the second light. When we indicate the second light that it is brighter than the first one. In the same manner the concept heavy has no meaning without lighter. The concept larger has no meaning without the concept smaller, so also the concept longer has no meaning without shorter. One concept is drawn rationally with the help of another concept.

Distortion in Concept Attainment

In the real life we find that certain things are not having clear cut distinction. We find many things that are indistinct or unclear and also distorted to some extent. It is not an illusion. But it is a conceptual distortion. When you present a concept it is not specifically referring to a clear cut object. It is mainly occurring in conjunctive concepts. For example, the conjunctive concept 'reptile' refers to the cold blooded crawling vertibrate creatures, like the snake, the lizard, the turtle etc. This concept is referring to three different types of creatures. The only common attribute to them is that they are crawling creatures. Rosch (1977) and Eleanor Rosch (1973) have studied the distortion characters in the conjuctional concepts.

Let us see another example, the conjunctive concept of 'bird'. The common feature of birds is flying. They have two long legs, have feathers, they build nests, sing, change the places according to climatic conditions. But, for example see some of the conjunctive concepts like Sparrow, Eagle, Crow, Goose, Chicken, Penguin. They all classified as birds, but the quality of birdness is going on decreasing step by step. The first three are very good examples of conjunctive concept. The eagle and the chicken do not possess the same quality. The eagle has more of the attributes of the bird, but the chicken has lesser attributes of a bird. Therefore, some confusion is caused in children in concept attainment when they confront the conjunctive concepts like eagle, penguin, chicken, bat and stone. For a child if all birds are flying stones are also flying, if they are thrown in the air. In the studies made by Eleanor Rosch (1973) some children included bats in the list of birds.

EXPERIMENTAL STUDIES OF CONCEPT ATTAINMENT

Concepts are formed by identifying the common properties or factors possessed by a number of objects and assigning names which cover all the objects having the same attributes. It may be also necessary to find out two or more common attributes that are running through several items. There are not many experimental studies on concept attainment. Experimental study on concept attainment makes the process simple. One of the earliest studies was made by Kurt Goldstein (1935) in the clinical situation among the brain injured persons. He has shown that a person who was suffering from brain injury has lost the capacity to think in abstract terms, but he may group things together.

For example, object of different form, colour, material or use are presented to a brain injured person and he was asked to put together those things that belong to the same category. The person may be using some method of classification. This would reveal whether he is employing concrete or abstract concepts in classifying them. If he would classify and group the objects on the basis of form or colour it indicates that he is using abstract concepts (attitude). But on the other hand if the person classifies and groups them only objects that are identical or similar in characteristics or on the basis of some specific use, he is using concrete concepts (attitude). For example, a person was given a task. He was asked to group yarns of different colours, by putting all the reds together and all the blues together and so on. But the person finally put all the reds that were exactly the same together and the reds that are slightly darker and the reds that are slightly lighter did not use for the classification. The reason he stated was that they were dark red and light red. He could not abstract that all reds should be put together. He has used just concrete concept in classifying the colours.

To illustrate this point in another way, a person takes abstract attitude or uses abstract concept if he puts into one category of 'instruments' or 'tools', the spanners, the hammers, screw drivers, nails, hacksaws of various sizes. He may use concrete concepts if he chooses one spanner and one screwdriver stating that these two are enough to remove the nail from the box. The person takes abstract attitude or uses abstract concept when he puts all the animals like, elephants, tigers, lions, monkeys, hippopotamus and reptiles into one concept, zoo. He takes concrete attitude or uses concrete concept when he wanted to see only the hippopotamus.

The capacity of an individual's abstract and concrete thinking can be tested by a test. Vigotsky (1962) has designed and developed a test in connection with his study of thought and language and this test was adopted by Haufmann and Kasanin (1937) for the study of concept formation. This test has been used in clinical practices. Vigotsky (1962) has pointed out that the understanding of certain concepts take place prior to the development of relevant terms and expressions in the child. According to him, for example, children should understand the logical properties of dimensions before understanding such words as bigger, smaller, shorter, longer and so on become meaningful for them. He states that there are virtually unquestionable evidence of thinking among non-human animals and proverbial children. From this it is clear that language is not necessary for thinking. But language is extremely important in thought and communication in societies.

In concept formation the Haufmann and Kasanin test has been widely used. This test consists of twenty-two blocks in five colours, six shapes, two heights, tall and flat and two sizes, large and small. There are 5 yellow blocks, 5 orange blocks, 4 green blocks, 4 blue blocks and 4 white blocks. With regard to *six shapes,* there are 4 triangle blocks, 4 quadrilateral blocks, 4 square blocks, 4 circular blocks, 2 semicircular blocks and 2 hexagonal blocks, a total of 22 blocks. There are 11 *large blocks* and 11 *small size blocks* with a total of 22 blocks. With regard to *height,* there are 10 *tall blocks* and 12 *flat blocks* with a total of 22 blocks. At the hidden side (bottom) of the each block a *three-letter* non-sense syllable is printed. This serves to classify the blocks into categories. *All the 5 tall* and *large blocks* are having the non-sense syllable *LAG* at the bottom, *all the 5 tall and small* blocks are the non-sense syllable *MUR,* at the bottom, *all the 6 flat and large* blocks are having the non-sense syllable *BIK* and *all the 6 flat and small* blocks are having the non-sense syllable *CEV* at the bottom.

In this test the *colour* and *shape* do not serve any purpose as far as classification is concerned. These two are an extraneous variables and are *distractors.* The *size and height* are the independent variables, important in this experiment. The blocks are spread out roughly into a circle or half circle and the subject is told to arrange them into *four groups* using any principle that he finds suitable for this task. But he is not allowed to reverse the blocks and see the bottom. The performance is scored according to the *number of trials* and the *amount of time* the subject has taken to complete the task by classifying them into four groups. When one trial is finished the experimenter should check up whether it is correct or not. If it is not correct he should show to the subject how it is not correct by showing the bottom letter of two different blocks from the same group. Then the experimenter once again should assemble all the blocks and arrange them into a circle and ask the subject to try again. The test performance should be continued with the same direction until the subject classifies the blocks into four groups correctly. The experimenter should find out whether the subject has classified them into four groups using a logical hypotheses, or by trial and error method or by guessing procedure.

As the subject try grouping the blocks in one way or other the experimenter should note down his behaviour. He should tell the subject whether the arrangement is right or wrong. Actually any subject in this experiment may first try according to colour and shape. But what counts actually in this test is the height and size. Thus the four categories are *large and tall, large and flat, small and tall* and *small and flat.* The possibility of committing error in this experiment is always there as the concrete attitude is often dominating over the abstract attitude. Murphy (1962) says that even perfectly normal people often lose their mental sequence and slip into a simple method of using the colour and shape even on finding it out that it is not correct.

Haufmann (1951) in his study was able to observe two fundamental types of sets which he called perceiving set and conceiving set. The people who approach the problem with *perceiving set* look at the blocks for natural grouping putting more importance on the physical look, concrete approach without thinking of a common factor operating in it. Whereas people who approach the problem with *conceiving set* try to find out a rational basis, a common factor that may be operating in it for grouping. They seek some abstract principle by which they can solve problem.

The experimental study on concept attainment made by Kenneth Smoke (1932) is one of the earliest studies. He used figures in his test. The figures are presented to the participant to develop concepts such as DAX. The test consisted of a circle one dot inside and one dot outside. Smoke used figures with different features. These figures had different shapes and sizes and the number of dots also varied. The dots are located at different parts of the figures. Some of these characteristics or attributes are relevant to the location and some are not, except that a dot is put inside and another dot outside. For attaining this concept DAX the subject has to indicate which attributes are relevant and which are not relevant.

In this experiment of concept formation, Smoke made it clear a point that the task of the participant is not only to detect the relevant attributes (features), but also to ascertain the principle or rule he applied in it. According to Smoke the DAX refers to a figures that is circular in character and has two dots, one inside the circle and other outside the circle. These dots can be anywhere inside the circle and also anywhere outside close to the circle. Having this as the principle of this experiment the participant should identify different types of concepts. Smoke used different figures with dots in his experiment. But a figure other than a circle with two dots is 'Not DAX' in this experiment. These figures or stimuli are presented to the participant one by one and he has to give his response by stating, whether it is 'DAX' or 'Not DAX'. If the response is not correct the experimenter should provide feedback. That is, trials should be carried out till the participant provides correct response to all the figures. The figures are presented in a successive rows. It was observed that the participant was able to master the concept in the sixth trial.

The important aspect of this experiment as observed by Levine (1966) was that the participant was testing the hypothesis in the concept attainment. The participant in this experiment in mastering the concepts was going on testing the hypothesis one by one after seeing the previous hypothesis was wrong. For example, in the *first* trial the participant may hypothesize that the concept is a simple one, that a dot is on the left of the figure irrespective of other things. In the *second* trial he is testing this hypothesis and finds that the first hypothesis is wrong and develops a second hypothesis that the concept is two dots one inside and one outside irrespective of other things. In the *third* trial since he sees a different figure with two dots he rejects the second hypothesis, as there was no circle with two dots. In the *fourth* trial he sees a circle with two dots which he already rejected it as NOT DAX and he was told that his response was wrong. Now the same figure appears and the subject learns to respond correctly by saying it is DAX. Now he understands that anything other than a circle with two dots is not a DAX. In the *fifth* trial, a figure with two dots outside the circle was presented and he was told that it is wrong. Now the subject begins to think that the concept DAX is a circle and two dots, but where the dots should be? From the previous experience he understands that the DAX should have one dot inside and one dot outside the circle. In the *sixth* trial, a figure with one dot inside and one dot outside was presented and the subject at once responded with DAX. Now he has learned perfectly how to identify the concept DAX. In the subsequent two trials he confirmed it following the principle.

The subject has taken six trials in mastering the concepts. He was going on testing the hypothesis one by one after understanding that the previous hypothesis was wrong. Levine (1966) has observed two aspects in the behaviour of the participant in formulating the hypothesis after every trial. The participant's behaviour was analytical. In the analytical approach he tried to isolate the attributes of the materials and the rules that govern the concepts. This is what the experimenter expects from every participant.

The second aspect, Levine has noticed was that some participants were not analytical. Generally, children were found to be non-analytical. For example, for children the concept bird includes every creature that flies in the air including bats, penguin and the concept fish includes every creature that lives in water, all kinds of fish, dolphin and whale. Levine states that in concept learning experiments it has been observed that most of the participants adapted a strategy of 'loss-shift and win-stay'. Therefore, the correct response is gradually narrowed till the participant reach the right concept.

Several experimental studies have been made to illustrate the process of concept formation. The psychological processes of generalization and discrimination are used in concept learning. Experimental evidences show that language is not necessary on concept learning. It is observed that rats and monkeys learn concepts. Hilgard *et al.* (1975) state that the stimuli that have certain amount of similarity with the original or stimuli more nearly alike are to the original, the more fully they will substitute the original. This principle they called generalization. This is how our ability reacts to some new situations as they are similar to familiar things. But the process of generalization diminishes in a systematic manner if the second and third stimulus becomes more and more dissimilar from the original stimulus. A form of process that completes the process of generalization is discrimination. Thus the process of generalization is a reaction to similarities and the process of discrimination is a reaction to differences.

Hilgard *et al.* (1975) states that rats learn the concept of triangularity. They were able to select triangles of various shapes and sizes for rewards and they were not responding to other forms. Similarly, they state, that monkeys learn the concept oddity. Though the stimuli varied trial to trial the monkey is responding to the common feature oddity and not to specific object. In both cases, the animals exhibited the process of abstraction, responding to triangularity (though the shapes and sizes are varied) and to oddity (though the objects varied from trial to trial).

A child, who learns the concept, 'dog' may generalize the concept in the beginning to all four legged small animals. Any four legged animal, for example, if a cat is seen the child at once calls it dog. Gradually the child learns to make correct discriminations from the corrections made by his parents and others and also from his personal experience. The child learns still further to discriminate between the house or friendly dog and the street dog and learns various concepts of the breeds. Since man has invented the language he uses it as a tool to express his thoughts and easily communicates his thoughts to others. He can easily deal with all sorts of single concrete concepts such as dog to highly abstract concepts such as faith, justice and so on which he has been attaining through his experience.

Experiments have been conducted to understand what types of concepts are formed most readily and easily. Few of the earliest studies made after Smoke, were the studies made by Heidbreder (1947) which were found to be very interesting and worth discussing. Heidbreder's studies revealed that the concepts with concrete features were found to be easier to conceive than the concepts with more abstract freatures of form and numbers. She employed paired-associate method with different types of materials.

The test is called 'attainment of concepts', which is consisted of thirty pictures. The author of this test wanted to find out how concepts are formed with the use of these pictures. All the thirty pictures are put into five sets and in each set there are six pictures. Under each picture of concept one non-sense syllable is associated. One non-sense word is printed at the bottom of every picture. Stating generally, in Heidbreder's concept formation test there are only six pictures, namely (1) human face, (2) circles, (3) buildings, (4) number, two, (5) number five, and (6) tree. The same six pictures are drawn little differently in five trial sets and each picture is paired with the same response (non-sense) word. The important characteristic of this test is that the pictures in every trial varied little, but the response (non-sense) words are the same. For example, under every human face in all five sets of trials the response (non-sense) word, RELK is given. Similarly, under every circle the non-sense word FARD, under every building, LETH, under every two number items, LING, under every five number items, DILT, and every three item MULP is given. Therefore, in every trial the same response (non-sense) word is repeated.

A trial consisted of six pictures. In each trial one set of six pictures are presented to the subject one at a time, starting from the top of the column and proceeded to the bottom. As the picture with the non-sense word at the bottom is presented the subject is expected to associate the word with the picture. When all the pictures of the first trial is presented, the subject is expected to call out the non-sense word associated with each picture. After the subject responded to the experimenter's order, the experimenter reads out the correct non-sense word which is written at the bottom of each picture.

On the first trial the participant has no chance of knowing the response word associated with each picture. What word was paired with what picture? He has no means of knowing this. But after several trials he slowly learns to respond correctly to all the pictures. Here it is not learning specific response (non-sense word) to the specific stimulus (picture). The same picture is not repeated. Therefore, it is learning to associate the response (non-sense word) to a concept. In this test, for example, the concept, 'face' is indicated by the (non-sense word) response, RELK. But in all five trials five different faces appear. Therefore, the response word, RELK is not specifically referring to a particular face in a trial. Whatever type of face is presented, *i.e.* whether it is a man's face, or woman's face or a child's face, RELK is referring to the concept of human 'face'. In Heidbreder's experiment, RELK is a response word to *concrete concept* of 'face', 'FARD' is a response word to *spatial concept*, 'circle' pattern, LETH is a response word to *object concept*, 'building', and LING and DILT are response words for *number concepts* of 'two' and 'five'.

The materials used in this experiment are such that the numbers or objects of each conceptual group do not have identical features, but they have something in common — on verbalized concept, for example, 'face'. All faces are to be called with the response word RELK, all circles are to be called with the response word FARD and so on. In these types of experiments, it has been observed that concrete and object concepts like tree, book, bird, face are the easiest to learn, the next easiest are the spatial concept, like circle pattern, figures, and then the number and abstract concepts. Generally human thinking tends to be more for objects rather than for abstraction. It appears that Heidbreder has developed several types of experiments on concept formation. The speed of concept learning depends upon how easily the stimulus generalization takes place. There may be interference in the process of generalization and certain similarities may cause confusion in identifying the concept. Because of this there may be slower learning on concept formation. According to Piaget (1952) children learn first only object concepts and they develop abstract concepts when they become older. Brain injured persons respond only to object concepts and fail to respond to abstract concepts.

Various forms of concept formation tests have been developed to study experimentally as to how concepts are formed in children and in adults. One of the important test designed and developed for experimental purposes was the test with 81 cards with varied characteristics. Bruner, Goodnow and Austin (1956) have developed this test with different figures such as circle, square and cross, different colours such as black, red and green, different numbers such as one, two and three and with different borders like single, double and treble. The cards are presented to the subject one at a time and he is expected to find out the concept the experimenter has assigned. The authors have observed that the subjects followed various methods in concept formation. The experiment is conducted in three stages and the procedure is highly cumbersome and it is very difficult to conduct this experiment in ordinary situation without much preparation on the part of the experimenter.

Bruner, Goodnow and Austin (1953) in their experiments have noticed that some of their subjects adopted a wholist strategy and some of their subjects adopted partist strategy in their task of concept attainment.

LANGUAGE COMPREHENSION

Psychologists in recent times have developed greater interest in associating language with Psychology to the extent of even evolving a field of study called Psycholinguistic. The early psychologists did not show so much interest as the present psychologists are showing. The main reason is that it has been realized that language is considered to be one of the component tools of thought and human thought is mostly controlled by language. It plays vital role in controlling human behaviour directly and indirectly. No scientific discipline can claim that language is its exclusive privilege. It is a common medium through which ideas from every discipline are communicated in whatever manner they want. Language is a product of the evolutionary process and the development of human race. Man differentiates himself from other species and establishes his superiority over other species in several ways. Language is one of the advanced product of man's evolution. It serves several purposes for man. It is one that clearly separates from other creatures.

Language is the repository of all our knowledge and wisdom. The human race not only stored information in language but also transmitting it through language. It is a means of transforming our knowledge to any place to any period. Language is accomplished with the universality of meaning and the storing of human experience. It is the chief medium of instruction. The development of language is a social phenomenon. It has a main function of conservation of energy, which is one of the important reasons for its development. Wherever there are human beings living, some language is used there either by speaking and writing or by speaking alone. The utility of language of a civilized man may vary from the man who is still living in primitive style. However primitive he is, or uncivilized he uses a language and through which he communicates his feelings and thoughts to others and activates them. Therefore, the use of language whether it is in civilized society or in primitive society, it animates, activates and motivates and makes things to move. It plays important role in the behaviour of human race. When we say, language, it is a concept, symbolically representing the meaning of a sound when one raises his voice. Every type of sound raised by a man is attached with a specific meaning. This was how every sound was given a meaning in the process of language development. The science of language, linguistic was born from the study of sounds and the linguists called them as phenomes.

Therefore, language may be defined as a system of expression of human feelings and thoughts by the use of spoken sounds or through some symbols, which is distinguishing the characteristics of man compared to other creatures. Gardner Murphy (1964) states that language is symbolic use of words or other expressive symbols. Language responses need a great deal of verbalization. Language is first expressed only in verbal form. The written form of expression might have come only after very long time. Language in the verbal or oral form serves several purposes. A man can live successfully in the society by using his language only verbally. He need not learn to write the language to express his feelings and thoughts. For example a man can command a dog verbally and make him to obey his orders. He can call his dog by a name 'Tito' or 'Bhima' and the dog at once may respond. Similarly, language is used orally by animal breeders and circus masters to command animals like, tiger, lion, elephant, bear monkey and so on. We can teach a dog to respond to different verbal expressions of language. For example, the dog can understand from the sound of the master whether he is fondly calling him or in an emotional state of anger. The same type of expression can be used on human beings. The verbal expression of language can be in such a way that we can easily develop friendship with a person or show our affection on a child or on a girl friend. But a different verbal expression of abuse may easily inflict painful feelings upon another person, or may arouse his anger. Man can make use of language for such indirect and direct communication and this is having immense importance in his social life. Therefore, language is having primary social values and this becomes the basic importance for human verbal learning.

Some equate language with communication. But language is not synonymous with communication as animals, birds communicate without language. Many animals living in forests reveal the location of food, indicate their territory, announce sexual arousal, indicate the presence of their enemies. Karl Von Frisch (1950) an Austrian naturalist, who studied the language of bees state that bees are able to communicate to other bees the place where food is available. Seyfarth, Chaney and Marlar (1980) have studied many verbalizations of the chimpanzees and they were able to distinguish the meanings of the sounds. The sounds are uttered to express temporary emotional states and other chimpanzees responded appropriately. The oral sounds evoked social contacts. The verbalized sounds also expressed the fright or asserting superiority, complain of isolation, aggression against hostile companions. The oral sounds also expressed painful cries of frustration and urgent call for mating and so on.

Animals like pronghorn antelope give warning signal with their tails, white-tailed deers also raise alarm by their short tails, birds like pheasants draw the attention of rivals by arousing. The hen on seeing an eagle gives warning sound to her chicken to take shelter. Hebb (1966) states that these types of behaviours of the animals and birds as communication through reflexive activity. Communication here is transmitting and conveying a message. There is a sender of the message and also a receiver of the same. Thus animals and birds communicate a message through the use of some arbitrary sounds, gestures, or symbols in a purposive manner.

Chimpanzees brought up in human environment were tried whether they could develop human speech. The well known studies were those of Kellogg and Kellogg (1933) and Cathey and Keith Hayes (1951). These chimpanzees have learned to carry out the orders of the masters but they have not able to utter any English words. These psychologists have failed in their attempt to make the chimpanzees to speak. They were able to find out that chimpanzees lack development of certain physiological parts that require to produce speech sounds in their bodies. Two defects are possible for this failure, one is inadequate development of the central nervous system (the brain) and the other is deficiency in the vocal apparatus of the chimpanzees. More recently attempts have been made to teach language to chimpanzees. David Premack (1976) has observed that the animal had the inability to control lips and tongue to have the potentiality for the language verbalization. Apart from Premack several others like Duane Rumbaugh *et al.* (1974) tried to teach chimpanzees for speaking language. But the chimpanzees were only able to reproduce certain words which were taught several times and they were not able to speak anything other than those words. Whatever attempts were taken to make the animals to speak a language, since the animals lack adequate physiological development they can never acquire the ability to speak with the existing physiological make up. If some surgical operations are tried in the lips and in the vocal apparatus of the chimpanzees some change adequate for true language sound can be produced. This may facilitate the animal to emit correct sound of the language words.

All these preceding discussions show that animals and chimpanzees have shown ability to learn to use gestures and symbols in communication in situations where they are needed. Heyers (1951) states that however laboriously the animals are trained they can respond only in symbolic fashion. The ability to use vocal response voluntarily is severely limited. The evolution of human communication is a social evolution of man. Man has been using gestures and signs in communication for thousands of years till the gestures took the form of verbal expression. The gestures and signs that man used prior to the invention of language are still used by the civilized man to teach his dumb and deaf folks and they become universal. Therefore, what distinguishes man from other animal kingdom is the tremendous amount of the verbal responses that he uses and his ability in using them in conceptual and symbolic ways in his communication processes.

How the human child acquires language? This has baffled the scientists and the ordinary man for a long time, Many studies have been made only after the first quarter of this century to answer this question. Jean Piaget (1896-1980) in his work (1926) has described the evolution of thought structure in child from his birth to two years. This two years period fall under six sub-stages. The age ranges in all six sub-stages are given approximately.

The first stage ranges from birth to one month, the second stage is from one month to four months, the third stage is from four months to eight months. The fourth sub-stage ranges from eight months to twelve months, the fifth sub-stage is from twelve months to eighteen months and the sixth sub-stage ranges from eighteen months to twenty-four months. We can describe here only the first two sub-stages which are relevant here.

In the first stage the child does not learn anything, but makes only reflex action. The child at this stage can fix on objects, reach, suck and grasp. The motor activities are not coordinated. In the second stage the child's activities are referred to as primary circular reactions. Mainly there is babbling behaviour. The child says something and hears and repeats the same thing several times. There are mainly primary circular reactions. The initial reflex action is accidental and not intentional. It is the babbling behaviour and the child imitates it and he imitates his own babbling behaviour and not others.

The human child at the age of two months all over the world without any exception has developed the capacity to display in his babblings every sound used in all the world languages. Osgood (1953) states that in the babblings all phonetic characteristics can be found. At this stage the babblings are common to all the languages. But as the child grows and develops in a specific culture many sounds slowly diminish and only those sounds which are specific to his culture are retained and repeated and many more are added. But how these sounds emitted by the child become organized in a meaningful pattern of the language. The explanation is that while the child is babbling he/she emits word like sounds. These sounds are reinforced by the adults and the frequency of these specific sounds increase. Adults at home often repeat the verbalization of the child and this way they serve as models for the child. It is only through reinforcement that children learn to imitate the speech patterns of those elders around him. The frequency of expression and the variety of speech sounds may decrease considerably if there were not imitation and reinforcement by the elders.

Piaget (1926) states that there is remarkable communication or correspondence between the development of thinking and the development of language. The child understands certain concepts often before he develops relevant terms and expressions. But only through language children develop terms and learn to express them.

There are several other views of language acquisition which are based upon some assumptions. First of all language is learned by the child by direct experience. The child during the course of learning receives rewards for correct responses and receives punishment for wrong responses. A child in a culture is expected to learn and to make use of the language that is prevalent there. There is no other choice for the child. The child cannot learn everything all of a sudden. It is gradually acquired through his development, maturity and age. The child's physiological development should be adequate to withstand both the mental and physical strains. Eric Lenneberg (1967) states that the child has innate disposition to learn language. The ability to produce and understand human language is a specific attribute, an inherited one. If it is not inherited ability, animals for example, the chimpanzees also could learn to speak a language. The inherited ability is also based upon well shaped vocal apparatus for fine tuning — the larynx which is specific to human beings, mouth and the proper arrangement of teeth, a brain centre for speech and an auditory system which differentiates speech sounds from other sounds. Lenneberg states that language acquisition is confined to physiological maturation and only a small portion is exposed to language. The strong evidence for the innate disposition is that children all over the world irrespective of culture and civilization acquire various aspects of language in a fixed order and almost at the same rate, and children acquire language very rapidly between the age of two and six years. It is because that during this period the brain development is also rapid. This is another physiological importance in language acquisition.

But in some exceptional cases maturation may be abnormally slow or the child may be in an unstimulating situation. Even then the child acquires language in the same way but at a lower rate. It may take one or two years more for the child to acquire language. Children acquire language by direct experience. They directly imitate the words, phrases and sentences they hear and then they learn to combine and use them quite appropriately by repetition and reinforcement and generalization.

Therefore, we may state that both innate disposition and direct experience contribute to the language acquisition of the child. Man is biologically prepared by nature to learn language. But it is also important that he should be exposed to express language through which he imitates, reinforces and generalizes. In this process of acquiring language adults have an important role to play in promoting and developing the language ability of the children. While adults provide their children with a model for speaking, they do not speak to the children in the same manner that they speak among the adults. The adult's speech differs in several ways when they speak to the children. As far as possible adults eliminate words and phrases which are difficult for the children to speak out.

They use as far as possible few words, very short sentences and few tenses. The prefixes and suffixes of words are also omitted sometimes. The important aspect of the speech is that the word should convey the meaning. As the child grows in a specific culture many sounds which the child was making earlier disappear and only those sounds which are used in the culture are retained. For example, the child utters the sounds which are associated with objects and persons. Therefore, he understands the meaning of the specific sound. By uttering the correct sound he gets the reward. For him Mama, or Mamma sound bring his mother nearby him. Further, whenever the mother feeds him with milk, and water she tells him milk, water and so on. Hence the child raises such sounds whenever they were brought near him.

At which age the child speaks the first words? The results of the studies vary with different age levels. Therefore, Nelson (1973), Greenfield and Smith (1976) state that it may be on the average sixty weeks for the child to utter the first words, but the range may vary. But in whatever the language the child speaks the first words are likely to contain certain consonants for which the tongue is at the front of the mouth. For example, the consonants like m, p, b, d, t. For vowels the tongue should be at the back of the mouth. For example, while speaking the words which have vowels, like a, e, i, o, u, the position of the tongue and the movements of the lips vary. Therefore, it takes little more time to speak correctly such words, for example, words like Mummy, Daddy, pick me, put me. Nelson and others state that six months after the utterance of the first words children add further fifty words slowly. After this there is rapid increase of words. After twenty one months the words increase at the rate of one per day. It has been found that at any age children understand more number of passive vocabularies than active vocabularies. According to Nelson passive vocabularies are the words a child easily understands. These are contrasting with active vocabularies. Active vocabularies are the words the child actually uses in his speech.

When children's first words are spoken with complete sentence? This depends upon the situation and movements and emotional states. The first words sentences are true words sentences. The cat may be simple to speak out, "It is a cat" or "Bring the cat", "I want the cat". These are one-word sentences and there is meaning in them. Later on the child may use the words in a complex sentence.

Children when they become familiar with two words sentences they construct complex sentences. According to Limber (1973) there is rapid progress in children's two-word sentences to more complex sentences by the time they reach three years. In the construction of complex sentences they use morphemes and syntaxes. Unlike the two words sentences, the child now uses complete sentence. For example, two-word sentence "I want the cat" is spoken by the child in complex sentence as, "I want the cat, I like it". "Bring the cat, I want to play with it". There is also sequence in the utterance and the correct meaning also is conveyed. The more complex sentences the child constructs the more conjunctions the child uses in the sentences.

In the beginning children omit the connecting words, auxiliary verbs, pronouns, propositions. Words of grammatical functions are slowly added in complex sentences. In one intensive study conducted by Brown (1973) on three children it was observed that they added fourteen morphemes to their sentences almost in the same order. Brown also found that the age of the children in using the grammatical morphemes varies. Some reach this stage earlier and some reach this stage after few months. It was also found that children use grammatical rules in using verbs such as come, run, take and so on, for which they formerly used irregular past tense. Since they were hearing these words frequently they could acquire the correct past tense for such verbs and also for some of the auxiliary verbs such as have, be, do and so on. At the age of three, children construct sentences in which verbs follow nouns. There is also subject, predicate and object in every sentence. There are also pronouns and auxiliary verbs and the clauses are joined. "*While* I was looking at you, you were looking at the other side". "I called you and you did not see me". "Daddy is in the office room". This sort of conjunctions and prepositions are used very often in the sentences.

Children, as and when gain experience appear to follow a general rule in using plurals both in nouns and verbs. They see many verbs while changing from present tense to past tense require adding 'ed' like learn to be used as 'learned', like as 'liked', call as 'called'. But they experience difficulties in some verbs which do not require adding 'eds'. For example the verb 'fall' is present tense, for which the past tense is 'fell', similarly for 'come', it is 'came', for 'see', it is 'saw'. In these type of cases children learn to correct the irregular forms. In the case of plurals, there is difference between nouns and verbs. Most of the nouns require to add only 's' for the plural forms. For example, rule is singular, by adding 's' to it it becomes 'rules' in plural form. The child easily learns the plural nouns, like cats, dogs, books, benches, chairs, eyes, hands and so on. But certain nouns like foot, tooth, goose, sheep are difficult for the child to use the plural words. Therefore, children acquire the general rules in learning a language through time and experience.

Pronunciation

With regard to the pronunciation the child in the beginning is lacking behind knowing correctly what it should be? The ability to pronounce a sound correctly is exciting for the child. Even among the adult persons in pronouncing the English words there is considerable variation from culture to culture. The Americans pronounce differently from the British, the Australians and the South Africans. Similarly children in every culture pronounce certain words differently from others. But through experience and time children acquire correct pronunciation. The correct pronunciation is also demonstrated in schools by the teachers when children differ in their pronunciation. Parents and elders at home correct the pronunciation made by the child. Berko (1958) has observed that children in Kindergarten and in the first grade generally give correct ending of the words. For children words ending with voiced consonants are plurals. The voiced consonants should have hissing sound, for example, it should have the sound as if pronouncing the letter 'z'. Such words are dogs, bells, birds. Words ending with voiceless consonants have 's' sounding. Such words are socks, pups, seats, cats and so on.

STRUCTURE OF LANGUAGE

The function of language in the life of man is said to be verbal behaviour. The verbal behaviour has several important functions. It is only through language that communication between persons takes place. Thoughts cannot be expressed without using language nor communication at any level is possible among human beings. It has a system of symbols and rules and through which we think and express our thoughts. Language is a sound phenomena in the evolutionary process of species and human beings. It has direct and indirect influence upon the behaviour of human beings. In our everyday life we are having many direct and indirect communications. You can command verbally, your dog to lie down or to hold something or to chase your enemy. The circus masters command the animals in the circus and the animals carry out the verbal orders of the masters. The Army Officer commands a company of soldiers in the parade ground directly by some verbal orders. Thousands of soldiers directly carry out the verbal orders. The teacher in the classroom verbally orders the students to carry out the assignments. The master verbally orders his followers to carry out the particular work. In these ways language is having direct communication with human beings and animals. But many of our communications through language are indirect. It is in the form of written message or through some media. We do not directly receive the message from the person, but indirectly by some means.

In the beginning of human verbal learning the social aspects of language played an important role. Henry Clay Lindgren (1974) states that of all the forms of social interaction the language people speak is the most compelling and enduring source of cultural identity. Cultural identities tend to follow linguistic lines. According to Lindgren language has a social and psychological importance. Language need not be considered that it is having only a tool value. That is, that it is transmitting information from one person to another. Psychologists are now thinking of the psychology of language structure and language related behaviour. Though some psychologists do not agree with the idea that language plays vital role in thinking, they cannot disagree the relationship that exist between language and the reality they perceive or the subject matter they study. They describe the reality they perceived and the subject matter they have collected from their studies only by using the language. They express their thinking only through the medium of language and communicate to others.

Generally, communicating is to transmitting or conveying a message. This requires a sender and a receiver. If sounds, postures and symbols alone are used in the communication they must be universally accepted to produce the same meaning. We use mostly symbols and symbols as they are not having any meaning unless we attach meaning to the symbols. Language is the simplest form of tool with which meaning can be attached. Any symbol, sound or gesture can be indicated by a word. A word may be referring to a person, place, object or an event which has a definite meaning. The word which is having a meaning is symbolically standing for a person, place, object or an event. This person or object is not there, but word is referring to it and, therefore, the word is symbolic representation of the object. On mentioning the name of the person, place, object or event we can at once describe about the person or object. The function of language in thought process cannot be explained without any reference to signs or symbols, since signs and symbols represent the objects and events of the world.

There are two kinds of signs or symbols, one is the linguistics and the other is non-linguistics. For example, waving the hands, nodding the head, raising the eye-brow, showing fist — all these are signs or gestures expressing some meaning. There are sign posts and traffic posts in various places like highways, road diversions, check posts, cross roads, school and hospital sides and also in electric posts sign boards with a skull and two bones. Generally, sign boards in all these places sometimes do not have written caution words, yet they express some meaning and people know the meaning of these signs. They are universally made known and accepted.

We also use mathematical and statistical symbols or signs. Every symbol is used meaningfully in all our calculations. Similarly, we are using diacritical marks, apostrophe marks, quotation marks, inverted commas — all are symbols stand for some meaning. In writing, using language we also use short forms of some words, which are called abbreviations. For example, 'Anot' stands for anatomy, 'alt' stands for alternative, 'bot' stands for botany, *'i.e.'* stands for 'that is', *'e.g.'* stands for 'for example' and so on. The letter combinations in short forms stands as symbols for the real words. In all these every symbol is used meaningfully in all our usages. Thus language is another kind of sign. In language a set of letters and sounds are combined in some particular fashion to represent the objects, events and our experiences. When we see an object we indicate it, when we witness an event we explain it and when we have some new experience we express it. For example, when we mention the concept, 'cat' to the child it has no meaning for the child. But for the first time when the cat was shown to the child and introduced to him by saying 'it is a cat'. The child associates the concept of cat with the animals, cat. In the second time the child finds no difficulty in attaching meaning to the concept of cat even though the cat is not there. The concept of cat stands as a symbol to the real cat. Thus we use language to point out persons, places, objects when they were not physically present.

We make use of language in two important functions. We use it for our communication purpose with one another and secondly, we make the system of symbols and rules in our thinking and expressions. Therefore, we must know something about language. Language is structured by certain formal rules. The subject matter of linguistics is to study the rules of language which include speech sounds, their meanings and grammar. The psychologists wanted to study the psychological aspect of language. Therefore, there are certain interconnections between language and thought. When psychologists and the linguists wanted to know the interconnections a new field of study called psycholinguistics was born. Psycholinguists analyse language into its basic components and the basic components can be put into complex structures. Psycholinguists use linguistic principles to understand the ways in which people use to generate and comprehend language. The linguistic rules cover various levels of complexity of the language.

All the languages are based upon several components or units. There are four main components or units. The components are phonemes, morphemes, syntax and semantics.

Phonemes

The simplest component or unit is called phoneme. Phoneme is the basic speech sound and every language is based upon certain basic or elementary speech sounds. Phoneme deals with the rules of language sound. Phonatics and phonology are the two terms invariably used in the study of language. Phonatics is the science concerned with the study of speech processes, including the production, perception and analysis of speech sounds from both an acoustic and physiological points of view. This science though may be applied to language studies technically it excludes linguistic considerations. But phonology is said to be the science of the sound system of a language or languages in general. Phonology has to do with units of sound that make up flow of speech. Though many sounds are made, only certain particular sounds can be combined to make a word.

In English language there are 46 possible sounds or phonemes. These serve to distinguish one word from another. A phoneme may consists of several phonetically distinct articulations which are regarded as identical by the language speakers since one articulation may be substituted for another without any change of meaning. Articulation means to act or process of speaking or expressing a word. Thus p/and/b/are separate phonemes in English as they distinguish such words as pet and bet. Phonemes as they are have no meanings of their own. They must be combined with one another to make up into meaningful words. In English it is stated that there are 20 basic vowel phonemes of which 5 are long vowels, 7 short vowels and 8 dipthongs and there are 24 consonant phonemes Hauks, (1984). Therefore, in English there are many different vowel sounds to distinguish all vowel sounds.

Morphemes

Morphemes are the simplest semantic units of language. They are the smallest components or units of meaningful speech sound. In other words morphemes are the smallest unit of sound carrying meaning in any language. Phonemes are single smallest speech sounds, whereas morphemes are the smallest combination of phonemes or speech sounds having meaning. Phonemes are combined and made into morphemes. Morphemes are words and meaningful components of words. All words are morphemes, but all morphemes are not words. A meaningful word may have more than one morphemes. Because in using a language we may use free morphemes, and bound morphemes. There may be also multi-morpheme words.

For example, wall and bar are morphemes in the same manner as wallow and borrow are. These morphemes cannot be analysed into still smaller meaningful units or morphemes. Walls and barrows both contain two morphemes wall/s, barrow/s. Wall and barrow are free morphemes and the 's' after wall and barrow are bound morphemes. For example, the word book-sellers has four morphemes. The noun in these morphmes is 'book' and the verb is 'sell'. These two are called free morphmes. They are independent words, could express the meaning without any combination of any other morphemes. But the morphemes, 'er' and 's' cannot express the meaning independently. Therefore, they are called bound morphemes. The bound morphemes always appear in combination with other free morphemes. All the root words like wall, bear, book, seller are referred to as free morphemes. The prefixes such as 'un-', 'il-' indicating negative meaning and the prefix 're-' indicating the 'again' or 'a new' meaning and the suffixes such as '-er', '-s', '-able', '-ness' indicating agent, 'ability' and 'quality' are also called bound morphemes. Thus free morphemes (nouns) in combination with other free morphemes (verbs) and with bound morphemes make words in any language. Bound morphemes in combination with free morphemes change or add the meaning of the free morphemes to which they are attached in a sentence.

Syntax

The third component of a language is the syntax. The rules of syntax govern the arrangement of components or elements to produce larger linguistic structure. Syntax is a branch of linguistics that deals with the grammatical arrangement of words and morphemes in the sentences of a language. It is the totality of facts about the grammatical arrangement of words in a language. The syntax or the grammar of a language is the most important aspect of verbal behaviour. For Psychology the most important of these are the rules of grammar. The rules of grammar tell us *how sentences must be constructed.* But in psychology it is stated *how sentences are constructed.* Syntax studies the rules for combining words and morphemes into phrases and sentences. In other words sentences are made up of phrases and sub-phrases which in turn, are combinations of words. Therefore, we may state that a phrase is a sequence of words arranged grammatically but not having a subject and a predicate. A phrase is a set of words that represents a single idea or component and also a phrase serves as a subject, the main part of the predicate, a complement, an object of some kind. There are several different types of phrases. These are (1) noun phrases, (2) verb phrases, (3) verbal phrases, (4) prepositional phrases, and (5) absolute phrases.

In constructing a sentence the phonemes are governed by rules and/or sequence and also indicate how words are formed from morphemes. For example, the bound morpheme 's' is added to form plural of some nouns and how sentences are formed from words. That is, how the noun phrases and the verb phrases in combination make the sentence. For example, "The cat chases the rat" is a sentence. "The cat is eating the rat", is another sentence. Let us see how the words are related to one another and the role of each word in these two sentences (Figures 7.1 and 7.2).

Example 1 : The cat chases the rat.

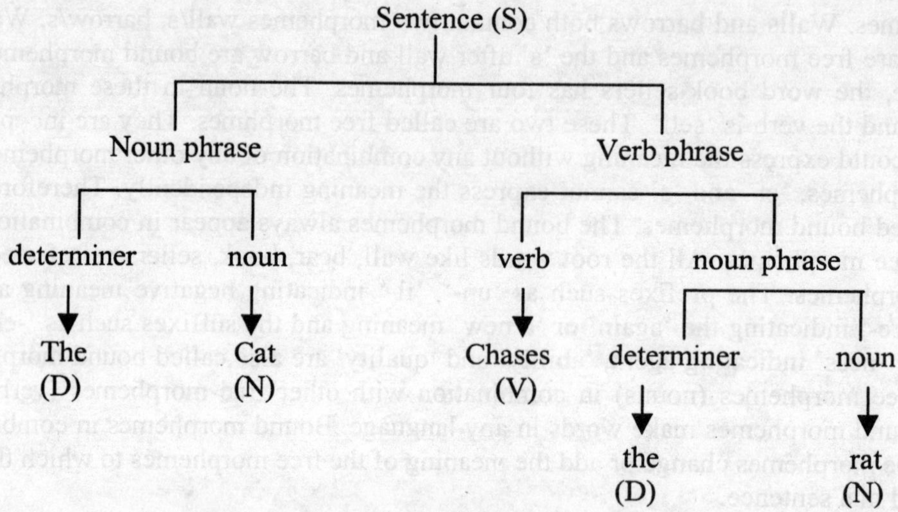

Fig. 7.1 : Noun Phrase and Verb Phrase (Illustration)

Thus a sentence (S) consists of a noun phrase (NP) following a verb phrase (VP).

Example 2 : The cat is eating the rat.

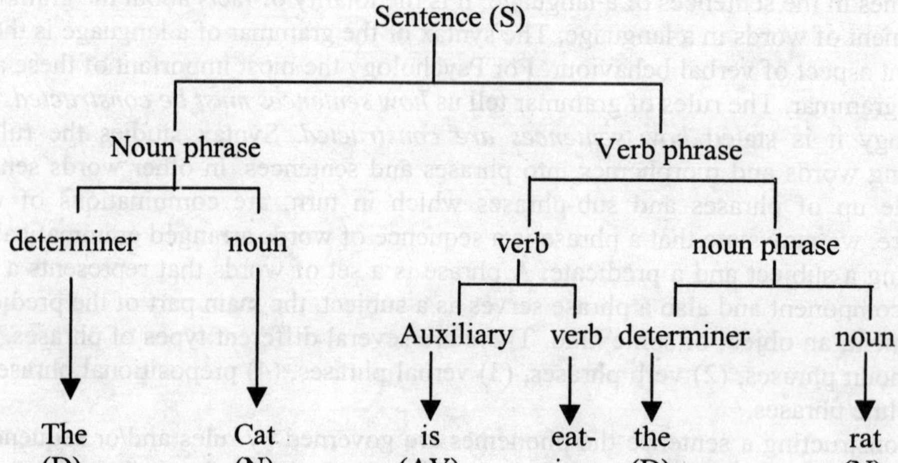

Fig. 7.2 : Noun Phrase and Verb Phrase (Illustration)

The analysis of the sentence into constituents is linguistic analysis. The constituents of this sentence has psychological meaning. Some lengthy passage can be broken in two ways. One is according to the constituent of the sentence. The other may be randomly broken. Richard Graf and Jana Torrey (1966) have conducted a test on two groups of subjects by exposing the passage for a short time through a machine. It was found that in the test the group which read the passage in constituent form did better than the group which read the passage broken randomly. Therefore, sentences which have constituent forms are psychologically important.

Let us illustrate the phrases by giving one example for each type of phrase.

Example 3 : Noun phrase : The large-sized television box with the rectangular design was made in England.

Example 4 : Verb phrase : The Prime Minister *was supposed to deliver* the convocation address. The verb phrase serves as the essential element of predicate.

Example 5 : Verbal phrase : In the verbal phrase the main component is a non-finite verb or gerund.

Emperor Asoka was happy *in embracing Buddhism* after the Kalinga battle (*in embracing Buddhism* modifies *happy).*

Example 6 : Propositional phrase : Propositional phrase functions as a modifier.

Ramesh bought the house *with all furnished materials* (with all furnished materials modifies house).

Example 7: Absolute phrase : This consists of a noun or pronoun plus a present or past participle phrase. Absolute phrase serves as modifier. It modifies the entire clause or sentence, i.e. *All things considered,* he would rather leave the place than to stay.

Semantics

The fourth component of language is semantics. The basic function of language is the semantic function, conveying meanings. Semantics is the study of the meaning of a language. Studies conducted in Psychology indicate that it is important to consider the types of meaning. Semantics mainly deal wi h the formal structure and the emotion of language employed in its use. In a particular context it depends upon how an expression sounds and looks. It may also vary culture to culture and person to person.

For example, the phrase, "I will see him" may be expressed in sounds in different context with different persons. "*I* will see him", "I *will see* him", "I will *see him*". The speech may vary in many attributes, such as pitch and amplitude. In the given phrase the expressions in the italicised words depend upon the intonation and accentuation of the words. The phrase has the same phonetic analysis, accentuation of the words. The phrase has the same phonetic analysis if the intonation (the manner of producing tones) and the accentuation (stress put on words) are disregarded. If you speak the italicised words of the phrase by putting more stress on the words in high pitch the meaning of the phrase differs in each one. The meaning of a word and its various associations among them and different possible occurrences constitute the structure of language. How extensively a language is used depends upon the structure of the language. There are mainly two kinds of meaning in a language. They are ***denotative meanings*** and ***connotative meanings***. The ***denotative meaning*** is also called ***extensional meaning*** and the ***connotative meaning*** is also called ***intensional meaning***.

A language may be thought of as a system of symbols that carry meaning which is universally accepted. Symbols possess meaning as they stand for some object or event. A word points out an object. For example, table, chair, pencil, book are concrete objects. This is a chair, this is a table, we can state pointing to them. The words chair and tables are external references to the words, chair and table. Children in every culture are taught like this, by pointing out, this is a chair, it is a table, that is a cat and so on. *This aspect of meaning, pointing out the concrete object* is termed as *denotative*.

The other type of meaning is derived from the cognitive associations people make with the objects by hearing and using. These meanings are not directly concerned with the objects that are specifically designated by the words. But these words imply and suggest. The connotative meanings are derived from the denotative meanings of many words. The connotative meanings of a word are often more significant than the denotative meanings. Connotative meanings are *emotional,* expressing some kind of *evaluation* and *preference*. This varies from person to person and from culture to culture. The person's possible responses are controlled by the choice of words which anchor the meaning of the words.

Connotative means an idea for an association suggested by a word or phrase. It is an idea implicated by a word. It is referring to other than the literal meaning. Therefore, connotative meaning in other words may be stated logically as intensive meaning. Connotation, logically, means intention. For example, the word 'hippy' (hippie) refers to the people who went against the traditional values in dressing, drinking and behaving in a society in 1960s in United States. They are called non-conformists. They rejected the traditional values and wished to live natural life. They moved all over the world dressed in rags, carrying their goods on their backs, sleeping on the roads. Decent people tried to avoid them. Therefore, the word 'hippy' created an impression among public that it is referred to people with unclean habits. But the word is also referring to something different meaning. The word 'hip' is also adds a connotative meaning. It is an exclamation used to introduce cheers. The word 'hip' is also having a connotative meaning of excited or emphatic cry. Another connotative meaning for 'hip' is following the latest trend or idea or fashion in music. Therefore, the word 'hippy', 'hip' is having two extreme meanings, involving emotional evaluation and preference. This is a good illustrative example for connotative meanings.

In the international level, formerly, various nations were using the phrase 'under-developed nations'. This was not liked by the underdeveloped nations. Now they are using the phrase 'developing nations'. Though both the phrases are referring to the same subject matter, in the former the language was emotionally involved, when we refer to some abstract words like peace and socialism. The words are so emotionally charged in some cultures that there is very little denotative meaning. The interpretations of these words are not accepted universally. These words are defined and interpreted differently in the different cultures. These words have more connotative aspects than denotative.

Another concept which causes some confusion in the meaning of the language is 'education'. Apparently there is no problem with this denotative word. But the meaning of the word varies from culture to culture. The word education generally is referring 'schooling' in English. But in Latin countries the word education is referring to 'upbringing'. The meaning of the word is more of connotative than denotative. Similarly certain words refer only to men and certain words to women. For example, the word 'handsome' is used to describe men, while, 'beautiful' is used for women.

Therefore the meanings of the words differ tremendously in emotions, evaluation and in preference. This causes misunderstanding because of the different connotations of the words of different people. Hence in order to attach firmly the connotative meanings in one position or make them more accurate Charles Osgood (1957) and his associates have formulated a method for measuring difference in connotations called ***'Semantic Differential'***. The method is called semantics, as it is dealing with meanings of the words, and differential because it provides several dimensions of meaning. They are of the view that even the simplest words, even those refer to concrete objects have an effective potential. This effective quality derives from the associations we make with them.

In determining the connotative meanings of a word by the method of Semantic Differential, the person is asked to rate a series of verbal concepts on a seven-point bipolar scale. At the two extremes of the scale there are two contrasting adjective words connected by a straight line divided into seven positions from left to right. The words to be rated may be referred to anything, for example, freedom, socialism, polite, handsome, China, Pakistan. The words to be rated on the scale may be twenty to fifty. The person has to decide and indicate the direction and the intensity of his judgment by making a mark at some point on the scale. Whether it should be on the extreme left or extreme right or somewhere in between the two extremes he should decide. The contrasting adjectives used for a concept may be, dirty-clean, sad-happy, weak-strong, light-heavy, slow-fast, passive-active, rough-smooth, good-bad, wet-dry and so on (Fig. 7.3).

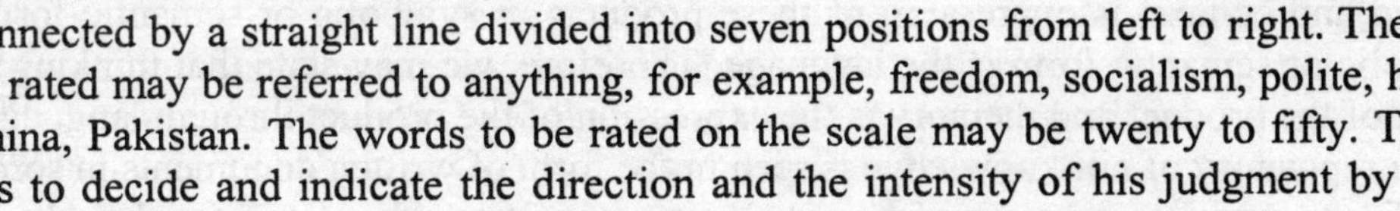

Fig. 7.3 : Semantic Differential Scale (Model)

In the beginning Osgood and his associates used some fifty pairs of contrasting adjectives for determining the semantic differential for various types of words. After analysing many words they have observed that connotative meanings of most of the words were stated in terms of three basic dimensions, such as *evaluative dimension* as good and bad, dirty and clean and so on, and *potency dimension* as weak and strong, small and big, heavy and light and so on, and *active dimension* like, fast and slow, active and passive, bright and dull. These three types of dimensions Osgood found that they provided clear connotative meanings. This type of study was found to be very much successful in various cultures. Osgood (1955) in one of his studies has found that terror on evaluation dimension, strong on potency dimension and active on active dimension were not precisely indicating the connotative meanings.

RATIONAL THINKING

Thinking is a higher mental process. Man alone is capable of thinking and what he is thinking is also expressed and communicated to others through the medium of language. Language is one of the most powerful instruments that man has invented and the most useful instrument in his life. There is close relationship between thinking and language. Whatever man thinks is revealed through his language either by orally or in written form. Thinking is very much influenced by experience and memory. When a man is required to think of something for the purpose of dealing with some problem he evolves some strategy for this by putting his experience and past memory relating to this in different combinations and arrives at some conclusion. Whether it is going to be successful or not he reveals his thought. Since thinking is a byproduct of experience and meaning, Jean Piaget (1961) states that thinking is a dynamic act. According to Guilford (1958) intellect consists of memory and thinking and all factors included either under memory or thinking and this can be classified under two kinds, the kind of *product* involved and the kind of *content* used. Products reflect what is remembered or known and content refers to the expression of these products in symbolic or semantic form. Thus product is what is remembered and content is expression of these products in symbolic or semantic form. This symbolic or semantic form is the language. Therefore, we may state that thinking is the content of the product and memory is the expression of the product through language. That all the repository of our knowledge is seen in the form of written documents in some language. From these store house of knowledge we can state what Socrates thought in B.C. 2400 and what Bertrand Russell thought in 1950 A.D. Therefore, we may also state that language is the manifested form of human thought process. It is uynamic in nature and has a structure. Whatever we speak it must make some sense. When it is not making any sense we say that it is not correct thing or it is not logical. Therefore, whatever we think and express sensibly is correct and accepted, because it is logical.

When we expect that we should think and speak logically we are expected to follow certain principles or rules. Therefore, logical thinking is a way of thinking in which the rules of reasoning are to be followed properly. At the higher level of thinking these rules are highly complex and rigid. Logicians have distinguished several kinds of thinking. In our everyday life we are applying some kinds of logical thinking even without learning the rules of formal logic. While we are engaged in some discussion with our friends we may easily fall into an argument. We would be very carefully presenting our points in a sequential manner, even involving some evidence to. support our side. Even though we do not follow the formal logic, we put up our points in an agreeable manner and evolve a correct conclusion. Here we apply reason, and rationally put up our argument which is also accepted by others. At such instances we do not bother whether we are following any logical principles. Because these are simple matters and we do not break our mind too much. But when we move from everyday logic to formal logic to establish the truth we will have to follow the rules systematically. In a formal logic we will have to use syllogism. Syllogism is the basis of formal logic.

What is Logic? Logic is the branch of Philosophy concerned with analysing the patterns of reasoning by which a conclusion is drawn from a set of premises, without reference to meaning or context. Logic is the science of thought process. Logic controls our thought according to certain principles. It is a normative science which deals with truth and right conclusion. It also deals with mistakes or fallacies in our reasoning. When a thinker reaches a conclusion which he accepts as certain he has been reasoning. When he reaches a conclusion following from some data he has been thinking or reasoning logically. When he reaches a conclusion that does not follow a data he is thinking illogically and his conclusion is also illogical. A conclusion follows from the data that have good judgment in the pattern of the data. If Devan is Arun's son and Hari's father, it follows that Arun is Hari's grandfather. This particular data can be drawn in a diagram with a vertical line with Devan in the middle, Arun at the top and Hari at the bottom : Arun → Devan → Hari.

Arun — Grandfather

↓

Devan — Father

↓

Hari — Grandson

Devan in the example, is a middle term standing in relation to both Arun and Hari and, therefore, it becomes possible to state the relation definitely between the two. Therefore, reasoning mainly depends upon the middle term.

There are two forms of reasoning in logic, one is ***deductive reasoning*** and the other is ***inductive reasoning***. The deductive reasoning is expressed in syllogism. In deductive reasoning a particular idea is brought under a general idea and a truth or conclusion is inferred from it. The syllogism is a sequence of sentences in which there is a major premise and a minor premise which together draw a conclusion. It is a classic device for classifying thinking. In a syllogism there are two sentences. The two are put together in such a way that a third statement called conclusion is inferred. Let us illustrate this point with well known correct syllogism.

All men are mortal (Major premise)
Socrates is a man (Minor premise)
Therefore, Socrates is mortal (Conclusion)

Another example may be stated:

All Anglo-Saxons are beautiful
All English are Anglo-Saxons
Therefore, all English are beautiful.

In these illustrative examples the particular idea, 'Socrates is a man', 'All English are Anglo-Saxons' are brought under the general idea 'All men are mortal' and 'All Anglo-Saxons are beautiful', from which inferred a conclusion or new truth, "Socrates is mortal", and 'All English are beautiful'. Thus in drawing conclusion here, we are going from a general idea to the particular idea. Therefore, this is deductive logic or reasoning.

Sometimes syllogism may be stated falsely when logical principle is not followed. The important point in deductive logic is that the object of the major premise should be the subject of the major premise and the subject of the conclusion should be the subject of the minor premise and the object of the conclusion should be the object of the major premise. In deductive logic any syllogism does not possess this principle may lead to faulty conclusion. Let us illustrate the point with a faulty premise and a faulty conclusion.

All A is B
All C is B
Therefore, all C is A.

This syllogism appears to be correct. But it is not correct. Because the rule is not applied. Let us apply the rule and set it correctly.

All A is B (Major premise)
All C is A (Minor premise)
Therefore, all C is B (Conclusion)

In inductive logic we go from a particular idea to some general idea. Taking into account several particular instances together we draw out a general conclusion from it. In concept formation we use the principle of generalization. Generalization in logical conclusions is not always correct.

This person is black – (S is M)
No European is black – (P is not M)
Therefore this person is not a European – (Therefore, S is not P).

In many cases this type of formal statement makes it easy to see whether the conclusion follows from the data. The formal statements show what is the middle term and what are its relations to other terms from which conclusion is to follow. We see that the conclusion is logical whenever an argument can be put up in this form. Here we go from a particular idea to some general idea that all black persons are not Europeans.

CREATIVE THINKING

Thinking is an intellectual cognitive activity. There is something going on in the mind. Thinking needs rational powers. The traditional view is that thinking is an obscure mental activity, Skinner (1968) states that thinking itself is not a behaviour. What is thought when given expression, it may lead to action. Skinner says that great thinkers rarely had any thought about thinking. A field of psychologists called cognitive psychologists study about the structure of expressed thought. Human thought may be expressed in various forms. We often come across three processes called ***creating, inventing*** and ***discovering***. These are performed only by human beings. Only man is capable of creating something new, can invent which is not in existence and discover which was not seen earlier. Here we are concerned with creative thinking. Creative thinking is studied as one of the dimensions of thinking. The thought or thinking associated or connected with creating means that we put our thought process in seeing something new by reconstructing the past. In creating something new we cannot say that everything is new. Gardner Murphy (1964) states that creation or initiation of new differs only in degree from the activity which is the reproduction of the past. Therefore, in creation there is something old and something new. The expressed thinking in the form of some activity has been presented in some new way.

Developing Creative Thinking

If thinking is analysed and taught separately to the students in schools and colleges the already known thought process may be transmitted with greater amount of efficiency. The teachers in schools and colleges have done some service to the desirability of developing creative thinking skills in students. But the school and college programmes are not deliberately designed to promote these skills. But there is every reason to belief that the abilities involved in creative thinking are mainly innate and inert. We also assume that a systematic exposure to greater amount of information of increasing difficulty and problems may activate to develop the ability of creative thinking. But these efforts may not produce the same amount of seriousness in everybody to develop the creative skills. Whatever the efforts are taken at the school and college levels and trainings are provided, at the end only few may exhibit fast proficiency and many may be struggling with the rudimentary principles or show slow progress in it. Probably they may be having some problem with regard to the development of some particular skills. If such skills in which they are lacking can be identified and helped in time it is possible to foster the creative behaviour in them. Though some who have innate tendencies and learn faster than others also can be taught to develop the creative skills. Some of the writers such as Rath *et al.* (1967) and De Bono (1976) have offered practical suggestions for fostering creative thinking and the same have been employed by the teachers in school.

Lateral Thinking

It has been observed that different types of thinking are involved in the development of creative skills (De Bono, 1970). If the person wants to discover more about something and accepted convergent solution to a problem then the appropriate approach is that he should carry out *vertical thinking.* But if one wants to find out the common or customary creative (divergent) solutions for problems then he should carry out *lateral thinking.* Lateral thinking according to De Bono is a different term from what has been traditionally called creative thinking. Generally lateral thinking means a way of solving problem by rejecting the traditional methods and employing unorthodox and illogical means. De Bono (1970) states that lateral thinking is a way of using mind that leads to creative thinking. Lateral thinking is closely related to insight, creativity and humour. Bono states that lateral (creative) thinking can be intentionally developed. According to Bono's (1976) programme the development of creative thinking requires encouraging the students to develop new ways of approaching the problems and they do not require to solve any specific problems.

Revolutionary Approach in Thinking

According Abraham Maslow (1908-1970) stereotype is a ready-made idea or previously determined conclusion. To justify this conclusion one is engaged in good deal of intellectual activity and finding evidence for this. People may simply believe without any thinking. Maslow (1970) states that it is possible for every person to live by a set of ready-made ideas that he has acquired during first decade of his life and this will never undergo any change even a small pit. This person with a high I.Q. spends good deal of his time in his intellectual activity and selecting pits of evidence to support his ready-made idea. This type of activity may be of some use to the world. But, it is worth having to make a sort of verbal differentiation between productive, creative thinking and rationalisation. Maslow states that ***rubricized*** (tending to confirm forcibly) ***thinking*** has a unique relationship with reproductive thinking. This has reference to the past. Problem solving is solving any new problem in the light of past experience. Therefore, this type of thinking is no more than a rearrangement of previously acquired habits and memories.

Here the discussion is, which kind of thinking may be used as the ideal model. The point put forth by the holistic-dynamic theorist is that thinking activity, if it transmits any meaning, has the meaning of creativeness, uniqueness, ingenuity and inventiveness. Thinking is the technique by which man creates something new. Then it becomes revolutionary because it is conflicting with what has been concluded already and also conflicts with what has been learned or with our existing habits. If the past learning and habits work together well, then we are responding automatically. Then we have no necessity to think from this point of view. This means thinking becomes opposite of learning. Maslow (1970) states that though it appears somewhat overstated, thinking may be defined as the ability to get rid of our habits and to ignore our past experiences. Maslow is of the view that since stereotypes and habits have ready-made ideas which are tending to produce conformity by violent method without having scope for thinking in new ways a definition of thinking naturally involve to disregard habits or stereotypes and past experiences. In true creative thinking one of the dynamic aspects which stands as an example is the great achievements of human history. The great achievements that man made in the world are the results of the characteristics of man's boldness, daring and courage. These are based upon true creative thinking disregarding habits and past experiences.

If man would have been wholly depending upon the ready-made ideas (stereotypes and habits) he would not have made so much progress in this world. All the achievements that we see today in the world are the results of the revolt that man made against the ready-made ideas in which he was living and boldly plunged into a type of radical creative thinking and going on farther away from the ready-made ideas or in Maslow's words from the rubricized thinking. Maslow compares the timid child who is clinging closer to his mother for safety and protection just like the thinking process is clinging to the habits. The bold child is free to keep himself in any venture and can go far away from his home just like the bold thinker who must be able to break the stereotypes, habits, past experience, learning, customs and conventions and to be free of anxiety. These are the characteristics of true creative thinking. The process of creative thinking as suggested by Maslow is quite similar to the lateral thinking suggested by De Bono. It is going ahead breaking the traditional way of thinking. If we go through the history of creative thinking or creativity we can observe that what suggested by Maslow and De Bono has been taking place in the world.

Modern Trend

We need not go to the remote past to understand this phenomenon. Let us just see and analyse what has taken place during and after the World War II. There has been increasing tendency among the developed nations to compete with each other in the scientific, artistic and technological inventions and discoveries. The real competition seemed to have been the utility of intellectual abilities and the development of human resources and talents. Those who were engaged in these areas were very much encouraged. It is an accepted fact that if we want to survive in this competitive world we need more brains. Therefore, it has become important to identify many kinds of talents and to tap the talents we require. Psychologists after the World War II have shown much interest in tapping of certain special talents in human beings. The existing intelligence tests were found to be inadequate and, therefore, they have developed tests that would identify the special talents, the creative skills.

Creative Thinking and the Creator

What is creative thinking and who is a creative thinker? Creative thinking may be defined as the ability to discover something new in the external world through past experience and imagination and the creation is the expression of such ability in some form so as to arouse the attention, curiosity, feeling and emotion of others in different degrees on the basis of their experience and knowledge. The forms of expression of creative thinking may be an epic poem, a play, painting, sculpture, a musical tone or even a design of beautiful building or bridge which was never in existence in the same form. Whoever makes out such form is a creator. Murphy (1964) says that we are all creators in a way. Under the pressure of some strong feeling some big idea in us may find expression is manipulating the material object in relation to the present situation not duplicating which is already in existence, but in new manner.

Under the stress of strong feeling to see something new, everyone of us may become a creator. Though some claim that it is a special gift bestowed on them, creative ability is not limited to artists and inventors. A child may work out a creative play. It is true that every child is trying. The child may create a new type of toy out of the clay that is available for him. A housewife may create a new kind of dish with palatable taste and flavour to please her husband. A dyer working colour dying may invent a new solution for colouring fabrics. A consumer psychologist may design a new type of advertisement to arrest the attention of the people. Thus everyone of us may become a creator in some way by producing something new or in strange manner which was not in existence previously.

Creative Process

The creative thinker is always trying to link the existing elements in some way, but not in the way that could naturally link with each other. The modern painters draw sketches in such a way that ordinarily we could not make out any meaning out of it. It needs certain explanation. The reason is that the painter sketches the picture in an ambiguous manner, just like the projective tests. He does not want to be a conventional painter. He wishes to see something new in his approach. This tendency has been in practice right from the ancient time onwards. The ancient sculptures and paintings were found with both human and animal features, sculptures with human head and animal body or animal head with human body were quite commonly found. For example we come across the Sphinx, the Egyptian statue, with a lion's body and man's head and a woman's head. So also we come across the Chimera a fanciful animal with a lion's head, a goat's body and a serpent's tail in Greek mythology, a mermaid, half woman's body and half fish's body, the outcome of the poetic imagination. In India many of the cave sculptures and paintings of the ancient period were full of strange combinations of human and animals, human and birds. Therefore, the tendency to see something in a new fashion has been existing in the minds and imaginations of artists and sculptors even in the ancient times and also in the middle ages.

Therefore, the ability to think and create new things in men is a continuous process and it is out of pressure of some idea that man indulges in the creative activities. Creative process involves greater amount of imagination and thinking based upon learning and experience and it is a continuous pressure – the pressure that John Stuart Mill (1806-1873) had during his adolescent period from his father as much as he could take and sometimes more pressure that he could take. The result was that Mill (1873) was able to master Greek, Roman, English literature and politics, mathematics, formal logic, economics and psychology at the age of sixteen. The more a person becomes sensitive to the environments, forms, colours, tones, shapes and contours the more he makes them as the roots of creativeness of his later life. It is the early joy in spatial and temporal relationships and mechanical laws that made Sir Isaac Newton (1642-1727) to become a great mathematician in his later life.

Murphy (1964) observes that an individual to become a creative is sensitive and selective to certain kind of materials in two ways. First, he directs his attention to them and secondly, he enjoys them for making new combination of them. McKinnon (1962) states that there are five stages in creative process. They are: (1) a period of preparation, posing a problem, (2) a period of concentrated effort to solve the problem, (3) a period of withdrawal from the problem, (4) a moment of insight, (5) a period of verification, evaluation, elaboration and application of insight. There is a tendency to see a problem and making enquiries constantly about it are the processes with which creative act starts.

The basic process of creativeness is awareness to the situation. The creative person is very sensitive to the situation. The creative process is not a unitary process, but it is a combination of several processes like cognitive, motivational and emotional processes. This combination of process involves psychological processes like perception, memory, imagination, appreciation, thinking, planning and deciding. All these processes are present in every human being, but only varies in degrees and the creator probably possess it to a greater degree.

Creativity and Intelligence

Psychologists studied creative thinking from various angles. According to Guilford (1959) creativity is a form of intelligence. He has associated a number of special abilities with creative process. They are flexibility, originality fluency with word associations and ideas. According to him these are associated with divergent thinking.

Apparently we assume that creativity and intelligence seem to be related to a greater extent. The psychologists who have studied this phenomenon reveal two kinds of views. Some are of the view that creativity is a form of intelligence and some are of the view that creativity is a special kind of ability distinct from general intelligence. These two concepts have attracted many psychologists and studies were conducted by them. All these studies have failed to establish that creativity is independent of intelligence. The study made by Getzels and Jackson (1962) is an important one in this context. The test was conducted by them in one of the private schools in Chicago. They have administered an intelligence test and a creativity test on a group of students. These students were from 5th grade to 12th grade. On the basis of test scores the students were divided into two groups, one those who scored high in the intelligence test and the other those who scored high in the creativity test. These two groups performed regularly in the academic achievements. There was no difference between these two groups in school performance. One personality inventory also was administered on them to see whether they vary in personality traits.

The investigators have observed that these two student groups varied considerably in personality traits. Those who had scored high in intelligence test were found to be conscientious, careful and self-controlled. Their goals appears to match with those of their teachers. The students who scored high in creativity test were found to be playful, expressive and independent and were not intelligent. They did not adjust well with their teachers, but worked harder than those students who had high I.Qs.

Wallach and Kogan (1965) have identified four groups of students in their study, classified as high and low on intelligence and creativity tests. It was found in their study that highly creative, but less intelligent students were most frustrated with school and highly intelligent and less creative students were addicted to school and well liked by the teachers. The studies made by Torrance (1962), Thorndike (1963) conclude that there is no high correlation between intelligence and creativity. Meer and Stein (1955) state that there is high relationship between intelligence and creativity. Crophy (1965) and Pribram (1963) state that there is no strong justification to treat intelligence and creativity as distinct and separate faculties.

CONVERGENT AND DIVERGENT THINKING

The two thought processes, convergent and divergent thinking are always associated with Guilford's model of intellect. Guilford (1959) after ten years of rigorous research work came out with a plan of intellect. This serves the purpose of illustrating the processes involved in both creativity and intelligence. Guilford's plan of intellect fall under two major heads such as memory and thinking. According to him memory has limited abilities while thinking is much more complex and has many abilities.

The model is organized around three main aspects of human functioning such as 5 operations, 6 products and 4 content. Any specific ability involves a combination of all these three. Therefore, according to Guilford there are (5×6×4) 120 abilities.

Guilford states that there are three, types of thinking abilities such as ***cognition, product*** and ***evaluation***. Of these three, cognition deals with knowing or discovering and recognizing information, production deals with the use of known information and also generating new information. The production type of thinking may be also stated as ***convergent thinking***. This convergent thinking is leading to a single correct answer. It is a process of thinking in which a common attribute is abstracted from many different ideas. In convergent thinking everybody may be thinking alike, but there is only one correct response. All others are wrong.

The other aspect of production type of thinking ability is ***divergent***. Since this is an offshoot of thinking ability it is also called divergent thinking. It is a process of thinking in which many different ideas or solutions are generated from a single idea or problem. This may take a number of directions and also it consists of generation of multiple responses. Anyone or more than one of which may lead to correct answer. In divergent thinking there is no single correct response, the value of responses depends upon its suitability, usefulness and meaningfulness.

The term convergent thinking was employed by Guilford to describe the type of thinking which results in finding out a correct solution to a problem. All the conventional intelligent tests measure convergent thinking abilities, whereas the term divergent thinking is used to describe the type of thinking that yields results in the production of several different solutions for one problem. One important characteristic of divergent thinking is that it is closely related to creative behaviour and some authors use it in place of creative thinking. Divergent thinking is the central point in the creative process. Convergent thinking and divergent thinking are the two operations that have stimulated for several research studies and many were interested in associating these two phenomena with other variables like intelligence and personality traits.

PRODUCTIVE THINKING

Productive thinking is one of the important processes of thinking mainly involved in solving a problem in which the past experience is repatterned and restructured. Productive thinking may be defined as the process of thinking which may be producing or which may be having the power of producing a favourable or effective results just by seeing. In psychological studies the part played by the past experience in human problem solving has been taken into account and it has been a subject matter for experimentation.

There has been diverse views with regard to past experience in problem solving behaviour. Some identified problem solving with trial-and-error learning and some were of the view that in problem solving behaviour past experience provides the raw materials and the role of past experience is only an invention and some have even neglected the role of past experience in problem solving behaviour. Gestalt psychologists disregard the past experience in problem solving behaviour.

The main reason for the diversified views with regard to past experience is that problem solving is not unitary process. There are several processes involved in problem solving behaviour. Maier (1945) has clarified this issue by showing the distinction between productive thinking and reproductive thinking. His clarification was in terms of the role of past experience in each type of activity. According to Maier reproductive thinking is characterised by the solution of the problems by means of the existing stimulus in the problem and in the previously mastered situations. Therefore, for Maier reproductive thinking and transfer of training are quite similar. Whereas productive thinking is not merely the process of arriving at a solution through direct application of previous learning, but by repatterning and reconstructing the past experience to meet the present demands. The studies of Maier (1945) and Duncker (1954) have identified the ways in which past experience may affect the problem solving behaviour.

Experimental studies on animals have been conducted to see whether animals are capable of making use of the past experience in problem situation. Birch (1945) has observed this in chimpanzee and has shown that the background of past learning represented an essential store house of behaviour and the same was restructured when any situation demanded. Birch and Rabinowitz (1951) state that productive thinking is impossible if the individual was restricted to past experience. According to them past experience may become a hindrance and an obstacle which may block productive thinking. Therefore, according to Duncker past experience is primarily having negative effect in problem solving experiments.

Max Werthimer (1880-1943) was working on thought process for many years. His first edition on productive thinking was appeared in 1945 and the enlarged edition in 1959. He has summarised a number of experimental studies in his work. His qualitative analysis of simple situations shows the differences between his approach and other's approach. He did not approve other's approach. The Gestalt had adapted two alternatives to thinking and problem solving. One is said to be formal logic and the other is the association theory. But it is said that both approaches are too limited when an individual meets with a problem.

Productive thinking is identified with solutions which have not been learned. Max Wertheimer has observed in his experiments that how a student thinks productively in discovering the area of a parallelogram. He disapproves the method of leaving the student to blind trial and error method. It is an inefficient method and the student learns nothing. He learns only how to try. Further, he states that the student does not learn very much from being taught to apply a formula.

According to Wertheimer productive thinking takes place when the student sees that the bulge on one side of a parallelogram fills the gap on the other. It should not be pointed out to him to see it. The solution must come as an insight. That is an idea or response should not be traced to past or previous conditions or experience. What

Wertheimer means is that the student should not start from scratch. He may have extensive stock of experience earned under similar situations. He can understand the problem and can calculate the area of a rectangle and he also knows something about triangles and how they differ in size and shape. He is more likely to have this insight if he has solved similar problems by cutting and arranging pieces or by drawing lines to divide areas into parts. With regard to productive thinking whether he has been taught this or learned in an occasional situation, it is not a matter for consideration. He must not have learned earlier this particular solution.

The acts of seeing should bring solution as a surprise. It is this aspect of productive thinking which Wertheimer expects actually what he calls an *original idea*. But there is an element of surprise in the discharge of any operant response. A stimulus is effective as a part of a situation or condition which builds up a moment and arouse a response. There is a transitory period of time in every such emission of response. We may be listening to an event that is occurring in a performance or to a musical note for some time and name it suddenly though it might not have been witnessed earlier. We may look for some time at a distant object moving towards us before seeing what it really is or we may react to it effectively in some ways. We are studying some material for some time before reproducing it correctly. So also in problem solving we produce conditions which may result in occurring a solution. But it will not be possible for us to say exactly when it occurred. According to Wertheimer solution to the problem is a pre-behavioural event. A response of this type is all the more an inspiring event when it appeared in a novel situation.

What is happening when a student sees that a protrude or bulge fills up a gap? It is very difficult to explain. Skinner (1968) says that it will not help us if we call it as insight. According to Skinner there is some features of visual perception and environmental variables relevant in it. Hedamard (1945) is of the view that productive thinking is to be understood only in terms of its structure and its origins are inscrutable. Hadamard states that thinking is an unanalyzable process of intuition. Wertheimer emphasizes the structural approach by stating that solutions appear in a regular fashion in line with true structure of the situation and children are prepared to hold such structural situations unless they are taught badly to blind repetitive practice.

EXPERIMENT 19 : CONCEPT FORMATION — VIGOTSKY TEST (MODIFIED FORM)

INTRODUCTION

We see innumerable number of objects everyday in the outside world and also we deal with them directly or indirectly. Most of the time we deal with them successfully. Sometimes there may be some confusion in dealing with them due to improper classification of categories of objects in our perceptual process. Garrett (1975) says that a man may identify a number of things by touch alone and solve everyday problem by muscular cues. But he cannot solve abstract problems in mathematics and science unless the various possibilities like the acceptance or rejection are expressed in words, numbers, certain symbolics, blue prints and so on. Much of confusions and misunderstanding in our perceptual world are caused due to the fact that we do not fully comprehend the concepts presented. Every object has certain characteristics. They are also called properties, traits or distinguishing qualities. In the psychological studies of imagination and thinking or in the studies of language and communication we often come across the word concept and concept formation. The word concept is derived from the Latin word 'com' and 'cepere' which mean 'together take' or 'take together', 'to become pregnant with', 'to form in the mind', or 'imagine to understand', or 'to form an idea'. Thus the word concept fundamentally refers to an idea or the general notion. It is conceiving something. It also refers to an idea representing the meaning of a universal term. It also comprehends the essential qualities of a class of species. Concepts are categories for classifying specific subjects, persons, animals and events on the basis of common elements.

In the light of these varied meanings, concept may be defined. According to Woodworth (1952) a concept is word which has a meaning and is a tool of thought. He states that one may know the meaning of a word, but he cannot define it correctly. A concept may be also stated as a response made to some aspects of a situation which can be later on generalized for a new situation which contains the same aspects.

The world in which we live is full of objects of different kinds. There are both animated and inanimated objects. These objects possess certain common qualities that make them different from other objects. When a person learns to respond to these common qualities of an object and learns to ignore other aspects he is said to have learned a concept. For example, a child is learning slowly to understand persons, animals, objects and events. He sees a small cup and understands that it is a cup in which he is given milk to drink. He learns that it is used for a specific purpose. He differentiates it from other types of materials used for drinking milk. That is, he sees the difference between the porcelain cup from the glass cup and the metal cup. Secondly, when he sees the porcelain cup he sees it as white in colour and differentiates it from a yellow cup and from other colours. In this context the child has learned also the colour concepts of yellowness, the quality and also has learned the properties of the same object, that cups may have different colours.

There are concepts of not only persons and objects, but also qualities, actions, interactions. There are concepts of the height of buildings, depth of seas, distance between places and also speaking and walking. According to Murphy (1964) a concept is a symbol which indicates a specific quality commonly possessed by a number of objects. It is necessary to find out more than one common characteristics running through some items.

According to Kimble *et al.* (1985) a concept is a symbol that represents for the common properties objects, events or ideas that are distinguishable in all other respects. Day by day we are coming across quite a number of new concepts. Some of them are simple and objective concepts. For example, dog, car, book, shirt, television, computer and so on are simple concepts. The other type of concepts are abstracts and complicated. For example, belief, habit, ambition, sympathy, justice are abstract concepts. The concept 'dog' is included in the common character of animal and it may be also included in a particular breed 'Eskimo'. Similarly the concept belief includes belief in God, belief that the sun rises in the morning.

CONCEPT FORMATION

We perceive characteristics which are common to a large number of objects and associate these with symbols and apply these symbols to other similar objects. This process of applying and using some symbols to objects for classification is called concept formation. Concept formation is a process of thinking. It is incubating for certain idea or formation of some idea. Concept formation is the process of discovering some common characteristics to series of objects. It refers to the description of these common elements. It differentiates the quality of one object from another.

In the illustrative example we used the concept of white cup. Cups may be white, blue, yellow. But all these cups are made of porcelain material. This common characteristic differentiates cups from glass tumblers and other similar materials used for drinking milk. Thus concepts are formed. For example, the concept, chair, it is included in the category of furnitures. A person who has learned the concept, chair is dealing with it as furniture. All those objects that have a seat, four legs, a back and two arms are called chairs. If a chair is introduced to a child in this way the child associates or relates all these aspects to the object and at once understands it as a chair. It does not matter whether it is made of wood or steel, it is a chair. Children learn colour concepts and number concepts easily. Concepts function to generalize, to discriminate and to abstract in our thought process.

There are two important psychological processes, operating in the concept formation. They are generalization and discrimination or differentiation. For example, a cow may be assigned to the general concept as animal or mammalian. Just like other animals the cow has four legs, a tail and two horns. Secondly, the concept, 'cow' facilitates us to differentiate this animal from other animals. From our personal experiences, from the knowledge gained through reading and some other sources we understand that cow is a harmless animal. Therefore, we are not reacting to a cow with any alarm that we may generally show to a bull or an elephant.

Thirdly, we may use this concept 'cow' for abstracting without making mention of any specific animals. For example, if the cow is introduced to a child for the first time and when he sees the cow for the second time he at once understands that it is a cow. He never mistakes it for a horse or for a donkey because it has four legs and a tail. In this process the cow has been discriminated from other animals that have certain common qualities. This is called abstraction. Abstraction is a process of isolating certain characteristics of an object and disregarding all other characteristics of the same object at the moment.

Concept formation is a complex process. But it takes place from simple form of learning the names of the objects to the forming of symbolic representation. The child first slowly learns only the names of the objects and their properties. In the example we used the white cup for illustration. The child understands how a cup looks like. Here the child learns the concept cup. Whatever looks like that and used for the same purpose it is called cup for the child. Here the cup is a general concept and white cup is specific concept. By saying white cup a kind of discrimination is taking place. Therefore, the child learns to discriminate white from other colours and differentiates it. The child may be forming concepts everyday in the same manner.

Thus there is formation of abstract ideas in the minds of the persons and they are attributed to objects which are having the same characteristics. In dealing with objects of different types with different characteristics the process of abstraction facilitates the person to classify into different categories. In the industrial work situation when the workers are dealing with spatial relations the process of abstraction helps them to solve the problem. They should find out the common characteristics that run through the objects with which the objects may be classified into certain types. This is how the process of abstraction works in concept formation. Candland (1968) equates the development of concept formation with operant learning. According to him it is similar to the techniques to facilitate the speed with which operant learning takes place. He considers that this concept formation is a problem solving task.

Kinds of Concepts and Fuzziness

Concepts as we use in our daily life are normally quite definite. An object is a genuine concept and also not a genuine concept. Therefore, concepts are classified into three categories. They are conjunctive concepts, disjunctive concepts and rational concepts.

The conjunctive concepts classify all objects, persons and events that have one common characteristics. For example, bird is a conjunctive concept. Rosch (1977) has studied the fuzziness of concepts. The common characteristics attributed to birds are, they fly, they are of a standard size, they have two legs, have feathers, they build nests, lay down eggs, they sing, move to warmer climate during cold season. But what about the birds like crow, sparrow, robin, eagle, goose, chicken, penguin. We can see that the fine quality of the concept bird are decreasing. Crow, sparrow, robin, eagle are very good examples of the concept of 'Bird'. But goose, chicken and penguin possess less of the birdness. These birds have the least qualities of a bird.

The concepts that include several characteristics are called *disjunctive concepts*. These concepts may have either this type of features or that type of features. For example, when a person has applied for some job, the management may consider his application either on the basis of his qualifications and grades he obtained or the athletic achievements he made, if the management is interested in sports or on the basis of letters of recommendations he has presented to the management. Either one of these or a combination of these may put him into the category of qualified candidate for the post. The candidate himself may not be knowing sometimes, which one of these brought him the job. There is something most considered and something least considered. But all of these qualifications cannot be considered as equal or equally important. Therefore, the three concepts qualification, athletic achievement and letter of recommendation have decreasing order of 'fitness' for the post. Hence there is conceptional fuzziness.

The rational concepts refer to comparative characteristics of persons, objects and events. In comparing persons, objects or events there may be first, second, third, fourth and so on. Even ranks may be given if the numbers are many. But in our everyday life we make comparative statement, like good, better and best. For example, there are three lights illuminating a place. We may state that the first is bright, the second is brighter and the third is the brightest. These are all relative concepts. We may attribute the quality of the second light as brighter than the first. In the same way the concept of size, heaviness, height and so on. Thus the concept of heavy has no meaning without another related concept light. Rosch (1973) has demonstrated the importance of different concepts.

Studies on Concept Formation

Several experiments and tests have been devised to study this process of concept formation. Since the process of concept formation passes through generalization and differentiation it is essentially a process of learning. Goldstein (1935) in his studies while comparing the normal persons with brain injured persons found that there was something in common between perceiving and conceiving. He was able to see this distinction. He observed that persons with brain injury or diseased have grouped things together though they are not capable of genuine abstract thinking. Goldstein states that a person who has brain injury is not capable of abstract thought. But he has ability to take a concrete attitude.

In one of the tests Goldstein wanted to test whether a subject is capable of arranging those objects that belong to the same category by putting together the objects of different form, colour, material and use. The subject should reveal his method of classification whether he is employing abstract or concrete concepts in this task. Grouping the material on the basis of form or colour indicated the subject's abstract attitude — it is abstract concept. But grouping identical and near identical objects on the basis of some specific use indicated concrete attitude. One example of the task performed by a subject is that he has taken abstract attitude when he put hammers, screw drivers, saws and nails of different sizes and shapes into one category of 'tools'. He takes concrete attitude when he selects one hammer and one nail on the idea that one hammer is enough to drive one nail. In another type of materials he takes abstract attitude when he puts materials like pipe, cigar, matches and ash tray under 'smoking materials'. But when he selects one cigar and one match it is concrete on the idea that one match is enough to light one cigar. This is the function of concept formation illustrated by Goldstein through his studies.

In concept learning the psychological processes of generalization and discrimination have been utilized. Hilgard *et al.* (1975) state that a process that is complementary to generalization is discrimination. The process of generalization is reaction to similarities and the process of discrimination is reaction to differences. According to them rats are capable of generalization but if there is no reward the response is extinguished.

Human beings, because of their language ability they are able to deal with all sorts of concepts from simple concrete objects such as cat to highly abstract concepts such as faith, belief, justice and so on. Experiments have been conducted to see what types of concepts are formed most readily and easily. The studies made by Heidbreder (1947) are very interesting and worth discussing. His studies revealed that concepts with concrete characteristics were found to be easier to conceive than the concepts with more abstract characteristics of form and numbers. She used paired-associate method with different types of materials.

In the Heidbreder's concept formation test there are *five sets* of pictures and in each set there are *six pictures*. On the whole there are 30 pictures. Under each picture of concept, one non-sense word is paired, printed under it. On each trial one set of six picture are presented one at a time. The participant is required to look forward the response (non-sense word) paired with each other. The experimenter has assigned a response word (non-sense word) to each concept (picture). For example, RELK is a response word to the *concrete concept* 'face'. FARD is used to the *spatial concept* 'circle' pattern. LETH is used to the *object concept*, the 'building', DILT is used as a response to the *number concept* of 'two' and 'five'. In the second trial none of the pictures are shown the same form as shown in the first trial and in the same order. All are little different pictures in some way but the same concepts were paired with the same response words used in the first trial.

In the first trial six pictures are presented and in the second trial another set of six different pictures of the same pattern in different order are presented and in the third trial another set of six little different type of pictures but in different order are presented. Thus in all the five trials six pictures with little different shapes but with the same response word in varied orders are presented. It is to be noted that none of the picture concepts are repeated in the same form in the same order for the second time. The experimenter continued the trials till the participant responded correctly to all the stimuli (pictures) in a trial. For example, LETH is used as an object concept, building. LETH is used in all the five sets but the shape of building varied from set to set.

The set of pictures are presented to the participant top to bottom one by one and as soon as one is presented he is asked to state the response word paired with the concept. On the first trial the participant has no chance of knowing the response word. But after several trials he slowly learns to respond correctly on all pictures. Strictly speaking in the Heidbreder's concept formation test there are six pictures, namely (1) human face, (2) circle, (3) building, (4) number, two (5) number five and (6) tree. The same pictures are, drawn differently in five trials and each picture is paired with the same response word (non-sense word).

In these types of experiments it has been found that object concepts are easiest to learn, the next easiest are spatial concepts and then number concepts. Generally human thinking tends to be more for objects rather than for abstraction. Piaget (1952) states that the child first learns only object concepts and the abstract concepts he develops when becomes older. Individuals with brain damage respond to only object concepts and lose their ability to respond to abstract concepts.

Several years earlier to Heidbreder, experiments on concept formation were conducted by Kenneth Smoke (1932). He used figures in his experiment. The figures consists of a circle with one dot inside and another dot outside the circle. He presented these figures to his participants to help them to develop concepts such as DAX. The figures used by Smoke for his experiment had certain characteristics such a different shapes, sizes, varied number of dots and locating the dots in different places of the figures. Some of these features' shape, number of dots and location of dots were relevant. Others are such as size, aspects of dot location, other than being inside or outside of the figure. For the purpose of attaining the concepts it should be stated what are the features relevant and what are not relevant. In this experiment of concept formation Smoke made it a point that the participant's task is not only to detect the relevant attributes (features) but also to ascertain the rule of principle applied in it.

According to Smoke, DAX refers to a figure that is circular and has two dots, one inside and one outside the circle. Having this as the governing principle the participant should identify different types of concepts. A figure other than a circle with two dots is a "Not DAX" in this experiment. The figure (stimulus) is presented one by one to the participant and he has to give his response whether it is 'DAX' or 'Not DAX'. If the response is not correct the experimenter provides the feedback. The trials should be carried out till the participant responds to all the stimuli correctly.

Levine (1966) has observed in this experiment that the participant in mastering the concepts was going on testing the hypothesis one by one after seeing the previous hypothesis was wrong. He has noticed two aspects in the participant's formulation of hypotheses and for each trial. First, the participant's behaviour was analytical. In this approach he tried to isolate the attributes of the materials and the rules that govern the concepts. This is what expected from every participant. But some participants use non-analytical approach. This is more common as many of the participants are not analytical. For example, children are not analytical. For them the concept 'fish' includes all kinds of fish, dolphin, whale and the concept bird includes, bat and every creature that flies in the air. Levine states that in concept learning experiments it has been observed that most of the participants follow a strategy of 'lose-shift and win-stay'. The correct hypothesis is gradually narrowed till the participant achieve the correct concept.

There are several types of tests designed by the experimenters to investigate the capacity of concept formation in persons. One of the tests very widely used in psychological laboratories and in the clinical practices is the test designed by Vigotsky. This test has been later on revised by Haufmann and Kasanin (1937). The participant is given blocks which differ from one another in size, height, shape and colour. He is told to put them into four groups. The experimenter notes down, the type of behaviour of the participant as he is trying to group. The experimenter tells the participant whether the group is right or wrong.

What is to be actually accounted in this test, is the size and the height. Combining these two criteria size and height there are four categories such as large tall, large flat, small tall, small flat. The person who groups the blocks on the basis of colour and shape or arranges according to any other form commits the error. The experimenter tells him that he is wrong. The participant himself should learn by some method to use the two criteria in grouping, the size and height.

Murphy (1952) says that it is not only the brain injured persons are incapable of forming such categories in their minds, but also many normal persons lack the capacity to solve this type of simple problem, simply by looking into the colour and shape. Haufmann (1941) has observed that there are two fundamental types of sets, one the perceiving set and the other conceiving set. Persons who are perceiving set, see at the blocks for natural grouping and persons who use conceiving set try to find out some kind of rational approach for grouping the blocks. The first is concrete attitude and the latter is abstract attitude, Several studies have been conducted comparing the brain injured with the normal people in this task of grouping the blocks (Goldstein, 1935) and found that many normal persons look at the blocks squarely without having any rational thinking.

EXPERIMENT

Problem

Study the process of concept formation that takes place in an individual person and state whether he/she had the perceiving attitude or conceiving attitude.

Method

Participant

Choose a person who has no knowledge about this experiment and the apparatus. Describe his characteristics, like age, sex, qualification, occupation and whether he/she has any deformities.

Materials

Vigotsky test modified and as used by Haufmann and Kasanin, stop clock.

Fig. 7.4 : Vigotsky Test: Thinking : Concept Formation : Experiment 19 (Before Experiment)

Fig. 7.5 : Thinking : Concept Formation : Completed Form : Experiment 19

Description of the Materials

The concept formation test was originally designed by I.S.Vigotsky to study the process of concept formation in a person. Later on this test was modified by Haufmann and Kasanin (1937). The test consists of 22 wooden blocks. The blocks are of five different colours, six different shapes, two different heights (tall and flat) and two different sizes (large and small). The blocks are having some symbols or non-sense syllables or numbers at the bottom. These are used to define the block type. That is, whether the grouping is right or wrong. For example all blocks having 'lag' at the bottom are tall and large, all blocks having 'mur' at the bottom are tall and small, all blocks having 'bik' at the bottom are flat and large and all blocks having 'cev' at the bottom are flat and small. Colour and shape are not relevant as far as grouping is concerned. These are distractors. They are packed in a wooden box in the appropriate places. This test is intended to test whether the subject is able to abstract on seeing many types of materials and able to classify them taking into account the common characteristics found in the objects. The test calls forth several types of reactions. Apart from forming symbolic idea there may be also trial and error method to solve the problem. It also involves reasoning in solving the problem. To some extent it is also involving insightful learning to solve the problem.

Procedure

The experimenter should remove all the wooden blocks from the box and check whether there are 22 blocks of different sizes, shapes, colours and heights. Then he should shuffle them for mixing the varied categories and then arrange them roughly into a circle on a table. He should take the subject to sit comfortably in front of the blocks and the following instructions should be given to him.

> "Here there are four kinds of blocks and you will have to arrange them in four groups on the basis of a principle which you discover suitable for the task. You will have to only move them and should not lift or tilt them in your performances. After classifying them into four groups tell me what idea or principle you have used for this. There is no time limit for this. You can take your own time. When I say, ready go, you must start the task. Now ready go."

As soon as the subject starts the task, the experimenter should start the stop clock to run and should note down the time in seconds taken by the subject for each classification. When the subject has completed one trial of classification the experimenter should check it, whether it is right or wrong by lifting the wooden block and seeing the non-sense syllables or symbols at the bottom of the blocks. If it is wrong then he should lift two blocks belonging to two different categories and show the bottom to the participant that the grouping is not correct.

The subject should be made to understand that in a group there should not be two different symbols or numbers, but all of them should have the same symbol or number. If it is wrong, the experimenter should collect all the wooden blocks and shuffle them once again and arrange them in a circle and ask the subject to start classifying them into four groups. This procedure should be continued till the subject classifies the wooden block into four groups successfully and for each trial the time should be noted down. Every time what principle or idea the subject applied in the classification also should be asked and noted down. At the end the introspective report should be obtained from the subject and he/she may be allowed to go away.

Results

The experiment has yielded three types of scores, one is the trial and the other is time and the third is principles or idea the subject has used in classifying the blocks. The scores should be presented in Table 7.1. Find out the number of trials and the amount of time he has taken to solve the problem. What are the ideas he progressively used. Collect the group data of the experiment from other experimenters and present them in Table 7.2. Work out the mean score and analyse the data.

Table 7.1 : Showing the individual data in the concept formation experiment

S. No.	Trials	Time taken in seconds	Criteria used	Remarks Right/Wrong
1.				
2.				
3.				
4.				
5.				
6.				
Total				

State how many trials the participant has taken to complete the task. State whether there was any progressively decreasing time for each trial or increasing time. Has he/she a perceiving attitude towards the problem or conceiving attitude. Has he/she followed trial and error method, or insightful method or reasoning method in solving the problem? State the mean trial of the group. How many of them below the mean and how many above the mean. Who has taken more trials to solve the problem and who has taken less number of trials. Where your participant stands? What is the mean time of the group? How many of them above the group mean and how many below the group mean in time? Where your participant stands?

Table 7.2 : Showing the group data of the concept formation experiment

S. No.	Name	Total No. of Trials	Total amount of time in seconds	Criteria used to solve it
1	R.V.			
2	M.S.			
3	K.R.			
4	B.V.			
5	K.P.			
6	A.L.			
7	T.S.			
8	V.S.			
9	M.T.			
10	T.V.			
Total				
Mean				

Discussion

Discuss the individual's performance. Whether he is slow or fast. How many trials he has taken to complete the task successfully. What are principles he used? How much time he has taken for the first trial and how much for the last trial? Whether he used trial and error method or insightful learning or abstraction. What was your participant's reaction when he failed to solve the problem? Discuss the data taking into account his introspective data. Discuss the general performance of the group. How many of them taken least amount of time and few trials? How many above the group mean in time and trials ? Where your subject stands in the group? What are the criteria used by the group. Which criteria were used by many? Whether the process of abstraction was there in the experiment? Is there any difference in the performance due to sex or age. What was the strange behaviour you have observed in your participant? Compare your results with the results of earlier studies and state whether there is anything new findings in your experiment.

Conclusions

How much time your subject has taken to successfully complete the experiment, and how many trials he has taken? What are the principles he progressively used?

Who has taken the least amount of time to complete the work and in how many trials? Who used only few trials? Is there difference in performance due to sex and age or other factors?

EXPERIMENT 20 : CONCEPT ACHIEVEMENT (PROF. RAJAMANICKAM'S ACHIEVEMENT TEST)

Man is a rational animal and also a thinking animal. He reveals his thinking in various forms of his everyday life experiences. This makes others to understand what the person is conveying in his process of communication. The basic elements of thinking must be man's sensations and perceptions. With his sense organs he is able to see, hear, touch, taste and smell every object he comes across and understand what the object is. The perceived objects sometimes may not be there, but he is able to indicate it by means of some symbols.

The primitive man in the beginning might have used some signs and signals to indicate an object. He might have used also some sounds though it might not have specifically indicated any meaning to others. But gradually he might have learned to associate some sounds with some objects. In the pre-language period some signs were used to indicate the objects and one's ideas or thinking. In the prehistoric period when there was no language to communicate one's ideas and thoughts, men were using pictures as symbols of the objects. The Egyptians used pictures for communication. The Egyptians drew small pictures of the objects on pieces of reed called papyrus. Papyrus is a kind of paper made from the stem pith of cyperaceous plant, used by the ancient Egyptians for writing. The ancient Greeks and Romans also used this plant and ancient documents were written on this paper. Only from the word papyrus we get the word paper (Horniblow, 1955).

DEVELOPMENT OF THOUGHT AND EXPRESSION

The ancient Egyptians carved the pictures on their buildings and monuments and on tablets of clay. For example the picture of a crane was indicative of bird, wavy horizontal lines were indicative of water, wavy vertical lines were indicative of rains. Similarly for wind, for eating and drinking pictures were drawn to indicate the object or event. So also the Sumerians used picture writing in tablets of clay to express their thinking. Some of the Semites learned how to write from the Sumerians. This, they have done by cutting little signs in damp clay with a stick. The clay tablets were then baked and in this way the signs of the idea were fixed.

All over the world the archaeologists have unearthed many such carved pictures and signs which were the thoughts of the ancient people. Even in India many archaeological discoveries illustrate what type of media the ancient Indians were using for expressing their ideas and thoughts. They were only in the form of signs and symbols. The unearthed ancient carved stones and clay tablets provide us with ample evidences about the means of expression of the ancient people's thinking. Modern man has very little to think as most of his life aspects by the process of evolution are functioning on stereotypes and habits as Abraham Maslow (1970) has stated. But the ancient man had to think about quite lot as he had to find out some means for his living. Since things around the ancient man were rough and uncultured he had to think and toil for getting his food either by hunting or by cultivation, and other requirements like making out his abodes, designing and manufacturing arms for fighting with his enemies. Therefore, necessities made the ancient man to think about many things and also he had invented many things at his level.

Even with regard to modern man it is the necessities that made him to think or some event or incidence made him to think. What incidence made Archimedes (287-212 B.C.), the Greek physicist to think of discovering the principle regarding the loss of weight of the floating materials in a liquid. Similarly what incidence made the English mathematician and physicist and philosopher, Sir Isaac Newton (1643-1727) to invent the law of gravitation. Something might have made Cook (1728-79) and Columbus (1451-1506) to think of discovering the new lands, Australia and America. Thus if we go into the details of the discoveries and inventions of the scientists and explorers some event or incidence might have made them to think of something new in existence and to go in search of them. Therefore, there is something stimulating man to think and work for it.

It is true that sensations and perceptions and learning are basic elements in thinking. But, in addition to these basic elements some specific event or incidence that is associated with the individual impels him to think in terms of putting him into action. It is not that all people who meet with some events and incidence think of them raising the question "Why it is like this?" Therefore, it is only very few people intend to think and evolve something out of their thinking and others get the benefits of them. From most of the process of thinking we can state that man normally do not show the expression of his thinking. If he would have thought, something would have occurred and which might be expressed in some form. If such things are not taking place man is not engaged in thinking. Solving some small problems might have been due to his experience, knowledge and learning. Man begins to think if only his experience, knowledge and learning and his stereotypical and habitual ways of life fail to help him to get his things done. Thinking whether it is simple, or complex emits something new.

This emission or discharge of something new may occur by rearranging or restructuring the old ones and also by adding something new to the old ones. Every human being might have experienced this in his life at least to some extent which perhaps he might not be remembering it. We put ourselves in thinking if only some necessity arises, whether it is very simple type of thinking or complex type of thinking. Sometimes, it is very difficult to express our thinking. Only when we express something it indicates that we are thinking. When there is no expressed form no thinking has taken place. Joseph Addison (1672-1719), an English easayist and poet, had founded a journal called *The Spectator* in 1711 in collaboration with Richard Steela and made notable contributions. All his essays were considered to be classic and he expressed his thought through his essays. Once it was so happened that he should address his club members and started to speak, "I conceive... I conceive... I conceive... but conceive nothing", by saying like this he sat down. But in the case of some people we could see that their thought preceding their expression. Therefore, thinking is a mental process that arises to meet some critical situation or to provide solution to some problem or to meet a need or necessity. At sometimes this may be quick and sudden at sometimes it may be slow. In most complex situations there may not be any thinking at all.

Children brought up in highly stimulating environments might have inherited highly stereotyped ideas and habits and they may meet difficult problems and complex situations with their stereotyped ideas very successfully whereas children brought up in unstimulating or underdeveloped environments could not face the difficult problems and critical situations successfully. They may try by trial and error methods. Max Werthiemer (1926) says that this is a blind method of approaching the problem.

Thinking is an internal mental process which is inexplicable unless the individual himself expresses it what he has been thinking. We cannot say by seeing a person whether he is thinking or not unlike other activities. Therefore, the study of thinking is a very difficult task. But man has discovered some methods of apprehending the process of thinking. The methods cannot be said as direct methods, but they are only indirect methods. In apprehending the process of thinking man makes use of language mainly. Language is a symbolic representation of objects and events.

Language is a universal phenomenon. In the evolutionary process human beings all over the world have intelligently invented some method of attributing some specific meaning to every human vocal sound. Every vocal sound produced had a meaning which was referring to some object. The object may be human beings, animals, birds, objects and events. Since human beings in the pre-historic period lived in groups or small communities, each community had its own meanings to their vocal sounds and this has become ultimately their vocal expression, in other words their language. Therefore, the spoken language had its origin in attributing meaning to the vocal sounds which were denoting some object or event. These vocal sounds denoting some meaning may be stated as symbols. In the course of evolutionary process the symbols received written forms. In the beginning only pictures were written to indicate the objects. Every object received a symbolic form and when the language was developed with letters the objects and events were indicated by words which were the symbolic representations of the objects and events.

Symbols were used either by vocal sounds or in words spoken or written forms in the absence of the objects, persons or events. When the full fledged language came into use this has become very easy, that is the use of symbols. The symbol became the internal stimulus of the objects, persons and events. Thus, thinking takes place through symbols which is expressed through the medium of language. Hilgard *et al.* (1975) states that language provides a system of symbols and rules that makes our thinking easy. As we look seriously on thinking and language we can even go to the extreme end by saying that there is no thinking without language. This is the reason that many textbook writers in Psychology associate thinking and language together in the same chapter. Symbols may be of different types having different meaning and, therefore, they require classification. This may fall under two types as generalization and discrimination.

Though the processes of generalization and discrimination are the components in learning they are also taking place in thinking. There are two forms of thinking. One is directed thinking and the other is associative thinking. The directed thinking is aimed at solving a particular problem. This form of thinking is limited in its scope. The associative form of thinking is unlimited in its scope. The associative thinking mainly refers to imaginations. Fantasies, daydreams, dreams and imagination are all associative type of thinking. All these are not directed to solve any problem.

Man is a thinking animal and also doing animal. In the execution of some activity whether it is simple or complex, it depends upon the person's previous thinking. The person has planned something earlier and now he is executing it. Therefore, planning is a form of thinking. Thirdly, man is a speaking animal. He speaks out what he has thought little earlier or few seconds ago. Man's thinking activity also caused the invention of language, through which he communicates his thoughts to others. John B. Watson (1878-1948) (1919) states that talking also is a kind of thinking. Talking is a louder thinking and thinking is silent talking. Therefore, we could see that man ultimately directs his own behaviour by his thinking. Man plans his own behaviour and anticipates the consequences of his own actions. At every stage man expresses his thinking whether his actions are successful or failure.

SYMBOLIC PROCESS

The process of thinking takes place step by step from simple symbols or signs through formation of concepts and development into language. From the discussion we have made in the preceding section this has been stated briefly. Therefore, there are three important aspects in the process of thinking. They are also called tools of thinking. According to Hilgard *et al.* (1975) thinking is considered as cognitive process described by the use of symbols. These symbols represent objects and events.

We are using quite a number of symbols in our everyday life, but we do not analyse them as to how they are symbols. In the absence of the real objects we use symbols. But every symbol is attached to an object and the idea is formed in our mind. Thus symbols become the internal representation of the objects and events. You might have described to your friends an event of eating a dinner. You also might have mentioned the names of the items served in the dinner and the dishes you have enjoyed. A specific kind of pudding you ate also might have included in your description of the items. But now you do not have the pudding in front of you nor you are eating it. But they are referred here symbolically. This symbolic reference may be stated as an idea or thought. Thoughts deal with objects and events of the past. It is also dealing with recollecting and reproducing the past events. In describing the dinner, you ate, to your friends it is only this process that is taking place. What you are describing to your friends giving the names of the items and the way in which you enjoyed or relished the dinner may be stated as symbolic process or thought process.

Modern textbook writers in Psychology invariably use the phrases symbolic processes, symbolic behaviour, ideational process, symbolic thinking and so on. But what all these refer to? After explaining everything they arrive at the conclusion that it is a process of thinking. According Moskowitz and Orgel (1969) symbolic process is a process of utilizing of symbols in making decisions, solving problems or adjusting to the environment. According to them we speak through symbols. What we have done in the past, what we are doing now, and what we are going to do next — all explained only through symbols. "Yesterday by this time I was in New Delhi Golf Club enjoying a nice drink. Today I am here at Madras eating lunch with you. Tomorrow by this time I will be witnessing a tennis match at Bombay". In all these three instances the person has been put in various situations, at various places and engaged in different types of activities. By making use of several symbols he conveyed several informations. Using several concepts several informations have been communicated. Symbolic processes are often used by human beings. According to Moskowitz and Orgel man has the ability to learn and also ability to utilize what has been learned. By using symbols in his activities a process called symbolization is taking place and this symbolic process enlarges man's capacity for learning and his ability to utilize what has been learned. Therefore, according to them symbolic processes become the basis for the development of language and the symbolic processes in turn are influenced by the development of language. By analysing what has been stated above we can arrive at an idea that there are certain amount of interdependence among learning symbolic processes, communication, development of language and problem solving.

Piaget (1926) says that there is clearly noticeable correspondence or communication between the development of thinking and the development of language. Piaget (1896-1980) made extensive studies on various aspects of the developmental process in children and their thinking and language development. A symbol is an internal representation of an object or event. It is a representation of a stimulus or event. Therefore, it may be also stated as a response. Symbols are only ideas of some objects. The process of identifying the response or idea is called symbolic process or symbolic thinking. Sometimes our vivid sensory experiences also may serve as symbols. But we call these as images — visual or auditory images. In our process of imagination we make use of such images. Since they are also a kind of symbols we may call the process as symbolic imagination or thinking. All our symbolic process or thinking activities are expressed through language and language is also constituted of symbols, language is very much involved in all our symbolic activities.

Concept

In the absence of an object we use the symbol to indicate the object. It is in the form of a word. The word is a language symbol. Therefore, a symbol is an idea. When an idea is expressed through a word we call it a concept. A concept is a word, having certain specific meaning. The term, concept is defined in various ways. But one important characteristic of the term concept is that the meaning of which is universally accepted.

Basically, concept is a philosophical term. It refers to some abstract idea or thought. Concept may be defined as a general idea of something formed mentally by combining all specific and general characteristic features. We have stated that a symbol stands for an object or a group of objects or events with some common characteristics. In this way a chair is a symbol and it is also a concept. Similarly, students, fruits, cart, tree, building, square, circle, five, eight, small, big, slow, fast, red, colour, justice, faith are all concepts. These include various types of concept, like concrete, object, spatial, number and abstract concepts. All these concepts have certain common properties by which we can classify them into certain categories.

Certain concepts are simple which can be comprehended by children easily and certain concepts are complex, very difficult for children to comprehend correctly. Concept, generally, means a word having some meaning. It has certain qualities by which it is identified. Concept may be referring to concrete objects or abstract ideas. The concrete objects are identified by their qualities and named. It is a very difficult task by finding out the common properties of certain objects which are having certain variations and categorise or classify them. It is easy to classify certain objects having simple or few qualities or objects that have clear concrete differences. But if abstraction is required for classification, however concrete the objects are very difficult. Human tendency is to go for concrete properties and, therefore, classifying objects on the basis of concrete properties becomes easy for any subject. Since there are concepts of various types, which are required abstraction, more thinking is needed to classify them on the basis of their abstract qualities.

Types of Concepts

Concepts may be classified on the basis of their properties. They are simple concepts, conjunctive concepts, disjunctive concepts and rational concepts. Concepts which are having one or two qualities are simple concepts. It is very easy to classify such concepts. For example, black colour has only one attribute. It is blackness. Black as a quality can be found in many things. For example, we use black ink, black paper, black coal and so on. There is no other quality associated with black colour.

The conjunctive concepts classify all objects, persons, and events that have more than two qualities. For example, dog is a conjunctive concept. It has more common characteristics. First of all dog is an animal, having four legs and a tail. Dog is a faithful animal to his master and watches the master and his house. There are varies of breeds in dogs. There are also street dogs. Dogs usually bark. Any other animals which do not possess these qualities are not dogs.

In *disjunctive concepts* several qualities are included. For example, the concept couple implies several meanings. Couple means a pair of persons or anything joining two things. It may refer to any two persons. A couple of persons coming at a distance. We could see that they are only human beings. As they come little closer we come to know that it is a man and a woman. A couple of persons — a man and a woman may be having any relationship. They may be husband and wife, or engaged and not married, or partners in a dance or game or they may be father and daughter, or mother and son, or may be brother and sister or they may be friends or neighbours or may not be having any relationship. Thus there are so many attributes. On verification anyone of the relationship may be correct.

The *rational concepts* are based upon comparative characteristics. It is to be decided by comparing one with the other objects. For example, there are three weights. We decide by lifting one by one and state which is light, which is heavy and which is heavier. Similarly in the case of objects of different sizes. Among four objects the first is small, the second is large, the third is larger and the fourth is the largest. If there are two objects we may state that the second is double size of the first. Children learn this type of concepts early. Because the difference is perceived physically. It is in concrete concepts. The concept heavier or larger, has no meaning without the concept lighter, or small or large.

Concept Attainment

Concept attainment is the process of discovering some quality or property that is common to a series of objects. Concept refers to the description of some common element. It is differentiating the quality of one object from another. Concept attainment or formation takes place in children when they begin to understand objects and learn to differentiate them. Concept attainment may also take place in adults. In these days when new instruments and tools are invented and scientific and technological advancements are in rapid progress, everyday some new concept attainment becomes a continuous process even in the case of adults. Adults attain quickly while children take time to learn and understand concepts.

Generally we teach children to learn things and events. When the child is introduced with a simple object, pencil, he understands what is a pencil and also pencil is green in colour. He learns two concepts, pencil and green. Later on he may understand that pencils may be green, red, blue and yellow and it is used for writing and it is made of wood and lead. These are common characteristics of pencil and he can differentiate them from other writing materials like, pen, ball-point pen, sketch pen. Thus, here the concept that the child has learned are pencil, and pencils are made of wood and lead and pencil is used for writing. Children learn easily colour concepts and number concepts, concrete and object concepts. It takes longer time for children to learn spatial and abstract concepts.

The functions of concept attainment in our thinking process are generalization, differentiation and abstraction. Therefore, there are three important psychological processes, operating in concept attainment. They are generalization, discrimination and abstraction. On seeing a horse in the premises we may introduce it to a child, "this is a horse", used for draught and riding. The child now understands a new animal called horse which is different from other animals having four legs and a tail.

Now, let us see how the psychological processes, generalization, discrimination and abstraction are applicable to the horse. The horse can be assigned to the general concept, 'animal'. The horse belongs to the domesticated perissodactyl mammalian. Just like all other animals the horse has four legs and a tail. This is generalization. Secondly, the concept horse helps us to differentiate it from other animals. From reading materials related to the horse and personal experience and also from some other sources we understand that horses are harmless animals since they are domesticated. Therefore, we are not reacting to a horse with any sense of fear or alarm that we normally show to an elephant, or tiger or bear. Thus the horse is discriminated.

Thirdly, we can use this concept horse for abstracting without making mention to any specific animals. In this process the horse has been discriminated from other animals that have common characteristics. The horse is isolated from other animals in certain characteristics. The horse is used for draught and riding. No other animals are used for draught and riding as the horse is used for. Therefore, the horse is isolated. This process is called abstraction.

Generally abstraction is a process of isolating certain characteristics of an object and disregarding all other characteristics of the same object at the moment. For example, when you mention that your friend is very 'thin' you isolate this thin characteristic of your friend from other characteristics. The adjective 'thin' indicates your friend's characteristic and also how other persons are.

But the process of generalization in the beginning appears to be very crude and not precise. A little child on seeing a donkey for the first time might call it a horse or a cow, if he has already learned something about a donkey and a cow, having four legs and a tail. Because this new animal donkey is related to the old animal (horse) in some manner. But when he comes to know that it is not the same animal that he saw earlier he is ready to learn the name of the new animal and differentiate it from the other. There are not many experimental studies on concept attainment as in other themes. Unlike other psychological processes experiments on thinking are very difficult.

Studies on Concept Attainment

The earliest study in this area was made by Kurt Goldstein (1935) in the clinical situation. His subjects were brain injured persons. He observed that brain injured persons have lost capacity to think in abstract terms. A few years after Goldstein's study Haufmann and Kasanin (1937) conducted another study on concept formation. They used the test material designed by Vigotsky (1962) in connection with his earlier study on thought and language in children. They used the test in such a way that the procedure attracted many researchers all over the world and the same procedure was used in concept formation experiments. More information about this test is provided elsewhere in this book.

This test consists of twenty two wooden blocks. These blocks are in five colours such as green, blue, yellow, orange and white. All fall under two sizes as large and small and also two heights as tall and flat and they have six shapes. At the bottom of every block a three letter non-sense syllable is printed. These three letter non-sense syllables serve the purpose of classifying the blocks into categories. There are only four three-letter non-sense syllables such as LAG, MUR, BIK and CEV. The subject is expected to classify the 22 wooden blocks into four categories using any criteria. The important variables are size and height and all other variables are distractors. The amount of time and number of trials are taken into account to classify the blocks.

Kenneth Smoke (1932) used figures in his test. This is supposed to be the earliest study on concept attainment. These figures had different shapes and sizes and the number of dots also varied. The figures are presented to the participants to develop concepts such as DAX. The test consisted of a circle one dot inside and one dot outside. All the dots are located at different parts of the figures. The participant should attain the concept DAX, for which he/she has to indicate which characteristics are relevant and which are not. A circle having one dot inside and one dot outside alone is DAX. All other forms are not DAX. The participant should indicate this. The experiment is testing five hypotheses one by one. The participant can perfectly identify the concept DAX only in the sixth trial.

A different type of experiment has been designed by Heidbreder (1947) employing paired-associate method for concept attainment. According to Heidbreder's study the concepts with concrete features were easier for the subjects to conceive than the concepts with abstract features and numbers. The concept attainment test consisted of thirty pictures. They were put into five sets and in each set there are six pictures. Under each picture one four letter non-sense syllable is given. This is the response word. There are mainly six pictures such as human face, building, tree, number two, number five and circle. The same six pictures were drawn in five different forms but with the same four-letter non-sense syllables at the bottom of each picture. The important characteristic of this test is that the picture in every trial is little differently drawn, but the response words (non-sense syllables) are the same and they were not changed. Therefore, in every trial (5 trials) the same response words (non-sense syllables) are given with little varied picture.

A trial consists of six pictures and in each trial one set of six pictures are presented to the participant one at a time, starting from the top of the column to the bottom. As the picture with non-sense word is presented to the participant he/she is expected to associate the word with the picture. When the six pictures of the first trial are presented the participant is expected to call the response word of each picture. After the participant's responses for the six pictures the experimenter reads out the correct response word, written at the bottom. In the second trial the same pictures are not presented, pictures are little varied, but belong to the same categories. The same category of pictures are shown in all five trials. It is tested whether the participant has learned to associate the response words to the pictures which are having certain common characteristics and whether the concepts are attained.

For example RELK is the response word for human face and it is given under every human face. The participant is expected to give the same response word to all the human faces presented in all five trials. The same thing should be followed for all the pictures responding with same response word. The author of the test has introduced in this test, concrete concepts, or object concepts, spatial concepts and number concepts. Heidbreder states that generally human thinking tends to be more for objects rather than for abstraction. This is a general conclusion drawn from every experiment in concept formation. We could not get more information other than these in Heidbreder's experiment. The scoring procedure and discussing the individual's performance by comparing with other members of the group have no place in this test.

EXPERIMENT

Problem

Assess an individual's process of thinking by administering a suitable test to him/her and state the nature of thinking demonstrated by him/her in the performance of the test.

Method

Participant

Find out a person who has no knowledge about this test and experiment and describe his/her personal characteristics such as age, qualifications, sex, socio-economic status and his/her other personal interests.

Materials

Concept Achievement Test designed by Prof. Rajamanickam, Stop clock, scoring tables (Appendix-I and II).

Description of the Materials

The concept achievement test consists of forty-eight pictures. This test is designed and developed by Prof. M. Rajamanickam on the pattern of Heidbreder's concept attainment test with some changes and modifications in it. They are put under six sets and in each set there are eight pictures. The eight pictures in each set belong to eight different categories. Under each picture a four letter non-sense syllable is given. These four letter non-sense syllables are the responding words. The eight pictures are (1) human face, (2) animal, (3) bird, (4) jug, (5) number three, (6) number six, (7) square pattern, and (8) circle pattern. The eight pictures which are given in set one, are drawn little differently in every set. Thus in all six sets the same picture is found in a varied form, but the response words (non-sense syllables) are the same and they are not changed. For example, the human face is seen in six different forms and similarly every other picture varied. The picture of an animal found in set one is not the picture of the same animal in set two. In every set different animals and birds are given. Similarly the jugs are having different shapes. The square and circle patterns vary in every set. But the response words (four letter non-sense syllables) found under every picture in the first set also found under corresponding pictures in other five sets. The pictures are arranged vertically from top to bottom in every set.

Scoring

The eight pictures of the first set are presented one at a time for ten seconds with the response word at the bottom of the picture. The participant is expected to associate the response word (four letter non-sense syllable) with the picture. After presenting all the eight pictures one by one to the participant with the response word the experimenter, now presents to the participant the same pictures one by one *without the response word*. The experimenter asks the participant to produce the corresponding response word of every picture. Pictures without response words are shown for ten seconds and the participant has to state the response word for the picture. The experimenter states, 'correct', if the same response word is produced by the participant. If it is not correct he says 'not correct' or 'wrong'. When the participant responded to all eight pictures, rightly or wrongly, to a set, the experimente reads out the correct response word for every picture in the same order to the participant. There can be twenty seconds interval between the first set and the second set.

Then the experimenter presents the second set of eight pictures drawn little differently from the pictures of the first set with the same response words one by one for ten seconds and when all eight pictures are over the same pictures without response words are shown for ten seconds and the participant is asked to report the response word for each picture. Then the experimenter reads out the correct response words given under each picture to the participant. The same procedure is followed in all six sets. The correct response for each concept is analysed.

Procedure

The experimenter should have the concept achievement test materials ready for the experiment. He should have six sets of pictures. In each set there should be eight pictures vertically arranged top to bottom and under each picture one response word (one four-letter non-sense syllable) should be written or printed. He should have also the same type of test materials *without response word at the bottom of each picture.* He should also check the stop clock. The pictures should be kept concealed before and after the presentation. They should not be within the visibility of the participant. The two types of pictures, one with response word and the other without response words should be kept separately.

The experimenter should make the participant seated comfortably in front of him in a well illuminated room. There should not be any visual or auditory distractions in the room. Then he should give the following instructions to the participant.

"I will show you a series of pictures of various categories. In each set there are eight pictures arranged top to bottom. Under each picture a four-letter non-sense word is given. This is called response word. I will show you each picture one by one for ten seconds. You will have to associate the picture and the response word. After showing you all the eight pictures one by one I will show you another type of the same pictures in the same order one by one for ten seconds but without the response words. Your task is, on seeing the first picture, to call the response word of each picture which you saw in the first type of pictures. Then I will show you the second picture and you should call the second response word you saw under the second picture. In the same manner I will show you the same eight pictures and on seeing each you should call its response word. If you call it rightly I will say 'correct'. If your report is not correct I will say 'wrong'. When you finish the report of the response words rightly or wrongly for the eight pictures I will read out the correct response word for each picture for your reference. This is the first set."

"Then I will start the second set in the same manner. But in the second set the eight pictures are drawn little differently from the first set of eight pictures, but the response words are the same as you saw in the first set. But there is no change in the response words. Your task is to see each picture and associate the response word with it. On presenting all the eight pictures one by one I will show you the same pictures without the response words and you will have to call the response word for each picture as you did in the first set. When you call the right response word I will say 'correct' and when you call a wrong word I will say 'wrong'. Then I will read out the correct response word for each picture for your reference. There are going to be six sets and the same procedure is followed in all the sets. In all the six sets your task is only to call the correct response word. Now get ready".

The experimenter arranges his test material in such a way that the participant could see only one picture at a time with its response word for ten seconds. The experimenter presents the first picture from the first set with the response word for ten seconds. As soon as the experimenter starts showing the picture he should start the stop clock to watch the time. Then the second picture is shown for ten seconds and so on. Then the same pictures without response words he presents and ask the participant: "Now call the response word for each picture". The experimenter records both correct and wrong responses. Then he starts the second set and carries out the same procedure till all six sets are over. When all the six sets are over the participant may be asked to give him/her introspective report and may be permitted to go away. The experimenter presents the responses of the participant in Table 7.3. If the participant could not attain the concepts at the end of sixth set the first set may be repeated as the seventh set and then the experiment may be stopped.

Results

The participant's reports are recorded in Table 7.3 as *R* and *W*, where *'R'* stands for right and *'W'* stands for wrong in each trial. The total rights are entered at the bottom of every set. The hypothesis is that the participant's right responses increase as he/she goes from set to set. In the first set there is very little chance for the participant to give the correct response to the pictures. Set to set he slowly learns to associate the response words to the pictures. But some participants may provide one or two correct response words even in the first set. Since different pictures of the same category are appearing in the coming sets it may cause little confusion to the participant as to which response word is to be associated with which picture.

Generalization is the important process to be adopted in this experiment in concept attainment. The response word is used generally to all objects irrespective of some small changes in the pictures. The pictures, human face, animal and bird are alive concrete concepts, the jug is object concrete concept. Numbers three and six are number concepts and the square and circle patterns are spatial concepts.

The assumption is that the participants can easily identify concrete concepts, then number and spatial concepts. But it is little difficult for them to identify abstract concepts.

Table 7.3 : Showing the Set and the response of the participant in Concept Attainment Test

S.No.	Picture (Stimulus)	Response word	Set 1	Set 2	Set 3	Set 4	Set 5	Set 6	Set 7
			Res. R or W	Res. R or W	Res. R or W	Res. R or W	Res. R or W	Res. R or W	Res. R or W
1	Bird	LAGH							
2	Human Face	MUKR							
3	Pattern (Square)	BIKT							
4	Number-3	CEVT							
5	Animal	GALB							
6	Pattern (Circle)	FARH							
7	Number-6	JISP							
8	Jug	KEMG							
	Total	R's							

R = Right; W = Wrong

Abstraction is finding out certain *specific quality* that would isolate object and *disregarding all other qualities* and identifying it and this specific quality is not found in other objects. This is how abstraction is made. In this experiment MUKR is the response word to the concrete concept human face, LAGH is the response word to 'bird' and GALB is the response word to 'animal', KEMG is the response word to concrete object 'jug'. There are two number concepts, three and six, and CEVT is the response word to the number concept 'three' and JISP is the response word to the number concept 'six'. The other two are spatial concepts. BIKT is the response word to the spatial concept 'square' pattern and FARH is the response word to the spatial concept 'circle' pattern.

Table 7.4: Showing the correct response words of the group for each set in the concept achievement test experiment

S.No.	Name	Set 1	Set 2	Set 3	Set 4	Set 5	Set 6	Set 7	Extra	Total
		Correct response	Correct response	Correct response	Correct response	Correct response	Correct response	Correct response	Correct response	
1	P.V.									
2	S.K.									
3	R.M.									
4	K.R.									
5	P.K.									
6	M.S.									
7	B.V.									
8	S.N.									
9	T.M.									
10	V.K.									
11	M.R.									
12	T.A.									
	Mean									

The experimenter can collect the group data from other experimenters. This is the total correct responses for each set of all participants in this experiment and may be presented in Table 7.4. The total and the mean for the group data may be worked out.

Though at the end all participants may report correct responses for all eight pictures the number of correct responses may vary from set to set. The individual differences can be also observed in this Table. Who is achieving correct response in set one. While completing sets 6 and 7 how many of the subjects arrive almost complete concept achievement stage. It can be also observed whether any of them did not achieve all the concepts at the end of 6th or 7th set.

The mean response for each set for all the subjects will show at what stage the mean responses are higher. The data may show whether the mean of the correct responses is going on increasing or goes at the same level till the end and at the end whether all of them have achieved correct concepts.

Discussion

The materials used in this experiment do not have identical features. But they have something in common. For example, when we express verbally the concept, human face, it is referring to the face of man or woman or boy or girl. But all faces are called with the response word MUKR, all birds are called with the response word LAGH and all animals are called with the response word GALB. In this type of experiment it is observed that the live concrete concepts like human face, bird and animal and object concept like the jug are learned easily and the next easy to learn are the spatial concepts like square pattern and circle pattern and then the number concepts like three and six and finally the abstract concepts are learned.

In this experiment it is expected that the participant should achieve all the concepts correctly maximum, in six or seven sets. If any participant could not achieve all the eight concepts in six or seven sets the experimenter should find out the type of concepts he/she has achieved and discuss the results. The experimenter should find out where the participant struggles to identify the concepts. He must also observe whether the person is abstracting or trying to find out any specific quality in an object and disregarding all other qualities of the same object at the moment.

This experiment of concept achievement on the pattern of Heidbreder appears to be more difficult than the Vigotsky test of concept formation experiment after Haufmann and Kasanin procedure. In the Haufmann and Kasanin procedure the size and height are used as criteria in classifying blocks into four categories. These criteria are conspicuously seen by the participant though the colour and shape stand as distractors. But in this concept achievement test the participant is required to observe in many pictures the common quality for generalization. For example, in the case of animals, there are six animals. All are different animals. Similarly there are six different birds. The person should not take each animal as separate object and try to find out certain common quality in them. For example, the participant should not isolate the animal cow with its specific characteristics, similarly the animal cat, monkey, pig, camel and dog. All these animals should be put into one category called 'animal', similarly the 'bird'. The six jugs are of different patterns. But in every shape it is only a jug. The experimenter, therefore, should observe whether the participant makes the process of generalization on seeing the same response word under the picture of all animals and under the picture of all birds and jugs.

In this experiment the process of generalization seems to be more difficult than the processes of discrimination and abstraction. The experimenter should discuss the results in terms of number of right response words (concepts achieved) in the first set, then in the second set, third set and so on. He should state whether the number of correct response is going on increasing or decreasing set to set. After how many sets the participant achieved all eight concepts. He should state which responses are easily reported and which are taken several in sets.

Discuss the individual's responses on the basis of the types of concepts. Which concept the individual has easily attained. What procedure the individual adopted in attaining the concepts. State whether the group as a whole was able to attain the concepts within the limited sets or needed extra set to attain the concepts. State the order of achieving the type of concepts of the most of the group members. Whether the members of the group learned easily the live concrete concepts like the human face, animal and bird, or they have learned the object concrete concept the jug easily or they have learned the number concepts like three and six easily or the spatial concepts like the square pattern and circle pattern. State the reason.

Conclusions

State whether all the subjects achieved all the eight concepts or any of them have failed to achieve. What are the concepts attained easily? Whether the process of generalization was adopted by the subjects? Or they followed trial and error method?

APPENDIX - I

Response Set

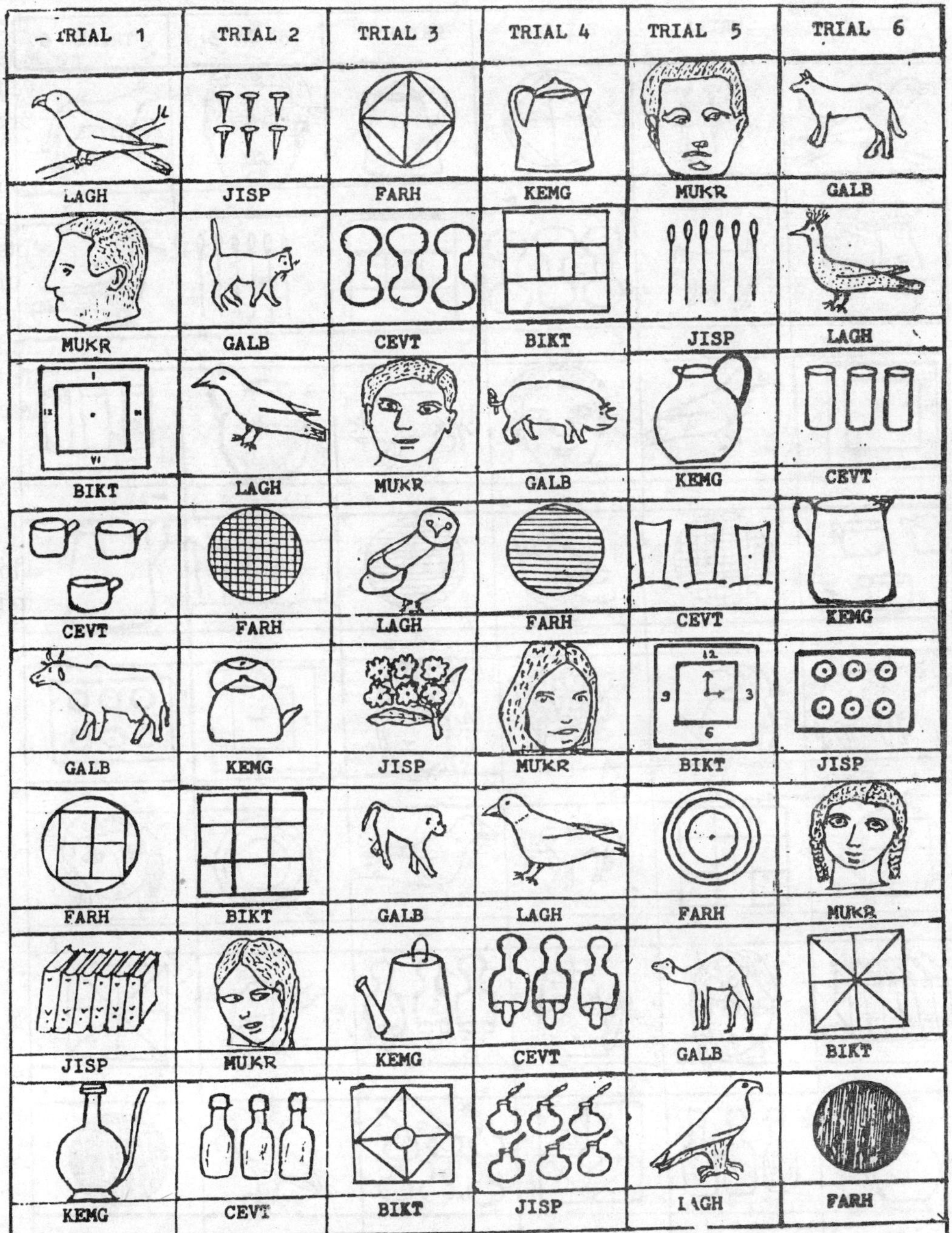

Fig. 7.6 : Response Set

APPEDIX - II

Stimulus Set

Fig. 7.7 : Stimulus Set

EXPERIMENT 21 : THE SEMANTIC DIFFERENTIAL
A Study of Meaning and Communication

Man has differentiated himself from other species by his ability to speak and express his thought and to communicate his thought to other human beings. Therefore, man has been using speech, whatever language he speaks, for communicating information from the time he has learned to speak. In the communication process the information conveyed should have the same meaning. But in the communication of information there are several alternatives. The stimuli used in the communication may be having several alternatives. In the communication process the amount of information is not depending upon a single stimulus that is used. It is better that one must use more alternatives or stimuli for better understanding and for more information.

For example, if one were to answer to a number of questions only by nodding his head, which means 'yes', there is very little information we are getting from it. Instead if there is scope for more alternatives like, true, false and unable to decide, there is chance for getting more or variety of information with varied meanings. Therefore, the hypothesis is that the more alternative stimuli we are using in our communication the more information we are receiving. According to the basic principle it may also be stated that in the communication we can get more information and also better meanings for the stimuli if the information contains the three important characteristics. Shannon (1948) states that the amount of information in the communication may be increased by (1) prolonging the length of speech, (2) by giving larger range of frequencies in the speech and (3) by varying the amplitude and pitch. With these characteristics we can make more distinctions in the meaning and information.

It is not always necessary to analyse the sound wave in a speech to evaluate the amount of information. It is better to evaluate the simplest component, the phonemes of the language. It is possible to work out the number of alternatives having in view the number of phonemes of the language. In English language there are 46 possible sound waves or phonemes. These are used to distinguish one word from the other. Therefore, it is assumed that each alternative has a definite chance of occurring like all other alternatives in a speech.

Studies have been made to measure speech perception. Lane (1962) has developed a procedure for this. Lane, Catania and Stevens (1961) have studied the difference between the speaker's own perception of his speech and the perception of the listener. The results of the study suggest that there is some deficiency for perfect communication between the physical world and the sensations of physical stimuli the speaker experienced. The speaker judges that the loudness of the vowel sounds in his own speech is increased as a power function. Therefore, changes in amplitude and frequency were perceived by the speaker as greater than the listener. This is the reason that sometimes the speaker complains that "I shout so much, but you are all not listening to me".

WORDS AND MEANING

In the communication we use language and through language we convey our thought to others. Therefore, language is a medium through which we exchange our thoughts. In expressing our thought we use the words of the language we speak. The simplest unit of a language is the phoneme or sound wave. But a word is one of the units of speech of a language. The speaker of a language usually regards it as the smallest isolate meaningful element of the language. Linguistics who analyse language state that all words are morphemes. The phonemes or speech sounds are combined and made into morphemes. Therefore, morphemes are the smallest meaningful units of speech. But, a meaningful word may have more than one morphemes.

The concept, 'meaning' refers to a sense. It is referring to the significance of a word, phrase, sentence and symbol. The characteristic of a word is to express a meaning. Every word used in a language has a meaning. If a word of a language fails to convey a meaning it is not a word. It is only combination of letters which is called 'non-sense syllable'. A word may have or express more than one meaning. For example, the word 'mean', which is having the meaning of 'mean', 'meaning', and 'meant'. It has also another meaning, 'miserly' or 'ungenerous' and 'petty'. The third meaning of the word mean is 'the middle point', 'average' in mathematical sense.

Apart from the meanings attributed more than one to a word, the word in sounds has varied connotations and denotations when expressed by different individuals. Words as they are in spoken sounds received different meaning. People, who learn their language also learn their meanings with different sounds which are assigned different meanings. For example, an object itself may give meanings and also create different feelings and emotions. The word 'cat' when mentioned is referring to a small animal. This arouses a feeling of liking or affection for some. But it may not be liked by many and hence may arouse a feeling of dislike. In other persons it may arouse a feeling of pity or sympathy for it even though they do not like to have it in their houses. Thus the word cat referring to an animal may arouse three different types of feeling in human beings. Though individuals differ in using the word, the word has acquired or the word has been assigned with liking, disliking and pitying. Therefore, words in a language when expressed were found to be having acquired different meanings for different persons.

This is causing some confusions in societies. It is also causing many misunderstandings and misconceptions as there are individual differences in the usage of words and the meanings. Individuals belonging to the cultural group may assign some meanings to words which may give different expressions in another cultural group. The thoughts of human beings and other rational conceptions are controlled by the meanings prevalent in a language. Douglas K. Candland (1968) says that personal word meanings are sometimes exaggerated that they are determining the behaviours of the individuals.

WORD ASSOCIATION

Generally in any language words have certain sequences. It is not occurring randomly. One word when uttered may prompt another word, which may prompt a third word. It is because the meanings are related. Francis Galton (1822-1911) calls this as word association. Galton (1879) was the pioneer in this field. He developed the word association test for personality assessment. His assumption was that the meanings of certain words are related. When one word is uttered, it prompts its related words in a chain like manner. When a word is suggested to a person and if he is asked to state the related words he would utter those words with which he is associated and thus reveals his personality traits. The word association primarily developed for personality assessment by Galton was continued by Carl Jung (1875-1961) in his studies. Jung (1919) used the word association test in the experimental situation for assessing personality. We are not concerned here about personality assessment. But we are interested how one word prompt other words in a chain-like manner as the meanings of those words are related. In what manner a person is related to some words is revealed by this word association test. A few years earlier to the study of Jung's word association test, Grace H. Kent and A.J. Rosanoff (1910) have published a list of words. This is most commonly used list of words in many word association experiments. The responses to this list of words from various groups were analysed.

For example, two persons were asked to give response words to the word 'tiger'. One person's response words were: Wild animal, forest, chases, deer, kills, eats, man, hunts. Whereas the other person's responses were: Circus, cage, master, whip, tone, dance. The first person associated the tiger with forest settings. But the second person associated it with the circus where he saw the tiger. Thus two persons perceive the same word with different meanings and, therefore, the types of associations vary. Since words have different meanings and individuals make use of these meanings in varied situations and the meanings may vary group to group, culture to culture it has been realised that it is important to know the amount of feeling attached to the meaning of a word and avoid certain unpleasant feeling, that could emerge from it.

SEMANTIC DEFFERENTIAL

Semantics

Language is an important medium of communication of our thoughts and information. Its main function is social. Therefore, we make use of language for two important functions. One is for communicating our thoughts to others and the other is making use of the symbols and rules in our thinking and expression. Therefore, it becomes very important for us to know more about language. Linguistics is the science which studies the rules of language. This includes speech sounds, their meanings and the grammar. The linguistics covers various aspects of complexity of language. Therefore, four important components by which linguistics is constituted. They are phonemes, morphemes, syntax and semantics. Here we are interested in semantics. The basic function of language is the semantic function. The function is conveying the meaning of the words of the language. Psychologists are of the view that it is important to know the types of meaning. Semantics is concerned with formal structure and the emotion of language used in its words. In a particular context how an expression of a word sounds and what type of meaning the expression conveys.

The meaning of word may vary in its sounding expression and thus it may have more than one meaning depending upon the expression or usage. For example, the sentence, "I will speak to him may be expressed in such a way that in every type of expression it will have different meaning. In the expression the *intonation* and the *accentuation* may change the meaning. The *pitch* and *amplitude* in speech may produce several attributes.

Therefore, the meaning of a word and the associations it has and its different possible occurrences constitute the structure of language. The structure of language determines the extensive usage of a language. In every language there are two kinds of meaning, the ***denotative meaning*** and ***connotative meaning***.

Denotative

The meaning of a word that indicates concrete objects is denotative. For example, table, chair, book, pencil, pen are concrete objects. "This is a book", "'That is a pencil". Thus we point out the objects. The words book, pencil and pen are indicating external references. It is indicating something explicit and overt. The word is the symbolic representation of an object.

Connotative

The other kind of meaning is *connotative*. The connotative meanings are not directly concerned with the objects. These words are implied and suggested. The connotative meanings are derived from denotative meanings. The connotative meanings are more significant than denotative meanings. These are more of emotional type, expressing some kind of preference and judging the worth. Connotative meanings vary from person to person and culture to culture. The responses a person makes are based upon the choice of the words which hold strongly the meanings of the words.

Connotative is an idea strongly affected by a word. Connotative meaning is referring to other than the literal meaning of an object or person. A word is associated with some other meaning which is having emotional charge. Therefore, connotative meaning is sometimes logically stated as intensive meaning. For example, the words 'democracy' and 'socialism' are described differently in different political cultures.

The socialists say that democracy at the ultimate level is the dictatorship of a single person. It is the exploitation of the wealth of the nation by the persons who are in power in the name of Government of the people, and by the people. On the other side the democrates say that in socialism people are deprived of their fundamental rights and they are made slaves to the state. In both statements, meanings other than the original meanings are attributed and also in every expression the meanings of the words vary tremendously in emotion, evaluation and preference.

Differential Method

Therefore, different connotations of words of different people cause misunderstanding. Hence Charles Osgood (1957) and his associates, in order to attach strongly the connotative meanings of the words in one position or make the meanings more accurate, have formulated a method called 'Semantic Differential'. This method is used for measuring differences in connotations. Since it is dealing with the meanings of the words it is semantic and since it provides several dimensions of meaning of the words it is called differential. Osgood and his associates have thought that even the simplest words, referring to concrete objects have effective potentials.

The semantic differential method has been used by several authors to study the connotative meanings of words. In determining the connotative meanings of a word by this method an individual is asked to judge and rate a series of words. The semantic differential scale is a seven-point bipolar scale. At the two extreme ends two adjective words of contrasting type are given. These two words are connected by a straight line. This straight line is divided into seven positions from left to right. Any word can be used for the rating on this scale. The word may be referring to anything. Candland (1968) calls this word as *stimulus* word as it evolves a response of rating. For example, democracy, communism, world, life, Pakistan, China, culture, chastity, pity, and so on may be used as words for rating. The individual has to judge and indicate his position on the scale. This position will show the direction and the intensity of his judgment of the word. Whether his rating should be on the left extreme or right extreme or somewhere in between the two extremes is to be decided by the individual. The two bipolar adjectives or the contrasting adjectives used for rating the meaning of a word may be good-bad, kind-cruel, representative-autocratic, generous-mean, large-small, heavy-light, clean-dirty, slow-fast, active-passive, smooth-rough, and so on.

Osgood and his associates (1955) have used many pairs of words in their studies. On analysing the responses of the subjects they found that the ratings fall under three dimensions. Therefore, there were three dimensions of meaning the rated words commonly had. The three dimensions are evaluation, potency and activity. The adjectives such as good-bad, clean-dirty, kind-cruel belong to ***evaluative dimension***. The adjectives weak-strong, big-small, heavy-light belong to ***potency dimension*** and the adjectives, fast-slow, active-passive, bright-dull belong to ***active dimension***. Osgood believed that this type of studies were providing with clear cut connotative meanings and also were found to be very much successful in various cultures.

A model of Osgood's Semantic Differential Scale can be drawn here to illustrate the meaning. In this, let us use the concept, democracy, communism and contrasting adjectives, rough-smooth, weak-strong, good-bad, slow-fast.

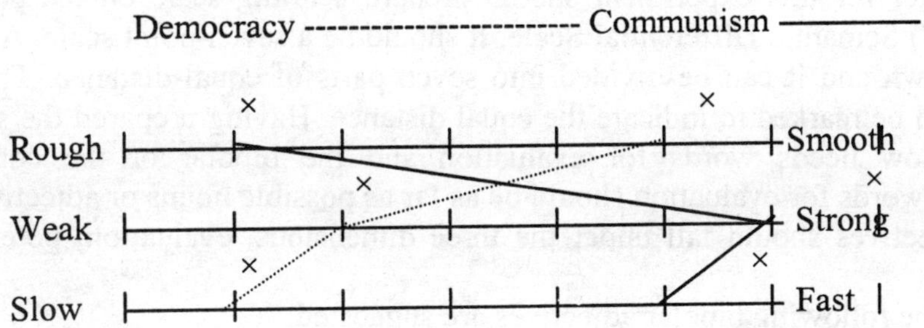

Fig. 7.8 : Semantic Differential Scale (Model)

The person who rated democracy and communism on the seven-point scale has indicated that communism is rough, strong and autocratic as he rated it on the left and right ends and also on the right beyond the middle position. The same person has rated democracy as smooth, weak and representative as his decisions are falling on the end of fifth position, on the end of second position and at the end of first position. But, from the rating we can conclude that both of his ratings are not extremely favourable to communism and democracy. The person is favourably disposed of towards democracy and accepts that democracy is a weak political philosophy. He is also favourably disposed of towards communism and accepts that communism is rough, strong and autocratic. This is how the person has attributed the meanings to democracy and communism. In the ratings the three dimensions suggested by Osgood are involved. The contrasting adjectives rough-smooth indicate the evaluative dimension, the adjectives weak and strong indicate potency dimension and the adjectives slow and fast indicate the activity dimension.

EXPERIMENT

Problem

Estimate and analyse the type of meanings held by your participant about certain words by the method of Semantic Differential.

Method

Participant

Choose a person who has no knowledge about this experiment and collect his/her personal information, such as age, sex, qualifications, occupation if any, socio-economic status, religion, personal interest, geographical area and so on.

Material

The experimenter for this experiment should prepare a rating scale on the pattern of Osgood's (1957) Semantic Differential Scale. It should be a seven-point scale. A straight line can be drawn and it can be divided into seven parts of equal distance. The seven positions should be marked to indicate the equal distance. Having prepared the scale the experimenter now needs words for evaluation and the bipolar or the contrasting adjectives. The words for evaluation should be as far as possible nouns or adjectives. The contrasting adjectives should fall under the three dimensions, evaluation, potency and activity.

Therefore the following bipolar adjectives are suggested:

1. *Evaluative dimension:* Fair-Unfair, Good-Bad, Happy-Sad, Loving-Unloving, Valuable-Worthless, Clean-Dirty, Beautiful-Ugly, Sacred-Vulgar, Wet-Dry, Fresh-Stale, New-Old, Lawful-Unlawful.
2. *Potency dimension*: Light-Heavy, Strong-Weak, Powerful-Powerless, Hard-Soft, Rough-Smooth, Large-Small, Brave-Cowardly, Bossy-Docile, Sweet-Bitter, Young-Aged, Rugged-Delicate, Dim-Bright.

3. *Active dimension*: Active-Passive, Fast-Slow, Sharp-Blunt, Tense-Relaxed, Hot-Cold, Energetic-Sluggish, Rapid-Restful, War-Peaceful, Efficient-Inefficient, Imposing-Elusive, Torrential-Breezy, Aggressive-Friendly.

With regard to the evaluation of the words to be rated by the participant on the seven-point scale there is no hard and fast rules as to what type of words should be used. Osgood (1957) who has formulated the method of semantic differential has used adjectives and participles in his studies. He used the words like, Polite, Eager, Burning for rating. But Lindgren (1966) used nouns like, France and U.S.A. Another author, John Parry (1977) also suggests nouns. Kimble *et al.* (1985) state that in semantic differential a series of verbal concepts that can be almost, anything, may be used for rating. Therefore, the words can be nouns or adjectives or even verbs of present participles can be used. The meanings of the words are anchored by contrasting adjectives either on the favourable side or on the unfavourable side. Therefore the following words can be used for rating:

(1) Russia, (2) Great Britain, (3) Justice, (4) Law, (5) Man, (6) Freedom, (7) Poverty, (8) Elegant, (9) Nasty, (10) Success, (11) Robbery, (12) Lady, (13) Father, (14) Slippery, (15) Chastity, (16) Modesty, (17) Alchohol, (18) Hippies, (19) Nomads, (20) Aristocrat. Nine test sheets and Stop Clock (See Appendix-IV).

Procedure

The experimenter should first of all prepare the test materials in advance for his experiment. The semantic differential scale for rating on seven-point scale should be prepared with sufficient number of copies. Each word should be rated in a separate sheet. It is better that each subject should rate a minimum of three words and each word is to be rated in a separate test sheet. The experimenter should administer the same three words on three individuals separately. Therefore, the experimenter should collect three persons of the same cultural group. He should also note down the time, the participant has taken for each test.

Since this is going to be conducted by the students of a class and each student is an experimenter, each one should use three separate words for rating. But the contrasting, bipolar adjectives can be the same for the experimenters. Each scale prepared by the (student) experimenter should have the bipolar adjectives from all these dimensions, such as (1) evaluative dimension, (2) potency dimension and (3) activity dimension. At the top of each test sheet the word to be rated and the name of the participant should be written. Since all the three dimensional words are given separately in the material section it may be easy for any experimenter to select three bipolar adjectives from each dimension. But to make the total ten in the first test sheet there can be 4 bipolar pairs of adjectives from evaluation dimension, in the second test sheet 4 bipolar pairs from potency dimension and the third test sheet 4 bipolar pairs from activity dimensions. All the experimenters can follow the same procedure. (See three tests of the semantic differential scale with bipolar adjectives — Appendix-I, II and III).

Having the materials ready, the experimenter should make the participant to sit in a well-illuminated room in front of him. Then he/she should give the following instructions. "I will give you three tests, one by one. In the test sheet on the top you will find a word, and at the two extremes of lines you will find contrasting adjectives. The straight line is divided into seven positions. If the left extreme is a positively worded adjective, on the right you will find a negatively worded adjective and *vice versa*. The centre division is supposed to be the middle position. When you read the word at the top *what sort of meaning you can assign* to it. You can just think over it for few seconds and make a *small cross mark* anywhere on the line in one of the seven positions in between the two bipolar adjectives. So your main task is to rate the word. Where do you like to place the word on the range left to right, on the seven positions? You will have to make a cross mark on the line anywhere. Remember, what sort of meaning you have decided for the 'word' at the top in between the two contrasting words at the two ends."

> "There are ten pairs of bipolar words in each test sheet. You will have to rate the word at all ten pair of bipolar words by a small cross mark on the straight line somewhere. As soon as you finish the first test I will give another test with different words at the top and different bipolar adjectives at the ends. Then I will give the third test. You will have to carry out the same thing in all the three tests."

After giving the instructions the experimenter should hand over the first test sheet to the participant and ask him to go through the test material once more. Every time when the experimenter gives the test sheet to the participant he should check whether the word to be rated and the name of the participant are written at the top of the sheet. He can permit one or two minutes for this. Then he should ask the participant to start the work: "Ready go". The experimenter should also start the stop clock.

Then the participant should start responding to the experimenter's order, rating the word with the each pair of the bipolar adjectives. When all three test sheets are rated the experimenter should obtain the introspective report from the participant and allow him/her to go away. The same thing the experimenter should carry out on two more persons and thus he conducts this experiment on three persons.

Results

The experimenter now has nine test sheets. The same three semantic differential scales have been administered on three different persons. Those three persons have rated three concepts (stimulus words) in terms of their own meanings to the three words. The hypothesis is that everyone may have his own meaning to the word and he/she might have assigned the place to each word. The meaning each one had for the word may be the same or vary in small level or even vary at larger level. This shows what type of meaning each one had developed for the word.

This should be assessed and the hypothesis should be tested. Therefore, the responses of all the three persons to the three words should be plotted on the semantic differential scale. The responses of the three persons to the first word are plotted in a semantic scale. Similarly the responses of the three persons to the other two words also may be plotted on two separate semantic scales.

The cross marks made by the three persons to each word with all the ten bipolar adjectives are connected. This figure will indicate each participant's assessment of the place for the stimulus word on the response continuum. When the responses of all three persons for one word are plotted on the semantic differential scale, it is easy for the experimenter to examine the placement of the cross marks by the three participants. This will indicate the meaning that each person has attached to the stimulus word.

Now the experimenter can analyse the distances between the responses of the participants. A comparative description of the placements among the three persons can be made. Whether there is much individual differences in the meanings for the same word. The experimenter should see whether there is any agreement among the three persons with regard to the meaning of the word.

The experimenter should also see whether the responses of the participants have optimum agreement among the three with regard to the various dimension, such as evaluation, potency and activity. Osgood *et al.* (1962) state that the choice stimulus word determines the participant's responses. The word is anchored by the scale either to the left end or to the right end or whatever it is, depends upon the stimulus word.

Discussion

The semantic differential scale in the beginning was used only to measure the meanings of words. Since each word is understood differently in different context and was causing confusion it has become essential to ascertain the extensity of the meaning of the words. Words are having cognitive meaning. But Osgood uses semantic differential to measure the meaning. Some felt that it appears to be incompatible. Therefore, Osgood made it a principle that the meaning of a word refers to what a person has had in his mind. The meaning of a word may vary from group to group. What one cultural group conceives of it is different from what the other has conceived. The meaning of a word may arouse different emotional reaction in a person or in a group if it would have been viewed as connotative.

If we want to know the correct meaning of a word we refer to dictionary. But the same meaning is not attributed to a word in a society. Therefore, the experiment is designed to ask the person what the meaning he is having in his mind and assigning to it and how he is going to rate it. We have discussed in preceding section that words have emotional temper. Individuals may attach emotional meaning to certain words. Therefore, the experimenter can examine whether any of the participants has been emotionally influenced in making the cross mark on the semantic differential scale. In certain cultures even ordinary words are having emotional tinge in expression. Therefore, what the individual feels about the word, positively or negatively, depends upon how he has been taught within a culture. The experimenter should discuss the results having in mind all these factors. The discussion can be also made from the results whether there is any differences in the meanings of the word due to age, sex, religion, socio-economic status or study of some subject.

Conclusions

State whether the three participants agree in most of the ratings or differ. If differ state the major factors contributed for this. State whether the differences in meanings are due to age or sex or study of subjects or culture.

APPENDIX – I TO IV

The following are the Bipolar Adjectives used for judging the meaning of the words. There are three dimensions. The Experimenter may select *three or four Bipolar Adjectives from each Dimension* to have a symmetrical scale.

APPENDIX - I

Bipolar Adjectives suggested for Evaluative Dimension

(1) Fair-Unfair, (2) Good-Bad, (3) Happy-Unhappy, (4) Loving-Unloving; (5) Valuable-Worthless, (6) Clean-Dirty, (7) Beautiful-Ugly (8) Sacred-Vulgar, (9) Wet-Dry, (10) Kind-Cruel, (11) Fresh-Stale, (12) New-Old, (13) Lawful-Unlawful, (14) Pleasing-Annoying, (15) Useful-Useless, (16) Unique-Common

APPENDIX - II

Bipolar Adjectives suggested for Potency Dimension

(1) Light-Heavy, (2) Strong-Weak, (3) Powerful-Powerless, (4) Hard-Soft, (5) Rough-Smooth, (6) Large-Small (7) Brave-Cowardly (8) Bossy-Docile, (9) Sweet-Bitter, (10) Young-Aged, (11) Rugged-Delicate, (12) Dim-Bright, (13) Cozy-Roomy (14) Private-Public, (15) Full-Empty, (16) Energetic-Lazy.

APPENDIX - III

Bipolar Adjectives suggested for Activity Dimension

(1) Active-Passive, (2) Sharp-Blunt, (3) Tense-Relaxed, (4) Hot-Cold, (5) Energetic-Sluggish, (6) Rapid-Restful, (7) War-Peaceful, (8) Efficient-Inefficient, (9) Imposing-Elusive, (10) Torrential-Breezy, (11) Fast-Slow, (12) Aggressive-Friendly, (13) Complex-Simple, (14) Ordered-Chaotic, (15) Formal-Casual, (16) Started-Finished

APPENDIX – IV

Any of the following words or concepts may be used for Rating

(1) Russia, (2) Great Britain, (3) Justice, (4) Law, (5) Man, (6) Freedom, (7) Poverty, (8) Elegant, (9) Nasty, (10) Success, (11) Robbery, (12) Lady, (13) Father, (14) Slippery, (15) Chastity, (16) Modesty, (17) Alcohol, (18) Hippies, (19) Nomads, (20) Aristocrat, (21) Rights, (22) Mother, (23) Religion, (24) Sex, (25) Marriage, (26) Death, (27) Birth, (28) Luck, (29) Diamond, (30) Heaven, (31) Hell, (32) Life, (33) School, (34) Love, (35) Prayer, (36) Communism, (37) Democracy, (38) Mind, (39) Heart, (40) Imagination, (41) Sacrifice (42) Dictator, (43) Service, (44) Tour, (45) Cinema, (46) Examination, (47) Race, (48) Election, (49) Family, (50) Husband/Wife.

8

PROBLEM SOLVING

INTRODUCTION

The concept 'problem solving' itself involves something that is not normally or easily performed task. It involves something more than what is required in performing ordinary task. Candland (1968) states that problem solving includes thinking, reasoning on creative behaviour. Therefore, it is concerned with higher mental processes. Gagne (1974) in his more recent work states that problem solving is classified as 'higher order rules' which is referred to 'thinking out' a solution to a problem by combining old rules to form new ones. According to him in problem solving the learning rules should be in the first place. He states that numerous examples can be quoted from everyday life for this. Problem arises whenever the previously learned rule is not appropriately helping us to deal with a problem successfully. Therefore, a condition necessary for problem solving is the presence of the appropriate rules. Gagne (1970) states three external conditions that are necessary for problem solving. First of all, the rules required for the solution of the problem must be active and must be following one after the other. Secondly, verbal questions should be used to recall relevant rules. Thirdly, the direction of thought processes should be also decided by verbal instructions.

Many of the scholars who deal with problem solving phenomenon associated fully with learning processes. There is every reason for them to associate problem solving with learning as learning is the prelude to problem solving. For example, educational psychologists, like Bruner, Jean Piaget, Ausubal and Gagne all deal with problem solving matter only through learning. Bruner (1959) has proposed elements of a model pertinent to learning, thinking, discovery and consequent problem solving. This model is derived both empirically and by observation. According to him one of the most important objectives in learning is to avoid future necessity of learning. It is by learning that we are able to recognize new problems. Such recognition is shown by the example of thinking which is the purpose of learning. Bruner is of the view that it is possible to learn without consequent thought. This type of acquisition of knowledge is taking place in a passive state.

It is true that most of the educational instructions are leading to passive learning. Bruner states that thinking is the reward for learning. Since much of educational instructions are the two common forms of passive learning, the student is deprived of this reward. The classroom teacher is responsible for this. The shortcomings of educational and cultural levels of the teacher should be overcome through experience. Bruner says that most of the educational systems make very little or no use of the experienced teachers in training new teachers. The ideal product for Bruner is the ability and desire of the students and adults to move from learning to thinking. Learning must make one to achieve thinking. To Bruner (1962) the purpose of education is to promote both. For him the most important function of a teacher is to save his students from the more dangerous consequences of learning.

Bruner, Oliver and Greenfield (1966) say that the most important single characteristic element is human beings that they are capable of learning complex skills. Therefore, learning is so deeply established in man in such a way that it is almost involuntary. All other species begin their learning fresh in every generation. But man belongs to a different species and his important function to conserve and transmit his past learning.

LEARNING TO SOLVE PROBLEMS

Every school expects the students to learn in an efficient way, so that they solve problems including skilled tasks. Human life is in a continuous process of readopting to changing environment. We do not see everyday everything is new, but, yet we see many things new. We always depend upon previously acquired habits that are quite relevant for adjustment in our everyday life. Our stock of habits is adequate to meet the everyday requirements and adjustments in the environment so long as we do not meet any unusual conditions or situations. If we do not have any habit or learned behaviour on which we will have to rely, then our life would be in a threatening state. Ramesh was asked by his Chairman who is far away to speak to him everyday at 10-30 AM and to explain the position of the company's work progress. He has been doing it everyday successfully for years. But on that day as usual he started dialling his telephone. There was no dial tone in his phone. It was dead. His Chairman would be waiting at his phone for the message from Ramesh. He tried again and again, but nothing happened. He cannot speak to his Chairman in his phone. Now he is faced with a problem. He must do something other than what he had been doing.

What has happened with this phone? He starts to think. Quite a number of questions piercing to his mind. What should I do? Should I go to the next house and speak? Or should I go to telephone booth and speak? It is a confidential matter, how can I speak it in the presence of others in public booth? Why he is thinking about such matters now? He ordinarily never used to think like this at 10.30 AM. The answer is clear. He is thinking in various ways because he was unexpectedly encountered a problem which he must solve. Everyday habit has failed to meet his demand and thinking has taken place in presence of a problem. Therefore, thinking here we mean not anything else, but deliberate application of an idea to the solution of a problem. Skinner (1968) says that thinking is often called problem solving.

What is a Problem ?

Generally, a problem means anything or a matter or a person that is difficult to deal with. But we are not satisfied with this definition in this context. We expect something more. Problems are not static events of our life. It is only an occurrence. What is a problem to one person is not a problem to another person. So also, what is a problem on one occasion for a person is not a problem on another occasion for the same person. It is also evident that problems vary in difficulty level. According to Wingo (1950) a problem occurs when an activity is blocked by an obstacle that cannot be removed by the stock of habits, because the situation is unfamiliar and confusing to us. The obstacle may be something on the way and must be changed. If you want to remove the obstacle you will have to think. Thinking is a form of behaviour which occurs when you face a problem. Therefore, in the absence of a problem no thinking is taking place. It may be also stated that a problem in any new situation or any new incident for which the individual is not having any readymade soluble or adaptive response.

Have you had any time problems for which you know the answers already? If you say, yes, then they were not problems for you. Because there were no obstacles and you did not think on them. A problem is there if only there is an obstacle and you fail to solve it by ordinary habits and learned behaviour. Therefore, a problem may be defined in a different way. A problem is the occurrence of an event in the form of an obstacle, blocking the goal seeking activity of a person which he cannot cope with his habits and learned behaviour.

An individual works to solve a problem only when he identifies and accepts it as a problem. Many of the social and emotional problems are not solved because people do not see them as problems. Like the problem of an individual nobody is held responsible for such problems. An individual will work to solve only the problems that are problems of his own. He thinks as best as he can when he confronted with something he sees it as a problem. Everyone may face somewhat different problems. The goals of the individuals may be the same, but every individual may be facing different types of obstacles and, therefore, the problems become different. What is the most difficult obstacle for one person, may be a simple, easily solvable for another person. Whatever the nature of the problem is, it may be found that it involves interference with the behaviour of the individual's goal seeking activity. A problem continues to exist as long as the goal or the demand is not achieved. An individual can solve a problem if only the obstacle between him and his goal is within his capacity that he can get rid of it.

How do we Solve Problems?

In problem solving, psychologists often conduct experiments on animals and compare the results with the problem solving activities of human beings. Sometimes it happens that when human subject confronts with a problem he starts looking at the roof for a few minutes and then provides the correct answer. Candland (1968) says that it is very difficult to explain or measure the learning process that has occurred while the person was looking at the roof. Because the observable and quantifiable behaviour was very small in this. It is a complex activity, but, because of it we cannot drop it from experimental investigation, the tendency involved in problem solving. From what appears to be a confused state of behaviour a response emerges which satisfies what the problem situation demanded and, therefore, it is called the solution for the problem. At this stage the problem disappears, the situation is modified and the problem is no longer there.

All those psychologists who conducted experiments on animals have taken much trouble to analyse the components of the learning process and the goal seeking or problem solving activities. Animals have been used in many problem solving experiments. In the psychological laboratories the presence of Thorndike's cats, Skinner's rats, Pavlov's dogs and Kolar's chimpanzees and Young's octopus need not be looked upon as surprising incidents. Till these authors analysed their animals' problem solving behaviour they looked upon as mysterious. The results of these studies were compared with human subjects in problem solving activities. Every author of general Psychology book on animal learning provides ample information on this.

Problem involves the presence of an obstacle in the goal seeking activity. Therefore, the problem can be solved by removing obstacles to reach the goal. First of all we are identifying the specific problem and start working towards removing the obstacle and then reach the goal. If the mode of approach to remove the obstacle does not work well then we try some other approach that is feasible and successful. Some obstacles may be removed easily by thinking but some needs the combination of thinking, observation and experimentation. John Dewey (1910) a reputed American philosopher has given a best known analyses of problem solving through reflective thought. Dewey has analysed it in five major steps from the beginning of the recognition of the problem to the final testing step. Dewey's work has become a source of inspiration to many scholars in this field.

The individual, the learner understands the presence of the problem situation. His goal-seeking activity is disturbed by the situation which prevented him from the goal. In such circumstances the individual begins to think in terms of getting rid of the obstacle.

The individual, in order to deal with the problem situation should organize the situation suitably to identify the problem. He must critically analyse the situation and see the relevant elements to the solution of the problem. The individual's ability to identify the problem depends upon his general intelligence and the previous experience. If the individual is not able to identify his problem and the obstacles to be removed then the only way is to use blind trial and error procedure. This means that he is not able to deal with the problem situation and direct the behaviour intelligently.

The third step is to formulate hypotheses for attacking the problem. In the example, Ramesh, when he was not able to speak over the telephone to his Chairman, has thought of several possible ways of getting his problem solved. Therefore, the hypotheses were the possible plans of action for achieving the goal sought.

The fourth step is to choose among the possible ideas and select the ideas that are most suitable and correct for solving the problem and reject those ideas that are not in line with the known facts or do not support the hypotheses. In the case of Ramesh, he has rejected the idea of going to his neighbour's house and telephone to this Chairman as inappropriate.

The fifth step in this process of thinking is to test the most appropriate hypothesis. This is by applying the hypothesis to the problem to achieve the desired goal. The formulated hypothesis is put into actions and the results are obtained to see whether the action based upon the hypothesis has brought out reliable information so as to accept the hypothesis. This type of problem solving seems to be primarily intellectual and does not involve experimentation or practical. Sometimes the formulation of several hypothesis are to be made and tested till the obstacles are removed. In the case of Ramesh one of the hypotheses worked well and he got his problem solved by going to his friend's house and using his telephone, informed his Chairman that he would speak everything in detail when his phone is set right. Therefore, in solving problem through thinking, ideas one after the other may be tried till the problem is solved.

Trial and Error Performance and Problem Solving

If an individual has learned to recognize various kinds of problems and try to apply the relevant methods he is not involved in thinking. Skinner (1968) states that the cognitive processes in his case seem to have vanished. For example, when a student has learned how to attend to his environment he has no need for any mental function. He acts with pre-current behaviour, with which he solves problems. Skinner says that this takes place thoughtlessly. Skinner is of the view that if previously learned technique of any sort does not apply to solve the problem the problem should be attacked by trial-and-error method. Skinner strongly asserts that trial-and-error is not at all a behavioural process. To him trial and error is at best a process of selection in which some of the successful responses are evoked by the situation which become effective in solving the problem. If the same final chance events are provided then the organism may respond with the same type of final behaviour without errors.

According to the law of multiple response formulated by Thorndike (1931) when individuals faced with a difficult problem for which they have no ready solution, they will engage in different types of responses until one produces a satisfactory result. This means that the individuals have solved their problems through trial-and-error method. Though trial-and-error is a form of instrumental learning Thorndike differs from others. In Thorndike's contribution there are two important aspects. First, his cat did not learn suddenly to make correct responses. His subjects did not show insight. If insight would have taken place the amount of time and the number of trials would have been reduced. Secondly, Thorndike's procedure stressed the split of complex behaviour into its components. This suggests that trial-and-error is a form of mental activity of the organism. Candland (1968) states that in certain way trial-and-error is not the correct term. It suggests that the organism consciously tries one motion after another until it finds a correct response to the problem. Candland says that trial and error behaviour is common among human beings in problem solving situations. For example, in learning to play a piano, learning to type a typewriter the initial movements are with trial-and-error in character. This usually consists of manipulation of instruments and it is often towards and away from the correct responses.

Insight Behaviour and Problem Solving

The concept 'insight' is the contribution the Gestalt psychologists made to the field, psychology in the context of their study of learning and problem solving behaviour. In opposition to the stimulus-response psychology the Gestalt presented two important elements, the phiphenomenon and insight. Insight was defined by Kohler (1925) as the sudden awareness of the relationship among various elements in a situation. Insightful behaviour brings out an ability in an organism to combine elements in such a way that a meaningful whole may be assumed. Kohler has stressed this insightful behaviour and demonstrated the presence of it in both animals and human beings. The parts of the patterns must be available to the learner for making the combination. If the key element of the pattern is missing then insight is difficult.

Gestalt psychologists are of the view that insightful behaviour is a common characteristic of human thought. It cannot be taught or learned since it is an inherent potentiality already present in human beings. The organism wishes to see things as whole, even by completing the gap, if any. The organism is not satisfied with incompleteness. The student faced with a problem in whatever situation is in a state of tension. He can relieve his tension only on solving the problem. Solution comes after a period of struggle or little or no progress towards the problem. The solution sometimes appears unexpectedly and suddenly than at other times. This sudden onset of the solution to the problem is described by the concept insight. The important characteristic of insight behaviour is that errors and excessive trials are reduced or disappear suddenly. An adequate adaptive behaviour to the solution of the problem remains. Insightful behaviour in some problem situations set in more gradually. The reason for gradual setting is that the organism might have mastered some elements of the pattern quickly and takes some more time to master other elements of the pattern to make the problem situation responded with correct solution.

Previous Experience and Problem Solving

The role played by past experience in human problem solving has been a subject matter for a long time for experimentation and discussion. There is diversity of view with regard to the role played by past experience in problem solving. It is viewed that problem solving is not a unitary process. In all problem solving a solution is arrived at. But the process or behaviour by which the solution is achieved is of different type. Maier (1945) has discussed in detail the part which past experience may play in each type of problem solving behaviour. Problem solving may be facilitated by immediate problem situation and the past experience. Birch (1945) states that the past experience may become an obstacle which blocks the productive thinking and reduces behaviour to fruitless attempts. Duncker (1945) has dealt with the past experience in his problem solving experiments and observed that past experience had exercised negative effect. Therefore, Birch and Rabinowitz (1951) suggest to remove the defects of interpretation involving the effects of previous experience on problem solving. But whatever the experimental evidences these two authors produce the individual's previous experience cannot be fully ruled out from problem solving task.

Stages in Problem Solving

Problem solving involves several basic psychological processes such as learning, perception, motivation and sensation. It is not a unitary process and no independent theory of problem solving can be evolved. We must first understand the behaviour involved in problem solving task. Woodworth (1952) states that man begins to incubate if the problem is a difficult one. After going round and round in a circular way he finds it not able to reach the solution and he lays down the matter for some time. This, was found to be working very well with many creative thinkers, scientists, inventors and in art and literature. Patrick (1935, 1937, 1938) and Rossman (1931) have studied many problems of creative thinking very extensively. On the basis of the studies, they have classified the whole process of creative thinking into four stages.

Therefore, the problem solving behaviour according to them occurs in four stages. They are (1) preparation, (2) incubation, (3) illumination, and (4) verification.

The first stage is preparation. During this period the person who solves the problem acquires the necessary basic skills for solving the problem and becomes aware of the particular problem and makes himself ready to face the problem. *The second stage* sets in when he finds that all his efforts to solve the problem were useless. He stops at this stage for the time being and engages himself in some other kind of work. He may even forget that he left a problem unsolved. This period is called incubation.

But when the person keeps himself engaged in some other kind of task he may have sudden insight into the solution which left unsolved. Now the answer is quite clear to him. This is the *third stage* and this period is called illumination. It is often thought that incubation leads to illumination and illumination need not be thought as an accidental phenomenon. It is rooted in the past experience of the problem solver. Illumination comes in the form of an idea that looks so sure that it at once occupied the entire attention of the thinker and leads him to vigorous thinking (Rosmann, 1931).

It often occurs in the case of some problem solvers that they do not check their solution after the period of illumination. But it is important to verify whether the method of solution really solves the problem. Therefore, the problem must be attended once again with the new method. This is the *fourth stage* which is called verification. All the four stages in the process of problem solving behaviour are taking place in our everyday life in such a way that we do not easily distinguish them. Modern psychological studies try to distinguish them in these ways.

MALADJUSTMENT AND PROBLEM SOLVING

Persons engaged in problem solving behaviour often feel satisfied and enjoy the pleasure of their achievement of the goal. But if they fail to solve their problems in spite of many attempts of trial and error or after a long period of incubation they are emotionally disturbed. This emotional disturbance may be seen usually in the development of the feeling of restlessness, anger or disgust or frustration.

The feeling of restlessness and anger may make the individual emotionally imbalanced. Gradually he develops the loss of coordination in his actions and this increases tension in his bodily parts and increases the feeling fatigue. He also feels loss of hopes. He falls prey to unnecessary irritations. This state of affair may take place due to a number of factors the individual faces in the problem situation. The factor include first of all the problem itself. It is a very difficult problem that he has faced and he has spent long period of time on it which he felt waste, put more efforts which have become fruitless.

Sometimes persons may be highly frustrated and become aggressive. When a person succumbs to frustration? Woodworth (1952) says that when a person is surmounted by obstacles or simply blocked by obstacles in all his attempts to solve a problem or has responded to his failure by anger, worry and anxiety, he is frustrated. But some individuals continue their problem solving attempts longer than others. Such people are called frustration tolerant individual.

This sort of emotional maladjustment may vary person to person in the same problem situation. Some may tolerate the situation and are not disturbed by the failures. They may give it up and seek another similar goal which he attained by his potentialities.

CREATIVE PROBLEM SOLVING

Strong criticisms are often coming from industries that there is less attention on the development of creative men than or production of creative things or ideas. Though this statement seems to be true to some extent it is not fully correct. Men engaged in industrial works have developed quite a number of techniques for solving creative problems.

One of the most important techniques is the Osborn's (1957) brain-storming technique. The brain-storming groups comprise of small number of people who are encouraged to think many ideas as to solve problems of creative nature. Wide variety of solutions may be evolved. Judgments on the suitability of solutions not allowed. It is difficult for inexperienced problem solvers. But leads to highly creative production. It encourages delayed evaluation. Because evaluation during production has reducing effect on the members. Osborn has described brain-storming more as a principle than a technique. The brain-storming session may have two or more hours. It involves five to twelve people from wide variety of backgrounds. It is a free-wheeling affair with ideas coming very rapidly from all sources. Evaluation comments and non-verbal expressions are forbidden. Checklists are prepared by researchers and they must stimulate the production of ideas.

The Gorden technique is a modified form of brain-storming technique. Gorden (1961) has developed a process called 'operational creativity'. In Gorden technique an abstraction of the problem is presented to the group members instead of detailed specific problems as given in Osborn's brain-storming. Gorden's argument is that by presenting an extreme abstraction we may get many ideas that are not ordinarily brought out.

The third technique for creative problem solving is the morphological analysis. This technique originally was developed by Osborn (1957) and Arnold (1962) and was assigned to Fritz Zwicky of Aero-Jet Corporation. According to this technique a problem is divided into a number of separate variables and for each variable the members were asked to think as many solutions or ideas as possible and then the results are combined in all possible ways. Arnold states a problem of developing a new type of vehicle. The problem has three different aspects and they are (1) type of vehicle, (2) type of power, and (3) the medium in which the vehicle is used. Every aspect of these yields itself to several solutions. For example, the type of vehicle may be a sling, a cart, a rocket, a box and so on. The type of power may be air, water, electric, horse and so on. The medium may be air, water, oil and so on. Arnold used 6 types of vehicle, 6 types of power and 5 types of medium, which yielded 180 solutions for the problem.

The fourth technique of creative problem solving is the *Collective Notebook Method.* This method was designed by Hafele (1962). This technique, unlike other techniques, presents both the problem and some possible solutions. These are presented on the first page of the notebook. This problem solving technique may be used among the factory members and also anywhere in a suitable environment. These notebooks are distributed among the members of the factory or any organization and they are to carry out the direction and find out solution for the problem, they will have to write down the solutions descriptively or diagrammatically in their books. After the given time is over the notebooks are collected and the solutions are scrutinized and evaluated. The members who have given the valuable ideas are then rewarded. The notebook method allows the members to take more time as to hatch for new ideas. This is one advantage of this technique.

The fifth creative problem solving technique is *The Collective Bulletin Board Method.* This method was devised by Lefrancois (1965). He has combined the brain-storming method and collective notebook method and redesigned to suit to tne classroom situations. The collective bulletin board method allows the individuals to have the benefit from the ideas of other members and allow complete freedom to express their own ideas. In this method the description of a problem and the possible solutions are presented on the bulletin board in the classroom and not in a notebook. Members or students are asked to add their solutions to the board and adapt the ideas written already there. The method was carried out for a period of ten weeks with two different classes. Lefrancois has found that after the period of ten weeks the members who participated in the collective bulletin board programme performed exceedingly well on tests of creative thinking problem solving than the control groups.

EXPERIMENTAL STUDIES

One of the experiments conducted by Haufmann and Kasanin (1937) to the study the concept formation using the test developed by Vigotsky is as good example for a complex kind of thinking activity. The capacity involved to deal with a problem of this sort is depending upon the kinds of set which the individuals utilize in their attack upon the problem. Actually the characteristics that consider in this test are height and size. This makes four categories, large tall, large flat, small tall and small flat. But the colours of the blocks may be misleading one. It is not only the brain injured and handicapped people lapse back but also normal people often commit the mistake of looking for colour and shape. The main reason for this kind of error in thinking that the concrete attitude is dominant over the abstract attitude. Haufmann (1941) states that there are two types of sets in this test, one perceiving set and the other conceiving set. The perceiving people do not look at the blocks for natural groupings. But the conceiving group try for a rational basis for groupings and they seek abstract ideas for their task. Murphy (1951) says that this abstract idea is fundamental for providing the mental set for the solution of the problem.

Experiments have been conducted on problem solving task through trial and error behaviour and insightful behaviour. Durkin's (1937) experiment has revealed that insight may appear in a very high or very low degree. According to his study methods of work may vary insightful behaviour from just appearance to almost pure insightful reorganization. Durkin observed that in the long run experience reveals larger integration and more insight. In this type of experiments reliable difference was found between persons who made their first attempt and persons who have been making attacks for a long time. The capacity for insight at any given time on a problem situation and the capacity to make one more capable of achieving such insight on problem situation are not determined by the amount of previous experience alone. The general intellectual level, good physical and mental health and alertness are important factors seem to be operating in it.

Students face varied type of problems in their course of study. When a student faced with a given kind of problem he begins to learn to behave in various ways which may increase the chance that he finds a solution. We cannot say that no effective response or solution is available. A solution is not available at present. But the problem is soluble. The student must first recognize the type of problem he has faced. Then he must select appropriate technique to attack the problem. Skinner (1968) states that a particular difficulty arises when a problem is to be solved only through successive steps. It is only the successive steps that one learns a response appropriate to each step and reach the ultimate solution for which many steps may be required.

According to Skinner non-verbal problem solving, like, inventing mechanical devices is not easy to describe and because of this it is also not easily taught. There are both verbal and non-verbal problem solving stocks of things which may go into hidden level and analysis of them may become very difficult. Therefore, they are taught at the overt level with more experimental evidences. The teachers while giving instruction in problem solving, they are often committing the mistake called the *formalistic fallacy*. To make the student to carry out the problem solving behaviour it is only needed simply to show him tempting what to do. But, instead, the student imitates what the teacher said and by doing so keeps himself in such a behavioural way which results in solving the problem. But the chances of keeping himself engaged in similar behaviour in future is not necessarily increased for the student. There may be different problem in different situation and the student may find it difficult to face the problem of this type. A different type of technique is needed for solving the problem for which the student has to equip himself with such technique.

Problem Solving and Decision-Making

Problem solving is sometimes associated with decision-making. These two are higher mental processes. Psychologists have been showing much interest in these for many years. Problem solving has been conceived by psychologists as discovery of correct response to a problem or situation that is new and difficult to the individual. But decision-making refers to selection of a correct responses out of several correct responses already brought out and discovered. Empirical studies have been made relating problem solving ability and age and it has been found that as and when people advance in age and become older and older their problem solving ability is reducing. Wiford's (1958) study was involved in comparing young and old in problem task and the results showed that old persons performed the task as accurately as the young persons, but the older people taken more time for the same task.

Problem solving is a complex process. It involves the study of many other components. Each one may have some influence with age. Therefore, if we accept the idea that problem solving ability diminishes with increasing age we must take into account the other components also. It is also true that in some instances problem solving depends upon the amount of information the individual is endowed with. Since problem solving situations may be arising in an individual's life span, he may be familiar with certain elements and possess reliable information with which he may solve the problem efficiently in old age much better than young and young adults.

But, if the problem involves factors like learning, experience, knowledge, information, intelligence the older people may do much better that the young people. Instead, if the problem emphasises more on memory, perceptual capacity and psychophysical the younger persons may perform better than older people. Birren (1964) states that age brings into the problem situation certain changes in the components of mental abilities of the individual. The other important factor is the adaptation of the individual to the problem solving situations. He states that because of the lack of adaptation the problem solving orientation process is also lacking in the individual as a function of age. But it will not be possible to say whether the problem solving ability increases or decreases due to advancing years of age as the problem solving task is a very complex in nature.

What has been stated in the case of problem solving also may be stated for decision making with regard to age. But we do not have much experimental evidences to state whether man's capacity in decision-making is changing with age. But the general observation is that there is a slow process in decision-making by the older people. There are some studies made on choice reaction time and in some simple tasks. In all these the older people have taken longer time for decision making than the young adults. In decision-making process some investigators include risk-taking behaviour. This means that an element of risk is involved in making a decision. For example, in races a man may bet a huge amount on a horse. In this the man is taking a risk in making the decision of betting. Hence this type of decisions may be stated as risk-taking decisions or risky decisions.

PROBLEM SOLVING

Problem solving has been defined in various ways. It is finding out some answer or solution to a question for which the answer is not ordinarily known nor an answer is in existence already. Problem generally refers to any thing or any matter or any person or any situation that is difficult to deal with, solve or overcome. A problem may be also referring to a puzzle or a maze or a question set for solution. A problem is a challenge to a person. Anyone who faces this challenge is under some kind of tension until the problem is solved. The person who faces the problem searches through reference books and also do much concentrated thinking until the tension is released. Finding out a solution to a problem is a discovery of correct response which is new to the person.

Man in his everyday life faces several problems. There may be simple problems or complex. He may solve some problems easily and some he may find very difficult and has to struggle very hard on them. If the problem solving activity is followed by some motive it receives importance and man however wants to achieve it. But if his skills and talents do not reach beyond certain limit after many trials he may give it up. Man solves his life problems in several ways. There is great amount of learning and certain amount of motivation in problem solving activity. Problem solving also includes the behaviour called thinking, reasoning or creating. Therefore, it is supposed to be a higher mental process. It is also called complex learning by some scholars. There is difference between learning and problem solving though problem solving is also a kind of learning. For example, in rote learning the responses are indicated to the participant and his goal is to reproduce them faithfully. But in problem solving the participant himself must discover the correct response and eliminate the incorrect ones. However, in problem solving task, it is not solved without any trial and error.

In problem solving situations, particularly in the case of motor skills, the goal is stated and the general mode of attack also is indicated to the participant. He must somehow, find out for himself the particular way and the integrated set by which he can achieve the goal. In the problem solving behaviour of human beings all the methods of learning are involved. For example, trial and error is common in human problem solving behaviour. Similarly the motor and verbal learning and transfer of training are also essential components in problem solving behaviour. The rat learns to run the maze, the cat opens the puzzle box, the monkey reaches its goal and man invents rocket – all in a way problem solving behaviour in which learning is involved. Apart from learning, problem solving involves tasks far more difficult than running the maze, opening the puzzle box and reaching the banana. It involves more complex tasks which are not observable nor measurable correctly than simple tasks like learning. Problem solving behaviour has two important aspects. Firstly, there is a low probability of observing the correct response or solution. Secondly, the discovery process seemed to occur.

Problem solving is a goal directed activity. It may be defined as a goal directed multiple trial and error approach to new situations in which the individuals discovers new modes of response. For example, let us consider this problem. Here is a three-digit number 999. Handle this number in whatever possible manner you like and make it 10. How do you solve this problem? The problem appears to be simple, but the method of solving it is difficult. A common reaction to this problem is to make into suitable components that should constitute into 10. The problem and the behaviour required for solving it would reveal some of the basic characteristics of problem solving behaviour.

The first thing in problem solving is that the correct solution is not made immediately. If this takes place, then there is no problem. Therefore, there is a psychological gap between the problem and its solution. The gap can be filled up only by discovering the correct response to the problem.

The second thing is, that problem solving behaviour in human being mainly consists of covert responses. For example, in the case of the problem given here you may be trying to solve the problem in various ways taking to yourself. But nobody knows what method you are using till you make some overt responses either by speaking it out or by writing the solution on a paper. This would reveal your problem solving method to others. In this way, most of the process in problem solving method are not exposed.

The third thing in problem solving is that the correct solution to the problem sometimes occur suddenly without much efforts. We sometimes solve the problems by sudden insight. For example, the author, when he encountered the problem given here, tried to solve it in several possible ways by manipulating the three nines. Suddenly the idea struck in his mind that what is required here is to add one to nine to make the number 10. How to get one from the rest of the nines. It is only dividing one nine by another nine and adding it to one nine.

That is : 9 + 9/9=9+1 = 10

This is how we are trying to solve every problem in our life situation.

In the preceding sections we have discussed in detail how thinking is a form of behaviour and also can be experimentally studied. It is not a simple behaviour like seeing, hearing or testing, but it is a complex behaviour pattern. It is a mental function of verbal type. Writing is also a form of verbal and motor type of behaviour. Watson (1925) the founder of behaviouristic school said that thinking is a form of silent speaking and speaking is a form of louder thinking. Whether it is silent speaking or lower thinking it involves the verbal element of language. What a man thinks? We do not know and nobody knows except the thinker. That is why experimentation in thinking behaviour is very difficult. The thinking behaviour has to be expressed in some form. It has to be expressed in the form of words. Words are symbolic expression of objects, persons and places.

Experiments on thinking may be conducted in two ways. One way is to use fully the verbal elements – language, words and expression of the same in the form of writing or speaking. The other way is to use non-verbal activity and to reduce the verbal elements to the least minimum level. The processes of thinking or problem solving or reasoning are used synonymously to indicate problem solving.

There are a quite number of problem solving tests. Some of the problem solving tests in use are, Healey's Puzzle Box Test, Heidbrealer's Test, Wisconsin General Test Apparatus, Purdue General Test Apparatus, Haufmann And Kasanin Concept Formation Test. Maier's Pendulam Problem Test, The Tower of Hanoir Puzzle Problem, The Water Glass Problem Test, The Nine-Dot Problem Test, Cut-Out Puzzle Problem Test, Pyramid Puzzle Problem Test, Outline Puzzle Problem Test and so on.

All these problem solving tests are mostly manipulation tests. They may reveal the explicit component of thinking behaviour. We can record of such explicit behaviour also as far as possible.

EXPERIMENT 22 : PROBLEM SOLVING EXPERIMENT-1 (An 'E' Letter Outline)

CUT OUT PUZZLE PROBLEM

Problem

The purpose of this experiment is to study the nature of the problem solving activity in a given cut out puzzle by assembling the parts to make it into a whole pattern.

Method

Participant

Choose a person who has knowledge about the problem solving activities of this type and describe his/her personal characteristics such as age, sex, qualification, occupation, socio-economic status etc.

Material

An *'E'* letter outline, 9 cut out pieces of the letter *E*, and stop clock. The *'E'* letter cut out pieces test was designed by Tinker and Baker (1938).

Procedure

The respondents may be seated comfortably in a well illuminated room in front of a table. The 9 *'E'* letter cut out forms should be shuffled. The *E* outline should be placed in front of the respondent and the 9 cut out forms should be placed in a jumbled form. The following instructions should be given:

> "In front of you there is an *'E'* letter outline. Here there are 9 cut out pieces. Fit these 9 pieces into an *E* letter into the same form of the *E* letter outline. Adapt any method or procedure convenient to you. The 9 small pieces may be used either side up and down. Do it as fast as you can. Ready go."

As soon as the participant starts the work the experimenter should start the stop clock to run. The total time taken by the participant to complete the task should be recorded. The experimenter should keep the record of the piece (by number) with which the participant begins. The experimenter should observe and record which parts seem to him/her easy for assembling and which parts seem to give him trouble, which part he/she finally placed and to solve the problem of assembling. Then the experimenter should collect the 9 pieces and shuffle them and ask the participant to perform the task for another time in the same manner as he/she did previously. The experimenter should note down the time he/she has taken to complete the outline and also other characteristics. The participant may be asked to give his introspective report and he may be allowed to go away.

Result

The experimenter's observation of the problem solving task should be recorded systematically and they should be given in Table 8.1. The number of trials the participant has taken to successfully completing the assembling task, the piece number he/she first chosen for his problem solving task and other pieces he/she subsequently picks up and the time for each trial should be given in Table 8.1. Which of the pieces has given him trouble in placing at the appropriate places. While performing the task for the 2nd time what were the changes taken place. Find out the difference between the 1st and 2nd performance.

The group data should be collected and should be given in Table 8.2. The total time each individual takes to solve the problem should be given in seconds.

Table 8.1 : Showing the result of the problem solving in the cut out puzzle experiment

Trial No.	Time in seconds 1st time	Time in seconds 2nd time	Pieces number order of placing 1st time	Pieces number order of placing 2nd time	Observations
1					
2					
3					
4					
5					
Total					

Table 8.2 : Showing group data in the cut out puzzle experiment

Sl. No.	Name	Trial Nos.	Time in seconds 1st time	Time in seconds 2nd time	Remarks – Important observations
1	N.K.				
2	R.T.				
3	V.S.				
4	T.V.				
5	S.V.				
6	S.R.				
7	V.C.				
8	T.S.				
9	P.S.				
10	S.T.				
	Total Time				
	Mean				

Discussion

Analyse and discuss the individual's score and the way in which the participant handled the problem solving situation. State whether he/she has understood the instruction properly and carried out. Whether he/she was found to be nervous and struggling in handling pieces or felt very easy. How he/she behaved in the first trial? How many trials he/she has taken to solve the problem. Describe the nature of his/her behaviour till he/she solved the problem. What method the participant followed in solving the problem. How much time he/she has taken to complete the task successfully. While performing the task for the second time were there any differences in placing pieces from the first time. Discuss it relating the choice of pieces for placing in the pattern.

In the group who solved the problem in the first trial? State his/her personal characteristics and other reasons if any to solve the problem quickly. How much time he/she takes for that? What procedure he/she followed? State, how the solution came to him, either by trial and error or by insight. Who has taken the shortest time to solve the problem and who has taken longest time? Compare your participant with other individuals. How many of them below the group mean and how many above the group mean? Was there anybody who did not solve the problem at all? Who has taken more trials? Find out reasons whether age, sex, qualification, other experience, socio-economic status had any influence in the problem solving task. Was there any difference between the first performance and the second performance among the members of the group in time, choosing the pieces and placing them. State, whether was there any difference in the performance due to sex. State how many percentage of males and how many percentage of females have solved the problem in below mean time. Discuss the problem solving behaviour of your participants with results of the earlier studies.

Conclusions

1. Was there anybody who solved the problem in the first trial with shortest time? State how much time he/she has taken to solve the problem and state also other informations about him/her.
2. Who has taken longest time and more trials to solve the problem? State the reason, if any, and also other particulars about him/her.
3. Was there any sex difference in the performance?

APPENDIX – I

The Cut-Out Puzzle

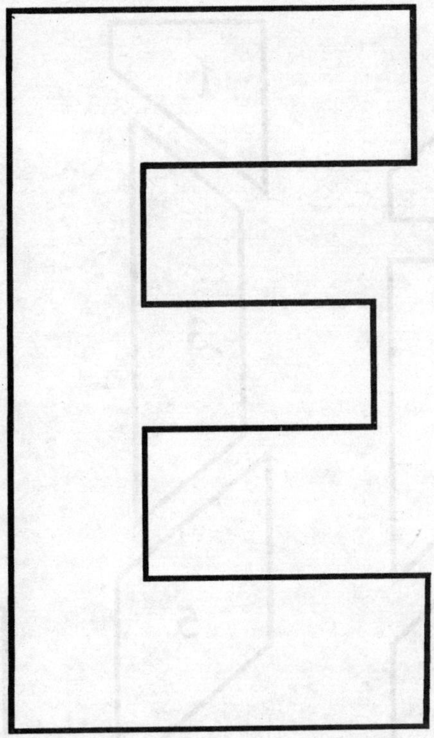

Fig. 8.1 : The E Outline : Experiment 22

APPENDIX – II

The Cut-Out Puzzle

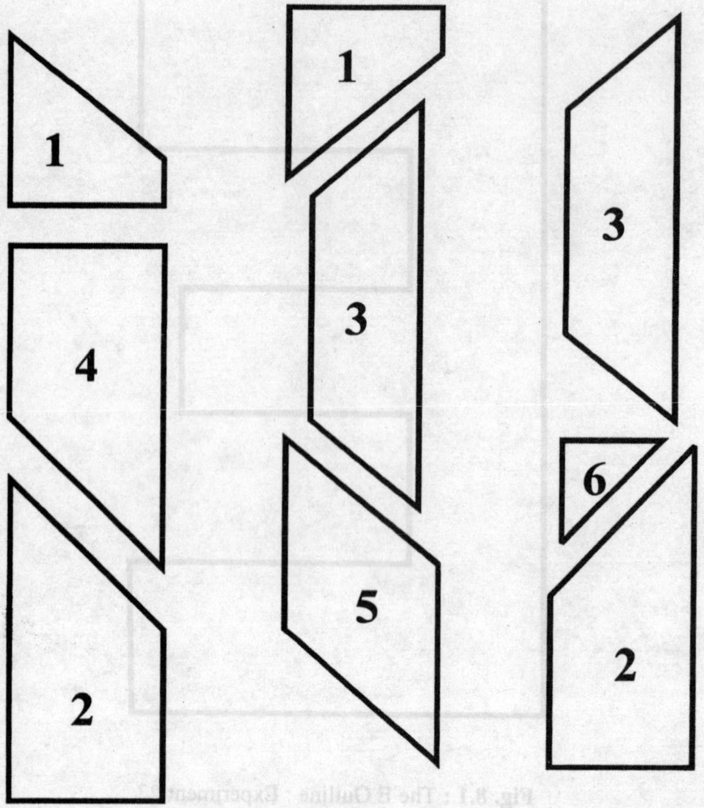

Fig. 8.2 : Cut-out pieces of the outline E : Experiment 22

APPENDIX – III

The Cut-Out Puzzle

Solution to the Problem

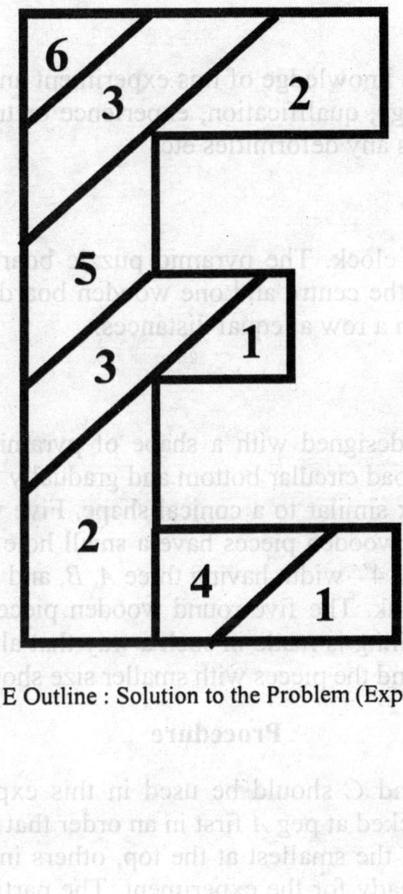

Fig. 8.3 : E Outline : Solution to the Problem (Experiment 22)

EXPERIMENT 23 : PROBLEM SOLVING EXPERIMENT – 2 (THE PYRAMID PUZZLE PROBLEM)

Problem

Study how the process of reasoning in a person is taking place by analysing his behaviour while solving the problems of increasing difficulty.

Method

Participant

Choose a person who has no knowledge of this experiment and describe his/her personal characteristic such as sex, age, qualification, experience or training in some job, socio-economic status, whether has any deformities etc.

Materials

Pyramid puzzle board, stop clock. The pyramid puzzle board consists of five wooden round pieces with a hole at the centre and one wooden board. The wooden board has 3 iron pegs *A*, *B* and *C*, fixed in a row at equal distances.

Description of the Material

This is a puzzle board test designed with a shape of pyramid and named after it. The pyramid board is having a broad circular bottom and gradually decreasing its width evenly to a point. It will have a look similar to a conical shape. Five wooden round pieces make this pyramid shape. The five wooden pieces have a small hole at the centre. There is one long wooden blank with 3′′ to 4′′ width having three *A, B*, and *C* iron pegs vertically fixed at equal distances on the blank. The five round wooden pieces of different sizes can be piled on the iron pegs. The piling is made in such a way that always the pieces with wider size should be at the bottom and the pieces with smaller size should be at the top.

Procedure

The three iron pegs *A*, *B* and *C* should be used in this experiment. The five round wooden pieces should be stacked at peg *A* first in an order that the largest piece should be at the bottom of the peg and the smallest at the top, others in between according to the size. Now the apparatus is ready for the experiment. The participant should perform the task correctly two times consecutively. As soon as he/she finishes it successfully first time, the experimenter should replace the same wooden pieces in the same peg once again and ask the participant to proceed as before.

The participant should be seated comfortably in a well illuminated room in front of a tea-poy or small table. Then the instrument should be placed in front of him/her with the 5 round wooden pieces piled at the peg *A*. The following instructions should be given to the participant.

"There are 5 round wooden pieces piled in an order at the peg *A*, the largest at the bottom and the smallest at the top. Now you will have to move the pile to the peg *B* or to *C* moving only one piece at a time. You should not place a larger piece upon a

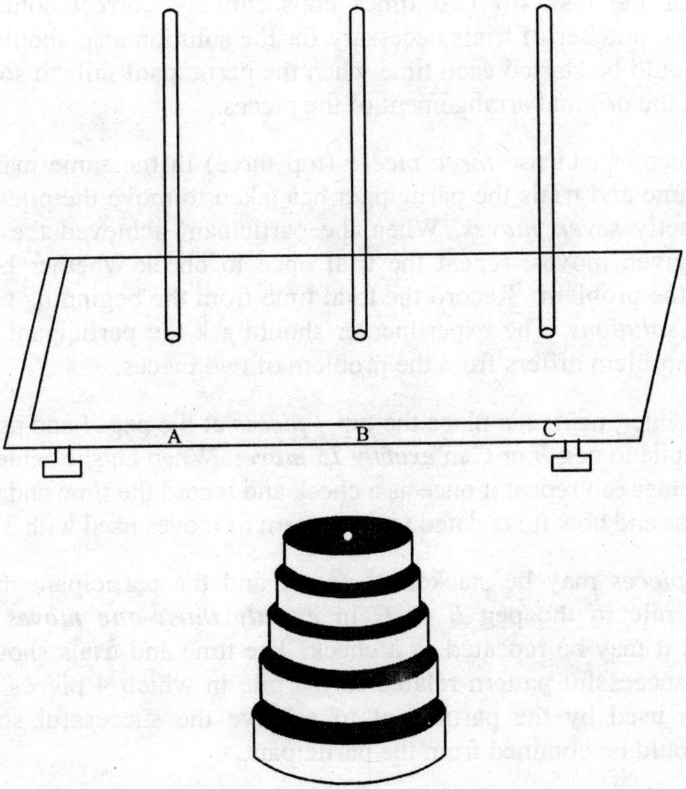

Fig. 8.4 : Problem Solving Experiment-2 : Pyramid Puzzle Problem Solving Test (Experiment 23)

smaller one. Use all the three pegs while making the moves. (The experimenter should demonstrate once). Before you start moving the pieces try to think ahead each time you make a move and formulate a plan of action. If you find it impossible to formulate a plan to move the pieces according to the procedure, you may start the task with trial and error and formulate a plan as you work out alone. *First I shall pile two pieces at peg A,* then three pieces, then four pieces and lastly five pieces. The *two pieces* should be transferred to the peg *B* or *C* exactly *in 3 moves*. You should do this correctly two times consecutively. (The experimenter should first pile the two *small pieces,* top 2 pieces at peg *A*). Ready go."

The experimenter should start the stop clock at once. Every time the experimenter should tell the participant the number of moves in the correct solution of the problem. The most important aspect of this experiment is to discover what plan of action or principle the participant employs in solving the problem. The time in seconds should be noted down for every stage. The experiment is performed in 4 stages.

1. In the first stage the experimenter should use the two small pieces at *A*. The pile is to be transferred to *B* or *C* in exactly ***three moves***. The stop clock should be started when the go signal was given. The clock should go until the proper solution is achieved twice in succession. The time the experimenter records is that from the

beginning of the task till two times consecutively correct solutions have been achieved. The number of trials necessary for the solution also should be recorded. A new trial should be started each time when the participant fails to solve it with given moves from the original arrangement of the pieces.

2. The experimenter can use *three pieces* (top three) in the same manner and should record the time and trials the participant has taken to move the pile from peg *A* to *B* or *C* in exactly ***seven moves***. When the participant achieved the correct solution within the given moves, repeat the trial once to check whether he/she has really understood the problem. Record the total time from the beginning till the end of the *two correct solutions*. The experimenter should ask the participant to indicate how the present problem differs from the problem of two pieces.

3. The experimenter, next, can place the top *4 pieces* at the peg *A* and ask the participant to move the pile to peg *B* or *C* in ***exactly 15 moves***. When he/she achieved the solution the experimenter can repeat it once as a check and record the time and trials. What plan brings success and how it is related to the pattern of moves used with 3 pieces.

4. Then all *5 pieces* may be stacked at peg *A* and the participant may be asked to transfer the pile to the peg *B* or *C* in ***exactly thirty-one moves***. On successful achievement it may be repeated as a check. The time and trials should be recorded. How is the successful pattern related to the pile in which 4 pieces were used. The general plan used by the participant to achieve the successful solution of these problems should be obtained from the participant.

Results

The experimental results should be obtained in terms of time, trials and observation and be given in Table 8.3.

Table 8.3 : Showing the data in the Pyramid Problem experiment

Number of pieces	Number of trials	Total time in seconds	Observations
2 (3 moves)			
3 (7 moves)			
4 (15 moves)			
5 (31 moves)			

Draw two graphs for the data. (1) in one, plot the number of pieces on the base line and the repetitions for successful solution on the vertical axis. (2) Draw another graph using the number of pieces on the base line and the ratio of the total amount of time to the number of moves for successful solution on the vertical axis. From the analysis of these curves show the relationship between number of pieces and the number of trials required by the participant. Collect the group data and present them in Table 8.4. Work out the mean trials and time.

Table 8.4 : Showing the group data of the pyramid puzzle experiment

Sl. No.	Name	Trials for each piece				Total	
		2	3	4	5	Number of trials	Time taken in seconds
1	P.S.						
2	T.V.						
3	S.V.						
4	S.L.						
5	S.T.						
6	S.R.						
7	V.C.						
8	S.K.						
9	S.B.						
10	T.S.						
	Total						
	Mean						

Discussion

State the individual's reaction on facing the problem. Was he able to find out the plan or principle in solving the problem or struggled or tried by trial and error method? Discuss the difficulty the respondent faced in moving the pieces towards the solution of the problem. At what stage the respondent was able to pile the pieces in the given moves? While repeating the successful solution was he taking the same time or less? Discuss it for every stage.

How many trials your respondent has taken to solve the problem at each stage? Was he/she committing errors? If so, state how they were committed? Who has taken more trials on the whole in the group in solving the problem and how much time? Who has solved the problems with least number of trials and lowest time? Where your respondent stands in the group? Discuss his/her performance in relation to other individuals. Compare the problem solving behaviour of your participants with the results of previous studies or problem solving.

Conclusions

State whether your participant was able to find out any general principle or plan that was successful for him/her in solving the problem at all four stages. State the difficulty he/she faced in the task.

EXPERIMENT 24 : PROBLEM SOLVING EXPERIMENT–3 (OUTLINE PUZZLE PROBLEM)

Problem

Study the progress of the problem-solving task of a person in an outline puzzle having directional habit in outlining.

Method

Participant

Choose a person who has no knowledge about this experiment and describe his/her personal characteristics such as age, qualification, sex, experience in any skilled activities, socio-economic status.

Material

A set of 15 outlines and stop clock. The outline is a square with dotted lines and within the square there are nine small squares in dotted lines. This outline was designed by Tinker and Baker (1938).

Procedure

The participant should be seated in a well illuminated room in front of a tea-poy or small table. Place the sheet of the outline. There should be 15 outlines. The following instructions should be given:

> "There are 15 outlines in this sheet. You are to begin with outline *'A'* (the experimenter should point out it). Your task is to trace over all the dotted lines in the drawing without removing your pencil from the sheet, without retracting or without going back on the line and without crossing any line you have already drawn. If you are blocked on one drawing leave it and at once you should begin tracing on the next outline. Like this you must do two successive errorless tracings. Now ready. Go."

The experimenter at once should start the stop clock to run and note down the participant's starting point on each outline. The problem is considered solved when the participant *can make two successive errorless tracings.* If time is taken by the participant for any reason at once stop the stop clock until the task is begun again. The experimenter should record the total time from starting the first outline to the end of the two successive errorless tracings. Also note down the number of outlines he/she has used in reaching the two errorless solutions. Note down the behaviour pattern of the participant and all the remarks that seem to be pertinent for the experiment. The method of attack on the problem his/her trial and error behaviour, expression of thinking habit etc. should be observed and recorded.

Results

Find out the time taken by the participant to make two successive errorless tracings in seconds and present it in Table 8.5. Also the number of trials he/she has taken to solve the problem. Collect the group data and present them in Table 8.6. Also state whether the participants have solved the problem or not by using 15 outlines. Work out the mean time and trials for the group.

Discussion

Discuss the results in relation to the time, number of trials and solving the problem. In the group how many of them have solved the problem and how many have not. Who are those taken the longest time and who are those shortest time. Who has solved the problem in minimum number of trials. What is the mean time and trial and how many were above and how many were below the mean.

Table 8.5 : Showing the individual data in the outline puzzle experiment

Time taken for two successive errorless tracings	No. of trials taken for solution	Observations

Table 8.6 : Showing the group data in outline puzzle experiment

S. No.	Name	Time for two successive errorless tracings in seconds	Number of trials	Problem solved or not
1	N.K.			
2	S.L.			
3	S.B.			
4	V.S.			
5	T.S.			
6	R.T.			
7	S.T.			
8	S.V.			
9	V.S.			
10	S.R.			
	Total			
	Mean			

Who are those solved the problem quickly and who are those did not. Find out the background of those persons who have solved it in a short time with few trials. Find out and state the influencing factors in the successful solution of the problem, whether it is qualification, age, sex or socio-economic status or experience in some other task. Find out and state the background of those who could not solve the problem. Compare your results with some of the earlier studies made on problem solving behaviour. State whether you are able to observe anything new in your results.

Conclusions

1. State how many in the group solved the problem and how many did not.
2. What was the shortest time in solving the problem? Who he was? Who has taken the longest time?
3. What was the minimum number of trials in solving the problem?

APPENDIX - IV

Problem

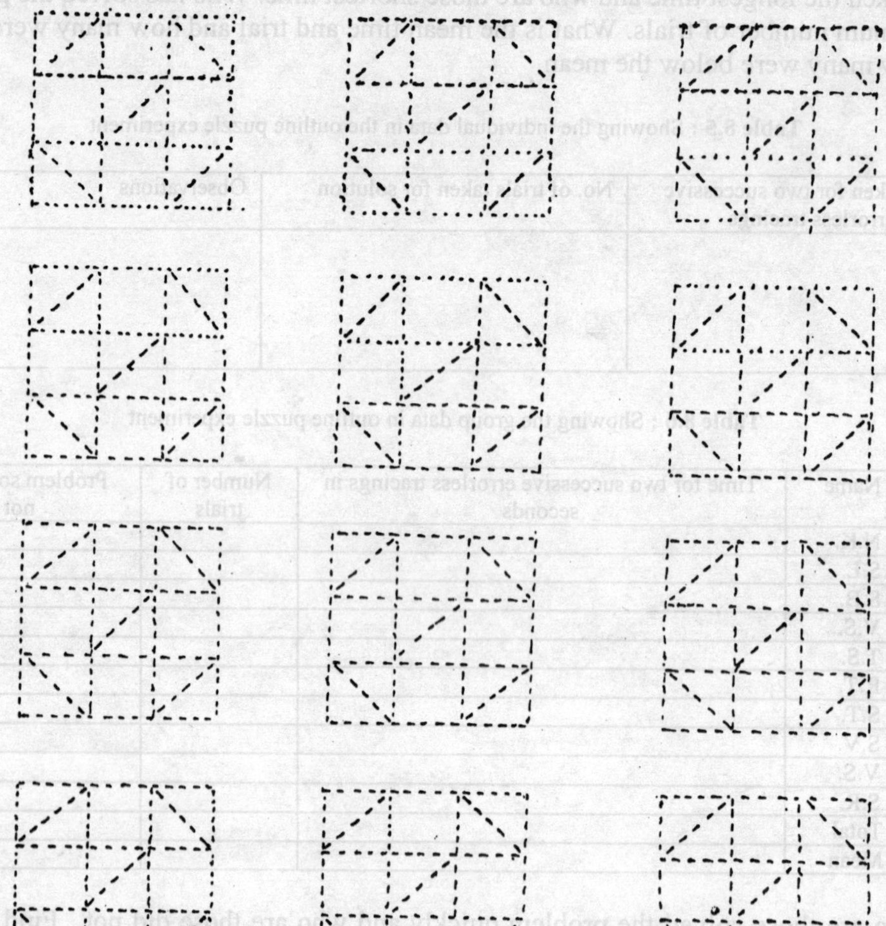

Fig. 8.5 : Problem Solving : Outline Puzzle Problem (Experiment 24)

APPENDIX - V

Solution to the Problem (Model)

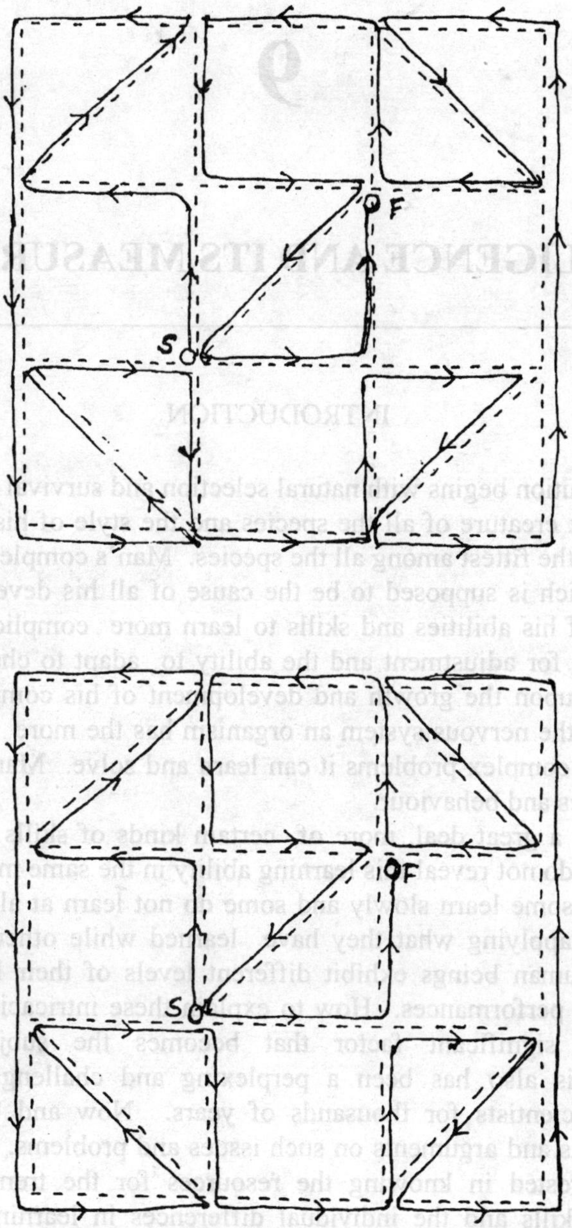

Fig. 8.6 : Outline Puzzle Problem – Completed Form (Experiment 24)

9

INTELLIGENCE AND ITS MEASUREMENT

INTRODUCTION

The history of evolution begins with natural selection and survival of the fittest and man is the most complex creature of all the species and the style of his survival is the best evidence that he is the fittest among all the species. Man's complexity is revealed in his neural structure which is supposed to be the cause of all his developmental processes. The development of his abilities and skills to learn more complicated matters and the increased flexibility for adjustment and the ability to adapt to changes in the environment all are based upon the growth and development of his complex nervous system. The more complex the nervous system an organism has the more complex situations it can face and more complex problems it can learn and solve. Man exhibits these in all his learning activities and behaviour.

Man can learn a great deal, more of certain kinds of skills than can any animal learn. But all men do not reveal this learning ability in the same manner. Some persons learn very quickly, some learn slowly and some do not learn at all. Some men exhibit great efficiency in applying what they have learned while others are inactive or act stupidly. Hence human beings exhibit different levels of their learning abilities and efficiencies in their performances. How to explain these intricacies? This is the most important striking significant factor that becomes the subject matter for the psychologists. This also has been a perplexing and challenging problem for the philosophers and scientists for thousands of years. Now and then there were hair splitting discussions and arguments on such issues and problems. But the modern man is very much interested in knowing the resources for the tremendous capacity for learning complex skills and the individual differences in learning abilities. Modern psychologists by way of exploring the resources of man's capacity for learning turned their interest and attention on the general construct — intelligence.

THE NATURE OF INTELLIGENCE

Psychologists' interest in the study of the individual differences in learning opened a new vista, named intelligence which they hoped would throw much better light upon the problem of individual differences in learning and also in other performances. It is extremely difficult to ascertain what part of the brain is related to learning activity. Researches in this area are going on and results may provide us the code to understand the changes that occur in the brain as the result of learning. The anatomical changes in the number and structure of synopses are accounted for our ability to learn and remember things. It is the neural structures that integrate the stimuli and the responses. The differences in the function of the neural structures cause the differences in learning ability. Probably the individual differences in learning may be due to the integrating process of the stimuli and responses.

The neural structural function is not directly measurable and, therefore, will not be possible to assess the learning ability. Learning ability is only inferred from the behaviour of the organism and intelligence is the inferred result of the organism's learning behaviour. In other words the learning ability of the organism is revealed in the form of intelligence. Therefore, intelligence refers to the nature of the function of the organism in a situation. Garrett (1975) says that intelligence may express itself in a number of ways. The school boy learns his lessons easily, the foreman skilfully handles his tools, the salesman deals with his customer successfully, the lawyer wins his case easily, the motor mechanic set right the car in few minutes. All these reports reveal the efficiency of the concerned persons – they may be stated as intelligent behaviour. Edwards and Scannell (1975) state that intelligence is a construct and nobody knows what intelligence is. It is inferred only from the behaviour. Intelligence may be also referred to varieties of skills and defined in various ways. Wickens and Meyer (1961) define intelligence as the degree of proficiency with which the individual solves problems, deals with abstract materials, verbalizes and learns new materials. According to Binet (1916) intelligence is a general capacity for comprehension and reasoning that manifests itself in various ways. Stagner and Karwoski (1952) state that intelligence is not a physical object. It is a complex characteristic of human performance, has two dimensions – speed and power.

The concept intelligence is viewed by Woodworth and Marquis (1952) little differently. According to them intelligence is a comprehensive term referred to observing, understanding, thinking, remembering and in all ways getting knowledge and it is related to intellect. They also state that intellectual activity produces knowledge, but intelligent activity does something more in addition to this. Kimble *et al.* (1985) state that intelligence is the ability to learn, to reason, to deal with abstract concepts, to adjust to new situations and to be friendly with society. We all know that everyone differs from others very much in these talents. All these definitions of intelligence are focused on different kinds of performances such as learning ability, ability to manipulate abstract symbols and ideas, ability to solve problems, ability to acquire new knowledge, ability to get along in society and so on.

Learning is a function that can be found both in animals and human beings. But there is tremendous amount of difference in speed and in complexity of learning. At the lower extreme there are feeble minded cases who are not mentally developed to learn the acts of normal human beings. At the other extreme there are persons who deal with extremely complex abstract functions. Gardner Murphy (1952) says that intelligence means the ability to learn and it increases steadily with the complexity of the nervous system. According to him the ability to profit from experience and environment is called intelligence.

Those definitions and descriptions of intelligence have academic basis. In every culture there is a tendency to evaluate the intelligence of the people from the academic activities. The academic activities and achievements of an individual provide us with substantial data to state what an individual is capable of. We could see that only people who can withstand mental and physical strain could make progress not only in academic life but also in every walk of life. How to describe these behaviour phenomena? Naturally they are intelligent people. They plan, execute and achieve. They are also responsible and dutiful people. Otherwise they could not make any achievement in life.

In the preceding section it is stated that intelligence is sometimes referred to learning ability. But some are of the view that learning ability is not fixed. It is the ability that a person possess at the time of performing a particular kind of task. After some time a person's learning ability may change and there may be a time in a man's life that he may not be having equal abilities in all walks of life. A man's successful performance is also depending upon his motivation and interest. A person who is very brilliant in language abilities and games may be lacking mechanical abilities and his acquired abilities in various areas may change when he advances in age. This may also change his motivation and interest. Therefore, learning ability is subjected to change. If intelligence is equated with learning abilities intelligence also is changing with age. Therefore, intelligence cannot be called a unitary and wholesome characteristics of an individual and he can solve all types of problems at all times. It is because of this variation in abilities that Guilford (1959) speaks of intelligences instead of intelligence.

Binet

Any study of intelligence is not complete and perfect without making some reference to the works of *Alfred Binet* (1857-1911) whose contributions were so great that he was the pioneer of the measurement of intelligence. Binet (1961) proposed first that memory and judgment were the essential elements in intelligence. Persons who have best memory and who could choose the most adequate among many are the most intelligent. Since he later on found that persons with best memories were not always most intelligent, he modified his theory and stated that intelligence declare itself in three specific manners as direction, comprehension and autocriticism or self-evaluation.

According to Binet an intelligent person can identify the proper *direction,* the most direct route towards his goal whereas a retarded person is usually lacking on this dimension. The ability of *comprehension* involves achieving ends, to reach a goal. An individual may encounter many stimuli in his goal directed way, some may be relevant and some may be irrelevant to his goal. A bright child and an intelligent adult are much more likely to take up the relevant stimuli whereas a dull child and a less intelligent adult are more likely to be diverted by the irrelevant stimuli and then they may lose the goal. The *autocriticism* is a self-evaluation, a rational appraisal of the actions of the individual. It is the highest level of intelligence. It is hard for the dull person to choose the direction and still hard to achieve comprehension and, therefore, he can never hope to achieve autocriticism. Persons who could evaluate their own actions objectively reveal this ability. It is observed that the persons who made greatest contribution to society possessed this ability to a higher level. Therefore, Binet asserts that highly intelligent adults reveal all the three abilities in their behaviours in varying degrees.

Terman

The method of measuring intelligence evolved by Binet was followed by Lewis Terman (1877-1956) who was mainly interested in the functional aspects of intelli-gence. Terman (1937) states that it is the language that has played an important role in the behaviour of man throughout the history with its complex and abstracts components. On this ground Terman asserts that intelligence is best revealed in the verbal and abstract expressive behaviour of an individual. Therefore, the best way to measure the level of intelligence of the individuals or groups is to provide them with verbal abstract tasks of increasing order. A person with highest verbal skills will also show high ability in other elements of comprehension and transfer.

Thorndike

While Binet and Terman emphasised their theories on practical effects of intelligence Thorndike (1874-1949) took a different turn to a more basic principle of associating intelligence to learning. Thorndike (1926) states that learning takes place through the formation of associations. Individuals are exposed to various stimuli in their everyday life and to which responses are made. When a response is made to a stimulus a connection is established. Connections of this type are neurophysiological processes which are taking place in the central nervous system. There may be as many associations or connections as many stimuli there are and responded. These connections are not taking place equally in all organisms as they do not show equal ability for connections. This difference to form connections is due to hereditary factor. Thorndike says that individuals with the greatest potentials for neural connections behave more intelligently than the individuals with less potentials for neural connections. The formation of connections is revealed through altitude or level and range or width. When altitude is combined with range a third factor area is obtained. Thorndike adds also the speed in behaviour as an important factor of intelligent behaviour. A bright person can do his task more quickly with less errors and less efforts than a dull person. Thorndike states that when certain kinds of behaviours function around the important factors the scope of intelligence is extended. Intelligence demands power and speed in solving problems and in meeting with situations All these can be met by only broad general intelligence. Since the general intelligence is very broad Thorndike classified its extension into three kinds such as abstract intelligence, concrete or mechanical intelligence and social intelligence.

Abstract Intelligence

Abstract intelligence is revealed in dealing with symbols, numbers, words, ideas and formulas. This is specially found in human beings. Animals are not capable of exhibiting abstract intelligence. Human beings with abstract intelligence discover the relationship between and among symbols and deal with logical syllogisms and draw conclusions. Since abstract intelligence deals with words and language the principles underlying conduct, truth, goodness, thought and the nature of the universe are expounded to the common man for the development of knowledge. Professionals like the lawyers, physicians, businessmen, statesmen, poets, novelists and literary men are successful in their professions because they possess the abstract intelligence at a greater level. The range of general intelligence in human beings is very wider.

Concrete Intelligence

The concrete intelligence is mainly dealing with machines and tools and mechanical comprehension. Generally men who possess this intelligence show more aptitude for machine works. They often keep themselves engaged in some mechanical or construction works. All their interests, we find, are mechanically oriented. The engineers, the mechanics, the technicians, technically trained workers, the trained industrial foremen and workers all exhibit in their works the ability called mechanical intelligence. Mechanical intelligence involve not only handling machines and tools successfully but also wide mechanical comprehension, manual dexterity and skill to coordinate various mechanical activities.

Social Intelligence

The third is, the social intelligence which is mainly concerned with the individual's ability to deal with personal and social matters. The person who is socially intelligent deals with social problems successfully. Wherever more human beings or groups are involved there may be conflicts or clashes and the person who is socially intelligent deals with them successfully and averts clashes. In professional or occupational areas also, generally the salesmen, auctioneers, brokers are socially intelligent persons. At higher levels, the diplomats, national representatives to high level committees must be socially intelligent persons. They must promote human relationships and friendliness.

This type of classification of intelligence is not very rigid and specific. There is considerable variation from person to person and also within the same person. A person, who is not educated and does not know any thing about the mechanical principles, can be a better mechanic than an educated person. A sales boy may become a successful businessman. It is not that everyone can do something better. In our day to day life we can see that some persons who have nothing to do with a business in the early years becomes very prosperous in some business. Therefore, there is individual difference in every ability, but sometimes practice or rigorous and continuous training can make even persons of mediocre ability to achieve something great.

Spearman

Since the individuals exhibit different levels of abilities in their performances in different tasks Charles Spearman (1863-1945) thought of general intelligence (abilities) and special abilities. Spearman (1927) a British psychologist after working on intelligence (individual differences) for twenty years came out with a conclusion that intelligence is more than an accumulation of specific skills. In 1904 he described the general intelligence which he called the 'G' factor as a kind of mental energy that flow into everything the individual does. The individual who is endowed with this (G) intelligence behaves intelligently in varieties of situations. He also found that some people with this intelligence do not act well in certain areas. Some are quick and successful while some are not. Spearman thought that it is a branch of 'G' factor in other activities in different degrees. When the general intelligence manifests in some specific activity it is called 'S' factor. There are many specific factors (S). Spearman observed that correlation between different test scores were never perfect. According to him every skill includes certain amount of the G factor and also different types of S factors. Spearman (1927) later on made changes in his two factor (G and S) theory and included intermediate factors. He observed that in between the general factors and the specific factors there are abilities which are less general than the former G factor and less specific than the S factor which are called group factors.

Thurstone

Thurstone (1887-1955), the American psychologist described intelligence in a different manner. Louis Thurstone (1938) called intelligence as the primary mental abilities. It consists of several abilities. According to him there are seven abilities. They are (1) Spatial ability, (2) Perceptual speed, (3) Numerical ability, (4) Verbal comprehension, (5) Memory, (6) Word fluency, and (7) Reasoning. He administered 56 tests on thousands of students and treated the data with advanced statistical methods. Thurstone used factor analysis for evolving these seven factors. Thurstone observed that these abilities are independent of each other. This means, a person who scores high in perceptual speed may score low in numerical ability. One or more abilities may be found in an intellectual activity. Thurstone's factor analytical study of mental abilities had tremendous influence upon subsequent studies of classification of abilities.

Following Thurstone's study of factor analysis which yielded several components of intelligence, other researchers attempted intensive study on complex intellectual abilities and came out with different components other than those evolved by Thurstone.

Guilford

After ten years of rigorous study Guilford (1966) developed a three-fold model of intelligence. Guilford felt that both Spearman and Thurstone have not described intelligence perfectly. He used the concept, 'the structure of intellect'. According to Guilford, intellect consists of memory and thinking abilities. Memory consists of limited number of abilities while thinking is complex and consists of many abilities. There are three types of thinking abilities such as : (1) Cognition, concerning with discovering of information, (2) Production concerned with use of information, and (3) evaluation concerned with determining the suitability of cognition and production.

Guilford postulated a threefold model of intelligence which yielded 120 potential factors or specific measurable abilities. They are : (1) *Operations,* (2) *Contents,* and (3) *Products.* The operations refer to the act of thinking (type of abilities), the contents refer to the terms through which we think such as words and symbols (expression), and products refer to the idea with which we arrive at (remembered or known). He states that his theory is a three-dimensional model. On the basis of these we may state that there is some kind of general intelligence and also special abilities depending upon the structure, development and knowledge of the individual.

MEASUREMENT OF INTELLIGENCE

Intelligence is what the intelligence tests measure. Here the measurement of intelligence becomes the primary matter. It was the evolutionary biology of nineteenth century that aroused the interest to measure intelligence. The evolutionary biology tried to explain the causes of individual differences. The psychologists were stimulated to measure the ability of man to learn by the Darwin's theory of the survival of the fittest. Sir Francis Galton (1822-1915) started an anthropometric laboratory in 1882 in London museum. He tested the visitor's visual acuity, sense of hearing and reaction time collecting a small fee from them and thus established the individual differences among the human beings. He believed that the sensitivity of the sensory systems was related to general intellectual ability. By the same time James McKeen Cattell (1860-1944) was interested in measuring simple abilities like sensory acuity, reaction time, memory, motor skills and head size on the idea that these must be the components of intelligence. He extended the use of tests. Mental tests, a term coined by him in 1890. Cattell has combined physical strength with measurement of intelligence. But, somehow the method and the results of his study did not attract many men in this field as the measures did not correlate significantly with intellectual activities. Wissler (1901) and others concluded that the essence of human intelligence is found in complex skills and not in simple activities.

While the dispute whether intelligence is fundamentally a general ability or a combination of several specific abilities was going on in various circles emphasis was made on development of workable intelligence test. During the same period the educational authorities in the city of Paris were very much concerned over the backwardness of the school children in the normal classroom performance and wanted to find out the causes and solve the problem. Alfred Binet (1857-1911) a French psychologist who was already interested in the matter of mental development undertook the responsibility of dealing with the problem and wanted to find out some method to measure intelligence. Binet's hypothesis was that the dull child is like the normal child, but backward in mental development and he would perform the task like the normal child, but like one or two years younger child. According to Binet the dull children could do better in their school performance when they grow older. Therefore, on this hypothesis Binet in collaboration with Simon (1873-1961) developed a successful mental ability test for different age levels from 3rd year to 7th year with increasing difficulty. Handwriting and palmistry were some of the variables included in the test. In the year 1905 the Binet-Simon scale with 30 items was brought out. Since the first version was in a crude form a revision of the scale in better form was brought out in 1908 and in 1911 enlarging the scale upto adult level. In the 1908 revision, Binet-Simon introduced a scoring concept called *Mental Age (MA)*.

Mental Age (MA)

The Binet-Simon scale consists of series of sub-tests arranged in the order of difficulty according to age levels. Upto 5 years there were two sub-tests for each year and from 6th year to 14th year one sub-test for each year. The experimenter while administering the test to each age group of children, starts first with each child with the easiest item at the top of the test and works out till the child answers no more questions. He goes from the easiest items to most difficult items of every year test. After collecting the data like this from a large number of children the experimenter should work out the norm. That is, in each age group large number of children were able to answer correctly almost at the same level — answered correctly the same number of questions. But a small number of children might have stopped not answering correctly some items at the bottom of that age level test and some children might have gone answering correctly beyond that age level. It is thus the average score or the norm of the normal children at each age level test and some children might have gone answering correctly beyond that age level is established. The norm or average score of a sample of six years children is six years norm. This norm or average score increases along with the age of the children.

For example, if a four years old child answered all the items of a fourth year test correctly his mental age is four and if the same child answered all the items of a fifth year test his mental age is said to be five. If a five years old child answered all the items of a fifth year test and also all the items of a sixth year test successfully, but none of the items of seventh year test his mental age is said to be six. The mental age is a measure of the absolute level of intelligence. The principle involved in establishing the age norms in the Binet-Simon scale is that the children of all age groups from 2nd year to 14th year should take up all the year tests. The 1937 and 1960 scales cover the years from two to fourteen with additional tests for average and superior adults.

If a child aged 9, passed all the items of a 9th year test and half of the items of a 10th year test then his mental age is said to be 9½. His overall performance equalled the performance of an average 9½ years old child. The mental age is obtained from the test scores and the chronological age (C.A.) is the individual's actual age. If a 9 years old child could perform only the average 8 years old child test then his mental age is 8. He is supposed to be a dull child. But if he could perform the average 10 years old child test then his mental age is 10 and he is said to be superior in intelligence.

Intelligence Quotient (I.Q.)

The intelligence quotient is an index of intelligence. The I.Q. is a comparison between the performance of two persons of the same age on a test. In normal children the mental age increases at the same rate as the chronological age. But in the case of a bright child who is far above the average the M.A. increases faster than his own chronological age and the chronological age of normal children. In the case of a dull child who is far below the average the M.A. increases slower than his chronological age. For example two children have the same mental age, nine, one is 6 years old and the other is 9 years old. This means that the 6 years old child is brighter than 9 years old child. This brightness of the children can be expressed quantitatively. At this stage we come across the intelligence quotient (I.Q.). This measure was first recommended in 1912 by a German psychologist Louis Stern (1871-1938) to divide the mental age by chronological age (M.A. by the C.A.). This is the brightness ratio, which later on developed into the Intelligence Quotient (I.Q.). This has numerical value obtained by multiplying the brightness ratio. The formula is :

$$\text{I.Q.} = \frac{\text{Mental Age (M.A.)} \times 100}{\text{Chronological Age (C.A.)}}$$

Let us apply this formula to the two children whose mental age was the same 9 and the chronological age was 6 and 9.

The I.Q. of the first child is : $\frac{9}{6} \times 100 = 150$

The I.Q. of the second child is : $\frac{9}{9} \times 100 = 100$

Let us assume that a 10 years old child also has the same mental age. Then the I.Q. of the 10 years old child is $\frac{9}{10} \times 100 = 90$

The first child is extremely bright, second child is average and the third child mentally retarded. The I.Q. indicated the rate of intellectual growth of the children. If the child performed the tests and obtained the M.A. to the level of his C.A. then he is an average child.

When the Binet-Simon scale was found to be measuring intelligence it got the attention of the American psychologists and in 1916 Terman (1877-1956) brought out a revision of Binet-Simon scale for administering it upon American children. This revision has become a model test for many intelligence tests. This was called Stanford-Binet Intelligence Scale. This scale was again revised and restandardised in 1957 by Terman and Merrill and this was most widely used and respected individual test in the world. Merrill again revised the 1937 revision in 1960 but it did not equal the 1937 revision.

The Wechsler Intelligence Scales

The Stanford-Binet Intelligence Scale did not fulfil certain requirements in United States. It did not help to measure the adult intelligence. There was a growing need for an adequate scale and David Wechsler, an American psychologist, developed several intelligence tests and one test specifically to measure adult intelligence in 1939 and revised it in 1955. The best known of the tests are the Wechsler Adult Intelligence Scale (WAIS), and the Wechsler Intelligence Scale for Children the (WISC). Wechsler's original plan was to devise a test to measure the intelligence of the patients at Bellevue Hospital in New York. The patients were mostly adults, from working class and were poorly educated. Since they were not progressing well in verbal tests Wechsler added more non-verbal items along with verbal items in his test. The WAIS consists of eleven sub-tests, six of them were verbal tests and five of them were performance (non-verbal) tests. This scale yielded three I.Q. scores, such as, verbal, I.Q. performance I.Q. and composite I.Q. (both verbal and non-verbal). He believed that the verbal scores should be supplemented by performance scores.

Group Tests of Intelligence

The tests discussed in the preceding sections are individual intelligence tests. These tests are administered on a single person. These tests need trained administers to score and interpret the scores and also very expensive and time consuming. But the tests that can be administered to a large group of individuals at a time are called group tests of intelligence. Since during the World War I thousands of men had to be tested for military service, there was an urgent necessity to develop group tests of intelligence. A committee of psychologists had developed two group tests of intelligence tests called the Army Alpha Test and the Army Beta Test. The Alpha test was a verbal test and the Beta test a performance and non-verbal group test for those who had language difficulty and not educated. The Army General Classification Test (AGCT) was developed by the American psychologists during the World War II for taking a large number of individuals. This was better than the Army Alpha Test. Following this several group tests of intelligence were developed to meet the needs of educational, industrial and counselling purposes. The Otis-Lennon Mental Ability Test and the Scholastic Aptitude Test were few of them.

Verbal and Performance (Non-Verbal) Tests

Most intelligence tests are verbal tests and they require the skills in using language. The verbal tests are designed to measure the general intelligence or some special abilities.

All the verbal tests are paper-pencil tests as they require reading and writing. The verbal tests are used to test the ability to deal with ideas by the use of words. Some of the important verbal tests are the Thorndike CVAD test. It has four sub-tests such as Comprehension (C), Arithmetic (A), Vocabulary (V) and following directions (D). The items are very carefully graded in all four areas. The items are so hard that everyone who takes up this test will have different level in scoring. The test mainly measures the individual differences in intelligence. The Army General Classification Test is another important verbal test. This bring out three primary functions of the individual. They are (1) Vocabulary, (2) Arithmetic reasoning, and (3) Block counting. Another verbal test widely used is the American Council of Education (ACE) test. This test emphasises the necessary skills dealing with words and mathematics.

The performance of non-verbal tests were developed for those who had language difficulty and those who could not take up verbal tests. These tests do not involve words or language. Culturally deprived groups, child of foreign language speaking parents, unschooled persons and those who do not fit for verbal tests would take up performance or non-verbal tests. A number of intelligence tests have been developed to reduce the use of verbal skills. The most widely used performance test of intelligence is the Draw-A-Man Test. This was developed by Goodenough and revised by Harris in 1963. The idea is that the ability to draw human figure in correct proportion with all details goes with age and intelligence. This test can be administered upon children up to the age of 15 years. Another important performance test is the Arther Adaptation of Leiter International Performance Scale. It is administered on children aged 3 to 8 years. There are blocks and they are randomly presented to the children and they will have to match them with the stimuli. This is widely used on deaf children and those who have language difficulty. The other performance tests are the Pintner-Patterson Performance Scale, the Healy Picture Completion Test, the Cornell-Coxe Performance Ability Scale, the Koh's Block Design Test, the Porteus Maze Test, Alexander Pass-along Test, Cube Construction Test, Block Design Test, Form Board Test and Matrices Test. All the performance tests are also called non-verbal tests. These tests can also be administered on educated children and on adults.

APPLICATION AND USES OF INTELLIGENCE TESTS

People who are educated and well informed in the common sense world know what intelligence is to some extent and they may also be appreciating a person who has acted intelligently in a situation. But they do not know that intelligence is measurable and stated in terms of numerical values. They may also be unaware or remain indifferent about the mental abilities and skills that a job or an occupation is required for efficient performance. The reason is that intelligence is regarded only an academic matter studied as a subject in colleges and universities in a course of study. They do not take into cognizance the inexplicable role that measurement of intelligence and intelligence tests are playing in various life and organizational situations. This is a very costly error that the people of our nation are committing when the services of a scientific technique is available for efficient handling of problems and issues.

There is a history behind the origin and development of intelligence tests and their uses. Alfred Binet has designed and developed intelligence tests only to solve the learning problems of the school children in the city of Paris on the request of the educational authorities. These tests helped the school authorities to identify and separate the dull children from the bright and to provide separate coaching system for the children who could not progress with normal and bright children. When this was published it has become highly sensational information and schools all over the world made use of these tests for improving the learning abilities of school children. Beginning with schools the benefits of the intelligence tests were extended to various areas.

Thus intelligence test has become one of the most useful tools available for a teacher, an educationalist, a counsellor, business and industrial organizations and government departments. Teachers and educationalists and others who are engaged in grading pupil, counselling clients, selecting and recruiting persons to various jobs should have some knowledge in the administration of intelligence tests, and in the interpretation of scores obtained from the tests. In using intelligence tests there is no trade secret. The procedures and keys are there with the tests for their use.

Diagnosis and Guidance

Intelligence tests are very widely used in most of the school systems. It serves as an aid in diagnosing the difficulties and problems of the pupil in schools. The most important purposes for which they are employed are counselling, career guidance, class placement, diagnosing for remedies and for enrichment.

A good teacher is interested in finding out whether the particular pupil is working to his potentialities. If it is not what is the real cause behind his low achievement. This particular pupil needs some help. The pupil's real aptitude and the level of his potentialities should be determined. Therefore, for determining his actual scholastic aptitude an intelligence test must be administered on him. The scores would reveal where he stands in comparison with other pupil. Then the teacher may find out the fact whether the pupil's low achievement is due to lack of capacity or something else or problems relating to his growth or rearing up.

Another important purpose for which intelligence test may be used is the educational and vocational guidance. The student may require guidance in various areas. The results of the scholastic aptitude test is very much useful in helping the student to select the correct course of study at high school and college levels. Guidance can be also provided to the students whether they can proceed further on their studies at college or take up a vocational course or a vocation after the completion of the high school studies. If the decision is to take up a vocation what type of vocation he should choose also may be decided.

Further, if the intelligence test is administered at the beginning of the school years and if the teacher knows the intelligence test scores of a group of students he can make use of this data along with other informations for several purposes. He can use them in grading and guiding the students, for assigning the standard of textbooks and other learning materials, type of exercises and home works. In addition to these he can set the goals and the expectations of the students. The planning of curriculum and grouping also can be made on the basis of the I.Qs. of the students.

Intelligence tests may measure apparently what the teacher wants. The test scores may be useful for him in predicting certain characteristics. For example, the student may be brighter in arithmetic, but he is lacking verbal skills. But verbal competence is highly related to achievements in academic subjects. If the teacher knows this in advance he can provide certain special care for the student to develop the verbal skills.

But there are various views expressed about the decisions taken on the basis of intelligence quotient. Lefrancois (1979) says that in predicting the academic success measured I.Q. is not on the same line as we generally assumed to be. According to Cohen (1972) the early success and later success correlate highly. But at the same time the measured early I.Q. and later I.Q. did not correlate to that level. Anderson (1939) states that intelligence is not a fixed and unchanging characteristic. In his study it was observed that the correlation between the test scores in the first two years of life and at the age of five ranged from (0) zero to 0.45. He observed that with increasing age the measured I.Q. found to be increasing steadily. Bloom (1964) also ascertained this from his studies. Therefore, Bloom states that the teacher should make a continual revival of the I.Qs. of the students from time to time and then make his decisions.

Placement and Programming

One of the important uses of intelligence tests has been the determination of placement of the students in schools and colleges. Which student should be admitted to regular liberal education and which student should be placed in the special course? In the United States many schools and colleges have their own standard of I.Qs. and mental age levels as a limit for admission to their institutions. Students from India or other foreign countries should take up an entrance examination or should pass a course of study prescribed by them. Intelligence test in this respect is of great help to the educational institutions in admitting students in various courses. Colleges and universities may find difficult time with programmes to executive if some of the students in every course of study fall below the average level. These institutions can progress well with their programmes if admissions are made strictly on the basis of merit and on the basis of the scores of some intelligence tests. Such institutions maintain certain prestige and status and the student also feels proud of his college or university. Some business or industrial organizations also prefer candidates from certain educational institutions. They believe that candidates from such institutions may be more efficient and honest and sincere in their performance than the candidates from other institutions which do not maintain any standard.

Further, certain intelligence tests are specially used for diagnostic purposes. These tests may help us to find out not only the strong and weak points of the individual in academic activities but also in his over all performances. Experts in testing programmes ascertain that people who are intelligent are also more responsible and dutiful in their assigned works and also maintain personal integrity at all levels. Intelligence tests may be used to measure one's present mental ability levels. People may be required in certain situations to learn many new skills for promoting efficiency. The measurement of mental abilities and the learning skills of the individuals may be of immense help to many business and industrial organizations who are selecting men to various trades and occupations. Selecting men to various jobs and placing them at different levels can be done easily and successfully on the basis of the scores of the intelligence tests.

The disadvantaged persons and persons who are lacking in certain specific skills may be traced out and proper care may be paid on them to develop their skills. Many of the business and industrial organizations in India select their employees at random and also by some conventional methods. They also spend enormous amount of money for providing training to these people before placing them in the respective places. But after a few years these organizations find it very difficult to run the organizations with these people. They not only lack efficiency but also not prompt in their works. They always find fault with the organizations and form associations and fight against the organizations. They always prefer less work, more salary, rapid promotion, more holidays, more fringe benefits, representations in the organizational administration and also shares in the profits. They do not demand all these things at the time of recruitment and till their jobs are insecure. Once they get stabilized in the organization they start the trouble. A systematic psychological testing programmes at the time of recruitment may solve such problems to some greater extent. There are few organizations in India which are free from employees strikes and shut downs because they are recruiting and placing employees according strict testing programmes. It is also important that a periodical revival of every employee's performance and his progress on the job should be made. The employee should show better progress in his work every year by inventing something new or finding out some flaw in the work or machine and report it to the management. This should be one of the conditions in the recruitment programmes.

EXPERIMENT 25 : MEASUREMENT OF INTELLIGENCE MODIFIED ALPHA EXAMINATION

INTRODUCTION

In psychological studies, generally the concepts are defined operationally. Such definitions are often applied to intelligence, motivation, learning and thinking. But the meaning of them often provoking disagreement. When an experimenter discusses his results obtained from different groups, he finds that the groups differ in intelligence. How do groups differ in intelligence ? This is the question raised by many scholars. The reply is that this is what the Simon-Binet test measures. Therefore, intelligence is defined in terms of what the Simon-Binet test measures.

The intelligence has been measured since social groups first came into existence in man's history. During the primitive period considerable emphasis had been placed on physical vigour and superiority. Those social groups, however, which chose their leaders only on the basis of physical strength, probably did not long endure. If the group was to survive foresight and planning were necessary. It had to know the proper time to move to new places. When war was necessary and likely to be successful, how to divide labour and utilize abilities for the maximum gain of the entire groups and so on. These are the kinds of decisions extending beyond physical strength and the groups with leadership capable of accurately anticipating the crucial situations, probably become dominant over those depending more completely on brutal strength for survival. As this type of leadership asserted itself and as customs and tradition allowed for more better selection of leaders, intellectual ability became more important and the assessment of such ability became critical to the successful function of the society. According to Woodworth (1952) the contribution made by Gall, a scientific anatomist who propounded a famous theory phrenology in the beginning of nineteenth century was a good example for all these though it did not receive scientific support.

In more scientific vein, Lombroso, an Italian anthropologist, believed that it is possible to recognize the genius from the physical signs. Attempts to measure intelligence as simply and precisely as possible persisted for some time after scientific study of human behaviour began. The tests designed in the beginning were almost of sensory and motor tasks, mainly based upon the works of Galton (1883). He had noted that persons in institutions who had been classified as in demands were relatively incentive to heat, cold and pain. They reacted slowly to stimuli of this sort. He reasoned, therefore, that persons with higher intelligence would react more quickly to such stimuli. Logically he then deducted that more intelligent persons would be superior in all kinds of sensory traits. He devised tests of a sensory and motor nature and began to administer them to as many people as possible. An anthropometric laboratory was established by him in the Kenisington Gardens, London, where for a small fee an individual goes out for a stroll in the Gardens could take up these tests. After devising means to measure pressure discrimination, weight discrimination, pitch discrimination, reaction time, visual and auditory acuity and so on, he collected a tremendous body of data and utilized statistical techniques from the field of mathematics to analyse them (1908). A person of superior intelligence differs from the subnormal not so much in eye sight or hearing as in superior used of the higher mental processes. Discrimination of weights can be measured with a fair degree of accuracy. Discrimination between two persons on traits such as judgment and reasoning is much more difficult. The measures are likely to be less precise and consequently there are more errors in measurement. At the same time it became apparent with experimentation that in order to measure intelligence one must deal with complex mental processes. The differences between persons in complex mental processes are much greater than in sensory and motor processes. This idea was first proposed by Alfred Binet. Early in his career, however, he became interested in individual differences in intelligence and their measurement. In 1895 Binet and Henri published an article dealing with individual differences criticizing the use of sensory tests in an attempt to evaluate intellectual level, they raised the question of the interrelation of different mental functions in an individual. Although some persons are slow in specific reaction it cannot be concluded that they are in general slower than those with faster reactions. The inference is that the individual high in general intelligence will also tend to be high in various specific intellectual endeavours. Binet and Henri also stated that certain tests could be selected which would have utility for practical purposes and be superior to sensory tasks for measurement and prediction. In order to discriminate between the sub-normal and the normal child, Binet (1916) realized that some kind of exact scale was necessary. He collaborated with Simon to construct such a scale, which was first published in 1905.

The Binet Scales are referred to as individual tests because there is a one to one relationship between examiner and examinee. The test can be given to only one person at a time and an administration may take about an hour. Since these things are true and since a trained skilled test administrator is necessary intelligence tests of this kind were not immediately accepted and utilised widely in school settings. The information on mental level could be helpful for school-programme planning but its use seemed impracticable from other points — one of the impracticabilities lays in the fact that the test was standardized on a French sample of school children. To overcome this limitation, Terman standardized the Binet Scales on a sample of United States school children in 1916. A third revision was made by Terman and Merrill (1960).

In the year 1917 the United States entered into the World War I and the armed forces were faced with the task of inducting millions of men and assigning them to the kinds of jobs which they could successfully and efficiently accomplish. A measure of intellectual ability would be very much helpful in this process particularly since the knowledge of intellectual ability was limited, but the use of an individual test was not practical. The armed forces, therefore, requested a group of psychologists, who were already active in intelligence testing, to devise a test which could be administered to a large group of people in a reasonable period of time and yield a general intelligence score of some validity. Otis (1954) was then in the process of developing such a test. His ideas were utilized and culminated in the Army Alpha and Army Beta tests. Each of these is a group test, the Alpha for literates and the Beta for illiterates. As the tests were used and research concerning their utility was made available, educationlists became very much interested in it. It was possible now to test groups in limited time period, the scoring was simple using the keys, the administration of the test and the interpretation of the score did not call for specialized training and the results provided the level of intellectual ability and by inference the learning ability. Such a tool would be of considerable use for schools. Ideally a child entering a school could be tested and then classified on the basis of the ability and the consequent progress in the school. There are a quite number of intelligence tests available. They are both individual and group tests. Some of them are Modified Alpha Examination Form 9, Alexander's tests, Raven's (1960) Progressive Matrices Test and so on. Modified Alpha examination Form 9 is one of the important tests and used in many institutions.

EXPERIMENT

Problem

The purpose of the experiment is to study how well the subject can think, remember and follow directions.

Method

Participant

Choose a person who has no knowledge about this test. Collect his/her personal information such as age, sex, socio-economic status, deformities, if any etc.

Materials

(1) Modified Alpha Examination Form 9, Test booklet, (2) Manual, (3) Ready made answer sheet, (4) Stop clock, and (5) Extra sheet for figure work.

Description of the Test

The Modified Alpha Examination Form 9 consists of eight sub-tests. Each one measures one aspect of mental ability.

Test-A : Test-A consists of a column of additions containing 24 items. The time required is one and a half minutes.

Test-B : It is a test to test the ability for following directions. It contains 20 items. The time required is 4 minutes.

Test-C : This test consists of arithmetical problems. There are 20 items in it. The time required is 5 minutes.

Test-D : This is the test for analogies. It consists of 40 items and the time required is 3 minutes.

Test-E : This is a test for the completion of number series. There are 20 items in it. The time required is 3 minutes.

Test-F : It consists of disarranged sentences. They have to be identified either as true or false. There are 24 items in it. The time required is 2 minutes.

Test-G : It is a test for finding out a largest common devices. There are 32 items in this test. The time required is 4 minutes.

Test-H : It is a test of synonym and antonym. There are 40 items in this test. The time required is half minute.

Procedure

The participant should be seated comfortably in a well illuminated room. The test booklet, the answer sheet and extra sheet should be distributed to the participant. Then the following instruction should be given to the participant.

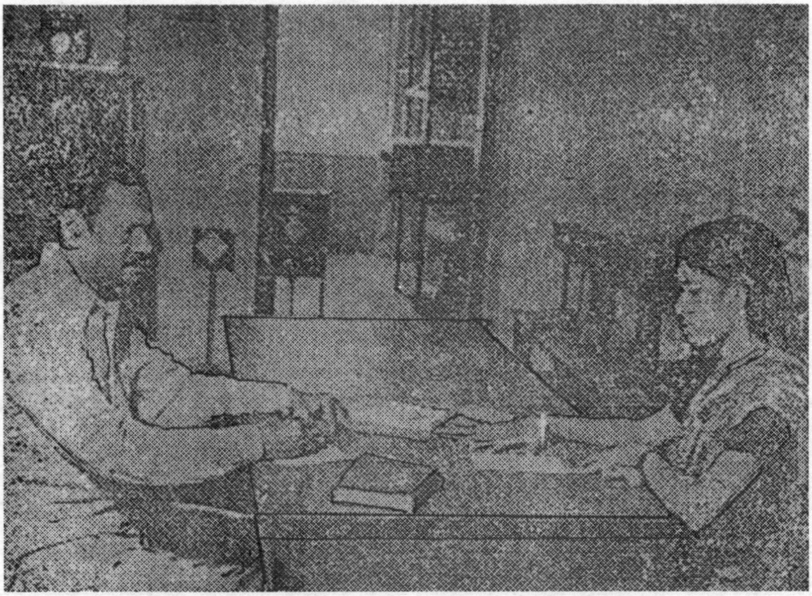

Fig. 9.1 : Modified Alpha Examination (Experiment 25)

"The purpose of this examination is to see how well you remember, think and carry out what you are told to do. Some of the things you will be asked to do may be very easy and some may be very difficult. You are not expected to give correct answer to every test nor to finish all the questions. Simply do the best you can in the time allowed. You will be told when to start and when to stop."

Specific directions for each test are printed on the left hand side of each test. They should be read aloud only once to the subject and no further assistance should be given. The following directions were given to the subject at the time of administering the test.

"You will have to write the answers on the answer sheet which you have put below the test booklet. Take care to see that the answers are written in the upper right hand corner in the answer sheet and the test sheet and the answer sheet meet correctly. Do your figuring work in the extra sheet of the paper provided to you. Make no marking on the test booklet. When I say stop, stop at once and no more work in the answer sheet."

As soon as the participant finished the first test, he should be asked to go to the next test and followed the same directions as in the first test. After the completion of all the eight tests the responses of the participant should be scored according to the key. The scores of all the eight tests should be tabulated which forms the raw scores of the individual. The scores may be classified into two groups on the basis of the test namely, numerical and verbal. Further the total scores of the numerical and verbal tests also may be calculated. The percentage scores may be noted separately for numerical and verbal tests from the manual on page 6 for the seniors. Then the group data may be collected and tabulated.

Results

The raw scores of the individual on the modified alpha examination Form 9 should be given for all the tests in Table 9.1. The percentage scores may be found out from the manual and entered in the respective cells. The group data may be given in Table 9.2 with the percentage scores. The numerical scores and the verbal scores should be worked separately and shown in Table 9.3 for the individual and for the group. Work out the mean score for each test and also for the group for all the tests.

A graph may be drawn for the numerical test scores and the verbal test scores for the individual.

Table 9.1 : Showing the individual score in the Modified Alpha Examination Form-9

S. No.	Test	Raw Score	Percentile Score
1	Test - A		
2	Test - B		
3	Test - C		
4	Test - D		
5	Test - E		
6	Test - F		
7	Test - G		
8	Test - H		

Table 9.2 : Showing the group data in the Modified Alpha Examination Form 9

S. No.	Name	Tests																Mean R. Score
		A		B		C		D		E		F		G		H		
		RS	PS	RS	PS	RS	PS	RS	PS	RS	PS	RS	PS	RS	PS	RS	PS	
1	P.V.																	
2	K.R.																	
3	M.S.																	
4	B.V.																	
5	R.S.																	
6	M.R.																	
7	P.K.																	
8	P.C.																	
9	R.A.																	
10	A.L.																	
11	S.D.																	
12	V.R.																	
	Mean																	

Table 9.3 : Showing the numerical and verbal scores for the individual and for the group

S. No.	Name	Numerical Test Scores				Total for N.T.	Verbal Test Scores				Total for V.T.	Grand Total
		A	C	E	G		B	D	F	H		
1	P.V.											
2	K.R.											
3	M.S.											
4	B.V.											
5	R.S.											
6	M.R.											
7	P.K.											
8	P.C.											
9	R.A.											
10	A.L.											
11	S.D.											
12	V.R.											
Mean												

Observing the data in Table 9.1 state in which test your participant has scored the highest and which test he/she has scored the lowest. What is the mean score for all the tests and what are the tests in which the participant has scored more than the mean and what are tests he/she scored less than the mean. Has he / she omitted any of the items in any of the test? State it.

Discussion

Discuss the individual's raw score in every test and state the percentage for the same. State the test in which he scored the highest and the lowest. In classifying the test scores into numerical and verbal do you see him proficient in numerical or verbal. Compare your participant's performance in relation to the group mean. In the group who scores the highest in numerical test and in verbal test. Is the same person who scores the highest in both the areas? Also state whether any of the personal factors like age, sex, socio-economic status, education of the parents has any influence in the test scores. Compare your test results with that of the test scores of earlier study. State whether there is any group difference in the test scores. Discuss the test scores on the basis of various cultural factors of the group.

Conclusions

1. In which test your participant is proficient?
2. What is his position in the group?
3. Who scored highest in numerical and verbal tests? State his backgrounds.
4. Find out and state whether the whole group was found to be proficient in Numerical test or in Verbal test. State the reason if any.

EXPERIMENT 26 : MEASUREMENT OF INTELLIGENCE

ALEXANDER SCALE : PASSALONG TEST

The concept, intelligence is generally referring to much broader practical value. Psychologists after working on it for several years distinguished its function in three areas. Each one may be said as intelligent activity. Particularly, the areas are abstract, mechanical and social. Most of the testings were made on abstract intelligence. The theorists in every field, for example, scientists and professional people may be said as abstract intelligent persons.

The mechanical intelligence is successfully applied in dealing with machines and mechanical comprehension. For example, the engineer, a skilful industrial worker, a well trained mechanic are said to be mechanically intelligent persons.

The social intelligence refer to the skilful dealings with men and policy matters. For example, a successful political leader who takes the people of the nation on his policy, a social scientist who deals with a social problem, for example, the family planning and birth control successfully, a diplomatic who can raise his nation's prestige in the international affairs may be said as socially intelligent persons.

Psychologists have been trying to identify the specific intellectual skill in persons and classify the types of intelligence. Though it is possible to make various classifications on the basis of matter dealt with they can be discussed under two broad categories as abstract intelligence and concrete intelligence. Why this type of classification is made? It is because abstractions do not always go with successful performances. Garrett (1968) says that an intelligent scientist who is dealing everything abstractly may not be a good skilful mechanic. Therefore, psychologists felt that there should be separate intelligence tests to test these various aspects as it has become important to identify men for various types of skilled jobs and practical works.

The Alexander scale is one of the intelligence tests designed to assess the intelligence and practical ability of person. This scale consists of three tests, each one measuring one aspect of intelligence of practical in nature. The first one is the Passalong test which measures the intelligence quotient, and the other tests, Koh's Block Design Test and Cube construction test are intended to assess the practical ability of an individual.

But, it can be observed that most of the performance tests are intended to assess the practical ability. They all non-verbal and involves manipulation and performance unlike paper-pencil tests. Since intelligence tests are required for various purposes and for different groups they were revised time to time, adding many things now and dropping that are not to the time. In this way the Simon-Binet test has underwent several revisions. The revisions made by Terman and Merrill (1937) in Stanford University served the greatest usefulness for different age groups.

On observing that these revised versions did not serve the purpose on adults, Wechsler (1958) developed an intelligence test for adults. This is called Wechsler Adult Intelligence Scale. This scale has two points, one as verbal tests and the other as performance tests. In each section there were five tests. Apart from other types of tests, there was also a block design test with sixteen small cubes painted with different colours on each face of the cube. There are seven design cards in this test. Wechsler has designed another intelligence scale for children on the same pattern as his Adult Intelligence Scale.

The Pintner-Peterson (1927) performance tests were one of the series of tests designed earlier. There are fifteen sub-tests each one designed to assess one separate task. There are seven tests of Form Board types and blocks are to be fitted to make the Form complete. All such tests are intended to assess the concrete intelligence or the practical ability.

ASSESSING INTELLIGENCE

Alexander Passalong Test

Psychologists and educationalists have been measuring intelligence in various ways for the past hundred years. Intelligence tests have been designed and developed time to time making improvements in the old tests. But it will not be possible to say that the particular test alone measures intelligence accurately and others are not. Accuracy in measurement is only relative. Test results are compared and the accuracy or the validity of a test is determined. Intelligence tests may be either performance tests or paper-pencil tests. The purpose of these two types are different. The performance tests are non-verbal whereas the paper-pencil tests are verbal tests. Which type is better? We cannot determine this. Both types have certain advantages and disadvantages. Therefore, both are good.

One main difference between the performance tests and the paper-pencil tests is that performance tests are individually administered and paper-pencil tests are administered to group of people. In performance test the individual personally manipulates the instrument or the test material. One intelligence test that is widely used in psychological and educational testing is the Alexander Passalong test. This test was developed by Alexander (1932) for measuring the intelligence of school students. Later on Alexander enlarged this test and added two more tests, Koh's Block Design and Cube Construction and called the test as performance scale. The 1958 revision of this scale is designated as Diagnostic Performance Tests.

EXPERIMENT

Problem

To assess the intelligence of a person by means of a performance using the Alaxander Passalong test.

Method

Participant

Choose a person who has no knowledge about this experiment. Collect the personal information such as age, sex, qualification, occupation, if any, experience, income, etc.

Materials

(1) Alexander's Passalong test with complete set, (2) Stop clock, and (3) Recording sheets.

Description of the Materials

Alexander Passalong Test : The passalong test consists on nine sub-test. There are four trays **with painted** ends. One end of the tray is red and the other end blue. There are thirteen **blocks.** Eight blocks are of one inch square, three with 2´´ x 1´´ rectangular and two 2´´ x 2´´ **square** blocks. Some are painted with red and some with blue. There are eight diagrams.

Procedures

The experimenter should keep the participant seated comfortably in front of a small table and he can establish proper rapport with him/her. There should not be any disturbance in the room. It is better that there should be only the examiner and the participant. The table and the seating arrangement for the participant should be so comfortable that he/she can move his/her arm and rest arms comfortably when he/she is working.

Fig. 9.2 : Measurement of Intelligence : Alexander Passalong test (Experiment 26)

While administering the test the experimenter should see that he/she keeps the test materials little away from the participant's sight and on any account the examiner should not keep two diagrams in a visible condition to the participant. Secondly, the participant should not see the preparation of the test, using the trays and colour blocks. When the participant is ready the experimenter should place the smallest tray No. 1 having two one inch square blocks in it and the design card No. 1. While placing the tray and the design card in front of the participant, the experimenter should see that always the red end of the tray is away from the participant. That is the blue end must be on the participant's side.

The experimenter can demonstrate once if the participant is not able to understand the verbal instructions. Then the experimenter should give the following instructions to the participant :

> "Look at the tray. It is painted. One end is red and the other end is blue. There are two coloured blocks, the red block is placed at the blue end and the blue block is placed at the red end. Now, your task is to move the blocks within the tray in such a way that the blue block should go to blue end and the red block should go to the red end as you see in the design (pointing out the design). You should only move the block and should not tilt or lift it. Do it as quickly as possible. Now, go ahead."

As soon as the participant started his task the experimenter should start the stop clock to run and count the time in seconds. The test is completed when the red block is at the red end and the blue block is at the blue end as shown in diagram. As soon as the first test is completed the experimenter can go to the second test and place the diagram and the tray with the blocks in front of the participant. He should simply say, "Now move the red blocks to the red end and blue blocks to the blue end as you see in the diagram". When the participant started his/her task the experimenter should start the stop clock and count the time in seconds.

On completion of the test 2, he can proceed to the test 3. This test is little different in setting up the blocks. The same tray as used for test 2, is used for the test 3 also. The experimenter should place two small blue blocks at the red end and two small red blocks at the blue end and one blue block in between the red and the blue blocks. But the final position should be the same as test 2. Each test should be performed in the same manner according to the specific prescription of each test. The experimenter should carefully note down the time in seconds. When the participant has failed to complete the task the experimenter can replace the blocks in the starting position and tell the participant to watch it carefully. The experimenter can demonstrate and he/she can move the red block towards the right side of the participant. There should not be any discussion. Simply should say 'go ahead'. If the participant completes the test in the second trial he can go on to the next test. If the participant has failed in two successive tests stop the experiment.

Table 9.4 : Showing the time limit for the tests

S. No.	Sub-Tests	Time in minutes
1	Sub-test 1	2 minutes (maximum)
2	Sub-test 2	2 minutes (maximum)
3	Sub-test 3	3 minutes (maximum)
4	Sub-test 4	3 minutes (maximum)
5	Sub-test 5	3 minutes (maximum)
6	Sub-test 6	3 minutes (maximum)
7	Sub-test 7	3 minutes (maximum)
8	Sub-test 8	4 minutes (maximum)
9	Sub-test 9	5 minutes (maximum)

This test should be completed within the specified time. The participant may lose his score if he fails to complete the test or take more time than the time assigned (See Table 9.4).

Where a participant has not successfully completed any sub-test at the end of these specified timings the experimenter should stop his task and make an entry in his record sheet as 'F' and start the next test.

Method of Scoring of the Sub-tests

The sub-tests are scored on the basis of the amount of time the participant has taken to successfully complete the test. There is deduction of scores for failure and for taking more time than the limit (See Table 9.5).

Table 9.5 : Showing the scores for each test successfully completed

S. No.	Sub-test	Time limit minutes	Full Score	Deductions	
1	Sub-test 1	2 minutes	2	No deduction if completed in one minute.	
				For completing in	
				2 minutes deduct	1
				more than 2 minutes	0
2	Sub-test 2	2 minutes	3	One minute : No deduction	
				1 to 2 minutes	1
				Above 2 minutes	0
3	Sub-test 3	3 minutes	5	One minute : No deduction	
				1 to 2 minutes	1
				2 to 3 minutes	2
				Above 3 minutes	0
4	Sub-test 4	3 minutes	5	One minute : No deduction	
				1 to 2 minutes	1
				2 to 3 minutes	2
				Above 3 minutes	0
5	Sub-test 5	3 minutes	5	One minute : No deduction	
				1 to 2 minutes	1
				2 to 3 minutes	2
				Above 3 minutes	0
6	Sub-test 6	3 minutes	5	One minute : No deduction	
				1 to 2 minutes	1
				2 to 3 minutes	2
				Above 3 minutes	0
7	Sub-test 7	3 minutes	5	One minute : No deduction	
				1 to 2 minutes	1
				2 to 3 minutes	2
				Above 3 minutes	0
8	Sub-test 8	4 minutes	7	1 ½ minutes : No deduction	
				Above 1 ½ minutes	
				for every 30 seconds	1
				2 ½ minutes	3
				3 minutes	4
				3 ½ minutes	5
				4 minutes	6
				Above 4 minutes	0
9	Sub-test 9	5 minutes	8	For 2 minutes : No deduction	
				2 ½ minutes	1
				3 minutes	2
				3 ½ minutes	3
				4 minutes	4
				4 ½ minutes	5
				5 minutes	6
				Above 5 minutes	0

The total possible score is 45.

A better and an easy method of scoring is also suggested by the author of the test. The experimenter may find it very comfortable to find out the individual score (See Table 9.6).

Table 9.6 : Showing the scoring procedure for Alexander's Passalong Test

Sub-Tests	Time taken in seconds and scores									
	0-30	31-60	61-90	91-120	121-150	151-180	181-210	211-240	241-270	271-300
1	2	2	1	1	0	0	0	0	0	0
2	3	3	2	1	0	0	0	0	0	0
3	5	5	4	3	2	1	0	0	0	0
4	5	5	4	3	2	1	0	0	0	0
5	5	5	4	3	2	1	0	0	0	0
6	5	5	4	3	2	1	0	0	0	0
7	5	5	4	3	2	1	0	0	0	0
8	7	7	7	6	5	4	3	2	0	0
9	8	8	8	8	7	6	5	4	3	2
Total Score	45	45	38	31	22	15	8	6	3	2

If all the nine tests are completed within one minute the individual score is 45 which is the highest possible in this test. If completed in 90 seconds the total score will be 38, and if it is within 2 minutes the total score is 31. But the scores may vary after every 30 seconds. The experimenter can assign the individual score for each test having this formula in scoring the tests

Results

The experimenter after scoring all the 9 tests of the individual should present the scores in Table 9.7. Then he/she can calculate the I.Q. using the formula $\frac{MA}{CA} \times 100 = I.Q.$ While calculating the I.Q. it should be presumed that persons above 16 years of age may be treated as 16. For example an adult has scored 38 in the test and his mental age is 17 years and ten months. Then his I.Q. is $\frac{204}{192} \times 100 = 106.3$

Find out the individual's I.Q. in the same manner and present it in Table. 9.7.

In calculating the I.Q. the total score of all the 9 tests should be taken and considered. The experimenter should keep in mind that all those who are above 16 years of age should be presumed that they are only 16 years old for the purpose of calculating the I.Q.

Table 9.7 : Showing the scores of the individual in the Alexander_Passalong tests

Test No.	Time in Seconds	Score	Remarks
1			
2			
3			
4			
5			
6			
7			
8			
9			
Total			

Then the experimenter should collect the group data from other experimenters and present them in Table 9.8.

Table 9.8 : Showing the scores and I.Qs. of the group in the Alexander Passalong test of intelligence (Maximum Score 45)

S. No.	Name	Total Score	I.Q.	Grade or Position
1	S.R			
2	K.S			
3	R.M.			
4	P.V.			
5	M.S.			
6	B.V.			
7	T.V.			
8	P.K.			
9	G.V.			
10	A.L.			
11	S.M.			
12	M.R			
	Mean			

Work out the mean score and mean I.Q. of the group.

Discussion

State what is the score and I.Q. of your participant. Where your participant finds it difficult to complete the test. Has he/she taken more than one trial in any test? What are the tests in which he/she has taken more than the limited time? How many tests he/she has completed below the time limit? Has he/she received 0 score in any of the tests?

Where your participant stands in the group? Has he/she scored more than the group mean or less than the group mean? What is the mean I.Q. of the group? How many in the group obtained the I.Q. above the group mean and how many less than the group mean? Whose I.Q. is the highest? Whose I.Q. is the lowest? State the reason, if any, for this. Is there any difference in the I.Q. between the sex, age and other factors.

Conclusions

1. What is the I.Q. of your participant?
2. What is the mean I.Q. of the group?
3. What is the highest I.Q. in this testing and who has obtained this?
4. What is the lowest I.Q. in this testing and who has obtained this?

EXPERIMENT 27 : ASSESSING INTELLIGENCE — PRACTICAL ABILITY (Alexander Scale : Koh's Block Design Test)

Problem

To assess the practical ability of a person by administering the Koh's Block Design Test.

Method

Participant

Choose a person who has no knowledge about this test. Collect his/her personal information such as age, sex, qualification, occupation, experience and so on.

Materials

(1) Koh's Block Design Test complete set with designs, (2) Stop clock, and (3) Scoring sheet.

Description of the Test

The Koh's Block Design Test is one of the tests in the Alexander's Scale. This test consists of ten sub-tests with ten design cards. This test has sixteen one inch (1′′ x 1′′ x 1′′) cubes. All the sixteen blocks are carefully painted with different colours. There are six sides in a cube. In all the cubes one side is painted with *red* colour, the other side with *blue,* the third side with *yellow* and the fourth side with *white* colour. Then the remaining two sides are painted respectively with *blue and yellow* and *red and white diagonally.* That is the *top side red and white are diagonally painted* and *bottom side blue and yellow are diagonally painted* in all sixteen cubes. Since each block is a one inch cube the top may become bottom and the bottom may become top. The ten designs are designed in progressively in increasing complexity. The simple design starts with four cubes, difficult design with nine cubes, and most difficult design using all the sixteen cubes. The time for block designs varies from one and half minutes to three and half minutes. In Koh's Block Design Test there are 17 sub-tests and design cards.

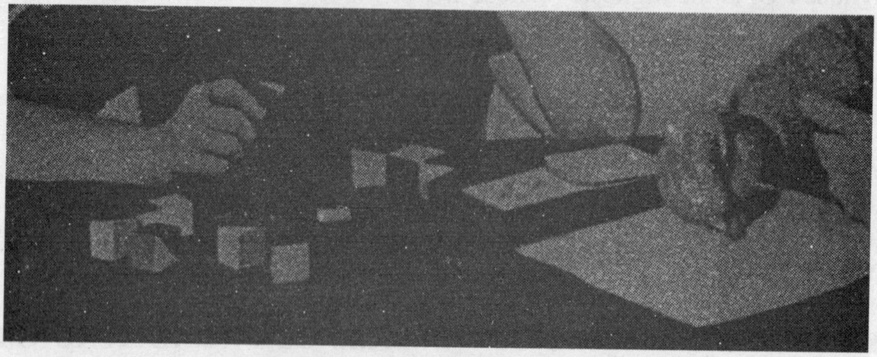

Fig. 9.3(a) : Koh's Block Design (Practical).

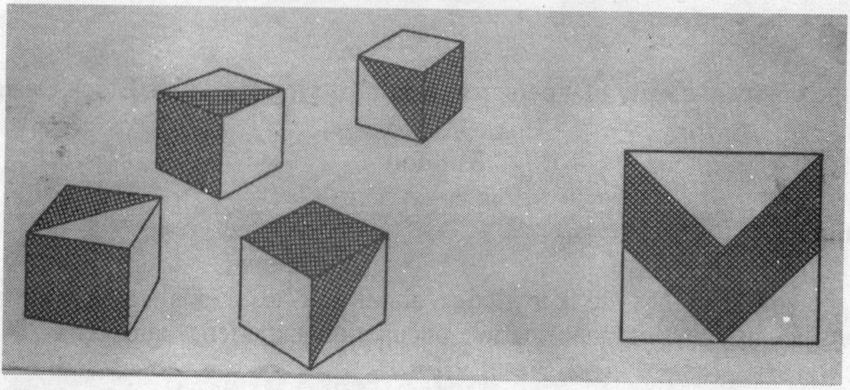

Fig. 9.3(b) : Measurement of Intelligence (Practical) Koh's Block Design Test (Alexander's) : Experiment-27

For the *Alexander Scale only ten* were selected. The designs are 1, 2, 4, 5, 7, 10, 11, 14, 15 and 16.

Procedure

The experimenter should keep the test box little away from the participant's visibility. The participant should be seated comfortably in front of a table in a well illuminated room where there should not be any disturbances visually or auditorily. Then he/she should remove four block (cubes) and place them on the table before the subject. The white side of the cube should be up in every case. Then the experimenter should turn the block showing the colours on the sides, blue, red, yellow and white. He must show to the participant that all blocks invariably are painted with these four colours on four sides and on the top red and white and at the bottom blue and yellow diagonally. Now take the design card No. 1 and place it before the participant. Out of four blocks one should have a white top and all the blocks should be scattered. The experimenter can demonstrate to the participant how to make design from the four blocks by placing them together in correct positions. Before this he should give the following instructions to the participant:

> "You just see, there are four blocks, all painted with four colours. Using these blocks you have to make out a design in the same manner as you see in this card (pointing out the card). You must make out this design as quickly as possible within 90 seconds. The moves of the blocks also should be as few as possible. Now, ready go."

The experimenter at once should start his stop clock and count the time in seconds and make entries in his record sheet. Also the number of moves he/she makes in making the design. If the participant finds it difficult, or if necessary further trials may be permitted. Because only in the first design the participant learns what he/she should do.

When the participant completed the first design successfully the experimenter can place the design No. 2 in front of him/her. The experimenter should shuffle the same four blocks and place them scattered having one white top.

> "Now carry out the task in the same manner and make out the design as you see in this card (showing card No. 2). The maximum time you can take is 90 seconds. But do it as fast as you can."

As the participant is performing the task the experimenter should note down the time he/she takes for completion and the number of moves made. If he/she is not successful he can show him how to make the design and no second trial is permitted. As soon as he/she completed the design No. 2, the experimenter can go ahead to design No. 3. The participant may be told to do the same thing as he/she did for design No. 2. But he/she must be told that time for design No. 3 is 2 minutes. When he/she completed the design No. 3, the experimenter can administer tests 4 and 5 (design 4 and 5) in the same manner as design No. 3. For designs 4 and 5 two minutes are allowed to complete each one.

Then the experimenter can go to design No. 6 and five more blocks should be added for this. This should be shown to the participant that these five blocks are quite similar to the four that he/she has been using for the other designs. Putting all the nine blocks on the table with only two white block up, the experimenter should scatter them. Then he should show the design No. 6 and ask the participant to complete the design within a maximum time of three minutes. When he completed design No. 6, design No. 7 also may be administered in the same manner as design No. 6.

On completion of the design No. 7 the experimenter should add the remaining seven blocks to the 9 blocks on the table and thus making the total blocks sixteen. Then he can present the design No. 8 to the participant and ask him to make out the design as seen in card 8 with these 16 blocks within a maximum amount of time three and half minutes. The experimenter should shuffle all the 16 blocks having only three blocks with white top and should scatter them.

When the participant successfully completed the design No. 8, then the experimenter should go for design No. 9 and design No. 10 in the same manner as he did for design No. 8. The time limit for design 8, 9 and 10 is the same, three and a half minutes.

If the participant has not correctly made the design the experimenter can point out the error and ask him to get it done correctly as in the design card. The experimenter should make it a point that the performing time should be always made in seconds and entered in his record sheet. The number of moves made by the participant should also be entered and they should be accounted while scoring. After completing all the designs the experimenter should get the introspective report from the participant and allow him/her to go away.

Method of Scoring

The scoring procedure varies for each sub-test on the basis of the amount of time and number of moves the participant made in making the design. For every excess of time points are deducted (See Table 9.9).

Table 9.9 : Showing the Scoring Procedure

Design or Sub-test	Time in seconds taken to complete the design and scores													Moves allowed	One Point deducted for more moves
	16-30	31-45	46-60	61-75	76-90	91-105	106-120	121-130	136-150	151-165	166-180	181-195	196-210		
1	2	2	2	2	2	0	0	0	0	0	0	0	0	6	1
2	6	5	4	3	2	0	0	0	0	0	0	0	0	7	1
3	8	7	6	5	4	3	2	0	0	0	0	0	0	10	1
4	8	7	6	5	4	3	2	0	0	0	0	0	0	11	1
5	8	7	6	5	4	3	2	0	0	0	0	0	0	11	1
6	12	12	11	10	9	8	7	6	5	4	3	0	0	22	1
7	12	12	11	10	9	8	7	6	5	4	3	0	0	22	1
8	14	14	14	13	12	11	10	9	8	7	6	5	4	31	1
9	14	14	14	13	12	11	10	9	8	7	6	5	4	31	1
10	14	14	14	13	12	11	10	9	8	7	6	5	4	31	1
Possible total score is : 98															

The scoring procedure for the Koh's block design should be carefully followed. For every 15 seconds one point may be deducted if the participant takes more than the limited time. The number of moves for each sub-test also are specified in the original Koh's block design test. If the moves are more than the limit points are deducted. But in

Alexander's scale only the time limit is specified and not the moves. If the experimenter wishes to follow Alexander's procedure points need not be deducted for moves.

Results

The experimenter then find out the individual's score according to the scoring procedure for all the ten sub-tests and present them in Table 9.10.

Find out the individual score for all the ten sub-tests after deducting the points for excess time and excess moves. Deduct this from the maximum possible score 98. Find out the individuals total score.

Table 9.10 : Showing the individual scores in the Koh's Block Design Test

Design or Sub-test No.	Score value	Maximum time allowed in seconds	Time taken to complete	Deduction	Moves permitted	Deduct 1 point for the above specified moves	Score
1	2	90			6		
2	6	90			7		
3	8	120			10		
4	8	120			11		
5	8	120			11		
6	12	180			22		
7	12	180			22		
8	14	210			31		
9	14	210			31		
10	14	210			31		
Total	98						

Collect the group data from other experimenters and present them in Table. 9.11.

Table 9.11 : Showing the group data in Koh's Block Design Test

S. No.	Name	Time taken	Moves made	Deduction of points	Score
1	P.K.				
2	M.S.				
3	R.V.				
4	K.R.				
5	R.M.				
6	S.T.				
7	T.V.				
8	B.V.				
9	A.R.				
10	T.S.				
Mean					

Work out the mean for the time taken and the moves made by the group. Find out the mean practical ability score.

Discussion

State the individual's score in Koh's block design test which is the practical ability score. Discuss the individual's scores in relation to the group mean in the time taken, number of moves made and the scores. What is the highest score in the practical ability score and who has obtained this? What is the lowest score and who has obtained this? Where your participant stands in the group? Who has taken longest time? How much? Who has taken the lowest time? How much? What is their corresponding practical ability scores of these persons? Is there any relationship between the duration of time to complete the task and the practical ability scores? Are the number of moves having any influence in the ability scores? If so, state how? How many in the group scored above the mean and how many scored below the mean? Do you find any other variables having influence in the ability scores of the individuals? Do you find age or sex causing any differences in the scores? State the number of performance of the group in this test as a whole.

Conclusions

1. State the practical ability score of your participant.
2. What is the highest practical ability score and lowest score?
3. How many scored above the group mean and how many below the mean?
4. State, whether age or sex had caused any variation in the scores.

EXPERIMENT 28 : ASSESSING INTELLIGENCE : PRACTICAL ABILITY (Alexander Scale : Cube Construction Test)

Problem

To assess the practical ability of a person by using cube construction test.

Method

Participant

Choose a person who has no knowledge about this test. Describe his/her characteristics such as age, sex, qualification, experience, whether he/she has any handicap, wear spectacles and so on.

Materials

(1) Alexander's scale-cube construction test, (2) Stop clock, (3) Scoring sheet.

Description of the Materials

The cube construction test consists of three sub-test. *The first test* is a block of 3˝ x 3˝ x 1˝. On all four sides it is red painted. But the top and bottom surfaces are not painted. There are nine small blocks of 1˝ cubes. If these nine blocks are so correctly put together then they make a painted 3˝ block. There is also a model block of the same pattern which can be shown to the participant.

The second test is comprised of a same type of block of 3˝ x 3˝ x 1˝. There are nine small 1˝ cubes. They are painted on all four sides and on the top, but not at the bottom. The nine 1˝ cubes are so painted that if they are correctly placed a 3˝ block is produced. Similar to this there is a model block which can be shown to the participant.

Fig. 9.4 : Cube Construction (Alexander's) Experiment 28 (Practical)

The third sub-test is a 2˝ x 2˝ x 2˝ cube. As a 2˝ cube it is not painted. It consists of eight 1˝ cubes. But the eight 1˝ cubes are painted red on three sides. If these eight 1˝ cubes are arranged correctly the 2˝ cube is produced which is not painted. There is a model block of the same pattern to be shown to the participant.

Procedure

The experimenter should make the participant to be seated comfortably in front of a table in well illuminated room where there is no disturbance of any sort. After establishing rapport, the experimenter should remove the first model of the cube construction test. The block is not painted on top and bottom, but painted red on all four sides should be shown to the participant and must be made clear. Then the experimenter should place the model flat on the table before the participant. The participant should see the painted surfaces and also the top not painted. Then the experimenter should show a demonstration to the participant as to how the model can be made from nine 1″ cubes. As soon as the demonstration is over the experimenter should lift the nine blocks jointly and should show that the top and bottom are not painted and only the sides are painted.

After the demonstration is over the experimenter should place all the nine blocks in scattered manner and give the following instruction to the participant:

> "Now, seeing the model in front of you, go ahead and make the same out of these nine blocks". The experimenter at once should start his stop clock and count the time in seconds.

The author of this test suggests that in cube construction test partially successful results also may be accepted if the participant is fully satisfied. At a stage if the participant states that he is satisfied then the test may be stopped and the time should be recorded. The experimenter should also note down the number of blocks correctly placed by the participant and the time taken for this.

The participant may be given a chance to correct his mistake. If he is satisfied then accept that the test is complete. The experimenter should check whether the bottom surface is correct. The time allowed to complete this test is *five minutes.* If the participant did not finish the test at the end of five minutes he should be stopped and the experimenter should note down the position of the blocks at this stage. Credit should be given to the blocks correctly placed. After making entries in the record sheet the performance of the participant, the blocks of model one may be removed and placed in the case. This 1″ cubes should not be mixed with other blocks.

After replacing the material of the test one, the experimenter can lift the model 2 and the corresponding blocks. Then he/she should show to the participant. The experimenter should explain to him/her that the model block of the test 2 is painted on sides and at the top and not at the bottom. He/she should place the unpainted surface of the block on the table and other painted sides are seen easily by the participant.

Then he/she should spread out evenly on the table the nine 1″ cubes. He/she need not give any demonstration for this model. The experimenter should just say, "Now you can go ahead to make just the same model in front of you. The experimenter at once should start the stop clock and count the time in seconds. The participant on completing the task if he/she has expressed that he/she is satisfied with it even if the performance is partially correct he/she should be stopped. Then the experimenter can note down the time and the number of blocks correctly placed. The time allowed is *five minutes*. Then the experimenter should replace the model 2 and the corresponding material in the case.

Now, the experimenter can go for the last sub-test with model 3. He/she should show this model 3 to the participant and explain to him/her that it is not at all painted. He should show to the participant all the four sides and the bottom and top that they are not painted. Then he/she should place model on the table. Then he/she should take the eight 1″ blocks and show to him/her that each one is the same as the other precisely. He/she must show that every 1″ block is painted red on three sides and the other three sides are not painted. Then all the eight 1″ blocks should be scattered on the table. Now the experimenter can give the following instructions to the participant:

> "Now, your task is to make out a block like this (pointing out the model) from the eight 1″ cubes. Go ahead."

As soon as the participant started the work the experimenter should start his stop clock at once and count the time in seconds. The time limit for this test is *five minutes.* If the participant has made only partial success and expressed that he is satisfied, then the experimenter should ask him/her, "Are you quite satisfied?" If he states that he/she is quite satisfied then the experimenter can stop the work and record the time and the number of blocks correctly placed. In this test generally the participants may offer partially correct answer. The form may be that all four sides and the top not painted. But they do not see the bottom. There may be one or two painted surfaces of the 1″ cubes. It is only for this reason they are asked to check whether they are quite satisfied. But, how they erred in this way is not pointed out to them.

Now the three sub-tests of the cube construction are over and the time and correctly placed number of blocks are also recorded. As soon as everything is over the experimenter should get the introspective report from the participant and may be allowed to go away.

Method of Scoring

The method of scoring of cube construction is somewhat more complicated than the other two tests we did. In this third test, two aspects are accounted for scoring. One is the level or degree of success and the other is the time. Therefore, the scoring system has two elements. They are called (1) Power-score, and (2) Time-score. The scores reveal the individual's practical ability score.

Model 1

(i) *In the case of model 1,* the *power-score* is referring to the *degree of correctness.* Thus one score for every block correctly placed. If all the nine blocks are correctly placed the total score is 9. If one block is wrongly placed then the score is (9–1) = 8. This is how the power-score is made.

(ii) *The time-score* is based upon the time limit given to this test. It is five minutes. For every minute 2 points are given. The time allowed is *five minutes*. Thus for five minutes the score is (5 x 2) 10. This is time score. For every half a minute or 30 seconds one point is deducted from ten. Suppose for example, a person completes the test in three minutes. Then the full time score is four, six deducted from ten (10 – 6) = 4. The more time a person takes the less his time score is.

(iii) *The actual time-score* is obtained by multiplying the actual power and full time score and by dividing the product by the possible power score (number of blocks = 9).

(iv) *The final score* is obtained by adding the actual power-score with the actual time score.

The formula to find out the *actual time-score* is:

$$\frac{\text{Full time-score} \times \text{Actual power-score}}{\text{Possible power score (Number of blocks 9)}}$$

The final score is = Actual power-score + Actual time-score

or

The following formula can be applied:

$$\frac{\text{Actual power-score} \times \text{Full time-score}}{\text{Possible power score (No. of blocks 9)}}$$

Let us work out one model to illustrate the scoring procedure of the model 1.

In a testing situation a person has placed correctly eight blocks and one is wrong. Then his actual *power-score* is 9 – 1 = 8.

The person has taken 3 minutes (or 180 seconds) to place the blocks correctly. Then this full time-score is 10 – 6 = 4.

Then the actual time score is $= \frac{4 \times 8}{9} = 3.6$

The final score is = 8 + 3.6 = 11.6

or

$$\frac{8 \times 4}{9} + 8 = 11.6$$

Model 2

(i) In the case of Model 2 the power score is worked by giving 2 points to each block correctly placed. There are nine 1″ cubes and thus the *possible power score is 18.* The actual power score is obtained by giving 2 points for each block correctly placed.

(ii) The time allowed for this test is *five minutes*. The full time score is one for each quarter minute (15 seconds) and thus *the possible time score is (5 x 4) 20.* The full-time score is obtained by deducting one from 20 for every 15 seconds taken by the participant. For example, a person has taken four minutes to complete the work, then his full time score is (20 – 16) 4. For four minutes, 16 quarter minutes. It is deducted from the total score 20.

(iii) The actual time score is obtained by multiplying the actual power score and full time score and by dividing the product by the possible power score (number of block = 9).

(iv) The final score is obtained by adding the actual power score with actual time score.

Let us illustrate the scoring procedure of this model with an example:

Suppose, in a testing situation a person has placed seven blocks correctly, for each correct block the score he/she gets is 2. Thus for 7 blocks 7 x 2 = 14. This is his/her *power score* (14).

The person has taken four minutes to complete the task. Then his *full time score* is (20 – 16) = 4.

The actual time score is $\frac{14 \times 4}{(9 \times 2)} = \frac{56}{18} = 3.1$

The final score = 14 + 3.1 = 17.1

This is how the scoring is made for Model 2 of the cube construction test.

Model 3

(i) In the case of Model 3 of the cube construction test the power score is obtained by giving 3 points to each block correctly placed. Thus the possible score will be (8x3) = 24. Since there are eight 1′′ cubes.

(ii) The time allowed for this test is *five minutes.* The full time score is obtained by deducting one from 20 for every quarter minutes (15 seconds) the participant has taken for placing the block.

(iii) The actual time score is obtained by multiplying the actual power score and full time score and dividing the product by the possible power score (8x3) = 24.

(iv) The final score is obtained by adding the power score with actual time score.

Let us illustrate the scoring procedure for Model 3 of the cube construction test with an example.

For example, in a testing situation a person has taken this test and successfully placed six 1′′ cubes correctly. Then his power score is (6x3) = 18. Three points to each correct placement.

The person has taken four minutes to complete the tasks. Then his full time score is (20 – 16) = 4. It is because for every quarter minute (15 seconds) one point is deducted. In four minutes there are 16 quarter minutes. Hence his full time score is only 4.

The actual time score is : $\frac{18 \times 4}{(8 \times 3)} = \frac{72}{24} = 3.0$

The final score = Actual P. Score + Actual time score = 18+3 = 21.

Results

On obtaining the scores of the individual on all three sub-tests they may be presented in Table 9.12.

Table 9.12 : Showing the practical ability scores of the individual as revealed by the Cube Construction Test

S. No.	Variables	Model-1	Model-2	Model-3
1	Time taken			
2	Number of Blocks correctly placed			
3	Actual power-score			
4	Full time score			
5	Actual time score			
6	Actual time score plus actual power score			
	Total Score			

These three tests bring out the individual's practical ability. What the individual is capable of in a problem situation? Therefore, state the scores of the individual in all the three sub-tests in terms of this. In which test he/she has obtained more and in which he/she obtained less. State whether the deductions are more for sub-test 1 or 2 or for 3. Are the durations are high for time score or for power score?

Collect the group data from other experimenters and present them in Table 9.13. Work out the mean score for all three sub-tests.

Table 9.13 : Showing the total practical ability scores of the group in 3 sub-tests of the cube construction

S. No.	Name	Model-1	Model -2	Model-3	Total
1	S.V.				
2	P.K.				
3	M.S.				
4	B.V.				
5	K.R.				
6	P.V.				
7	P.C.				
8	M.R				
9	D.V.				
10	S.V.				
	Mean				

Discussion

Discuss your results in detail taking into account the practical ability scores obtained by your participant in all three sub-tests and his/her place in the group. Which model is having high mean score and in what test the group as a whole scored higher? Compare the mean scores of the three tests. Who made highest score in the first test? Has he made correspondingly higher in other two tests? Who has made the lowest score in the first test? Has he/she made consistently lower scores in other two tests? Do you find any improvement in the performance of the individual from test to test? Whose practical ability is higher? In the group where your participant stands? How many in each test are above the mean score and how many are below the group mean. Do you find any difference between male and female subjects in these three tests? Is age having any influence in the scores? Discuss your results comparing with the results of earlier studies and state whether your results are in agreement with them or vary.

Conclusions

1. What is the practical ability score of your participant?
2. Who has scored the highest in all three tests and who has scored the lowest?
3. How many above the group mean and how many below the group mean?
4. Is there any variation due to sex, age, or any other factors in the practical ability scores?

10

MEASUREMENT OF INTELLIGENCE

HUNDRED YEARS OF STUDY

The measurement of intelligence is more than a century old in the history of psychological studies. The scientific spirit of the study in biological sciences and the study of the evolutionary processes of the species in the second half of the nineteenth century have aroused the curiosity in the minds of the scientists about the individual's adaptive abilities to the environmental situations and on the way in which the organisms survive, reproduce and decay in the physical environment. In the adaptive ability and survival there was great amount of individual differences. It was observed that the adaptive capacity was the important factor in the survival of the fittest. This process was systematically observed by many and Charles Darwin (1809-1882), an English naturalist became the champion of this phenomenological issue and formulated the theory of evolution by natural selection and expounded it in his work, *On the Origin of Species,* in 1859 and the same has been applied in his work, *The Descent of Man* in 1871. This has aroused interest also in measuring intelligence in man. He was the person mostly responsible for the study of human development from psychological point of view. The emotional expression on man and animals was one of his important studies. Applying Darwin's theory of man, the individual differences were found to be very conspicuous. Among the adaptive characteristics of man, the ability to learn was found to be the most crucial one.

Following this, around 1880s scientists like Francis Galton (1822-1916), a cousin of Darwin, has explored the theory of evolution, and carried out researches in heredity, meteorology and statistics and found the theory of eugenics and the theory of anti-cyclones. Galton's work *Hereditary and Genius* was brought out in 1869, in which he discussed the idea of biologically inherited intelligence. The present method of measuring individual differences in intelligence and personality are entirely based upon Galton's work. Galton was of the view that personality and intelligence were inherited from one's progenitors. In his scientific studies of human differences Galton believed that human intellectual abilities can be objectively assessed and for which he also invented one of the most important tools of measurement, the statistical treatment of the data called the 'correlation coefficient'. Galton strongly believed that humans differ from each other in intelligence and in abilities in measurable amounts. His strong emphasis was that individual differences in intelligence are determined by biological hereditary factors.

A little later, James McKeen Cattell (1860-1944), an American psychologist, who worked under Wilhelm Wundt (1832-1920) in Germany had developed qualitative methods in Psychology. Though he was basically interested, under the influence of Wundt, in the study of simple cognitive abilities like, the reaction time, sensory acuity, rote memory, and simple motor skills, later on he believed that these activities are simple components of human intelligence. Cattell has developed many measuring techniques and several complex statistical methods. Cattell as early as 1890 for the first time announced the concept 'test' in the context of the study of problems in relation to abilities. It was he, who first started the 'test norms' and standardization of test materials statistically. Then following Francis Galton and Cattell, quite a number of scientists started studying intellectual abilities of man, and the area of measurement of intelligence was added with enormous amount of test materials and the measurement of intelligence has gained more importance in the psychological studies. Thus the history of the measurement of intelligence is more than one hundred years old. Many intelligence tests were designed and developed to measure more complex mental abilities. Some of the important persons who made notable contributions in this area are Charles Spearman, Thorndike, Binet and Simon, Thurstone, Guilford, R.B. Cattell, Arnold Gesell, Piaget and Bruner.

INTELLIGENCE AND COGNITIVE FUNCTIONS

Intelligence is considered to be one of the higher mental functions. Any higher level of activity which is functioning in relation to the environment is a cognitive activity in which the central nervous system plays an important role. This higher level of activity is generally stated as the cognitive activity. It is the activity of the mind which is in contrast to motor activity which is a bodily activity. The classification of these two activities, mental activity and motor activity is only an arbitrary classification and not rigid. But both the activities involve the functioning of physical structure. Therefore, the classification is only one of varied degrees of neuromuscular activity. Generally, motor activities are controlled by cognitive functions and some motor activities are involved in cognitive functions. These functions vary according to the developmental process. The child's activity is fully motor or neuromuscular activity. But with time, the mental activity is gradually emerging in the child.

As the child matures in age and experience there is more awareness of the objects and persons and of the total environment. When there is increased level of the functioning of the central nervous system, there is more amount of manipulation of ideational image of the objects of the world. The inherent behaviour of this type is stated as the cognitive behaviour. The child also reveals his mental activity in his preference of motor activity. This aspect of the cognitive process continues in the child and expands as he grows and develops physically and mentally. The behaviour of the child improves time to time and the effectiveness becomes more and more valuable in fulfilling the needs. The cognitive awareness becomes more and more many sided and spreads through many facets and also manifests in various ways. The concept of intelligence is generally used to explain this whole area of the functioning.

THE CONCEPT OF INTELLIGENCE

The concept of intelligence is used by everyone of us in many of the life situations. It is used invariably in almost all contexts irrespective of achievement or failure. There are several words, used by people which would involve the functioning of intelligence or stupidity. The performance of an act or a task or solving a problem or passing a test or examination is referred to by using the word intelligence. Many people on seeing a child, at once make comments like, he is a bright boy, looks intelligent, clever, smart, thoughtful or looks dull, stupid, not active and so on. What all these indicate? All these are referring to a person's level of intelligence or the achievement he is going to make in the competitive world. Generally all these refer to the requirements of a person, the competence, capability and the efficient performance he is going to make in the life situations. These are the ways in which the concept of intelligence is defined, explained and demonstrated in the common sense world. These views held by the common sense world are not far away from the views held by the scientific world or in the field of Psychology. A student of Psychology may find all these are quite close to the subject matter what he studied in his subjects. Therefore, the meaning of intelligence held by the common sense world cannot be fully ignored on the ground that it is not systematically defined. What is spoken of intelligence in the common sense world is also correct, but the student of Psychology and education goes further from the place to make his knowledge more perfect, scientific and measurable.

Intelligence has been defined in various ways. Some have defined it theoretically and some defined it from operational point of view. Spiker and McCandless (1954) have defined operationally, stating that intelligence is what the intelligence test measures. The general definition of intelligence is that it is the ability to learn quickly and deal with abstract concepts, and solve difficult problems and adopt to difficult situations. It also involves adjustment with the other and with the society at large. Some are of the view that it is a single whole that makes the person to be successful in every aspect of his life. But some are of the view that it is combination of several abilities. Therefore, there were controversies among the experts with regard to the definition and function of intelligence. Woodworth (1952) states that intelligence is intellect put to use.

FACTORIAL STUDIES

However, there were differences of opinions with regard to the definition of intelligence, experts accept that intelligence consists of a number of different attributes or capacities. Thurstone (1941) and Guilford (1966) have made extensive studies independently and revealed the multi-model of intelligence. Prior to Thurstone and Guilford, Charles Spearman (1904, 1923, 1927) has demonstrated his two-factor theory in his works, the general factor (G-factor) and the specific factor (S-factor) and he explained how these two factors of intelligence function in the everyday life of man. According to Woodrow (1921) intelligence is an acquired capacity. It is the capacity with which a person acquires all kinds of knowledge. Stern (1914) states that intelligence is the general mental adaptability to new problems and conditions of life.

The multi-factor theorists have assumed that intelligence is not general, but it consists of a number of specific skills and also they have traced those skills in their works. A person may be stated as intelligent. But, is he an expert in every branch of knowledge or is he an expert in every field? It is not. Apart from the general intelligence which is common to all, there are also certain specific skills. By developing a particular skill, the person becomes an expert in that field. For example, in a society, everyone is not a musician or a painter or a poet or a writer. These persons have more skills in the particular area and they became experts. Though many people like music and painting they could not exhibit it as they lack the skill. Some say that it may be due to the training one receives in the particular area. But it is not the training alone that accounts for specialization. Primarily, it is the skill that one is endowed with and secondly, the training he/she received. The person who is intelligent in the specific area learns the art quickly and makes use of it, but the person who lacks the skill or not intelligent in that area, however efficient training is given to him/her, he/she cannot exhibit any proficiency in that area. Therefore, intelligence consists of a number of specific skills. The success in the field of music needs one type of skill, while success in another area requires another type of skill in that specific area. Spearman was the first to observe these specific skills in individuals.

Later on Thurstone and Thurstone (1946) applied the statistical technique of factor analysis by administering 64 tests and extracted seven factors such as perceptual speed, word comprehension, word fluency, space, number, memory and induction. Thurstone called these as primary mental abilities. They have also observed that the general factor which Spearman formulated was operating in every task and the analyses of other factors were very much needed for assessing the intellectual abilities. Similarly Guilford's (1959, 1967) model of the intellect involves 120 abilities, which are classified under three faces as operation, content and product. According to Guilford a specific ability involves a combination of all the three faces.

Intelligence, according to Hebb (1966), is the innate potentiality for cognitive development. Thus everyone gives his own definition for intelligence from the studies he made. But formulating theoretical definitions has become less and less since 1920 in favour of empirical definitions. Robinson and Robinson (1965) state that though there are several reasons for this, one important reason is that there is no general agreement based on theoretical model of intelligence. But how the existence of intelligence can be perceived or could be demonstrated? In what form or process it is present? Since intelligence is not a concrete entity, it cannot be adequately described nor the existence can be demonstrated. The presence of intelligence can be realized only from the product of some function. For example, in learning a subject matter or solving some problem one does it successfully and another fails to perform it successfully. Even in learning some material, there is no direct standard or measure for judging a person's learning ability. We could only infer a person's learning ability by observing his behaviour or from the result of his performance.

Similarly, in solving a problem, for example, a cut out puzzle may not be solved ordinarily by everybody. It requires certain amount of thinking, perceiving the whole pattern, spatial relations and so on. Out of ten persons only one or two can perform it successfully in a limited amount of time. Some may take longer time and the rest of them may fail to complete it. How to explain these three stages of performance? The performances of all these persons are observed and also we observe the successful product. From this we can infer one thing that in these learning and problem solving tasks, the ability to learn and the ability to solve the problem may be labelled as intelligence and the person who performed the task successfully is intelligent. Therefore, intelligence is not a matter or material substance and can be located in any part of the body of a person. It is the knack tactics, innate potentiality, or a quality or trait or characteristic inherent in an individual which is revealed in his behaviour or performance of a task in a given time at a specific situation. Any attempt to assess this by a suitable norm or criterion which may be conceptual or operational may be stated as the measurement of intelligence. It may be a test material (conceptual) or an instrument (operational) designed and developed comparing the performances of many individuals and evolving a standard.

DEVELOPMENT OF INTELLIGENCE TESTS

Thus many test materials and instruments have been designed and formulated during the course of one hundred years right from the time of Sir Francis Galton, Cattell, through Alfred Binet, Wechsler, Terman and Merrill to the present day experts. It is a difficult task to enumerate all the intelligence tests that were developed up to this time. The *Mental Measurement Year Books* may provide all these chronologically. The mental testing movement did not meet with any serious controversial situation like the model of intelligence.

Intelligence is measured by standardized mental ability tests. These tests are carefully validated and norms are fixed. It is also found that tests standardized for one culture or nationality is not fitting to another culture or nationality. Therefore, separate mental ability tests were designed and standardized for every culture. But there are certain tests which are culture free and common to all cultures or nationalities. Such tests are very rare. Mostly the performance tests may be treated as culture free tests.

A test standardized in one culture for a specific age group may be administered to another culture of the same age group. But the performance may not be of the same level. There are certain variables having influence in the performance of the subjects. For example, the language, socio-economic status, bringing up of the children, parental care, parents' educational standard, type of schooling, all may have influence in the performance of the intelligence test scores of the subjects. Therefore, such tests are required to be revalidated in the culture where the tests are administered. Further, there may be also differences in the test scores of the same age group. In the developed nations, because of the highly stimulating situations the children may perform the tests much better than the children of the developing nations. Children of the developing nations may do the same test little later, because of the slow learning ability. To avoid such problems every nation is interested in developing intelligence tests of their own to suit to their subjects. Many of the intelligence tests developed in the United States and in European countries have been revalidated in the native languages and administered successfully. But the tests should be constructed strictly according to certain principles. The selection of test items quite suitable to the purpose, the theory on which the test is based, establishing the norms, the reliability and the validity of the tests all should be carried out carefully step by step. All the intelligence tests that were widely used at present were standardized according to these principles.

Intelligence tests were designed and developed to various age levels, like the infant tests, right from the age of 3 months to 24 months, the children tests, from the age of two years to 5 years, and from the age of 4 to 7 years, 5 through 15 years, 16 through 18 years and then the adult intelligence test, 20 years through 60 years. The early intelligence tests were developed only for the infants, children, early adolescents and later adolescents. Later on around 1939 David Wechsler published the adult intelligence scale. Thus intelligence tests for different age levels were published.

Types of Tests

These tests and scales are both individual and group tests, some are individual tests and some are group tests. Some of them are performance tests and some paper-pencil tests. Thus there are two types of tests.

Individual Tests

The individual tests are administered individually, one at a time, whereas the group tests are administered simultaneously to a group of subjects. The group tests are mostly paper-pencil tests. There are many group tests. The group tests have certain advantages over the individual tests. The group tests are widely used and they are inexpensive. The individual tests are much more expensive, and time consuming. The scores are more reliable and provide some information about the subject. The individual tests are more useful in diagnosing specific problems of the individuals, particularly the learning difficulties of the children. The group tests can be administered and scored by any classroom teacher who knows the procedures. The individual tests require great deal of training and skill to administer them successfully.

Some of the individual tests most widely used are the Revised Stanford-Binet Test, the Peabody Picture Vocabulary Test (PPVT) and the Wechsler Intelligence Scale for Children (WISC).

The Standford-Binet test is one of the best known and most widely used individual intelligence test. This test has been translated in many languages soon after it was published. This test needs much training and competence in administering it successfully. It has different levels of difficulty to meet the various age levels. The score of this test can be converted into intelligence quotient (I.Q.). Terman and Merrill employed the notion of I.Q., the ratio of individual mental age to his chronological age, multiplied by 100 to eliminate the decimal figures. Thus the I.Q. is = $\frac{MA}{CA} \times 100$.

The Peabody picture vocabulary test can be administered very easily and the individual score can be elicited. This test normally required fifteen minutes time to complete, but the time is not limited. This test required the subject simply to point out one picture out of four when the test administer reads out the word. The words are arranged in the order of difficulty, from easiest to most difficult. The test is stopped when the subject makes six failures consecutively. The intelligence score of the individual is obtained on the basis of his/her chronological age and the level of the last correct responses.

The Wechsler intelligence scale for children has been designed and developed by David Wechsler in 1949 for using it on children aged five through 15 years. He developed another scale for children called Wechsler Preschool and Primary Scale of Intelligence (WPPSI) for children aged 4 to 6, and also a scale for adult called Wechsler Adult Intelligence Scale (WAIS) (1955). The WISC is similar to Stanford-Binet scale. The scores can be obtained on a number of specific sub-tests. The whole test consists of two sections and each section there are six sub-tests. The first is the verbal scale which includes general information, general comprehension, arithmetic, similarities, vocabulary and digit span. The second section is performance scale which consists of picture completion, picture arrangement, block design, object assembly, coding and mazes. Scores can be obtained in all these sub-tests. But the whole test scores fall under two categories such as verbal and performance (non-verbal). In getting the I.Q. the scores of these two categories are put together.

The adult version of Wechsler scale is a major contribution of David Wechsler. Wechsler adult intelligence scale consists of two sections like the children scale, verbal sub-tests and performance test. The verbal sub-test are, information, comprehension, arithmetic, similarities, digit span, vocabulary. The performance of sub-tests are digit symbol, picture completion, block design, picture arrangement, and object assembly. The items in each of these tests vary.

Group Testing

We can discuss also some of the group tests. Goodenough has first designed a test called, Draw A Man Test. The other tests like the Otis Quick Scoring Mental Ability Test, Lorge-Thorndike Intelligence Test and the California Test of Mental Maturity have been brought out. The Draw A Man Test is a group intelligence test, first developed by Goodenough in 1926, and later on it was revised by Harris in 1963. It is believed that children's drawings may reveal their thinking from simple to complex levels. In this testing the child is asked simply to draw a man as best as possible. There was no time limit. The scoring was made on the basis of the details, accuracy and perfection of the picture. The scores are converted into I.Qs. of the subjects from the age of three to fifteen. Goodenough has stated that the scores of the infants and some children do not have the predictive value as there was decrease in the scores in the successive testing of the same children. This means, those who scored high in the first testing and obtained high I.Q. have scored low in the subsequent testing and those children who scored low in the first testing scored high in the subsequent testing.

The Otis Quick Scoring Mental Ability Test was designed and developed by Otis. Otis and Lannon have jointly developed several forms of group Mental Ability Test. These tests are used at elementary level, at junior high school level and senior high school level. The Otis tests are widely used in schools as these tests cover throughout the school career and also the college level. This test consists of a number of forms and can be used at all age levels from grade one to sixteen. The forms are scored by machine. The test scores are obtained from both verbal and non-verbal tests and the I.Q. is obtained from the total scores.

The Lorge-Thorndike Intelligence Tests are multilevel paper-pencil test and are used on students from grade one to twelve. These tests were developed by Irving Lorge and Robert L. Thorndike and these are well constructed tests. These tests are very widely used in schools. There are both verbal and non-verbal forms in this test. The children were required to understand the verbal concepts and reasoning. The test user reads the items and the subject has to mark the correct pictures. In grade 4 the verbal and non-verbal tests are separately administered. The non-verbal items are mostly for general ability, not relating to verbal and reading.

The California Test of Mental Maturity (CTMM) was developed by three authors, Sullivan, Clark and Tiegs (1936). The test was revised in 1963. This test is administered from kindergarten to adult level. This is one of the most widely used tests. The test consists of variety of items. The full test, among school age students, requires one and a half hours to complete. There is also a short form of this test, but the result is not so much reliable. The I.Qs. are obtained separately for language and non-language parts of the test. The test provides eight specific sub-test scores such as logical reasoning, spatial relationship, numerical reasoning, verbal concepts, memory, total language, total non-language and the whole total.

Individual Differences and Binet-Simon Test

The efforts made by early scientists to explain the individual differences in performances resulted in exploring the area of the complex skill called as intelligence. The works of Francis Galton, J.M. Cattell and Charles Spearman have thrown much light upon the phenomenon, intelligence. Galton tried to explain it as the biologically inherited skill. He was of the view that intelligence is the basis for all the natural abilities. This is found in humans in varying quantities from idiots to geniuses and are also measurable. Thus Galton's work became the fundamental basis for further development of knowledge in the area of intelligence both for explaining and measuring intelligence. Galton, designed many laboratory instruments for studying sensory responses, but did not develop a workable intelligence test. Cattell, who was influenced by Wundt's techniques, designed tests and administered to the students of Columbia University for assessing the abilities.

The individual differences were looked upon from different angles in Paris. Alfred Binet started exploring the causes of the individual differences in schools. He advised school authorities to separate the dull students who were not benefited by the ordinary classroom function and perform poorly due to lack of motivational factors from those who were doing better in school performance. He stressed an objective rating method of intelligence, as the subjective method had very little reliability. Therefore, Binet (1905) brought the first successful test of intelligence which was called Matrical Scale of Intelligence in the year 1905. Binet prepared the test in collaboration with Simon. Binet, a leading psychologist of the time in Paris, had the brilliant idea that children increase in ability as they grow older. Binet, after conducting many experimental methods, came to a conclusion that a single task would not bring equally fair result in all the individuals. Therefore, he invented some new type of tests in which the child is to perform a variety of tasks and has many chances to reveal himself what he is capable of.

The tests were ranging from easiest to most difficult. Thus the tests were graded in difficulty. The easiest tests were successfully performed by three-year old children. The next tests were in the order of difficulty and were beyond the reach of the average three-year olds. They were within the reach of four-year olds. But few of the three-year olds also were able to reach a few of the items of the four-year old tests. Like this tests were designed for different age levels. The first Binet test consisted of thirty problems in ascending order of difficulty. The idea he had in mind was to identify children who were likely to fail in school, so that such children can be taken to special class. Thus a systematic comparison was made between normal children and mentally retarded children. Binet tried his test items on different age levels since it was not possible to judge by inspection the exact difficulty levels. The first test underwent several revisions and modifications. The fundamental basis was that the test designed for three-year olds means that all the children of average ability of three years should pass all the items of the three-year olds test, four-year old tests item should be passed by all the average four-year old children. In the same manner it went up to the age of thirteen years. Thus a revision was made in 1908. The test was shaped into an accurate measuring instrument. The arrangement gave rise to the concept of mental age. A child's mental age was the age of normal age of the child whose performance was equal to that age.

A further revision and modification was desired by Binet. The work was going on after the first revision of 1908. Before the second revision was over Binet himself died in 1911. By this time, Binet test was put in use in many countries all over the world. The 1911 revision of the Binet-Simon test was mainly based upon the classroom observations. The teacher's assessment as bright and dull and also the trial and error adjustment were taken into consideration. The scores of the tests and the teacher's ratings of intellectual ability were found in agreement strongly. The Binet-Simon test was again revised in 1916 by Louis Terman at Stanford University and it was called Stanford-Binet test. The Stanford-Binet test was again revised by Terman and Merrill (1937). This revision was made more elaborately by collecting thousands of items that were used in the previous tests and each item was administered to a minimum of 20 subjects of known ability levels. The scale was then constructed using these items that could best differentiate the children of different ages. The scoring was made very easy and it was done objectively. A norm was preparec by administering the scale to more than 3000 school students. To establish the validity and norm, Terman and Merrill constructed two alternative scales, called Forms L and M. The items were carefully matched at each age level and each form consisted of 129 items. The two forms brought out almost the same result.

The Terman-Merrill 1937 revision of the Stanford-Binet scale underwent another revision in 1960. This was done by Merrill by updating the items in both forms. Some of the older items were dropped and replaced with some new items. The scoring method also was changed. The revised forms of the Stanford-Binet scale became the basis for all other intelligence scales and it was highly reputed scale in the field of Psychology and education. Based upon the Stanford Binet Intelligence Scale many intelligence scales have been constructed in the United States and in England. These scales have tested various types of hypotheses.

INTELLIGENCE AND THE INFLUENCING FACTORS

There were several studies relating intelligence to various factors. Therefore, there were many studies of the correlates of intelligence. Though some of the factors had only limited influences they helped to explore several assumptions. Some of the variables correlated with intelligence were the family background, socio-economic status, racial differences and the ethnic variables. The early presumptions were that intelligence was genetically inherited and also was the cause of environmental stimulation. The early authors did not examine the factors related to intelligence. The development of several theories of intelligence lead to the assumption that there could be some factors having influence in the development of intelligence. Therefore, studies were made to find out the variables related to intelligence or influencing intelligence.

The existing information on intelligence have very little to do with the variables that are having influence, the family includes, both parents, siblings, the home, and other relatives and their educational status and the style of life. The child is born and brought up in the family and all the developmental process are according to the way in which the parents have designed. Whether the parents have brought up their children in an intelligent way or in a stupid way would have the impact on the child's development. But it is very difficult to classify and evolve a norm to measure the way in which parents deal with their children and the way in which a particular characteristic was inculcated in the child. Further it will not be possible to assess the type of behaviour pattern that is having influence in the performance and behaviour of the child. This needs enormous amount of research work of longitudinal type. In view of these problems only few studies have been made relating family variables to intelligence. Some studies have been made on infants, but they were not relating family variables to intelligence. In one of the studies made by Honzik (1967) on infants aged twenty-one months, the family variables were related to test performances between twenty-one months and thirty years. The findings showed positive relation between the I.Q. scores of the boys and the mothers' worrisomeness, tenseness concern and energy. These mothers were found to be more responsive to the wants and needs of their children. In the case of girls early closeness to mother was related to test scores. But during the adolescent period the father-daughter relation surpassed mother's early relation. The I.Qs. of the girls seemed related to the parental agreement.

The studies made by Bayley and Schaefer (1964) were longitudinal relating the sex to intelligence of the children. This study revealed that the hostility of mothers was related to high I.Q. scores in the case of boys in the infant period, whereas in the case of girls, loving and controlling of maternal attitude was related to high I.Q. scores during the infant period. In one of the studies the influence of discipline was related to intelligence. Baldwin, Kalhorn and Breese (1945) have found in their studies that children from democratic attitude homes revealed high I.Q. scores. But children from autocratic and rejecting attitude homes revealed low I.Q. scores. Kent and Davis (1957) have classified parents as demanding, overanxious, normal and unconcerned and related these attitudes of parents to their children's intelligence. These investigators found that the children of demanding families performed better than those children from normal families. The children from unconcerned families performed not so well.

The intelligence of an isolated mountain children was studied by Sherman and Key (1932). The children were from several hollows in a mountain area. The culture was completely different from plains and characterised by extreme degree of poverty, low literacy, poor educational facilities and isolated from other communities. These children's performances in the intelligence tests were related to the cultural levels of the individuals. It was found that these children have performed and scored less than the average scores. The families of these children were categorised on the basis of parental education, occupation, income, social status and combined these variables to get one consolidated score and then related to intelligence. The I.Qs. scores of these children fall into different categories of the socio-economic status. In the study conducted by McNemar (1942) it was observed that children from high status families achieved high I.Q. scores on the average level of 115 while those children whose fathers were holding skilled occupations had the I.Qs. on the average level of 97. Similar results were obtained in the study of Eells (1953). In this study it was observed that there was 15 to 25 points difference between the scores of children of professionals and the children of labourers. These results were obtained by using the Wechsler Intelligence Scale for children.

In one of the studies, father's occupation was related to the I.Q. scores of their children. The Wechsler intelligence scale for children was used to assess the I.Q. scores of the children. The parents were holding different occupations from professional workers to farm labourers. The occupational hierarchy was maintained in the I.Q. scores of the children. The children of professional workers scored on the average level of 110, the children of skilled tradesmen scored on the average level of 105, and the children of domestic workers scored on the average level of 97 and the children of farm labourers scored on the average level of 94. There are some evidences, available to show the relation of socio-economic status to intelligence. The results of these studies show that socio-economic status is having definite influence on intelligence.

RACIAL AND ETHNIC DIFFERENCES IN INTELLIGENCE

The racial and ethnic differences are not without any relationship with intelligence. The I.Q. scores of black and white children have been compared. Theorists of intelligence studies are having strong view that intelligence is genetically inherited. But the behaviourists did not accept this view. There were burning arguments and debates on either side. But the results of some studies confirm the view that black children perform in intelligence tests considerably lower than the white children. The studies made by Dreger and Miller (1960), Kennedy (1964) and Sherey (1966) have firmly established that the black children's I.Q. scores were considerably low. The difference in I.Q. scores between the white children and black children were ranging from fifteen to twenty points. The white children scored on the average 100, while the black children scored about 80 to 85. The low scoring of the black children may be due to several reasons, just like the isolated mountain children. In the case of the black children there were incidences of prematurity, poor medical care, food problems – all these were the characteristics of black children. Liebert, Poulos and Strauss (1974) state that one important reason for the lower performance of the black children in intelligence test has been that most of the intelligence test were standardised using white children as the subjects. A test standardised in white population without the representation of the black population may have white biasing. Such intelligence tests should be administered to both white and black children and scores should be compared. Investigators often make comparison between whites and blacks and announce that the black children score less than white children. Thus the superiority of the white is established.

A group of investigators wanted to establish a norm for the black children of the grades, one to six among the American black children. Kennedy, Van De Riet and White (1963) have conducted the investigation. The sample consisted of 1800 black children and the Stanford-Binet Scale was administered. The mean I.Q. score for the black children was 80.7. The children from big cities scored higher than those from urban and rural communities. Therefore, they were able to get the same level of I.Q. scores from the black children. It has been observed in many studies that the effects of race and socio-economic status did not bring out a clear-cut result. Mostly these were confusing. The main reason may be that a disproportionate size of blacks come under the lower social classes. These blacks were mostly characterised by variables such as poverty, low education, improper schools etc. These are quite common among the black population. Attempts were made to equalize the social status of the blacks. When tests were conducted they scored still lower than the white children. Therefore, the racial factors are said to be accounting for the low performance of the black children in intelligence tests.

Another important reason attributed for the racial and ethnic differences in scores of intelligence tests is that the tests are culture biased. Tests developed in one culture or racial group are used in another culture or race. Certain items may have different meaning in another culture. The language used to explain the matter may take more time to understand for the children of different cultures. Hence the performance is low.

But, however we cannot disregard or reject the standardised intelligence test and the valuable use of them in assessing the levels of intelligence. Particularly in the case of prediction of future prospects of students and the assessment of their academic performances are found to be more valuable. Some of the culture free intelligence tests are used all over the world.

EXPERIMENT 29 : MEASUREMENT OF INTELLIGENCE : RAVEN'S PROGRESSIVE MATRICES (Standard Form)

One culture free intelligence test is the Raven's Standard Progressive Matrices Scale designed and standardized by J.C. Raven (1960). This test consists of only designs or patterns and no verbal statements or indications. It is fully a non-verbal test. This test can be administered to both illiterates and literates. It is also very interesting task. The standard progressive matrices test is very much useful in the field of clinical Psychology. One important use of this test is measuring intelligence for selection and guidance work. Another important advantage of this test is that the instructions in administering this test to the children or to any subject including the illiterates are very simple and easy. The subjects themselves may find this test very interesting and they may like it. It can be administered to any language group easily. Therefore, it is not only culture free, but also language free and nationality free test. It may cover very wide range of age and ability. This test has three forms as, the Standard Progressive Matrices sets A, B, C, D and E, the Coloured Progressive Matrices, sets A, Ab and B, and Advanced Progressive Matrices, Sets I and II.

Raven's Progressive Matrices Scales have a long history. It was first designed in 1936 and revised, standardized and published in 1938. It was reprinted with extended norms in 1940. This underwent several reprints from 1940 to 1943. This test was again revised and published with norms for subjects from 6 years to 65 years of age. After a reprint in 1951, it was again revised with the ordering of the problems in 1956. This revision underwent several reprints from 1958 to 1971. Raven first designed and developed the Mill Hill Vocabulary Scale for assessing a person's ability to recall the acquired information, whereas the progressive matrices tests were designed to assess a person's immediate capacity for observation and clear thinking and accurate intellectual work. Raven used both of his tests to find out under what conditions his two tests, the Mill Hill Vocabulary Scale and the Progressive Matrices can assess a person's intellectual developments, specifically the trainability or mental impairment for practical purposes. Progressive Matrices were extensively used in recruiting men for Armed Forces in England and also in clinical works in Royal Hospital in Scotland.

Raven has based his progressive matrices tests on Carl Spearman's (1923) Principle of neogenesis. Both of his tests were standardized in British population and developed norms and the re-test reliability was highly significant for his Mill Hill Vocabulary test. The re-test reliability of the Progressive Matrices tests was found to be significant at .02 level for all five sets. All the three Progressive Matrices, mentioned above, were developed by Raven for different purposes, each one serves one purpose.

EXPERIMENT

Problem

Assess a person's immediate intellectual capacity to observe and for clear thinking and his/her accurate intellectual work using a standard intelligence test.

Method

Participant

Find out a person who can show some interest in taking up an intelligence test and collect his/her personal information such as age, qualification, sex, parent's qualifications, occupations, income, home town, whether urban area or rural area, or isolated farm area or mountain area.

Materials

Raven's Standard Progressive Matrices Scale with sets A, B, C, D, and E, Guide to the Standard Progressive Matrices sets A, B, C, D and E with scoring key or manual with marking key, norms etc., test booklet, answer sheet, stop watch and lead pencils.

Description of the Materials

The Standard Progressive Matrices, 1956 revision with sets, A, B, C, D and E is an intelligence scale designed to assess the individual's immediate capacity to observe and clear thinking. This was first developed by Raven in 1936. The scale underwent several revisions and reprinted and the final form is used here. The Progressive Matrices Scales have three forms such as the Standard Progressive Matrices, the Coloured Progressive Matrices, and the Advanced Progressive Matrices.

1. The Standard Progressive Matrices, Sets A, B, C, D and E

This was developed to assess a person's intellectual ability at the time of testing him with designs or patterns having no meaning. The patterns are presented for his observation and to solve the problems. The person has to find out the relation between the designs and perform the task. This scale consists of five sets such as A, B, C, D and E. In each set there are 12 problems. Thus there are (5 x 12 = 60) sixty problems on the whole. It may be stated that there are five sub-tests in this. In each set, the first problem is the easiest one. Then the problems are going on progressively increasing in difficulty. As and when the subject works on these problems he gains some training and he/she can understand the pattern of arrangement of the problems. All the five sets provide the subject chances to understand five progressive methods. It may be stated that there are five progressive assessments of the mental capacity of the person. The designs are presented in the test booklet in bold types to avoid the fatigue effect. The subject has to find out the missing piece of the design in every pattern to complete it. Thereby the person is able to grasp the idea of discovering the correct piece among the designs. Sets A and B have six pieces of designs from which the person has to find out the correct piece which can fit into the pattern. From sets C to E the pieces of designs are increased to eight. Out of eight pieces of designs the person has to find out the correct piece to fit into the pattern. This means there is a systematic order of difficulty in the sets.

This scale is administered to everyone irrespective of age in the same order and he/she can work at his/her own speed from the beginning to the end. This scale can be used as an individual test or as a group test. The individual's score is the index of his/her intellectual capacity, irrespective of one's culture, nationality, age and education. The author of the scale has checked its reliability for each age group and it varied from .83 to .93. The scale has the 'G' factor of intelligence and it is spread out to the level of .82. Therefore, it is a reliable scale to assess the general intelligence of a person. But the scale has certain limitations. The author of the scale states that young children, mentally defectives, and very old people are expected to solve the problems only in sets A and B, and the easier problems in sets C and D and nothing more. This scale is designed with a view of assessing the whole range of the level of intellectual development of a person and not mainly intended to assess individual differences.

2. The Coloured Progressive Matrices, Sets A, Ab and B

This scale is intended to test the intellectual capacity of young children and old people. It is mainly used in clinical works and anthropological studies. This is a language free test and persons who do not know English language, persons who suffer from physical deformities, or mentally sub-normal or feeble minded also can take up this. This scale has two main sets A and B and the Ab set is a combination of A and B, having intermediate difficulty level. Sets A and B include all the cognitive aspects of children aged 11 years. In order to reduce the verbal instruction, the designs are printed with coloured backgrounds and with sufficient illustrations. This scale is also designed with movable bits to make the task easy. In this form, the scale is easily administered by showing or indicating the designs. The coloured progressive matrices can be used to assess a person's level of capacity of observation and clear thinking that he has developed or the level of defects or impairment of his/her mental development. If a person is very much successful in sets A, Ab, and B, then sets C, D and E can also be followed immediately.

Advanced Progressive Matrices, Sets I and II

This scale is designed for the persons who are above 11 years old and also for those who are above the average in mental ability. It is used to assess an individual's maximum intellectual capacity for observation and clear thinking. It is also administered with time limit in assessing the speed and accuracy of the individual's works on this scale. The first set consists of twelve problems as in Standard Matrices. These problems are intended to make the person to get acquainted with the method of working. The second set consists of forty-eight problems. These problems are quite similar to the problems of the sets C, D and E of the Standard Matrices Scale. Thus the whole scale consists of 60 problems as in the Standard Scale. The response to Set I of this scale would reveal whether an individual is superior, average or dull. If the individual is found to be average or more than an average or bright in the intellectual ability, then, Set II can be administrated to find out the exact level of his/her ability. To assess the final level of his/her intellectual ability he/she may be asked to solve as many problems as possible in Set II within the time limit. For the Set I, *half a minute* time for each item may be allowed . In the case of Set II *one minute* for each item may be permitted. Thus the Set I takes 6 minutes and Set II 48 minutes. For the whole scale 54 minutes may be permitted, when it is used as a speed test. Originally this was developed for the Army Officer selection in the British Army during World War II in 1943.

While administrating the Raven's Progressive Matrices the age and type of persons should be kept in mind. It has been suggested by Raven that in assessing intellectual capacity of the individuals for anthropological, genetic and clinical studies the Standard Progressive Matrices Scale may be more useful if it is administered without time limit. But for the vocational guidance and occupational selection and for pursuing higher technical course of training the Advanced Progressive Matrices Sets I and II with time limit may be more useful.

In testing children's immediate intellectual ability for comparison and reasoning the Standard Progressive Scale may be employed without time limit on children below 11 years old. Since the progressive matrices are standardized on a sample of people aged from 6 to 65 years, all the three forms can be employed from the age of 6 years to 65 years, except the coloured matrices which may be used on young children. The standard matrices may be administered to little older children.

Procedure

The Raven's Progressive Matrices Scales can be used as an individual test and also as group test (self-administered). Whether it is an individual test or group test the procedure and the instructions are the same.

Individual Test

Keep the person, to be tested, comfortably seated in a quiet and well illuminated room. The experimenter should give out the answer sheet and a pencil to the person to be tested and ask him/her to fill up in the scoring sheet with the particulars asked for about him/her. Then the experimenter should give the test booklet to the subject and open the first page of the test booklet A-1. Then the experimenter should give the following instructions by saying, "look at this design (pointing out the large design). It is a large design with a bit of the design removed and the space is empty. The six bits below (pointing out the bits below the large design) are having the same shape to fit into the empty space. But all of them do not complete the design as every one is having different design". He should explain to the subject how the bits 1, 2 and 3 are not the correct bits to complete the design by placing in the empty space. "The bit number 6 appears to be suitable to fill up the space". (At this stage the experimenter should ask the subject), "can you find out any of the bits fully suitable to fill up the gap"? If the subject could not find out the correct bit for the space even at this stage then the experimenter should continue the instruction until the subject grasp the point and solve the problem. As soon as the subject traced the correct bit he should be asked to write down the number of the bit in the answer sheet against A-1.

Then the experimenter should continue the instructions by saying, "on every page like this (showing A-2) you will find in the large design a pattern of space is left out and your task is to find out the correct bit which can fit into the empty space to complete the design and write down the number of the bit in the answer sheet against A-2. In the same manner on every page in the large pattern a bit is removed and placed below along with other bits which do not fit into the pattern to complete it. You will have to do the same thing as you did in the first design A-1. As you go on from page to page you may feel difficult to find out the correct bit, but if you pay more attention as you did in the previous patterns you may find them not so difficult. You will have to find out the bit which can complete the pattern. You need not hastily do this. You can take your own time. You must go page by page serially. Try to get as many rights as possible. Remember, there is only one correct bit among the six or eight bits. Now ready go."

As soon as the participant starts solving the problems, the experimenter should start the stop clock. The experimenter is not supposed to give any other assistance to the subject in solving the problems. When the person completed the test, the experimenter should collect the answer sheet, the test booklet and the pencil and also should get an introspective report from him/her and allow him/her to go away.

Group Testing

Raven's Progressive Matrices Scale can be used as a self-administered or group test. The experimenter should collect randomly as many subjects as possible. Since it can be administered in several group settings, it is better to have a small group at each time for administrative conveniences. A small group of 15 to 20 school or college or university students is convenient for supervision.

The experimenter should keep the group of 15 students seated comfortably in a well illuminated classroom situation at tables. These 15 students can be put in three rows, five students in each row. Each row should be at a reasonable distance and there should be sufficient space between students, so that there may not be any chance for copying. There should not be any visual or auditory distractions where the test is conducted. All students should face the test administer only.

The experimenter should get ready with all the test materials required for the group. There should be 15 test booklets, 15 answer sheets (record sheets), 15 lead pencils and one stop clock.

The experimenter, on having fully satisfied with the arrangements, should distribute the test materials to the participants, the test booklets, answer sheets and lead pencils.

On distributing the test materials the experimenter should tell them not to open the test booklet till he orders to open. First of all the experimenter should ask the participants to provide the information asked for in the answer sheets. When everybody is ready, the experimenter should ask them to open the first page of the booklet. The experimenter or the test conductor should demonstrate this visually to the whole group. Then he/she should give the instruction: "Look at the top of the first page. It is set-A, and the problem is A-1. There is one large pattern at the top and six bits in the form of small pattern. Each one is numbered. In the large pattern there is a bit missing. The shapes of the six bits are the right shapes (pointing out the bits). All the six bits may correctly fit into the gap or empty space but they do not complete the pattern correctly. For example, look at number one (pointing to the bit and to the pattern). It is not correct. Similarly bit numbers 2 and 3 also are not correct. You must see that these bits fit into space, but do not complete the pattern. Therefore, bit numbers 1, 2 and 3 are not the correct patterns."

"What do you think about the bit number 6? Is it right bit to complete the pattern? It appears quite similar to the pattern. But it is not the quite right bit to complete the pattern. Therefore, "can you point out the bit number which is quite right by your finger?" If the test conductor is fully satisfied with the participant's ability to grasp the correct bit, then he can proceed further, by saying, "Yes, you are right, bit number 4 is the correct one to complete the pattern. Therefore, the answer to A-1 problem is 4, and write down 4 against A-1, in the column in your answer sheet. Do not go to the next page till I tell you."

The test conductor at this stage should see that the participants in room have finished the task in the first page of the test booklet and when he is satisfied he may continue his instructions: "as you have seen in the first page, in every page of the test booklet a bit is missing in the pattern. Your task is to think and decide which of the bits given below can fit into the gap of the pattern and complete it. When you have decided a particular bit is the most suitable to complete the pattern at once write down the number of the bit in your answer sheet. When you begin it, may appear simple and easy, but on going page to page you may find it difficult to decide and, therefore, you will have to pay more attention in deciding the right pattern. Start at the beginning and go through the pages to the end of the booklet. Do not omit any of them. You should not turn back the pages on any account. Try to answer all of them and take your own time. Now, you can go to the next page. Ready go."

Soon after giving this order, the experimenter should start stop clock and note down the time. In order to ascertain whether the participants have grasped the problem correctly, the experimenter can check once with the correct answer after allowing sufficient time for answering. When everybody has answered the A-2 problem the experimenter should say: "the correct pattern for the problem A-2 is number 5. Check whether you have written the number 5 against the problem A-2 in the column of your answer sheet. Now you can go on doing like this by yourself to the end of the test booklet."

The test conductor or the experimenter should go round and supervise the work and see whether everyone has written correctly in his/her answer sheet his/her own solution upto five problems of Set-A. If the person has clearly understood upto this level no assistance for the method of reasoning is needed, but he/she must make the entries in his/her answer sheet correctly in the appropriate places. It is quite possible that sometimes persons may skip over certain problems. Therefore, after fifteen minutes of the commencement of the test, the conductor should go round and see whether he/she entering the choices against the correct numbers in his/her answer sheets. The test conductor while supervising the work, if he has observed that a person has given more than one number in the answer sheet, he/she must be told to cross out the wrong and to retain the right one. After 30 minutes the test conductor should again check how far they have finished the test. It is expected that some should have finished the test. It is expected that some should have finished everything by this time taking half of a minute for every problem. Some may take one minute for each problem in sets C, D and E as they are more difficult problems. Therefore the appropriate time may be 48 minutes to 60 minutes. Others may be going through set-D or set-E. The test conductor should announce, "those who have finished all the problems at once can raise your arm and can return the test materials and go away". The time is accounted from the time the conductor said, "turn over yourself to the next problem, A-2" to the time he/she raises the arm after finishing the test. The experimenter at once should note down the time the person has taken to complete the test. When everybody has completed the test, on verifying it, the test conductor should stop the stop clock and account the time for the whole group. He should collect all the test booklets, answer sheets and the pencils and allow them to go away on getting the introspective report from them.

Method of Scoring

The experimenter himself can score the problem with the help of a key. The key for the Raven's Standard Progressive Matrices Scale has been revised in 1956 and is given in the guide to the Standard Progressive Matrices page 12. The numbers of the correct patterns are given in the key. The key as given in the manual or in the Guide to Standard Progressive Matrices sets A, B, C, D and E, should be used in scoring the answers. Sometimes, it is possible that a person might have mentioned more than one bit number, might be 2 or even 3. In such a case the number he mentioned as the final may be taken as his/her correct answer whether it is right or wrong. But if a person in a group test has written more than one number against any of the items of the scale and if the test supervisor has not discovered it and advised him to cross out the wrong answers and to retain one correct answer, that is, if the mistake was not observed by the supervisor till the test is over, then the number on the extreme right, that is the number lastly entered should be considered as his correct answer whether it is right or wrong. An individual's score on the scale is the total number of problems he/she has solved correctly from the beginning to the end. That is, the summing up of all the rights in five sets constitute the individual's score.

The author of this scale suggests to assess the consistency of the person's work. This is obtained by deducting the expected score from the score obtained in each set. If a person's score in anyone of the sets deviates more than two then his/her total score on the scale cannot be accepted as a reliable measure of his intellectual ability. But for most of the general purposes the total raw score may be taken as relatively valid as an index of the person's intellectual performances even if the discrepancies are more than two points. Therefore, the total score the individual has obtained in the whole scale (in all the 5 sets) may be used to assess the individual's intellectual ability. The I.Q. may be obtained by comparing the raw score with the appropriate age norms. But the norms on Indian population for this scale by administering on a large sample has not been worked out. Therefore, the raw score on this scale obtained by the individual can be used for interpreting the individual's intellectual ability.

Results

The experimenter after collecting the answer sheets from the participants should check first whether the participants have answered all the items of the five sets according to the instructions given to them. If there is any mistake in writing down the correct answer in the column, the experimenter should follow the procedure suggested above in the case of entering more than one number against any of the items. If everybody has answered according to the procedure, then the experimenter with the help of the key should find out the score for each set separately and also the total for the whole scale. The idea of giving the scores of each set separately is to see whether the scores are progressively decreasing as and when the subject goes from set to set. It is expected that the scores should go on decreasing from the highest to the lowest. If the experimenter had conducted the test on a single person, he should present the score in a separate table for each set and also the sum. The maximum possible score in this scale is 60 (5x12 = 60) provided if the participant has answered all the items correctly. Otherwise it will be less than 60 depending upon the number of items answered correctly.

Table 10.1 : Showing the individual score on the Raven's Standard Progressive Matrices Scale

S. No.	Sets	Total number of items attempted	Obtained score	Expected score (55)	Discrepancy score
1	A			12	
2	B			11	
3	C			11	
4	D			11	
5	E			10	
Total				55	

Present the scores of the individual, whom you tested, in Table 10.1. Arrange the sets vertically and enter the number of items the person has attempted and also the number of items correctly answered. Enter the scores against each set.

Find out the scores for each set separately and also for the whole scale. Mention the discrepancy for each set and also for the total. The discrepancy score is obtained by deducting the expected score from the obtained score. It can be minus (–) score or plus (+) score.

Present the group data in Table 10.2. The group data may be obtained either by administering the scale on a group of 15 participants or the teacher himself might have acted as an experimenter and administered the Raven's Standard Progressive Matrices

Scale to the students and scored according to the procedure or by collecting from other 15 experimenters who independently administered the scale to 15 individuals. The same can be presented as the group data.

Table 10.2 : Showing the group data on the Raven's Standard Progressive Matrices Scale

S. No	Name	Age	Score in Sets					Total score obtained	Expected score	Discrepancy Points – or +	Percentile points	Time taken in minutes
			A	B	C	D	E					
1	P.V.									55		
2	K.R.									55		
3	M.S.									55		
4	P.K.									55		
5	S.M.									55		
6	B.V.									55		
7	R.A.									55		
8	T.S.									55		
9	C.G.									55		
10	D.V.									55		
11	R.N.									55		
12	A.N.									55		
13	V.S.									55		
14	M.R.									55		
15	T.V.									55		
	Total											
	Mean									55		

Work out the scores in such a way that the Table 10.2 is completed with all the required scores for the purpose of discussion of the individual's and the group's performances on the Raven's Standard Progressive Matrices Scale. By deducting the expected score from the obtained score we get the discrepancy points. The author of the scale has given the expected score for each set on administering scale to a large population and the same can be accepted for the practical purpose. This is given both in Table 10.1 for each set, and the total expected score for the whole scale in Table 10.2. Therefore on obtaining the scores of the 15 persons the discrepancy scores can be elicited.

With regard to the percentile points, the age of these subjects must be between 20 and 25 years and the author (1960) has given in the manual the percentile points by administering the scale to a large sample. Therefore, the same percentile points can be accepted for our purpose of discussion as there is no percentile norm on Indian sample. The percentile points for the chronological age of 20 and 25 are the same. Therefore, if we know the score of the individual, the percentile points can be found from the Table V of the manual. Accordingly, we can find that for the scores of 55, 54, 49, 44, 37, 28 and 23 the percentile points are 95, 90, 75, 50, 25, 10, and 5 respectively. Therefore, for any raw score between 55 and 25 we can find out the percentile points.

Work out the total and the mean scores for each set and also for the whole total score. Find out the mean discrepancy score and the percentile points for the group. State the score and percentile points. State the range of the total score of the whole group from highest to lowest. State the range of the discrepancy score and the percentile points from highest to lowest. State the mean score of the group for the whole scale.

Discussion

The Raven's Standard Progressive Matrices Scale is designed to assess the individual's or the group's intellectual ability. The intellectual ability specifically refers to the immediate capacities for observation and clear thinking. The order of the five sets of this scale provides the standard training in the method of working. This scale is very much useful for the clinical psychologists and for those who wanted to measure intelligence for occupational selection and for vocational guidance work. Raven's Standard Progressive Matrices Scale has several advantages. The instructions to administer the test is easy and very simple. It is fully non-verbal in character and, therefore, it can be administered in any cultural or national groups. Another advantage is that it can be used on any age group ranging from 2 years to 65 years of age. It can be easily scored unlike other intelligence tests. All the five sets are arranged in the order of difficulty. This scale can be specially used for the guidance or selection of persons who are wishing to go for higher technical course of training.

The scale may be helpful to classify people into five grades on the basis of the scores and percentile. According to this scale, persons whose scores fall at 95th percentile and above are placed in Grade I and who are supposed to be intellectually superior of their age. Those whose scores fall at 75th percentile and above are placed in Grade II. These people are definitely above the average in intellectual ability of their age. Persons whose scores fall in between 75th and 25th percentile are placed in Grade III. These people are supposed to be on the average in intellectual ability of their age. Those, whose scores fall below 25th percentile are below the average in intellectual ability and placed in Grade IV. Lastly, those whose scores fall below 5th percentile are placed in Grade V, and they are mentally defective of their age level. Now, state according to this, where the person whom you tested falls? State, what is his score and the corresponding percentile points and also his discrepancies. State the grade to which he/she belongs to. If you can find out any reason for his high or low score, state it. Relate the score and the time the person has taken to complete the test.

State the mean score of the group for the whole scale, and state how many have scored above the group mean and how many below the group mean. State the group mean time and who has taken the least time and who has taken the longest time. State the group mean for each set and discuss whether the means are in progressive order or in descending order. State and discuss who has scored the highest and who has scored the lowest. From the personal information find out the reason, if any, for the high and low scores. Find out and discuss whether anyone in the group scored more than the expected score 55. If so give his/her personal characteristics like age, sex, specific training or experience and the family backgrounds.

Find out whether there is anybody whose age is above 20-25 or below 20. State his/her total score and discuss his/her performance on the scale. State and discuss the discrepancy scores. What is the meaning of discrepancy score and how many had above the group mean and how many had below the group mean? Discuss it. State those persons who have got the high discrepancy score and those who have got the low discrepancy score. Compare and discuss the total score, time taken to complete the test, and the discrepancy score of those who had high scores and those who had low scores. Discuss whether there is any relationship among these three variables. Find out and discuss whether the proportion of the high obtained score – the low discrepancy score – the shortest time for completing test exists in this study, or the proportion of low obtained score – high discrepancy score - the longest time for completing the test exists.

Grade your participant on the basis of the percentile score. Is he/she above average or average or below average. Give the ranking to the 15 persons, who have taken the test, from 1 to 15 on the basis of the percentile scores. State the persons who got the first rank, second and third ranks. Also state the person who got the last three ranks.

Grade the 15 persons on the basis of the obtained scores or percentile scores and discuss their performances and their intellectual ability. How many persons are classified into the Ist grade? Who are they? How many are placed in the IInd Grade by scoring at 75th percentile and above? Who are those in the IIIrd Grade, graded as average? How many are below average in the intellectual ability? Who are they? Are there persons who scored below 5th percentile? Who are the intellectual defectives? State them and discuss, mentioning their raw scores in every set and also in the whole scale. State and discuss who are those intellectually superior and intellectually defective as revealed by the Raven's Standard Progressive Matrices Scale. State the reason, if you find any, for some being intellectually superior and for some being intellectually below the average and defective.

Do you find any variation in the intellectual level on the basis of socio-economic status, age, sex, rural, urban backgrounds. If so, state who are those influenced by these variables? Discuss it.

Conclusions

1. State the grade and the intellectual level of the individual whom you have tested with the Standard Progressive Matrices Scale.
2. In the group who scored the highest and who scored the lowest?
3. How many in the group above the group mean score and how many below the group mean?
4. How many are intellectually superior and placed in Grade I and how many are placed in Grade IV and V, below the average, intellectually.
5. Do you find any relationship among the high score-short time-low discrepancy score.
6. Is there any difference in scores on the basis of sex and age?
7. Who scored high? Whether male students or female students.
8. State the percentage of males who have scored above the group mean and the percentages of females who have scored above the group mean?

REFERENCES
(Volume – I)

CHAPTER 1

Teaching of Psychology

Black, J.A., and Champion, D.J. *Methods and Issues in Social Research,* New York : John Wiley and Sons, Inc., 1976.

Bridgman, P.W. *The Logic of Modern Physics,* New York: Macmillian, 1927.

Burt, C. *The Factors of the Mind,* New York : Macmillan, 1941.

Candland, D.K. *Psychology : The Experimental Approach,* New York: McGraw-Hill Book Company, 1968.

Cattell, R.B. *Factor Analysis,* New York: Harper, 1952.

Comerey, A.L. *First Course in Factor Analysis.* New York: Academic Press, 1973.

Ebbinghaus, H. *Memory: A Contribution to Experimental Psychology* (1885) (H.A. Ruger & C.E. Bussenius Trans) New York: Columbia, 1913.

Fechner, G.T. *Elements of Psychophysics* (H.S. Langfeld Trans) in B.R. and (Ed) The Classical Psychologists, Boston: Houghton Mifflin, 1912.

Fisher, R.A., *The Design of Experiments* (6th Ed.) New York: Hefner, 1953.

Fruchter, B. Factoral content of right response and wrong response scores in a battery of experimental aptitude tests. *Ph.D. Dissertation.* University of Southern California Library, 1948.

Fruchter, B. *Introduction to Factor Analysis,* Princeton, New Jersey: D. Van Nostrand Company, Inc., 1954.

Galanter, E. Contemporary Psychophysics. In R. Brown, E. Galanter, E. Heff and Mandler (Eds), *New Direction in Psychology,* New York: Holt, 1962.

Galton, F. *Inquiries into Human Faculty and its Development,* London: Macmillan, 1883.

Gorsuch, R.L. *Factor Analysis,* Philadelphia; W.B. Sounders, 1974.

Guilford, J.P. *Psychometric methods,* New York: McGraw-Hill Book Co., 1948.

Helmholtz, H.Van. *Handbruch der Physiologisches:* Optic., Leipzig, Voss. 1st Ed., 1867.

Helson, H. Adaptation level theory. In S. Koch (Ed) *Psychology : A Study of a Science, Study I* Conceptual and Systematic, Vol. I. Sensory, perceptual and physiological, New York: Harper & Row, 1958.

Helson, H. *Adaptation Level Theory,* New York: Harper & Row, 1964.

Hotelling, H. Analysis of complex statistical variables into principal components, *Jl. Ednl. Psychol.* 1933, 24, 417-441.

Hull, C.L. *Behaviour System,* New York: Haven Con., Yale University Press, 1952.

Kelley, T.L. *Essential Traits of Mental Life,* Cambridge, Mass: Harvard University Press, 1935.

Maslow, A.H. *Motivation and Personality* (2nd Ed.) New York: Harper & Row Publishers, 1974.

McGuigan, F.J. *Experimental Psychology* (4th Ed.) Englewood Cliffs, New Jersey: Prentice Hall, Inc. 1978.

Selltiz Claire, Merie Johoda Morton Deutsach and Stuart W. Cook, *Research Methods in Social Relations* Rev. ed. New York: Henry Holt. 1959.

Sheridan, C.L. *Methods in Experimental Psychology,* New York: Holt Rinehart and Winston, 1979.

Spearman, C.E. *The Abilities of Man,* New York : Macmillan, 1927.

Stevens, S.S. *Handbook of Experimental Psychology* (Ed). New York: Wiley, 1951.

Stevens, S.S. *Psychophysics,* Ed., by Geraldine Stevens, New York: John Wiley & Sons, 1975.

Thompson, G.H. *The Factorial Analysis of Human Abilities,* Boston : Houghton Mifflin, 1948.

Thurstone, L.L. *Vectors of Mind,* Chicago: Chicago University Press, 1935.

Thurstone, L.L. *Primary Mental Abilities,* Chicago: Chicago University Press, 1938.

Thurstone, L.L. *Factorial Studies of Intelligence,* Chicago: Chicago University Press, 1941.

Thurstone, L.L. A factorial study of perception, *Psychometric Monograph No. 4.* 1944.

Thurstone, L.L. Factory analysis of body types, *Psychometrica,* 1946, *11*, 15-21.

Thurstone, L.L. *Multiple Factor Analysis,* Chicago: Chicago University Press, 1947.

Woodworth, R.S. *Experimental Psychology*, New York: Holt, 1947.

CHAPTER 2

Treatment of the Experimental Results : Statistical Methods

Black, J.A., and Champion, D.J. *Methods and Issues in Social Research,* New York: John Wiley & sons, Inc., 1976.

Candland, D.K. *Psychology : Experimental Approach,* New York: McGraw-Hill Book Company, 1968.

Ebel, R.I. *Measuring Educational Achievement,* New Delhi: Prentice-Hall of India Private Limited, 1980.

Garrett, H.E. *General Psychology,* (2nd Edition), New Delhi: Eurasia Publishing House (Pvt.) Ltd., 1975.

Guilford, J.P., and Fruchter, B. *Fundamental Statistics in Psychology and Education.* International Editor, (6th Edition), New York, McGraw-Hill Book Company, 1987.

Gullikson, H. *Theory of Mental Testing,* New York; Wiley, 1950.

McGuigan, E.J. *Experimental Psychology*, (3rd Edition), Englewood Cliffs, New Jersey: Prentice-Hall, Inc., 1976.

Moore, D.S. *Statistics : Concepts and Controversies,* Fourth Edition, New York: W.H. Freeman & Company, 1997.

Parry, Johny and Adiseshiah, W. *Experimental Psychology*, Bombay: Allied Publishers Private Limited, 1977.

Rajamanickam, M. *Statistical Methods in Psychological and Educational Research,* New Delhi : Concept Publishing Company, 2001.

Sheridan, C.L. *Methods in Experimental Psychology,* New York: Holt, Rinehart and Winston, 1978.

Spearman, A. "General Intelligence" Objectively determined and measured, *American Journal of Psychology,* 1904, *15*, 201-93.

Thorndike, R.C., and Hagan, E. *Measurement in Psychology Education,* (2nd Ed), New Delhi : Wiley Eastern Private Ltd., 1961.

CHAPTER 3

Sensory Experience

Bell, C. Idea of a new anatomy of the brain: Submitted for the observation of his friends. In W. Dennis, *Reading in the History of Psychology.* New York: Appleton-Century Crofts, 1948.

Candland, D.K. *Psychology : The Experimental Approach.* New York: McGraw-Hill Book Company, 1968.

Crocker, E.C. and L.F. Henderson, *Odour Classification Set.* Cargille Scientific, INC., 33, Village Park Road, Cedar Groove, N.J.

Easton, D.M. *Mechanisms of Body Functions* (2nd Ed.). New Delhi: Prentice-Hall of India Private Limited, 1978.

Geldard, F.A. *The Human Senses.* New York : Wiley, 1953.

Helmholtz, H. Van. *Handbrch der physiologischen optik,* Reipzig : Voss, (1st ed.), 1867.

Henning, H. *Der Geruch,* Leipzig: Barth, 1916.

Magnen, J. Etude d'une methode d'analyse qualitative de l'olfaction, *Ann. Psychol.* 1942-1943, *43-44,* 249-264.

Mueller, C.G. *Sensory Psychology.* New Delhi: Prentice-Hall of India Private Limited, 1969.

Muller, J. *Elements of Physiology.* Vol. 1 (W. Baly : Trans.) London : Tylor & Walton, 1838.

Pfaffmann, C. Gustatory afferent impulses. *J. Cell Comp. Physiol.* 1941, *17*, 243-258.

Sherrington, C.S. *The Integrative Action of the Nervous System.* London: Constable, 1906.

Winger, M.A., Jones, F.N. and Jones, M.H. *Physiological Psychology* . New York: Holt, 1956.

Woodworth, R.S. and Marquis, D.G. *Psychology.* London: Methuen & Co. Ltd., 1952.

Zwaademake, H. Profung der Geruchssinnes und der Geruche. In R. Abderhalden (Ed.) *Handbuch der biologisches Arbeits methoden,* Abt v. Teil 7, Heft 3, 1920.

References for the Experiment –1 : Determination of Two-Point Threshold

Candland, D.K. *Psychology : Experimental Approach.* New York: McGraw-Hill Book Company, 1968.

Epstein, W. and Shontz, F. *Psychology in Progress.* New York: Holt Rinehart and Winston, Inc., 1971.

Haber, W.B. Effects of loss of limb on sensory functions. *Jl. Psychology,* 1955, *40*, 115-123.

Haber, W.B. Reactions to loss of limb : *Physiological and Psychological Aspects*. Ann. No. 7, Acad. Sciences, 1958, *74*, 14-24.

Postman, L. and Egan, J.P. *Experimental Psychology : An Introduction.* Ludhiana: Kalyani Publishers, 1949.

Simmal, M.L. On phantom limbs. *Arch. Neurol. Psychiat.,* 1956, *75*, 637-647.

Simmal, M.L. Phantoms in patients with leprosy and in elderly digital amputees. *Amer. J. Psychology*, 1956, *69*, 529-545.

Simmal, M.L. The conditions of occurrence of phantom limbs. *Proceedings of the American Philosophical Society,* 1958, *102*, 492-500.

Simmal, M.L. Phantoms, phantom pain and denial. *Amer. J. Psychotherapy*, 1959, *13,* 603-613.

Simmal, M.L. The absence of phantoms for congenitally missing limbs. *Amer. J. Psychol.* 1961, *74*, 467-470.

Simmal, M.L. Phantom experiences following amputation in childhood. *Journal of Neurology and Neurosurgery,* 1962, *25*, 69-78.

Simmal, M.L. Psychological after effects of amputation, *Rehab. Couns. Bull.* 1963, *6,* 75-83.

Weddell, G. The pattern of cutaneous innervation in relation to cutaneous sensibility. *Jl. Anat*. 1941, *75,* 441-446.

Weddell, G. The anatomy of pain sensibility. *Jl. Anat.*, London, 1947, *81,* 374.

References for the Experiment – 2 : Assessing Kinesthetic Sensitivity

Candland, D.K. *Psychology : Experimental Approach.* New York: McGraw-Hill Book Company, 1968.

Dietze, A.G. Kinesthetic discrimination: The difference limen for finger span. *J. Psychol.,* 1961, *51*, 268-273.

Laidlaw, R.W. and Hamilton, M.A. A study of threshold in apperception of passive movement among normal and control subject. *Bull. Neurol. Ins*. New York: 1937, *6*, 268-273.

Leukel, F. *Introduction to Physiological Psychology.* Saint Louis: The C.V. Mosby Company, 1968.

Oberlin, K.W. Variation in intensive sensitivity to lifted weights. *J. Exp. Psychol.,* 1936, *19*, 438-455.

Pattern, H.D. In T.C. Ruch and J.F. Fulton, *Medical Physiology and Biophysics.* Philadelphia : Sounders, 1960.

Reid, R.L. An illusion of movement complementary to the horizontal-vertical illusion. *Quart. J. Exp. Psychol.,* 1954, *6*, 109-111.

Stevens, S.S., and Stones, G. Finger span : Ratio scale, Category scale and JND scale. *J. Exp. Psychol.,* 1959, *57*, 91-95.

Wenger, M.A.; Jones, F.N.; and Jones, M.H. *Physiological Psychology.* New York: Holt, 1956.

CHAPTER 4

Physiology of Emotions

Appley, M.H., and Trumbull, R. (Eds.) *Psychological Stress : Issues in Research.* New York : Appleton-Century Crofts, 1947.

Ax, A.F. The physiological differentiation between fear and anger in humans. *Psychosom. Med.*, 1953, *15*, 433-442.

Bard, P. On emotional expression after decortication with some remarks on certain theroetical views, Part I and II. *Psychol. Rev.* 1934, *41*, 309-329, 424-449.

Bard, P. and Mountcastle, V.B. Some forebrain mechanisms involved in expression of rage with special reference to suppression of angry behaviour. *Res. Publ. Ass. Res. Nerv. Ment. Dis.*, 1947, *27*, 362-404.

Bills, A.G. The influence of muscular tension on the efficiency of mental work, *Amer. J. Psychol.*, 1927, *38*, 227-251.

Brady, J.V.; Porter, R.W.; Conard, D.G. and Mason, J.W. Avoidance behaviour and the development of gastroduodenal ulcers, *J. Exp. Anal. Behav.*, 1958, *1*, 69-73.

Candland, D.K. *Psychology : The Experimental Approach.* New York: McGraw-Hill Book Company, 1968.

Cannon, W.B. *Bodily Changes in Pain, Hunger, Fear and Rage,* 1915, (1st Ed.), 1928, 2nd Ed.: New York: Appleton Century Crofts.

Cannon, W.B. The James-Large theory of emotions: A critical examination and an alternative theory. *Amer. J. Psychol.*, 1927, *39*, 106-124.

Cannon, W.B. Again the James-Large and the thalamic theories of emotion. *Psychol. Rev.* 1931, *38*, 281-295.

Coleman, J.C. *Abnormal Psychology and Modern Life* (Rev. Ed.) Bombay: D.B.Taraporevala & Sons Ltd., 1974.

Darwin, C. *Expression of Emotions in Man and Animals.* London : J. Murray, 1872.

Duffy, E. *Activation and behaviour.* New York: John Wiley & Sons, 1962.

Epstein, W. and Shontz, F. *Psychology in Progress.* New York : Holt, Rinehart and Winston, INC., 1971.

Eysenck, H.J. *The Dynamics of Anxiety and Hysteria: An Experimental Application of Modern Learning Theory to Psychiatry.* London: Routledge & Kegan Paul, 1957.

Eysenck, H.J. Conditioning, Introversion-Extroversion and the Strength of the Nervous System. In H.J. Eysenck (Ed.). *Readings in Extroversion-Introversion,* Vol. 3. London: Stables, 1966.

Eysenck, H.J. *The Biological Basis of Personality.* Springfield, Ill.: Charles C. Thomas, 1967.

Franken Lauser, M. Experimental approaches to the study of catacholamines and emotion. In L. Levi (Ed.), *Emotions : Their Parameters and Measurement.* New York: Raven Press, 1975.

Goltz, F. Der Hund Ohne Grosshirn. *Archiv fur die gesamte Psychologie,* 1892, *51*, 570-614.

Grossman, S.P. *A Text book of Physiological Psychology.* New York : John Wiley & Sons, Inc., 1967.

Herrick, C.J. The functions of the olfactory parts of the cortex. *Proc. Nat. Acad. Sci.* (Washington) 1933, *19*, 7.

James, W. *The Principles of Psychology.* New York: Holt, 1890.

Kluver, H. and Bucy, P.C. Preliminary analysis of functions of the temporal lobes in monkeys. *Arch. Neurol. Psychiat.* (Chicago). 1939, *42*, 979-1000.

Lazarus, R.S. *Patterns of Adjustment* (3rd ed.). Tokyo: McGraw-Hill Kogakusha Ltd., 1976.

Leukel, F. *Introduction to Physiological Psychology.* Saint Louis: The C.V. Mosley Company, 1968.

Levi, L. The Urinary output of adrenalin and noradrenalin during pleasant and unpleasant emotional states: A preliminary report. *Psychosomatic Medicine,* 1965, *27*, 80-85.

Lindsley, D.B. Psychophysiology and motivation. In M.R. Jones (ed.) *Nebraska Symposium on Motivation.* Lincoln, Nebr: University of Nebraska Press, 1957.

Magoun, H.W. *The Ascending Reticular System and Wakefulness*. Springfield, Ill.: Charles C. Thomas, 1963.

Malmo, B.R. Activation: A neuropsychological dimension. *Psychological Review,* 1959, *66*, 367-386.

Mason, J.W. Strategy in psychosomatic research. *Psychosomatic Medicines,* 1970, *32*, 427-439.

Masserman, J.H. *Behaviour and Neurosis.* Chicago : University of Chicago Press, 1943.

Moskowitz, M.J. and Orgel, A.R. *General Psychology.* Boston: Houghton Mifflin Company, 1969.

Papex, J.W. A proposed mechanism of emotion. In Isaacson, R.L. *Neuropsychology.* New York: Harper & Row Publishers, 1964.

Russell, B. *An Outline of Philosophy* (6th ed.) London: George Allen and Unwin Ltd.,1951.

Pavlov, I.P. *Lectures on Conditioned Reflexes*. New York: Liveright, 1928.

Selya, H. *The Stress of Life.* New York: McGraw-Hill, 1956.

Sherrington, C.S. *The Integrative Action of the Nervous System.* London: Constable, 1906.

Sternbach, R.A. *Principles of Psychophysiology.* New York: Academic Press, 1966.

References for the Experiment – 3 : Measurement of Physiological Changes in Emotion

Altman, P.L. and Dittmer, D.S. (Eds.). *Biology Data Book.* Washington, D.C.: Federation of American Societies for Experimental Biology, 1964, pp. 239-241.

Bard, P. and Mount Castle, V.B. Some forebrain mechanism involved in expression of rage with special reference to suppression of angry behaviour. *Res. Publ.* Ass. *Res. Nerv. Ment. Dis.,* 1947, *27*, 362-404.

Candland, D.K. *Psychology: Experimental Approach,* New York: McGraw-Hill Book Company, 1968.

Duffy, E. *Activation and Behaviour.* New York : Wiley, 1962.

Hull, E.L. *Hypnosis and Suggestibility : An Experimental Approach*. New York: Appleton-Century Crofts, 1933.

Lindsley, D.B. Emotion. In S.S. Stevens (Ed.). *Handbook of Experimental Psychology.* New York: Wiley, 1951.

Moskowitz, M.J. and Orgel, A.R. *General Psychology.* Boston: Houghton Mifflin Company, 1969.

Murphy, G. *An Introduction to Psychology.* New Delhi : Oxford and IBH Publishing Co. 1964.

Postman, L. and Egan, J.P. *Experimental Psychology.* Ludhiana: Kalyani Publishers, 1949.

Woodworth, R.S. *Experimental Psychology,* New York: Henry Holt & Company, W.C., 1938.

Woodworth, R.S. and Marquis, D.G. *Psychology.* London: Methuen & Co. Ltd. 1952.

Woodworth, R.S. and Schlosberg, H. *Experimental Psychology* (Re. Ed), New York : Holt, 1954.

References for the Experiment 4 : State Trait Anger

Cannon, W.B. The James-Large theory of emotions: A Critical examination and an alternative theory. *American Journal of Psychology,* 1927, *39*, 106-24.

Spielberger, C.D. and London, P. Rage Boomarange: A Key to Preventing Coronories is emerging: Learn to handle anger and avoid seven deadly responses. *Indian Journal of Community Guidance Service,* 1984, *1,* 15-19. Distributed by the Author.

CHAPTER 5

General References for Perception

Bergson, Henri. *An Introduction to Metaphysics.*

Berkeley, G. An Essay toward a New Theory of Vision, Doulin (1709). In G. Berkeley, *Works on Vision.* (Edited by C. M. Turbayne). Westport, Conn.: Greenwood Press, 1981.

Boring, E.G. and Holway, A. the moon illusion and the angle of regard. *Amer. J. Psychol.,* 1940, *53*, 109-113.

Bornstein, M.H.; Kessen, W. and Weiskolf, S. The categories of hue in infancy. *Science*, 1976, *191*, 201-202.

Bovard, E.W. Social norms and the individual. *Jl. Abnom. Soci. Psychol.,* 1948, *43*, 62-69.

Candland, D.K. *Psychology : The Experimental Approach.* New York : McGraw-Hill Book Company, 1968.

Cohen, W. Apparent movement of simple figures in the Gangzeld. *Percept. Mot. Skills,* 1958, *8*, 32.

Fantx, R.L. The origin of form perception. *Scientific American,* May, 1961, 66-72.

Fernberger, S.W. New phenomena of apparent visual movement. *Amer. J. Psychol.,* 1934, *46,* 309-314.

Franklin, J.C., Schiela, P.C., Brozek, J. and A. Keys. Observations on human behaviour in experimental semi-starvation and rehabilitation. *J. Clin. Psychol.,* 1948, *4,* 28-45.

Garrett, H.E. *General Psychology.* (2nd ed.), New Delhi; Eurasia Publishing House (Pvt.) Ltd., 1975.

Gibson, J.J. The visual perception of objective motion and subjective movement. *Psychol. Rev.,* 1954, *61*, 304-314.

Gibson, E.J. and Walk, R.D. The visual cliff, *Scientific American,* 1960, *202,* 2-9.

Gregory, R.L. and Wallace, J.G. Recovery from early blindness: A case study. *Experimental Psychological Social Monograph,* 1963, *2.*

Gregory, R.L. *Eye and Brain : The Psychology of Seeing* (3rd ed.), New York : World University Library, 1977.

Hockberg, J. E. *Perception.* Englewood Cliffs, N.J: Prentice Hall, 1964.

Hull, C.L. *Principles of Behaviour.* New York: Appleton-Century Crofts, 1943.

Hull, C.L. *Essentials of Behaviour.* New Haven Conn.: Yale, 1951.

James, W. *The Principles of Psychology.* New York : Holt, 1890.

Kendler, H.H. *Basic Psychology.* London; Methuen & Co. Ltd., 1963.

Koffka, K. *Principles of Gestalt Psychology.* London: Routledge and Kegan Paul, 1950.

Kohlar, W. *Gestalt Psychology.* New York: Liveright, 1929.

Korte, A. Kinematostopische untersuchungen. *Z. Psycho.*, 1915, *72*, 193-290.

Leukel, F. *Introduction to Physiological Psychology.* Saint Louis; The C.V. Mosby Company, 1968.

Levine, R.; Chein, I. and Murphy, G. The relation of the intensity of a need to the amount of perceptual distortion: A preliminary report. *J. Psychol.,* 1942, *13*, 283-293.

Locke, J. *An Essay Concerning Human Understanding.* London (1890) edited by P.H. Nidditch. New York, Oxford University Press, 1929.

McClelland, D.C.; Atkinson, J.W. The projective expression of needs. 1. The effects of different intensities of the hunger drive on perception. *Jl. Psychol.*, 1948, *25*, 205-222.

Morris, C.G. *Psychology : An Introduction.* Englewood Cliffs, New Jersey; Prentice-Hall, Inc. 1973.

Moskowitz, M.J. and Orgel, A. R. *General Psychology.* Boston: Houghton Mifflin Company, 1969.

Munn, N.L. *Introduction to Psychology*. Boston: Houghton Mifflin, 1962.

Murphy, G. *An Introduction to Psychology*. New Delhi: Oxford and IBH Publishing Co., 1964.

Postman, L. and Egan, J.P. *Experimental Psychology : An Introduction.* Ludhiana: Kalyani Publishers, 1949.

Riesen, A.H. Arrested vision. *Scientific American.* July 1950, 16-19.

Russell, B. *An Outline of Philosophy* (6th Imp.) London: George Allen & Unwin Ltd. 1951.

Senden, M.V. *Space and Sight.* (Trans. by P. Heath), New York: Free Press, 1960.

Sherif, M. *The Psychology of Social Norms.* New York: Harper and Brothers, 1936.

Siipola, E.M. Implicit or Partial Reversion-Errors : A Technique of Measurement and its relation to other Measures of Transfer. *Journal of Experimental Psychology,* 1940, *26*, 53-73.

Stagner, R. and Karwoski, T.F. *Psychology.* New York: McGraw-Hill Book Company, 1952.

Tolman, E.C. *Purposive Behaviour in Animals and Men*. New York : Appleton-Century Crofts,1932.

Tolman, E.C. A Psychological model. In T. Persons and Shils (Eds.) *Towards a General Theory of Action.* Cambridge, Mass: Harvard, 1951.

Wallach, H. The perception of motion. *Scientific American.* July 1959, 56-60.

Wertheimer Max. Experimentalle studien uber des Sahen von Bewegung, *Z. Psychol.*, 1912, *61*, 161-265.

Wertheimer Max. Psychomotor coordination of auditory and visual space at birth. *Science*, 1961, *134*, 1692.

Wickens D.D. and Meyer, D.R. *Psychology*. New York: Holt, Rinehart and Winston, 1961.

Woodworth, R.S. and Marquis, D.G. *Psychology*. London: Methuen & Co., 1952.

References for the Experiment - 5 : Perception of Space in Depth

Bartly, S.H. *Vision: A Study of its Basis,* New York: D. Van Nostrant Co., 1941.

Boring, E.G. *Sensation and Prception in the History of Experimental Psychology,* New York: D. Appleton-Century Co., 1942.

Hilgard, E.R.; Atkinson, R.C. and Atkinson, R.L. *Introduction to Psychology* (Sixth Edition), New Delhi: Oxford and IBH Publishing Co., 1975.

Kendler, H. *Basic Psychology,* New York: Appleton-Century Co., 1963.

Moskowtiz, M.J. and Orgel, A.R. *General Psychology*. Boston: Houghton Mifflin Company, 1968.

Woodworth, R.S. and Schlosberg, H. *Experimental Psychology*. New York: Holt Rinehart and Winston, 1954.

References for Experiment – 6 : Perception of Size Constancy

Bower, T.G.R. The Visual work of infants. *Scientific American,* 1966, *215*, 80-92.

Gibson, E.J. and Walk, R.D. The "Visual Cliff", *Scientific American,* April 1960, 64-71.

Hilgard, E.R.; Atkinson, R.C. and Atkinson, R.L. *Introduction to Psychology*. New Delhi: Oxford & IBH Publishing Co., 1975.

Hochberg, J.E. *Perception*. New Delhi: Prentice-Hall of India Private Limited, 1969.

Kimble, G.A.; Garmezy, and Zigler, E. *Principles of Psychology* (6th ed.). New Delhi: Wiley Eastern Limited, 1985.

Moskowitz, M.J. and Orgel, A.R. *General Psychology*. Boston: Houghton Mifflin Company.

Postman, L. and Egan, J.P. *Experimental Psychology : An Introduction*. Ludhiana: Kalyani Publishers, 1949.

Stagner, R. and Karwoski, T.F. *Psychology*. New York : McGraw-Hill Book Company, 1952.

CHAPTER 6

General References for Learning, Transfer of Training, Remembering and Forgetting

Archer, C.P. Transfer of Training in Spelling. *University of Iowa Studies in Education,* 1930, *5*, No. 5.

Bayles, E.E. An Unemphasized Factor in Current Theories Regarding the Transfer of Training. *Journal of Educational Psychology,* 1936, *27*, 425-430.

Beach, F.D. The descent of Instinct, *Psychol. Rev.* 1955, *62*, 401-410.

Bedell, R. and Nelson, E.S. Word Attack as a Factor in Reading Achievement in the Elementary School, *Educational and Psychological Measurement,* 1954, *14*, 168-175.

Brown, R. and McNeill, D. The "tip of the tongue" Phenomenon. *Journal of Verbal Learning and Behaviour,* 1966, *6*, 325-337.

Bruce, R.W. Conditions of Transfer of Training. *Journal of Experimental Psychology,* 1933, *16*, 343-361.

Bunch, M.E. The Amount of Transfer in Rational Learning as a function of Time. *Journal of Experimental Psychology,* 1936, *22*, 315-322.

Chaplin, J.H. Scientific Curiosity in relation to Science Achievement in School, *British Journal of Educational Psychology,* 1953, *23*, 205-206.

Davenport, J.W.; Chamove, A.S.; Harlow, H.F. The Semi-automatic Wisconsin general test apparatus. *Behaviour Research Methods and Instrumentation,* 1970, *2*, 135-138.

Dewey, J. *The Quest for Certainty,* New York: Minton, Balch, 1929.

Edwards, A.J. and Scannell, D.P. *Educational Psychology*. Bombay : Allied Publishers Private Limited, 1975.

Ekstrend, B.R. Effects and sleep on memory. *J. Exp. Psychol.,* 1967, *75*, 64-72.

Freud, S. Psychopathology of everyday life. *In the Basic Writings and Sigmund Freud.* New York: Random House, 1938.

Frutchey, F.B. Retention in High School Chemistry, *The Educational Research Bulletin,* 1937, *8*, 217-218.

Garrett, H.E. *General Psychology.* (2nd ed.). New Delhi: Eruasia Publishing House (Pvt.) Ltd. 1975.

Gladis, M. Grade Differences in Transfer as a Function of the Time Interval between Learning Tasks. *Journal of Educational Psychology,* 1960, *51*, 191-194.

Hardex, K.C. The Relative Efficiency of the Separate and Together Methods of Teaching Hormonious. *Journal of Experimental Education,* 1937, *6,* 7-23.

Hendrickson, G. and Schroeder, W.H. Transfer of training in learning to hit a submerged target. *Journal of Educational Psychology,* 1941, *32*, 205-213.

Hilgard, E.R. and Bower, G.H. *Theories of Learning* (4th Ed) Englewood Cliffs, New Jersey: Prentice-Hall, Inc., 1975.

Hinde, R.A. *Animal Behaviour : A Synthesis of Ecology and Comparative Psychology.* New York: McGraw Hill, 1966.

Hudgins, B.B. Effects of Group Experience on Individual Problem Solving. *Journal of Educational Psychology*, 1960, *51*, 37-42.

James, William. *Principles of Psychology,* New York: Holt, 1908.

Jones, R.L. Interaction and Applicational Transfer Following Small Group Discussion-Decision. *American Psychologists*, 1961, *16*, 315.

Judd, C.H. The Relation of Special Training to General Intelligence. *Educational Review*, 1908, *36*, 28-41.

Kant, I. *Critique of Pure Reason* (Translated by J.M.D. Meiklejohn) London: George Bell, 1887.

Kersh, B.Y. The Motivating Effect of Learning by Directed Discovery. *Journal of Educational Psychology*, 1962, *53*, 65-71.

Kingsley, H.L. and Garry, R. *The Nature and Conditions of Learning*. Englewood Cliffs, N.J: Prentice-Hall, 1957.

Kliene, L.W. An Experimental Study of Associative Inhibition. *Journal of Experimental Psychology,* 1921, *4,* 270-289.

Kohler, W.*The Mentality of Apes* (Translated by E. Winter). New York : Harcourt, Brace and World, 1925.

Maier, N.R.F. An Aspect of Human Reasoning. *British Journal of Psychology:* 1933, *24*, 144-155.

McGeoch, J.A. The Influence of Sex and Age upon the Ability to Report. *American Journal of Psychology,* 1928, *40*, 458-466.

Melton, A.W. and Irwin, J.M. The influence of degree of interpolated learning on retroactive inhibition and the overt transfer of specific responses. *Amer. J. Psychol.,* 1940, *53,* 173-203.

Muchl, S. The Effect of Visual Discrimination Pertaining on Learning to Read a Vocabulary List in Kindergarten Children. *Journal of Education Psychology,* 1960, *51*, 217-221.

Munn, N.L. Bilateral Transfer of Training. *Journal of Experimental Psychology,* 1932, *15*, 343-353.

Murphy, G. *An Introduction to Psychology.* New York : Harper & Row Publishers, 1968.

Osgood, C.E. *Method and Theory in Experimental Psychology.* New York: Oxford University Press, 1953.

Pimsleur, P. and Bonkowski, R.J. Transfer of Verbai Material across sense Modelities. *Journal of Education Psychology,* 1961, *52*, 104-107.

Russell, B. *An Outline of Philosophy.* London: George, Allen & Unwin Ltd., 1951.

Schroder, H.M. and Rotter, J.W. Rigidity as Learned Behaviour. *Journal of Experimental Psychology,* 1952, *43*, 141-150.

Schultz, D.G. Dose One Year of High School Algebra Improve College Achievement? *American Psychologists,* 1960, *15*, 427.

Siipola, E.M. and Israel, H.E. Habit Interference as Dependent upon Stage of Training. *American Journal of Psychology,* 1933, *45*, 205-227.

Siipola, E.M. Implicit a Partial Reversion-Error: A Technique of Measurement and its relation to other Measures of Transfer. *Journal of Experimental Psychology,* 1940, *26*, 53-73.

Stroud, J.M. *Psychology in Education.* New York: Longmans Green, 1956.

Thorndike, E.L. *Animal Intelligence.* New York: Macmillan, 1911.

Underwood, B.J. *Experimental Psychology.* (2nd ed.). New York: Appleton-Century Crofts, 1966.

Watson, J.B. *Behaviourism,* New York: Norton, 1924.

Webb, L.W. Transfer of Training and Retroaction. *Psychological Review Monographs,* 1917, *24*, No. 104.

Woodworth, R.S. and Marquis, D.G. *Psychology.* London: Methuen & Co. Ltd., 1952.

Yarrow, L.J. The Effect of Antecent Frustration on Projective Play. *Psychological Monograph*, 1948, *62*, No. 6.

Yonge, G.D. Structure of Experience and Functional Fixedness, *Journal of Educational Psychology,* 1966, *57*, 115-120.

Yum, K.S. An Experimental Test of the Law of Assimilation. *Journal of Experimental Psychology,* 1931, *14*, 68-72.

References for the Experiment 7 : Learning by Insight

Candland, D.K. *Psychology : The Experimental Approach.* New York : McGraw-Hill Book Company, 1968.

Garrett, H.E. *General Psychology.* New Delhi. Eurasia Publishing House (Pvt) Ltd., 1975.

Kohler, W. *The Mentality Apes,* Translated by E. Winter. New York: Harcourt, Brace, 1925.

Murphy, G. *An Introduction to Psychology.* New Delhi : Oxford and IBH Publishing Co., 1964.

Postman, L. and Egan, J.P. *Experimental Psychology.* Ludhiana: Kalyani Publishers, 1949.

Woodworth, R.S. and Marquis, D.G. *Psychology.* London: Methuen & Co. Ltd., 1952.

References for the Experiment 8 : Study of Habit Interference

Allport, F.H. *Social Psychology.* Boston: Houghton Mufflin, 1924.

Garrrett, H.E. *General Psychology.* New Delhi: Eurasia Publishing Co., 1964.

Guthrie, E.R. *The Psychology of Learning* (Rev. Edition) New York: Harper & Row, 1952.

Hall, J.W. Elements of General Psychology. In Morse, W.C. and Wingo, G.M. *Psychology and Teaching* (3rd ed). Bombay: D.B. Taraporevala Sons & Co. Private Ltd., 1970.

Lefrancois, G.R. *Psychology for Teaching* (3rd ed.). Belmont California: Wadswords Publishing Company, Inc., 1979.

Murphy, G. *An Introduction to Psychology.* New Delhi: Oxford and IBH Publishing Co., 1964.

Woodworth, R.S. and Marquis, D.G. *Psychology,* London: Methuen & Co., 1952.

References for the Experiment 9 : Tapping Test – Learning

Milton Metfessal and *et al. Students' Guide for Demonstration of Psychological Experiments.* New York: McGraw-Hill Book Company, Inc., 1932.

Woodworth, R.S. and Schlosberg, H. *Experimental Psychology* (Rev.Ed.) New York: Holt, 1954.

References for the Experiment 10 : Study of Bilateral Transfer

Bruce, R.W. Conditions of Transfer of Training. *Journal of Experimental Psychology,* 1933, *16*, 343-361.

Candland, D.K. *Psychology : The Experimental Approach.* New York: McGraw-Hill Book Company, 1968.

Edwards, A.J. and Scannell, D.P. *Educational Psychology*. Bombay: Allied Publishers Private Limited, 1968.

Frutchey, F.P. Retention in High School Chemistry. *The Educational Research Bulletin*. 1937, *8*, 217-218.

Munn, N.L. Bilateral transfer of learning *J. Exp. Psycho,* 1932, *15*, 343-353.

Postman, L. and Egan, J.P. *Experimental Psychology,* Ludhiana: Kalyani Publishers, 1949.

Webb, L.W. Transfer of Training and Retention. *Psychological Review Monographs,* 1917, *24*, No. 104.

Woodworth, R.S. and Marquis, D.G. *Psychology*. London : Methuen & Co., 1952.

Yum, K.S. An Experimental Test of the Law of Assimilation. *Journal of Experimental Psychology,* 1931, *14*, 68-82.

References for the Experiment 11 : Transfer of Training (Cup and Ball)

Edwards, A.J. and Scannell, D.P. *Educational Psychology*. Bombay : Allied Publishers Private Limited, 1975.

Munn, N.L. Bilateral Transfer of Training. *Journal of Experimental Psychology,* 1932, *15*, 343-353.

Postman, L. and Egan, J.P. *Experimental Psychology:An Introduction*. Ludhiana: Kalyani Publishers, 1949.

Renshow, S. The visual perception and reproduction of forms by tachistoscopic methods. *Jl. Psychol.,* 1945, *20*, 217-232.

Webb, L.W. Transfer of Training and Retraction. *Psychological Review Monographs,* 1917, *24*, No. 104.

Woodworth, R.S. *Experimental Psychology,* New York: Henry Holt & Co., 1938.

References for the Experiment 12 : Paired Associates Learning – Immediate Memory Span for Numbers

Atkinson, R.C. and Shiffrin, R.M. *Mathematical Models for Memory and Learning*. Stanford: Stanford University Press, 1965.

Atkinson, R.C. and Shiffrin, R.M. Human memory: A proposed system and its control processes. In K.W. Spence and J.T. Spence (Eds.). *The Psychology of Learning and Motivation,* Vol. 2. New York: Academic Press, 1968.

Atkinson, R.C. and Wickens, T.D. Human memory and the concept of reinforcement. In R. Glaser (Ed). *The Nature of Reinforcement*. New York: Academic Press, 1971.

Bernbach, H.A. Stimulus learning and recognition and paired associate learning. *Doctoral Dissertation*. Univ. Michigan, 1965.

Bower, G.H. Verbal learning. In H.H. Helson and W. Beven (Eds). *Contemporary Approach to Psychology*. Princeton: Van Orostrad, 1967.

Candland, D.K. *Psychology : The Experimental Approach*. New York : McGraw Hill Book Company, 1968.

Conrad, R. and Hille, B.A. The decay theory of immediate memory and paced recall. *Canad. J. Psychol.* 1958, *12*, 1-6.

D'Amato, M.R. *Experimental Psychology: Methodology, Psychophysics, and Learning.* New York : McGraw-Hill Book Company, 1970.

Ebbinghaus, H. *Memory : A Contribution to Experimental Psychology.* Trans. by H.A. Ruger and C.E. Bussenius, New York: Teachers College, Columbia University, 1913.

Falkanberg, P.R. Recall improves in short-term memory the more recall context resembles learning context. *J. Exp. Psychol.* 1972, *95*, 39-47.

Fechner, G.T. *Elements of Psychophysics* (Trans. by. H.E. Adler), New York: Holt, Rinehart and Winston, 1860.

Harriman, P.L. *The New Dictionary of Psychology.* New York : Philosophical Library, 1947.

Hebb, D.O. *The Organisation of Behaviour :* New York: Wiley, 1949.

Hebb, D.O. *The Organisation of Behaviour : A Neurophysiological Theory.* New York : Basic Books, 1961.

Hilgard E.R. and Bower, G.H. *Theories of Learning.* (4th ed.) Englewood N.J.: Prentice-Hall, 1975.

Kimble, G.A., Hilgard and Marquis *Conditioning and Learning* (2nd ed), New York: Appleton-Century-Crofts, 1961.

Kintsch W. and Morris, C.J. Application of Markov model to free recall and recognition. *J. Exp. Psychol.* 1964, *69*, 200-206.

Miller, G.A. The magical number seven plus or minus two: Some limits on our capacity for processing information. *Psychol. Rev.* 1956, *63*, 81-97.

Muller, C.E. and Pilzecker, A. Experimentable Beitrape Zur Zehre vom Gedachtrio, *Z. Psychol.,* Erglad, *1*, 1900.

Olson, G.M. Learning and retention in a continuous recognition task. *J. Exp. Psychol.*, 1969, *81*, 381-384.

Peterson, L.R. Immediate memory: Data and theory. In C.N. Cofer & B.S. Musgrave (Eds.), *Verbal Behaviour and Verbal Learning,* New York : McGraw-Hill, 1967.

Peterson, L.R. and Peterson, M.J. Short-term retention of individual verbal items. *J. Exp. Psychol.* 1959, *58*, 193-198.

Postman, L. and Egan, J.P. *Experimental Psychology.* An Introduction. Ludhiana : Kalyani Publishers, 1949.

Mumelhart, D.E. The effects of interpretation intervals on performance in a continuous paired associate task. *Unpublished Doctoral Dissertation.* Stanford University, 1967.

Runquist, W.N. Verbal behaviour. In J.B. Sidowski (Ed.) *Experimental Methods and Instrumentation in Psychology.* New York: McGraw-Hill, 1966.

Thorndike, E.L. Animal intelligence : An experimental study of the associative processes in animals. *Psychol. Monogr.* 1898, No. 8.

Tulving, E. and Medigan, S.A. Memory and verbal learning, *Ann. Rev. Psychol.,* 1970, *21*, 437-484.

Underwood, B.J. *Experimental Psychology,* New York : Appleton-Century Crofts, 1949.

Woodworth, R.S. *Experimental Psychology*, New York : Holt, 1947.

References for the Experiment 13 : Paired Associate Learning – Immediate Memory Span for Non-Sense Syllables

Atkinson, R.C. Ingredients for a theory of instruction. *Amer. Psychol.*, 1972, *27*, 921-931.

Ebbinghaus, H. *Memory.* Translated by H.A. Ruger and C.E. Bussenius. New York: Teachers College, 1913.

Kimble, G.A.; Garmezy, N. and E. Zigler. *Principles of Psychology.* New York: John Wiley & Sons, Inc., 1984.

Miller, G.A. The magical number seven plus or minus two: Some limits on our capacity for processing information. *Psychol. Rev.,* 1956, *63*, 81-97.

Postman, L. and Riley, D.A. A Critique of Kohler's Theory of Association, *Psychological Review,* 1957, *64*, 61-72.

Underwood, B.J. *Experimental Psychology.* (2nd ed.) Englewood Cliffs, N.J.: Prentice-Hall, 1966.

References for the Experiment 14 : Paired Associate Learning – Immediate Memory Span for Meaningful Words

Cook, J.O. and Spitzer, M.E. Supplementary report: Prompting versus confirmation in paired-associate learning. *Journal of Experimental Psychology,* 1960, *59*, 275-276.

D'Amato, M.R. *Experimental Psychology: Methodology, Psychophysics, and Learning.* New York: McGraw-Hill Book Company, 1970.

Gibson, E.J. A systematic application of the concepts of generalization and differentiation to verbal learning. *Psychological Review,* 1940, *47*, 196-229.

Hall, J.W. Elements of General Psychology. In Morse, W.C. and Wirgo, G.M. (Eds). *Psychology and Teaching* (2nd ed). Glen View, Illinois: Scott, Foresman & Company, 1969.

Hawker, J.R. The influence of training procedure and other task variables in paired associate learning. *Journal of Verbal Learning and Verbal Behaviour,* 1964, *3*, 70-76.

Jung, J.A. Cumulative method of paired associate and serial learning. *Journal of Verbal Learning and Verbal Behaviour,* 1964, *3*, 290-299.

Kendler, H.H. Motivation and behaviour. In D. Levine (Ed.) *Nebraska Symposium on Motivation.* Lincoln: University of Nebraska Press, 1965.

Lockhead, G.R. Methods of presenting paired associates. *Journal of Verbal Learning and Verbal Behaviour,* 1962, *1*, 62-65.

Postman, L., and Goggin, J. Whole *vs.* part learning of paired associate lists. *Journal of Experimental Psychology,* 1966, *71*, 867-877.

Runquist, W.N. Verbal behaviour. In J.B. Sidowski (Ed.) *Experimental Methods and Instrumentation in Psychology.* New York : McGraw-Hill, 1966.

Sidowski, J.B. Response Prompting. The practice time variable. *Psychonomic Science,* 1968, *13*, 311-312.

References for the Experiment-15 : Serial Learning – Non-Sense Syllables

Atkinson, R.C. A stochastic model for rote serial learning. *Psychometrika,* 1957, *22*, 87-95.

Glanzer, M. and Peters, S.C. Re-examination of the serial position effects. *Journal of Experimental Psychology,* 1962, *64*, 256-266.

Morris, C.G. *Psychology: An Introduction* Englewood Cliffs, New Jersey: Prentice-Hall, Inc., 1973.

Young, R.K. Serial learning. In T.R. Dixon and D.L. Horton (Eds.) *Verbal Behaviour and General Behaviour Theory.* Englewood Cliffs, NJ: Prentice-Hall, 1968.

References for the Experiment-16 : Serial Learning – Meaningful Words

Glanzer, M. and Peters, S.C. Re-examination of serial position effect. *Journal of Experimental Psychology,* 1962, *64*, 258-266.

Hilgard, E.R. and Bower, G.H. *Theories of Learning* (Fourth Edition). Englewood Cliffs, New Jersey: Prentice-Hall, Inc., 1975.

Jenson, A. Spelling errors and serial position effect. *Journal of Educational Psychology,* 1962, *53*, 105-109.

Megeogh, J.A. and Irion, A.L. *The Psychology of Human Learning.* (2nd ed.) New York: David Mickay, 1952.

References for the Experiment – 17 : Free Learning - Meaningful Words

Candland, D.K. *Psychology : The Experimental Approach.* New York: McGraw-Hill Book Company, 1968.

Craik, F.I.M. The fate of primary memory items in free recall, *Journal of Verbal Learning and Verbal Behaviour,* 1970, *9*, 143-148.

D'Amato, M.R. *Experimental Psychology, Methodology, Psychophysics and Learning.* New York: McGraw-Hill Book Company, 1970.

Deese, J.E. *Psychology as Science and Art.* New York: Harcourt Brace Jovanovich, 1972.

Feigenbaun, E.A. and Simon, H.A. Performance of a reading task by an elementary perceiving and memorising program. *RAND Corp. Paper,* p. 2358.

Glanzer, M.S. Storage mechanism in recall. In G.H. Bower, ed., *The Psychology of Learning and Motivation : Advances in Research and Theory,* Vol. 5, New York: Academic Press, 1972.

Hall, J.W. Elements of General Psychology. In Morse, W.C. and Wingo, G.M. (Eds). *Psychology and Teaching* (3rd edition) Glenview, Illinois: Scott. Foresman & Company, 1969.

Murdock, B.B. *Human Memory: Theory and Data.* Potomac, Md. : Lawrence Eribaum, 1974.

Postman, L. Experimental analysis of learning to learn. In Bower, G.H. and Spence, J.T. (Eds). *The Psychology of Learning and Motivation*, New York: Academic Press, 1919.

Simon, H.A. and Feigerbaun, E.A. An information-processing theory of some effects of similarity, familiarization and meaningfulness in verbal learning. *Journal of Verbal Learning and Verbal Behaviour*, 1964, *3*, 385-396.

Tulving, E. Subjective organization and effects of repetition in multitrial free recall learning . *Journal of Verbal Learning and Verbal Behaviour*, 1966, *5*, 193-197.

Underwood, B.J. The Representativeness of Rote Verbal Learning in A.W. Melton (ed). *Categories of Human Learning*, New York: Academic Press, 1964.

References for the Experiment – 18 : The Effect of Mnemonic Strategies on Memory

Atkinson, R.C. Mnemotechniques in second language learning. *American Psychol.* 1975, *30*, 821-28.

Bobrow, S.A. and Bower, G.H. Comprehension and recall of sentences, *J. Exp. Psycho,* 1969, *80*, 455-61.

Bower, G.H. Analysis of mnemonic device. *American Scientist,* 1970, *58*, 596-610.

Bower, G.H, and Hilgard, E.R. *Theories of Learning* (5th ed.), New Delhi : Prentice-Hall of India, 1986.

Cermak, L.S. *Improving Your Memory.* New York: McGraw-Hill, Co. 1975.

Mandler, G. Organization and Memory. In K.W. Spence & J.T. Spence (eds.), *The Psychology of Learning and Motivation : Advances in Research and Theory,* Vol. 1. New York : Academic Press, 1965.

Mandler, G. *Association and Organization : Facts, Fancies and Theories.* In T.R. Dixon & D.L. Horton, (eds.) Verbal behaviour and general behaviour theory. Englewood Cliffs, N. J.: Prentice-Hall, 1968.

McQurie, W.J. A multiprocess model for paired-associate learning. *J. Exp. Psychol.,* 1961, *62,* 335-47.

Raugh, M.R. and Atkinson, R.C. An application of the mnemonic key word method in the acquisition of a Russian Vocabulary. *J. Exp. Psychol. Human Learning and Memory,* 1979, *104,* 126-133.

Tulving, E. and Madigan, S.A. Memory and Verbal Learning. *Annual Review of Psychology,* 1970, *21,* 437-484.

CHAPTER 7

General References : Thinking, Imagination and Language Comprehension

Berger, R.J. and Oswald, I. Eye movements during active and passive dreams. *Science,* 1962, *137*, 601.

Berko, J. The child's learning of English morphology, *Word,* 1958, *14,* 150-177.

Binet, A. and Simon, T. Upon the Necessity of Establishing a Scientific Diagnosis of Interior States of Intelligence. In James D. Jenkins and Paterson, D.G. (Eds.) *Studies in Individual Differences.* New York : Appleton-Century Crofts, 1961.

Birch, H.C. The relation of previous experience to insightful problem-solving, *Journal of Comparative and Physiological Psychology,* 1945, *38,* 367-383.

Birch, H.C. and Rabinowitz, H.S. The negative effect of previous experience on productive-thinking, *Journal of Experimental Psychology,* 1951, *41,* 121-125.

Brown, R. *A First Language : The Early Stages.* Cambridge, Mass : Harvard University Press, 1973.

Bruner, J.S., Goodnow, J.J. and Austin, G.A. Fixed choice strategies in concept attainment. *Amer. Psychol* , 1953, *8,* 314.

Bruner, J.A., Goodnow, J.J. and Austin, G.A. *A Study of Thinking,* New York : John Wiley, 1956.

Calkins, M.W. Statistics of dreams. *Amer. J. Psychology,* 1893, *5,* 311-343.

Cartwright, R.D. and Monroe, L.J. Relation of dreaming and REM sleep. The effects of dream deprivation under two conditions. *J. Pers. Soc. Psychol.,* 1968, *10,* 69-74.

Dropley, A.J. Originality, intelligence and personality, unpublished Ph.D. dissertation, University of Abbesta Edmonton, Alberta, Canada, 1965.

De Bono, E. *Lateral Thinking : A Text Book of Creativity,* London : Ward Lock Educational, 1970.

De Bono, E. *Teaching Thinking.* London : Smith, 1976.

Dement, W. and Wolpert, E.A. The relation of eye movements, body motility and external stimuli to dream content. *J. Exp.Psychol.,* 1958, *56,* 543-553.

Derment, W. An essay on dreams. In T.M. Newcomb (ed.), *New Directions in Psychology II,* New York : Holt, Rinehart and Winston, 1965.

Duncker, K. On problem solving. *Psychological Monographs,* 1945, *58,* 5.

Ellson, D.G. Critical conditions influencing sensory conditioning. *Journal of Exp. Psychol.,* 1942, *31,* 333-338.

Erikson, E.H. The dream specimen of psychoanalysis. *Journal of American Psychoanalytic Association,* 1954, *2,* 5-56.

Foulkes, D. Dream reports from different stages of sleep. *J. Abnorm. Soc. Psychol.,* 1962, *65*, 14-25.

Foulkes, D. and Vogel, G. Mental activity at sleep on set. *J. Abnorm. Psychol.,* 1965, *70,* 231-243.

Freud, S. *The Interpretation of Dreams* (1900), Standard ed. 1953. London : Hogarth Press.

Freud, S. *New Introductory Lectures on Psychoanalysis.* New York, Norton, 1933.

Frisch, Karl Von. *Bees : Their Vision, Chemical Senses and Language,* Ithaca, New York : Cornel University Press, 1950.

Gagne, R.M. *The Conditions of Learning* (2nd ed.). New York : Holt, Rinehart and Wiston, 1970.

Galton, F. *Enquiries into Human Faculty and its Development,* New York : Macmillan, 1883.

Garrett, H.E. *General Psychology,* New Delhi : Eurasia Publishing House (Pvt.) Ltd., 1975.

Getzels, J.W. and Jackson, P.W. *Creativity and Intelligence,* New York : Wiley, 1962.

Goldstein, K. *The Organism,* New York : American Book, 1935.

Graf, R. and Torrey, J.W. *Perception Phrase Structure in Written Language.* American Psychological Association. Convention Proceedings, Washington, D.C. American Psychological Association, 1966.

Greenfield, P.M. and Smith, J.H. *The Structure of Communication in Early Language Development.* New York ; Academic Press, 1976.

Guilford, J.P. Three Faces of Intellect. *The American Psychologist,* 1959, *14,* 469-479.

Haber, R.N. Eidetic Images, *Scientific American,* 1969, 36-44.

Hadamard, J. *An Essay on the Psychology of Invention in Mathematical Field.* Princeton : Princeton University Press, 1945.

Hall, C.S. A cognitive theory of dream symbols. *J. Gen. Psychol.,* 1953, *49,* 69-74.

Hall, C.S. *The Meaning of Dreams,* New York : Harper & Row, 1953.

Hall, C.S. and Van de Castle, L. *The Content Analysis of Dreams,* New York: Appleton-Century Crofts, 1966.

Haufmann, E. and Kasanin, J. A method for the study of concept formation, *J. Psychol.,* 1957, *3,* 531-540.

Hauks, P. (ER) *Collins Dictionary of the English Language,* London, Collins, 1984.

Hartmann, E. *The Biology of Dreams,* Springfield Ill. : Charles C. Thomas, 1967.

Hartmann, E, Bernstein, J. and Wilson, C. Sleep and Dreaming in the Elephant, *Psychophysiology,* 1968, *4,* 389.

Hayes, C. and Hayes, K. *The Apes in Our House,* New York : Harper & Row, 1951.

Hebb, D.O. *A Textbook of Psychology* (2nd ed.), Philadelphia : W.B. Saunders, 1966.

Heidbrader, E. The attainment of concepts III. The Problem, *Journal of Psychology,* 1947, *24,* 93-138.

Hilgard, E.R., Atkinson, R.C., and Atkinson, R.L. *Introduction to Psychology.* (6th ed.), New Delhi, Oxford & IBH Publishing Co., 1915.

Horton, L.H. Lavitation dreams : their physiology. *J. Abnorm Psychol.,* 1919, *14,* 145-172.

Hunt, J. Mav. *Intelligence and Experience,* New York, Ronald Press, 1961.

Kellogg, W.N. and Kellogg, L.A. *The Ape and the Child,* New York, McGraw-Hill, 1933.

Kendler, H.H., and Kendler, T.S. Effect of Verbalization on reversal shifts in children. *Science,* 1961, *141,* 1919-1920.

Ladd, G.T. Contribution to the Psychology of visual dreams, *Mind,* 1892, *1,* 299-304.

Lefrancois, G.R. *Psychology for Teaching* (3rd ed)., Belmont, California, Wadsworth Publishing Company, Inc, 1979.

Lenneberg, Eric. *Biological Foundations of Language,* New York, Wiley, 1967.

Levine, M. Hypothesis behaviour by humans during discrimination learning, *Journal of Experimental Psychology,* 1966, *71,* 331-338.

Limber, J. The genesis of complex sentences, In Moore, T.G. (ed.), *Cognitive Development and the Acquisition of Language,* New York, Academic Press, 1973.

Lindgren, H.C. *An Introduction to Social Psychology,* (second edition), New Delhi, Wiley Eastern Private Limited, 1974.

Mackinnon, D.M., and Dukes, W.F., Repression, In L. Postman (ed.), *Psychology in the Making,* New York, Alred A. Knopf, 1962.

Maier, N.R.F. Reasoning humans: III. The mechanics of equivalent stimuli and reasoning, *Journal of Abnormal and Social Psychology,* 1945, *35,* 349-360.

Maslow, A.H. *Motivation and Personality* (2nd ed.), New York, Harper & Row, Publishers, 1970.

Meenes, M. and Morton, M.A. Characteristics of eidetic phenomenon, *J. Gen. Psychol.,* 1936, *14,* 370-391.

Meer, B., and Stein, M.L. Measures of intelligence and creativity, *Journal of Psychology,* 1955, *39,* 117-126.

Mill, J.A. *Autobiography,* New York, Holt, 1873.

Morgan, C.D. and Murray, H.H. Method for investigating fantasies, *The Thematic Apperception Test, Acta Neurol., Psychiat,* 1935, *34,* 289-306.

Morris, C.G. *Psychology : An Introduction,* Englewood Cliffs, New Jersey, Prentice-Hall, INC., 1973.

Morse, W.C. and Wingo, G.M. *Psychology and Teaching* (3rd Ed.), Bombay, D.B. Taraporevala Sons & Co. Private Ltd., 1970.

Moskowitz, M.J. and Orgal, A.R. *General Psychology:* Boston : Houghton Mifflin Company, 1969.

Murphy, G. *An Introduction to Psychology,* New Delhi, Oxford & IBH Publishing Co., 1964.

Nelson, K. Structure and strategy in learning to talk. *Monographs of the Society for Research in Child Development,* 1973, *38,* 149, 1-2.

Osgood, C.E., *Method and Theory in Experimental Psychology,* New York, Oxford University Press, 1953.

Osgood, C.E. and Suci, G.J. Factor analysis of meaning. *Journal of Experimental Psychology,* 1955, *50,* 325-338.

Osgood, C.E., Suci, G.J. and Taurenbaum, P.J. *The Measurement of Meaning,* Urban, Ill, University of Illinois Press, 1957.

Osgood, C.E. Studies on the generality of affect meaning system, *American Psychologist,* 1962, *17,* 10-18.

Osgood, C.E. Semantic differential technique in the comparative study of cultures in Jakobovits, L.A. and Miron, M.S.(eds.) *Readings in the Psychology of Language,* Englewood Cliffs, N.J., Prentice Hall, 1967.

Perky, C.W. An experimental study of imagination, *Amer. J. Psycho.,* 1910, *21,* 422-452.

Piaget, J. *The Language and Thought of the Child,* New York : Harcourt, Brace & World, 1926.

Piaget, J. *Judgement and Reasoning in the Child,* New York : Harcourt, Brace & World, 1928.

Piaget, J. *The Moral Judgement of the Child,* London : Kegan Paul, 1932.

Piaget, J. *The Child Conception of Number,* New York : Humanities Press, 1952.

Piaget, J. *The Origins of Intelligence in Children,* New York : International University Press, 1952.

Piaget, J. How children form Mathematical Concepts, *Scientific American,* 1953, *189*, 74-79.

Piaget, J. *The Construction of Reality in the Child,* New York : Basic Books, 1954.

Piaget, J. The Genetic Approach to the Psychology of Thought. *Journal of Educational Psychology,* 1961, *52,* 275-281.

Piaget, J. The stages of the intellectual development of the child, *Bulletin of the Menninger School of Psychiatry,* March, 1961..

Premack, B. *Intelligence in Ape and Man,* Hillsdale, N.J. : Lawrance Erilbaurn Associates, 1976.

Pribrunn, K.N. The new neurology : Memory, novelty, thought, and choice, *in EEG and Behaviour* (ed) C.N. Glaser, New York : Basic Books, 1963.

Rath, L.F. Jones, A. Rothstein, A. and Wassermann, S., *Teaching for Thinking : Theory and Application,* Columbus, Ohio : Charles E. Merrill, 1967.

Rechtschaffen, A., Verdone, P. Wheaton, J. Reports of Mental Activity during Sleep, *Canadian Psychiatric Association Journal,* 1963, *8,* 409-414.

Rechtschaffen, A. Dreams reports and dreams experience, *Experimental Neurology,* 1967, Supplement 4, 4-15.

Reik, Theodor, *Listening with the Third Ear,* New York. Farrer, 1949.

Rosch, E. On the internal structure of perceptual and semantic categories, In T.E. Moore (Ed.) *Cognitive Development and the Acquisition of Language,* New York : Academic Press, 1973.

Rosch, E. Human categorization, In N. Warren (ed.) *Advances in Cross Cultural Psychology,* Vol. 1, London : Academic Press, 1977.

Rumbaugh, D. M., Glaserfeld, E. Von Warner, H., Pisani, P. and Gill, T. V. Lane (Chimpanzee) *Learning A Language : A Progress Report, Brain and Language,* 1974, *1,* 205-212.

Seyfurth, R.M.; Cheney, D.L.; and Marlar, P. Monkey responses to three different alarm cells : *Evidence of Predator Classification and Semantic Communication Science,* 1980, *210*, 801-803.

Skinner, P.F. *The Technology of Teaching,* Englewood Cliffs, New Jersey : Prentice-Hall, Inc., 1968.

Smoke, K.L. An objective study of concept formation, *Psychological Monograph,* 1932, 42.

Thorndike, R.L. The measurement of creativity, *Teacher's College Record,* 1963, *54*, 422-424.

Titchner, E. B. *Lectures on the Experimental Psychology of the Thought Processes,* New York : Macmillan, 1909.

Torrance, E.P. *Building Creative Talent,* Englewood Cliffs, N.J.: Prentice Hall, 1962.

Vigotsky, L.S. *Thought and Language,* Translated from Russian and edited by E. Hamsman and G. Verkan, Cambridge, Mass : M.T. Press, 1962.

Wallach, M.A., and Kogan, N. *Modes of Thinking in Young Children : A Study of the Activity-intelligence Distinction,* New York : Holt, Rinehart and Winston, 1962.

Watson, J.B. The Unverbalized human behaviour, *Psychological Review,*1924, *31*, 273-280.

Weed, S. and Hallan. A study of dream consciousness, *Amer. J. Psychol.,* 1986, *7,* 405-411.

Wertheimer, M. *Productive Thinking,* New York : Harper & Row, 1945.

Wertheimer, M. *Productive Thinking* (enlarged ed.) Michael Wertheimer (ed.), New York : Harper & Row, 1959.

Woodworth, R.S. and Marquis, D.G. *Psychology,* London : Methuen & Co., 1952.

References for the Experiment-19 : Concept Formation

Candland, D.K. *Psychology : The Experimental Approach* : New York : McGraw-Hill Book Company, 1968.

Garrett, H.E. *General Psychology,* New Delhi : Eurasia Publishing House (Pvt.) Ltd., 1975.

Goldstein, K. *The Organism,* New York : American Book, 1935.

Haufmann, E. and Kasanin, J. A method for the study of concept formation, *Journal of Psychology,* 1937, *7,* 521-540.

Haufmann, E. A study of personal patterns in an intellectual performance, *Character and Personality,* 1941, *9,* 315-325.

Heidbreder, E. The attainment of concepts : III The Problem, *Journal of Psychology,* 1948, *24,* 93-138.

Hilgard, E.R., Atkinson, R.C. and Atkinson, R.L. *Introduction to Psychology,* New Delhi : Oxford and IBM Publishing Co., 1975.

Kimble, G.A.; Garmezy, N. and Zigler, E. *Principles of Psychology* (Sixth edition), New Delhi : Wiley Eastern Limited, 1985.

Levine, M. Hypothesis behaviour by humans during discrimination learning. *Journal of Experimental Psychology,* 1966, *71,* 331-338.

Murphy, G. *An Introduction to Psychology,* New Delhi : Oxford and IBM Publishing Co., 1964.

Piaget, J. *The Origin of Intelligence in Children,* New York : International University Press, 1952.

Rosch, E. On the internal structure of perceptual and semantic categories, In T.E. Moore (ed.) *Cognitive Development and the Acquisition of Language,* New York : Academic Press, 1973.

Rosch, E. Human categorization, In N. Warren (Ed.) *Advances in Cross-cultural Psychology,* Vol. I, London : Academic Press, 1977.

Smoker, K.L. An objective study of concept formation, *Psychological Monographs,* 1932, 42.

Woodworth, R.S. and Margins, D.G. *Psychology,* London : Methuen and Co. Ltd., 1952.

References for Experiment-20 : Concept Achievement

Goldstein, K. *The Organism,* New York : American Book, 1935.

Haufmann, E. and Kasanin, J. A method for the study of concept formation, *J. Psychol,* 1957, *3,* 531-541.

Heidbreder, E. The attainment of concepts III. The problem, *Journai of Psychology,* 1947, *24,* 95-138.

Hilgard, R.H., Atkinson, R.C. and Atkinson, R.L. *Introduction to Psychology (Sixth Edition),* New Delhi : Oxford & IBH Publishing Company, 1975.

Horniblow, E.C.T. and Sullivan, J.J. *The March of Time,* Book I, London., The House of Grant Ltd., 1955.

Maslow, A.H. *Motivation and Personality* (2nd ed.), New York., Harper & Row Publishers, 1970.

Piaget, J. *The Language and Thought of the Child,* New York, Harcourt Brace, World, 1926.

Smoke, K.L. An objective study of concept formation, *Psychological Monographs,* 1932, *42.*

Vigotsky, L.S. *Thought and Language,* Translated from Russian and edited by E. Hamsman and G. Verkan, Cambridge, Mass : M.T. Press, 1962.

Watson, J.B. *Psychology from the Standprint of a Behaviourist,* (2nd ed.) Philadelphia : J.A. Hippincolt, 1919.

References for the Experiment-21 : The Semantic Differential

Candland, D.K. *Psychology : The Experimental Approach,* New York : Mc-Graw Hill Book Company, 1968.

Galton, F. Psychometric experiments, *Brain,* 1979, *2,* 149-162.

Jung, C.G. *Studies in World-Association,* (M.D. Eder, Trans.) London : Heinemann, 1919.

Kent, G.H. and Rosanoff, A.J. A study of association in insanity, *Amer. J. Insanity,* 1910, *67*, 37-96, 317-390.

Kimble, G.A.; Gusmezy, N.; and Zigler, E. *Principles of Psychology* (Sixth edition), New Delhi : Wiley Eastern Limited, 1985.

Lane, H. Psychophysical parameters of vowel perception, *Psychol. Monogr.,* 1962, *76,* No. 44.

Lane, H.L.; Catania, A.C.; and Stevens, S.S. Voice Level : Autophonic Scale, perceived loudness and effect of sidestone, *J. Acoust. Soc. Amer.,* 1961, *33,* 160-167.

Lindgren, H.C. *The Semantic Differential as a Measure of Perceptual set towards various Countries,* unpublished research, San Francisco State College, 1966.

Osgood, C.E., and Suci, G.J. Factor analysis of meaning, *Journal of Experimental Psychology,* 1955, *50,* 325-338.

Osgood, C.E.; Suci, G.J.; and Tannenbaum, P.H. *The Measure of Meaning,* Urbana Ill, University of Illinois Press, 1957.

Osgood, C.E. Studies on the generality of effect meaning system, *American Psychologist,* 1962, *17*, 10-18.

Parry, J. and Adiseshiah, W. *Experimental Psychology,* Bombay : Allied Publishers Private Limited, 1977.

Shannon, C.E. Mathematical theory of communication, *Bell. Systtech. J.,* 1948, *27,* 379-423, 623-656.

CHAPTER 8

General References for Problem Solving and References for the Experiments 22 : Cut Out Puzzle Problem; 23 : Pyramid Puzzle Problem and 24: Outline Puzzle Problem

Arnold, J.E. Useful creative techniques in *A Source Book for Creative Thinking,* Eds Sidney J. Parnes and Harold F. Harding, New York : Scribner's & Sons, 1962.

Birch, H.G. The relation of previous experience to insightful problem-solving, *Jl. Comp. Psychol.,* 1945, *38*, 367-383.

Birch, H.C. and Rabinowitz, H.S. The negative effect of previous experience on productive thinking, *Journal of Experimental Psychology,* 1951, *41*, 121-125.

Birren, J.E. *The Psychology of Aging,* Englewood Cliffs, N.J. : Prentice-Hall, 1964.

Bruner, J.S. Learning Thinking, *Harvard Educational Review,* 1959, *29,*184-192.

Bruner, J.S. The Functions of Teaching, in William C. Morse and G. Max Wingo (eds.) *Readings in Educational Psychology,* Chicago : Scott, Foresman, 1962, pp. 162-169.

Bruner, J.S.; Oliver, R.R.; and Greenfield, P.M. *Studies in Cognitive Growth,* New York : Wiley, 1966.

Candland, D.K. *Psychology : The Experimental Approach,* New York : McGraw Hill Book Company, 1968.

Dewey, J. *How we Think,* Boston : Heath, 1910, 1933.

Duncker, K. On problem-solving, *Psychological Monographs,* 1945, *58* (5).

Durkin, H.E. Trial and error, gradual analysis and sudden recognition : An experimental study of problem solving, *Arch. Psychol.,* 1937, No. *210*.

Gagne, R.M. *The conditions of learning* (2nd Ed.), New York : Holt, Rinehart and Winston, 1970.

Gagne, R.M. *Essentials of Learning for Instruction,* Hinsdale, III, Dryden Press, 1974.

Gordon, W.J.J. *Synectics : The Development of Creative Capacity,* New York : Harper & Row, 1961.

Hafele, J.W. *Creativity and Innovation,* New York : Reinhold, 1962.

Haufmann, E. and Kasanin, J. A method for the study of concept formation, *Journal of Psychology,* 1937, *3,* 528-540.

Haufmann, E. A study of personal patterns in an intellectual performance, *Character and Personality,* 1941, *9,* 315-325.

Kohler, W. *The Mentality of Apes,* New York, Harcourt, Brace, 1924.

Lefrancois, G. R. Developing creativity in high school students, Unpublished M.Ed. Thesis, University of Saskatchewan, Saskatoon, Saskatchewan, Canada, 1965.

Maier, N.R.F. Reasoning in humans : III. The mechanics of equivalent stimuli and of reasoning, *Journal of Abnormal and Social Psychology,* 1945, *35*, 349-360.

Murphy, G. *An Introduction to Psychology,* New York, Harper & Brothers, 1951.

Osborn, A. *Applied Imagination,* New York, Charles Scribner's & Sons, 1957.

Patrick, C. Creative thought in Poets, *Arch. Psychol.,* 1935, No. *178*.

Patrick, C. Creative thought in Artists, *Jl. Psychol.*, 1937, *4*, 35-73.

Patrick, C. Scientific thought, *Jl. Psychol.,* 1938, *5*, 55-83.

Rossman, J. *The Psychology of the Inventor,* Washington : Inventors Publishing Co., 1931.

Skinner, B.F. *The Technology of Teaching,* Englewood Cliffs, New Jersey : Prentice-Hall, Inc., 1968.

Thorndike, E.L. *Human Learning,* New York, Appleton-Century Crofts, 1931.

Tinker, M.A. and Baker, K.H. *Introduction to Methods in Experimental Psychology,* New York, I. Appleton-Century Company, 1938.

Watson, J.B. *Behaviourism,* New York, Norton, 1925.

Welford, A.T. *Aging and Human Skill,* London : Oxford University Press, 1958.

Wingo, G. Max. Methods of Teaching in Chaster W. Haris (Ed.), *Encyclopedia of Educational Research,* New York : MacMillan, 1960.

Woodworth, R.S. and Marquis, D.G. *Psychology,* London., Methuen & Co. Ltd., 1952.

CHAPTER 9

General References : Intelligence and Its Measurement

Anderson, L.D. The predictive value of infant tests in relation to intelligence at five years. *Child Development,* 1939, *10,* 210-212.

Binet, A. and Simon, T. *The Development of Intelligence in Children,* Baltimore : Williams and Wilkins, 1916.

Binet, A. and Simon, T. "Upon the Necessity of Establishing a Scientific Diagnosis of Inferior States of Intelligence". In James J. Jenkins and Donald G. Paterson (eds.). *Studies in Individual Differences,* New York : Appleton-Century Crofts 1961(a).

Binet, A. and Simon, T. New Methods for the Diagnosing of the Intellectual Levels of Subnormals. In James J. Jenkins and Donald G. Paterson (eds.) *Studies in Individual Differences,* New York, Appleton-Century Crofts, 1961(b).

Bloom, B.S. *Stability and Change in Human Characteristics,* New York : John Wiley, 1964.

Cattell, J.M. Mental tests and measurements, *Mind,* 1890, *15,* 373-381.

Cohen, D.K. Does IQ matter? *Current,* 1972, *141,* 19-30.

Edward, A.J. and Scannell, D.P. *Educational Psychology,* Bombay : Allied Publishers Private Limited, 1975.

Galton, F. *Inquiries into Human Faculty and Its Development,* London : MacMillan, 1893.

Garrett, H.E. *General Psychology* (Second Edition), New Delhi : Eurasia Publishing House (Pvt.) Ltd., 1975.

Guilford, D.P. The Structure of Intellects, *Psychological Bulletin,* 1956, *53,* 267-293.

Guilford, J.F. Three Faces of Intellect, *Amer. Psychologist*, 1959, *14,* 469-479.

Guilford, J. F. Intelligence, *Amer. Psychologist*, 1966, *21*, 20-26.

Harris, D. *Children's Drawings as Measures of Intellectual Maturity,* New York : Harcourt, Brace and World, 1963.

Kimble, G.A., Germezy, N. and Zigler, E. *Principles of Psychology,* New Delhi : Wiley Eastern Limited, 1985.

Lefrancois, G.R. *Psychology for Teaching* (3rd ed.), Belmont, California : Wordsworth Publishing Company, Inc., 1979.

Murphy, G. *An Introduction to Psychology,* New Delhi, Oxford & IBH Publishing Co., 1964.

Spearman, L.E. General Intelligence, Objectively determined and measured, *Amer. J. Psychol.,* 1904, *15,* 201-293.

Spearman, C.E. *The Nature of Intelligence,* London : MacMillan, 1927.

Stagner, R. and Karwoski, T.F. *Psychology,* New York : McGraw-Hill Book Company, 1952.

Stern, W. *General Psychology from the Personalistic Standpoint,* New York : The Macmillan Company, 1938.

Terman, L.M. *The Measurement of Intelligence,* Boston : Houghton Mifflin, 1916.

Terman, L.M., and Merrill, M.A. *Measuring Intelligence,* Boston : Houghton- Mifflin, 1937.

Thorndike, E.L.; Bergman, E.O.; Cobb, M.V.; and Woodyard, E. *The Measurement of Intelligence,* New York : Teachers College, Bureau of Publications, 1926.

Thurstone, L.L. Primary Mental Abilities, *Psychometric Monograph,* Chicago : Chicago University Press, 1938.

Weachsler, D. *The Measurement and Appraisal of Adult Intelligence,* (4th ed.), Baltimore : Williams and Wilkins, 1958.

Wickens, D.D. and Meyer, D.R. *Psychology,* New York : Holt, Rinehart and Winston, 1961.

Wissler, C.D. The correlation of mental and physical tests, *Psychol., Rev. Monogr. Suppl.,* 1901, *3,* No. 6.

Woodworth, R.S. and Marquis, D.G. *Psychology,* London, Methuen & Co. Ltd., 1952.

References for Experiment-25 : Modified Alpha Examination

Binet, A. and Simon, T. *The Development of Intelligence in Children,* Baltimore : Williams and Wilkino, 1916.

Galton, F. *Inquiries in the Human Faculty,* London : MacMillan, 1883.

Galton, F. *Memories of My Life,* London, Methuen, 1908.

Otis, A.S. *Otis Quick Scoring Mental Ability Tests,* Beta, Yonkers-on-Hudson, New York : World Book, 1954.

Raven, J.C. *Guide to Standard Progressive Matrices,* London : H.K. Lewis & Co. Ltd., 1960.

Terman, L.M. and Merrill, M.A. *Stanford-Binet Intelligence Scale Manual for the Third Revision Form.* L.M. Boston : Houghton, Mifflin, 1960.

Woodworth, R.S. and Marquis, D.G., *Psychology,* London : Metheun & Co., 1950.

References for Experiment - 26 : Alexander Passalong Test, Experiment 27 : Block Design, Experiment 28 : Cube Construction

Alexander, W.P. *Passalong Test, British Journal of Psychology,* 1932, *23,* 52-63.

Brown, A.W. The development and standardization of the Chicago Non-verbal Examination, *Journal of Applied Psychology,* 1940, *4,* 36-47, 122-129.

Garrett, H.E. *Great Experiments in Psychology*, 3rd ed., New York : Appleton-Century Crofts, 1951.

Garrett, H.E. *General Psychology,* (2nd ed.), New Delhi, Eurasia Publishing House (Pvt.) Ltd., 1968.

Pintner, R. and Peterson, D.G. *Pinter-Paterson Scale of Performance Test,* New York : Psychological Corporation, 1927.

Terman, L.M. and Marrill, M.A. *Measuring Intelligence,* Boston : Houghton Mifflin, 1937.

Wechsler, B. *The Measurement and Appraisal of Adult Intelligence,* (4th ed.), Baltimore : Williams and Wilkins, 1958.

CHAPTER 10

Measurement of Intelligence

Baldwin, A.L.; Kalhorn, J.; and Breese, F.H. Patterns of Parent Behaviour, *Psychological Monograph,* 1945, *25,* 106-108.

Bayley, M. and Schaefer, E.S. Correlations and Maternal and Child of Behaviour with the development of mental abilities, Data from the Berkeley Growth Study, *Monograph of the Society for Research in Child Development,* 1964, *29,* (6 Whole No. 97).

Binet, A., and Simon, T. The Development of Intelligence in Children, *Annu. Psychol.*, 1905, *11*, 163-90.

Dreger, R. and Miller, K.S. Comparative Psychological Studies in Negroes and Whites in the United States, *Psychological Bulletin,* 1960, *57*, 361-402.

Eells, K. Some implications for school practice of the Chicago Studies of Cultural Bias in Intelligence Tests, *Harvard Educational Review,* 1963, *23*, 284-297.

Goodenough, F.L. *Measurement of Intelligence by Drawings,* Yongers. World Book Co., 1926.

Guilford, J.P. Three Faces of Intellect, *Amer. Psychologist,* 1959, *14*, 469-479.

Guilford, J.P. Intelligence, 1965 Model, *American Psychologist,* 1966, *21*, 20-26.

Guilford, J.P. *The Nature of Human Intelligence,* New York, McGraw-Hill, 1967.

Hebb, D.O., *A Testbook of Psychology* (2nd ed.), Philadelphia : W.B. Saunders, 1966.

Honzik, M.P. Environmental Correlates of Mental Growth : Prediction from the family setting at 21 months, *Child Development*, 1967, *38*, 337-367.

Kennedy, W.A.; Van, De Riet V.; and White, J.C. (Jr.), A normative sample of intelligence and achievement of Negro elementary school children in south-eastern United States, *Monographs of the Society for Research in Child Developmet*, 1963, *28*, No. 6.

Kennedy, W.A. A follow-up normative study of Negro Intelligence and Achievement, *Monograph of the Society for Research in Child Development,* 1969, *34*, Serial No. 126.

Kent, N.A. and Davis, D.R. Discipline in the home and intellectual development, *British Journal of Medical Psychology,* 1957, *30*, 27-33.

Liebert, R.M.; Poulos, R.W.; and Strauss, G.D. *Developmental Psychology,* Englewood Cliffs, New Jersey, Prentice-Hall, Inc., 1974.

McNemar, Q. *The Revision of the Stanford-Binet Scale. An Analysis of the Standardization Data,* Boston : Houghton Mifflin, 1942.

Raven, J.C. *Progressive Matrices, Sets, A, B, C, D and E,* London : H.K. Lewis & Co., Ltd., 1938.

Raven J.C. *Guide to the Standard Progressive Matrices, Sets, A,. B, C, D and E,* London : H.K. Lewis & Co. Ltd., 1960.

Robinson, H.B. and Robinson, N.M., *The Mentally Retarded Child,* New York : McGraw-Hill, 1965.

Sherman, M. and Key, C.B. The intelligence of isolated mountain children, *Child Development,* 1932, *3*, 279-290.

Shuey, A.M. *The Testing of Negro Intelligence,* (2nd ed). : Lynchburg, Va. Bell, 1966.

Spearman, C. General Intelligence, objectively determined and measured, *Amer. J. Psychol.,* 1904, *15,* 210-293.

Spearman, C. *The Nature of Intelligence and the Principles of Cognition,* London : McMillan & Co., 1923.

Spearman, C. *The Abilities of Man,* New York : McMillan, 1927.

Spiker, C.C. and MaCandless, B.R. The concept of intelligence and the philosophy of science, *Psychol. Rev.,* 1954, *61*, 255-266.

Sullivan, E.T.; Clerk, W.W. and Tiegs, E.W. *California Test of Mental Maturity,* Monterey : California Test Bureau, 1963.

Stern, L.W. The psychological methods of testing intelligence, Trans. G.M. Whipple, *Educational Psychology Monographs,* 1914, No. 13.

Terman, L.M. and Merrill, M.A. *Measuring Intelligence,* Boston : Houghton Mifflin, 1937.

Thurstone, L.L. Primary Mental Abilities, *Psychometric Monograph,* 1938, No. 1 Chicago : Chicago University Press, 1938.

Thurstone, L.L. and Thurstone, T.G. Factorial Studies of Intelligence, *Psychometric Monograph, No. 2,* 1941, Chicago : Chicago University Press, 1941.

Wechsler, D., *Adult Intelligence Scale, Manual,* New York : Psychological Corporation, 1955.

Woodrow, H. Intelligence and Its Measurement : A Symposium, *Journal of Educational Psychology,* 1921, *12,* 207-210.

Woodworth, R.S. and Marquis, D.G. *Psychology,* London, : Methuen and Co. Ltd., 1952.

EXPERIMENTAL PSYCHOLOGY
WITH
ADVANCED EXPERIMENTS

VOLUME - II

By the Same Author

1. Contemporary Fields of Psychology and Experiments

2. Statistical Methods in Psychological and Educational Research

3. Modern Psychophysical and Scaling Methods and Experimentation

4. Psychological Perspective of HIV & Aids

About the Author

M. Rajamanickam has made many contributions to the field of Psychology both in English and Tamil. Some of them were used as Text-Books. He was the first to obtain the Doctorate Degree of the Annamalai University in Psychology. He taught psychometry to the Post-Graduate students for more than two decades and the experience he had gained inspired him to write this book. Generally, to the students of Psychology, Psychometry is a nightmare. But this author has simplified all the intricacies of the field and made the students to like the subject. He has published quite a number of papers in foreign and in Indian journals. With his knowledge in Psychometry, he has designed and developed about thirteen test materials. His important publications are, *Contemporary Fields of Psychology and Experiments; Statistical Methods in Psychological and Educational Research; Modern Psycho-Physical and Scaling Methods and Experimentation; Modern General Psychology;* and *Advanced Experiments in Psychology,* in two volumes. There are also a number of books in Tamil on Psychological themes. He is very much interested in guiding research, and produced twelve Ph.D., and about 40 M.Phil. Degrees. He started the Community Guidance Service Centre in 1980 in the Department to extend the Psychological Services to the rural people also founded the *Journal of Community Guidance and Research* in 1984, of which he is the Editor and it is run for the past 20 years without any break or delay.

EXPERIMENTAL PSYCHOLOGY WITH ADVANCED EXPERIMENTS

VOLUME - II

M. RAJAMANICKAM

M.A., M.Litt., Ph.D.

Formerly Dean, Faculty of Education and Professor and Head of the Department of Psychology, Annamalai University

CONCEPT PUBLISHING COMPANY, NEW DELHI-110059

Cataloging in Publication Data—DK

Courtesy: D.K. Agencies (P) Ltd. <docinfo@dkagencies.com>

Rajamanickam, M. (Muniappa), 1925-

Experimental psychology with advanced experiments / M. Rajamanickam.

v. cm.

Includes bibliographical references.

Includes index.

ISBN 8180691101 (set)

1. Psychology, Experimental. I. Title.

DDC 150.724 21

ISBN 81-8069-113-6 (Set) - P.B.
ISBN 81-8069-114-4 (Vol. 1) - P.B.
ISBN 81-8069-115-2 (Vol. 2) - P.B.

First Published 2005

Published and Printed by
Ashok Kumar Mittal
Concept Publishing Company
A/15-16, Commercial Block, Mohan Garden,
NEW DELHI-110059 (India)

Phone : 25351460, 25351794
Fax : 091-11-25357103
E-mail : publishing@conceptpub.com

PREFACE

Experimental Psychology had its origin in Fechner's Psychophysical Theory and Psychophysical Methods in the third quarter of nineteenth century. Gustav Theodor Fechner (1801-1887), a German Physicist and Philosopher, was very much interested in studying the relationship between mind and body and the problems involved in it. He was working on it for more than ten years and at last was able to evolve a theory. Even before this time when Fechner was concentrating on mind-body problems, there were also other scientists working on the relationship between the body-soul or spirit (mind). Fechner has perceived this problem from different angle with his experience and knowledge in Physics and Philosophy.

The problem, he specifically studied was the amount of physical energy that could have corresponding effect on mental energy. From this idea he was able to formulate a hypothesis relating physical energy and mental energy. What amount of physical energy that could cause changes in the mental energy? He was able to succeed on this attempt and also could evolve a formula which was called Fechner's Law. Later on he has modified it.

There were also many contemporary scientists during Fechner's time, like Weber, Wundt and others, but they were not working in the same line as Fechner did. Psychologists at this time strongly believed that experiments conducted according to the Psychophysical and Scaling methods yielded reliable and accurate information. Psychophysical methods keep one actively involved in the laboratory works to watch the new results. Any experiment conducted in the laboratory situation provides exciting experience whether it is based on Psychophysical or Scaling methods or an expeiment in any other field, yields new and accurate information.

Experimental Psychology is as much important as any other subject in a course of study. Since experimental results provide new information to every area of the subject experiments should be considered as the foundation for the area of any subject or discipline.

The discipline of Psychology was very much nourished by rich laboratory works in European countries during the last quarter of nineteenth and first quarter of twentieth centuries.

Indian Psychologists in their efforts to build up the discipline of psychology are very much anxious about the performance of psychological experiments in psychological laboratories. They expect that the experiments should go according to the correct procedures in the same manner as experiments conducted in Physics laboratories, Chemistry laboratories and other biological laboratories with all seriousness. Unless proper seriousness is shown in the laboratory situation in the performance both by the teacher and the student, psychological experiments could never fulfil the requirements. Indian psychological laboratories have come up very slowly. They are still with very few instruments and many universities never thought of having a Department of Psychology with a laboratory. Consciously or unconsciously, an idea is set up in the minds of the teachers who handle the practical classes and the students who take up practicals that *it is a leisure time activity. This attitude is prevailing in many Indian Universities.* This is a very dangerous attitude which should be thoroughly avoided.

This can be done only by giving more importance to practical classes with all seriousness.

In this book, '*Experimental Psychology with Advanced Experiments*' the experiments included and discussed are from various areas and these experiments can be conducted with different types of tools and instruments. The experimenter may find not only interest in it but also some kind of inspiration to conduct similar experiments more and more. The experimenter may feel that he/she has discovered something new and also may feel happy that he/she has made some new contribution to the field of Psychology and got the reward for what he/she has perspired. Therefore, it is immanent that any student of Psychology who conducts experiments with all seriousness and sincerity according to the procedure can certainly discover something new and may feel very proud of it.

There are twenty chapters with sixty experiments in these two volumes, which are covering the areas of sensory experiences, physiology emotions, and mainly the cognitive areas like perception, learning and remembering, thinking, language comprehension, problem solving, nature of intelligence and its measurement, creativity and its measurement, measurement of human aptitudes and special abilities, motivation and achievement, assessment of personality types and traits, social behaviour and environment and behaviour.

In certain areas where suitable test materials were not available I, myself had designed, developed and validated the tests and included in this book with examples and interpretation of the results. Most of the experiments included in this book are such that could reveal the skills of the participant and the efficiency of the experimenter. Therefore, I am quite confident that the experimenter may feel ease and grace with this book in all his experimental works. The Psychophysical Experiments are not included in this book. All the Psychophysical Experiments and the Scaling experiments are included in my another book, '*Modern Psychophysical and Scaling Methods and Experimentation*'.

This book is written with utmost care and untiresome efforts after analysing the syllabuses of Experimental Psychology of many Indian Universities with the view of making it as a comprehensive guide, and as an alternative to the existing study materials available for Experimental Psyschology. Therefore, I strongly believe that this book would certainly meet the requirements of students of Psychology at all levels which may serve as a textbook as a manual, as a self-guide and as a reference work to the teacher who handles the Experimental Psychology classes.

I am extremely grateful to Mr. Ashok Kumar Mittal of Concept Publishing Company for having envisaged so much interest as to bring out this voluminous book, *Experimental Psychology with Advanced Experiments* in two volumes in an excellent manner. Mr. Mittal, an outstanding Indian publisher, a man of fertile imagination and lofty thoughts with creative spirits, has already published quite a large number of books for the benefits of Indian intellectuals and for the colleges and universities. He deserves all admiration for the wonderful services rendered for the cause of education of this Great Nation by way of publishing valuable scientific and creative works with international standard.

Chennai-40 **M. Rajamanickam**
January, 2004

CONTENTS

Volume - II

LIST OF EXPERIMENTS

Volume - II

LIST OF FIGURES

Volume - II

11

CREATIVITY AND ITS MEASUREMENT

INTRODUCTION

Parents of the present generation expect their children to be all-round in their achievements. Their children's intellectual maturity though is imminent and also important, their children's progress in all other areas should also be taking place side by side through maturity. As one matures through years of his life, hundreds of big and small things he meets with. Many may attract him, some may simply stimulate him and some may neither attract him or stimulate him. He may pay more attention on those which attract him very much or in which he can make progress easily and some of those which stimulate him may catch his attention if he finds that it would facilitate his programmes. Though a student in the higher secondary or college level has his own academic activities as important, he may be also paying some special interest on certain extra curricular activities.

These extra curricular activities may be anything, which may be an activity in which he might have casually paid some interest. But later on he may find that it is more successful for him and also bringing him great reputation and he might start paying more attention on it. This was how many of the men in the world made outstanding contributions in particular field. Now-a-days, both at State and Central level children in schools are encouraged by making them to take part in some competitive activities and they are sufficiently rewarded. On an important occasion competitions in painting, handicrafts, essay writing, designing scientific and electronic moppets and instruments, various sports activities and so forth. These programmes at some level appear to be simple or ordinary in the case of some individuals. This might have been a big opportunity to exhibit his talent which may be rare and draw the attention of many that it is a thing of beauty and brings joy for many. The person himself gets excited at his extraordinary talent in which he has been toiling and moiling for days. This task may result in creating something new and he may be developing it at a large level which may bring him much bigger costly rewards and fame. Thus creating something new may result from making use of given opportunities and exhibiting one's skills and talents. Till the skills and talents of a person put into some kind of activity and bring out something strange or new he is an ordinary person like others. When he shows to the world that he is capable of creating a strange thing which can be of a rare one, he becomes a creator.

If we go through the life histories of some greater inventors or creators there are interesting incidents we can find in them. Psychologists made extensive studies on intelligence show the distribution of intelligence in human population or in some societies. They illustrate it by drawing a curve which they call it normal curve. It is not lopsided curve. The meaning of which is that a small percentage of the population falls either side of the summation of the curve indicating that on one side people below normal intelligence and on the other side above normal intelligence. The first category people are called feeble minded or individuals who have I.Qs considerably below the average level and the people on the other side to the distribution are called very intelligent or individuals who have high I.Qs. People who have very high I.Qs are called gifted persons or genius. In between these two divisions most of the people fall who are called the normal in intelligence. If a creativity test is conducted on a population the same type of distribution we may get. At one end we could see people with very poor in creative ability, at the other end a small number of people who are gifted for creation or considered to be the creative genius and in the middle most of them with average or moderate creative abilities. This may be natural distribution as in intelligence is.

One factor often stressed is the maturity in the developmental process. It involves the use of one's capabilities. Therefore maturity implies the fulfilment of various potentialities. These are one's physical and mental abilities to the fullest possible extent. People who are at the lowest range of intelligence are least capable of thinking, particularly thinking creativity. If the younger people, particularly those who are students to become mentally and physically mature and to become creative and responsible citizens they need to be provided with much help during their school days for developing their capacities. They must equip themselves with abilities for analyzing interpreting, evaluating ideas, formulating hypotheses and go in search of satisfactory answers and take up responsibilities. They must also be ready for a creative thinking with the view of emerging with something new. It is true that such things often depend upon the amount of encouragement one gets. The school programmes, the teacher and the school environment all must have a positive attitudes for this. If one is to take part creatively in an environment he is required to adjust himself to the situation and this adjustment is through the process of maturity. A person who is mentally, physically and emotionally matured, is also capable of adjusting himself to a creative situation. There are experimental studies that creative persons are lacking in adjustment and also often frustrated.

A person participating in a creativity activity means, not that he should create something materially. His creative thinking and participation may be also in the fields of politics, economics, social, scientific, artistic and so forth. This may also means a good background in mathematics, physical and social sciences, in art and literature and in other technical skills. Therefore a mature person in a society is one who must do anything more than what he knows about his environment. In this way a matured mind is a creative mind, and also matured to put up himself in any situation successfully.

Therefore, one to become a creator, he should be matured mentally, physically, emotionally with a good background knowledge of the situation and development of knowledge and skill in some physical or social sciences. The individual must be readily adapt to the situations which are having favourable atmosphere for creative activities. Thereby some pressure is exerted on the individual for creative thinking and activity. Sometimes the individual might have been put into heavy pressure in the environment more than what he can hold.

WHAT IS CREATIVITY

To define creativity is a difficult task. It is defined in various ways. A definition of creativity that can be universally accepted is impossible. Therefore it is defined in whatever manner acceptable to a situation and persons. Generally creativity is seeing something in a new relationship. it is one's abilities or skills expressed in a new combination. Creativity can be seen in thinking, in expressed activities, in manipulating objects or materials. According to Morse and Wingo (1970) creativity is also possible in modes of social interaction, in studying, in working and in playing. Whatever you do, you can remove the old pattern or relationships and set a new type of arrangements, a new shape and look to the old objects and set-up. All the old ways of dealing with the objects, persons, places can be done creatively in a new way of thinking.

Creativity mostly involved in problem solving. Therefore whenever you solve a problem you are doing a creative work. Newell *et al.* (1962) say that you cannot solve a problem without creative thinking. Because the old habits and customs will not help you to solve a problem. You will have to add something new. According to them creativity is nothing but problem solving which is carried out under difficult and novel conditions. If this is creativity it is not confined to genius. Morse and Wingo (1970) state that only few of us might have the originality which Leonardo da Vinci (1452-1519), an Italian painter, sculptor, architect and engineer the most versatile talent of the Italian Renaissance had. But all human beings are having new ideas and insights in varying degrees from time to time. Many people might have had the same idea many times and expressed the same idea many times. But for the creator every new idea or expression is original and an example of creativity. Creativity is an investigation of new ideas. It is the determination of some unique characteristics. It is a requirement. It is an urge in the person to reveal in an object something new. It is also learning to estimate and to deal with the new situations on their unique characteristics.

Right from the ancient times onwards man has been creating something new out of stress or pressure or out of necessity. In the case of the ancient man it is more appropriate to say that it is out of bare necessity that he invented many things new by the method of trial and error or by accident. Modern man, we can say, is continuing inventing new things. Many artistic, scientific and technological achievements have been made in modern times. Whether it is solving a new problem or development of a theory of relativity or creation of a new musical note or an invention of a new type of railway engine or an artistic painting or composing a poetic play, or sculpture — all have the same psychological process.

Creations are made by human beings — individuals who have put the selves in the field of work. Human being brings out creative products after working on it for a very long time of vigorous work. They might have made lots of unsuccessful attempts and received ineffective responses. Several times they might have failed to bring out a correct response or solution to a problem. Therefore, it implies correct response or solution is not readily achieved by one or two attempts. There might have been several trials and errors occurred before the incident of invention or creation emerged.

Creation means an act or process of creating something new brought into the world. It is something brought into the existence or created out of human intelligence and imagination. In the task of creation the individual himself is required to judge his own skill whether he can make any creation or invention. The painter must judge whether a particular shade of colour could make out something new in the whole product. So also a scientist may be making of some judgment about his exploring activity with an ambition or hope of getting some promising result in the particular direction he is working. A writer of fiction might have put all his creative imaginations in his forthcoming creative work with the expectations that the fiction might create thrill and emotional excitements in the minds of the readers.

Modern industry provides ample scope for creative thinking and creative products. We are seeing in the market some new materials produced out of raw materials brought out for consumption. Whether it is food products, textile products, household articles or furniture, or fancy goods are all packed in attractive packages and displayed in attention getting attractive manner with new type of advertisement techniques in the market. No one who goes for shopping can never escape from such attractions and even the person who is not accustomed to use such goods may be tempted to buy atleast once. The creative imagination is very much put into action in the modern industrial products. What was considered as luxury a few years ago has become necessity. Particularly the Indian textile industry has made wonderful progress by the creative imagination of the colour specialists, the design engineers and the chemical engineers. The new attractive designs not only in the varieties of cloths but also in varied designs of garments which make the present younger generation to become fanatic and crazy for the fancy materials. This appears to be easy and simple for a spectator but how much time and mental work the manufacturers might have spent over them is a matter of creation.

The process of creation and invention are most complex activities which require much greater degree of schematic and sustained activities and critical abilities. In every creative activity much of our imagination and images are put into action. The mental images one makes in the case of a creation involve many things. He must have close contact with materials, his past experience, his knowledge in various fields, and bringing to his memory the various types of materials with which he has been associated in the past and also think of what type of product may be easily marketable and make people to like it and buy. Therefore, it is a most complex activity involving much critical abilities.

Based upon this discussion we can define creativity, as the capacity to discover something new, a mixture of realistic and imaginative thinking often resulting in new solutions. Gardner Murphy (1964) says that creative ability is the capacity to produce some original work through thought and imagination. According to him there are two kinds of mental activities in creations, one is that which reconstructs the past and the other is that which initiates the new. Therefore according to Murphy in every creation there is something old and past and something new and recent. Maslow (1970) states that there is a universal characteristic in all people and there is no exception in it. Every person in one way or another shows a special kind of ability which may be called creativeness or originality or inventiveness. Maslow says that the so called geniuses display ability which we do not understood. What we could understand about them is that they are endowed with a capacity which is different from all others. The person is born with a rare talent. Maslow is of the view that the creative ability of the self actualized man is related to the universal creative ability of the person who is not influenced by the vices of the world. It appears that the creative ability is a potential common human nature with which every man is born. But, Maslow says, most individuals lose this potentiality as they grow in a culture and influenced by the cultural factors. Some persons keep in possession of this potentiality and look at the life directly. These people even if they have lost it as most of the people they later on regain it and make use of it.

This universal creative ability normally in most of the human beings emerge in the form of creative products like books, producing musical notes, in the form of intrinsic products like painting, sculpture and so forth. But some people do not exhibit their creative potentialities in any form of product. They are very modest. Because of this, the special type of creative ability is the sign of healthy personality and such people can throw it in whatever activity in which they are interested in. They can emerge as creative persons. Whether one does an artistic or scientific work or some kind of physically skilled works like carpentry or shoemaking or a salesman or clerical work he does the work with the creative spirit, an attitude that is natural characteristic of the person to perform the task. He does his work creatively in an inducement manner modestly. These people seem to perceive the creativity and truth very easily. It is because of this humble nature they are looked up as men of only limited creative potentialities. Maslow (1970) is of the view that it is because of the inhibition that man's spontaneous nature is not given expression. The superficial and powerful set of inhibition is checking the spontaneous expression of one's potentialities. If there was no such checking every human being may become a creative person.

The so called creative persons or creators are less inhibited, less constrained, less bound and less encultured. They are freely expressing their natural potentialities without inhibition. They also develop a sense of privilege in the community to act freely. This we can see in the world of art. Maslow is right in expressing this idea. The people of the world of art are such that they are till recently were very ordinary people. Once they started giving expression to the healthy personality–a special type of creativeness, they become creative artists. As we have stated above, others are humble with their natural potentialities and not exhibiting it as there is no pressure on them.

What prompts man to make something which should appear attractive and beautiful to him and to others and how do we judge things that they are beautiful. It is the driving force that everyday one is experiencing that makes him to do something new or strange. It is not the intuition that is the cause of it. There is an impulse in man to search for something new in art and science or in the environment or in the society. The impulse or we can say pressure, to create or invent something new was also present in the modern man. The cave man was the first creator and inventor. If he was lacking, what Maslow says, the natural common potentiality, he would not have carved out the caves in the mountain nor painted the wall with beautiful colours and written his thoughts. The impulse or the natural common potentiality, to create new is also based upon necessity. This is also an impulse or urge to discover something which is not in existence.

The concepts, creation, invention and discovery appear to be similar in meaning. But they are not, they are different. The concept discovery refers to finding out something from the existing things, which were till now not known or seen. The object of the discovery is already there. The person who discovers only finds out where it is and what it is. The British, the French, the Spanish and the Portuguese and other Europeans were once the Empire builders and discoverers of new nations. Every European nation in fifteenth and sixteenth centuries was in keen competition in discovering new lands and made big expeditions and explored the new lands on earth. Captain James Cook (1728-79), an English navigator, made expeditions and discovered Australia in eighteenth century. Similarly, the Spanish navigator, Columbus (1451-1506) made big expedition and discovered America in 1492. Australia and America as nations were already there. Similarly in the case of invention, the inventor is not finding out something out of nothing. He finds out something out of something in existence. He puts or assembles the existing materials to different combinations and finds out a product which was not existing previously. For example, James Watt (1736-1819) a Scottish engineer invented the steam engine in 18th century. Graham Bell (1847-1922) a U.S. scientist, born in Scotland has invented telephone in 1876. In the case of these inventions the basic materials for invention were already there. Therefore, in discovery and invention there were nothing personal and effective. If Cook and Columbus did not discover Australia and America somebody also would have discovered them. Similarly if James Watt and Graham Bell did not invent the steam engine and telephone some one would have invented the steam engine and telephone. In the biological sciences there is only discovery, in physical sciences there is both discovery and invention. Charles R. Darwin (1809-1882), an English naturalist and biologist formulated the theory of evolution by natural selection, expounded in *On the Origin of Species* in 1859 and applied it to man in *The Descent of Man* in 1871. It was he who discovered how the animal and the plant species were originated.

But creation is something different from these two phenomena. In creation the person creates something out of nothing. It is from his own thought, experience, imagination and intelligence that he makes out something new which did not exist previously. For example, a poetic play like William Shakespeare's (1564-1616) *Macbeth* and the epic poetry like John Milton's (1608-1674) *Paradise Lost* or a lyric poem, *Ode to a Nightingale* of the English poet John Keats (1795-1821) or the famous English novelist Charles Dickens' (1812-1870) *Great Expectations*, the most famous painting *The Virgin of the Rocks* by Leonardo da Vinci (1452-1519), an Italian painter, scultor, an architect – all are creations.

The act of creation is the same in science as in art. As Maslow said that it is the natural common potentialities in all human beings. It varies only in the way in which one is executing it. Some may execute it out of extreme pressure or impulse some may execute it out of necessity.

But how does a poem or a play or a painting or sculpture differ from a theorem? All are creations, brought out from nothing. But the units in both differ because they match human experience in different ways. Every child can rediscover the Pythagorean theorem in the same form. His experience is intellectual and can be matched exactly. But in art it is not happening. Many people, both painters and common man are painting women, but not like Leonardo da Vinci. Many people write play, epic and poetry, but not like Shakespeare, Milton and Keats. It is impossible to match the experience of one person with that of another in arts.

Therefore, it is said that creativity in certain areas is a multifaceted phenomenon. It involves developing something in unique manner. Every one may become a creator under stress of strong feeling. Therefore, creativity is not limited to few artists and inventors. If a child is allowed to play freely he can design a creative play. He makes use of the tools or toys and devise new type of play. The clay may be easily available for the child and he may manipulate the clay in whatever manner he wanted and it may result in the creation of a new toy out of it. A house wife may create a new kind of taste in cooking to please her husband.

Therefore, creativity may be defined as the sense of ability to perceive something new in the external world through past experience and imagination and creation is the expression of such ability in some form so as to arouse the attention, curiosity, feeling and emotion of others in different degrees on the basis of their experience and knowledge. The form of expression of creativity may be a poem, play, painting, sculpture or even a musical tone which was never in existence in the same form at any time.

Creativity is not subjected to any context. If it is to be original and creative it should be free from any restriction. Skinner (1968) states that creativity is stated in terms of trait character. It is the freedom of the mind and inquiring spirit. In other words it is creativity. Skinner says that we can measure the trait and compare people's creative ability with this and also test the presence of the traits associated with it. Skinner states that man is a genetic product and the environment in which he is born and brought up becomes his nurturing factor. The education one receives may change the characteristics he has developed in an environment. The behaviourists have strong view that the genetic or hereditary factors can be changed by education and environmental factors. Therefore, creativity cannot be due to hereditary factors. Human beings can be taught to become creator. If the person is allowed to function freely without any restriction he can reveal originality and creation.

Every definition of creativity is viewing the end result placing things in new ways and seeing connections that were not existed previously. Kimble *et al.* (1985) states that creativity requires originality, but originality is not always creativity. According to them mentally ill persons, particularly the schizophrenic is said to be original and unique, but not creative. The same thing can be applied to most of the thoughts. The thoughts of many people may be original and unique, but they are not creative people or their thoughts may be used for any creative products.

An idea or product must be original and quite suitable to be creative. The original idea must fit to the context and make meaning to the demands and situation. Research studies in the field show that environmental factors heightened creativity in creative people. Murphy (1964) states that we are all creators in a sense. Under the stress of strong feeling, when some big idea emerge we may state that something may come out as something original. An ordinary person is not very much different from a poet. George Thomson (1946) says that an illiterate woman with no artistic claims suddenly expressed her feeling without any previous thought which had all the characteristics of poetry. This was out of inspiration she revealed the characteristics of a creative imagination. According to Murphy creativeness emerges out of sensitiveness to the environment. There are, therefore, quite a number of ways in which creativity can be defined involving factors within and factors from the outside world. Kimble *et al.* define creativity as the ability to generate many new and usual ideas.

Mednick (1962) defines that creativity is forming of relative elements into new combinations which meet the specified requirements. The more the distant elements mutually shared in the new combination the more it is creative process of a solution. For Guilford (1959) creativity requires fluency, flexibility and originality.

Though there were many studies on creativity the important points relating to creativity research and speculation remain not answered. Stephen Winsten (1949) the biographer of George Bernard Shaw (1856-1950) the famous English dramatist indirectly has stated that the distance between the genius and madness is only a hair's breadth. A hair's breadth separates a genius from a madman. He means that there is very small difference or no difference between a genius and an insane. Winsten is of the view that man likes to consider the creative man as defective or quite close to madness. When Winsten referred this to Bernard Shaw, Bernad Shaw said that most of these are like, but he is an exception to this.

While discussing about creativity and creative persons Guy R. Lefrancois (1979) states that though he is not afraid of creative persons, as he was once, he states that there is certain uneasiness and uncertainty in this matter. He puts the question, "Are the creative people non-conformists, eccentrics, radicals and fools or ordinary people?" The answers to this is that there are some with both the characteristics. He states that there is nothing magical quality in creativity. Creativity is a quality of humans just like intelligence. This is a quality present with everyone. But it may vary in degree. There may be very low creative ability in some and few may possess it at a very high level. But a man is not thoroughly devoid of the creative ability. Some may develop a particular skill right from childhood and become experts when they become adults. But others may be interested in something else which may not be having any relationship with art or creation. But every one, as Maslow states is having the common human potentialities.

CREATIVE PROCESS

Creative process mainly refers to the functioning of creative ability. In what manner creative ability functions to indicate that the particular person is a creator. The creative person or the creator always tries to link or connect the elements with some other elements with which they do not in any way link naturally. A painter can imagine to combine the parts of two elements or the parts of two species and like to see how it would look like. Since his imagination is free and not controlled he can even imagine any uncommon species.

For example, a painter could create an uncommon species which may be a combination of both woman and fish and he could call it a mermaid. It is a new combination, not in existence. It never existed in the past and it could never exist in future. But it exists in the poetic imagination of a poet and in the artistic imagination of a painter. It is only an imaginary combination and not existing in reality. A sculptor also can make out a sculpture in stone or bronze metal a funny combination of human body and snake head or a snake body and human head. We see all these funny sculptures in most of the Hindu temples carved out of stones and kept in the pillars around the temple walls. Therefore, this type of creative imagination was in existence right from the ancient times onwards and it existed in stone sculptures in the medieval times and in paintings in modern times. Therefore, there were generations of creators making out creative products in various forms. The type of creation of combining two different unnatural elements imaginatively was in vogue all over the world. It is because of this, authors who deal with creativity, state that it is a universal tendency to see and to produce new and strange things. In the same way we come across in Greek mythologies, the animal 'chimera' is often spoken of. It is a fanciful animal with lion's head, a goat's body and a serpent's tail. It was never existed at any time in the world. It existed only in human imagination. Some one wished to see such a creature and he has created it in his imagination. In Greek mythology, Sphinx, a monster with a woman's head and a lion's body is described. It is said that she lay outside the city of Thebes asking travellers a riddle, a guzzling question and killing them when they failed to answer it. Oedipus answered the riddle and the Sphinx then killed herself. A huge statue of Sphinx lays near the pyramids at Gisa in Egypt. In Hindu mythology there are several references with animal, animal's head and human body and human head and animal's body. These are ancient people's creation based upon the creator's imagination. Even modern writers attributed human characteristics to deities in animal forms. This is called anthropomorphism in classics. Human creative products thus taken various forms.

Creative process involves greater amount of imagination and thinking based upon learning and experience and also continuous pressure–the pressure that John Stuart Mill (1806-1873) had during his adolescent period from his father. It is the environmental pressure that sometimes makes one a creator. The more sensitive a person is to the environment, the forms, colours, shapes and contours the more he makes them as the roots of creativeness in his later life. The joys in spatial and temporal relationships and mechanical laws in his early days that made Sir Issac Newton (1643-1727) to become great mathematician and physician in his later life. Murphy (1964) observes that man's sensitiveness helps him in two ways to become a creator. First is by way of directing his attention on them and secondly by way of enjoying them and making them into new combinations. His keen observation makes him to understand the characteristics in the objects that could be combined in a new fashion. Then he enjoys himself on knowing these characteristics in the object and then he puts them into a new combinations. This, he is not achieving suddenly. He might have tried in different combinations to arrive at a right solution. Thus the creator sometimes struggles to get the solution what tried.

Creative process involves continuous enquiries. Mackinnon (1962) states that there are **five important stages in the creative process.** The first is the period of preparation assuming a problem. The person must first satisfy himself that there is a problem. The second period is the concentrated effort to solve the problem. This means the efforts must be continuous and also he must fully keep himself engaged in it. The third period is withdrawal from the problem. This takes place when the person finds that it is difficult for him to achieve the solution. Instead of spending his time more and more he feels that it is better to withdraw from it. The fourth period is the moment of insight. When the person withdraws himself from it, he is not fully given it up his efforts. It is going on working in his mind and there is sudden striking in his mind the solution to the problem. The period of withdrawal seemed to have been a period of incubation. After this he gets the insight, a way to solve the problem or the manner in which the problem can be handled. Then, the fifth period which is indicated by a period of verification, evaluation, elaboration and application of insight. Thus the tendency is to observe a problem and constantly making enquiries about it and looking out for a correct approach and applying it to get it solved are the processes with which a creator starts his creative work.

The person's sensitiveness to the environment is one of the important characteristics that makes him to work out on the creative line. The other important characteristic of the creator is his awareness to the situation. This awareness is the basis of creativeness. The awareness is to find out what is lacking in the situation. There is something lacking in the situation and, therefore, there is a problem. This is another important characteristic of the creator. Therefore, a creative person is supposed to be too sensitive to the environment and also aware of the situation. Creative process is not a unitary process. It is a combination of several processes like cognitive, motivational and emotional processes. This combination of processes involve a group of psychological processes such as perception, learning, memory, imagination, apperception, thinking, planning and making decisions. These processes, generally, may be present in everybody, but it varies in degrees. It may be present to a greater degree in a creator and lesser degree in ordinary persons. But these characteristics can be developed by suitable training and practice.

Some authors relate creativity with intelligence. Guilford (1959) considers that creativity is a form of intelligence. Guilford combines a number of special abilities with creative process. Therefore, according to him a number of factors associated with the creative process. We have discussed a few of the important characteristics in the preceding passage. The special characteristics that are associated with creative processes are flexibility, fluency and originality with word associations and ideas. These are associated with Guilford's divergent thinking. The divergent thinking is one of the five operations in his thinking aspect of his theory of intellect. The convergent thinking emphasises one single solution whereas divergent thinking emphasises a number of different solutions. According to Guilford there are also some other abilities found outside of divergent thinking.

Creative process is looked upon from various angles. Authors like Gagne (1965); Bruner (1961, 1963, 1964, 1965); Bruner, Goodman and Austin (1956) stress that creative process is an educational process. Creative process is an approach for teaching children. Gagne assumes that a scientist in his scientific exploration acts as a creative. To make out a creative or to produce such creative acts in a person training should be given to children to become creative. According to Gagne creativity is wider than science in its applicability. Creative expression needs a very broad form of training in a number of diverse fields. The training for the creative expression, or to make one a creator, should consist of encouraging originality in thinking, and there should be sufficient reward for it subsequently. Gagne's process of approach is inclusive of creative dimensions and content. Content, here he means the curricular materials. Children should be trained on intellectual processes and this training should be used for the application of science. By this means a process of transfer of training takes place. The children must be explained what these processes are and they should be rewarded far every successful application of the method on science. This principle involves that cognitive processes are taught, learned and transferred as other behaviour. Gagne, therefore, is of the view that creativity as a cognitive process can be trained and transferred for application in a number of fields.

CREATIVITY AND EDUCATION

Creativity is expressed in various forms and it is judged only from the products. It is also judged by social acceptance. At times, it may appear to be sudden, fancy for the people and people may be very much excited and show much interest in such creative fancy products. The mental process is the cause of the products. It is resulting from the mental content and the contents are recombined in the process and revealed as a new product which is fanciful to the people. Edwards and Scannell (1975) state that the mental process does not always cause a product. Suppose if the individual is not sensitive to the environment and not aware of the conditions whatever the mental content is, the mental process does not bring out a product. The mental process can be identified as what it is and it can be also measured. If we know these two aspects it is possible to mention the nature of creativity. This view of Edwards and Scannell seems to concord with the view of Gardner Murphy (1964). Murphy says that what the individual is capable of achieving in creativity can be ascertained if the individual is given opportunities in the early days and if we give importance to the early experiences.

Educational psychologists and educationalists have been giving more importance to creativity in recent years since Gagne's stress that school children should be given chances to know more about creativity and the application of creative process. Creativity has been used in association with Guilford's divergent thinking and sometimes it is also used synonymously. It is because Guilford (1950, 1959, 1962) has made extensive work on creativity. This has led to some kind of wrong understanding about creativity. The product of creative ability has been converted into test scores and the presence of creative process is informed from the test scores, but the nature of creative process could not be inferred from the test scores. Therefore, Guilford has avoided using the term creativity while discussing about factors of productive thinkings. Many research works have been carried out in school situations with creativity tests. We are discussing this aspect under measurement.

Textbook writers in the field of psychology now-a-days include a section or a chapter on creativity in their books after the outcome of research work of Guilford on creativity. But we must know that even many years before Guilford's work appeared, psychologists stressed the importance of creativity in schools. Great authors like Stern (1924), Cowell (1926), Spearman (1931), Rossman (1931), Patrick (1934), Hollingworth (1942), Hadamard (1945) have spoken much about creativity and the creative skills in children.

Modern authors do not agree with the view that creative skills are based upon genetic factors. Many of them are holding strong views that it can be taught and learned and produced. That is the reason many of the educationalists stress that creativity should be combined with curriculum. In this task the importance of a teacher in the school in inducing the creativity is questioned. The question raised by them is whether a person, the teacher can make another person more creative? The answer in short to this question is 'yes'. There are evidences that the teacher in school can play a vital role in inducing creative spirit in children and thereby develop their creative skills to make them creative person. In one of the studies made by Haimowitz and Haimowitz (1960) two hundred scientists were interviewed and fourteen hundred great men and women were examined with their childhood historical events in schools. These people repeatedly reported the influence of the teachers in their careers. They did not refer to all the teachers but to a particular teacher who had much influence in their lives.

The classroom climate and the characteristics of the teachers and their influences on the school students were studied with special reference to flexibility and originality. The study made by Spaulding (1963) on the style of two teachers revealed the fact that flexibility and originality were negatively related. One teacher has put more emphasis on emotional and social behaviour of the students and not much on their intellectual activities. Whereas the other teacher put more emphasis on discipline. The students had different type of influence from the teachers. One of the assumption is whether group participation may stimulate creativity or thwart it. This is a debatable question. The effect also depends upon individual differences. One view held by many is that when several people discuss the same subject matter there is possibility of getting more new ideas and, therefore, there is likelihood of inducing new ideas on creativity. Osborn (1953) in his study of creative thinking has asserted this point.

There has been difference of opinions on this issue and several researchers had opposite views from their studies. The group was not actually stimulating creativity. Morse and Wingo (1970) while reviewing the studies made by various authors ascertain that in some studies, group participation fostered higher creative production, whereas in some studies the results were on the reverse. It is observed that there were several factors having influence in school students' creative production for such variations. The varied results show that the type of task assigned to the students, the pressure from classmates towards creative work and away from creative work, the leadership methods, the desire of some students to work alone and also work in a group were found to be the influencing causes. One great disadvantage some individuals find is that the group is a deterrent force by threatening, ridiculing, and demanding conformity with the group performance. Osborn (1953) observed in his study that the group can provide stimulation and encouragement, but also work as a threatening force against those who perform better. With all these results encouraging and discouraging, Osborn from some pilot projects has observed that it can be concluded that a teacher can increase creative products in school students. He suggests seven ways by which creativity can be increased in school students by a teacher. The seven ways are (1) creating an encouraging climate, (2) placing value on variety, (3) encouraging creative perceiving, (4) teaching productive thinking and evaluation, (5) encouraging self trust, (6) children tend to behave in rewarding ways. Hence Torrence's (1965) six principles for rewarding creative behaviour in the classroom should be followed, and (7) being a creative person oneself. A brief discussion of them may throw some light on this problem.

1. Creating an Encouraging Climate

It is one way of creative interest in the minds of children to prepare them for the creative work. The teacher should invite new ideas from the child. He may be allowed to express himself in new ways. By this method the teacher creates a sense of confidence in the child. He should think that he is given importance in such a way that he is accepted by the teacher and also appreciated by him. It is true that many children are not bold enough to bring out their imaginations or inner thoughts in their assigned works fearing that their teacher may criticise them and show to the whole class and put them into shame. Therefore, in order to get rid of such situations they remain silent without exposing their creative skills. Children who have developed strong repression have faulty expressions. Such children are not feeling free to express their impulses. If the assigned work is difficult the child is not courageous enough to come out with his original work and he becomes very rigid. Such children should feel free to express their thoughts whatever type of imagination they have. Therefore, the teacher should create a sense of confidence in the child and encourage him to go ahead with his creative thinking.

2. Placing Value on Variety

The teacher should make the students to think in terms of varied opinions in their assignments. There is not only one correct answer, but there are different types of right answers to the problem and, therefore, everyone may try with his own idea to find out the correct solution. Further, the teacher should bear in mind that he is dealing with children and he cannot expect the work imposing adult standards of creative products or images. Instead if the teacher encourages, values and interprets and appreciates differences of opinions and expressions in the children's work, they may carry out their work without fear or discouragement.

The teacher and other students of the class should value the work of the child, and approve that it is original, and creative work. The students of the class should be made to accept and tolerate the differences of other students. If the students come to know that even though their works are not upto the standard they are not threatened or discouraged by other students of the class they may get a confidence that they can do better next time. Such students should often be helped to establish a standard. The teacher should not create a situation in which student should be in a position to decide the creative work or approve the works of others. To foster creativity among a group of children and to confirm certain standard in achievement is really a difficult task. But it is true that while dealing with large mass of students fixing some uniformity of standards is important. But if uniformity becomes a goal and if it is stressed in all areas then students are subjected to penalty for want of originality. Then the policy of teaching to be more creative or inducing students to become creative is at stake. Therefore, pressing for a uniformity of standard or insisting on conformity in creativity in schools may jeopardize the creative spirit. Whatever every child produces from his own imagination is original and creative for the child. This should be encouraged by saying, "yes, it is correct, you are putting something original, but try more in this line", by saying like this he should be encouraged and guided in a new and little more difficult line.

3. Encouraging Creative Perception

A creative person should perceive objects around him or in the outside world more vividly and also listen everything with keen interest. A person's creative imagination should be expressed in some form substantially to show that he is a creative person. For which he should receive and register the impressions of the objects in his mind, look at objects clearly at close range in the field. Then he should look critically at the objects, analyse, interpret and evaluate the important elements of the objects or the toys with unusual arrangements and form mental images and make new changes in the parts of the objects perceived. Every person perceives the given objects, but particularly the creative person perceives it with different outlook. Willingly or unwillingly we look at certain objects selectively on certain parts and we do not see certain parts, when we look at an incidence or occurrence on the road we look curiously only on those who are involved in it. We do not see anybody else other than this. Even if our friend is standing and watching the incidence we do not see him. Later on we may recognize him. Helson (1959) states that even in some simplest perceptual process the person makes some selection in the objects or events. There is always a tendency in mean's perceptual process to seek certain things and to avoid certain things. Our activity of seeing out is having very wider range of area. But the creative person, Helson says that unlike other persons he is more active in the situation than most of the people and also he is more flexible than most of the people. Both the activities, the perception and flexibility can be fastened by teaching or can be inhibited. Sometimes it is quite but actual for some persons to conceal or misrepresent certain information in agreement with their beliefs and cultural values or may be without any intention add in their memory which is lacking with some information. By adding this new information they feel that their memory is complete with adequate information. In a sense this is also a process of creativity. This involves changing or adding something to the stored information.

4. Productive Thinking and Evaluation

Teaching generally, involves inducing new information to the taughts. In fostering creativity the task is to induce more productive thinking and evaluation. Whatever has been learned earlier and whatever the content from it is required the matter can be recalled for varied intellectual functions by asking different type of questions which may be having some relationship. Productive thinking, Skinner (1968) says, is identified with solutions which have been learned. According to him productive thinking is generated by problem solving techniques that have been learned already. These processes occur because they have good form or the structure corresponds to the structure of the problem in some way. Therefore, the correct subject matter has to be drawn from the past learning by suitable and relevant questions.

5. Promoting Self Trust

Several studies have been made on the factors that foster or promote or restrict creative thinking in the schools. The results of the studies made at the university of Minnesota show that there are several factors which are discouraging creativity. The studies made by Torrance (1960) involved rewarding creative behaviour. Torrance has observed six factors such as (a) premature attempts to eliminate fantasy, (b) restricting manipulation and curiosity, (c) over emphasis on prevention of errors, (d) fear, timidity and pressure from age-mates, (e) adult pressure on verbal skills, (f) lack of resources for working out ideas. These factors are discouraging the confidence of the children in their own ideas.

When children were encouraged with confidence in their own ideas and also recommended deliberately to reduce the effects of discouraging factors there was uncritical acceptance to it. This shows children wish to have confidence in their own ideas and they trust themselves. In fostering creativity in children it is also important that they should be taught to develop self-confidence and self-trust on themselves. The teacher should also know that more than what he teaches to foster creativity, the development of self-trust and self-confidence seem to be more important. No child can develop his creative skill when he has no confidence in himself and his activities. Therefore what Osborn has suggested self-trust as one of the ways of fostering creativity is correct.

6. Children's Tendency for Reward

Researchers believe that any assignment with children is successful if it is associated with rewards. This hypothesis is true also in the case of fostering creativity in children. After making intensive study Torrence (1965) suggests some six principles for rewarding creative behaviour to the children in the class room. The six principles are to (a) be respectable of children's questions, (b) be respectable to imaginative and unusual ideas, (c) show children that their ideas have values, (d) give chances for practice for experimentation without evaluation, (e) encourage and evaluate self-initiated learning of children, (f) combine evaluations with causes and consequences.

These six principles are mainly concerned with rewarding children's activities. The rewarding is referring to honouring children's questions, ideas, showing interest in their ideas, allowing them for practice and experimentation, honouring the self-initiated work, and combining the causes and consequences with the evaluation. To all these the children ideas and activities were given importance. This may not only create interest in children but also may develop their self-esteem. All these may facilitate fostering creativity in children.

7. The Teacher as a Creative Person

The teacher himself must be a creative person, if he is interested in fostering creative skills in children. A person who is devoid of creativity may not be dealing with the matter with proficiency. If the teacher who is engaged in different type of routine work, may not arouse the creative spirit in children to the expedited level. He is putting his students in great doubts. The teacher who is himself still learning may be secretly clearing his doubts in his subject cannot be a model for students in creative ability. He may not be in a position to direct the students for evolving new ideas and images for the creative products. The teacher's demonstrations, the descriptions of the same and the explanations, his answers to the students' questions all may speak much loudly than what he speaks about the subject matter by way of teaching. Therefore the teacher himself should be an expert creative person if he is entrusted with the responsibility of fostering creative abilities in children.

CREATIVE EDUCATION

There should be some definite rules or principles laid down which can be of some use to the individuals by giving chances to them in the early years to develop their creative skills whatever their creative potentials are. The individuals may be capable of learning and achieving a level which can be developed later on if they are more sensitive to the environment and aware of the conditions around them. Murphy (1964) says that providing chances to individuals to develop the creative skills does not mean that we are setting magical power that it is very important and every child must try to become outstandingly creative and that creativeness is the only way of life for fulfilling the needs and happiness of one's life. The central point is that a creation can provide a person great deal of satisfaction. There are evidences that while creative products providing sense of satisfaction and happiness on one side, on the other side great deal of creative potentialities are wasted without being properly guided and utilized. Creative potentialities are wasted also due to the teaching with ineffective methods and foolish approaches. Every society needs creativity for finer display of people's skills. It is not that only great geniuses should fulfil the requirements of the community. The creative needs can be fulfilled by all the members of the community.

These requirements of creativity can even be fulfilled if the individuals in early years were left out without developing the creative skills. In this respect there are certain practical difficulties. It is providing opportunities to the children. Children at the tender age level may be given freedom to react to whatever in which they are interested. They may be sensitive to certain objects or activities in the outside world. There should not be any rigidity or compulsion that they should do only such things and not the other things. Whatever the child responds in a situation, to certain objects it is only dealing with his play — which may be creative for him. He wishes to see something new with the materials easily available to him. He assembles them in a different combinations and feels happy that he succeeded in creating something new. The child is very much interested in reaching every finer object and wanted to dismantle it and reassemble. This is a universal tendency in children all over the world. If the child is left freely in his own way he is happy in dealing with materials. The child need not be taken to a school where creativity is taught. The home environment is the school for the child to practice creative skills of his own. Therefore the important principle should be helping the child to become a creative person and allow him to move and act freely without any restriction and indirectly providing him certain tools and materials which are attractive for him and allow him to manipulate in whatever manner he wished.

The child may be interested in objects with bright colours, contours, shapes, tones, rhythms, paints, brushes, hammers, screwdrivers, moving toys and wanted to do experimentation with them. If he would have left certain activity incomplete sometimes ago, or on the previous day he wanted to continue the same on the next day. The child tries to express the primitive instinctive or spontaneous response to objects in the outside world. He is always having endless desires for them since they give him the satisfaction and happiness when he has achieved them. The more sensitivity the child shows to the objects and the vivid perception he has on the events and things the more he develops the creative skills in his later years. This type of experience the child gains in manipulating objects as he desired and handling everything to his satisfaction facilitate to develop his creative skills. Any creative activity on similar lines become, easy for him and he can even independently attempt to create something new. This we may say creative education at home environment. If parents are keen enough in making rewards for each activity the child performs creatively, it may even foster the creative activities of the child. It is better that parents also should put some indirect pressure on the child when he gets into some difficult problems or situations. It need not be the same type of pressure which J.S. Mill had experienced in his early years from his father John Mill.

Apart from rewards or gifts from the parents for every successful creative work the child has executed, there may be also recognition of the child's performances from the society. Slowly as and when he grows into adolescence and adulthood there is social recognition, which brings him encouragement and excitement apart from happiness. He feels very proud for making such achievements in the creative field. In this manner every one can be a creative person and he can make contributions to the society.

Promoting Creativity

Can we increase our creative ability? It is true that everyone possesses some amount of creative ability. No human being is fully devoid of creativity. What one possesses is also subjected to increase. There are a quite number of materials available to increase our creative ability. Articles, books, training programmes are available on creativity. People can learn to be creative by practicing the lesson prescribed on creativity. Some authors doubt whether this faculty can be increased or we can only exercise the existing potential ability to its fullest extent. Whether creativity can be learned or not, but its existing ability can be fostered by activities or exercise or by practice. Dorothy and Jerome Singer (1977) have studied this aspect of creativity and found that people who demonstrated creative achievements in childhood period, engaged in imagination as children developing imaginary playmates and playing only gesture game, without words. In children's symbolic games creativity is expressed. Creativity is very common among children, but rare among the adults. What amount of creative ability one possesses in childhood is probably subdued when he becomes an adult person due to the process of general education and socialization.

Since children all over the world have the same type of creative skills they exercise the same in various activities. But adults do not exercise creativity unless his work is related to it. Society encourages and promotes creative abilities of the individuals. In some societies it is not encouraged. Silvano Arieti (1976) states that a society can promote and suppress creativity of its people by its values and resources. He observed in his studies several factors that foster a creativogenic society. The cultural factors are made available to all citizens without any discrimination, with openness and tolerance. The society emphasised the people to become creative person for immediate gratification, pleasure and comfort. Therefore, the society takes special interest in promoting creativity among its citizens.

EXPRESSION OF CREATIVITY IN ARTS

Any art is an expressed form of creativity. Art means the creation of works of beauty or other special significances. It is the exercise of human skill distinguished from nature. Art is the imaginative skill as applied to representation of the imagination. It is the product of natural world or figments of the imagination. It is the product of man's creative activities. The work of art collectively, referring especially to visual arts, also music, drama, dance and literature. In all these it means creativity and is expressed in some form. In all these areas imagination is easier to foster in some areas. Arts are of two types, the fine arts, and the practical or applied arts. The fine arts are produced chiefly for its aesthetic value as opposed to applied arts. The fine arts are also called *beaux arts*. Any of such arts produced are painting, sculpture and engraving, and architecture. Art is applied in various areas mainly for commercial purposes. In the advertisement of various commodities, foods, manufacturers apply the art with the view of attracting people towards their goods. Motion picture producers use various techniques in advertisement by using the art of painting, printing, and electrically displaying their productions. Everything in these are presented artistically. In all these imaginations and creative expressions of art, are applied in all possible attention getting attractive manner.

Creativity is encouraged in all these areas. Creativity is a universal and fundamental, common human potentiality, is the view expressed by Maslow (1970). Therefore, creativity is needed fully in almost all the areas curriculum. A young child's artistic expressions are found in his overall behaviour. In each and every movement he expresses his artistic spirit. Whenever any object is put before him, he sees curiously, reaches, touches, feels and then manipulates it. If he finds that it is freely available he tries to explore it. Exploration of objects is his primitive artistic characteristics. Rhode Kellogg and O'Dell (1967) have analysed more than one million drawings made by children all over the world and have found that children everywhere follow the same method in their approach to art and the development of their skills.

According to Kellogg's study the child at the age of two starts scribbling on the paper all his imaginations which may be anything else. It say be a human figure, or an animal like a cat or a toy. Then he moves on to the next stage making shapes only in the form of an outline. Then at the age of four and five he makes out a picture, a man with a coat and hat. He makes out identifiable pictures. Now he is a young artist. But the parents and other adults even at this stage expect from the child much more than what he is capable of. Kellogg was worried on the attitude of the adults and cautions them. It is not wise on the part of the over-eager adults to find fault with the child's drawings and then expect that it should be still more accurate reproduction of the real object. Such attitude of the adults may suppress the creative spirit of the child and may kill his interests in creative art. Instead of encouraging the child's creative work they criticise and discourage him. They must instantly encourage the child and allow him to go ahead with his free imagination and creative work. In Maslow's words, the child's expression of creativeness is the expression of healthy personality and it is projected out upon the world.

Children's creativity is encouraged in some schools is making them to express it in arts. They provide materials in which the child's artistic impulses can be expressed. Schools not only provide materials, but also space and sufficient time, ample freedom to explore the materials through which the child's creativity can be expressed. Materials like wood, plastic, metal, clay all touched, handled, manipulated, formulate first something out of them without knowing what he is going to produce. In handling materials some children may be slow, but explore sensitively, some may be fast in forming out something.

In visual arts children will have to explore space, colour and designs to express their creativity. Similarly artistic expression may find place in dance, play, singing, planning for displaying the creative products. A real or genuine artist feels that every aspect of his life is an expression of his creativity. Designing and making out a fancy dress, preparing a new type dish with a different type of taste, or mixing out a kind of juice, showing his skill and talent in a game or play, dressing with attractive ornaments for a dance programme, making out a new toy with clay or wood and painting with fascinating colour, designing an advertisement poster with some new ideas – all these are also creative expressions through art.

As we have stated in the preceding passage that the teacher himself should be a creator. Such teacher's task is to arouse interest in the students in creative art. The teacher should stimulate them to search and find out a suitable means for the expression of his creativity. The teacher can make his suggestions in such a way that the students themselves should find out some theme of their own and protect themselves into it. The teacher at the same time should maintain the essential qualities of artistic creation such as originality and individuality. It is sometimes dangerous to compare and judge one's creative expression with public stands. This will jeopardize the individual's critical thinking and creative imagination. The teacher should encourage constructive self evaluation though it is subjected to public criticism in the outside world.

The critical ability that one develops to criticise his own creative works in a normal developmental sequence should be from the point of view of Lowenfeld (1957). Students who pass from preadolescence period to adolescence period often develop a kind of vague notion about their creative works. As result of it they become ambivalent and critical of their own images. This feeling becomes more severe when they reach fullfledged adolescence period. They speak of a standard in artistic expression which they can never attain. Therefore, his attitude becomes a matter of satisfying the judges or others instead of expressing his real creative art. Lowenfeld states that this is the general attitude of the adolescents because of their age which is highly self critical and also a stage of pleasing others. But a really sensitive person to creative products never falls prey to the adolescent's ambivalent attitude.

Whatever type of attitude one develops, a creative person at a time, his creative impulse and the urge to create something new can never be suppressed. Having developed so much and produced many artistic materials his aspirations may become stronger and stronger and he may plan even for a very wider scope for his creativity. At this stage many creative persons isolate themselves and find out a solitude place and plan their creative works, keeping away all the criticisms. He gives explanations and descriptions to the shades of his creative works. As he gain more experience he may even name the type of artistic style he is following and how his artistic creations need more thinking to understand the meanings of the expressions. Many famous painters seem to have followed this kind of policy when their artistic creations were in the transitory stage.

CREATIVITY AND INTELLIGENCE

Any discussion on creativity is not complete and perfect without making reference to intelligence. Psychologists often relate creativity and intelligence in their studies. But there are two kinds of views about this. Some are of the view that creativity is a form of intelligence while some are of the view that creativity is a special kind of ability distinct from general intelligence. These two concepts have attracted many psychologists and right from the time of Francis Galton this has been receiving greater attention among the psychologists. Several studies have been made and all these studies have failed to establish that creativity is independent of intelligence. It would be quite relevant in this context to discuss one of the important studies made by Getzels and Jackson (1962). This is supposed to be the classical study which highlights the distinction which has been assumed sometimes existed. They conducted the test in one of the private schools in Chicago. Getzels and Jackson have administered one intelligence test and one creativity test on a group of students. These students were from 5th grade to 12th grade. On the basis of this testing the students were divided into two groups, as those who scored high in the creativity test and those who scored high in intelligence test. These two groups of students were performing regularly in their academic achievements.

The authors of this study have also administered a personality inventory on the group and these student groups varied in their personality traits. It was found that those who had high I.Qs. were found to be conscientious, careful and self controlled. Their goals appeared to match with those of their teachers. Those who scored high in creativity test were playful expressive and independent. But they worked harder than the students with high l.Qs. and they did not adjust well with their teachers. They had some adjustment problems. The results of this study were interpreted that there is relatively low relationship between creativity and intelligence. The mean I.Qs. of the school was 132. The mean I.Qs. of the highly creative and low I.Q. group was 127. This is a very limited range in intelligence test scores. Therefore, it will not be possible to state that there is any relationship between creativity and intelligence. Therefore, this study did not generalize beyond this selected group.

Another interesting study in this field has been made by Wallach and Kogan (1965a) relating intelligence and creativity. The subjects were 5th grade students in one of the sub-urban schools. These authors felt that the conventional intelligence test would not serve their purpose and so they developed tests consisted of game like activities to evaluate creativity. There was no time limit for this test. The students were divided into four groups on the basis of their scores in intelligence and creativity tests. The students who scored high on both the tests performed well in both free and controlled situations and well with child like and adult behaviours. The high creative and low I.Qs. were frustrated in the school environment, but they have performed well in a free context. The low creative and high I.Qs. valued school achievements highly and worked very hard to excell other students. The fourth group, the low creative and low I.Qs. wanted to award the school work. They were involved in social activities and expressed regressions. This study suggested that there is different modes of thinking in creativity and general intelligence.

Wallach and Kogan (1965 b) also worked on personality correlates of intelligence and creativity in children. They have observed that children with high creativity and low intelligence were angry, conflict with themselves and with their school. They were with feeling of unworthiness and inadequacy. They were in stressful state in free context. The children with low creativity and low intelligence were basically confused themselves helplessly. They were engaged in various defensive activities. They plan skilfully from useful intensive social activities to regressions such as passivity or psychosomatic symptoms.

But those who were high in creativity and high in intelligence can exercise both control and freedom. Those who were low in creativity and high in intelligence addicted to school achievements. They were desperately striving for academic success and they were liked by their teachers.

From the available studies it will not be possible for us to make any conclusion regarding the correlation between creativity and intelligence. Torrence (1962) states that if we classify students as gifted on the basis of intelligence test scores we would be missing about seventy per cent of the most creative persons. Thorndike (1963) has analysed all the studies made in this area and summarised the findings. According to his observations the correlation between creativity and intelligence is not appreciable as it ranges from 0 to 0.4. But at the same time quite a number of investigators stated that there is high correlation between creativity and intelligence. In the study made by Meer and Stein (1955) it was observed that at the level of 95th percentile alone the I.Q. scores fail to correlate with creativity. Others like Wallach and Kogan (1965), Cropley (1965) and Pribram (1963) claim that the creativity and intelligence were distinct and separate as there was very small evidence for justification. But, however the highly creative persons may be also scoring high in intelligence test, the eccentricity of the creative persons cannot be ruled out from the personality correlates point of view.

MEASUREMENT OF CREATIVITY

Psychologists have been engaged for many years in the task of identifying the talents of school children by using intelligence tests. But the conventional intelligence tests were found to be inadequate to exhibit certain special talents that human beings are endowed with. At present there is very big need for mathematicians, scientists, and linguists all over the world. Even though these abilities are included in the intelligence tests probably these professions require little different abilities. If the psychologists search for these kinds of talents they should not search them through their existing intelligence tests. Therefore, it has become a challenging task for the psychologists to develop tests that would identify the special talents of creative type.

Who can make something original in artistic and scientific worlds? Who are the potential creators? Persons who are making extraordinary achievements and if their achievements are very much useful they are called genius. The concept of creativity, invention and discovery are always associated with genius or creator, inventor or discoverer. If we want to know who are the creators and who are inventors we must go deep into the origin of the drives or impulses towards creativity to identify the conditions that seem to induce creativity or impede it. Psychologists have identified two contrasting theories as the cause or origin of creativity. According to the conflicting theory, creativity is the consequence of conflict. The founder of this theory is Sigmund Freud (1856-1939). Creativity is the outcome of one's inner conflict of motives. The individual who fails to fulfil his desires in the real world due to the developmental process, suppresses them to the unconscious mind, since his desires are not socially acceptable in real life. They lay dormant as latent content. These desires become motives and dynamic and they try to come out as manifest content in the sublimated form, which are socially acceptable. The person can express all such unfulfilled and socially unapproved desires in the sublimated form of art. He puts his whole personality in this task of artistic creation and he becomes an expert in the field he has chosen for expressing his dynamic of motives. This is how a creative person emerges in the society. Morse and Wingo (1970) stated that some of the maladjusted persons with much creative talents are afraid of resolving their problems by some therapeutic techniques. But we cannot apply this theory to children who develop their creative abilities.

The other theory of creativity is conflict-free creativity and is the consequence of the expression of the healthy personality. This is Maslow's (1930) view. Creativity is the fundamental human potentiality which may be expressed in any artistic form. It is common to all human beings and universal. Everybody can express certain amount of creativity if tested with suitable materials. Creativity is a complex phenomenon. It is a product of several diverse forces and conditions and there is no single factor operating in this. There are also factors inhibiting creativity. Studies have been made correlating creativity and intelligence and the researches state that the study is not the same thing as intelligence. Frank Barron (1958) has intensively studied by investigating people who were considered to be highly original scientists and artists. Among those studied, there were painters, writers, biologists, economists, anthropologists, physicists and physicians. All these people were given some tasks. These tasks were interpreting ink-blots, creating mosaics from coloured squares, creating stage designs on a miniature stage completing unfinished drawings and expressing an artistic preference for various figures and designs.

They were asked to suggest new uses of common place, objects, and to struggle hand to hand to the consequences of improper events. And also they were asked to create a story using as much words as possible at randomly drawn list of nouns, adjectives and adverbs. The results of these studies were that the creative person, whether he is a scientist or an artist preferred painting and drawings that are not symmetrical, complex than those balanced. Their own mosaics and drawings were the same asymmetrical and complex qualities. The ink-blot test they described was what they saw in outlines. Barron's study is one the earliest in this area. Unlike other creativity tests Barron's test consisted of a combination of varied types of tests including the Rorschach Ink Blot test. This appears to be a very difficult method of testing creativity.

Another creativity test called 'The Remote Association Test' was developed by Medrick in 1962. This test was devised to meet certain specific needs. This test consisted of 30 items. They are in the form of sets and each set contains three words. The subjects are asked to bring out a link among three words by giving a fourth word. The time allowed to complete the test is 40 minutes. The number of right links is taken as the score of the subject. This test is very widely used for various purposes.

A battery of creative tests were designed and developed by Wallach and Kogan (1965). These authors followed Mednick in defining creativity. This is one of the creativity tests very widely used for research purposes. In India this test was used in many research works by Prof. Paramesh (1972). Torrance (1966) measured creativity among the elementary school children. He was very much impressed by the wonderful single aspect of the intellectual aspect of man. Many significant functioning was observed in children's mind even when it is limited to the examination of the qualities of thinking on fluency of ideas, spontaneous flexibility, originality and elaboration. Some children were found to be exceptionally fluent in the product of ideas expressed in words, but unable to express ideas in figural or auditory symbols. Others were extremely fluent in expressing ideas in figural form, but when asked to express them in word, they appeared to be mentally paralysed.

Researchers tried with several ways of identifying creative talents of the students. One way is that the teacher rating the creative talent of the students. Lefrancois states that it is very simple method but most unreliable. Gallagher (1960) states that in teacher rating the creative talents of the students about 20 percent of the most highly creative students would be missed. Another method of rating creative talent is to have the individual themselves to rate. Taylor and Holland (1964) have observed that this method is one of the most effective methods of measuring creativity.

There were tests, just like intelligence tests, developed to measure creativity by several researchers. Investigators who made studies in this field assume that the nature of creative process is diversified. Some of them were of the view that the projective tests were not found so much useful in testing creative skills of the individuals. Therefore, group tests, though they have certain limitations, were developed by several research experts like Guilford (1950), Torrance (1962), Getzels and Jackson (1962), Yamamoto (1964), Wallach and Kogan (1965). Guilford included factors like word fluency, expressional fluency, association fluency, ideational fluency, semantic flexibility and originality in his test. But Thorndike (1963) while reviewing Guilford's creativity test has stated that 'Guilford's test did not measure the real creativity. Based upon the more popular method advanced by Guilford, the Minnesota Test of Creative Thinking was developed in 1964 and Torrance developed his test of creative thinking in 1966. According to the test developed by Torrance the subject is asked to produce a variety of responses to the task designed. These responses are scored on the nature of responses and other factors. The most cited example of such a test is the unusual uses of the test. The subjects were asked to think and report as many uses as they can for an ordinary object. The objects such as brick, nylon stocking etc. The responses for each object are counted and an index of fluency is prepared. Flexibility is measured by counting the number of shifts between classes of response. For example, brick is used for building a house, planter, or a compound wall, a road and so on. Each response is scored for fluency here and not for flexibility. A shift or move from one category of uses to another involving moving objects rapidly is flexibility. Originality is scored on the basis of the number of responses judged as unusual by the experimenter and less than five percent of the total responses.

The Minnesota test of creative thinking also includes several non-verbal measures of creativity. These consisted of a picture construction task and a figure completion task. This is a battery of test which also consisted of creative writing as opposed to creative ability. All these tests tasks are having time limit. Peel (1968) states that the tests on writing tasks which are having time limit depress the scores of some students.

EXPERIMENT 30 : STUDY OF MENTAL IMAGERY

Thinking is the process of thought and man alone is capable of thinking. Sensation, perception and learning serve as the basic components of thinking or thought process. We perceive the objects and world through our senses. A person who is deprived of all his senses is like an inanimate object. Imagine for a moment if you have lost all your seeing, hearing, tasting, smelling, touching, kinesthetic, organic and bodily sensations, you are nothing but a man of lifeless object, capable of not moving or acting and doing nothing. Therefore, it is the senses that make man worth living and make him to perceive and know the world. Similarly man's ability to learn makes him to react quite appropriately to the situations and events. When confronts some problem which he could not solve it in the course of time he begins to think with all his experiences he has acquired through his senses and learning abilities. He could do many things without much effort through his normal course of life experience.

In certain extraordinary situations or in dealing with certain difficult cases, man puts himself with all his thought processes as how to get over the situation or deal with the cases. In most of the situations man brings back into his consciousness all those events, ways and means he had in the past relating to the problem situation. The ideas relating to the events may appear in the form of images. He could recall or recollect all these related to the present situation and may apply one by one to solve the problem. Thus the thought processes appear in the form of mental images. In all our mental images there are unsuspectable elements of reproduction of the past experience. Our memory is not lost. By some suitable cues it is possible to trace out certain idea in our memory related to the situation or problem. Thus the memory images, traced out from the past experience, come into play in every form of mental activity.

While discussing about mental process, apart from thinking we always come across the concepts like imagery, images, mental images, memory images, after-images, eidetic imagery, associated imagery, synesthesia and chromesthesia and so on. In order to avoid certain confusion involved in this, it is important to clarify all these terms at this stage. Strictly speaking in a broader sense all these are related to the process of thinking in one way.

Imagery

Imagery generally refers to images collectively. The collection of images is called imagery. It may be also stated as the general processes of imagination or materials of imagination. Imagery is the characteristic kind of mental images formed by a particular individual. A series of mental images also may be called as imagery. We have various mental images through sensory experiences. Thus we have visual images, auditory images, olfactory images, gustatory images, tactual images, kinesthetic images. The auditory and tactual images, are more dominant over other images in a blind man. Men can never have any mental imagery without sensory experiences.

Images

The general meaning of image in the common sense world is that it is a representation. It is also stated as the likeness of a person on a thing especially in sculpture. It is also referred to a person or object that resembles another closely or a copy of the other. But in the scientific world, image refers to a mental representation or a picture or an idea produced by the imagination of a person. Image psychologically refers to the mental experience of something that is not present immediately to the senses, often involving memory.

Mental Images

We are having every day many mental images of some type. But it is very difficult to define it correctly. Generally, mental image is a mental picture. We are forming images of objects persons, places, events and experiences. Therefore, an image is a personal experience of an object, person, place, an event or an experience a person has formed. These images may be visual, auditory, olfactory, gustatory or even a sensation of touch.

We might have seen a beautiful picture drawn by a famous painter, or a sculpture designed by a renowned sculptor, or a scene from a famous play. All these may make us at some time to think of them. Or when you narrate it to your friend you make a mental picture of it and enjoy. This is an image of your past experience of the object. Though the object is not there in front of you, you make it to appear before your mental eyes and you enjoy it. In the same way a musical performance, vocal, or instrumental played by a famous musician, or a violinist, witnessed by you, often may make you to have the image of the musical notes in your ears hearing and you may enjoy it.

Even with regard to the taste we go back to our past experience to the days when we ate a dinner with several delicious dishes. The taste of the food often make us to think of it and have the image of the incidence. Sometimes we may even have the images of the tastes which we disliked. In this respect of enjoying the sensations, our nose is not falling behind other senses. The sense of smell is often associated with taste. Smell and taste are called chemical senses. The fragrant smell we got from flowers or the ethereal smell we get from fruits or the smell of a delicious dishes prepared by a skilful cook, the smell of chicken roast or the turkey biryani and the food smells which can make you to salivate in your mouth — all can make the images of the scenes. The garden of the east and scents or perfumes of the Arabs are noted for the smells. Shakespeare in his play *Macbeth*, makes reference to the perfumes of Arabia and makes Lady Macbeth to speak of it, 'even all the perfumes of Arabs could not change the smell of my hand'. Thus Shakespeare had the image of the smell of the scents manufactured by the Arabs.

Our sensation of touch also makes us to have the image of some object that we touched and enjoyed. The touch sensation is felt by our skin. Our skin is capable of realizing four sensations such as cold, warmth, pleasure and pain. The sense-organs associated with these sensations are called spots. These spots are located below the outer surface of the skin. Some smooth touch of some object gives us pleasure, but a deep pin prick gives us pain. Therefore, we may have the images of pleasant touch and also a painful prick when we think of cutaneous sensations.

Apart from these five senses we also have organic sensations, the sensations felt in our internal organs. The sensations like hunger and thirst are called organic sensations. Hunger is the sensation aroused by the contraction of the stomach muscles. The low position of water in the body and the consequent dryness of the throat and mouth create the sensation of thirst. Men who starved without food for several days often have the images of food and food materials. Similarly caravan travellers often get the images of water springs as they were feeling thirsty for long time.

We have a sensation called kinesthetic sensation. Persons who have lost their skin sensation often find this sensation. In judging weights we realize the sensation of pressure. An object when lifted is felt by a person whether it is light or heavy. This is felt by the muscles.

The vestibular sense is also an internal sensation. This sensation is related to the maintaining of bodily balance and the three semi-circular canals in the inner ear are mainly responsible for this sensation. When a person is seated in a rocking chair, or in a merry-go-round, moved swiftly he is made to move and his semi-circular canals are affected and the person may complain of motion sickness. In normal situations he may have the images of these experiences even though they are not pleasant experiences. Therefore all our sensations sometimes may make us to realize our past experiences and have the images of our experiences whether they were pleasant or unpleasant.

From all these, therefore, we can evolve the idea that human beings acquire knowledge through the processes of sensation and perception. The formation of image is not possible without keen sensitivity to objects, places, persons and events. All of us might have had vivid visual, auditory and other experiences of something at some occasions which might have occurred even without our will and we might have utilized such images for our symbolic thinking. Moskowitz and Orgel (1969) state that symbols are ideas and symbolic processes are referred to ideational processes and ideas are equated with images.

Memory Images

Our visual perceptions of objects, persons or events in the outside world are not completely disappeared from our memory. It is true that we do not remember some, but we remember many of them. What we have learned and experienced earlier are stored in our memory. But it all depends upon our ability to recall and identify the object or the person or the event. When we restate our past experience after few days it is restated in the form of images. This image is the mental picture. It is clear sometimes and sometimes not clear. But some persons have very vivid, sharp and clear picture of the objects and events. Some may have only some vague impressions and they cannot clearly describe them. Therefore, the images of the people may differ significantly from person to person.

If two persons happened to witness the same event on the road and if they were asked to explain we would not get similar statements from both the persons. We would certainly get two different versions of the event. This depends upon the kind of image each has formed of the object or event and the formation of image also depends upon several factors such as the individual's past experience, motivation, learning, age and other social contacts. Further, each one may use different style of language to describe the image he has formed of the event. The verbal description of the image one has made is not the same as that of the other. The two descriptions given by two persons of the same event vary from each other. Therefore, the images of the two persons vary even though there is no chance for changes in the memory. Therefore, all the images are mental images and also memory images.

When we perceive something, fixing our eyes on the object in the outside world the object persists for few seconds. The object makes impressions in our senses and these impressions do not fade away from the brain centre at once. The impressions fade away gradually and ultimately disappear. It is because it takes few seconds for the impressions to stimulate the brain centre and so also it takes few seconds to fade away. The image of the object remains in the brain centre for few seconds. This is sometimes called as after-image. Everyone of us might have had such experiences. Either the face of the person with whom we were having conversation or his voice we were hearing from him may be continued for few seconds even after the person has left us. Murphy (1964) calls this as the "mind's eye". This type of after images are also called memory images. This process takes place without any efforts on the part of the individual.

The after image is also called sometimes as sensory impression. This after image may be differentiated from the memory image. The sensory impressions are the first stage and the memory images are the second stage. The sensory impressions when rearoused they are called memory images. What is that rearoused? It is the impressions of the object perceived. In this way it also becomes mental images. Francis Galton (1822-1911), the British psychologist, had made notable contributions to the development of experimental psychology. Galton's (1883) achievements were unique in the study of imagery and of individual differences in imagery. Boring (1969) states that Galton has originated the theory called ideational type. He was the first to conduct experiments on images. The nature of his work was to collect information about the frequencies and vividness of imagery. His study was mainly on memory images. According to him memory images depend upon the individual's voluntary recall. In Galton's study the subjects were asked to recall the appearance of 'this morning breakfast table' to know about the picture of 'mind's eye'.

The results of the study put Galton into great surprise, because some of his subjects had no imagery at all. Some had very vivid imagery. In Galton's study both the memory images and after images were involved and, therefore, it is very difficult to bring out a clear cut distinction between these two images.

After-Images

In the preceding passage we have made mention about the after-image. This after image has something to do without senses and sensory impressions. While we are looking at an object or a scene for few minutes the brain centre is stimulated and the image of the object is registered there. This process is not taking place quickly. It forms slowly and, therefore, when we turn our eyes away from the object or stop seeing the object any more we are not fully devoid of the object. We retain the image of the object in our perceptual process without any voluntary effort. It is because the impressions registered in the brain centre are slowly diminishing. Till the impression fully diminish we see the object even though the object is not there. This is called after-image. The cones and rods in the retina area retain the image of the object for few seconds. Since the impressions of the object are retained in the retina area for few seconds Garrett (1975) calls it as the *prolongation of sensory experience*. The sensory experience continues even after the object is disappeared. Murphy (1964) calls this after-image as *the persistence of sensation*. According to Murphy (1964) the visual after-image may be positive and also negative. If the visual after-image is of the same as the original stimulus with all its colour and shape then it is positive, but if there is any change in the colour of the object then it is negative visual after-image. The positive after-image is the reproduction of the original sensation of the object after the stimulus ceased to exist.

Therefore, an after-image may be defined as the sensory experience that occurs or persists even when the stimulus has disappeared from the scene. Since only the experience persists in the after-image, Woodworth (1952) calls it as *after-sensation* which according to him is a better name for this. Therefore, images may be also stated as sensory images. The visual after-images are applicable only to the sensation of sight or seeing. But sensory after-images are applicable to all other senses also. For example, in hearing we have after-images, in smelling and in tasting. But, after-image one should experience immediately after the stimulus object disappeared. There cannot be any after-image after that incidence. Morris (1973) says that his subjects had recurrent after-images. He says that they have reported ghosts and apparations in their after-images. Persons who have abnormal symptoms of persecution often report of hearing recurrent threatening voices.

There were many recent experimental studies on after-images. Investigators like, Ellson (1952), Reik (1949) report that among the after-images the visual after-images are very vivid and quite common. Those who are high in one type of after-images also high in other types of after-images. There are individual variations in reporting sensory after-images. It is quite, but natural that painters and sculptors have very vivid visual after-images then others. Similarly musicians and instrumentalists have vivid auditory after-images than others. It has been observed that children have better vivid images than adults. Objects which are concrete and strange produce better vivid images than normal objects and abstract ideas. The reason for the painters and sculptors and musicians to have better vivid sensory after-images has been that they were put into rigorous training in the respective field and also they have developed more interest and strong motivation in their respective fields. These factors have been found much influence upon their sensory after-images.

Persons who are deprived of food or those who are starving often report of images of food. Their sensory after-images were found to be much more vivid and clear than the normal people. Sex starved youths often go into their past experiences and have the vivid images of all those events and activities in which they were involved. The present day motion picture producers often make it a policy of introducing the past events as important scene in their pictures to show the sensory after-images of the heroes and the heroines. Therefore the sensory after-images mostly the visual and auditory receive greater importance in the every day life of human beings.

Eidetic Images

Human beings have the capacity to develop images that could be as vivid and as clean as the objects themselves. These images or imagery are quite similar to photographic picture. These we call eidetic images. Therefore, an eidetic image is a clear vivid and identical image of an object or event evoked after some time of its experience. It may be also stated as the ability of an individual to retain visual images of objects that are quite similar to photographic picture in clarity. Normally such images are described in far greater detail than memory. Persons who are endowed with such images are called eidetics. Woodworth (1952) calls the eidetic images as primary memory images.

Much has been said about various types of images in the preceding passages. The eidetic images we are stating here are different from those images. The important characteristics of eidetic images are that they are unusually clear, vivid and sharp. It is true that same people have eidetic imagery. Those people are said to possess photographic memory. Persons who have eidetic imagery are capable of describing the features of an image with minute details. Children between the age of ten and fourteen exhibit eidetic imagery very often. It is quite common in the later childhood period and the eidetic imagery is fading away when they become adolescents.

Studies have been conducted on eidetic imagery. Meenes and Morton (1936) jointly and individually conducted investigation among eidetic persons and non-eidetic persons. They made comparative study among these two groups of people. It has been observed by them that the after-images of the eidetic persons were found to be much superior to those of non-eidetic persons in many respects. In the case of eidetic persons the images that were aroused suddenly remained for longer duration and they were more positive and with more details than those persons who were non-eidetic. After a week these two groups of persons were again tested with the same matter and the eidetic persons were found to be much superior to non-eidetic persons. The study conducted be Haber (1969) on twenty school children who were eidetics out of 500 brought out several interesting results. One of the girls who was a subject in this study was able to reproduce the images after several weeks of the study. Haber has observed that in the memory test the eidetic children were found to be in no way better than the non-eidetic children. Therefore, Haber is of the view that there may be several factors that determine eidetic images in individuals. However there was individual differences in the quality of eidetic images.

Associative Imagery

While discussing about images and imagery we come across another type of imagery called associative imagery. We can have imagery of every sensory experience. An imagery of one sensory experience may be associated with another sensory image. In this way it is easy for us to associate visual sensory imagery with auditory sensory imagery. For example, on seeing famous musician, visually, we can have the images of his musical vocal performance. Here we see the person in flesh and blood, but we are able to have the image of his voice. In the same manner we may associate visual, olfactory and gustatory images when we see some delicious dish colourfully combining it with pleasant smell and taste. The image of some unforgetful sensory experience or seeing or hearing or tasting may bring the image of other sensory experiences in a chain like manner.

It has been observed that some people could convert sounds heard into visual pictures. A sweet sound heard in a radio programme could arouse the image of the person who had performed the song. There are people who have the tendency to associate colour with musical notes. In every culture this tendency prevails. What colour is to be associated with what musical note needs more experience and elaborate studies. These are called cross-modelity associations.

Cross-modelity Associations

This is a phenomenon in which psychologists have shown special interest. They wanted to discuss this with different type of terminology. Synesthesia and chromesthesia are the two important terms they have evolved to discuss the cross-modelity associations.

Synesthesia is a general term. It has both physiological and psychological meanings. Physiologically synesthesia means that it is a sensation experienced in a part of the body other than part stimulated. Psychologically synesthesia means that it is a subjective sensation of a sense other than the one being stimulated. For example, a sound may evoke sensations of colour. Synesthesia indicates the form of imagery in which a person experiences one sense in terms of another. Synesthesia refers to the vivid image aroused by a stimulus of one sense in a different sense modelity.

Chromesthesia is a term, referring to the occurrence of the images of colour in association with non-visual stimulus. In chromesthesia state the person sees colour when sound is heard. It is converting colour into sound. Chromesthesia is also called 'colour hearing'.

Among these two forms of imagery, synesthesia images are widely found among children and they are also called eidetics. There are people who are colour hearing and they often associate high tones with bright colours and low tones with dark shades of colours. Psychologists who have studied this aspect are of the view that creative artists have synesthesia image to a greater extent than the average persons. Studies were made in this area by Marks (1975), Heinz Werner (1957) among children and it has been estimated that about 40 to 50 per cent of children reported of colour hearing. They heard "light and dark-red whistiling". They also merge colour and smell. One child reported that her mother smelled purple when she appeared as wearing grape perfume.

Hypnagogic Image : This is a type of image, experienced by a person just before fully falling asleep which often resembles hallucination.

Hypnopompic Image : Hynopompic image refers to the state existing between sleep and full waking state. It is characterized by the persistence of dreamlike imagery.

IMAGES AND IMAGINATION

The various types of images are the primary elements of imagination. Imagination is not possible without the formation of mental images and mental images are not possible without keen sensitivity to objects, places, persons and events. Therefore images are collection or reconstruction of sensory experiences which are mostly visual, auditory, gustatory, olfactory, touch and bodily. All images are the result of sensory experiences, and those sensory experiences one might have had sometimes back. It might have been thirty minutes ago, or one hour back, or one day ago or one week ago or even one or two months ago. But man can have the image of the experiences irrespective of duration. Certain images may be clear and vivid, certain images may be somewhat clear, certain images may be faint, certain images may be very faint and certain experiences do not form the correct image. But all images are mental images as a result of mental activity.

The mental images are the reproductions of our earlier perceptions. These reproductions in the form of images vary in a very wide range. In some people the past sensory experiences are reproduced in the form of images very sharply with minute details like the original stimulus. Individuals differ very much in providing the images of the past perception. Psychologists have studied the nature of mental images of various sensory experiences in children in adolescents and in adults. Experiments have been conducted to determine the strength of reproduction of the sensory experiences by making the subjects to reproduce the mental images in whatever form they can present.

Hypothesis

(1) It is assumed that human beings have very vivid mental imagery in the visual sensory area. (2) Persons who have vivid mental images in one area are also supposed to have in other areas.

EXPERIMENT

Problem

Determine the strength of mental images of a person in various sensory areas and compare them with the mental images of a group of persons.

Method

Participant

Choose a person who has no knowledge about this experiment and collect the personal information like, his/her age, sex, qualification, occupation if any, his/her personal interests, socio-economic status and so on.

Materials

Prof. Rajamanickam's Mental Imagery Questionnaire with 6 sub-tests, stop clock, white sheets of paper.

Description of the Material

The Mental Imagery Questionnaire consists of six areas of experiences. This Questionnaire with six sub-tests was designed and developed by Prof. Rajamanickam for estimating mental imagery of individuals.

In each area there are ten items referring to an event of experience which everyone can have. The areas of experiences are: (1) visual, (2) auditory, (3) gustatory, (4) olfactory, (5) tactual and (6) bodily. Each area is a sub-test.

In every item one phrase referring to some event, characteristic, quality, features of a person, name of some object and its nature and characteristics, type of experience one can have from an object are given. Each item is not a complete sentence. The participant is required to provide the information for each item by writing a number against each item.

The participant on seeing an item in an area, he/she has to close his/her eyes and search for such experience he/she had in his/her life and form an image of the past experience and judge the vividness of it. On the basis of the mental picture of the experience he/she must decide how far it is clear and make out the range of clarity of the image of the experience. If it is very clear and vivid like the original experience he/she should write number 5 against the items on the right where space is provided. If the image of the experience is fairly vivid then he/she should write down 4, if just clear 3, if somewhat clear 2, if dim 1 and if there is no image then should write down 0. Thus the image is described in a six-point scale.

The experimenter should not hand over all the six test sheets at a time to the participant. He should give the test sheet one by one. As soon as the first sub-test of an area is answered the experimenter should receive it from the participant and then hand over the second test sheet, then the third and so on. The experimenter should start the stop clock to run as soon as the participant started answering the items for each area and record the time the participant has taken for each area of ten items. The experimenter should inform the participant that there is no time limit to answer the items and he/she can take as much time as he/she wants. In every sub-test ten items relating to the area are given. They are selected out of 15 items of the original test to make out a test for conducting this experiment. The item numbers may vary from 6 to 10. Here the items are limited to ten. Therefore, all the six tests shown here are having only ten items. The experimenter can use the same ten items for each test. Thus he/she finds 6 sub-tests. The same can be *written in 6 separate sheets* for administering on the subjects.

Procedure

The participant should be made to sit comfortably in a well illuminated room where there should be no visual or auditory distractions. The experimenter should have already prepared the six sub-tests, from the test materials of six areas of experience in sheets of paper, each with ten items of varied experiences. At the top of *every sub-test* the six response categories with the respective points should be given like this: 5 for very clear and vivid image, 4 for fairly vivid, 3 for just clear, 2 for somewhat clear, 1 for dim image and 0 for no image. Then the experimenter should give the following instructions to the participant.

"This test is with regard to assessing your mental images in various areas. There are six areas of sensory experiences and there are ten items in each area. I will give you the six test sheets one by one. Your task is to answer all ten items. Whatever type of answer it is going to be, you will have to write it in the space provided.

"When you hold in your hand the first test sheet with ten items read the first item. Then for a moment close your eyes and go into your past experience recall it as much as possible to get a clear picture of the event or experience or object, stated in the item and retain the mental picture of the specific sensation.

"For example, if the event was, meeting your school teacher after a long time and having some conversation with him. You can form a mental picture of this event, how he looked like, spoke to you and your reactions to his querries. If it is an object, a fountain pen, how it looked like and how you felt. If it is a soft drink, how its taste was, if a song, how you enjoyed it and so on.

"Having had a mental picture of it, just judge how clearly and vividly it is presented in your mental picture or image. Judge the degree of its clarity and then write down the clarity of the image as follows: If the mental image is very clear and vivid write 5 against the item in the space provided. If the mental image is fairly vivid then write down 4. If it is just clear write down 3, if somewhat clear write down 2, if it is dim write down 1 and if there is no image write down 0. Carry out this procedure while going through every item and you should not omit any of them.

"As soon as all the items of an area are answered return the test sheet to me and I will give another sheet. Your task is to carry out the same in all the areas. Now, hold this first test sheet". The experimenter hands over the first test sheet to the participant. "Now you can start, Ready' go".

As soon as the participant started responding to the items the experimenter start the stop clock to measure the time the participant takes to answer the first sub-test of ten items. When the participant has completed it, the experimenter should stop the stop clock, receive it, and issue the second test sheet at once and start the stop clock. In the same manner all the six sub-tests should be conducted. When the participant has completed all the six sub-tests the experimenter should obtain the introspective report from the participant and he/she may be permitted to go away.

Results

The responses of the participant for the sub-test in visual area are worked out and presented in Table 11.1. The scores ranged from 5 to 0. There are six response categories on which the participant has answered each item. The time taken by the participant was also recorded. The mean score for visual area was worked out and presented in Table 11.1.

Table 11.1 : Showing the scores of the participant in the visual area on the Mental Imagery Questionnaire

Sub-test-1 : Visual area

Time taken to complete the test : minutes

Write against the item under the column score on the right 5 for very clear and vivid image of the object, 4 for fairly vivid image, 3 for just clear image, 2 for somewhat clear image, 1 for dim image, and 0 for no image of the object.

S.No.	Items relating to the sensation of seeing	Score
1.	Your College or University Tower Building	
2.	Your final year School Teacher's face	
3.	A famous dance performance you witnessed	
4.	The first day you saw the classroom in your College/University after getting admission	
5.	The face of your favourite film Hero	
6.	The face of your favourite film Heroine	
7.	A Temple Tower or a Church Tower excited you	
8.	The Republic Day Parade	
9.	The first time you saw the sea	
10.	A tiger you saw in a zoo or circus	
Total		
Mean		

State the items which were responded with the lowest scores and the items responded with the highest scores. State the mean score by working out from the total and state how many items are above the mean score and how many below the mean score. What are the items which had very clear and vivid mental images having score of 5 and 4 and the items which had somewhat clear and dim images having the score of 2 and 1. State also the items which had no image.

If all the items were responded with very vivid mental images then the highest score is 50 and if all the items were responded with no image the lowest score is 0. Therefore, the scores in this test ranges from 0 to 50. But every individual's score should fall between 0 to 50.

The high score in the visual area indicates that the individual is having vivid visual mental imagery and also he/she is having high creative images. High creative people are supposed to have high mental imagery. The low score indicates the low visual mental imagery and also indicates low creative image. It is also presumed that artists, painters, sculptors and some writers and poets have very vivid visual mental imagery.

Table11.2 : Showing the scores of the individual in the auditory area on the Mental Imagery Questionnaire

Sub-test : 2 Auditory area

Time taken to complete the test : minutes

Write against the item under the column score on the right 5 for very clear and vivid image of the object, 4 for fairly vivid image, 3 for just clear image, 2 for somewhat clear image, 1 for dim image, and 0 for no image of the object.

S. No.	Items relating to the sensation of hearing	Score
1.	The voice of your favourite musician	
2.	The voice of your father	
3.	The noise of a take off aeroplane	
4.	Your favourite Television News reader's voice	
5.	The noise of a steam engine	
6.	The thunder sound	
7.	The sound of your College/University bell	
8.	The voice of your favourite teacher	
9.	The clattering noise of the typewriter	
10.	The sound of a Temple or Church bell	
Total		
Mean Score		

The responses of the participant for auditory items on the Mental Questionnaire are presented in Table 11.2. The scores are given under the column score for each item according to the procedure ranging from 5 to 0 on the basis of the strength of the image of the object or event the participant has experienced or dealt with in his life. Work out the mean and state how many items are above the mean and how many items below the mean. State the items having very vivid and clear image and the items having no image. Find out and state whether the pleasant sounds or noises received high score and unpleasant sounds or noises received low score from the participant.

Table 11.3 : Showing the responses of the participant for gustatory items on Mental Imagery Questionnaire

Sub-test-3 : Gustatory area
Time taken to complete : minutes

Write against the item under the column score on the right 5, for very clear and vivid image, 4 for fairly vivid image, 3 for just clear image, 2 for somewhat clear image, 1 for dim image, and 0 for no image of the object.

S. No.	Items relating to the sensation of taste	Score
1.	The taste of mango fruit	
2.	The taste of Jankery (sweet meat)	
3.	The taste of coffee	
4.	The taste of orange fruit	
5.	The taste of boiled milk	
6.	The taste of toothpaste (colgate)	
7.	The taste of lime juice	
8.	The taste of soda water	
9.	The taste of plantain fruit.	
10.	The taste of ghee	
Total		
Mean Score		

The participant's responses to the items relating to the sensation of taste on Mental Imagery Questionnaire are presented in Table 11.3. These items are referring to various types of tastes that any person might have had on some occasion. Work out the mean score for the mental imagery of tastes. State what are the tastes above the mean score and what are the tastes below the mean score. What are the tastes responded with the scores of 5 and 4 and what are tastes responded with the scores of 0 and 1. The highest possible score for all the items should be 50 for this area. But what is the total score obtained for all the items. Is this total score above the previous two areas or below the two areas.

Table 11.4 : Showing the responses of the participant for olfactory (Smell) sensation on the Mental Imagery Questionnaire

Sub-test-4 : Olfactory area
Time taken to complete the test : minutes.

Write against the item under the column score on the right 5 for very clear and vivid image of the object, 4 for fairly vivid image, 3 for just clear image, 2 for somewhat clear image, 1 for dim image and 0 for no image of the object.

S. No.	Items relating to the sensation of smell	Score
1.	The smell of jasmine	
2.	The smell of camphor	
3.	The smell of petrol	
4.	The smell of burning rubber	
5.	The flavour of tea	
6.	The smell of bathing soap	
7.	The smell of raw fish	
8.	The smell of peppermint	
9.	The smell of roasting onion	
10.	The smell of fine apple juice	
Total		
Mean Score		

Table 11.4 carries the responses of the participant to the items relating to the sense of smell. It is true that there are innumerable number of smells that man can produce or nature can produce. Every material that is used in cooking is producing a smell. The participant has given the strength of his mental image of the smell he has smelt and experienced in his life. Work out the mean score and state that in how many items the score is above the mean score and in how many below the mean score. State the names of the materials that have received the high score 5 and 4 as having vivid and clear image of the smell and also state the names of the objects received the score 0 and 1 as no and low image. What is the total score for all the items? Is it high or low compared to other areas of sensory images ? Has the participant taken more time or less time compared to other areas ? State it.

Table 11.5 : Showing the participant's responses to the items relating to the cutaneous (touch) sensation on the Mental Imagery Questionnaire

Sub-test-5 : Cutaneous (touch) area
Time taken to complete the test: minutes

Write against the item under the column score on the right 5 for very clear and vivid image of the object, 4 for fairly vivid image, 3 for just clear image, 2 for somewhat clear image, 1 for dim image and 0 for no image of the object.

S. No.	Items relating to the sensation of touch	Score
1.	A touch of a hot iron piece	
2.	Your toes strongly hitting against a stone	
3.	A blow by your friend on your back when you were not aware of it	
4.	Shaking hands with a V.I.P.	
5.	Touching of an ice bar	
6.	A pinprick in your finger tip	
7.	The gentle breeze on your body when you were sweating due to hard physical work	
8.	An electric shock you received while dealing with a switch	
9.	The touch of a rough stone surface	
10.	The touch of a highly polished marble stone	
Total		
Mean Score		

The participant's responses for the items relating to the touch sensation on the Mental Imagery questionnaire are presented in Table 11.5. Work out the mean score for the touch sensation and find out the items having the scores above the mean score and the items having the scores below the mean score. Name the objects which were responded by the participant with the scores of 5 and 4, and the objects responded with the lowest scores of 0 and 1. State whether the total score is high or low compared to other sensory areas in forming the images.

Table 11.6 : Showing the responses of the participant to the bodily sensation on the Mental Imagery Questionnaire

Sub-test-6 : Bodily sensation by movement
Time taken to complete the test : minutes

Write against the item under the column score on the right 5 for very clear and vivid image of the object, 4 for fairly vivid image, 3 for just clear image, 2 for somewhat clear image, 1 for dim image and 0 for no image of the object.

S. No.	Items relating to bodily sensation	Score
1.	Sitting in a rotating chair and moving fastly.	
2.	While fastly going over the steps to upstairs.	
3.	Pulling forcibly a jammed drawer of a table	
4.	Riding fastly a motor cycle or bicycle	
5.	Making a long jump and falling forward	
6.	Jumping high to catch a cricket ball or something Lifting a heavy	
7.	weight	
8.	Your feeling of very hungry	
9.	After a long time desk work the way you are yawning your body	
10.	Your feeling of very thirsty	
Total		
Mean Score		

The participant's responses to the bodily sensation by movement on the Mental Imagery Questionnaire are presented in Table 11.6. Work out the mean score of the test and state that in how many items the scores above the mean and in how many items below the mean score. State the items in which the scores are very high with 5 and 4 and also state the items in which the scores are very low with 0 and 1. Compare the total score of all the items with other areas whether it is higher than the other areas or low. What you infer from this result? State how much time the participant has taken to complete the items of the questionnaire and compare the time with other areas of sensation. Do you find any significant difference between the bodily sensation and other sensations ?

Table 11.7 : Showing the mean scores for all the six sensory areas and the time taken by the participant to complete the items

Measures	Sensory areas					
	Visual 1	Auditory 2	Gustatory 3	Olfactory 4	Cutaneous 5	Bodily 6
Mean Score						
Time in minutes						
Rank						

Compare the mean scores of the six areas and state the areas which are providing the most clear and vivid mental images. State the order of the vivid images from the highest to the lowest. State how much time on the whole the participant has taken for all six areas to complete the questionnaire and to which area he/she has taken shortest time and to which area he/she has taken longest time.

Collect group data from other experimenters, the mean score for each area and. also the time taken for each area and present them in Table 11.8. Work out the mean for the group for both the strength of the imagery and for the time. Compare the scores of the individuals in all six areas. Who are those obtained the highest scores in all six areas and who are those the lowest. Are there significant differences in the mental images of the group members? Is there any difference in the time between the individuals in completing the whole test? What are the sensory areas in which the group as whole has formed very vivid imagery and also the areas in which formed poor imagery?

Table 11.8 : Showing the group data for all six areas on the test of Mental Imagery Questionnaire and also the time taken for each area

S. No	Name	Sensory areas												Total Score
		1		2		3		4		5		6		
		Visual		Auditory		Gustatory		Olfactory		Cutaneous		Bodily		
		Mean	Time	Mean	Time	Mean	Time	Mean	Time	Mean	Time	Mean	Time	
1.	P.V.													
2.	S.M.													
3.	B.S.													
4.	K.R.													
5.	T.S.													
6.	V.R.													
7.	R.M.													
8.	V.L.													
9.	C.M.													
10.	M.G.													
11.	D.K.													
12.	S.V.													
	Mean													

Discussion

State and discuss the nature of your participant's mental imagery in various areas. Stat the sensory area in which he/she had very vivid mental imagery. Verify the hypothese one and two whether they are applicable in your participant's case: that human being have very vivid mental images in visual sensory area. Test this hypothesis with th obtained data from your subject. Secondly, persons who have very vivid mental image in one sensory area are supposed to have also in other areas. Test this hypothesis wit obtained data from your participant and state whether they are accepted or rejected in hi case.

Compare the scores of your participant with the scores of others in every area an state where he/she stands. In the whole group who has scored the highest in all area showing very vivid mental images. Is he/she consistently high in all six areas or only i certain areas? Apply the two hypotheses to the members of the group and state whethe the hypotheses are accepted or rejected.

Find out whether individuals differ very much in their mental imagery generally an then see specifically the sensory area in which they have very vivid mental imagery an discuss why such differences have taken place. State whether individuals differ in thei mental images on the basis of age, qualifications, subjects they study, specific interests socio-economic status. State and discuss whether the males scored high or the females i their vivid mental images test. Is there any relationship between mental imagery an mental creativity and mental imagery and imagination? Discuss in detail.

Do you find that the time factor has anything to do with the mental imagery of th individuals in any of the areas? They might have taken much time in certain areas. Wha are those areas? Why did they have taken more time to respond to the items? Explain. I the time factor do you find individual differences. What factor was found to be operatin in it? Is there difference between sexes? Who took more time, males or females?

Conclusions

What are the sensory areas in which your participant was having very clean and vivi mental images? Do the vivid mental images vary among the group members or all are o the same level? In the group how many of those were having very vivid mental image and how many had very poor mental images. What are the areas in which most of th subjects had very vivid mental images and what are the areas they had poor menta images.

APPENDIX-I

MENTAL IMAGERY QUESTIONNAIRE

SUB-TEST- 1 : VISUAL AREA

Direction

This is with regard to visual sensation, seeing. There are 15 statements. Read each statement carefully. Try to recollect your experience with the object. Indicate the degree of the image of the event. Against each item, under the heading 'Score' on the right, write 5 for very clear and vivid image of the scene or event, 4 for fairly vivid image of it, 3 for just clear image of it, 2 for somewhat clear of it, 1 for dim image and 0 for no image of the object.

Time taken to complete the test in minutes :

S. No.	Items relating to visual sensation	Score
1.	The first day you saw the classroom in your College / University after getting admission	
2.	Your College or University Tower Building	
3.	The Republic Day Parade	
4.	The first time you saw the sea	
5.	A Tiger you saw for the first time in a zoo or circus	
6.	A beautiful painting you saw in an exhibition or museum	
7.	A long express train moving fast at a distance	
8.	A huge and unprecedented public meeting you witnessed	
9.	A steep Mountain Valley you saw	
10.	Your final year school teacher's face	
11.	A famous dance performance you witnessed	
12.	The face of your favourite film Hero	
13.	The face of your favourite film Heroine	
14.	A Temple or Church Tower that excited you	
15.	A big dam on a river you saw	
Total		
Mean		

SUB-TEST - 2 : AUDITORY AREA

Direction

These 15 statements you are finding below, are with regard to the sensation of hearing. Read each statement carefully and try to recollect your past experience of hearing of the specific voice. Indicate the level of the image of the voice. Against each statement under the heading 'Score' on the right, write 5 for very clear and vivid image of the voice, 4 for fairly clear image of it, 3 for just clear image of it, 2 for somewhat clear image of it, 1 for dim image of it and 0 for no image of the voice.

Time taken to complete the test in minutes :

S. No.	Items relating to auditory sensation	Score
1.	The voice of your father	
2.	The voice of your favourite musician	
3.	The sound of a roaring lion	
4.	The mill siren near to your house/loudly moving vehicle	
5.	Your favourite Television News reader's voice	
6.	The noise of a take off aeroplane	
7.	The voice of a roaming street fruit seller	
8.	The barking of street stray dogs	
9.	The noise of the steam engine	
10.	The thunder sound during rainy season	
11.	The sound of your College/University bell	
12.	The voice of your favourite teacher	
13.	The clattering noise of the typewriter	
14.	The sound of a Temple/Church bell	
15.	The speech sound of your favourite political leader	
Total		
Mean		

SUB-TEST - 3 : GUSTATORY AREA

Direction

The gustatory area is with regard to the experience of taste. Read the following 15 statements carefully. Try to recollect your experience in tasting of the particular item. Indicate the level of the image of the taste, whatever it was. Against each statement under the heading 'Score' on the right, write 5 for very clear and vivid image of the taste of the object, 4 for fairly clear image of it, 3 for just clear image of it, 2 for somewhat clear image, 1 for dim image of it and 0 for no image of the taste of the object.

Time taken to complete the taste in minutes :

S.No.	Items relating to the sensation of taste	Score
1.	The taste of mango fruit	
2.	The taste of jankery (sweet meat)	
3.	The taste of coffee	
4.	The taste of honey	
5.	The taste of buttermilk	
6.	The taste of orange fruit	
7.	The taste of pepper	
8.	The taste of onion	
9.	The taste of sea water	
10.	The taste of tooth paste (Colgate)	
11.	The taste of lime juice	
12.	The taste of soda water	
13.	The taste of plantain fruit	
14.	The taste of ghee	
15.	The taste of Ice cream	
Total		
Mean		

SUB-TEST - 4 : OLFACTORY AREA

Direction

The olfactory is referring to the sensation of smell of an object. Read the following 15 statements carefully and try to recollect your experience of smell of the object and get the image of it. Against each statement under the heading 'Score' on the right, write 5 for very clear and vivid image of the smell of the object, 4 for fairly clear image of the smell, 3 for just clear image of it, 2 for somewhat clear image of it, 1 for dim image and 0 for no image of the smell of the object.

Time taken to complete the test in minutes :

S. No.	Items relating to the sensation of smell	Score
1.	The smell of jasmine	
2.	The smell of camphor	
3.	The smell of petrol	
4.	The smell of burning rubber	
5.	The smell of roasting onion	
6.	The flavour of coffee	
7.	The smell of pineapple	
8.	The smell of bathing soap	
9.	The smell of raw fish	
10.	The smell of peppermint	
11.	The smell of mango fruit	
12.	The smell of garlic	
13.	The smell of sulphur	
14.	The smell of sandal paste	
15.	The smell of rose	
Total		
Mean		

SUB-TEST - 5 : TACTUAL AREA

Direction

The tactual area is referring to the sensation of touch. Read the following 15 statements carefully and recollect your experience of touch in similar incidence. Form the image of it whatever level it is. Indicate it in numbers. Against each item under the heading 'Score' on the right, write 5 for very clear and vivid image of the experience, 4 for fairly clear image of it, 3 for just clear image of it, 2 for somewhat clear image of it, 1 for dim image of it and 0 for no image of the touch experience.

Time taken to complete the test in minutes :

S. No.	Items relating to the sensation of touch	Score
1.	A touch of a hot iron piece	
2.	A touch of an ice bar	
3.	Hitting your toes against a stone	
4.	A blow on your back by somebody when you were not aware	
5.	of it	
6.	Shaking hands with a V.I.P.	
7.	A pinprick in your finger tip	
	The gentle breeze on your body when you were sweating due	
8.	to hard physical work	
9.	Some cut injury you had in some part of your body	
10.	The feeling of a strong scratch in your knee	
11.	An electric shock received while dealing with a switch	
12.	The touch of a rough stone surface	
13.	A touch of boiling milk	
14.	Bath in chill water	
15.	Touching of a highly polished marble stone	
	Touch of boiling water	
Total		
Mean		

SUB-TEST - 6 : BODILY MOVEMENT

Direction

This is with regard to the bodily experience in movements. Read the 15 stateme carefully and get the image of that particular experience and indicate the level of Against each statement under the heading of 'Score' on the right, write 5 for very c and vivid image of the bodily movement experience, 4 for fairly clear image of experience, 3 for just clear image of it, 2 for somewhat clear image of it, 1 for dim im of it and 0 for no image of the experience.

Time taken to complete the test in minutes :

S. No.	Items relating to the bodily sensation	Score
1.	Sitting in a rotating chair and moving fastly	
2.	While fastly going over the steps to upstairs	
3.	Pulling forcibly a jammed drawer	
4.	Riding fast in a motor bicycle/bicycle	
5.	Making a long jump and falling forward	
6.	After a long time desk work, the way you are yawning your body	
7.	Diving in a swimming pool or similar incidence	
8.	Jumping high to catch a cricket ball or something	
9.	Lifting a heavy weight	
10.	Muscling arm your and clenching your fist	
11.	Going on in a winch from one end to another end	
12.	Munching your mouth vigorously something	
13.	Running fast and suddenly halting	
14.	Your very hungry feeling	
15.	Your very thirsty feeling	
Total		
Mean		

ANSWER SHEET

MENTAL IMAGERY QUESTIONNAIRE

ease find in the following pages certain statements relating to various sensing periences, like seeing, hearing tasting, smelling, touching and bodily movements. ere are 6 sub-tests, each one having 15 statements relating to a particular sensing perience. You might have experienced all these senses. Now, try to get the image of the ticular experience mentioned in the statement. You may have different levels of image the experience in your mind. Whatever level of the image of the experience you could m, please express it according to the direction given in the pages. This is not an elligence test nor a personality test. This is only to assess the level of the mental image your past experience. The information provided by you are used only for the academic research purposes and will be treated strictly confidential. Therefore, you can be te free and frank in answering this questionnaire. Please provide the following ormation accurately.

1. Age :
2. Gender :
3. Qualifications :
4. Religion :
5. Married or Single :
6. Your or Your parent's occupation :
7. Approximate monthly income to your family :
8. Number of members in your family :
9. State whether your Home Town is an Agricultural or Industrial or Metropolitan area :
10. Are you interested in Sports and Games ?
11. Have you any interest in Fine Arts, like Painting, Singing, Dancing, Sculpture and the like ? State it :
12. Are you interested very much in seeing Motion Picture or Television Programmes? If so. What types of programmes attract you very much? State in detail :

ANSWER SHEET

MENTAL IMAGERY QUESTIONNAIRE

e write your answer *in this answer sheet*. There are six sub-tests to be answered and are 15 statements in each test. The number of statements or items are given from 1 . Against each item write down your answer according to the direction given in the f the test booklet. Write your answer either as 5 or 4, or 3, or 2, or 1, or as 0 (zero), e basis of the level of the imagery you could form in your mind about the particular tion. After finishing all the 15 items of the first test seeing (visual), go to the second test hearing (auditory), then the third test, fourth test and so on. Please do not omit any them.

Item Nos.	Test-1 Seeing (Visual)	Test-2 Hearing (Auditory)	Test-3 Tasting (Gustatory)	Test-4 Smelling (Olfactory)	Test-5 Touching (tactual)	Test-6 Bodily (Organic)
1.						
2.						
3.						
4.						
5.						
6.						
7.						
8.						
9.						
10.						
11.						
12.						
13.						
14.						
15.						
Total						
Mean						

12

MEASUREMENT OF HUMAN APTITUDES AND SPECIAL ABILITIES

INTRODUTION

hile the individuals are very much anxious about the jobs available for them, the siness and industrial organizations are very much worried about the availability of table persons for various types of jobs in their organizations. Men may be available t they must be suitable for the jobs and jobs may be available but suitable men are not ailable. Thus in the personnel programme the important function is the matching of lividuals to jobs for which they were qualified. Finding out individuals aligned with propriate jobs at the time of employment is very important. When someone makes a cision to place an individual on a job, he is making a prediction. He is predicting, ticularly, that the individual will perform on the new job very effectively. But edictions made by human beings are not always correct. However, whatever one edicts, his prediction is supposed to be more accurate as he has more practical and tinent information to use in his prediction. In the same way an employment officer in edicting the job performance of applicants, is able to make better predictions if he is lowed with correct information. He needs, in particular, the information about what man characteristics are related to success on the present job and also which applicants ssess such characteristics.

Personnel officers, Psychologists in organizations and employment officers often use ious pychological tests for identifying the specific characteristics or abilities the plicants possess. The range of human abilities, however, fall under two groups. The st group of abilities mainly refer to basic human abilities. These are general class of man abilities that most of us have in common, but in varying levels. For example, the ntal abilities and the psychomotor skills belong to this type. The second group of lities mainly refer to the specific abilities on job performance. These abilities are que to the individuals who have learned to perform a particular job or activity. Here, are mainly concerned with the first group of abilities.

Human beings are able to acquire certain specific skills and knowledge to perform tain jobs and activities and they differ in the level of acquisition of skills, because of ir varied capacities. Capacities are potential abilities and achievements are actual lities. These capacities are generally referred to as aptitudes and they are actually the sic human abilities. Individuals have marked differences in these basic abilities. These basic human abilities are measured and used as criteria in predicting job success in individuals who claim as competents for certain jobs.

Since the concept, 'basic human abilities' has certain ambiguity, it is important to define it operationally. According to Fleishman (1967), the basic human ability is "a more general trait of the individual which has been inferred from certain response consistencies (*e.g.* correlations) on certain kinds of tasks". Farina (1969) states that "these basic abilities are fairly enduring traits which have both learning and genetic components underlying their development".

According to Woodwoeth and Marquis (1952) the word 'ability' is having two meanings though closely related. "It may mean actual ability or it may mean potential ability". A person who has the actual ability to do the task can do it now, while a person who has the potential ability to do it can learn and do it. To avoid some confusion caused in the meaning Woodworth uses the word achievement to mean actual ability and the word capacity to mean potential ability. Achievement is limited and capacity is unlimited. Therefore, a man's capacity always exceeds his achievement since he can learn and do the task, which he could not do it at present. He is capable of learning many things and can perform them later on and also can make improvement over his present achievements. But a man's capacity to learn certain things also is limited. A man can never make thirty feet long jump, however training and practice he made. His performance is limited to certain extent.

A man's achievement can be measured as he could demonstrate it by his performance whereas capacity cannot be measured directly. There is lack of demonstration or evidence in the case of capacity. How much capacity one has in learning to speak German ? It is very difficult to answer this question directly. It can be stated only in terms of levels or degrees. However the same type of intensive training is given to a group of students athletic and literary fields each comes out with unequal achievements. This means that each one started the activities of various fields with unequal capacities and as a result of it they could not perform the task better than what they demonstrate.

Psychologists believe that achievement can be measured by taking into account the performance of an individual in a given task. In measuring achievements psychologists have formulated certain standard for every task which may be called as tests. If the performance of an activity is to be measured, it is compared with standard already set up by means of test scores. In the performance of an activity various components of the activity are taken into consideration. For example, while testing the driving of a vehicle, the test consists of driving up and down the hill, driving in a narrow lane, backing according to the specifications, parking, slow driving, reaction to the traffic signals, overtaking and so on. In the practical situations achievements are measured very well and it was also noticed that individuals differ in their speed of performance. There are also standardised tests to measure the clerical speed and accuracy of the workers.

Psychologists believe that if the capacity of a person could be measured or estimated before he is recruited for a job or before he begins a training much time and energy could be saved or wasted efforts could be eliminated. But unfortunately capacity cannot be measured directly. Still one best way of solving this problem is to devise aptitude tests which are nothing more than achievement tests which can be used to measure the present performance and used as index of future performance for other tasks. The assumption is that the present performances in the aptitude tests will predict the future performances. But one should bear in mind that predicting future performance on a special task is much more difficult than to measure the present achievement.

It is possible to estimate the aptitude of a group of students in a particular work. The students must be ignorant of the present work and they are given certain amount of training. We can find that some of them may make rapid progress in the task than others. For example, memorizing certain passages from a technical book some students learn much more faster than others. It is observed that those who made better progress in the beginning continue to do well in the later stages also. But at the same time those who made poor progress in the beginning found to be poor at the later stages also. But this cannot be made as general rule in all the tasks. There might have been certain intervening variables or certain advantages or disadvantages may become the cause of the success or failure in the start. Some students who might have got real aptitude for this work might have become nervous and failed to do the justice and some students who might have been very intelligent in learning scored high in learning though they had no real aptitude for the special kind of work. That is why aptitude cannot be measured accurately.

Gardner Murphy (1964) says that the term aptitude means "simply the readiness of an individual to profit by future training". According to him "a great many types of abilities which are important in relation to success in a given vocation or in a specific job can be tested one at a time". This is very advantageous because a person might have one of the required aptitudes, but not all of them, and he might consequently do better to find a job which requires his own special pattern of ability. The abilities which we test today in a person may reveal different results if he gets further practice and experience after some days. Therefore, we should differentiate between his fundamental aptitude and the achievement he reveals to us today and what he could be capable of tomorrow.

Suppose, for example, a man has applied for some job in an organization and he is at 22 years of age. He is found strong, shows much interest in learning semi-skilled industrial job, moving fast and of moderately intelligent in general. A test was conducted on him and he was found to be having certain aptitudes for jobs which require eye-hand coordination, speed of response, accurate performance and so on. He has never done this job and found to be of worthy of something. Another person of the same age did not show so much of success in the aptitude tests. But he had a high level of achievement on the job. Here we can see the difference between these two persons. The first man scored high on the aptitude tests, but on the job did not exhibit much success, whereas the second man did not score high on the aptitude tests, but exhibited high level of proficiency on the job. The reason was that the second man had some special training on similar job. If the first man also had some special training he could have done much better than the second man on the job test. Therefore, an achievement test will reveal with how much proficiency the two men will actually carry out the operation and an aptitude test may provide us with a general view of how much we can expect from each of them after each one had the required training or practice.

We all know that individuals differ widely in various ways, like intelligence, knowledge and skills. Therefore, if we want to determine whether a person has the skills for a particular job or he is intelligent, will be benefited by the college education we must have reliable methods of measuring his present abilities and potential abilities. The present society is highly technological and complex. Therefore, it has become very important to match the unique talent of each person to the requirements of the job. If we could perform this task successfully it will be very much advantageous to both the individual and society. What a person can do now and what he might do when appropriate training is given are not going to be the same. For example, a first year medical student cannot do the operation of opening the stomach of a person and remove an abnormal growth nor a first year engineering student can assemble the parts of a motor car. Both of them cannot perform the task at present, but both have the potential ability to learn the required skills. Now they cannot do, but after few years on completing the course of study they can perform the tasks.

It is also important tö appraise the distinction between a capacity to learn and an accomplished skill. This is done through suitable tests. There are tests, designed to measure capacities or potential abilities to predict what one can acquire with training are called aptitude tests. These tests are generally called tests of general intelligence and tests of special abilities. The tests that what one can do now are called achievement tests. For example an intelligence test that predicts how well a student will do in college is an aptitude test and at the end of the course, the examinations which are given to see how much the student has learned are achievement tests. In general both the aptitude tests and the achievement tests are ability tests. In business and industrial situations achievement tests are given on the job. How well one performs the job. This may be assessed by more than one supervisor and the performance may be compared.

Psychologists try to relate human abilities to see whether there are any relationship between abilities or any similarity of skills. It is found that most of the abilities have some association with one or more abilities. But some human abilities are quite independent of other abilities as they do not have any association with other ability tests scores. It is very difficult to make mention about all the basic human abilities. Fleishman and others have enumerated thirty seven human abilities. McCormick and Tiffin (1979) have put all the human abilities into four broad classifications. They are : (1) Mental abilities, (2) Mechanical and related abilities, (3) Psychomotor abilities (4) and Visual abilities. We shall discuss here briefly the various abilities and the corresponding ability tests.

1. MENTAL ABILITIES AND MENTAL ABILITY TESTS

The mental abilities, in other words may be stated as cognitive or intellectual abilities or intelligence. Studies on this area were undertaken in the last quarter of nineteenth century by Sir Francis Galton in England and by Alfred Binet in France. Galton was the first to design and develop tests to measure intelligence. He developed a battery of tests to measure variables like reaction time, visual acuity, memory for visual stimuli, strength of hand grip, head size and so on. He collected data from several thousand people who visited an exhibition at London in the year 1884. His conclusion was that superior intelligence was associated with superior physique and vigour. But the type of intelligence test with which we are familair today was designed and formulated by the French Psychologist Alfred Binet. Binet in collaboration with Theodore Simon, another French Psychologist published a scale to measure intelligence in 1905 and revised it first in 1908 and again in 1911. The present intelligence or mental ability tests are based on this. Even though Binet-Simon Intelligence tests were first intended to identify the mental deficiency in school children they served as model for all other intelligence tests developed later on.

Binet tests consisted of many items referring various aspects. But, he believed that all items tap the general intelligence. This was the view of some psychologists that intelligence as a general ability for comprehension and reasoning that may manifest in various patterns of behaviour. But psychologists like Spearman and Thrustone were of the view that intelligence is a collection or an orderly arrangement of certain special abilities having positive correlation with one another in varying degrees. They ascertain this view by applying the statistical methods of correlation coefficient and factor analysis and these methods provide precise information about the special abilities concealed in intelligence. Charles Spearman (1927) said that all individuals are endowed with a general intelligence factor called the 'G' in varying degrees. A man is described as very intelligent or dull only on the basis of the amount of the G factor. The G factor is the major determinant of performance on intelligence of the individual on the test items. Further, there are also special factors which are called specific factors, the small 's' referring to particular abilities. According to him an excellent performance in mathematics is a function of a person's general intelligence (G) and a person's good performance in painting or music is a function of the person's special ability (s).

Thrustone (1935), after working on this factor analysis with large number of tests for many years came out with a view that intelligence can be broken down into a number of primary abilities and published all his findings in his psychometric monographs under the title "The Primary Mental Abilities". Thrustone in his studies was able to identify seven factors which he called as the primary abilities. These abilities are (1) Verbal comprehension, (2) Word fluency, (3) Number, (4) Space, (5) Memory, (6) Perceptual speed, and (7) Reasoning. He developed battery of tests to measure each of these abilities and they have predictive power and still they are widely used all over the world. It was found later on that Thrustone's primary mental ability tests are not fully independent. They have significant intercorrelations among them. This seems to support Spearman's theory of general intelligence factor.

The studies of factor analysis of Spearman and Thrustone accelerated further studies in this line of thinking. One of the profound scholars, who made a major development in this area by many years of study was Guilford. He has published several monographs, papers and books on the basis of his series of studies and also formulated a theory called the 'structure of intellect'. Guilford (1959) did not accept Spearman's factor theory of general intelligence and Thrustone's theory of Primary mental abilities, but suggested the concept called, "Three faces of intellect". The three faces are contents, operations and products. Guilford (1967) was of the view that intelligence test items should be distinguished not only in terms of content, but also in terms of operations made on the contents and also in terms of the product, that is, the results. Thus his structure of intellect consisted 4 contents, 5 operations and 6 products yielded 120 components which he called 120 unique intellectual abilities.

According to Guilford's 'structure of intellect', the content includes four aspects such as, F–figural (size, form, colour etc). S–symbolic (letters, digits etc.) M–semantic (verbal meanings and ideas), and B–behavioural (dealing with social situations).

The second face, operations include five aspects, such as, C–cognition (rediscovery of recognition), M–memory (remembering what is learned or cognized), D–divergent production (going in search of different ideas or solutions to problems) N–convergent production (processing information to arrive at correct answer or solution) and E–evaluation (judging adequacy of one's thinking) .

The third face, the products refer to the nature of responses of people, consisted of six aspects such as U–units, C–classes, R–relations, S–systems, T–transformations and I–implications.

The combination of these three aspects, content, operations and products add up to (4×5×6) 120 varied aspects or facets of intellect and each one is an ability stated as a triagram indicated by the letters as CSU, of which C stands of cognition, S for sematic and U for units.

There are quite a number of mental ability tests available and most of them are in use in various situations. All the mental ability tests are not of the same pattern. They differ in various respects. There are individual tests and group tests, paper-pencil tests and performance tests, children tests and adults tests and so on. These tests also differ in content. For example some tests are designed to assess the general intelligence and some are designed to assess certain specific aspects of mental ability. There are certain mental ability tests which can cover very wider area of mental abilities and identify the individuals with more specific abilities. Mental ability also differ in the administration procedures, method of scoring, method of analysis of scores. Further, some of the mental ability tests are designed for certain specific purposes. There are mental ability tests indented for educational counselling and guidance, for vocational selection and placement in business and industrial organizations for a wide variety of jobs.

The mental ability tests that are widely used: The Adaptability Test, Army General Classification Test, Alpha Examination Modified Form 9, California Test of Mental Maturity, Differential Aptitude Test, Otis Self-Administering Test of Mental Ability, Thrustone's Primary Mental Ability Tests, Revised Beta Examination, SRA Non-verbal form, SRA Verbal Form, Thrustone Test of Mental Alertness, Wonderlic Personnel Test, Alexander Passalong Test, SRA Typing Skill Test, SRA Clerical Aptitudes Test, Thrustone Typing Test, Minnesota Clerical Test and so on.

2. MECHANICAL ABILITIES AND MECHANICAL ABILITY TESTS

The importance of mechanical ability in the present day world has been increased tremondously. The rapid development of science and technology resulted in the invention of new industrial machines in every type of work. Machines are replacing men in all manual works. Since men will have to operate the machines men with mechanical abilities are now on enormous demands. The invention of electronics and computer machines have simplified many of human works in the everyday life. Therefore there is a search for men with mechanical abilities everywhere and the importance of the mechanical ability has been increased. Mechanical work has something more to do with the cognitive or mental aspects than with physical or motor aspects. Tyler (1965) says that mechanical ability has two aspects and the ability is generally called mechanical aptitude.

One aspect of mechanical aptitude is (1) comprehension of mechanical relations, the recognition of instruments and tools to be used in various jobs and also the related cognitive abilities. These factors of the mechanical aspects are based upon the practical knowledge gained from using mechanical instruments and tools. Friedman (1954) and others call this as general mechanical aptitude. The other aspect of mechanical aptitude is with regard to the (2) perception and manipulation of spatial relations. This is with regard to the understanding the parts to fit together as a whole. In Thrustone's Primary mental abilities also this factor is included. Many of the mental ability tests have some factors related to the comprehension of mechanical relations and also spatial relations. Generally we could see that mental abilities and mechanical abilities are inter-related. There can be no mechanical work without the use of mental abilities. But the utility may vary in degrees.

The mechanical ability tests most commonly used for various purposes are : The D.A.T. Space Relations, D.A.T. Mechanical Reasoning, Bennett Mechanical Comprehension Test, Minnesota Spatial Relations Test, O'Connor Wiggly Block Test, Purdue Mechanical Adaptability Test, Purdue Mechanical Performance Test, and SRA Mechanical Aptitudes Test.

3. PSYCHOMOTOR ABILITIES AND PSYCHOMOTOR ABILITY TESTS

Many of the human activities in the everyday life involve psychomotor abilities. Any accurate or correct performance of some work or activity involves psychomotor ability. Therefore, the concept psychomotor ability refers to characterizing movements of the body and bodily parts associated with mental activity in some given task. All psychomotor skills involve mental abilities. Measurement of simple motor performance goes back to the beginning of experimental psychology to the times of Wilhelm Wundt, Fechner, Cattell and Galton. Some of the psychomotor abilities tested in the psychological laboratories are the reaction time, steadiness, speed and dexterity, two hand coordination, arm-hand steadiness, postural discrimination, fine psychomotor coordination, multiple limb coordination, response integration, manual dexterity, finger dexterity and muscular coordination. Thus psychomotor abilities cover very wide range of abilities that involve frequently some amount of visual control.

Some are of the view that psychomotor abilities are relatively independent. But how far they are independent? This has to be experimentally verified and established. Vernon (1950) has made several studies on psychomotor abilities. The test scores of his study have shown some relationship with each other test scores. Though the intercorrelations were found to be very low the relationship cannot be ruled out. But most of the psychomotor abilities were considered to be predominantly specific in nature. But, however such abilities are required to be tested with suitable tests in a given situation and revalidated for the job in question. The psychomotor abilities are very pertinent in engineering field in connection with physical output responses made with the body members especially arms, hands, fingers, legs and eyes. The human abilities to perform various types of motor actions can have relationship to three aspects of human factors in engineering. The first deals with selection, design and location of control device systems. The second is with design of hand tools which are movable to perform the work and the third is with work methods involving specific procedure and sequences of carrying out the work activities and materials. The psychomotor skills are very important to a number of occupations.

There are good psychomotor ability tests in use. They are: Minnesota Rate of Manipulation Test, MacQuarrie Test for Mechanical Ability, O'Connor Finger Dexterity Test, O'Connor Tweezer Dexterity Test, Purdue Pegboard Test, Purdue Hand Precision Test, Purdue Grooved Pegboard Test, Bennett Tool Dexterity Test and Rajamanickam Target-Dart Test.

4. VISUAL ABILITIES AND VISUAL ABILITY TESTS

Every job needs the use of visual ability or the vision. We must see the object if it is to be handled or manipulated. But there are certain jobs in which the visual skills of various types are specially important. Visual skills very much needed for office jobs, some inspection operations, handling some machines, operation of vehicles and for understanding the signals at a distance. The visual skills which are relevant to job performance are visual acuity, depth perception, colour discrimination and postural characteristics of the eyes.

Visual Acuity

Visual Acuity is the ability with which we could discriminate black and white. This involves discriminating the minimum separable areas. Particularly in some of the psychophysical experiments the variable stimulus is moved in small distances and which could be discriminated. Visual acuity is measured both for distant view and nearby view depending upon the nature of job performance.

Depth Perception

The Depth Perception is referring to seeing the objects accurately at a distance. This is also known as perception of space. The performances of both eyes are involved in this. Depth perception is a three dimensional phenomenon. Both eyes are fixed on an object at a distance and a triangular is formed among the two eyes and the object away from the eyes.

Colour Discrimination

This colour discrimination is different from the discrimination of black and white by visual acuity. Therefore, colour discrimination refers to identifying various colours by discriminating between red and green, between blue and yellow and orange and pink. People who are suffering from colour blindness may not be able to differentiate one colour from the other. But even people who are not suffering from colour blindness find it difficult sometimes to discriminate colours if the colours vary in degrees and shades.

Postural Characteristics of the Eyes

While seeing an object at a distance with both eyes in normal situation the eyes make movements in relation to each other so that the sight of both eyes thrown upon the object symmetrically. Because of this we could see one single object eventhough both eyes are projected upon the object. The eyes take a postural position by convergence or divergence process called Phoris in such a way as to cover up the distance between the eyes and the object. The adjustment process or the deviation may be vertical and lateral and this can be measured by suitable instrument on either direction. But some individuals who have some difficulty in discriminating objects or colours take pains and strain the eye muscles and direct both eyes on the object. This is done in a strained posture and take the risk of muscular fatigue if the object is to be seen for longer period. Sometimes the posture of the head also has to be changed depending upon the visual acuity of the individuals.

There are not many vision tests. The most widely used test is the Snellar Letter Chart. The physicians and the eye specialists use this test for testing the vision of the individual who are deficit of normal vision. This test consists of several rows of block letters of decreasing size. It is placed normally at a distance of 20 feet from the subjects. This test is administered separately to each eye to determine the smallest possible letter the individual can read. It is determined that the smallest letter the individual can read the greater the visual acuity he has. The test scores in the Snellen Letter Chart is always in the form of fraction. In the fraction the numerator 20 is constant which represents the distance between the subject and the chart. Therefore, the standard visual acuity score is 20/20. A score of 20/40 means that the person can read at 20 feet distance only letters twice as large as the standard letter and a person with standard eye can read it from a distance of 40 feet. The smaller the fraction the poorer the vision is.

The American Medical Association has simplified the Snellen rotation of acuity scores by converting it into percentage scores. This is called (Visual Efficiency). The difference between the standard scores 100 percent and the individual's visual efficiency is stated as loss of vision. The AMA's percentage notation normally is not carried out above l00 percent and below 20 percent for the placement purposes. It is also stated by some authors that the Snellen Letter Chart test and the American Medical Association test do not provide satisfactory visual acuity for placement purposes in certain industrial jobs.

Since these two testing procedures have certain limitations and fail to fulfil certain specific needs in personnel selection and placement procedures, more efficient instruments have been devised. These instruments are the Ortho-Rater, the Sight-Screener, The Telebinocular and the Vision Tester. These instruments measure various visual skills in addition to visual acuity. Among these instruments the Ortho-Rater is a prevision, stereoscope with long focal length that has certain advantages over other instruments. It has seprarate control of test stimuli for for each eye. There are twelve tests incorporated in the Ortho-Rater instrument and they fall under two categories as far-distance and rear-distance. They are: (1) far-distance test of vertical phoria, (2) lateral phoria, (3) acuity of both eyes, (4) acuity of left eye, (5) acuity of right eye, (6) rear-distance test of vertical phoria, (7) lateral phoria, (8) acuity of both eyes, (9) acuity of left eye, (10) acuity of right eye, (11) depth perception, and (12) Colour discrimination.

The Purdue University's Occupational Research Centre has conducted extensive research work. The Research Centre was able to show that different jobs demand, different visual skills quantitatively and qualitatively. They have demonstrated that crane operators and truck drivers require the ability to see things at a distance. Watch makers radio and television mechanics require the ability to see things at a very close range. There are jobs requiring accurate colour discrimination. For example, in printing colour pictures, painting in different colours, at traffic signals, in dyeing factories the ability to discriminate colours is much needed. In selecting persons for such jobs the Ortho-Rater may be of very much useful in testing candidates.

EXPERIMENT 31 : MEASUREMENT OF BASIC ABILITY

INTRODUCTION

We could see that people differ from one another in various ways. They differ in their abilities, aptitudes, interests, motivations, personality, skills, hobbies and education. One of the most important aspects of the individual differences is the abilities. Many of the problems we face today are based upon some estimates of the individual abilities. Parents send their sons and daughters to schools to do painting lessons or to college to do engineering or medicine on the hope that he has the ability to profit for such training. An organization appoints a person whom it considers that he would do best on the job. Like this we are making some estimates of the abilities of some individuals. This is a very crude way of estimation. A careful way of the estimating the abilities of the individuals is by the use of psychological approaches.

The psychological approaches to deal with various problems such as recruitment, selection, placement and training of personnel to business and industrial organizations, to deal with educational problems of children in guidance situation and to deal with personal problems of individuals in adjustments have been applied very successfully in the second quarter of twentieth century after the World Wars I and II. The excellent services rendered by the pscychologists in various parts of the world during the time of wars, in industries when there were depressions and prosperity and in community when there were social and emotional problems, have contributed to the development of various applied areas in the field of Psychology. One of the important areas to be mentioned in this context is the psychological testing and measurement. This is the most important contribution psychology has made to the practical guidance of human affairs.

Psychological testing and measurement need suitable instruments and tests. Instruments and tests should be designed and developed to suit the requirements. To design a suitable instrument and a test which should serve the purpose accurately, needs much effort, time and knowledge. The psychologists relentlessly faced the needs and designed and developed innumerable number of instruments and tests. Tests are classified according to the function they are designed. A general classification is (l) Intelligence, (2) Aptitude, (3) Achievement, (4) Personality and interest, (5) Sensory and motor skills. One of the areas, the psychologists wanted to study extensively is the area of ability. Excellent tests to assess the abilities of the individuals and groups have been developed in recent times and they are used for several purposes in areas like educational, vocational, guidance and counselling situations.

In guidance situation several types of test are used at present. One important test is aptitude test, designed to measure the individual's aptitudes.

Generally, aptitude refers to the quality of being fit for a purpose or position. It also refers to fitness and suitableness. It is a natural tendency or propensity of a person, his natural capacity for any pursuit. It is also defined as the ability to profit by training. Psychologists use this concept parallel to intelligence or quick-wittedness.

The concept "aptitude" was elaborately dealt with by Bingham (1937) in his book "Aptitude and Aptitude testing". According to him aptitude is a condition symptomatic of a person's relative fitness of which one essential aspect in his readiness is to acquire proficiency in his potential ability and another is his readiness to develop an interest in exercising that ability. This definition refers to general and technical usage.

Thorndike and Hegan (1970) define aptitude as the "readiness for learning". This readiness depends on unknown properties on the genetic characteristics of the individual and his whole life history of physical surroundings and previous learning.

Psychologists and educators are interested in measuring what a person can do in a situation and what he will do. In both cases they are measuring the ability of the individual. In the first situation they measure the aptitudes and in the second they measure achievements. Woodworth (1952) says that the word ability has two meanings. It may mean actual ability and also may mean potential ability. As the meaning may cause some confusion, he differentiates these two abilities by referring actual ability to achievement and potential ability to capacity. Woodworth equates potential ability or capacity to aptitude. Since a person's capacity cannot be measured directly in recruitment and selection situations his aptitude can be tested as an index of his future performance in related tasks. But Thorndike says that ability may be divided into aptitude and achievement. Aptitude refers to what a person could learn to do in a job and achievement refers to what he has learned to do in job. The test designed to measure aptitude or capacity is called aptitude tests and the test designed to measure achievement is called achievement test.

A person's future performance in a job or school work can be assessed by using suitable aptitude tests. Thurstone (1948) states that his primary mental ability tests can be used for guidance purpose, because these tests could predict where a persons could make a greater success.

Psychologists have observed that while appraising the individual's ability in one area the test often produces two or more scores of a specialized nature. They also provided more specific information about the individual. Psychologists are also interested in some specific information which would provide various phases of the ability area. This interest led to the development of test batteries. To measure special aptitudes single tests also were developed.

One of the areas in which psychologists are interested is the vocational area where they can guide young people for various types of works in which they are interested and also could be successful and happy. Another important concern the psychologists are having is to help the employers in selecting right people for right job. The psychologists, while going through various aspects of vocational guidance in selecting persons to jobs, analysing and evaluating jobs, discovered that different jobs require different special abilities and different levels of general mental abilities. For example a watch maker requires finger dexterity and steadiness of the hand to handle small parts of watches, a lawyer needs verbal comprehension, a mechanical engineer needs understanding mechanical comprehension and principles and the clerk needs combining things appropriately. Therefore, different jobs require different abilities and every human ability is a specialized one to some degree.

A few of the aptitude test batteries are the differential aptitude test battery and the general aptitude test battery. The DAT is a power test and GATB is a speed test. For some tests, time limit is set and the person who takes these tests can make best scores if he could do things quickly. These are speed tests. The other tests are power tests. They are designed to test the ability of a person to answer difficult questions or solve difficult problems without any time limit. But sometimes, time limit is fixed for power test also for certain conveniences. There are also single aptitude test. They are the Bernnett Mechanical Comprehension Test, the Minnesota Vocational Test for clerical works.

A group of aptitude test is also called prognostic test. They are designed to predict readiness to learn or the degree of success in some subjects. A group of such tests are also called "reading readiness" tests. These tests are administered on young children after their entry into the first grade to find out how they could progress in the school performances.

There are certain aptitude tests, specially designed and developed for selecting persons for professional training in certain particular jobs. They are also known as vocational aptitude tests intended to assess the ability to learn a specified vocation. For example clerical aptitude is the ability to learn a clerical vocation. Professional organizations are using these tests in test programmes for selecting their students. These tests are very much in use in the fields like engineering, medicines, law, veterinary medicine, dentistry, nursing and accounting. These tests are generally called professional school aptitude test batteries and are varied from those scholastic aptitude tests. The scholastic aptitude tests assess the ability to succeed in some specified type of formal schooling. A number of tests have been developed in the field of music and arts and they are called music and artistic aptitude tests.

EXPERIMENT

Problem

The purpose of this experiment is to measure the fundamental or basic abilities of the person/persons.

Method

Participant

Choose a person who has no knowledge of this experiment. Describe the person's characteristics like age, class, sex, socio-economic status etc. State whether the person is wearing eye glass or not.

Materials

A battery of eight differential aptitude tests : 1. Verbal Reasoning, 2. Numerical Ability, 3. Abstract Reasoning, 4. Space Relations, 5. Mechanical Reasoning, 6. Clerical speed and accuracy, 7. Language Usage : I. Spelling, and 8. Language Usage : II. Sentences. Test manual, Answer sheet, Scoring Key, Stop clock, pencil are also required.

Description of the Tests

The Differential Aptitude Tests (DAT) consists of eight tests. These are developed to measure the fundamental intellectual abilities of the persons, particularly school students and to give them Vocational guidance. The authors of these tests are George K. Bennet, Herold G. Seashore, and Alexander G. Wesman (1959). The tests were published by the Psychological Corporation, New York. They have been widely used all over the world. The Indian edition is available with Manasayan of Delhi.

Every test is intended for specific purpose and a knowledge in these tests may help an Indian student who wishes to appear for competitive examinations in various areas like Union Public Service Commission, Bank Probationary Officers, Bank Clerk, Engineering, State Service Commission Examinations and Executive in Business and Industrial Organisations.

1. Verbal Reasoning

The verbal reasoning test measures the individual's ability to understand concepts composed of words. The aim of this test is to measure the student's ability to abstract and to think constructively. The test item in analogy form is suitable for the measurement of reasoning ability. The analogy item is specially devised for this test. It has three characteristics. The item is reliable, versatile and measures reasoning of complex type. Academic success is one of the fields in which this test is successful. The test also tests the occupational aspirations of the students. It consists of 50 items of multiple choice type. In each item the first and the last word is omitted with blank spaces. Each item words are given and the subject has to fill up the blank with suitable letter (word).

2. Numerical Ability

This test is intended to test the individual's ability to comprehend the numerical relationship and to handle the numerical concepts. The problems are called arithmetic computation. It tries to avoid language element which involves reading ability. Most of the items requ understanding of numerical relationships. The items computationally appear simple l complex as problems. It is tested among high school students. It involves quantitati thinking as the students who could like to progress in the fields like mathematics, physi chemistry and engineering. There are also other occupations such as statistician, laborat assistant, salesman etc., require numerical ability and this test may be useful to test the abili This test consists of 40 items with all types of calculations like addition, subtracti multiplication and division. All these types of calculations are with multiple choice items.

3. Abstract Reasoning

It is also a reasoning test but non-verbal type. Hence, the subject or participant expected to perceive the operating principles in the diagram that causes change and fi out the principles operate the change of figures. His understanding should be logica presented. The abstract reasoning test may be an alternative to verbal and numerical te with regard to general intelligence. But the ability of word reasoning cannot be trea with the ability of abstract reasoning. In this case abstract reasoning cannot be treated an alternative to verbal reasoning. There are 50 items in this test.

In the problem figure there are four figures and in answer figures there are five. T subject should understand the pattern shown in the problem figures and should indic how the problem would continue in the fifth figure on the answer sheet.

4. Space Relations

The space relation test is a pattern test. It has 40 items. In each pattern five figures given. The participant has to judge which of these five figures can be made from pattern. The pattern shows only the outside of the figures. So, for each pattern there five figures in a row and in every row there may be at least one correct figure. But th may be also more than one correct figure.

It is a test to judge one's ability to understand the constructed figure from the pattern it is a test of one's imagination as to how an object could appear if it is placed in diffe manners. It is also a test of one's space perception correctly. The important factor in this is that the participant is required to manipulate his mind in a three dimensional space.

5. Mechanical Reasoning

The mechanical reasoning test consists of pictures of various mechanical situati There are also questions relating to the pictures. There are 68 pictures in this test and subject has to answer the questions asked for relating to the pictures. It is a form mechanical comprehension test. It does not require any special knowledge to answer questions, but require reasoning. The mechanical reasoning test is regarded as one of intelligence tests. The person who scores high on this test can easily learn the mechan principles for operating and repairing of complex type of machines. A student wh studying physical sciences or engineering sciences is expected to score high on his otherwise he may find his job very difficult.

Clerical Speed and Accuracy

; a test of speed and accuracy and tests how fastly and correctly a person works on it. ere are letter and number combinations. In this test there are one hundred items. In h test item there are five letter groups. In some, among the five there is one number one letter combination and in some two numbers are combined and in some two ers are combined. They are, either both are capital or one capital and one smaller. The e combinations are given in answer sheet, but in different order. In each test item one he five combinations is underlined. The subject has to find out the same combination ach answer sheet and has to underline it. In clerical jobs there are situations in which person may encounter with similar elements. It does not involve much of intellectual ities. Generally this test measures perceptual speed, momentary retention and speed onse. Students who score very low in this test may lack standards of neatness, speed accuracy in classroom work. A bright student who scores low on this test but scores on other tests reveals lack of ability to work rapidly. A retest can be made on a ht student if his score on this test is very low.

anguage Usage

test consists ot two tests, the first one being (I) Spelling and the second (II) ences. The spelling part consists of 100 words. Some of them are correctly written some of them are incorrectly written. The subject has to indicate in the answer sheet her the particular word is correct or incorrect by putting a cross mark (×) in the circle v the right and wrong headings.

nguage Usage : II. Sentences

second part of the test is sentence section. It is to measure the subject's language y. It is by distinguishing between good and bad grammar, punctuation and word :. There are 50 sentences in this test. Each sentence is divided into five parts as A, B, , and E and indicated by these letters. There may be mistakes in any part of the nce. The mistakes may be in one part or two parts in grammar, punctuation or ng. When the subject finds a mistake in any part of the sentence he has to mark a (×) in the answer sheet in the circle just below the letter indicating the part. Some of ntences are fully correct and some are fully incorrect. The test as a whole is a test of y to differentiate the right from the wrong. This ability is mainly required in ess correspondence, journalism, printing, proof reading and steno-typing.

Procedure

e administering the test, it is to be noted that a planning is necessary to avoid certain larities. It is important that all these tests should be administered within two weeks There should not be any interruption while the testing is going on.

ne student should look fresh and alert. Therefore, it is better to administer it in the ng hours. These tests should be arranged in different order to create some interest oid boredom. The order suggested by the authors of these tests is verbal reasoning, language usage, numerical ability, clerical speed and accuracy, abstract reasoni mechanical reasoning and space relations.

The teacher who administers the Differential Aptitude Tests should make extens preparation. The best preparation is that the teacher himself should take the tests. The administrator should stick to time fixed for each test. Otherwise, the scores will have meaning.

It is important that all test materials and other relevant things should be collected the experimenter before the session starts. The administrator should check all the materials one day before the session. The number of test materials, answer sheets, k should be counted and kept separately in envelopes.

The room should be properly lighted and ventilated. The temperature of the room shc be comfortable. Students should be seated comfortably at a reasonable distance, so that administrator can pass through all the rows. The students should not see each other's face.

There should be advance preparation on the part of the test administrator before test is administered. The students should be put at ease and should be instructed, to forth their best efforts on this as it is relating to their personal values and they shoulc told what each test is dealing with.

The test administrator should see that all the students are having two sharp pen and erasers. If one student is not in possession of these the administrator should sup them. This is to be followed specially when clerical speed and accuracy tes administered. Then the respective answer sheets should be distributed to the students they should be asked to provide necessary information required for this. The follow instruction should be given before distributing the test booklets : "please do not open booklet until I tell you to do so".

Then the direction given in each booklet should be read aloud by the administr and should tell the subjects to read the direction to themselves. Each example also shc be read aloud. At this stage the administrator should ask them whether they have doubts or questions. If the group is young it is important that the instruction and reac should not be done in haste in case of verbal reasoning. After giving all the direction administrator should say "ready, begin, and at once should start the stop clock. At the of the time he should say "stop" and should stop the writing and stop the stop clock once, separate direction should be given for the other test as given in the test booklet.

Table 12.1 : The correct time allowed for each test is given in Name of Test and its Time

S. No.	Test Name	Time
1.	Verbal Reasoning	30 minutes
2.	Numerical Ability	30 minutes
3.	Abstract Reasoning	25 minutes
4.	Space Relations	30 minutes
5.	Mechanical Reasoning	30 minutes
6.	Clerical Speed and Accuracy :	
	Part I	3 minutes
	Part II	3 minutes
	Language Usage :	
7.	Part I – Spelling	10 minutes
8.	Part II – Sentence	25 minutes

Then the answer sheets and test booklets should be collected. The collection should done carefully. As soon as this is over write down a note about the name of place, time d date and the type of the group.

Method of Scoring

ıe scoring procedure should be done according to the key supplied by the authors. ıere are two types of keys as right keys and wrong keys. The procedure for the tests ıere right keys are only to be used is to go through the scoring sheets and recording the mber of rights on the answer sheet and then wrong keys and recording the number of ·ongs on the answer sheet.

The next step is to subtract the wrong responses from the right responses on the basis the proportions and assign marks as given in the key. It is important to check each oring sheet because there is every chance to mark more than one. In such cases only e is permitted for scoring.

It should be borne in mind that the clerical speed and accuracy test should be scored ly for Part II. For both the forms *A* and *B* the score key should be used.

Results

ork out the individuals score according to the scoring procedure for all the tests and esent them in Table 12.3. From the manual (if available) find out the percentile scores the test scores and enter them in the respective column. Explain where your rticipant is strong and where he is weak. In which test he made the highest score and ıich the lowest.

Table 12.2 : Showing the scoring information of the test in DAT

No.	Test	Maximum possible score	Formula	Key	Notes
1	Verbal Reasoning	50	R	Rights only	One mark for each item
2	Numerical Ability	40	R-1/4 W	Rights and wrongs	One mark for each item number
3	Abstract Reasoning	50	R-1/4 W	Rights and wrongs	One mark for each item number
4	Space Relation	100	R-W	Rights and wrongs	Multiple marks are permitted for each item number
5	Mechanical Reasoning	68	R-1/2 W	Rights and wrongs	One mark for each item number
6	Clerical Speed & Accuracy	100	R	Rights only	One mark for each item number. Only Part II scored. Same key for form A and B
7	Language usage Part I – Spelling	100	R-W	Rights and wrongs	One mark for each item number.
8	Part II – Sentence	95 (94)	R-W	Rights and wrongs	Multiple marks are permitted for each item number

Table 12.3 : Showing the individual's scores in the DATs

S. No.	Tests	Raw score	Percentile score
1	Verbal Reasoning		
2	Numerical Ability		
3	Abstract Reasoning		
4	Space Relations		
5	Mechanical Reasoning		
6	Clerical Speed & Accuracy		
7	Language usage Part I – Spelling		
8	Part II – Sentence		

Table 12.4 : Showing the group data in the DATs

S. No.	Name	Name of Tests															
		V.R.		N.A.		A.R.		S.R.		M.R.		C.Sp & A		LU–I & II		LU–II	
		R	P	R	P	R	P	R	P	R	P	R	P	R	P	R	P
1	R.V.																
2	M.R.																
3	B.V.																
4	S.R.																
5	P.V.																
6	P.K.																
7	R.M.																
8	M.S.																
9	S.V.																
10	C.S.																
	Mean																

R = Raw Score, P = Percentile Score. Mean may be worked out for the new scores only.

Collect the group data for all the tests and present them in Table 12.4. Work out the mean scores and explain them. State the tests in which the scores are high and low.

DISCUSSION

First of all state the purpose of conducting this test. Since the DAT can be used for more than one purpose the specific purpose should be stated clearly. On the basis of this the discussion should be made. How many of them in the group scored high to fit themselves for the purpose. What are the abilities in which the group scored high and low. How many of them are above the mean in each test and below the mean. What is the position of your participant? What is the rank he secured in the group. State whether the individual's personal factors are having any influence in the test scores, such as age, sex, socio-economic status etc.

CONCLUSIONS

1. State the position of your participant in every ability test in the group.
2. What are the abilities in which the group is proficient.
3. State the abilities the group scored very low.
4. What are the personal factors influenced the proficiency of the participants.

EXPERIMENT 32 : MANUAL DEXTERITY TEST

INTRODUCTION

Human aptitudes though potential in nature, may be made to manifest in performing certain activities. The performed activities may reveal what a person can perform in these activities. If several people perform the same activity the level, nature and quality of performance may not be the same. The same activity, but the outcomes vary. The reason might have been that the level of input in the activity was not the same in all the individuals who have performed the task. Why this variation in the input? Every person applies his mind, experience and knowledge in performing the task. What amount of mind, experience and knowledge applied by *'A'* was not applied by *'B'* and what amount of this applied by *'C'* was not applied by *'D'*. In this way every one's input in the task is what he/she is capable of. One person may be proficient in one type of task, another may be in another type. Thus individuals vary in their capabilities in performing the same task. These capabilities may be stated as aptitudes and abilities. What is inherent in relation to the task is called apptitude and what is manifested in the task is called ability. All individuals are not endowed with same amount of apptitudes and abilities. They vary considerably from person to person.

Psychologists believe that whatever one is endowed with genetically or by practice or learning or both genetically and practice can be assessed and his/her future level of performance can be predicted. They have demonstrated this in various areas. In the preceding section we have described few types of abilities like, mental abilities, mechanical abilities, psyshosomatic abilities and visual abilities. These abilities are not absolutely independent of each other. They are interrelated to some extent. That is why we are performing all our activities in our everyday life successfully. There is certain amount of coordination of all these abilities in every human performance.

Psychologists wanted to assess every ability independently of each other and tried to devise methods and procedures, and designed tests and instruments. Thus many instruments and, paper-pencil tests have come into psychological studies. These tests were used for various purposes such as selection and placement in the employment situation, counselling and guidance in dealing with individual problems, diagnosing and identifying individual's abnormalities and treating them, identifying individual interests in educational situations and finding out occupational interests and so on.

Some tests are designed to assess certain specific skills required for certain industrial jobs. It is true that machines are invented to perform the industrial works easily and in certain areas machines are replacing men. But if anything goes wrong in a machine it is man who sets it right. Therefore, for certain industrial or mechanical or electrical jobs certain special skills are required. It is not possible to identify the special skills by a single psychological test. A battery of tests are required to identify the special skills in man. These tests are in one way or other interrelated. Series of attempts must be made to design and develop a battery of tests. For example, Thurstone's (1938) primary mental ability tests consisted of fifty six tests which identify seven abilities, which he called primary mental abilities. The general aptitude test battery, used by the United States Employment service measures ten abilities such as intelligence, verbal, numerical, spatial, clerical, form perception, aiming, motor speed, finger dexterity and manual dexterity. There are also tests measuring spatial relations, speed of perception, mechanical principles, mechanical comprehension, tracing, tapping and hand dexterity. A fector of manual dexterity has been located in tests which require routine quick movements. Manual dexterity involves quick and accurate movement of one's bodily parts.

In certain small industries parts of maichines or tools should be moved and placed accurately without comitting errors in the performance. It may also involve spatial relations and perception of space. For such jobs persons having the particular skill are selected and placed. This also involves time, speed, accuracy or placing. There are psychological tests designed to asess these special abilities. It is also possible to assess these aspects in the laboratory situations.

EXPERIMENT

Problem

The purpose of this experiment is to measure the speed with which a person is able to : (1) pick up the cylindrical wooden blocks and place them in holes and to (2) turn the blocks and place them in the same type of holes.

Method

Participant

Choose a person who has no knowledge about this test and describe his/her personal characteristics, such as age, sex, qualification, experience in some job, personal deformities if any, whether wearing eye glasses etc.

Materials

Minnesota Rate of Manipulation (Test) Board with 58 holes, 58 wooden blocks and stop clock, two thin plywood board to the size of the test board.

Description of the Materials

The Minnesota Rate of Manipulation board is a rectangular wooden Board of 75 cm in length, 25 cm in breadth and about 4 cm in thickness. There are 58 holes in 4 rows. In two rows there are 15 holes and in other two rows there are 14 holes. The holes are about 2.5 cm deep. The distance between any two holes or any side is 1.6 cm. There are 58 wooden blocks which can go into the holes of board and about 1/3 of the wooden blocks may be projecting outside of the holes. The top of the wooden block is red in colour and the bottom is green. Two kinds of test may be conducted with this board. One is placing test and the other is turning test. The diagram of the test material should be drawn with names of the parts.

Procedure

Placing Test

The experiment should be conducted in a well illuminated room. In placing test the rectangular wooden test board should be kept along the edge of the work table. A thin plywood board to the size of the test board should be kept at the bottom of the board. Then all the 58 holes should be filled with the wooden blocks, red side on the top and green at the bottom. The test board should be slided forward gently with the wooden blocks leaving the plywood board in the same place. Then lift the test board gently leaving the wooden blocks and place it on the plywood board in the original place from where it was slided. Now the test is ready for the experiment. This should be done by the experimenter himself/herself.

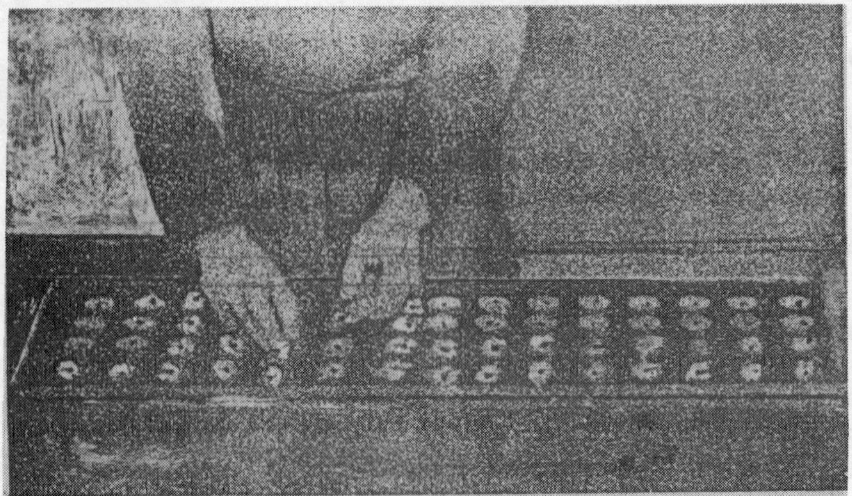

Fig. 12:1 : Manual Dexterity Test. Minnesota Rate of Manipulation Board. Experiment 32

The following instruction should be given to the participant. "Stand in front of the work table, facing the board. This is a speed test and you will have to go as fast as you can. Use only one hand. The blocks are in four long rows and in 16 columns. In each column there are four blocks except in the last column. You will have to lift the wooden blocks with your fingers and put them in the holes of the board. Pick up the block nearest to you, that is from the front row on the left side of first column and put it in the hole farthest to you, that is in the last row. Then pick up the next block from the same column and put it in the next nearest hole in the same column. You will have to move towards you while filling up the column. After finishing the first column go to the second on the left and do the same thing and fill up the second column. In the same manner you will have to fill up all the 15 columns one by one, going from left to right. You should not skip or cross over the column. You must go column by column. Do it just what I do here." (The experimenter should demonstrate the test for two or three columns). "If you are right handed begin the task on the left side, if left handed begin it on the right side.

Now you understand this. Carry out the work as fast as you can when I say ready go." As soon as the participant starts the task the experimenter should start the stop clock to run.

The participant may be allowed to fill up the board with the wooden blocks once as a preliminary trial. Then four trials should be given. After every trial a rest for 30 seconds should be given. The time taken by the participant to complete each trial should be noted down and recorded in the record note book in seconds. The score is the amount of time taken by the participant for each trial. The result should be given in a tabular column-I in the respective place.

Turning Test

After completing the placing test the turning test should be given to the participant. The experimenter should make the arrangements for the turning test. Everything should be done in the same manner as he did for the placing test. The test board should be placed at the edge of the work table towards him. One plywood board should be kept below the board and all the wooden blocks should be placed in the holes with all 58 red ends at the top as in the placing test. Now the test board along with the wooden blocks should be gently slided (moved) forward sufficiently till the test board fully leaves the plywood board kept below. Then gently lift the test board leaving all the wooden blocks on the work table and place the test board in the original place on the plywood board. Now the arrangement is over for the turning test. The experimenter should give the following direction to the participant.

> "This time you are going to perform the task little differently. You will have to pick up the wooden block with your right hand farthest from you on the left first column and turn the wooden block by twisting your right hand till you see the green colour at the bottom and transfer the wooden block to your left hand and place it in the hole farthest from you on the left hand side. Then pick up the next block from the same column turn it by twisting your right hand and pass it on to the left hand and place it in the next hole in the same column. When you complete the first column with 4 wooden blocks start the next column and do the same thing. Carry out the task in the same manner till you arrange all the 58 wooden blocks in the board holes. You should do this column by column and on any account should not skip or cross over the column. When you complete you can see all green ends of the wooden blocks. You can go as fast as you can. You will have to do four trials in the same manner."

The experimenter can demonstrate the procedure by picking up the wooden blocks and turning and placing them in two or three columns. He can ask the participant to do the task in one or two columns. The right handed participant should begin on the left end of the board and the left hand participant should begin at the right end of the board. The time the participant takes for each trial should be noted down in the record note book in seconds. Therefore, as soon as the participant starts the task the experimenter should start the stop clock to run. The results in seconds should be given in the respective columns of Table 12.5.

Results

The time taken by the participant for each trial in placing test and in turning should be given in a tabular column for all the trials. The individual score in the time taken by the participant for all the 4 trials in placing test and in turning test and this should be given in Table 12.5. The group data should be collected as a whole for placing test and turning test and should be given in Table 12.6. The data should be treated statistically and analysed.

Table 12.5 : Showing the time taken by the participant in placing and turning tests of Minnesota Manual Dexterity Test

Trials	Time in seconds	
	Placing Test	Turning Test
1. 2. 3. 4.		
Total		
Mean		

Table 12.6 : Showing the total and mean scores of the group in placing and turning tests in Minnesota Manual Dexterity Test

S. No.	Name	Mean time (Scores) in seconds			
		Placing test		Turning test	
		Total	Mean	Total	Mean
1.	G.R.				
2.	S.R.				
3.	V.C.				
4.	P.S.				
5.	V.S.K.				
6.	T.S.				
7.	G.S.				
8.	P.T.				
9.	V.S.				
10.	S.T.				
	Total				
	Mean				

The experimenter should analyze the scores which are in seconds. He/She should compare the first trial with the other.

The experimenter should analyze the scores which are in seconds. He/She should compare the first trial with the other three trials. He/She must observe whether there was any change in scores. It is expected that trial to trial the time should go on reducing. Similarly in the turning test the experimenter should compare the score of the first trial with other trials and see whether there was any reduction of time in the subsequent trials. Then by totalling the scores of all four trials the mean score should be worked out for both the placing test and for turning test. Give the mean scores of all the participants in Table 12.6 and work out the group mean. Compare the mean scores of all the participants.

Discussion

In selecting persons for various jobs it is important to know whether the person has any aptitude for the job or not. Whether it is big industry or small industry or simple mechanical work or office work unless one has real aptitude for the job he will be causing loss for the organization or damage or even get into accidents. Therefore, psychologists have devised various types of tests to assess one's aptitude and they are administered on them and his future performance on the job is predicted. The Minnesota Rate of Manipulation test is designed to assess how far one can accurately manipulate the instruments. In certain works, instruments or parts are handled manually. Some could do such tasks accurately and some may struggle to perform the task. This test, therefore, is also called Manual Dexterity test. It is to measure the speed with which a person is able to pick up and place the wooden blocks of the same size in the holes of the test board and how fast he/she can turn the blocks and place them in the holes of the board.

Ths scores obtain in this test may explain the speed of performance of a person in a job which involve similar task. Observe whether the person is making any improvement in his performance from trial to trial. Is he/she becoming nervous in performing the task. In turning and placing test, was he/she carrying out the task accurately or reveal any nervousness.

Compare your participant's mean score with that of other participants. The lower the score the higher the dexterity is and the higher the score the poorer the dexterity. Therefore, find out who scored the lowest and who scored the highest. Find out where your participant stands. Analyze the scores taking into account the participants' personal information. Discuss what has contributed to one's low scores and high scores. Discuss from the group mean, how many of them above group mean and how many of them below the group mean. Where your participant stands in the group mean. Discuss whether age, experience, sex or any deformities or other personal factors are having any influence in the scores of placing test and turning test. Compare your results with any of the earlier experimental results and state whether your results vary from them or in agreement with them.

CONCLUSIONS

1. State who has the highest dexterity?
2. State who has the lowest dexterity?
3. How many of these are below the group mean?
4. How many of them are above the group mean?
5. What are the personal factors having influence in the dexterity ?
6. Do the individuals differ very much in this test.
7. Is there any specific factor found to be operating in the performance of the individuals.

EXPERIMENT 33 : FINGER DEXTERITY TEST

INTRODUCTION

Human beings always live in competitive societies whether they are in primitive societies or civilized societies. History is full of instances of conflicts, aggression, violence and wars. To achieve desired goal or object man strives hard with all his skills and talents. But men vary in skills and talents. The early man might have recognized the individual differences in arm strength, physical strength, speed of running, accuracy of aiming the arrow in hunting and so on, while the modern man recognizes the individual differences, apart from physical strength, in the intelligent way of dealing with men and matters, solving problems and making achievements in his everyday activities. One of the main studies of Psychologists is the individual differences in various abilities and achievements. In order to make their studies more accurate and standard they have devised several test materials to assess human aptitudes, abilities skills and talents, personality characteristics, interests and attitudes. Thus there are practically hundreds of psychological tests in use.

These tests are classified into various categories for purpose of conveniences. There are individual tests which can be administered on individual persons, separately, in the laboratory situations and his performance in it is observed by the experimenter or the examiner whereas the group tests are given to a group of people at a time and data are collected. The other classification is the performance tests and paper-pencil tests. The performance tests are to be manipulated by the individual and these tests are generally instruments and apparatus. The paper-pencil tests require writing and the subject will have to write the answer in the answer sheet whatever he thinks correct.

These four types of tests (individual tests, group tests, performance tests snd paper-pencil tests) are designed to measure some human characteristics which we want to know in a particular situation for a specific purpose. These are mainly the aptitudes, achievements, personality traits, interests, beliefs and attitudes. The concept aptitudes is used here in a very wider sense which include the abilities of various types. There may be mental abilities (intelligence) mechanical abilities, psychomotor abilities and visual abilities.

Since this is the age of science and technology, it is important that we must know how individuals differ in skills and abilities. When there were no scientific developments and industries men were depending upon only physical or manual work and there was no necessity to test a man's skill of ability as the demand of men were not so much. But in the present industrial world there is much demand for men to be put into various kinds of work. Since the population is increasing very rapidly too many men compete for the same job. Therefore, who is to be selected and who is the fit person to perform the job efficiently is the problem facing the organizations and job seekers. In this context the work of the psychologists becomes not only important, but also crucial. They believe that the psychological tests may solve the problem.

There are organisations which need men who can do the job with more precision. The job may be manipulating small instruments accurately and efficiently. In addition to accuracy, precision, and efficiency the job should be performed intelligently. A stupid way of handling a machine tool may cause not only accident but also losses. Therefore, if the person's skills and abilities are carefully assessed at the time of selection and placement losses may be avoided and efficiency may be accelerated. The type of abilities required in the said job must be psychomotor abilities. In psychomotor ability there is coordinated function of bodily parts like hands, fingers, legs, and the whole body. Psychomotor ability may be defined as the ability that is relating to the movements of the body or bodily parts in a coordinated manner associating with mental ability.

Psychomotor abilities are applied in various ways in many situations. Extensive studies were made by Fleishman and others and extracted quite a number of psychomotor abilities from their studies. The term psychomotor covers the range of abilities that include what we generally call dexterity, manipulated ability, motor ability, eye-hand coordination and other muscular performances. Thus, there are various kinds of psychomotor abilities. Friedman and others say that there are three factors operating in a psychomotor performance. These are : (1) Psychomotor coordination – this is the ability to coordinate the movements of the large muscles of the body. (2) Psychomotor precision – this is the ability to carry out the activity accurately. (3) Psychomotor ability – this is the ability to carry out the activity quickly. It is also observed that psychomotor abilities are relatively independent of mechanical abilities. There is also appreciable association between psychomotor abilities and anthropometric measurements, such as length of fingers and other bodily measurements. There are a number of psychomotor abilities tested in the psychological laboratories and tests have been designed.

Fleishman and others have identified a number of psychomotor activities. They are : (1) Choice reaction time – the ability to select suitable stimulus out of many and react quickly. (2) Reaction time – the ability to react speedily to a single stimulus. (3) Speed of limb movement – the ability to make discrete movement of arm or leg speedily. (4) Wrist finger speed – the ability to make discrete movements of the fingers, hands and wrists speedily. (5) Finger dexterity – the ability to make skilful coordinated movement of the fingers in a manipulative ability. (6) Manual dexterity – the ability to make skilful coordination in the movements of hand and arm. (7) Rate control – the ability to make motor adjustments in anticipation to changes – the continuously moving object according to the time. (8) Control precision – the ability to control and make necessary muscular movements and adjust to the position of a machine.

The tests designed to measure the finger dexterity are : Hays Pegboard test, Purdue Pegboard test, Stromberg dexterity test, O'Connor fingers dexterity test.

EXPERIMENT

Problem

The purpose of this experiment is to measure the finger dexterity of a person or how accurately a person handle small instruments.

Method

Participant

Choose a person who has no knowledge about this test and describe his/her characteristic such as age, sex, qualification, occupation, married or single, if employed salary for a month or parents, income, technical experience if any, whether he/she wearing spectacles and so on.

Materials

O'Connor Finger Dexterity Test Board, 300 cylindrical steel pins and stop clock.

Description of the Materials

The O'Connor Finger Dexterity Test is a wooden board of 12´´ long and 6´´ breadth and 1´´ thickness. On the upper surface of the board on left half of the board there is metal plate fixed and on the right half is a shallow tray. In the tray there are 300 cylindrical steel pins of 1 inch long and 1/10 of an inch diameter in size. In the metal plate there are 100 holes. Each hole is having 3/16 of an inch diameter and about ½ of an inch depth. These 100 holes are in 10 rows and in 10 columns. Each hole is spaced half of an inch from the other hole. The task of the person is to pick up three pins at a time and insert in one hole. The score is the time the person takes to complete the 100 holes with 300 pins.

Procedure

The test should be conducted in a well illuminated room. The participant should be seated comfortably in a chair in front of a table of about 30´´ in height. The O'Connor Finger Dexterity Test Board should be placed on the table one foot away from edge of the table. The tray of the board should be on the right side of the participant if he is right handed and to the left side of the participant if he is left handed. It should be at an angle of 90 degree with the participant's working hand, but may be changed if so desired by the participant. The 300 cylindrical steel pins should be placed in the tray of the test board. The experimenter should keep his stop clock ready for work and also he/she should keep the table in his/her record note book ready to record the readings of the performance of the participant.

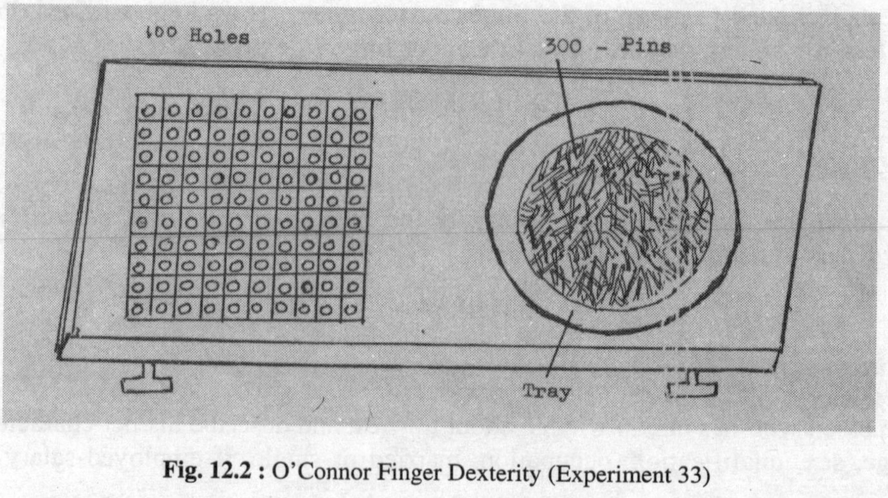

Fig. 12.2 : O'Connor Finger Dexterity (Experiment 33)

Then the experimenter should give the following instructions to the participant.

"Here is a board with holes that have room for three pins in each hole. Pick up three pins from the tray (showing the pins in the tray) at a time and fill up the holes by placing three pins in each hole as fast as you can. Use only one hand and only your fingers. Start from the left top corner and work towards the right. Fill up the first row, then start the second row (the experimenter can demonstrate filling up two rows) and then go down one by one to the bottom row. Instead, if you go filling up from up and down your sleeves or fingers may catch the pins. Be sure to fill up each row completely before you start the next row. Do not skip over the holes or rows. If you miss or drop down one or two pins, do not stop to pick up them. There are extra pins in the tray."

The experimenter should show by gesture that the holes should be filled from left to right of the top row if the participant is right handed. When he/she has filled up the top ten holes the experimenter should remove the pins and allow a moment of rest. Then the experimenter should tell the participant, "When I say, 'ready go', you must start the task in the same manner as fast as you can."

Now the experimenter can give the order, 'ready go'. At once he should also start the stop clock to run and make note of the time and then should make entry in his record about the time the participant has taken for the first five row and then for the second five rows. When the participant has completed all the ten rows the experimenter should remove all the pins and put them in the tray as they were before and make ready for the second trial. In the same manner four trials should be performed and the time should be noted down. The time in seconds is the score. This should be given in a tabular column. The participant may be asked to go away.

Results

In Table 12.7, the time taken by person to fill up the first five rows and the second five rows should be given and the total time taken for each trial should be worked out. In the same manner for all the four trials the time should be worked out. Then the mean time for first half and for the second half of the trials should be worked out.

Table 12.7 : Showing the time taken by the participants in the Finger Dexterity Test

Trial No.	Time taken for first half in seconds	Time taken for the second half in seconds	Total time taken to complete the Board in seconds
1.			
2.			
3.			
4.			
Total			
Mean			

The group data should be given in Table 12.8. This should be the mean time for the first half and for the second half and the total time each person has taken for all four trials and also the mean. The mean score for the whole group should be worked out.

Table 12.8 : Showing the time taken by the group in the Finger Dexterity Test

S. No.	Name	First half	Second half	Total	Mean
1.					
2.					
3.					
4.					
5.					
6.					
7.					
8.					
9.					
10.					
Total					
Mean					

Analyze the data given in the tables. What the low score indicates and what the high score indicates. Indicate the high and the low scores in the group data. How many of them scored more than the group mean and how many less than the group mean.

Discussion

Write the discussion taking into account the individual's score. State how the participant performed task. Was he/she nervous in the task or performing freely? State how much time he/she has taken for the first trial and also for the subsequent trials. How much time he/she has taken in the last trial? Was there any difference between the first trial and the last trial? How much time he/she has taken for the first five rows and for the second five rows in the first trial and also in the last trial? Was there any difference between the first half and the second half in the first trial and in the last trial? State the reason.

Discuss the group data taking into account the scores in the first half and in the second half of the individuals. Who scores the lowst in the first half and who scores the highest and also who scores the lowest in the second half and who scores the highest. State, how many of them scored more than the group mean and how many of them scored less than the group mean. State the reason from the personal information. State whether the person wearing eye glasses scored high or low.

CONCLUSIONS

1. State whether your participant was benefited by the trials (training) in the Finger Dexterity Test.
2. Did he/she show any difference between the first half of the first trial and the first half of the last trial.

3. What specific characteristics did you observe in your participant during the course of trials?
4. Where your participant stands in the group data?
5. Who scored the lowest and who scored the highest?
6. Was there anybody not benefited by the trials?

EXPERIMENT 34 : ASSESSING HAND-TOOL DEXTERITY

INTRODUCTION

Vocational opportunities are increasing enormously day by day, owing to the rapid development of various types of industries and technological and chemical organizations. This has created keen competition among the unemployed youths for various jobs. But organizations are interested in recruiting only competitive men for their purpose. Certain jobs require handling instruments and tools accurately and precisely to avoid damage or accidents and loss. Therefore, persons are to be tested whether they possess the required skills and talents. Certain jobs required certain specific skills for increasing the production.

In mechanical works, generally, the bolts and nuts are to be used in a sequential order with much proficiency. Assembling, dismantling and reassembling are to be made while attending repair and servicing works. In a machine or vehicle manufacturing industry assembling is a major work which needs handling of various parts, bolts, nuts and screws to be done rapidly and correctly. Unless one is proficient in all these he cannot be a successful worker. Even in handling ordinary mechanical tools one should possess the proficiency or the natural abilities. In the modern industrial world the day-to-day work whether it is at home, office, a firm or a small industry servicing is becoming inevitable as some falts may occur in various home or office appliances. For all these we need persons who are familiar with such works and who can be proficient in handling the tools and equipments or instruments with ease and grace.

Psychologists in vocational counselling situations find it necessary to assess whether the candidate is endowed with the aptitude essential for the specific job situation. They are of the view that the candidate should not feel nervous in handling the tools and should also move fastly and accurately. This ability is called tool dexterity. This type of skill is found now-a-days very important in many different types of factory jobs, home and office situations. A test devised and constructed to measure this aspect of skill is called Hand-Tool-Dexterity Test. This was designed by an American Psychologist George K. Bennett. The ability measured by this test is a combination of aptitude and of achievement based on one's past experience and knowledge in handling tools. This test is designed and administered in such a way that it minimizes the advantage that might have gained by a person by other ability. This measures the manipulative skills of a person independent of intellectual factors. This is not used to measure one's intelligence. However, it is observed that some brief practice will not seriously affect the test scores. This is also used as a Psychomotor Ability Test. (McCormick and Tiffin, 1979).

EXPERIMENT

Problem

The purpose of this experiment is to assess one's hand-tool dexterity.

Method

Participant

Choose a person who has no knowledge about this experiment and describe his personal characteristics such as age, sex, qualification, socio-economic status and has job experience if any.

Fig. 12.3 : Manual Dexterity : Bennett Hand-Tool Dexterity : Experiment 34

Materials

Bennett Hand-Tool Dexterity Board with bolts and nuts in three rows, wrenches, one adjustable wrench, stop clock.

Description of the Material

The Bennett Tool Dexterity Test was designed by George K. Bennet (1943) for the purpose of securing proficiency in handling mechenical tools and named after him. The test board is a thick one foot long and about 9″ width of wooden plank at the bottom with two uprights at the ends with three rows of holes of different sizes. The upright one about 8″ in height. Normally the bolts are in the holes on the left upright with heads on the bolts on the inside.

Procedure

The test should be conducted in a well illuminated room and should be free from noises. The test frame should be mounted firmly on a work table of 4′×2′ and of 34 inches height. There should be sufficient space on all sides of the test frame. The bolts should be on the respective holes on the left upright with the heads of the bolts on the inside. The nuts should be so tightened that they cannot be removed with fingers and can be removed only with the help of wrenches. Once the nuts loosened with wrenches, can be removed easily with the fingers. The jaws of the adjustable wrenches should be closed. The tools should be placed between the uprights. When the arrangement is ready the experimenter should give the following instructions :

"There are three rows of bolts with nuts on the left upright and you will have to place them on the corresponding rows on the other (right) upright. The heads of the bolts should be on the inside as you see here. First of all you should remove all the bolts and nuts from the top row and place them on the work table. Then remove the middle row and place them on the work table. But while removing each bolt of the bottom row straight away you can fit the bolts in the respective holes of the bottom row of the other upright. First tighten the nuts with fingers and then with the help of appropriate tools (wrenches).

"After mounting the smallest bolts and nuts in the bottom row of the right upright of the frame then mount and tighten the middle size bolts and nuts in the middle row in the same manner. Thirdly, the largest bolts and nuts should be mounted and tightened in the top row holes of the right upright. All the bolts should be so tighten with the help of appropriate wrenches that the bolts and nuts should not be removed with fingers.

"Lastly, when you start removing the bolts and nuts from the left upright, since they are very tight, use two wrenches, one for holding the bolts and one for loosening the nuts. You can first loosen all the nuts with wrenches and then you can use your fingers to remove them and to mount on the other upright. When they are sufficiently tightened with fingers then you can use the wrenches for tightening strongly.

"Now you can go ahead. Work as fast as possible."

As soon as the participant started the task by picking up the first wrench the experimenter should start his stop clock to record the amount of time. As soon as the participant complete the task the test frame should be turned to 180 degree for facilitating for another participant.

Scoring

The score in this test is the amount of time the participant has taken to complete the task once. The time is in between when the participant picks up the first wrench and completes the task by tightening all the three rows of bolts and nuts as directed. The experimenter can have four trials, (1) from left upright to right upright, (2) from right upright to left upright, (3) from left upright to right upright and (4) from right upright to left upright to see whether there is any improvement in his performance.

RESULTS

The results as recorded by the experimenter may be given in tabular columns. The amount of time taken by the paticipant for each trial may be given in minutes and seconds in Table 12.9. The mean time may be worked out. The group data may be collected and given in Table 12.10.

Table 12.9 : Showing the scores of the individual in Bennett Tool Dexterity Test

Trial No.	Time in minutes and seconds
1.	
2.	
3.	
4.	
Mean	

Table 12.10 : Showing the scores of the group in Bennett Tool Dexterity Test

S. No.	Name	Trials				Mean
		1	2	3	4	
1.	P.V.					
2.	K.P.					
3.	S.M.					
4.	K.R.					
5.	C.G.					
6.	M.R.					
7.	S.V.					
8.	A.C.					
9.	R.V.					
10.	M.K.					

Explain the results of your participant. How much time he has taken for the first trial and for the subsequent trials. In the group data where your participant stands. Who has taken too much time and who has taken too little time.

Discussion

Discuss the performance of your participant. Was he feeling nervous in handling the tools first or he was handling them easily or with any familiarity ? Explain his behaviour. Has he committed any mistake in the task? Has he made any improvement in his performance. Compare your participant with others in the group. Discuss the performance of the group in terms of first and last trials and the mean scores. State whether any of the participants made rapid improvement in the task. Who completed the task in the shortest time and who has taken the longest time. Compare these individuals. Also see whether the personal factors of the individuals such as age, qualification, sex, socio-economic status previous experience had any influence in the task. How many of them above the group mean and below the group mean.

Conclusions

1. Is your participant proficient in this task.
2. Who completed the task in the shortest time. State his personal factors.

EXPERIMENT 35 : ASSESSING HAND-STEADINESS OF A PERSON

INTRODUCTION

The development of industry and technology have increased the job opportunities of people who are specialized in certain areas of skills. Apart from the educational qualification required for a job it is also important that the person must possess certain specific training or skill for the successful performance of the job. Certain skills are learned by practice and certain skills are innate in the individuals. Whatever it is, it is to be known who possesses the particular skill for the job. It is known by assessing the skill in a person by certain tests.

There are quite a number of psychological tests designed to assess the aptitudes or natural abilities of men for selection and placement in organizational and industrial situations. The mental ability tests, mechanical, psychomotor, visual and other special ability tests are intended only to fulfil the needs of the demands arising often in industries. Apart from the general aptitudes required to certain occupations, certain jobs require certain special abilities for efficient performances. For example, in certain jobs speed, accuracy and dexterity are required. The manual dexterity needs the eye-hand co-ordination, and certain jobs require the finger dexterity with eye-hand co-ordination. In handling certain small machines and tools the hand and finger steadiness are inevitable. It is very difficult to identify such persons who possess the hand steadiness among the group of people without any systematic assessment.

Unless a well designed test is available it will not be possible to assess such skills in human beings. There are several tests available to test one's skills in this line like manual dexterity, finger dexterity tool dexterity, eye and hand coordination, arms and eye cooridnations and so on. But the skill required in watch making, radio television, electrical and telephone wire connections is little different from the former ones. This may be called steadiness test. Steadiness refers to the skill to hold a small instrument or tool at a particular point for a long time without any vacillation. Whipple has designed a test for assessing one's skill of steadiness and it is named after him. It is Whipple's steadiness tester. This is a very familiar test often used in psychological laboratories (Woodworth, 1955).

EXPERIMENT

Problem

The purspose of this experiment is to assess the nature of skill of steadiness of a person with a suitable test material.

Method

Participant

Choose a person who has no knowledge about this experiment and the test material and describe his characteristics such as age, sex, qualification, trade skill if any, socio-economic status etc.

Material

Whipple's steadiness tester, counter, stop clock, electric bulb, 6 or 9 volt battery, battery box, connecting wires, and a metal stylus.

Description of the Instrument

The Whipple's steadiness tester is a small wooden box with a brass plate mounted in the front of the box with 9 circular holes of different sizes. The box is made to stand on four washers. The box is a three sided one.

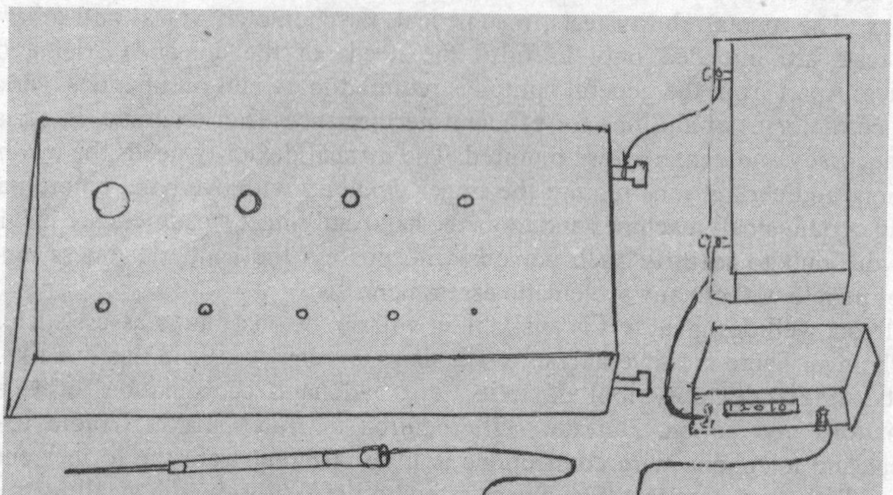

Fig. 12.4 : Assessing Hand Steadiness Whipple's Steadiness Tester (Experiment 35)

There is a bottom and above it there are two sides in 45° slanting position. The metal plate is fixed in the front. The size of the metal plate is 9″×5″. There are 9 holes of varying size. The biggest hole is ½″ in diameter. All others are going on decreasing in size step by step and the last one is the smallest with .1″ diameter. The first four holes are in the top row and the other 5 holes are in the bottom row. At the bottom of the metal plate, at one corner there is a binding screw with which a battery box can be connected. The stylus is a long needle can be inserted even in the smallest hole of the box without any contact with the side.

Procedure

First of all the experimenter should set up the instrument for the experiment. The Whipple's steadiness tester should be connected serially beginning with the battery box, to the bulb, to the counter and ending with the stylus. The circuit is complete when all the parts are connected serially with box by wires. The experimenter can test whether it functions with all responses. That is, when stylus touches the metal plate the bulb should glow and numbers in the counter should run. When the experimenter is fully satisfied, he/she can make the participant to sit in front of the instruments and also should see that there are no interferences and distractions at the time of conducting the experiment. Then the following instructions may be given :

> "This is an experiment to test the steadiness of your hand. Hold this stylus in your favoured hand with the first three fingers firmly. Your fore hand must be in an angle of 100°. You will have to insert the stylus slowly into the first biggest hole without touching the side and hold it there for 15 seconds without any support to your hand at any level. When I say, 'out' you will have to remove the stylus from the hole. You will have to begin with the biggest hole and move to the smallest in the order of sequence. Any touch of the stylus with the side of the hole is counted as an error. Therefore, carry out the task carefully. When I say 'ready, go' you must start."

The experimenter himself can give one or two demonstrations as to how it is to be done by the participant. When the participant has understood the procedure and got ready the experimenter can start the experiment. As soon as he starts the experimenter should also start the stop clock and be ready to record the time of the light burning and also the numbers run in the counter. When the participant complete all 9 holes once, it is one trial. There should be four trials. There should be rest pauses between trials for about 30 seconds. In the scoring, the contact time in seconds and the errors indicated by the counter important point are the important points that the experimenter should note down carefully. The contact time is obtained by recording the amount of time the bulb has glowed.

Results

A common score should be obtained by using the variables of contact time and number of errors and the duration of the steadiness time. The formula used for this is :

$$\text{Weighted score} = \frac{\text{Number of errors} \times \text{Contact time}}{15 - \text{Contact time} \times \text{Number of errors}}$$

Here 1. The number of errors refer to the numbers run in the counter.

2. The contact time is the amount of time the stylus had contact with the side of the metal plate, which is recorded from the bulb and stop clock.
3. 15 refers to the duration of steadiness time. The amount of time the participant was holding the stylus inside the hole of the metal plate.

The experimenter should present the errors and contact time scores for all 9 holes separately in Table 12.11. He should give this for all the four trials and find out the mean error and contact time. Then he should work out weighted scores with the mean errors and mean contact time for all 9 holes and present them in Table 12.12. The lower score indicates less errors and less contact time. This indicates the participant's steadiness of the hand. The hypothesis is that this score should be low for higher steadiness. This is tested in this experiment.

The group data should be collected and given in Table 12.13. Work out the mean score for the individual and also for the group.

Table 12.11 : Showing the errors and contact time for the individual in the Whipple's Steadiness Test

Trial No.	Errors and contact time in seconds for each hole																	
	1		2		3		4		5		6		7		8		9	
	E	CT	E	CT	E	CT	E	CT	E	CT	E	CT	E	CT	E	CT	E	CT
1.																		
2.																		
3.																		
4.																		
Mean																		

Table 12.12 : Showing the weighted scores for the individual in the Whipple's Steadiness Test for all 9 holes

S. No.	Dimension Size	Mean Error	Mean Contact Time	Weighted Score	Formula
1.	1				
2.	2				
3.	3				
4.	4				
5.	5				$\frac{\text{Means errors} \times \text{Mean contact time}}{15 - \text{Mean contact time} \times \text{Mean errors}}$
6.	6				
7.	7				
8.	8				
9.	9				

Table 12.13 : Showing the Weighted scores for the group for all 9 dimensions

S. No.	Name	1	2	3	4	5	6	7	8	9	Mean
1.	P.V.										
2.	S.R.										
3.	R.M.										
4.	K.R.										
5.	M.S.										
6.	V.P.										
7.	B.V.										
8.	M.R.										
9.	T.V.										
10.	A.M.										
Mean											

Discussion

Discuss the performance of the individual for each dimension. State whether he commits more mistakes in the beginning or in the end. Was there any progress in his performance trial after trial? What is the difference between the first trial and the last trial. State also the nature of contact time for each dimension. Was the weighted score for the last dimension the highest compared to other dimensions. What is the nature of steadiness of your participant? Is it in agreement with the hypothesis.

Discuss the steadiness score (weighted score) for the group. Who scored the highest and who scored the lowest? Is the mean for the group is increasing order for the dimensions (largest to smallest).

Compare the individual's mean weighted scores with the mean score of the group. Whose score is the lowest? Who scored the highest? How many of them above group mean and how many are below group mean. State who are those maintained higher steadiness in the test. Discuss the scores relating to individual's personal factors. Find out whether age, experience, sex, socio-economic backgrounds had any influence in the individual's hand stediness.

Conclusions

1. Has your participant maintained steadiness (scoring less than group mean) or not?
2. How many had maintained steadiness and how many had not.
3. Had any personal factor influence in the steadiness.

EXPERIMENT 36 : STUDY OF ACCURACY OF PSYCHOMOTOR SKILLS IN A GIVEN TASK

INTRODUCTION

Psychology as an empirical science supports its stand through experimental evidences. Psychological tests are the instruments, designed to study various aspects of human behaviour. All the measurements in the field of Psychology and Education are done through standardized tests only. The measurements of psychophysical and other sensory responses, intelligence, personality traits, motivation and emotion, perceptual process, interests and attitudes, aspirations, social motives and achievements are all made through tests only. These tests mainly fall under two types as performance tests and paper-pencil tests. But there is no hard and fast rules that a particular human character should be measured only by a performance test or by a paper-pencil test. Any one of these two types may be designed to measure any aspect of human behaviour. In this way psychologists have designed innumerable number of tests for their purposes. One type of test generally used in psychological laboratory is the performance test. This needs certain explanation to avoid some confusion caused in using tests like the projective type.

The concept 'performance' in the psychological laboratory is meant, the act of performing, the execution or accomplishment of work. It refers to a particular action, deed. It is the manner of performing or functioning. The concept 'performance test' refers to a test designed to elicit manual or behavioural responses rather than verbal ones. Therefore, performance tests are referring to something designed which can elicit a particular kind of behavioural response. The response is seen in the result of the act or deed whatever it is. It may be a successful response or unsuccessful response. But the instrument or apparatus used in the experiment elicits a particular act of response which may reveal the person's motives, ambitions or aspirations or skills. Cronbach (1966) says, a performance test is an observation in a standard situation designed to elicit a particular type of response. Performance tests are also designed to provoke exhibitions of truly critical behaviour. The features of such tests are these : (l) The stimulus situation is made as uniform as possible for all the subjects. (2) The situation is designed to create difference in the behaviour in the subjects which the tester wants to observe. (3) The observer makes records of the subject's method of performance.

The performance test is having several advantages over other types of tests. One important advantage is that it comes closer than techniques to compare subjects in the same type of situations. But performance tests vary greatly in design, purpose and administration. Sometimes they are used as psychometric instruments to measure a narrowly defined constructs. Sometimes they are used to evaluate the person's total life style. They are also used as work samples for predicting success in a specific task.

The performance tests are also designed to test the psychomotor abilities of the person. These tests are used to test the simple speed, accuracy, or dexterity measures. There are some psychomotor performance tests for measuring more complex abilities. Some tests are confined to industrial situation and some in the military situations. All psychomotor tasks are not purely manual. They are based on intellectual abilities. Measurement of simple motor performances goes back to the beginning of experimental psychology to the times of Wilhelm Wundt, Fechner, Cattell and Galton who were the pioneers in the field. Some of the psychomotor abilities tested in the psychological laboratory are the Reaction time, Steadiness, Speed and dexterity, Two hand coordination, Arm-hand steadiness, Postural discrimination, Fine psychomotor coordination, Multiple limb coordination, Rate control, Speed of response to a signal, Response orientation, Response integration, Manual dexterity, Muscular strength. The so called aiming test involves precise eye-hand coordination. Aiming may be measured by using any suitable test which involves eye-hand coordination and steadiness. Lloyd Humphreys (1939) and others have made several studies in which they found that there was very high correlations between aiming or steadiness tests and rifle markmanship. A rifle man must hold his sight steadily on a target while resting in an elbow in a prone position. A pistol markman must hold his sight steadily on a target while standing. A dart aimer should stand steadily without leaning his bodily parts.

The psychomotor abilities are very pertinent in engineering field in connection with physical output responses made with the body members especially arms, hands, fingers and legs. The human abilities to perform various types of motor actions can have relationship to three aspects of human factors in engineering. The first deals with selection, design and location of control devices of systems. These devices are contro-lling mechanisms which perform work. The second deals with the selection and design of hand tools which are movable and are used to perform work. The third deals with work methods involving specific procedures and sequences of carrying out work activities and work materials and inner space.

There are other kinds of skills which are relevant and important in the industrial and business environment. They are important to a number of jobs and were subjected to considerable amount of study. Among those the abilities which are of concern in a large number of occupations are those concerned with the actual physical manipulation of subjects in an environment, the psychomotor skills. The motor skills are quite relevant to a number of occupations. Unlike intellectual abilities, in manual or psychomotor abilities, a factor observed is that some people generally perform well whereas others generally perform poorly. Fleishman (1962) has suggested eleven skills. These are, control precision, multilink coordination, response orientation, reaction time, speed of arm movement, rate control, manual dexterity, finger dexterity, arm-hand steadiness, wrist-finger speed and aiming.

Human performance can vary across different types of psychomotor activities. We are concerned with average differences in human performance. There are several criteria of psychomotor activities. They are speed of activity, accuracy of activity, force applied during activity (strength), energy cost of activity, biomechanical analysis and other methods.

There are difference of opinions with regard to the functions of psychomotor abilities. McCormick (1974) is of the view that psychomotor abilities are relatively separate type of abilities. If they are really independent how accurately they would be functioning. Vernon (1950), after making several studies with psychomotor tests has found psychomotor tests have some relationship with each other. But the inter-correlations scores between them are very low. In his study the inter-correlations were ranged from 0.13 to 0.30 which are not significant. Vernon says that though some psychomotor test scores are correlated with each other, correlations are very low. Therefore, the various psychomotor abilities may be considered as specific or independent predominantly. When such tests are used in any situation, they should be revalidated.

Another study has been conducted by Wolins and Mackinney (1961) in a handicraft company among the packers. The Purdue Pegboard and Minnesota Rate of Manipulation tests were administered on twenty seven packers. The performed job was packing handicraft kits in boxes. The Packer's performance was assessed and rated by their supervisors. He classified them under five categories. The correlations of various parts of the two tests with the ratings of the supervisor were made. The correlations for the Minnesota rate of manipulation test were negative. In the case of Purdue Pegboard test the correlations were very low except one test (assembly).

Drews (1961) has developed a validation method of psychomotor ability tests specifically for the Purdue Pegboard test components. After analyzing the results he has developed one test for each job that was in line with the motions of the job. They were found to be more valid for predicting job success.

EXPERIMENT

Problem

The purpose of the experiment is to study the level of accuracy of psychomotor skills of the participant in a given task.

Method

Participant

Choose a person who has no knowledge about this experiment and collect certain personal informations such as, sex, age, qualifications, whether wears spectacles or having any deformities, socio-economic status, height, and handedness.

Materials

Prof. Rajamanickam's Target-Dart test-grouping method. The Target Board, Test Sheets, a set of five Darts, Meter Scale, chalk piece, and a design to measure the grouping.

Description of the Material

It is a test designed and fabricated by Prof. M. Rajamanickam to measure the accuracy of psychomotor performance of a person in a given task. The Target is a board, made out of rubber sheet/card board of 2′ × 2′ size and fixed on a 3′ × 3′ wooden block board which is to be fixed on a wall 3′ above the ground. The surface of the board is painted with white colour on which four circles of different sizes are drawn. The innermost circle is of 3″ diameter. It is called 'Bull'. It is divided into two parts, the upper half is black in colour and the lower half white. The second circle is of 6″ diameter. It is called "Inner". The third circle is of 9″ diameter. It is called "Magpie". The fourth circle is of 12″ diameter. It is called "Outer". It is the same as the target used in firing ranges to give practice to the soldiers and to test their psychomotor skills in the Army (Appendix-I). The same may be printed to the same size in number of copies and used as 'Test Sheets' (Appendix-II).

The Dart is of 3″ in length, having a brass metal stem of $1\frac{1}{2}$″ length with a sharp needle of $1\frac{1}{2}$″ at the head. At the back of the Dart a feather head made up of plastic is attached. It is detachable. The feather tail is to facilitate the Dart to move fastly in the direction it is aimed. The Dart weighs about 100 grams. There are five Darts in a set. Each Dart makes one aim and five aims constitute one trial.

Method of using material

The method used to measure the accuracy of the psychomotor abilities of a person is *grouping method.* The person is to make the Darts into a group on the target by aiming. The Darts should fall quite close to each other to form a group to the size either smaller or bigger. The idea is that the grouping should be as smaller as possible. The smaller the grouping the greater the accuracy of the psychomotor abilities and the bigger the grouping the poorer the accuracy of the psychomotor abilities. The *grouping is assessed in terms of the size of* the circle. The smaller grouping is supposed to have a smaller circle. Thus the smallest grouping is to have 2″ circle. When all the five Darts are aimed at the Target all the five Darts should form a circle of two inches (2′′). The circle should have 2′′ diameter. *The 2′′ diameter circle is the smallest* grouping and most accurate performance of psychomotor abilities. The next bigger grouping is 3′′ circle, the third 4′′ circle and the fourth 5′′ circle which is the biggest one, having 5′′ diameter in this experiment, designed to measure the accuracy of psychomotor abilities. The observer is always expceted to aim for the smallest grouping to exhibit the accuracy of his psychomotor abilities. Therefore, there may be 2′′, 3′′, 4′′ and 5′′ groupings in this experiment and the scores are assigned according to the size as shown in Table 12.14. The grouping Darts may fall any where in the Target, but within the outer circle of the Target the grouping may be formed.

Table 12.14 : Showing the size of the grouping and corresponding scores

Grouping	Score
2′′ circle (Diameter)	20
3′′ circle (Diameter)	15
4′′ circle (Diameter)	10
5′′ circle (Diameter)	5
Outside the Target	0

Procedure

(a) Preparation for the Conduction of the Experiment

The experiment should first of all make some preliminary arrangements for the conduction of this experiment. Since the Target is already fixed on a wall the experimenter should make the area congenial for his purpose. The area should be properly illuminated. The room must have sufficient length (at least 18 feet length). The experiment should be conducted minimum at five positions of varied distances. The shortest distance should be 8 feet minimum and the positions may be varied with a difference of one foot between positions. Thus the five positions may be 8 feet, 9 feet, 10 feet, 11 feet and 12 feet. The experimenter with the help of a metre scale can measure first 8 feet from the place where the Target is fixed towards the open space and make a mark and draw a long thick line with a chalk piece across on the ground. He should write as 8 feet at the two ends of the line. Similarly at 9th feet, 10th feet, 11th feet and at 12th feet by measuring the distance, thick lines on the ground should be drawn and marked with number at every position. Now the experimenter should ascertain and satisfy himself that there are five positions with 8 feet, 9 feet, 10 feet, 11 feet and 12 feet distances (see the layout of the experimental situation – Appendix-III).

The next step is to put a small table or teapoy on the 8 feet position, putting the two front legs along the line. That is, the lengthy end of the table is 8 feet away from the Target. At the other end the participant should be made to stand close to it. After the task is over at 8 feet position, the table or teapoy should be moved to the 9 feet position. Then it should go to the next positions one by one till the task is over. This is called increasing order of the experimental procedure. In this procedure the participant is allowed to have some practice and he should be asked to make five aims 7 the darts at the target by which he could understand the nature of the task and the amount of energy should be exercised in this task.

(b) Conduction

The subject should be asked to stand in front of the table or teapoy steadily. He should place his left foot on the lower frame of the table or teapoy and place his left hand on the table or teapoy and stand steadily. Then the experimenter should explain to him, what is to be done. The Target should be shown to the subject and the 4 circles ranging from the smallest to biggest should be shown. Then the following instructions should be given :

> "Look at the Target. The Target area is within the area of the biggest circle line. Beyond that line the Target ceases to exist. You will be given five Darts. (The Dart should be shown to the subject). You will have to aim the Darts always at the centre of the Target where there is the black point. But it does not matter if the Darts fall somewhere in the target. All the five Darts should be aimed in the same manner. It is important that all the five Darts should hit at the Target almost at the same place within the target area so as to make all the five Darts fall in 2′′ diameter circle for forming 2′′ grouping. If all the five Darts fall within a circle of 2′′ diameter anywhere within the target you get 20 marks. If all the five Darts fall within a circle of 3′′ diameter it forms 3′′ grouping and you get 15 marks. If all the five Darts fall within a circle of 4′′ diameter it forms 4′′ grouping and you get 10 marks. If all the five Darts fall within a circle of 5′′ diameter it forms 5′′ grouping and you get 5 marks. But if all the five Darts fall in a circle bigger than a 5′′ diameter you get 0 mark as it does not form the required grouping. Therefore, you must carefully aim all the five Darts with the same amount of accuracy so as to make them into a smallest grouping. The smaller the grouping of five Darts the higher marks you get. This you will have to perform at five different places of different distances. Please see to it that you move only your arm and wrist and not the body. Aim the Darts at the target." (At this stage the experimenter hands over all the five Darts into the hand of the participant). Now you can have 5 practice aims before starting the trials. "Now start the trial aims one by one." "Ready, go".

After the trial aims are over the participant should get ready for the test aims. Then the experimenter should fix *one test sheet on the Target pattern* with pins on the four corners. The test sheet is of the same size as the Target put up on the wall. When the testing is over at all five positions the experimenter should remove this sheet and he can attach it in his record note book as an experimental data.

The experimenter should watch whether the subject follows the instructions correctly. If he/she commits any mistake the exprimenter should correct him at once. *When all the five Darts are aimed at the Target* from the first position the experimenter should take the participant near the Target for *measuring the size of the grouping.*

It will not be possible to measure the size of the grouping when the Darts are on the Target. Therefore the experimenter should *remove each Dart gently by rolling the Dart.* Otherwise the *sharp ends would be broken.* As soon as one Dart is removed from the Target *a small circle around the hole should be drawn* with a pencil. This should be done as soon as every Dart is removed from the Target. When all the five Darts are removed, and a small circle is made around the each hole the Dart needle has made, the *five small circles should be joined by a pencil mark.* Then the experimenter should measure the size of the grouping these five Darts have made with the help of *a design for measuring the grouping* (Appendix-IV), or with a one foot scale and tell the participant the size of the grouping he has made and the marks he obtained in that trial. He must find out whether it is 2″ grouping or 3″ grouping or 4″ grouping or 5″ grouping by assessing the circle by its diameter. The experimenter at once should enter this result in the columns of Table 12.16 he has already prepared for this. The experimenter should make it a point that the Darts should be with him and only at the time of the task they should be given in the hands of the participant.

Then the experimenter and the participant should go back to the position and move the table or teapoy to the next position (9 feet) which is one foot longer than the previous one and continue the same procedure. He should ask the participant to be ready for the task and should hand over the Darts. As soon as it is over he should move to the 3rd position (10 feet), 4th position (11 feet) and 5th position (12 feet) to do the same thing. *Five Trials should be performed at each position.* At every position the experimenter should do the same thing with regard to removing the Dart, making small circle by pencil around the Dart hole and joining all the five small circles to make the grouping circle, and measuring the size of the grouping circle. On the whole, in all the five positions the individual makes 125 (5×5×5) aims in the Target-Dart Test.

At the end of the task at all five positions the experimenter should get the introspective report from the participant. He should also ask him whether he was benefited by knowing the result of his performance in the next trial and what sort of experience he has gained from this and what he has exhibited in this task.

Method of Scoring

As stated in the preceding section the scoring of this test is based upon the performance of the participant. One Trial means five aims with five Darts. The participant's task is to throw all the five Darts at the Target almost close to each one so as to form a grouping. Five trials should be performed at each position. The grouping may be small or big and they may vary in sizes. The grouping may be in the form of a circle when all the five Darts holes are connected. It may be 2″ circle, having 2″ diameter. This is the smallest circle for which 20 marks are assigned. The next bigger grouping is 3″ circle, having 3″ diameter for which 15 marks are assigned. The next bigger grouping is 4″ circle, having 4″ diameter for which 10 marks are assigned and the last is the biggest grouping 5″ circle, having 5″ diameter, for which 5 marks are assigned. But if the Darts fall away from the Target or if the circle is bigger than 5″ diameter 0 mark is assigned. If four Darts fall close to each other within the Target and one Dart falls outside the Target, that is, outside of the outer circle it does not form a circle. Hence no marks.

Table 12.15 : Showing the marks and grades

One Trial = 5 aims. Five aims with 5 Darts should make one grouping	
Maximum marks for a grouping	20
Maximum marks for 5 (Trials) grouping	100
Minimum marks for a pass	40

Score Board	Marks
2″ grouping (2″ diameter circle)	20
3″ grouping (3″ diameter circle)	15
4″ grouping (4″ diameter circle)	10
5″ grouping (5″ diameter circle)	5
No grouping (More than 5″ diameter circle)	0

Scores and grades

Below 40	:	Poor (Fail)
40 to 54	:	Fair (Pass)
55 to 69	:	Good
70 to 84	:	Very Good
85 to 100	:	Outstanding

RESULTS

The results of the experiment presented in Table 12.16 should be treated statistically. The total marks for each position should be worked out and then the mean for all five trials. Determine the mean score for all five positions. The minimum marks required for a pass in one position with five trials is 40. If less than 40, it is a fail. Grades should be assigned to the individual in every position. The total score for all five trials is 100. Hence the grades may be assigned to the individual on the basis of the marks in all five positions in Table 12.16.

State whether the individual has obtained a pass mark in every position. State the groupings your subject has made. What is the size of the smallest grouping? In how many positions he made measurable groups, that is circles with 2″ diameter, circles with 3″ diameter, circles with 4″ diameter and circles with 5″ diameter. State whether he/she was making any improvement trial to trial. Mention this in the remarks column. Collect the group data from other experimenters and present them in Table 12.17 and work out the mean.

Discussion

This is a test of individual's psychomotor skills by using Target-Dart Test with grouping method. This involves the individual manipulating his bodily parts like, fore-arms, hind-arm, wrist, fingers and eyes and also his/her accuracy of aim. The accuracy of his/her aim and the manipulation of bodily parts are the tests of his/her psychomotor skills. If one is more skilful in this he/she can achieve better score by getting the smallest groupings in aiming the Darts. Therefore, state the nature of your subject's psychomotor skills. First of all whether he/she has passed in all five positions by scoring more than 40 marks. Then state on the basis of the norm what was his performance, whether his grade is fair, or good or very good or outstanding. From the grade you can judge his/her psychomotor performance.

Table 12.16 : Showing the individual's scores on the test of psychomotor abilities using the grouping method of the Target – Dart Test

One Trial = 5 aims.

Maximum score for each trial	20
Maximum score for five trials	100
Minimum marks for a pass	40

Score Board	Marks
2″ grouping	20
3″ grouping	15
4″ grouping	10
5″ grouping	5
No grouping	0

Scores and grades

Below 40	:	Poor
40 to 54	:	Fair
55 to 69	:	Good
70 to 84	:	Very Good
85 to 100	:	Outstanding

Trials	8 ft.		9 ft.		10 ft.		11 ft.		12 ft.		Remarks
	Size of Group-ing	Marks	Size of Group-ing	Marks	Size of Group-ing	Marks	Size of Group-ing	Marks	Size of Group-ing	Marks	
1. (5 aims)											
2. (5 aims)											
3. (5 aims)											
4. (5 aims)											
5. (5 aims)											
Total Marks											
Mean Score											
Pass or Fail											

Table 12.17 : Showing the group data on the test of psychomotor skills using grouping method of the Target-Dart Test

S. No.	Name	Total Grouping Score in 5 trials (Each trial 5 aims)					Total Score	Mean Score	Grade Pass or Fail
		8 ft	9 ft	10 ft	11 ft	12 ft			
1.									
2.									
3.									
4.									
5.									
6.									
7.									
8.									
9.									
10.									
11.									
12.									
Total									
Mean									

Was your participant able to achieve groupings in the trials of the first position at 8 feet ditance? In all the five trials at 8 feet distance, did he expose his/her psychomotor skills in thrwing the Darts? Was the skill consistently displayed by him in the subsequent trials and in the other positions? Compare his/her score in the first trial and in the last trial and analyze the performance. At what position the person was able to achieve the smallest grouping to get the highest score in a trial? Compare the mean scores of the first position with other positions and state whether his experience of the previous trials and the scores of the previous trials have helped him to improve his performance. If there was no improvement state the reason for the same from the personal informations collected.

State by assessing his/her performance in this experiment, the psychomotor job in which he/she may be successful. Explain it as a counselling procedure. From the group data state and discuss who are those made the highest score? Are there any outstanding performances? State whose individuals and the nature of performance in all five positions. If there are cases of failure, discuss them relating the mean scores of all five positions. That is, whether the person was consistently scoring the fail marks/failed to make groupings in his/her performance or his/her capacities are limited in this type of task. State and discuss the performances of those individuals who made better grades based on the scores. State and discuss the performances of those whose aiming abilities were very accurate and also very poor.

Discuss the individual differences in the scores and state the reason and the characteristics of the person. Were there differences on the basis of gender or physique, or height or experience in similar task? The way in which the person handles the instrument and the way in which he performs the task may reveal the nature of his/her psychomotor skills. The experimenter should observe this and write in the discussion.

Conclusions

State the nature of psychomotor abilities of your subject. Has he passed the test and what grade he/she has achieved in every position and in the whole test. Where he/she stands in the group.

How many in the group have passed the test and how many have failed in the test. Who has made the outstanding achievement? How many have made poor performance? State your general conclusion regarding the performance of the group.

APPENDIX – I

TARGET – DART TEST (GROUPING METHOD)

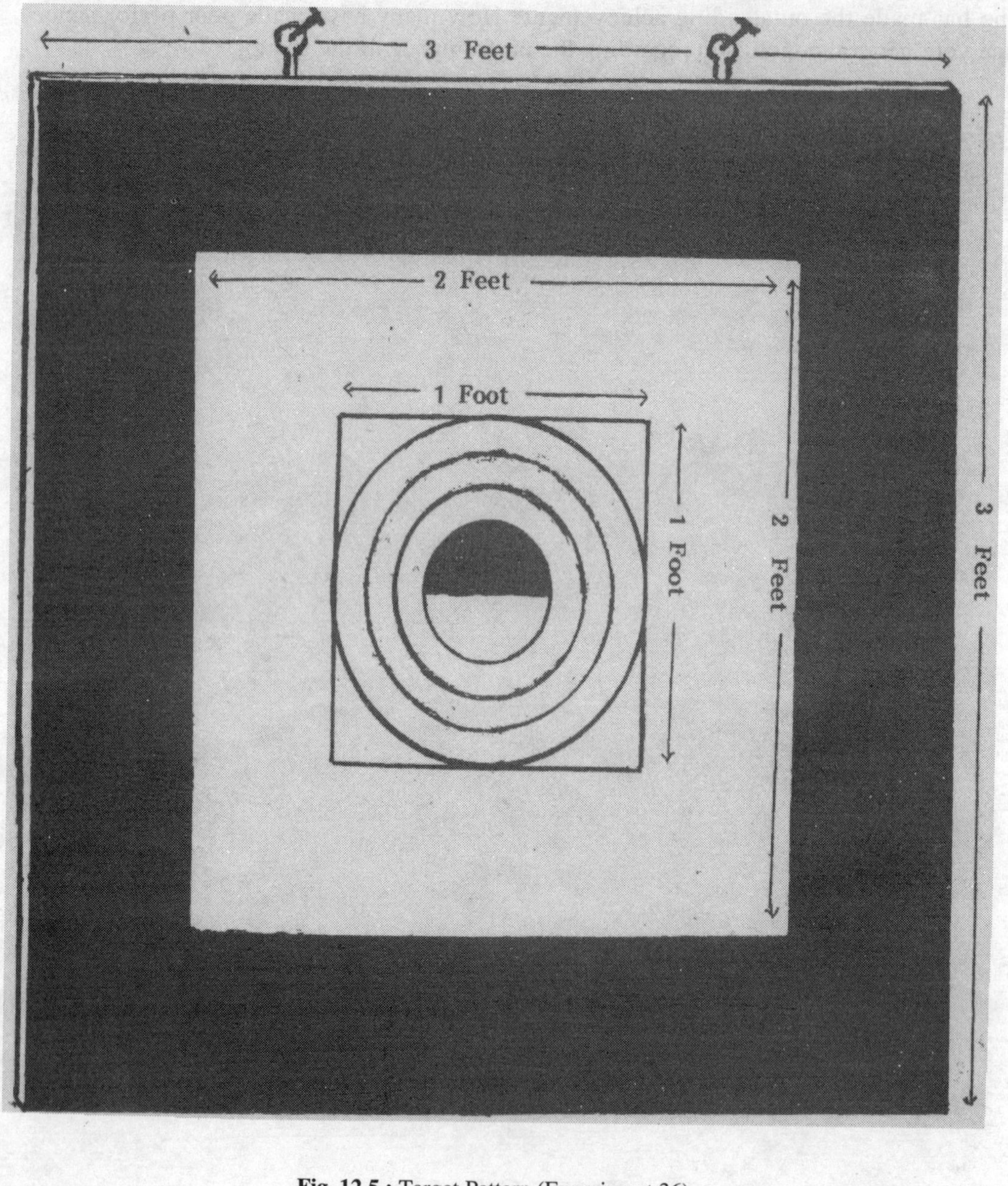

Fig. 12.5 : Target Pattern (Experiment 36)

APPENDIX – II

LAYOUT OF THE TARGET – DART (GROUPING METHOD) EXPERIMENT

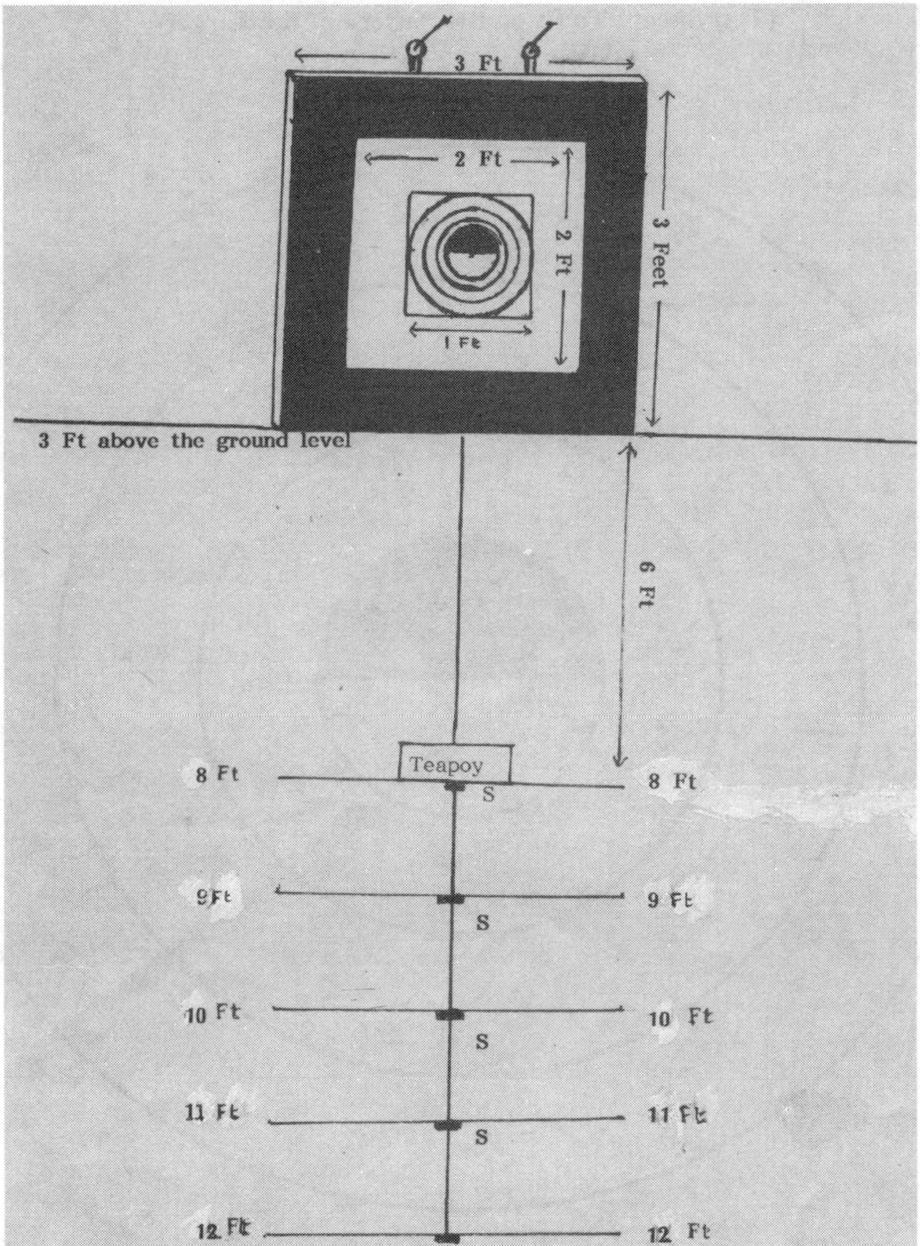

Fig. 12.6 : Layout of the Experimental situation
S = Positions of the subject.

APPENDIX – III

PROF. RAJAMANICKAM'S TARGET – DART TEST
(Test Sheet : To fix on the Pattern of the Target)

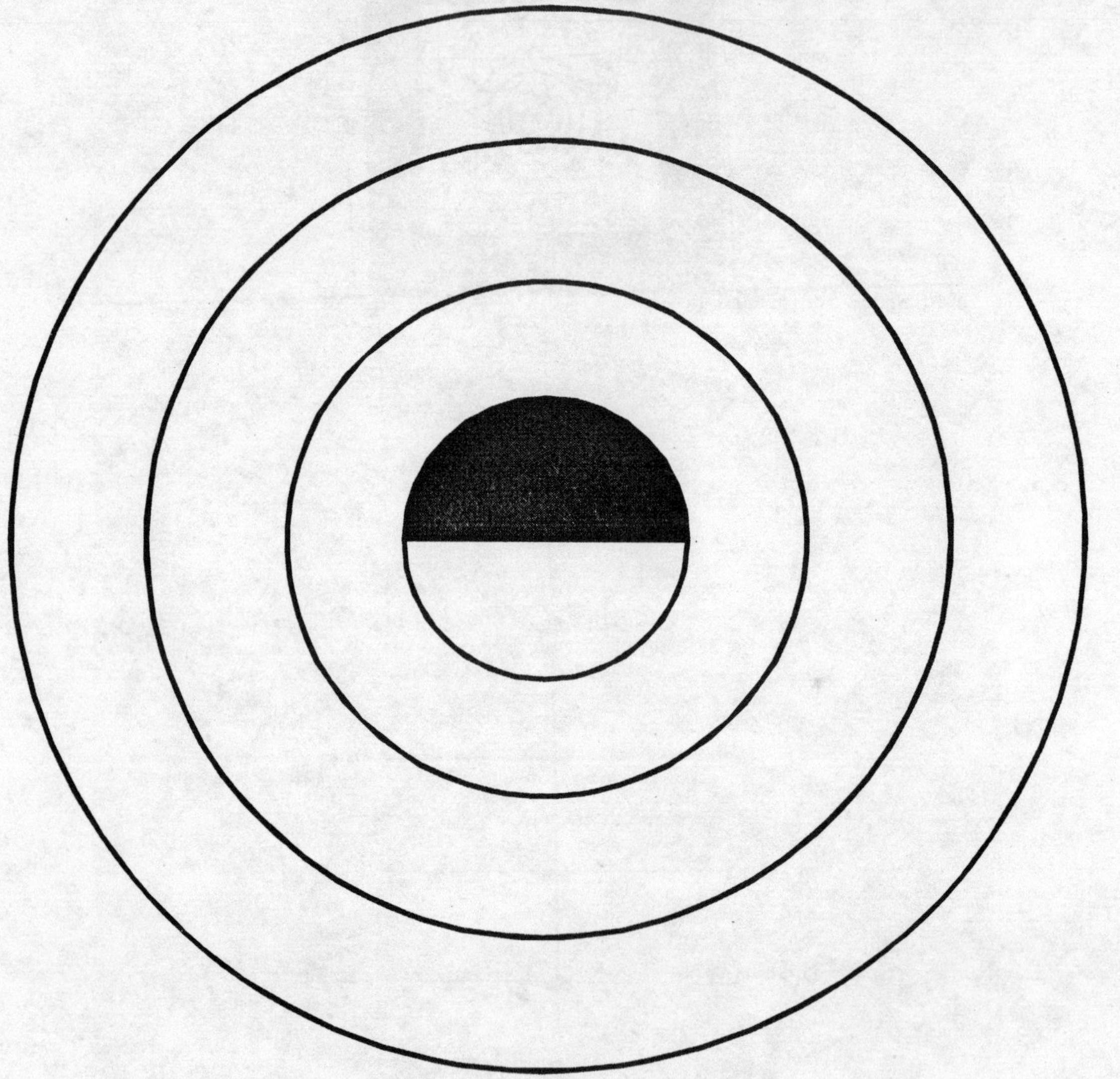

Fig. 12.7 : Test Sheet (Experiment 36)

APPENDIX – IV

TARGET – DART TEST (GROUP METHOD)

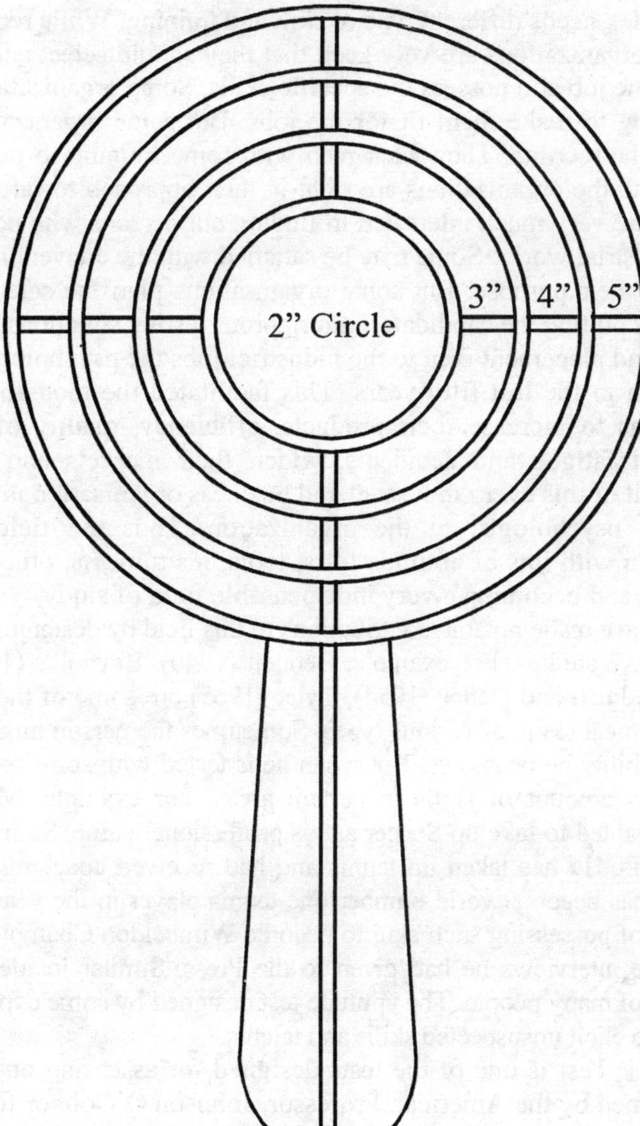

Fig. 12.8 : Target-Dart Test : Pattern for measuring Grouping. This Pattern may be prepared in steel or copper wire (Experiment : 36)

EXPERIMENT 37 : ASSESSING MECHANICAL SKILL

INTRODUCTION

Modern industry and technology have grown so vastly that every industry is becoming different from the other in some respect and as such every industry whether it is big or small simple or complex needs different type of skill and training. While recruiting men for such industrial jobs, organizations are very keen that they should select and palce persons who are suitable for the job and possess the specific skills. Some organizations recruit men and give them training to make them fit for the jobs. But some organizations could not provide training for the recruits. They want men with some training to put them straight away on the jobs. Both the organizations are right in their approach towards selection and placement. But they are very much interested in finding out persons who possess the skills required for their industrial works. Some may be satisfied with the conventional methods of selecting persons for their purpose. But some organisations plan the selection procedure very systematically by putting the candidates into rigorous testing situations.

In the selection and placement men to the industrial jobs the psychologists have made very rich contribution in the last fifty years. This facilitated the modern industrial and business organizations to increase their products, efficiency, quality of the products, reduce the industrial fatigue and accidents, widen their markets and increase their consumers. As a result of this every industrial and business organisation is becoming very keen in having one psychologist in the organization. Thus the field of industrial psychology has grown with lots of aptitude tests, tools, instruments, other test materials and research findings and becoming a very indispensable field of study.

Several scholars have made notable contributions to this field by designing tools and tests and also made extensive studies. For example, Bennett (1940), Reynolds (1961), Friedman and Ivens (1954), Friedman and Detter (1954), Tyler (1965) are some of those who studied particularly the mechanical skills of various types. Sometimes the person himself may not be aware of the skill or ability he possesses. But when he is tested with some tests and tools he may exhibit enormous amount of skills in certain areas. For example, Micheal Stich, a German young man wanted to take up Soccer as his professional game. Some one suggested to him to go for tennis. He has taken up tennis and had received coaching systematically. Within few years he has become world number one tennis player in the year 1991. He said that he never thought of possessing such skill to become Wimbeldon Champion. This he had revealed in one of the interviews he had given to the Press. Similar incidents might have taken place in the life of many people. The aptitude test designed by some experimenters may make the individuals to elicit unsuspected skills and talents.

The Wiggly Block Test is one of the tests designed for assessing one's mechanical skills. This was designed by the American Professor, Johnson O'Connor for testing skills for industrial purposes. This involves mainly assembling of parts. There are many tests designed for testing assembling and other mechanical skills. But one specific feature of this Wiggly Block Test is that it has three dimensions. The third dimensional spatial relation is supposed to be highly taxing for the person who performs the task.

This test is very widely used in industrial situations where visualization and spatial relations particularly drawing mechanical parts for the development of machines are needed. The test reveals the ability and aptitude that is required for the occupations like draftsman, mechanical engineer, machine architect, tool and discmaker. O'Connor says that the administration of this test is very complicated and one must receive proper training in it. The teacher or experimenter who administers this test should practice himself and get familiarized with the test before he administers it.

EXPERIMENT

Problem

The purpose of this experiment is to assess the mechanical skill of a person.

Method

Sample

Choose a person who has no knowledge about this test and describe his characteristics such as age, qualifications, sex, occupational experience if any, socio-economic status, physical deformities if any etc.

Materials

O'Connor Wiggly Block test and stop clock.

Description of the Material

This test was designed by Johnson O'Connor for testing one's mechanical skill and it is named after him as O'Connor Wiggly Block test. This test gives the look of a rectangular wooden piece. It has 9 inches in length, 5 inches in width and 5 inches in depth. This size is made up of 9 pieces. There are two horizontal cuttings in lengthwise which make three separate pieces horizontally and two vertical cuttings in depthwise thus making 9 pieces all lengthwise. The cuttings are made in a zig-zag manner and the cuttings may not be noticed unless the pieces are separated. Otherwise it may look like one solid piece. So it has three vertical piles and three pieces in each pile and has three layers with three pieces in each layer. When all the nine pieces were once disassembled it is difficult and also takes more time to reassemble them in the same manner.

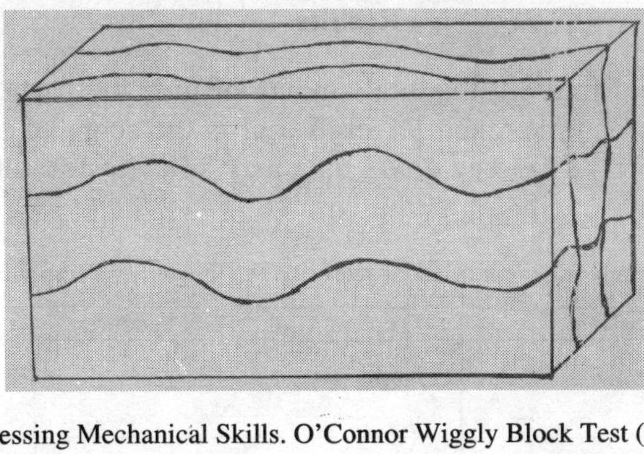

Fig. 12.9 : Assessing Mechanical Skills. O'Connor Wiggly Block Test (Experiment 37)

Procedure

This experiment should be conducted in a well illuminated and ventilated room. There should not be any visual or auditory disturbances for the participant. He should be seated comfortably in front of a work table. The O'Connor Wiggly Block test in the assembled form should be placed in front of the participant, length-wise. Then the following directions should be given :

> "This wooden block is made up of 9 pieces. Each piece is like this" (One top front piece may be removed and shown to him and should be replaced). "I shall remove all the 9 pieces separately and mix them together and you will have to arrange them in the same manner what you see here."
>
> "It is cut through into three piles and there are three pieces in each pile and thus making them into nine pieces." (The experimenter at this stage should separate the blocks into three vertical pieces). "There are two vertical cuttings and two horizontal cuttings and the cuttings are done in a zig-zag way." (This also must be shown to the participant. The three layers should be lifted one by one, first the top layer with three pieces, then the middle layer with three pieces and lastly the bottom layer with three pieces). "You must arrange them in the same order, first the bottom layer, then the middle and lastly the top layer. Now I dismantle the whole structure and lay the nine pieces on the work table." (By saying like this the experimenter should mix all 9 pieces randomly)." When I say, 'ready, go' you must start assembling all the nine pieces into one structure you saw earlier. You must assemble these three times and you can take your own time. Yes, now ready, go."

As soon as the participant started assembling the pieces, the experimenter should start the stop clock to run and must note down the time. When the participant completed the assembling task successfully, the experimenter once again should dismantle them and mix the pieces and spread them for the next trial. After completing the second trial the experimenter should arrange for the third trial. The time must be noted for all the three trials. When all the tirals are completed the exprimenter can ask the participant to give his introspective report as to how he felt while assembling the pieces. Then the participant may be allowed to go away.

Results

The experimenter should find out the total amount of time the participant has taken for each trial. The amount of time taken for each trial is the score of the individual. This should be given in Table 12.18. The group data may be collected for all the three trials and given in Table 12.19.

Table 12.18 : Showing the individual scores on the O'Connor Wiggly Block Test

S. No.	Trials	Time in Minutes and Seconds	Remarks
1.	First Trial		
2.	Second Trial		
3.	Third Trial		
Mean			

Table 12.19 : Showing the group data of the O'Connor Wiggly Block Test

S. No.	Name	1st Trial	2nd Trial	3rd Trial	Mean
1.	B.V.				
2.	M.S.				
3.	K.R.				
4.	M.R.				
5.	S.V.				
6.	P.K.				
7.	V.R.				
8.	R.S.				
9.	T.V.				
10.	C.P.				
Mean					

State what amount of time your participant has taken for the first trial and for the subsequent trials. Was there any difference in the time among the trials. In the group data, state who has taken shortest time and who has taken the longest time. State the mean score for the group and how many scored below the group mean and above the group mean. Analyze your participant's score in relation to the other members of the group. What the test actually assesses and how many have the specific skill.

Discussion

Discuss the general performance of your participant. Was there any improvement in his performance in trial to trial? What type of skill he possesses? In the group data where your participant stands? What is the rank he occupies? Discuss his mechanical skill in relation to the other individuals. Also examine whether age, sex, and other personal factors have any influence in the performance of the individuals. Compare your result with the results of earlier studies. Discuss whether the individual's previous experience in similar task had any influence in the present performance. State also what type of mechanical works the individuals can perform efficiently. What do you understand from your participant's introspective report.

Conclusions

1. What was the general performance of your participant in this test.
2. What was the level of the performance of the group?
3. How many of them may be proficient in the mechanical type of work?

13

MOTIVATION AND ACHIEVEMENT

INTRODUCTION

The study of motivation is becoming more and more difficult and writers in this area are getting confused as no correct definition of motivation has been evolved. The meaning of this concept also is often varied. Candland (1968) states that for want of suitable definition of the concept and for want of suitable techniques for the study of motivation, researchers in this field are very much thwarted. Further, the recent theories of emotion equate motive with emotion and also with activation. Therefore, doubts arise whether motivation is a separate concept fully free from emotion and activation in its meaning and definition or to be considered similar to emotion or activation. Authors like, Duffy (1934, 1948, 1957, 1962), Leeper (1948), Webb (1948) and Cofer and Appley (1964) have made extensive research work on this and stressed for a motivational theory of emotion.

Definition

Motivation is the cause of human behaviour and behaviour is a motivated function in an organism. The human organism, Maslow (1970) states, is an integrated organized whole. In motivation, this means many specific things. The whole organism (individual) is motivated and not just a part of him. An organism's behaviour is manifesting in the form of responses. The responses are both internal and external. There are internal physiological needs and also there are external stimuli. The internal physiological needs and the external stimuli determine the organism's responses. Therefore, the organism is made to move and act towards the stimuli to meet the internal needs of the organism. The organism is caused to behave. Hence motivation may be defined as the cause of behaviour of the organism. Motivation, according to Lindsley (1959) is a combination of forces which initiate, direct, and maintain behaviour to achieve a goal. Moskowitz and Orgal (1969) state that, why does an organism respond in a particular fashion, because there is continuous interplay between the internal conditions of the organism and the immediate environmental situation of the organism. For them a response is an observable change in the behaviour of the organism.

The motivational study according to Hall (1970) is based upon the problems derived from two observations. We see on some occasions an organism is more active and it is aroused and its actions are surcharged with more energy than on other occasions. Secondly, we see that much of its behaviour appears to be directed towards the attainment of some goals. This is an observational description of the behaviour of an organism in a motivated situation. The organism is moving towards achieving some goal.

Hence there is a need for the organism. It may be internal physiological need and the goal may be the stimulus, the food that fulfils the need of the organism. It is because of this type of behaviour of the organism Kimble *et al.* (1985) define motivation as the process that provides behaviour its energy and its goals. According to them an organism experiences motives as feelings of wanting, needing and desiring. Motives cannot be seen directly. We can only infer them from the behaviour of the individuals.

Authors like Garrett (1975) state that organic needs, attitudes, interests – all these are motives. It is because they urge the individual to actions and motives and operate from within the organism. There are specific motives for specific needs. A person who is feeling hungry goes in search of food. The motive is a drive. It directs the organism internally for actions or to fulfil the needs. Woodworth (1952) defines a motive as a readiness or preparation of the organism for certain kind of behaviour and, therefore, for Woodworth a motive is practically the same as a set. According to Murphy (1952) motive is a basic tendency to activity which is initiated by the shifts of physiological balance that is accompanied by sensitivity to some particular types of stimuli, resulting in a consummatory response. For Murphy motives and drives are the same.

In their attempt to locate the seat of emotion in the brain and other nervous system, the experimenters have discovered the *limbic system* and the *reticular activative system* as the important parts in'the arousal of emotion in an organism. This has become the source of information for some activation theorists to make radical changes in the names, definition and interpretation of some concepts. Emotion and motivation were the two concepts received greater attention of the activation theorists. The old definitions and interpretations are now slowly changing into some new names and interpretations.

CLASSIFICATION OF MOTIVES

Motives are classified in various ways. An organism is to behave because of the motives of various types. The behaviour of the organism depends upon the strength of motives or drives. When an organism's motive or drive becomes stronger? When the need is increased? For example, if the organism is deprived of food for a long time, the length of time and the deprivation make the organism restless and the internal stimulation becomes stronger and the organism begins to act vigorously to get its goal, i.e. the food. Similarly the other motives are determined by the strength of stimulation. The motive to avoid danger or pain of the physical punishment, the motive to get rid of anger or irritation causing stimulus, the motive to have the feeling of love satisfied by meeting the object of love are the clear example, Thirdly, the experiences the individual had gained in the past also may be contributing to the creation of motives. The degree of affiliation the object had produced in an organism motivates it to move in search of the goal. Therefore, motives may act at various levels and classification of motives is made on the basis of the nature of function and the needs and goals of the organism.

Motives are generally classified as primary motives and secondary motives. Woodworth (1952) classified motives into three main classes. They are : the motives which depend upon internal bodily states, which he calls organic needs, the second is the emergency motives which are aroused by environmental situations, and the third objective motives, which are to deal with objects in the environmental situations. Authors like Garrett (1975) classified motives as physiological motives, common social motives, unconscious motives and conflict of motives.

The Primary Motives

The primary motives, organic motives, physiological motives are all referring to the same thing, but in different words. The primary or physiological motives or drives are hunger, pain, sex, rest and sleep which are all bodily needs. All the primary motives are capable of reducing drives. They also become the potential reinforcers and can act as incentives. Drives are resulting from some physiological disequilibrium, for example, hunger and thirst. The physiological or bodily disequilibrium is a temporary organic need. It results from a general increase in arousal. The organism is energized for action. All the primary motives have the same characteristics. When the goal is achieved the drive is reduced temporarily.

The Secondary Motives

The secondary motives are supposed to be learned or acquired motives. In our every day life we can see that many events that arouse human behaviour seem to have no direct connection to physiological or organic motives. Learning and experience have influence on motives indirectly. Human beings are not always feeling hungry or feeling thirst or always go in search of food. After the hunger and thirst drives are fulfilled man does not keep quiet. Man is motivated by many other factors other than the primary drives.

Man needs money for many things. He has to purchase many things other than food, like clothing, must have a building to live in and furnitures to have comforts. He paints the walls and beautifies them. He must send his children to schools to educate them. For all these he needs money. Therefore, he must do some work which can fetch him money. He moves and acts vigorously. Thus money becomes an incentive because of its relations to various things he needs. In a society a man achieves social prestige or status through the association of money. He might have developed interest in going to various clubs to spend some time with friends or may go to races. These acquired motives are also physiological and psychological.

Psychological Motives

The psychological motives are also called emergency motives. These motives mainly depend upon environment. Garrett (1975) treats the psychological motives with emergency character and states that all emotions act as powerful drives and direct human behaviour. The unpleasant emotions stir up the organism and increase physiological force and mobilizes body's resources and attacks the object if the emotion is anger and takes flight if the emotion is fear. According to Garrett these emotions do not possess emergency character by themselves. But they are potential drives of human behaviour. Similarly the emotions of grief derive the sorrowful characters from the feelings of discomfort or acts of punishment. The emotion of love acts as a drive in an organism for the feeling of affection. In the same way, the emotion of joy is a drive for excitement.

Social Motives

The learning experience and social conditioning initiate a number of urges in an individual which may be described as social motives. The order of dominance can be observed right from birds through animals to human beings. An individual aroused by his ***mastery motive*** wanted to dominate over others. He wanted to assert his self in the society. Towards which he may be involved in various kinds of activities. The individual can assert his self in competitive games, in intellectual performances, or by making good oration in public and political meetings, by organizing various social activities, by leading some expeditions, by doing service to the needy people, by occupying high position or rank and so on.

Activity Drives

There are certain drives lie inactive under some vital organs. Generally there is a craving for activity. The basis is found in the central nervous system, large muscles of arms, legs and the trunk. These muscles need always contraction and relaxing and when this is not taking place they get tension and the individual is craving for some activity. There is a need for a continuous action and longing for new activities. Wenger (1943) has studied the level of muscle tension and this tension was found to be existing even after relaxing, because it is related to the total personality of the individual who is longing always for some new activity.

We could have observed in some situations that some individuals out of extreme drive for activity might be playing some game or doing some exercise alone secretly. Children are full of energy and they wanted to spend the energy by keeping themselves constantly involved in some useless activities. All these are out of the activity drives only.

Sensory Drives

Man is a behaving animal. He needs some kind of activity for the bodily parts. If the activities are not taking place in some form there is tension in the vital organs. In the case of sense organs they are often stimulated and activated by the brain. Since the sense organs are connected with the brain when they are stimulated they keep on functioning for a quite long time. There is tendency to see beautiful colours, to hear sweet musical tones, to eat tasty food, to smell flagrant smells and also to touch soft and sensuous things. These tendencies arise out of sensory drives since the sense organs kept inactive for a long time. There may be individual differences in reducing the drives of the sense organs. In the case of seeing beautiful things, one child may need similar one for him, another may simply see and go away, a third may want to create similar one. So also in the case of hearing music one person may enjoy hearing by nodding his head and moving his hands, another may want to learn similar music and want to give such performances. Therefore, the sensory drives create tension for activity or for achieving the goal. Lerner and Murphy (1941) studied these aspects among young children.

Unconscious Motives

We come across the concept of unconscious in the study of abnormal personality. But here it is applied to motives. The unconscious motive is not fully separated from consciousness or normal personality. For example, there are instances in which a person may at one time be dominated by one motive and may be acting on it and on another time may be dominated by another motive. Sometimes when one motive is dominating another may unite and cause confusion. The old motives and the motives that are not reviewed lie dormant are unconscious motives.

Achievement Motivation

In recent years this has been receiving greater attention from psyochologists. The basic idea in achievement motivation is that what motivates one to achieve and to master his environment. It may be an exploratory drive or an unlearned and learned curiosity that may impel a person to make better achievement in an environment. The works of McClelland (1953) and his co-workers have been often discussed in the contemporary psychological studies and experiments have been designed to study various aspects of achievement motivation. According to them motivation is the learned expectation that a goal may produce a positive or a negative reaction.

According to the assumption made by this theorist, motivations are learned, there may be also unlearned motives. But motives which are important for human behaviour are learned and they must be studied by suitable tests, particularly, the projective test to know the level of an individual's imaginations and day dreams. The results of the test may alter or change this level of motivation. It means that the test has revealed that the individual's aspirations are too high or unrealistic and, therefore, it needs to be changed.

As stated above the learned motives are very important for human behaviour. The goals of such motives are also known. It is assumed that the goals previously learned may provide pleasant feeling and which may motivate the individual more strongly than a goal which is not known. Further the feelings of pleasure and pain are attached to learned and unlearned goals respectively. The phenomenon of achievement motivation has a long developmental process, having its beginning with Murray's (1938) explorations of personality. He called this first as need achievement which is one among his twenty manifest psychogenic needs. His next contribution was the development of an instrument called, the 'Thematic Apperception Test' (TAT). This test is supposed to uncover 'coverts and unconscious complexes'. This test was accepted universally and used to assess ths achievement needs.

A systematic study of achievement motivation was made by McClelland, Atkinson, Clark and Lowell (1953). It also resulted in making necessary refinement in the TAT for measuring achievement needs.

Out of twenty psychogenic needs, why was achievement chosen for the study? McClelland states that he and his colleagues were primarily interested in the laboratory studies and it so happened that this motive was often found to be arousing in it. Further, McClelland was not satisfied then existing motivational theories of Freud and Hull. For McClelland, it seemed that the learning together anxiety and food deprivation led to stimulus balance. But on the other hand if emphasis is made on achievement needs it readily produces stimulus attack – this is the functioning of motivation. On the basis of this experimental result McClleland evolved his theory of motivation. Some are of the view that McClleland's theory of motivation was influenced by Hebb's idea of optimum level of stimulation.

On publication of the work, the Achievement of Motive McClleland went back to laboratory work in 1953 and his colleague Atkinson carried out the experimental studies on achievement motivation. He validated the TAT which has paved way for Atkinson (1964) for formulating a theory of achievement motivation. Therefore, the theory of achievement motivation is always mainly associated with the names of Murray, McClleland and Atkinson. Atkinson in his later studies associated the need of achievement with level of aspiration.

The kernel or the central points of this theory are that the desires of achievement needs are to accomplish something difficult to master or handle or organize physical objects or human beings. It is also to do things as rapidly and independently to overcome obstacles and to attain high standard. It is to excel everybody. It is to increase self-esteem and successful exercise of skills. It is assumed that high need achievers may tend to be internally oriented whereas low need achievers may tend to attribute their performance to external factors. White (1959) states that individuals are driven by need to achieve competence. A successful performance of this need may manifest in positive self-concept and higher degree of achievement motivation may result. An unsuccessful performance may cause a negative self-concept and lower degree of achievement motivation.

HILGARD'S CLASSIFICATION OF MOTIVES

Hillgard (1962) classified motives under three heads, as survival motives, social motives and ego integrative motives.

The survival motives refer to the physical needs such as food, water, maintaining temperature, rest and sleep. The social motives refer to sex, parental motives, affiliative motive. It also refers to dominance, submission, aggression and abasement. The ego integrative motives refer to self-esteem, mastery of the social environment, construction, invention and destruction.

MASLOW'S CLASSIFICATION OF MOTIVES

The motivational theory puts a few questions and tries to find out answers for the same. What make an organism to act or behave? What directs the behaviour? Why there is a specific kind of behaviour? Why is behaviour learned? Why does the behaviour stop? The organism acts or behaves, because there is an urge in it to move and act in a particular fashion. There is a need in the organism. The need in the organism is a general state of deficiency. When there is deficiency in the organism, it loses its energy and efficiency to function. It becomes restless. Therefore, the organism must re-equip the body with energy which was spent. The organism is aroused by the needs and it reacts to it. This reacting tendency to the needs of the organism is the drive. If there is a need for food then this activity ceases to exist after food is consumed. The organism in the state of need is unpleasant and when the need is satisfied it is pleasant. The restlessness and the muscular tension in the organism are reduced. These are physiological needs which include food, water, sleep, rest, activity and sex. There are psychological needs which include the need for love, affection, achievement, independence, social recognition and self-esteem.

Murray (1938) has enumerated twelve viscerogenic (physical) needs and twenty eight psychogenic (psychological) needs. But Abraham Maslow (1970) proposed two kinds of needs, the basic needs and the meta-needs. The basic needs consists of first physiological needs such as the need for food, water, temperature regulation, someastbetic, minerals, hormones and vitamins.

Physiological Needs

Maslow says that as soon as physiological needs are satisfied at once the higher needs emerge and these dominate the organism. When these needs are satisfied again still new higher needs emerge. After fulfilling these needs still higher new needs are emerging. This process takes place step by step because the basic human needs are organized into a hierarchy of relative greater force. This means whichever motive is more powerful begins to dominate the organism. If the needs are obstructed they may emerge again and dominate the organism. The behaviour of the organism is dominated by unsatisfied needs.

Safety Needs

When the physiological needs are fully satisfied then there emerge new set of needs which may be next in the heirarchy of needs. These are safety needs. The safety needs include security, stability, dependency, protection, freedom from fear, freedom from anxiety, and chaos, need for structure, need for law and order, limits and strength for protection. These safety needs seem to be as important as the physiological needs. The organism seems to be euqually well dominated by these needs. The organism almost functions like a safety-seeking mechanism. The organism makes use of the receptors, effectors, the intellect and other capacities as safety seeking tools. These safety needs are very important and urgent in social situations when there is threat for life and property.

Love and Belonging Needs

On fulfilling the physiological and safety needs there emerge new needs which are next in the hierarchy of needs, love and affection and belonging needs. What are all said about physiological and safety needs may also apply to these needs. The organism feels very much from the absence of friends, a lover of the opposite sex, family and children. The person longs for affection with people and may strive very hard to achieve them. He may need all these more than anything else in the world. Though sex is considered as purely a physiological needs, when it is associated with the needs of love and affection it involves both give and take – a reciprocal affection which may make a man or woman to ponder over their present loneliness and strongly strive for the company.

Self-esteem Needs

In every society individuals have strong desire or need for evaluating the self highly. Consciously or unconsciously man develops this tendency and wanted to place himself at a higher level and wanted to be respected by others. Maslow calls these as esteem needs. These esteem needs emerge next in the hierarchy of needs when all the said needs are fulfilled. This is with reference to self-respect, or self-esteem. This is discussed under two parts. The first part deals with need for strength, need for achievement, need for competency and mastery. This mainly deals with equipping oneself with competency or capabilities. The second part deals with need for reputation, status, fame, glory, dominance, recognition and dignity. These refer to the consequent of the first part.

These physiological needs, safety needs, love and belonging needs and self-esteem needs are hierarchical in the sense that the higher level needs are strived only when the lower level needs are satisfied. When a man is feeling hungry he needs immediately food and not love or self-esteem.

Maslow calls the basic needs as ***deficiency needs***. These needs motivate the organism for activity. Since the organism is deficient of food or water these needs first should be fulfilled. When all the basic needs are fulfiiled then there emerge the meta-needs which are called the ***growth needs***. The meta-needs motivate behaviour in an organism not out of deficiencies. Behaviour is motivated from instinctual tendencies towards growth. Only the basic needs have hierarchy in the organization. The meta-needs are not hierarchical. They include aesthetic urges, cognitive urges and they are associated with ethical principles like truth, beauty, goodness and justice. According to Maslow the highest is the need for self-actualization. Man has a tendency to make himself capable of achieving all his needs. ***Self-actualization*** is a tendency to develop the self to the full level of attainment of which one is capable of. This also may mean to put all the potentialities of a person into action and achieve all the goals he needed.

Maslow, in his treatise of the theory of motivation not only speaks of a series of increasing gratifications of the basic needs, but also stresses a series of increasing degrees of ***psychological health***. A man who is safe and loved is healthier than a man who is safe but unloved and rejected. A man who is psychologically healthy wins honour admiration, achieves self-esteem status, prestige and also love and affection from all to be still more healthy and attains full humanness.

Strength of Motives

Having discussed Maslow's hierarchy of motives, right from the physiological needs to the needs for self actualization we might have formed some idea regarding the strength of motives. Only after fulfilling the basic for the deficiency needs one thinks of the higher needs. Therefore, the basic needs are supposed to be more stronger than the higher needs. In the hierarchy of motives the physiological motives are gratified first because of their importance and strength. They are so strong as to overpower all other motives and dominate the organism and direct for behaviour.

If a person is given with a long list of human motives and asked to state which are the stronger motives and which are the weaker? It is true that he will hesitate for sometimes to decide the strong and weak motives. But finally he may state the motive that impels him for action or for some kind of behaviour. It may be for a short period but still it is stronger. Sometimes we may be required to estimate the strength of motives of a person in an interview to know whether he likes the job or if appointed will he stick to the job or quit in search of some other job. In such instances we are waiting from him strong expression for accepting that job and his wish to continue in it and also his likes and dislikes of the organization. Answers to questions relating to these aspects depend upon the individual and circumstance. There may be personal motives and also social motives. Suppose, if there are more competitors in the interview his reaction may be different. Some individuals make it a combat motive of fighting for the post and develop strong motivation. Sometimes the individual himself may find it very difficult to decide which is strong and which is weak though he is involved in it.

Psychological tests have been designed and developed to measure motives. More information are available on strength of motives from animal experiments (Richter, 1927, Wardon, 1931). Many studies have been made on motivation relating learning (Seelamon, 1940; Clark, 1940). There are learning theories relating motivation. Educational psychologists have done much in this area. But the researchers who work on animal motivation do not work on human motivation. It will not be possible to attribute all characteristics of animal motivation to human beings. The factors of human motivation are completely different from the factors that motivate animal behaviour. The precision of the data obtained from the experiments conducted on animals in a controlled situation could not be compared with the strength of human motivation in various situations.

PSYCHOLOGICAL HEDONISM AND THE INSTINCT THEORY

If an organism has performed the task quite successfully the result is gratifying the organism and it feels pleasant. Instead if the result is not gratifying, the organism is not feeling pleasant. Hence it does not perform it appropriately. In the case of human beings it may be accounted that human beings act to enjoy the pleasure and avoid pain. This is referred to as psychological hedonism. The instinctual behaviour pattern in animals and birds can be also used to explain human behaviour. The earlier authors tried to explain the problem of human motivation in terms of instinct theories. McDougall (1908) the doughty champion of the instinct theory and Bernard (1924) who followed McDougall have explained any type of human behaviour through the instinct theory. The behaviours that are ever explained are those for which an instinct is named. But certain behaviours are not explained and predicted by instincts. Therefore, Thorpe (1963) correctly applied the instinct theory more to animals than human behaviour. Further the instinct has been redefined by Lorentz (1952) and Tinberger (1951). According to them instinct is a complex, species-specific, relatively unmodifiable behaviour. But it can be observed that the human behaviour patterns are instinctual as many animal behaviours are. The instinct theorists may be correct in saying that we are born with certain potential instincts but they are modified through learning and experience. The psychological hedonism seems to approve this view.

Though the instinct theory was strongly criticised by Watson (1953) and others in 1930s a time has come for the psychologists to re-examine the instinct theories formulated by Freud and McDougall. The dynamic and humanistic psychologists have come to a conclusion that it is impossible to define the whole human activity by stimulus-response mechanism alone. Maslow (1970) states that the instinct theory formulated by Freud and McDougall had certain virtues. The instinct theory advocates that man is a self-mover, his own nature and environment determine his behaviour. The instincts supply man readymade framework of ends, goals and values. There is a need for good and a need to avoid illness. All human beings belong to a single biological species and the organisms are functioning on their resources. They often display biological efficiency which is the function of the instincts. The humanistic psychologists find some mistakes in the instinct theory which Freud and McDougall have failed to explain. However time may come that others may be making another re-examination of the instinct theory which may bring the instinct theory much still closure to the present theories.

MOTIVATION AND LEARNING THEORIES

All learning theories have strong association with motivational variables whether the variables have any relevance to human learning or not. The learning theorists in their studies have attempted to measure the motivational variables. Hull (1952) was one who had studied the problems associated with measurement of motivation. According to him it is impossible to predict the performance of the organism by identifying the relevant, variables. Since the variables are associated with the drive level of the performance, prediction can be made. The environmental variables act upon the organism and the organism behaves in predictable way.

The other theorist, Tolman (1932, 1951) differs from Hull and put more emphasis on the purposive behaviour of the organism. According to Hull organisms anticipate events and are aware of the consequences. The rat for example is hungry and expects food. When it gets its food the expectation is confirmed. For Tolman this conformation of expectation is the basic property of motivation. Another contemporary theorist to Hull and Tolman was Guthrie (1952). All the three theorists are behaviourists in a sense but formulated different learning theories. According to Guthrie all learnings take place in the first trial itself and all other subsequent trials are intended to eliminate incorrect and inappropriate responses and also to strengthen the correct responses. The repeated trials or the reinforcements are for reducing the possibility of learning other responses. To Guthrie motivation is not important. Guthrie states that the hungry rat runs restlessly to be away from painful hunger for food. The animal is kept active for external stimulus whatever motivating variables are there. Another contemporary learning theorist, Skinner (1938) came out with his theory called operant conditioning in which he has given some importance to motivation. In operant conditioning the primary concept is ***reinforcement***. If a reinforcement makes an organism to repeat the response already made then that response is learned response. In Skinner's experiments the rats are deprived of food or water and they are rewarded with food and water when they make appropriate responses. A reinforcement is made whether the same response is made by the animal. Skinner wanted to see the effects of reinforcement by using schedules of reinforcements with different effects. This means using different forms of motivation for eliciting the response.

Motivation, just like intelligence cuts across all topics in the field of psychology. There are various types and kinds of motives and they are related and interchangeable terms such as needs, drives incentives and motives. To study the role of motivation in learning and to put it in proper places is a difficult task. In recent times systematic study of motivation in learning and their effects have been made. One of the early studies made by Hurlock (1925) is still valued as important. She used verbal incentives like praise and blame to determine the effects on learning in the classroom. The results showed that praise as an incentive had greater effect on learning. Praise was found to be an efficient motivating force in learning. Following Hurlock's study of the role of motivation on learning, Dadvids and Sidman (1962) conducted a study of dynamic of stability and instability in achievement using variable which they believed reflected motivational dynamics. With a view to investigating motivational properties and the effects in learning Taylor and Spence (1952) conducted an investigation among the students using Manifest Anxiety scale. Studies on motivation in learning were also made by Mandler and Sarason (1952), McClelland (1955) and Olds (1953). These studies reveal that motivated students are inclined to learn more efficiently than not motivated students. These results put more emphasis on the importance of the role of motivation in learning. Since human motivation is a very complex phenomenon it is very difficult to identify the correct variables that exert influence upon the performance of the organism.

EXPERIMENT 38 : ASSESSING LEVEL OF ASPIRATION AND ACHIEVEMENT

INTRODUCTION

Every man is interested in making his self more successful in this competitive society. He wanted to achieve his goals for a happy living. Several psychological processes are involved in this phenomenon of achieving the goal. Some of the processes are interests, level of achievement, degree of motivation and emotional factors. Each of these are inter-related in some way. But what are the specific goals, the attainment of which will make the individual to assert as the great achiever. According to Murphy (1951) there are four such goals of great importance. They are physical attributes, possessions, prestige or status and power.

The physical attributes refer to one's physical strength, skills, which can facilitate him to achieve what he wishes. The other goal, possessions may be referring to many things. It is the larger or extended self. Man in a society is very much honoured if he possesses property enormously. It is a distinction for him. Right from childhood period this tendency of possessing something of one's own is growing in an individual. The child collects toys, stamps, gifts, coins and many things and claims that they belong to him. As and when the child grows this tendency also is growing at a larger level. Mainly this possession refers to the possession of wealth, because wealth can bring many things. Therefore, this possessive nature is a common trait found in every human being unless one is a saint or one who has renounced the worldly pleasures of life.

The third goal to be achieved is prestige or status. It is quite, but nature that the achievement of second goal may bring one the third goal. One's physical attention, the reputation he builds up on this, or the love and affection one shows to his fellow beings may all fetch him prestige or status. Winners of competitions, laurels, medals, titles and distinctions also may add prestige or status to a man. Veblen (1899) states that the Kwakiutl Indian chiefs destroy huge amounts of wealth in fire for the sake of vain display and putting the other chiefs into shame who could not destroy so much wealth. The fourth goal, power is also one of the gratifying goals that human beings try to achieve. Power is one that man enjoys it. Power also corrupts the man who handles it. Power as a goal man achieves it in two ways. One way of getting the power is through hereditary method. That is by way of ancestral inheritance. The other way is by struggle. In the present democratic society there is neck to neck struggle for power. The man who is politically powerful in the present society is supposed to be all powerful and he can do and undo anything. Therefore, man is easily motivated to struggle at all costs to achieve the power. The man who achieves power is satisfied by asserting his power successfully against others. It is true that in some societies one can enjoy status without having any possessions and power. Great scholars, poets, philosophers and religious leaders enjoy respect and status. In all these, the goals achieved fetch to the concerned man self-enhancement, the goal of status.

There are several ways of putting the man in goal of status in which he is likely to achieve. In an experiment conducted by Hoppe (1939) on the aspiration level, the persons who were working on a task were told how well they have done, and also were asked, "How well would you do next time?" It was found in this study that some persons set a goal so low that they were sure to succeed and thereby achieved very cheap victory.

Every one of us is aspiring for some goals. Some of the goals we must meet reasonably. Some goals may be of immediate need and some goals may be of long ranged. Long ranged goals are less sure to attain. One has to buildup slowly the feasibilities of achieving them. What goals the individual aspires? What are the feasibilities that exist for reaching them? What are the inconsistencies or discrepancies between the goals aspired and to those to be achieved? A person may have several goals in his mind or may wish. But what he expects to achieve? What the goals one wishes may differ from the goals he expects to achieve and these two may differ from what he has actually achieved. English (1962) has conducted a survey of the level of aspiration relating child development. Level of aspiration was compared with achievement, socialized goals, affiliation and other needs.

A study on level of aspiration was conducted by Child and Whiting (1949) among 151 under graduate students. In this study each student was asked to describe three incidents of his life. In the conclusions it was found that one described a frustrating event without goal attainment. The second described a frustrating event finally with goal attainment. The third was goal attainment with little or no frustration. Then each student was asked to state the effects of each incident on his level of aspiration to see whether the incidents have raised the level of aspiration or lowered. There were five generalizations, observed by Child and Whiting when they consolidated and analysed the descriptions.

1. It was found that success or attainment of goal has increased the level of aspiration and failure lowered the level of aspiration.
2. If the success was stronger then the rise of the level of aspiration was greater. When failure was stronger the lowering of the level of aspiration also was greater.
3. There was change of confidence in student's ability to achieve the goals. As a result of this there were shifts in the level of aspiration.
4. If the student has failed to attain his goals then he was inclined to avoid announcing ths level of aspiration.
5. The failure had greater varied effects than success in setting up the level of aspiration.

Generally when we are successful in achieving the goals we raise our head and feel that our self-esteem is raised. When we fail to make attainment of goals we shrink ourselves. The hypothesis that performance influences the level of aspiration has been tested by Church, Millward and Miller (1963) in their study among sixty college students. The result of this study indicated that the level of aspiration had influence in the result of the performance.

ACHIEVEMENT AS THE LEVEL OF ASPIRATION

In the classroom situation the teacher provides opportunities for success and failure for the students. The success and failure may affect the level of aspiration of the students. One measurement upon which this influence may be reflected is in marks the students expected and achieved. Rosenfeld and Zander (1961) have defined the level of aspiration as the level of achievement. The pupil realistically expect the marks to be high. Rosenfeld and Zander have administered a questionnaire on 400 junior high school tenth grade students. The questions were centering on achievement in mathematics and the relations of students to their mathematics teachers. The investigators have found that there was a strong relation between mathematics grades aspired to and the grades received at the end of the semester. This result suggested that there was considerable amount of truth on the part of most of the students. The students generally have set their grade aspirations a little higher than the marks obtained in the previous examination. Thus the authors have defined level of aspiration as the level of achievement.

Rosenfeld (1964) has also defined level of aspiration equating it with the difficulty level of the goal. He discriminates between aspirations wished for and expected. In another study Weigard (1953) has examined the relation between occupational choice and college success. The definiteness of a vocational choice and a strong desire for that choice are strongly related and led to the success in his study. Greater success is expected if the student sees that what he is doing is related to the goals that are actually aspired.

LEVEL OF ASPIRATION AND ACHIEVEMENT

All aspirations are not always achieved. Some may be aehieved as soon as the performances are over and some may be taking longer time to achieve – these are long ranged goals. Therefore, aspirations have differential effects. This may result from the respective achievement. Why aspirations have differential effects? It is because, there are other variables which are related to the level of aspiration. People, who set up level of aspirations are not of the same nature. They may have certain personal factors related to their level of aspirations. These factors may interfere in their level of aspirations. Every one does not approach the problem in the same manner. For example, there are individuals who have relatively high sense of ***fear for failure*** of the goal as opposed to low sense of fear for failure. Some have relatively high ***need for achievement*** as opposed to some who have low need. There are persons who have high ***need for affiliation*** as opposed to low need for affiliation. There may be also individuals with different combinations of these variables. Moulton (1965) has studied ninety-three senior high school male students with respect to their need achievement and test of anxiety. They were divided into three groups. One group had higher need achievement and low test anxiety, the second group showed higher need achievement and higher test anxiety, and the third group showed higher test anxiety and low need achievement. Each student was asked to choose among the tasks of three degrees of difficulty. Then they were dealt with in terms of success or failure. The hypothesis was tested. The results showed that there was abnormal shifts in aspiration level as a result of success or failure for the third group. The result confirmed the hypothesis. This shows that there are certain underlying variables in aspiration which determine the goals and also make shifts in goal setting.

The aspiration levels may also influence the roles played by the persons in their goal achieving task which may lead to differences in performance. Medow and Zander (1965) have conducted an investigation among 180 high school junior and senior students and verified this idea. The students were made to work on a designed task in groups of three. One student was made as the leader of the group. The other members of the group are not supposed to go on the task solution until the leader had completed each step. The investigators were studying the effects of this procedure on the leader and the followers. In the task there were various levels of difficulty. In half of the groups the leader named the difficulty level of the task to be performed. In the other half, the members secretly named the difficulty level, but the leader has selected and specified the task. The completed works were compared with several measurements.

The analysis of the results indicated that the leaders took their responsibilities seriously and over-estimated their importance. The leaders expressed that they had greater role in the task of the group. The other members were more dependent on the leaders than on themselves. The leaders said that they were more responsible for the quality of the product. The effect was stronger on the leader's evaluation of his performance when the condition of failure was imposed experimentally. The leadership roles assigned to the students seemed to have both positive and negative effects.

Generally, if goals and level of aspirations are set by others and persons are made to perform to achieve the goals then it will have different effect from the goals and level of aspirations set by the persons themselves of their own. Many classroom learning goals are of the latter type. Diggory and Morlock (1964) have suggested this phenomenon.

LEVEL OF ASPIRATION AND INTELLIGENCE

It is also posed to the investigators who are interested in the study of level of aspiration, whether level of aspiration has anything to do with intelligence. This has been tested by Davids and Sidman (1965) in their studies. These researchers have used in their studies bright high school male students who were academically successful and those who were not successful academically. The students had to perform on a motor task and their performance was predicted with knowledge of results in two trials. The students who were not successful academically performed lower than the academically successful students in the first trial and revealed initially a lower level of aspiration. But under conditions the under achievers exceeded their aspiration they set up a new goal similar to the group which was academically successful. The authors state that so far as the degree of brightness is concerned variability in an unsuccessful group may be reduced.

Level of aspiration is highly concerned with one's self. It is one way of asserting the self. The level of aspiration is one of the means of raising a person's self-esteem. Therefore, a person by putting himself in a situation in which he is likely to win the goal, he can raise his self-esteem. In setting up level of aspiration one must be quite aware of his abilities or competence, the feasibility of achievement and also the value of the goal. According to Murphy (1951) aspiration level means, the level which one strives to attain. In some cases the failure of which may lead to frustration if the person puts too much of importance on the goal. Some may strive many times in spite of failures and at last discard the goal saying that it is not good. Or the person may change the mode of approach and put up fresh efforts to attain the original goal or he may try for another similar goal. Sometimes the failures may make him to realize that he is aiming at too high and he may lower his level of aspiration to make himself confident of winning the goal.

In choosing life goals many individuals do not take into account the factors influencing the feasibility of achieving the goal. Further individuals differ considerably among themselves in choosing their life goals and setting up their levels of aspirations, because individuals differ fundamentally in their interests, motives, intelligence, physical abilities, socio-economic status and so on. Some may he aware of their limitations and some do not. Therefore, individual's expectations and their accomplishments always vary. Suppose, if we ask two persons separately how quickly and how successfully you can complete a particular task, one may over-estimate his ability and set up his aspiration at a higher level, but when he performs the task he may fail to complete it successfully. Another may under-estimate his ability and set up his aspiration at a low level and when he performs he can complete the task successfully as he desired. In academic life many students find it difficult to complete certain courses after one or two years of study and wanted to change over to some other course. If one is successful then he increases his level of aspiration and when he fails he lowers it. But level of aspiration and winning the goal are influenced by several psychological variables.

Psychological studies on level of aspiration have brought out the factors underlying the process of success or failure of the goals. The individual's interests, attitudes, motivation, physical abilities, personality, intelligence, sense of fear, feeling of confidence, self-esteem, ego-involvement and achievement are the influencing factors in the level of aspiration or goal aspiration of the individual. It is possible to assess the level of aspiration of an individual in a given task. There are quite number of experiments conducted in the laboratory situation for assessing individual's level of aspiration. Instruments have been designed and paper-pencil tests have been developed and validated.

Rotter (1954) who has invented the instrument called 'Level of Aspiration Board' has studied this extensively and explained in his Social-personality Theory. Rotter's (1954) locus control theory assigns responsibility for life's outcomes. According to this theory one's assumption about good and bad events of life influences his behaviour. This is the basic personality trait. An individual who thinks that he is able to act and increase the possibility of good results and reduce the possibility of bad results is said to have internal locus control. These are internally oriented people. In contrast, the individual who assumes about his ability to control events that he is not able to do it and is at the hands of luck, fate, and other uncontrollable outside forces is said to have external locus control. People of this type always look for external explanations for their successes and failures. Rotter (1966) has proposed individual differences in this regard. His subsequent studies on externality-internality dimensions has established that his classification of people on the basis of their reactions to their own behaviour was found to be very much useful and valid. It is true that some people are highly dependant on others and function at varying degrees of ***'felt powerlessness'*** – externally oriented, whereas others are more independent and function at varying degrees of ***'felt powerfulness'*** – internally oriented. Heider (1958) has elaborated Rotter's external-internal dimensions relating it to motivation. His argument was that there must be some interaction between 'personal casuality' and motivation. According to him ***intention*** is the motivating factor of behaviour. Internal or external forces cannot motivate an individual for behaviour unless the individual has got the intention to aim for the goal. Therefore, ***intention, expectancy*** and ***value*** are the important concepts in this explanation of behaviour which is goal oriented.

It is also observed that level of aspiration is necessarily involved in the individual's motives. Persons who are well-adjusted tend to have correct evaluation of themselves and correctly relate themselves to the world around them. Therefore, they have a good realistic level of aspiration. On the other hand persons who are maladjusted tend to be unrealistic and set up their level of aspiration at too high or too low. When an individual realizes that he cannot attain the original goal he is aiming, he may seek for a best substitute goal. This means that he is lowering his level of aspiration to the level of his achievement. Coleman (1974) says that regression enables one to lower his level of aspiration and makes him to depend upon others.

The level of aspiration one sets up also depends upon his potentialities and his previous achievements. An individual is only what he is capable of. The level one strives to achieve may be his life goal, set up by him and he may be strongly motivated by the factors around him. As Heider states, unless one has the strong intention to keep himself engaged in the goal seeking activity whatever possible chances exist and however he is capable of goal cannot be achieved. Therefore, the individual should keep himself intentionally ready to win the goal at whatever level he sets up his goal aspiration.

EXPERIMENT

Purpose : To measure the level of aspiration and the corresponding achievement of a person in a level of aspiration experiment and to find out the factors influencing the level of aspiration in his performance.

Method

Participant

Choose a person who has no knowledge about this experiment and describe his characteristics such as age, sex, qualifications, occupation, any physical handicaps, whether wears a spectacles etc.

Materials

Rotter's Level of Aspiration Board, Wooden rod, Iron ball and Recording sheets.

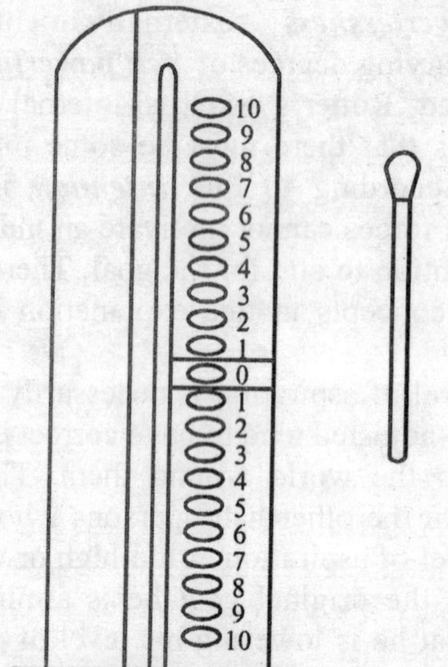

Fig. 13.1 : Rotter's Level of Aspiration Board (Experiment 38)

Description of the Level of Aspiration Board

The Rotter's level of aspiration board is a wooden board, having one metre long and about 6′′ width. In the middle of the board there is a long groove end to end with one inch width. At the centre of the board in the groove there is a grading like a scale with equal spacing and there are numbers written in white colour which are very much visible. The numbers are starting increasing from the bottom end from 0, 1, 2, 3, 4, 5, 6, 7, 8, 9,

10 and then going on decreasing from 10 to 9, 8, 7, 6, 5, 4, 3, 2, 1 and 0 at the upper end. There is a small hallow space for the iron ball to be placed. With the help of the wooden rod the iron ball can be pushed upward through the long groove and the ball can easily move from bottom end to the upper end. The ball can stop at any of the graded place numbered depending upon the force exerted on the ball. The position of the board should be parallel to the ground. This can be adjusted with the help of a clamp attached to the upper portion of the board. The wooden rod is of one foot long and about 1/2″ diameter. The iron ball is of normal size of 3/4″.

Procedure

The experimenter first should set up the apparatus quite appropriately for the experiment. Particularly with regard to the inclination at which the board should be placed, he must try and test whether the ball moves through the groove easily and also stops at a point steadily. He should prepare his record sheet for making entries of the trials. After making all these preparations the experimenter should make the participant to sit comfortably in front of the apparatus. Then the experimenter should give the following instructions to the participant.

> "Now, please look at this board. There is a long groove and there are also numbers. You will have to push the ball with this rod along the groove, so that the ball moves through the groove and stops against a number. If the ball stops at the number 10 you score 10. If it stops at 8 your score is 8. If the ball stops at zero or where there is no number or if it bounces off then your score is zero. I shall show you some demonstrations."

The experimenter should demonstrate some trials and hand over the rod to the participant.

> "You have understood now, what your task is. First you can have some practices. You can have 20 practice hits."

For every practice hit the experimenter should tell the participant the score he has made. On completing 20 practice hits the experimenter should tell the participant :

> "Now you can begin the test. I shall give you twenty trials. A trial constitutes five hits. Thus you will hit the ball 100 times in all. If you hit the ball every time at 10 then your score for a trial is 50 (5 × 10). If you hit the ball every time at zero, then your score is zero. Before you start every hit call for a number you going to aim at and tell me and I will tell you the score you have achieved in that trial of five hits. After every trial of five hits I will tell you the score."

The experimenter should note down at once the number called for by the participant and also the number he hit and at the end of five hits he should work out the score for each trial and tell the participant. In the same manner when all the twenty trials are over the experimenter should get the introspective report from the participant and allow him to go away.

Table 13.1 : Showing the scores of the participant in the Rotter's Level of Aspiration Experiment. One trial means five hits. The average points of 5 hits is the score for one trial. Thus 20 trials consist of 100 hits

Nature of Score	Trials (Each Trial consists of Five hits)																				Total	Mean
	1	2	3	4	5	6	7	8	9	10	11	12	13	14	15	16	17	18	19	20		
Aspiration score for each Trial (AN)																						
Performance score for each Trial (PN)																						
Goal Discrepancy score (GDS) (AN – PN-1) PN-1 : Performance in the previous trial.																						
Attainment Discrepancy score (ADS) (AN – PN)																						

Results

The experimenter, then should work out the scores for all twenty trials performed by the participant and present them in Table 13.1. What the score participant aspired in each trial and what he performed also should be given in Table 13.1. The goal discrepancy score (GDS) is the difference between aspiration for a trial and the performance in the previous trial. The formula is :

GDS = AN (Aspiration) – PN-1 (Performance-1). It is positive if the aspiration is higher than the performance on the previous trial. PN-1 refers to the performance in the previous trial.

There may be also difference between aspiration and the performance in the given trial. This score is called attainment discrepancy score (ADS). The formula is : ADS = AN(Aspiration) – PN (Performance). It is positive (+) if the performance score is lower than the aspired score tend if it is negative (–) the aspired score is lower than the performance score.

For 20 trials there are 19 goal discrepancy scores (GDS) and 20 attainment (ADS) scores. The mean score for all the 19 GDS and for all the 20 ADS should be worked out for the participant and should be presented in Table 13.1.

Find out the mean aspiration score of the participant for all 20 trials and his mean performance score and present them in Table 13.1.

Collect the group data and present them in Table 13.2. In Table 13.2, present the mean aspiration scores, performance scores, Mean GDS and Mean ADS.

Table 13.2 : Showing the group data in the Rotter's Level of Aspiration Experiment

S. No.	Name	Mean Aspiration Score	Mean Performance Score	Mean GDS Score	Mean ADS Score	Aspiration Positive (+) Negative (–)
1.	R.V.					
2.	M.S.					
3.	K.R.					
4.	P.V.					
5.	P.K.					
6.	T.S.					
7.	K.S.					
8.	R.M.					
9.	P.S.					
10.	V.G.					
11.	T.G.					
12.	M.R.					
Mean Score						

Discussion

Discuss the individual's goal discrepancy scores (GDS) and the attainment discrepancy scores (ADS). Is he influenced by success. Does the level of success increase his level of aspiration or decrease his level of aspiration? Do you find any rigidity in your participant? That is, inspite of lower performance, is he making the statement of the same aspiration level? State the main factors influencing the level of aspiration of the participant.

What is the mean level of aspiration of your participant in the group. Is it above the group mean or below the group mean? Where your participant stands in the performance of the group mean? Is he above or below the group mean? What is the rank of GDS of your participant? Is his ADS is below or above group mean? Whose GDS is the highest and whose GDS is the lowest? Whose ADS is the highest and whose is the lowest. Is there too much individual differences in the ADS in the group. State the reasons for all these. How many of them in the group are having positive aspiration and how many negative aspiration? How many in the group have positive attainment scores. State the reasons for all these. Discuss your results with the results of earlier studies given in the introduction.

CONCLUSIONS

State, whether your participant has positive or negative aspiration. Has he attained the positive ADS or negative. How many in the group have positive GDS and positive ADS. State the factors having influence in the level of aspiration.

EXPERIMENT 39 : STUDY OF GOAL ASPIRATION AND ACHIEVEMENT

INTRODUCTION

Modern society is a highly competitive one and everyone must strive hard to achieve something he needs. The child even before going to school is asked what he is planning to do when he is grown into an adult. Parents and others are very curious about the child's future prospects. Every culture provides ample opportunities for its members to achieve what they want, and everyone is subjected to some pressures to create aspirations for the future. It is true that aspirations motivate a person to make use of the opportunities provided by the parents and the society. In planning the future the social pressures in the individual play important roles by creating competitions with members of the peer group, in various situations. The social pressures add new meanings to the aspirations of the individual, put more emphasis to develop aspirations that are satisfying the individual's ego and bring admirations from the members of the society. In this way formulation of aspirations becomes an important aspect of an individual's future plans and self betterment .

Aspiration and Ambition

The concept, 'aspiration' is used synonymously and interchangeably with ambition. Ambition to make achievements in life and to live happily, is not an unnatural feeling for a man. Man desires to get money, power and honour. Therefore, ambition may mean a desire for fame, power, honour, to get ahead of all superiority. But aspiration may mean a longing for advancement, a desire to improve one's status or a desire to rise above all in a competitive society. It is a desire to be higher than the present position or status.

In the process of aspiration and ambition motivation plays an important role. But there is difference in the function. In ambition, once the goal is achieved motivation ceases to exist. In aspiration, motivation is long enduring, making improvement in the achievements one makes. Since aspiration is the function of improvement for better and better achievement it does not always satisfy the individual. The individual often aspires for better and no end. In aspiration the individual sets up his goal for himself in the tasks in which he is ego involved. The individual's self-esteem is raised by way of success. The failure leads to disappointment.

Characteristics of Aspiration

After setting up the goal by himself individual strives hard and confidently expects to achieve his goals in the ego involved tasks. If he achieves, he regards the achievement as success. If he fails to achieve his goals, he treats his achievement as failure. If the individual did not expect any success in the task he would not have the feeling that he has failed or disappointed. Hurlock (1974) states that there are three aspects in aspiration. The first is what aspect of it the individual considers as important and what he wants to do. The second is how well he reports to perform the important aspect of the activity. Thirdly how important the performance as a whole or partially is for him.

There are three kinds of aspirations. They are (1) positive and negative, (2) immediate and remote, and (3) realistic and unrealistic. The positive aspiration on the goal is towards achieving the success while the negative aspiration on the goal is to avoid failure. The immediate aspiration is to reach a goal at once, while the remote aspiration is not for achieving immediately. It is for the future and sometimes not realistic. In a person some aspirations are realistic and they are satisfied that he expected to achieve the goals. The unrealistic aspirations are mainly wishful estimate of his skills and ambitions. Coleman (1974) states that in an individual's motive pattern his level of aspiration is involved. Well-adjusted people have accurate evaluation of their relationship with the world and develop fairly realistic level of aspiration whereas maladjusted people tend to be unrealistic and set up their goals at too high or too low. The operational definition of level of aspiration based upon the nature of the task is quite suitable and recognizable. Frank (1935) has offered a definition which was adopted by Kurt Lewin. According to Frank the level of aspiration is the level of future performance of an individual in a familiar task announced by him on the basis of his level of past performance in that task.

All aspirations in achieving the means are not of equal in strength. It is possible to predict the strong and weak aspirations. It is observed that remote aspirations are stronger than the immediate aspirations. The positive aspirations are stronger than the negative aspirations, and the realistics are stronger than the unrealistics. The level of aspiration may be measured by a suitable test.

Discrepancies in Aspiration

What motivates a person for best performance? It is the individual's goal-setting behaviour which motivates one for better performance and it is called the level of aspiration. According to Monte (1977) an individual sets up his goal in situations involving different levels of difficulty, which is called level of aspiration performance. The concept, *level of aspiration* was invented by Dembo (1931) and later on it was used by Lewin (1944) in his works. An associate of Lewin, Hoppe (1930) conducted experimental studies on the level of aspiration. He used various types of tasks and employed different types of instruments like, dart throwing, ring tossing and target shooting. He employed several terminologies in his studies of level of aspiration.

According to Hoppe, the difference between what a person can actually do at the present aim or throw and what he is going to do in the next aim or throw becomes his goal discrepancy. If the person sets up his aims on a difficult goal (higher level of aspiration) than his previous achievement it is said to be the *positive goal discrepancy*. On the other hand if the person sets up an easy goal (lowering the level of aspiration) than one he has achieved previously, it is said to be the *negative goal discrepancy*. For example, a person has now scored 6 points and he announces that he is going to achieve 8 points. He has set a positive goal discrepancy of +2. Instead if he announces 5, then he set a negative goal discrepancy of –1.

Hoppe also speaks of another discrepancy called ***attainment discrepancy***. The difference between a person's announcement of his new level of aspiration and the actual achievement is called ***attainment discrepancy***.

Attainment discrepancy also may be positive and negative. It depends upon the relationship between the level of aspiration set by the person and his achievement. For example, suppose a person has announced that he is going to achieve 9 points, but actually he has attained only 7 points. His attainment discrepancy is – 2. Instead if he has announced his level of aspiration as 7 points and has attained 8 points, then his attainment discrepancy is +l. He has achieved above the level of his aspiration.

An important doubt that may impel any one in this context is, what is the condition under which a person sets up his level of aspiration above his past performance (positive) and what has influenced him to set up his level of aspiration below his past performance (negative)? Lewin and his co-workers have answered these two questions by formulating a conceptual model of the person. The person is exposed to forces that push him in various directions. Lewin and his colleagues to explain this phenomenon employed several terminologies such as force, regions, locomotion, valance (value) and life-space.

Aspiration does not indicate how close the person is to achieve his goal. It is what he hopes to reach and what he aims for. The gap between the goal the person has already reached and the goal he hopes to reach may be small. Under this condition his level of aspiration is realistic because the choices are very good that he will be able to reach his goal. When the gap is large the level of aspiration is unrealistic. The choices of reaching the aspired goal become rare as the gap between the achieved and the hoped for goal becomes wider. When the gap between the person's achievement and his hope for goal is large, his self-concept will be severely damaged if he fails to reach the goal. He thinks of himself as a failure. As a result, his subjective judgment may be reinforced by the objective judgment of others.

It is not good to encourage high aspirations without knowing how failure in particular situation would affect a person. Failure to reach any goal is ego-deflating and damaging. Goals vary in importance according to the degree of personal satisfaction. A person's former achievements are a good indication of future performance in the same areas. Some authors design tests in this line. But it need not be always like that. It is possible to determine aspirations by Laboratory studies.

There is considerable amount of value in knowing the level of aspiration of the individual. It focuses the person's attention in the kind of aspiration he has set and makes him aware that he might have developed the habit of aspiring unrealistically high or unrealistically low. It may help him to understand why he is often dissatisfied with himself to the level of self-rejection. Hilgard (1962) is of the view that it is also important to know the strength of his motivation and other factors that compensate the limit and contribute to achievement.

Aspiration and Achievement

Mastering the developmental tasks leads to success which in turn brings happiness, social recognition to the person. Aspirations are the beginning to the chain of activities that lead to successful or unsuccessful achievement. Success must be won by the person's own efforts. Help and guidance are also needed for continuous success. Ability alone is not enough to achieve success. The person must use his ability in the right way at the right time. There are several elements contributing to the complex process that lead to the successful achievement the ***achievement syndrome***. The person must have training under the guidance of an experienced person to know how to make the best possible use of the abilities.

It is important that to achieve success in any activity at any age the person must be flexible – willing to adjust to new roles and undertake new activities. The person must be independent of others that he can make his own decisions. Another syndrome close to independent is that *the person must take risk.* The central element in achievement syndrome is motivation. He must be willing to sacrifice time, effort and immediate pleasures to achieve higher goals and greater success in future. Achievement can be judged objectively, by comparing the person's achievements with those of his peers or subjectively by comprising the person's achievements with his level of aspiration. That is, the person can be objectively a success or subjectively a failure in other way.

STUDIES ON LEVEL OF ASPIRATION

Several studies have been conducted on level of aspiration following the first experimental study made by Hoppe. Lewin (1944) and his co-workers found in their experiment that their subjects in the beginning of the experiment set up only moderate positive level of aspiration. In the presence of the experimenter the subjects have gained some experience and this increasing experience played important role in the success and failure of their attempts. It was observed that the success has induced to raise their aspiration level in most of the subjects. At the same time the continuous failure compelled them to lower their aspiration level.

Level of aspiration experiments were also conducted among the children. Jucknat (1937) used the mazes in his experiment with increasing difficulty among the children. Successful completion of the series of mazes as the attainment of the level of difficulty set up by the children resulted in raising their level of aspiration by 51 percent of the subjects. About 41 percent of them kept their aspiration level at the same levels and only 8 percent of the children lowered their level of aspiration. Jucknat's study was theoretically analysed and the shifts in the level of aspiration of the children was interpreted by Fetinger (1942).

An experiment on aspiration level was conducted by Gould (1939). In this experiment the subjects who were working on a task were asked, how well they have done their work and they were asked how well they would do the task next time? Gould had observed that some persons set up their goal so low that they were sure to succeed.

Studies on level of aspiration were made in school settings. Rosenfeld and Zander (1961) conducted a study among 400 junior high school students studying in tenth grade. It was found in this study that the students generally set up their grade aspirations higher than the marks they have obtained in the previous test. Rosenfeld (1964) has made also another study equating aspiration level with difficulty level of the goal chosen by the person.

The relationship between occupational choice and college success was investigated by Weigard (1953). In his study it was found that in both they had definite success. The goal choice was an effective motivator for the students' success in the college. Worrell (1959) made a study among a group of the college students for a period of four years. He found support for using the level of aspiration for predicting the academic performance of better students and for predicting the inability to complete the college degree course of the unrealistic students. Worrell had observed in his study that the 400 students who were unrealistic have attained poor grades continued to attain poor grades. For Worrell, unrealistic means the student who sets up his level of aspiration higher than his earlier performance, who aspired to get a grade higher than he has attained earlier, and who believed that he could achieve much better than he has made in the past.

There is nothing wrong in aspiring for a better achievement next time. One can work hard and improve his capabilities to achieve a better goal. But with limited capabilities and repeated failures, one should not set up a goal at a too higher level. Studies have been made relating level of aspiration and achievements. It has produced varied effects on aspirations and on achievement.

The effects of success and failure on level of aspiration relating achievement motives was investigated by Moulton (1965). He administered the measures of need achievement and test of anxiety on 93 senior high school male students. These 93 students were divided into three groups on the basis of the scores on those two tests. One-third of the students had higher need achievement than in the test of anxiety. The second one-third of students had shown high scores on both the tests. The third one-third of students had shown high scores in test of anxiety than in need achievement. Each group was asked to choose three degrees of difficulty among the tasks. Then they were handled variedly on the basis of success or failure. Moulton's hypothesis was that abnormal or unrealistic shifts in aspiration level should occur most often for the third group following the success or failure.

The results of this study supported the hypothesis. The variables in aspiration, determined not only the goal setting, but also the degree and the type of shifts in goal setting. There were different degrees in goal setting by this group and also the type of shifts were either too high or too low. Similar study was also made by Medow and Zander (1965) and Diggory and Morlock relating level of aspiration to achievement.

Studies were also made relating level of aspiration to the levels of mental abilities whether level of aspiration has the same meaning and results for different levels of intelligence. Davids and Sidman (1965) tested this phenomenon among the academically successful and unsuccessful groups of high school male students. Mecoy (1963) studied the level of aspirations of educable retarded children.

EXPERIMENT

Study of Goal Aspiration and Achievement

Problem

The purpose of the experiment is to study the goal aspiration and achievement and to determine the level of attainment discrepancy (ADS) and the goal discrepancy (GDS) of a person through co-ordinated skills in a given task.

Method

Subject

Choose a person who has no knowledge about this experiment. Collect his personal information like his/her age, educational qualification, gender, whether he/she wears spectacles or has other deformities, and also his/her socio-economic status.

Material

Prof. Rajamanickam's Target-Dart Test : Application Method. The test consists of a Target Board, a set of 5 Darts weighing 100 grams of 3″ length, metre scale and chalk piece.

Description of the Materials

This is a test designed and fabricated by Prof. M. Rajamanickam to measure the goal aspiration and achievement of a person through co-ordinated skills and accuracy of judgment. The Target is a board made out of rubber sheet/thick card board 2′ × 2′ size fastened on a 3′ × 3′ black plywood board. This board is fixed on a wall, 3 feet above the ground. The surface of the board is painted with white colour on which four circles of different sizes are drawn. The inner most circle is of 3″ in diameter. It is called "Bull". It is divided into two equal parts, *the upper half is black* in colour and *the lower half is white*. The second circle is of 6′′ diameter. It is called "Inner". The third circle is of 9′′ in diameter. It is called "Magpie". The fourth circle is of 12′′ in diameter. It is called "Outer". It is the same as the target used in firing range to give practice and to test firing skills of the soldiers in the Army (Appendix-I). The same Target, fixed on the wall, is also printed to the same size in sheets of paper and used as 'test sheet' (Appendix-II). This 'test sheet' is to be pinned on the Target board on the pattern for performing the experimental task. After the experiment is over the test sheet is to be removed from the Board and attached in the Record Note Book. The Dart mark holes should be indicated in the sheet by drawing a small circle around the holes. The number of Bulls, Inners, Magpies and Outers be counted again. Every experimenter should be supplied with the test sheet and it should be pinned on the Target board. Darts should not be thrown at the Target board when there is no test sheet fastened.

The Dart is 3′′ length, having a brass metal stem of 1′′ length with a sharp needle of 1 ½′′ at the head. At the back of the Dart a feather tail made up of plastic is attached. It is detachable. The feather tail is to facilitate the Dart to move fastly in the direction it is thrown. The Dart weighs about 100 grams. There are five Darts in a set. Each dart makes one throw and five throws constitute one trial.

Method of Using the Materials

The participant's goal aspiration and achievement should be assessed by the ***Application method*** of the Target-Dart test. In achieving the goal the person tries to throw all the five Darts at the inner most circle the 'Bull' of the Target to get the highest points 20. A Dart thrown at the Bull gets him 4 points. In case if he could not throw the Darts at the 'Bull' and if they fall in the 'Inner' or 'Magpie' or on the 'Outer' he gets the points 3, 2 and 1 respectively for each Dart thrown. But if the Dart falls away from the Target (outside the four circles) he gets 0 point. Therefore, the person's goal aspiration should be to throw all the Darts at the 'Bull' to score the highest points out of 20. But what he achieves actually may not be what he aspired. It depends upon how accurately he throws the Darts at the Target and the points he gets. In the application method the person does not get any point if his throw becomes washout, that is the Dart falls away from the Target (outside the 4 circles). The scores are assigned according to the following procedure.

Score Board

A throw at Bull	:	4 Points
A throw at Inner	:	3 Points
A throw at Magpie	:	2 Points
A throw at Outer	:	1 Point
A throw outside (washout)	:	0 Point

Procedure

(a) Preparation for the Conduction of the Experiment

The experimenter should first of all make some preliminary arrangements for the conduction of this experiment. Since the Target is already fixed on a wall the experimenter should make the area congenial for his purpose. The area should be properly illuminated. The room must have sufficient length (at least 18 feet long). The experiment should be conducted minimum at 5 positions of varied distance. The shortest distance should be 6 feet minimum and the positions may be varied with a difference of 2 feet between two positions. Thus the five positions may be 6 feet, 8 feet, 10 feet, 12 feet and 14 feet (see layout Appendix-III).

The experimenter with the help of a metre scale can measure first 6 feet from the place where the Target is fixed, towards the open space and make a mark and draw a long thick line with a chalk piece across on the ground. He should write as 6 feet at the two ends of the line. Similarly at 8th feet, 10th feet, 12th feet and at 14th feet by measuring the distances, thick long lines on the ground should be drawn and write the respective numbers on the two ends of the lines at every position. Now the experimenter should ascertain and satisfy himself that there are five positions with 6ft, 8 ft, 10 ft, 12 ft and 14 ft, distances.

The next step is to place a small table or teapoy on the 6 feet position, putting the two front legs of the table or teapoy along the line. It should be in such a way that the lengthy end of the table/teapoy is 6 feet away from the Target. At the other end the participant should be made to stand close to it. This arrangement is made to prevent the participant from leaning forward at the time of performance. When the task is over at 6 feet position the experimenter should move the table or teapoy to the next position 8 feet and then to the next positions one by one till the task is completed at the last position. This experimental procedure is called ***increasing order***.

(b) Conduction

The experimenter should fix the places by measuring the distances with metre scale. There should be five positions as 6′, 8′, 10′, 12′, 14′. The positions should be marked by the chalk piece with a long line. The subject is to be made to stand in front of the table or teapoy first at a distance of 6 feet. He should stand close to the table or teapoy placing his left hand on the table or teapoy and his left leg on the lower bar of the table or teapoy and stand steadily. Then the experimenter should place the Dart in the right hand of the participant and ask him to hold it with his three fingers supported by other fingers just like holding a pen. The following directions should be given.

> "Hold the Dart steadily with your fingers. In front of you there is a Target at a distance. There are four circles in it from smallest to biggest. Try to throw the Dart at the centre, that is at the smallest circle "the Bull". If the Dart is thrown at the centre you get 4 points. If your throw falls away from the centre in the second bigger circle (Inner) you get 3 points. If your throw falls in the third bigger circle (Magpie) you get 2 points and if your throw falls at the last circle – the biggest (Outer) you get 1 point. If the Dart falls away from the Target you get zero point. You must put sufficient accuracy and force in all your throws, so that the Dart thrown is firmly gripped at the Target. Try to score the highest point by throwing the Darts always at the centre – that is at the 'Bull'. By throwing all the 5 Darts at the Bull you get 20 points – the highest and this is your goal. This should be your aspiration in all the trials. When

you throw all the five Darts you are making one trial. For each trial there are five throws. Before you start throwing the first Dart of every trial you must tell me *what is the goal you aspired.* That is how many points you are going to obtain in all five throws out of the maximum 20 points."

Now the experimenter should test whether the subject has understood the task he is going to perform. After testing the subject verbally the experimenter may permit the subject to have one trial of 5 throws. Then, if the experimenter is satisfied he should ask the subject ***the goal he aspired*** and enter the points in his record book, and give the order, "Ready, go".

The subject should throw all the five Darts on the target one by one and when all the five are thrown, the experimenter and the subject should go near the Target and examine the places where the Darts are thrown and the experimenter should make entry of the points in Table 13.3, which he has prepared already. Then the Darts should be removed from the Target gently by rolling the rod, otherwise the sharp end may be broken. At once the experimenter should make ***a small circle around the hole*** in the Target test sheet, the Dart has made. It is important that the experimenter himself should do this and *the Darts should be with the experimenter.* The experimenter should pass on the Darts to the subject only at the time of the Task. At the *end of every trial the subject should be informed of the points he has scored in that trial.* Then the subject should begin his/her next trial. Before that the experimenter should ask the aspiration of the subject. That is, how many points he/she is going to achieve in this trial. The subject should tell the experimenter his goal aspired. At once the experimenter should make entry in the Tabular column.

When all the five trials are over in that position the experimenter should vary distance to 8 feet and then to 10 feet, 12 feet and 14 feet, and the subject should be asked to perform the same task, at different distances. While lengthening the distance from 6 feet to 8 feet and then to 10 feet, 12 feet and 14 feet, the experimenter should warn the subject that he should put more force in his throws as the distance is getting longer and longer. Since the subject may not be familiar with such throwing task, in order to avoid more unrealistic throws it is better to caution the subject at each position to put more force in his throws. The same procedure should be followed in making entries in Table 13.3. At every position the subject should stand in front of the table or teapoy steadily. *The table or teapoy legs should be placed on the lines drawn to indicate the distance.* Five trials at each position should be made. This means that at every position (5×5) 25 throws the participant makes. The aim of this experiment is to test what the subject aspired and what he/she achieves and to see the level of attainment discrepancy. Aspiration is made first and this is tested in the task assigned whether he achieves it or not. The procedure of relating the present aspiration to the previous achievement for getting the goal discrepancy score (GDS) as stated in some experiments (Hoppe, 1930; Lewin, 1944) is also followed here. This experiment is designed to conduct at 5 positions, each with 5 trials and to assign grades for the individual's achievements in all the 5 positions. There are five grades on the basis of the scores the individual achieves. One must obtain a minimum of 40 points for a pass in the test. Therefore, attainment discrepancy is determined first for every position of 5 trials. The simple idea whether the subject achieves the goal he aspired every time or not, is tested. The goal aspiration is there and the subject is made known the task and he is to achieve and he is also informed of the scores he had obtained in the previous trial. Therefore, the individual may set up his next aspiration (goal) having in mind the scores of the previous trial. Since the performance of the experimental task in this experiment involves several achievement syndromes such as bodily posture, psychomotor skills, co-ordination, manual dexterity, Arm-hand steadiness, wrist-finger movement, speed and accurate aiming, it brings happiness and social recognition on achievement. After all the trials are over the experimenter should get the introspective report from the observer. Finally the experimenter should ask the participant whether he has taken any guidance from the results of the previous trials and benefited by them in the subsequent trials or not. If benefited how he was benefited, if not, why?

Method of Scoring

As given in the preceding section, a throw at the Bull is given 4 points, at the Inner 3 points, at the Magpie 2 points, at the Outer 1 point and for Washout 0 point.

One Trial = Five Throws.
Maximum Points for a Trial : 20

Maximum Points in one
Position for 5 trials (5 × 20) : : 100

Minimum Points to be scored for a pass
in one position : 40

Mean Achievement Scores and Grades

Below 40	-	Poor (Fail)
40 to 54	-	Fair
55 to 69	-	Good
70 to 84	-	Very Good
85 to 100	-	Outstanding

Results

After every trial of five throws the experimenter should enter the results in Table 13.3. The experimenter should ask the participant the goal he aspired before he starts every trial and enter the points in the respective place (in column 4).

Table 13.3 : Score Sheet : Showing the scoring of the Individual's Performance in all five Trials and in all five Positions in the Application Method of Dart-Target Test

S. No.	Position	Trial No.	Goal Aspiration Point	Score for each throw (attained)					Total Score for each trial	No. of Bulls	No. of Wash-outs
				1	2	3	4	5			
1	6 Feet	1									
		2									
		3									
		4									
		5									
2	8 Feet	1									
		2									
		3									
		4									
		5									
3	10 Feet	1									
		2									
		3									
		4									
		5									
4	12 Feet	1									
		2									
		3									
		4									
		5									
5	14 Feet	1									
		2									
		3									
		4									
		5									
Total											

Then the achieved score for each throw in the respective column should be entered. There is one column for each throw (*i.e.,* 1, 2, 3, 4, 5). The experimenter as and when he conducts the experiment can complete most of the columns in Table 13.3. The goal aspiration points he can enter by asking the subject before he starts every trial. Then the points for the throws can be entered as soon as one trial is over. The number of Bulls and Washouts should be given in the respective column. This is to assess the subject's accuracy or the aim in every trial of the test. The maximum score in all 5 trials in one position is 100. Then the achievement scores of the subject in all five trials and in all 5 positions should be worked out. The total number of Bulls and Washouts in 125 throws the subject has aimed should be entered in the respective columns.

Then the aspired scores and achieved scores for all the five trials should be entered in Table 13.4 under each position.

The total aspired scores and achieved scores for all the five positions should be entered separately. From these two, the *attainment discrepancy score* (ADS) for each position should be determined. This discrepancy score may be positive (+) or negative (–). If the attainment score is higher than the aspired score then the attainment discrepancy is positive (+). Instead if the attainment score is lower than the aspired score then the attainment discrepancy score is negative (–). Determine this for every position and for all five positions and enter in Table 13.4.

Table 13.4 : Showing the Scores of the Goal Aspiration and the Achievement and the Attainment Discrepancy in all 5 Trials at all 5 Positions in the Application Method of the Target-Dart Test. (The Attainment Discrepancy Scores should be worked out from the entities made in the **Score Sheet**) (Table 13.3 : Score Sheet)

S. No.	Trial No.	Variables	Positions									
			6 Ft. 20 Pts.	Att. Dis. Ss. + or –	8 Ft. 20 Pts.	Att. Dis. Ss. + or –	10 Ft. 20 Pts.	Att. Dis. Ss. + or –	12 Ft. 20 Pts.	Att. Dis. Sc. + or –	14 Ft. 20 Pts.	Att. Dis. Ss. + or –
1	1	Aspired										
		Achieved										
2	2	Aspired										
		Achieved										
3	3	Aspired										
		Achieved										
4	4	Aspired										
		Achieved										
5	5	Aspired										
		Achieved										
Total Goal Aspired Score												
Total Achievement Score												
Attainment Discrepancy + or –												
Mean Achievement Score												
Number of Bulls												
Number of Washouts												
Pass or Fail												
Remarks (Grade)												

From the total find out the mean achievement score for each position by dividing it by five. State the mean achievement score, whether it is above 40 or below 40. On the basis of the marks, the grade the subject has achieved are determined.

State the number of Bulls and number of Washouts in 25 throws for each position.

Similarly the *goal discrepancy score* (GDS) should be determined. It is the difference between the scores of the present goal aspiration and the score attained in the previous trial. If the present goal aspiration score is higher than the score attained in the previous trial then the goal discrepancy score (GDS) is positive (+). Instead, if the present goal aspiration score is less than the score attained in the previous trial then the goal discrepancy score (GDS) is negative (–). Determine the GDS for every position and for all five positions and present the scores in Table 13.5. Find out the total discrepancy score for each position and also work out the mean.

Table 13.5 : Showing the scores of the Goal Aspiration and Achievement and the Goal Discrepancy in all Five positions. Work out the Goal Discrepancy Score from the Score Sheet (Table 13.1).

S. No.	Trial No.	Variables	Positions									
			6 Ft. 100 Pts.	Goal Dis. Sc. + or –	8 Ft. 100 Pts.	Goal Dis. Sc. + or –	10 Ft. 100 Pts.	Goal Dis. Sc. + or –	12 Ft. 100 Pts.	Goal Dis. Sc. + or –	14 Ft. 100 Pts.	Goal Dis. Sc. + or –
1	1	Aspired										
		Achieved										
2	2	Aspired										
		Achieved										
3	3	Aspired										
		Achieved										
4	4	Aspired										
		Achieved										
5	5	Aspired										
		Achieved										
Total Goal Discrepancy in each position												
Mean G.D.S.												

N.B. : The *Goal Discrepancy Score* is the difference between the Achieved Score of the Previous Trial and the present Trial's Aspired Score. Therefore, in determining the Goal Discrepancy Score discord the Aspired Score of the FIRST TRIAL and the Achieved Score of the LAST TRIAL. Hence there will be only FOUR Goal Discrepancy Scores for FIVE Trials.

Collect the group data from other experimenters and enter them in Table 13.6. In Table 13.6, column three enter the total achievement score for all 25 trials the subject has performed in this experiment. Also find out the total aspired score far all the 25 trials. Find out the attainment discrepancy score (ADS) and also the goal discrepancy score (GDS). State whether it is positive or negative. State the total number of Bulls each subject achieved in 25 trials and also the total number of Washouts of every subject. State the how many passed the test and how many failed.

Discussion

The discussion part should form the answers to the question what the subject hoped to reach and what he has achieved. While discussing experimental results the experimenter should observe the scores at every stage and make a critical note of the performance. Since the task of this experiment involves several skills to be manipulated in accordance with his/her goal aspiration the score obtained in every trial is important to assess his/her skills.

Table 13.6: Showing the Total Scores of the Goal Aspiration and Achievement and the *Attainment Discrepancy and Goal Discrepancy* of the Group

S. No.	Name	Total Achievement Scores in all 25 Trials	Total Aspired Scores in all 25 Trials	Attainment Discrepancy Scores in all 25 Trials + or –	Goal Discrepancy Scores in all 25 Trials + or –	Total No. of Bulls in all 25 Trials	Total No. of Wash-outs in all 25 Trials	Pass or Fail and Grade	Remarks
1.									
2.									
3.									
4.									
5.									
6.									
7.									
8.									
9.									
10.									
11.									
12.									
Total									
Mean									

Discuss the results, taking into account of the various aspects of the individual's scores. Discuss whether he/she scored the minimum marks for a pass in every position. What was the nature of his/her aims? Was he/she really interested in making progress in the trials or apathetic in his/her approach.

At every distance 6 ft, 8 ft, 10 ft, 12 ft and 14 ft, the individual's scores should be examined to see whether the individual was progressing forward or backward. It is also to see whether distance had any adverse effect in his throws and achievement. Compare the scores the individual has attained at every position.

Observe, state and discuss whether the achievement of the previous trial had any positive effect or negative effect in the subject's next trial either in setting up his goal aspiration or in making his achievement. Discuss whether the total aspiration score and the total achievement score varies from trial to trial and distance to distance.

Discuss the subject's attainment discrepancy score for every distance. Whether it is positive and negative. Discuss whether the varied distances had any effect in the attainment discrepancy score. If so whether it is positive or negative?

Discuss the mean attainment score of the subject of all the positions. Whether it is pass or fail? What is the grade your subject has obtained? State the reasons if any if he has failed. Discuss what were the factors that contributed for better achievement and poor achievement. Has your subject set his goal aspiration unrealistically? If so, what must be the reason for this? At what distance your subject has achieved more number of Bulls and at what distance he made more number of Washouts?

Discuss the goal discrepancy score (GDS) for each trial, and each position. Whether it is positive or negative. State the reason if any. Compare the goal discrepancy scores between positions. Whether the level of goal aspiration set by the individual was realistic or unrealistic. Discuss the mean goal discrepancy score of all the positions comparing the mean attainment score. State whether it is positive or negative?

State the reason, if any for the subject to have unrealistic goal aspiration. Was it taking place even after, knowing the previous attainment score? Did you observe any frustration-aggression in the subject in his/her failure to achieve the aspired goal. How it was expressed by the subject?

Discuss the scores of the individuals in comparison with the group mean. Whether the individual was above the group mean or below the group mean in his achievement scores. The individual's accuracy of aim or lack of accuracy in his achievement is to be discussed in terms of the number of Bulls and Washout he achieved. How many subjects in the group get pass marks and how many get fail marks. State the reasons. Find out what type of subjects scored better grade and what type scored poor grade. Find out the reasons and discuss.

Discuss in how many trials the attainment discrepancy score (ADS) is positive and in how many it is negative? Discuss in how many trials the goal discrepancy (GDS) is positive and in how many it is negative. Critically analyse the positive and negative ADS and GDS. Discuss the observer's introspective report in relation to his achievement. Discuss whether the individual was benefited by knowing the score of the previous trial in the subsequent trials. Compare your results with the results of earlier studies in this area. State and discuss how individuals differ in their goal aspirations and in achievements. Whether the differences were based on sex, physique, socio-economic status? What type of subjects made outstanding achievement and what type made poor.

Conclusions

State whether the individual has achieved the goal aspired. Whether the individual was benefited by knowing score of the previous trial? What was his grade achievement? Pass or fail? Whether he/she is below the group mean or above the group mean. How many in the group get pass and how many fail? Are there any outstanding achievements? State them.

APPENDIX – I

TARGET-DART TEST (APPLICATION METHOD)

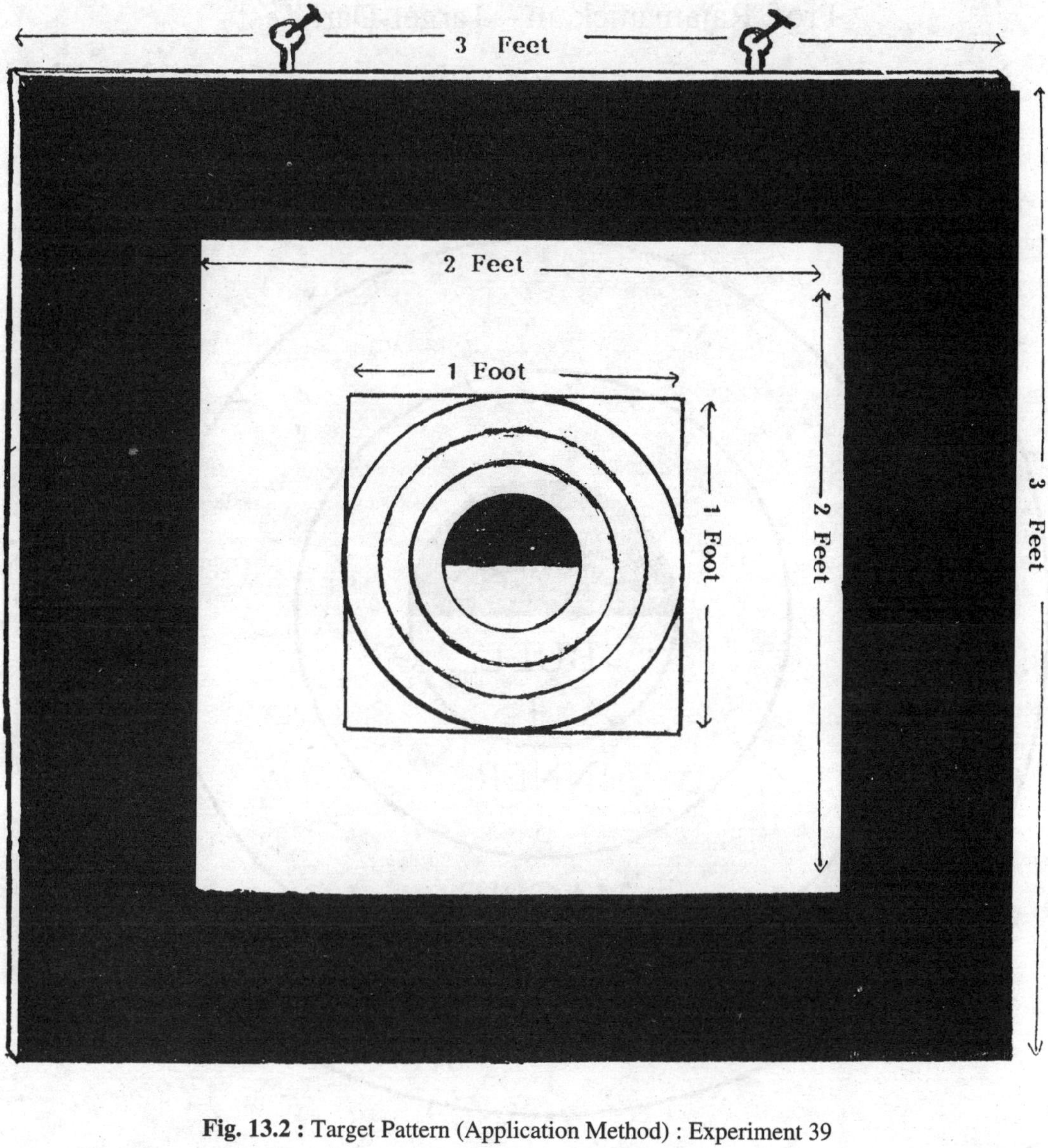

Fig. 13.2 : Target Pattern (Application Method) : Experiment 39

APPENDIX – II

TEST SHEET : TO BE FIXED ON THE TARGET (APPLICATION METHOD)

Prof. Rajamanickam – Target-Dart Test

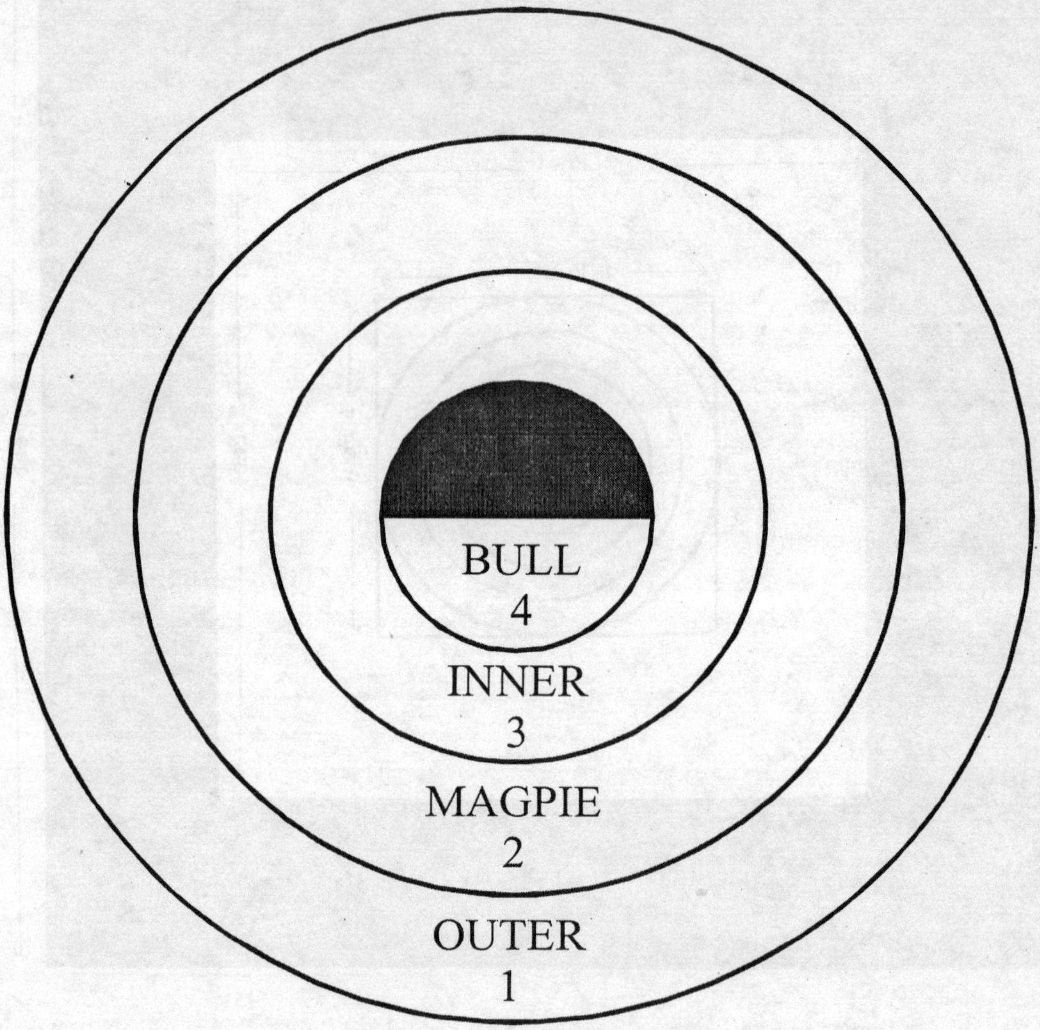

Fig. 13.3 : Test Sheet (Rajamanickam – Target-Dart Test)

APPENDIX – III

LAYOUT OF THE TARGET-DART (APPLICATION METHOD) EXPERIMENT

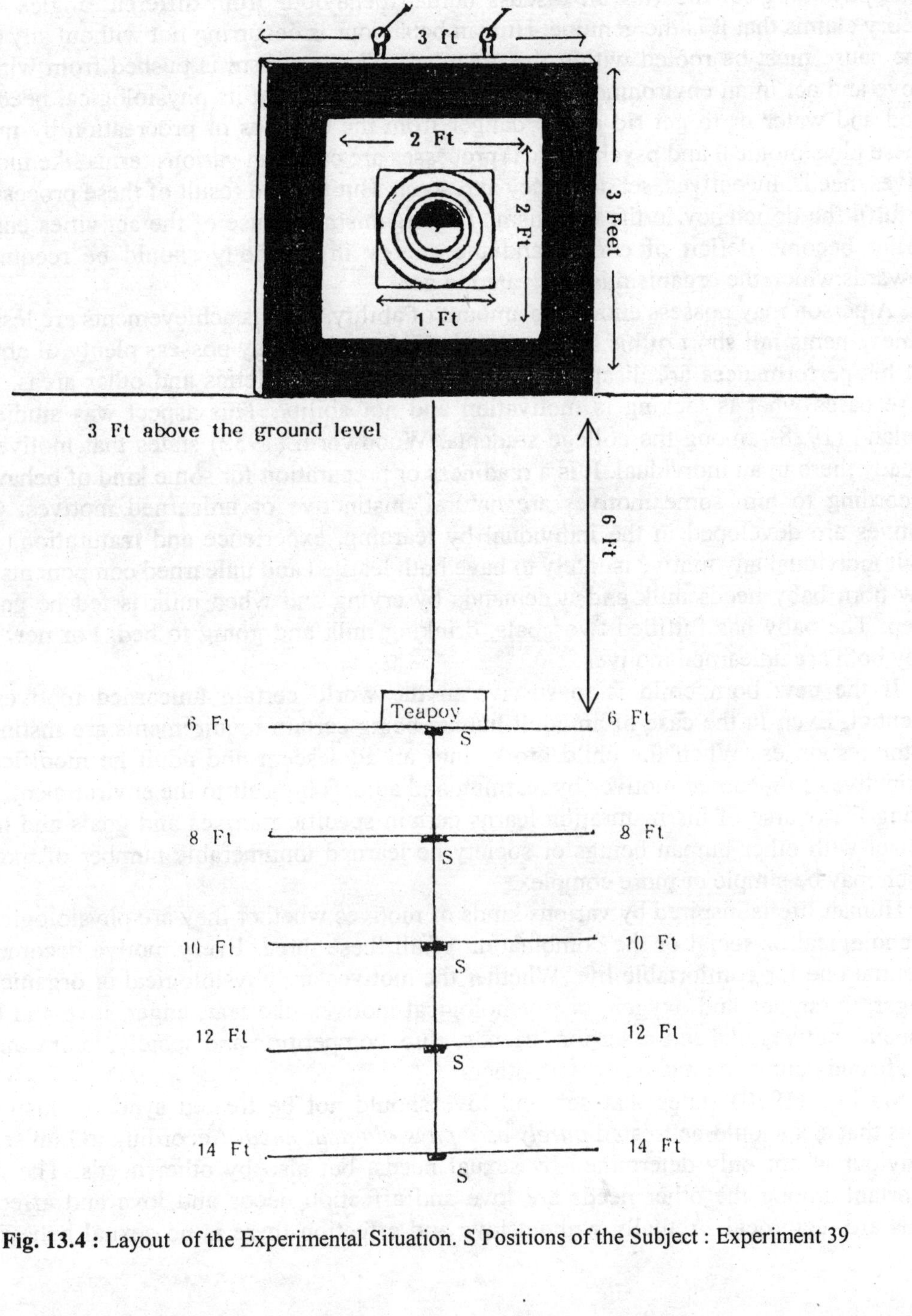

Fig. 13.4 : Layout of the Experimental Situation. S Positions of the Subject : Experiment 39

EXPERIMENT 40 : ASSESSING THE STRENGTH OF VARIOUS MOTIVES

Human behaviour is explained through different psychological theories. The instinct theories, the behaviouristic theories, social learning theories, cognitive theories and neurophysiological theories all discuss human behaviour from different angles. Every theory claims that it is the genuine. Human behaviour is occurring not without any cause. The cause must be rooted within the organism. The organism is pushed from within to move and act in an environment. Whether it is for achieving its physiological needs like food and water or to get rid of the danger from the enemies or procreation by mating. These physiological and psychological processes are called in various terms like motives, drives, needs, incentives, set, tendency and so on. But the end result of these processes is to fulfil the deficiency in the organism. The organism because of the activities engaged earlier become deficit of energy and the energy in the body should be reequipped. Towards which the organism is motivated to act.

A person may possess enormous amount of ability, but his achievements are less. His achievements fall short of his ability. Similarly, a student may possess plenty of abilities but his performances are disappointing in his study, in athletics and other areas. In all these cases what is lacking is motivation and not ability. This aspect was studied by Troland (1928) among the college students. Woodworth (1952) states that motives are already there in an individual. It is a readiness or preparation for some kind of behaviour. According to him some motives are natural, instinctive or unlearned motives. Other motives are developed in the individual by learning, experience and maturation. In an adult individual any motive is likely to have both learned and unlearned components. The new born baby needs milk and it demands by crying and when milk is fed he goes to sleep. The baby has fulfilled two goals, drinking milk and going to bed. For new born baby both are unlearned motives.

If the new born child is to survive in the world certain unlearned motives are essential. Even in the case of an adult human being certain requirements are instinctive motor responses. When the child grows into an adolescent and adult he modifies his instinctive or unlearned motives by learning and adjusts himself to the environment. Man during the course of his maturation learns certain specific motives and goals and in his contact with other human beings or society he learned innumerable number of motives which may be simple or more complex.

Human life is inspired by various kinds of motives whether they are physiological or psychological or social or the combination of all these three. Every motive becomes an essential one for comfortable life. Whether the motives are physiological or organic like hunger, thirst, sex and oxygen, or psychological motives like fear, anger, love and hate, or social motives like self-assertion, friends, club, competition and society, man wants to be safe and happy and make gay with others.

Maslow (1970) states that sex and love should not be treated synonymously. He wants that sex should be treated *purely as a physiological need*. According to him sexual behaviour is not only determined by sexual needs, but also by other needs. The most important among the other needs are love and affection needs and love and affection needs are reciprocal. Actually without love and affection there is no sexual behaviour.

Maslow states that sexual behaviour is ***multi-determined*** because it is also determined other needs, *the most important are* ***love and affection needs***.

Strictly speaking, unlike other physiological needs (hunger, thirst, oxygen, rest and sleep) sex is determined more than one component. Therefore, sexual behaviour is a multi-stage and multi-factor operation. There are more psychological and social components. Normally sexual behaviour involves physical maturity (physiological), mutual attractions, awe, piety, tolerance, love and affection (psychological). Sex is also a part of group or social life. Unlike physiological needs like hunger, thirst etc., the needs cannot be fulfilled without the company of the opposite sex. Meeting takes place in a social function, there is self attraction, it results in the development of love and affection which finally leads to sexual behaviour. Hilgard (1962) has classified sex as a social motive since it involves also social factors. Therefore, sex cannot be purely a physiological need.

While presenting his heiarchy of motives Maslow makes self contradiction in explaining certain concepts, of which sex is one. Having said that sexual behaviour is *multidetermined and the chief determinants are love and affection* needs, Maslow stresses that sex may be studied purely as a physiological need. How can these two ideas can go together? There are more reasons to state that sex is a psychological and social motive. Sex is more associated with feelings and emotions and group life. Hence it may be considered as a psychological motive or social motive.

Strength of Motives

Man is not always seeking to fulfil one type of needs or motives. When one, which is very essential is achieved he is satisfied and goes to fulfil other important motives. Sometimes man may come into contact many motives and he may not be in a position to decide which is more important and which is less important and which is more needed and less needed or which is to achieve and which is to be dropped. In such situations he meets with conflict of motives, either he has to give up the first or the second. Therefore, man is sometimes put in a dilemma with regard to his choice of motives. Conflict of motives may take place if only two or three motives are of the same kind or strength. Motives of varied kinds having different values may not attract the individual with the degree.

Any strong motive may dominate the organism and may have influence in his behaviour. It may be for the time he needs. In another occasion, in another situation some other motive may be dominant and may have influence in his behaviour. Therefore, no one motive may be dominating in an individual and influencing his behaviour all the times. Woodworth (1952) says that the combat motive of fighting against suppression and slavery may be the strongest of all motives.

In normal situations people may be spending more money on the motive which may be dominating in their life. For example there is always some craziness for entertainment and events associated with motion pictures and movie stars. The people of the United States are not exempted from this type of craziness. Thorndike (1940) states that in 1920s the people of United States spent twice as much money for entertainment of all kinds than for education. The entertainment motives at this time were very much stronger than education for the people of United States. But after sometimes it was found that many students were spending more money, time and energy on the motives of education than on entertainment motives. At this time entertainment motives were considered as social motives.

Assessing Strength of Motives

The strength of motives may be assessed by designing suitable test. Psychologists have designed and developed tests for assessing the strength of motives and experiments have been conducted most among the college students. One way of assessing the strength of motives is by collecting a long list of motives and asking the person to arrange them in the order of his preference of the motives in the list. The motives which are dominating and strong in him may receive the top preferences and the motives which are weak, having less influence in him may receive least preferences in the list.

Further he may be also asked how much he is ready to spend for the satisfaction he is going to have from the motive and how long he can work for it? In order to secure the particular motive what other comforts or pleasures he is going to give up? Answer to these questions may reveal the strength motive which is dominating in the individual and the nature of influence it is going to have in him. In addition it will also reveal the comforts and pleasures he is ready to sacrifice in order to enjoy the satisfaction he is going to have by the particular motive. But it also depends upon the individual. The answers to such questions may vary according to individual differences. The same motive may not attract persons of different sex, age, qualifications, social status, self-esteem, his earlier achievements socio-economic status, racial and communal stocks.

EXPERIMENT

Problem

Assessing the relative strength of physiological, psychological and social human motives such as hunger, thirst, fear, sex, self-esteem and self-actualization.

Method

Participant

Choose a person who has no knowledge about this experiment. Collect his/her personal informations such as age, sex, qualifications, occupation, his/her main entertainment interest, socio-economic status of his/her family.

Material

Six list of stimulus words. Ten words from each area of the motives. A list of five stimulus words for trial, stop clock, paper and pencil.

Description of the Material

The material for assessing the strength motive consists of six lists of words. Each list is having ten words. These words are called stimulus words. But they are also the response words since the participant has to report the same words. The six lists of words are from three areas like physiological, psychological and social. From physiological motives two motives are selected. They are hunger and thirst. From psychological motives two motives are selected. They are sex and fear. From social motives, self-esteem and self-actualization are selected for this experiment.

In the case of hunger and thirst, the food materials that can satisfy hunger and thirst are used. In the case of sex and fear, components related to sexual behaviour and components can cause fear are used as stimulus materials. In the case of self-esteem, all those related to assert the self are used as stimulus materials. And in the case of self-actualization the components of self-actualization as suggested by Maslow are used as stimulus materials. In certain motives more than ten words are given and the experimenter can select any ten words as stimulus words (Appendix-IV).

Procedure

This is a word list method. The participant is to be seated comfortably in the presence of the experimenter. Then the experimenter should give the following instructions to the participant.

> "Now I will read out a list of words to you. As soon as I finish the list you must at once tell me ***those words which occur to your mind*** without taking much time: First I will give you a preliminary trial by calling some words and you can tell the words that strike in your mind".

Then the experimenter should call out a few stimulus words and at once should start the stop clock to run. The time should be noted down in seconds when the participant start reporting the words. The experimenter should make the note of the response words and the time he has taken to report. If the experimenter is fully satisfied with the performance of the participant he can start the test. If the participant commits any mistake the experimenter can correct him as to how to respond.

The experimenter should keep a Table prepared for making entries of the response words and response time. Then the experimenter should begin with first list of words by reading out all the ten words in the list one by one clearly audible to the participant. He should give a caution, "ready" and start reading the words and simultaneously should start the stop clock. Since the response time will be in seconds the experimenter should watch the stop clock carefully to get the correct time. As soon as the participant reported a word the experimenter should enter the word and also the time he has taken for this in the respective columns of Table 13.7.

When the first list is over with ten words, then the experimenter should go to the second list of words. The experimenter should tell the participant the following instructions: "The first list of words is over. Now we go to the second list of ten words. You must do the same thing what you have done previously. Now ready." Then the experimenter should start the second list and do the same thing by noting down carefully the response words and response time. The same procedure should be followed for other four list of words. After completing all the six list of words the experimenter can allow the participant to go away. Sometimes the participant might not have reported all ten words of the list. This means whatever words left out are not important for him. Therefore, those words may be left out from the response words.

Table 13.7 : Showing the Response Words and Response Times of the individual in the experiments on Strength of Motives

I. PHYSIOLOGICAL

S. No.	Stimulus words for HUNGER (1)	Response word	Response Time in seconds	Stimulus words for THIRST (2)	Response words	Response Time in Seconds
1.						
2.						
3.						
4.						
5.						
6.						
7.						
8.						
9.						
10.						
Mean						

II. PSYCHOLOGICAL

S. No.	Stimulus words for SEX (3)	Response word	Response Time in seconds	Stimulus words for FEAR (4)	Response word	Response Time in seconds
1.						
2.						
3.						
4.						
5.						
6.						
7.						
8.						
9.						
10.						
Mean						

III. SOCIAL

S. No.	Stimulus words for SELF-ESTEEM (5)	Response word	Response Time in seconds	Stimulus words for SELF-ACTUALISATION (6)	Response word	Response Time in seconds
1.						
2.						
3.						
4.						
5.						
6.						
7.						
8.						
9.						
10.						
Mean						

Table 13.8 : Showing the group data in the experiment on Strength of Motives in three areas

S. No.	Name	Mean Response time in seconds					
		1 Hunger	2 Thirst	3 Sex	4 Fear	5 Self-esteem	6 Self-actualisation
1.							
2.							
3.							
4.							
5.							
6.							
7.							
8.							
9.							
10.							
Mean							

Results

The response words and response times, the experimenter has already noted down in Table 13.7. Then the mean response time for each list of words should be worked out by dividing the total time by the number of words reported by the participant. Compare the first mean response time the participant has taken with that of the second, third, fourth and so on. State for which list of words he/she has taken the least amount of time and for which he/she has taken more amount of time. Collect the group data from other experimenters and present them in Table 13.8. Work out the group mean for all six response times.

Discussion

Discuss the individual's scores in terms of time taken by him to respond to the various stimulus words. State and discuss the motive for which he/she has taken shortest time to respond and the motive for which the longest time he/she has taken. What does it indicate? State and discuss the individual's strength of motives in the order from highest to the lowest of the six motives. State the strongest and the weakest motive in each list and state the reason for it. In the group mean where the individual stands? Is he/she above the group mean in all the motives or below the group means. In the group who has taken the shortest time and who has taken longest time to report the total response words.

State and discuss which word was responded with the shortest time in the group? How many of them reported the same word? What it indicates? State the order of preference of the group mean and discuss it with reference to the nature of each motive. Do you find any variation in the score on the basis of age, sex, qualification, socio-economic status of the participant? Who has taken the least time for self-actualisation? Discuss it in relation to his personal factors. Generally which motive dominated the individuals and which motive dominated the group. In the group state the strongest response word in each list and discuss the reason for the same.

Conclusions

State the strongest and the weakest motive of your participant. What is the strength motives of the group? Which motive dominated the group? State the individual differences if any in the strength of motives.

APPENDIX – IV

The following six lists of words are from three areas like Physiological, Psychological and Social. There are two motives from each area.

I. Physiological motive Stimulus words for Hunger (1)	II. Psychological Motive Stimulus words for Sex (3)	III. Social motive Stimulus words for Self-esteem (5)
1. Vegetable briyani	1. Love	1. Self-assertion
2. Lime-rice	2. Meeting	2. Mastery
3. Curd-rice	3. Glamorous	3. Protest
4. Chicken briyani	4. Marriage	4. Demonstration
5. Sappathi-Guruma	5. Embarrassing	5. Strike
6. Mutton-briyani	6. Sweet heart	6. Revolution
7. Pepper-rice & vadai	7. Sexual union	7. Social prestige
8. Puri-potato/Guruma	8. Jilting	8. Winning award
9. Thosai roast	9. Strong affection	9. Winning prize
10. Bread and egg	10. Kissing	10. Unrestrictedness
		11. Aggressive
		12. Title and diploma
Words for Thirst (2)	**Words for Fear (4)**	**Words for self-actualisation (6)**
1. Water	1. Stabbing	1. Self-acceptance
2. Milk	2. Murder	2. Spontaneous
3. Lime Juice	3. Tiger	3. Self-determined
4. Orange Juice	4. Enemy	4. Appreciative
5. Apple Juice	5. Killing	5. Peak experience
6. Badam Juice	6. Cobra	6. Humanistic
7. Soda	7. Wild elephant	7. Friendliness
8. Ginger	8. Death	8. Democratic
9. Butter milk	9. House breaking	9. Sense of humour
10. Ice-cream	10. Burning alive	10. Creative
		11. Culturally detached
		Justice

EXPERIMENT 41 : ASSESSING THE INFLUENCE OF KNOWLEDGE OF RESULTS IN LEARNING

Knowledge of Results as Motive

In the preceding sections we have discussed motivation and motives and we can find that motivation can cut across almost all the topics in psychological studies. The importance of motivation has been felt very much by the educational psychologists and they have studied extensively relating it to learning.

Human behaviour is goal oriented and every behaviour pattern is the cause of some goal seeking activity. Attaining the goal is to fulfil needs that have aroused in the organism internally. As some deficiency has occurred because of the previous activities of some type, the organism must go in search of its goal. It moves and acts as there is a drive that impels it to action. Psychologically it may be stated that it is a motive or a set that predisposes the organism for some activities, through which to attain its goals. Motivation, therefore, involves certain conditions activate the organism and also regulate the behaviour constantly. The behaviour is directed towards the goal and the goal is highly related to the condition that drives the organism for action. Edwards and Scannell (1975) state that motivation, just like intelligence, is a broad inclusive term may cut across many things to play its role.

The role of motivation in learning has been studied by all the learning theorists. Thorndike (1932) one of the early theorists of learning worked with simple motivating drives like, hunger, thirst and avoiding pain. These elementary drives are used in animal learning because it is very easy and simple to control the experimental situation and measure the motivating effect on learning. Candland (1968) states that by understanding clearly the laws that govern the simple motivational states in learning, the learning theorists believed that the complex human motives can be understood. But some are of the view that the motivating factors in animal learning cannot be correspondingly applied to human beings as animals have only one drive at a specific occasion whereas human beings may have many. Therefore, the motivating factors in human learning may vary considerably from the factors applied on cats and rats.

Since human motives are complex, the values of them are not always recognized and understood by the learner. The learner's motives may change from time to time. Garrett (1975) illustrates this point from the behaviour of a hungry cat. The cat is a very docile animal. If it is fed very well it will lie calmly inside the cage. At the same time a hungry cat, however docile and tamed it might try to run away rashly to escape, even when food is placed outside the cage. The hungry cat is highly motivated to escape from the cage more than the food in front of it. The hunger drive starts the activity and food is the goal and the organism on attainment of it keeps itself quiet. In the activity of the attainment of the goal the organism learns to win over several hazards. Thus motives activate the organism for behaviour and maintain the behaviour by making the organism to learn certain things and learning becomes possible. But the learning must be done by the learner (organism) himself.

The motive to be effective depends upon several factors like intelligence, interest, personality traits, age, sex, emotional state, experience and other situational factors of the learner. Human beings learn certain things quickly if they find that it would bring them certain pride or honour or advantage to make their business more prosperous, or powerful. Therefore, if pride or honour or wealth or power is the motive, learning certain skills or knowledge can accomplish a person, he would learn it by all means. Certain person was interested in playing football. A tennis coach wanted him to take up tennis. He reluctantly took up tennis and he found himself very much proficient in that game and made lots of money by winning the matches. This money made him to further learn and improve the games. This person was Michael Stitch of Germany who won the Wimbeldon trophy in 1991. What made him to learn and proficient of the game was money. Hence money and fame motivated him to learn and to develop his skill in tennis. It has also earned for him pride, ambition, prestige and place among world renowned persons. Garrett says that social motives are derived motives and they are important in maintaining the behaviour which leads ultimately to learning. Therefore, motives play important roles in human learning though they are considered to be very complex. But once the learner understands the value of it he puts enormous amount of energy in it and exhibits his talents in achievement.

Set or attitude is one of the important components associated with learning. With a favourable set or attitude the learner can show better progress. Since learning always involve selection of the material, if the learner likes the material then learning becomes easy and rapid. Further, the degree of interest with which the learner approaches the material may facilitate learning. Understanding the values of the material to be learned may also make the learning task easy and worthy.

Man, unlike animals is interested in learning complex skills. Learning skill is for efficient way of doing. Anything we do might be done skilfully or ineptly. But usually a skill learned can be improved and developed for better performance. There are two results in learning a skill. Firstly, a new skill is learned and secondly a habit of using it appropriately where it is necessary. This principle may be applied to any skill. We know that many human skills are combination of several skills. There may be certain components of skills common to more than one skill. These skills can be easily learned and once developed can be applied in some new situations.

In certain activities the motor skills are involved and there may be motions of varied types. Merely making the motion does not involve learning. For example a simple motor activity like writing involves certain motions or movements. Right from the period of childhood one is engaged in writing. Writing the exercises, writing letters, writing examinations and so on. With so much practices one has learned the skill to make improvement in his writing. In improving the writing skill the student may make a mistake each time and may recognize the mistake and then he may make a correction. Thus he has chance for self-correction. If the student fails to recognize his mistake then he will continue in practising his errors. But if the student wanted to improve his writing it helps him to learn by making him to pay more attention in the practice. Unless there is such motivation, learning to improve his writing will never take place.

Improvement, whether it is in hand writing, or some other skills, will not take place unless the individual recognizes his mistake and understands what mistake he has made. In developing skill like handwriting the individual himself gets the information directly. There are tasks in which the individual needs the help of others. The student needs the help of the teacher. The teacher helps the student to identify the mistake and also to correct the mistake. The student is made to know his mistake and also to correct it. Thus in a task the student gets the knowledge of right and wrong and also with the help of right knowledge he corrects his mistakes. This process is called feed-back system. If the feed-back is clear without any hesitation on the part of the student, also carried out in the task then there occurs more efficient learning. After every response or act of performance the individual comes to know the result of his performance and also how far it is correct and incorrect. The learner, thus, is testing his own performance and gets an image of the act he should perform next. This type of motivation makes the learning more accurate. Hence the knowledge of the results of the individual's previous act motivates him/her for better performance. Having this as a hypothesis experiments have been devised.

Studies of the Influence of Knowledge of Results

In the field of education and psychology experiments have been conducted using certain incentives for better performance and also providing the results of the previous act to the subjects better performances are expected. It is presumed that if the student is providing with the knowledge of his performance his further performance is having influence in his attitude or approach. Knowledge of results, it is believed, induce some kind of attitude and degree. Edwards and Scannell (1925) state that knowledge of results is a kind of reinforcement though it appears to be facilitating in its effect. The phenomenon of knowledge of results facilitates learning is not a new one. This has been used in studies for very long time. In the first quarter of twentieth century Judd (1907) used knowledge of results in his studies. Simultaneously McGaoth and Irion (1952) studied the effect of knowledge of results in the classroom. The students, in their studies received information about their progress in an unsystematic manner. Therefore, whatever influence knowledge of results caused were not properly analysed.

A different type of study has been made by Spitzer (1939) using the knowledge of results as determining factor in learning and retention. The test was conducted on a large sample of sixth grade students. They were divided into eight groups. A pre-test was conducted first and the subsequent tests were matched. The post-tests were conducted with varying intervals, right from immediately after the pre-test to two groups of students, after one day for one group after seven days for one group, after fourteen days for one group, after twenty-one days for three groups, after twenty-eight days for two groups and after sixty-three days for three groups. Groups one and two have taken three post-tests, groups three and four have taken two post-tests, groups five and six have taken two post-tests, group seven and eight have taken one post-test.

The results indicated that the post-tests conducted immediately after the pre-test had better effect on retention. The longer the duration between the pre-test and post-test the moderate the influence of the knowledge results had. Post-tests after three weeks time had poor influence. Knowledge of results have been used in the experiments by Angell (1949), Sassenrath and Garverick (1965), Patterson and Brewer (1963). Wallace (1964) assessed the effects of feed-back on stating expectations of what comes next. Brackbill, Wenger and Wilson (1964) studied the effects of delayed assessment on young children. Blaker (1965) did not agree with the results of Brackbill's study. English (1969) found differential effects of time of delay among college students.

The effects of knowledge of results were also studied in programmed instruction. Knowledge of results were used by Stolurow (1961) in machine teaching and related it with some passage. Morrill in his study of using knowledge of results in programmed instruction has observed that many facts have to be investigated though he accepts the procedure and knowledge of results. Researchers like Krumboltz and Weisman (1961), Seagoe (1956) and Bruner (1964) have studied the influence of knowledge of results in programmed instruction.

EXPERIMENT

Problem

Assess the influence of knowledge of results in the task of reproduction of constant length of lines and compare it with the reproduction of the same line without knowledge.

Method

Participant

Choose a person who has no knowledge about this experiment and collect his/her personal information such as age, sex, qualifications, whether wearing eye-glasses, deformities if any and so on.

Materials

A white cardboard of one square foot with a straight line of 20 centimetres (200 millimetres) at the centre, a metre scale, stop clock, paper and pencil, one inch width and one foot long plain cardboard.

Procedure

This experiment is to be performed in two series. The first series is performed without the knowledge of results of the performance. The second series is performed with the knowledge of results of the performance.

The participant should be seated comfortably in a well illuminated room in front of the experimenter. Then the experimenter should give the following instructions to the participant: "Listen here. I show you a white cardboard in which you will see a straight line at the centre. You will have to look at it for one minute and draw twenty lines of the same length. After drawing the first line you will have to fold the sheet of paper behind and you should not see that line drawn previously. You should do like this every time you draw a line. In the same manner you will have to draw twenty lines.

As soon as you finish drawing twenty lines you should hand over the paper to me. You can use this cardboard bit for drawing the straight line. After giving this instruction the experimenter should give one long sheet of paper and one pencil and the one inch-width one foot long cardboard to the participant and asking to get ready. Then the experimenter should show the cardboard with a straight line of twenty centimetre to the participant with a caution, "now ready". At once the stop-clock should be made to run. Exactly after one minute the cardboard should be withdrawn from the sight of the participant. When the participant has finished his task the experimenter should collect the paper in which the lines are drawn.

For the second series the experimenter should get twenty bits of paper cut from the large sheet of paper and hand over them to the participant. Then the experimenter should give the following instruction for the second series.

> "Now, again I will show you a line drawn on a cardboard for one minute and your task is to draw the same line in a bit of paper and give to me. I will measure the length of the line and tell you the length of it. Then you will have to draw the second line in another bit of paper. In the same manner you will have to draw twenty lines in twenty bits of paper."

Then the experimenter should hand over the twenty bits of paper and ask him, "ready, go". At once he should show the cardboard line to the participant and also start the stop clock. After one minute the experimenter should withdraw the board and the participant should carry out the task. As soon as one line is drawn the experimenter should measure it with the metre scale and tell the participant whether the line is longer or shorter or equal to the line shown. This he should do for all twenty lines. As soon as the participant finished the task he/she should be asked to give the introspective report and also a statement whether he/she was benefited by the knowledge of results of the previous performances. Then he/she may be allowed to go away.

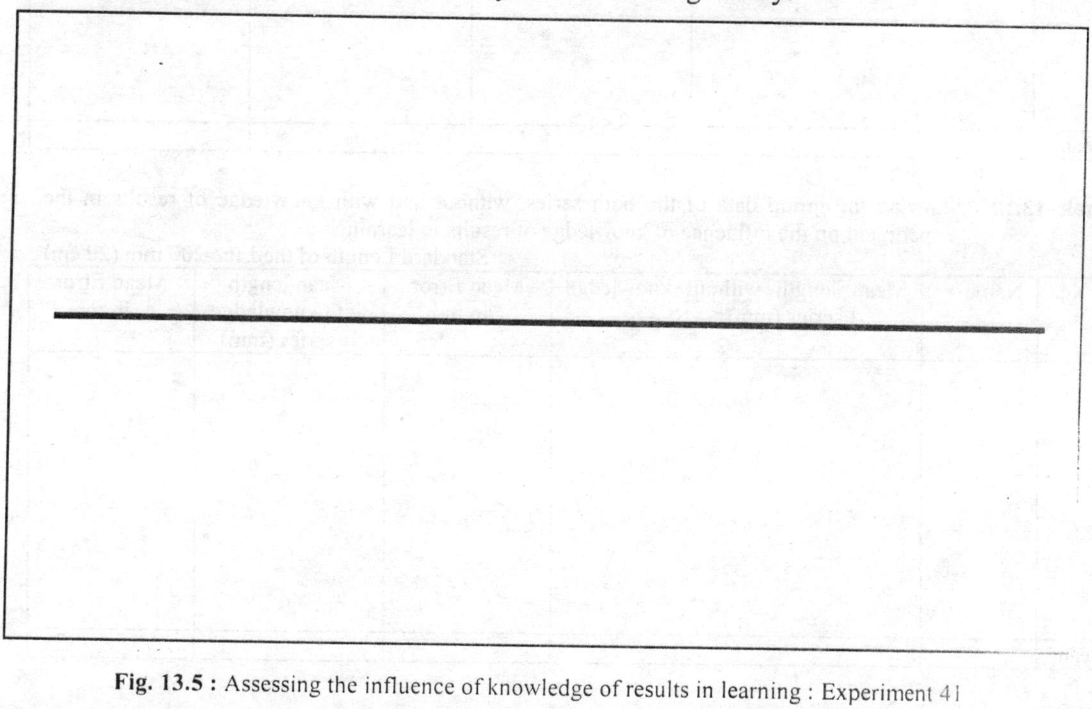

Fig. 13.5 : Assessing the influence of knowledge of results in learning : Experiment 41

Results

The experimenter should measure the length of the twenty lines drawn without knowledge and with knowledge, with the metre scale and present them in Table 13.9. The standard length of the line is 20 centimetres (200 millimetres). For accuracy millimetre measurement can be used.

Table 13.9 : Showing the individual scores in both the series, without and with knowledge of results

S. No.	I series without knowledge length drawn in mm	Error in mm + or –	II series with knowledge length drawn in mm	Error in mm + or –
1.				
2.				
3.				
4.				
5.				
6.				
7.				
8.				
9.				
10.				
11.				
12.				
13.				
14.				
15.				
16.				
17.				
18.				
19.				
20.				
Mean				

Table 13.10 : Showing the group data of the both series, without and with knowledge of results in the experiment on the influence of knowledge of results in learning

Standard Length of the Line 200 mm (20 cm)

S. No.	Name	Mean length without knowledge I series (mm)	Mean Error in mm	Mean length with knowledge II series (mm)	Mean Error in mm
1.					
2.					
3.					
4.					
5.					
6.					
7.					
8.					
9.					
10.					
Mean					

The experimenter should work out the mean length and error for both without knowledge of the first series and with knowledge of the second series. Then he/she should collect the group data from other experimenters and present them in Table 13.10. The mean score for both the series should be worked out for the group. State the positive error with plus signs and the negative errors with minus signs.

Discussion

Discuss the nature of errors committed by the individual in the first series when he/she had drawn the 20 lines without knowledge. State whether any of the lines were close to the standard line or always longer or always shorter. Discuss the nature of error the participant has committed in the second series and had drawn the lines with knowledge of results. State and discuss whether the subsequent lines drawn with the knowledge of results were close to the standard line or longer or shorter. Discuss the nature of commission of errors. Has he/she learned anything from the knowledge of results. Compare the mean length of lines and the mean error of the both series for the individual. State whether the knowledge of the results had any motivating effect upon the individual's performance. Was it positive or negative influence?

Discuss in the same manner the group data. State, whether any of the members of the group had drawn lines close to the standard line and had minimum amount of error. State and discuss whether the group members committed more errors in the first series or in the second series.

State the group means of the both the series and discuss whether the results supported the hypothesis or not. The knowledge of results should have a positive influence in the performance by committing the least amount of errors. State whether this has taken place. State and discuss whether group mean lengths are more away from the standard length of the line or close to it. State and discuss whether the errors are more positive or negative. How many in the group are above the mean length and how many below the mean length in the second series. State as a whole whether the group was benefited by the knowledge of results in learning in this experiment or not. If not state the reason for the same. State whether there is any individual difference in the result due to sex, age and other factors.

Conclusions

State whether your participant was benefited by the knowledge results in learning or not. Has he/she did better in the first series or in the second series. State whether the group members were benefited in the second series or not. If not state the reason.

14

STUDY OF FATIGUE

INTRODUCTION

The study of fatigue in psychology has become one of the important subject matters as it mainly concerned with some performance, activity, work, response, reaction, attention and perception in which the psychologists are interested. We often hear this concept fatigue in all kinds of works which involves muscular activity or physical activity. Therefore, in whichever the above said phenomena the bodily parts or the muscles are involved or put into action for a long time the consequence is referred to as fatigue. Whether it is industrial work in some organization or a personal work at home we put our muscles and bodily parts into action. After performing the work for a long time there is a decrease in the work output. The decrease of the work output is known from the production. The production is the perceivable one of the work output. Why there is a reduction in the production and in the work output? The employees are not continuing their work with same speed as they started in the beginning. It is because of some impairment in the bodily parts, the employees could not continue the work with the same speed. What sort of impairment has taken place? Outwardly we do not see any impairment. But, they say, because of the fatigue they could not continue the work rapidly.

When the production is reduced all those who are concerned with the work output and production become very much perturbed and worried. They do not want to meet with any loss. Therefore, they are interested in it and wanted to avoid the reduction causing phenomenon in the work. It is because of the feeling of fatigue, a physical impairment that the output has gone down. The physical impairment of the employees is not a perceivable variable. There is no any such sign outside. We do not know what is going on inside the body. But we see the work decrement in the production.

Since the work decrement is a clear evidence there must be something physiologically and mentally set in the organism. That something we say fatigue. The individual is experiencing the feeling of fatigue physically or mentally. The work is of physical type involving muscular activities we call it physical fatigue. Instead, if it is mental work it is called mental fatigue.

WHAT IS FATIGUE

Fatigue is a state of weariness or tiresomeness on the part of the person who is involved, in some kind of physical or mental work. It is a kind of reduction in zeal for further work.

The nature of fatigue is generally revealed in the reduced readiness or efficiency in the employee to do further work and in the increased feeling of strain.

Fatigue is an inevitable incidence in the work situation and it occurs in almost all works. It is also described as a sort of negative aptitude for work. Fatigue is also a kind of activity decrement which means reduced capacity for further work. As a consequence of previous activity in which the person has put himself as hard as he could he is physically hampered for further continuance of the work with the same speed. Fatigue exists in any forces. They may be called physical fatigue, or muscular fatigue and mental fatigue. There is also central or nervous fatigue as there are fatigue effects in the central nervous system. The sensory apparatuses also sometimes are subjected to fatigue effect. Particularly the ear and eye get fatigued when exposed to intense stimuli. This fatigue includes both physical and mental reactions and also the phenomena of monotony and boredom.

In the common sense world the concept fatigue is confused with monotony and boredom. Therefore, these three concepts need certain clarifications.

Monotony

The concept, monotony refers to the state of mind caused by performing repetitive tasks. The same type of work is performed with the same type of movements. There is no chance for any variation in this work or movements. This causes tiresome as one has to go on in the same tone for a long time. The person becomes dull and tedious because of the repetition. Even when we are listening to some musical performance if it is of repetitive type, lacking of variation in the pitch we become monotonous.

Boredom

In certain works people feel a sense of boredom. It is the feeling aspect of the individuals towards some work. Generally, it is referring to lack of interest in something we do or witnessing. Boredom is characterized by depression and a desire for change of activity. It is tinged with emotional distaste and is accompanied by a corresponding negative attitude to work. Thus boredom is more heavily affected by factors such as personality, attitude and interest patterns than monotony. Further boredom can be differentiated from fatigue, because it is a desire for change in activity whereas fatigue needs rest pauses and relief from the work temporarily in order to regain the lost energy to continue the work.

PHYSIOLOGICAL CHANGES AND PHYSICAL FATIGUE

Physiologists say that some physical changes have taken place in the organism during fatigue time, which may be physical impairment as the consequence of prolonged physical works. It is because of the physical work the organism has developed a kind of mental attitude. There is a kind of disinterestedness for further work.

In all physical works the muscles are mainly involved. In physical works the muscles are stimulated fastly and repeatedly. This eventually results in loss of ability in muscles and further work becomes slow and difficult. The physical fatigue takes place when the muscles receive pressure greater than what they are capable of. When we expect the muscles to perform more than their abilities the muscles lose their strength and the organism experiences fatigue. It is a state of exhaustion and the decrement of work is the perceived result.

The bodily conditions impose certain organic needs. The needs are for rest after continued activity and for sleep at interval after work break. Continued bodily activity causes muscular fatigue. It is partly because of using up of the fuel stored in the muscles and partly due to the accumulation of waste products of the muscular activity in the muscles.

The most important physiological change is the accumulation of lactic acid in the blood caused by the breaking down on the glycogen or sugar in the blood because of the continued activity. There are also other physiological changes present in the junctions of the nerve cells and in the muscle cells. Harrell (1951) states that these do not necessarily accompany changes in the production or changes in the feelings. These two are not measurable as the changes in the production. The activity may become painful because of the soreness of the muscles and joints. This painful feeling causes strong intention to stop the activity and take rest though the muscles have plenty of energy and strength.

Prolonged physical work causes fatigue and physiological changes. The physical fatigue is mainly due to insufficient oxygen supply to muscles. During work the muscles consume more oxygen for activity than that the lungs, heart, and the circulatory system can provide the muscles. At the same time the waste products of the muscles and the lactic acid accumulate faster than the circulatory system can remove them. The accumulated lactic acid can cause contraction of the muscles without the nerve cell excitation and the nerve endings are stimulated and the pain is caused. Consequently there is an impairment of the activity of the muscles in performing the task. Ultimately it results in work decrement. But under many conditions the central nervous system is not much affected by this oxygen deficiency.

Fatigue, whether it is due to physical work or mental work cannot be fully eliminated. But, it can be reduced by providing rest pauses and work breaks. If rest is provided for few minutes the individual can resume the work with renewed energy. The rest pauses should be sufficient to re-equip the muscles for the lost energy.

There are experimental studies on physical fatigue. In psychological textbooks the instrument called ergograph is often mentioned. This instrument is commonly used to measure the output of energy of the human finger. In this experiment the person is asked to pull a weight through a pully with his fore finger at the rate of one pull for every beat of the metronome as adjusted by the experimenter. The subject should carry out the task to a specified time. Individuals may vary in this task. It can be observed in this experiment that as time goes on the speed of stretch and flex of the finger are reduced. Ultimately the person stops the task. The task of stretching and flexing of the finger in pulling the weight makes out a curve in the paper affixed on the drum of the instrument. It can be noticed that the amplitude of the curve would be going on reducing in its size as time goes on in the task. The reduction in the size and the amplitude is the indication that fatigue is setting in the bodily parts *i.e.* in the finger. Similar to this there are also other tests to measure physical fatigue.

SENSORY FATIGUE

Man's efficiency in behaviour depends upon the nature of sensitivity of his sensory apparatus. If man's sense organs do not function efficiently he becomes dull, slow, sluggish, sometimes even inactive and loses the pleasures of his life. Therefore, man must know the values of sensory apparatus and take proper care of them and maintain without any impairment to his sensory apparatus such as eyes, ears, nose, tongue, skin and kinesthesis. Any impairment to any of these sense organs causes many inconveniences and makes him difficult to respond properly in this competitive world.

Exposing the sensory apparatus to extreme situations which have nothing to do with our work or unwillingly and unintentionally when we are forced to endure hazardous environments often cause fatigue and impairment to our sense organs. It may be some extreme and hazardous light or sound, heat or cold. The eye muscles should not be exposed to fatigue causing situations like extreme bright lights. The pains we experience in the eyes are due to strain of the eye muscles. This strain is the fatigue of the eye muscles. When we keep ourselves engaged in reading in very dim light situation in which the letters are not clearly seen or when we read some passages printed with very small letters eye muscles get fatigued. Sufficient information on sensory fatigue are not available as experiments on this are very difficult.

There are some studies made only on auditory fatigue. The ears should not be exposed to unwanted sounds and noises. The sensitivity of the ear may be impaired when sudden horrible sounds attack the parts of the ear. According to Rawdon-Smith (1934) most common intense stimuli the ear has to tolerate are the loud noises of the transports, high powered machines, loud chord music. In our country the commonest horrible sounds to which we are forced to expose our ears are the sounds of the street loudspeakers alarming in various places in various forms. Hearing often these horrible sounds causes auditory fatigue which was not realized by people. But in the long run it results in losing their sense of hearing. Workers from boiler and metal industries retire with deaf ears. Probably they are helpless to control these auditory disturbances. Rawdon-Smith states that the ear is often subjected to fatigue by severe acoustic stresses.

FATIGUE IN PSYCHOPHYSICAL EXPERIMENTS

In psychophysical experiments and in the study of learning we discuss about the practice or training effects. In practice or training what is occurring, is repetition. Psychologists believe that practice or repetition will increase the efficiency. Candland (1968) says that the subject while making several determination of the stimuli he/she acquires the practice effect which is supposed to increase the efficiency with repeated trials. It is true that some subjects perform better when the number of trials are increased. But it is not true with all. Many subjects' efficiency is decreased because of the fatigue effect. The more the number of trials the subject performs there is less accuracy in the performance. The subject gets tiresome and it causes decrement in his efficiency. This is how the fatigue effect takes place in the accuracy of sensory perceptions and judgments.

In sensory experiences the processes of masking and adaptation are quite commonly taking place. Stimuli of various intensities are combined in sensory stems. This is more common in auditory sensation. The human auditory sensation alters the basic sensation for determination. In auditory sensory experience it is said that fatigue is masked and adaptation takes place mostly. In the fields applied psychology all these have been tested. Wegel and Lane (1924) have conducted experiments on a single subject. The data obtained by them in their experiments have excellent model for this hypothesis. These experimenters wanted to determine the masking effect of fatigue in auditory sensation. They used a kind of secondary tone to mask the primary tone. They have measured the primary tone, two times. One measurement was made while the secondary tone was on. Another measurement was made in the absence of the secondary tone.

In this experiment, the strength of the primary tone was 1200 Hertz (Hz) at 80 decibel (dB). They have observed after conducting several trials in which the high tones were affected more than the low tones. Secondly when the primary and the secondary tones were presented to different ears masking was occurring still. But the severity was not so much when the tones were presented to both ears and heard by both the ears. Thirdly the experimenters have observed that the low tones were more effective in masking the high tones than the high tones effect on low tones in masking.

This kind of experiments on masking fatigue and adaptation in audition were conducted in applied areas for some specific purposes. Fatigue and adaptation generally are referring to the effects of the stimulations made continuously. When the sensory apparatus is made to respond repeatedly to the same stimulations the sensory nerves of the sense organs lose their efficiency and become less sensitive to the continued stimulations because of the fatigue effect.

Human beings may come across such insensitive effect in some parts of the body if it is put into continued contact. Candland (1968) quotes a good example of wearing a ring in our fingers. In the beginning we are aware of the presence of the ring in the ring finger. We are very sensitive to the cutaneous sensation of the touch of the ring. But after many days we do not feel the presence of the ring in the finger. The continued stimulation of the touch causes less sensitive to the touch and we do not feel it. Therefore, the masking or fatigue effect because of the continued stimulations may be caused not only in auditory sensation, but also in other sensory areas. The same type of masking effect can be demonstrated in olfactory and gustatory and other sensory areas. Experimental information are seldom available in these areas. But in these areas adaptation effect are taking place easily except in the area of audition.

In one of the experiments conducted by Wood (1930), Geldard (1953) it was observed that there was a strong desire or tendency to return to the normal sensory system in auditory sensation when a person's both ears were subjected to an experimental situation in which a 1000 Hertz (Hz) tone at 50 decibel (dB) was presented for a period of ten minutes to both the ears. The recovery to normal sensitivity was very rapid. This is because to get rid of the effects of fatigue.

FATIGUE IN INDUSTRY

In industrial situation the employees are motivated to do more and vigorous work and increase the production. There are several types of incentives to motivate the employees for hard work. There may be also competition among the employees to become booster and get the reward. Some employees may develop self-competition. The output of muscular work can be increased considerably by this. Motivation from the visible goal and knowledge of results motivate employees for more work in spite of prevailing fatigue. This was tested in the study conducted by Crawley (1926). It was observed that the employee was motivated by saying that if he could reach a particular goal level he is surpassing his previous achievement. The employee under this condition performed 20 to 40 percent more work at the cost of some extra fatigue. The result of this study shows that employees take risk to have self-competition and exhibit better performance.

Therefore, when there are keen competitions employees even though are loaded with fatigue they are ready to bear some extra fatigue and continue the work because of the strong motivation for work to get the extra benefits. The organism is ready to endure more tolerance for painful experiences. Studies on fatigue show that the output was often found to be increased near the end of each work period, before coffee breaks, lunch breaks or at the end of the day. At the time when we are expecting more fatigue and work decrement the output is increased at the end of the work period. This shows the motivation of the employees towards the end of the work or day's work that the monotonous work is going to stop for the day.

Some of the authors who made studies in industrial situations state that the accumulation of waste products and lactic acid and consequent physiological fatigue is very remote in modern industries. Brown (1966) says that industrial fatigue is not causing the accumulation of toxic products from the muscles in most of the modern industries. In modern industries the fatigue causing factors are controlled. Therefore, even after eight hours of continuous work employees do not show any sign of fatigue.

Studies relating to fatigue and wage payments were made. Wyatt (1938) found in his study that plants with good wage incentive system had higher production level with lower level of fatigue expression.

FATIGUE AND ACCIDENTS

It is not incorrect to assume that fatigue leads to accidents in work situation, particularly in the industries. Several reasons have been attributed to accidents by accident specialists. There are certain physiological and psychological factors and conditions cause the individual to be involved in accidents. In this event of accident, fatigue, emotional upset, attitudes, alcohol, drugs, experience and training all play their parts in the total picture of accidents. When, particularly industrial work is concerned, fatigue is one of the important causes of accidents.

By performing the work for long periods of time without sufficient rest pauses or breaks the employee naturally becomes weak by spending more energy from his body. Hence, he is likely to miss a part of the machine, or fail to move appropriately in an assembly work which may result in causing an accident, resulting in cutting his fingers or damaging the parts of the machine, or causing some fire if the working is done by using electricity.

Thus an accident may be defined as an event that is taking place without any foresight for expectation causing some type of personal injury and damage to life and property. This definition may be applied to any type of accident. Every day large number of people are involved in industrial accidents. The number of industrial workers (including transports) killed and disabled runs in several lakhs in our country. Accidents not only costs industrial economy but they also result in anguish, pain, and often death to the unfortunate individuals.

Therefore, in industrial situation accidents may be caused by fatigue. Fatigue is caused by many factors, like long hours of work, work days, extreme hot temperature. By making improvement in all these, fatigue effects can be reduced to the minimum possible extent. A study was conducted by Vernon (1945) in England during World War II about the accidents rates among the woman workers in a shell factory. The reasons for the accidents was the long work hours and the woman workers become very much tired. The accident rate was reduced by more than 60 percent when the work hours was changed from a twelve-hour-work to a ten-hour-work for a day.

EXPERIMENT 42 : ASSESSING MUSCULAR FATIGUE IN A CONTINUOUS PHYSICAL WORK

INTRODUCTION

Fatigue is a subject matter that interests both the employer and the employee. Since a decline in the production is often attributed to the workers fatigue, the employer is interested in eliminating this decline by reducing the fatigue. The employee recognizes fatigue as a feeling of tiredness or pain, he regards it as intrinsically unpleasant, and so he too is interested in its decrease or elimination. Industrial fatigue, despite its elusive nature, has been the subject matter of many investigations from both physiological and psychological points of view. One of the goals of time and motion study has been to reduce the effects of fatigue in workers by introducing more efficient work methods. Evidences indicate that proper training methods and correct work methods can do much to prevent the onset of fatigue. It is also recognized that effective motivation can be successful in eliminating fatigue or at least in reducing its effects. In some of the studies on fatigue a relationship has been found between accident rate and fatigue. Therefore, it is assumed that fatigue leads to accidents.

All those physiologists and psychologists who have been working on fatigue recognize the diverse factors involved in this phenomenon. The word fatigue is used and misused very widely. Sometimes fatigue of a physical nature is differentiated from fatigue of a psychological nature. This leads to the use of the term "fatigue" as muscular fatigue and "mental fatigue". Monotony, boredom and other similar phenomena in industry are used in some respects similar to fatigue as far as reduced production is concerned. Although these terms are sufficiently differentiated, anyone when in certain physical work situation is fatigued or tired as compared with when he is bored in some other situation, there is often confusion between these two concepts. The psychologists' study of fatigue may include both the physiological and psychological aspects as there are changes in both the aspects due to continued or prolonged performance of a work. The concepts used synonymously in the common sense world need certain clarification.

Generally fatigue is referred to physical exhaustion. Irrespective of its technical meaning people use this word to express their feeling of tiredness incurred in some physical work and also in mental work. Therefore, we may state that fatigue is the weariness from bodily or mental exertion. It is the cause of continued labour or exertion. The physiological meaning is that fatigue is the temporary dimension of the irritability or functioning of organs, tissues or cells after excessive exertion or stimulation (Viteles, 1932). From the mechanical point of view, fatigue is the weakening or breakdown of the material subjected to stress. Fatigue is also referred to thc labour of a non-military type of work done by soldiers. For example, it may be stated that the soldiers are engaged in such labour, as the fatigue. It is also generally referred to weary with bodily or mental exertion. Fatigue from short bursts of activity whether by the whole body or by isolated muscular groups is characterized by increase in lactic acid and temporary inability to continue the work. In man the chemical analysis of the blood reveals a low level of blood sugar. In a hot environment fatigue has several manifestations. The most simple to measure the fatigue is the increase in heart rate. In the instance of two individuals doing the same task, one becomes more fatigued than the other because of the poor nervous coordination of the unskilled man. The unskilled man spends more energy than the other who is skilled and trained in the work.

In an intensive study called "Fatigue and hours of service of Inter-State Truck Driver", the term "fatigue" is defined as an altered psychological and physiological state in relation to the states of recovery or normal capacity. Mosso defines fatigue as a condition caused by activity in which output produced by that activity tends to be relatively poor and the degree of fatigue tends to vary directly with the poorness of output. He, himself dissatisfied with this definition and he concludes that the term fatigue should be banished from scientific terminology, whatever fatigue is, it is safe to say that any muscular work even that involved in sitting in a chair will result in fatigue provided the work of the muscles and the resulting expenditure of energy are at a faster rate than recovery is. The major difficulty in understanding the true nature of fatigue is that it is caused by the fact that many factors, in addition to muscle use, contribute to its onset. The length of the work period, the speed of work, the extent of the musculature involved and the tension accompanying the task are a few of these factors. Further complications are, a few, caused by the fact that industrial fatigue rarely concerns a single muscle but rather the individual as a whole. Depending on the strength, stamina and preparation of the individual fatigue is manifested at different rates in different people. Psychologists can help to eliminate much of the unnecessary fatigue that is caused by unfavourable conditions on the job. According to Tiffin (1952) insufficient work spaces, inefficient work methods, inefficient tools and inefficient people to do the work cause much unnecessary fatigue. These causes can be attacked and the conditions may be changed so as to eliminate or reduce much fatigue on the job. A simple approach to an understanding of fatigue has been the study of it in the laboratory.

The process of fatigue in driving has been studied by Crawford (1961). He was able to see several stress factors were inherent in the driving situation. Heavy traffic, unfavourable weather conditions and so on may cause stresses and result in stress fatigue. This may lead to more difficult performance than normal driving fatigue.

In another study Heimstra (1970) observed the effects of stress fatigue on performance in a driving situation. The measures were, tracking, speed control, reaction time, and vigilance in each subject. The results revealed that the subjects who experienced stress performed at a lesser level than the subjects who did not experience.

Safford and Rockwell (1966) studied the effect of fatigue on driving performance in an actual vehicle. The measures were recording the vehicle velocity, accelerator position, steering wheel reversals and brake pedal depressions. The subjects were told that they would receive different amount of money depending upon how long they drive up to 24 hours. The speed was limited to 70 mph. The subjects were given 15 minutes rest after every 8 round trip of 160 miles. Out of seven only three have completed the test. The result of this study did not show any decrement in driving performance after long driving. The important finding was that the drivers showed great variability in performance and some were affected very much by fatigue than others.

Fatigue is a factor that is causing accidents in industrial works. This has been studied by Vernon (1945) in England during World War II. In a steel factory the accident rate among women workers was reduced by more than 60 percent when factory work hours changed from twelve hours to ten hours. Extreme fatigue leads to increased frequency and accidents.

Length of work period is one of the causes of fatigue. The output is reduced because of this. This has been studied by Vernon (1920). This was one of the earliest studies made by Vernon during the World War I. When the work week was shortened in a munitions factory from 58.2 hours to 51.2 hours his production was increased about 20 percent and when there was further reduction to 50.4 hours there was another increase of production by 17 percent.

An early study of this nature was conducted by Mosso in 1890. He constructed an ergograph to record and measure the work done by the muscles in flexing a finger. By attaching a weight at one end of a string and a strap with a loop at the other end and making it to go through the groove of the pulley which was fixed on a wooden plank, was the device. The weight was made to hang down from the pulley. The hand was placed parallel on the plank and a finger was put into the loop of the strap. The finger was made to move by flexing and stretching, thereby the weight was made to pull up and down. The finger was put into work and it was possible to obtain a sample of work for which a fatigue curve can be drawn. Using this technique Mosso demonstrated the rate and extent of the work that could be done by a single finger lifting a specific weight. He was also able to demonstrate that if a fatigued muscle was forced to work, the period of recovery was longer. It was observed that if the work was done for a period of 10 minutes then few minutes were required for the recovery of fatigue. But if the work was done for a period of 15 minutes then the time required for the recovery of fatigue was one and a half times of the recovery time of 10 minutes work. Although Mossos ergograph does study fatigue it has the obvious disadvantage of doing this under extreme unnatural conditions. Furthermore, no job in industry is likely to be done with the movement of only one finger.

Vernon developed a variation of the dynamometer which enables one to study the effects of work when the larger muscles of the body are involved. He also demonstrated that the introduction of rest pauses increases the capacity for work, and further that a change of posture during the rest pauses helps recovery to a greater extent than merely stopping work.

EXPERIMENT

Problem

The purpose of this experiment is to measure the level of muscular fatigue in a given continuous physical work of the subject.

Method

Participant

Choose a subject who has no knowledge of this experiment and describe his/her characteristics such as age, sex, qualifications and whether he/she is wearing spectacles or not and other physical aspects like, short and fat, thin and tall etc.

Apparatus

Ergograph with automatic recording device (Anand Agencies, Poona), weights like 1 kg. and 500 gms., Metronome and stop clock. The motorised ergograph may also be used. The motorised ergograph is available with all instrument suppliers. It has certain advantages.

Description of the Apparatus

The Anand Agencies Ergograph is a modified one and it has a paper feeding system. It has certain advantages over the conventional models. The paper movement is automatic and, therefore, it needs no power. There is no necessity to fix the paper over recording drum. The record markings are evenly spaced and they are neat in appearance. The recording can be made for any length of time without changing the paper.

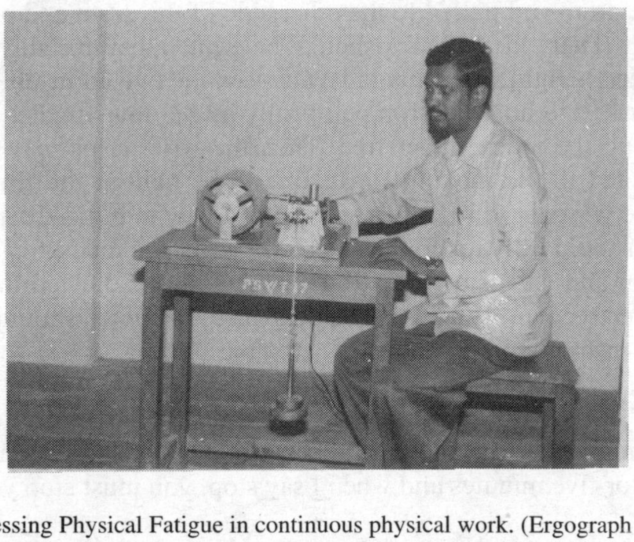

Fig. 14.1 : Assessing Physical Fatigue in continuous physical work. (Ergograph Experiment 42)

The apparatus is fixed on a rectangular wooden board. The recording arrangement is fixed at one end and a pair of wrist grips and an iron peg for hand gripping at the other end. The paper feeding system consists of a paper roll stand, a drum with gear wheel, attached to a lever and a spring loaded clutch. The clutch prevents the drum from rotating on the reverse side. It consists of a single pawl ripping over the gear wheel. A pair of rubber rollers on either side of the drum are provided in the holders for holding the paper over the drum. There is a spring loaded pencil holder mounted on the moving carriage. The moving carriage has two sides one on the subject side and other on the weight side. The carriage moves on two guide bars and a string is attached to the weight side of the carriage and the string ends in a small 'S' hook for the purpose of holding the weights. This string is passing, through the groove of the metal pulley and the pulley is connected to the pawl. The counter will count one number whenever the string over the pulley makes to and fro motion to the length of more than two inches. This should be borne in mind during the operation. The carriage has a small finger loop on the subject's side which is attached to the system consisting of a metal collar and wingnut with which the distance between the subject's finger tip and the moving carriage can be adjusted.

Before operating the instrument a paper roll of suitable size should be inserted by unscrewing the spindle. This should be tighten again after the paper roll is placed properly in position. Then the free end of the paper should be gently passed under the rubber rollers of the folder and over the drum. The holders of the rubber rollers should be gently lifted for passing the paper through them. A sharp soft pencil should be placed on the spring loaded pencil holder. Now the apparatus is ready for operation. (If motorised ergograph is used, it may be described).

Procedure

The Ergogragh apparatus should be placed at the edge of the table and clamped firmly. The weight pan is to be attached to the 'S' hook and 1.5 kg. of weights be placed over it. Guide the string through the groove of the pulley and the weight should be hanging free from any touch of any thing. The paper roll is to be inserted into the apparatus under the holder and over the drum and pulled to the other side of the drum and placed it under the grip of the holder. Then the subject should be seated comfortably in front of the apparatus. The subject's right wrist should be between the two adjustable wrist grips. The subject should be asked to hold the iron peg firmly by his four fingers (thumb and other three fingers) leaving the index finger free. Then the wrist grips may be adjusted. The leather-loop connected to the moving carriage is to be held at the first segment of his index finger and the wingnut to be adjusted in such a way that the sharp blade at the left end of the carriage should fully touch the lever, whenever the index finger is extended. In order to rate flexion and extension a metronome should be adjusted in such a way that there is one beat per second. The initial reading in the counter should be noted down. Then the following instructions should be given to the subject :

> "When I say 'ready, go' your task is to flex your finger once at every beat in the metronome and to extend it to another beat. You should carry on the task continuously for five minutes and when I say stop, you must stop your task."

The metronome and the stop watch should be started simultaneously and at the same time "ready, go" caution also should be given to the subject. For every minute a mark is to be made on the graph paper till the subject is asked to stop at the end of the fifth minute and he may be left free. An introspective report may be obtained from the subject. Final reading in the counter should be noted down. The initial reading is to be subtracted from final reading and that is the number of strokes made by the subject. The amplitude of the pencil mark is related to the amount of work done at a time. The decrease in the amplitude indicates the onset of muscular fatigue.

Results

In Table 14.1 give the total time of the task and number of revolutions per minute the subject has made. This is to be elicited by deducting the initial reading in the counter from the final reading. The number of amplitude made by the pencil mark also should be given in Table 14.1. The height of the amplitude should be measured. What is its height in the beginning and what is its height at the end? After how many minutes the height of the amplitude is reduced in height. Is there gradual decline in the height of the amplitude? Observe this in the graph paper and note down. State whether the subject has continued the task for full 5 minutes or stopped it after few minutes. State what he has experienced from the introspective report. Work out the mean rotation and mean height of the amplitude. Collect the group data from other experimenters and present them in Table 14.2.

Table 14.1 : Showing the result of the performance of the subject in muscular fatigue test in an Ergograph

Time in minutes	No. of rotations	No. of amplitude	Height of the amplitude in cm
1st minute			
2nd minute			
3rd minute			
4th minute			
5th minute			
Mean			

Table 14.2 : Showing the group data of the muscular fatigue experiment in Ergograph

S. No.	Name	Mean No. of rotations	Mean No. of amplitude	Mean Height of amplitude cm
1.	P.K.			
2.	S.M.			
3.	R.V.			
4.	K.R.			
5.	B.V.			
6.	T.V.			
7.	M.S.			
8.	A.L.			
9.	C.P.			
10.	G.R.			
11.	E.R.			
12.	M.R.			
Mean				

Discussion

Discuss the nature of performance of the subject. State how many rotations the subject has made in the first minute, in the second minute, in the third, in the fourth and in the fifth minute. State whether the number of rotations are increasing or decreasing. If decreasing at what time it is decreasing. Is the decreasing gradual or sudden? State after how many seconds of work decreasing takes place.

State the number of the amplitude in the beginning. After how many seconds the number starts reducing.

Discuss the height of the amplitude in the beginning, after one minute, after two minutes, at the end of the task. After how many seconds the height of the amplitude begins to reduce in height. Discuss the onset of fatigue in your subject in terms of seconds.

Discuss the group data. Who are those get fatigued early and who are those continued the task for longer time. Compare the individual's mean with the group mean and discuss the result for all the three variables, *i.e.* rotations, amplitude and height of the amplitude. State the nature of experience of the subject from the introspective reports.

Conclusions

State, after how many seconds the muscular fatigue sets on the subjects. What type of subjects prolonged the task. Does muscular strength facilitates late onset of fatigue?

EXPERIMENT 43 : ASSESSING THE EFFECTS OF CONTINUOUS MENTAL WORK

The concept, fatigue is used to explain the consequences of performing various types of physical and mental activities for a long time. It describes a general state of an individual, a state resulting from continuous activity or work. It is conceived as a reduction of power to perform or to react.

The development of fatigue is mainly due to long periods of activity, lack of opportunity for rest, heavy work, and environmental tensions. Fatigue, being a general state of the individual it shows its effects in all aspects of behaviour and in all its manifestations. In short it may be defined as a sort of negative attitude for activity or work. According to Tiffin and McCormick (1965) the study and control of fatigue is difficult because it has three aspects. First, it is defined as decline in quantity of work for a given level of effort. The second is a feeling of tiredness – accompanied by reduced capacity for work. The third aspect is the physiological changes.

Thus fatigue includes both physiological and mental reactions as well the phenomena of boredom and monotony. Boredom is lack of interest in work. It is a kind of emotional dislike for work and it is subjective and from within. Monotony is the state of mind caused by repetitive task. Boredom is a desire for change of activity, whereas fatigue is a state for rest or relief from work. Therefore, fatigue includes all these three aspects.

Mental Fatigue

Though we speak of physical fatigue and mental fatigue most of the characteristics of fatigue apply to both types of fatigue, except the certain physiological changes like the accumulation of lactic acid in the muscles. There is work decrement in both and also mental tension, increased heart beat, increased blood pressure and a feeling of depression of reduced tendency to do further work. Therefore, long periods of mental work results in work decrement due to the onset of fatigue. We do not see the direct cause of the fatigue, but it is seen in the work decrement which is also attributed to conflict of motives or boredom. Persons doing office work show no sign of physical impairment at the end of long day's work, though they often make complaints of fatigue.

Mental fatigue is experienced only` by human beings where verbal tasks are involved. It may be in the form of reading, writing, doing mathematical works, copying, typing and other office works. It is said that writers or authors of books get fatigue of their imaginations involving complex situations and problems. Though the task is of verbal nature the increased repetition of the same matter causes fatigue. It is true that if the repetition is for practice it may increase the efficiency in performance. But the repetition itself is a task which results in mental fatigue.

Fatigue, whether it is physical or mental has detrimental effect on work. There are various factors which are having influence in the efficiency of the work. Wyatt and others (1929) and Anastasi (1964) studied the influence of rest pauses and work breaks in fatigue.

In prolonged mental activity there may not be so much loss of energy. But there is a feeling of disinterestedness for work. It is called boredom. Boredom also is a kind of mental fatigue. There is a loss of interest in the work on the part of the individual. The person develops a feeling of distaste for work. Either the person wanted to stop the work and take rest or to do a different type of work.

If the mental fatigue is severe the person goes in a state of drowsiness and develops a tendency to go to sleep. This is one of the most powerful physiological motives a person wanted to fulfil. This is how mental fatigue causes physical fatigue. Vitales (1932) says that this type of fatigue can be overcome by strong desire for activity. After taking sufficient rest the organism may be again motivated for activity.

Since fatigue is associated with several factors in the industrial situation workers' attitude towards the work, organisation, and their attitude towards the work group may cause fatigue in certain situations. Some writers associate fatigue with motivation, some associate with job satisfaction of the employees. Writers like Vitales (1953) associate with motivation and morale. If there exists good morale among the industrial employees motivation also would be high and the onset of fatigue becomes remote. Since morale refers to an attitude of an individual towards a group situation he is satisfying his own motives through the group cohesiveness. When there is strong group cohesiveness (morale) the feeling of fatigue seem to be having the least effect.

After working out a lengthy assignment university students often take a stand that they cannot do anything more than that as they are very much exhausted and must go to bed for sleeping. But at that particular moment if the telephone bell rings and a particular student's girl friend wants to meet him out somewhere then the fatigue caused by the strenuous mental work goes out of the door with the student. Therefore, it is no doubt that the major component of mental fatigue is the attitude one has developed.

From the foregoing discussions we can also understand that it is very difficult to make any distinction between mental and muscular fatigue. There is interaction between different forms of fatigue. In the fatigue state there is certain alteration in the individual both in physical and mental aspects. This is because the individual is put under severe stress. Stress is caused due to mental work and the environmental tensions during the activity.

Studies of fatigue involving various factors have been made. Kossoris and Kohler (1947) found in their studies that fatigue was caused by hours of work and nature of work. Production was reduced because of these factors. In the study made by Nelson and Bartley (1968) it was observed that the female workers were very much fatigued in the first few hours of the work period. Griffith (1950) and his co-workers state that the workers felt greatest feeling of fatigue during the fourth .and eighth hours of the eight-hour-work-shift.

In an experiment on mental fatigue three subjects worked for a period of 12 hours on four successive days. The work was, multiplying four place numbers by four place numbers. Each problem was solved without the aid of pencil and paper and only the answer was recorded. The effect of continued mental work was seen with an increase of errors. The increase of errors may be attributed to the increase of mental fatigue. Apart from such objective evidences, there are also feelings of intense dislike, ennui, boredom to the extent of headache and illness, dizziness, nervous instability and physical and muscular weariness and some of other states accompany the decrease of mental efficiency. Geldreich has conducted an experiment on mental task. He found that there were increases in heart beat, respiration rate, blood pressure and skin condutance during the performance of this task. Thus mental tasks produce not only work decrement but also physiological changes. Tension is the largest component of mental fatigue if the task is considered by the individual as a challenge. Tension is not only a psychological component but also a physical.

The first experimental study on fatigue in the laboratory situation was made by the Italian physiologist, Mosso. He devised an apparatus called 'Ergograph'. This apparatus actually recorded the work performed. After Mosso's Ergograph several types of Ergographs have been devised. One type of Ergograph that is very much used in the laboratory situations is the 'Finger Ergograph', in which only the index finger is used. Experiments conducted on the phenomenon of fatigue using this apparatus in the laboratory situations revealed several factors associated with fatigue like rest interval, nature of work, speed of work, interest and motivation. On the basis of these factors simple experiments are devised to study the mental fatigue of individuals in the laboratory situations.

EXPERIMENT

Problem

Measure the effects of continuous mental task in an individual.

Method

Participant

Choose a person who has no knowledge about this experiment and describe his characteristics. State his/her age, sex, qualifications, his/her body built, whether there is any physical deformities etc.

Material

Sheets of papers with vertical columns and some prepared numbers for cumulative adding, pencil, stop clock, (Appendix-I).

Procedure

This experiment is to be performed in two postures, one the restricted posture and the other unrestricted posture. The participant should be made to sit comfortably in a well illuminated room in front of a wall. There should not be any disturbance by noise to the subject. The experimenter in advance should prepare the experimental material in columns. He must choose any 30 members and add to each number 2, 3, 4 and 5 cumulatively. For example if the number is 9, add 2 to it, it is 11; and add 3 to it, it is 14. By adding 4 to it, it becomes 18; and by adding 5 to it, it is 23. So, the cumulative numbers after adding, 2, 3, 4 and 5 to 9 are 11, 14, 18, 23, 25, 28, 32, 37 and so on. After adding the four numbers once the same four numbers should be again repeated by adding to the cumulated number as shown above. The experimenter, therefore, should prepare a long cumulative numbers to each number he has selected. He should select 30 such numbers and prepare a series of cumulative numbers with 2, 3, 4 and 5 by adding. Thus there should be 30 columns with 30 numbers with cumulative additions. Then he should give the following directions to the participant. First he can begin with the restricted posture. After 30 trials for 30 numbers he can go for the unrestricted posture.

> "Please listen carefully. This is with regard to adding some numbers continuously to a number I call for. You will have to add first 2 to it. Then add 3 to the total, add 4 to the total and add 5 to the total. After adding 5, you will have to begin again with 2 and add to the total, then 3, 4, 5 and so on till I call for another number. As soon as I call for another number you will have to apply the same adding procedure to the new number. Please make it a point that you will have to add only the numbers 2, 3, 4, 5 in rotation. You can rotate any number of times in adding till I call for a new number. For example I call the number 8. You must loudly pronounce the added number 10 after adding 2 to 8. Then after adding 3 to it you must say 13, then add 4 and say 17 and then add 5 and speak aloud 22, then repeat the adding numbers 2, it is 24, then 27, and so on."

Now the experimenter can verify whether the participant has understood the procedure by giving trial with a number. If he is satisfied, then he can start the restricted posture. In this posture the participant should sit erect, looking at a point on the wall (The experimenter can make a mark on the wall to pin down the participant's attention) throughout the work he should not vary the bodily posture, with feet flat to the ground, head and neck and arms in the same erect position. Then he can call out the first number which he has already worked out with cumulative numbers by saying to the participant, "ready, go". At once the experimenter should start the stop clock. As the participant goes on giving the cumulative numbers then experimenter should check them. If the number is wrong, he should correct the number and ask him to proceed on and should make an entry of the error in his tabular column. *After exactly 30 seconds* the experimenter should say the next number and the participant should begin to add the numbers, 2, 3, 4, 5 to the new number without any stopping. At once the experimenter should put a bar after the last number to indicate how many numbers the participant has given. This procedure should be continued for *15 minutes* continuously till the 30 trials are over in the restricted posture. The number of errors and the number of additions for each number should be entered.

As soon as the restricted posture was over the experimenter should begin the trials with unrestricted posture. The participant may be allowed to sit at ease and can move his body as he liked and can give his responses.

The experimenter on giving a number the participant ***should loudly give the added numbers*** as he was giving in the previous procedure for 2, 3, 4 and 5. ***After every 30 seconds a new number should be entered*** in the tabular column. This should be continued ***for 15 minutes to complete 30 trials***. When all the trials are over the participant should be asked to give an introspective report. Then he should be left free.

Results

The experimenter should workout the number of errors for each number used for adding and also the number of additions for each number the participant has made. This should be worked out for both the restricted posture and unrestricted posture for all the 30 trials and should be given in Table 14.3. This is the individual score in the mental fatigue test. Work out the mean score. The mean of the number of errors and the mean of the number of additions for the whole group may be collected and given in Table 14.4. Work out the mean score for the group.

Table 14.3 : Showing the individual scores in the Mental Fatigue Test

Trial Nos.	Restricted posture		Unrestricted posture	
	No. of additions	No. of errors	No. of additions	No. of errors
1.				
2.				
3.				
4.				
5.				
6.				
7.				

Contd.

Table 14.3 *Contd.*

Trial Nos.	Restricted posture		Unrestricted posture	
	No. of additions	No. of errors	No. of additions	No. of errors
8.				
9.				
10.				
11.				
12.				
13.				
14.				
15.				
...				
...				
...				
...				
...				
30.				

Table 14.4 : Showing the group data in the Mental Fatigue Test

S. No.	Names	Restricted posture		Unrestricted posture	
		Mean additions	Mean errors	Mean additions	Mean errors
1.	K.S.				
2.	P.V.				
3.	K.R.				
4.	M.S.				
5.	B.V.				
6.	T.V.				
7.	R.N.				
8.	S.V.				
9.	M.N.				
10.	R.V.				
Total					
Mean					

Discussion

Discuss the individual's performance in the mental fatigue test in terms of number of additions he has made in each trial. State whether the number of additions are increasing after every new number or decreasing. State whether the number of errors are increasing after every new number or decreasing. Find out the reasons for the decrease of the additions if any. Also find out the reasons for the increase of errors, if any. Is there any number having no errors. State the reason. At what state fatigue sets in the participant.

Compare the number of additions of the restricted posture with that of unrestricted posture. Is there any substantial difference between these two postures. State the reason for the same. Compare the errors between the two postures. Do you find any difference between the two postures. Find out the reason and discuss. See whether the mental fatigue is indicated by the gradual decrease of the number of additions and increase of errors. Find out whether there is any difference between the big numbers given at the end and the small numbers given at the beginning in additions and errors.

Do you find much variations in the group data among the individuals. Who gets fatigue much earlier by committing more errors and gives less number of additions. Who scores better with more additions and less errors. Is there any body free from fatigue?

Is there any difference between the restricted posture and unrestricted posture in the scores of the group. Discuss the individuals place in the group by taking into account the group mean. How many are above the group mean in additions and below the group mean in errors. State the nature of mental fatigue of the group in general. State whether the individual's personal factors have any influence in the mental fatigue. Relate the introspective report to the scores of the individuals and discuss the results.

Conclusions

After how long time the number of additions are reduced and errors increased in the individuals? Who is more fatigued in this test and who is less fatigued? Whether there is any difference between the restricted posture and unrestricted posture in the performance. Are there any individual variations in the degree of mental fatigue in this test?

APPENDIX – I

Work sheet for Mental Fatigue test : Adding the numbers 2, 3, 4, 5 to a number assigned by the experimenter

S. No.	Number for Trial assigned	Cumulative additions : Restricted Posture
l.	6	8, 11, 15, 20, 22, 25, 29, 34, 36, 39, 43, 48, 50, 53, 57, 62, 64, 67, 71, 76, 78, 81, 85, 90, 92, 95, 99, 104, 106, 109, 113,118.
2.	8	l0, 13, 17, 22, 24, 27, 31, 36, 38, 41, 45, 50, 52, 55, 59, 64, 66, 69, 73, 78, 80, 82, 85, 89, 94, 96, 99, 103, 108, 110, 113, 117, 122.
3.	10	12, 15, 19, 24, 26, 29, 33, 38, 40, 43, 47, 52, 54, 57, 61, 66, 68, 71, 75, 80, 82, 85, 89, 94, 96, 99, 103, 108, 110, 113, 117, 121, 126.
4.	12	14, 17, 21, 26, 28, 31, 35, 40, 42, 45, 49, 54, 56, 59, 63, 68, 70, 73, 77, 82, 84, 87, 91, 96, 98, 101, 105, 110, 112, 115, 119, 124.
5.	14	16, 19, 23, 28, 30, 33, 37, 42, 44, 47, 51, 56, 58, 61, 65, 70, 72, 75, 79, 84, 86, 89, 93, 98, 100, 103, 107, 112, 114, 117, 121, 126.
6.	16	18, 21, 25, 30, 32, 35, 39, 44, 46, 49, 53, 58, 60, 63, 67, 72, 74, 77, 81, 86, 88, 91, 95, 100, 102, 105, 109, 114, 116, 119, 123, 128.
7	18	Work out in the same manner as shown above for other numbers : l8, 20, 22, 24, 26, 28, 30, 32, 34, 36, 38, 40, 42, 44, 46, 48, 50, 52, 54, 56, 58, 60, 62, 64.
..		
..		
..		
..		
30.		

APPENDIX – II

S. No.	Number for Trial assigned	Cumulative additions : Unrestricted Posture
1.	7	9, 12, 16, 21, 23, 26, 30, 35, 37, 40, 44, 51, 53, 56, 60, 65, 67, 70, 74, 79, 81, 84, 88, 93, 95, 98, 102, 107, 109, 112, 116, 121.
2.	9	11, 14, 18, 23, 25, 28, 32, 37, 39, 42, 46, 51, 53, 56, 60, 65, 67, 70, 74, 79, 81, 84, 88, 93, 95, 98, 102, 107, 109, 112, 116, 121.
3.	11	13, 16, 20, 25, 27, 30, 34, 39, 41, 44, 48, 53, 55, 58, 62, 67, 69, 72, 76, 81, 83, 86, 90, 95, 97, 100, 104, 109, 111, 114, 118, 123.
4.	13	15, 18, 22, 27, 29, 32, 36, 41, 43, 46, 50, 55, 57, 67, 68, 69, 71, 74, 78, 83, 85, 88, 92, 97, 99, 102, 106, 111, 113, 116, 120, 123.
5.	15	17, 20, 24, 29, 31, 34, 38, 43, 45, 48, 52, 57, 59, 62, 66, 71, 73, 76, 80, 85, 87, 90, 94, 99, 101, 104, 108, 113, 115, 118, 122, 127.
6.	18	19, 22, 26, 31, 33, 36, 40, 45, 47, 50, 54, 59, 61, 64, 68, 73, 75, 78, 82, 87, 89, 92, 96, 101, 103, 106, 110, 115, 117, 120, 124, 129.
7.	19	Work out in the same manner as shown above for other numbers : 19, 21, 23, 25, 27, 29, 31, 33, 35, 37, 39, 41, 43, 45, 47, 49, 51, 53, 55, 57, 59, 61, 63, 65.
..		
..		
..		
..		
30.		

EXPERIMENT 44 : TESTING ENDURANCE : MUSCULAR FATIGUE (TAPPING TEST)

Endurance refers to the hardship one undergoes in a situation. It is also stated as the strain one is experiencing in a work situation. The strain or hardship continues to exist without yielding. It is also called forbearance or long suffering by performing some task. In some training programmes, particularly in the army the soldiers are expected to pass the endurance tests. How much difficulties one could withstand in emergency situations? For which the soldiers are put into very difficult tasks continuously for a long time. Endurance is equated with fatigue. Since both endurance and fatigue involve physical and mental exhaustion due to exertion or some tiring activity both lead to decrement of products or efficiency.

Physiologically, fatigue may be defined as the temporary inability of an organ or part of a body of an organism to respond to a stimulus because of over activity. The stress put on the organism makes it weak for further continued activity. If the activity is of repetitive in nature the effects of it may yield satisfaction provided the operations are few. At the same time some jobs permit large number of operations for longer duration. This involves the physiological process of fatigue. Physiologists have long been recognized that the continuous activation of a muscle results in a decrease in the probability and amplitude of its response. This type of reduced responsiveness is called fatigue. This is associated with an accumulation of lactic acid in the muscles.

It seems reasonable that performing the same task or operation over and over again would be causing more fatigue than performing different tasks or operations. If fatigue is a state of aversion it means that the persons engaged in the tasks would prefer varied performance to repetitive work. The muscular activity involved in the performance of a single operation is not that it is so difficult or not that the rate of work is so fast that it cannot be performed without fatigue continuously. In fact workers do not describe such works as fatiguing. But it is boring for them to continue. The job is not interesting with varied activities or movements. The work actually does not demand more amount of physical effort. It is with regard to the likeness of the job. The workers point is that the work is a source of displeasure, which makes them a feeling of discomfort and disinterestedness. Thus repetitive jobs generally lack variety of stimulation in the work which the workers expect in the performance.

According to the studies made by Hebb (1949) McClelland *et al.* (1953) the effective consequences of a stimulus pattern are dependent on the amount of disparity between the stimulus and adoption level established in a person as a result of past experience which he acquired in some other job. If there is mild disparity then it is tolerable or pleasant. But if the individual enters with greater disparities then there is disruptive and unpleasant consequences. If the person is performing a highly repetitive operation then the sources of stimulation are highly constant and not changing from the kinesthetic and sense modalities. He can foretell what it is going to be. The consequence would be high degree of displeasure. Since it is a feeling aspect of human characteristics though there would be no fatigue, a fatigue like feeling sets on the individual in the physiological sense.

An experiment was conducted by one of the students of Lewin, Anitra Karsten (1928) in the University of Berlin. It was considered to be a classical experiment in this area. He demonstrated the aversive qualities of a highly repetitive task. The experiment was conducted on college students. They were asked to perform a very simple task of drawing lines in a particular rhythmic way. As soon as the students completed the task in one sheet of paper they were given another sheet of paper to carry out the same task, and then another sheet was supplied and so on. It was observed that gradually the students began to draw the lines with variations in their task. The lines were made heavier, lighter, tilted, curved, larger, smaller and so on. Finally, the students reached a stage that they were unable to continue the task further. What is reason for this type of performance? Lewin (1935) states that this is not the result of fatigue. It is called psychological satiation. They have been repeatedly performing the same activity and they get satiated for the activity. It is the expression of the feeling that what they have done is more than sufficient.

In the beginning though the work might have been pleasant the repetition of an activity resulted in reduction of performance indicating the desire to vary the activity. There was large amount of intolerance for continuing the activity. Hebb (1949) states that this kind of intolerance for continuing the same activity occurs in a person even in re-reading a novel which he read a few days ago. The repetitive task is unattractive. There is much greater demand on the worker's muscular system without any scope for better rewards.

Fatigue is an occurrence of physiological state which often results in reducing the speed of performance or efficiency of the work performed by a worker. Particularly, if the work is of repetitive in nature whether it is simple or complex task the consequences are there. Since, every physical work involves muscular system of the body of the organism muscular fatigue is imminent. Generally after a period of time in the operation of any work the person is subjected to certain mental and physiological changes. These changes constitute operation fatigue. This type of fatigue, it is generally assumed, will impair performance ability. But it is very difficult to prove with many evidences that a decrement of performance is taking place when the person is fatigued.

When we use the term fatigue in certain situation, we know what it means. But it is not as simple as we think to know about it. Actually fatigue is a complex process. It results from a variety of causes and it is manifesting in a variety of ways. It is difficult to measure fatigue effect and present it in qualitative terms. In many of the studies fatigue is defined in terms of the amount of time a worker has been engaged in a work. Some studies define fatigue in terms of movements the worker has performed and the production rates during the course of work, the number of errors he has committed as and when the time is increased and the number of accidents, if any, involved by the person.

Therefore, it may be also assumed that a person who performs the task for a period of three hours non-stop is more fatigued than a person who does it only for 15 minutes. A number of studies have been conducted to study the fatigue effect in driving. Crawford (1961) says that there are a number of stress factors inherent in the driving situation. Heavy traffic, unfavourable weather conditions, lighting conditions may serve as stresses and result in a condition called stress fatigue which may lead to more impairment of performance than normal driving fatigue.

The relative terms of fatigue are boredom and monotony. Sometimes these three terms are used interchangebly, though these three indicate varied aspects. All the three are of great importance to business and industry. All have detrimental effect upon production. But when there are decrement effects in production in an industry it is the indication of fatigue or boredom or monotony. There is no one accepted definition of fatigue though it is widely used. According to Brown *et al.* (1966) fatigue is a condition of a person brought about by work and resulting in a work decrement. It is also a tired of feeling and physiological changes in the person, whereas boredom refers to a psychological condition or feeling of a person as a result of performing activities without interest. Boredom may also mean an aversion to continue the same activity. Monotony refers to the psychological condition resulting repetitive task. Hence work decrement may be due to fatigue, boredom and monotony and all the three are closely associated.

Physiological Changes in Fatigue

Any work fatigue involve physiological changes. Brouha (1954) states that when a person goes to work from resting certain physiological changes take place. There is increase in heart rate, as it is related to activity. If the work puts more physiological load upon the body, the heart rate increases correspondingly and it slowly returns to normal level on completion of the work. This heart rate increase also depends upon the physical health condition of the worker. Temperature also was found to be influencing the heart rate. Brouha states that due to increased temperature the load becomes heavier in the cardiovascular system and fatigue sets in soon and exhaustion also becomes faster. The factors such as age, sex, nutrition states, effect of chemical substances also influence the fatigue level. Advancing age in work is easily accessible to greater stress and fatigue sets in faster.

As and when the work duration is increased the muscles may become rapidly fatigued to the state that no longer work can be performed. In such cases rest pauses are introduced to counteract the fatigue effect. Brown (1954) has pointed out that the accumulation of waste products due to fatigue effect is very rare in modern industry. The onset of physiological fatigue is quick in heavy muscular work. But physiological fatigue in most of the modern industries is not the result of the accumulation of lactic acids in muscles. Because according to Brown most of the modern industries are free from uncomfortable conditions like, heat, cold, dust, illumination and so on. Therefore, physiological or industrial fatigue is rare. But fatigue in modern industry is due to psychological condition. That is the feeling of aversion the worker develops during work. In this way Brown equivates fatigue with boredom. A person who has no interest in the work becomes bored in his work and there is work decrement. He is tired of the work and not happy in continuing the work.

It is also possible to avoid the feeling of boredom by changing the activity. The individual's efficiency in the work can be increased by changing him from one activity to another. Boredom is not from outside. It is within the worker. What is interesting at one time is not interesting for him on another time. Therefore, it is the psychological reaction the worker projects at the time of work. The work done more repetitively may result in monotony and boredom to some workers. This is indicated by work decrement. There are instances in which workers with highly repetitive tasks show increased production for longer period of time. On the other hand some workers in complex and non-repetitive works reveal boredom and monotony and consequently decrement of productivity. What must be the cause behind these state of affairs? It is the motivation that plays an important role in these states. If the person is highly motivated he will not succumb to monotony or boredom easily than a person who is less or not motivated for work. Therefore, the attitude and motivation, of the individual worker are having influences upon the physiological and psychological effects of the work of the person.

EXPERIMENT

Problem

The purpose of this experiment is to measure the endurance or the onset of muscular fatigue as affecting the strength and as affecting the speed of the participant in his/her performance.

Method

Participant

Choose a person who has no knowledge of the purpose of this experiment. Note down his/her age, sex, qualifications, handicap if any and other particulars.

Materials

Whipple tapping board, tapping stylus, electric counter, stop clock.

Description of the Material

The whipple tapping board is a wooden board of 55 by 10 cm in size. ***It has two metal plates of 10 cm square at each end.*** The two metal ends are connected to screw at the centre of the board by copper wires. The board and the stylus are in circuit with battery and electric counter.

Procedure

The subject is to sit comfortably in front of the whipple tapping board. The board is placed horizontally within his hand-reaching distance. The elbow should be kept at rest without any movement upon table and the movement should be made with hand and forearm only. One end of the metal plate should be on the right and the other on the left of the participant. He is to tap each plate alternatively with a stylus as fast as possible for 160 seconds by each hand. First he should tap the plate on the right and then on the left. This may be done with each arm first with his right arm for 160 seconds and secondly with his left arm for 160 seconds. The following instructions should be given :

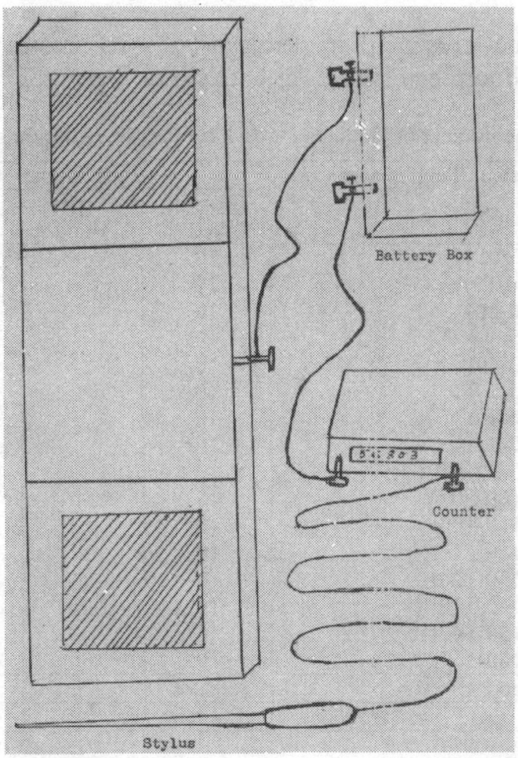

Fig. 14.2 : Whipple's Tapping Board (Experiment 44)

"In front of you on the table, there is a board with metal plate on either side (The board should be pointed out the subject). You would be given with a metal stylus. You should hold it first with your right hand, and tap the metal plate on the right and on the left alternatively as fast as possible. When I say "ready, go" you must do it and when I say "stop" you must stop tapping instantly. After few seconds you must do the same thing with your other arm, make it a point that you must react quickly." At the end the person should be asked to give his/her introspective report.

Precaution

The experimenter should watch very carefully with a stop watch that the task is performed only for 160 seconds with each hand. Every 10 second the experimenter should watch the counter and tell the subject, "now, your tapping number in this" and the reading on the counter at that point is to be taken as closely as possible and recorded in his observation note book. The readings are cumulative and the total must be taken at the end. The number of taps for each 10 seconds is taken by subtracting from the other. This should be done for every 10 seconds for both the hands separately. The subject should not experience any discomfort nor there should be any distraction to his task. It should be observed that the counter runs for each tapping made by the person.

Results

The readings should be presented in Table 14.5. The number of tapping for each successive 10 seconds should be elicited for both hands separately.

Table 14.5 : Showing the number of tappings for each successive 10 seconds for right and left hands

Sl. No.	Successive intervals Time in seconds	Number of tappings			
		Time in Sec.	Right Hand	Time in Sec.	Left hand
1.	First 10 seconds	10		10	
2.	2nd 10 seconds (20-10)	10		10	
3.	3rd 10 seconds (30-20)	10		10	
4.	4th 10 seconds (40-30)	10		10	
5.	5th 10 seconds (50-40)	10		10	
6.	6th 10 seconds (60-50)	10		10	
7.	7th 10 seconds (70-60)	10		10	
8.	8th 10 seconds (80-70)	10		10	
9.	9th 10 seconds (90-80)	10		10	
10.	10th 10 seconds (100-90)	10		10	
11.	11th 10 seconds (110-100)	10		10	
12.	12th 10 seconds (120-110)	10		10	
13.	13th 10 seconds (130-120)	10		10	
14.	14th 10 seconds (140-130)	10		10	
15.	15th 10 seconds (150-140)	10		10	
16.	16th 10 seconds (160-150)	10		10	
Total 160 seconds + 160 seconds		160 Sec.		160 Sec.	
Mean					

What is your participant's fatigue index for tapping? Work it out and indicate the index using the formula. It is the number of taps for the last 10 seconds, subtracted from the number of taps for the first 10 seconds and divided by the number of taps for the first 10 seconds. The fatigue index for tapping should be less than one. Collect the group data of the tappings and fatigue index from other experimenters and present them in Table 14.6.

Table 14.6 : Showing the mean score of the group for right and left hands

Sl. No.	Subjects	Mean Score		Fatigue Index
		Right hand	Left hand	
1.	K.R.			
2.	P.V.			
3.	S.V.			
4.	R.M.			
5.	P.K.			
6.	G.G.			
7.	R.V.			
8.	M.S.			
9.	B.V.			
10.	T.S.			
11.	B.A.			
12.	R.V.			
	Total			
	Mean			

$$\text{Fatigue Index} = \frac{\text{Number of taps in first 10 sec.} - \text{Number of taps in last 10 sec.}}{\text{Number of taps in first 10 seconds}}$$

Draw the fatigue curves for the respondent's data for right and left arms. Find out likenesses and differences. Compare the right hand and left hand curves and point out any interesting differences. Compare the group data for right and left arms.

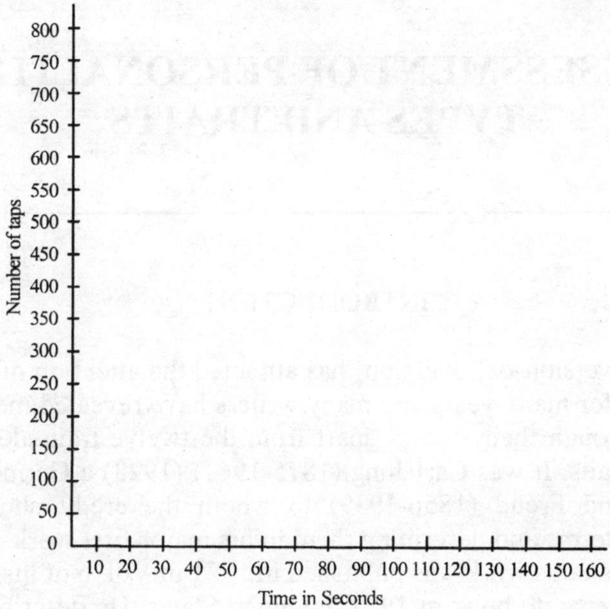

Fig. 14.3 : Graph showing the individual's tapping and time of right and left hands

Discussion

In the original experiment the amount of time for tapping is only 5 seconds. Since there are certain practical difficulties in observing, recording and informing the participant the number of tappings the time is extended to 10 seconds and total amount of tapping is 160 seconds for 15 successive steps for each hand.

There are 15 successive steps for each hand in this experiment. Observe whether there is any difference step to step and interesting change in the successive steps. What is the tapping speed? Which hand is faster? Where fatigue takes place in your subject? Whether fatigue effect continues at the same level, increases or decreases? In what way the subject's right hand task differs from left hand task? Which hand exhibits more efficiency? See whether fatigue is the same as the feeling of weariness?

Use your judgment and find out the factors that had effect upon your subject's muscular endurance (a) for increasing it; and (b) for decreasing it. See whether a complete rest, restlessness, sleep, heavy meal, high room temperature, high humidity, coffee, sad news, news of success has any effect.

Conclusions

What is the fatigue index of your subject? Where he stands in the group? Which of his hands exhibits more endurance? What is the tapping speed? Which hand is faster?

15

ASSESSMENT OF PERSONALITY TYPES AND TRAITS

INTRODUCTION

The concepts, 'introversion-extroversion' has attracted the attention of researchers in the field of psychology for many years and many writers have revealed more comprehensive personality traits through their studies apart from the twelve traits described by Cattell (1946) as primary traits. It was Carl Jung (1875-1961) (1923) a German psychiatrist, an associate of Sigmund Freud (1856-1939) to whom the credit should be given for inventing these two terms and describing them in his major first work, the *Psychological Types*. Though this famous work was published in 1921 in Vol. 6 of his collected works it was published as a separate book in 1923 in United States. He described the personality traits through these two concepts. According to Freud and Jung the key element in an individual person is his conscious and unconscious relationship to people and things in the external world.

Alfred Adler (1870-1937) an associate of Freud and Jung was of the view that the individual's focus of attention is more of subjective (personal) with an intention of striving for inner security and also to compensate his personal inferiority.

JUNG'S ATTITUDE TYPES OF INTROVERSION AND EXTROVERSION

Jung had difference of opinion with Freud and Adler on some major issues of their types of personality. After some time Jung (1917) came out with some fresh thinking and raised the issue, there are only two human types, one of them is more interested in the external objects and the other is interested in himself. In his *Analytical Psychology* published in 1917 Jung described his two psychological types with some more experimental information. For Jung the introversion is described as hesitant, reflective, retiring, shrinking from objects, always on the defensive side and prefers to keep behind. On the other hand the extroversion is described as an outgoing, unprejudiced, accommodative, adapting easily to situations, developing attachment quickly and so on. But Jung has stated that his classification of these two types and the descriptions of characteristics of the introverts and extroverts is not complete and perfect.

It is also true that the introverted person may be more interested in the external world of things when his inner life is affected. Similarly an extroverted person may be more interested in the subjective world when the objective world has affected his nature of life..

Then the extrovert may develop temperamental moods and becomes the self-oriented. Jung was also aware of all these changing phenomena of personality traits and, therefore, he was not fully satisfied with the two simple types of personality. Therefore, Jung thought of four functional types such as (1) sensation, (2) intuition, (3) thinking, and (4) feeling in addition to the attitude types of introversion and extroversion. Thus it makes eight combinations of attitude functional types of introversion and extroversion. Therefore, attitude types of introversion and extroversion that Jung has postulated are not rigid type of classification of individuals.

DESCRIPTION OF EXTROVERT CHARACTERISTICS

The extrovert is under the influence of the external object forced to subjective experience by environmental calamities. The extrovert under the powerful influence of social forces is moulded by the cultural factors like moral and ethical principles. The extreme extrovert, Jung (1923) says, gets into the great danger of dissolving into the objects and he is lost completely in them. The extreme extrovert is a typical neurotic and his specific type of neurosis is hysteria. The hysterial reactions are primitive, depending on others in the environment. The extrovert is interested in visible tangible world. The extroverts likes to make actions and decisions quickly and easily.

The four attitude-functional types of extroversion are : (1) sensation extrovert, (2) intuitive extrovert, (3) thinking extrovert, and (4) feeling extrovert.

The Sensation Extrovert

This is one type of personality description according to Jung. The sensation extrovert goes in search of new sensory experiences and collects actual experiences of concrete objects. This person cultivates his sensory ability and becomes not merely a sensational type but could make very good discrimination of acts. He seeks always good company and always favours external objects. He is a well adjusted person to realities and interested in the welfare of others. He is a more realistic and sensational type of person.

Intuitive Extrovert

To put it precisely about the characteristics of this person, he is more a visionary, changeable and creative. Since he is changeable it is very difficult for him to maintain interest in every one of things for a long time. In him the other characteristics are acting only to a minimum extent. The intuitive extrovert makes decisions without conscious thought. But his decisions are good. He is in close touch with the wisdom of his own unconscious world. This type of persons do not take interest in the welfare of others but capable of inspiring others with confidence for some new causes.

Thinking Extrovert

This type of persons are very much objective and rigid in their convictions. The thinking extrovert is fascinated by external ideas and objects. He is not able to draw any logical conclusions as he does not want to go away from the objective ideas he has made. He refuses to accept any other principle for himself. He is mostly rigid and intolerant. He is for universal truth and is dogmatic. For the thinking extrovert, thinking is the superior function and hence he develops a suppressed feeling of life.

Feeling Extrovert

This type of person like the thinking extrovert finds place in the external world. With regard to his feeling he is like a woman. He has a tendency to make friends easily and also influenced by the emotions of the society. But his thinking function keeps him suppressed and puts him in strong display of emotionally extravagant.

DESCRIPTION OF INTROVERT CHARACTERISTICS

The introvert individual is aware of the external conditions. But the introvert stresses his subjective responses to them. The introvert's attitude to his life is to have his ego independent of the external influences and towards it he struggles. He is interested in primarily in his own thoughts, feelings and ideas. He will engage himself in dreaming and planning for the future and respects his own standards and sentiments. He is often interested in invisible forces and laws of nature. The typical extreme introvert gives way for neurosis and he suffers from psychasthenia. The symptoms of this neurosis is intense anxiety reactions, chronic fatigue and exhaustion. The introvert attitude in combination with the functional types constitute four attitudes functional types of introversion. These four are (1) sensation introvert, (2) intuitive introvert, (3) thinking introvert, and (4) feeling introvert.

Sensation Introvert

The sensation introvert's chief traits are passive, calm and artistic. He is often changing to external situations. His personal reaction is mainly to objective sensual events. He expresses his thinking and feeling primitively. The sensation introvert is interested in traditional and poetic description of the world. His sense impressions are in terms of what is good and what is evil for him. Sometimes he wrongly interprets the reality and keeps himself away from the external world. But he is not disturbed by this isolation. He remains quite and calm when others criticise him.

Intuitive Introvert

The intuitive introvert's prime traits are mysticism, dreaming and in these traits he is singular. If the intuitive tendency is in its height the introvert keeps himself isolated and not at all concerned about the realities of the world. In this stage he reveals his peculiar artistic ability by drawing pictures of conventional types. His creative efforts are strange and beautiful. He claims that he is a creator. Perceiving things in a different way is the main problem of the intuitive introvert. That is, he satisfies himself. He will become unfriendly from those who are around him suddenly. He claims that he is full of spiritual truths and obscure thinking and play miracles. With these traits sometimes he may collect large number of disciple. In India there have been many such mystics in the past. Even now there are many who claim that they are good inspired, full of spiritual wisdom, can play miracles and hundreds and thousands of disciple worship them.

Thinking Introvert

The prime traits of thinking introvert are intellectual, theoretical and impractical. He has strongly motivated desire for privacy. If there is any disturbance for his privacy and solitary state he becomes very furious. The thinking introvert since he is concerned with abstract thinking and formulating theories he develops a tendency to ignore the practical aspects of every day life. He makes his decision without any feeling. He is stiff, absolute and without compassion or pity.

Feeling Introvert

The feeling introvert's important traits are silent, childish and indifferent attitude. Actually he has no regard for the feelings and opinions of other people. It is generally observed in the feeling introverts that there is a tune of superiority and critical indifference in the relationship of the feeling introvert with other people. There is a minimum emotional expression outwardly. But the inward emotions are strong and troublesome for the feeling introvert. This type of intense emotions are sometimes expressed by the feeling introvert in writing of poetry that is kept very carefully from the eyes of others. Feeling introversion is more common among women and it is observed that they show much indifference to others.

Jung's important work, *Psychological Types* was no sooner appeared in 1921 than many writers in United States started studying introversion-extroversion traits experimentally and brought out enormous amount of information and critical views about them. The early work of Heidbreder (1926) and Prince (1929) and a host of other investigators studied introversion and extroversion. Woodworth (1952) states that we should not classify everybody either as an introvert or an extrovert type of person. When large sample of people are investigated through suitable methods it has been found that large number of people fall in the middle between the two extreme traits. Heidbreder was able to recognize an 'ambivert type' in his study. The ambivert type is a fixed type and most individuals were found to be belonging to this type. Prince (1929) states that in any dimension of personality the peak of the distribution falls near the middle of the distribution and in between the two extreme traits. The same type of distribution may also take place in Jung's introversion-extroversion testing in a large population. But Jung's the four functioning types may make the distribution evenly among the eight attitude-functional types of introversion-extroversion. He has also observed that the two fundamental types of introversion and extroversion had associated with various forms of mental illness.

EYSENCK AND INTROVERSION AND EXTROVERSION

Jung's typology of personality dimensions received a new turn when Hans Eysenck (1916 - 1998) undertook the study of introversion-extroversion and neuroticism from a different angle. Eysenck's (1987) professional background made him more curious to find out reliability of the psychiatric diagnoses made among the soldiers and airmen who were coming to the Mill Hill Emergency Hospital where he was working as a research psychologist. The patients were diagnosed and classified. Eysenck's doubt was that if two or more psychiatrist examined the same patient do they agree among themselves. If their diagnoses do not agree what sort of treatment would be prescribed to the patient? Though this was not the work of psychologist in a psychiatric hospital Eysenck's account raised some important issue. Though the situation was not favourable for the issues he has raised, he was able to collect information that the professional judgments had very little similarity in the diagnosing psychiatric patients. What the second psychiatrist diagnosed was different from the first psychiatrist of the same patient even when the second already know the result of the first.

Eysenck (1952) preferred the hypothetico-deductive method for measuring quantitative aspects of personality. He often criticised the existing methods of study of personality as not scientific, less empirical and less factual. Psychoanalysis was his main target of attack. He charged that Freudian approach is fictional and not testable (1953). Eysenck's (1957) prime aim of psychological formulation has been that it should be empirical, observable, testable and should have more source for new hypotheses. He was following these qualities in all his studies. Eysenck did not approve the experimental approaches made by Freud and Jung. Eysenck (1972) has criticised Freud and Jung unsparingly in one of his later works, Freud as the great story teller and Jung as the brilliant myth-creator. Eysenck's whole dimensions of personality had its beginning only in the study and introversion-extroversion and neuroticism. Since he was associated with London Psychiatric Hospital to deal with patients who were suffering from neurosis, Eysenck had to explore the personality types and various forms of neuroses or mental illness. At this time the most attractive personality types are introversion-extroversion and the mental illness was neurosis. Freud was studying the various forms of neuroses and Jung was studying the personality types of introversion-extroversion. Eysenck in his early works and in his later works used only these concepts. Neuroses, introversion-extroversion were used by Freud and Jung as their own discoveries. Eysenck's works without these three concepts become null and void.

The Maudsley Personality Inventory designed by Eysenck and its later revision were intended to study introversion-extroversion and neuroticism. Many years prior to Eysenck's entry into the London Psychiatric Hospital for starting his research work, Freud has published quite number of papers and books on neurosis and Jung has published many works on the basis of introversion-extroversion. All these works become basis for Eysenck to develop his personality dimensions on the basis of introversion-extroversion and neuroticism. Instead of paying encomium to Freud and Jung for their decades of study on human nature Eysenck has made open attacks on them. The most unrealistic one is that Jung was not the first writer to use the concepts introversion and extroversion, even though he was the first to use for studying human nature. Eysenck has stated that there were other writers like, Hippocrates (464-351 BC) and Galan (AD.130-200) as early as 400 BC and 200 AD used similar terms. But neither Hippocrates nor Galan nor Kant had used the same terms introversion and extroversion in their writings. Eysenck simply quoted these authors in order to reject Jung's invention.

Hippocrate's concepts are the four humours, such as blood, black bile, yellow bile and phlegm. According to him the balance physiological processes is essential for normal mental health. Any adverse mixture of these humours cause mental diseases. Galan developed a temperamental theory of personality based upon the four humours. The dominance of blood causes sanguine temperament, the excess of black bile results in melancholic temperament, the dominance of yellow bile produces the choleric temperament and the excess of phlegm results in the phlegmatic temperament. Emmanuel Kant (1724-1804) a German philosopher has also discussed four temperamental traits. He made verbal description of the four temperamental types. Kant tried to state that how various separate traits cluster together in a particular type of temperament.

WUNDT'S DIMENSIONAL THEORY OF TEMPERAMENT

After quite long time subsequent to Kant, one dimensional theory of temperament was devised by Wilhalm Wundt (1832-1920) (1897) the founder of the first psychological laboratory. He recognised four types of temperament and each has two variable truth, one the degree of emotionality and the other degree of changeability. Thus Wundt suggested two continuously variable dimensions – the emotionality and the changeability. Each dimension has extreme opposite ends. Thus one end stated as Emotional and the other end Non-emotional. In the same way the other dimension is anchored at one end as Changeable and the other end as Unchangeable. It is also stated in other words as Strong *versus* Weak emotional types and Quick *versus* Slow reaction types. According to Wundt's temperamental theory the phlegmatic and sanguine (Galan's types) personality types are emotionally weak and at the other end the melancholic and choleric types are emotionally strong temperaments. While anchoring with the changeable variable the melancholic and phlegmatic personality types are unchangeable while the sanguine and choleric personality types are extremely changeable. But each one of these forms of temperamental personality traits is independent.

We have discussed the Kernal points of the earlier studies with regard to dimensions of personality by the early physicians, philosophers and psychologists. It is clear that none of them used the concepts introversion, extroversion and neurosis except Jung. They all used different concepts to describe their temperamental traits of personality. If we state that there are concepts in the studies of early authors quite similar to introversion, extroversion and neurosis they are only derivatives and it cannot be construed that they used the original concepts, introversion-extroversion as Jung has used in his studies. Though these concepts are found in dictionaries that the words were in usage as early as 1656 and 1866 they were not used to describe human characteristics till Jung made use of them. Therefore Eysenck's contention is not correct. It may be stated that Eysenck has continued his studies further on dimensions of personality from the place where Jung has left or he built his personality types on the edifice laid down by Jung and others. He has taken all those aspects of the earlier authors acceptable to him and has built up his personality theory combining Wundt and the biological aspects of Pavlov (1927, 1928) and Teplov (1964).

EYSENCK'S APPEARANCE ON THE SCENE

Right from the very beginning of his career as psychological researcher in the Mill Hill Hospital, Eysenck had been dealing with persons who had mental disturbances, particularly neurosis or war neurosis. One of the earliest studies in which Eysenck involved was the administration of personality, psychomotor, and performance tests on a large number of neurotics and soldiers and airmen and collecting data from about ten thousand subjects and treated the data with sophisticated statistical method like factor analyses and cluster analyses and isolating the personality factors having association with various types of neuroses.

POSTULATIONS OF JUNG AND EYSENCK

Jung was of the view that an ***introvert*** is likely to develop ***psychasthenia*** if he yielded for mental illness and an ***extrovert*** may develop ***histerical disorders*** if he yielded for mental illness. Jung had come to this conclusion and postulated the two attitude types, namely introversion and extroversion only on the basis of numerous observations he made and personal experiences he had gained.

Eysenck's study of factor analysis of thirty-nine items yielded two bipolar factors consisting of number of traits. One factor that could be best described was general neuroticism. A second factor emerged from the factor analysis seemed to accept Jung's hypothesis. Eysenck (1947) felt that the ***general neuroticism*** could be divided into two groups, one ***neurotically maladjusted*** and the other may be characterised as ***hysterical disorder***. Eysenck thought whether the *highly neurotic group* could be further divided into ***anxiety neurotics*** and ***hysterical neurotics***. He further thought whether he could find out empirical evidence for introversion-extroversion associating his two neurotic groups and make it as the second dimension. Because Jung's neurotic group the psychasthenia (extreme introvert) and another neurotic group the hysterical disorders (extreme extrovert) and Eysenck's two neurotic groups, the neurotically maladjusted and the hysterical disorder (which are derived from his high general neuroticism) seem to be almost the same. What Eysenck laments is that there is no empirical evidence for Jung's postulations and also they are not testable. But, since Jung's postulations are quite close to Eysenck's factor analytical studies Jung's postulations may be accepted as correct or near correct. When Eysenck (1947) collected experimental literature on the problem of introversion-extroversion he found very wide sources of evidences in the literature that introversion was associated with a variety of neurotic tendencies and extroversion was associated with hysterical neuroses.

While searching for empirical evidences on the issue of introversion-extroversion whether this dimension could be associated with neurotic illnesses, Eysenck found in his own study among the large sample of soldiers and airmen that the personality dimension of introversion-extroversion can be used to differentiate the two groups of neurotics. Eysenck (1947) has invented a new term, 'dysthymia' to describe the neurotic symptoms such as depression, anxiety, phobia and obsessive-cumpulsive neuroses. These are the same neurotic tendencies which Jung has stated. He used this term, *dysthymia* as equivalent to Jung's *psychasthenia*. He used this term also to describe his introvert personality types. But Eysenck could not invent any equivalent term to Jung's *hysterical disorders* to describe the extrovert and, therefore, retained the same and stated as *hysteric* and approved the characteristics of extrovert personality types as stated by Jung.

EYSENCK'S DIMENSION OF INTROVERSION AND EXTROVERSION

Having in our minds the above stated information on the dimension of personality traits of introversion and extroversion, let us know what Eysenck has originally contributed from his empirical studies other than what others stated. Eysenck's postulations of the introversion-extroversion dimensions are mainly based upon variety of experimental investigations using several types of tests and questionnaires. He has applied statistics very extensively on his data for identifying the variables operating on the dimensions of personality traits.

Extreme Introvert

According to Eysenck the ***extreme introvert*** reveals symptoms of the development of anxiety and depressions. These are revealed in obsessions, irritability and apathy. There is always autonomic instability in the neurotic introvert's behaviours. The neurotic introvert is easily hurt in his feelings. He is self-conscious, having the feelings of nervousness. He often suffers from inferiority feelings and becomes moody suddenly. He is often involved in day dreaming. He keeps himself away from social activities. He often complains that he is not getting sleep. With regard to his body development there is more vertical growth than horizontal growth. He makes poor response to efforts. He often suffers from salivery secretion. His mental ability level may be higher than the average. He may have very good language ability. He has a tendency to repeat everything constantly. The neurotic introvert is accurate in his judgments, but reacts slowly. He can exhibit excellent performance in tweezer Dexterity works.

The introvert neurotic puts, his level of aspiration at a very high place, may fail to achieve and under rate his ability to perform the task. He is very rigid in his decisions. He is interested in the traditional arts and ancient things. He does not show any interest in funs and jokes including sex. He may have clear handwriting. He does not show much interpersonal variations.

Eysenck states that the typical introvert is quite, may retire into loneliness. He engages himself in introspection. He likes books more than people. He is friendly only with close friends. He is a serious person, shows interest in well defined mode of life. He always keeps his feelings under control and rarely behaves aggressively. He is reliable and does not lose his temper. He shows interest in great ethical values and has pessimistic attitude to life.

Neurotic Extrovert

The ***neurotic extrovert*** according to Eysenck (1947) has a tendency for hysterical disorders. He is interested in social activities, in making many people as friends. He expects that people should speak to him more. He is not interested in study or reading books. He is interested in excitements and also acts impulsively, like practical jokes. He is carefree, easy going, keeps himself moving, acting, behaving aggressively. He does not keep his feelings under control, becomes temperamental quickly. He is not a trustworthy person. His bodily development is more of horizontal than vertical. He may be suffering from stammer or stutter. He may often involved in accidents. He is often getting pains and aches. But his salivery secretion is normal. His level of mental ability is below average and his language ability is poor.

The neurotic extrovert according to Eysenck (1947) may react quickly, but erroneously. His tweezer dexterity is poor. He puts his level of aspiration at a low level. He has a tendency always to over estimate his ability. He is not rigid in his decisions, but shows more interpersonal variations. He likes modern art and interested in creative art. He likes funs and jokes of all types. He shows optimistic attitude towards life.

EYSENCK'S BIOLOGICAL BASIS OF PERSONALITY DIMENSION

We have discussed in the preceding section the introverts and the extroverts differ in their extreme levels and the varied types of behaviour patterns both the arousal and the inhibitory characteristics. Eysenck (1967) states that the differences are taking place by the intervention of the *central nervous system*. Eysenck has been revising his theory of introversion-extroversion on the basis of the results of the studies he was making time to time. This has enriched his personality theory. Though his theory was basically biological, his testing theory was behavioural. He is forced to have this behavioural testing as such testing is not possible biologically.

Eysenck claims that there is relationship between his two personality dimensions and the *ascending reticular activating system* (ARAS) of the brain. On the basis of this he has revised and extended his basic theory of introversion and extroversion. According to Eysenck (1967) as quoted from Thomspon (1967) that the ARAS accounts for the cortical function in learning, conditioning wakefulness and attention. This ARAS appears to settle down the levels of arousal, ranging from sleep to the utmost level of excited behaviour. A part of ARAS has inhibitory effect on the cortex which reduces the cortical excitation. In this way Eysenck finds physiological excitation and inhibition in the ARAS and makes it the basis of introversion-extroversion behaviour pattern. By associating the excitation and inhibition of the ARAS to personality types, Eysenck hypothesised that the introverts are aroused by the higher level of ARAS and the extroverts are by lower level of ARAS.

Actually the ascending reticular activating system (ARAS) is a system of network of many short fibres originate in the central part of the medulla, mid-brain and diencephalon carries impulses through multiple synaptic pathways up to the cortex. It is a grey matter surrounded by some white material. Much of this grey matter is composed of short, many branched interconnecting cells called brainstem recticular formation (BSRF). The part of

BSRF makes up the ARAS (Leukel, 1968). The ARAS cells are excited through many synapses and the excitation is passed from cell to cell through the centre of the brain and reach all areas of cerebral cortext. The ARAS excites the cells of the cerebral cortex and reduces their threshold from other sources to incoming stimuli. Easten (1978) states that the cerebral cortex is maintaining receptive awake condition when the impulses come steadily into the cortex. Thus the important function of the ARAS is to keep the brain awake. If the ARAS is more active then the organism is more aroused and alert. If the ARAS is inactive the cortical activity is low. This means the organism goes in a state of sleep.

Another stage in this BSRF is that the cells of the BSRF is regulating the muscle tone. They can excite and inhibit stretch reflexes. This system is called descending reticular activating system (DRAS). It plays the role of sensory feed-back to maintain the state of wakefulness. When the BSRF is active the ARAS stimulates the cerebral cortex and the DRAS increases extensor muscle tone. In an awaken state the BSRF stimulates the cortex through ARAS. Moruzzi (1954) states that the BSRF is capable of functioning from its original place by activating the brain.

The functional organisation of the human brain is so complicated that it will not be possible to ascertain how a particular function is caused in human behaviour. All the research findings are elicitated mostly from monkeys and cats. A man's reaction to cortical stimulation may vary in person to person depending upon the function and structure of the body built and the level of sensitivity of the sensory nerves and apparatus.

Thus Eysenck (1967) made his introversion-extroversion dimensional theory of personality more biologically oriented and his transition from 1957 to 1967 theory of biological functions became the basis for finding four extreme types of personality. According to this theory a person may be said to be normal or neurotic, introverted or extroverted and in between these extremes there are various degrees. Thus an individual may be normal introvert, or neurotic introvert, or normal extrovert or neurotic extrovert. According to Eysenck (1967) these four extreme types of personality dimensions are caused by the excitation-inhibition of the ARAS and the activation of the visceral brain.

A THIRD DIMENSION : PSYCHOTICISM

Eysenck has continued his laboratory experimental studies on the treatment of neurosis. He has been constantly reviewing his personality dimension of introversion-extroversion, and normality-neuroticism to see whether they have any genetic basis. He found that both the dimensions had strong genetic dependency and there were hereditary components operating in these dimensions.

The introvert, Eysenck has observed, inherits an emotionally reactive autonomic nervous system which leads him to high neuroticism and anxiety response and he is afraid of being in the presence of complex social environment. In contrast the extrovert inherits the nervous system that is not sensitive to stimulation and not afraid of being in the presence of complex social environment. It is also suggested that the extroverts may be of two types. The first type is highly sociable, outgoing and excitable behaviour. He looks like normal personality but may get into hysterical disorder. The second type of extrovert is impulsive, has anti-social urges. He may turn into a criminal, a psychopathic personality.

But there are certain criticism made on Eysenck's introversion-extroversion and neuroticism by Claridge (1967). Eysenck has revised his personality inventory several times. The last revision he made was in 1975 which he named it as Eysenck Personality Questionnaire. This inventory yielded a third independent personality dimension called *Psychoticism*. Based on this data Eysenck (1975) clarified how Psychoticism is a dimension of personality. This is Eysenck's recent discovery and it is an additional dimension of personality disorder. The person who scores high on Eysenck's Personality Questionnaire seeks to be alone, cares very little for company of others. He is more hostile to others and disregards even dangers. He makes fools of others and he may go to the level of extreme cruelty. After 1975 there was no further revision of his Personality Questionnaire. His continued research may result in adding another dimension, a fourth dimension of personality. Probably he may be completing it.

Thus Eysenck (1967) justifies that his personality dimensions have both biological and social component. Both these factors shape the individual's personality make up. The biological stimulation predisposes the individual's responses. But the stimulation may excite or may not excite depends entirely upon the environmental or social situations. Thus Eysenck was able to put a bridge between the two different strategies that we have discussed in the preceding section.

Personality measurement is not pertaining to the whole personality make up. It may be referring some aspects of personality. There are a quite number of personality traits. Each test designed has got his own hypothesis and strategies. For example Cattell (1946) first has designed a Personality Questionnaire to identify twelve primary personality traits. Later on he extended it to 16 personality factors (1972). Guilford (1936) combined introversion with three dimensions like, liking for thought as against action, liking for solitude as against society and proneness to find trouble in life. For Guilford introversion-extroversion is a very broad and comprehensive personality description. Heidbreder (1926) found out a mixed type, a middle type called ambivert which falls in between the two extremes of introversion-extroversion.

Around 1950s Hathaway and McKinlay (1951) have devised the Minnesota Multiphasic Personality Inventory (MMPI). This was designed for clinical judgments but could measure several normal personality traits. It consists of 550 statements with three response categories, true, false and cannot say. The items cover wide range of topics such as physical conditions, psychosomatic symptoms, social attitudes, family and marital factor, psychiatric symptoms. There are fourteen scales in this commonly used for personality analysis. Out of these, four are validity scales, remaining ten are clinical scales. Out of ten, eight are for clinical diagnosis one scale is for masculinity-faminity and one is for social introversion. Based upon this MMPI there are about 200 scales have been empirically devised. This Inventory has been translated into twenty-six languages.

The three dimensional personality Inventory is a recent development that Eysenck has made. Any personality inventory can be used to assess the personality traits of a person or a group. But if we need a specific dimension like Psychoticism, extroversion and neuroticism and the other extremes of these three let us go for Eysenck's Personality Questionnaire. The historical development of this inventory and the characteristics the inventory can identify reveal the real contributions that Eysenck has made in the field of personality measurement. This is what we have described in the preceding section. It is important that every student of psychology should know to what extent Eysenck has perspired to establish his dimensional theory of personality. Any study on personality in these days is imperfect and unconvincing without making some valuable references to Eysenck's study of personality measurement.

EXPERIMENT 45 : ASSESSING DIMENSIONS OF PERSONALITY (Eysenck Personality Questionnaire)

Personality is one of the most interesting part of psychological studies. Both in the common sense world and in the scientific world the concept personality is invariably used in conversations, discussions or in describing a person. In a conversation between two persons, one may ask the other, 'You know Mr. X, please tell me what sort of a person he is'. The other person only may describe about him in terms of his characteristics. He may say that Mr. X is a pleasing and forceful person. He is energetic, attractive, cheerful, self-reliant, cooperative and intelligent. Or he is a highly persistent, hot tempered, selfish, uncooperative, hostile, suspicious and so on. Therefore, personality refers to the type of behaviour pattern of a person. The behaviour pattern may be pleasing or insulting, right or wrong, good or bad for others. It may be agreeable or disagreeable to his fellow beings.

There are thousands of adjectives and adverbs that can be used to describe a personality. We may be using some of them when we speak or write about a person favourably or unfavourably. Personality is defined in various ways based upon the person's physique, conduct, character, personal mannerism, social mannerism, and so on. Garrett (1961) states that personality is an organization of characteristic behaviour traits which serves to describe a person. According to Woodworth (1952) it is the total quality of an individual's behaviour as it is revealed in his habits of thought and expression, his attitude and interest and his manner of acting. Kimble and others (1984) state that personality is a unique organization of permanent characteristics that puts the individual apart from other individuals. Thus there are several such definitions of personality.

All these definitions include many aspects of the individual's personality characteristics. It will not be possible to include all aspects in a study of personality. As stated in the definitions it involves broad areas like behavioural traits, or total quality of behaviour or organization of characteristics. Therefore, psychologists have simplified the problem by classifying the personality into some traits or dimensions. These traits or dimensions similar to the dimensions of ability as stated in terms of verbal ability, numerical ability and spatial ability and so on. Now the problem is whether any outstanding traits or dimensions of personality can be identified. Any personality inventory cannot reveal all the traits or dimensions sharply for our purpose. Since there are hundreds of names of dimensions, many of them may be of similar type and also there may be overlapping one trait over the other.

An extensive statistical and correlational method with factor analysis may identify personality dimensions sharply to some extent. Studies of this type were made earlier on the area of mental abilities by Thurstone (1936) and others. In the field of personality assessment Cattell has identified 16 personality dimension using factor analysis. But a systematic correlational and laboratory experimental studies with factor analysis were made by Eysenck. Eysenck's studies were in continuous process right from 1940s. His studies on clinical patients in the Mill Hill Hospital made him to develop curiosity on the reliability of the psychiatric diagnosis made by the psychiatrists. He has raised an important issue that if two or more psychiatrists do not agree in their diagnosis what kind of treatment can be prescribed? But his work was only to collect data and do the required statistical correlations. Further he has made huge collection of data to test some particular hypothesis on personality, Eysenck also tested the validity of Jung's psychological types, the introversion-extroversion and he found that the description of the characteristics introvert and extrovert did not sharply conform the Jungian definitions, but fairly close to them.

The assessment of introversion-extroversion is made by questionnaire of self-report measures. Eysenck's (1957) measure of introversion-extroversion is made by the Personality Inventory (EPI) designed by him.

Eysenck has established his dimensions of personality empirically, by factor analysis. It is a mathematical technique of classifying traits on the basis of relative strengths of inter-correlations among them. Apart from other types of technique used for assessing personality dimensions, introversion-extroversion inventory was also sometimes used.

Eysenck considered that extroversion is one of the three dimensions of personality to be of prime importance. The other two are neuroticism and psychoticism. For sometimes Eysenck (1947) has used tough minded, tender-mindedness in place of psychoticism. These three dimensions are independent of each other and provide a variety of normal and abnormal personality characteristics. With respect of extroversion dimension it was found that most people are actually ambiverts rather than extreme extroverts or introverts. They may be more or less stable or neurotic irrespective of their position in extroversion. The third dimension, psychoticism characterises the type of behaviour that is eccentric, impersonal, hostile and anti-social. An individual person can be located anywhere in the three-dimensional space. This model of personality is to be greatly appreciated because the concept extroversion is partly related and independent from other personality traits.

The modern researcher faces a dilemma whether to accept Cattell's sixteen dimensions of personality and carry out the research or to accept Eysenck's three dimensions of personality and carry out the research. But apparent difference between Cattell and Eysenck is not very sharp. The differences seem to be only at the surface level. When you go deep into these two theories of dimensions the differences disappear. It is clear that Cattell's sixteen personality dimensions are not independent of one another. They are all interrelated and they can be factor analysed and clustered into few. Probably they may fall under two or three major dimensions as Eysenck has suggested. Further the names are given to factors arbitrarily. They can be modified. Further with very large number of factors the description will not be grisp and accurate. There may be several overlappings of one factor over the other. Cattell's primary factors give detailed picture of the personality. But it is doubtful to what extent they are reliable. It is very difficult to verify experimentally from biological point of view.

In the later researches Eysenck has demonstrated that his three dimensional systems of personality can be enlarged to include several forms of psychiatric disorders. He has revised his personality inventory from time to time adding his findings. The final revision of his personality Inventory was made in 1975 under the title of Eysenck's Personality Questionnaire-R (EPQ-R).

The three important personality dimensions, Psychoticism, Extroversion and Neuroticism are measured by means of ninety questions. These 90 questions he has selected from a large number of items after making factor analysis. The extroversion dimension is opposed to introversion. Hence, by stating extroversion the opposed traits also are involved. Extroversion, therefore, refers to outgoing, uninhibited, sociable. Neuroticism is characterised by general emotional liability of a person. The person is emotionally over responsive or stressful state. This is his liability to neurotic tendency. The third dimension, psychoticism is an independent one. This is characterised by solitary, troublesome, cruel, lacking in feeling of sympathy, hostile to others. The person is sensitive seeker. He likes strange and unusual things.

Eysenck has made his earlier studies only by paper-pencil tests, like (1) The Maudsley Personality Inventory (1959), (2) The Eysenck Personality Inventory (1968) and (3) Eysenck Personality Questionnaire-R, (1975). Since he has been consistent in his laboratory based and empirically testable hypothesis for the treatment of neurosis Eysenck (1957, 1969, 1973) has several times looked into his description of personality with genetic basis of the dimensions of introversion-extroversion and normality-neuroticism. The conclusion he has made was that both of these two dimensions are depending upon genetic component and hence he stated that both dimensions have inherited factors.

According to Eysenck the introvert inherits a nervous system which can make him conditioned responses easily and if an introvert inherited emotionally reactive ANS he will be acquiring strong anxiety responses (high neuroticism). The introvert will be sensitive to stimulation, fearful in a complex and demanding social situation. On the other hand an extrovert inherits the types of nervous system that is easily conditioned, not sensitive to stimulation, not fearful of complex and demanding social environment.

The first personality inventory revealed that the introvert is more closely related to high neuroticism, but the extrovert (hysteric) resembled normal on the neuroticism dimension. Eysenck (1969) was not satisfied with this result as the inventory had more biased items. To set right this he formulated a new instrument called Eysenck Personality Inventory. Through this inventory he was able to confirm the independence of neuroticism and introversion dimensions. This inventory revealed that extreme introversion is by nature more neurotic personality than extroversion. This confirmed the independence of these dimensions. Eysenck (1975b) was still further driven to the necessity of designing a third instrument called Eysenck Personality Questionnaire in which he included more items to measure a third dimension of personality disorder called psychoticism. Eysenck believed that this third dimension like the other two is also inheritable. It is independent and also consistent with neuroticism and introversion-extroversion. This is how briefly we can state Eysenck's development of personality dimensions.

The personality dimensions, (1) psychoticism, (2) neuroticism, and (3) extroversion-introversion as conceived by Eysenck can be briefly summarised for the purpose of exploiting the results of the ensuing experiment.

(1) Psychoticism (P)

Psychoticism is an additional dimension of personality. It is a recent addition to Eysenck's (1975b) personality dimensions. It is a disordered personality dimension inheritable like extroversion-introversion and neuroticism. This is supposed to be consistent with the other two dimensions and also independent. An individual who scores high on this scale is a psychotic. Eysenck (1975b) has described him as a loner. He prefers to be lonely and wishes to be free from the company of others. He is overtly hostile to others, including close relatives. The psychotic is afraid of dangers. He makes fools of others and enjoy. He may be sometimes taken to extreme cruelty. The psychotic has a peculiar characteristics of enjoying unusual things.

(2) Neuroticism (N)

The individual who scores high on this scale is considered to be a neurotic person. He is reactive emotionally. He is highly anxious, moody and depressed. He is often disturbed in his sleep. He often feels fatigue experiences and various kinds of bodily pains. According to Eysenck (1975a) he is always preoccupied with things that may go erroneously. He will have strong emotional reaction to all kinds of thoughts.

(3) Extroversion (E)

This is one of the three dimensions of personality as conceived by Eysenck (1980). The other two are Neuroticism (N) and Psychoticism (P). These three according to Eysenck are independent of each other. The three are supposed to provide all the characteristics of normal and abnormal personality. The person who scores high on extroversion scale is one who likes people very much. He is sociable, he is strongly attracted to exciting and novel situations. He is easy going and very optimistic. The extrovert likes very much practical jokes. He may sometimes become impulsive and aggressive and go to the extent of even assaulting others. He is unable to control his temperaments.

Introversion (I)

This is the other extreme of extroversion. Persons who score low on the Extroversion scale are supposed to be introverts. The introvert may be described as shy person. He is quite. He often withdraws from situations. He is isolated from social situations. But he is a thinker and planner. Eysenck (1975a) says that he may be also expecting some problem before they occur. The introvert avoids overt activities, excitements and stimulations. He strictly controls his emotions and keeps under check. He desires a quite and well ordered life. He is highly pessimistic.

Eysenck thus by presenting a biologically based dimensions of personality has been completely successful in demonstrating a strategy that has bridged the gap between observable phenotype and inferred or casual genotype.

EXPERIMENT

Problem

To assess the personality dimensions of Extroversion, Neuroticism and Psychoticism as conceived by Eysenck.

Method

Participant

Choose a person who has no knowledge about this test. Describe his/her characteristics like age, sex, qualification, experience, whether he/she has any handicap, wear eye glasses and so on.

Material

Eysenck's Personality Questionnaire 1975 revision Manual, Test-booklet, Key, Scoring Sheet.

Description of the Material

The Eysenck Personality Questionnaire-R is the result of several years of work by Eysenck (1960). The 1975 revised edition has been designed by H.J. Eysenck and S.B.G. Eysenck. This Questionnaire measures three personality dimensions, such as Psychoticism, Neuroticism and Extroversion. This questionnaire consists of ninety questions carefully selected out of many items after factor analyses. According to Eysenck Psychoticism is an independent variable. There are four scales in this. They are the Psychoticism scale, Neuroticism scale, Extroversion scale and the Lie scale. The Psychoticism scale consists of 25 questions, the Neuroticism scale 23 questions, the Extroversion scale 21 questions and the Lie scale 21 questions. The responses to these questions are made by 'Yes' or 'No'. There is a box against each Yes or No answer and the subject has to put a cross (×) mark in the box against the answer.

Procedure

The Eysenck Personality Questionnaire may be used both as an individual test or as group test. Keep the participant seated comfortably in a well illuminated room and hand over the questionnaire. Give the following instructions :

> "Please answer each questions by putting a cross mark (×) in the box against 'Yes' answer if you agree with the idea asked in the question. If you do not agree with it put a cross mark in the box against 'No' answer. There is no right or wrong answers or tricky questions. Answer quickly and do not think too long over the meaning of the questions. Now, 'ready, go'."

As soon as the respondent answered the questionnaire the experimenter should collect it and allow the respondent to go away.

Results

Then the experimenter should scrutinize, whether all the questions are answered. Then he/she can score the Questionnaire according to the key and find out the individual score on all four scales, P, N, E and L. If the *Lie score for male exceeds seven it should be rejected* and in the case of *female if it exceeds eight it should be rejected.* If the individual scores on the Lie scale for male and female are 7 and less or 8 and less than it should be accepted and the raw scores should be presented in Table 15.1.

Table I5.1 : Showing the raw scores of the individual in the Eysenck Personality Questionnaire

S. No.	Scale	No. of items	Highest possible score	Score obtained in each scale
1.	Psychoticism	25	25	
2.	Neuroticism	23	23	
3.	Extroversion	21	21	
4.	The Lie Scale	21	21	
	Total	90	90	

State the scores of your participant in each scale. State the highest score in each scale and the significance of it. What your participant has obtained and describe his characteristics, as hypothesised by Eysenck.

Collect group data from other experimenters for each scale and present them in Table 15.2. Work out the mean for the group in each scale. Work out the inter-correlation between the scores of the scales using the group data and present them in Table 15.3. State what is the group mean for each scale. That is the correlation coefficient between the scores of the scales.

Table 15.2 : Showing the group data in Eysenck's Personality Questionnaire

S. No.	Name	Psychoticism	Neuroticism	Extroversion	Lie	Total
1.	S.V.					
2.	P.K.					
3.	M.S.					
4.	K.R.					
5.	T.V.					
6.	S.L.					
7.	P.V.					
8.	R.M.					
9.	P.M.					
10.	B.V.					
	Mean					

Table 15.3 : Showing the Inter-correlation between the scores of four scales P, N. E and L of Eysenck's Personality Questionnaire (R)

Scale	Psycho.	Neu.	Extro.	Lie
P	×			
N		×		
E			×	
L				×

Make entries in the columns and rows in the correlation matrix with the correlation coefficients and test the significance of them.

Discussion

Discuss the scores as revealed by the EP for the individual and for the group. In the psychoticism scale where your participant stands? Is he/she identified with Eysenck's description of psychoticism or varies? In the neuroticism scale what your participant has scored? Is it low or high? What it describes? State whether your participant is more of extrovert or introvert or ambivert, where he/she falls in the range of scores. What is his/her Lie score? What do you consider about this?

Describe the characteristics of the group in the light of the scores of the four scales. Describe each person in the light of the characteristics of each of the three dimensions of personality. State where your participant falls on the three dimensions. Do you find psychoticism as an independent dimension in this study. How many are extreme psychotics? Explain how? How many in the group are normal and how many are extreme neurotics? How many in the group are extreme extroverts and how many are extreme introvert? How many fall in between these two ?

In the inter-correlations between dimensions, are the correlation coefficients higher or lower, state them. What low correlation indicates and what high correlation indicates?

Conclusions

1. What personality dimensions your participant belongs to?
2. How many in the group are Psychotics, Neurotics and Extroverts?
3. Are there any extreme cases?
4. Do you find any variation in personality dimensions of male and female persons?

Simplified Scorings
KEY for Eysenck Personality Questionnaire – R

S. No.	Psychoticism	S. No.	Neuroticism	S. No.	Extroversion	S. No.	Lie Score
2.	No	3.	Yes	1.	Yes	4.	No
6.	No	7.	Yes	5.	Yes	8.	No
9.	No	12.	Yes	10.	Yes	13.	Yes
11.	No	15.	Yes	14.	Yes	16.	No
18.	No	19.	Yes	17.	Yes	20.	Yes
22.	Yes	23.	Yes	21.	No	24.	No
26.	Yes	27.	Yes	25.	Yes	28.	No
30.	Yes	31.	Yes	29.	No	35.	Yes
33.	Yes	34.	Yes	32.	Yes	39.	No
37.	No	38.	Yes	36.	Yes	44.	No
43.	Yes	41.	Yes	40.	Yes	48.	No
46.	Yes	47.	Yes	42.	No	51.	No
50.	Yes	54.	Yes	45.	Yes	55.	Yes
53.	No	58.	Yes	49.	Yes	59.	No
57.	No	62.	Yes	52.	Yes	63.	No
61.	No	66.	Yes	56.	Yes	69.	No
65.	Yes	68.	Yes	60.	Yes	73.	No
67.	Yes	72.	Yes	64.	Yes	78.	Yes
71.	No	75.	Yes	70.	Yes	81.	No
74.	Yes	77.	Yes	82.	Yes	85.	No
76.	Yes	80.	Yes	86.	Yes	89.	Yes
79.	Yes	84.	Yes				
83.	Yes	88.	Yes				
87.	Yes						
90.	No						

EXPERIMENT 46 : ASSESSING INTROVERSION-EXTROVERSION
(Neymann - Kohlsted Diagnostic Test)

The results of psychological studies on human beings are presented and interpreted in various ways by researchers. All these are concerned with human behaviour. Behaviour is the revelation of human nature or human mind. The human mind is manifesting in various characteristic ways which may show what type of person one is, what are his ambitions, aspirations, interests, attitudes, likes and dislikes, aptitudes, intelligence and personality traits and so on. Since every psychologist is interested in the study of one of these aspects enormous amount of information about human nature have been collected and stored in volumes of works. Certain informations are accepted readily, certain informations are accepted after discussion, still certain after verification and certain information become controversial and also subjected to severe criticism. The study of personality traits is one such phenomenon that has often been raising much controversy among the psychologists. This resulted in giving birth to various theories of personality.

Just like, it is said that "all roads lead to Rome", all personality theories are only studies of human nature. Every theorist follows one method of study which is most suitable for him. He may claim that he alone is right. But it will not be appropriate for one who studies personality and personality traits to state that a particular theory alone is correct and others are not correct. Every theorist brings out something new which are not found in others and what he states also seems to be convincing and agreeable. As Jung (1936) has said, whoever looks into the mirror of the water will see first of all his own face, the mirror does not flatter, it faithfully shows whatever looks into it, namely the face. What Jung has stated is not agreeable to some though outwardly they do not express it.

The important problem that psychologists confronts is whether it is possible to study a person thoroughly and attribute the same characteristics to everybody or is it possible to generalize them? Many have tried with all their inventions of instruments and test materials to evolve a universal law or theory that could be common to all. The individual differences, the personality psychologists observed among the people became the fundamental concern and all personality researches and theories had this as the centre of their problem.

Two research traditions in psychology side by side independently have been formulating strategies in research. The experimentalists and the correlationists have been going on parallel to each other like the railway lines. Experimentalists used control as the essence of their experimental method. They compared the performance or test scores of the experimental group with that of the control group. The experimental group is exposed to the influence of some independent variable whereas the control group is kept away from the influence of the independent variable. These two groups of subjects are given some task to perform and then it is measured. If the independent variable has exerted any influence in the performance of the experimental group then it would be differing from the performance of the control group. Here the experimentalist is interested only in the difference between the two groups in the performance. He is not interested in the individual performance in the group. The average performance of the two groups is taken and compared. In the laboratory study the experimentalists wanted to be more precise and accurate and make abstraction.

But the correlationists are interested in the difference that exists among the individuals of their study. The individual's reaction to the stimulus is revealed in the difference among the subjects of the group. How an individual has reacted to a stimulus and how another individual reacted to the same stimulus. This has been tested and the level of relationship between the individual the differences is measured. The other variables that caused difference in the individual's relationship also assessed. Thus one discipline, the experimentalists are interested in the difference between and among groups while the correlationists are interested in the individual differences within a group of subjects.

The correlational method was found to be more practical and suitable for testing many problems. This was revealed in one case when Alfred Binet faced the problem of the different mental abilities of the school children in Paris. Binet's problem, then, was to determine the most suitable school environment for the problem child and not to find out the type of school for the average child.

Binet has designed intelligence tests in varying degrees of difficulty and tested the children's mental abilities by the correlational method. Binet has faced with unusually difficult problems. It was to deal with a serious practical problem of classifying the children studying in the same class, some as brilliant and some as dull or feeble minded. Therefore, he has taken into consideration the teacher's biases in identifying the students as dull and bright, the intellectual and motivational factors of the children and the properties of the school situations. Thus on one side accuracy and abstractions by the experimentalists, on the other side the level of relationship in the differences between individuals. Two different strategies directed towards the assessment of human nature on various characteristics. Why these two should function independent of each other? Cronbach (1957) has stressed the merger of these two strategies for a better assessment of human characteristics, whether it is intelligence or motivation, aptitude or personality traits.

Generally investigations on intelligence, motivation, aptitude or personality are concerned with only individual differences. Normally these investigators do not go beyond the observation of individual differences. Once an individual difference is established by the experimental method either by applying intelligence test, or aptitude test or personality inventory investigators can go one step further by associating the performance or the scores obtained in these tests with that of the performance or scores of other laboratory experiments like learning, perception or judgment and so on to see whether there is any relationship between the two types of test scores. This is one way of synthesizing the experimental method and the correlation method. This type of study does not involve much difficulty. But the only problem is that both types of testing must be conducted on the same subject.

Cronbach has strongly emphasized the merging of these two methods. But it can be done only on a small sample. Having accepted this as a healthy suggestion investigators must strive for inventing better ways of implementing the suggestion on larger samples. From the Cronbach's statement it is understood that till 1957 no investigator has tried to merge the experimental and correlational strategies in his studies. But the studies made by Eysenck (1957, 1967) and presented by him in his works seem to be very close to the suggesion stressed by Cronbach. Monte (1977) states that Eysenck has made the true merger of the experimental and correlational strategies.

Let us not go into the details of Eysenck's theory of personality dimensions, but let us see how he has merged the experimental and correlational strategies in his works. Three of Eysenck's (1947, 1952, 1953) early works were mainly based upon his theory of personality. The important concepts he used to describe his personality dimension are introversion and extroversion. The personality type of an individual can be established by a paper-pencil personality inventory. Eysenck himself has designed one personality inventory called The Maudsley Personality Inventory in 1959. Later on in 1975 he has revised it and called it as Eysenck Personality Questionnaire in collaboration with B.G. Eysenck (1975 a and 1975b). The individual score on this instrument is compared with norm of the group and the individual's degree of introversion or extroversion is assessed and his personality type is established. Then his performance on the other laboratory tests was correlated. Eysenck was able to see that parsons who scored high on the introversion personality test learn verbal material more quickly than the individuals who scored high on extrovert test. This means according to Eysenck, the introverted persons are very sensitive and in them the central nervous system is highly aroused in the processes. In the case of extroverted persons, they are less sensitive and have more inhibitory brain processes. These ideas are derived from the correlation made between the individual's test scores on personality and their subsequent laboratory performance. This type of study, merging the experimental method and the correlational method can be adapted in studies wherever feasible.

Introversion-extroversion was the prime theme of study for Eysenck during 1950s and also thereafter. Almost around the same time in America attention was drawn to the problem of introversion-extroversion. It was felt that in order to eliminate certain experimental problems there should be accurate differentiation of introversion from extroversion.

It was generally thought that introversion and extroversion were normal personality traits. But it was assumed that certain types of insanity were associated with introversion and extroversion. For example, the behaviour patterns the schizophrenics seem to be typical introversion and that of the mania depressive psychotics seem to be typical extroversion. Neymann and Kohlstedt assisted by Morgan have formulated a test to verify the hypothesis. The test is designated by Neymann-Kohlstedt introversion-extroversion diagnostic test.

The important characteristic of this test development is that large number of questions were administered on one hundred schizophrenics and one hundred manic depressive psychotics and the best 50 questions which yielded highly significant reliability were selected. Later on when the test was administered on one thousand professional students the scores yielded a normal probability curve. Therefore, the test is accepted as one of the best diagnostic tests for introversion-extroversion. But the test developers intention was not to merge the experimental and correlational methods in their studies even though the development of the test was based upon identifying the introverts and extroverts among the persons who are on the border line of abnormality. This test has been used very successfully in psychiatric hospitals for long time. Investigators who made personality measurement have found this test was very much reliable in identifying introversion-extroversion traits in United States. But now it is used widely in many universities.

EXPERIMENT

Problem

To assess the personality traits of an individual and to identify whether he/she is introvert or extrovert type of person.

Method

Participant

Choose a person who has no knowledge about this test. Describe his personal characteristics such as age, sex, qualifications whether he/she wearing eye glasses or having any deformities.

Materials

The Neymann-Kohlstedt Diagnostic test which identifies introversion-extroversion personality traits.

Description of the Test

The Neymann-Kohlstedt Diagnostic test was developed by these two authors and it is named after them. J.B. Morgan of the North Western University has assisted them in making the test more of a diagnostic type. It was first considered that introversion-extroversion is a normal personality trait, but later on it was found that certain abnormal characteristics were associated with introversion and extroversion. Particularly Schizophrenia was associated with introversion and manic depressive psychosis was associated with extroversion. This test is capable of identifying these characteristics.

This test consists of 50 items, each one is answered with 'Yes' or 'No' words. The person who takes up the test is asked to express his/her personal like or dislike for each statement. If all the items are correctly answered then the result is pure or extreme extroversion.

Procedure

This test is used to assess the personality traits that are required for various types of vocational success. Persons who are temperamentally suitable for some particular jobs are identified by this test. It is also used in the clinical situations. This test should be handled or administered by well trained psychologists or psychiatrists. This can be used as an individual test and also as a group test. The participants should be made to sit comfortably in a well illuminated room where there should not be any distractions. The experimenter should hand over the test booklet to the participants and ask them to provide the personal information asked for at the top of the booklet. As soon as they did this the experimenter should give the following instructions to them.

> "Please listen here. There are 50 statements in this booklet. Read each statement carefully and state whether you like or dislike the idea expressed in the statement. You will have to express your own reaction to the idea. You should not think in terms of what it ought to be or might not to be. State your immediate reaction to the idea given in the statement. Do not take time to think over it. If the idea is agreeable to you answer with yes, circle the word 'Yes'. If the idea is not agreeable to you answer with 'No', circle the word 'No'. Your quick judgment is needed. There is no time limit, but you will have to finish it as quickly as possible. Please answer all the items. Don't omit any of them."

As soon as the test booklets are answered the experimenter should collect them and allow the participants to go away. Then he/she should scrutinize the answered test booklets and score them according to the procedure.

Method of Scoring

The answers should be checked keeping the key of the test. See Appendix I. Every correct answer according to the key should receive one point. Since the test is fully an extrovert test, a pure extrovert should have answered all the items according to the key with 'Yes' answer and his score should be plus 50. Instead a pure introvert should answer all the items with 'No' word and his score should be minus 50. Thus in this test the scores range from minus 50 to plus 50. The individual's score is: number of items answered correctly minus number of items answered incorrectly.

Statements answered with both 'Yes' and 'No' words are not counted. They are eliminated. Suppose, for example, one person has answered forty items correctly (extrovertly) and three items not answered. Then his score should be forty minus seven or number of items not answered correctly. It is seven. Since he has not answered three items they are eliminated. Hence forty minus seven or plus thirty three is his score, The scoring of this test is straight by the formula: number of items correctly answered – (minus) number of items incorrectly answered. But if there is any item unanswered it should not be accounted in scoring.

Results

The experimenter should score each answered booklet according to the procedure described above. The individual score should be obtained by deducting the incorrect answers from the correct answers and it should be presented in Table 15.4. Examine whether the individual's score is plus or minus. The extrovert should have plus score and the introvert should have minus score.

Table 15.4 : Showing the individual score in the Neymann-Kohlstedt Diagnostic Test for answering the personality type

No. of items correctly answered	No. of items incorrectly answered	Difference (Plus or Minus) (Score)	Personality Type

Find out your participant's score. State whether it is plus or minus. If plus or minus state the range of scores whether it is in between plus ten and minus ten. Individual's score between these two points (+10 and –10) indicates neither extroversion nor introversion. In this type of cases the experimenter once again should examine the answers and check up the score. The experimenter can collect group data from other experimenters and present them in Table 15.5.

Table 15.5 : Showing the group data on Neymann-Kohlstedt Diagnostic Test for answering the personality type

S. No.	Name	No. of correct answers	No. of Incorrect answers	Difference (plus or minus) (Score)	Personality Type
1.	A.S.				
2.	B.V.				
3.	K.R.				
4.	K.P.				
5.	M.S.				
6.	P.K.				
7.	S.T.				
8.	R.S.				
9.	S.P.				
10.	T.V.				
11.	R.A.				
12.	T.R.				
	Mean				

Work out the mean scores for the group and compare individuals with mean score of the group. How many obtained plus score and how many minus score? Is there border line cases.

Discussion

Discuss the individuals scores. State whether his/her score is plus or minus? Is it falling on the border line or high on plus dimension or on minus dimension. State whether he/she is an extrovert or introvert? State also the degree of the extroversion or introversion. What type of personality he/she is?

What is group mean score? Is it plus or minus? How many of them fall above the group mean and how many of them below the group mean? How many of them are falling on the border line between plus ten and minus ten. Who scored the highest plus score. What does it indicate? Who scored the highest minus score? What does it indicate? Where your participant falls in the group? Whether the group as a whole is an extrovert type or introvert type? State the reason, if any from the personal information. Is there difference in the personality type due to sex? If so, explain who are those extroverted males and females and introverted males and females. Discuss your results comparing with the results of some earlier studies.

Conclusions

1. State whether your participant is extrovert or introvert by giving his/her score.
2. State the group mean score. Is it plus score or minus score?
3. How many in the group are extroverts and how many are introverts?
4. Is there much sex difference.

APPENDIX - I

KEY FOR NEYMANN-KOHLSTEDT DIAGNOSTIC TEST OF EXTROVERSION-INTROVERSION TYPE OF PERSONALITY

1.	NO	26.	NO
2.	YES	27.	YES
3.	NO	28.	NO
4.	YES	29.	YES
5.	NO	30.	YES
6.	NO	31.	NO
7.	YES	32.	YES
8.	NO	33.	NO
9.	YES	34.	YES
10.	NO	35.	NO
11.	YES	36.	YES
12.	NO	37.	NO
13.	NO	38.	NO
14.	YES	39.	YES
15.	NO	40.	YES
16.	YES	41.	NO
17.	NO	42.	YES
18.	YES	43.	NO
19.	NO	44.	NO
20.	YES	45.	YES
21.	YES	46.	NO
22.	NO	47.	YES
23.	YES	48.	NO
24.	YES	49.	YES
25.	NO	50.	YES

EXPERIMENT 47 : ASSESSING PERSONALITY TRAITS

INTRODUCTION

Personality is defined in various ways. There is no one single method by which we can study personality accurately. Personality finds its origin in the consistency of an individual's behaviour over a period of time in different situations. People tend to behave consistently because they adapt what they learn in one situation to another situation. An individual acquires a characteristic and a uniform system of behaviour. Whatever consistency takes place in one's behaviour his personality traits are not permanent like his name. Time to time when one gains experiences, old patterns of behaviour are modified and new patterns are tested. Further personality is influenced by situational variables. This may have influence in one's motivation. A shy boy may develop into a gregarious and an aggressive adult.

Generally man is interested in his own and other people's personality. The main theme for the literary, dramatic and for the art of painting is the portrayal of personalities and their interactions. Despite the long usage of the concept 'personality' by the philosophers from the ancient time the systematic study of personality and its measurements were not taken place until the beginning of twentieth century. The first systematic work on personality carried out by psychiatrists like Kraeplin (1905), Freud (1925), Jung (1923), Adler (1935) and his followers using clinical methods.

Personality may be defined as the relatively stable organization of a person's motivational dispositions, arising from the interaction between biological drives and the social and physical environment. The term implies both cognitive and physical attributes and also refers chiefly to the affective cognitive traits, sentiments, attitudes, complexes and unconscious mechanisms, interests and ideals which determine one's characteristics or behaviour and thought. From the social point of view personality has been applied to the impressions made on others rather than to the inner self. Psychometrically specialized scholars like Thurstone (1954), Cattell (1949), Guilford (1959) and Eysenck (1947) accept in reality of stable personality traits of varying degrees of generality. They state that personality traits cannot be directly observed. They must be inferred from consistencies in speech and behaviour. They try to isolate the main dimensions of personality through factor analysis of correlations between tests. Therefore, personality is conceived as a person's scores on all the measurable factors or common traits. Emotionally instability or anxiety and extroversion-introversion are widely accepted. Some stress the need for intuitive understanding of the unique organization of each individual personality as against measurement of his attitudes. This is subjective approach. Allport (1937) defines personality as the dynamic arrangement in the individual of those psychophysical systems which determine his unique adaptation to his environment. Guilford (1959) considers that an individual's personality is a unique structure of traits. For Murray (1938) personality is the controlling organ of the body effecting change from birth to death.

Character in the psychological sense is equated with personality. The concept of character has emphasised the static element more than personality which has stressed the functional dynamic self-evolving and processful aspects. Therefore, personality is that which has emerged from his character to date under the influence of his environment. By and large it can be stated that the American scholars view personality in terms of external behaviour, superficial traits, motor components, inter-human relations and modifiability, and investigate mainly the interaction between the individual and the social environment (culture). But the European scholars on the other hand tend to favour concepts such as profound disposition, determination by constitution, fixed structure, relative independence from society, and relative unchangeability.

Personality Inventories play important role in the measurement of personality characteristics. They measure several different aspects of personality at one time. It saves both time and cost in the administration. The Bernreuter Personality Inventory is of very immense use for those who are interested in assessing personality characteristics of industrial employees and other employees of different types of organizations. The advantages of this Inventory is that it consists of six scales and for all the scales the same statements are used with different scoring keys. The Inventory is a bipolar type, having both the plus scores and minus scores. The six scales are Bl-N, B2-S, B3-l, B4-D, Fl-C and F2-S. These scales measure neurotic tendency, self-sufficiency, introversion-extroversion, dominance-submission, confidence in oneself and sociability.

EXPERIMENT

Problem

Assess the personality traits of your subjects using Bernreuter Personality Inventory.

Method

Subjects

Describe the subject's characteristics, whether they are students or adults, their age, other physical conditions, etc.

Materials

Bernreuter Personality Inventory, Manual, test booklet, scoring sheet, and key and norms.

Description of the Test

This inventory was developed by Robert G. Bernreuter (1938) for measuring various personality traits. This is very widely used test with more accurate assessments. It consists of the following 6 scales. The Inventory consists of 125 questions and every scale is scored on all 125 questions.

Scale B1-N

This scale the measure of *neurotic tendency*. Persons scoring high on this scale tend to be emotionally unstable. Those who are scoring above 98 percent would benefit from psychiatric or medical advice. Those who are scoring low on this tend to be very well balanced emotionally.

B2-S Scale

This scale the measure of *self-sufficiency.* Persons who are scoring high on this scale prefer to be alone, rarely ask for sympathy or encouragement and tend to ignore the advice of others. Those who are scoring low on this dislike solitude and often seek advice and encouragements.

B3-I

This scale is the measure of introversion-extroversion. Persons who are scoring high on this scale tend to be introverted. They are imaginative and tend to live within themselves. The scores above the 98 percentile bear the same significance as done in similar scores on the B1-N scale and would benefit from psychiatric advice. Those who are scoring low are extroverted, that is why they rarely worry, seldom suffer from emotional upsets and rarely substitute day dreaming for action.

B4-D

This scale is the measure of *dominance-submission*. Persons who are scoring high on this scale tend to dominate others in face-to-face situations. Those who are scoring low tend to be submissive.

F1-C

This scale is the measure of *confidence in oneself.* Persons who are scoring high on this scale tend to be hamperingly self-conscious and have feelings of inferiority. Those who are scoring 98 percentile would probably benefit from psychiatric or medical advice. Those who are scoring low tend to be wholesomely self confident and to be very well adjusted to their environment.

F2-S

This scale is the measure of sociability. Persons who are scoring high on this scale tend to be non-social, solitary or independent. Those who are scoring low on this tend to be social and gregarious.

Procedure

This is a self administering Inventory. It does not require any instructions. To ensure careful reading, the administrator of the Inventory should read the instructions on the first page aloud while the individuals are reading them silently. There is no time limit. It can be completed in 25 minutes. The examiner should give the instructions in such a way that it is not an intelligence test nor requires right or wrong answer. The following instructions may be given to the subjects.

> "There are 125 statements in the test booklet and you will have to answer all of them. Please do not omit any of them. Make no mark on the test booklet. You are to read each question in the test booklet and mark all your answers on the separate answer sheet.
>
> "The questions on the test booklet are intended to indicate your interest and attitudes. For each question there are three types of answers given in the answer sheet with "Yes", "No", ?".
>
> "If your answer to the question is 'Yes', mark a circle around 'Yes' to the question you are answering. If your answer is 'No' make a circle around 'No' . If you could not answer with either 'Yes' or 'No' make a circle around the question (?) mark".

Scoring

The test booklets should be collected as soon as they were answered by the subjects. There are six keys one for each sub-scale. Each scale should be scored separately by checking the answers with the key. The first scale is Bl-N. For Bl-N all the answers for all 125 questions should have a score. Write down on the score sheet against the number of the question the corresponding score as given in the key. The same answer either it is 'Yes' or 'No' or '?' is having a separate numerical value for each scale in the key. Therefore, there are six scales and six sets of scores. There may be plus score and minus score. The individual score is the difference between the two. Therefore, the individual score may be either + or –. This score can be treated statistically for your purpose. Find out the percentile scores of the individual for each scale from the table.

Results

The individual score for all the six scales should be given in Table 15.6 and state how your participant has scored in each sub-test. Whether your participant score is plus or minus. State what plus indicates and what minus indicates. In how many tests your participant scored plus score and in how many minus score?

Table 15.6 : Showing the individual score and the percentile score on the six scales of the Personality Inventory

Category	B1-N	B2-S	B3-1	B4-D	F1-C	F2-S
Plus						
Minus						
(Score) Difference						
Percentile Score						

Collect the group data from other experimenters and present them in Table 15.7. In the group how many had obtained plus scores and how many minus scores ? What are the tests in which the group mean is plus and in what tests the group mean is minus ? How many in the group scored below the group mean in both plus and minus ?

Table 15.7 : Showing the group data on the six scales of the Personality Inventory

S.No.	Subject	Raw Score Difference						Percentile Score					
		B1	B2	B3	B4	F1	F2	B1	B2	B3	B4	F1	F2
1.	V.R.												
2.	P.K.												
3.	S.M.												
4.	K.R.												
5.	P.V.												
6.	A.R.												
7.	A.S.												
8.	R.M.												
9.	S.T.												
10.	E.N.												
Total													
Mean Score													

Discussion

Give your interpretation on the basis of the mean scores and on the basis of the percentile scores. How many of them require psychiatric attention. How many are emotionally well adjusted and how many are emotionally balanced. State who are those self-sufficient and who are those need advice and encouragement. Who are those introverted in the group ? Do they need advice ? Who are those extroverted in the group? State their characteristics. Who are those dominant in the group? Who are submissive? How many of them in the group self-confident? Who are those developed inferiority feelings? How many of them non-social and how many of them socially inclined? Do you find variation in sex and in age. State the reason if any for differences among the individuals.

Conclusions

State whether your participant is emotionally balanced and also possess all the desirable characteristics. How many in the group have problems? How many are free from problems?

EXPERIMENT 48 : ASSESSING TEMPERAMENTAL TRAITS

INTRODUCTION

Observers of human nature right from early days have noted that there were conspicuous variations in human energy level, mood, and in general style of life. Scholars in the field of literature and men of science have devised certain systems for classifying temperaments. Hippocrates, the Greek physician, who lived in fourth century B.C. had classified men into four categories. They are sanguine who is energetic and cheerful, choleric who is energetic and irascible, phlegmatic who can be sluggish and placid and melancholic who is characterised as sluggish and sad. He made these classifications on physiological bases. Temperament generally is referred to the individual's characteristic mood, activity level, excitability, and focus of concern. It also includes personality dimensions like, cheerful-gloomy, energetic-lethargic, excited-calm, introverted-extroverted, and dominant-submissiveness. Therefore, temperament is an aspect of the individual's personality.

Temperament is the basic characteristic of personality. Personality differences may appear right from early infancy through adolescence to adulthood as temperaments, vary through developmental process. Scarr and Kidd (1983) have stated that as infants identical twins tend to be more alike in temperament than fraternal twins. When they become adolescents and adults, identical twins indicate a moderate degree of similarity on measures of personality traits. In terms of specific traits, introversion-extroversion is often cited as the aspects of personality for which genetic basis becomes fundamental. This has been verified in a study conducted by Slater and Cowie (1971).

From the time they are born, children differ from one another in many traits, and some of these differences appear to be inborn. Because of such inborn differences, even children who are exposed to similar circumstances and child-rearing practices do not turn out to be the same.

The study of temperament has come into effect during the nineteenth century. The famous dichotomic type theory of temperament has given rise to so much speculation and to so many efforts at experimental validation that it is perhaps the most universally known psychological theory in the field of personality study.

In the first major review of the topic in 1941, Calvin Hall (1941) defined temperament, "as consisting of the emotional nature, the basic need, structure, and the activity level of an organism. Such traits as timidity, aggressiveness, sexuality, spontaneity, variability, speed of reaction, and activity are examples of temperamental traits". Although the view is often expressed that culture is the major determinant of individual differences in humane, Hall (1941) pointed out that individuality in animals is largely determined by heredity factors and that culture plays only a minor role. To support this view, he noted that differences in the social environment of a group of animals can be reduced to a minimum (e.g. isolated, cage-reared animals), and yet experimental manipulation of the genetics of the group can produce a large number of variations in temperament as well as characteristics.

In a further study of temperament, during 1969 Birns, Barten and Bridger (1969) tried to determine whether such temperamental characteristics as sensitivity and strength of response remained stable over the first four months of life. In their investigation, 24 infants were tested and rated at three days of age, at one month of age, at three months of age, and at four months of age. At each age the infants were exposed to several conditions: a visual following task: exposure of skin of abdomen to an ice-cold metal disk; feeding; soothing, and smiling and stroking of the infant's abdomen. After observation of the infant's responses to each of these conditions, ratings were made on eight dimensions using six-point scales. The scales were labelled irritability, soothability, activity level, alertness, vigour of response, sensitivity, tension, and maturity level.

The results demonstrated significant stability for the characteristics of irritability, tension, sensitivity, soothability, and activity level, but not for alertness, vigour, or maturity. The authors concluded that certain individual differences among neonates (presumably due to constitutional or hereditary factors) probably play a major role in the subsequent development of personality.

The role of the child's temperament in personality formation has been examined. In 1968, a major clinical study has been reported by Thomas, Chess and Birch (1968). These authors began by pointing out that parents frequently behave quite differently to their different children and that such variations were probably related to the specific temperamental characteristic of each child. In addition, the development of behaviour problems may be related to the specific interactions between the parents and each particular child.

Another study of temperament in infants was reported by Stern, Caldwell and Richmond (1969). These investigators were interested in relating the mother's personality and behaviour to the child's personality and development. Their analysis showed that there was a reciprocal interaction between the temperamental characteristics of the infants and the behaviour styles adapted by the mothers.

Jung (1923) classified temperament as extroversion and introversion. The extrovert's attention and interests are turned mainly outwards – to persons and things in the world around. The introvert's interest is directed inwards to the individual himself. Jung himself, however, admits that these types apply only to a mixed type.

TEMPERAMENT STYLES

From birth infants show individual differences. One baby curls quietly in an adult's arms; another squirms and kicks with great energy. One baby feels almost limp when picked up; another is always tense and rigid. Some researchers have suggested that these individual characteristics express the child's inborn temperament. For example, Braselton (1969) distinguishes three temperamental types : the quiet baby, the active baby, and the average baby. In another study based on 14 years of research, Thomas, Chess and Birch (1970) concluded that most children retain the same general temperament from birth to adolescence. They described babies as easy, difficult, or slow-to-warm-up. They defined easy children as those who relaxed and adapted well from birth. In later life such children find schools quite agreeable and learn rather easily how to make friends. Difficult children, on the other hand, are moody and tense. They react violently to new people and new situations, sometimes withdrawing from them, at other times protesting until the well-meaning adult gives up. Slow-to-warm-up children are comparatively inactive, withdrawn, and slow to react.

Although new born babies may look like to the unpracticed eye, they differ greatly in general activity level, sensitivity to external stimuli, social responsiveness, and sleeping and feeding patterns. Many of these individual characteristics seem to persist as the infant grows older. Thomas and Chess (1977) and Thomas, Chess and Birch have identified some of these persistent traits which constitute an individual's temperament or temperament style. In a longitudinal study these investigators observed children from infancy through adolescence, in an effort to determine the relation between temperament and mental health.

They identified three types of temperament in infants, although many babies were not of a pure type. The majority had the temperament of the easy child, a pleasant mood, regular patterns of hunger, sleeping and excretion, and little fear of new objects or persons. The intensity of their responses was generally low or moderate. A second group of children were slow, weary of new situations and of strangers. They were somewhat variable in biological functions, and the intensity of their responses was generally low. In the third group about one in ten babies was a difficult child, who often became the tyrant of the household. These children had irregular feeding and sleeping patterns and were slow in accepting new foods or in adjusting to new activities. They tended to cry often and quite loudly; they seemed generally unhappy, unfriendly, and unpleasant, frustration usually sent them into a tantrum.

These three temperamental styles appeared to be unrelated to sex or intelligence, but they were associated with the development of behavioural problems in childhood or adolescence. Only 18 percent of the easy children developed behavioural problems judged serious enough to warrant psychiatric attention, whereas 40 percent of the slow-to-warm-up children and 70 percent of the difficult children needed such help in later years.

There may be several reasons to account for the differences in temperament of children. Evidence that inheritance may help to determine temperamental style was obtained by a Norwegian investigator Torgersen, who found that identical twins were more likely to have similar temperaments than fraternal twins. The study by Sameroff (1977), Sameroff and Zax (1973) has suggested that temperament may be fashioned by physiological and clinical influences on the brain as it develops before birth. These investigators interviewed women in their later months of pregnancy and then studied their children soon after birth and at four, twelve, and thirty months of age, using a variety of neurological and psychological tests. Many of the children who are labelled difficult at four months of age had mothers who had suffered high levels of anxiety in late pregnancy.

These temperamental differences observed in early infancy imply that behaviour patterns once thought to be the outcome of poor child-rearing practices may actually be constitutional traits of the child. According to Thomas, Chess and Birch (1968) a child who stands at the periphery of the group in nursery school may be anxious and insecure, but he may also be expressing his normal temperamental tendency to warm up slowly. An infant with irregular sleep cycles who cries loudly at night may possibly be responding to an hostile, rejecting mother, but he may also be expressing his temperamental irregularity. A six-year old who explodes with anger at his teacher's commands may be aggressive and oppositional, but he may also be showing the frustration reactions of a very persistent child when he is asked to terminate an activity in which he is deeply absorbed. A mother's guilt and anxiety may be the result of a deep-seated neurosis, but they may be the result of her problems and confusion in handling an infant with a temperament pattern of a very difficult child. Such reasoning suggests that by knowing an understanding a child's temperament, parents may be better able to guide the development of their children.

The parents' behaviour towards a difficult child may become as negative as the child's behaviour. Thus a vicious circle begins. As the parents scold, frustrate, ignore, or speak their child, the child grows even more difficult. And the difficult baby turns into a child with serious behavioural problems in school, in public and elsewhere.

Many psychologists are cautious about categorising babies as quiet or active, easy or difficult. These psychologists point out that the same baby may go through all these states in a single day. While one style or another will usually dominate, it should not be thought of an unchanging. Other psychologists flatly reject the belief that temperament is inborn, or predetermined. Brofenbrennder (1977) suggests that infants behave the way they do because they are part of the social system in which various people – the infant, the mother, the father, siblings – influence each other's behaviour and temperament.

EXPERIMENT

Problem

Assess the level of your participant's temperamental traits and find out whether the person is having stable and consistent temperament or unstable and inconsistent temperament.

Method

Participant

Choose a person who has no knowledge about this test and collect his/her personal information such as age, sex, height, weight, past ailment if any, health history, parent's economic position, number of family members, his/her order of birth, etc.

Materials

Thurstone Temperament Schedule, Manual, Scoring Key, Answer Sheet.

Description of the Material

The Thurstone Temperament Schedule (1950) is one of the important temperament schedules widely used in the class room works and for research purposes. There was a need for a schedule that emphasizes important, stable traits which describe how normal, well adjusted people differ from each other. Thurstone Temperament Schedule was devised for this purpose. It is designed to assess those traits which are relatively permanent for each person. It has broad coverage. Seven areas of temperament are apprised in a short questionnaire. They are 1. Active (A), 2. Vigorous (V), 3. Impulsive (I), 4. Dominant (D), 5. Stable (E) (Emotionally stable), 6.Sociable (S), and 7. Reflective (R).

1. Active (A) : A person who is scoring high in this area works and moves rapidly. If he has to be quite he becomes restless. He is tending to be hurry always and "on the go". The person usually speaks, walks, writes, drives and works fast even when these activities do not require speed.

2. Vigorous (V) : A person who scores high in this area takes part in physical sports. He is interested in works requiring the use of his hands and use of tools and he is in outdoor occupations. This area refers to physical activity using large muscles and needs expenditure of energy. This is masculine character, but girls and women also may score high in this area.

3. Impulsive (I) : The person who scores high in this area is a happy-go-lucky, daredevil, carefree, acting-on-the spur-of-the moment characteristic. He makes quick decisions, enjoys competition, and changes easily from one task to another.

4. Dominant (D) : Those who score high in this area think themselves as leaders, capable of taking initiative and responsibilities. They are not domineering even though they have leadership qualities. They organize social activitics, enjoy public speaking and promote new projects. They attend situations of accidents.

5. Stable (E) (emotionally stable) : Those who score high in this area are cheerful and have even disposition. They can relax in a noisy room and remain calm in crises. They can disregard distractions while studying. They are not irritated if interrupted. They are not annoyed if the work is not finished.

6. Sociable (S) : Those who score high in this area enjoy the company of others, make friends easily, are sympathetic, cooperative and agreeable with others.

7. Reflective (R) : Persons scoring high on this like meditative and reflective thinking and enjoy dealing with theoretical rather than practical problems. Self-examination is characteristic of reflective persons. They are quite, work alone, enjoy work that requires accuracy. They plan a job rather than carry it out.

The schedule consists of 140 items covering these seven areas, each area having 20 items. There is a scoring key. Those who score high are stable and consistent in temperament. There are three response categories such as 'Yes', 'No', '?'. The person will have to choose any one of the three responses.

Procedure

This temperament schedule is self administered. It can be administered on individuals or on a group of people without any supervision. The subjects may be seated comfortably in a well illuminated room. The schedule may be distributed among them. They may be given a short instruction.

> "There are 140 questions given below. Please read each one carefully and answer all of them. There are three possible answers given at the end of each question. If you can answer the question with Yes put a circle around 'Yes'. If you can answer with No put a circle around 'No' and if cannot decide to answer with Yes or No put a circle around the '?' mark. There is no time limit."

As soon as the subject completed the questions the experimenter can collect it and can score it according to the procedure. If the answer agreed with the key one mark is given. If the answer did not agree zero (0) is given. The individual's score is the total marks obtained by him for all the 140 questions. Separate scores can be assigned to each area.

Results

The individual's score is obtained using the key and suitable treatment should be given to the scores of each area and to the whole score and presented in Table 15.8. There are 20 questions for each area and the highest score may be 20 and the lowest score may be zero. The scores may be ranged from 0 to 20 for each area.

The group data may be collected and given in Table 15.9. Give the scores of each area for the whole group and compare them with the individuals. Who scores high in which area and who scores low. Work out the mean for the group for each area and for the whole scores. Compare the individual's scores with the group mean.

Table 15.8 : Showing the scores for all 7 areas on the Thurstone Temperament Schedule for the individual

S. No.	Areas	Scores obtained	Total possible scores
1.	Active (A)		20
2.	Vigorous (V)		20
3.	Impulsive (I)		20
4.	Dominant (D)		20
5.	Stable (E) (Emotionally stable)		20
6.	Sociable (S)		20
7.	Reflective (R)		20
Total Score			140

Table 15.9 : Showing the group data of the test of Thurstone Temperament Schedule

S. No.	Name	Area							
		Active A	Vigorous V	Imp. I	Domi. D	Stab. E	Soci. S	Refl. R	Total
1.	M.S.								
2.	R.V.								
3.	P.K.								
4.	S.V.								
5.	K.R.								
6.	M.R.								
7.	P.V.								
8.	A.M.								
9.	T.V.								
10.	S.R.								
11.	M.T.								
12.	R.M.								
Total									
Mean									
Rank									

Work out the total score for each area and also the mean score. Find out the total score for each individual. Analyse the scores in terms of total score obtained by the individual for the whole areas and also for each area, who scored the highest and who scored the lowest in each area. On the whole who scored the highest? In which area the group data is high and which area it is low. Give the rank order for the areas for the group mean.

Discussion

Discuss the individual score for each area. In which areas the individual scored high and in which areas scored low. State the reason in terms of his personal information. State the trait in which he is highly stable and consistent. State the area in which he is not stable. State the reason. Discuss the group data in terms of the mean score for each area. State in which area the group as whole is highly stable and consistent and the area in which the group is not stable. How many of them were above the group mean and how many below the group mean. State the rank of the areas and discuss them. Who is the most stable person temperamentally and who is the most unstable person. Find out the reasons for the same and discuss. How many of them moderately stable. Discuss the traits in which the group as a whole scored high and scored low. Compare the results of this study with the results of earlier studies and discuss.

Conclusions

1. State whether your participant is highly stable or moderately stable or unstable.
2. State the areas in which he/she is highly stable and the areas in which unstable.
3. How many of them in the group highly stable and how many unstable.
4. State the level of stability of the whole group.
5. Where your participant stands in the level of stability in the group.

EXPERIMENT 49 : ASSESSING PSYCHOSOMATIC DISORDERS

INTRODUCTION

Modern medicine, with all its advancements in diagnosis and treatment, finds certain cases to be beyond its scope and treatment. Medicine is concerned with physical illness and concentrates upon the control of the organic factors of diseases, whereas psychopathology is interested in uncovering the pathological and emotional factors in mental illness. But today it has been realized that both the fields are having certain limitations. Because an illness may be primarily physical or psychological or may be both. However it is a disorder of the whole body of the person. It is not of his heart or lungs or his nerves or psyche. Whether a man in a healthy state or in the illness state an emotional upset may reduce his resistance to physical disease and a bad cold or fatigue may reduce a person's sense of tolerance for psychological stress. Taking into account all these it can be perceived that man functions as a psychobiological unit and is put into continuous interaction with his environment. The incidence of illness or disorder is also having socio-cultured interaction because of the disorders or illness found in different groups and thus the environmental factors affect him. It is because of this that an interdisciplinary approach has been devised by combining biological, psychological and socio-cultural and it was found to be more coherent to deal with disorders or illnesses and it is referred to as the ***psychosomatic approach*** and the treatment is also called ***psychosomatic medicine***. Dunbar (1943) the champion of the Psychosomatic medicine has stated that it is often "more important to know what kind of patient has the disease than what kind of disease the patient has". It is because that one out of every ten persons who seeks medical treatment suffers from a kind of illness based upon emotional stress. Unlike patterns of abnormal behaviour, psychosomatic disorders involve changes in the structure and function of internal organs. The psychosomatic disorder is also stated as Psychophysiological disorder covering a very wider range of disturbances in which emotional factors play important roles. Inspite of many other disorders, the most common types of psychosomatic disorders are peptic ulcers, migrains headaches, asthma, eczema, hypertension.

In the psychological study of adjustment we must look into two kinds of demands, one is internal which is primary, arising out of biological make up of the person. Certain conditions should be met for the individual's comfort and survival. One must be protected from dangers and must get sufficient food, water, warm and shelter. The other basic demand is external which is arising from external physical or social environments. Of these the social demands are considered to be the most important for explaining the emotional life of human beings. The emotions are toned positively and negatively. The positively toned emotions are love, joy, thrill, excitement and the negatively toned emotions are fear, anger, anxiety, guilt, depression.

Adjustment appears to be essentially a matter of problem solving. The person must find out the characteristics of his environment and also how it can be dealt with in living. This learning begins at birth and continues throughout the life. In adjustive activity strong emotions, particularly the stress emotions such as fear, anger, anxiety, guilt and shame are also generated. This results when the environmental demands involve high stakes and entail frustration, threat and conflict. Therefore, it may result in ineffective solution of the problem and production of stress and bodily disease. Lazarus (1976) states that stress is a bodily tension, occurs when there are demands on the individual which tax or exceed his adjustive resources. The study of stress is very important in the health sciences because it can lead to bodily illness and maladjustment. It was observed that two-thirds of medical patients reveal symptoms that are psychological in origin. They are in fact stress linked. The connection between psychological problems and bodily illness is called psychosomatic disorders. It may be also called stress disorders, diseases adaptation, and psychophysiological disorders. This statement implies that a problem that begins psychologically may produce organic disturbances which may even end in permanent damage of the tissues.

Psychosomatic illness refers to certain bodily disorders. To mention some of the most important are, stomach ulcers, intestinal colitis (diarrhoea), hypertension (high blood pressure) and migraine headache. There are also other forms of illness which psychological stress is capable of causing such as chronic indigestion, hives, allergies, susceptibility to infection, accident proneness and many others. Many thoughtful professionals think that all illnesses have psychosomatic overtones. They are of the view that most of the illness is not the result of an attack by viruses or micro-organisms or by conditions external to the body. Illness in part arises because the person's bodily resistance to it has been impaired by stress. It is also observed that emotional disturbances lie at the core of the causation of psychosomatic illness. The physiology of emotion will explain how particularly the negative emotion affect the body.

Stress is having special importance in adjustment, because it implies a change from single problem solving to a situation having emotional overtones. Strong emotional situation often results in irrational way of coping with adaptive problems. Stress increases potentials for maladjustment. Stress will impair the person's handling of life tasks. It produces bodily diseases. Stress emotions serve as distractions that interfere with productive thinking. A person's anxiety is an important stress emotion arising where harm is anticipated. Life stress can result in maladjusted behaviour. Therefore, a stress is an adjustive demand that requires coping behaviour on the part of an individual or group. It occurs when there are demands on the person which tax or exceed his adjustive resources. The idea is that chronic emotional status and their bodily reactions may result in change in certain organic and body functions. Many disorders have a psychosomatic basis in some people like stomach ulcer, high blood pressure, and asthma. Therefore, emotional disturbance lies at the core of the causation of psychosomatic illness. Physical disorder is caused, and also suitably can be influenced by the emotional state of the patient. Under stress some individuals develop bronchial spasms and some hypertension and still others migraine headaches. Therefore, life stress is the underlying cause of all these disorders. Chronic emotional tension elicited by life stress can cause profound changes in the physiological functioning of the human body. These changes can play in turn important causal role in the various psychosomatic disorders.

PSYCHOSOMATIC DISORDER

The term psychosomatic is derived from the Greek words psyche and soma. Psyche in ancient times means soul or mind and more recently has come to mean behaviour. Soma typically refers to the physical organism, the body. The term psychosomatic, therefore, indicates relationships between psychological processes or behaviour on the one hand and the somatic structures or bodily organs on the other.

During the middle of nineteenth century psychologists had concentrated on the bodily functions. in the year 1950, Wolff (1950) had proposed a theory of psychosomatic disorders which is based on the physiological theory of emergency reaction. Alexander (1943) was one of the leading psycho-analysts who had also investigated about the problems of psychosomatic disorders. Halliday (1943) had defined psychosomatic as a bodily disorder whose nature can be appreciated only when emotional disturbances, that is, psychological happenings, are investigated in addition to physical disturbances like somatic happenings.

Stalker (1950) had classified psychosomatic problems into three groups. The first group contained the functional disorders. The second, emotional factors even though the organic disease was present. The third, physical disease has to do with the autonomic nervous system. Fenichel (1939) differentiated between two categories of functional disturbances. One of them is physical in nature and consists of physiological changes caused by the inappropriate use of the function in question. The other one is an expression of a fantasy in a body language and directly accessible to psychoanalysis in the same way as a dream. In the same period psychosomatic was also defined by Glover (1937) as the organic changes consequent on effective changes in the organic manifestation of dammed-up instinct and physical consequences of unconscious attitudes or unconsciously determined behaviour patterns.

Psychologists after the first quarter of the twentieth century have concentrated more on the emotional tension status of individuals, which is a psychosomatic illness, the cause of which is primarily psychological. Different types of illnesses like ulcers, asthma, headache, high blood pressure and skin eruptions are related to emotional stress. Modern psychologists stressed the importance of long-term emotional stress because this will influence individual's physical health and mental efficiency. One of the earliest theories of emotion proposed that emotion is nothing but perception of the physiological changes.

Many psychologists have argued that emotions are the chief sources of disturbance and maladjustment. A mild level of emotional arousal tends to produce interest in the task at hand. When emotions become intense they usually result in some decrement in performance.

Reiser (1953) said that the word "psychosomatic" has become so firmly established in common usage that it is not feasible to discard it. We prefer to speak of "psychosomatic approach" and "psychosomatic mechanisms" rather than of "psychosomatic illness or "psychosomatic medicine". Psychophysiological disorders have been known as psychosomatic disorders which imply that the mind determines the disease or disorders of the body. It does not imply a dichotomy between mind and body.

Characteristics of Psychosomatic Disorders

Psychosomatic disorders show the disturbance of function together with damage in the organs of the body. In this they differ from mental disorders. In psychosomatic disorders emotional disturbances play an important role in precipitating the onset, and recurrence of symptoms. This distinguishes them from purely organic disorders. They are chronic disorders with a phasic course. They tend to be associated with other psychosomatic disorders. This may occur in the family or at different periods of life in one patient. They show a great difference in the incidence between sexes. Thus asthma is twice as common in boys as in girls, before puberty, after, it is less common in men than in women. Duodenal ulcer is much more common in men and thyrotoxicosis is common in women. Psychosomatic disorders are classified on the basis of organ system affected and no part of the body is exempted. The American Psychological Association (1974) uses the term Psychophysiologic disorders synonymously to Psychosomatic disorders. The APA has classified it into ten categories.

They are such as (1) Psychosomatic skin disorders or neurodermatosis, atrophic dermatitis, eczema and cases of acne and hives; (2) Psychosomatic musculoskeletal disorders or backaches, muscle cramps, tension headaches and some cases of arthritis; (3) The respiratory disorders or bronchial asthma, hyperventilation syndromes, hiccoughs and recurring bronchitis; (4) The cardiovascular disorders or hypertension, paroxysmal tachycardia, vascular spasms, heart attacks and migraine headaches; (5) Heamic and lymphatic disorders or disturbances in the blood and lymphatic systems; (6) Psychosomatic gastrointestinal disorders are peptic ulcers, chronic gastritis and mucous colitis; (7) The genito-urinary disorders are disturbances in menstruation and urination. (8) The endocrine disorders are hyperthyroidism, obesity and other endocrine disorders are those in which emotional factors play a causative role; (9) Chronic conjunctivitis are the disorders of organs of special sense; and (10) The other types of disorders are disturbances in the nervous system in which emotional factors play a significant role, such as multiple sclerosis.

All of these disorders are caused by life stress. In essence, chronic emotional tension elicited by life stress perceived as threatening can cause profound changes in the physiological functioning of the human body; these changes, in turn, can play an important causal role in the various disorders we refer to as psychosomatic.

It has clearly been stated by medical and psychological scholars that emotional upheavals may aggravate or produce psychosomatic illnesses like stomach trouble, indigestion, metabolic disturbances leading to possible emaciation, heart diseases, insomnia, increased susceptibility to cold and dysentery. Many somatic abnormalities or psychosomatic disorders have a psychogenic origin and it is important to note that these kinds of states may affect one's mental set-up and thereby create problems of adjustment.

The satisfactory adjustment and resolution of the individual's basic psychosomatic needs and urges are prime conditions for the attainment and preservation of mental health and sound physical health. The individual's failure to achieve inner harmony in himself, and to reconcile himself with the folkways and mores of society, will tend to produce in him personality modifications or distortions in varying degrees. This depends upon the nature and severity of the emotional stress. It is clear from the above statements that psychosomatic problems of any kind will affect the individual's normal life. When an individual is having some problems he finds it difficult to adjust himself and with his environment.

Selection of the Instrument

The need has been felt at present for an instrument for evaluating the rapid development of psychiatric and psychosomatic symptoms among large numbers of persons in a variety of situations in our society. It has been ignored for a long time. But the development of community psychology made the psychologists to realise the value of such instruments and the measurement of neuropsychiatric and psychosomatic problems. The test consists of series of questions relating to neuro psychiatric and psychosomatic symptoms. The test would serve as a standardized psychiatric guide to the interviewers in the clinical field and it may differentiate persons statistically. Persons suffering from severe psychosomatic disturbances are identified from the general population. It is divised as a supplement to the interview. But when the interview is not possible this test is used as substitute.

The index does not function as an instrument for measuring and comparing various personality characteristics. The Index is able to identify the deviates from the average or normal and these are reflected in the index scores. The Index may be very much useful in identifying the problems among the adolescents and young adults. In the clinical situations it may be applied on patients complaining of somatic problems, depressions, gastrointestinal problems, defects in adjustments etc.

1. Neurologic and psychiatric wards and out-patient departments often have large numbers of patients than can be cared for adequately by the available staff. One phase of investigation of such patients that is time-consuming is the accumulation of a sufficient body of historical data and information about present complaints. This material can be collected by means of the ***Cornell index*** and placed at the disposal of the physician with no expenditure of his time. He can, then, attach diagnostic and therapeutic problems promptly.

2. Medical and surgical wards and out-patient departments also have many persons with structural disorders and also may have emotional disturbances which interfere with rapid convalescence. These disturbances are sometimes not clear and may remain unrecognized as factors contributing to prolonged convalescence. Physicians desiring to check on the presence of personality disturbances in medical and surgical patients may be referred to for consultation with a psychiatrist and will find the Cornell index helpful in achieving the end.

3. In Industry this test can be of very much useful to physicians in the medical departments in determining employees who need, guidance for emotional disturbance or lowered morale associated with poor work adjustment. The index may be applied in the placement of the employees in business and industrial organizations.

4. This test offers a simple and expeditious method for obtaining standardized data on personality adjustment which may be subjected to statistical analysis in the implementation of research projects.

5. This test can be administered in a community where there are emotional maladjustment, physically handicapped and other social problems like delinquency and criminal tendencies, aggression, violence, alcoholism, unemployment, poverty, overcrowding, social unrest, communal tensions, seriously disturbed children, not going to school and so on.

This test is an effective time saving device in providing information to the interviewing examiners concerning psychosomatic and neuropsychiatric tendencies in a person.

EXPERIMENT

Problem

Find out the Psychosomatic/Psychophysiologic disorders of your subject using a diagnostic instrument.

Method

Subject

Find out a person who has no knowledge about the purpose of this test. Collect his/her personal information such as age, sex, height, weight, his past health history, occupation, marital status, income, present ailments, etc.

Materials

Cornell Medical Index Health Questionnaire (CMI) Manual, Key and Scoring Sheet.

Description of the Materials

The Cornell Medical Index Health Questionnaire originally was developed and standardised by Arthur Weider, Harold G. Wolff, Keave Broadman, Bela Mittleman and David Wechsler (1949) jointly for the study of psychiatric illness. The original form consists of 195 questions. The questions are of four kinds: (1) those relating to bodily symptoms, (2) questions relating to past illness, (3) those relating to family history, and (4) those relating to thought, behaviour, mood and feeling. The questions are arranged in sections, headed by letters of the alphabets. It has two forms, one for men and the other for women. They are identical except for six questions in the Genito-urinary sections. This was modified and revalidated later on by Broadman and others. The modified version consists of 101 items with ten areas. The following is the modified version and it is called in a short form as Cornell Medical Index (CMI).

S. No.	Name of the area	Items (from to)	No. of Items
	Introductory neutral question	1	1
1.	Defects in adjustment expressed as feeling of fear and inadequacy.	2-19	18
2.	Psychological mood reactions especially depressions	20-26	7
3.	Nervousness and anxiety	27-33	7
4.	Neurocirculatory psychosomatic disorders	34-38	5
5.	Pathological startle reaction	39-46	8
6.	Other psychosomatic disorders	47-61	15
7.	Hypochondriasis and asthenic	62-68	7
8.	Gastrointestinal psychosomatic disorders	69-79	11
9.	Excessive sensitivity and suspiciousness	80-85	6
10.	Troublesome psychopathy	86-101	16

There are two response categories to answer the questions such as 'yes' or 'no'. There is no time limit to answer the questionnaire.

The Cornell Medical Index (Questionnaire) test has been revalidated at Annamalai University (1980) in connection with the study of Psychosomatic symptoms in athletes and non-athletes. In 1980 it was administered on 350 University students and the split half method reliability was used. The reliability was found to be 0.94 and the validity was 0.97. The same version was again revalidated at Annamalai University (1990) for the study of psychosomatic disorders among the adolescent girls. The Cornell Index was administered on 270 adolescent girls students. The reliability was 0.80 and the validity was 0.89. It can be administered on student population and on adults.

Procedure

This test may be administered as an individual test or as a group test. As a group test it can be administered on any size. The test is given to the subject and he/she is asked to complete it. The subject should first of all provide the personal information such as age, sex, marital status, past health history, qualifications, occupation, income, present ailments, height, weight, size of the family etc. After filling up the questions with personal information the subject can start answering the test questions. The experimenter should read out the instructions loudly to the subject:

> "There are 101 questions in the booklet. You will have to read each question carefully and understand the meaning of it. There are two answers. 'Yes' or 'No' at the end of each question. If you can answer the question with 'Yes' put a circle around 'Yes'. If you can answer the question with 'No' put a circle around 'No'. If you are not sure about the answer, guess and answer with Yes or No. Please answer all the questions and do not omit any of them."

When the subject has answered all the questions the experimenter should collect the questionnaire from the subjects and also make a note, how much time the subjects have taken to answer the questionnaire. The subjects may be allowed to go away.

Method of Scoring

With help of the key the Index test is to be scored. If the answer given by the subject is found to be in agreement with key then one mark is given. Otherwise no mark is given. The individual's score is the total marks given to the items. The high score indicates more psychosomatic disorders, the low score less psychosomatic disorder in the area.

Results

After scoring all the items according to the key, the experimenter should present the score of the individual for each area in Table 15.10. He should also work out the percentile score and give it in the columns.

Then the group data should be collected from other experimenter and present them in Table 15.11. He should explain the individual score in each area, where his score is high and where it is low. Similarly state the high score and low score of the individuals. Work out the mean score of the group for each area. State how many above the mean and how many below the mean. Give the rank order to the subjects.

Table 15.10 : Showing the individual's scores on the Cornell Medical Index Test

S. No.	Name of the disorder	Score	Percentile Score	Rank order
1.	Fear and Inadequacy			
2.	Depression			
3.	Nervousness and anxiety			
4.	Neurocirculatory psychosomatic disorder			
5.	Startle reaction			
6.	Other psychosomatic disorders			
7.	Hypochondriasis and asthenia			
8.	Gastrointestinal psychosomatic disorder			
9.	Sensitivity and suspiciousness			
10.	Troublesome psychopathy			

Table 15.11 : Showing the group data on the Cornell Medical Index Test

S. No.	Subject	Scores in areas of disorders										Total Score	Rank Order
		1	2	3	4	5	6	7	8	9	10		
1.	P.K.												
2.	M.S.												
3.	B.V.												
4.	R.S.												
5.	K.R.												
6.	P.V.												
7.	M.R.												
8.	R.M.												
9.	K.M.												
10.	R.V.												
Total													
Mean													

Discussion

Discuss the individual's problem in relation to each area and the whole test taking into account the score. State the disorder which is found to be widely prevalent in the subject. Also state the area in which least disorder is observed. Discuss whether the results observed in various areas are in line with the subject's past health history or present ailments. State whether it is hereditary or new disorder.

Discuss also the group data in the same manner. Compare the group mean in each area with the individual's score and state how many have scored above the group mean and below the group mean. Discuss the widely prevalent disorders in the group. State the areas of severe disorders, moderate disorders and less disorders. State the reasons if any. Also state the rank order of the subject and discuss it. In the group who is in the first rank of the disorder and who is in the last rank. While writing the discussion part relate your data to the subject matters given in the introduction. State whether there is any difference between men and women in their psychosomatic disorders. State any other variable having influence in the participant's disorders.

Conclusions

1. State the highly prevalent disorders in the group.
2. State the least prevalent disorders of the group.
3. Who is more subjected to disorders and who is least subjected to disorders.
4. State the important reasons for the disorders as revealed by the study.
5. State whether the group as a whole positively inclined to disorders or negatively.
6. State where your subject stands in the group.
7. State how many of them in the group are in need of any counselling.
8. State the areas in which your subject is showing high disorders.

16

COUNSELLING AND GUIDANCE

WHY COUNSELLING IS A MUST ? : AN INTRODUCTION

Counselling has been regarded as a part of the broad discipline of psychology from its earliest time. It has been essentially a field of application of the basic psychological principles and experiments have been conducted to identify the problem persons. It has taken several years for formulating hypothetical structure for counselling. Several developments have taken place to achieve the status of the present position. But, since contemporary world is characterized by rapid change in all walks of life it is very difficult to ascertain the nature and characteristic of human problems and the methods of dealing with them. What has been considered as current trend of action is found to be no more adequate as new ideas and testing procedures are emerging from time to time.

Society is often facing many types of psychological problems due to the various political, social, environmental changes and also due to the rapid development of science and technology. The more money the nation spends on microbiology and for its research the more micro problems humankind is facing today and the micro problems become macro problems next day. The micro causes of human pathological conditions are thoroughly neglected and millions of unemployed and semi-employed youths are left in the blind alley without proper guidance. There are many problem areas such as drop-outs, delinquents, under achievers, intelligent and creative youngsters not sticking to any specific work, minority youth, the non-academically oriented youth. The strength of population in these areas is constantly increasing, which means increase of more psychological problems.

If proper attention is paid in time on these problem causing population many of the psychological and social problems would have been solved. The proper attention, the psychologists have been stressing to alleviate such problems is counselling and guidance in the schools where the problems emerge at micro level. It will not be possible to judge apparently by the teachers who handle the classes. Suitable tests must be administered by the school psychologists to identify the problem individuals. Psychologists have designed and developed tests and inventories of various types for various purposes. Mostly school students often face adjustment problems in the beginning which may lie dormant, but develop slowly through adolescence and adulthood.

An individual may be having some problems with regard to his/her personality make-up. His/Her reactions in the family situation or in the social situation may not be pleasant and he/she may be finding it very difficult to cope up with others very successfully. His/Her emotional states and moods may be putting him/her in a most disadvantageous position. The individual may be living in his/her own world of imagination and confining himself/herself with it and accepting the reality of the world may be a very difficult task for him/her. One need not be abnormal in his/her characteristics. But any normal human being also sometimes expresses certain unpleasant and uncompromising attitudes and takes a very unusual way of life. This sort of behaviour may cause concern to teachers, counsellors, and others who are working with youth. More than these professionals, the parents are very much worried and the person himself/herself is more interested in knowing something about the self.

Adjustment is a psychological process which could encompass the entire span of life of a man. Man sometimes finds himself helpless when got into the grips of adjustment problems. His problems may arise from within himself or from outside situations. But there is always a tendency in man to get rid of his problems by some means. If his personal efforts are not successful in this attempt he may seek the help and services of the people who are qualified and trained in dealing with some problems.

It is important for a man to know what he is capable of and what are his assets and liabilities. Problems arise in a person only when he is not aware of these things. Some understand about themselves by some trial and error methods and by experience. They are intelligent in a way. But many of us neither learn by trial and error nor gain by experience. Such people often face problems and they need to be guided. This sort of guidance may take place in educational and vocational situations. In some cases it may take place in dealing with personal problems of adjustment.

Generally, psychologists, counsellors and teachers take up this responsibility of guiding people after assessing the characteristics of the 'problem people'. So long as problems lie dormant nobody bothers. But when they manifest and become active then it is to be attended to with all seriousness. Generally the individual's problems, whatever the nature of type they are, may manifest and become active in the later adolescent period or in the early adulthood period. During these stages the individual must hold certain responsibilities for which he is not ready. Zudick (1971) states that there is general agreement in the guidance literature that many of the educational vocational and *personal maladjustments* we see in later years have *their genesis during the elementary school years*. Zudick is of the view that a well-conceived guidance programme designed to help students may develop their understandings about both the self and the society which may deter the feelings of discouragement and problems in the later life. In this programme the counsellors should hold a key position to help the teachers in schools. This may prevent many of the adolescent and young adulthood problems.

WHAT IS COUNSELLING?

Counselling is not a new concept. It has been in existence from time immemorial in different names. When a person is in trouble or in difficulty seeks someone in whom he has confidence and gets his advice and guidance. Younger persons seek the advice of older persons. Children get advice from parents. Less successful persons seek the advice of more successful persons. In all these situations counselling is in existence where one seeks the guidance of another in sorting out his problems.

Therefore, the situation for counselling is very common, where one person seeks the advice of another person. But the expertise giving appropriate advice is not common. This requires special knowledge, skill and competence of a counsellor for understanding the problems of a person and then lead him through the process of discovery to a solution. This should be possible and acceptable to the person. Counselling should not be considered like the common act of advice performed in society by teachers in schools, parents at home, doctors of dispensaries. Counselling has become a field of scientific psychology. It is an art and also a science. It is an art in dealing with a person to discover his problems and science in studying the problem systematically and finding out solution for the problems.

Therefore, the aim of counselling is to help persons to overcome many of their present and future problems. Since the nature of problems vary time to time the counsellor's approach to the individual problems also may vary. For example the second and third quarter of twentieth century caused many new types of problems. The rapid development of science, technology, industry and medicine and the new type of nuclear and chemical destructive weapons have contributed not only new human problems but also become the cause of uncertain future.

It is very strange to see in the present generation the two uncompromising phenomena, one causing rapid destruction and the other increasing the longevity of man by 30 to 40 years. On one side men go in search of job and seek the advice of experts to tell him the job for which he is fit and on the other side organizations do not find competitive men for various jobs. Therefore, in these contexts the nature of counselling cannot be the same what was practised 40 or 50 years ago. New techniques and methods are evolved to deal with the problems that arise from time to time.

At present it has been realized that counselling should start early in schools and should continue to enable the individuals to meet educational, vocational and personal problems of adjustment during the time and thereafter in life. Modern medicine and better ways of living had reduced the rate of early death and increased the span of life. This creates one more problem of old age adjustment. Therefore, since problems can arise at any age of man's life, counselling also must become necessarily a continuing process at all age levels.

The purpose of counselling, in whatever form it might exist, it should help individuals to become self-confident, self-dependent, self-sufficient and self-directed and learn to adjust themselves successfully to the demands of a meaningful happy life. Individuals should be assisted by suitable means to enhance primarily their emotional, personal, social and intellectual development. In this way the task of the counsellor is manifold, such as perceiving the problems and preventing them, assessing the capabilities and developing them and administering therapeutic techniques to resolve the problems.

Counselling may be defined in several ways and there are a quite number of definitions available in scientific literature. But the central point in any definition is the interacting process between the counsellor and the individual with problems which he/she cannot resolve them (Rao, 1981). Therefore, counselling indicates a professional relationship between a trained counsellor and a client. The relationship usually is person to person, but sometimes it may involve more than two people.

COMPONENTS OF COUNSELLING

According to Jones (1979) there are four basic assumptions. They must be accepted if counselling is to be successful. (1) First of all the person must be willing to cooperate and participate in the counselling process. He may wonder about what is going on around him and may have the desire to know what he is going to experience. Since the process may be new and not familiar to him he may resist at the moment. Therefore, the problem person should willingly cooperate with the counselling situation. (2) Secondly, the person who practices counselling should possess efficient professional training, experience and knowledge. He must be in a position to create in the person a sense of confidence and assist him in identifying the objectives and goals. (3) Thirdly, the environment should be congenial as to provide the person a sense of confidence and a mood of expectation. (4) Finally, there should be a kind of relationship in counselling for meeting the immediate and long-term needs for the person. The counsellor should work cooperatively with the person in identifying his problems and alter his behaviour according to the needs of the environment or organization.

CHARACTERISTICS OF A COUNSELLOR

Counselling techniques have become more sophisticated and the procedures are becoming more accurate. The benefits of the refined techniques of counselling go to only the counsellees. A counsellor with all his expertise tries to modify human behaviour. What sort of behaviour needs modification? A counsellor is interested in modifying the behaviour of the problem student. A student in school or a drop-out or an unemployed youth may be a problem person. If he is given counselling in proper time he can be rescued from getting more problems. Therefore, the counsellor's task is first to identify the problems of the individual and design counselling techniques to deal with them. The counselling may be continued for a day or two or may be continued for a week. This counselling may bring some change in the individual's behaviour and the individual may be free from the problem and may be ready to take up some responsibility.

But the individual may not be ready for any change in his behaviour. There is resistance to change and it is a natural consequence of the tendency of every individual to maintain his habits and the perceptions he has learned. Therefore, the counsellor comes to know that every counsellee is reluctant for a change. The counsellor should know the specific development that is active and that is not active. The counsellor should use many of his expertise and other resources like the interview, assessing the traits through tools, and records, through his teachers, parents and friends.

Apart from attitudes and skills, practice could be more fruitful for the counsellor. If greater attention is paid as a counsellor the success will be more consistent. A counsellor should understand that he cannot put on himself more skills by putting on the clothes of an expert. He should realize that it is he who is the person should shape the course of counselling efficiently. The counsellor influences other people in a number of ways. It is by way of his looks, qualifications, training, techniques, approaches, his socio-economic status, customs and manners, his aspirations and success in his profession.

A counsellor should be a person who is actually possessing all the desirable and agreeable characteristics. A counsellor should easily promote interpersonal skills and everyone should have such impression about him. He should have extensive professional background knowledge and also good knowledge of dynamics of human behaviour. He should work with other counsellors, teachers, parents, administrators and other specialists harmoniously.

A counsellor should have sound intellectual aptitude and a sound knowledge of career and occupational information. He must have been put himself in rigorous testing situations and should have developed the ability to easily administer tests, inventories and questionnaires upon various types of groups and able to analyse the scores critically by himself and identify the characteristics of each person. He should be quite conversant with library works. Though research is not his full time work, when necessity arises he should be in a position to easily trace out the information required from the books and journals. According to Zeran and Riccio (1962) the main work of a counsellor is dealing with school students and conducting research on them to the extent of determining the local needs. He should be able to evaluate the available information and make use of them for counselling and guidance programme. In order to avoid or minimize mistakes the counselling should show interest in deep research and the interest for knowledge should be continued. The type of research should be for stimulating intellectual activities. The counsellor must be familiar with the changing trends and the new developments in various areas. There are always the changes in the population trends, rapid changes in the fields of science and technology, introduction of new concepts on the nature and function of education, new looks into the nature of human beings, emergence of new vocational and occupational trends due to new type of industrial developments. The counsellor is required to renew his knowledge in all these areas now and then to keep his professional efficiency at a standard. Sometimes, even very small fraction of the ideas may have the chance of serving for a big cause.

The counsellor should have the sense of humour in dealing with counsellees. Fullmer and Bernard (1972) state that he should have a non-hostile sense of humour which is a characteristic that would be a saving device for the counsellor. A counsellor who is sincere and competent in his work always considers his work as vitally important, may get frustrated when he comes across persons who are very indifferent in the counselling situation. Such students are very inactive in school performances and are beneath the estimation of the teachers. In such situations the counsellor's task becomes highly critical. A sense of humour free from aggressive attitude may serve some useful purpose. Maslow (1970) suggests that a sense of humour could facilitate the counsellor in his task. The sense of humour is humble and not hostile. This needs making a fun at the weakness of the character of the people in general. Maslow is for philosophical and unhostile sense of humour. Some do not consider a fun as funny what the average man considers to be funny. Therefore, they do not laugh at hostile humour. For Maslow, hostile humour means making people laugh by hurting someone else. They also do not laugh at the superiority humour which means laughing at someone else's inferiority. Another humour, Maslow states is authority – rebellion humour which is very indecent talk. Maslow calls it as oedipal or smutty joke. What they call humour consists of making fun at human beings when they are foolish or forget their place in the universe. They try to be big when they are really small and foolish.

Maslow (1970) quotes, Lincoln's humour as an example for this. He states that Lincoln never made any joke that hurts another person. It is said that most of his jokes had something to do with a function beyond causing a laugh. His jokes were often found to be education, similar to parables. Therefore, humour or joke should not be at the cost of another man's self-respect or by hurting his feelings.

Bernard (1961) states that the image of the counsellor is that of a person who is aware of his specialized responsibilities. Thus a counsellor is an educator and a psychologist. He is also a specialist in the study of student behaviour. He always sees the student as a dynamic changing personality. A counsellor is a specialist in promoting human relations. He must be a person who is able to live with the awareness to things and events.

The counsellor should promote creative people which is one of the ways of helping people. Who is a creative person? A creative person is not one who is free from problems. He is one who is anticipating excitement and enjoying it. According to MacKinnon (1962) a creative person is interested in perceiving life events and not interested in judging them. In fostering creative people the counsellor pays more attention in the development of feeling in people rather than the development of thinking.

In fostering creative people the counsellor is having an open mind. A closed system of counselling may lead to aggressive attitude which will discourage the creative aptitude. Gary (1961) states that it will develop undemocratic system, may result in perversions. The counsellees should have open mind, remain free from any dominance and also should be open to experience. The counsellor should be a self actualizing person, in which he may find himself very difficult sometimes. It is true that some may find opportunities to project the self with all his skills and execute the responsibilities, while others may not find opportunities to put their skills in action. There are several factors leading to self-actualization. Therefore, the counsellor should strive towards self-actualization which is the highest level of need satisfaction. He is the model for practising counselling and guidance.

GUIDANCE : VOCATIONAL

Prosperity opens new avenues for vocation, sometimes even making scarcity of men for occupations. Depression closes the avenues and throws men out jobs. These two phenomena often play successive roles in the life of human beings and create anxieties and uncertainties in the minds of men for centuries. Now, time has come for man to realise certain values of life and rearrange his thought process and reorganise the whole set-up of his life. The development of Science and Technology, Industry and Organisation are the conditions creating and increasing job opportunity in various public and private sectors. This is on one side. This, we may call as prosperity. On the other side, the rapid increase of population, drought, scarcity of food, the neck to neck aggressive competition between men outwitting each other, having some employed and others unemployed make life very miserable and unhappy. This, we may call depression. Therefore, we could face both prosperity and depression not on different occasions but at the same time in India. At this juncture the phenomenon of vocational counselling and guidance should receive greater importance for better understanding about job opportunities. Therefore, not only the student of vocational and Industrial Psychology should know this subject matter more scientifically, but also all those who are undergoing a course of study in any discipline and those who are looking out for some job and those who are unemployed for want of qualification and skills and those who are not satisfied with present job should know something about this.

Therefore, it becomes an important task for the psychologists to educate people who are qualified or unqualified, trained or untrained, skilled or unskilled, talented or untalented with some kind of knowledge about vocational psychology which can be useful for them in some situation when they face some problems. Education for vocation, we meant, helping a person to find out a job for which he is fit, irrespective of his qualifications, by testing his aptitudes, interest and personality traits and guiding him to make his choice with same confidence. This process involves both counselling and guidance.

THE IMPORTANCE OF VOCATIONAL SELECTION

(a) What is a vocation?

The word vocation has been used both in the common sense world and in the scientific world. Therefore, it needs certain clarification. In the common sense world it is used with different meanings. Generally, it is referred to a job, occupation, business or profession one does. The Oxford Dictionary states that vocation is an action of a person to exercise some special function. It is the direction towards special work in life. It is the particular function or station to which a person is called for. It is also stated as the natural tendency to fitness for such work. One definition for vocation is that it is a function of a station in life to which a person is called for by an organization.

It is also stated in the meaning of a trait of a person. It is referring to a strong impulse or inclination to follow up a particular activity or career.

In the scientific studies the concept, "vocations" is referred to a job, occupation or a life career. In other words, a job one does and earns for his livelihood. Therefore, the word vocation is used in a wider meaning.

(b) Vocational Selection and Malnutrition

Vocational selection means choosing a life career, the job one chooses for keeping himself engaged and earning for his living. Every educational system prefers people for better life and provides them training for the needs of the time. There are some technical institutions in our country which squarely face the problem of training people for vocation. Some institutions are dedicated to culture oriented education for the purpose of propagation of certain principles and leave the person for acquiring his technical skills and knowledge on the job he chooses for his life career. At the higher educational levels particularly in the colleges and universities there are conflicting philosophies regarding vocational information. The Engineering colleges view this little differently from the Arts and Science colleges. The student who gets admission to Engineering college thinks that he has made his vocational selection correctly. But when he is about to complete his course in the Engineering college or after completing the course he feels that he had made a wrong choice of his vocation and wanted to go on in a different line. There were several such instances we did come across in our counselling and guidance centre of the Department of Psychology. In the counselling interview it was observed that the candidate had confessed that he had to go into the Engineering Course soon after the Higher Secondary education because of the pressure of his parents and relatives and not of his own.

Another engineering graduate had stated: "It was highly exciting in the beginning of the course. Later on as and when I go higher and higher it was highly mechanical and life was becoming tensional and frustrating. I would have been happy if I would have studied literature or philosophy. At this stage I can't do anything. Therefore, now I am seriously thinking of going into some desk-work job. Can I be guided in such lines." There are hundreds of cases like this and we cannot enumerate all of them here. At present hundreds of Engineering graduates are turning to managerial type of occupations where what they studied in their colleges become useless. Whereas the Arts and Science colleges provide liberal education and leave the candidate to choose a vocation where he can apply his knowledge suitably.

There are various sources where a job is available. Some youngsters worry about their vocational objectives and often ask their teachers and verify about their future prospectives. Some institutions employ school psychologists to provide vocational guidance. There are also Government employment services and other ourside and private agencies for guiding people for better jobs. The contribution of psychology to education also includes vocational guidance. The individual is the centre of the entire educational process and the individual should be helped in all his problems and choices.

Vocational choice is the central problem facing many youths. Getting into wrong vocation is a very serious affair. If a man enters a job for which he does not have any aptitude he has to struggle in it. He may not be successful to support himself. This failure sets up an emotional under current. He finds that others are more successful in their jobs and he is not and, therefore, he develops an attitude of indifference and a feeling of inferiority complex. He projects his difficulties and failures on someone else. He seeks other emotional outlets and may become a delinquent. On the other hand if he takes up a vocation for which he is well adapted then, there is satisfaction in doing the job well.

Many workers, sometimes, are dissatisfied with their work. This is because of inadequate information and unwise vocational choice. After making a choice out of too many jobs a worker may feel that he has made a wrong choice of his job and others made wise choice. There have been numerous studies of job satisfaction in India and in many of the foreign countries.

A simple question to study this is: Are you satisfied with your job? If you were going to live your life again, would you take the same job? This method of counselling leads to an elaborate clinical analysis, but we can get some information about the candidate's aptitude and interest.

Some investigators are interested in finding out how many people are dissatisfied. But others are interested in making detailed analysis of the sources of dissatisfaction. They are looking toward making improvement in the morale of the employees by doing something in the light of the analysis.

A review (1984) of various studies reported over a period of 21 years found to be 343. The percentage of dissatisfaction varies from 1 to 92 percent, but had an average of 13 percent. There were varieties of causes of dissatisfaction. It is a challenge for the industrial personnel department to improve morale and for the vocational counsellor to get people in the right job to start with.

It was found that those who were dissatisfied with their jobs also dissatisfied with their life. There was overall dissatisfaction in some problems. If a wise vocational choice is made earlier the dissatisfaction need not arise at all.

VOCATIONAL APTITUDE

Aptitude is the natural ability to profit by certain types of training and to do the work required efficiently in a particular situation. Vocational aptitude refers to the ability to learn the skill involved in a specific job. For example, clerical aptitude is the ability to learn clerical skills. There are tests to assess the ability to learn the skills of a particular job. The vocational aptitude test is a predictor of job success. By administering such tests on a person it is possible to find out to what extent a person is capable of performing a task. Aptitude is also defined as a set of characteristics indicative of ability to learn. This implies that an aptitude is a combination of entities – the set of characteristics which enables a person to learn something, may be even different that which enables another person to learn the same thing. A characteristic or a trait may be developed into broader vocational skills.

Different types of traits and abilities may be combined in a person for success in a given field. One teacher may be successful because of his scholarly ability, interest in his subject and clear presentation of a wealth of material. Another teacher may be a mediocre in his scholarship and academic ability, but equally successful because of his genuine interest in students' welfare and in expressing warm and friendly feelings. In the same manner differences could be pointed out among successful professional men like the lawyers, physicians, foremen, salesmen and others. The varying combination of special aptitudes and trait makes a person to be successful in a given occupation or vocation. Aptitude involves specificity, unitary composition and the facilitation of learning of some activity and self actualization to certain extent. But aptitude is relatively constant in its nature and composition. It is the innate capacity of an individual.

The term aptitude is generally used in conjunction with the name of some occupation or activity. Therefore, Super (1971) defines aptitudes as psychological factors, each relatively stable, unitary and independent which contribute in varying degrees to success in various occupations. Aptitudes are to some extent based on inherited characteristics. Aptitudes are potential abilities or capacities. They enable one to learn. A person may be ignorant of sciences, but may have a high degree of aptitude which is important in scientific work.

According to Garrett (1961) aptitude is a potential capacity to learn, inferred from performance or from tests. When we say that a person shows aptitude for languages, for mechanics, for sports or for teaching we mean that he has given evidence of such aptitude through superior performance in one of these areas. Aptitude is revealed in performance or inferred from performance. It sets up a limit what a person will achieve when given opportunity or training. Achievement is actual performance, that is what a person does. But aptitude often exceeds achievement, because a person never achieves what he is capable of and achievement never exceeds one's aptitude. Aptitude shows a high degree of transference among related activities. For example, a boy who plays tennis well is likely to be good in base ball and badminton though he is not equally competent in all these.

Psychologists have devised various tests for assessing varieties of aptitudes of the people while selecting for various jobs. Aptitude tests measure the individual's capacity or latent ability to learn a given job. They are used for prediction and they measure basic human characteristics or ability related to capacity to develop proficiency on specific job. These basic characteristics are thought of as aptitudes. Great deal of research work has been done to determine the basic human characteristics. According to MaCormick and Tiffin (1984) aptitudes can be grouped into five classes as mental abilities, mechanical abilities, psychomotor abilities, visual abilities and other special abilities. Any vocation needs any one of these abilities, to certain degree and a combination of more than one ability to some extent. Psychologists discuss these things in detail under the heading of psychological tests in industry.

DEVELOPMENT OF VOCATIONAL INTEREST

Interest and ability are related to some extent. A person will do better in his job and be happier if he is interested in his job. Vocational psychologists have paid much greater attention on interest during the past few decades than on any other higher characteristics like intelligence, aptitude and personality traits. Studies show that interest is related to all these characteristics in some way, because there is some empirical evidence that there is positive correlation between expressed interest and graded achievement. But here also there is the question of cause and effect. One may try harder at a job in which he is interested or he is well in a job. Sometimes success in his job may create in him some kind of interest. Interest is often synonymously used with aptitude and motivation. Sometimes it is defined that motivation is interest. It is also observed that if a person is interested in an object or an activity he would be motivated to carry out the same. If a person is very much interested in business he is said to be having aptitude for business. But, since aptitude is referred to capacity or innate ability and motivation is referred to impetus or to a stimulus incentive and interest is referred to the expression of one's preference for an object, person, place or work which he proposes to have in life situation, interest is different from the other two processes. There are four types of interests. They are expressed interests, manifest interests, inventoried interests and tested interests.

The expressed interests refer to the expression of specific interest. It is the preference a person is having for something. A person may say that he is interested in playing tennis. Manifest interest is not expressed in words, but in action. It is expressed through participation in some specific activity. A person may express his interest by joining a group of music party. The inventoried interests refer to responses made by a person to a large number of questions with likes and dislikes or giving the order of preference for a number of things or activities. These responses were treated statistically to find out the level of preference one made for vocation or activity.

The Strong Vocational Interest Blank and the Kuder Preference Record are some of the inventories used to assess one's vocational interest. It is based upon subjective estimation. But the tested interest refers to objectively measured interests. It is hoped that the interest one has for a vocation may manifest itself in action and may result in providing suitable information about the vocation.

Carter (1969) was the first to synthesize the findings of the vocational interest and to formulate a theory of the development of vocational interests. According to him the individual derives satisfaction from identification with some group by which he gets a status. According to Carter, the interest patterns of adolescents tend to become increasingly practical. The individual is trying to adjust to a complex culture and the individual's experiences in this provide him basis for the integration of personality. Therefore, the pattern of vocational interest which forms gradually becomes identified with the self. Findings show that there is relationship between general aptitude and interest, and in some cases aptitude comes first resulting in approval, and then satisfaction and the development of interest. It is probable that aptitude plays a part in the development of personality traits and, therefore, in the development of interest it should be recognised that there is relationship between interest and the deeper layers of personality. Interests are the products of interaction between inherited aptitudes and endocrine factors on one side and opportunity and social evaluation on the other.

Most young people at adolescent period have opportunity to explore social, linguistic, mathematical, technical and business activities to some extent. They identify themselves with parents, other adults and schoolmates and self-concepts find a definite form. Therefore, it is generally believed that the vocational interest patterns of adolescent period form substantially into permanent state and manifest in the later life to some extent. Therefore, there is some evidence to say that there is good correlation between expressed interest and graded achievement.

Is the interest genuine or does a person merely like the hours of work? When a client expresses a vocational interest one of the first problems for the counsellor is to determine whether that interest is genuine. Generally it is not. A youngster thinks banking would be a nice career because he likes the hours of work. Sometimes interest is out of line with opportunity. To quote one instance is: a group of young men came for vocational guidance and out of which 612 wished to enter professional work. None were interested in unskilled labour. Strong's Vocational Interest Inventory was administered on them and it got into their real interest. It was found that only 16 percent of them were to do professional work.

The interest inventories reveal unsuspected interests of the individuals. The clients might have never thought about the kind of work or might have never heard of it. To quote one instance for this; a youngster who came from a farm, enrolled in a course of dairy technology. His parents thought that it was nice for him. He was interested in occupational opportunity service. But interest inventory indicated that he was fit for salesmanship for dairy equipment. He was found to be very much successful in it in his later life.

TECHNIQUES IN SELECTION

A person must be assigned to specific job or he should be trained for specific job. Basically this function means matching the individuals to jobs considering on the potential or actual qualifications of individuals or requirements of jobs. Every organisation needs qualified persons for every job, for which they must make personnel selection. In technical sense, this refers to choosing one or more persons to be employed from a number of available candidates. The selection decision is made on the basis of the satisfactory performance of candidate in the interviews and on the basis of the fulfilment of the required qualifications. Personnel selection is made for placements, training, promotion and transfer to similar jobs. In the personnel selection procedure, certain personnel specifications are taken for consideration. They are educational qualifications, work experience, training, biographical data, physical data, information on various abilities, personality, interests and other characteristics.

The most common techniques used at present for evaluating individual differences for personnel selection for industrial organizations are: (a) Interview, (b) Application Blank, (c) Personal Data Sheet or Biographical Data.

(a) Interview

All these three techniques are used sometimes to supplement each other. The personal interview along with application blank is used almost by every employer. There is a universal desire to see and talk with any one who is being considered for some job.

Every individual who faces an interview has an optimistic faith in his ability to size up people. A survey has been made among 325 companies to see what techniques they use in personnel selection and the study revealed that 98 percent of the companies used interview.

Some authors like McCormick (1984) stated that the interview is a kind of communication process, used for various purposes in industry and other organizations, such as employment counselling, attitudes and opinion surveys and for marketing and consumer researches. Here we are interested only in employment process and it is aimed at deciding whether to offer an employment to an individual candidate or not.

The personnel manager in any organisation may interview the candidates who are considered for the jobs. There are employment interviewers or recruitment boards, executives, departmental heads or supervisors who can interview the candidates. In some instances candidates may be interviewed by two or more than two members.

(b) What is an Interview?

An interview is a conversation between the employer and the candidate with a definite purpose. There are three purposes in the interview such as: (l) obtaining information from the candidate, (2) giving information to him, and (3) finding out the candidate's motivation. The employment interview should serve all these purposes. The interview should provide an appraisal of personality by obtaining relevant information about the employee's backgrounds. The interview should supplement the application blank to obtain additional information. Probably interview is the most widely used source of information and decision making material in which an expert can collect a wealth of information and impressions. This technique assumes that a given stimulus has the same meaning to every respondent.

The interview may be either formal or informal. But these two are often combined for assessment of an individual. In the formal interview, the interviewer wants to get definite information from the subject. He makes entries in a form. The individual's suitability for college, for professional training for certain job is determined from his academic record, his work history, his character and personal references and other pertinent information. In a vocational situation the demands are good health, physical stamina and temperament. The interviewer with these requirements in his mind examines the candidate. The salesman who must meet the consumers in face-to-face situations needs to possess the personality traits which should convince the consumers. Success in the formal interview depends on how readily the interviewer can identify the personality traits sought in an individual. But in an informal interview the interviewer engages in conversation with the interviewee and permits him to speak mostly. A skilful interviewer can get valuable information about the individual's personality traits from his expressed interest, ambitions, his dress, his manner of expression, hesitation, emotional states and confusions. The informal interview is often used as means of aiding people with personality problems of those who are worried about themselves, their school work, their jobs, their social or personal relationships. In personal interviews the subject is encouraged to 'talk out' his problems and he is let to discover himself where his difficulties lie.

(c) Patterned Interview

This is also called standardised interview or diagnostic interview. It is possible to get good result with a careful interview. McMurry (1958) calls it a patterned interview. By this he meant that this interview had a definite plan. It is a handy combination with application blank and a careful reference check. McMurry designed it to measure the personality traits that are wanted for employees. Such traits are 1. Stability, 2. Industry, 3. Ability to get along with others, 4. Self-reliance, 5. Willingness to accept responsibility, 6. Freedom from emotional immaturity, 7. Motivation. The basis of this approach is that an applicant's future behaviour can be judged by his past performance. If he was industrious he will be the same in future. If he was unstable he will be the same in future. Patterned interview contains no questions relating to the job skills. It is designed to appraise personality, motivation and interests. The validity of patterned interview was tested in several studies. In one study the correlation between supervisors ratings and interviewers judgment was 0.68. This indicates the definite plan of patterned interview. Patterned interview requires careful preparation of questions, use of printed form, uniform method of recording information and making judgments. One of the instruments called diagnostic interviewer's guide was developed by Havland and Wonderlic as part of the selection programme. It consists of applicant's background such as 1. Work history, 2. Family history, 3. Social history and 4. Personal history.

(d) Non-directive Interview

It is the study of merits of the applicants by the use of non-directive questioning. The basic is the minimum use of direct questioning in which the "Yes" or "No" answers are avoided. Broad general questions are substituted. For example, questions like, what sort of work that interests you most? are mostly asked. The more the applicant is allowed to talk freely about himself the more he reveals his personality.

The non-directive interview is essentially a client-centred interview. The technique was formulated by Carl Rogers (1939) as an outgrowth of his experience in personal counselling. It is based on a theory that individual has the capacity and drive to grow and develop so that he can meet the conditions found in reality. Therefore, the counsellor is mostly passive here. The individual is fully accepted for what he is. He is free to express any attitude except destruction of the objective or violence toward the counsellor. But it is not as deep as psycho analytic type.

(e) Probing Interview

Psychologists use questions of probing nature in this interview. They use them for two reasons. One reason is that it is used as a tool in the clinician's attempt to discover the underlying drives in the candidate's mind. It is used as a kind of performance test in the selection of men to various jobs in industries. For example, if your former principal of your college were to ask you to address the students in the college, what do you think of your subject or topic would be? or if you have children of your own, how do you plan to bring them up in comparison with the way you were brought up ? Answers to questions of this type will reveal certain fundamental reactions to your college, home and situations.

The second type of probing interview as practised in industry involves asking related questions so as to force the applicant to reveal his attitudes, judgments as applied to job satisfaction. The interviewer may ask how the applicant's previous supervisors treated him. The main object is what the applicant reveals about his adjustment tendencies. The interviewer also notes down the emotional reactions of the candidate. This technique is relatively new in industry.

(f) Advantages and Limitations of Interview

The interview is having certain advantages over tests. Abilities and traits cannot be accurately measured by tests. Many psychologists think interview is much better than paper-pencil tests. On the positive side we see whether the candidate is good looking or not and how he reacts to the situation and conversation. The candidate's mannerism and alertness can be observed in the interview.

The interview is not a fool-proof method of selection. It has its own limitations. The interviewer cannot judge from a man's face such as personality and character traits, honesty, etc. A shifty gaze does not mean dishonest character, conduct and physiology cannot agree with contours of face. There is no special immunity to this weakness.

(g) Halo Effect in Interviews

There is the tendency to judge the total worth of the person on the basis of a few specific characteristics closely connected to stereotyping. It is a dangerous attitude. The interviewer may like one or more traits not connected with the job. Therefore, the interview should be more objective in nature and the recognition of human errors in judgment is a step for more objectivity. The interviewer can be trained to escape the halo effects and pitfalls.

(h) Finding Interest

The interviewer wants to determine vocational interest of the candidate for which he can use tests. But one approach is to ask the applicant just what he wants to do. What is interested in getting out of job? His responses will reveal his interest. The strength of his interest is also revealed. A complete description of the job may be presented so that the interviewer will learn whether he is really interested in this kind of work. Many college graduates find it very difficult to visualise job duties unless they are provided with correct information about the job, Therefore, the interviewer sometimes provides the candidates with information to find out the interest of the candidates.

APPLICATION BLANK

Application blank and interview are the two mostly used selection methods. They are often used as the supplement to each other. A recent survey showed that 99.5 percent of the companies used application blank. Application form may contain many pages with hundreds of questions about the applicant's history. The main items are: name, address, age, qualifications, marital status, schooling, experience etc. Most of the items in application blank can be answered by a short one or two word statement or simple 'Yes' or 'No'. The most impressive use of application blank is the prediction of success.

PERSONAL DATA SHEETS: PERSONAL HISTORY

Personal data sheet is associated with turnover. Men who stayed longer on a job were over 30 years of age and had more than one dependant and were married. The Bell Telephone System conducted an investigation in which salesmanship is associated with executive success. Men with highest scholarship in college made highest salaries. School achievements were associated with success of 150 executives employed in a member of companies. There was little difference between two top groups out of three in scholarship and qualifications. College education and high scholarship was a test of motivation and intelligence.

Physique

Physique of the employees is not without any effect in the successful performance of any job. Executives have been found to be taller and heavier than non-executives.

On the basis of personal history a weight of application blank was used. Being under 28 years of age had an advantage. It was weighted high. If college education is unnecessary, it is weighted low. Although application blank cannot be weighted successfully for jobs weighted application blank is found to be highly promising.

VALIDITY OF SELECTION

There is wide margin of errors in ordinary employment procedures. In one study twelve sales managers rated 57 applicants for sales jobs. Each interviewer conducted interview the way he saw fit. At the end the sales manager rated the applicants according to his overall suitability. The ratings were placed in rank order from 1 to 57 for salesmen. This showed that there was low agreement among the sales managers in rating the applicants.

In another study of interviewing errors, there were six sales managers and a psychologist. They interviewed 12 candidates for a job as salesman. After each interviewer's judgment it was compared with a group judgment. The correlation was 0.47. This was below the demands of a satisfactory validity.

The interviewing technique can be improved. (1) It can be accurate if it is carried out with sufficient care. (2) The interviewer need to be trained in definite procedure. (3) The interviewer must maintain an objective attitude towards applicants.

ORIENTATION

Orientation begins in early stages. It is intended to make one to study himself in order to understand better about his abilities and interests. He may have a better awareness of how he may make better out of job he does. The real point is that many employees need a great deal of retraining themselves in their prospective fields of work. This can be provided by various means like lectures, field trips, films, reading and discussion. Occupational roles are of central importance in our society. Much of an individual's role playings are vocational in content. It may even begin in High Schools and may continue throughout life. After cessation of formal education some orientation to the total life span should be provided. Organisations, personnel policies and activities should move in this direction with their executive development, management succession, supervisory training, employee counselling and refrainment counselling plans. Orientation programmes may be narrow in scope and limited in perspective perpetuating the *status quo* in serving the interests of small groups or it may be also broad in scope serving the interests of society as a whole. Orientation programmes help able young people to find out appropriate ways of using their abilities in meeting society's needs.

APPRAISAL TECHNIQUES

Psychology has provided vocational guidance with a variety of analytical and predictive devices. The test of intelligence, aptitudes, proficiency, interests and personality traits were the forms of devices. These came into widespread use as a part of appraising vocational promise and fitness. The formal programmes of vocational guidance in schools and colleges are either very slow or absent. Atleast some-short term vocational counselling at critical chord point in careers should be emphasised. Few Indian Universities have introduced diploma and certificate courses in guidance and counselling. Appraisal, that is analysis and diagnosis, is one part of counselling process. Data about a person are collected, synthesized, understood, so that others may assimilate them and use them in their thinking. The personal data should cover psychological and social data and the situational data should include the social and economic data. These are to be assessed along with his intellectual and special abilities, his interests, his personality and skills. The counsellor needs to know his client's role models, identifications, self-concepts, motivation, concept of work and values he seems to observe. Appraisal for counselling involves the use of a variety of pertinent techniques like tests, inventories, interviews, observations and projective methods. The process of appraisal is done with orientation and counselling. The best appraisal work is to be done in school situations. New appraisal should go with orientation. The effective appraisal is a part of reality testing procedure.

VOCATIONAL COUNSELLING TECHNIQUES

We have seen that appraisal is to be done with orientation and counselling as they do not constitute different categories. They are blend into or overlap each other. A good cooperative appraisal work may emerge with counselling. Vocational development consists of implementing a self concept and self concepts often need modification. It is important that the client or patient puts his self concept into words in counselling process first. He needs to clarify his actual role and his role aspirations, so that the counsellor may understand his nature of the vocational counselling problem confronting him. If testing is to be helpful, more than one interview is often needed.

The vocational counselling is best proceeded if it is handled non-directively at first. Interviews may be shifted sometimes to directive handling when reactions to reality are to be explored. Thus it calls for the cyclical use of non-directive and directive methods. Counselling is a service needed by most persons as they move through life, face new choices and make new adjustments.

IMPLICATIONS FOR VOCATIONAL GUIDANCE AND PERSONNEL PROGRAMMES

Vocational development is a major responsibility of schools and colleges. Pupils and students should acquire an orientation to life and to work in schools and colleges. The adolescents should prepare to earn a living and to lead a way of life and from there the young adults should enter the world of work and careers. This programme is successful in foreign countries. But in India we are not even thought of it. The Indian educational policies should provide scope for this kind of education: Vocational development in schools and colleges is different from job oriented education. We in India often confuse vocational development with job orientation. Even students who are trained under job orientation educational system need vocational development. The confusion can be avoided if only we appoint counselling psychologists in every school and college and by providing right knowledge about vocational development. In short, vocational development is a major concern of society and right persons should be appointed in schools and colleges to disseminate the correct knowledge.

PROGRAMMES IN EDUCATIONAL INSTITUTIONS

Education is a continuous process over a period of nine to twenty years. Before the boys and girls develop a degree of self understanding and orientation to society to seek and make their own places in the world, the schools and colleges serve as the most important single agent of vocational development. The school serves as a unique place to guide vocational development bringing resources of society to bear on the individual.

It is important that always a vocational programme should be planned. Because continued contact alone will not result in individual growth and development. Therefore, schools and colleges need to develop and carry out educational programmes having the objectives of the development and self concept in students and the translation of the self concept into occupational terms.

It is also important that any educational programme should have staff and facilities. The instructors who were considered best qualified for this particular task should be assigned or recruited. If any educational programmes were to make any contribution to vocational development, training in vocational counselling must be assigned to specially trained staff members. A teacher in history or biology or an athletic coach may be good in teaching the subject or coaching a game, but not orienting young men to careers and helping them to develop vocationally realistic self concepts. The training in this line is primarily psychological supplemented work in education and sociology. In schools it requires one year training, but in colleges and universities, most counsellors have to complete a doctorate or atleast a Master's Degree in one of the programmes of study in counselling psychology.

The time is an important factor in guidance programmes. The counselling for guidance of development may be personal and vocational. It is done in series of interviews lasting about an hour with spacing. From several studies it is understood that counselling of this type is usually not effective. Counselling requires time for consultations. Further, it requires continuous contacts throughout the educational process. It is important that the counsellor should confer with each student from three to six times per year for about half an hour to one hour.

GUIDANCE FOR VARIETY OF CAREERS

Men and women enter various professions, managerial occupations clerical, and sales occupations, and skilled trades. These are described as careers in popular sense. Super (1957) has classified careers into three categories, as ***stable careers, conventional careers*** and ***unstable careers***. Men and women enter such professions on leaving colleges and universities and remain in them for ***longer period***. Therefore, they have been characterised as having ***stable careers.***

There are also conventional career patterns. People before entering stable occupations tend to enter in some other type of occupations and progress through ***trial jobs*** to life work. These are called conventional careers. Clerical and sales occupations, skilled craftsmen, semi-skilled workers pass through varieties of conventional careers.

The third type is called ***unstable career*** patterns. Some people in every occupation stabilize early and may have stable careers. But many settle down in an occupation after a series of trials. Some move from one type of trial to another type of trial taking risk and time. But some uproot from one occupation and go through another ***series of trial works*** till they settle down. This we may describe as unstable career pattern. Particularly people with personality or character problems and unskilled workers often fall into their unstable or multiple trial career patterns. Therefore, we may state that vocational development means not the unfolding of a career in the popular sense, but it is rather movement towards and through a life of occupational change, of multiple trials and instabilities.

At first, career patterns of this type appear to be very undesirable because the concept of instability, the idea of trying a number of times without stable occupation is not compatible with our lower and middle class social values. It is always thought in our society that the aim of education and guidance should help everyone to develop a career rather than to have a career pattern. Vocational development and guidance should eliminate trials and help everyone to occupy an appropriate life work on the completion of formal education. This laudable objective is psychologically possible for persons qualified for professional careers if the professions are so planned as to make it practical. But stability may not be appropriate for certain types of occupations. Most of the managerial positions require experience in other types of works before the person is ready for executive responsibility. In certain jobs, experience on the job is considered more important than formal training. Therefore, occupational stability as conceived by professionals may not be the universal desideratum. Because for the semi-skilled workers occupational mobility, flexibility, versatility may be more appropriate objectives.

The professional worker is going wherever he finds the best opportunities to pursue his chosen occupation. He tends to be mobile geographically and institutionally. But the semi-skilled worker is on mobile occupationally working whatever job for whatever company offers him the best work. A good job for a professional is one in which he can use his knowledge, improve his understanding and skills, exercise his own judgment and associate himself with interesting and intellectually stimulating people and he is called work-oriented man. A good job for a semi-skilled worker in business or industry is one in which he can work in a pleasant environment with agreeable associates and supervisors for good pay at suitable hours and he is called outcome-oriented worker. In the case of professional work the work itself is the reward. In the case of semi-skilled, work is rewarding as it is making life possible. For the professional, stability is living centred. Therefore, vocational guidance programmes in schools and colleges have to be taken in account. Regarding the aim for further vocational development some students can look forward to stable career patterns (a life work) and some other students can look forward to unstable and multiple trial career patterns (a life of changing work). Under these circumstances, what vocational orientation appraisal and counselling can do? The answer is that for career or stable career, group orientation appraisal and counselling may have clear occupational objectives on which focus can be made.

In the orientation programmes, managerial, clerical, sales and semi-skilled fields can be surveyed. The nature of work, the patterning of careers, the way of life involved in each field can be understood. The task of appraisal is a more difficult one. But the kind of abilities, skills, interests and values, the possession which characterises persons in those fields can be assessed by usual methods.

Thirdly, counselling sometimes baffles the counsellors in making the decisions with the group's vocational objectives. It sometimes leaves the counsellor and the client with a feeling that the counselling is ending with insufficient planning. According to Canthier the semi-skilled workers need a sort of multi-potential technical education for fulfilling their vocational objectives.

PROGRAMMES IN COMMUNITY AGENCIES

The vocational guidance programmes planned for schools and colleges are also applicable in community agencies. It is to be understood that many schools and colleges do not have formal programmes of vocational guidance. Even if they had such guidance as they go from school to college or technical schools or as they enter labour market many boys and girls need more help than school or college has given them.

Vocational development does not stop with getting a satisfactory job. New problems may arise with regard to vocational choice and adjustment during adult years. The needs were not met by educational institutions and new needs may occur in later life. This has led to the growth of many community guidance services.

Community agencies like child and family welfare agencies, rehabilitation centres, hospitals, neighbourhood centres or fraternal organisations have provided some of these services. They have found that their clients, patients and members are helped to lead more satisfactory, constructive and effective lives when they are helped to find appropriate outlets for their abilities and interests and they move from dependency to independency.

PROGRAMMES IN BUSINESS AND INDUSTRY

Business and industry have recognised their state in vocational development. Even though it had its own limitations it has been visible in some form. Robert Owen in nineteenth century had demonstrated that educating the workers could be profitable. After the World War I, it was felt that there should be programmes for better selection and training of executives, supervisors and other types of personnel. In 1930s, the Hawthrone experiments of the Western Electric Company demonstrated that employee counselling could produce more effective workers and set up a pattern for the expansion of personnel services of World War II. From these instances, a new emphasis has stemmed for executive and supervisory in the form of training and counselling.

The concepts executive development and management succession were widely used. It fostered rapid industrial and business expansion which put premium on talent. Talents at all levels were made to seek. Further the need, continuity in management is felt and this is to be achieved by continuing and planning of vocational development. Personal planning and personnel development were done companywise. But when it is applied to individual it manifests as in development and planning of individual career. It is the vocational development. The individual's potential must be assessed early in his relationship with the company and this may facilitate his growth and advancement so that there may be no dissatisfaction and his talent elsewhere can be used. He must be helped to understand his own capacities for development so that he may aspire and plan realistically, whether for security or advancement in position. This is vocational development counselling.

CAREER PATTERN THEORY

It is important that one must plan for choices and adjustments before the need arises, for which knowledges of career patterns and life stages help him.

The widespread mobility of semi-skilled workers and the different mobility of professionals, the pathways to executive positions, the floundering and trial of the early employment years, finding the motivation for innovation without pressure, decline and retirement, finding satisfying roles – all these problems encountered in personal work are the problems of vocational development, and career patterns and life stages. An understanding of these concepts and knowledge are essential to the effective planning and conducting of a personnel programme.

STAFFING

The planning of personal programmes, which takes into account the individual differences, vocational development, career patterns cannot be done by men who are not familiar with the fields of vocational psychology and occupational sociology however able they may be at management. Vocational development programme should not be entrusted to men and women who do not have the skills of vocational orientation, appraisal and counselling however competent they are at supervision and other works. The training called for is training in personnel psychology, group dynamics and counselling psychology, measurements, social psychology and educational philosophy.

COUNSELLING AND GUIDANCE

Counselling is the component of guidance. The task is to assist the students in schools to examine the conflicting values that are the root cause of many personal problems. But in the beginning, guidance was dealing with vocational and related problems. It is concerned with getting job for young persons. The most important reason for this was to reduce problems like delinquency and criminal activities.

It is observed that many of the teen-aged boys and girls who are not in schools have nothing to do with anything. They are free and have the whole time in their hands. Similarly the young adults have more time in their hands. These situations, if not controlled, would certainly breed delinquency. The college drop-outs and the college completed youths when unemployed, find more time for unethical and delinquent activities. Therefore, counselling and guidance according to Hughes (1971) become imperative and essential component of school curriculum in order to prevent the emergence of the present and future problems among the young students in schools.

Psychologists, having realized the importance and the values of counselling and guidance have designed and developed various tools and tests and formulated methods to deal with such problems. The first step is to identify the problems through suitable tests. Some of the test materials widely used are the Minnesota Counselling Inventory, Thurstone Interest Schedule, Brainard Occupational Preference, Mooney's Problem Check List, California Psychological Inventory and so on.

CONCLUSION

The fusion of various trends in the field of counselling psychology has infused a new life into vocational guidance. Counselling psychology has been given broader focus. It has acquired new meaning and has developed new methods. In formulating a theory of vocational development, developmental psychology, occupational sociology, psychology of personality have been drawn for supplementing the field. The field of vocational guidance has demonstrated new vitality with the emergence of theories of vocational development.

EXPERIMENT 50 : ASSESSING THE PROBLEMS IN VARIOUS AREAS FOR COUNSELLING
(Minnesota Counselling Inventory)

Counselling has been receiving considerable attention in various situations. But it is placed strongly in broader social and scholastic perspective in recent times. Counselling is defined in various ways according to the situation. Generally it refers to the advice or guidance given by a person who is in better perspective to a person who is in need of it. Since counselling has become a separate field of study in psychology, its perspective has also become wider, specialization in the field has become very rapid, professional position becomes very rigid and expertise is available not only in United States and United Kingdom but also all over the world. Therefore, everyone defines counselling in his own way. Once, counselling has been monopolized by a group of specialists called clinical psychologists who were dealing with mental patients and maladjusted persons. Now it is treated as a component of guidance and its scope is widened and it is also applied in the three commonly accepted aspects of guidance – vocational, educational and personal. Therefore, counselling can be defined as the type of interview that is more pertinent to the needs of the client rather than to the needs of the adviser in guidance.

Counselling is one of the ways in which young people are enabled to grow in independence and confidence. It is a substitute for what has not been provided in other ways in the school or college. According to Wall (1968) the school is pre-eminently the place where boys and girls should be able to get unemotional and balanced answer to the questions, "What sort of person am I? What do others really think of me?" Wall states that how successfully it fulfils the function does not depend upon counselling, but on the whole issue of interaction between the individual student and his/her school.

One of the objectives of counselling is clarifying the helping process. The counsellor with his training and experience is able to do this by providing the student correct understanding about himself – who he is and what he is capable and what he expects from the student.

Another objective is that the counsellor works assisting the student to examine the psychological dimensions of what he is thinking and telling. This is quite different approach and opposite of a person's work in a social conversation.

The third objective is that the counsellor works to develop a second relationship with the student. When the client enters the guidance room the counsellor should receive him cordially and greet him and should open a topic of common interest. These three objectives should be carried out in concert.

Williamson (1950) is of the view that counselling is one of the personalized and individual processes designed to aid the individual to learn school subject matter, citizenship traits, social and personal values and habits, and all other habits, skills, attitudes and beliefs which can make up a normally adjusting being. Therefore, the function of counselling is to assist the student at school or college to learn efficient method of achieving the desired goals. There are several studies of counselling effectiveness in educational situations. Kranzler's (1966) study was mainly based on Elementary School children. Rothney's (1958) study was on secondary school students and Williamson and Bordin (1940) made their counselling study on college students.

Any counselling, whether it is for school students or college students or for a person with maladjustment, vocational or educational or personal guidance, needs the administration of some suitable psychological test for understanding the person's personality characteristics, though interview is also conducted. One will supplement the other in getting information about the person or the client. In counselling situation one of the most widely used inventory is the Minnesota Counselling Inventory. This inventory provides information required by the counsellor for his purpose. This inventory is also useful for teachers, educationalists and also to those who deal with school students, college students and youth who are not employed. This is also useful to persons working in community clinics, youth centres and other non-school sittings.

In whatever situation the counselling is taking place it should be systematic and should serve some purpose. The counsellor should assess the individual's personality characteristics. On the basis of the personality characteristics the counsellor designs his counselling process. The person need not be abnormal or maladjusted. Any normal person may be put to this situation. The individual's reaction to various family and social situations has to be assessed. This cannot be done only with the counselling interviews. Suitable tests or inventories designed for this purpose should be administered on the individual and the type of reaction should be assessed. The Minnesota Counselling Inventory is considered to be the most suitable test for this purpose as it has been specially designed for assessing various levels of adjustment. Further, the variables of this inventory are based on the needs of the adolescent boys and girls. The test administers can easily identify the needs of the adolescents. The inventory may help the counsellor or the test administrator to acquire further information about the personality structure, personality dynamic and personality problems of the young people.

The Minnesota Counselling Inventory has certain specific purposes apart from what is stated earlier. It sensitizes the counsellors and teachers to the relevant personality characteristics differentiating students. It helps to identify the students who are in need of therapeutic attention. This inventory should not be used with children less than eighth grade. It will not provide any information about children who are psychotic or who are severely disabled because of major psychological disturbances. This inventory can be administered by any teacher or by persons who are experienced in test administration. But the scores should be interpreted only by persons with technical knowledge in testing, test construction and analysis of the answers. Persons who interpret the scores of MCI should have good academic and practical background in the personality theory of the adolescents.

EXPERIMENT

Problem

Assess an individual's various aspects of personality make up and his/her the level of reactions to various situations by using Minnesota Counselling Inventory.

Method

Participant

Choose a person who has no knowledge about the test and describe him/her personal characteristics such as age, qualifications, sex, socio-economic status etc.

Materials

Minnesota Counselling Inventory Manual, Test Booklet, Key and Answer Sheet.

Description of the Material

The Minnesota Counselling Inventory was designed and developed by Ralph F. Berdie and Wilbur L. Layton and it was published by the American Psychological Corporation in 1957. But it is not fully original. It is one of the descendants of the Minnesota Multiphasic Personality Inventory. The MCI was specially designed for relatively normal high school and college students. It consists of 355 statements with True or False answers. There are seven areas in this inventory. They are (1) Family relationship (36 items), (2) Social relationship (60 items), (3) Emotional stability (43 items), (4) Confirmity (35 items), (5) Adjustment to Reality (55 items), (6) Mood (46 items) and, (7) Leadership (35 items). These items and areas are similar to Bell's Adjustment Inventory and California Test of Personality.

The MCI has two other scores. One is question score which is the number of omitted items used to determine whether the student has answered enough items to justify the scoring and the other is the validity score which helps to identify the student who might be over anxious to display socially acceptable characteristics.

The scores of the first three areas (FR, SR and ES) assist in identifying the areas in which students may be adjusting particularly well or poor. The remaining four scores (Confirmity, Adjustment to Reality, Mood and Leadership) are more directly related to the methods students employ in making adjustment. The remaining 45 items in the MCI are not referring to any scale. They are included for possible further use in the development of new scales.

Procedure

The participant should be made to sit comfortably in a well illuminated room. If there are more participants in the room or hall they should be made to sit at a reasonable distance so that there may not be any consultation among them. The following directions should be given:

"The purpose of this investigation is to help you to learn more about yourself. This booklet (showing the test booklet to the participant) contains 355 statements. You will have to read each statement carefully and state whether the content in each statement is true or false as applied to you. There is separate answer sheet (showing the answer sheet to the participant) in which you will find only the item numbers. Against each item there are letters 'T' and 'F' which indicate True or False. If the item is true as applied to you make a check (✓) mark on 'T' otherwise make a check mark (✓) on 'F' in the answer sheet. Answer the items as quickly as possible and try to answer all the items without omitting any of them. When I say, ready, go, you must start answering the items. First read the instructions at the top and then answer. Ready, go."

The experimenter can make a note of the time the participant has taken to answer all the 355 items. The approximate time may be 55 minutes to 60 minutes. Then the experimenter should collect the answer sheet and the test booklet and the participant may be allowed to go away. The scoring should be done according to the procedure with the help of the scoring key of the inventory.

Results

The experimenter should find out the individual score for the whole inventory and also work out the scores for all the seven areas and should give in Table 16.1. The high score indicates more problems and the low score less problems in this inventory. Similarly in every area if the individual scores high he/she is having problem in that area and a low score less or low problem.

A group data may be collected for all the areas and may be given in Table 16.2. The individual's total score for all the areas should be worked out. And also the total score for each area for all the participants and the mean score for each area should be worked out.

Table 16.1 : Showing the individual scores for all the seven areas in the Minnesota Counselling Inventory

S.No.	Area	Score
1.	Family Relationship	
2.	Social Relationship	
3.	Emotional Stability	
4.	Conformity	
5.	Adjustment to Reality	
6.	Mood	
7.	Leadership	
	Total Score	

Table 16.2 : Showing the group data for all the seven areas in the Minnesota Counselling Inventory

S. No.	Name	FR	SR	Est	Conf	Ad. Re.	Mood	Leader	Total all areas
1.									
2.									
3.									
4.									
5.									
6.									
7.									
8.									
9.									
10.									
Total									
Mean									

Discussion

Analyse, first, the individual participant's score in each area whether he/she scored high or low. Wherever the individual scored high it is the indication of more problems. Discuss indicating the specific area and state his/her nature of problem.

Discuss the group data in the light of total score for all the areas. In the group who scored high and what are his/her problems and who scored low. How many of them scored high and how many scored low. How many of them scored above the group mean in all the areas and how many below the group mean. Find out the reason for the same from the personal information. State in which area the mean score for the group is high and in which area it is low. State what are the problems widely prevalent in the group. Discuss whether age, sex and socio-economic status are having any influencing effect upon the individual or group problems.

Conclusions

1. State what sort of problems your participant is having.
2. In the group whose score is the highest in which area and what sort of problem he/she is having.
3. Who scored the lowest and in which area ?
4. How many of them in the group scored above the mean in the Inventory.
5. How many of them are free from problems.

EXPERIMENT 51 : UNDERSTANDING VOCATIONAL INTEREST

INTRODUCTION

Vocation is the job or occupation in which one is engaged to earn his livelihood. There are various sources where a job is available. Some youngsters worry about their vocational objectives and often ask their teachers and verify about their future prospectives. Some institutions employ school psychologists to provide vocational guidance. The contribution of psychology to education also includes vocational guidance. The individual is the centre of the entire educational process and the individual should be helped in all his problems, interests and choices. At present vocational choice is the central problem facing many youths. Getting into wrong vocation is a very serious affair. Many youths of the present generation get into wrong vocation accidently or by choice and feel sorry later on. This is because of inadequate information and unwise vocational choice. Studies on job satisfaction have revealed such problems (Super, 1971).

It is also observed that interest and ability are interrelated. A person will do better in his job and be happier if he is really interested in his job. Therefore, educational and vocational psychologists have paid much greater attention on interest of the candidate who seeks some job. Interest is often synonymously used with aptitude and motivation. Sometimes it is defined that motivation is interest. It is also observed that if a person is interested in an activity he would be motivated to carry out the same. It is stated that there are four types of interests (McCormick, 1984). They are expressed interests, manifest interests, inventoried interests and tested interests. The ***expressed interests*** refer to the expression of specific interest. It is the preference a person is having for something. ***Manifest interests*** are not expressed in words, but in action. It is expressed by participating in some specific activity. The ***inventoried interests*** refer to responses made by a person to a large number of questions with likes and dislikes for a number of activities and ***tested interests*** refer to the statistical treatment given to the responses to find out the level of preference one made for a vocation.

There are a number of interest inventories designed and developed by researchers for testing the vocational interests of adolescents and young people. The inventories reveal unsuspected interests of the individuals. They were found to be very much useful in assessing the nature of vocational interests of the individuals and also they are helpful in making a person to systematically clarify his/her understanding of his/her vocational interests.

The Thurstone Interest Schedule is one of the instruments available for assessing the understanding of the individual's vocational interests. It is also used as a counselling instrument. The purpose of this instrument is not concealed. The client is expected to make honest expression of his/her choices. It is more suitable for secondary school students and unemployed youths.

EXPERIMENT

Problem

The aim of this experiment is to provide the participant a systematic understanding of his/her vocational interests.

Method

Participant

Choose a person who has no knowledge about this experiment and describe his/her personal characteristics such as age, qualifications, sex, socio-economic status, etc.

Material

Thurstone Interest Schedule and Manual.

Description of the Materials

Thurstone Interest Schedule is designed as a counselling instrument at the secondary school level. It is used in a situation where the client-counsellor relationship is straight forward and the purpose of the instrument is not disguised. The schedule consists of ten vocational areas. They are Physical Science (PS), Biological Science (BS), Computational(C), Business (B), Executive (E), Persuasive (P), Linguistic (L), Humanitarian (H), Artistic (A), and Musical (M). Each area is compared twice with each of the other areas. Thus, ten scores of various areas available. They represent the interest in ten vocational fields.

The Thurstone Interest Schedule requires only ten minutes to answer and can be scored in 2 or 3 minutes. The lowest possible score for an area is 0 and the highest possible score is 20. The scores may be represented in a profile. The schedule consists of a sheet of 4 page folder of demy size. There are ten rows and ten columns. Thus there are 100 boxes and a pair of occupations are printed in each box with numbers 1 and 2. The individual will have to make his/her choice by drawing a circle around the number, 1 or 2 which represents an occupation.

Thurstone first published this schedule under the title of Thurstone Vocational Interest Schedule in 1937. Later on in 1941, he published it under the title of the Thurstone Interest Schedule to replace the former one.

Procedure

The participant is to be seated comfortably in a well illuminated room and the Thurstone Interest Schedule should be given to him. He may be asked to provide all the information asked for at the top of schedule. Then the following instructions should be given:

> "In this schedule there are 100 cells and the occupations are given in pairs in each cell. You will have to express your preference for different occupations. Among the two occupations there is no difference in income or prestige. Therefore, you can indicate which one you prefer. Draw a circle around 1 if you prefer the first of a pair of occupations given in the cell. Draw a circle around 2 if you prefer the second of the pair of occupations. Draw a circle around both the numbers if you like both the occupations. If you dislike both the occupations cross out both the occupations. Now turn the next page and mark your own preferences for occupations. Start making marks from left to right and complete all 100 cells. You can take your own time."

The experimenter may also note down the time to know how much time the participant has taken to complete the schedule. As soon as the schedule is completed the experimenter can collect the schedule and may allow the participant to go away.

Method of Scoring

The experimenter now can score the schedule according to the procedure devised by the author of the instrument. There are ten areas in the schedule and, therefore, ten interest scores. The interest score for Physical Science (PS) is obtained by, counting all the circles around 1's in the column PSl and all the circles in around 2's in row PS2. Write down the total in the box PS at the end of the row PS2.

The interest score for Biological Science (BS) is also obtained in the same way by counting all the circles around 1's in column BSl and all the circles around 2's in row BS2. Write down the total in the box BS at the end of the row BS.

Follow the same procedure to score all other interest areas. Then sum up the scores entered in all the boxes at the end of the row and write it in the box T. Count all the 1's and 2's in all the columns and rows and verify whether your score entered in Box T is correct.

Results

Find out the score for each field and for all the ten fields for the individual and present them in Table l6.3. Also arrange the fields in the order of preference, giving number l to the first preferred and 10 to the last preferred. Collect the group data and present them in Table 16.4. Work out the total or the mean score for the group data and find out the order of preferences of the whole group and present them in Table 16.4.

Draw a profile for the individual scores of ten areas. Draw a profile for the total score for each area of the whole group for all the ten areas. Explain the scores of your participant and also the whole group. Explain the nature of the two profiles.

Discussion

Discuss the individual's preferences on the Thurstone Interest Schedule with the help of the scores he/she has obtained in all ten fields. State the order of preferences from highest to the lowest. State what type of occupation your participant prefers most. Discuss this in relation to his/her personal characteristics as to why he/she prefers it or rejects the last occupation. Also explain the profile.

Table 16.3 : Showing the individual scores for all 10 fields of preferences on Thurstone Interest Schedule

S. No.	Field	Scores	Rank
1.	Physical Science - PS		
2.	Biological Science - BS		
3.	Computational - C		
4.	Business - B		
5.	Executive - E		
6.	Persuasive - P		
7.	Linguistic - L		
8.	Humanitarian - H		
9.	Artistic - A		
10.	Musical - M		
	Total		

Table 16.4 : Showing the group data for all 10 fields of preferences on Thurstone Interest Schedule

S.No.	Name	Fields of Preference										Total	Mean
		PS	BS	C	B	E	P	L	H	A	M		
1.	K.R.												
2.	M.S.												
3.	B.V.												
4.	P.K.												
5.	R.M.												
6.	S.L.												
7.	M.R.												
8.	P.V.												
9.	V.P.												
10.	S.M.												
Total													
Mean													

Interest Profile for the Individual

Score	PS	BS	C	B	E	P	L	H	A	M	Score
20	—	—	—	—	—	—	—	—	—	—	20
19	—	—	—	—	—	—	—	—	—	—	19
18	—	—	—	—	—	—	—	—	—	—	18
17	—	—	—	—	—	—	—	—	—	—	17
16	—	—	—	—	—	—	—	—	—	—	16
15	—	—	—	—	—	—	—	—	—	—	15
14	—	—	—	—	—	—	—	—	—	—	14
13	—	—	—	—	—	—	—	—	—	—	13
12	—	—	—	—	—	—	—	—	—	—	12
11	—	—	—	—	—	—	—	—	—	—	11
10	—	—	—	—	—	—	—	—	—	—	10
9	—	—	—	—	—	—	—	—	—	—	9
8	—	—	—	—	—	—	—	—	—	—	8
7	—	—	—	—	—	—	—	—	—	—	7
6	—	—	—	—	—	—	—	—	—	—	6
5	—	—	—	—	—	—	—	—	—	—	5
4	—	—	—	—	—	—	—	—	—	—	4
3	—	—	—	—	—	—	—	—	—	—	3
2	—	—	—	—	—	—	—	—	—	—	2
1	—	—	—	—	—	—	—	—	—	—	1
0											0

Interest Profile for the Group

Score	PS	BS	C	B	E	P	L	H	A	M	Score
20	–	–	–	–	–	–	–	–	–	–	20
19	–	–	–	–	–	–	–	–	–	–	19
18	–	–	–	–	–	–	–	–	–	–	18
17	–	–	–	–	–	–	–	–	–	–	17
16	–	–	–	–	–	–	–	–	–	–	16
15	–	–	–	–	–	–	–	–	–	–	15
14	–	–	–	–	–	–	–	–	–	–	14
13	–	–	–	–	–	–	–	–	–	–	13
12	–	–	–	–	–	–	–	–	–	–	12
11	–	–	–	–	–	–	–	–	–	–	11
10	–	–	–	–	–	–	–	–	–	–	10
9	–	–	–	–	–	–	–	–	–	–	9
8	–	–	–	–	–	–	–	–	–	–	8
7	–	–	–	–	–	–	–	–	–	–	7
6	–	–	–	–	–	–	–	–	–	–	6
5	–	–	–	–	–	–	–	–	–	–	5
4	–	–	–	–	–	–	–	–	–	–	4
3	–	–	–	–	–	–	–	–	–	–	3
2	–	–	–	–	–	–	–	–	–	–	2
1	–	–	–	–	–	–	–	–	–	–	1
0											0

Discuss the group data in terms of the total scores in all ten fields. What the group as a whole prefers? What are the second, third preferred occupations by the group? Discuss the rank order of the occupations. State the reason if any, why the group preferred the particular occupation and rejected the last one. Is your participant's occupational preference in relation to the group preference varied? Is your participant able to clarify his/her understanding of his/her vocational interests systematically? Or is he in a confused state? Explain

Conclusions

1. What are the occupations your participant preferred? Give the first three occupations and the last preferred.
2. What is the order of preference of the 10 fields of occupations for the group. State it in the form of rank order.

EXPERIMENT 52 : STUDY OF OCCUPATIONAL INTERESTS

Generally the word occupation refers to a person's regular work or job. It is the principal activity in which one is engaged. It may also refer to any activity on which time is spent by a person. It is the act of occupying or the state of being occupied by a person. A person is engaged in something or that in which one is engaged in a course of action. Occupation refers to an employment or business. It may also mean for taking possession of a place or land. But here we are using the word occupation in the sense of employment. The concept of occupation seems to have certain similarity with vocation and career. The concept of career is also having similar meaning. Career is a profession or occupation chosen as one's life's work. The concept of vocation refers to a specified occupation or profession or trade. Broadly speaking occupation is the job or profession or position in which a person is engaged and earns for his livelihood to support him and his family. The job may be of anything. It may be physical work or mental work or both physical and mental, skilled or unskilled. In this way there are hundreds and thousands of jobs or occupations. But one must choose only one as his job or occupation. This is one serious question that arises often in counselling at any level.

The individual may not be in a position in making the choice. He/She may be vacillating between two occupations or among many. In such case the individual may seek the help of some person whom he/she considers as right person or expert. Since occupational choice is a single event, at some point in life the individual decides what he should do for a living. Then the individual himself could not make any decision, he must make use of guidances available in making a choice. The social and economic forces in which the individual is having little or no control over them guide his vocational development. Giving the individual's personal history such as his qualification, competence, socio-economic backgrounds, his race, religion he may be asked to say how much choice he can have and how freely he can make the choice among many jobs. Neither the client nor the counsellor could predict what the person is capable of and for what job he is fit. Therefore, the trait and factor theory of vocational development has a long history and it still constitutes as the rationals for most vocational counselling in school or college or occupational situations. This theory puts more emphasis on test development so that traits and collection of traits or the personality factors could be measured (Blum and Balinsky, 1961).

One way of making the individual to make decision about his occupation is to test his aptitudes and abilities and provide him the list of jobs for which he is fit. The other way is to help the individual to find out what kinds of occupations would be most interesting for him among many of the jobs. The second approach may be agreeable and interesting to the individual. With the view of assessing the occupational preference of the individuals and provide them with vocational counselling and guidance several tests and inventories have been designed and developed by the psychologists. Some of the tests and inventories widely used in the vocational counselling situations are Thurstone's Interest Schedule, Strong's Vocational Interest Blank. The Dailey Vocational Tests, The Kuder Preference Record (Vocational) and Brainard Occupational Preference Inventory.

The Brainard Occupational Preference Inventory yields scores in six broad occupational fields for both sexes. The six fields are (1) Commercial, (2) Mechnical, (3) Professional, (4) Esthetics, (5) Scientific, (6) Agricultural (for boys) personal service (for girls).

The Inventory may help the counsellor to test one's abilities and interests. In testing the abilities the raw scores must be converted into normative system of ranks. In measuring interests the raw scores and the ranks may be useful. The raw scores reveal the strength and choice of a person's interest. For comparing individuals the normative system is needed. The counsellor must use both the raw scores and ranks to help the person to arrive at a decision.

Sometimes, it is possible in this inventory for an individual to score in an unusual way. The person may be highly optimistic in everything and he may have high scores in all the fields. Another may express his decision naturally in all items. These unusual scores may reveal the aspects of personality into which the counsellor should probe further. These scores may also represent either carelessness or non-cooperation of the respondent. Such problems may be reduced in a class group by proper orientation of the group to the purpose of the inventory and test. The six fields of occupations are associated with 120 occupations, each items referring to one occupation. Each field has four sections and each section has five occupations. The occupations range from unskilled to highly skilled. In making the decision socio-economic status of a person may play important part. The author of the inventory tested its reliability in this test. The test-retest study yielded the coefficient of correlations ranging from 0.73 for commercial to 0.91 for scientific and the reliabilities for all the fields were highly significant. The correlation coefficient between odd-even items ranged from 0.88 for commercial to 0.95 for mechanical and scientific fields. All these were highly significant.

EXPERIMENT

Problem

Study the occupational interest of a person in a systematic manner.

Method

Participant

Choose a person who has no knowledge about this Inventory and describe his/her personal characteristics such as age, sex, qualifications, socio-economic status, whether employed or unemployed.

Materials

Brainard Occupational Preference Inventory Form R Manual, Test Booklet, and Answer Sheet.

Description of the Material

The Brainard Occupational Preference Inventory Form R, was designed and developed by Paul B. Brainard and Ralph T. Brainard. This was published by the American Psychological Corporation, New York, in the year 1945. This Inventory permits systematic study of a person's occupational interest. There are six fields in this Inventory. They are, (1) Commercial, (2) Mechanical, (3) Professional, (4) Esthetic, (5) Scientific,

(6) Agricultural (for boys) and Personal Service (for girls). It can be administered on both boys and girls except the last field. It is very much useful for the counsellor in making the counsellee to make his/her decision with regard to the occupational interests. There are 120 items in this Inventory, each field having 20 items. There are five response categories such as strongly dislike, dislike, neutral, like and strongly like. This can be administered on an individual person and also on a group of students.

Procedure

The respondents of this Inventory should be seated in a well illuminated room. If there are more persons, there should be reasonable distance between persons, so that they will not have any conversation or discussion about the idea expressed in the items. Then the examiner should distribute the test booklets and the answer sheets to the respondents. They also should have black lead pencil. If any one is not having, the examiner should supply it. The first page of the test booklet consists of only instructions as how to answer the test booklet. When the respondents have the test booklets in their hands the examiner should say, "Look at the instructions on the booklet while I read them with you." After reading the first page the examiner should give some time for the respondents to ask questions and to clarify doubts if any. Then the examiner should say, "When I tell you, begin, you will have to open the booklet, read the first question and write your answer in the separate answer sheet. Answer all the questions in the same manner. But be sure that the serial number of the answer as you give in the answer sheet is the same as the number of the question in the test booklet. In the answer sheet there are two dotted parallel lines with five dots from left end to right end and above the lines there are letters: SD, D, N, L, SL. You will mark your answers by drawing heavy straight lines of different lengths on the answer sheet. If you strongly dislike the idea relating to the occupation a short heavy line to the SD be drawn. If you dislike it somewhat a heavy line from the beginning to the letter D is to be drawn. If you are neutral or indifferent about it stretch the line to N. If you like somewhat extend the heavy line to L and if you strongly like it extend the line from the beginning to the end SL. Be sure to make a mark for every question in the answer sheet. The marking should be always in between the two dotted lines. The area should be blackened in the answer sheet. See the model below.

SD — Strongly dislike

D — Dislike

N — Neutral

L — Like

SL — Strongly like

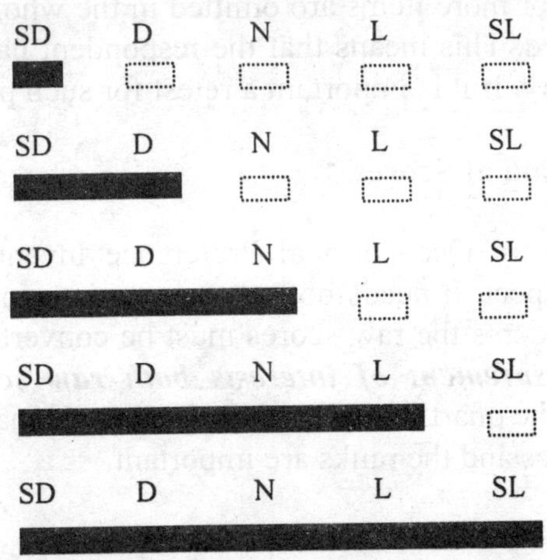

When you go to part 2, beginning with question number 101, you will notice one set for boys and one set for girls. If you are a boy answer the boy's set and if you are a girl answer the girl's set.

Work fast. There is no time limit. But do not spend too much of time on any one of the question and on any account do not skip over or omit any question without any answer.

Before starting to answer the questions provide the personal information asked for in the separate sheet. Now, "ready, go".

As soon as the respondents answered all the questions the examiner can collect the test booklet and the answer sheets and he can allow them to go away. Then the examiner can scrutinize whether the respondents have answered all the questions as directed.

Method of Scoring

The respondent has answered all the 120 questions by drawing strong black line below the response category to indicate his/her preference for the job, The strong line may be below the SD, or S, or N, or L or SL for very question number. The scoring is to be done according to the Manual. The weights for various responses are 1 for SD, 2 for D, 3 for N, 4 for L and 5 for SL. Score for each field may be got by summing up the weights of the appropriate items. The individuals total score is the sum of all the fields. The lowest possible score for any field is 20, which means that the respondent has marked SD for all the 20 items of the field. The highest score for any field is 100, which means that the respondent has marked SL for the 20 items of the field. The 20 items of every field are not given in the serial order in the test booklet. The items for Commercial are 1 to 20, Mechanical 31 to 40 and 41 to 50, Professional 31 to 40 and 51 to 60, Aesthetic 61 to 70 and 81 to 90, Scientific 71 to 80 and 91 to 100, Agricultural (for boys) Personal Service (for girls) 101 to 120. High score indicates liking and low score disliking.

Omitted Items

If the respondent has omitted only one item it may be treated as Neutral and a weight of 3 may be given to it. But two or more items are omitted no scoring should be made for that field. If six or more items are omitted in the whole answer sheet the answer sheet should not be scored. This means that the respondent has not answered the items to the best of his/her ability. If it is important a retest for such persons should be conducted.

Interpretation of Scores

In the Brainard Occupational Preference Inventory the interpretation of scores is an important aspect. It needs both the raw scores and the percentile scores. In comparing a group of students the raw scores must be converted into normative system of ranks. But ***in the measurement of interests both raw scores and the ranks may be useful.*** Sometimes the quartile rank is found to be adequate. For the purpose of counselling both the raw scores and the ranks are important.

Results

Work out the individual scores for all the six fields of occupations and give them in Table 16.5. Explain the fields in which respondent scored low and the fields in which he/she scored high. Mention the jobs included in the field which the respondent preferred.

Collect the group data for each field separately and give them in Table 16.6. Work out the total score for all the fields for the individual and also the mean score for the group for each field. Indicate the fields in which the whole group scored low and high.

Table 16.5 : Showing the individual scores in six fields of the Brainard Occupational Preference Inventory

S.No.	Field	Raw Score	Percentile Score	Order of Preference
1.	Commercial			
2.	Mechanical			
3.	Professional			
4.	Aesthetic			
5.	Scientific			
6.	Agricultural or Personal Service			
	Total			

Table 16.6 : Showing the raw scores and percentile scores of the group for the six fields in the Brainard Occupational Preference Inventory

S. No.	Name	Fields of Occupations												Total	
		I		II		III		IV		V		VI		R	P
		R	P	R	P	R	P	R	P	R	P	R	P		
1.															
2.															
3.															
4.															
5.															
6.															
7.															
8.															
9.															
10.															
	Total														
	Mean														

Discussion

Explain the meaning of the scores of the individual respondent for each field. What the low score indicates and what the high score indicates.

State the jobs included in the field in which the individual scored high. Indicate his/her first preference, second preference of his/her field and the occupations. State the reasons if any.

While interpreting the group data make out a rank order from the total score of all the fields of the individuals and state who scored the lowest and who scored the highest.

State the fields in which most of the individuals scored high and their interests. Find out the field mostly preferred by many of the respondents and state the reasons.

State which field received the lowest mean score and which the highest. Find out the reason for the same. Compare the results of your study with any of the previous studies and state whether your study agrees with the previous studies or disagrees. State the reasons for the same.

Conclusions

1. State your respondent's interest as revealed by the Brainard Occupational Preference Inventory.
2. State his highest preference and the lowest preference.
3. State the occupational preference of the whole group individually.
4. Which are the occupations preferred by many and which are not preferred.

17

SOCIAL BEHAVIOUR

INTRODUCTION

The study of social behaviour is highly interesting and also is very valuable as there are many social problems we are facing today. The most serious problems are interpersonal relation, group prejudices, friction between nations, adjustment among social groups, violence, over population, environmental pollution and so on. Social behaviour is revealed in various ways. Some psychologists seek to understand the causes of social behaviour. Every social group is having certain norms and values and they influence the individuals with their standard values and stereotypes. Stereotypes are fixed standard conventional notions. This may vary from group to group. But it is the individual persons who practice the stereotypes of the specific groups. The individual carries with him his group norms and values, develops negative feelings towards others, whose norms and values are different and excludes them from neighbourhoods, jobs, and schools. But if the individual person who provides help to others who are helpless, he fights for their rights, and negatively he commits acts of violence against such groups. In this way social behaviour is performed by the individual persons in groups ultimately. Therefore, social psychology studies the bulk of individual's action and thoughts in a social situation and this constitutes the study of social behaviour of the group. Since the individual persons constitute groups the study of individual behaviour in a group setting and studying his potentialities with which he is acting and reacting to certain specific situations all include social behaviour.

INDIVIDUAL AND THE GROUP

The primary concern of social psychology is the analysis of social behaviour of the individual. While describing the social behaviour of the individual we will have to describe his world of social situation in relation to the characteristic of the world. The term group according to Krech and Crutchfield (1948) refers to two or more people who have an explicit psychological relationship to one another. This means each member of the group acts in such a way that it is more or less having immediate effect upon another member of the same group. These members have something in common. For example, the members of a political party belong to a social group. What reaction a member of that party has at Delhi, is also held by members of the same party at Mumbai, Chennai, Kolkata and in other places. It is because they have something in common i.e. common ideology. Hence they behave alike wherever they are. These individuals who are reacting in the same manner to a common cause may have different age, qualification, income, occupation, marital status, personal traits, religious affiliation and so on. But still it is called social group. Since they have one common interest, they are made to act similarly.

SOCIAL GROUP AND HUMAN CROWDS

A human crowd should not be taken for a social group. A crowd has no common interest and it will not react to any incident or idea. Every individual in a crowd is independent and does not bother about what the other person does.

If the crowd is tense it will have negative consequences. This is true not only in the case of human being, but also in the case of animals. Over crowding will have negative effect on social behaviour (Baron and Byrne 1987). Calloun (1962, 1971) conducted some experiments on rats and found extreme negative consequences in putting the animals tensely. Though the animals were well fed and provided with plenty of water in a tensely packed situation they reacted in different types of negative ways. The negative behaviour ranged from extreme aggression to cannibalism or physical sickness. Anderson *et al.* (1977) Massey and Vanderbegh (1980) also found similar results in their studies.

What about human being? Will the overcrowding have the same effect on human beings? Overcrowding will have some negative consequences, but not in the same manner as it cause in the case of animals. Zultnick and Altman (1972) state that there is a positive correlation between population density and social problem. But the problems arise among the people of lowest income. Though there are no experimental evidences there were incidents occurring now and then that due to overcrowding people get stampeded and died. Paulus (1980) says that crowding refers to a negative psychological response. The effect of crowd is influenced by the type of density. The social density seems to have stronger effect – the negative effect than the spatial density.

The people who are coming out of a cinema house as long as they sit in the theatre they are called audience but when come out they become crowd. The people who get down from a plane, train and the people gathered in a market – all may be classified as crowds.

SOCIAL GROUP AND ORGANIZATION

A social or psychological group is different from a class of people or assemblage of people who may be seen as group because they are spatially in close proximity. Krech and Crutchfield (1948) state that whether an individual constitutes a social group or not, is determined by the way in which he perceives and reacts with other members of the social group and the dynamic interaction of the members of the group. The social group need not be too big. The size of the group may be small, may meet for short duration and the meeting may be informal.

SOCIAL GROUP AND SOCIAL ORGANIZATION

Social grouping may be also referring to the social organization. In social organization people have many things in common. The term social organization has been used by many social scientists synonymously with social institutions. But there are strong criticisms to use this concept of social institution as it lacks precision in its definition. This is mostly used by Sociologists referring to family, marriage, education, economics and religion. But the term social organization, according to Allport (1927) refers to specific groups of people which are described in terms of cultural products, norms and values, collective names, patterns of action, common beliefs and attitudes and so on.

Every social group or organization has its effects upon the behaviour of the individual. Certain generalization can be made in the behaviour of the individuals. Though certain distinctions can be made between social/psychological groups and social organizations they have much in common. For example the cultural products, eating habits, dressing patterns, religious activities and many of the generalizations apply to both social groups and to organizations. This facilitates to study small groups experimentally.

SIZE OF THE GROUPS

The smallest group is the family consists of the husband and wife. Therefore, the size of the group may range from 2 members, the husband and wife to several members. For example, in a small shop, there may be two or three members. There is potential relationship between members. For example with four members the potential relationships are six, a-b, a-c, a-d, b-c, b-d, and c-d. The larger the group grows the potential relationships become complex. The individual as a member of a group has certain limitations in his perception and reaction to the members of the group. Therefore, there is a limit to the size of the social group. This limitation is not no bar with social organization.

THE ROLES OF INDIVIDUALS IN GROUPS

Just like the husband plays his roles in the family, members in small organizations play their roles. When a person takes up a job in a company, his relationships with his superiors, subordinates, colleagues, are predetermined to him. How he should behave with others in the company is already determined. But he behaves towards others of the group partly depends upon his personality traits and the nature of the group and also the roles he plays within the company members. In social organizations there is always the tendency to formulate interpersonal relations in a regular manner.

Generally when people come together for the first time in a social group bring with them the beliefs and attitudes that they have developed from the previous group experience. Their interpersonal relations are mainly based upon them. These beliefs and attitudes and interpersonal experience and skills determine their social behaviour according to the present context.

SUB-GROUPS AND THEIR RELATIONSHIPS

Every social group is made up of some sub-groups. Generally all members of a group do not maintain their interpersonal relationships so closely as to judge that there is homogeneity. The group may be divided into several sub-groups and within the sub-groups there may be homogeneity. One sub-group may differ from other groups. What factors determine the major groups also determine the sub-groupings. The members of sub-groups perceive other sub-group members in a somewhat different manner. This will show that the members of one sub-group perceive members of other sub-groups and interact with them in a little less than their own group members. Thus there exists complex interrelationship in the structure and function of social groups.

STUDY OF INTER-GROUP RELATIONSHIP

It is exceedingly difficult to study the type of inter-relationships that exist within social groups. Moreno (1934) developed a method called 'Sociometric methods'. The objective of this method is to assess the existence of the type of feelings of acceptance and rejection or likes and dislikes that exist among members of the group. The sociometric method as designed by Moreno involves just to ask a member of a group to state which are the members with whom he wishes to act, work or live besides and which are the members whom he dislikes or wishes to avoid. When such informations are obtained from all the members it may facilitate to make out a sociogram. The sociogram is a pictorial illustration of the type of relationship every member of a group is having with every other member. The type of relationship will reveal the mutual likes, dislikes and indifferences. There may be also isolates who are not liked by any other members and also there may be a star member who is liked by many.

PERSONAL CHARACTERISTICS OF THE GROUP MEMBER

In small group studies Candland (1968) states that certain type of characteristics are most often examined by the social scientists. They are mainly (1) abilities, skills and experience (2) attitudes, values and interests, (3) demographic attributes such as sex, age, education etc., and (4) personality traits.

1. Abilities, Skills and Experience

In a small group researches it has been observed that the abilities of the members are based upon three variables such as group performance, individual's interaction within the group and the position of the members in the group structure. Heslin (1964) while making the survey of earlier studies found that the general ability of the members was found to be positively related with the effectiveness of their group. Heslin states that it cannot be the measurement of ability, but the skills that are relevant to the group task. It is also found in some studies of specific abilities that greater ability means better task performances. Rohde (1958) found that the group's performance reliably correlated with individual scores of all the members of the group. It is demonstrated in this that the ability of each member of the group contributed to the whole group performance. In Rohde's study only one skill was involved in the performance.

Investigators like Thalen (1949) states that if there are more than one type of performance then the effectiveness of the group is increased, particularly when the group is heterogeneous. Torronse (1954) states that the group may possess the required skills but this does not make sure that they will be allowed to make use of them as several factors may prevent them from using. It was also observed from the studies made by Triandis, Hall and Ewin (1965) that if heterogeneous attitude prevails in a group then it leads to social stress which breaks the communication among the members and hinders their abilities. Triandis and his colleagues found in their studies that a group with heterogeneous attitudes make very poor performance than a group of homogeneous attitudes.

In one of the studies made by Goldman (1965) it was observed that performance differs between individuals working alone and working in the group. An individual's performance is better when he works with other members of the group than when he works alone. It was also found that persons with low ability showed greatest increment in the work. These persons performed poorly on a problem on an earlier occasion. Goldman also observed in his study that when a person works with a person of low ability he shows better performance than working with person of his own ability and also the performance is greater when he works with persons of greater ability.

2. Attitudes, Interests and Values

These processes depend upon the attraction of group members on one another. The members of the group should develop a feeling of oneness that they belong to one group. This psychological process is called group cohesiveness. The type of feeling one develops may have effect upon the group cohesiveness. It may be positive feeling or negative feeling among members. The stronger the cohesiveness the group has the greater the efficiency with which the group will perform. Cartwright and Zander (1960) have found in their studies that group cohesiveness was an important determining factor of group's performance. Newcomb (1961) studied the importance of homogeneity among 17 students. Newcomb measured their attitudes, values and interest in beginning when they were stranger and after putting them together in a room and made them to live for fifteen weeks. It was found in this study that persons who had similar attitudes, values and interests have become friends easily and the scores were positively correlated. They agreed mutually on a number of issues on which they differed earlier.

The homogeneity of attitudes in a group behaviour was tested by Fiedler, Meuwese and Oonk (1961) and found that groups having homogeneity in religion showed low interpersonal conflict and performed better than groups which showed heterogeneous attitudes. Therefore, it was observed by these authors that to remove potential obstacles for effective group behaviour the homogeneity of attitudes, values and interests among the members of the group should prevail. This condition in a group promotes better group performance. But homogeneity in a group has also one disadvantage. Festinger and Thibant (1951), Spitzer (1964) observed that homogeneity and the group conformity suppress the creative ability of the individual members who can contribute their ability to successful performance of the task. Numerous studies on attitudes, values and interests were made in various situations. Bogardus (1925) developed a classical technique for measuring attitudes towards various nationalities and to compare the attitudes with various nationalities. The scale he designed is called Social Distance Scale. In social distance scale the person has to express his acceptance or reaction of any national group. Each nationality's likes and dislikes are indicated by seven types possible contacts a person can have towards a person belonging to other nation ranging from a visitor to any country to close kinship by marriage on a five-point scale with response categories, any, must, some, few and no. Beliefs and attitudes are measured on a very wide variety of ways. Apart from Bogardus's social distance scale, Thurstone's Psychophysical method of equal appearing interval, Likert's summated rating scale and Guttman's scalogram are some of the techniques often used for attitude scale construction and for measurement attitudes. Adorne *et al.* (1950) developed the F scale to measure the Authoritarian personality syndromes.

3. Demographic Characteristics

These refer to a person's physical and social attributes. The physical demographic characteristics are the person's age, sex, height, weight and race. The social demographic characteristics are religion, educational level, employment, socio-economic status and area of residence. These characteristics have varied effects on group's behaviour. Members of a group may vary in their age very much and this may have effect in the real communication process. The young may show some indifference to new ideas as old people initiate ideas. Similarly sex difference can have the same effect. Men always initiate ideas. As result of it ideas from women are not honoured.

The study made by Willis (1966) using the demographic variable brought out interesting results. His study was referring to speaker's relationships and the distance from which they speak to each other. Willis recorded the physical distance between people when started their conversation first. He related these distances to their age, sex, race and social relationship. The results indicated that generally people stand close to their good friends, than to their general friends, and casual friends (acquaintances). This type of behaviour was found to be much more common among women than among men. Willis further observed that parents stand far from their children, people stand close to people of their own age than they do with their younger persons. The speaking distances indicate the type of familiarity people have with others in communication. The Negroes maintain a greater speaking distance than the Caucasians do. Willis's study revealed that age, sex and race have definite effect in group behaviour apart from other variables.

4. Personality Traits of the Members in Group Behaviour

This variable also is having greater effect on group behaviour. Studies have been conducted to find out which personality characteristics have greater influence on the interactions of the members of the group. Mann (1959) has reviewed 150 studies in which the personality characteristics and the performance of small groups were involved. These studies show that adjustment, extroversion, dominance, masculinity and conservatism were all related to certain specific types of behaviour. It was also found that adjustment and extroversion are positively related to the leadership as perceived by the members. Conservatism was positively related to popularity, but negatively related to leadership. Dominance was positively related to the institution of task contribution and leadership. Masculinity is slightly related positively to leadership and popularity. Mann says that many of the correlations between personality characteristics and performance have low level of significances.

The relationships between the personality characteristics of leaders and the personality characteristics of the followers in group behaviour were studied by Sandford (1952), Haythorn, Couch, Haefner, Langham and Carter (1956). Haythorn and his colleagues compared in their studies the performance of groups with varying level of authoritarian leaders and followers. The authoritarian leader is one who is very much interested in power, wishes either to control or to be controlled. It was observed that the homogeneous groups have high morale, better cooperation less interpersonal conflict and effective communication than the heterogeneous groups. Haythorn *et al.* stated further that there was less friction between leaders and followers in homogeneous groups. Studies relating to democratic leadership and group behaviour were made by Lewin and Lippitt (1938), Lippitt (1940), Lippitt and White (1947) and found that the democratic leadership produced higher group morale, higher satisfaction among the group members and better quality and performance than the autocratic or authoritarian leadership.

What we call authoritarian leadership is more like the dictatorship whereas the democratic leadership is like the egalitarian. The authoritarian leader exercises more absolute power than the democratic leader. The authoritarian leader alone determines the policies of the group, he alone makes major plans, he alone dictates the activities of the members and he is the ultimate agent to judge reward and punishme t. The democratic leader does not show any difference from the authoritarian leader with regard to the amount of power he exercises in the group structure.

AUTHORITARIANISM AND SOCIAL GROUPS

The concept authoritarianism has been receiving greater importance in the study of personality traits, leadership, ethrocentrism, anti-semitism, attitudes and prejudices from 1940s when Erich Fromm (1941) first introduced it to describe the German character structure of that time. The root of this concept lies in the German Fascism and Nazism developed in Germany in 1920s and 1930s. Authoritarianism provides connection between personality and ideology and between character and prejudice. Fromm says that it is a new character structure developed from the new social and economic development that took place in Europe and the consequence of it is seen in personalities of the people in Western Europe and particularly in the lower middle class people of Germany. It was the authoritarian character structure. Fromm selected this concept, authoritarian because it was functioning authoritatively in the social and political affairs of the Fascist. The authoritarian character is the same as personality structure of Fascism. Authoritarianism mainly applied to the lower middle class people of Germany of the 1930s and it is accounted at the German national character.

The Barkeley studies by Adorno and his colleagues (1950) were made among the Americans and published in 1950 under the title of Authoritarian Personality in a volume consisted of 900 pages. This study was mainly based upon anti-semitism and nothing specific with the German character structure. Because the anti-semitism was practised against the Jews which was found not only in Germany but also in most parts of Europe.

According to Adorne *et al.*, the authoritarian is one who rigidly adhere to conventional middle class values, submit himself uncritically to idealized moral authorities of the in-group, has aggressive tendency to condemn, reject and punish people who are against the conventional values, pre-occupied with the dominance-submission, strong-weak, leader-follower dimension, identified with power figure, hostile towards out-group, projecting emotional impulses outwards and sexually deviant.

Another behaviour pattern associated with the authoritarianism is the group membership and group behaviour. It is the act of affiliation within the group and expressing the norms of the group through what one does explicitly. The relationship takes place within the group because of the socialization. Seaman, Michel and Dillchay (197l) state that once persons become members of the group they become more authoritarians. One becomes authoritarian either by attraction to the group or by socialization of the group members. There is also difference between the membership groups and a reference group. A reference group may not be a membership group for an individual. The reference group has certain important functions. The reference group is a normative body which has sanctioning power and controls the individual's actions for whom it serves. Secondly the reference group is used by individuals as a point of reference for evaluation of beliefs, attitudes and behaviour.

EXPERIMENT 53 : MEASUREMENT OF SOCIAL DISTANCE

INTRODUCTION

Much of human behaviour throughout life is one way or another conditioned by our fellow men. We may be stimulated by them and directed towards them. In other words we may be attracted by them and drawn towards them in some ways. This takes place in an environment in which man lives, acts, interacts and carries out his mission. This may be called a social environment and social relations with others are important determinants of a individual's behaviour. The general laws of human behaviour manifest themselves in the interaction of individuals in the social environment (Bird, 1940). The field of psychology which deals with this aspect of human behaviour is social psychology. Any experiment in which an experimenter and a subject face each other constitute social situation. The motivation of the subject, his co-operation in the experiment, accepting the instructions and his communication all could be taken as social variables.

Man is a social animal and he desires to be in the company of other human beings. Psychologists call this as gregarious nature. But this gregarious nature is not general. It is selective (Brown, 1965). Here the nature of social situation determines the behaviour of the individual. An individual's standards of perceptions, and judgments are affected by the social influences, perceptions, norms and judgments.

Experiments can be conducted to study how an individual's standards of judgment are acquired in a social context and how they represent in accordance with a social norm. Social norms vary from group to group (Sherif, 1936). Individuals born and brought in one group perceive the members of other groups with their norms. Social norms have strong cultural impact. Therefore, individuals are always making judgments of other individuals as members of different groups through their social and cultural norms. In some situations we wait and judge others with their performance. We may enter into certain situations with some expectations as to the behaviour and attributes of other people and belongings. Since we are going there with our social and cultural attributes and norms we make judgments of them in our own way. We judge other men and matters without any verification. What we make is prejudgment. Most of the pre-judgments or prior judgments evolve from the past experience and, therefore, past learning affects the present action and also the future (Murphy *et al.* 1947). Prior judging of men and matters without any verification or fact with our own cultural and social norms are developing a sort of ***apathy is called prejudice***. The word prejudice comes from the concept pre-judgment.

The consequences of prejudice may range from simple verbal abuse or feeling of hatred to severe physical attack to the level of extermination by lynching. In social situation even simple part learning is involved in the development of prejudice. We always carry into our social situation attitudes and expectations gained from our past experience and also from the experiences of others that are reported to as disappointments and emotionally tinged instances. Any report based on tact or rumour cause us anxieties and tension. According to Krech and Crutchfield (1948) every society is generated by prejudices of some kind which may be determined by cultural variations but also by some kind of physical attributes. The nationality, skin colour, the shape of the eyes, colour and length of one's hair, race, religion, community, caste all may contribute to the development of prejudice irrespective of the ability and achievements the people with those varied attributes may show in any situation.

Psychologists believe that whatever persists in a society can be measured. Similarly social conflicts, group tensions, prejudices and attitudes can be measured. One of the important studies made in this area has been the study of prejudice towards different nationalities. Bogardus (1933) the American Sociologist has designed and developed a scale to measure the amount of prejudice that a nationality could project towards another nationality. Bogardus called this scale as social distance scale. The concept ***'Social Distance'*** used by Bogardus (1925) referred to the amount of likes and dislikes that a person displays towards members of different group, race or nation.

EXPERIMENT

Problem

The aim of the present experiment is to assess and measure an individual's level of prejudice towards the ten different nationalities.

Method

Subject

Choose a person who has no knowledge about this experiment. Collect his personal information, like, name, sex, age, educational qualifications, socio-economic status, religion, nationality and complexion.

Materials

Bogardus Social Distance Scale and Scoring Sheets.

Description of the Scale

The scale has ***seven types of possible contacts*** a person can have towards a person belonging other nations. Each contact is put into ***five response categories***. The person who takes up the test will have to express his first feeling of reaction on all seven contacts through five response categories, The scale is given as follows :

According to my first feeling of reaction I would willingly admit :

(a)	Any: Most: Some: Few: No:	Canadian as visitors to my country.
(b)	Any: Most: Some: Few: No:	Canadian to live and work in my country.
(c)	Any: Most : Some: Few: No:	Canadian to full citizenship including right to vote in my country.
(d)	Any: Most: Some: Few: No:	Canadian to my school or University for my profession or occupation.
(e)	Any: Most: Some: Few: No:	Canadian to my club.
(f)	Any: Most: Some: Few: No:	Canadian to my home as my personal friend.
(g)	Any: Most: Some: Few: No:	Canadian to close kinship by marriage.

The ten nationalities towards whom the subject will have to express his feeling of reaction on the above pattern are as follows :

1. American, 2. Pakistanis, 3. Japanese, 4. British, 5. African (Negro), 6. French, 7. Russian, 8. Chinese, 9. German, 10. Ceylonese.

Procedure

Make the subject to sit comfortably. Give the test sheet, Bogardus Social Distance Scale in the hands of the subject. The test sheet for all ten nationalities should be in the same pattern as shown above with ***seven types of contact*** and ***five response categories***. The following instructions should be given to the subject.

"The object of this enquiry is to ascertain your own attitude towards different nationalities. It is not for any political or personal purpose. It is simply to assess your views for research purpose. You can be absolutely frank in your replies, concealing nothing about your feeling towards the nationalities. There are ten nationalities given in the test sheet and you will have to express your first feeling of reaction in seven different types of contact through five response categories as given in the test sheet. Underline any one of the response words given in front of each statement which expresses most nearly your view towards the particular nationality."

Method of Scoring

When the subject has answered the Questionnaire for all ten nationalities the experimenter should collect it and score the responses. Marks are given to the responses according to the nature and degree of the responses to the statements. The key for scoring the responses to the statements is as follows:

	Any	Most	Some	Few	No
(a)	+2	+1	0	–7	–14
(b)	+4	+2	0	–6	–12
(c)	+6	+3	0	–5	–10
(d)	+8	+4	0	–4	–8
(e)	+10	+5	0	–3	–6
(f)	+12	+6	0	–2	–4
(g)	+14	+7	0	–1	–2

According to the key the plus (+) mark indicates the absence of prejudice and the minus (–) mark presence of prejudice. It should be noted that the number of marks given to the responses vary according to the nature and degrees of the statements enclosed. That is, the willingness to admit any member of a nation as a *visitor* to my country receives +2 marks as this relationship does not involve great amount of racial friendliness. But on the other hand unwillingness to admit *any* member of a nation *as a visitor* to my country receives –14 marks as this relationship indicates extreme social hostility and prejudice. But on the other hand for the statement, I would willingly admit *any* member of another nation into a *close kinship by marriage* receives plus (+) 14 marks as this shows very high degree of good racial friendship and absence of prejudice. At the same time the response unwillingness into close kinship by marriage receives minus (–) 2 marks as it does not involve high degree of racial prejudice.

In this type of scoring system the highest positive score in this scale for a nationality is +56 indicating high degree of tolerance and absence of prejudice. The lowest possible score is –56 indicating high degree of hostility and presence of prejudice. This scale can be applied to any nationality for assessing social prejudice and to any community or caste for assessing communal or caste prejudice.

Results

As soon as the subject answered the questionnaire for all ten nationalities collect it and score it according to the key. Give the data in Table 17.1 showing the individual score for all ten nationalities. Rank the nationalities on the basis of the scores by giving rank 1 to the highest positive score and 10 to the lowest score. Get the group data, find out the mean score and give the rank order on the basis of the scores.

Table 17.1 : Showing the individual's scores for the ten nationalities on the Social Distance Test

S. No.	Nationalities	Scores
1.	American	
2.	Pakistanis	
3.	Japanese	
4.	British	
5.	African (Negro)	
6.	French	
7.	Russian	
8.	Chinese	
9.	German	
10.	Ceylonese	

Table 17.2 : Showing the group data and the ranks given by the subjects on the Social Distance Test for the ten nationalities.

S. No.	Name	1 Ameri		2 Pak		3 Jap		4 Bri		5 Afri		6 Fre		7 Rus		8 Chin		9 Germ		10 Cey	
		Sc	Rk	Sc	Rk	Sc	Rk	Sc	Rk	Sc	Rk	Sc	Rk	Sc	Rk	Sc	Rk	Sc	Rk	Sc	Rk
1.	R.V.																				
2.	M.S.																				
3.	T.V.																				
4.	B.S.																				
5.	P.K.																				
6.	K.R.																				
7.	E.S.																				
8.	B.V.																				
9.	D.U.																				
10.	S.V.																				
	Mean																				

Discussion

Analyse the data and state the individuals stand on the Social Distance Test. What are the nationalities against which the subject projects high degree of prejudice and what are the nationalities against whom he shows high degree of friendliness. Who is preferred first and who is rejected to the last? Find the possible reason and state it. How many of them in the group show high degree of prejudice and against whom, state the reason. What are the nationalities preferred for friendship? How many of them above the group mean and below the group mean? Discuss the ranks given by the group on the basis of the scores. Compare the ranks of the nationalities of the present study with one or two of the studies made earlier.

Conclusions

State the position of your subject regarding the degree of his prejudice and preference. Which nationality he/she preferred first and which he/she rejected to the last? State the group's preferences and the ranks. State the reason, if any.

EXPERIMENT 54 : SOCIOMETRY

INTRODUCTION

Man is a social animal. Much of man's environment is made up of other human beings and much of man's behaviour is influenced by other human beings. It is very difficult to say whether there is any human behaviour that do not function in a social context. Social behaviour may be defined as the behaviour of an individual in response to social stimuli. Social stimuli may be people or acts of people. Social behaviour is a kind of social interaction which may be the response on both the sides. The social psychologists are generally concerned with the study of social stimuli, social behaviour and social interaction. This falls under two types of analysis. One is focussing its attention on the individual's behaviour in a social context and the other is his concentration on group behaviour. Behaviour in a social context involves the consideration of social stimuli and social behaviour. It may be describing the type of behaviour like frustration, aggression, affiliation, affection, domination, submission and so on that occur in social situation. Psychologists wanted to discover the nature of differences if any between objects and persons in a social context. They believe in a broader sense that differences exist between objects and persons in degrees.

Eventhough social psychologists give more importance to the study of a particular social stimulus and a psrson's response to it and their attention is more focussed on social interaction. In other words the type of mutual response or the cyclic pattern of stimulus response sequences between persons receive greater attention. This shows that much of their interest is on the analysis of small group behaviour. What is a small group? A small group is one in which every member is aware of the individuality of every other member. Therefore, every member's likes and dislikes of the members of the group may vary in degrees. The type of the relationship may be interpersonal attraction.

The social motives or derives of the group may be the same. Each one's association with others may be for some practical reasons. It may be for mutual help, protection, security, sex, reassurance etc., dealing with important involvement. Each one's association with other may identify and define his relationship with social and physical world. The fact that we wish to associate with others implies that we are attracted to them or that we wish to attract them to us.

It is generally felt that appearance is the main basis for attraction between strangers. If attraction is to result in any kind of positive relationship it is necessary to maintain appearance. In our everyday life the word attraction or attractiveness is often referred to the surface aspects of a person.

The idea that appearance is the main basis of attraction, is somewhat repugnant to psychologists. In the experiment conducted by Walster (1962) and others it was observed that as far as social attraction is concerned physical attractiveness was the only significant variable. Other measures like intelligence, self acceptance, extroversion etc., were found to take no relationship to liking.

Physical appearance also becomes basis for social attraction. Persons who contest for American Presidency were taller than normal persons. They were also taller than the unsuccessful candidates. These findings tend to support the physical attractiveness theory of social attraction. Similarity is another factor that may promote interpersonal attraction. Physical appearance tend to be somewhat not strong explanation of social attraction. The cultural or serecultural groups have certain standard appearance and a member of such group possesses certain standard of appearance, he is liked by the member of other cultural group. The similarity in cultural norms promote interpersonal attraction. The closer a person approaches this standard in his appearance he becomes more attractive to those who use the standard as a basis for their judgments.

Studies on inter-personal attraction show that similarities may be equally important in the function of interpersonal attraction. It is quite but natural for the winners in any matches or games likely to seek the company of each other. But it is also true that losers also are more likely to seek the company of each other. According to Baron and Byrne (1987) physical attractiveness has a strong effect on the assumptions we make about other people and in the degree of attraction we feel toward them. However it is we who attribute more positive and desirable qualities to those who are attractive than those who are not. The fact that an individual who met with disappointment may lead him to perceive similarities between himself and others who also had similar disappointments. The losers of defeated players are likely to be attracted by other losers and recognise them. The experience of loss in both cause sympathy on each other.

Persons who have similar attitudes in various social or political issues or parties are attracted towards each other and share their similar attitudes and values. Researches on these lines have been conducted to measure the relationship between attractiveness and shared similarities in attitudes and values. Newcomb (1963) in his studies tested a general theory of interpersonal attraction dealing with the attempts of individual, involved in interpersonal situation to develop the feeling of affection and harmonious relationship. Another study has been conducted by Newcomb with 17 male university students.

These 17 students lived in a home provided by the experimenter for a period of 10 weeks. They were all new to each other before the study. The type of friendship patterns were studied by asking each student to fill up a questionnaire each week having their friendship choices attitudes. Newcomb (1963) observed that room mates tend to be attracted to each other from the beginning irrespective of similarities in attitudes and beliefs. Also when two students shared similar attitudes about other student of the group the liking was becoming stronger between them.

The study of interpersonal attraction was also related to racial prejudices following the studies of similarity in attitudes and values. In one of the studies the relationship between attitude similarity and race prejudice was studied. In this study the subjects who rated high or low in prejudice were asked to fill up an attitude scale. The experimenter then filled up a scale for each member of the group. This was presented to the group members saying that the answers were given by a Negro stranger and they were asked to rate him as a possible work partner. The questionnaire were filled up in varying amounts of agreement from 0 to 100. It was found that there was increase in the attractiveness of the Negro stranger. The increase in the attractiveness was depending upon the extent to which his attitudes were in agreement with those of the subjects expressed earlier. It was found in this study that even the highly prejudiced student stated that the Negro stranger was more attractive work partner. This was because that the Negro's opinion was in agreement with that of the students. Actually this Negro was a fictitious subject created by the experimenter. Experimental results on similarity of attitudes show that we are more attracted to people whose opinion resemble ours than those whose opinions vary.

It was also observed that frequent exposure promote interpersonal attraction among the persons who do not share anything in common. Homan's (1950) has stated that if frequency of interaction between two or more persons increased then the degree of their liking for one another will increase. Homans's view was experimentally verified by Festinger (1950) and others. It was shortly after the World War II young veterans and their families were housed in colleges and universities throughout the United States. One such housing area was a complex of seventeen buildings, each building consisted of ten apartments, five in each floor. It was observed by the researchers that there was spontaneous group formation among the families. Proximity was the main factor in determining who become friends with whom. Further it was found that functional distance was more effective than physical distance between apartments. People who used the same staircase and the same utility rooms were having more social contacts and become friends than the persons who lived in another apartment house. The usage of common places making the people to see each other often resulted in closer contact functionally than those who live in next apartments.

In one of the Whyte's (1956) studies it was observed that families who participated together in various social gatherings in a sub-urban area near Chicago were mostly neighbours. The families whose children played together were spontaneously drawn into interaction without any initiatives. In all these studies it is noted that proximity plays an important role in the interpersonal attraction as it provides more chances for interaction. The other factor was that they had something in common. There is a degree of similarity as they share that same building, same staircase, same utility rooms. Further, in all these studies it was found that the groups were homogeneous in age, education, economic status and interests. These factors contributed for building up a feeling of similarity which made it easier for them to find out some common interest and frames of reference than those people whose backgrounds were heterogeneous.

In view of these citations it is possible to study the interpersonal attraction experimentally in a group of people. The easiest method is to ask the member of group whom he likes very much or dislikes. But to ask such a question directly is not liked by people. Therefore, the question is put indirectly involving certain recreational activities and interests and the like. The number of times each member of a group is preferred by others is the indication of his/her attractiveness. In the same way how many of them in group disliked or ignored a member can be also rated. This method of measuring interpersonal attraction is called ***sociometry***. This sociometric method was developed by J.L. Moreno in 1953. The method was extensively used in studying the social structure of work groups, play groups, student groups in classrooms. The method was also used for ***measuring inter-personal attractiveness*** of individual group members. This method of sociometry had certain similarity with Bogardus (1933) Scale of Social Distance in the beginning and later on radical revision was made. The main difference of this method from the earlier technique is that it is chiefly concerned with attractions and repulsions between individuals within a small group.

Helen Jenneings (1948) an associate of Moreno stated that sociometry may be described as a means of presenting simply in scores and graphically the entire structure of relationships that existed at a particular time among the members of a group. The important lines of communication, or the model of attraction or rejection in various degrees are readily understandable at a glance.

Goode and Hatt (1952) state that this sociometric technique has been used in varied group situations like informal groups, school classes, prisons, organizations and in industries. Since there is no restriction in the application of sociometric technique, researchers who use this technique should bear in mind the three characteristics mentioned by Jennings (1943). The three important characteristics are, first a specific number of choices should be allowed, the choices should be varying according to the number of members in the group to be tested. Secondly, a specific standard for choice should be used and it should be varying with functional activity of the group. Thirdly, different degrees of preference should be designated for each choice in number.

While applying this method to measure interpersonal attraction in any situation three choices were allowed for each person and the situation must be defined as one in which the criterion for selecting the other person in the group must be like to play with them on a play group. The levels of choice were designated as number l, the person they like most, then the second choice with 2, and then the third choice with 3. But there are a number of ways in which the situation can be structured. For example, in a school it might be "with whom would you like to play, like to eat, share a room, spend leisure time". The important point here is that the situation must be a real one to the group being tested.

The choices in situation may be varied. There may be mutual choice, choosing each other among the two in a group. It is a mutual acceptance, indication of cohesiveness. There may be star attraction, one is liked by many in the group. He is called natural leader in the group. There may be isolates among the group members — the one not liked by any one in the group. There may be cliques, few members choosing one another among themselves and no one else.

In social psychological researches sociometric methods are used. They used to measure any personal dimension on which group members have an opinion. It is also used to study the leadership potentials, disruptive tendencies, personal warmth, ambition, dominance and political liberatism-conservatism. The interpersonal attraction is designated in choices as number 1, 2 and 3. These choices of each member of a group may be shown diagrammatically. The diagram is called sociogram. The sociogram is drawn to the data in the form of preferences of each member of the group. The preferences are presented in table, called sociometric matrix. Sociometric data are reported in terms of number of choices (attractions or rejections) made by each member of the group tested. The data become the basis for a diagram called sociogram. According to Lindgren (1974) sociogram has provided with one additional advantage to the experimenter that he can have an overall view of the preferences of the social group and also see the interpersonal and social attraction prevailed in the group.

EXPERIMENT

Problem

Study the interpersonal attraction of a small group or study the spontaneous group formation in a given small group.

Method

Participant

Choose a small group or a class consists of 10 to l6 students. Collect personal information of those students like, age, sex, religion, occupation of parents, wear spectacles or not, other physical deformities. The participants are named as A, B, C, D, E, F, G, H, I, J, K and L.

Materials

A slip contains one question, three situations.

Procedure

The test may be conducted by the class teacher himself as the experimenter as the class consists of only 12 students. If it is a big class it may be divided into 2 or 3 or 4 groups and each group may be assigned with one situation. Since the test is conducted in a 12-member group the experimenter gets ready 12 slips in which the following question may be written and the following instructions are given.

> "You are arranging your *birth day dinner* for your friends in a Hotel. Who are the three students from this class you will choose and invite for the dinner? Give the names in the order of preferences as lst, 2nd and 3rd. Answer the question confidentially and fold the paper and hand it over to me."

Then the experimenter (teacher) distributes the slips to all the 12 students. The students as soon as they answered the question, fold the paper and return to the experimenter. If necessary two more situations may be assigned. For example, going for picnic, as play-mates, room-mates, going to a picture and so on may be assigned and student may be asked to choose any three in order of preference as first choice, second choice and third choice.

Here, every student should think that he as an experimenter who conducts the experiment.

Results

Prepare sociometric matrix and have the names of the students as the chooser and chosen. It should be arranged in rows and columns. This is called sociometric matrix. Enter the choices of all 12 students (*A* to *L*). Each student makes 3 choices in order of preference. For example *A* makes 3 choices. Whom he preferred first? Write l, below the name whom he preferred at first, similarly, the second and also the third. Enter the 3 choices made by all 12 students in the matrix. Work out the total of the first choices, second choices and third choices in the row and lastly the sum total of all the three choices. A model of sociometric matrix with a source data is given in the Appendix-I.

From the matrix find out who has received the highest first preference and who has received the lowest preference. Who has received the highest second preference and who has received the lowest? State, who has received the highest third preference and who has received the lowest third preference?

Draw the diagram, the sociogram presenting the data from the Matrix in the figure. Indicate who are the stars and cliques, isolates and actually attracted. A model of sociogram drawn to some data is given in Appendix-II.

Discussion

Discuss the results as given in the matrix. What frequency each student shows of being chosen? Who is having the highest frequency of being chosen by the group members. State if there is any reason for being chosen with highest frequency. Who is the second best in the group chosen. State the reason, if any for the same. Who is the third best chosen in the group? State also the reason for the same. Is there any natural leadership? How it occurs? Is there any cliques? Who are they? How many isolates are there ? Whether there is any interpersonal attraction ? Discuss all these in the light of the earlier studies given in the introduction.

Conclusions

Is there any star ? How many ?
Is there any clique ? How many ? What it indicates ?
Is there any isolate ? How many ?
Are there mutual attractions ? Who are they ?

APPENDIX – I

SOCIOMETRIC MATRIX (A MODEL METRIC)

	Chosen												
Chooser	**A**	**B**	**C**	**D**	**E**	**F**	**G**	**H**	**I**	**J**	**K**	**L**	**Total**
A			3		1			2					6
B	3			1							2		6
C		2				1	3						6
D													–
E			3				1			2			6
F	1				2		3						6
G				3					1			2	6
H							1		3		2		6
I		1	2				3						6
J	1	2		3									6
K						2	1		3				6
L							1			2	3	–	6
1st Choice	2	1	–	1	1	1	4	–	1	–	–	–	11
2nd Choice	–	2	1	–	1	1	–	1	–	2	2	1	11
3rd Choice	1	–	2	2	–	–	3	–	2	–	1	–	11
Total	3	3	3	3	2	2	7	1	3	2	3	1	33

APPENDIX – II

A MODEL OF SOCIOGRAM

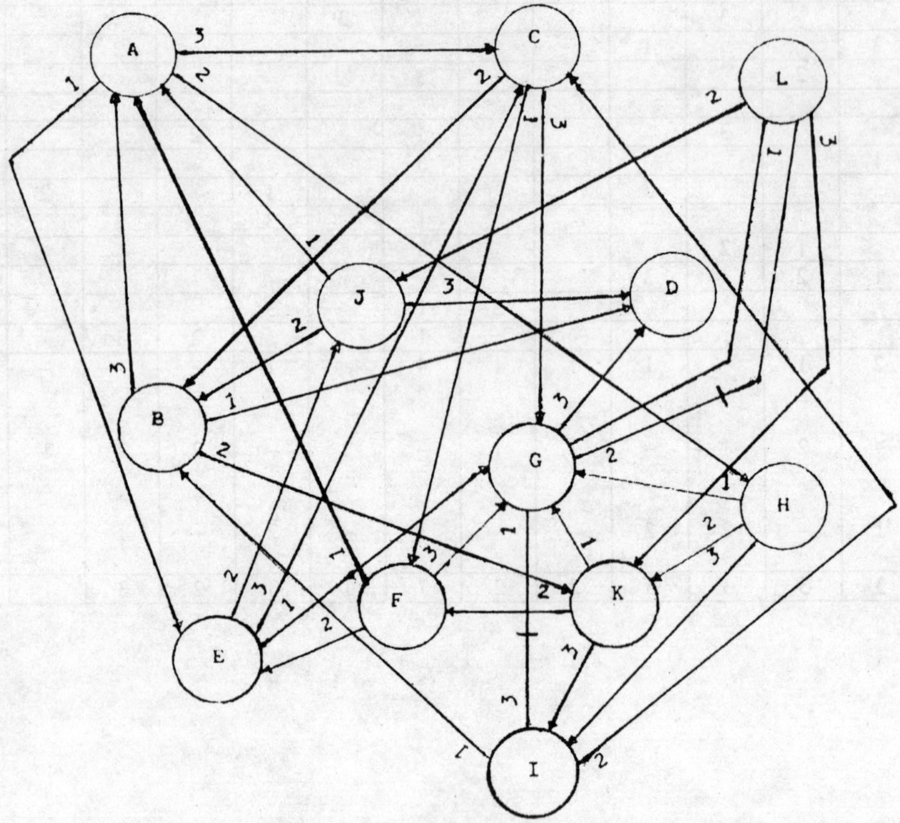

One Way Choice

Mutual Choice

18

ATTITUDES, RELATED TRAITS AND MEASUREMENT

INTRODUCTION

Man is a social animal and he lives always in a society in the company of his fellow beings. The social nature of man is such that he builds up contacts and relationships with other human beings. A person may be a member of various organisations and social groups and he builds relationships in all those situations. Though man's social behaviour is learned to some extent there is an innate tendency in man to be always in the company of other human beings. William McDougall (1936) the doughty champion of the instinct theory stated that man wants to be always in the company of other human beings because of the gregarious instinct, an innate nature in man to be in the company of other human beings. Though it is stated as primitive instinctive tendency it is this which promotes social pattern of behaviour like gregariousness, cooperativeness, competitiveness in which the individual's relationship with other is observed. There are certain permanent features of social behaviour. Garrett (1975) calls them as permanent social traits. They are generally described with the terms of attitudes, value, sentiments, and public opinion. We are here concerned with attitudes.

The concept, attitude is an old concept, generally used in social psychology. But in Germany in the beginning of this century it was an important concept in general psychology. The origin of this concept has been assigned to two German scholars, Muller and Schumann. The nineteenth century philosophers Alexander Bain, Herbert Spencer and Wilhelm Wundt (1832-1920) have used this concept to refer to mental processes. Bain (1868) has pointed out that, the forces of the mind might have got into a set track or attitude. Thomas and Znaniecki (1918) were the two sociologists first used this term very comprehensively. Giddings, an American sociologist, and Thomas, an American psychologist used this concept in their works. Warren was the first to use this concept in a textbook, *Human Psychology*. After this the concept gained more importance and invariably, it has occupied more pages in social psychology books and experimental studies were made on various aspects. Gordon W. Allport (1897-1967) wrote in his book (1935) that attitude is the central concept in social psychology. Today attitudes have become an inseparable personal and social values. Hence the psychology of attitudes has to be discussed only with the nature of human values. As result of this attitude and social values are expressed in some form in all social contexts.

If social psychological phenomena are to be experimentally studied and expressed in quantitative terms one has to relate the values of his study to attitudes. It has received so much importance in social psychological researches. It is because of the importance that Allport (1935) has stated that the concept, attitude is probably the most distinctive and indispensable concept in the contemporary American Social Psychology and no other term appears more frequently in experimental and theoretical literature as the concept attitude. Garrett (1975) states that it is very difficult to define the concept, attitude, since it often escapes from correct understanding. The reaction is that it primarily functions from inner state rather than expressing it overtly. But, however it is a function of inner state, attitude is expressed overtly in the personal or group behaviour in social contexts.

There is no one accepted single definition of attitude. Quite a number of definitions are available in every social psychology book. One can evolve his own definition of any concept if he would have some experimental study on the subject matter. The results of his experimental study may suggest some ideas from which one can formulate his own definition. Therefore, there are innumerable number of experimental studies on attitudes and also so many definitions.

The social behaviour of human beings can be explained in terms of motivational, emotional, perceptual and learning processes. Man's social behaviour is evolved from the dynamic nature of these four processes functioning together. Human behaviour is neither momentary nor immediate. It is a long-term arrangement. Therefore, it is better to see man's social behaviour from some higher level. This higher level of arrangement of motivational, emotional perceptual factor is called beliefs and attitudes. Only on the basis of this process Krech and Crutchfield (1948) have formulated a definition. According to them an attitude is an organization of the motivational, emotional perceptual and cognitive process of some aspect of the individual's world. Allport (1935) says that it is a mental and neural state of readiness established through experience exerting a direct influence upon the individual's response to all objects and situations with which it is related. According to F.H. Allport (1924), an attitude is a preparation in advance of the actual response setting up an important determinant of the ensuing social behaviour. Bird (1940) defines it as an adjustment of an individual toward a selected aspect of his environment or of his own conduct.

Attitude has been defined by Petty and Cacioppo (1985) as lasting, general evaluation of people, objects or issues. According to them lasting means it tends to persist across time and general means that it involves some degree of abstraction. A feeling existing for a moment is not an attitude. According to Breckler (1984), an attitude has three components such as *affect, behaviour*, and *cognition*. The affect component refers to emotional aspect. It may be positive or negative emotions. The behaviour component refers to our intention to act in certain ways. The cognitive component refers to thinking and interpreting the use of an attitude. Every attitude is a combination of feelings of likes and dislikes, behaviour and intentions and thoughts and ideas. Since every attitude has three components and function in different degree of combinations it is named as *ABC model* of attitudes by Breckler.

While carefully examining these definitions there is one central point running through all of them. It is a sense of readiness of a person towards something outside of him. On this basis an attitude can be defined as positive or negative responsive tendency of a person towards a person, object or situation. The important factor implied in an attitude towards an object or person is a response or reaction to it or him. It need not be necessarily a favourable response. However it involves same action overtly or inherently. Attitudes themselves are not actions though the possessors of them are often lead to action. Attitudes are not directly perceived. What type of attitude one possesses is revealed in his actions or overt social behaviour. The overt behaviour of an individual is the manifestation of his attitude. A man may reveal his attitude verbally on his own accord or in a response to a question or when a feeling of irritation is aroused in him.

SOCIAL AND PERSONAL OR PRIVATE ATTITUDES

There may be many issues involved in an individual's adjustment to the day-to-day problems. In his dealing with various social groups a man may develop a large number of attitudes in varying degrees towards various groups, individual persons, objects and situations. Some of his attitudes are found to be common to a group of people. Every individual of his group may find the same interest in it. These attitudes may be called social attitudes. But there are also certain attitudes prevailing in a group of people which are not having any common interest. These attitudes may be functioning around an individual's personal or private affairs. Such attitudes of the individuals are called personal or private attitudes. This kind of attitude mostly prevail among the members of various ethnic groups.

When various social and communal groups come into contact and if the members had occasions to live together for a short period there is likelihood of the development of personal or private attitudes. This personal attitude may be friendly or hostile in nature. Allport says that a person possesses two sets of attitudes. One set of attitude is kept confidentially as private and personal and the other set since which is socially determined it is publicly acknowledged. There may be some conflict between the social attitudes and personal or private attitudes. William James (1842-1910) quotes an incidence in his work (1890) to illustrate this point. A judge in the court finds his two attitudes in conflict when he is sentencing a criminal. He states, "as a man, I have pity on you, but as an official I must show you no mercy".

Every individual is supposed to possess these two attitudes and they are revealed in the respective situations. A person's attitudes towards his parents, children, cat, job, car are personal and private. These attitudes are not shared by others. A person's personal or private attitudes are very important for the individual's personal adjustments. Social and public attitudes are shared by others. They are common to the members of a group. For example, attitudes towards school, college, Government, road, airport, election, office, industry, Pakistanis, Srilankans, Americans, Muslims, Hindus all social or public.

ATTITUDES AND OTHER STATES OF READINESS OR TRAITS

There are certain psychological processes that are functioning quite similar to attitudes. In the common sense world all these are used synonymously. Though the function of certain concepts appear similar to attitudes they are not the same. Therefore, it becomes important to show the similarities and differences between attitude and other concepts. Some of them are, in traits, interests, motives, beliefs and opinions.

Attitudes and Traits

Garrett states that an attitude is a trait. It is a tendency to act in a particular fashion. It is a mental set or readiness to act. An attitude fundamentally is an implicit response. Hence it may be also defined as an implicit response towards or away from an individual value or social value. It is a potential adjustment towards a person or object or situation. There are certain common features in these definitions such as intention or purpose, anticipation and duration. An attitude like a trait differentiates one from another. It determines an individual's tendency to respond in a particular situation towards a person or object. Murphy (1964) states that traits are the means of distinguishing one person from another. Similarly attitude also differentiates its possessor from another. A person reacts favourably or unfavourably to certain persons. In this way an attitude functions like a trait. But a trait is a permanent personality characteristic of a person, possessed in more or less degree. Therefore, all traits are not attitudes.

Differentiating these two more specially, an attitude has always a specific reference. It is a learned responsive tendency either positively or negatively. It has a motivational component. Attitudes are effected by reinforcement and also subjected to modification. But traits are more or less permanent and general and may manifest in a variety of situations. According to Allport (1935) trait has no definite reference to an object or person whereas an attitude has definite object of reference. A person's views towards liquor, marriage, birth control, divorces, Fascism or Democracy are attitudes, but a person's reactionism, radicalism, introversion, extroversion, rigidity are traits. A trait may be narrowly limited to a state of readiness for response whereas an attitude may be specific as well as general. An attitude indicates the acceptance or rejection of an object or person with which it is related. But a trait has no such definite indication. For example a man may love his dog. He has a loving attitude towards the dog. But if the person has a sympathetic outlook towards the animals and human beings he has that trait of kindliness.

Attitude and Interest

Generally, interest is a feeling goes with attention. Murphy (1964) says that it is a disposition towards an object which one easily or freely attends to. Interests are distinctive states of readiness of the individual towards an object, whereas attitudes are states of readiness projected towards some persons or objects or situations. Interests are objective and attitudes are subjective. All social adjustments made by an individual involve both attitudes and interests. MacIver (1959) quotes an example to explain this point. A burglar, a citizen, a policeman and a judge are interested in law. But their attitudes to law are quite different. The burglar wants to violate the law and cheat the citizen, the police and the judge. The citizen wants to safeguard his property by the help of law and the policeman and the judge try to maintain the law. Similarly a person may not be interested in witnessing a cricket match, but he may have a strong attitude towards it as it maintains his nation's prestige. A communist does not agree with the principles of democracy, but at the same time he is interested in knowing how far the democratic party functions and progresses. Therefore, interest is concerned with one's likes and dislikes whereas attitudes are concerned with one's approval and evaluation.

Attitudes and Motives

An attitude is a state of readiness for motivated behaviour of performance, perception and thought. This means that there is some relationship between attitudes and motives: Motive often implies a need and direction of behaviour towards a goal. Motive is often used synonymously with need or drive. Motivated behaviour is directed to one goal only at a time. But an attitude may have many goals to be fulfilled.

Attitudes and Beliefs

An attitude is acquired through learning and experience, but it does not involve a higher mental process. A belief is a statement or proposition accepted on some evidence. It requires some amount of thinking. Krech and Crutchfield (1948) distinguish these two concepts. According to them a belief is a pattern of meaning of a thing. It is the total cognition of the individual about a thing. It refers to the conviction that certain things are true. We believe that the sun rises in the morning and sets in the evening. It is based on some evidence. Whereas an attitude is a state of readiness, organized through experience putting some influence upon the individual's towards persons, objects favourably or unfavourably.

Attitudes and Opinions

Opinion falls in between attitude and belief. An opinion like a belief is an acceptance of some proposition or statement. A person who holds an opinion does not have a strong desire to prove it. An opinion is neither a verified fact like knowledge or unverified proposition like a faith. Therefore, an opinion is a probability between verified knowledge and unverified faith. On verification opinion becomes knowledge. Otherwise it is faith. In the same way one cannot say that attitudes always be verified. When there are facts and evidences we have faith and when there are no facts and strong evidence we have opinions. But this distinction is not always correct. Thurstone and Chave (1929) clarify the concepts attitude and opinion. According to them attitude is a complex affair and it is not easy to describe it fully in few words. Attitude denotes the sum total of a man's inclinations, feelings, prejudices, perceived notions, ideas, fear, threats and convictions about any specific topic. It is purely subjective and personal affair. They state that opinion is a verbal expression of attitude. An attitude is symbolized in an opinion. For example, a person states: "In a democratic state there is a scope for all sorts of irregularities. This statement is his opinion, expressed verbally, which reveals the person's attitude towards democracy. Therefore, here attitude is implicit and opinion is explicit.

ATTITUDE FORMATION AND CHANGE

The psychology of attitudes has not only scientific and academic interest, but it also has a practical value. It is more concerned with regard to prediction of human behaviour in various areas. Some are interested in controlling and changing of human behaviour. From these points of view it is important to know how attitudes are formed and changed. Any attempt on the part of the psychologists to foretell or to regulate human behaviour involves understanding of the formation and development of attitudes. Political leaders, manufacturers, businessmen, and advertisers wanted to implant a new attitude in the minds of the people concerning their ideas and goods.

There is no human being who is born with attitudes nor attitudes are formed in them on their own accord. Attitudes are formed in relation to a person or place or object. Can we say that children enter the world with their political preferences, communal and racial hatreds, religious affiliations already fully formed. In fact these attitudes are acquired over a long period of time. Then how to account for their formation and development. Researchers in this area suggest several processes. Garrett (1975) states that there are two determinants that are playing important role in the formation and development and change of attitudes. They are psychological determinants such as motivation, emotion, need, thinking, dominance and submission and cultural factors such as social status, family environment and education.

One of the most important factors that operates in the formation of attitudes is selectivity. If an individual is interested in certain issue he willingly tries to acquire some attitude, favourably or unfavourably. Sherif and Cantril (1949) state that the first and the primary stage in the actual formation of an attitude is the perceptual stage. In this stage the person comes into contact with objects. Contact is possible only with perceptual situations. According to them there is always *subject-object relationship* in attitudes and attitudes are related to stimuli situations. The stimulus situations may be home, eating place, persons such as mother, father, brother, friends, teachers, playmates, community, school, college, flag, constitution and democracy.

It is stated that *culture* exerts *greater* influence in the formation and development of attitudes. Kretch and Crutchfield (1948) Sargent and Williamson (1948), Murphy *et al.* (1937), Lapiere and Farnsworth (1948), Macrone (1937) all put more emphasis on cultural factors for the formation and development of attitudes. According to them the home, parents, school and teachers are the cultural forces that play important part continuously on the child's life in the formation and development of attitudes.

Researchers suggest that an individual's *personality traits* have considerable influence in the formation and development of attitudes. Personality traits found to be highly correlating with social attitudes. Dexter (1939) in his study found that women students who were radical in their attitudes were more introvert, more self-sufficient, more dominant than women students who were conservative. But Morgan (1934) found the opposite results in his studies in another culture. A positive correlation between personality trait and radicalism in one culture may turn to be negative correlation in another culture. Murphy (1954) states that many personality traits seem to be related to prejudice. Rokeach (1960) in his investigation has observed that hostility towards out groups goes along consistently with certain personality traits.

Learning and Experience

Recent researches in the field attitude studies put more emphasis on two important factors for the formation and development of attitudes and change of attitudes. Baron and Byrne (1968) state that attitudes are formed by social learning and by direct experience. The formation of attitude by learning is largely a part of socialization. This is a process by which the new born child is transformed into a responsible member of human society. Children acquire their attitudes everywhere from parents, teachers, friends, media. Hence children acquire attitudes both from the schools and from outside the schools. This learning determines their likes and dislikes, favourable and unfavourable responses to persons, situations, objects, institutions and values. According to them there are three main processes which play important role in the formation of attitudes. They are classical conditioning, instrumental conditioning and modelling. ***Classical conditioning*** means learning by association. A mother's negative emotions will be noticed by the child and the child develops negative attitude to the object. ***Instrumental conditioning*** refers to learning in which responses are made to positive outcomes and eliminate negative outcomes. Children want to have right view of parents as it brings reward and eliminate wrong views of parents. ***Modelling*** refers to learning by observation. It is the limitation of the parents actions even if parents are not influencing them. Ruch (1958) states that attitudes are also learned behaviour pattern of the individuals. The existence of an attitude towards something is quite impossible without having some knowledge about that thing.

The child generally learns most of the things by the process of imitation and identification. The child grows up in the immediate presence of his father and mother and slowly learns to imitate and identify himself with them. Later on in the school and college he tries to imitate and identify his teachers and professors in some manner. It is because of the regard and respect he has for them. Experimental evidences show that learning of this type had some influence in the formation and development of attitudes. Vetter and Green (1932) have studied the formation and development of anti-religious attitudes of 350 members of the American Association. Many of them stated that they acquired their attitude from the study of history and science, some of them from the study of materialistic philosophy. Other traced their anti-religious attitudes to the feelings and experience of grief and disgust. A few of them traced their attitude to the influence of their friends who were atheists.

People form their attitudes from their own personal experience. They draw conclusions and make generalization from the results of their life events. Social cognition mainly contributes to the formation of attitudes. The formation of attitudes help individuals to reduce certain strain for getting more information about certain things. If you have already formed some idea about the place you visit, the food you wish to eat, the person you wish to meet your work is simplified. Thus formation of attitude by experience determines future activities and reduces the efforts in some activities. In this way attitudes become stereotypes.

Change of Attitudes

The attitudes once forned and developed are not permanent. They undergo change according to the experience one gains from the circumstance he faces. As and when one's experiences increase at various levels the attitudes already formed may appear to be not suitable or need modification. Allport states that the most favourable occasion for the change of attitudes is the period of emotional disorganization. Young (1931) says that the emergence of some new attitudes in place of old may resolve some crisis, may be a solution of a difficulty, end of a period of chaos and also the termination of some disorganization.

An attitude is very strong when it suits to the conditions and fulfils the desires of the individuals. It becomes weak and yields to change when it lacks these characteristics. Allport (1935) says that the shock of grief, economic disaster and falling in love are the times favourable for new attitudes to come into being. It is generally an orientation to contexts and current events. The change of attitudes attempts to systematise old experience with the emergence of a new attitude. Sherif (1948) states that attitude change means change of the individual's stand on a given issue.

Attitude change takes place in relation to a person, an object or some values. Several factors both internal and external work actively when an individual is exposed to change. Attitude change is found to take place in various areas of human values. Economic depression in the country, unemployment problems, insufficient income, social inequality and poverty can be stated as some of the factors exerting influence upon the individuals for change. The individual's experience and knowledge, social contacts, personality traits, active participation of the individual in interacting situations, prestige, mass media of communicating forces like, newspaper, magazines, motion picture, radio and television programmes are some of the factors constantly contributing to attitude change.

Experimental studies show that knowledge and experience in varying degrees have some effect in changing the attitudes of the people. Cherrington (1934) used nine different groups for his study of international attitudes. Attitudes were measured before and after putting the group into various kinds of experience and knowledge in varying duration. The experimental factor consisted of a three-day conference in Geneva and Switzerland, a summer school course, a series of lectures fo. two years. The group members' responses to various general and specific items for all nine groups were tabulated separately. The groups' average responses for the items before the test were 84 and 63 and 89 and 72 after the test. This shows that the experience and knowledge they have gained in the conference, summer school and lecture class have changed their attitudes.

Psychologists have observed in their experiments that attitudes can be changed by inducing certain factors. Controlling the experience of the people and providing them a systematic education and using the modern mass media, and the methods of communication were some of the successful methods discovered. Making efficient propaganda, screening motion pictures with new ideas, delivering inspiring public speeches in person, through radio, television, printed pages are some of the forces that can change attitudes. Attractive advertisements, political changes, prolonged interaction, majority opinions, expert opinions are also other influencing factors.

Persuasion

Psychologists for a long time have been studying effects of persuasion in changing the attitudes. Persuasion is a process of influencing people to change their attitudes. People engaged in trade and business urge the people in various ways to buy their goods. Political leaders plead with all their resources to get your vote. Humanistic service organizations warn people with collected evidences against smoking, drinking, speed driving, over eating and environmental populations. To be precise, at every turn we are encountering attempts to change our attitudes. Sometimes these attempts succeed and sometimes they fail.

A group of social psychologists at Yale university studied the effects of persuasion in attitudes. They measured some subjects' attitudes before and after their experiments and compared the effects of different ways of dealing with the attitudes. The findings of these studies are still valued. Hovland and Weiss (1952) found that experts attempts had better effect than non-experts. Walster and Festinger (1962) observed in their studies that if people think that the matter is not purposely intended to persuade them than they are easily persuaded. This is the reason that we do not fall prey to every commercial advertisement in the Television. Janis (1954) found in his study the influence of personality trait in persuasion. People with low self-esteem were more easily persuaded than people with high self-esteem. In the study of Kiesler and Kiesler (1969) it was found that popular and attractive means of communication had better effect than unpopular and unattractive ones. Messages from known and liked people had better effect than from unknown and disliked people. Allyn and Festinger (1961) noticed that people when they were distracted, were easily influenced by persuasion than when they were observing fully and when the persuasive message was simple. Hovland, Lumsdaine and Sheffield (1949) in their studies observed that when the audience had opposite attitude to what the persuader was trying on them persuasion become very tough. In such cases the persuader had to present more effectively the attitudes of both sides than just presenting one side. Miller *et al.* (1976) discovered that fast speaking had better persuasive effect than slow speaking. People think that the person who speaks fast knows his matter better. Laventhal, Singer and Jones (1965) in their studies have observed that persuasion had better effect by the information that could arouse fear in the audience. People should be persuaded to stop smoking by creating a high sense of fear of dying from lung cancer and by showing them picture of a smoker's lungs.

People sometimes resist persuasion and when persuasion occurred people deny it. Bem and McConnell (1970) observed in their studies that when people fall prey to persuasion they forget very easily what the original attitude they had previously.

Theories of Persuasive Processes

The study of persuation for attitude change has been a continuous process evolving one theory after another. More recently, studies in this area questioned the validity of the earlier Yale University Studies. Several researchers like Petty Ostrom and Brook (1981) have promoted an approach called Cognitive Response Analysis.

Cognitive Response Analysis

This has been further influenced by some ideas from social cognition. This theory stresses the point, "what makes you to change your mind?". It is how and what you think about a persuasive information or message. This means that people do not passively accept the persuasive message. They actively think about them. This thinking are thought results either in attitude change or resistance to change. People while thinking about the message often *elaborate* the contents of the persuasive message. That is, they go beyond the original message with their ideas and arguments. This is what the cognitive response analysis theory states.

Elaboration Likelihood Model

But, of late a theory has been evolved by Petty and Cacioppo (1985) on the basis of the elaboration that people often make on the persuasive message. The theory is called ***Elaboration Likelihood Model.*** Elaboration means thinking about and analysing the ideas and arguments relevant to the persuasive messages that are left out in the persuasive messages are included here. According to this theory there are two kinds of persuasion processes such as (1) the Central Route and (2) the Peripheral Route.

The Central Route

The central route to persuasion involves careful thought about the issue and about the ideas and argument, so that persuasion is most likely to take place if there are strong and convincing ideas and arguments. In this the attitudes in question are at the centre of the person's attention at the moment. Therefore, it is called central route.

The Peripheral Route

The peripheral route at the same time involves convincing the person without his full attention and careful thinking. Therefore, the strength of argument is less important and it is persuading the person who is distracted and it leads to attitude change.

The central route to persuasion should be used when the persuader has strong, good and convincing ideas and arguments. The peripheral route to persuasion should be used when the persuader cannot make strong cases as he is persuading a person who is distracted and in a confused state. People sometimes resist persuasion if they come to know in advance that some one is going to force them to range their beliefs and attitudes.

COGNITIVE DISSONANCE AND CHANGING ATTITUDES

The cognitive dissonance theory was propounded by Leon Festinger in 1957. This is one of the most important concepts in social psychology and it has stimulated a great deal of research in a short period. According to this theory an individual attempts to establish internal harmony, consistency or congruity among his beliefs, opinion, attitudes, knowledge and values. According Festinger (1957) an individual wants to maintain internal harmony or consistency among the cognitive elements. But the individual may find that pairs of cognitive elements are in irrelevant consonant or dissonant relationships with each other. The relationship is consonant if one element follows from the other. If the two elements have nothing to do with each other, then the relationship is irrelevant. But the relationship becomes dissonant if there are only two elements and if the obverse (counterpart) of one element follows from the other. In other words, dissonant relationship occurs when one of the two cognitive elements function disharmoniously to the other element. This dissonance can be resolved or reduced by changing one of the cognitive elements or by adding new elements or by reducing the importance of the elements. Dissonance results when others disagree with us. This can be resolved either by changing our opinion or by making others to change their opinion or by coming to a decision that there is disargument with us.

This cognitive dissonance theory is applied to attitude change also. Disharmonious feelings are generated by conflicts between a person's attitudes or beliefs and actions. Generally cognitive dissonance – 'an unpleasant feeling' – results from the conflict between an individual's thought and action. We might face simple conflicts even in our day-to-day life events. But we do not give more importance to it and try to resolve it by some means without much conscious efforts. But conflict may arise between more important and complex attitudes and behaviour. For example, a person might have formed many attitudes regarding his university education, professional careers and his further higher studies in a foreign country. But there was a sudden breakdown in his father's lucrative business. As a result of it he lost all the financial supports from his father. He has to take up a job in a firm as an assistant executive. The person's action did not concord with his attitudes. Hence he met with dissonance. This person's action-behaviour did not concord with many of his attitudes, such as desirability of a good programme. But, in a man's life it is often difficult to satisfy all his attitudes. It is because of this dissonance, which is a common experience in a man's life. Every one has many attitudes in his mind which do not agree to the best course of action or behaviour. Dissonance is the struggle the individual puts forth with personal inconsistencies. Several new looks have been tried to Dissonance theory recently and Cooper and Fazio (1984) have paid considerable attention on it.

The psychologists who worked in this area were able to trace out many factors that are contributing to dissonance. The personal choice as stressed by Linder, Cooper and Jones (1967) is the cause of dissonance. Generally, dissonance takes place only when people feel that their inconsistent action or behaviour had happened by their own choice. Firstly the inconsistent behaviour should have some foreseeable bad consequences. Goethals, Cooper and Naficy (1979) stated that if there was no harm in our behaviour we do not feel dissonance. Secondly, the inconsistent behaviour involves the self in some important way. Baumeister and Tice (1984) said that the action was performed by the self out of his free choice which creates an inner feeling that the individual should hold personal responsibility for the behaviour. Since the behaviour was witnessed by others the individual's personal reputation is involved. Thirdly, dissonance appears to depend also on physiological aspects. Croyle and Cooper (1983) stated that there was mild bodily upset as a part of dissonance.

THE MEASUREMENT OF ATTITUDES

The measurement of beliefs and attitudes has become one of the major studies in social psychology. If we know the beliefs and attitudes of the people it is possible to do something towards the prediction and control of their behaviour. Many business organizations, political parties wanted to know what people think about their products and ideologies. Unless a systematic measurement is made it will not be possible to say anything about these. The Gallup poll, the Fortune poll were some of the public opinion surveys conducted in United States for various purposes. Attitudes cannot be directly assessed. They are inferred from behaviour. It is assessed by observing the individual's responses to persons, objects and views by evaluating his statements or by the verbal expressions. Generally, it is easier to assess the individual's beliefs and attitudes through his spoken or written statements. This can be done by interviews. Since interviews are time consuming job a more reliable and efficient and the quick method is using paper-pencil scales and questionnaires. This method has several advantages. It is easy to administer the paper-pencil scales or questionnaires on a large number of people at the same time. If anonymity is assured more reliable information can be obtained and the data can be easily analysed statistically.

Dimensions of Attitudes

Attitude measurement is not directed to a single source. Attitudes have several dimensions. They are, direction, intensity, centrality, salience and consistency.

Direction

The scale used to measure attitudes should indicate the *direction*. First of all attitude measurement is concerned with the direction of the attitude. It may be for or against the person, object or ideas or events. Generally, in an attitude scale the direction is indicated on a like-dislike, approve disapprove, agree-disagree dimension by making a tick mark on 'Yes' or 'No' or True or False options.

Intensity

In attitude measurement the *intensity* of an attitude is also important. Is the individual expressing his feeling strongly about the object of the attitude or expresses a weak or neutral feeling. Intensity of the individual's attitude can be measured in various ways. There can be graded scale of values and one's position can be indicated. Also there can be statements expressing the view with different degrees ranging from most favourable to most unfavourable and the individual may be asked to select the statement with which he agrees. There can be also varied response categories ranging from strongly agree to strongly disagree and the person can choose any one of the response categories by which he can indicate the degree of his acceptance of the idea expressed in a statement. Direction and intensity are affective or feeling components of attitudes. The component 'direction' indicates the individual's feelings for or against. The 'intensity' component shows the degree of acceptance or the strength of feeling of acceptance.

Centrality

The third component of attitude measurement is the position of the attitude. Is it occupying an important key position or the end or marginal position of the attitude? An attitude which occupies a central position is likely to motivate for action. The *centrality* of an attitude implies cognitive, affective and behavioural aspects. An attitude if it is concerned with the self then it is felt more strongly for action. Therefore, centrality functions as an important motivating factor for action.

Salience

The importance given by the holder to the attitude is stated as *salience*. This refers to the level or degree of importance. An attitude may find a central position but may not be salient. Salience is the importance an individual or a group of people give to an attitude apart from its central position.

Consistency

To what extent an attitude measured is in harmony or agreement with other attitudes and attitude systems. This aspect is called *consistency*. Vernon Jones (1970) tested this component 'consistency' among the college students. It is a 37-year period study. The students who had favourable attitude toward religion and the church also had favourable attitude towards service to society, personal values and ethics. There are methods to measure each of these dimensions independently.

Types of Measurements of Attitudes

There are wide variety of ways in which attitudes can be measured. Attitudes are mainly measured by scaling methods, rating methods and by Questionnaire methods. In scaling methods there are different types of scales and they have been developed by Thurstone, Likert, Guttman, Bogardus and others. One method of attitude measurement is the scaling method. The measurement of the individual's attitude is based upon the judgments made by expert members, clinicians and interviewers. This type of measurement is based upon the ratings by a single person or an average ratings of a group of persons.

The earliest attempt to measure attitudes was made by Sumner (1898) who measured attitudes by using Questionnaire method. His study in 1898 for studying beliefs consisted of 25 questions with varying degrees such as "Will the death penalty for murder always be held justifiable among civilized people?" The subjects were asked to rank the questions on the basis of their answers. He also used another method of relative ranking which is known as order-of-merit method. This method has become the fundamental basis for many attitude scales developed later on. He used both the indirect and direct methods for measuring attitudes. In the indirect method a group of experts were required to act as judges for arranging the statements in an order from strong agreement to strong disagreement. After the arrangement, the subjects were asked to mark a statement which is nearest to their own opinion. In the direct method the subjects were asked to rank the statements which express their own attitude toward the problem from most favourable to most unfavourable and also to state the intensity and certainty of their opinion towards the issues. All the scales developed by psychologists and sociologists for the measurement of attitudes fall under these two heads. All the methods now used for measurement of attitudes are scaling methods.

There was some anxiety both in Europe and United States to know the attitude of people toward religion and church. Thurstone (1877-1955) and his colleagues have published some carefully designed methods of attitude scale construction. They have also published a number of refined scales for the measurement of people's attitudes or opinion towards church, war, capital punishment, evolution, the Negro, birth control, censorship, the Chinese and towards many other social problems. Thurstone and Chave (1929) have published a scale to measure attitudes towards church.

There was a feeling among the British people in the second quarter of this century that the British people were becoming less and less religious. Various explanations were given for this. Argyle (1958) states that the peak period of the Britishers' religious activity was around 1880. This was attributed to the Victorian way of life and to the arousal of guilty feelings over sex. Children were put under strict discipline and adults were expected to take in various aspects of formal religious activities irrespective of their personal inclination. Argyle (1958) stated that the development of scientific knowledge was one of the contributing factors for people getting less religious. In the United States it was found that there was decline in adult church membership, Sunday school membership and voluntary contributions to church during 1900-1940. Argyle states that this decline of religious activity in the United States during this period was due to social change and the change of attitude towards religious authorities. Therefore, Thurstone and Chave have first published a scale to measure attitudes towards the church in 1929. The technique used by them to develop this scale was called 'Psychophysical method of equal appearing intervals'. Following this several scales on the same technique were developed.

For the development of an attitude scale Thurstone and Chave have prepared hundreds of statements ranging from most positive to most negative about certain problem or issue. Out of which 130 statements were selected and mimeographed one statement on each slip. Thus a set of 130 small slips were given to 300 persons and were asked to sort out the statements into eleven piles, ranging from most favourable to most uniavourable. The most favourable statements were placed in the first pile that is in number one pile (1) and the most unfavourable statements were placed in the last pile, that is in number eleven pile (11). The middle is the sixth pile (6) should contain the neutral statements and all other statements were sorted out according to the degree of intensity in the respective piles. If a person put more than 30 statements in a pile out of 130 that judgement was eliminated from the valuation. Then the statements were tabulated into the form of cumulative frequencies. A graph was prepared for each statement with the cumulative proportions and the scale value of a particular statement was determined. Thus 130 graphs were prepared to find out the scale value of 130 statements. The scale values ranged from 0.0 to 11.0 representing the opposite extremes.

A different approach from Thurstone has been made by Likert (1932) for designing an attitude scale. Likert developed attitude scales to measure towards imperialism, internationalism, and toward the Negro. The technique Likert followed to develop his scale involved, first the collection on a number of statements referring directly to the object in question. There were both positive and negative statement towards the object in question. Under each statement five response categories such as (1) strongly agree, (2) agree, (3) undecided, (4) disagree, and (5) strongly disagree were given. Then the statements were mimeographed into a number of copies and given to a group of people and they were asked to indicate their reaction to each statement the degree of agreement or disagreement through the response categories by making a mark over the response. The five responses for the positive statements were assigned scores as 5, 4, 3, 2 and 1 and the individual score was the total score for all the statements the subjects answered. The scores were assigned reversely for the negative statements as 1, 2, 3, 4 and 5. After having determined the individual scores for all the statements, correlation between the score of each item and the total item was made. Any item fails to have a significant correlation with the total score was eliminated. All the items of this scale measure the same attitude or complex of attitudes. This scale was called summated rating scale. Many researchers prefer to use Likert scales in their researches as the technique to develop the scale is easier than Thurstone and Guttman scales.

A new technique has been followed in designing an attitude scale by Guttman (1950). His scaling technique is called Scalogram. The important objective of this scaling technique is to determine whether the attitude is scalable. It is determined by examining whether a larger proportion of the subjects have answered the same item with the same response categories. The response categories were (1) agree, (2) undecided, and (3) disagree. The scale is a measure of unidimensionality. The set of items of the scale should show consistency. If a set of items of the scale was found to have inconsistent responses, then the assumption is that there is more than one dimension in the scale. In such case the items which received inconsistent responses were eliminated. Thus the scale is consisting of only those items which were responded consistently by all the subjects. The determination of consistency and the unidimensionality is a laborious task though it appears to be simple.

AUTHORITARIAN ATTITUDES

Authoritarian attitudes have been receiving greater importance in social psychological researches during the end of second quarter of this century in United States. Quite a number of studies have been published clustering authoritarian attitudes. People who are interested in law and order are generally wish to express other attitudes related to law and order. The 37-year survey made by Vernon Jones (1970) states that students who were favourable to preparedness for war were also strongly inclined to favour the severe punitive approaches in dealing with crime and delinquency. Attitudes of this type were found to be interrelated and clustered around authoritarianism and, therefore, called as authoritarian attitudes by the researchers who were engaged in research works in the fields of personality, social psychology and in related fields. Studies of this type have become more important for different kinds of social behaviour such as communal and racial prejudices, child-rearing practices, emergence of leadership and political affiliations.

Authoritarian attitude involve much of ego-defense. Hence the ego-defensive attitudes are associated with ethnic prejudices. People do not accept their deep sense of inferiority feelings, impulses to violence and aggression and extreme sexual urges. But they never fail to project these consciously unacceptable characteristics upon an individual or on antegonistic social group. By this means the individual is able to escape from the blame that he is in possession of these traits. Thus the attitude of prejudice involves attributing undesirable traits to some minority groups which were not liked by his own group. Feshbach and Singer (1957) state that prejudice is an ego defensive attitude and it is aroused in an individual or among the group members by an increased frustration of sexual and aggressive feelings, threats to status and security caused by the failure of experiences.

The development of prejudice against the minority group, by a person who suffers from severe insecurity about his own self and property is quite common in ethnic groups. It is because that this attitude permits him to establish his superiority over the members of the prejudiced group. The expression of attitudes in hostility and prejudice may be satisfying the individual emotionally. The expression of hostile attitude may help him to release the held-up aggression and hostility that he has received and stored from his aggressive employer or from his nagging wife on an innocent outgroup member who cannot retaliate it effectively. Experimental studies show that there is also relationship between prejudice and ego-defensiveness.

The type of relationships that exist in the families, specifically the parents' exercise of authority over their children is a good example as a source of the development of prejudiced attitudes. An authoritarian family structure with a strong punishing father may produce authoritarianism – orientation children towards the world. The authoritarian thinks that there is continuous conflict and struggle in the world in which the strongs dominate the weaks. The authoritarian is concerned with power and rigidity in his thinking. The deeper levels of the authoritarian attitudes are extreme hostility toward the outgroups and extreme submissiveness toward the ingroups. He establishes sharp boundaries between the groups in which he is a member and against all other groups. For an authoritarian submission means uncritical attitude towards idealized moral authority of the ingroup. Obedience and respect for authority are the most important virtues the children should learn. Every person should have complete faith in some supernatural power whose decisions he must accept and obey without question. What this country needs more than the laws and political programmes is a few courageous, devoted and tireless leaders in whom the people should put their faiths.

Thus an authoritarian attitudes are expressed in behaviour in various ways. He is a dogmatic, rigid, strong conviction for traditional values and status oriented. An authoritarian's attitude basically is to be submissive to higher power and status and to dominate the lower status individuals and to use drastic methods in dealing with those who go against the authority. Social psychologists have been referring to the authoritarian personality as a distinct type of trait that is not common in the general population. But in reality it is found to be a scalable behaviour that can range from extreme authoritarianism to extreme democratic. Therefore, social behaviour and authoritarian attitude can be determined on a scale whether one is authoritarian or not and if he is authoritarian whether he is more authoritarian or less authoritarian. The study of authoritarian traits conducted by Adorno (1950) and others in the University of California in Berkeley between 1940 and 1950 brought many facts about authoritarianism and also stimulated many investigators in this direction of research.

The most important aspects of the Berkeley study is the development of four basic types of questionnaire for the study: These are the measures of anti-semitism, political and economic conservatism, ethnocentrism and antidemocratic trends. The two scales that attracted the psychologists are the E-scale and the F-Scale. The E-scale measured the ethnocentrism and anti-democratic attitudes. The F-scale measured the anti-semiticism and political and economic-conservatism attitudes.

Further researches on authoritarian attitudes have thrown some more light on this. According to the studies made by Christie (1956), people who were found to be rigid, demanding absolute and unquestionable obedience to higher authority and people who are intolerant of ambiguity and believe in drastic solution of the social problems will score high on the F-scale. The Berkeley researchers also show that these people score high. But these people must be politically reactionaries and conservatives. People who are rigid and demand absolute obedience with leftist political views, score at low and middle ranges on the F-scale. Therefore, it appears that the F-scale is a measure or right wing or reactionary and conservatism-authoritarianism and not a measure of leftist or radical authoritarianism.

EXPERIMENT 55 : MEASUREMENT OF ATTITUDES (RELIGIOUS)

INTRODUCTION

Man has enriched his life with material, mental and spiritual wealths through thousands of years of accumulated hard work and pious living. Time to time some valuable content is always added to human life which makes life more perfect, pleasant and happy. Every action of man is to enrich some aspect of his life. Religion is one such value which has given man solace and comfort in life. In whatever form and name it exists in the life of man the end result of it is the same in all religions and in all cultures. Religion is the main governing influence with man. Man in the beginning was superstitious and has consequently remained backward in certain areas when people in other areas have progressed by leaps and bounds. Some religious groups may fall back behind other races in efficiency test.

But the hard labour of our ancient stalwarts resulted in the evolution of the marvellous system of *yoga*, the art of contacting Reality and in the actual realization of Truth. The *Yoga Sadhanas*, the spiritual practices are the outcomes of the deepest psychological research carried out for thousands of years. This psychology is not merely that of the mind but is something more. It is a supramental psychology or it may be termed as spiritual psychology. Spiritual psychology finds its root in the basic conceptions of Indian philosophy. The spiritual psychology has been practiced through *sadhanas* and salvations have been attained. This has become a methodology of spiritual psychology.

Now modern psychologists are going to explain the outcome of the spiritual psychology which has been neglected all the time in preference to deceptive phenomena. Therefore, there is nothing wrong in attempting to synthesize the spiritual heritage we have accumulated through several centuries into the feasible modern techniques to assess the nature of transcendental traits which the Indian saints have been preserving and speaking about them in their discourses.

What makes man a sacred Being ? Is it the inner soul of the mortal that graves for identifying with the Divine Soul ? There is something which makes man sacred, religious minded by imbibing some kind of supra-perceptual faculty.

Religion has become a way of life for man from time immemorial and man has been cultivating his faith continuously through some changes, and modifications have been taking place time to time. But the concepts and the ideas have been receiving moral importance and they were propagated in inspiring and devotional words by the saints and seers, whom people respected as the prophets and as the representatives of God. If there is any element of truth in it nothing wrong in adapting the way of life which may bring peace and happiness to the people of the present century which is in the midst of anxieties and uncertainties.

Therefore, what type of attitude prevails among the people of different religious denominations of the present day is to be explored by developing suitable instrument to assess the nature and characteristics of their religious affiliations. The study of religious behaviour in India is very interesting since India has been the birth place of several religions. Her wonderful religious diversity and also the unity in diversity provides interesting information to the student in this field.

Religion is a very widespread significant social and cultural force in the life of human beings. We cannot say at what stage man has become religious. He is religious from the time immemorial. From the very beginning of human history man is found himself submitting to some unknown powers. Rituals, sacrifices, ceremonies, singing and dancing were some of the expressions of his submission. This unknown power was named variously as gods, spirits, ghosts etc. Religious practices have coloured both the social and economic life of man. Arts and literature owe much to the ecstasy of religious emotions. It is very difficult to formulate a definition of religion, which can be accepted universally. There are many definitions available in religious literature. According to Clark (1958) religion is the inner experience of the individual when he senses a 'beyond'. This experience has some effect on his behaviour when he actively attempts to harmonize his life with this 'beyond'. The word 'Beyond' here stands for a supernatural or supersensible force which is felt and known. This experience which defies verbal expression however influences him to move and act on a significant manner. An important characteristic of this inner experience is to make the individual submit himself to the supernatural force. The individual at such activities like prayer, worship and rituals tries to harmonize his life with this force.

Religion may be defined as a way of life revealed in emotional expressions and inspired by faith in God or in a supernatural power who is believed to be controlling and guiding the actions of man and the destiny of the world. Man often feels that his life is incomplete and imperfect. He longs for perfection. This longing makes him to seek the help of some supernatural power which can do for him things that he cannot do for himself. Psychologists tried to study the human behaviour from different angles. Their studies to explain man's religious behaviour resulted in formulating different theories. Some of the psychological theories of religious behaviours are : (1) The instinct theory, (2) The social learning theory, (3) The frustration theory, and (4) The conflict theory.

1. The Instinct Theory

This theory tries to explain many aspects of human behaviour including the religious. It was very popular in the early decades of this century. When behaviour was labelled as 'instinctive', it was deemed that no further explanation was necessary. The word instinct has been used in a variety of senses. McDougall (1948) the doughty champion of the instinct theory defines it as an inherited or innate psychophysical disposition which determines its possessor to perceive and to pay attention to objects of a certain class, to experience in emotional excitement of a particular quality upon perceiving such an object and to act in regard to it in a particular manner, or at least, to experience an impulse to such 'attention'. Jung (1932) thinks that the commonly understood meaning of the term instinct is an impulse toward certain activities. This impulse is caused by an outer or inner stimulus and results in the discharge of psychic casualty. Every psychic phenomenon is instinctive impulsion irrespective of whether such impulsion has its origin directly in organic sources or effected by the purpose of will. Therefore Jung (1939) says all those psychical processes over whose energies the conscious has no disposal, come within the concept of instinct.

The traditional meaning of the term 'instinct' as behaviour not learned or acquired by the individual but innate, requires certain modification when applied to religion since the element learning is also present. Religious behaviour is instinctive not in the sense of unlearned behaviour but in the sense that it is moved by instinct. Leuba (1912) states that religious behaviour is the outcome of a learning process, itself made possible by the presence of an instinctive foundation. Vetter (1932) said that any behaviour found universally in man must of course be instinctive.

Sex

The relation between religion and the instinct of sex has been widely discussed by the student of psychology of religion. A school of American writers calling themselves as Erotogenists (1956) also affirm that religion is based on sex instinct. The early religious writers on this subject claims that religion was a form of psycho-sexual perversion. Hendric (1957) says that the term, psycho-sexuality comprises, "all aspects of love and pleasure-seeking and their mutual inter-relationships; it emphasizes unconscious wishes for sensual gratification and their conscious de-erotogenic derivates, normal and abnormal, as well as wishes which culminate in complete and mature hetrosexual union". The desire for hetrosexual union is expressed due to disapproval, disappointment or fear and the individual turns aside from the normal course (a morbid alteration of function which may occur in emotional, intellectual or spiritual fields).

Sex perversion is also very conspicuous in the field of asesthetics, in music, painting, dancing and literature. Hadfield (1956) has stated that it is indeed a matter of astonishment to the pervert himself to find out within himself tendencies so alien to his refinement of character and taste. Those who acquired with these perversions can also detect morbid perversive tendencies in the works of some artists, whose sadism, masochism, anal-erotic or exhibitionistic tendencies are freely displayed in their artistic works. They turn to these creative imaginations as they have failed to obtain perfect sexual satisfaction in the real world. Similarly a love-sick person who is unable to fulfil his sexual desires biologically, replaces the loved object by God. Freud (1921) finds the source of religion in sex. He finds evidence for his theory in various primitive cultures. He says that all religions attempt to cope with the problem of a sense of guilt. This guilt is partially derived from the Oedipus Complex. This complex is the situation in which the male child 'falls in love' with his mother and rebels against his father. Later on as an expiation for his act of rebellion and as an expression of a tender feeling towards his father he tries to reconcile himself with him. As part of the process of reconciliation the boy tries to appropriate the father's good qualities. This leads to the search for the "father image". This father image is later on identified with God. Trotter (1916) says that the Herd Instinct is the root cause of man's religious behaviour. The herd instinct is a biological necessity. It reveals man's sense of incompleteness and compels him to go in search of some longer existence than his own. The herd instinct makes us to seek a remedy for physical loneliness and intellectual isolation. The company of his fellowmen gives men some comfort but it is only in fellowship with God that he attains complete satisfaction.

Weiser (1922) says that emotional thrill is the core of religion. An emotional element of this kind is the most constant factor in the religious complex. The emotional experiences awakened during the religious life of man depend upon time and situation. In primitive religion the emotion of fear and awe were dominant. Fear has received the attention of more writers than any other emotion. Lucretus said, fear begets religion. Some theologians opine, 'the fear of God is the beginning of wisdom'.

Animism is the belief that there is a spirit in every object which enables it to move and act. Disasters are caused by these spirits. The animistic view suggests that the emotion of fear caused by spirits and ghosts made the primitive man to worship any object in which he thought there was a spirit. Spencer's view of ancestor worship has no other significance than the fear for the ghosts of ancestors. The primitive man believed that if he failed to worship his dead ancestors they would cause diseases and disasters to his family. Ancestor worship is prevalent even now in many Hindu families in India.

2. The Social Learning Theory

Religious behaviour, experience and attitudes are only a part of culture. These are communicated from generation to generation like other customs. Religion is learned by a process of socialization. Man is a social animal. Religion is both an individual and social matter. When does an individual become religious? Becoming religious is a slow and gradual process. It is the result of the individual's interaction with society. Pratt (1934) says that the adult world had entered into a conspiracy to force the child's attitude into approved social grooves. Religious belief becomes part of cultural conformity. Socialization includes also certain negative practices. According to Briddle (1956) giving religious sermon about the wrath of God, hell-fire, and expression of anger is a negative type of religious practice. In primitive society all social gatherings and religious significance, law, morality, art and religion went together. The gods of Roman religion bore social characteristics. Mars was the god of war, Venus the goddess of love, Creses the goddess of growth of crops. Consus and Ops of the harvest. Religious beliefs colour every aspect of human life, food supply, livelihood, trade, war and Government. In ancient China the relation between religion and political organization was on a national footing. The Emperor of China was the son of Heaven.

3. The Frustration Theory

According to this theory religion is a kind of response to frustration. When a man fails to resolve his state of frustration directly he tries to do it indirectly. Some of the methods the individual adopts to resolve his frustrations, conflicts and anxieties are the identifications of displacement, sublimation and other defence mechanisms. Frustration is reduced by turning towards religious behaviour. Karl Marx says that it is the economic frustration that drives the working class to adopt religion. Argyle (1958) says that there was an increase of membership in the church of England and Roman Catholic church in Great Britain during the period of economic depression in 1930-35. The religious activities of ex-servicemen who had been on active service were more intense than their religious life before their service. Wilson (32) says that incurable diseases, severe illness cause frustration. Investigations reveal that many deformed people in Britain were highly religious.

4. Conflict Theory

One of the psychological sources of religious behaviour is conflict. Religion according to this theory is an outcome of conflict between two opposed tendencies, for example, good and evil, activity and passivity, humility and pride. The religious behaviour of man that arises out of conflict has a special reference to the pain caused by it. According to Flugal (1945) religion is a product of conflict between two parts of the personality, particularly between super-ego (consciousness) and the instinct. The super-ego is projected on God. Therefore, conflict is a psychological source of man's religious behaviour.

FORMATION OF RELIGIOUS ATTITUDE

Since the study is religious attitude, it is also important to say something about attitude, directly. Enough has been said about religious aspects. But religion and attitude should be related. There are six related areas included in religious behaviour. They are nature of God, prayers and worship, formal religion, priests, future life and spirits and spirit world. All these may be functioning through the individual's attitude. Whether one is highly religious or moderately religious or anti-religious, he expresses it through his attitude and action. There are about seventy different definitions of attitude available in psychology literature. Therefore, we can quote one standard definition given by Allport (1935). He states that an attitude is a mental and neural state of prepared act established through the experience stretching out a direct and forceful influence upon the individual's response to all objects and situations with which he is related. Taking into account the central point in this definition we can formulate a definition of our own religious attitude. A religious attitude is a positive or negative response tendency towards various aspects of religion, like nature of God, Prayer and Worship, Formal Religion, Priests, Future Life, Spirits and

Spirit World. An important function implied on an attitude is response or reaction to the value. It may be favourable or unfavourable response involving some kind of action inherently are overtly towards God etc.

LEARNING

Every human being learns many things in his everyday life. He learns both in school and outside the school. This learning is not without any effect in our behaviour. It determines our likes and dislikes, favourable and unfavourable responses to persons, situations, institutions and values. According to Ruch (1958) attitudes are also learned behaviour of the individuals. We learn our attitudes towards person, institutions situations and social issues. The existence of an attitude towards something is quite impossible without having some kind of knowledge about the things.

The child grows up in the immediate presence of his father and mother and slowly learns to imitate and identify himself with them. In high schools and colleges students try to imitate or identify their teachers and professors in some of their ways of life. It is because of the regard and respect they have for them. Experimental evidences show that learning has some influence in the formation of attitudes. Vetter and Green (1932) have studied the formation and development of anti-religious attitudes of 350 members of the American Association. Many of them stated that they derived their attitudes from a study of history and science, some of them from a study of materialistic philosophy. Others traced their attitudes to the feeling and experience of grief and disgust. A few of them traced them to the influence of friends who were atheists.

Attitude studies were made very widely relating important values like religion, social, economic, artistic etc. Allport was of the view that the attitude concept, is probably the most distinctive and indispensable concept in contemporary American social psychology. The reason is that attitudes are of greater interest to social psychologists. It is thought that attitudes are having effective cognitive and behavioural components involving feelings and emotions, beliefs and action. Some attitudes are heavily charged with effective components and do not require any action beyond the expression of feelings.

Generally attitudes are inferred from behaviour and cannot directly be observed. Attitudes are assessed from the individual's verbal, written or spoken statements. This is through interviews, paper and pencil tests and through questionnaires. The paper-pencil test method though has certain disadvantages, it is easier to collect information from large number of subjects and analyse the results statistically. Attitude measurement first of all is concerned with the direction of the attitude being assessed. It may be for or against a race, community, object, person, place or event or values. In a typical attitude, scale directions may be indicated as like or dislike, yes or no, agree or disagree, favourable or unfavourable. In this, the strength or intensity of the expressed view or attitude is also important. A respondent's attitude may be strong, or weak or ambivalent. Therefore, intensity of the attitude may be checked.

Measurement of attitudes is also concerned about the centrality of an attitude. It is close to the centre of the values that are highly significant to the welfare and goals of the individual. Attitudes that are strongly held are likely to be central and are likely to provide motives for action. To what extent an attitude is given importance by its holder is referred to as salience. It is also observed that all central attitudes are not salient. The measured attitudes should also have certain amount of consistency. That is, a devout and loyal member of a particular religious denomination should also have high regard for other members of that sect. If we discover that he has low opinion about members of his own denomination, such an attitude is consistent with his attitude towards his religion. The attitude that an individual has towards his religion is generally consistent with his attitude towards persons who are associated with the religion. In one of the surveys conducted by Vernon Jones (1970) among the student it was found that attitudes favourable towards religion and church were associated with personal ethics rather than social ethics.

MEASUREMENT OF RELIGIOUS ATTITUDES

Attitude measurements were made using the questionnaire methods. Sumner (1898) was the first psychologist to study attitudes. In 1898 he conducted an investigation to study the beliefs. He framed only 25 questions first. Will the death penalty for murder always be held justifiable among civilized people? This was the type question he used. He asked the subjects to mark the questions on the basis of their opinions.

Study of attitudes towards religion and church were made when there was decline of feeling. There was a feeling among the British people in the second quarter of this century that their country was getting less and less religious. Various explanations were. given for this. Argyle (1958) says that the peak period of the Britisher's religious activity was around 1880. This was attributed to the Victorian way of life and to the arousal of guilty feelings over sex. Children were put under strict discipline and adults were expected to take part in various aspects of formal religion irrespective of their personal inclination. In U.S.A. there was a decline in adult church membership, Sunday School membership and voluntary contributions to church during 1900-1940. The decline of religious activity in the United States during this period was due to social change and the change of attitudes towards religious authorities.

Thurstone and Chave (1929) were the first to measure attitudes towards church by a systematic procedure. The method they used was psychophysical method of equal appearing intervals. It was also called *a priori* approach. The method and the scale developed by them were very widely used by psychologists. Newcomb and Svehla (1937) studied the attitudes of 540 families towards the church. The mean score towards church was arranged for classification on the basis of age, sex, religious affiliation and occupational status. But there was not much difference in the scores of these groups. The females were more favourable towards the church than the males. But there was steady decrease in the favourable attitude towards the church between the age groups of 17 and 30, particularly among females. The favourable attitudes towards the church of fathers and mothers increased steadily with advancing age.

Shuttleworth (1927) studied the family influence upon religious practices of the children. The subjects were sophomores. Religious attitudes were measured in terms of both interest and degree of acceptance of statements. The religious trainings of the students were measured by self-ratings of their own homes in six different respects and correlated with interests and beliefs. It was found that there was definite relationship between home influence and present religious practices of the students.

Katz and Allport (1931) have studied students' attitudes towards religion and church at Syracause University. The majority of the students had some belief in religion and the church at the time of entering the university. After spending few years at the university the upper class students were found to have more faith in an impersonal God and less belief in a personal God and prayer than those who had spent less time there.

The study of Sappenfield (1942) centred round the denomination of the attitude characteristics which the typical catholics, the typical protestants and typical jews were thought to possess. This study was conducted to find out the relative attitude positions of groups having different religious affiliations. It was also to see how far different religious groups were conservative and radical in their attitudes towards communism, war and birth control. The important findings of his study were that attitudes expressed by individual student were more highly correlated with estimated attitudes of members of their own religion than with members of other religions. Sappenfield observed that the catholic group favoured communism the least, the Jewish group favoured it most, while the protestants were in the intermediate position. On the issue of war the protestants were most favourable, the Jewish least favourable and the Catholics in an intermediate position. The Jewish group was least opposed to birth control, the Catholics most opposed and the Protestants choose the middle position.

Kuhlen and Arnold (1944) have tried to find out the kind of religious beliefs of school students. There were about 547 students and they were given statement consisted of various beliefs. The students were between 12 and 18 years of age and were asked to respond to each statement in terms of their belief such as 'belief', 'not belief' or 'wonder about'. It was found that matters like Heaven and Hell, the actuality of God, the truth of every word in Bible were of serious concern to the older adolescents. The degree of uncertainty in religious beliefs of the students increased with advance of age. They found that this was due to their study of natural sciences and social studies or discussions with adult associates.

A study has been conducted to verify the hypothesis whether religious education has any marked influence upon the persons' life. Allport and his colleagues attempted this study among the college students. It was found that religious training influenced the students' upbringing. There was only 6 percent of the men students and ten percent of the women students not influenced by the religious training. The students' present felt need for religion was compared with the religious influence on the students' upbringing. It was found that the need was great among the students who received the religious training. Almost all the students who were brought up in the Catholic faith felt the need for religious orientation while 40 percent of the students who were trained in Judaism or liberal Protestantism did not feel the need for religious sentiment. Garrison (1959) remarks that it has been observed in several studies that the students of Roman Catholic faith found to be more loyal to their church than the students of other faiths in United States.

It is widely believed in India that the modern youth is becoming increasingly irreligious. The education he receives in colleges and universities and the study of sciences and technology are believed to have a secularising effect upon him. It is also believed that the east is becoming increasingly materialistic due to growing industrialization. The beliefs were experimentally examined in the study conducted by Rajamanickam (1976). The university and college students 750 in number were the subjects in this study. Among the things attempted were assessing the students' attitudes towards God, immortality of soul, opinions about priests, the spirit world, attitude towards personal religion (prayer) and attitude towards the institution of religion. The study also aimed to find out the effect social, economic and cultural backgrounds on students' religion. It was also tried to find out the relationship between religious attitudes and radicalism, conservatism and personality traits.

The religious attitude scale was developed by Rajamanickam (1962) was used in this study. It was a summated rating scale, developed according to Likert technique and validated with a criterion scale developed by him according to Thrustone (1929) scaling technique of psychophysical method of equal appearing intervals. The other instrument used in his study was modified version of Vetter's attitude scale.

One of the important treatments of the data was the factorial analysis. Thrustone's (1947) complete centroid method yielded three factors, such as Religionism, Overt religionism and Spirit world in the student religious attitude. There was a positive correlation between religious attitude and conservatism. The correlation co-efficient for the Arts students was higher than the Science and professional students. The rural and urban backgrounds had some effect upon the student religious attitude. Moral and religious instructions received in high schools and colleges had definite influence upon the students' religious attitudes.

In one of the earlier studies made by Rajamanickam (1962) the student attitudes toward religion, the spiritual and supernatural were assessed. The sample was 300 university students studying in various areas like Arts, Science and Professional courses. A religious attitude scale developed by the author on the basis of Likert technique was used. It was found that there was a positive correlation between religious attitudes and conservatism. Students who were conservative were found to be more pro-religious than the radical students. Women students were found to be more religious than the men students. The science students were found to be more religious than the arts students.

It was observed that God, prayer, sacred books, Temple or Church, religious festivals received more support from the students than Heaven, Hell, Future life, Priests, Spirits world and Devils. The city students were found to be more religious than the village students. About 60 students out of 300 said that God does not exist. These students belonged to Art section. They come from rural areas and from low income group.

EXPERIMENT

Problem

Assess the religious attitudes of a person with various components and state the components in which he/she is extremely pro-religious, moderately pro-religoius and the components in which he is anti-religious.

Method

Participant

Choose a person who has no knowledge about this test and collect his/her personal information like age, sex, educational qualification, religion, socio-economic status, geographical area etc.

Materials

Prof. Rajamanickam's Religious Attitude Scale, Hand-book, Test material, Key and Answer sheet.

Description of the Material

Prof. Rajamanickam's (1989) Religions Attitude Scale consists of 60 statements relating to various areas of religion like (1) Nature of God, (2) Prayer and worship, (3) Formal religion, (4) Priests, (5) Future life, and (6) Spirits and spirit world. There are ten statements for each area. Out of 60 statements there are 30 positive statements and 30 negative statements. In each area there are 5 positive and 5 negative statements. Under each statement there are five response categories, like strongly agree, agree, unable to decide, disagree and strongly disagree. The participant will have to underline any one of the responses with which he agrees most.

Procedure

The participant is to be seated very comfortably to have free thinking on the subject matter and to express his views freely. After getting the personal information in the answer sheet he should be given the following instructions : "This booklet consits of 60 statements, each expressing one idea on some religious value. You are quite familiar with these. You may strongly agree with certain ideas and strongly disagree with certain ideas. With some ideas you may moderately agree or disagree. Your agreement or disagreement may vary in degrees. Whatever it is you can express it here. There are five responses given under each statement. They are strongly agree, agree, unable to decide, disagree, and strongly disagree. You will have to underline any of the responses with which you agree. Read each statement carefully and underline the response in the answer sheet provided to you. Please do not make any mark on the test booklet. Answer all the statements without any omission. When I say, 'ready, go', you can start answering questionnaire. There is no time limit. You can take your own time. Now, 'ready, go'."

As soon as the participant completed the questionnaire the experimenter can collect it from him and allow him to go away. The experimenter can also note down the time the participant has taken to complete the questionnaire.

Method of Scoring

The experimenter can score the questionnaire by giving numerical values to the responses in the column provided in the answer sheet with the help of the key. The experimenter is advised to study first the key sheet and understand how the key sheet is prepared. The key sheet provides certain information about the statements. It is, whether the statement is positive or negative and also to which value the statement belongs to. The experimenter should check whether the participant has answered the statement by underlining any one of the responses. Then as given in the key sheet 1 mark may be entered on the answer sheet in the score column if strongly agree response is underlined for the positive statement, if agree is underlined, 2 may be entered, 3 for unable to decide, 4 for disagree and 5 for strongly disagree. If the statement is negative, the strongly agree response will receive 5, agree 4, unable to decide 3, disagree 2 and strongly disagree 1.

In this method of scoring, the strongly agree response of the positive statements and strongly disagree of the negative statements receive the same score 1. Similarly the strongly disagree of the positive statements and strongly agree of the negative statements, receive the same score 5. This means that the lowest score indicates the pro-religious attitude and the highest score anti-religious attitude. For the whole scale the lowest possible score is 60 which is the extreme pro-religious attitude and the highest possible score is 300 which is the extreme anti-religious attitudes.

Results

The individual's score may be obtained by totalling the scores assigned to each response of the 60 statements. It may fall in between 60 and 300, 60 being extremely pro-religious and 300 being extremely anti-religious.

The individual's scores may be sub-divided into six components on the basis of value areas like (1) Nature of God, (2) Prayer and worship, (3) Formal religion, (4) Priest, (5) Future life, (6) Spirits and spirit world. These scores may be presented in Table 18.1. Since each component is having ten statements the extreme pro-religious score may be 10 and the extreme anti-religious score may be 50.

Table 18.1 : Showing the area-wise scores of the individual in the Religious Attitude Scale

S. No.	Area/Component	Statements	Total Scores for 10 statements
1.	Nature of God	1, 7, 13, 19, 25, 31, 37, 43, 49, 55	
2.	Prayer and Worship	2, 8, 14, 20, 26, 32, 38, 44, 50, 56	
3.	Formal Religion	3, 9, 15, 21, 27, 33, 39, 45, 51, 57	
4.	Priests	4, 10, 16, 22, 28, 34, 40, 46, 52, 58	
5.	Future Life	5, 11, 17, 23, 29, 35, 41, 47, 53, 59	
6.	Spirits and Spirit World	6, 12, 18, 24, 30, 36, 42, 48, 54, 60	

Table 18.2 : Showing the range of scores and the level of attitude for all 60 statements (Enter the individual's total score in the last column

Range of Scores	Level of Attitude	Individual's total score and place
60 – 90	Extremely Pro-Religious	
91 – 150	Moderately Pro-Religious	
151 – 210	Neutral	
211 – 270	Moderately Anti-Religious	
271 – 300	Extremely Anti-Religious	

Table 18.3 : Showing the individual's scores in each area and the Level of Attitude

S. No.	Area/Component	Scores	Level of Attitude : Extremely religious Moderately religious or Anti religious
1.	Nature of God		
2.	Prayer and Worship		
3.	Formal Religion		
4.	Priests		
5.	Future Life		
6.	Spirits and Spirit World		

Table 18.4 : Showing the score for each area and the total score for the group on the Religious Attitude Scale

S.No.	Name	Nature of God	Prayer and Wor-ship	Formal Religions	Priests	Future Life	Spirits and Spirit World	Total Score in all 6 areas
1.	P.K.							
2.	V.M.							
3.	R.S.							
4.	K.R.							
5.	M.S.							
6.	T.V.							
7.	S.L.							
8.	E.K.							
9.	G.G.							
10.	H.T.							
11.	J.R.							
12.	A.L.							
Mean								

Work out the individual's score for all 60 statements and give it in Table 18.2. Work out the individual's score for each area and present it in Table 18.3. Collect the group data for each area and also for the whole scale and work out the mean and present them in Table 18.4.

Discussion

Write the discussion taking into account the individual's score for all 60 statements. State his level of attitude and whether he is extremely pro-religious or moderately pro-religious or anti-religious. State the individual's scores in various areas and discuss the area where he scored high or low and what it indicates? Discuss the individual's score taking into account the group mean. State whether he is above the group mean or below the group mean. State the reason. Relate his/her personal information and discuss his/her religious attitude.

Discuss the individual's score in various areas comparing them with the group mean. State the areas in which the group has scored low and scored high. State the reason. In the group who are those extremely religious persons? Who are these extremely anti-religious persons? Discuss their stand relating their personal information such as their religion, community and socio-economic status and other variables that seem to have contributed to this. Compare the group result with the result of the earlier studies given in the introduction.

Conclusions

State whether your subject is a strong thiest or athiest. State the area in which he scored very high and very low. What does it mean? How many in the group scored below the mean score and above the mean. State the backgrounds of the strong thiests and atheists.

APPENDIX - I

RELIGIOUS ATTITUDE SCALE

NOTES FOR RESEARCHERS/TEST USERS

India is a nation of multiplicity of races, religions, cultures caste and communities, where a very wide diversity of views exist side by side and also they are put into active operation which sometimes resolve the differences for unity and universal brotherhood and sometimes leads to acute rivalry and serious disintegration. Since people of diversified groups live in the same social environments side by side for thousands of years minor differences do not cause much. But all the times religious differences never come to any permanent resolution. This means that it is a very powerful social force which cuts across all the values and stand as a unique cause of the differences. But what psychological factors are operating in this social force is a matter for psychologists to comprehend and explain to those who are interested in it. Studies should be designed and conducted in the areas where problems disturb the normal peaceful life. If all the psychological studies confine within the academic and laboratory walls psychological contribution to the acute social problem becomes rare. Conferences and seminars do not show any serious concern over the perennial psycho-social problems. The nature of religious affiliation is such that a call could congregate in no minute to all those who belong to a particular denomination. What sort of inner force is functioning in the minds and hearts of those people is an interesting phenomenon for study. The studies, the author has made for many years in this field is not very big, but it has aroused the curiosity of many young men to study some of the aspects of religion which may throw much better light on this problem. This scale may be of some use for those who want to the study of religious behaviour of people of all categories. Researchers and test users are advised to read the Hand Book thoroughly once and then make use of the scale. It is always better to use the designed answer sheet to avoid certain possible mistakes.

Designed and prepared in	1958
Revised with some changes in	1962
Rewritten and Revised with additional items in	1988

DIRECTION FOR FILLING IT UP

There are some statements given below about various aspects of religion. Your opinion or view may agree or disagree with the idea expressed in each statement. Please read each statement very carefully and express what you feel about the idea in the statement by underlining one of the five responses, *i.e.* Strongly agree, Agree, Unable to decide, Disagree, Strongly disagree, according to the strength of your faith on the idea of the statement. This is used only for academic purposes and your identity will not be revealed at any cost. Please write your answers in the *Answer Sheet* and do not make any mark in this booklet.

Examples

1. God is the prime mover of human history.
   Strongly agree : Agree: Unable to decide: Disagree: Strongly disagree
   Here Mr. *X* expresses his view by underlining the response Strongly agree.

2. God is the prime mover of human history.
   Strongly agree: Agree: Unable to decide: Disagree: *Strong disagree*
   Here Mr. *Y* strongly disagrees with the view and, so he has underlined it.

3. God is the prime mover of human history.
   Strongly agree: Agree: *Unable to decide*: Disagree: Strongly disagree
   Here Mr. *Z* is not able to decide the idea. Therefore, he has underlined the middle view. Whatever your view is on each statement please indicate it through any one of the five responses as shown above.

STATEMENTS

1. God is the Supreme Being and His Supremacy was always unquestionable and indisputable.
   Strongly agree: Agree: Unable to decide: Disagree: Strongly disagree
2. Prayer is a solemn and humble request to God or any object of worship.
   Strongly agree: Agree: Unable to decide: Disagree: Strongly disagree
3. Temples, churches and mosques are not the real places where we can find solace and happiness.
   Strongly agree: Agree: Unable to decide: Disagree: Strongly disagree
4. Priests have played an important role in the religious life of people by making them to understand religious principles.
   Strongly agree: Agree: Unable to decide: Disagree: Strongly disagree
5. The human soul attains its liberation by taking a series of rebirths with continually of the same self.
   Strongly agree: Agree : Unable to decide: Disagree: Strongly disagree
6. Spirits are supernatural being imperceptible, troublesome terrifying and hostile to mankind.
   Strongly agree: Agree: Unable to decide: Disagree: Strongly disagree

7. God is the Creator and the Destroyer of the Universe.
   Strongly agree: Agree: Unable to decide: Disagree: Strongly disagree
8. Prayer is a religious observance sacredly performed in private or public.
   Strongly agree: Agree: Unable to decide: Disagree: Strongly disagree
9. When we enter sacred places (temple, church, mosque) we feel inspired by some kind of faith and our pride dissolves.
   Strongly agree: Agree: Unable to decide: Disagree: Strongly disagree
10. Priests by their simple and pious living have set up a good tradition of true religious life.
    Strongly agree: Agree: Unable to decide: Disagree: Strongly disagree

11. The soul in its journey of births and rebirths reaps the fruits of its actions of the previous birth.
Strongly agree: Agree: Unable to decide: Disagree: Strongly disagree
12. Spirits and Spirit world are like Heaven and Hell. All foolish imaginations.
Strongly agree: Agree: Unable to decide: Disagree: Strongly disagree
13. Nobody has seen any God anywhere in this world. Therefore, there is no God.
Strongly agree: Agree: Unable to decide: Disagree: Strongly disagree
14. Prayer and worship are waste of time. and energy. There is no use of praying or worshipping.
Strongly agree: Agree: Unable to decide: Disagree: Strongly disagree
15. The Sacred books are full of ethical and moral values and people should read them everyday to be free from evil idea.
Strongly agree: Agree: Unable to decide: Disagree: Strongly disagree
16. The priests serve as the middle men in Sacred places, misguiding and cheating the people.
Strongly agree: Agree: Unable to decide: Disagree: Strongly disagree
17. Whatever a man sowed in the last birth reaps the fruits in this birth.
Strongly agree: Agree: Unable to decide: Disagree: Strongly disagree
18. There is no spirits and there is no spirit world.
Strongly agree: Agree: Unable to decide: Disagree: Strongly disagree
19. God reveals Himself through incarnations on the earth whenever there was increase of diabolic evils.
Strongly agree: Agree: Unable to decide: Disagree: Strongly disagree
20. Prayer is a means of establishing contact between God and Man.
Strongly agree: Agree: Unable to decide: Disagree: Strongly disagree
21. Sacred books are good for nothing. They are not correcting the behaviour of the people.
Strongly agree: Agree: Unable to decide: Disagree: Strongly disagree.
22. Priests have made use of religion for their personal gains.
Strongly agree: Agree: Unable to decide: Disagree: Strongly disagree
23. Rebirth, future life are all only imagination. There is no future life after death.
Strongly agree: Agree: Unable to decide: Disagree: Strongly disagree
24. The Devil, Spirits, Ghosts, Demons are not true.
Strongly agree: Agree: Unable to decide: Disagree: Strongly disagree
25. God's presence in any form is not really felt by even great saints. It is only an Illusion.
Strongly agree: Agree: Unable to decide: Disagree: Strongly disagree
26. Prayer and worship never brought any tangible benefits for man.
Strongly agree: Agree: Unable to decide : Disagree: Strongly disagree
27. Sacred books are always guiding us on righteous path.
Strongly agree: Agree: Unable to decide: Disagree : Strongly disagree
28. Priests have been interpreting and practising religious principles with all devotion and guide people on righteous path.
Strongly agree: Agree: Unable to decide: Disagree: Strongly disagree

29. People who do good will enjoy their life in Heaven and those who commit sins will suffer in Hell.
Strongly agree: Agree: Unable to decide: Disagree: Strongly disagree
30. Spirits can never come to earth as they do not exist anywhere.
Strongly agree: Agree: Unable to decide: Disagree: Strongly disagree

31. God has created man in His own image.
Strongly agree: Agree: Unable to decide: Disagree: Strongly disagree
32. Prayer relieves mental unrest, resolves tensions and crises of human beings.
Strongly agree: Agree: Unable to decide: Disagree: Strongly disagree
33. Religious festivals and processions are highly exciting and creating faith in God. People find pleasure in taking part in them.
Strongly agree: Agree: Unable to decide: Disagree: Strongly disagree
34. When the religious functions are performed by the Priests in all ceremonious way, certainly we find some kind of mental satisfaction.
Strongly agree: Agree: Unable to decide: Disagree: Strongly disagree.
35. Nobody knows anything about Heaven or Hell. It is invented by the religious propagandists for their own purpose.
Strongly agree: Agree: Unable to decide: Disagree: Strongly disagree
36. It is possible to communicate with the spirits of the dead people.
Strongly agree: Agree: Unable to decide: Disagree: Strongly disagree

37. God sees every good and bad deeds, but waits to reward.
Strongly agree: Agree: Unable to decide: Disagree: Strongly disagree
38. The more prayer and worship we perform the more we develop anxieties and restlessness. Therefore, there is no use of praying.
Strongly agree: Agree: Unable to decide: Disagree: Strongly disagree
39. Religious festivals, processions, congregations are only waste of time and money. They do not serve any real purpose. They must be stopped.
Strongly agree: Agree: Unable to decide: Disagree: Strongly disagree
40. The hereditary priesthood is good for nothing. It is high time that it should be abolished.
Strongly agree: Agree: Unable to decide: Disagree: Strongly disagree
41. There is no life after death. Our life ends once for all after the death.
Strongly agree: Agree: Unable to decide: Disagree: Strongly disagree
42. The spirits are invisible and can bring good or evil.
Strongly agree: Agree: Unable to decide: Disagree: Strongly disagree

43. It is from the man's fear of crimes and sins that God was born. Otherwise there is no necessity to think about God.
Strongly agree: Agree: Unable to decide: Disagree: Strongly disagree
44. Prayers are rewarded by God only in imaginary stories, but not in real life.
Strongly agree: Agree: Unable to decide: Disagree: Strongly disagree
45. Religious discourses performed by some religious scholars are intended only to make money and improve their personal life.
Strongly agree: Agree: Unable to decide: Disagree: Strongly disagree

46. It is the priests who submit the grievances of the people to God, praying Him to redeem their sins and troubles.
Strongly agree: Agree: Unable to decide: Disagree: Strongly disagree
47. We cannot simply reject the religious truths of future life, rebirth, Heaven or Hell. Our knowledge is inadequate to understand these religious truths.
Strongly agree: Agree: Unable to decide: Disagree: Strongly disagree
48. The evil spirits which are called demons or devils may cause dangers, disturbances, sickness and death.
Strongly agree: Agree: Unable to decide: Disagree: Strongly disagree
49. There is no evidence that God takes any Interest in the affairs of man.
Strongly agree: Agree: Unable to decide: Disagree: Strongly disagree
50. God is in Heaven. That Heaven is in your heart. Pray and search for Him in your heart and you will find Him there.
Strongly agree: Agree: Unable to decide: Disagree: Strongly disagree
51. There are too many religions in the world with too many troubles. People can live happily without these religions which are illusions.
Strongly agree: Agree: Unable to decide: Disagree: Strongly disagree
52. In some religions the priests pretend to be the true representatives of God and enslave the people and treat them badly. They should be punished severely.
Strongly agree: Agree: Unable to decide: Disagree: Strongly disagree
53. Let us not worry about the last and the next birth. Let us be happy in this birth which is true.
Strongly agree: Agree: Unable to decide: Disagree: Strongly disagree
54. It is true that some people control the evil spirits and perform some miracles.
Strongly agree: Agree: Unable to decide: Disagree: Strongly disagree
55. It is man who created God in his own image.
Strongly agree: Agree: Unable to decide: Disagree: Strongly disagree
56. It is sure that even a man's life time praying and worshipping will not bring anything good.
Strongly agree: Agree: Unable to decide: Disagree: Strongly disagree
57. Man by nature is a religious creature and he is all the times worshipping some God and getting his problems solved.
Strongly agree: Agree: Unable to decide: Disagree: Strongly disagree
58. People should perform the rituals by themselves as they wish. There is no need for a priest in Temple, Church or Mosque.
Strongly agree: Agree: Unable to decide: Disagree: Strongly disagree
59. Our joys and sufferings of this birth never really make us to think that it is all the results of our actions of the last birth. Therefore, there cannot be any rebirth or future life.
Strongly agree: Agree: Unable to decide: Disagree: Strongly disagree
60. It is not possible to see the actions of the spirits in this world. All we hear about the spirits are false.
Strongly agree: Agree: Unable to decide: Disagree: Strongly disagree

---

Please write here if you wish to make any comments or remarks:

APPENDIX-II

RELIGIOUS ATTITUDE SCALE

ANSWER SHEET

PROFESSOR RAJAMANICKAM'S RELIGIOUS ATTITUDE SCALE

1. Age: 2. Sex: 3. Religion

4. Qualification: 5. Occupation: 6. Monthly Income:

Underline the response with which you agree

Score	St. No.	Response
	1	Sty. Agree: Agree: Unable to decide: Disagree: Sty. Disagree
	2	Sty. Agree: Agree: Unable to decide: Disagree: Sty. Disagree
	3	Sty. Agree: Agree: Unable to decide: Disagree: Sty. Disagree
	4	Sty. Agree: Agree: Unable to decide: Disagree: Sty. Disagree
	5	Sty. Agree: Agree: Unable to decide: Disagree: Sty. Disagree
	6	Sty. Agree: Agree: Unable to decide: Disagree: Sty. Disagree
	7	Sty. Agree: Agree: Unable to decide: Disagree: Sty. Disagree
	8	Sty. Agree: Agree: Unable to decide: Disagree: Sty. Disagree
	9	Sty. Agree: Agree: Unable to decide: Disagree: Sty. Disagree
	10	Sty. Agree: Agree: Unable to decide: Disagree: Sty. Disagree
	11	Sty. Agree: Agree: Unable to decide: Disagree: Sty. Disagree
	12	Sty. Agree: Agree: Unable to decide: Disagree: Sty. Disagree
	13	Sty. Agree: Agree: Unable to decide: Disagree: Sty. Disagree
	14	Sty. Agree: Agree: Unable to decide: Disagree: Sty. Disagree
	15	Sty. Agree: Agree: Unable to decide: Disagree: Sty. Disagree
	16	Sty. Agree: Agree: Unable to decide: Disagree: Sty. Disagree
	17	Sty. Agree: Agree: Unable to decide: Disagree: Sty. Disagree
	18	Sty. Agree: Agree: Unable to decide: Disagree: Sty. Disagree
	19	Sty. Agree: Agree: Unable to decide: Disagree: Sty. Disagree
	20	Sty. Agree: Agree: Unable to decide: Disagree: Sty. Disagree
	21	Sty. Agree: Agree: Unable to decide: Disagree: Sty. Disagree
	22	Sty. Agree: Agree: Unable to decide: Disagree: Sty. Disagree
	23	Sty. Agree: Agree: Unable to decide: Disagree: Sty. Disagree
	24	Sty. Agree: Agree: Unable to decide: Disagree: Sty. Disagree
	25	Sty. Agree: Agree: Unable to decide: Disagree: Sty. Disagree
	26	Sty. Agree: Agree: Unable to decide: Disagree: Sty. Disagree
	27	Sty. Agree: Agree: Unable to decide: Disagree: Sty. Disagree
	28	Sty. Agree: Agree: Unable to decide: Disagree: Sty. Disagree
	29	Sty. Agree: Agree: Unable to decide: Disagree: Sty. Disagree
	30	Sty. Agree: Agree: Unable to decide: Disagree: Sty. Disagree
	31	Sty. Agree: Agree: Unable to decide: Disagree: Sty. Disagree
	32	Sty. Agree: Agree: Unable to decide: Disagree: Sty. Disagree
	33	Sty. Agree: Agree: Unable to decide: Disagree: Sty. Disagree
	34	Sty. Agree: Agree: Unable to decide: Disagree: Sty. Disagree
	35	Sty. Agree: Agree: Unable to decide: Disagree: Sty. Disagree
	36	Sty. Agree: Agree: Unable to decide: Disagree: Sty. Disagree

Score	St. No.	Response
	37	Sty. Agree: Agree: Unable to decide: Disagree: Sty. Disagree
	38	Sty. Agree: Agree: Unable to decide: Disagree: Sty. Disagree
	39	Sty. Agree: Agree: Unable to decide: Disagree: Sty. Disagree
	40	Sty. Agree: Agree: Unable to decide: Disagree: Sty. Disagree
	41	Sty. Agree: Agree: Unable to decide: Disagree: Sty. Disagree
	42	Sty. Agree: Agree: Unable to decide: Disagree: Sty. Disagree
	43	Sty. Agree: Agree: Unable to decide: Disagree: Sty. Disagree
	44	Sty. Agree: Agree: Unable to decide: Disagree: Sty. Disagree
	45	Sty. Agree: Agree: Unable to decide: Disagree: Sty. Disagree
	46	Sty. Agree: Agree: Unable to decide: Disagree: Sty. Disagree
	47	Sty. Agree: Agree: Unable to decide: Disagree: Sty. Disagree
	48	Sty. Agree: Agree: Unable to decide: Disagree: Sty. Disagree
	49	Sty. Agree: Agree: Unable to decide: Disagree: Sty. Disagree
	50	Sty. Agree: Agree: Unable to decide: Disagree: Sty. Disagree
	51	Sty. Agree: Agree: Unable to decide: Disagree: Sty. Disagree
	52	Sty. Agree: Agree: Unable to decide: Disagree: Sty. Disagree
	53	Sty. Agree: Agree: Unable to decide: Disagree: Sty. Disagree
	54	Sty. Agree: Agree: Unable to decide: Disagree: Sty. Disagree
	55	Sty. Agree: Agree: Unable to decide: Disagree: Sty. Disagree
	56	Sty. Agree: Agree: Unable to decide: Disagree: Sty. Disagree
	57	Sty. Agree: Agree: Unable to decide: Disagree: Sty. Disagree
	58	Sty. Agree: Agree: Unable to decide: Disagree: Sty. Disagree
	59	Sty. Agree: Agree: Unable to decide: Disagree: Sty. Disagree
	60	Sty. Agree: Agree: Unable to decide: Disagree: Sty. Disagree

Individual's Total Score on Religious Attitude Scale …

Area-wise Scores on Religious Attitude Scale

S. No.	Area/Component	Statements	Total Scores of 10 statements
1.	Nature of God	1, 7, 13, 19, 25, 31, 37, 43, 49, 55.	
2.	Prayer and Worship	2, 8, 14, 20, 26, 32, 38, 44, 50, 56.	
3.	Formal Religion	3, 9, 15, 21, 27, 33, 38, 45, 51, 57.	
4.	Priests	4, 10, 16, 22, 28, 34, 40, 46, 52, 58.	
5.	Future Life	5, 11, 17, 23, 29, 35, 41, 47, 53, 59.	
6.	Spirits and Spirit World	6, 12, 18, 24, 30, 36, 42, 48, 54, 60.	

APPENDIX-III

RELIGIOUS ATTITUDE SCALE

PROFESSOR RAJAMANICKAM'S RELIGIOUS ATTITUDE SCALE

Key for Religious Attitude Scale

Statements	Weight for Responses				
	Strongly Agree	Agree	Unable to decide	Disagree	Strongly Disagree
1 – GOD – P	1	2	3	4	5
2 – PRY – P	1	2	3	4	5
3 – FOR – N	5	4	3	2	1
4 – PRI – P	1	2	3	4	5
5 – FUL – P	1	2	3	4	5
6 – SPR – P	1	2	3	4	5
7 – GOD – P	1	2	3	4	5
8 – PRY – P	1	2	3	4	5
9 – FOR – P	1	2	3	4	5
10 – PRI – P	1	2	3	4	5
11 – FUL – P	1	2	3	4	5
12 – SPR – N	5	4	3	2	1
13 – GOD –N	5	4	3	2	1
14 – PRY – P	5	4	3	2	1
15 – FOR – P	1	2	3	4	5
16 – PRI – N	5	4	3	2	1
17 – FUL – P	1	2	3	4	5
18 – SPR – N	5	4	3	2	1
19 – GOD –P	1	2	3	4	5
20 – PRY – P	1	2	3	4	5
21 – FOR – N	5	4	3	2	1
22 – PRI – N	5	4	3	2	1
23 – FUL – N	5	4	3	2	1
24 – SPR – N	5	4	3	2	1
25 – GOD –N	5	4	3	2	1
26 – PRY – N	5	4	3	2	1
27 – FOR – P	1	2	3	4	5
28 – PRI – P	1	2	3	4	5
29 – FUL – P	1	2	3	4	5
30 – SPR – N	5	4	3	2	1
31 – GOD –P	1	2	3	4	5
32 – PRY – P	1	2	3	4	5
33 – FOR – P	1	2	3	4	5
34 – PRI – P	1	2	3	4	5
35 – FUL –N	5	4	3	2	1
36 – SPR – P	1	2	3	4	5
37 – GOD –P	1	2	3	4	5
38 – PRY –N	5	4	3	2	1
39 – FOR –N	5	4	3	2	1
40 – PRI – N	5	4	3	2	1
41 – FUL –N	5	4	3	2	1
42 – SPR – P	1	2	3	4	5

Statements	Weight for Responses				
	Strongly Agree	Agree	Unable to decide	Disagree	Strongly Disagree
43 – GOD – N	5	4	3	2	1
44 – PRY – N	5	4	3	2	1
45 – FOR – N	5	4	3	2	1
46 – PRI – P	1	2	3	4	5
47 – FUL – P	1	2	3	4	5
48 – SPR – P	1	2	3	4	5
49 – GOD – N	5	4	3	2	1
50 – PRY – P	1	2	3	4	5
51 – FOR – N	5	4	3	2	1
52 – PRI – N	5	4	3	2	1
53 – FUL – N	5	4	3	2	1
54 – SPR – P	1	2	3	4	5
55 – GOD – N	5	4	3	2	1
56 – PRY – N	5	4	3	2	1
57 – FOR – P	1	2	3	4	5
58 – PRI – N	5	4	3	2	1
59 – FUL – N	5	4	3	2	1
60 – SPR – N	5	4	3	2	1

Note : P – Positive Statement; N – Negative Statement
GOD – Nature of God; PRY – Prayer and Worship
FOR – Formal Religion; PRI – Priests
FUL – Future Life; SPR – Spirits and Spirit World.

EXPERIMENT 56 : ASSESSING REACTIONARY AND RADICAL TENDENCIES

INTRODUCTION

Psychology till the first quarter of twentieth century was non-experimental and any attempt to measure any psychological phenomenon was looked upon with suspicion and hostility. However there were some measurements in quantitative terms in the areas like reaction time, learning and forgetting, the level of intelligence and aptitude were going on though they were ignored as unworthy. After 1900 psychology had the impact of physical and biological sciences and began to function as an independent science forming new alliance with biological and other social sciences and extended its basic measurement techniques in various areas. Early psychologists were mainly interested in seeing, hearing, feeling and speed of response and the latter psychologists extended their studies to sensory and perceptual areas. Many tests and techniques have been developed to appraise the individual and they have been applied to many aspects of human abilities, personality and other aspects.

What we measure in a person and how we describe it? We cannot measure everything in a person and describe every thing about him. Having measured certain aspects of a person we describe him in quantitative terms the levels of the attributes that were manifested and perceived. Psychologists and educationalists are mainly interested in measuring two aspects. What a person is capable oı doing and what he would do? The first is referring to the measurement of individual ability. Ability may be potential ability and actual ability. The potential ability is aptitude and the actual ability is achievement. In other words, what we measure is what a person could learn and what a person has learned to do some task. To measure these two aspects psychologists and educationalists have developed aptitude tests and achievement tests. What the individual will do or capable of doing correspond to the area of personality measurement. A rough definition of personality is, what a person will do. It is also referring to how he feels and his inner motives and conflicts, what we study in a person today will be helping us to predicts something about him in the future.

We may be interested in seeing a person whether he will go for a degree course or a professional course or whether he will be happy with a teaching job or an executive job. These are the inner aspects of the individual and we make inferences from outer observations of behaviour. What we observe is a series of reactions. A person is happy or tearful, sociable or introverted. This may refer to behaviour consistent over some time and place. But we see in a person a behaviour consistency of a unified aspect. This may be stated as the syndrome or cluster of some related behaviour and this may be stated as trait. Always a trait is a construct which means putting together substances or parts. It is a model devised on the basis of observations to be related to what is observed in theoretical framework. There are a number of terms referring to the various aspects of personality that we wish to evaluate. Some of them may be character, adjustment, temperament, interest and attitudes etc. In each of these aspects of personality several things may be clustered.

Character

Educational and vocational organisations are very much interested in the behaviour of an individual which is specifically designated as conduct and character. This character is a cluster of traits to which some social importance is attached. For example, loyalty, sincerity, honesty, co-operativeness, kindliness, thrift are social virtues attached to this trait and character. A number of tests and evaluation procedures have been developed to assess character.

Adjustment

It is one of the important areas of study for the psychologists and educationalists. Organisations are interested in well adjusted persons in the recruitment to various occupations. The unhappy person is an unproductive and fails to fit into the situation. The degree of adjustment varies from culture to culture depending upon what is normal for that culture. Adjustment may mean the behaviour pattern that enables a person to get along well in his social situation.

Temperament

Temperament include the mood and general style of life. Mood refers to the type of momentary reaction of a person in a particular situation. A repetition or constant occurrence of this may develop into temperament. This construct has been used widely in the scientific and literary world. Hypocrates had stated that men could be classified into four categories on psychological basis as sanguine, choleric, phlegmatic and melancholic. According to Thorndike and Hagen (1970) temperament generally refers to an individual's disposition, energy levels, excitability and social orientation and it includes dimensions of cheerful-gloomy, energetic-lethargic, excitation-ca...n, introverted-extroverted and dominance and submissiveness. In short it is an aspect of personality.

Interest

Interest refers to the choices made by an individual with respect to his activities in which he is engaged. He may show preference to some and aversion to some and may be indifferent to some. In interest measurement the individual's preference and avoidance can be assessed by interest inventories.

Attitudes

In psychological studies we often come across the concept attitude invariably in almost all the areas as it covers a very wider area of human experiences. Therefore, it has become a major field of study. The concept attitude has been employed to explain the manner of thought of a man towards some place, person or thing. A person's stand on any of the political or social issues is often referred to as his attitude. A person's or group's approach or avoidance to certain problem or subject is explained in terms of this word.

Around 1920's and 1930's it became a fashion for the social scientists to make attitude studies. Bogardus (1933) studied social distance, Thurstone and Chave (1929) published monographs on attitude measurement. Allport (1935), Newcomb (1932), Murphy and Likert (1933) and others have made extensive studies in this direction. A large number of articles, reports, pamphlets and books were published on this subject.

A man's social behaviour is explained in terms of motivational, emotional, perceptual and learning process. It is better to see a man's social behaviour from some higher level. This higher level may be called beliefs and attitudes. According to Krech and Crutchfield (1940) an attitude is an enduring organisation or motivational, emotional and perceptual cognitive process with respect to some aspect of the individual's world. It does not demand emotional excitements. Attitudes themselves are not directly perceived. They are perceived in actions and overt behaviours. Therefore, an individual's overt behaviour is the manifestation of an attitude. The important factor implied in an attitude towards an object or person is response or reaction to something outside. It may be a favourable or unfavourable response. However it involves some action overtly or inherently. Attitudes may be social and personal. Allport (1935) says that a person possesses two sets of attitudes, one private and personal and kept confidentially and the other is publicly acknowledged and socially determined. An individual's attitude towards his parents, children, wife, dog etc. is not shared by anybody and, therefore, it is a personal and private attitude. It is with regard to individual's adjustment. The social attitudes are concerned with objects, persons and issues common to a group of people and to the society in which they live.

ATTITUDES AND OTHER AREAS

Attitudes are compared with other areas of study like interests, faiths and opinions etc. Attitudes are subjective while interests are objective. The terms such as love, hate, sympathy, pride may be grouped under attitudes while the terms friend, enemy, family, play, games can be grouped under interests. Everybody is interested in law. But the burglar wants to violate the law and cheat the citizens. A citizen wants to safeguard his property by helping the law. The policeman and the judge want to maintain the law. A conservative person may not agree with the principles of communism, but still he is interested to know how the communist party functions. Therefore, attitudes are associated with approval and evaluation whereas interest is with one's likes and dislikes.

Attitudes and Traits

It is possible to show how these two processes function in different ways. Allport (1935) differentiates these two. A trait has no reference to any object, but an attitude has a definite object of reference which may be material or conceptual. An individual's views towards family planning, birth control, marriage, liquor, divorce, democracy, communism or Fascism are attitudes whereas a person's conservatism, radicalism, liberalism, introversion, extroversion, asceticism are traits. Secondly a trait may be narrowly limited to a state of readiness for response whereas attitude may be specific as well as general to the world at large. Thirdly an attitude indicates acceptance or rejection of an object with which it is related. But a trait has no such definite indication. A man's preference for a particular race or group is an attitude, but a man's hostility is a trait.

Formation of Attitudes

Attitudes are formed in relation to person, place or thing. Attitudes are formed in an individual in his interaction with human beings and with social institutions. The most important factor in the formation of attitude is selectivity. If an individual is interested in certain issue he willingly tries to acquire some attitude which may be favourable or unfavourable. According to Sherief and Candril (1947) the first stage in the formation of attitude is the perceptual stage. In this the internal factors come into contact with external factors of the stimulus situation. There is always a subject-object inter-relationship in attitude formation. The stimulus situation may be anything outside like a home, a sex, parents, friends, teacher, temple, club, etc. This description therefore, implies that attitudes are formed.

An attitude is also an expressed opinion of an individual with regard to some issue. A man may oppose some change in the society or he may welcome it very strongly. Some people react swiftly objecting any change and some may accept it on principle. All these stands are attitudes. One may be conservative, the other may be radical, a third one may be reactionary and so on. Attitude scales measure all these aspects. A man's expressed opinion may fall into five categories, such as reactionism, conservatism, neutralism, liberalism, and radicalism. These five categories need certain explanation.

REACTIONISM

A person's social or political attitude may be expressed in various forms which may reveal his characteristics in some way. In attitude and interest measurements, the response categories strongly agree, agree, neutral or unable to decide, disagree and strongly disagree are generally used to indicate the type of response of an individual towards some problem. Instead of response categories the quality itself may be straightaway used to find out the person's nature of attitude and if the person functions with such quality or characteristics he is said after the name of the quality. It may be also his personality characteristic. The most widely used concept in this line of thinking are conservatism and radicalism. But there are also other traits used to widen this range of conservatism and radicalism. They are reactionism, neutralism and liberalism. Reactionism is the extreme form of conservatism and liberalism the mild form of radicalism.

Reaction is a response to some foregoing action or stimulus. It refers to reciprocal action. Reactionism refers to the characteristic of some function based upon some philosophy. Reactionism is opposed to change. It is a desire to return to a former condition or system.

Reactionism refers to repulsion or resistance exerted by a person on a body opposed to the impact of pressure or another person or body. Reactionism is a movement towards the reversal of an existing tendency or state or things in position or in society.

A person who functions on the philosophy of reactionism may be called a reactionist. He is a professed reactionary and is pertaining to or characterised by reaction. Therefore, the reactionary is a man in motion refuses to acknowledge that whatever has been settled must henceforth be considered good or atleast tolerable and he is willing to erase some laws, scrap some institution, even amend his nation's constitution, so that he can roll back the social process to the time at which his country men first went foolishly astray.

The main theme of reactionism is that the past life was very glorious in all its social and cultural values and any change in these will be a decline from the crowning glory and spoil the resplendence and the ancient traditions which were guarded by the ancestors for generations. Therefore, the reactionist's response to any change in the present set-up is sudden and violent. The reactionist mainly concentrates on protecting the values of the past. It may be social, cultural, religious, artistic and also political. The cultural heritage is the monument to be preserved forever to establish their past glory. The reactionist is the lover of ancient things and wanted to maintain them to speak about their past glorious life. Modern politicians never fail to quote their past glories in their speeches on the public platforms to gain support for their parties.

The much guarded value is religion. The people who follow great religions of the world put strong protest for any change in their religions. Thus the function of reactionism is manifesting in man as a trait and it depends upon how a man respects an existing value. Therefore, a reactionary is a person who is having strong faith in the antiquity of values. The existing culture and civilization, customs and traditions in which he lives are very precious for him and he strongly protest, even a small change in any of the values he has been preserving for all the times.

CONSERVATISM

Conservatism means opposition to change or innovation. It is a political philosophy advocating preservation of the best of the established order in society and opposing radical change. It refers to policies, doctrines, practices of the conservative party favouring the preservation of established customs, values etc., and opposing innovation. A conservative is a person who is reluctant to change the old ideas and he is tending to be moderate and conventional in style. Therefore, conservatism is the characteristic relating to the preservation of established order in the society. In Britain and Canada the major right wing political party is the conservative party which was developed from the Tories in 1930. It advocates a mixed economy, encourages property owning. In Canada, there is a progressive conservative party opposing changes. Toryism generally, conservative, advocates the principle in politics and theology. Its pattern of thought and actions are real and enduring and since no substitute is likely to be accepted conservatism will have a long life as a handy tool for social sciences. Conservatism is classified into four types. They are : (a) temperamental conservatism, (b) situational conservatism, (c) political conservatism, and (d) philosophical conservatism.

(a) Temperamental Conservatism

It describes a temperament which is a cluster of traits that are on daily display by most men in all societies. There are four important elements in the conservative temperament. They are (l) habit, (2) inertia, (3) fear and (4) emulation.

1. The habit according to William James is the enormous flywheel of society and it is the most precious conservative agent.

2. The inertia is referred to a force that often seems to be as powerful in the social world as in the physical world.

3. The fear is the fear of the unexpected, the irregular and the uncomfortable.

4. The emulation is the product of both fear of alienation from the group and a craving for its approval. All these four traits of character flourish with particular vigour among some of the most unfortunate and insecure and expendable groups of society.

(b) Situational Conservatism

It is not confined to the well placed and well to do persons at all levels of being. The possessor may lament for a change in the *status quo*. These people express a highly personal sense of satisfaction and identify with the established order. One happy home of situational conservatism is the upper reaches of society.

(c) Political Conservatism

One may find political conservatism in every country and culture. Since conservatism is essentially in defense of changing society, the leaders of Soviet Union are conservatives. The Tories of Great Britain, the Republics of USA, the Gaullists of France and the Democratics of the European countries are conservatives in the real sense. The liberal democratics of Japan and the Swatantra Party in India were conservatives in a less meaningful sense. The conservative is generally well adjusted psychologically to a world he never made. The conservative is above all, a man of order and he has no place in a sheltered society.

(d) Philosophical Conservatism

It is an important political force in most of the countries in the west. It is a philosophy dedicated to the defence of an established order. During the World War II, the conservative thinkers have been more active than at any other periods in the past two centuries. Conservatism exists in the realm of ideas and one must recognise the existence of a core of principles. The existence of a universal moral order is sanctioned and supported by organised religions. The imperfect nature of man in which unreason and sinfulness lurk always behind the curtain of civilised behaviour. The primary role of conservatism is to protect the private property in the pursuit of personal liberty. The conservative has preference for liberty over equality. The conservative always insists that *education must discpline before it can liberate*. The philosophical conservatism as a school of political thought is identified with Edmund Burke. It is a pattern of personal behaviour. It is a cluster of social attitudes and prejudices.

NEUTRALISM

Neutralism is a state of character of being neutral in dispute or contest. It is not siding with any party to war or dispute. The neutral is one who remains between two parties or sides, not assisting either party. It is a kind of indifferent attitude towards the problem or the crisis. It is occupying a middle position between the two extremes. The term neutralism was used synonymously to non-alignment in 1960's to refer to the attitude of the Asian and African nations to avoid military alliances with either side in the cold war between the big forces. In specific conflict even a major power might choose to be neutral. Smaller and weaker powers sometimes declare themselves to be in a state of neutralisation. Switzerland is the most frequently cited nation for example of neutralization. Neutrality as a policy or as the strategy appears to be safe. When two extreme ideas put forth before a man to decide which one he prefers he might not be in a position to decide either way and he might choose to be in between the two ideas. He is said to be neutral. Neutral also refers to a state of undecided or unable to decide.

LIBERALISM

The term Liberalism is generally known from political point of view. A man's stand on some problem may be stated that 'he is very liberal'. 'He viewed the matter liberally'. On this way, the concept 'liberal' is used in different contexts. In England there is a political party called 'Liberal Party'. The concept liberalism is having an important place in the history of European political philosophy. The word 'liberal' is pertaining to a 'free man'. The etimological meaning of the word 'liberal' is being free from any restraint from outside. The liberal is one who wants to have a change or reformation in the existing conditions. As and when new ideas develop one may be tempted to have a change. It has been taking place in social, economic, political and religious institutions from time to time. There have been instances that ideas were put forth for and against changes all the times. Therefore, liberalism refers to liberal opinion, liberal practice or liberal politics. Liberalism is a movement in modern protestantism that rejects the Biblical authority.

Liberalism is the political tenets of a liberal. Liberals are favourable to changes and reforms, tending in the direction of democracy. Therefore, liberalism may be stated as a system of philosophy that is relating to social, political theological and religious views, that favour progress and reform. It is also relating to politics or views advocating individual freedom. Liberalism also refers to the generous temperament or behaviour of an individual, a behaviour tolerant of other people which is not strict but free. It is free in giving and it is generous and open hearted. It is also free in speech and action. It is not rigorous, but free from narrow prejudice.

RADICALISM

Radicalism is relating to the characteristic of basic, inherent constitution of a person favouring or tending to produce extreme or fundamental changes in political, ecnomic, social conditions, institutions and habits of mind. There were radical movements in political and social issues. A radical is one who holds most advanced views of political reform on democratic line and thus belongs to an extreme section of the liberal party. The term 'radicalism' always points to some analytical or revisionist function. It concentrates on a particular principle of change at the expense of the traditionally sanctioned issues.

In its positive sense, this tendency implies a projection of a completely new version of human life. In its negative sense it implies a threat to all aspects of on-going life. Radicalism need not have logically derived conclusion. Although radicalism is characteristically rational or rationalized it is inconsistent with reasonableness. In principle pure radical thought and action is entirely devoid of practical wisdom of sensitivity for the occasion. The ethics of radicalism is invested with tensions. In the long run radicals succumb to worldly pressures.

Modern radicalism was founded largely upon academic discourses of Locke and Montesquieu. The doctrines of modern democratic radicalism was formulated by Rousseau and a host of writers. Their argument urged that all existing conditions and customs are unnatural and must be destroyed and replaced by a new and traditional order. But the religiously inspired radicalism of the reformation period was wholly secular.

EXPERIMENT

Problem

Assess a person's reactionary and radical and other tendencies by using suitable scale.

Method

Participant

Choose a person who has no knowledge about this test and collect certain personal information like his age, sex, educational qualifications, socio-economic status, religion etc.

Material

Prof. Rajamanickam's Reactionism-Radicalism Scale consists of Hand-Book, Test Material, Key and Answer Sheet.

Description of the Scale

Prof. Rajamanickam's (1989) Reactionism-Radicalism scale consisted of 60 statements each one expressing one idea about some issue. There are 30 positive statements and 30 negative statements. There are 30 issues belonging to six areas like (1) Political, (2) Economic problems (3) Social, sex and family, (4) Race, community and tradition, (5) Progressive education and (6) Individual freedom. There are two statements for each issue, one positive and one negative. For each area there are ten statements, five positive and five negative. The scores may be treated as a whole for all the 60 statements as an individual score or may be treated for each issue separately. It may also be treated for the

6 areas separately to assess an individual's place in each area. The score for the specific issue may be obtained by matching 30 positive and 30 negative statements and adding the scores. These 30 issues may go into 6 areas. The individual score is the total scores obtained for all the 60 statements. Under each statement there are five responses like strongly agree, agree, unable to decide, disagree and strongly disagree. The participant is to underline any one of these responses by way of answering the questionnaire.

Procedure

The participant is to be seated comfortably in a manner that he should have free thinking on each issue. If the personal information are needed he/she may be asked to provide all such particulars and the following instructions should be given.

> "This is a questionnaire, consisting of 60 statements. In each statement one idea is expressed. You may agree or disagree with it. Whatever you feel about the idea in the statements you will have to express what you feel about it. Your agreement or disagreement may vary in different degrees. There are five responses given under each statement such as strongly agree, agree, unable to decide, disagree and strongly disagree. Read each statement carefully and express your views about the idea by underlining any one of the five responses in the answer sheet. Please do not make any mark in the test booklet. Answer all the statements in the same manner. When I say, 'ready, go', you can start answering. There is no time limit. You can take your own time".

After giving the instruction, the experimenter may hand over the test booklet to the participant. As soon as he/she starts answering the questionnaire, the experimenter may note down the time, just to find out how much time he/she takes to answer the whole questionnaire. The experimenter can collect the questionnaire as soon as the participant completes it and allow him/her to go away.

Method of Scoring

The scoring is made with the help of the key. There are five response categories. Each is assigned with a numerical value. For the positive statements the strongly agree response will be given 1 mark, for agree 2, for unable to decide 3, for disagree 4 and for strongly disagree 5. For negative statements, strongly agree will be given 5 marks, for agree 4, for unable to decide 3, for disagree 2 and for strongly disagree 1 mark. The strongly agree response of the positive statement indicates Reactionism, agree Conservatism, unable to decide Neutralism, disagree Liberalism and strongly disagree Radicalism. The strongly agree of the negative statement indicates Radicalism, agree Liberalism, unable to decide Neutralism, disagree Conservatism and strongly disagree Reactionism.

The individual's score may always fall in between 60 and 300 and may not fall on the terminal points like 60 or 300.

Results

There are 30 issues and 6 areas in the scale. Find out the scores for each issue by matching the positive and negative statements and give the scores serially area-wise in Table 18.5.

Table 18.5 : Problem-wise scores of the participant on Reactionism - Radicalism Scale

S. No.	Problem areas	Items	Scores for Items	Reactionary or Radical
I.	**Political Issues**			
1.	The Right of the Government	1, 31		
2.	Power of Labour Organization	2, 32		
3.	Communism	3, 33		
4.	Democracy	4, 34		
5.	International Affairs	5, 35		
II.	**Economic Problem**			
6.	Duties on Import Goods	6, 36		
7.	Hereditary Wealth	7, 37		
8.	Minimum Wage Laws	8, 38		
9.	Equality of Income	9, 39		
10.	Distribution of Wealth	10, 40		
III.	**Social, Sex and Family**			
11.	Social, Morality, Sex and Family	11, 41		
12.	Divorce	12, 42		
13.	Marriage with Blood Relations	13, 43		
14.	Abortion	14, 44		
15.	Inter-Caste and racial marriage	15, 45		
IV.	**Race, Community and Tradition**			
16.	Respect for Customs and Traditions	16, 46		
17.	Attitude towards Foreigners	17, 47		
18.	Achievement of White Race	18, 48		
19.	Caste System	19, 49		
20.	Birth Control	20, 50		
V.	**Progressive Education**			
21.	Preference for Science to Religion	21, 51		
22.	Religious Education in Colleges and Schools	22, 52		
23.	Co-education in Colleges and Universities	23, 53		
24.	Admission to Colleges and Universities	24, 54		
25.	Compulsory Education	25, 55		
VI.	**Individual Freedom**			
26.	Observance of Law	26, 56		
27.	Freedom of Speech against Nation	27, 57		
28.	Compulsory Military Training	28, 58		
29.	Medical Care of the Individual	29, 59		
30.	Discipline in Colleges and Universities	30, 60		

There are 10 statements (5 positive and 5 negative) for each area. Work out the scores for all 6 areas and give the scores in Table 18.6 area-wise. State scores of your participant in each area and state whether he/she is a reactionary or radical. That are areas in which he/she is reactionary and what are the areas in which he/she is radical. Collect the group data for area-wise items, and work out the mean scores and give them in Table 18.7. In the group who are those politically reactionary and who are those radicals? What are the areas in which more participants are reactionary and radical? Who are those scored above the group mean and who are those below the group mean?

Discussion

Discuss the individual scores in each problem/issue and state where he/she is reactionist, conservative, neutral, liberal and radical. Check whether the participant's scores in all these issues are consistent or having any variation. Discuss the participant's scores area-wise and state where he/she is reactionist, conservative, neutral, liberal and radical. State his trait in every area. Is he/she varies in his/her attitude from area to area.

Discuss his/her trait on every area relating his/her personal information like age, educational qualification, socio-economic status, religion, community, geographical area and so on. Discuss the group data. How many of them are above mean and below mean? What is the trait of the group as a whole?

Table 18.6 : Area-wise scores of the individual on Reactionism - Radicalism Scale

S. No.	Areas of the issues	Items	Total Scores	Reactionary or Radical
1.	Political Isssues	1, 2, 3, 4, 5, 31, 32, 33, 34, 35		
2.	Economic Problem	6, 7, 8, 9, 10, 36, 37, 38, 39, 40		
3.	Social, Sex and Family	11, 12, 13, 14, 15, 41, 42, 43, 44, 45		
4.	Race, Community and Tradition	16, 17, 18, 19, 20, 46, 47, 48, 49, 50		
5.	Progressive Education	21, 22, 23, 24, 25 51, 52, 53, 54, 55		
6.	Individual Freedom	26, 27, 28, 29, 30 56, 57, 58, 59, 60		

Table 18.7 : Group data for the 6 areas on the Reactionism-Radicalism Scale

S. No.	Names	Political	Economic	Social etc.	Race etc.	Progressive Education	Individual Freedom	Total score in all areas
1.	R.M.							
2.	M.S.							
3.	B.V.							
4.	K.R.							
5.	P.V.							
6.	A.R.							
7.	B.C.							
8.	P.V.							
9.	R.M.							
10.	S.R.							
Mean								

If any one deviates very much from the group, compare his score with other individuals of the same personal information and discuss it. State whether age, religion cause any difference. Do the members of the group react in their tendencies in line with the prevailing political or social current trends? Are members consistently reactionary or conservative or radical in all the areas? Discuss it critically. State the traits mostly prevalent in the group. Compare the individual scores and the group scores with any one of the earlier studies in each area.

Conclusions

State the trait of your participant taking into account his/her whole score.
State the trait of the group.
How many of them Reactionist and how many of them Radicals.
State whether there is any variation due to sex, religion, age, socio-economic status.

Guidelines for determing the individual traits from the whole score and from each area

Range of Scores (Whole score)	Trait	Range of scores for each area
60 to 90	Reactionism	10.0 to 15.0
91 to 150	Conservatism	15.1 to 25.0
151 to 210	Neutralism	25.1 to 35.0
211 to 270	Liberalism	35.1 to 45.0
271 to 300	Radicalism	45.1 to 50.0

APPENDIX - I

REACTIONSIM-RADICALISM SCALE

INSTRUCTIONS FOR RESEARCHER/TEST USERS

The present scale has been designed and developed and revalidated with the view of providing quantitative measure of an individual's stand on various issues and certain problem areas. It is very difficult to say that a person is very rigid, another is flexible and a third is very lenient and so on unless and until a systematic study is made. There is no material instrument to measure human feelings, views, opinions and various types of reactions. Psychologists have devised certain methods by which some test materials could be constructed and the views and opinions of human beings can be assessed as far as possible. This scale is intended to assess the stand of a person and also of a group of people in various issues and problems. It is very carefully developed and validated for research purposes and, therefore, researchers should take proper care in administering the scale on any population. This scale can be administered on the general population, student population and on professional population. The subjects should be asked only to answer with the responses, that is, by underlining one of the five responses. The answer sheet is designed in such a way that the individual scores may be collected easily and can be treated area-wise and problem-wise. It is always safe for the researchers and test users to use the designed Answer Sheet to avoid the possible mistakes by the subjects while answering the questionnaires.

DIRECTION FOR FILLING IT UP

There are 60 statements given below, each expressing one idea about certain social issue. You may have different opinions about these social issues. You may agree with certain ideas and disagree with certain ideas. There are five responses *i.e.* strongly agree, agree, unable to decide, disagree, strongly disagree given under each statement. Indicate your agreement or disagreement by underlining any one of the five responses according to your first reaction. This is only for the academic purposes and your identity will not be revealed at any cost.

Examples :

1. War is a necessary factor in the struggle for survival and supremacy in the world.
   <u>Strongly agree</u> : Agree : Unable to decide: Disagree: Strongly disagree
   Here the response strongly agree is underlined by Mr. *X*. This indicates that Mr. *X* strongly agrees with the idea expressed in the Statement No.1

2. Every citizen should join the Army if there is any danger from the neighbouring country.
   Strongly agree: Agree: Unable to decide: Disagree: <u>Strongly disagree</u>

Here the response strongly disagree is underlined by Mr. *X*. This indicates that Mr. *X* strongly disagrees with the idea expressed in the Statement No. 2

If your agreement falls in between the two extremes underline response with which you agree among the five. Answer all the 60 statements correctly without omitting anything. Please do not write any thing on this booklet. Write only in the *Answer Sheet*, supplied to you.

STATEMENTS

1. Government should have all the rights to take over any properties, lands or industries belonging to individual persons.
   Strongly agree : Agree: Unable to decide: Disagree: Strongly disagree
2. Labour organisation and trade unions should be prohibited totally as they are dangerous to individual freedom and industrial progress.
   Strongly agree: Agree: Unable to decide: Disagree: Strongly disagree
3. Communism is absolutely not suitable to the cultural temperament of our country.
   Strongly agree: Agree: Unable to decide: Disagree: Strongly disagree
4. Democracy is the only form of Government in which every man can find his freedom, equality and happiness.
   Strongly agree: Agree: Unable to decide: Disagree: Strongly disagree
5. We can easily solve certain international problems by being friendly with other nations.
   Strongly agree: Agree: Unable to decide: Disagree: Strongly disagree
6. There should be high taxes on all import goods in order to encourage our own industries and products.
   Strongly agree: Agree: Unable to decide: Disagree: Strongly disagree
7. All wealth should go to the State at the death of the person who owned it.
   Strongly agree: Agree: Unable to decide: Disagree: Strongly disagree
8. The State should not interfere in the determination of wages or salaries of the workers.
   Strongly agree: Agree: Unable to decide: Disagree: Strongly disagree
9. Whatever the income one gets should not be subjected to income taxes.
   Strongly agree : Agree: Unable to decide: Disagree: Strongly disagree
10. The standard of living of the working class should be raised by equal distribution of wealth and status.
    Strongly agree: Agree: Unable to decide: Disagree: Strongly disagree
11. We must remove all the sex restrictions and permit complete freedom in sex relations.
    Strongly agree: Agree: Unable to decide: Disagree: Strongly disagree
12. Divorce should not be granted on any account to either party as it would destroy the foundations of our social life.
    Strongly agree: Agree: Unable to decide: Disagree: Strongly disagree

13. We must remove all restrictions and social laws regarding mating and marrying with blood relations.
Strongly agree: Agree: Unable to decide: Disagree: Strongly disagree
14. A woman should be the master of her own body and have the unlimited right to decide whether she would have child or abortion.
Strongly agree: Agree: Unable to decide: Disagree: Strongly disagree
15. Every true citizen of this country should do his best for the biological inter-mixture of castes and races by inter-marriages.
Strongly agree: Agree: Unable to decide: Disagree: Strongly disagree
16. All our customs and traditions are outgrown. We must discard them and introduce progressive ways in our life.
Strongly agree: Agree: Unable to decide: Disagree: Strongly disagree
17. We must preserve the purity of our land and heritage by preventing all foreigners setting in our country.
Strongly agree: Agree: Unable to decide: Disagree: Strongly disagree
18. The white race has to be appreciated for its intellectual superiority, because modern science and civilization are entirely the products of the white race.
Strongly agree: Agree: Unable to decide: Disagree: Strongly disagree
19. Caste system is a way of living, very ancient in origin. We must maintain it forever to safeguard our prestige and respect.
Strongly agree: Agree: Unable to decide: Disagree: Strongly disagree
20. We must take all possible measures to control the birth rate in every family to save the country from poverty and other problems.
Strongly agree: Agree: Unable to decide: Disagree: Strongly disagree
21. All religions are highly superstitious and irrational. Therefore we must adapt a scientific way of life and develop progressive thinking.
Strongly agree: Agree: Unable to decide: Disagree: Strongly disagree
22. Man finds peace and happiness only in religion. It must be taught in all schools and colleges as a compulsory subject.
Strongly agree: Agree: Unable to decide: Disagree: Strongly disagree
23. Men and women should not be admitted to the same Colleges and Universities since co-education spoils the study and the behaviour of the students of both sexes.
Strongly agree: Agree: Unable to decide: Disagree: Strongly disagree
24. We will certainly lose the benefit of education if we do not consider merits in admitting students in Colleges and Universities.
Strongly agree: Agree: Unable to decide: Disagree: Strongly disagree
25. The state should provide free and compulsory education for all the children up to their capacities and until they are placed in some jobs.
Strongly agree: Agree: Unable to decide: Disagree: Strongly disagree
26. If a law exists it must be observed always and any person breaking it should be punished severely as a criminal.
Strongly agree: Agree: Unable to decide: Disagree: Strongly disagree

27. Freedom of speech should have limitation in the interest of Nation. Every Government must watch and apply strict censure on the speeches made by persons against the Nation.
Strongly agree: Agree: Unable to decide: Disagree: Strongly disagree
28. The State should have the right to enlist compulsorily any person for military service during war-time to train him as an efficient soldier.
Strongly agree: Agree: Unable to decide: Disagree: Strongly disagree
29. The State should take entire responsibility of providing medical care to all the citizens.
Strongly agree: Agree: Unable to decide: Disagree: Strongly disagree
30. Strict measures should be taken to maintain discipline among the students in Colleges and Universities as indiscipline everywhere leads them to all sorts of criminal activities.
Strongly agree: Agree: Unable to decide: Disagree: Strongly disagree
31. Government should not have any right to own, control, and operate the whole economic resources of the country.
Strongly agree: Agree: Unable to decide: Disagree: Strongly disagree
32. The problem of modern society can be solved only by labour organization and Trade unions when they capture the government.
Strongly agree: Agree: Unable to decide: Disagree: Strongly disagree
33. Communism alone can provide best form of Government. We must go for such a Government to solve our problems quickly.
Strongly agree: Agree: Unable to decide: Disagree: Strongly disagree
34. Democratic ideals appear to be attractive, but in reality it is very badly practised leading to all sorts of irregularities and corruption by few party leaders. Democracy as practised in many countries is worst than Dictatorship.
Strongly agree: Agree: Unable to decide: Disagree: Strongly disagree
35. It would be dangerous for our country to make friendship with other nations who are aggressive and unfriendly.
Strongly agree: Agree: Unable to decide: Disagree: Strongly disagree
36. There should be no duties on any import goods and we must allow everything to flow freely into our markets.
Strongly agree: Agree: Unable to decide: Disagree: Strongly disagree
37. Government should not interfere with hereditary wealthy on any account. It will destroy individual's interest in work and savings.
Strongly agree: Agree: Unable to decide: Disagree: Strongly disagree
38. Salaries and wages should be the same for all men and women from President to factory worker according to their needs.
Strongly agree: Agree: Unable to decide: Disagree: Strongly disagree
39. There should be heavy income taxes so as to remove the big difference in income.
Strongly agree: Agree: Unable to decide: Disagree: Strongly disagree

40. The existing property rights and status should not be destroyed by distributing the wealth to the people of the working class who are lazy and dishonest.
Strongly agree: Agree: Unable to decide: Disagree: Strongly disagree

41. We must teach and practice perfect morality and prevent all unlawful sexual activities.
Strongly agree: Agree: Unable to decide: Disagree: Strongly disagree
42. Individuals should be allowed to divorce themselves at their own free will.
Strongly agree: Agree: Unable to decide: Disagree: Strongly disagree
43. We must prevent and punish severely all marriages between blood relatives.
Strongly agree: Agree: Unable to decide: Disagree: Strongly disagree
44. The act of abortion is a murder and it deserves severe punishment.
Strongly agree: Agree: Unable to decide: Disagree: Strongly disagree
45. All the possible efforts should be taken to prevent the inter-caste and inter-religious marriage.
Strongly agree: Agree: Unable to decide: Disagree: Strongly disagree

46. Our customs and traditions are very much valuable for us and they must be preserved forever.
Strongly agree: Agree: Unable to decide: Disagree: Strongly disagree
47. Foreigners of any race should be permitted to come and settle in our country without any restrictions.
Strongly agree: Agree: Unable to decide: Disagree: Strongly disagree
48. The white race has not done anything specifically for the development of science and modern civilization to establish its superiority over other races.
Strongly agree: Agree: Unable to decide: Disagree: Strongly disagree
49. Caste system is a barbarious way of living. We, the civilized men must put a drastic end to it and build up a casteless society.
Strongly agree: Agree: Unable to decide: Disagree: Strongly disagree
50. Birth control is a sin against the wishes of God. We must give birth to as many children as we can.
Strongly agree: Agree: Unable to decide: Disagree: Strongly disagree

51. Science serves only our material comforts and increases mental tensions. Religion alone can relieve us from mental tensions and provide us with happiness.
Strongly agree: Agree: Unable to decide: Disagree: Strongly disagree
52. All religions are based upon irrational and superstitious beliefs. Young minds should not be spoiled by such religious education in schools and colleges.
Strongly agree: Agree: Unable to decide: Disagree: Strongly disagree
53. Co-education provides proper understanding in men and women about each other for a right living. It must be encouraged at all levels.
Strongly agree: Agree: Unable to decide: Disagree: Strongly disagree

54. There should not be any restrictions in admitting students in Colleges and Universities. All students must be admitted.
Strongly agree: Agree: Unable to decide: Disagree: Strongly disagree
55. Education is a private matter of the parents and their children. Going to school and learning should be left with the individual person and the State should not interfere in educational matters.
Strongly agree: Agree: Unable to decide: Disagree: Strongly disagree

56. A man must have the right to break any law, because it prevents his freedom of action and makes him a slave to the state.
Strongly agree : Agree: Unable to decide: Disagree: Strongly disagree
57. A man must have the right to express his opinion. Government should not make any restriction on it.
Strongly agree: Agree: Unable to decide: Disagree: Strongly disagree
58. The State should not have the right to recruit compulsorily any person to military service whether there is war or not.
Strongly agree: Agree: Unable to decide: Disagree: Strongly disagree
59. Medical care is an individual affair and State should not interfere in it.
Strongly agree: Agree: Unable to decide: Disagree: Strongly disagree
60. It is not only the student but also the teacher is responsible for the indiscipline in Colleges and Universities. Therefore, there should not be any interference in the freedom of action of the student on the name of discipline.
Strongly agree: Agree: Unable to decide: Disagree: Strongly disagree

APPENDIX - II

ANSWER SHEET

PROFESSOR RAJAMANICKAM'S REACTIONISM-RADICALISM SCALE

1. Age : 2. Sex : 3. Religion :

4. Qualifications : 5. Occupation : 6. Monthly Income :

Underline the Response with which You Agree

Score	St. No.	Responses
	1.	Sty. Agree: Agree: Unable to decide: Disagree: Sty. Disagree
	2.	Sty. Agree: Agree: Unable to decide: Disagree: Sty. Disagree
	3.	Sty. Agree: Agree: Unable to decide: Disagree: Sty. Disagree
	4.	Sty. Agree: Agree: Unable to decide: Disagree: Sty. Disagree
	5.	Sty. Agree: Agree: Unable to decide: Disagree: Sty. Disagree
	6.	Sty. Agree: Agree: Unable to decide: Disagree: Sty. Disagree
	7.	Sty. Agree: Agree: Unable to decide: Disagree: Sty. Disagree
	8.	Sty. Agree: Agree: Unable to decide: Disagree: Sty. Disagree
	9.	Sty. Agree: Agree: Unable to decide: Disagree: Sty. Disagree
	10.	Sty. Agree: Agree: Unable to decide: Disagree: Sty. Disagree
	11.	Sty. Agree: Agree: Unable to decide: Disagree: Sty. Disagree
	12.	Sty. Agree: Agree: Unable to decide: Disagree: Sty. Disagree
	13.	Sty. Agree: Agree: Unable to decide: Disagree: Sty. Disagree
	14.	Sty. Agree: Agree: Unable to decide: Disagree: Sty. Disagree
	15.	Sty. Agree: Agree: Unable to decide: Disagree: Sty. Disagree
	16.	Sty. Agree: Agree: Unable to decide: Disagree: Sty. Disagree
	17.	Sty. Agree: Agree: Unable to decide: Disagree: Sty. Disagree
	18.	Sty. Agree: Agree: Unable to decide: Disagree: Sty. Disagree
	19.	Sty. Agree: Agree: Unable to decide: Disagree: Sty. Disagree
	20.	Sty. Agree: Agree: Unable to decide: Disagree: Sty. Disagree
	21.	Sty. Agree: Agree: Unable to decide: Disagree: Sty. Disagree
	22.	Sty. Agree: Agree: Unable to decide: Disagree: Sty. Disagree
	23.	Sty. Agree: Agree: Unable to decide: Disagree: Sty. Disagree
	24.	Sty. Agree: Agree: Unable to decide: Disagree: Sty. Disagree
	25.	Sty. Agree: Agree: Unable to decide: Disagree: Sty. Disagree
	26.	Sty. Agree: Agree: Unable to decide: Disagree: Sty. Disagree
	27.	Sty. Agree: Agree: Unable to decide: Disagree: Sty. Disagree
	28.	Sty. Agree: Agree: Unable to decide: Disagree: Sty. Disagree
	29.	Sty. Agree: Agree: Unable to decide: Disagree: Sty. Disagree
	30.	Sty. Agree: Agree: Unable to decide: Disagree: Sty. Disagree
	31.	Sty. Agree: Agree: Unable to decide: Disagree: Sty. Disagree
	32.	Sty. Agree: Agree: Unable to decide: Disagree: Sty. Disagree
	33	Sty. Agree: Agree: Unable to decide: Disagree: Sty. Disagree
	34	Sty. Agree: Agree: Unable to decide: Disagree: Sty. Disagree
	35	Sty. Agree: Agree: Unable to decide: Disagree: Sty. Disagree
	36.	Sty. Agree: Agree: Unable to decide: Disagree: Sty. Disagree
	37.	Sty. Agree: Agree: Unable to decide: Disagree: Sty. Disagree
	38.	Sty. Agree: Agree: Unable to decide: Disagree: Sty. Disagree
	39.	Sty. Agree: Agree: Unable to decide: Disagree: Sty. Disagree
	40.	Sty. Agree: Agree: Unable to decide: Disagree: Sty. Disagree

Score	St. No.	Responses
	41.	Sty. Agree: Agree: Unable to decide: Disagree: Sty. Disagree
	42.	Sty. Agree: Agree: Unable to decide: Disagree: Sty. Disagree
	43.	Sty. Agree: Agree: Unable to decide: Disagree: Sty. Disagree
	44.	Sty. Agree: Agree: Unable to decide: Disagree: Sty. Disagree
	45.	Sty. Agree: Agree: Unable to decide: Disagree: Sty. Disagree
	46.	Sty. Agree: Agree: Unable to decide: Disagree: Sty. Disagree
	47.	Sty. Agree: Agree: Unable to decide: Disagree: Sty. Disagree
	48.	Sty. Agree: Agree: Unable to decide: Disagree: Sty. Disagree
	49.	Sty. Agree: Agree: Unable to decide: Disagree: Sty. Disagree
	50.	Sty. Agree: Agree: Unable to decide: Disagree: Sty. Disagree
	51.	Sty. Agree: Agree: Unable to decide: Disagree: Sty. Disagree
	52.	Sty. Agree: Agree: Unable to decide: Disagree: Sty. Disagree
	53.	Sty. Agree: Agree: Unable to decide: Disagree: Sty. Disagree
	54.	Sty. Agree: Agree: Unable to decide: Disagree: Sty. Disagree
	55.	Sty. Agree: Agree: Unable to decide: Disagree: Sty. Disagree
	56.	Sty. Agree: Agree: Unable to decide: Disagree: Sty. Disagree
	57.	Sty. Agree: Agree: Unable to decide: Disagree: Sty. Disagree
	58.	Sty. Agree: Agree: Unable to decide: Disagree: Sty. Disagree
	59.	Sty. Agree: Agree: Unable to decide: Disagree: Sty. Disagree
	60.	Sty. Agree: Agree: Unable to decide: Disagree: Sty. Disagree

Individual's total score for all 60 items ....

Area-wise Scores on Reactionism-Radicalism Scale

S. No.	Area	Items	Total Scores
1.	Political Issues	1, 2, 3, 4, 5, 31, 32, 33, 34, 35	
2.	Economic Problems	6, 7, 8, 9, 10, 36, 37, 38, 39, 40	
3.	Social, Sex and Family	11, 12, 13, 14, 15, 41, 42, 43, 44, 45	
4.	Race, Community and Tradition	16, 17, 18, 19, 20, 46, 47, 48, 49, 50	
5.	Progressive Education	21, 22, 23, 24, 25, 51, 52, 53, 54, 55	
6.	Individual Freedom	26, 27, 28, 29, 30, 56, 57, 58, 59, 60	

APPENDIX - III

PROFESSOR RAJAMANICKAM'S REACTIONISM-RADICALISM SCALE

Key for Reactionism (RE)-Radicalism (RA) Scale

Statement	Weights for Responses				
	Strongly agree	Agree	Unable to Decide	Disagree	Strongly Disagree
1 – RA –	5	4	3	2	1
2 – RE –	1	2	3	4	5
3 – RE –	1	2	3	4	5
4 – RA –	5	4	3	2	1
5 – RA –	5	4	3	2	1
6 – RA –	5	4	3	2	1
7 – RA –	5	4	3	2	1
8 – RE –	1	2	3	4	5
9 – RE –	1	2	3	4	5
10 – RA –	5	4	3	2	1
11 – RA –	5	4	3	2	1
12 – RE –	1	2	3	4	5
13 – RA –	5	4	3	2	1
14 – RA –	5	4	3	2	1
15 – RA –	5	4	3	2	1
16 – RA –	5	4	3	2	1
17 – RE –	1	2	3	4	5
18 – RA –	5	4	3	2	1
19 – RE –	1	2	3	4	5
20 – RA –	5	4	3	2	1
21 – RA –	5	4	3	2	1
22 – RE –	1	2	3	4	5
23 – RE –	1	2	3	4	5
24 – RE –	1	2	3	4	5
25 – RA –	5	4	3	2	1
26 – RE –	1	2	3	4	5
27 – RE –	1	2	3	4	5
28 – RA –	5	4	3	2	1
29 – RA –	5	4	3	2	1
30 – RE –	1	2	3	4	5
31 – RE –	1	2	3	4	5
32 – RA –	5	4	3	2	1
33 – RA –	5	4	3	2	1
34 – RE –	1	2	3	4	5
35 – RE –	1	2	3	4	5
36 – RE –	1	2	3	4	5
37 – RE –	1	2	3	4	5
38 – RA –	5	4	3	2	1
39 – RA –	5	4	3	2	1
40 – RE –	1	2	3	4	5
41 – RE –	1	2	3	4	5
42 – RA –	5	4	3	2	1
43 – RE –	1	2	3	4	5
44 – RE –	1	2	3	4	5
45 – RE –	1	2	3	4	5

Statement	Weights for Responses				
	Strongly agree	Agree	Unable to Decide	Disagree	Strongly Disagree
46 – RE –	1	2	3	4	5
47 – RA –	5	4	3	2	1
48 – RE –	1	2	3	4	5
49 – RA –	5	4	3	2	1
50 – RE –	1	2	3	4	5
51 – RE –	1	2	3	4	5
52 – RA –	5	4	3	2	1
53 – RA –	5	4	3	2	1
54 – RA –	5	4	3	2	1
55 – RE –	1	2	3	4	5
56 – RA –	5	4	3	2	1
57 – RA –	5	4	3	2	1
58 – RE –	1	2	3	4	5
59 – RE –	1	2	3	4	5
60 – RA –	5	4	3	2	1

EXPERIMENT 57 : ASSESSING AUTHORITARIAN PERSONALITY

INTRODUCTION

One of the important areas of research the Psychologists are very much interested is to assess the personality characteristics of the individuals. The type of personality characteristic on which they paid more attention after the World War II was the Authoritarian Personality. The reason has been that it was very much associated with prejudice and anti-semitism. Research studies state that prejudiced and unprejudiced individuals differ basically in their outlook and personal philosophy of life towards their life and their attitudes towards others. The origin of the concept, Authoritarian Personality is found in the works of Erich Fromm (1941). An extensive research works on Authoritarian Personality were conducted by Adorno, Frenkal-Burnswick, Levinson and Sanford (1950) and they have published their important work, *The Authoritarian Personality* in 1938. Their earlier attempt was to understand the Fascism and Nazism that developed in Italy and Germany in 1920s and 1930s. Their aim of the study was also to describe the psychological variables involved in anti-semitism. Anti-semitism is a tendency to discriminate against the Jews. The study has been described as the most important one because of its attempt to examine the political ideology of anti-sematic individuals and relate them to the personality structures of such individuals. Several studies have been brought out subsequently. Some of them experimentally studied this problem were Shils (1954), Titus and Hollander (1957) and Kisscht and Dillchay (1967).

Adorno and his co-workers developed the 'F' scale, a scale for measuring the Fascist's attitude (F). Fascism is an ideology. It was a movement inspired by Italian Fascism, like the German Nazism – National Socialism. Both the Fascism and Nazism practised right wing nationalist ideology or movement, with an authoritarian and hierarchical structure that is fundamentally opposed to democracy and liberation. Fascism was a political movement under the regime of Benito Mussolini. Fascism encouraged militarism and nationalism, organizing the country along hierarchical authoritarian lines. Thus the scale has become the standard instrument for assessing the fascistic tendencies of the individuals. This has resulted in the formulation of an important area of study in psychology and authoritarianism provides a link between personality and ideology, between character and prejudice. This construct has been used in a wide variety of psychological, social, political and economic phenomena in recent times.

When Adolph Hitler came to power in 1933, he started thinking of the wanderings of his earlier falls. This has affected the world for many years to come. That is the atrocities of Nazi Government under his leadership. The atrocities required partnership in doing wrong things and cooperation of vast number of the German people. The German Government's actions against the Jews in Germany, Poland and France were very cruel and inhuman. These actions of the German people urged psychologists, sociologists and other behaviour scientists to study the personality characteristics of the German people. At this time the German people were seen passive and accepting the violent actions of their Government. The people of Germany who were long educated in a developed democratic country accepted the propaganda of Nazi Government against the Jews. The Nazi Government's propaganda was that the Jews were the source of Germany's economic, social and political fight. The Jews were identified as the evil to be won over to save the country from the difficulties. But it was true that Germany at that time had many difficulties and the Jews were scapegoated and targeted.

Erich Fromm (1941) felt that a historical analysis was necessary to understand modern social character. He believed that the German character structure of that day was the consequence of age old long changes in the religious experiences of the individuals since Reformation that took place in Eruope in Sixteenth century and the rise of capitalism. These changes provided potential strength and independence and set man free. But it produced a sense of isolation and fear. The social and economic changes with intensified religious doctrines resulted in the formation of new character structure. This has become an important factor in turn in making further changes in social and economic spheres. The consequences, a century later was the authoritarian character structure in the personalities of the people in the lower middle class people of Germany. This character structure is identified as the idealization of authority, fear and submission. According to Fromm, the term 'authoritarian character represents the personality structure which is the human basis of Fascism'. This was what Fromm thought of German national character.

Berkeley Studies

When we think of authoritarianism at once the famous publication, 'Authoritarian Personality' and the 'F scale' brought out by Adorno and his colleagues strikes in our mind. The study made by Adorno and others was called 'The Berkeley studies' as it was conducted at Berkeley California University. It was originally planned to study anti-semitism. The main reason for starting this study was the incidence of persecution of Jews all over the world. Millions of Jews and some ethnic minority groups were exterminated by the Nazi Government. This was the outcome of the prejudice that existed from fourteenth century in many countries around the world. One of the reasons for encouraging this study was that prejudice was one of the serious problems that existed at that time. Adorno and his colleagues tried to analyse this problem. One of the important hypotheses was that under anti-semitism, there lay characteristics deep in the structure of personality.

The procedure Adorno and others followed was to identify the strong anti-semitists and to assess the level of their prejudices and those who are now in such prejudices and to compare these two groups of people and to identify their personality structure and the dynamic function of the personality characteristics. To complete this work they must use a measure to assess the prejudice against the Jews. The ideology that prevailed in Europe and America about the Jews was that they are of separate class of people to be persecuted. Adorno and his colleagues developed a scale that could assess the personality structures of the individuals from psychoanalytic perspectives. The aim of developing this scale was that it must assess the level and degree of anti-semitism and also the extent of negative sentiments individuals could express. The scale was called *Anti-Semitic Scale*. The original scale consisted of 52 items, each item expressing prejudice against the Jews.

The Anti-Semitic Scale has been used by Adorno and others to measure the ideology about the Jews and to study most prejudiced and least prejudiced individuals by comparing the scores of the individuals in the upper 25 percent with that of the lower 25 percent. Later on the authors of the A-S Scale expanded their study to deal with ethnocentrism. Their conclusion was that if we remove the outer skin of the anti-semiticist we can find that he is an ethnocentrist. Therefore, they expanded their study to examine ethnocentric ideology using a topic for this and the groups included for the study are the Negroes, Okies (an impoverished migrant farm workers in Oklahome State in U.S.) women, criminals and other ethnic groups in U.S.A. It was called *Ethnocentrism Scale (E-Scale)* developed after the A-S Scale. Ethnocentrism means belief in the intrinsic superiority of the nation, culture or group to which one belongs to, often accompanied by feelings of dislike and contempt for other groups. This type of feeling was prevalent at that time and manifested in American patriotism.

Sometimes the authors used the E-Scale as a substitute to the A-S Scale to find out the prejudiced and unprejudiced individuals and against whom an intensive follow-up study was conducted by using the A-S Seale, because the scores of E-Scale and A-S Scale were highly correlated. In the development of ethnocentrism the political and economic forces have played greater roles both in individuals and in institutions against the Jews and other ethnic groups. Adorno and his colleagues have found that political and economic forces were behind ethnocentrism. Therefore, they thought of a scale to study the political and economic ideology and developed the *Politico-Economic Conservation Scale (PEC)* in the same fashion as they have developed the A-S and E-Scale. But the scores of this scale did not correlate with the A-S Scale as much as the E-Scale. This has high relationship with conservative ideology. In the light these studies Adorno and his co-workers have stated that Authoritarianism is a driving force in a person towards prejudice and acceptance of anti-democratic propaganda. The anti-democratic trends in the personality were implicit which was also prefascistic attitude.

The original scale developed by Adorno and his co-workers was the F-Scale to measure Auhoritarian Personality. This scale was having reference to certain specific groups. But they wanted a scale that would measure prejudice without making any reference or mentioning the name of any specific ethnic or minority groups. They wished that the scale should elicit honest self-disclosing in a person. The original F-Scale, therefore, was revised and items referring to ethnic groups were dropped and finally they had 34 items. These items are designed in such a way that each one describes the authoritarian personality.

Authoritarian and Democratic Personalities

The authoritarian person is basically a weak person. He relinquishes his personal initiatives and gains strength by surrendering to a powerful leader. His weakness is due to the upbringing, when he was a child. The authoritarian personality sees the world is divided into strong and weak, in terms of dominance and submission. The idea of equality is basically alien to him. His weakness will prevent him from questioning the authorities as such doubts would endanger his own security. He uncritically obeys the customs and traditions of his society. For an authoritarian, the individual is on the higher level because of the virtue of his innate superiority. A person is considered as either totally superior or totally inferior.

The people in authority should be regarded as both all powerful and wise because they are regarded as protectors by those who are weak. The authoritarian personality is predisposed towards prejudice. The predisposition originates in his failure to recognise the basic equality of man irrespective of their status in life. His basic concept about the world is that it is a dangerous place where security can be attained only by submission and conformity to the superior power.

The democratic personality, in contrast to the authoritarian believes in the basic equality of all men. He recognises that every one has a good and bad sides, superior in some aspects and inferior in some other aspects. For a democratic, personality status line is not rigid. It is fluid and subjected to change as a result of the individual's efforts. He obeys the temporary leader as he has more knowledge and experience at the time, not because he has more power. He feels that he is equal to those who are higher in the society. The democratic person feels secure and need not submit to blind authority. He is fully aware of his own worth and looks at the world as place where he is master of his own destiny. The world is a friendly place for him. There is give and take in human relationship with warm affection and love rather than struggle for power. But for the authoritarian person, love, affection and sympathy are sign of weakness. For him to speak of love, affection, sympathy and kindness is senseless and unrealistic. Intolerance is the general state of mind. In Hitler's Germany societies emphasised submission to an absolute authority. But on the contrary in democratic society there is encouragement for equalitarian tendency and self sufficiency.

The authoritarian person is against all out groups and highly prejudiced towards them. The ideals of Nazism and Fascism functioned in the same manner. Nazism was the National Socialist party in Germany founded by Hitler in 1922 after the World War I and it ruled Germany under the leadership of Hitler till 1945. Similarly, Fascism was founded by Binito Mussolini as a nationalist movement with the view of rebuilding the lost Roman Empire in 1919. Both the organizations were against the parliamentary system and anti-communistic ideology. They have strong faith in an economic system based on the State controlled capitalism. For them the State is the supreme authority and man is respected if only the State is powerful. According to the German and Italian nationalistic movement, in a democratic State the individual is selfish and self progress is encouraged and nation is neglected and sometimes the individual even betray the nation. An individual without national feeling is dangerous to the society and to the nation.

The F Scale

The F Scale was designed to assess the individual's fascistic and anti-democratic tendencies. Adorno and others attempted to validate their scale with interview results and projective test results. They concluded that the person with high F Scale score was rigid, prejudiced towards minority groups, submissive to authority and did not examine or reflect on his own life. In fact person who secured high on F Scale seemed to have an authoritarian personality and, therefore, the scale is called scale for the authoritarian personality. Hoffer (1951) has discussed that the general authoritarianism as a personality style may manifest itself in any political pressure. This has also been discussed by Rokeach (1960) extensively in his work in *The Open and Closed Mind*. In the studies of these authors there is some mixing of dogmatism with general authoritarianism. Kirscht and Dillchay (1967) in their studies have demonstrated that authoritarianism as measured by the F Scale is different from dogmatism. But ideologically dogmatism (General authoritarianism) and authoritarianism as measured by Berkeley studies are different. For example, a communist may be dogmatic, but he is not a pro-fascist.

In the authoritarian personality Adorno and his colleagues were able to identify nine variables. They called these variables as Syndrome of authoritarianism. They are (1) *Conventionalism* which means rigid adherence to conventional middle class values. (2) *Authoritarian Submission :* this refers to submissive uncritical attitude toward idealized moral authorities of the ingroup. (3) *Authoritarian Aggression*: this is a tendency to be on the look out for to condemn, reject and punish people who go against the conventional values. (4) *Anti-intraception* : this is referring to opposition to the subjective, the imaginative and the tender-minded. (5) *Superstition Stereotype*: this is the belief in the mystical powers which determine the individual's fate, thinking rigidly. (6) *Power and Toughness*: this is pre-occupation with the dominance and submission, strong-weak, leader-follower, identification with power figures, egoistic, asserting strength and toughness. (7) *Distinctiveness and Cynicism* : this is referring to hostility, defaming of the human. (8) *Projectivity*: which believes in wild and dangerous things to go on in the world, projecting unconscious emotional impulses. (9) *Sex*: this is referring to extreme sexual concern.

EXPERIMENT

Problem

Assess the authoritarian personality traits and the level of anti-democratic tendencies of a person by using the F-Scale.

Method

Participant

Choose a person who has no knowledge about this problem. Collect all his personal informations such as sex, age, educational qualifications, religion, community, socio-economic status, place of his birth, contact with other religious and community people.

Materials

The F-Scale test, paper, pen. The F-Scale consisted of 38 statements, each one expressing one idea.

The participant is required to indicate his agreement or disagreement of the statement through any one of the responses. There are seven response categories, each having certain weightage. The weightages are assigned from +3 to –3. For the response agree very much, +3 mark is given, agree pretty much +2, agree a little +l, neither agree nor disagree 0, disagree a little –l, disagree pretty much –2, disagree very much –3. Plus score indicates presence of prejudice and minus score absence of prejudice. All the statements fall under nine areas. They are (1) Conventionalism, (2) Authoritarian Submission, (3) Authoritarian Aggression, (4) Anti-intraception, (5) Superstition and Stereotype, (6) Power and Toughness, (7) Distinctiveness and Cynicism, (8) Projectivity, and (9) Sex.

Procedure

The participant should be made to sit comfortably and the following instructions should be given.

> "In the test sheet there are 34 statements. You may agree with certain statements and disagree with certain statements. Your agreement and disagreement may vary in degrees. Whatever it is, read each statement carefully and express your first reaction to the idea expressed in each statement. Write your answer in the space provided for each statement in the answer sheet. A number is given to every degree of response. Answer all the statements using the following code by writing the number assigned to the respective response. There is no right or wrong answer. You should not omit any of them. As soon as you complete all the statements please handover the test sheet to me."

If you agree very much with the statement	put +3
If you agree pretty much with it	put +2
If you agree a little with it	put +1
If you neither agree nor disagree with it	put 0
If you disagree a little with the statement	put – 1
If you disagree pretty much with it	put – 2
If you disagree very much with it	put – 3

When the subject has completed the questionnaire the experimenter should collect it from him/her and allow him/her to go away.

Method of Scoring

The scoring of the answers should be made according to the procedure as given by the authors of the scale. In order to find out the individual score the numbers that subject has given to the responses are to be transformed into positive numbers ranging from 1 to 7. Thus individual's response for each item is converted into numerical value and then summed up to get the total score. A high scale score indicates the presence of prejudice, which reveals the individual's high fascistic and anti-democratic attitude, while a low score indicates the absence of prejudice which reveals the non-fascistic and democratic attitude. For authoritarian items the conversion of score is made as follows:

+3	+2	+1	0	–1	–2	–3
7	6	5	4	2	2	1

For non-authoritarian items the conversion of scores is made as follows:

+3	+2	+1	0	−1	−2	−3
1	2	3	4	5	6	7

The individual's score is summed on the basis of this conversion.

Results

Find out the individual's score on the F-Scale using the above mentioned conversion formula. Then find out the score for each area and put it in Table 18.8.

Collect the group data for each area and work out the mean score for each area and also for the whole scale and put them in Table 18.9.

Table 18.8 : Showing the individual scores obtained by the F-Scale in nine areas

Name	Con.	A.S.	AA	Anti I.	Sup. St	P&T	D&C	Proj.	Sex	Total

Table 18.9 : Showing the group data as obtained by the F-Scale in nine areas

S. No.	Name	Con.	A.S.	AA	Anti I.	Sup. St	P&T	D&C	Proj.	Sex	Total
1.	R.V.										
2.	P.K.										
3.	S.V.										
4.	M.S.										
5.	K.R.										
6.	P.V.										
7.	M.R.										
8.	T.V.										
9.	S.M.										
10.	T.C.										
11.	R.M.										
12.	K.M.										
Total											
Mean											

State your participant's score in each area and state whether he is having prejudice or not. How many of them in the group are highly prejudiced or anti-democratic and how many of them unprejudiced and democratic.

Discussion

Discuss the results with reference to the characteristics of authoritarian personality and anti-democratic tendencies. State the level of your subject's prejudice in the whole scale and in respective area. Discuss whether the group as whole is authoritarian or democratic. Using other independent variables discuss the level of authoritarian and anti-democratic tendencies of the group. Also state the specific areas in which the subjects scored highest and lowest. What these indicate. State also the reasons if any from the personal information of the individuals. How many of them remained neutral. Compare the results of your experiment with that of the earlier studies and discuss.

Conclusions

State whether your participant is authoritarian or democratic type of person on the whole. What is his/her position in all the nine areas of the traits? How many of them in the group are authoritarians and how many democratics? State the variables that determine the type of personality. Do you find any difference between the two sexes? State whether males are more authoritarians or females. State, whether age has any influence in determining the types. State, whether any other factor having impact upon the participants in determining the types of personality.

19

DEVELOPMENTAL PSYCHOLOGY

WHAT DEVELOPMENTAL PSYCHOLOGY IS?

Developmental Psychology is one of the fields of Psychology mainly dealing with the developmental processes and changes of behaviour at different levels of development in humans. The changes in behaviour occur in relation to growth development, maturation and learning of the organism. It is also interested in the factors that shape the human behaviour from the time of birth, through childhood, adolescence, adulthood and old age. Developmental Psychology studies the behaviour of the individual, in relation to his physical development, motor development, mental development, speech development, emotional development, social development, his relation to religious activities, aesthetic sense, his interest in developing various skills and talents, his interest in play and the development of personality.

Though human behaviour is accessible only after birth, when man is born into this world, some development psychologists started the study of behaviour even before birth right from the time of conception. Physiologists say that information is available for study from the foetus eight weeks after the conception. The physical changes, movements and the growth of the foetus (unborn human infant) the downward movement of the head of the foetus all can be observed. Whether the foetus is going to be an underweight or normal weight can be assessed. The foetus comes into the world after 270 days or 300 days. At this stage we call the new born as neonate or infant. Now the behaviour of the new born infant is assessable for observation and assessment. Developmental psychologists, particularly the child psychologists begin to study every aspect of this human infant from the stage of neonate through various developmental stages.

The process of development of the human organism begins with the cause of pregnancy and eventuates in birth of a human infant (neonate) and comes to an end at death. Therefore, the period at which developmental psychology is generally very much interested is from the time of birth and ends at the old age and at death. Developmental Psychology basically tries to describe and explain systematically the changes that occur in a human organism during the course of the life cycle of the organism. Developmental Psychology primarily deals with the physical, motor, cognitive, emotional, social and personality changes, separately. But these are not independent functions. All these developmental processes interact with one another in a coordinated manner. The normal form of developmental processes in all these areas lead the individual to maturation. Though the normal physical development may take place in an organism, one of the factors, responsible for maturation is the development of brain and nervous system. Maturation is a process innately occurring exercising influence over the physical and psychological development. All the developmental processes cited above are continually occurring in coordination with each other in an integrated manner making the person physically, cognitively, emotionally and socially matured.

While describing the functions of developmental psychology we come across three important concepts like growth, development and maturation. The concepts need to be explained adequately before we proceed to the subject matter in detail.

Growth, generally, refers to a seed germinating into a plant. In this case, it is the spurting of the seed or anything that is suddenly developing. Growth means an act of growing in the living organisms including humans. This growth is the process in organisms, following the assimilation of food. Growth here means increase in the size or enlarging in all respects significantly. It is the stage of development. The concept, growth is used in different contexts. The plant is growing, we say, there is growth of industry or there is growth of population and so on. In the case of humans, growth specifically refers to physical growth in height, and weight and size. Lefrancois (1979) states that growth refers to quantitative physical changes, such as increase in weight and height. In the case of humans, we say physical growth, and we also say physical maturity and physical development, over growth or under growth. Thus development and maturity are also associated with growth.

But we may differentiate development from growth for clarity. Development generally means the process of growing or progressing, a fact of an event that is changing in a situation. We may sometimes speak what is the development of the problem, issues, or matter, or an agreement. This refers to the progress of the issue or matter. But in the case of humans, development refers to the total growth of the organism. We sometimes, speak that the person is grown taller, but there is no bodily development or proportionate physical development. What does it mean? This means that the individual has not attained the total process of growth in all respects so as to make him to adapt to the situation. Here development includes growth, physically and mentally.

The concept, maturation used in various areas, wherever growth and development are involved. Maturation mainly involves human's growth and development, specifically an advanced stage physically, mentally, emotionally, and socially. Maturation also involves personality development. We often hear, people speaking like these, as, "he is not physically matured' (developed), if a person is physically undergrowth, or 'he is not mentally matured', when he fails to deal with a situation successfully or if one is mentally retarded, sometimes, we also speak that, 'he is not emotionally matured', if he is ill tempered. A person is socially immatured if he fails to conduct himself properly in a social situation. Therefore, maturation involves all these aspects. Maturation may be defined as the competence the individual has developed physically, mentally, emotionally and socially to deal with the situation successfully in a psychologically meaningful manner.

In view of this developmental process, developmental psychology may be defined as the study of behaviour through all the developmental changes right from the period of infancy, childhood, adolescence, adulthood and old age. Stevenson (1968) states that developmental psychology is alternatively used with child psychology, adolescent psychology and genetic psychology. Developmental psychology is a very vastly developed area that it has both research and applied areas. The applied area at present is mainly concentrating on emotionally disturbed mentally retarded and disadvantaged children.

PHYSICAL DEVELOPMENT

Human development is a many sided affair and it takes place on a unitary manner. There is no single aspect that can be isolated and state that it is taking place independently. Therefore, human development should be considered not only as unitary process, but also as a holistic function. But human development is studied in different stages on the basis of various levels of growth, development and maturation. At every stage the growth and development have similar characteristics everywhere in humans except the cultural variations and the type of environmental stimulations. For the purpose of studying and comprehending the subject matter clearly human development is conceived and classified into three main stages, as childhood period, adolescent period and adulthood period. But each period is also divided into two or three stages. For example the childhood period is studied under different stages like the period of neonate, infant, early childhood period and later childhood period. The adolescent period is classified as early adolescent period, later adolescent period or preteenage, teenage and later teenage. The adulthood is classified into early adulthood period, early middle age, later middle age. The old age is classified as early old age and later old age. At every stage remarkable changes are taking place. Therefore, developmental psychologists are interested right from the period of neonate to the later old age period. Since their interest is studying the human behaviour throughout the span of life their particular interest in developmental psychology has been to find out answer to the question, where does the human organism start to behave.

The developmental psychologists in order to find out answer for the question; they carried out rapid studies and carried out many investigations by regressing back into the prenatal development, birth and neonate stages. Some studies were started right from the time of fertilization of the ovum. Evans (1971), Tanner (1970), Knobloch and Pasamanick (1958), Landreth (1967) and others have studied the prenatal development. It is observed that the rate of growth is rapid during the early foetal period. There are changes in bodily proportion in this period. The development of hair, nail and external sex organs takes place later on. Newberry (1946) states that as the foetus approaches the birth period the kicking decreased and the head movements strikingly increased. Watson and Lowrey (1954) also observed those functions. Landreth (1967) states that during the fifth month reflex (automatic) responses develop and they become important for survival.

Birth Neonate

The human foetus is fully developed after 270 days of conception, and is ready to come into the outer world. As and when the day of the birth of the child is approaching, the head of the foetus moves downward. This makes the birth to bring out the head first at the time of birth. Then there is a relief of pressure in the mother's abdomen and it makes her to breathe more freely as the head has moved downward. The pressure for the mother is lightened. This lightening is the indication that the foetus is going to be expelled from the mother's womb within few hours to few days, depending upon the age of the mother and the number of children she had already delivered. In the study made by Prechtl (1963) it was observed that the labour involved in delivering the child also depends upon certain types of shortages in the physical condition of the foetus. The birth of the child is an important event, it is a dramatic one and is having psychological importance. But however, it is a biological event.

In some cases, birth becomes difficult if there is any complication in the placenta. Then it may result in some damage to the child or even death. There may be also respiratory or circulatory sudden changes. The impairments that are taking place in the child are not noticed at this stage. As the neonate grows and develops the impairments would be traced. It may be sensory deprivation or some physiological in nature. But generally this is very rare.

The neonate arrives into the outer world any day after 270 days of conception. The period of development in the mother's stomach may be 9 to 10 months. But there might have been children born before this period either by natural means or by operation. Some neonates may take over more than 300 days, and some neonates may cause trouble to the mothers for few weeks before birth. Such children may have certain eccentric characteristics. Whatever the number of days the foetus takes to come out from the mother's womb, the neonate or the new born of human infant is a creature generally born with twenty inches long and five to nine pounds in weight.

The new born human child is called an infant or neonate. The concept, infant is a Latin word which means 'very young' or 'the youngest' or literally 'speechless'. The new born child is not speaking, but it makes sounds or noises, it cries. The neonate's skin is very thin and also dry. The infant cries and the skin has wrinkling appearance. It is true that all new born human infants have flat nose, protruding forehead, with a small chin and black or blue eyes. But when the neonate grows and develops, there will be changes in all these. The changes take place rapidly. Though the child is helpless and not having good appearance, he is capable of growing and developing with all kinds of skills into a successful young person. But the child, who is born prematurely from his mother's stomach, has certain deficiencies much more than a child who is born after full time development in mother's uterus. The impairments of sensory and motor skill make the child with more debilities. Mortality is a common occurrence of the prematurely born infants. Caputo and Mandell (1970) state that the mortality rate of prematurely born infants has been considerably reduced in the United States during 1933-1955 and more premature infants survived. Whether the newborn infant was sufficiently matured in the mother's uterus or not, it was judged in most of the studies by finding out the weight of the infant at the time of birth.

Birth Weight

The birth weight of the infant has been used as one of the criteria in judging his future developments. Many longitudinal studies made in this area, related birth-weight and intelligence in the later life. It was observed that the prematurely born infant weighted less than 3.3 pounds at birth and these infants in the later life did less than normally born infants in the measures of intelligence. There was also difference in the degree of impairment between infants prematurely born with heavy weight and low weight. The heavy weight premature infants showed less impairment than the low weight premature infants. Babson, Kangas, Young, and Bramhall (1964), Churchill (1965) Willerman and Churchill (1967) have used several variables in their studies and compared with the development of intelligence. In these three studies it was observed that infants born with high birth weight showed higher IQs in their middle childhood period than the infants born with low weights. The investigators have controlled all other variables in their studies. Wiener (1968), Caputo and Mandell (1970) studied the black children who were drop-outs in high schools. These children had the lowest birth weight and they were also slow learners.

Studies were made comparing low birth weight with other physical developments. In the study made by Drillien (1961) it was observed that more than fifty percent of the low birth weight children in his sample had a minimum of one physical or neurological defect. There were also 37 percent of the cases with visual defects and 12 percent with hearing defects. These defects occur both at the time of birth and in the later life. There were several studies associating physical defects with low birth weights. The studies made by De Hirsh, Jansky and Langford (1966), Wiener, Rider, Oppel, Fischer and Harper (1965) and Robinson and Robinson (1965) confirmed these results that neurological damage bas been found with low birth weight infants.

The infants, low birth weight has been associated with adjustment and social behaviour in the later life. Caputo and Mandell stuuies have related these variables and established that the low weight infants had problems in adjustment and in their social behaviour. The difficulties found in the low weight infants, though were occasionally identified earlier, they were also found in their later life. But the advancement of science and medicine has brought out dramatic changes in the life of low weight infants. These studies were not continued till their adolescent period to see whether the defects observed in the early years, were rectified or not and the low weights have made improvement and can go with the normal born or not.

Reflex Action

During the first few months, the new born child is making only some reflex movements. The world is a new environment for the new born and it has to adapt to the new situation in all his activities like breathing, blinking and sneezing. He makes also some feeding reflexes like, sucking and swallowing. The searching reflex is taking place in response to some stimulations. Followed by searching reflexes there are also grasping reflexes. The infant firmly holds any thing that is available within his grip. The ***Babinski reflex*** is moving out the toes to the gentle touch of the sole of the foot. The infant also makes swimming reflex by laying on his stomach and moving his arms and legs in a swimming manner. The infant makes a type of reflex in response to loud noises, or when he feels loss of support, which is called ***Moro reflex***. These reflexes that are appearing in the new born infant slowly disappearing as he advances in his age and at the end of one year all the reflex movements are disappearing.

The new born infant's activities are mostly unidentified mass activities. These activities are not voluntarily coordinated motor activities. Though they are mass unidentified activities, the reflex activities are individual and coordinated specific responses. All the reflex responses are controlled by the spinal cord. Many of the reflexes are vital and important for survival.

Stages of Growth

The human infant grows very rapidly during the period of early childhood, the first three years. This is very much surprising for the parents and biologists. But considering the general bodily growth of humans, the ***first twenty-one years of life*** is the important period of growth. In some cases growth may take place even upto the ***age of twenty-four***. Generally, the physical or bodily growth stops at twenty-one and in some case at eighteen and nineteen. This varies due to environmental and nutritional factors. Liebert, Poulos and Strauss divide the physical growth period into three stages. The infancy and early childhood period are said to be the first stage, which is extended from birth to five years. The second stage is from the middle childhood to late childhood period. This stage is from the 5th year of life to 12th year. The third stage is the period of adolescent. This is from the age of thirteen to twenty years of life. According to these authors, there is a rapid development during infancy to early childhood period and during adolescent period. These stages of classification are not rigid. They are based upon the bodily development arbitrarily.

There are also other classification of stages, the growth is not taking place stage by stage. Growth is a continuous process right from the time of birth up to twenty-one years. There is no standard measurement of growth at every stage. In some children the growth and development is slow. In some cultures the standard height of persons may be six feet and above and in some cultures it is between five feet and five and a half. The chromosome variations also are there. Therefore, it is not possible to fix the growth of body or height and make it a standard.

The growth of body is associated with other developmental maturational processes. In developmental psychology a single aspect is not studied independent of other aspects. Jean Piaget, classified the developmental stages in various ways. Though Piaget made his studies very extensively, his main object was to study the mental development. Therefore, most of the Piaget's (1926, 1928, 1930, 1932, 1951, 1952, 1969) studies were on cognitive development, and mental development. Therefore, he classified the intellectual development period into five stages as sensorimotor, which takes place from birth to two years, the perceptual stage, which is from two years to four years, the intuitive stage which ranges from four years to eleven years and the formal operational stage which is from eleven years to fifteen years.

There are also other authors who classified the child's developmental stages from different points of view. In the works of Hurlock (1968) we could see that the developmental stages are six, right from prenatal stage to later adolescent stage. According to Hurlock the prenatal developmental period is about nine months or 270 days, the infancy period is from birth to 14 days, the babyhood period is from two weeks to two years, the childhood period is from two years to twelve years. The adolescent period is from thirteen years to twenty-one years. According to Hurlock the adolescent period for girls begins at 13 years and for boys it is fourteen years. But Hurlock had classified the adolescent period into early adolescent period and later adolescent period though there are many differences between the early adolescent and later adolescent of boys and girls. Stott (1974) classified the developmental period of the adolescent into three stages. According to him the early adolescent period is from ten years to thirteen years, the middle adolescent period is from fourteen years to sixteen years and the later adolescent period is from 17 years to twenty years. This classification seems to be more practical. Because different developmental changes are taking place during these three stages. The first stage is the period of puberty which is called pubescence or pubescent. Since many changes are taking place collectively during this period it is called pubescence. The changes are mainly appearance of pubic hairs, change of voice in boys, several bodily changes in girls. There is sudden spurt in the physical growth. Other physical changes are the endocrine glands, increase in height and weight. But in some boys and girls these developments may take place in the middle stage due to certain impairments in development.

The other stages of development are adulthood, middle age and old age. Lefrancois (1979) states humourously that the adulthood period is from 21 to 25 years, the middle age is from 25 to 30 years and the old age is from 30 years and above. But this classification appears to be superficial. A man does not become old at the age of thirty, nor the middle age begins at 25 years and ends at 30 years. The developmental stages have been viewed variously by Erikson (1956) from psycho-social point of view, and Freud (1949) viewed it from psychosexual point of view.

The bodily growth is taking place in humans systematically along with increase of age. The various parts of the body and the organs, tissues are developing as the individual grows. The body size includes height, weight, skeleton, muscles, internal organs and the nervous systems, in head and body, all develop fully at the age of six or seven. In some cases, the development of these parts may take one or two years more. But the reproductive parts develop only when the individual attains the adolescent stage. It is not possible to fix the rate of physical growth at each age level as there are large individual differences in the rate of growth. The bodily development also depends upon the development of thyroid glands. The hypothyroidism or thyroid deficiency affects the growth from the age of eight to twelve which causes symtoms of cretinism. The cretin fails to develop physically and mentally. The person becomes dwarf because of inadequate bone growth. He becomes feeble minded because of inadequate development of nervous system. The pituitary gland is having more influence in the bodily development of this individual. The over secretion of the hormone of this gland causes rapid growth of the bones and muscles and the person becomes giant with seven or eight feet height. The under secretion of the hormone greatly affects the growth and the person becomes dwarf or midget. The person may have normal growth of the body if only these endocrine glands secrete the hormones normally.

The physical development generally refers to bodily growth and maturation in developmental psychology. It also includes emotional and social development, and cognitive development. As the child grows and develops physically and mentally his behaviour becomes more and more mature, comprehensive and meaningful. The physical growth is taking place in a dimensional manner. If we analyse the overall physical growth, each aspect's assessment may differ quantitatively. The growth or change is to be stated as the addition of some more quantity to the overall assessment. For example, the growth of change in height and weight, the increase is added at every stage in unit of measurement. Krogman (1956) states that the total weight and the total height are increasing with age or time. This is a dimensional change that takes place in the human organism. This dimensional growth process is a continuous one and it ceases to increase at certain times.

According to Krogman (1956) height is a vertical dimension of growth of human organism and weight is a horizontal dimension of the growth. In fact it refers to segment of a total bodily assessment. Therefore, with regard to bodily size, the important dimensions commonly used are the total height and the total weight of the human organism. Since these two are in relation to each other to any bodily assessment or body size or body type they are assessed together. In recruiting persons to defence and police services the height and weight dimensions are measured and it is expected that these two should be in certain proportion. Therefore, these two dimensions are as indexes of body size and body type for various purposes. The dimensions of height and weight are also considered as the indexes of healthy physique and healthy growth. In schools, the physical education departments classify students on the basis of a standard measurement using the height and weight of the students. They classify the students on measuring the height and weight as sub-junior, junior, senior, and super-senior and accordingly they assign the individual to a group for playing the games. They do not put the senior boys with junior boys group for playing a game as it involves more physical strength. Dubos (1968) states that the change of height and weight is attributed to better nutrition. The individual's potentiality for tallness is genetically determined. But it is also observed that tall parents always do not give birth to tall children. Therefore, Falkner (1966) is of the view that the final potentiality to tallness can be attained only from the environmental factors.

A person's body weight is used as an index for the purpose of comparison of body size and body type in the every day world. We also learn many things about human growth from the study of groups and populations and the measurement of weight of people and structure. Any longitudinal study in developmental psychology of individuals and children relating the environmental factors involves the weight of the persons. In these studies, weight is considered as the important measuring variable as it is highly influenced by physical health, nutritional and emotional factors.

While comparing the importance of height and weight, Stott (1974) states that though weight is biologically an important indicator, it is not as important as the height is for several purposes. Weight is not a constant or permanent dimension of an individual's unit of measurement. It is a changeable or reversible variable unlike height. A person's weight may increase and also decrease, but the height once increased will never decrease. It is a constant dimension of measurement. It is because the skeletal body or the bony framework of the body is not subjected to reversible in height. A person may lose his weight or reduce his weight for health reasons or for building up his structure, but he cannot lose or increase his height. But height always gives a man a towering look. It has certain advantages over weight. However, a height-weight proportion is needed for a man's body type. Watzal (1941), a pediatrician has worked out a standard measurement relating to height and weight. He called it ***a grid***. It is a method of presenting the change in a child's body, particularly the size and shape throughout the period of developmental stages. The Iowa study made by Jackson and Kelly (1945) is a classical study which provides valuable information about the growth of children. The Iowa study set up a norm for average weights and heights. The norms are based upon the study made on 1500 boys and 1500 girls. They were mostly from higher socio-economic-education groups. According to the Iowa norm the average six and a half-year-old boy should be 46.8 inches tall and 48.0 pounds in weight. This is the norm worked out for the growing children of the United States. Accordingly any child aged six and a half years old should have 46.8 inches height and 48.0 pounds weight. It is also important that every nation should have a norm like this. Since the Iowa norm is based upon the high socio-economic-education groups, it had to be verified whether the same norm would be applicable to other socio-economic-education statues groups. The same number of boys and girls should be studied from other cultures also.

The progress in growth is sometimes enhanced by certain related factors and also affected by certain factors. The enhancing factors are mainly the types of feeding the child, the number of feedings as the child grows may be decreased and the pattern of feeding also may be changed from breast-feeding to cup and spoon-feeding. Stuart and Prugh (1960) have observed that the child naturally loses his weight when there was changes in feeding. When the child grows into pre-school stage certain changes are taking place in the growth. The growth rate is altered in some children. The child puts himself in more activities of energy expenditure and, therefore, there is need for increasing the energy. The child as he grows, his world is widened and more attention is needed with regard to his health. As he gets into the nursery school, joins the play group, and comes into contact with other children. The common health problem is the communicable diseases. This health hazard occurs more often between two and a half years and six years. It is more frequent during the school age period. These developmental aspects are quantitative in characteristics and also measurable. These are mostly physical growth and bodily changes.

MATURATION

But there are certain processes that are not accessible for measurements. These can be only felt and observed and may be called qualitative changes or manifestations. Psychologically we may call them as maturational processes. Maturation goes with growth and development of the human individuals. There can be no maturation without growth and development. Humans grow when they are children. The increase is in various dimensions like height, weight and body structure. As years increase the child grows into adolescent and adulthood and also old age. During these growth periods there are progressive development in various aspects other than the physical growth. These child's ability to manipulate the environments, objects and persons increase. He gains experience more and more, learns more and more. These gains are expressed in his behaviour. Whether one's gains of experience is in agreement with his people at home his peer groups at school, play ground, and other people whom he meets depends upon the process of maturity.

Maturation is defined in various ways. A person's physical growth and development is not called maturation. Along with the physical growth and development the person develops also mentally and socially. Unlike physical growth, maturation is not coming to an end or ceases to develop. According to Stott (1974) maturation is a life long process. Maturation involves several aspects. Therefore, maturation may be stated as relatively advanced physically, mentally, emotionally and socially grown up state. It is the quality of being mature of full development. We hear, people often speaking about maturation or maturity. They say physical maturity, mental maturity, emotional maturity and social maturity. Therefore, a correct definition of maturity should involve all these aspects. A person who is said to be matured is supposed to possess all these qualities in equal proportion. Suppose if a man remains as midget he is said to be not attained suitable physical maturity. Similarly if a person fails to learn to solve some problem which is normal or average to his age, he is said to be mentally not matured. When a person acts impulsively in a situation and picks up quarrel with everybody, he is emotionally immature and a person, who is not successfully interacting with other in a social situation is a person socially not matured. Therefore, a definition of maturation or maturity is incomplete if it is not involving all these aspects. But the development of maturational qualities depends upon several factors, like the way of bringing up by the parents, their treatment, the interaction with siblings, and other visitors, the physical development, physical health, learning mental development, motor development, development of perceptual process, the school playground, peer groups all contribute to the individual.

MOTOR DEVELOPMENT

Motor development generally refers to the ability the child gain to make movement of his bodily parts by coordinating his muscles of the body for performing the required overt activities. The motor activity is the basic mode of response of the child to the external world. Most of the behaviour patterns of the new born child is only a mass action. The child makes a sweeping motion involving the arm and shoulder to make contact with an object by his hand. This we may state as the grasping response. Then the trend is changing from the mass action. The child begins to walk slowly maintaining balance. During the period of growth, development and maturation, the motor development is one of the prime functions the child makes. In the motor activities specific type of muscular coordination is very much involved. Though the muscular coordination is not very accurate and erroneous in the beginning, later on the child makes it more accurate and correct. In the motor development or development of behaviour, the development of nervous system plays an important role, because the nervous system is a coordinating system.

The child's behaviour in total is a motor function and it is continued by the child in various forms to complex motor activities. Then the child develops motor activities in many ways. All the motor activities involve the sequences of controlling muscular movements. For example in grasping, holding, walking, writing, talking, the motor skills are involved. The development of walking of the child is progressed from laying, sitting, crawling, standing, walking with assistance and then walking without any body's help or assistance and then running. This may take place in normal children from 18 months to 24 months. Like all other skills motor activities are also highly reinforceable. It is for making the skill more accurate and rapid and also evolving perfection. As the child grows older and older many motor functions are reinforced. Not only the child or the adolescent, but also an adult finds pleasure in seeking that his activities are accurate and perfect. Therefore, the motor functions are pervading in different areas. Once the child is conscious of his ability and skill in performing a motor function, he always tries to repeat the function and also carries over to other functions. He applies his skills and motor abilities in all the plays and games he wishes to play.

The child's grasping and holding activity involves cognitive awareness. Activities of these types are visually directed with awareness. At this stage the child's mental development can be also judged objectively. This can be from the type of motor activity the child displays. If the child approaches the object with an intention, it is the indication of his mental awareness. But does the child approach the object with object vision? McGraw (1943) has studied this aspect. According to her observations the young child at this stage in not having the power of accommodation and convergence of the eyes. The child is not able to focus the object in the retinal images clearly. The objects at varying distances are not focussed properly in the retinal area. Therefore, the vision of object is not clear and specific for the child though the child appears to be looking at the object. This aspect has been studied by other researchers like Fantz, Ardy and Udelf (1962), Gorman, Cogan and Gellis (1957). These are some of the recent studies to confirm McGraw's observations.

But there are contradictory views with regard to the child's object vision. According to an ophthalmologist, Walsh (1957) the fusion of visual images from both eyes is not achieved by a child until he is about six years old. It is stated that upto the age of preschool period the child sees every thing as double. But the earlier authors state that the stereoscopic vision is possible in children as early as at 2 years old. Therefore, there is no strong evidence to confirm any view on visual perceptual ability of young children. There is also clear evidence that the newborn child possesses the cognitive awareness or some level of cognitive ability.

The child's ability to walk is called ***locomotor development***. The child gains this ability to walk in upright manner. This stage has something to do with visual ability of the child. He is now free from manual exploration for his walking or standing. He is interested in exploring everything visually. This is another psychological development in the young child. Unless the child gets erect posture and stand without and hold, walking is not possible without falling down. If the child fails in his attempts to stand and walk independently he resorts to crawling. Therefore, the crawling and slow walking may take place at the same time. Parents provide all the facilities for the child to stand erect and walk. Walking is considered to be the most important development in the life of the child. Stott (1974) states that the presence of locomotor ability in the child is traced in the prenatal period. Whether the child is capable of walking soon or takes time can be judged by observing the child's muscles of the trunk and limbs. These parts of the body are involved in the exercise given to the legs for developing ***bipedal locomotion***. But some of the complex and voluntary activities, and maintaining balance for upright posture require the development of neuro-muscular functions. The nervous system and the muscular system should function in a coordinated manner.

The development of locomotor functions in the growing child are mainly of two types. First the locomotor function takes place on fours. The child uses, in this function, two legs and two hands. We call this motor function crawling. Prior to crawling the child makes the movements by his body which is called creeping. When creeping takes place, it means that the child is getting ready for making locomotion. After this, the child gets sufficient strength in his limbs and trunk, slowly raises up putting his two hands and legs on the ground and tries to raise up his head. If this becomes easy for the child he slowly moves on his chest and then raises his body pressing his hands and legs on the ground and makes movements. He may fail in this effort many times, but he continues his attempts and begins to crawl. After few weeks or months he begins to crawl as fast as he can. While crawling he also learns to sit erect and examines the objects whatever available to his hands. Then the child tries to stand up by holding something or with some assistance. The parents at this stage pay much attention and care to make the child to stand for more times and make the child slowly to put his steps forward. The steps in the beginning are very much erroneously placed on the ground. He does not put his foot with any amount of firmness. When he feels unsafe, he sits back. This type of motor activity serves as an exercise and soon he gets sufficient strength to his legs and puts his legs without any bending and makes himself erect. The erect posture of the body in standing position provides confidence to the child to put his steps forward. In the child's life this is one of the most important locomotor functions, because, the child passes from the stage of moving on fours to the stage of moving on twos. He passes from the stage of crawling to the stage of walking.

The developmental stage of locomotion on twos is more exciting stage for the child. He reinforces his movements in twos often and becomes aware of his ability to walk with help and later on tries to carry out the same without any assistance. When the child attained the stage of walking on his own, he makes it as fast as he can and also runs. In his attempt to run he may fall down many times. Sometimes he cries or weeps and sometimes he smiles. The locomotion or the power of moving fast by running leads the child to climb on anything above the ground level. First he tries to climb on furnitures and stands and claps his hands and smiles that he was successful in getting over them. The child does not stop with this his locomotor activities. Climbing over the trees, walls and ladders and swinging holding something are some of the exciting motor functions of the child. Child is very much interested to get into the car and he tries to manipulate the steering and also handles the starting key on seeing his father. Once the child has developed certain amount of locomotion skills he tries to imitate the elders in all respects. When he sees his father wearing the shoes for walking, he too tries to put his feet into the shoes and tries to walk.

When the child reaches the school age period he gets mastery over many motor skills. These motor skills are pertaining to locomotor and manipulating and examining. Manipulation and examination may also result in breaking and damaging some materials at home. He tries to open and examine everything and when he could not open it, he breaks it by hitting it on the ground or on something. The school going years for the child are physically very important period of years. These years are full of activities and he uses his body in coordinating activities. He makes use of tools and toys. Any costly material also is a toy for him to play with. Since he is full of energy, he spends it mostly in physical activities like running, jumping, swinging, climbing, swimming, riding bicycle and in various exerting activities. These exerting activities may vary from culture to culture. But generally the child's activities are motly gross bodily activities, which exhaust him physical energy.

As the child increases in his age the nature of activities are going on changing from simple to complex, from individualistic to competitive. His association with peer groups are increasing when he reaches the high school age. Further there is more development of complex activities. More challenging type of activities become common events. He takes part in various game activities which require highly developed skills and practice. He is also willing to practice any game and also takes risk.

The early adolescent period is a critical period. Since this is the period in which the physical maturity begins in both boys and girls, it is the period that determines the individual's future prospectus mainly. The individual's development of motor skills and proficiency in them, the cooperative spirit, the peers acceptance on the basis of the skills or rejection are the main changes in this period. Most of the individuals decide what he/she is going to be in the latter stage. His proficiency in motor skills decide his participation in games and events in the play ground. The development of motor skills in this stage is increasing steadily. When the individual reaches the later adolescent period, the level of the motor skills attain certain proficiency which qualifies him for becoming a member of a team to play a game like football, cricket, hockey, basket ball, throwball or baseball and so on. Individuals who fail to reach a level of proficiency are not getting chances to get into any team which means they lack in the development of motor skills. Therefore, this stage is a more significant one in the life of an adolescent.

The development of motor skills is also the development of physical strength. It is a period of great excitement for the adolescent, because he is a person of much vigour, or strength, speed, action, good neuromuscular coordination, strong determination, self conscious and with all adolescent characteristics. The motor skills may make him a victorious or conquering hero and may bring him fame and glory. Playing games not only helps one to develop his motor skills, but also shapes his body structure. Development of motor skills and participation in games and athletic events are mostly individual matters and the peer group or play group has nothing to do with them. The importance of motor activities can be realized that even crippled children by constant training and practice of some specific activities and movements get relieved of their handicaps. Therefore, the motor activities right from infant stage to the end of adolescent stage make one with good physique and able bodied person in this competitive world.

COGNITIVE DEVELOPMENT

The concept, cognitive development came into usage when a group of psychologists in the United States tried to study the knowledge getting process in disagreement with motor development and the S-R theory and Watson's Behaviourism. Watson's Behaviourism did not speak much about thought, language and intellectual functioning. Watson did not accept these phenomena fit for research as they are mainly internal functions. The cognitive psychologists mainly depend upon human mental processes, the internal processes which come in between the child's revelation to the external world and his reaction to the events and objects. The cognitive psychologists are of the view that the internal mental processes alone can work to explain the behaviour of man. The classical cognitive psychologists stress the fact that people have intentions, purposes, ideas or cognitive expectations, etc., and because of these they make decisions which may be revealed in the actions of various types. Therefore, cognitive psychology has played a strong role in the field of American Psychology for more than four decades. The champions of this field of psychology are Bruner (1964), Lawrance and Festinger (1962), Tolman (1896-1959) (1958), Hilgard and Bower (1975), Piaget and Inhelder. Bower and Hilgard have brought out all theories under two categories as Behavioural-associated theories and cognitive-organizational theories.

Cognitive theorists state that man is not a passive receptor of stimuli. The mind (Though it is a universal concept, cognitive psychologists see it), is an active process and it processes the information received, and transforms it into a new form and category. Cognition, according to the cognitive psychologists is a mental process. The other meaning of cognition is knowledge. Therefore, cognition is a knowledge getting mental process. When a stimulus is presented, it is seen by our eyes. It is a sensory input to the mental process. The mental process transforms the sensory input in various ways. It is coded and is stored in the memory and retrieve it for the use whenever it is needed. According to Neisser (1967) perception, learning, remembering, imagery, problem solving, thinking and the use of language are all supposed to be the stages of cognition.

People may have images, and think, plan and make decisions from the stored information or from the remembered information of the earlier learning. People also make decisions and plan selectively choosing the stimulus from the environment. Out of many, only few, which would serve the specific purpose draw the attention of the individual. Therefore, cognitive psychology may be defined as the science which attempts to investigate the internal mental processes objectively. Cognitive psychology is of the view that the mental process or cognition may be compared to an electronic computer. The information which needed or presented to the mental process in terms of sensory process, are processed in various ways. The information are selected, and compared with other information already stored in memory and then they are transformed and rearranged in whatever manner required for the specific function. The nature of the response or output of the information depends upon the nature of the internal processes after the input.

Cognitive psychologists state that cognitive development is the same as the intellectual development. According to them there are various stages of intellectual development. Jean Piaget (1896-1980), the Swiss psychologist classified the cognitive development into four stages. They are (1) Sensorimotor stage, (2) Pre-operational stage, (3) Concrete operational stage, and (4) Formal operational stage. According to Piaget (1968) while observing the sensorimotor stage one could see that there is an inter-play between motor activity and perception in the child.

Sensorimotor Period

The approximate age of this stage is from birth to 2 years. The infant develops motor skills by reacting to the environment. This reaction is not taking place internally. It is a response of seeing and just reponsing to it. For example, a bright light is shown and the infant is very much attracted by it and when it is hided, he looks around for it. Until the child mentally ready for representing the world he is responding only with the sensori-motor functions. Objects exist for the child only when he sees, touches, hears, tastes and smells them. When the objects are removed from his sensory experience they do not exist for him. During the sensorimotor stage one of the important achievements the infant acquires is the object concept. The infant has developed an identity of objects of his own even when the object is not within the infant's immediate presence. Piaget (1954) conducted an experiment on an infant to find out the development of the object concept by showing strange bright objects and then hiding them. Piaget states that there is a continuous progress from natural reflex actions to habit acquisition and a type of behaviour which reveals intelligence.

During the sensorimotor stage there is no sign of language development. But symbolic representation of the objects begins at this stage. Imitation may bring of the object in the infant. Thought is an internal process. The process of internalization takes place relating the activity to the object or event that were presented earlier to the child. The infant then begins to imitate in the absence of the object or event. Piaget called this as deferred imitation. Any imitation is an expression of the internal process, which is a symbolic representation of the objects, persons and events. This stage, Piaget, calls as the beginning of language, following this, words will come out later on and replace the images and symbolic representations and concrete objects. The more variety of stimulations the child receives the more they become the source of instrumental for his greater capacity to cope with the environment. This process of development is stated as accommodation and assimilation during the sensorimotor stage and Piaget discribes these developments in six sub-stages.

The first stage is the period from birth to first month. The new born infant makes most of the time only in reflex responses and is waking and moving his hands and legs. The reflex exercises slowly getting improved. The sucking response becomes perfect at the end of the month by recognizing the nipple from other similar objects. But there is no see and reach, and reach grasp reflex movements. The motor activities are independent of each other and there is no coordinated motor activities.

The second stage of the sensorimotor period is from the end of first month to fourth month. In the second stage the child gains some new responses. Mainly there are more of primary circular reactions. These reactions centre round the child's body. The child is responding to his own responses. In other words the child imitates his own activities. Thumb sucking becomes voluntary activity and the child reinforces it.

The third stage of the sensorimotor period is from four to eight months. During this stage the circular reaction continues and it is called secondary circular reaction. This is a transition period in which the child learns to coordinate between vision and motor activities. During this stage the child starts to grasp the objects seen and wanted to manipulate. For example the shaking of a rattling object caused some sound and the child is able to associate object and the sound and even if the object is not held in his hands the same shaking movement the child is making. By this means the child is not able to differentiate the movement and its effect. Therefore, there is no intentional behaviour in the child.

The fourth stage starts at eight month and ends at twelfth month. At this stage the child learns to coordinate isolated activities. The child in this stage can reach the object he sees and can grasp and manipulate. Since the child can crawl now he can move and grasp the object he wants. The child at this stage behaves intelligently by reaching the object, recognizing that it is wanted for him. He even removes the obstacles if any, on the way and gets it. But however he is not able to differentiate two similar objects. He can identify two different objects that are having association. For example, bottle means milk, water tub means bath. He understands anything that is put in a continuous process in the same environment.

The fifth stage goes from twelfth month to eighteenth month. The child at this stage begins to examine everything that is available within his reach in the environment. If the child comes to know that the object he wants is hided behind something, he pulls the object behind which the object he wanted is hided and searches it. Some amount of cognitive ability at this level he has developed. Piaget calls this type of activity as tertiary circular reaction. The child dicovers the object he wanted to grasp and examine it. If the child is able to find out what he wanted in his search, he repeats the same activity. For example, if the child accidentally hit a metal plate with a spoon, he hears a bell like sound and he repeats the same activity to hear the same sound. The child may repeat the same activity on another metal plate to hear the sound. Now he discovered that with the spoon he can make metal plate to sound. This accidental behaviour of the child is to explore the environmental situation. This we may call as the instrumental behaviour.

The sixth stage of the sensorimotor period in the child's life falls between eighteenth month and twenty-fourth month. The child completes two years at the end of this stage. Piaget states that this is a transitory period and the child passes on from motor activity or motoric intelligence stage to mental activity or symbolic intelligence stage. The child has been dealing with environment in early life in terms of his motor behaviour, like looking at objects, touching, grasping, manipulating, throwing, and even by biting. When he reaches his second year there is a process of internalization in the child's actions. The child tries to coordinate sensorimotor activities and integrate them by a trial and error method. There is some sign of using his intelligence in all his activities. In solving a problem, the child makes use of his cognitive abilities, like perceiving, motor activity of trial and error or exploring the situation and coordinating and integrating them. The child reveals also his mental images or internalization of the situation.

Pre-operational Period

The pre-operational period starts at the end of second year or at the beginning of third year of the child and goes up to the age of seven years. During this period the child only prepares himself for the concrete operation at the age of seven. Till then the child could not develop the cognitive ability of thinking for actual operation. The period of pre-operation is not conceptual in real sense. Because the child does not make use of concepts. In the pre-operational period the child develops cognitive ability to use symbols. At end of second year the child is able to distinguish various symbols like words and images, the actual events and actions. The child is able to associate the name and the object and in the absence of one it becomes possible to manipulate one without the other. One represents the other. The child is able to discriminate symbolically the alternatives. This ability the child acquires by learning. Piaget calls this ability as signifier or learning words.

In the pre-operational stage the child's cognitive process is explained by Piaget (1960) and his co-workers through the development called conservation. Piaget's conservation theory is a test of intelectual growth of the child. Actually the quantity of a mass of substance is not changing if the shape is made to change. Similarly the same quantity of milk, if poured in two beakers of different sizes, one to vertical in shape and the flat in shape, is not changing. For an adult person, there is no change in the quantity in both cases, in the case of change of shape and in the case of change of beaker. But for a child there is a change in both the cases. Piaget has conducted experiments with objects varying in size, shape, height, length, weight or mass, number, area, and volume. In all these he observed that the children below the age of seven had the difficulty in conservation of concept. These children still react only through their sensorimotor abilities. Thinking is very much influenced by visual perceptions or impressions. The ability to differentiate objects or conservation is necessary for all logical operation. The child below the age of seven is not developing this ability, as the cognitive maturity is not attained by the child. On attaining the age of four the child is becoming intuitive in his dealings and he is learning more egocentric in problem solving activities. The child is more dominated by visual perception and the ability to apply logical rules or classify object yet to develop. The child is slowly passing away from the sensorimotor stage through the processes of pre-operational stage.

Concrete Operational Period

The concrete operational period ranges from 7 years to eleven or twelve years. The important development during this period is the relation between activity and the environment. The child at this stage learns to manipulate objects. There is also more of physical and mental development in this period. What the child was lacking in pre-operational stage is attained in the concrete operational stage. It is the ability to conservation of problems. The child understands the logical principles which are needed for solving the problems. According to Piaget there are nine different logical classes which the child begins to manipulate which includes the limitations of the child's thinking ability (Flavell, 1963) For example, the child is lacking in his ability of reversing and identifying in the pre-operational period. The ability to reverse the operation is internalized and for every operation there is an inverse operation to cancel it. For example, the adding and subtracting of the numbers may be used to illustrate the operations, reverse and identity. For example 6+4=10. This is an operation. The inverse operation is 10–4=6; the identity operation for addition is 6+0+0+0=6; the identity operation for subtraction is 6–0–0–0=6. There is no change in the number 6, with whatever number of zeros are added to 6 and whatever number of zeroes are subtracted from 6, the number 6 is not changed. Thus any type of conservation problem can be used to test the operational rules which the child has comprehended in relation to this thiking in the concrete operational stage of cognitive development. At this stage the child can demand objects on the basis of dimensions such as height, weight, round, square and so on. The child, now understands the relationship between objects and among objects and events in the environment. But he is not in a position to formulate a hypothesis and explain it. He can relate a balloon and water, an environmental state that the balloon is floating on water. But he cannot explain why it is floating. The child can say that a coin dropped in water will not float, but it will immerse in water. But he cannot explain, why. In this way the child in the concrete operational stage can maintain the quantity attributed to objects. The child's conservation may be extended to various things like numbers, volume, length, breadth, distances, area, continuous object, discontinuous object, liquid materials, solid material, floating substance, immersing substance and so on. At this stage the child can classify objects and arrange them in serial order on the basis of the dimensions such as size, height, volume and in relational terms. For example *X* is longer than *Y,* football is larger than tennis ball and so on.

Formal Operational Period

The formal operational stage is the final form of the developmental stage of the child. The age of which is approximately from 12 years and above. It is the period of pre-adolescent stage and there is display of the ability of reasoning. This is the stage in which the evolution of the cognitive ability of thought structure takes place and it is stated as the formal operation. There is more of hypothetic dealing of subject matter and children at this stage apply formal sets of logical rules. The child isolates the elements of a problem and tries to explore systematically all possible ways of solving the problem. The classical experiment Piaget used was the pendulum problem. The formal operational thinking was tested by Piaget by using the pendulum. The subject has to discover what determines the number of oscillations of a simple pendulum. The experimenter presented to the subject with long strings suspended from a hook and several weights are attached to the lower end of the strings. The subject can vary the length of the string and change the weight attached to it. The height may be altered from which the weight is released. The pendulum is pushed with varying degrees of force. The problem is which of these factors alone or in combination makes the pendulum to swing rapidly. The length of the string is the critical variable and so the shorter the string the faster the pendulum swings. The experimenter is not giving any indication of it. The way in which the problem is solved will indicate much about the cognitive operation and ability of the person. At the concrete operational period children generally have great deal of difficulties in solving the problems. But at the formal operational period they become adolescents and they proceed in the task of solving the problems through three steps in determining the formal conclusion of the problem. Before proceeding to meet the problem they design their task of solving the problem, secondly they observe and systematically record the results, and thirdly they draw the conclusions on the basis of logical principles. Inhelder and Piaget (1958) describe the case study of a fifteen years old girl. In the case of the pendulum problem, swingings were made with three weights 20, 100 and 200 grams. These three weights were made to swing with a long string and a short string. Each weight was made to swing both with a long string and a short string by pushing with varied degrees of force. The girl was able to say that the shorter the string and the greater the force the faster the swinging of the pendulum.

While analyzing Piaget's theory, it was observed that there were sequences in the cognitive development of the children of different ages and backgrounds. This means, the increasing age was having increasing effect in the cognitive development of children. The socio-economic status and the environmental conditions had increasing effect in the cognitive development. Middle social class children get through consevation concepts at early age than lower social class children. Children from urban area master the conservation concepts much better than rural children. With regard to formal operational thinking of the cognitive ability, a very bright ten years old child can systematically analyze the problem and solve it. But an adult was not able to achieve the formal operational thinking of cognitive ability.

Some were of the view that training can improve the cognitive ability and the children who were backward can be raised in the cognitive stages. With regard to conservation of concepts varied techniques were used to shift the children from one stage to the next higher stage. There were conflicting results in these attempts. There were some studies reporting that the attempts were not successful while some other studies reported that the techniques of teaching conservation principles were successful. From the studies made by Glaser and Resnick (1972) it is understood that though progress of cognitive ability from one stage to the next stage depends upon maturation, training was found to be the most effective device for speeding up the cognitive development in children, if they are ready with normal growth of the average children of their age.

OTHER THEORIES OF COGNITIVE DEVELOPMENT

Cognitive theories had their origin in the first quarter of this century. The cognitive process was already in usage in Gestalt theory, Tolman's theory of sign learning, the information-processing theories of behaviourism. The recent cognitive theorists are, Piaget, Bruner, Gagne and others. But a full-fledged cognitive psychology has emerged only around 1950s. One of the contemporary American psychologist, Bruner (1966) has formulated and developed a theory of cognitive development. Though Bruner did not base his theory on Piaget's theory, he owes much to the works of Piaget. Because it was Piaget's works that stimulated Bruner to proceed in this work. Bruner's cognitive theory is mainly based upon the interaction of the child with the social and culture conditions. The child, according to Piaget, will respond to the object only when it is present in the immediate environment. When the object is removed or hided the child will not respond to it. This is one of the aspects of cognitive growth. Bruner states that cognitive growth is a process that depends upon the child's internalization of events or objects or persons of the culture. The child is developing some system of reflecting the world around him. He finds the world around him for manipulating it cognitively. According to Bruner, man is a self-conscious creature which makes him to go beyond the simple process of adaption. He keeps himself in logical and rational behaviour. In this process he should have been increasing his capacity to deal with others by means of words and symbols. This development of increased capacity to deal with others by means of words and symbols may be called intellectual development. According to Bruner intellectual development is an outcome of the systematic interaction between the teacher and the student. Bruner's cognitive psychology puts more emphasize on the development of mental abilities.

Because of this teaching the child had developed some theorems regarding the instruction. Later on, Bruner (1966) stated a theory of instruction, called *prescriptive*. In this theory of instruction he formulated rules for achieving knowledge or skill and provided methods for assessing the outcomes which he called as the *normative*.

The cognitive development occurs mainly due to the interaction between child's innate capacities such as sensory, motoric and reflective capacities and culturally provided information which are from his environment. Every culture provides the child with different types of amplifier at different times. Therefore, we cannot expect that there will be parallel growth of cognitive development in all the cultures. Every culture may put importance in certain specific aspect and also impairments. But there are many aspects attributed to human culture in general. For example, for certain jobs the same level of qualifications are needed from everybody irrespective of cultural variations. In the case of proficiency in games and sports the same level of motor skills are needed from every culture.

Language is a cognitive skill which is important for communication purposes and also for the individual learner, to comprehend the instructions. One important test of cognitive theories is to explain the development of language to which Bruner is in agreement with other theorists. It is the language that is used as a medium of teaching. There are innumerable number of languages in the world. But the ability to learn any language is inherent in every individual. The cognitive development to learn two languages may vary person to person. The inherent capacity that is developed to deal with several matters at the same time and to attend several things in succession during the same period of time are the expressions of intellectual development in the individual. The multiple demands are met in appropriate manner by the individual whose level of intellectual development is rated as higher than the other. The world for Bruner is extraordinarily complex and in dealing with it there is great distance between the young children and the ten years old children. According to Bruner the world is represented in three different manners such as enactive, iconic and symbolic.

The enactive mode of representation of the world involves action. Enact means to make into act or represent. Therefore, the enactive way of representing the world involves action and the world is represented through action for the child. The child during the sensorimotor period of cognitive development does this. Bruner (1965) states that the child is converting the visual image into three turns, which he explains through Piaget's work.

The iconic mode of representation involves the use of images. The child sees and stores the visible characteristics of the objects as aspects for dealing with them. In this, the cognitive process of perception is very much involved. Icon means an image or picture, a symbolic representation of an object. Iconic also means having the characteristics of the image. Bruner uses this concept to explain the child's cognitive development of symbol. Therefore, for the child, the imagery becomes an important cognitive instrument. As the child grows older and older there is much change of improvement in the imagery, more towards perfection. For example, an incomplete letter or figure drawn with dots is not correctly understood by a young child. But when he grows older he is able to make out the dotted lines as letters or as figures. The ability to recognize the forms and objects increase rapidly as the age of the child is increased.

Symbols stand for objects and things according to Bruner. Symbols can be manipulated much better than actions and images. In the manipulation of symbols, alternatives are eliminated quickly. Conclusions can be drawn by induction and deduction methods. This shows one's level of cognitive growth.

TOLMAN'S THEORY OF EXPECTANCIES AND COGNITIVE MAP

The original theory of Tolman (1886-1959) was called the 'Purposive Behaviourism'. His important first work, purposive behaviourism in animals and men, was brought in 1932. This was later on called a sign Gestalt theory or an expectancy theory. In the whole work the later terms put emphasize on cognitive processes. Though originally Tolman was a behaviourist, he has given completely a different shape to his theory from the theory of Watson, Pavlov, Guthrie and others. He has also distinguished himself from the stimulus-response theories of Thorndike, Hull and Skinner. He used the concept sign learning. Tolman was very much concerned as to how to deal with knowledge, thinking, planning, decision, purpose and intention. He was only describing the animal behaviour in the same way a person described another man's behaviour, the motives, cognitions (Knowledge) expectations and purposes. He was a critic of the S-R reinforcement theory of his time. He firmly believed that human behaviour should be studied in terms of actions, and not a simple movements. For him, behaviour is a goal directed action and it is a purposive action. Therefore, Tolman strongly advocated that behaviour is a goal directed action with a definite purpose.

A man keeps himself in many activities in his day's work. He writes a letter and posts it in response to a letter received this morning. He attends to a phone call and speaks, in return, he dials a number and speaks to a person. He drives his car to market and buys some stationeries. He goes to a restaurant, orders for a cold drink and drinks it. In all these actions a goal is immanent, a purpose is there. The goal may be going toward something or going away from something, or getting rid of some danger. The organism's behaviour of the time is directed by what the organism wished to be its result. The immediate result expected from the behaviour of the air passenger is to catch the flight for which he was waiting, to win the match for the player, to interview the leader for the Television reporter, to express the emotion of extreme joy for the Television actor, to get out or a Pazzle box for the cat. Thus in all these behaviour is directed to achieve some goal or to fulfil some purpose. It is taking place in the form of some action or movement and the same is stated as the molar or large scale movements and not molecular, or small scale movements. According to Tolman, the descriptions of behaviour involved in molar movements may be stated as the assumed purposes. The comparative psychologists take this in the sense, that human traits are attributed to animals and the purposes are inferred from simple principles. Therefore, Tolman's purposive behaviourism was criticised by many behaviourists.

The fundamental principle Tolman postulated was that individuals acquire knowledge of their environment where their important goals are discovered. The part and parcel of knowledge acquired is from the relation between the stimulus and the corresponding response followed to it and another stimulus followed by a response. This leads to the rise of three-term unit instrumental learning, which Tolman called, 'expectancy'. The three-term expectancies are S_1-R_1-S_2. This indicates the individual's or organism's learning that in the stimulus situation S_1 there is R_1 response. This is followed by the stimulus situation S_2. Tolman explains, the three-term instrumental learning with the door bell. This can be explained in various ways. When you go to see a person, the door is closed and you see a door bell. Seeing the door bell is a stimulus (S_1). Then you press it. The bell as a stimulus demands an action, that is pressing the bell (R_1). The bell is supposed to ring. When it rings some body will open the door. Therefore, the ringing of the bell is stimulus (S_2), for which there is a response, which is immanent which may be termed as (R_2). The stimulus (S_2), may arouse a response and also may not arouse any response (the opening of the door). But, the door bell as the stimulus (S_1) makes the person to press it. Therefore, pressing is a response (R_1). For further action or behaviour on the part of the organism, the pressing (R_1), should raise a sound (S_2) and a response (R_2). This is how Tolman explains the three-term unit of instrumental learning or expectancy. The stumulus S_2 creates an expectation on the part of the organism. The organism gains a bit of knowledge without performing anything about it. Knowledge is acquired by the organism simply by exposing and attending to the environmental occurrences.

Tolman states that the three-term unit of instrumental learning (S_1 - R_1 - S_2) or expectancy can be strengthened when the S_1, R_1 and S_2 events occur every time. But it will be weakened if the S_1- R_1 are not followed by S_2 in sequence. If the S_2 is absent then there is loss of the specific expectancy. This Tolman calls as ***extinction***. Tolman's expectancy is the device applied to describe the results of learning, stimulus-generalization, transfer, response and competition. One important characteristic of Tolman's theory is that the successive expectancies may be synthesized in a bigger unit of knowledge. Inference is another thought process that Tolman has recognized. The animal is already taught for an expectancy (S_1 - R_1 - S_2) with the three-term unit of instrumental learning. At this stage another event connected to the stimulus S_2 is taught. If this is occurred the animal infers a new (response) expectancy for the old S_2. The animal infers a new reward (R_{2n}) or response for the stimulus S_2 at the goal point. This sequence (S_2- R_{2n}) if works, then other response selections will be affected.

An animal in its exploratory activity of its environmental area may gain quite a number of bits of knowledge like, S_1 - R_1 - S_2 association. Thus the animal according to Tolman gains knowledge about the whole environment, and the bits of knowledge become structured into a kind of *cognitive map* of the particular area and it is not simple unassociated or independent bits of knowledge of the particular place or area or with stimulus-response units. A wise or knowledgeable animal, according to Tolman, can make use of the knowledge of the objects and of the spatial arrangements, and the routes in discovering the path to the goal which was sought. The animal's behaviour is quite similar to a person who makes use of a map to find a place in a big city. All other types of behaviour of the animal is to use the shortest way to the desired goal. Tolman states that one good way for the animal to learn the spatial relations in the maze is non-rewarding environment condition.

In the maze, there is a dead end for every unit and another end is the door or the way. The animal takes the last trial to go through the door to the desired goal. The way recently or lastly discovered is stregthening the cognitive development of the animal, which guides the animal that the door is the way from one segment to another. But the animal, under non-rewarding situations does not show all that what it has learned. Now the substitution is food which is highly needed goal object at the end of the maze. It is this situation that led the animal to make use of its ***cognitive map***, in taking the correct turns which are leading the animal from one segment to another. This cognitive map also makes the animal to reduce the errors abruptly. Tolman has used the fascinating term *cognitive map* to describe the knowledge, the animal has acquired in the spatial area of the maze.

The cognitive psychologists study the cognitive development and cognitive developmental stage through two approaches. One is the psychometric approach and the other is the developmental approach. The psychometric approach puts emphasis on the development and measurement of intelligence and the individual differences. Those who studied cognitive development through psychometric approach tried to classify people on the basis of the performance of the children in the tests conducted by them. The tests are called intelligence tests. They classified persons into three categories as normal, mentally retarded or feeble minded, and genius. According to them most of the people in culture are normal because whose scores on the intelligence test are found to be on the average level. The feeble minded are those who score much below the normal level in the intelligence test and the genius are those who score much above the average level. In this way intelligence also is a cognitive variable.

The other approach, the developmental approach, puts emphasis on the development of perception, thinking, problem solving, use of lauguage and other related sensory, motor and mental processes in the average individual. Investigators of this approach are of the view that the intelligence test scores provide certain information with regard to the level of intelligence of the individuals and groups, but these test scores are not providing any information about the whole processes of the cognitive development. The results of the test scores do not tell us why some problems or items are rightly answered and why some are wrongly answered? How and why these children arrive at two types of answers? The first approach is mainly dealing with another aspect of cognitive ability. But both the approaches are contributing to the development of intelligence and cognitive abilities.

The cognitive developmentalists put more emphasis on perception, thinking, imagination, learning, remembering, problem solving, use of language and other related processes. But intelligence cannot be explained or described without these processes. An intelligent activity involves perception, thinking, learning, and problem solving abilities. Therefore, by studying these functions we are also studying intelligence. Further, cognitive developmentalists also emphasize the process of maturation of the individual and maturation goes with age and experience in the environmental situation. The psychometric approach does not say anything about perception, thinking and maturation. According to them, maturation is a developmental acquisition of skills by normal children through physical and physiological growth and development.

COGNITIVE STAGE

The child becomes what is required for an individual and gains more amount of intellectual skills through cognitive development. But there is some difference of opinion, whether the child's cognitive skills are achieved through gradual increase of intellectual potency or thorugh the succession of stages of growth with each stage qualitatively different from the other. But whatever difference of opinions exist, the theories that advocate the various developmental stages, have brought out much reliable information about cognitive development. One way of verifying the existence of cognitive development is to ask the children of different ages, carefully designed questions pertaining to their culture. For example we are presenting a long fountain pen and a short pencil to the child and ask him to tell us the similarities and differences between them. The child may tell us that one is long and other is short. The child's answer is relating to the physical qualities of the two objects. The way in which the child has thought about the objects is concrete mode of thinking. This answer is based upon functional aspect, that is what can be done with them. But if the child says that both are writing materials, then the child uses abstraction, a high level of thinking. The child attains this stage only when he grows older to the next stage. Thus the child's mode of thinking passes on from concrete to functional and to abstract. In the process of cognitive development, the abstraction stage comes only later on.

Cognitive development is a continuous process and it is taking place in a sequential manner over time. What development has taken place in the first stage is carried to the second stage. Then the third stage is followed. The cognitive development is always moving from concrete to abstract through functional and there is reverse moving. It is not the time or age that determines the cognitive development. Time or age is used only as a background, based on which the cognitive development is rated.

Cognitive development is unavoidable since the individual passes on from one stage to another in a sequential manner. The internal development is expressed in every stage. At each stage there are some new changes in the life of the child. The cognitive stages are occurring in an unaltered sequential manner. The bringing up of the child, and the educational methods and the culture are having influence in the successive process in the child's cognitive development.

Learning

In the cognitive psychology, learning is associated with behavioural development and maturation. Hence behaviour at any developmental stage is determined by what the child has learned up to that time. The cognitive stages of development take place in an unchanged successive manner. This is because the methods of teaching the children put more pressure on sequence upon what the children are taught. Progression, for the environmentalists, in learning is moving from concrete to abstract thought through functional processes.

The child, as a result of learning, is able to differentiate the simple dimensions like shape and size and later on learns to discriminate more abstract dimensions like the functions and classes and divisions. Thus the environmentalists state that children as they learn more and more move from one stage to another when they were given suitable training. The maturationalists state that children learn more and more as they develop and mature from stage to stage in succession. The environmentalists state that children learn more as they pass from simple concrete stage to complex abstract stage through functional process as a result of teaching.

But the cognitive theorists do not accept both these extreme views. The cognitive theorists prefer a stand in between these two extremes. According to them, the stages of cognitive development emerge in an unaltered order of the interactional process between the individual and the environment. Learning, according to the cognitive theorists, is a *process of information*, the child makes in a certain way. The mind of the infant or new born child is furnished with primitive or crude or simple basic construction. As the child grows, develops and matures he finds more contacts with environmental conditions. In this process the primitive or simple, basic construction is restructured and reconstructed and new stages are emerging. In the process of contacting the environmental conditions, children pass through various stages of development. The stages of development may have varied or different rates depending upon the pressure of the environmental conditions. Hence the stages of development might have attained different limits.

In the developmental stages there may be individual variation in the rate of intellectual growth and variation in the final level of attainment. The intellectual level may vary at each developmental stage such as dull, average intellectual level, and bright. The actual number of intellectual stages vary with particular cognitive process. Every child may show different levels of developmental sequences. Any child may show a genuine combination of continuous developmental process at any one stage if appropriate training is provided in a cultural setting and social environment. For example a child at the age of two might have learned to speak well and also handle objects, but may struggle to walk better. Another child might have learned to walk and run well, but has not learned well to speak, utters only few words. A third child has learned to walk and run well and also manipulates objects successfully and also be able to speak well with many words. These stages of children may be stated as below average, average and bright. These developmental processes depend upon the interaction of the individual with the environmental conditions and the pressure of the culture upon the individuals.

We have stated earlier that the child's mind is primitive, and simple and the child on receiving the sensory impressions makes information processing in certain way. The basic components of information processing are sensory receptors, effectors units, memory store and central processor. The human brain is a kind of computer. Just like the computer, the mind has all the characteristics of a computer. There are separate levels of organizational processes like feature analyzers, pattern of recognizers, sensory memories and conceptual memories. It may be also treated at some level as a symbol manipulating system. Thus mental symbol system may represent the environment internally and construct patterns of symbols. The patterns of symbols may be of any complex type to represent the places and events of the world. These scenes or places and events of the world in the forms of symbols and symbolic structures are stored in the memory. The places and events are stored in the memory with some indicators, with which the symbols and symbolic structures can be brought back. In other words, the human can bring back the stored materials to the conscious level and can remember them. The indicators or the designators are just like the key, can help to bring back the structure and can handle and transform the symbols and symbolic structures and engage them in the activities of the organism, and the process is called thinking.

The sensory receptors, mentioned above, as one of the components of information processing, receive inputs from the environment. The computer scientists use different concept in information processing. What their machines do are stated as stimuli, data, instructions, and the whole class of names is called information. When all these are fed in the computer, it is called input or reads in. After making the computer to move fast or slow some specific end result is displayed either in the form of figures or pictures or in the form of printing in a ray tube. This is called outputs or reads out.

What is taking place between the input and the output? The computer is said to be performing a series of manipulations as instructed or ordered on the input information or data. The computer cannot do anything on its own. It is to be instructed to do the desired processing. It is because of this that it is necessary to specify the sequence of operations that the computer should perform to solve a data processing problem (Ray and Dastidar, 1993). In the manipulation process in the computer, the input mass data are undergoing into several processes, like altering, transforming, calculating, and searching for something stored earlier and comparing with them evaluating and making decisions on them. In the computer program every manipulation is required a series of short instruction on which the computer functions. This is called ***subroutine***, which means regular or customary procedure. Any doubt arises, or any information requires may be referred to the routine in the information processing machine.

When you speak of the functions of computer, it is in terms of program. Programing is writing a long sequence of instructions carried by some input information or information already stored in the memory of the machine. The total sequence of the instructions is generally called the program. Any such programs should be written in a standardized language, precisely on a format. It should be in the language of the computer. There are certain computer languages used in the program to carry out numerical calculations.

The environmental stimulus events are get registered in the visual, auditory and tactual sensory receptors. The receptors are structures specialized for irritability (excessively sensitive) in response to various forms of energy or stimuli, like light, heat, cold, sound or touch. Receptors consist of tissues or organs that are specialized to respond to some specific form of excitation. The tissues are the collection of cells and they are specialized for a common function. These are muscle tissues, nerve tissues and so on. These tissues are said to have very low threshold to energy or stimulation for which they are specialized. The receptors may be stimulated with a minimum energy. There are five types of stimulus energies for which the receptors are specialized. They are mechanical, thermal, chemical, acoustic and photic. In the cognitive field we are interested in ***three types of stimulus energies*** such as ***thermal receptors*** relating to skin, ***acoustic receptors*** relating to audition (hearing) and ***photic (visual) receptors*** responding to light. These receptor tissues are quickly reacting to the changes in the environment.

All the cells of muscle tissues are specialized alike for contraction, the receptor tissues for irritability and the nerve tissues for conduction, and the glandular tissues for secretion. The receptor tissues are more sensitive to the events of the environment. The receptor cells are directly connected with muscle cells which cause rapid responses to events. Now, the nerve tissues spread out to connect the receptors and the effector cells of the muscles and the glands for more coordinated responses to stimuli. The effectors are involved in producing responses. In this process, five sets of components are involved. The receptors receive the stimulus events from the environment and the stimulations are taken by the sensory nerves to the central nervous system. Thus the receptors are connected with the central nervous system. Then the information from CNS reaches the association area (neurons), which are the parts of the cortex. These areas are associated with sensory input and with motor output. The association neurons carry the information from cerebral cortex to the motor neurons. These motor neurons carry the message from the central nervous system (cerebral cortex) to the effectors. Thus the stimulation received by the receptors (input) are led to sensory afferent neuron impulses. These impulses excite the association neurons in the central nervous system (cerebral cortex area) and the association nerves excite the effectors (the muscles and glands) and the effectors produce responses (output). Thus an object providing a visual sensation or a sound providing an auditory sensation is realized by an individual stating. "Here is the person whom we searched for", or "This is the song I wished to hear and enjoy". This type of response will take place if attention is paid to the event. This is an internal description of the event stored in short-term-memory. When a stimulus character entered the short-term-memory, there is an active program of manipulating the event, judging and making decision by referring to the person or to the event or to learned material if relevant to the matter and disregarding it if irrelevant.

Thinking and Problem Solving

Thinking is defined in various ways by the theorists. Generally, thinking means the process of thought, it is putting one to ponder over a problem or matter. Thinking is exercising the mind in order to make a decision. In making a decision or judgment one has to think. It will be known, whether one has thought or not, only from the action. In the common sense world thinking has many meanings. In psychological studies also, thinking has been stated with different meanings. Psychologists view thinking as a behaviour pattern in terms of ideas or symbolic processes, or an ideational problem solving mental activity, different from the solution of a problem through overt activity. However, we explain or define thinking and demonstrate the process of thinking it is inaccessible and intangible directly. It can be assessed only indirectly. It is not an observable behaviour. According to Watson (1924) the founder of behaviourism, thinking and feeling are intangibles. He states that thinking like talking involves the movement of throat muscles. All our mental activities are in the form of speech-sound symbols. Therefore, there is close relation between speaking and thinking. Talking or speaking according to Watson is thinking aloud, and thinking is an implicit or a silent speech.

The basic functions that are important to human life are very much based upon cognitive and emotional activities. There are quite a number of mental activities revealed in an individual's intelligence. The mental activities, that are intelligent are not, separate from man's overt motor activities. Actually a person's early development of intelligence is assessed only in terms of motor behaviour. It is also visual that learning is one aspect of the total unitary process that is taking place continuously in the individual's development. Therefore, learning, in a broader sense, is a functional behaviour of development. Looking from this cognitive developmental point of view, intelligence also is an outcome of the unitary process of development. The learning of language or the development of the verbal expressions of sound which represent symbols of objects, acts, situations and events not accessible to the immediate senses, constitute the outcomes of thinking, the higher cognitive process. Thus thinking is one of the most differentiating qualities of human kind from other creatures. It is an implicit or inherent cognitive process which generate further mental processes. Thinking is a broader functional process which includes the mental activities generally stated as reasoning, imagination day dreaming, remembering and problem solving. Thinking activity is different from motor activities and all other cognitive activities like perception, speech etc.

Thinking activity is not bound by time and space. It may be referring to any time, past, present and future and it may get into any place. Stone walls and iron bars cannot prevent it. Thinking activity may deal with any object not present in front of us and also any situation not accessible for perception. The thinker can create a world of his own in any normal manner. In thinking any idea can be manipulated in any manner with the help of the knowledge of learning that took place earlier. The common features in thinking process are some kind of new mental representations for the earlier cognitive experiences. While feeling our past experience from the memory what taking place is, giving representation to our past experience. What is remaining from the previous learning is manifested in memory recall. The memory, is also a form of thinking, makes some kind of revival of the past experience or learning or behaviour. Therefore, memory involves learning of past experience or knowledge, retention, recall and recognition. Memory may be referred to the effects of the past experiences that are assumed to have stored in human mind.

Memory, according to Ausubel (1960), is the ability to remember. Remebering is a function which determines whether the new material can be disassociated or separated from the existing material structure. After learning, the newly incorporated material also gets formed into the structure, already existing cognitive structure. It means that the material loses its characteristics of disassociability. When it reaches its zero point of disassociability it cannot be recalled and it results in forgetting.

Retention generally refers to the registration of the learned material in the mind. It is a process of storage system of the learned material. It is an interval period between learning and retrieval. The cognitive view of retention is that if a behaviour is to exert any influence upon the learner's behaviour on a later time, then retention as a stimulus event has to be set as a signal to represent symbolically. Till then it is retained over that period of time. Observational learning represent in two ways in humans, one is imaginal and the other is verbal. When the stimuli are presented to the organism, sequences of sensory experiences occur and integrated at the same time. Later on when it is needed there is revival of the integrated sensory experiences which may guide the learner by imitating behaviours.

Recall and recognition are the two active aspects of memory. The second representative system, "the verbal coding" is functioning in varying levels of efficiency. A simple verbal learning may take place, for example, the names of persons or places or events. The person can describe to himself the characteristics of the names and events, and the person can rehearse or repeat and learn the verbal descriptions of the names and events. When the person wanted to recall the names and events, the verbal descriptions may serve as cues to guide the person by imitating the characteristics. Bandura, Grusea and Menlure (1968) have conducted an experiment to test the recalling methods. They have used these methods and found that the persons who have learned by describing the words recalled and recognized the names and events best.

Development of Concepts

Concepts formation is a process of thinking. It is the collection of perceptual experiences or ideas which are associated with the properties they possess. It refers to the properties common to a class of objects or ideas. There are concrete concepts like animals, birds and abstract concepts like right, freedom, justice and so on. Concept development in the child, forms as he grows and comes into contact directly with objects and gains more sensory experiences. Concept development or formation is the ability the child develops to think out of various direct contacts and sensory experiences he had with his outside world or in his environment. It is an ability to represent mantally, the experiences of the objects when they are not in his immediate presence.

The cognitive view with regard to concept development is that it is a more internal state of the organism. Concept development mainly depends upon the level of development of the organism and the background in which the new learning is taking place. Therefore, the individual's cognitive structure reveals the stages of learning. The cognitive investigators do not take into consideration the specific earlier conditions of new learning. They take into account the individual's whole history of learning. According to Ausubal (1963) the present cognitive structure that is existing in the individual is the important factor which is having impact on the learning and retaining the new meaningful materials. Bruner and Olver (1963) also emphasize the same view. People, while learning, associate the learned materials with each other. They often associate the newly learned materials by extending and combining them in the form of grouping with the materrials already learned and structured in the cognitive style.

The study made by Bruner and Olver were mainly concerned with rules and procedures used by the people in associating and grouping the materials. A number of procedures were described by them. These procedures or strategies were concerned with how for the subject matter is important at that time, and the procedures have nothing to do with the contents in the grouping. The procedure the child uses in grouping depends upon the mental development, and the previous grouping experience or cognitive structure. This development is based upon the age and the level of intelligence of the child. One important finding made by Bruner and Olver (1963) is that, how the young child distinguished himself from the older child. The young child in changing and classifying the procedure in his developmental stages passes from complex to simple. That is, when he is young, the strategy is complex and when he grows older it is simple.

The child's ability increases with age and the child perceives objects and events to classify and to categorize them in relation to the earlier groupings, and to categorize the objects progressively, taking into consideration other than those aspects used earlier, and bases his classification on the actual quality of the objects, thus the efficiency in thinking ability is facilitated. Hunt and Hovland (1961) have evolved a model of human concept formation. Later on Hunt (1912), Hunt, Marin and Stone (1966) evolved an information processing model that could solve the concept formation problems. Hunt used a holistic strategy. According to him the learner is an algorithm, a special way of solving certain kind of mathematical problem. According to this strategy the concept learner searches for a common characteristic of the objects by which they can be classified.

Human beings are constantly dealing with the realities of life in the world which are so numerous and cause utter confusion and, therefore, it becomes necessary for man to apply his mind to classify, categorize and conceptualize them to reduce the load or tension. Concept learning, therefore, becomes a function of an adaptive mechanism by which man can successfully deal with the realities of the world (Sigel, 1964). But we must know at the same time that the problem of concept formation in thinking is a joint product of maturation and learning.

Types of Thinking

Thinking has been studied in variety of ways. Merrifield (1966) has specified five types of thinking such as cognitive thinking, memorative thinking, divergent thinking, convergent thinking and evaluative thinking. The divergent and convergent are the two aspects of productive thinking. The divergent thinking is more concerned with creativity. Therefore, it is also stated as creative thinking. Since we are mainly concerned with cognitive development in this section, only the types of thinking are mentioned here.

Cognitive Thinking and Learning

Cognitive thinking is mainly referring to the knowing aspects, or comprehension, awareness, conceptualization, perception, learning, intellectual abilities, problem solving and so on. In the cognitive thinking there is more mental representation and manipulation of the ideas. The cognition of any new situation, like the child facing the school for the first time, not only involves the awareness of the situation, but something more. There is also some kind of classification, and discrimination. The mental process of identifying or placing objects into different classes or categories on the basis of certain inherent common qualities is stated as concept formation or concept learning. This is one of the important aspects of cognitive thinking according to Bruner and Olver (1963). Learning takes place only in relation to cognitive structure. Cognitive structure is one's source of knowledge. It results mainly from genetic learning. Hence learning involves in these terms, the process of associating the new learned materials with this structure. The new knowledge and the items of knowledge are integrated into the existing structure meaningfully.

Thinking is not taking place spontaneously or in a vacuum. It is not the expression of one's innate ability. Though it is a function of the intellect, it is the result of learning process that has been taking place for a long period of time. According to Harlow and Harlow (1949) the brain is very indispensable for thinking or for thought process. But however good brain one is having if it is not taught or tutored it is of no use for thinking and solving the problem. An untutored brain may try any problem by trial-and-error method or in a clumsy manner. Learning and training alone can make an individual to think and to develop concepts and ideas. Therefore, learning is the source for the development of thinking. Even in the case of children, they only learn to develop their thinking. Children can be helped to acquire their thinking approaches and skills through learning and solving their problems.

The potentiality for learning and thinking varies widely among the human beings. Man possesses many types of abilities to function in the world. He develops all these abilities only within the limit of his innate potentialities. Therefore, all the development of abilities are also matters of learning and whatever the level of thinking ability one has developed, is also gained through learning and the training.

Problem Solving

There are stages of increasing differences in children's learning abilities. This is a problem that has been frequently observed in problem solving experiments. Gagne (1970) in his recent works has stated that problem solving is a class marked as 'higher order rules'. According to him it is 'thinking out', a solution to a problem by putting together the old rules to form new ones. It is because of this function in problem solving that learning rules become more important. Problem solving occurs when the rules learned previously are not suitable to the solution of the present problems. From the learning point of view problem solving is important as it brings out the transfer of the individual's skills to the problem situation in a perceptual manner. In our everyday life we are facing many problem situations. Some of them may be simple and some of them may be complex. Children in playing situations are often asked to find out the hidden objects. They go in search of the objects in all possible known ways or with the strategies they have known through learning.

One of the productive thinkings, the convergent thinking is generally called problem solving. Convergent thinking is more of a controlled thinking in nature. It is a way of thinking in the direction of the solution to a problem. In a structured problem there may be few limited number of correct solutions. Convergent thinking is a mental activity with a purpose of finding out a solution to the existing problem. In convergent thinking, objects, situations and relations are manipulated through symbols and ideas for the solution of a problem. It may be also resolving a conflict situation. The problem objects may be physically present, and also difficult situations in the environment, may be faced directly. Such problems may be solved by the use of images and symbolic representations of some kind. In solving certain problems, the images and the symbolic representations and the learning may be modified or reorganized to suit to the situation in solving the problem. For example, in solving some arithmetical problems, the child might have solved them mentally. On seeing that a calculating machine if available, the child can make use of it and try to solve the problems quickly with the help of learning set. The child surveys the situation perceptually and takes use of the situation for his purpose. Perception of the situation and the mental images are used by the child. Harlow and Harlow (1949) state that a cognitive structure is already set by learning and the same has been used by the child to solve the problem. Thus problem solving is a type of cognitive thinking as the concerned activity is controlled and directed.

In problem-solving, the child requires several cognitive abilities. As stated above the child is required to solve some arithmetic problems, the abilities like perception, learning, and mental ability. Many research works have been conducted in the area of problem solving. Bourne, Dominowski and Leftus (1974) and Duncker (1945) have traced three areas of importance in their studies. According to them, the ***first area*** involves, how a person observes, interprets or represents a problem, is asserted to himself and how he understands what the problem is about. The ***second area*** involves the strategies (Procedures) applied in finding out a solution on understanding the problems. The ***third area*** involves the way in which the person compares an evolved solution rule, and decides whether a solution would occur so as to stop the work. The problem solving procedure is well-structured and interpreted in this programme. All these three stages are essential to deal with any problem situation.

All the problems are not faced objectively with physical objects, events or environmental situations. There are problem situations in which there may be no direct sensory images. The problem may be theoretical, and abstract symbols need to be combined in various ways to find out a solution. The symbols are mentally manipulated unlike the physical materials involved in concept formation. The abstract symbols are rearranged and reorganized in solving such problems. For example, children face verbal symbols, words and phrases. Symbols are used to the pronunciation of work. Mathematical symbols like the numbers and other diacritical marks, the notations are used in physical and chemical sciences. All these may be required in problem solving. Though this involves a process of reasoning they are used in abstract problem solving. There is progressively increasing difficulty in cognitive system of problem solving behaviour. Thus the ability to understand and solve problems increase with the age of the child.

LANGUAGE AND COGNITIVE DEVELOPMENT

Human beings differentiate themselves from other organisms in several ways. One of them is by their competence of using the language. Language, they use in various ways, like by reading it, listening to it, understanding it, using it for thinking, and using it in writing, in speaking it and also for the purpose of communicating their ideas. Human beings gain almost all their needed and essential information by being told by some and not by acting. Children's schooling and formal education are the outcomes of imparting information by teaching and telling them to the children. We acquire all our knowledge of everyday matters only from reading news reports and from listening to the television and radio transmissions. The medium that is used in all these is the language that one speaks. Language is the safe treasury chest of all our knowledge and wisdom. What is stored in the individual's mind is also in the form of language. All the knowledge and information are stored in languages and also transmitted whenever they required. Language is the medium of instruction in schools and colleges and the medium of communication all over the world among humans. Language is not synonymously used with communication, because communication is taking place also among the animals and birds. Therefore, language is the most clearly distinguishable form of human behaviour from other living creatures. Man alone is capable of verbal behaviour. Psychologists, therefore, interested in knowing how humans acquire language and how they use it in their verbal behaviour, becomes essential for an understanding of human learning. Skinner, the American psychologist, has long been studying verbal behaviour. In 1936,

Skinner has produced a phonograph record made up of speech sounds. The record is called verbal summator and it was used to study the sounds of words by the listener. According to Skinner (1968) verbal instruction first stresses the ears while the teacher and student speak to each other. But the printed materials and other visual materials change the stress to the eyes.

Children's various abilities are growing with their age and maturation. The ability to use language provides them with an indication of knowing the objects in the world. The child uses the language to recognize and to remember the object around his home. The child may speak by using the language, 'it is my toy', "it is mother's ear ring", "it is father's umbrella" and so on. On knowing that the child is capable of using the language, parents and other elders also may be asking the child: "speak out in words, what am I doing?" Studies have been made with regard to the importance of the relation between language and cognition. Some of the studies indicated that when there was any difficulty in applying language (words), then the task of problem solving was not successful. Keeney, Cannizzo and Flavell (1967) have observed in their studies that for spontaneous production of verbal behaviour and to make the behaviour more effective cognitive development should take place in children. Children who lack active development do not use the words spontaneously during the time when task was assigned.

There were deep studies, conducted by several Russian psychologists like Alexander Luria (1961), Vigotsky (1962) and Sokolov (1959) into the role of language and the direction of behaviour. Luria's study was with regard to the control of behaviours by language. Luria has observed that children begin their task without any difference between motor functions and language functions. Children have demonstrated that in the absence of language, motor responses may be made. When they found language, the children believed that there are two systems for their purpose. They believed that the verbal and motor behaviours are two separate systems and the problem was with regard to the integration of these two systems. Luria observed that there were progressive developmental stages in the verbal behaviour with increasing of age. Language may stimulate action in a two years old child, but the action is not quite suitable to the situation.

A more difficult procedure was used in an experiment by Luria. A child was made to sit before a board of instrument which was containing a light bulb. A rubber balloon was given to the child for squeezing and the squeezing of the balloon was noted. Children of different ages from two years and above were used in this experiment. Orders were given to the children such as, "When the light is on, you will have to squeeze the rubber balloon, and when the light is put off, do not squeeze the balloon". Here the order was to squeeze balloon when the light was put on or not to squeeze the balloon. There were two years old children. These children looked at the light whenever they heard the word "light' in the command, and whenever they heard the word "squeeze" in the command they squeezed the rubber balloon. They were not able to comprehend the full meaning of the command. They were carrying out the task to the meaning of each word. At the time when the children were carrying out the command to squeeze the balloon they were also told, "That is enough". To this instruction the children did not stop the squeezing, instead, they intensified the squeezing of the balloon.

In this experiment, Luria has observed that the two years old children's simple activities were directed by the language referring to the motor system, was prevailing over the language system. The children while acting on one command, they do not ordinarily stop the action nor change from one action to another. This is the behaviour pattern of two years old, who have not developed the cognitive learning. But the three years old children, mostly, were able to follow the command to squeeze the balloon and stop squeezing it and were waiting till the light was on. But the three years old children were not responding successfully to complicated commands. When they were told to squeeze when the red light was on, and not to squeeze when the green light was on, it was found that they squeezed for both red and green lights. In another command when they were told to squeeze twice when the light was on, they squeezed more than twice. These children like the two-year olds could not stop the activity once they started. Acquiring language for correct direction on behaviour, children attain only when they reach five or six years of age. In the case of squeezing the balloon these children acted appropriately on the command. In Luria's study of language-action relationship it was found that there were several stages of cognitive development in carrying out the instruction to the correct meaning of the words.

Learning to Talk

The child, thus, slowly makes use of language in his behaviour, either by responding to the verbal commands from others or by himself using the words in the appropriate situation in appropriate manner. This development takes place with age and maturation. Another variable that is having influence in the child's speech or language acquisition is the stimulating social environment. In the stimulating social environment the most important function of language is the communication. Language is a means of communicating one's ideas and feelings among individuals. Communication in a language is taking place by the use of highly evolved system of articulation of vocal and related sounds. The speech communication is a two-way process of interaction between persons. A person must first understand the meanings of the words expressed by another person and secondly must communicate the meanings to another person through a commonly understood verbal symbols. A higher order of mental functioning is expressed in verbal communication. In short, it is the manifestation of mental activity and verbal behaviour.

Verbal behaviour, according to Skinner (1957), is the behaviour that is strengthened in another person by the use of language. It is also a mode of psychological interaction among persons (Stott, 1974). The verbal behaviour or the functioning from the behaviouristic point of view is an acquisition through learning. It is like learning of any other observable and measurable behaviour. Therefore, like any other learning, verbal learning also can be analyzed cognitively.

Learning to speak a language of verbal behaviour is a complex process. The child's learning to talk in a language has been an important developmental process for the psychologists. In recent years a new science called psycholinguistics has emerged. The important problem of this new science is to find out the specific tasks involved in speech learning in childhood period. The number of words the child used in his speech is considered as the indication of speech learning. This has been studied more elaborately by Brown (1968).

It has been found that profound changes have been taking place in the child's speech behaviour at the end of the second year. Bruner (1964) was surprised to see that the child has acquired the use of productive grammar and also learned the rules of language formulation. The child gradually learns to follow the rules without any planned efforts. He starts first like a parrot, imitates what others speak. Those who studied speech learning in children have found that there were two levels in the problem of language composition, the rules of grammar and the rules of syntax. The rules of the grammar are to deal with the arrangement of morphemes meaningfully. The rules of syntax are to deal with the arrangement of words as elements in sentence to show their relationships. It can be also stated that grammar is the abstract system of rules in terms of which a person's mastery of his native language can be explained, whereas syntax is the systematic statement of rules governing the grammatical arrangement of words and morphemes in a language.

The child gradually develops his own grammar and syntax, though at the beginning not following the elaborate adult grammar. The child begins with two word sentences in his talking and shows gradually the semantic relationship. He learns to use words and phrases, parts of speech and also comes to identify the class of new words and placement of them.

Children, it seems, assimilate and regularise the syntax of the language in the conversation with their mothers and fathers and otiier elders. Brown and Bellugi (1964) state that the child's speech development is not only a matter of learning to speak word by word or sentence by sentence, but what others speak about him, and his speaking. Speech development involves the use of the principles and rules of language structure that could induce the child to produce sentences in his speech. In the speech learning there must be something of the total acquisition of speech skills. Experts who studied speech learning of children suggest this important aspect. Theorists did not make any reference to this. But the cognitive theory offers some explanation for this. Bruner (1964) says that cognitive growth is the abstract summary, or the thing that typifies the whole class of the interaction between the child and his culture and language. According to him language is a technique created by the culture and provided to the child. The child is constantly interacting with the language and the experience of which occupies the whole life of the child. If there is any change in the cognitive structure, it is because of the development of expert skills exercised by the culture. In such expertise development everyone including the child contributes for the creation of new ideas.

SOCIAL AND EMOTIONAL DEVELOPMENT

Emotional Development

We have been so far dealing with various aspects of physical, motor and cognitive developmental processes. Man is a social animal. His interaction with other individuals in the society is based upon the nature of his social and emotional development right from his childhood period. Man becomes social only after his birth in the society and after learning to be social. The child is always in the company of others and he is shaped and groomed by his society and culture. During the course of the physical growth and development and maturity the child learns many things. He realizes that he is one among the people live in the society and in the world and must learn how to be happy with them. Therefore, his social life is combined with his emotional life. In the developmental process the social responses of the child become more specific. The early emotional reactions also are going on changing.

The social and emotional responses of an individual are more important matters of study for the psychologists along with the other developmental processes. The scientist, who was the first to investigate and to study about the emotional behaviour of children was Charles Darwin (1809-1882). Darwin in his classical work, *The Expression of the Emotions in Man and Animals*, which was brought in 1872, has stated that human emotions are innate. Human beings are experiencing several types of emotions in varied situations. The study of human emotions is always an interesting matter. Because we can not imagine human life in this world without experiencing the feelings and emotions. Early textbooks in psychology speak about quite a number of feelings and emotions. The instinct theorists like McDougall (1908), Bernard (1924) speak about several feelings and emotions. McDougall, the doughty champion of the instinct theory, states that there are fourteen instincts and each one generates one feeling. According to Bernard there are six thousand human instincts. Any type of behaviour that seemed to be human is based on some instinct which is also an expression of an emotion. Later on Watson and Morgan (1917) have studied emotion experimentally and stated that there are only three basic emotions like fear, anger and love and the other emotions are the combination of these three primary emotions. All these emotions are occurring through learning. According to Watson, the founder of behaviourism, fear is caused by painful experience, hence by inflicting pain in a child fear can be aroused. Similarly loud noise, sudden loss of support may cause fear in a child. Anger or rage can be caused by interfering in the child's play by removing his toys, by restricting his movements. Love can be aroused by gently touching him. The behaviourists are of the view that all the emotional reactions may be conditioned.

Some investigators refuse to confirm the existence of the primary emotions such as fear, anger and love. According Bridges (1932) children are born with one basic emotional reaction called generalized excitement. As the children grow and develop the general excitement is changing into more specialized forms like pleasure and displeasure, delight and distress, elation and depression, love and hate, joys and sorrows, anger and disgust fear and courage. As the child matures more and more specific emotions are emerging. This is quite similar to cognitive and motor development. Bridges states that when the child becomes three years old he expresses subtle emotions like anger, envy, jealousy and disappointment. Emotions are classified into two categories as unpleasant emotions or emergency emotions and pleasant emotions. Fear, anger, envy, jealousy are unpleasant emotions. Love, affection, joy, happiness, feeling of thrill are pleasant emotions.

The area of emotionality is a very difficult one to study experimentally. While viewing it objectively, it is a behaviour pattern that has common characteristics with animals. The cognitive processes are more involved in the emotional behaviour of man.

But it is observed that man's feelings and some of the emotional experiences are partly controlled by the spinal cord and mostly by the cerebrum. A well developed person displays all the emotional behaviour patterns. Emotionality is a complex and multifaceted phenomenon which may be disorganized in certain situations and reflect the primitive functions. Emotion is defined in varied ways. There is no unanimity with regard to the difinition of emotion. In the early textbooks it is stated that emotion is a stirred up state of the organism, making internal physiological changes like increased heart rate, blood pressure, mobilization of energy to face the situation either for flight or for attack. Kimble and others (1985) define emotion as an excited state of mind based on physiological changes from homeostasis.

The level of emotional state may vary from mild form to very intense state. Every emotion is expressed in some form. The child expresses his emotion of joy by smiling, fear by crying and anger by kicking and rolling. Smile appears in a child at the age of two months. Laughter is a later development. According to Freedman (1964) there are three stages in the development of smiling. At the very early stage, after birth, there is only reflex or spontaneous smiling and is not in response to any external stimulus. The second stage of smiling takes place after two weeks and it varies in children upto eight weeks. This is not for any specific stimulus, but to some visual and social stimulus. Freedman calls this as non-selecting social smiling. It is more of general type of smiling. The third stage of smiling is a more selective social smile expressed by the child at the age of five or six months. This is the stage that the child learns to smile at the familiar and known faces. There are several theories of smiling formulated by investigators. The three important theories of smiling are, ***reinforcement theory***, ***evolutionary theory*** and the ***cognitive theory***.

The reinforcement theory was formulated by Brackbill (1958). According to this theory the smiling of the children depends upon how they are reared up. Children smile where the adults smile back at their smiling. Their smiling is reinforced by the adults' smiling. The adults smiling is a reinforcement or a reward for the child's smile.

When the child's smile is reacted by other people, the child continues his smiling. This reaction forms the basis for the theory of ***evolutionary smiling***. Crying serves as a survival value for the child to get food and smiling also serves as a survival value by the reward of the parents of feeling pleasant. By smiling, the child gets more attention of the parents. Then the child's needs are met by the smiling. The attachment between the child and parents is increased. Freedman (1967) had formulated this theory of evolution of smiling.

The cognitive theory was designed by Shultz and Zigler (1970). This is a very recent theory of smiling. The theory states that the child expresses his innate joy by the act of smiling. Children when they learn certain cognitive tasks, feel that they have gained some new experiences and the same give them the feeling of pleasure and it is expressed in the form of smiling. The new experience for the child is very much perplexing when it is not easily fitting into the cognitive framework. The child finds the new experience away from his cognitive field and as result of it he is not in a position to absorb the experience. In such case the child has to reconstruct his internal stand of the object. When this has taken place the child can absorb the object and can be satisfied with this and can smile. Cognitive theorists are of the view that the child take little more time to absorb simple objects to smile. But in the case of complex objects he takes longer time to absorb them and express his feeling of joy by smiles. Shults and Zigler (1970) have demonstrated this in their experiments.

Fear

Fear is one of the primary emotions. The child develops the emotion of fear when he suddenly hears a loud noise. It is also true in the case of adults. The loss of support also causes fear to the child. These views of Watson (1917) was not accepted by many psychologists. According to Hebb (1949) fear is the result of maturation and not due to learning. For example, a child upto the age of five is not afraid of snakes. At the same time an adult person gets startled at the sight of snake. Psychologists are of the view that many of the fears are learned and conditioned. Human beings always associate events and experiences. A painful event causes fear. Apart from dangerous situations, the physician's injection needle, the teacher's cane all cause fear to the child. A steep valley from the top of the mountain may cause fear even to adults. Children develop fears to objects not causing painful experience. For example, school children are afraid of heavy home works given by the teachers.

There is rapid development in children's cognitive abilities by which they are acquiring abilities to recognize dangers and also objects and situations that threaten them with painful consequences. Children generally develop fears to darkness even though there is no danger in it. The child seems to have very few fears. When the child grows and matures he develops all sorts of real and imaginary fears. Jersild and Holmes (1935) have illustrated a number of common fears of children. The fears developed by children for loud sounds, strangers, darkness, unexpected movements are decreased during the first five or six years of their age. After this period children develop new fears in the later childhood period of life, such as fear of animals, fear of dreams, natural events like thunder and lightning. These fears also are getting declined when they grow into the adolescent stage.

Generally younger children develop fears for concrete objects, and events that are associated with painful feelings. Older children develop fears more for abstract things, dreams, ghosts, school assignments and failure in examinations. But the widely studied children's fears are separation, anxiety and fear of strangers.

Fears may vary in intensities, ranging from mild form of fear called worry or anxiety to the level of disorganization. Anxiety is a feeling of discomfort. It is associated with real or imagined or anticipated dangers or unpleasant events. If the child did not return home even after six O'clock in the evening, the mother develops severe anxiety and worry and imagines all sorts of dangers to her child. This is one of the common anxieties occurring in our everyday life. Phobia is always associated with anxiety. Phobia is an unrealistic fear of some particular events, objects or circumstances. The child is afraid of darkness, imagining that there may be ghosts and devils. Anxiety and phobia are the reactions the children develop during the period of their upbringing. In some cultures phobia is associated with abnormal reactions and supernatural beings. As the children grow older and older they understand that these are all imagined and they dispel such fears.

Anger or Rage

Anger is an unpleasant and emergency emotion. Anger and fear have something in common. Because both are called unpleasant and emergency emotions. In both physiological changes are taking place when aroused. According to Bridges (1932) anger is not found to be existing at the time of birth. The situations that cause anger in the early child period are painful and discomfort situations and interferences in the free movements. The child expresses his emotion of anger in various ways. Morgan (1934) states that the child expresses his anger in four characteristic ways. The child is fully ***egocentric***. He becomes self-centered when the outside conditions cause him painful experience and bodily discomfort. Followed by egocentric, the child is very ***intolerant*** of any bodily discomfort. He expresses it in various ways like crying, kicking, bending and twisting his body. Thirdly, the young child is making his demands in a very urgent manner. This demand is for his ***immediate relief*** from his discomforts. He cannot wait in his demands, and needs immediate satisfaction. Fourthly, the young child becomes a noisy outburst and ***explosive*** in his demands. He expresses his anger in a violent manner. These characteristics of anger are common in young children. The child's anger may be controlled by the process of socialization. Levin and Turgeon (1957) state that parents were found to be successful in socializing their children and controlling the children's emotional outbursts. It is found that environmental influences often cause emotional problems in children. These problems can be solved by suitably manipulating the environment and by protecting the children from such problematic environmental conditions.

Social Development

The human infant from the time of his birth till he can exist independently has to be under the care of his parents and other elder persons. Of all the living creatures man is the most difficult creature from the developmental point of view. We know that certain creatures are left free soon after the birth or from the time they come out of the shell. Some animals will have to be under the care of their mothers for few days, some for few months and some young animals for one or two years. But man alone must be under the care and protection of his parents for the longest period. A new born child till he grows and matures into an adult person he is a liability to his parents. The parents' cares are so strainful and difficult that they will have to put all their attention on the child to bring him up and make him successful citizen in the competitive world. The child's various developments such as physical, perceptual, motor, intellectual, emotional and social–all fall under the care of the parents. Any impairment at any stage of development results in physical or mental deformities. These conditions cause much more strain for the parents. Therefore, the child is constantly under the care of the parents and continually shaped by his parents and those who are around him and the society. In the social development of the child many people are involved right from the time of his birth. The social development of the child is also responsible for the development of one's personality characteristics.

Social Attachment

The processes involved in social development of the child are making him a socially acceptable creature in the society. Parents, siblings, other members of the family, peer groups, school, playground, all contribute to the social development of the individual. The individual is made socialized. Social scientists agree that socialization involves all these processes through which a young child tends to feel and act similarly like most others in the society and culture. The family, school, religious and other institutions play important role in socialization processes. The young child is very sensitive and egoistic in enjoying the comforts from the mother. The physical contact and attention and people sitting close to him are the needs of the child. The combination of all these is called social attachment.

The small child is emotionally and physically attached to his mother. The mother is the source of comfort and pleasure and the absence of pain for the child. Studies have been made on the child's early attachment. Bowlby (1969) states that children seek contact with their mothers when they were frightened, and demand protective responses from their mothers. According to Bowlby, punishing attachment is not inhibiting it, instead, it is intensifying it. Punishment is a threatening stimulus for the child for demanding greater attachment from the mother. Schaffer and Emerson (1964) state that there is an innate demand or need in infants to be close to other elder persons. The infants' need for others company is manifesting in many patterns and this varies from child to child, and from one developmental stage to another developmental stage.

Studies have been made with regard to the development of social attachment of children during their childhood period. Ainsworth and Bell (1973) have observed three types of attachments the children have exposed to their mothers. When mothers are separated for a short period of time and reunited, most of the children wished to be in close contact with their mothers at once on seeing them. This is called *security attachment* with their mothers. Another type is called *ambivalent attachment*. The children wanted to have attachment with their mothers, but since their mothers were absent for a period of time they show some sort of resentment by resisting the attachment. The third group of children do not want attachment with their mothers as they were away for a long period of time and hence they avoid their mothers. The authors state that this type of *avoidance-attachment* is quite common among children who were away from their mothers for a long time. Children who were deprived of social attachment for a long period may face severe developmental disturbances.

The father of the child has his own share in the social development of the child. Most of the studies have described only the mohter's role in the social development of the child. Though the father plays only a minor role, still it is very important in the life of the child. It has been observed that the father's personality characteristics and the type of behaviour are considered to be more important in the emotional and social development of the child. In the research studies conducted by Lamb (1977) it was observed that children develop social attachment with both fathers and mothers. The type of cares provided by the father and mother vary. Mother's cares are mainly with regard to feeding, cleaning, making them to sleep and so on. Father's cares are mainly stimulating the children to play. The father-child interactions are more physically stimulating than for play and games, and the type of social experiences are different. Peterson and other (1959) have found in their studies that children's maladjustments mainly depend upon their parents' social attachments. Fathers can extent the social experiences of their children. Generally father's absence at home for a long time may affect the behaviour pattern of the sons and daughters. The studies made by Ainsworth and Bell (1970) on the absence of fathers at home showed that there were much adverse effects on children. It has been found that boys from fatherless homes or father's absence at home for long time on foreign countries, were found to be having more adjustment problems and were less skilful than those boys whose fathers were at home. Further, children develop sex identification with their parents. Children imitate their parents in various aspects of life styles, like interests, attitudes, like and dislikes and other behaviour patterns. The boys appropriate their fathers qualities and the girls appropriate their mothers qualities. Sheldon and Glueek (1958) state that more delinquents were bred from fatherless homes. They have observed in their studies that more than forty percent of adolescent boys were found to be delinquents from homes where there were no fathers. According to Rutter (1971) boys become anti-social elements in families where there were continued conflicts and quarrels between parents and from fatherless families. Fathers' absence in families has its impact very much on girls'. The girls' problems are many in fatherless homes. One of the more serious problems is that adolescent girls tend to have premarital promiscuous sex relations with men. The girls take liberty in enjoying moral holidays in fatherless homes. Hatherington (1972) has enumerated the problems that are emerging from fatherless homes. Thus the role of the father in child rearing, in social attachment and social development of the children have become very big issues of social development.

In the socialization process of children, other than parents, there are several other forces operating significantly. They are the culture to which the children belong to, the school, teachers, peer groups and the mass media of communication. Children as they grow up and go to schools, they are with their friends and peer groups most of the time. One of the important sources of social development is the interaction with peer groups and playgroups. Children develop many social characteristics only in the peer groups. For example, the development of competitive spirit, aggressive and altruistic tendencies, moral characteristics and discipline are all the results of peer group contacts and interactions. Children explore the physical and social environments only when they play with peers and adults. Even when children are alone they imagine about playmates and have social interactions with them.

When children grow, develop and mature, physically and mentally they come into contact with the mass media of communication like newspapers, magazines, radio, television and motion pictures. All these have profound influence in the development of social behaviour of the children. Allowing children to have too much of time in watching the television also causes some problems. The television and motion picture violences have caused more harms and problems to the children than the knowledge they acquire from them.

Parents imbibe some cultural and social values in the minds of the children. Teaching certain cultural and social values, that are respected by their society, become one of the duties of the parents of some cultures. Parents believe that this is one of the child-rearing practices of the parents. In some cultures, parents teach their children with the values of economic needs of their culture. The child rearing is very much affected by the economic conditions of a culture. There are cultures economically surplus and find no problem in child rearing. There are also cultures economically deficient or impoverished who face difficulties in child-rearing. Studies have been conducted worldwide by Barry, Child and Bacon (1959) for assessing the nature of the effects of economy on child rearing. Nations or societies that accumulated more quantity of food by their improved methods of agriculture and animal husbandry were greatly insisting upon their children, to be more altruistic, responsible, obedient and also to have a tendency to give to others who are in need.

People who tend to feel and think alike share certain common attitudes and interests in the society. The common attitudes they have about their children, tend to resist the influence from outside of their social class for change. There are wide social class differences in child rearing practices and attitudes about children. Ericson (1946) states that there is considerable amount of difference between middle social class and lower social class in their attitudes and expectations about their children. The middle class begin their training on children earlier than the lower social class families. They put more emphasis on social responsibility, individual achievement and also put supervision on the activities of children. Thus the processes involved in social development in children are very complex and there are multiple factors that determine the socialization processes.

PERSONALITY DEVELOPMENT

In the developmental process, the development of personality is an important component of the person. Personality is defined in a number of ways in the textbooks. It mainly refers to the total quality of a person.

In the interpersonal relations a man may be meeting many people, dealing with different situations and meeting with many problems. In all these how the person places himself depends upon the type of personality characteristics he/she has developed during the course of these physical and mental developments. There are different theories of personality. Each theory desecribes personality in its own way.

The nature and sources of motivation is one of the main issues on which the theorists of personality vary. According to them the central points in understanding personality are the motives, drives, wishes, wants, needs, inclinations and the purposes for which one is activated. These are considered as sources and the origin of human conduct. For example, conduct, character, moods, temperaments, ambitions, aspirations, dominance, submission, emotional stability and personal appeal are viewed as components or characteristics of personality. In personality development any one of these characteristics may become dominant or a particular quality may be more effectively functioning in a person. A person is often given to a particular type of mood constantly, may develop a permanent quality in him, which may be getting annoyed quickly or irritated, dominance assertive, quarrelsome and so on.

Development of personality is explained mainly involving the physiological factors, the physical environmental factors and the social factors. The physiological factors that influence personality development are the drugs of various types, blood sugar, diet, disease and the endocrine glands which by secreting more hormones into the blood, the damage of thyroid glands, the under secretion and over secretion of adrenal glands, the under development of gonads, the over activity and under activity of the pituitary gland all have influence in the development of personality. It is important that the endocrine balance must be normal. It is possible that the extreme over activity or under activity of any one of these glands may result in affecting the normal development of behaviour and personality. The abnormal function of the endocrine system determine the individual differences in personality. Thus psychologists like Ingle (1935) and others are of the view that the endocrine glands that secrete the hormones directly into the blood vessels are the basic biological factors in personality development.

Environmental Influence

The physical environment in which humans live may be having influence in the development of personality. Environments may be physical environment and social environment. The very recently emerged environmental psychology stresses much the influence of the physical environmental factors in the behaviour of humans and in the development of personality. The physical environments such as the landscape, forest, trees, sounds or noises, temperature, heat, cold, humidity, air, wind and man made environments such as buildings, roads, dams and industries, all have influence on the behaviour and personality development of the individual. Personality development also takes place due to individuals activity. But the individual's activity depends upon the stimuli received by the individual from the environments. The social environment alone need not be emphasized for the development of personality. The physical environment like the tropical climatic conditions may cause the development of an easy going personality disposition. A moderate climatic condition with the change of seasons may bring forth a more energetic temperament.

Cultural Influence

Since the individual is born and brought up in a society, he cannot be free from the influence of the society. The social environment is having several important influences on the individual's personality development. The individual receives a code of conduct from the society and he/she adopts the code of his/her group. The individual himself may build up a code of bahaviour in his group and may play his role. The code of the group includes manners, conducts and morals. It also includes mores and folkways apart from what an individual is acquiring from the family. The individual as a child picks up many of the rules easily and happily. If he resists any of them he may be made to accept by criticisms, ridicules, punishments and by restrictions. Finally he accepts the cultural norms and the standard of behaviour by adopting them. The parents and other elders determine the roles, the child has to play in the family. Some parents are over solicitous and protective. In such cases the child may develop a dependent personality. But some parents may be harsh and punitive type and the child may lack affection from the parents. In such families the child may develop a defensive negativistic personality.

Adults' roles in the family generally vary on the basis of occupation and income which form the social class and status. Children in such families understand their social status and develop personality traits which are neither aggressive or dominance, but may be docile and accept disciplined way of life, determined by their families. Therefore, the development of personality is very much influenced by physiological, physical environmental and social factors. The personality development does not mean the combination of or sum of these factors. It is interaction and integrative product of all these factors in a harmonious manner.

Therefore, personality refers to the total quality of a person. A person as an individual is unique and different from every other person. According to Stern (1938), a person is a living integrated whole organism, is unique in every respect and striving toward the goals. He is a self-contained, but open to the world around him. A person's personality is, therefore the total quality, a complex combination of qualities that make one a unique and reveal an individual identity. Therefore, in the development of such unique whole several factors are involved.

EXPERIMENT 58 : ASSESSING THE LEVEL OF MOTOR ACTIVITY IN THE DEVELOPMENTAL PROCESS

INTRODUCTION

Man, like all other creature is born in this world, grows and develops physically and mentally, moves and performs activities in his environment. But man does not stop with this. He goes many steps further beyond from the physical growth and development of the animals of all kinds. Man's physical growth also goes with his mental growth and development. He matures physically and mentally to perform his activities, moves and reacts in response to other human beings in the environment, learns many things to the needs and requirements of himself and others. Man takes longer time for his physical and mental growth and development than any other creatures in the world. For example, the chick picks up small grains within few minutes of its coming out of the egg cell. All the four legged animals can move about within few hours of their birth and remain under the protection of their mothers only for few months. But the human child will have to be under the care of his parents for many years. The human child has to be a dependent under the care of some elders till he can be independent by himself. This takes normally sixteen to twenty-two years. In some cases, particularly in Indian society it even exceeds 25 years.

The reason might be that the growth and development take place very slowly in human. His physical and mental growth and development make him to mature in all aspects of his life. Man's developmental processes should be such that he should face all kinds of situations successfully by himself without the aid and help of others. Therefore, it is very difficult to state when the developmental process ceases in man.

Human Development

This is a very wider field with many facets which involve a combination of a number of scientific disciplines. Development, in general, includes the total functioning of the individual. According to Stott (1974), development refers to the growth of the elaborate system of organs and parts, maturation and the development of man's intellect, the coordinated overt activities and skills, the development of his emotions and the patterns of interaction with others and the functions of bodily structure and also learning.

Therefore, human development includes physical, functional and psychological developments in a unified manner. The functions of human physical organism constitute psychological phenomena. Therefore, it becomes important for us to examine the functional aspects of human development which involve the physical and psychological development in a broader sense. The physical development mainly includes growth and maturation of nervous system and the brain. The psychological development includes the development of feelings and emotions, temperament and personality traits. The functional aspects of development include mainly learning, the influence of culture and environment and mental or competitive developments. The functional development also involves the vital functions like, the ingestion of food, elimination of waste, activity, sleep and rest. These are important for fulfilling the basic biological needs regularly and to maintain life and health.

Intelligence

In human development, the development of intelligence is an important factor. This begins with the emergence of the cognitive function in the child. Cognitive function or cognitive activity generally is referring to the higher level of activity. It is the activity of the mind. This is in contrast to bodily activity, the motor function. The concept, intelligence is not defined satisfactorily that can be accepted by all. The general, simple definition of intelligence is the ability to deal with a problem situation successfully. Individuals vary considerably in this ability. This ability mainly depends upon the development of the central nervous system, the brain. The central nervous system is a very complex system, the more complex the central nervous system is the more one can successfully deal with the problem situation. With regard to the development of the central nervous system there is no so such difference in humans unless one is mentally handicapped or mentally retarded. Otherwise, the development is almost the same for all humans. The development of the brain in animals vary considerably from humans. Man's superiority over animal kingdom mainly depends upon his complex mental development. Man is inferior to animals and birds in all other respects like physical strength, sense of seeing, hearing, smelling, tasting and in predicting events, seasons and so on, except in mental activity. Therefore, the mental development is specific to human for his survival in the world.

Maturity

In the process of maturity of humans, the development of speech is an essential component for communication. Communication may take place in all living creatures, but in the case of man it is language, that is used as a medium of communication. Communication can be made also through signals, signs, gestures and postures, but it is language that makes the communication more meaningful and only through speech that man communicates all his needs and wants. In the process of interaction in the society, language plays an important role. The speech acquisition and language development take place right from childhood onwards. These take place through learning. The ability to learn a language, to speak and to communicate is a great achievement that man makes to differentiate himself from other living creatures. In the developmental process language development is intended for verbal communication and interaction with others. It is also called verbal behaviour in psychological researches. Verbal behaviour and communication are the manifestations and expressions of higher mental functions. Therefore it is considered as one of the cognitive functions.

MOTOR FUNCTIONING IN THE CHILDHOOD PERIOD

The individual in this developmental process right from the period of childhood is going on changing constantly in every respect in response to the external world, the environment. The mode of direct contact with the stimulating environment is through his body activity. In the body activity the motor functioning is the primary function in the external world. The infant's mode of function during the period of eighteen to twenty-four months, Piaget (1952) states, is the sensory motor period. In this period the infant's activity becomes progressively more complicated. The infant finds it difficult to control the process of activity. For example, reaching an object, the child does not reach in the first attempt itself. He takes more than one attempt in reaching an object. Sometimes he may extent his arm only near the object, but not reaching it. Sometimes he may extent his arm more than what is required. He takes few trials to reach the place where object exactly is. His bodily development and sensory development are involved in it. The child's reaching for an object and grasping it that is placed in front of him is called cognitive awareness.

How a cognitive awareness has taken place in a child? How to judge this? This can be judged only in terms of his mental development which is revealed in the quality of his motor activity. The child's body performance more and more is becoming accurate and efficient. The infant's emotions also can be judged from his overt behaviour. During the early months and years, the infant's primary developmental tasks are getting control of this body and body parts. When this is achieved and becomes more accurate and efficient the child gets control of the external environments. He comes to the stage of reaching the objects and manipulating it and also exploring it. He wants to know the nature of objects by collecting everything by his side. In all these the child's motor performance is involved. He not only perceives and explores the objects around him, but also puts himself 'in action' vigorously. As he grows older and older he is full of energy and he wants to spend his energy by his motor activities and exhausts it. When he gets tired of his activities he goes to sleep. According to Caven *et al.* (1949) and Lansing (1952) in the waxing years of life, the motor dexterity, habits of physical activity and manipulating skills, that the child has acquired, have much to do with the mobility, vitality and the sense of personal adequacy. Hence the early years in the life of the child is the period which makes him what he is going to be in the later years. The motor skills he acquired, the spirit and the energy he developed for the mobility make him whether he is going to be an active and creative person or a docile and take life as it comes to him.

The child as he grows physically, he gains more ability to manipulate his surroundings. Such child also makes rapid growth. These characteristics when a child has acquired indicate that he has developed perceptual motor skills. The perceptual motor development is relating to two important principles such as differentiation and hierarchical integration. The principle of differentiation deals with the physical development of the child. This is the description of the increasing degree of specific control and its motor functions. The growing child quickly starts showing the development of motor coordination. It is first seen in the control of arms movements, hands movements, and then the finger movements. When there is increase of coordination then comes the increase of differentiation in the development of the child.

In this way the child gets mastery in every movement and then puts them together in a more complex pattern of behaviour or movement. Werner (1948) calls this process as hierarchical integration means that the child's new motor skills that are learned, are put together to make them in an integrated pattern of a large and coherent whole unit of motor activity. The child gets more and more control over his arms, legs, neck and other bodily parts and begins to put these differentiated skills together in simple activities and then tries to perform complex and integrated motor behaviour.

In the process of the perceptual motor development two aspects seem to be involved. One is the function of bodily maturation and the other is the role of experience or training. It is very difficult to say how much of each variable is involved in the perceptual-motor development. Generally maturation refers to the type of physical capacity revealed by an organism in the performance of an activity accurately, in which there is no environmental influence and experience or training. According to Collins English Dictionary (1979) maturation means relatively advanced physically, mentally, and emotionally grown-up. Stott (1974) states that maturation is the most significant physical changes that takes place along with age in relation to the psychological functioning. It is a change in quality which is not accessible for direct measurement in quantitative terms. Therefore, according to him this growing aspect of the human's development is called maturation. Maturation is not ending with the physical growth and development of the child. It is a continuous process. In other words, as Stott states, maturation is a life long process. Even after the growth and development of the physique, some changes in the internal structure of the body are taking place in various parts like the tissues and skeletal muscles and bones of the body. The bodily structures include the organs, muscles, nervous tissues, blood vessels and so on.

In the perceptual motor development apart from maturation, experience or training also plays an important part. But the amount of experience in motor development cannot be isolated and correctly assessed, since these two always go together. According to Todd (1937), development is a unitary process of change emerging from the interaction the organism makes in the environmental situation. The environmental influences also are brought in. Therefore, for the growth and development of an organism maturation and experience or training should not be considered as the independent processes. But, however, investigators tried to study this aspect, whether experience or training can be independent of maturation. Gesell and Thompson (1929) and McGraw (1935) tried to verify this phenomenon of maturation-experience issue.

A pair of identical twins who were having common hereditary qualities were used in this study to see whether these two twins who are supposed to have genetically equal in maturation and physical ability differ in their ability by learning and experience. One of the twins was given training in motor skills like walking, climbing, rolling and skating. The other twin was not given any training. It was observed that the child was not ready, that is when he is not sufficiently matured, experience or training was not effective. The twin who was given training have shown some gains in the beginning. When the twin who was not given training previously was able to take up the task very quickly on introducing the training when he was ready by maturation. Therefore, it is understood that it is the maturation that determines training. The level of experience and gains depend upon the level of maturation in a person. Therefore, the authors suggest that as the child is ready maturationally to perform any task, then teaching or training him any task is going to be a waste of time and effort.

The basic pattern of motor skills become the foundation for further development of motor skills. Objects are grasped to handle, examine and to manipulate. New skills the eye-hand coordinations are built upon the basic pattern of the motor skills. Children at this stage show great amount of variation in the number of motor skills developed. In this process, maturation plays small determining part. The child learns to thumb materials that are available to him. He now learns to write, learns to hold brush for painting, constructs pyramids with blocks and makes use of all the materials available within his reach. He learns to run, jump, climb, swing within the limited environmental situations. In the nursery or kinder garten school and in the play area he exhibits a variety of motor skills which he has already acquired or learned. In this there may be wider individual differences among the children.

When the child reached the school age he seems to have learned a great variety of motor skills. These skills are mainly manipulative and locomotor type. During this period most of the activities, plays and games of the child are involving various types of simple and complex activities. There may be various types of combinations, coordinations and integration of skills already developed in him.

Having learned many new types of motor skills the child enters the elementary school with all his skills. The elementary school period is going to be full of physically active years. He sees in the elementary school vicinity many children of his own age with full of energy and moving fast. He is very much interested in making use of physical energy in exercising of many bodily activities. He learns more to coordinate his bodily features with the skills he has already gained. He handles and manipulates new types of tools and play materials which are more related to his bodily activities. In his activities there are more muscular coordinations. The child is full of energy and he has to spend it by putting himself physically in more coordinated activities. These activities the child chooses both at home and also among the peers and they are mainly involved in fast bodily movements like running, jumping, climbing, diving, swimming, riding bicycle, skating and so on, which involve brisk movements.

The children are very much interested in achieving greater motor skills by performing acts which may surprise their parents and others, mainly the peers. In all these the child shows that he is a highly courageous boy. In this way there are more complex process of motor skills which are steadily and rapidly developing in the child. A large scale of muscular coordinations and manual dexterity are the characteristics of the play. Children in India, in the school age level show more interest in games such as football, cricket, basketball, judo (the Japanese game), swimming, and also tennis. Varieties of skills are involved in these games and the superiority of the skill also is established. Gutteridge (1939) has conducted a study with regard to the development of skills, involved in playing the base-ball in the United States. In her studies it was found that children were learning to throw the ball at the age of two or three. They start practicing and were able to throw the ball accurately only at the age of five or six years. About 75 to 85 percent of children were able to throw the ball well. This game of throwing the ball involve a complicated task. In India also the throw ball game is played at the schools but it is not a very familiar game.

Several authors like Gesell, Helverson, Thompson, Ilg, Castner and Ames (1940) have studied the processes involved in the game of throw ball. This game mainly involves visual localization, the way one stands, displacing the whole body, reaching, release and maintaining the static equilibrium. They have observed that the throwing of the ball involved a good sense of static and dynamic balance of the body, accurate aim, delivery and release, good eye-hand coordination, correct functioning of the fingers, quick arm movement, turning of the trunk, head, legs and controlling and throwing of the ball. Therefore, the throw ball game needs the development of complex motor skills of many types.

PREADOLESCENT AND ADOLESCENT PERIODS

In the preadolescent or preteen age period along with throw ball skills, there are also other skills to be practiced and developed. These skills are mainly fielding. This skill is involved in cricket and baseball. The other skills to be developed are catching, batting, running without getting disqualified from the game and other fine skills. Particularly in cricket game several motor skills are involved like bowling, batting, wicket keeping, fielding and so on. In all these activities lots of training are required. In bowling there are various techniques of delivering the ball. Each technique involves a particular kind of motor skill. This involves greater amount of force in bowling the ball. Similarly in facing the ball the player follows certain tactics, depending upon the speed of the ball. In playing this game one needs great amount of stamina and practice as it is played for hours together standing in the field. Like the game cricket, every game is required certain specific motor skill. This is achieved by continuous training and practice. However, experience one is having at the time of playing a match of the game one is required to practice it.

Apart from the development of motor skills for playing game, a person must also withstand the fatigue for long time. At this age one of the important factors involved in playing the game, efficiently is the motor skill. This we could see in the case of almost all the games. For example, in the game of tennis the factor of age plays a greater role. Since it is a combat or challenge between two individuals whoever is young and energetic with the skills and practice wins the game. Experiments and researches were conducted in the field of development of skills for each game time to time and it was found that the skills and techniques were changing and whoever was proficient of the skills at the time became the champion of the game. Whatever amount of skills one had developed he cannot be champion all the times or for longer period. The record one person sets this year is challenged and better record is established next year by another person. For example in tennis, the world number one rank is going on changing often, among three or four persons in a year. The same conditions prevail in other games such as cricket, football, hockey, basket ball, baseball and also in all athletic events. All these depend upon the development of motor skills, performing the task accurately and proficiently and also possessing great amount of stamina. Stott (1974) states that motor skills in children are developed gradually and steadily with the increase of age. Motor skills are not developed spontaneously, but a person at every stage should put himself in action and take training and exercise regularly to achieve them.

Generally, children as they appear are the same all over the world. But variations arise only from the manner by which they are brought up and trained. This is also true in the case of motor activities. Children on reaching their adolescent period are very much engaged in rigorous gross bodily activities like, doing everything fast, moving fast, running, jumping, climbing, pushing and doing many things without any aim simply to keep their body in action. These activities, particularly, the gross bodily activities also depend upon the environmental stimulations. Children in highly stimulating situations are very active and they engage themselves both in constructive or creative and destructive activities. In the preadolescent and adolescent periods it is the gross bodily activity that is important. They do not see whether it is going to be constructive or destructive. When one is motivated for an activity, the motivated activity normally springs from tensions. This is what Murphy (1964) says, while discussing about the concept of tensions. These tensions are some sort of accumulated tensions within the body. The tensions are aroused for action from outside stimuli and are arising from the vital organs, that are from the striped muscular system. During this period great deal of tensions are quickly accumulated due to outside pressures, and they are the activities performed by others. Murphy associate motivated activity with accumulated bodily tensions, outside stimulation, excitation of the body and activity. According to him maturation plays a large part in motor development as a whole.

Cultural Influence

The type of stimulations the individuals face for activity, vary from culture to culture. Particularly children and the preadolescents and adolescents are highly stimulated by environmental forces for activity. In some cultures there may be continuous over stimulation and the children may be highly tempted for activity. But in some cultures there may be understimulation and children may become docile and inactive and the development of skills may be delayed or they may be lacking in certain motor skills. In some cultures the cricket may be highly exciting game and children from childhood period may start practicing the same. For example some of the nations in the world, like England, West Indies, India and Pakistan are noted for their skills in cricket. All other nations either show little interest in this game or do not play. Similarly hockey was played long back only in India and Pakistan.

Now many nations in the world show interest in this game and also developing the skills. Some of the European nations, though started recently, they have developed much better skills in this area. Football is the game causing more excitement among the people in European countries. The excitements and tensions caused by this game among the European people were such that it has become a national pride for a nation to win the game in the world soccer matches. Italy is one of the nations giving more importance to develop the skills in winning the soccer. Once, when Benito Mussolini (1883-1945) was the dictator of Fascist Italy, there was a football match between Italy and Romania, in which Italy was defeated and Mussolini got wild and said they have lowered the dignity of the nation and ordered the players to go and play once again with Romania and come back with victory. Otherwise they all will be beheaded. Then the players had vigorous practice and played the football match against Romania and won the match. Mussolini was happy and praised the players and said that they have brought back the lost glory of the nation. Such was the importance given to the game, football in European countries. Italy is still maintaining the fame in the world soccer game. Therefore, culture plays an important role in developing certain physical skills and motor activities right from the childhood onwards. The Russian and the Chinese cultures pay more importance to gymnastics. The Negroes (both the American and African) give more importance to athletic activities. The Indian and Chinese cultures show proficiency in wrestling. The ancient Indian culture was also noted for the skills in shooting arrows from the bow and sword fighting. Therefore, right from the time of ancient culture to the present time man has been giving importance in developing some type of motor skills and these skills varied from culture to culture depending upon their physical needs and social and environmental stimulation.

Family Influence

Though culture is demanding some type of physical and motor skills in the individuals, the family is the primary source of inspiration for the children to develop the motor skills. In some families the father may be showing more interest in his children's physical activities and development of skills. The father, himself might have been once a champion of some games or athletic events and he wanted his sons also should be trained in the same lines and himself may be training his sons in various activities to find out the specific skill in which they may become proficient or experts. The mother also may be providing all the facilities for her children to train them in some physical or motor skills. At present this is taking place all over the world. To become proficient in game or in an athletic event is not only intended to show one's physical or motor skill, but also it has become a vocation or career to make money. For example in the games, tennis and cricket, the family influence plays more important role in developing the required motor skills. In India this attitude is very much encouraged. One Ramanathan from Madras was an expert tennis player once and when he retired from this game, he trained his son Krishnan in the same game. Krishnan has developed his skills in tennis and became an internationally reputed tennis player. When Krishnan retired, he trained his son Ramesh in the same line and he also became very proficient and played tennis at the international levels, representing India. Here we could see that the same game was played for three generations in the same family. Therefore, the family environment was a stimulating factor for the development of the skills of the individuals. The more winnings one makes in the area in which one is proficient, the more interest he develops to increase his skills for making further winnings. The development of such physical skills with enormous amount of stamina and proficiency in muscular and motor skills, should start from the time of preadolescent period.

Sex Variations

In developing motor skills the sex sometimes is causing cleavage at the adolescent and preadolescent period. Till certain years there is no difference between boys and girls in the development of motor abilities and physical and muscular skills. In one of the studies made by Govatos (1959), it was observed that up to the age of ten no differences were noticed between boys and girls in motor development. But when they reached the preadolescent stage the changes in physical growth caused differences gradually between the boys and girls and the boys develop greater physical strength and established their superior position in the development of physical skills and motor activities. In the bodily speed of movement, activities involving large muscles, in fatigue and endurance, the boys excel the girls. There were separate games and events for boys and girls, so also for men and women. But during the last two decades enormous amount of changes have taken place in the participation of games and athletic activities.

Now-a-days, girls in schools and colleges play the same games as the boys are playing and also the girls take part in the same athletic events as the boys are taking part. In school and college sports activities there is no event in which the girls are not taking part. The same level of activities have been extended from school and college to state level, national and international levels. The girls now-a-days play all the men games such as football, hockey, cricket, basketball, volleyball, tennis, table tennis, badminton and so on. In the same manner the girls are taking part in all the athletic events, like all types of running, jumping, throwing, swimming, gymnastics and so on. In fact in the international sports activities the girls excel men in gymnastics. In all these games and athletic activities the motor activities, physical and muscular skills should be developed to the required level of proficiency. The women who participate at the international sports events should have been developing their motor activities and other physical and muscular skills right from childhood or preadolescent period onwards. In all these the culture and the family environments play as the stimulating forces.

Peer Group influence

Apart from culture and the family environment, the influence of peer group is having greater amount of influence in the development of motor activities in the life of later adolescent boys. The boy at this stage must participate in the activities of his age group. If one is lacking in the required motor skills he will not be permitted to participate in the activities. Then he will be isolated and teased by other boys. If one is found below the level of the average boys, he should be given proper guidance and special training to develop his skill to be with others. In the prime adolescent period there is significant development in the speed and accuracy of motor activities and in physical strength.

ADOLESCENT CHARACTERISTICS

But the characteristic way the adolescent behaves sometimes is highly embarrassing. The behaviour is without any aim. His activities are very hasty, meaningless, more of self-conscious, a tendency to ignore others. The reason for this type of behaviour among the adolescents is that this period is full of energy, vigour, speed and great amount of muscular and bodily coordination and activity. The adolescent's motor activities are generally characterized by agility, that is, quick and easy movement.

The individual's developmental process further, is marked by high degree of temperamental characteristics. At various developmental stages it has been observed that the adolescents have been exhibiting contemplative, stable, delineative, persistent, shyness and socially not responsive characteristics. These characteristics are quite common only among the adolescents who are fast moving in their activities. The muscular training also is more involved in this. The muscles at this stage are trained along with the pperceptual process. The development of muscular activities and temperamental traits make the person more quick and accurate in his/her motor skills. Some temperamental persons solve certain difficult problems involving motor skills more quickly than non-temperamental persons.

Once a person has developed one type of motor skill and becomes efficient in dealing with the activities related to that skill, he may also develop certain other motor skills easily. Since the person has trained his muscles and other bodily features to meet the motor activities in one type, he can also appl; the same practice and experience in another type of motor activity successfully. For example persons who use fingers and arms in some motor skill like typing in a typewriter can correspondingly apply the same skill in electronic typewriters and in computers. A typist who is skilful in a manual typewriter, can efficiently handle the computer in feeding the matters. All motor skills involving the use of hands and fingers are interchangeable. Therefore, if a person is proficient in one type of motor activity, can perform successfully in another type of motor activity. In certain industrial job the level of motor skills are tested before one is employed. If he is lacking the skill, sufficient training is given to him to master the skill.

However one is efficient in some motor skills the business and industrial organizations provide some training to the newly employed persons to make them familiar with the work. Thus the development of motor skill right from the childhood period becomes very essential while the individual grows and matures physically and mentally.

In psychological studies, the developmental psychology deals with various developmental process. Psychologists study the developmental process not only theoretically, but also experimentally. They have designed and developed certain tests and instruments to test a person's level of motor skills in the laboratory situations. These tests, some of them are simple and some complex. They use such tests and instruments mainly for guidance and counseling purposes.

EXPERIMENT

Problem

Assess a person's level of developmental process in a simple motor activity through the speed and accuracy of his/her performance.

Method

Participant

Choose a person who has some interest in knowing his/her accuracy of motor performance and collect his/her personal information such as age, class, sex, games in which he/she is interested and the games played and other activities performed by hand.

Materials

The materials used in this experiment are, two circles with one inch radius and the centre of these two circles with six inches apart in a sheet of paper. The central point of these two circles are not marked. Ten sheets of paper of 11″×8 ½″ size, compass box, pencil, papers and stop watch / clock.

Preparation of the Material

Since this test material is a simple one, the experimenter himself can prepare it in advance and take it to the laboratory where he/she is conducting this experiment. The experimenter should get 10 sheets of paper to the size of 11″×8 ½″. Make the sheet one folding lengthwise and one folding breadthwise. You get the central point of the sheet. At the centre, where the two foldings meet, make a small mark with pencil. Keep the sheet lengthwise or up and down and measure 3″ apart from the central point on the upper portion and also on the lower portion on the folder line and make small dots in pencil at the two points. These two points make a distance of 6″ apart. Draw two circles of 1″ radius by putting the compass needle at one point and then at another point. Now you have two circles of 1″ radius with 6″ apart. Put your compass needle on the points marked on the folded line when you draw the circles. Now you find the two 1″ radius circles, one on the upper portion and one on the lower portion of the central point of the sheet of 11″ × 8 ½″ size. This is the test material for this experiment. Make out ten sheets like this. When you have prepared the ten sheets, your test materials are ready. If printed material is available it can be used (Appendix-I).

Procedure

The experimenter should first make ready the test materials. There should be ten sheets, with two cricles of one inch radius drawn, 6″ between the two central points of the circles. The distance between the centre of the two circles should be six inches. The experimenter should make the participant seated comfortably in front of a table in a well illuminated room. Then he/she should give some instructions about the test and the task the participant is going to perform. Showing one test sheet, the experimenter should ask the participant: "Do you see two circles at two places in this sheet?" The paticipant should approve of it and should give an affirmative answer. Then the experimenter should place one test sheet on the table in front of the participant vertically. This means, the two circles on the sheet should appear to the participant up and down, one circle should be above the central point and the other should be below the central point.

Then the experimenter should hand over a sharp lead pencil to the participant and give the following instructions:

> "There are two circles, one above and one below. Your task is to put dots by this pencil on both the circles by moving your hand up and down. You should move your hand up and down or forward and backward. You must put the dots in both circles in the *middle of the circles* as far as possible and as fast as you can. But see that every time you move your hand to the circles a dot is marked in the middle of the circles. When I say stop. You must stop it at once. Now keep the pencil ready to do the task and 'ready, go'."

As the participant stared the task, at once the experimenter should start the stop watch/clock and carefully watch the time. For every trial a period of ***ten seconds*** is permitted. After ten seconds the experimenter should ask the participant to stop it. At once the experimenter also should stop the stop watch. The experimenter should collect the test sheet from the participant and give him another test sheet for the second trial. Then the second trial should be started in the same manner as the first trial was done. On completing the second trial the experimenter should give another test sheet for the third trial to the participant and give him/her the signal to start and the experimenter should carefully start the stop watch for counting the time. In the same manner ten trials should be completed. When all the ten trials are over the experimenter should collect the last test sheet and the pencil from the participant and he/she may be allowed to go away.

Then the experimenter should scrutinize the ten test sheets and count the number of dots placed by the participant in each circle of every test sheet and enter in a table. Then the experimenter with his compasse draw over every one inch circle, three small circles to the size of ¼″, ½″ and ¾″ radius. Then he/she must count the number of dots within ¼″ radius circles, ½″ radius circles and ¾″ radius circles and enter in a table (Appendix-II).

Results

The results in this experiment are the responses made by the participant by placing dots in the middle or near middle of the two circles in all ten test sheets. The number of dots placed in the whole one inch circles should be counted and entered in Table 19.1. Then the number of dots fallen within the smaller circles of ¼″, ½″ and ¾″ radius should be counted and entered in Table 19.2. This is the individual data in this experiment. The task performed by the participant is an assessment of the development of his motor skill. The number of dots in each trial placed in both the upper circle and lower ciecle by the individual are presented in Table 19.1. The total for each of the circle of the ten trials, the total for all trials and the mean for the ten trials should be elicited.

Table 19.1: Showing the individual data, the number of dots in each circle and for each trial and mean of all 10 trials in the test of motor skills of the person.

S. No.	Trials	Number of dots in 1 radius		Total
		Upper Circle	Lower Circle	
1.	1			
2.	2			
3.	3			
4.	4			
5.	5			
6.	6			
7.	7			
8.	8			
9.	9			
10.	10			
Total				
Mean				

State the total number of dots your participant has marked in both circles for every trial. Do the dots in the circle vary from the other. In the series do you find any change in the number of dots as he/she progressed through the trials. Are the dots going for on increasing or decreasing? Find out the total number of dots for each trial and for all ten trials and work out mean and state it.

Find out the total number of dots in ¼′′, ½′′ and ¾′′ radius circles for each trial and enter them is Table 19.2 for all ten trials.

Table 19.2 : Individual data: Showing the speed and accuracy of the individual's Motor Development in the test of motor skills

S. No.	Trials	Number of Dots in each circle						Total
		¼′′ Circle		½′′ Circle		¾′′ Circle		
		Upper Circle	Lower Circle	Upper Circle	Lower Circle	Upper Circle	Lower Circle	
1.	1							
2.	2							
3.	3							
4.	4							
5.	5							
6.	6							
7.	7							
8.	8							
9.	9							
10.	10							
Circle Total								
Grand Total								
Mean								

State the total number of dots in the ¼′′ radius circle of the first trial and check the dots in the other trials 1 to 10, and state whether the dots are going on increasing or decreasing. Similarly find out whether the dots in ½′′ radius circles are going on increasing or decreasing as the trials increased. Find out the total number of dots falling in ¾′′ radius circles. Are the dots going on increasing or decreasing trial to trial.

Work out the total for all the ten trials for each of three sizes of circles and find out the mean. State the size in which the mean is high and in which the mean is low. Check whether the total dots of each trial are in agreement with the total dots as given in Table 19.1. If it is not, scrutinize once again and correct the mistake. State the trial in which the total number of dots are highest and the trial in which the dots are lowest in each size of the circles.

Collect the group data from other experimenters for all the ten trials and present them in Table 19.3. Work out the total and the mean score for the whole group for each trial, one to ten. Find out the mean, for each individual for all the trials. State the highest score and who scored it and state the lowest score and who scored it. Give the rank order for the subjects from one to 12 or 15 whatever the number of subjects in the group. Enter in Table 19.3, the number of dots in the ¼′′ radius circle of the first trial and the last trial for all the subjects in the respective columns.

Table 19.3: Group data showing the mean number of dots in all the trials for all the subjects and the level of accuracy of motor development in the test of motor skills

S. No.	Name	Trials and Dots										Total	Mean	Rank	Mean Dots in ¼′′ Circle	
		1	2	3	4	5	6	7	8	9	10				First Trial	Last Trial
1.	P.V															
2.	S.V															
3.	T.N															
4.	G.R															
5.	P.K															
6.	B.V															
7.	K.R															
8.	M.S															
9.	A.L															
10.	R.M															
11.	S.V															
12.	R.A															
13.	T.S															
14.	S.C															
15.	D.V															
Total																
Mean																

State the highest mean score for all the ten trials of the group, who has obtained this. State the lowest mean score in all the trials and who has obtained this. Who has obtained the first rank by getting the highest mean score and who obtained the lowest score and the last rank. With regard to the ¼′′ radius circle, who has obtained the highest number of dots and who the lowest. State the number of dots in both cases. State whether the mean dots are higher for the first trial or for the last trial. What in the group mean for each trial? What is the group mean for all the trials. How many have obtained above the group mean and how many below the group mean.

State the trial in which, the group as a whole obtained the highest score and in which the lowest score. Are the scores for the group progressing trial by trial, going on increasing or decreasing, state it.

Discussion

Explain the development of motor skills in an individual and how this skill can be developed by an individual. State and discuss the factors that are mainly contributing to the development of motor skills. How do you relate the development of motor skills in a person to the experiment you have conducted. State it in terms of the hypothesis, what the results of this study indicate? Discuss this stating the scores obtained by the individual in the test of motor skill.

The assumption is that if a person is proficient in motor development than his performance in this test should be revealed. He should score high in the task assigned to him/her. If he/she is lacking in motor skills, the performance would be lower or poor. This is tested in this experiment in individuals. Therefore, discuss the individual's performance in terms of his scores. State, what the individual has obtained in the first trial, and go through all the trials whether the scores are going on increasing or decreasing. In how many trials he/she scored more than the mean and in how many less than the means. What does it indicate? Is he/she progressing in the performance?

The task assigned to the participant is to put dots in the middle of the two circles, moving his hand forward and backward and as fast as he can. In this task where he/she put most of the dots. This is tested by drawing three circles to the radius of ¼′′, ½′′ and ¾′′ and counting the number of dots in each circle. Putting most of the dots within ¼′′ radius circle indicates that the person is highly skilful in motor activity, if most of the dots fall within ½′′ radius circle he/she is moderately skilful in motor activities and if most of the dots fall within ¾′′ radius circle the person is somewhat skilful in motor activity. But if the dots fall away from the ¾′′ radius circle the person lacks in motor skills. This is how a person's motor skill is discussed in this experiment. Discuss it in terms of the individual's score in these areas. Is your participant highly skilful in motor activities? State the number of mean dots in all these three areas and also the grand total and discuss his/her proficiency in motor skills.

Compare your participant with the group members. In the group where your participant stands. State the mean of the group for all ten trials. State whether your participant has obtained the score above the group mean or below the group mean. State his rank in the group and assess his motor skills as revealed by this test. State and discuss who is the person above the group mean and those who scored below the group mean.

Discuss about the performance of these persons. What the scores above the group mean and below the group mean indicate? Who are those more proficient in motor skills as revealed by this experiment, who are those developed the skills moderately, and who are those poor in their motor skills.

With regard to the accuracy of motor skills, who are those put the largest number of dots in the ¼′′ radius circle in the first trial and in the last trials, in the same size circle. Discuss the performance of these persons. Who are those put the lowest number of dots in the ¼′′ radius circle. What does it indicate? Discuss and compare their dots in all the three areas of ¼′′, ½′′ and ¾′′ radius circle. State and discuss who are those persons performed the task more proficiently and more accurately and who are those performed the task poorly and less accurately in this test of motor skills. Do you find any reasons for better and accurate performance in such task? If so state and discuss it. Is there any difference between males and females in their performance. State the reasons. Do you find that the participant's experience or training in some motor task had any influence in the performance? If so, discuss it. Are there any differences in the performance of those who play some games or do some exercises similar to this task? Do you find that age, and the subject one studies are having any influence in this task? Discuss it.

Conclusions

1. State the nature of performance of your participant. Is he/she highly proficient, or moderately proficient or poor? Where he/she stands in the group? State the rank.
2. How many of them in the group highly proficient in the task? And how many moderately proficient and how many poor?
3. How many in the group more accurately and efficiently performed the task and how many were not accurate and not proficient in the task.
4. Is there any difference in the performance of this task due to sex and age, experience or training in similar tasks.

APPENDIX – I

TEST OF MOTOR DEVELOPMENTAL PROCESS

Fig. 19.1: Motor Ability: Developmental Process : Experiment : 58

APPENDIX – II

THE TEST AFTER TESTING THE MOTOR DEVELOPMENTAL PROCESS

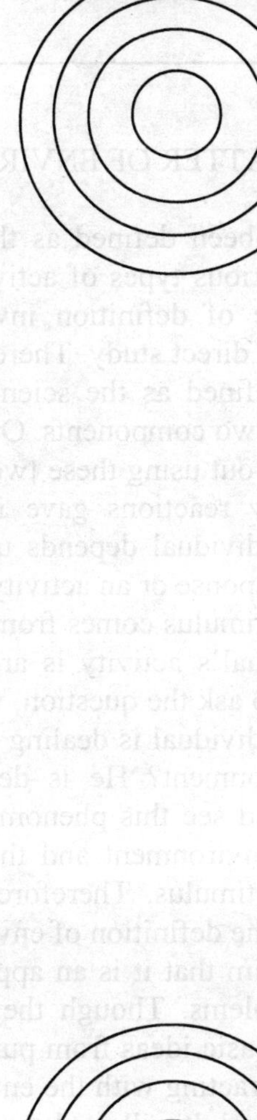

Fig. 19.2 : Motor Ability : Developmental Process : Experiment : 58

20

ENVIRONMENT AND BEHAVIOUR

INTRODUCTION : SUBJECT MATTER OF ENVIRONMENTAL PSYCHOLOGY

In the earlier days psychology has been defined as the science of the mind. Mind is revealed in behaviour, mainly in various types of activities such as cognitive, conative and emotional activities. This type of definition involving the mind raised several controversies as it is inaccessible for direct study. Therefore, the definition went through several changes and at last it is defined as the science of human behaviour. Human behaviour is always associated with two components. One is the stimulus and the other is the environment. Researches carried out using these two components. The psychologists who were concentrating in sensory reactions gave importance to stimulus and the environment. The activity of the individual depends upon the stimulation. A stimulus arouses a response or activity or a response or an activity is aroused by a stimulus. Where does the stimulus come from? The stimulus comes from the environment. Therefore, we may state like this, that an individual's activity is aroused by the stimulus from the environment. In this case we can also ask the question, what does the individual do? The answer to this question is that the individual is dealing with the environment. How does the individual deal with the environment? He is dealing with the environment by responding to the stimulus. We could see this phenomenon like this. The individual is aroused by the stimulus from the environment and the individual is dealing with the environment by responding to the stimulus. Therefore, Woodworth's (1952) formula, W-S-O-R-W is rightly applicable to the definition of environmental psychology.

Environmental psychologists claim that it is an applied field of study, dealing with the pressing of the real world problems. Though they are not concerned with basic researches, they have drawn all the basic ideas from pure or basic researches. The basic principle is that the individual is interacting with the environment through stimuli which are constantly influencing the individual. It will not be possible for us to define or explain environmental psychology without this basic principle. Therefore, we must accept the idea suggested by Heimstra and McDonald (1973) that both kinds of researches are essential if psychology is to advance as science.

WHAT IS ENVIRONMENTAL PSYCHOLOGY?

Environmental Psychology has been defined in various ways. Each definition gives importance to one aspect of the environment or to the organism. Heimstra and

McFarling (1974) bring relationship between human behaviour and man's physical environment. Proshansky (1976) in his definition brings relationship between man's behaviour and experience and his built environment. According to Kantowitz (1977), it is the study of man in the physical setting. Holahan (1986) relates physical world with human behaviour. All these definitions do not bring out the kernel point in these definitions. Therefore, a definition which involves all the salient elements in the definition may be sought.

Environmental Psychology may be defined as the study of the effects of built or non-built or physical and natural environment on human behaviour and the consequent human reaction toward the built or physical and natural environment. In environmental psychology the individual's or the group's behaviour is assessed both in the man-made environment and in the physical and natural environment and also in some unusual environment. Therefore it becomes necessary to involve every type of environment where human beings function with some kind of behaviour. Man interacts with the physical environment in a number of ways and in return the physical environment effects human life in a number of ways. The physical environmental factors determine how man should behave in a particular situation. Man's sensory experience, his perceptual processes, his intelligence, personality development, motives, emotions, interests and aptitude, likes and dislikes, prejudices and attitudes were studied in detail by the psychologists and they invariably associated the influence of environmental factors with human behaviour. What has been conceived as environmental conditions by the psychologists include the physical and natural environments like the roads, forests, mountainous areas, rivers, the tropical areas, the sea and sea shore, under the sea, the space, the arctic and antarctic regions. The built or man-made environmental conditions are the urban settings with buildings, work places, offices, industries, dams, and roads. The social environmental conditions are the population, the communities, the crowds and so on. The aim of environmental psychology is to understand the manner in which the physical and natural environment, the built or man-made environment and the social environment affect the human behaviour and to make use of this result and knowledge to make improvement or change or modify the environment to suit to the human conditions.

Further, environmental psychologists and others have observed that there are environmental problems steadily increasing all over the world like the rapid growth of population, overcrowding, environmental pollutions, air pollution, noise and heat in urban settings, the destruction of forests and construction dust, smoke and the toxic materials producing industries which are affecting the behaviour of the people and their health. People are constantly put into environmental stress.

Environmental Psychology is also interested in the problems of interpersonal environments. The personal space of the individual, the effects of sex and culture in personal space, the situational effects on personal space, intrusions into the personal space, territorial behaviour, territorial dominances, territoriality and family relationship, territory and architectural designs, overcrowding and its effects on human behaviour. Thus environmental psychology has a very wider scope in its study and it has become one of the important applied areas of psychology.

Environmental psychology had its origin around 1960 and within a short period, lots of researches have been conducted and enormous amount of materials in the form of books, pamphlets, journals have been brought out. Studies have been made and analysed the psychological consequences of environmental problems. The business and industrial organizations and other public and private sectors are already aware of the potential environmental dangers. The scientists of various disciplines are also dealing with the environmental problems. When psychologists attempted to deal with the environmental issues a new field of psychology called the environmental psychology was born. What psychology is going to offer and how the environmental psychologists are going to deal with the environmental hazards? Psychology has plenty of things to offer and psychologists can formulate strategies to solve the pressing real world problems.

METHODS OF ENVIRONMENTAL PSYCHOLOGY

Experimenters in the field of psychology use several methods or approaches in conducting research. Though it is possible to use all these methods in the field of environmental psychology to some extent, environmental psychologists use some selective methods. They mainly apply the experimental method, the correlational method, obtrusive method and unobtrusive method.

Experimental Method

In experimental method, generally, the experimenter directly controls the variables and causes of changes in the behaviour. The experimenter manipulates the independent variable and finds changes in the dependent variable, the behaviour in the organism. This is possible only in the laboratory situation. In environmental psychology, laboratory experiments were conducted by making changes in the environmental components like noise, heat and air or wind. The dependent variable in an experiment should be in some measurable response from the respondents who are being studied. In these situations the responses may be verbal, physiological or bodily changes or motor responses. Suppose, we are studying the effect of heat or noise on some persons, the intensity of heat in a room may be raised from normal level to some abnormal level which the respondents could not bear it and the reactions may be recorded. Similarly in the case of noise the independent variable noise can be presented from very feeble noise to louder noise with varying intensities in a room and the responses of the respondents may be recorded. A wind tunnel may be constructed to study the reaction of the wind of varying intensities in some persons who serve as subjects in this experiment. Broadbent (1956), David Glass and Singer (1972) have studied the effects of noise in various types of tasks. The effects of varying temperature in a room was studied by Griffitt (1970) in a laboratory situation by Page (1978) and in a field by Cunningham (1979). The effects of air pollution was experimentally studied by Blot *et al.* (1975), Mendelsohn and Orcutt (1979) and others.

But in many of the natural situations we see some typical reactions of the people. Mostly in extreme hot places in summer season, and in extreme cold places in winter season the reactions of the people may be recorded only by observation. Similarly, people who live by the side of railway stations, by the side of aerodromes, metal and steel industries are constantly experiencing the horrible noises. What type of reactions and adaptive mechanisms are exercised by the people, can be studied only by observation or by contacting them. How do people face air pollution in certain environmental situations can be studied only by conducting some investigations. The effects of torrential wind cannot be studied whenever we want. This is an occasional natural event takes place due to depression of the atmosphere. Therefore, the effects of the consequences can be studied only by observation on such occasions. But a wind tunnel experimental effects may provide accurate information, whereas the effects of torrential wind must be observed occasionally. Thus experimental method in environmental psychology is very difficult as people may not be willing to face the severe effects the environmental components.

Correlational Method

Correlational method involves comparing between the results. It is comparing the results occurred in two different situations. It may be comparing the result of some natural event with similar event occurred in some other situations. These results are obtained by careful observations of both the events and collecting information. The effects of any one of the environmental components may be compared. For example, the temperature of heat may be causing the same effect in city area. But how people are behaving or reacting to heat in the city roads, shopping places in comparison to the behaviour of the people in suburban or the slum areas. The types of reaction of the people in one area may be changing from the other area. The experimenter can observe and record whether the changes in behaviour in the crowded city area are having any relationship with the changes in behaviour in the suburban or slum areas.

In these situations the experimenter is not manipulating the situation or the variables. The experimenter observes specifically the relationship between the changes in the reaction between these two groups of people in two natural situations, which are having different characteristics. But the experimenter's aim is to find out the effects of heat in the changes of reaction of the two groups of people. But this correlational method is based upon the observation of the effects of natural events. For example, if the intensity of heat is severe, the city people may adopt all the artificial devices to face the heat. But the suburban or the slum people directly face the heat effect without any adopting devices. In this way for the environmental psychologist the whole city area including the suburban and the slum areas constitute the natural laboratory, since he observes the reaction of the people in a natural situation and not in an artificial laboratory situation. Thus the environmental psychologist can observe the reactions of the people for a specific environmental variable in the fields, in the forest, in mountainous regions, in tropical areas, in desert area, in the sea and compare the changes occurring in the situations. But the correlational method is not as reliable as experimental method. Actually the data collection made for correlational method is only by observation in the natural setting. The experimenter, apart from observing and recording the information, can use also the self-reporting method for collecting the data from both the areas and compare the results. Persons may be asked to express their views by assessing the situation and the reports may be collected from both the areas and correlated.

Questionnaire Method

The questionnaire method can be also combined with the interview method in collecting data. The situational or environmental characteristics may be worded with both positive and negative ends and the person may be asked to assess and rate the situation on a five-point or seven-point scale, ranging from most negative end to most positive end. Scores are assigned to the level of rating and the individual score is the total score for all the items rated. This scale was designed first by Fisher (1974) for judging the quality of the environment. Bell *et al.* (1978) call this as obtrusive measure. The same type of scale can be used in an interview and data can be collected on the qualities of the environment. This method is having certain disadvantages. Since the person, from whom the experimenter is collecting the information, comes to know that his/her judgments are collected, there is every chance for the person to behave differently. He/She knows that his views are watched and hence the person may give dubious judgments and not his/her real judgments. This type of self reports may spoil the aim of the study.

Indirect or Unobtrusive Method

The indirect or unobtrusive method was found to be more reliable in collecting reports from the persons in various environmental situations. This method is free from the suspicion caused by the questionnaire method in the minds of the persons from whom informations are collected. This method was developed by Web and others (1966). They conducted an investigation by using this method from high and low density dormitory residents. Webb and others used this unobtrusive method extensively. Brickman and others (1973) wanted to study the effect of various levels of density of population in 'helping behaviour' among the high and low density dormitory residents. They send stamped addressed envelopes with self-report scale to the residents. By this means, the respondents may remain unidentified and hence they may express their judgments freely. Many investigators used this method in various situations.

RELATIONSHIPS BETWEEN ENVIRONMENT AND BEHAVIOUR

The relationships between environment and behaviour are built upon extensive information obtained from many experimental results. There are several mediators evolved from the empirical results and they try to explain. In this way there are several theories which attempt to explain the relationship between the environment and human behaviour. Each theory uses one or more mediators to explain the relationship. The mediators serve as antecedent causes, and the consequences are the coping responses or protective devices. Therefore the cause and effect hypothesis is applied here. There is a cause behind every act in this work. An event when occurred, there is another event followed by it. Thus there is a cause behind every effect. It means something determines an effect. Psychologists, hence accept the idea of determination. Something determines the human behaviour.

The fundamental assumption on which the psychological studies are made, is that human behaviour is determined by the environmental events. Environmental events may determine human behaviour cognitively, conatively and emotionally. Therefore psychological researches are directed in finding out the antecedent causes of various behavioural events or acts. In every scientific research the hypothesis is verified, whether the specific effect could be caused by making an event to occur. The verification of a hypothesis yields some new knowledge and on the basis of the knowledge some hypothesis may be formulated and the knowledge can be tested and further verified. In this way knowledge is made observable empirically. On several such attempts toward empirical observation and on gathering several facts of the same type we can go for evolving the abstract idea of formulating the theory. This is how theories are framed in environmental psychology, just like all other fields. Though the field is just beginning, valuable theories have been framed with regard to the determination of behaviour. There are no elaborate theoretical formulations in the field of environmental psychology as it is a young science. Hence there are very few theoretical systems which can be used for the prediction and generalization of the effects of environmental components on behaviour. Generally in the environmental situations the organism is put into stress, aroused, experiences over stimulation or under stimulation.

Stress

Stress may be physical or mental. Stress generally refers to mental or physical tension, force that causes strain physically or mentally. Psychologically, stress is a response to an external threat that involves fear, anxiety and irritation or anger. Physiologically stress refers to physical discomforts or physical strain or some physical impairment. The environment component if it is strong or severe it causes physical stress. Man experiences stress when there are demands on him, which exceed his adjustive resources. Man's adjustive resources are limited and when the pressure on him exceeds he faces stress, a kind of disorganization of the body. The environmental conditions are noxious or harmful to the tissues of the body. They cause harm to the normal integrated functions of the tissues.

In the environmental conditions there are physical stressors such as heat, cold, humidity, natural disasters like floods and earthquakes, environmental pollution caused by dumping of toxic wastes and garbage in public places and polluting the air, water, which may result in producing microorganisms or virus which may cause infectious diseases. All cause stress to humans and they are termed as environmental stressors.

Noise is another environmental component, which causes much stress and strain in its extreme level. Stress is a reaction to these environmental stressors. They serve as stimuli in an environmental condition and make the organism to respond with the stress to the condition. Every environmental component in its extreme levels high or low causes stress to the organism. Thus according to environmental psychology in a stimulus-response situation what we call stressors are all environmental components. They may be natural or man-made components.

According to Lazarus (1961) there are certain environmental social conditions, which may cause damage to the organism at present or in future. Such conditions are called psychosocial stressors. The overcrowding is a social environmental condition which causes much stress and strain. This may result in aggression and violence in a situation, also causing stampeding and death and criminal activities. Though stress originates in an environment it is not always there. Stress also depends upon the weakness of the individuals and the strength of their system of defense mechanisms built physiologically and psychologically. Man learns to withstand certain stressful situations. For example, the military operations, imprisonment, captivity, natural disasters like heavy rains, floods, storms, earthquakes, fires, explosion, disabling injuries, incapacitating illness, terminating diseases like the cancer and aids, loss of loved ones are some of the stressors depend upon the defense mechanism built by the individuals. Individuals react to the same stressors in varied ways according to their adequacies. The reaction also varies in the environmental conditions according to the severity of the environmental components. For example the heavy rains and floods which may wash away everything on the route. Those who have lost their properties and dear ones may be disorganized by the stress while some may consider it as an adventurous event which they rarely come across in their life.

Thus every environmental component in its extreme level may cause stress to humans and stress is a reaction to the stressors. Stress is an alarming reaction in which the physiological processes are accelerated and there may be increased heart rate, blood pressure, and glandular secretions. The other state of reaction to stressors is residence. The body automatically puts up certain amount of resistance to the stressors. But this is possible only to certain extent. For example, in case of environmental component, temperature is tolerated only to certain level. If the environmental temperature is increased above the normal level, then there is severe sweating all over the body and if the same condition continues for a long period, then the body get dehydrated and it results in disorganization and death. Some of the environmental stressor are polluted air and water, extreme temperatures of heat and cold, humidity, weather condition, wind, noise apart from the natural physical environmental conditions like the storm or tempest or violent wind, earthquakes, rains, floods, fires and so on. All these are environmental stressors cause stress to humans. Experiments have been conducted on certain stressors to study the effects of them on humans, the type of reactions or the type of resistance to the stressors in laboratory situations.

Arousal

This is supposed to be one of the reactions to the stressors. It is an elevating response from an organism by a stimulus or a stressor. When an organism is exposed to stressors the organism is made to awake swiftly or vigorously. Thus arousal is an increased reaction effect to the stressors. Arousal is an abstract concept. It can only be inferred from empirical evidences. Evidences can be inferred from the increased physiological changes, in the autonomic system. When an organism is aroused, then there are several autonomic changes, like increased heart rate, blood pressure, increased respiration rate, constriction of blood vessels, galvanic skin response, palm sweating, urine secretion, muscular tension, adrenal secretion and so on. Hebb (1972) and others who have studied this aspect from neurophysiological perspective state that arousal is an increased brain activity by stimulating the brain centre called reticular formation.

Arousal is also considered as an intervening variable in various types of situations while studying the behaviour. It serves as a mediator in many situations. Therefore, though arousal is considered as a reaction to the stressors in the environmental psychology, in many of the situations arousal is used to explain behaviour. In this case the arousal is not considered as the effect of stressor or reaction to stressor. The arousal is a response to the ordinary stimulus and the stimulus is not causing any stress or causes increased arousal nor the stimulus is of a threatening type. But the organism is aroused for action. Any normal event in the street, an unusually designed building, new type of an automobile, a couple with disporoportionate height, are also the incidents causing arousal.

Arousal may have varied effect on behaviour, depending upon the level of arousal. It may be lowered arousal and increased arousal. The lowered arousal does not cause instant behaviour whereas makes the organism ready for immediate action which may cause also stress. For example, a loud noise at a distance may have lowered arousal on behaviour, whereas the same loud noise may have startled or heightened arousal on behaviour to the extent of becoming also a stress. Arousal of both the extremes, either too low or too high will have disadvantage on the behaviour.

Every environmental component is a stimulation in certain situation if any organism reacts to it. Human beings expect the optimum or intermediate level of stimulation. Broadbent (1971), Hebb (1972) and Kahneman (1973) have observed in their studies of environment-behaviour relationship that when the environmental stimulations from various sources like the noise, temperature, air pollution or crowding were increased, the arousal, the behavioural performance or the tasks were also increased, or reduced. They found that this type of behaviour depended upon the level of response of the person in a particular task. The response might have been below or above the optimum arousal level for performing the task. These investigators have observed that low arousal was facilitating for maximum performance of the task and extremely heightened arousal was hindering from paying more attention on the performance of the task. Arousal serves as an intermediate variable between stimulus and behaviour. It also functions as a mediator in arousing other behaviour like aggression in certain situations. According to Berkowitz (1970), Geen and O'Neal (1969) the increased arousal will be conducive for aggressive behaviour, if the situation could arouse the response of aggression. For example, a noisy situation may cause irritative behaviour, and if the noise is of extreme type, then it results in arousing aggressive behaviour.

OVER STIMULATION IN ENVIRONMENT —BEHAVIOUR RELATIONSHIP

The environment-behaviour relationship is explained through various theoretical approaches. We have described in the preceding passage, the stress and arousal. Since human capacity is limited man cannot put sufficient efforts to receive all the stimulations that are attacking him in a particular situation and respond to them. Cohan (1977) studied this aspect of over stimulation in environment-behaviour relationship. In a complex environmental situation there may be many environmental stimulations which a person with his limited abilities cannot process them and make out meaning out of them. In such situations the environment behaviour relationship cannot be adequately explained through stress and arousal. Therefore Cohan (1977) and Milgram (1970) have proposed a different theoretical approach called environmental overload approach to explain the environment-behaviour relationship.

What is Environmental Overload ?

Man's behaviour is determined by environmental stimulation. In an environment there may be many environmental conditions or components operating. Thus it becomes a complex environmental situation and too many types of stimulations from too many sources are attacking the man and man cannot respond to all the stimulations and process them meaningfully. Thus the amount of information from the environmental sources exceed the man's capacity to receive the relevant stimulations and process them meaningfully. This process is called environment overload or information overload.

When there are too many stimulations from several sources or when there is environmental overload, how humans deal with them. One way of coping with such situation is to respond to relevant stimuli as far as possible and to ignore the stimulations which are not relevant to the immediate need. This is what humans do in their behaviour. In such situations man may be positively reacting to certain stimulations and negatively reacting to certain stimulations. Hence man's coping up plan is negative behavioural effect to excess or over environmental stimulations. The stimuli which are important for the immediate purpose or for the task are given more attention and the behaviour-environment relationship is positive and the stimuli which are not very important for the immediate purpose or for the task are ignored without paying any attention toward them and hence the behaviour-environmental relationship is negative.

Sometimes, it is also possible that the less important stimuli may interfere with the main task when ignored. Ignoring such stimuli may facilitate and improve the main task. But if the interference of the ignored stimuli are strong, then it may reduce the efficiency of the main task. If it is a strong distractor of attention, then the overcoming of the distraction is to make it as an advantageous stimulus which may even increase the efficiency. This can be done by putting more attention and effort in the main task. For example, the piano player while playing piano with his right hand often experiences that his left was a constant source of distraction and spoiling his performance on the piano. He cannot remove his left hand and play piano. Therefore, he tried to overcome the distraction in various ways. As he found a way of making use of both the hands in playing the piano, it has also increased the efficiency. The environmental overload of distracting stimulus was used as one of the positive stimulus and the excess load was avoided. In many situations humans themselves find out ways and means to overcome the distracting environmental overload.

The environmental overload may persist even if we ignore the stimuli which are less or not important for the main task and it may result in distracting our attention in the main performance. The excess stimuli in the environmental situations are causing distraction, which refers to the failure of concentration of attention. The excess stimuli are the forces that are impairing our performance and turn our attention from main task. These distracting forces can be overcome in four ways. (1) The distracting or the excess environmental forces can be overcome by putting more attention and effort in the work.

(2) If the performance was continued for many days with the distracting forces, the work output can be increased by putting more energy in the work and by developing negative adaptation. This means, the worker has learned to ignore the distracting forces or the overload. (3) There are certain environmental excesses or distracting forces or stimuli that could neither be ignored nor overcome by any means. Sometimes the excess stimuli or distracting forces themselves become the centre of interest and the worker forgets his main work. Woodworth (1952) says the music, which was in the background, was a source of distraction for the work, later on became a source of inspiration for the same work. (4) One best way of avoiding or overcoming the excess stimuli or the distracting forces is to avoid them physically by removing them from the place or shifting work place to some other places.

In the study of environment-behaviour relationship it was also observed that if work is performed in the environment overload situation for a longer period of time the worker's capacity get exhausted by putting more effort in the work and even a very small need for work is considered as overload. Further the environmental overload results in several other behavioural negative after effects such as frustration, more errors in mental works, less interest in the welfare of others. All these occur because of reduced capacity due to exhaustion and there is lack of attention to even relevant stimuli. Milgram (1970) states that when there was increased urgent requirements for every day activities, there was disintegration of social life in big cities and urban areas. People fail to attend the welfare of others because of the environmental overload. Thus the normal behaviour pattern is very much affected by the environmental overload as man will have to spend much energy in it.

ENVIRONMENT-BEHAVIOUR RELATIONSHIP IN UNDER-STIMULATION

Human capacity for work is not unlimited. Man has only limited capacity for work and he tries to make use of it to meet all his demands in a short period of time. When this is not possible he get exhausted, vexed and frustrated. It also involves more emotional disturbance. When there are too many stimuli in the environmental situation, man is not able to respond to all of them with his limited capacity. This is one type of problem man faces due to overload of environmental stimuli. This is one extreme. The others extreme in which man faces environmental problems is the under-stimulation or inadequate source of stimulation. Human behaviour can be also related to environmental under-stimulations. What we call under-stimulation is referring to, mainly inactions or non-response on the part of man when the environmental stimulation is weak or feeble and fails to arouse response in an organism.

We are discussing understimulation in this context, referring to the characteristics of environmental stimulation and not in the context of sensitivity of human sensory apparatuses. We know that human sensory apparatuses are limited in their capacities compared to some animals and birds. Man needs an optimum stimulation to respond. The stimulation should be to the level of arousing response in an organism or in a normal human being who is endowed with normal sound sensory apparatuses and not defective. If a normal human with a sound sensory apparatuses fails to respond to a stimulus either by seeing, or hearing or smelling or in any other form, the stimulus is said to be too weak stimulus and there is too small or little stimulation. Persons who have sensory deprivation may not be responding to stimulations however big or strong they are. We should not call this as understimulation.

From the environmental psychological point of view, as some authors like Parr (1966) has suggested that understimulation may also mean opposite of overstimulation, referring to the environmental conditions which can cause excitement with few stimulations. Parr (1966) is of the view that the urban areas are with overstimulating social environment and people get disgusted with it and wish to have non-overstimulating physical environment. According to him the mountainous regions, the forest areas, the fields have limitless variety of changes with natural sceneries stimulating visually, whereas the urban areas have the same type of streets and buildings in some particular structure resembling each other. Man feels as if he is shut in all directions without any chance of free expression for sensory aspiration. This type of environment is not causing any interest in life. Man develops a feeling disinterestedness or boredom in life aspects which becomes the cause of some sort of illness. Persons or young boys become disgusted and turn to delinquency, criminal activities and turn to damage the properties.

This type of social environment which fails to provide sufficient stimulation for the expression of sensory aspirations and sense of well-being make people to go in search of natural physical environment. Therefore, understimulation, according to urbanologist means the environment which is not providing sufficient stimulation for happy living and to have exciting experience in life. Hence environmental understimulation is not defined in terms of intensity of stimulation so as to arouse sensory responses. Thus environmental understimulation is having relationship with human behaviour in this way.

OPTIMUM LEVEL OF ENVIRONMENTAL STIMULATION–BEHAVIOUR RELATIONSHIP

In the preceding section we have discussed about environmental overstimulation and understimulation and behaviour. It is observed that both the conditions are having harmful effects on behaviour. These levels of stimulations stimulate also other behaviours like feeling of frustration and anger. Therefore, any stimulation of intermediate or medium type may be the most suitable for behaviour. Halson (1964) evolved a theory called adaptation level theory. Whohlwill (1974) believes that environment-behaviour can be explained through this theory. Adaptation means making suitable changes to the situation. This phenomenon can be applied to environment-behaviour relationship. Man is reacting to the situation through his sensory responses, social responses and his movements. He adapts himself to various types of situations. The optimum level of stimulation is most desirable. In this situation he is not finding any problem. His reactions and behaviour are normal.

Man in the environmental world is exposed to all the three types of stimulations, such as over, under and optimum levels. Though the optimum level is desired, it is not always possible to be in the midst of optimum level of stimulation. The level of stimulations in the environment may be going on changing. In a situation where there is overstimulation and if it is not possible to avoid it man learns to live in the situation by making changes in himself or in the environment. By this means he learns to adapt himself to the overstimulation or understimulation. This is also a kind of adjustment.

There may be environmental conditions of varying levels of stimulation. There may be different levels of intensity in the environmental components. For example the physical environmental components sound or noise, temperature, low level of oxygen concentration and wind all may be varying place to place and time to time. Though man is experiencing the loud sounds constantly, after few hours he adopts to the noise and he does not find it a disturbing force. Or if possible he reduces the tone of the volume. But in a place where there is sound or noise man feels unnerved or threatened. A place where there is absolute silence for long time, it is also a disturbing condition. Similarly the temperature too high or too low is not avoidable to us. In case if it is not avoidable man feels uncomfortable for sometimes, but later on leans to adapt himself to both extreme heat and extreme cold by making changes in himself and in the environment.

The environmental stimulations may have varied diversities. Most of them may be of sensory stimulations like seeing, tasting, smelling and touching. But too much and too little diversities also are not desirable. This can be seen mostly in lighting arrangements with different colours, ranging from very bright to dull lighting. Similarly in eating and smelling humans experience extreme levels. For example, too hot, too sweet, too salty, and also too cold, too sour and no salt may cause unpleasant feeling. But in these cases there will be individual differences. Individuals differ in tastes. Therefore, there may be diversity in matters relating to likes and dislikes of colours, feelings and tastes.

The adaptation level theory is not as strong as the other theories in explaining the environment-behaviour relationship. Unlike other theories we cannot generalize the consequences nor predict the relationship correctly between the environment and behaviour. It is because of the variations in individual's level of adaptation to situations. What is agreeable to one person is not agreeable to another person and what is acceptable in one place is not acceptable in another place. Therefore, it is not a strong variable through which environment-behaviour relationship can be satisfactorily explained.

ENVIRONMENTAL POLLUTIONS

Environmental Psychology is mainly interested in the study of the environmental effects on human behaviour and the behavioural effect on the environment. For environmental psychologists environment is perceived with a very wider perspective involving the physical natural environments like the fields, forests, mountainous regions, the tropical areas, the rivers, the sea and sea-shore, the arctic and antarctic regions, the space and all unusual environmental conditions, the built in environments like the residential areas, organizational buildings, offices, courts, prisons, roads, dams, bridges, tunnels, breakwells, harbours, wall between countries, the railway lines, aerodromes and so on are all built in or man made environments. All these are environments for the environmental psychologists. These environments are closely connected with humans in one way or the other and they are influencing human behaviour. In return man reacts to these environments in various ways. Human reactions to the physical natural environments results in making changes according to their requirements. Man makes changes and modifications in the natural environments. In this venture man destroys the natural environments like the forests, fields and mountainous areas for constructing industries, buildings and houses to live in and so on. When man makes his establishment in these areas, it results in producing chemical substances, toxic materials, throwing lots of wastages, smokes and gases. In due course these toxic materials and wastages accumulate in heaps. Sometimes these wastages if they are in liquid forms channalized into the nearby rivers and also in seas. These are causing health hazards. In this way a clean fresh air area is polluted which causes health problems to humans.

When this type of behavioural activities increase more and more beautiful green field and forests, mountain valley areas which were once healthy areas with fresh air and green sceneries becomes polluted by the toxic materials, smokes, dusts, and wastages. Thus there is fast deterioration in the life style of man due to environmental pollutants. Environmental psychology tries to investigate the effects of various types of pollutants like the littering environment, the effects of environmental pollution on behaviour, the effects of noise, extreme temperatures, the effects of air pollution and so on. In urban areas air pollution is mostly caused by automobile transports. The automobiles run by oils, petrol, and diesel produce smokes and pollute the air. The pedestrians who have no means of escaping from these automobile smokes, but to breath and spoil their health.

The environmental pollution in urban areas not only caused by the automobiles, but also by other means. There are large industries, factories and workshops around the city areas which are functioning round the clock producing various types of materials, throwing and dumping enormous amount of wastages, toxic materials, smokes and dusts which are mixed in the air and cause bad odours around the residential areas. There are also all sorts of unwanted noises and glaring lights. People are made to live in the midst of garbage, wastages, excrements, noises, glaring lights. In certain areas people keep all the times awake and experience inconveniences. People in India, generally do not bother about these environmental pollution problems they are experiencing, till they get affected personally. People do not know what are the ways they are affected by the environmental pollutions. Since they are immuned to the environmental pollutions they do not normally try to find out remedy for the problems.

The urban environmental problems include the design of buildings and houses. The modern designs do not take into the consideration whether the building or the house or the flat is having proper fresh air circulation or ventilation and natural lighting conditions. By having these facilities, the rooms in the buildings do not get hot temperature. There is no need to use the artificial light for all the twenty-four hours and the glaring of the artificial light can be avoided. The residents of the houses should have all these facilities and also should have easy social relations. There should not be any chance for friction or conflicts among the residents who are living in flats in large number. In due course the residents may develop much strain which may lead to quarrel, aggression and violence. Therefore, it is better that only like minded people should occupy the flats. The great disadvantage in the modern flat system of construction is that people are crowded in a small space. Any damage to the building like the fire, it is very difficult for all the people to run away quickly from the building. There may be stampedes and deaths.

Thus serious environmental problems cause much disastrous consequences in human life and behaviour. Environmental pollutions particularly in urban areas cause serious health hazards like cardiovascular diseases, infectious and epidemic diseases and cause deaths in large number. It is because of the negligence of the public health departments in big cities, where the toxic materials and dumped garbage were not removed. The dumped garbage in large quantities in every nook and corner cause stink and strong unpleasant smell which mixes in the air of the surrounding areas and cause health hazards. Sometimes the sewage water goes into the corporation's drinking water pipes and pollute the fresh water. These things are routinely happening in Indian city areas. Environmental psychology studies all these aspects and wanted to bring an awakening among the general population about the environmental pollution. In this way it becomes more a practical science.

NOISE AND HUMAN BEHAVIOUR

Among the environmental components noise is one of the strong forces that interferes in the behaviour of humans. There are also noises or sounds caused by natural events like the thunder and lightning. Though this natural event is more threatening it is not happening all the times. It is happening only during the rainy seasons. But the man made noises are happening all the times causing impairing effect on humans and interfering with behaviour. Noise is caused everywhere in various ways. Whether we want it or not it is sounding and interferes with our work performance and other activities and reduce the efficiency in performance and also the production.

What is Noise?

Noise is considered to be sound which causes irritation or annoyance and provokes anger. Generally noise is a sound that is louder and disturbing. It is also loud shouting. It is referred to periodic disturbance. The simplest definition of noise is that it is an unwanted sound. The concept noise is used in several situations. We say, why he is making noise? When we do not like a person's talk we may say like this. The machine is making some noise. Noises are produced by the moving vehicles in the urban areas. We can hear these noises round the clock without any break in big cities. People who live by the side of railway stations hear the noises of incoming and the outgoing trains. The aeroplanes while take-off or landing produce horrible noise till it disappeared or landed. Persons who play the rock music in a concert generally produce intolerable noise to the level of 120 decibels for hours together. They are of the view that the more louder noise they produce the more they are excited in dancing. Some people enjoy more louder noises under certain circumstances. Some religious denominations also produce louder noises by reciting religious hymns. They enjoy, but it is a disturbing event for others.

There are certain industries producing continuous noise night and day and people who work there often suffer from loss of hearing and retire as deaf persons. The metal and steel industries, the textile industries cause sensory impairment to the people who work there, because they are continuously exposed to high decibel levels of noises. Thus noise is unwanted, disturbing, intolerable, annoying, causes damage to our sense of hearing when it is produced at high level of decibles and when we are continuously exposed to it. Therefore, it should be avoided or controlled.

Noise is an unwanted sound. This sound can be produced by the vibration of an object when it is hit strongly. Noise, sound and voice, all refer to the same things. These can be also differentiated. When a sound is produced loudly and when we do not like it we call it a noise. Sound is a sensation produced by anything that can be heard. Sound may be also stated as that which is heard, resulting from stimulation of auditory nerves by vibrations in the air. Sound is having volume or quality. It is used in the communication process. But voice refers to the sound made by the vibrations of the vocal cords. Specifically, voice is the sound produced by human beings. All sounds are not called voices, because sound produced by animals and birds have different names. We say that the lion is roaring, the dog is barking, the crow is crowing and so on. Thus voice is applicable to only humans.

We reject noise only when it is extremely loud and of damaging type. We do not reject totally all the noises. It is only from sounds that humans have evolved rhythmic notes to compose music. Both the vocal and instrumental music notes compositions are made by producing the sound in different levels of decibels. However sweet a musical note is, whoever performs it, only when the sound is at an intermediate or optimum level we are enjoying it. If it is overstimulation with extreme sound or understimulation with very low sound we do not enjoy the music. When the music is transmitted through loud speaker very loudly we hate it. But some want it very loudly. Therefore, it becomes subjective. What is music for one person is merely noise for another person. Therefore, noises and sounds can be tolerated only to certain decibel level of stimulation. Noise is both a physical component and a psychological component. The physical component of sound is perceived and evaluated by our ear and the brain structures, whether it is wanted one or unwanted one. But the psychological component of sound is liking or disliking, accepting or rejecting. When it is liked and accepted it is of intermediate level of stimulation and if music we even enjoy it. When disliked and rejected it is overstimulation and we hate it since it is intolerable causing damage to our senses of hearing.

THE PHYSIOLOGY OF HUMAN AUDITORY MECHANISM AND SOUND

When we know noise as a physical component and we perceive it with our ears and brain structures, it is also important to know something about human auditory mechanism. Physics is the science which studies about sound components. Any solid body when struck vibrates. This vibration varies according to the force with which the solid body is struck. These vibrations of the solid body are produced consistently. The vibrations are regular and periodic. Sound is heard by our ears. Sound when produced pass through the sound waves. Sound vibrations through sound waves give pressure on the air molecules. The air molecules reach the eardrum. The air molecules are forced to go together, create positive and negative pressures on the eardrum. The positive pressure are the peaks of the waves and the negative pressures are the volleys of the waves. Thus sound waves reach the ear with peaks and volleys.

The eardrum, a membrane, vibrates to the pressure of the sound waves and in return makes the parts of the middle ear which are called the *ossicles* which are functioning on a lever systems are vibrated by the rapid movement of the eardrum. The three parts of the middle ear are the hammer (malleus), anvil (incus) and the stirrup (stapes). The hammer receives the vibrations from the eardrum and vibrates the anvil which in turn moves the innermost part of the middle ear, the stirrup. The stirrup transmits the vibrations to the oval window of the inner ear. The important parts of the inner ear are cochlea, vestibule and the three semicircular canals. The important part of the inner ear is the coiled, cochlea. This is filled with fluid, divided into three major compartments. One of them is the basilar membrane. The fluid in the cochlea is disturbed by stirrup of he middle ear which is made to vibrate in different parts of the basilar membrane. These vibrations cause impulses in the nerve endings of the hair cells which are sent to the temporal lobe of the brain and the sensation of hearing is produced (Fig. 20.1).

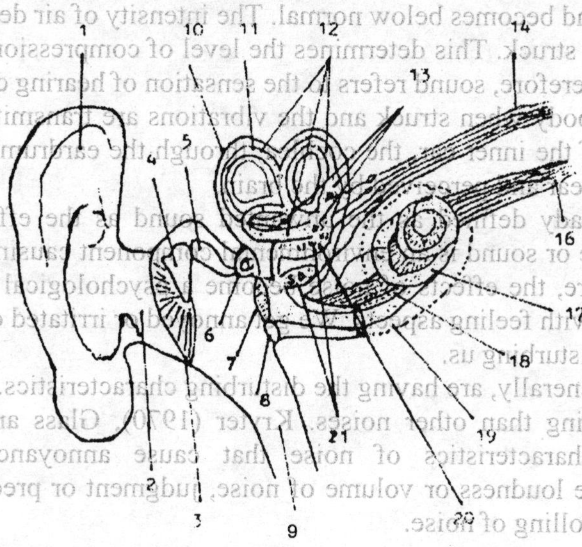

(1) Outer ear (Pinna), (2) Ear canal, (3) Tympanic Membrane (Ear drum), (4) Malleus (Hammer), (5) Incus (Anvil), 6. Stapes (Stirrup), (7) Ovel window, (8) Round window, (9) Eustachian tube, (10) Perilymph, (11) Endolymph, (12). Semicircular canals, (13) Cupula in 3 Ampulas (in the semicircular canals), (14) Vestibule nerves, (15) Auditory nerves, (16), Cochlear nerves, (17) Cochlea, (18) Scala vestibuli, (19) Scala tympani, (20) Vestibule, (21) Utricle and Saccule.

Fig. 20.1 : Showing the various parts of the outer, middle and inner ear

PSYCHOLOGICAL ANALYSIS OF NOISE AND SOUND

Sound of greater vibration or amplitude causes more impulses or pressures at each cycle and a louder sound is experienced. We are hearing varieties of sound or noises in a just few minutes time in our everyday life. There are loud noises, mild noises, medium noises, high pitched tones, low pitched tones, and so on. Sounds or noises are not always regularly occurring at definite frequencies. Most of the noises are with mixed vibrations with varied frequencies. There is no consistency in the noises we hear now and then.

Sounds or noises draw the attention of psychologists as they often interfere in the behaviour of humans, and also they are interested in the aspects hearing as sounds or noises stimulate humans to react. They wanted to know how the hearing sensory apparatus, the ear, converts the sounds into action potentials or how the functions of the ear makes man to react to a sound. Secondly, how the ear is able to discriminate the low pitched sounds from the high pitched sounds.

A sound or noise consisted of many types of pitches. This can be illustrated in a classroom situation with a tuning fork. The tuning fork can be made to vibrate with varied tunes by striking it with a hammer. When the tuning fork is struck the vibrations are very deep and louder and then slowly they are reduced. The motions of the tuning fork strike the air molecules and cause pressure in the air. The air molecules clash with one another and move apart regularly and the change in air pressure of the waves produce the sound. The air pressure is not uniform always. The intensity of the air pressure changes. At first the air pressure is above normal and then the intensity of the pressure is slowly reduced and becomes below normal. The intensity of air depends upon how hardly the tuning fork is struck. This determines the level of compression and rarefaction of the air molecules. Therefore, sound refers to the sensation of hearing caused by the vibrations of a solid metal body when struck and the vibrations are transmitted to the coiled fluid-filled structure of the inner ear, the cochlea, through the eardrum and the ossicles of the outer and middle ear and perceived by the brain.

Noise is already defined as the unwanted sound as the effects of noise are very unpleasant. Noise or sound is an environmental component causing unpleasant feeling to humans. Therefore, the effects of noise become a psychological phenomenon. Noise is often associated with feeling aspects. We get annoyed or irritated or get angry if the noise is continuously disturbing us.

All noises, generally, are having the disturbing characteristics. However, some noises are more disturbing than other noises. Kryter (1970), Glass and Singer (1972) have identified the characteristics of noise that cause annoyance to humans. These characteristics are loudness or volume of noise, judgment or predictability of noise and checking or controlling of noise.

Loudness of Noise

Noise or sound is having a range of intensity. If sound is too feeble then it is possible to hear the sounds of internal body functions like the heart beats, blood flow. These can be heard through our ears. But too loud sound may cause damage to the parts of our ear. Noise is measured in terms of unit. The decibels (dB) are the basic unit of sound or noise. If sound intensity is stated in terms of decibels, then we must know that there are two important factors. One is that decibel is stated in the form of ratio. The ratio always would indicate the relationship between two varied properties. In the same way when it is used for sound or noise the decibel is a ratio, which indicates the relative difference between two noises or sounds. Since the intensity of noise or the loudness of sound depends upon how forcefully it is produced by heavily striking some solid material, the force or energy that strikes the air puts more pressure in the sound wave which produces louder sound experience. The normal human ear that can ordinarliy detect is about .0002 dynes or microbars as per square centimetre. The other factor, that the decibel unit is a logarithmic scale. A scale of sound ranges from .0002 to 1000 dynes. The highest range 1000 dynes cause more painful experience.

Judgment of Noise

Noises may be regularly produced or irregularly produced. The irregular noise is not predicatable. It is produced intermittently and cause annoyance or irritation. The irregular noise also causes stress and we will have to put more effort in our work. We can adopt any one of the avoidance mechanisms and ignore it if it is regular noise. But we cannot do this in the case of irregular noises.

Checking of Noise

Certain noises we can check and control and avoid the disturbances caused by them. But we cannot check and control all the noises, for example, in sound proof rooms we don't hear noise and we are not disturbed, however louder noises are outside. The aeroplanes while flying in the air cause horrible noise, but those who are travelling in the plane do not hear any noise of the planes. Similarly, while travelling sitting in the airconditions compartment of the express trains we do not hear the noise produced by the train. In these situations noises are controlled and we are not disturbed. But we can check and control certain disturbing noises if it is within our control. Noises in boiler and metal industries cannot be stopped and men who are working there retire with permanent deafness. These noises since cannot be checked and controlled, workers and people who live a close to those industries have to endure the hardships and meet the consequences.

The Effects of Noise

Human beings are quite familiar with noises and sounds. They know what a noise is in the man-made world. The man-made world is full of noises. In other words the world is noisy. With all these unpleasant experiences man lives in the noisy world. It is because that he cannot avoid or eliminate noise from the man-made world. It is an unwanted sound. But it is produced constantly everywhere. Nobody wants it yet it is there. The reason is that modern man has associated the noise with his life events. When he wanted to move form place to place he was to find out a vehicle which may be a motor car or a city transport bus or a train or an aeroplane . All these cause noises. When he reached his work place, if it is an industry, it causes much louder noises in a non stop way round the clock. Therefore, man has to endure the effects of noise. No man is willing to lose his sense of hearing because it is very big loss for him. But his life situations are such that he has to experience all sorts of noises and must even lose his sense of hearing. Apart from the roads, industries, shopping place, crowds, there are also places like clubs, restaurants concert halls producing louder noises on the name of rock or pop music. Modern youths who constantly exposed to the louder rock music for hours together everyday lose sense of hearing. Louder noises of unpleasant type cause stress and when continued it results in hypertension, ulcers, heart rate, excess secretion of adrenal glands, deafness and so on.

Investigators like Cohen (1973) and Jansen (1973) have found in their studies that exposing to louder noises for longer duration was having relationship with cardiovascular disorders, allergic reactions, sore throat and digestive disorders. Specially young and new workers in the noisy work environment suffer very much due to noises.

The effects of noise is also found to be causing the symptoms of mental disorders. Investigators like Cohan *et al.* and Harridge (1974) have combined physical health with mental illness and found the symptoms of mental diseases due to louder noises.

Just like noise is having effect on physical and mental healths, it is also having effect on work performance and decrease the efficiency and reduce the production. Though there are different views prevailing on this issue the effects of noise on work performance cannot be ruled out. The interference of noise in the process of communication is causing much concern for person who are dealing with commerce and trade. Noises also interfere with human social relations and behaviour. Noise decreases the interpersonal attraction and increases the distance between the persons. Further noise causes annoyance in the extreme level and the person becomes aggressive and even shouts loudly and makes noise. We often see people in society that aggressive people are always noisy. Thus noise is causing damaging effects physically, mentally and socially. Therefore, noises has to be avoided or controlled or reduced as far as possible.

TEMPERATURE AND HUMAN BEHAVIOUR

Temperature is one of the important environmental components which is having influence in human behaviour. Temperature may be referred to hotness or coldness. The temperature may be hot or cold in an environment and man is reacting to these variations in different ways. Therefore, temperature may be varying in degrees from extreme cold to extreme hot. Human beings are accustomed to normal temperature. The human body temperature is 98.6° F or 37° C. If there is any increase or decrease in the atmospheric temperature the human body core temperature (98.6° F) is affected. The earth on which we live is covered by gases and smokes. The air or climate in a particular place is called atmosphere. The lowest atmosphere layer is called troposphere. In this atmosphere the air temperature is decreased. The troposphere is the atmosphere from the earth's surface to about 6 to 12 miles above the earth. The atmosphere may be also thick with smoke. But life on earth can exist only in normal atmospheric condition. The behaviour and experience of the people may vary according to the atmospheric conditions. Since climatic condition of the earth varies considerably form region to region, the nature of life of the people also varies. Temperature is one of the physical environmental components having effect upon human behaviour.

Human Reaction to Temperature

It is very difficult for man to live in extreme cold climatic regions. But certain birds, animals and plants exist in the extreme cold regions like the arctic and antarctic regions. Man cannot think of living in such regions. But man can make expeditions to such areas for a temporary period. Human bodily make up is such that he needs moderate temperature between 115° F and 77° F. The human body temperature should be always at

98.6° F and any change in this core temperature will have serious consequences. Any increase above 115° F and any decrease below 77° F will cause death. In the extreme temperatures the body's adaptive mechanism fails. Therefore, man finds out artificial methods as adopting devices. In summer seasons due to extreme heat there are instances of deaths by sun strokes. People from cities temporarily migrate to cold places or hills resorts to get rid of the extreme heat of their places and when temperature comes down they go back to their respective places. In the case of extreme cold condition the body gets freezed and it becomes motionless. Though the hypothalamus part of the brain is working as body's adaptive mechanism, in the extreme temperatures it fails to function to save the body.

The effects of heat and cold cause various changes in human behaviour and in adjustment problems. Man's feeling aspects thoroughly change and he develops negative attitudes toward the environmental conditions that are harmful. A person's reaction to the environmental condition reveals his moods and temperaments, likes and dislikes, approaches and aversions. If an individual is asked to describe an environmental condition, he would certainly bring out all his personal experiences he had in the past mostly the pleasant and unpleasant events. He might give more importance to the extreme heat and cold temperatures he has experienced in the places and the consequences of the same.

Apart from the natural extreme climatic conditions that man is experiencing there are man made conditions in which heat is raised by 10 to 20° F. Man's artificial adaptive devices to heat and cold also increase the temperature of the surrounding areas. Man's constancy is to modernize everything with his scientific materials. The rapid increase of industrialization and urbanization results in cutting trees and destroying the forests are constructing buildings and industries. When the forests are destroyed we lose the natural resources for fresh air, rains and shades. This results in increasing atmospheric temperature. Man makes artificial cooling devices in rooms of the buildings in houses and offices. The air conditioner throws out the heat from inside of the buildings, and the heat thrown out by the air conditioning devices is not going away. The heat is within the surrounding areas and increase the heat of the atmosphere by 10 to 20° F.

According the studies made by Hurt (1979) the atmospheric temperature in the urban areas is always higher than the temperature of the rural, agricultural areas. That is the reason the workers in rural and agricultural areas work in the fields continuously for a longer period of time, whereas in urban areas the workers getting fatigue within few hours of starting the work and they want rest pauses and tea breaks. The rest pauses and tea breaks are due to the effects of hot temperature only. Further, heat reduces the efficiency in performance and reduces the production in organization. Many of the Indian cities became more hot at present time in 1990s than they were 40 years ago. The reason is that the green plantations in the city areas are destroyed and in the suburban areas or in the nearby areas the forests are destroyed and lots of heat and dust producing industries have come up in the surrounding areas. All these added the urban heat.

Extreme Tempratures

Temperature according to human perception falls under various levels of degrees. They are extreme heat, moderate heat, extreme cold, moderate cold, humidity and so on. These levels of temperature cause different types of reactions. Man's reaction to temperature is to avoid the extreme levels and try to maintain the normal temperature of the body.

Heat is one of the physical components of the atmosphere and it is measured and perceived in terms of Fahrenheit and Celsius scales have been devised to measure the heat. The clinical thermometer is used by the physicians to test the temperature of the human body. There is also gas thermometer, resistance thermometer, thermocouple, pyrometer and so on, all used only to measure the heat. The barometer is used to measure the atmospheric pressure and climatic conditions. Thus heat received so much importance in the life of human beings as it is affecting the comforts and behaviour. The extreme heat that man is experiencing is during the summer season. The human body and the skin are very sensitive to temperature. Heat is a physical component. The receptors of the skin are the psychological components in human body. It is the skin receptors that receive the heat from outside and do not transmit the heat to the body. The change in the temperature is felt between the skin and the environment. If the atmosphere is hotter than the body then the body perceives the outside temperature as hot and if the atmosphere is cooler than the body, the body perceives the outside atmosphere as cold.

When the atmospheric temperature is extremely hot, above the body temperature (98.6°F) then the adapting mechanism of the body begins to function. Because of this the body is made to throw away the heat. Some changes are taking place at this time in the body. The breathing takes place with some noise. There is also dialation in the blood vessels which are near the skin surface. This process takes place fastly. Blood in large quantity is sent from the heart to the skin surface area. In this process the excess heat sent through the blood from the central part of the heart to the skin surface is thrown out by the outside air and the body is sweating. The body sweats profusely and returns to the normal temperature.

Another reaction of the body to extreme hot temperature is supplying more water to the skin surface for evaporation by sweating. This is taking place when there is heat wave due to extremely hot temperature. This is called emergency measure of the body. The urine formation is reduced and the body tissues are squeezed and more water is supplied to the skin surface for evaporation and the body get cooled. These two processes take away the body water to meet the emergency situation of reducing the heat. But the body is put into a state of dehydration and the person feels extremely thirsty. The throat and mouth get dried up and the person feels dizziness. The immediate remedy is that the body should be again filled with significant water. This part of the function is performed by the hypothalamus part of the brain. These processes are taking place automatically when the body is exposed to extremely hot climatic condition. Sometimes this adaptive mechanism fails to take place in the body. In such a situation many physiological changes are taking place in the body. Some of the serious consequences are heat exhaustion, heat stroke, heat asthenia, increase of blood pressure and heat attack. Particularly when there is heat stroke there is rapid increase of blood pressure and then it falls off. This is the indication of deep unconscious state which may prolong for a longer period and may end in the death of the person. In certain areas in India during summer months there are deaths by heat strokes and heart attacks. This takes place also in big cities as the cooling system of air conditining throw much heat outside which is added to the atmospheric heat by 10 to

20° F. Thus the summer extreme heat is further increased by 10 to 20° F which becomes fatal to human beings.

Just like the extreme heat, the extreme cold also causes serious consequences. In winter seasons there will be extreme cold in some places. People who live in such places are reacting to it to maintain the bodily equilibrium. The reaction should be to maintain the body heat at normal level. The adaptive mechanism, as in extreme hot climatic condition, functions in such a way that the body heat is not reduced. This is taking place in the body by increasing the metabolic rate.

When the body is exposed to extreme cold condition, persons who are not accustomed to such a situation are subjected to certain physiological changes if they did not take any precautionary measures. They are put to severe shivering of the body, there is vasoconstriction, the blood vessels become narrow, there is pilo erection, and also certain cardio-vascular symptoms are also taking place. The vasoconstriction process maintains body temperature inside. This supplies more blood in the internal organs which increases the metabolic rate and thereby increases the body temperature.

The atmospheric temperature is not always extremely hot or extremely cold causing physiological inconveniences. More than fifty percent of the time in a year is free from extreme climatic conditions. The climatic condition is moderately high and people easily adapt to this climatic condition. But only persons from cold climatic conditions may find the moderate climatic condition as somewhat hot, but can adapt to it without much discomfort. Persons from hot climatic conditions can also easily adapt to this moderate climatic condition without any physiological changes. The tendency of people to accustom to new environmental condition of moderately high temperature is stated as acclimatization. People in India mostly live in moderately high temperature environment which may facilitate them to adapt any climate conditions.

Humidity

This is another primary environmental condition in which the air is mainly saturated with water vapour. Humidity means moisture in the air. The air is filled with water. This is very uncomfortable climatic condition and many people find it difficult to adapt to this condition. The main reason is that the air will not absorb the sweating water of the body as there is already water in air. Therefore, persons who are having high eosmophil counts in blood would have breathing difficulties. Normal people would have more perspiring if the humidity level is high in an environmental situation.

AIR POLLUTION AND BEHAVIOUR

Till very recently, it is obvious that little attention has been paid to the study of physical environment and human behaviour. The emergence of environmental psychology has brought an awakening in the scientific attitude of the social scientist and they hurriedly started studying various components of the environment on knowing the values of this relationship with behaviour. Among the other physical environmental hazards the air pollution is considered to be most vital one. Air pollution is becoming the most important primary problem all over the world. Though the phenomenon of air pollution is a century old it has drawn the attention of the social and physical scientists only in the second half of this century. Since air pollution affects very much the health of the people it has to be dealt with all seriousness.

What is Air Pollution ?

Air pollution means that air is made unclean, impure and contaminated with poisonous and harmful substances or mixed with garbage, wastages and dusts, causing very bad smell. Air pollution is the main environmental pollution which is the cause of all kinds of physical diseases to humans. It has been observed that air pollution is accounted for 50 to 90 percent of cases of asthma and cancer in human beings. Since air is the vital physical component for our survival it is supposed to be pure and should not be mixed with any odourous material. Air pollution may be also defined as the mixture of toxic components such as gases, smokes, dusts, industrial wastages, chemical substances, smells, fogs in the atmospheric air making the air unclean, impure and contaminated so as to cause damage to the physical health of human beings and other living organisms on the earth.

Air is mostly polluted in the urban areas where there are lots of roads and industrial dusts and the smokes of automobiles. The urban people produce artificial air movement by the system of air conditioning. But outside they breathe only the polluted air. People often aspire for fresh air and go to rural areas and hilly places. It is not only for breathing the fresh air, but also to feel the effect of fresh air on the body.

How Air is Polluted ?

This question is briefly and correctly answered in the definition given above. When air is mixed with toxic and chemical substances it gets polluted. In the definition and description of air pollution Hammel *et al.* (1973) make more mentioning of smoke and smog (mixture of fog and smoke) than other components. The smoke and smog are highly air polluting components. These are produced in the atmosphere and are mixing with air. In the modern scientific world there are innumerable things which are producing air polluting components. The modern industries like the iron and steel industries, cotton industries, metal industries, cement and leather industries are functioning night and day, throwing heavy load of dusts smokes and toxic materials in the atmosphere.

There are some other industries like the sugar industries, paper industries, tanneries (leather industries) throw tons of waste products from the industries everyday. These industries channelize the waste products which are in liquid form in the nearby flowing rivers and pollute the river water and air. They also store the liquid form of waste products in large reservoirs. These waste materials are toxic chemical substances and cause intolerable smell and pollute the atmospheric air. Since they store these substances for a long time without recycling them, they also pollute the water in the nearby wells and agricultural plantations. Agricultural farmers in these areas often complain that their wells are getting spoiled and their crops have failed. The smells of these toxic chemical substances spread out in the atmosphere and pollute the air. The air pollutions, specifically caused by the sugar industries and the leather industries, are intolerable.

There are industries causing very heavy smokes. The fertilizer industries, the power producing industries by burning the coal release enormous amount of smokes in the atmosphere which are polluting air in surrounding areas. Some of the common pollutants are the carbon monoxide, sulphur dioxide, nitrogen dioxide, hydrocarbons and photo-chemical pollutants. Just like the industrial smokes there are also automobile smokes mostly in urban areas which pollute the atmospheric air by releasing the smokes and gases everywhere. The local transports like the two wheelers, autorickshaws, motor cars, light and heavy weight lorries, transport buses all ply the whole day throwing enormous amount of smokes and gases in and around the areas of the cities. These smokes and gases mix in the atmospheric air and pollute it. Now-a-days since there are roads connected to the semi-urban and rural areas and transport buses in large number ply in both in semi-urban and rural areas. The increase of population and the increased mobility of the people to various places for various purposes like trade and commerce, people wanted more transports everywhere and the state governments provide these facilities. Thus the automobile smokes are spreading out everywhere polluting the atmospheric air. The industrial and the automobile smokes, gases and dusts apart from polluting the atmospheric air increase the temperature of the atmosphere by 10 to 20°F.

The Effects of Air Pollutions

In the surrounding areas where the air is polluted there is no other go for people except to breathe the polluted air. The urban pedestrians breathe the automobiles smokes and gases everyday when they go out of work places, shopping centres and returning home. There are some unwanted odours, smells and gases in the air which cannot be ignored. But people when they go out do not directly perceive the air pollutants. They only breathe in without any knowledge about them. There are some pollutants very harmful. For example the carbon monoxide has neither smell nor colour. Therefore it cannot be directly perceived, but it is perceived indirectly from the harmful effect it has caused.

The air pollutions caused by the sugar industries and tanneries cause stink (most unpleasant smell) and the consequences of dehygienic effects are not yet diagnosed correctly . The people who live in the surrounding areas of these industries succumb to various types of respiratory ailments and skin diseases. But if these industries make provisions for recycling the waste products quickly then the air pollution can be avoided and people can be saved from health hazards. Many people when they breathe in the polluted air often get headache, nausea, giddiness, sometimes even fever. If the air smells stink, then some exhibit symptoms of allergic reaction. Prolonged exposure to polluted air also causes tuberculosis. Thus the consequences of breathing the polluted air are many and they cause serious health hazards to people who are exposed to it.

Perceiving Air Pollution and Human Olfactory Function

Air pollution is perceived through smell and visual acuity. The chemical or the gas mixed air can be differentiated from the pure air. We breathe the fresh air and any change in the smell is at once realized and we understand that it is not a pure air. We can easily identify the smell of a rose or jasmine and a bad odour. Nose is the sensory apparatus by which we smell and identify the type of smell and the olfactory membrane is the exact part of the nose which distinguishes the smells.

It is important at this contest that we must know something about the olfactory mechanism. The individual can judge the type of odour he breathes. We breathe all the times, but we do not speak about the smells we are breathing either this smell or that smell. Whenever we breathe, we are smelling something. But we are not aware of the accompanying smell. We become aware of the smell only when some unusual odour in quality and intensity is smelt. We breathe and smell only through the nose. It is the most common way of smelling the odour.

The smelling apparatus is physically associated with the breathing apparatus, i.e. the nose. There are receptors for olfaction in the nose. Therefore, in the apparatus nose, there is some other method of perceiving the odour. The stimulus for the sensation of smell should be from the nasal cavity to the receptors. The organ of smell is the olfactory membrane. The olfactory membrane is located at the top of the nasal cavity. It is just behind the outer nose. The smell is realized only when the receptors receive the stimulus (Figs. 20.2 and 20.3). Since smell is not a distant sense like audition or vision, only when there is stimulus, the odour is realized. It is because of this that we take things near to our nose, smell and find out what it is. Experimental studies show that there are six elementary smell sensation.

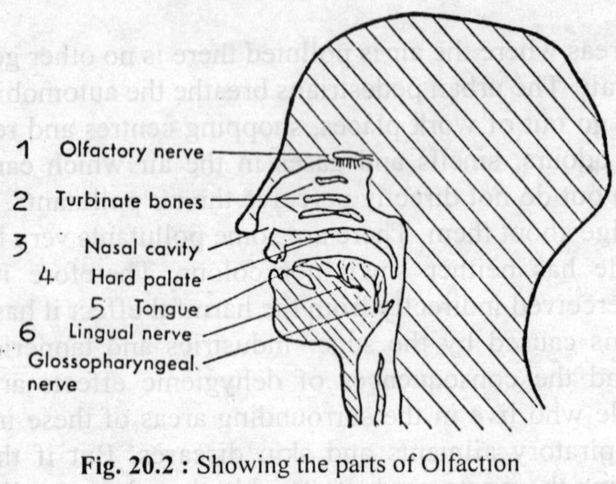

Fig. 20.2 : Showing the parts of Olfaction

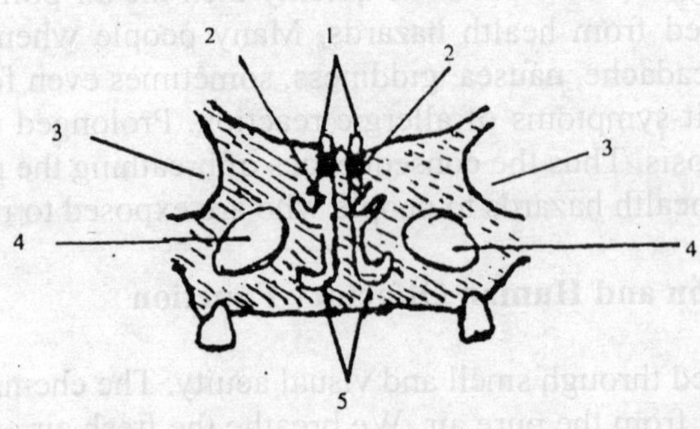

1. Olfactory bulb, 2. Olfactory epitheliam 3. Orbit, 4. Mazillary sinus, 5. Nasal passages.

Fig. 20.3 : Showing the Frontal section of skull in the region of Olfactory Mucous

The olfactory membrane is stimulated by pollutants and man's sense of smell can detect whether the smell of the air is polluted or not. Therefore, the air pollution is perceived through the sense of smell. Air pollution is also perceived by vision. The smokes and the smog can be perceived as air pollutants. If we perceive the atmosphere with a hazy appearance then we consider that it is with the presence of air pollution.

OTHER MEANS OF PERCEIVING AIR POLLUTION

There are pollutants which are not perceived either by smell or by vision. Since the pollutants have no colour and smell they are not ordinarily perceived. There are instrument designed to detect chemical pollutants. The police in our country are using some instruments to detect the amount of smoke or gas released by the automobiles. Hummel *et al.* (1975) state that in urban areas about 50 per cent of air pollution is caused by automobiles. The non-experts are judging the air pollution only by visual perception, whereas experts perceive air pollution by the instruments. People may be educated to reduce air pollutions. Though this method may have some effect, there are also other ways to detect air pollution. According to Hummel (1977) air pollution can be detected by the information available in the environmental situations. For example, the tall buildings prevent the easy flow of air and winds. The air is unclear and polluted which finds no means of clearing the area since the buildings prevent the easy flow of air in urban areas. The traffic post, where the automobiles were made to stop, they produce more smokes and gases and pollute the air more than the moving transports.

The smokes are also produced by the smokers and pollute the air in places where they assemble. The smokers smoke various types of materials for smoking like, beedies, cigarettes, cigars and pipes. In all these the tobacco leaves are used for smoking. The smokers produce smokes where there are also non-smokers. The smokes of the beedis, cigarettes and cigars, pollute the air and the non-smokers have to breathe the smokes. The smokers pollute the air by their smokes in all the public places like clubs, picture houses, shops, canteens, hotels, bus stops, offices and also in the trains, buses and planes, while halting and also while moving. The smoke is more disturbing in moving vehicle than on the ground.

Air pollution is also made by throwing the garbage on the streets. When large amount of garbage are collected in one place on the roadside it produces very bad odour and it mixes in the air and pollute the whole area. In some big and small cities in India garbage are not removed at once. They lay there for weeks together and all sorts of animals, cows, buffaloes, pigs and dogs search for their food in the garbage and spread out everywhere. Further, there are men who collect certain materials from garbage for recycling also thrown in the surrounding areas. In certain places the Corporation and Municipality, the workers of the public health department set fire to the garbage and huge smokes are produced and it is spreading out the whole area and pollute the air. So in two ways the garbage pollute the air, one by smells and the other by smokes. People who are living in such areas know how the air is polluted in various ways, but they are helpless.

There will be high concentration of pollutants if this type of situation continues. This may have adverse effects on health of the people. The garbage and burning of the garbage, rubber, the oozing of drainage pipes and sewage pipes are occurring in many places in cities and immediate action is lacking. When this condition continues for many days then the air is polluted around the city. If the pollution is not controlled for long time then the toxic effects in the air increase which may cause serious damage to the health of the people. Epidemic disease may spread out rapidly, virus may spread out and infectious diseases leading to deaths of people become quite common.

It is observed that the common pollutant, carbon monoxide (CO) prevents oxygen from going to bodily tissues including the brain and heart which may cause several health problems. Some of the common health problems are visual and hearing impairments. There are also other health problems like epilepsy, fatigue, headache, insomnia, depression, back pain, and certain psychotic symptoms. Garland and Pearce (1967), Rose and Rose (1971), Schulte (1963) who studied these aspects state that old people and sick persons become victims to these situations very easily. Hart (1970) states that physicians who studied these patients have identified certain disease symptoms caused by air pollutions such as headache, fatigue, irritability, burning of the eyes, certain gastrointestinal problems and so on. LaVerne (1970) states that the number of ailments aggravated by air pollution are found to be numerous. It is also believed that 50 to 90 percent of cancer cases were due to air and water pollutions.

PERSONAL AND INTER-PERSONAL ENVIRONMENT

People live and work only in an environment. People come into contact with other people in an environment and have interpersonal contact. In our everyday life our reaction to other persons and the type of relationship we have is of great importance in the study of environmental psychology. Generally we have an immediate environment which is an area surrounding each one of us. This is an integral part of every individual. The immediate surrounding area of a person is called his personal space. This personal space of an individual is extended on all the four sides. It is longer in the front and shorter on the sides and at the back. Since we are moving toward it is longer at a time. The personal space is also stated as personal distance. Hence personal space or distance may be defined as the area around each person into which most of the people are not possibly go beyond the limits and trespass.

Personal Space and Interpersonal Distance

The personal space is a person's exclusive privilege. Therefore, entry into a person's space is opposed, but it is not thoroughly opposed or restricted to all. The entry or intrusion depends upon the individuality of the person who intrudes. There is a general tendency to stand more close to a friend of ours than to a stranger in a place. This can be observed when we go out for shopping, club, picture house, bus stops, railway stations and so on. A male member wishes to stand more close to his friend, and not close to a person not known to him. This is also happening among the females. We wish to stand close to our relatives and friends and not with strangers. But in the case of opposite sex, man may stand close to a female friend, and a female may stand close to a male friend, but not with female stranger or male stranger. Who can intrude into one's personal space, is determined by culture and sex. Ashton, Shaw and Worshan (1980) have conducted experiments on this aspect and observed that females interact at close distance with the same sex than males.

The interpersonal distance has been studied by Hall (1966) in various countries. The results were more of culture oriented. According to his study, people of North America, Scandinavia and Great Britain expect greater distance between persons, while the people of France, Greece and Middle East are not for greater distance between persons. This type of cultural differences cause confusion and embarrassment for the visitors and tourists from foreign countries who are not aware of the cultural norms of those countries. The personal distance maintained by the people of various countries seems to be the result of the socialization process specific to the country.

In Indian community the personal distance in the interpersonal environment is a very big problem and issue. The interpersonal distance is maintained on the basis of community or caste one belongs to. In the rural environment the interpersonal distance is more severe, and more than in urban environment. The main classifications of community are, the forward communities, backward communities and scheduled caste communities. In each community there are many sub-communities or castes. The forward community maintains certain distances with the backward and scheduled caste communities. The backward community maintains certain distance with the schedule caste communities. The schedule caste community people are always kept at a distance. They were once treated as untouchables. The Britishers who ruled India have given them some benefits in education and in providing jobs in Government services, and also representations in the councils by which the distance between them and other communities was reduced to some extent. The freedom fighters under the leadership of Mahatma Gandhi, Rajagopalachari and others have realised the enigma of untouchability and tried to eradicate it by all means.

After independence, the Government of India have taken all possible steps to reduce the distance between the communities. It may be stated as community space instead of personal space. The community space, means the distance between the schedule caste communities and other communities, which was demanded by the so called forward community people. This distance once was considered to be the respect and honour which the lower castes community people pay to the higher caste community people. This type of community space and caste space is specific to Indian communities. Though in urban areas the distance is reduced, in rural areas the distance is still maintained. Any intrusion into their space or distance may cause communal clashes. However, in urban areas this distance is reduced due to economical, educational and social developments among all communities, especially in big cities. Therefore, the personal space or distance varies from culture to culture based upon the socialization process.

Studies have been conducted on the personal space or distance. According to Hall, there are four types of distances between persons in the personal space. The first one is the intimate distance which is very short, very close between persons. The second is the personal distance which is little longer than the intimate distance. The third is the social distance which is longer than the personal distance that exists in interpersonal relations such as business, office, and other organizations. This exists between masters and followers. The fourth is the public distance which is a distance between the leader and the followers or the speaker and the audience. Therefore, the personal space or distance varies according to the nature of the function and interaction in which individuals are involved. Personal space and distance has more of cultural and social significances.

Invading into one's Personal Space

Invading into one's personal space is also stated as intrusion into one's personal space. It is getting into a person's space without any welcome or being asked for. Invading into one's personal space without being asked for is a forcible entry. Intrusion may be also stated as encroachment into one's space. Spatial intrusion causes negative experience or feeling. Intrusion may also cause positive feeling or experience. The negative feeling is caused if the intruder is a stranger. There is a positive experience if the intruder is a friend or an opposite sex. An intruder's entry into one's personal space causes positive feeling if the intruder helps the persons. This takes place mostly with children. Children accept help from strangers and develop positive feeling. Generally when a person is in a distressed state he accepts help from strangers and develops positive feeling.

Studies have been conducted to find out what type of behaviour takes place in a person towards persons standing at different distances. Konecni *et al.* (1975) in their study had observed in persons who were waiting to cross a road, at a distance of one foot, two, five and ten feet from the subjects who were of the same sex associates. The associates were strangers to the subjects. It was observed that the subjects who were closer to the associates crossed road fastly than the subjects who were standing at a longer distance. This means the subjects felt that the associates were intruders into the personal space of the subjects. Hence they avoided their nearness as quickly as possible. This type of personal space is invisible, has no definite area or boundary line. Therefore, by physically avoiding the strangers, intrusion can be averted and the personal space can be protected.

TERRITORY AND BEHAVIOUR

Territory here is referring to one's personal space. But there is difference between personal space of a person and the territory of a person. Personal space has known area or defined area in space. Personal space is relaxed on certain occasions. But territory is not like that. Territory has a wider meaning. It is an area of the jurisdiction of a person, ruler, nation. It may also mean any large tract of land owned by a person or an area assigned to a person or an organization. Generally territory refers to any tract of land. It is a geographical domain under the jurisdiction of a political unit, specifically, of a sovereign state. Business and industrial organizations have the salesman's territory. It is an area inhabited and defended by an individual person or an organization, or an individual animal or a breeding pair of animals or birds. Territory is an area defended by a federal state. Thus the meaning of territory is wider, applies not only to an individual person, but to an organization, state, animals and birds. Territory may be also defined as an area that is marked with boundaries, guarded or protected.

Human beings indicate the territory of an area which is under their protection by putting stones at the boundary lines, by putting fences, raising walls, name boards, sign boards or employing watchman. Animals mark their territory by the smells of urination.

The territorial behaviour was maintained by animals and birds for the purpose of food and shelter, mating behaviour and building nests. These are purely instinctive behaviour pattern. But the territorial behaviour of human beings` is much different from the animals and birds. Humans make use of territory for the purpose of maintaining privacy and control. There are three types of territories in human behaviour. They are the primary territory, secondary territory and public territory according to Altman (1975).

The primary territory is with regard to holding ownership of a house, flat or apartment, a piece of land or a field etc. The primary territory may be owned by a single person or more than one person or an organization. The primary territory is having a definite boundary line with name and marks. There is no encroachment into this area. It may be owned permanently or may be also used on a contract basis for some years, for example, occupying a house or building for rent for ten or twenty years. The person who occupies it is having the same right as the person who owns a house or building with regard to its territorial right. These territorial persons are more interested in maintaining privacy and any invasion or intrusion or encroachment by some intruders or person is strongly protested and intercepted. They also employ watchman to guard and prevent any act of encroachment.

The secondary territory is not based upon any ownership right. It is with regard to using of a space for certain period of time. The secondary territory is also referring to the space, but it is assigned to a person temporarily only for using for a limited time. For example, in a picture house the particular seat is reserved for a person. Other persons cannot occupy this seat. In schools and colleges students regularly everyday occupy the same seat in the classroom and students do not occupy others' seats. Though a student is not having any personal right over the seat, it is left for him, and no one will occupy it. There is mutual understanding among them. Similarly seats are allotted to the members of Parliament, in Parliament House, in Assemblies and offices. This type of territoriality for behaviour is temporary.

The public territory refers to the seating arrangements made for sitting and waiting in railway stations, aerodromes, in a sports pavilion, stadium, auditorium. This is only for a temporary period. These are all public places and people wait there only for a short period till the train or the plane arrives or till the function is over. In public places, like the library or the sports pavilion, the person who goes there first may indicate that the particular seat is for him and others do not occupy it. Though it is a public place and everyone has the right to occupy any seat, the person who goes first is preferred first. In secondary public territories there is no chance for privacy and personal space and distance.

Territorial Conflicts

Conflicts arise with regard to territory among the neighbours. The owners of neighbouring houses, lands or fields often quarrel on their territorial limits. There is likelihood of one owner encroaching into another person's area by few inches or feet. This type of conflict, quarrel, claim are quite common among the Indian landlords in rural areas and also among the owners of the neighbouring houses in urban areas. The conflict and quarrel for the territoriality in Indian communities, the encroachment of one's territory by another is a common occurrence. It may take place with regard to the boundary line between the houses or between the agricultural lands. The conflict may take the form of violent action even leading to murder. They may also go to the court claiming that their right of ownership being questioned.

OVERCROWDING AND BEHAVIOUR

Crowding means large number of people or things are collected in a small space. Overcrowding refers to the collection of large number of people or too many people are collected in a too small space more than the spatial capacity. We have now-a-days see that there are crowds everywhere and many people try to avoid the crowds, the crowds in the picture house, shopping centres, clubs, railway stations, aerodromes and in bus stops. Particularly these crowds are moving fast in mornings and evenings on the roads. The people who go to offices in the city and in the suburban areas wanted to catch the particular train or bus move fastly in crowds and in the evening these people return to their homes at the same time. In these crowds everyone is going on his own business and does not see each other's face or talk to each other. This means that they wanted to avoid the crowds and thus express their negative feelings. Each one wanted to be away from the other, but he/she cannot do it as everybody wanted to go by the same train or bus. Therefore, these crowds create a sort of negative feeling in everybody as there are too many persons in a too small space, packed very close against the wishes of the persons. There is high density in such space which is interferring in the goals of the people.

In the study of population they use the concept density, the population density. This means, the number of people living in a particular given space at a particular time. The density may be high or low. We often speak that a city is high densely populated. Most of the metropolitan cities are densely populated. There is a standard fixed for a person, how many square feet are required for a person to live in comfortably. That is the reason, that in lodges, the size of the room should be mentioned. The capacity of the room and how many persons can be lodged there, should be stated. This refers to the capacity of the room. But crowding is referring to the type of response a person develops in a situation. The crowding in the place is such that there can be no more people get into the space. Therefore, there is negative or aversive attitude toward the crowding. People's bheaviour is much more affected by the crowds. Generally, crowds cause stressful feeling, and people try to avoid it as too many people are occupying too small space.

Overcrowding is, occupying a space more than its capacity by too many people to the extent of finding out a space only to put one's foot. There would not be any space to move about. Suppose, if some mishap has happened, then every one in the crowd wanted to escape from the situation and try to take a flight. Since the space is highly packed, it will not be possible for the people in the middle to move about and they try to rush with all their anxieties. This type of behaviour results in stampede and people get into the head-long rush and even die. There were instances that took place in many places that overcrowds resulted in stampedes and many people died. It is because of collecting too many people into a too small space.

But crowding cannot be fully avoided and labelled it with negative reactions and aversive feelings. We cannot enjoy certain concerts without sufficient crowd. For example, the football match, or cricket match when played between nations there will be strong reaction of much excitement, thrill and pleasure among the audience. If there is no crowd, there is no excitement or pleasure. The player himself will not find interest if there is no crowd. Except such occasions, crowding or overcrowding interferes with behaviour and performance.

Studies have been conducted to see how far the crowding affect the behaviour of people in various situations. Worchel and Teddle (1976), Worchel and Brown (1984) have studied this phenomenon experimentally. It was observed that high density always impaired individual's performance on complex task. Therefore, there is relationship between crowding and people's behaviour and the relationship is negative on the performance and also there is development of aversive feelings toward the crowding.

EXPERIMENT 59 : MEASUREMENT OF ATTITUDE TOWARD ENVIRONMENTAL POLLUTION

INTRODUCTION

The environment in which we live determine our behaviour to the greater extent. This environment need not be only the natural physical environment. It includes social environment, and man-made environment. Whatever type of environment in which we live we learn to adapt ourselves to the situation. Man has been experiencing the environmental stimulations in various ways. Some may be simple and tolerable, but some environmental stimulations are too much to deal with. They cause harmful effects on our bahaviour. Both understimulation and overstimulation may have undesirable effect in human behaviour. But what is required is neither understimulation nor overstimulation. A medium environmental stimulation alone would be congenial for human behaviour.

While discussing about the level of stimulation Wohlwill (1974) incorporates the adaptation level theory of sensation and perception of Helson (1964) in his environmental stimulation. Wohlwill states that human beings on certain occasions dislike the crowds. For example when we go for some shopping or to attend a function or dinner too much of crowds in the street, in the shopping centres, in the dinner hall make us disgusted and sometimes we even withdraw from those places. The environmental crowds sometimes become tense and people may get stampeded suddenly which may cause even deaths. Such environmental situations are quite common in these days. This type of excess tense situation we do not like. But at the same time we do not like also complete social isolation. For example, we cannot enjoy a football match or cricket match sitting alone in the stadium. We wish a big crowd there to share the feelings of other people in the winning or in the defeat of the match.

We expect some crowds in certain functions, meetings, particularly the political meetings. In some meeting if there were no crowds the speaker himself may not be inclined to continue his speech. The speaker expects a big environmental crowd should give him a big clapping. The environmental crowds may cause both favourable and unfavourable consequences. The same type of situations say arise for all types of environmental stimulations. For example, in a situation noise may be causing damaging effect to our ears and we dislike such noises. But in an auditorium or stadium if there is no noise there is no enjoyment or thrill for the audience. The audience express their pleasure or joy when their team scored a goal in a football match. In an auditorium the play or concert is enjoyed by the audience by clapping loudly and cheering the artist. In such situations if there is no noise there is no joy or life. Thus the environmental situations are having effects upon human behaviour. Similarly, the temperature should not be either too low or too high. Both the conditions are having effect upon the behaviour of human beings.

Another environmental problem situation is the transportation or road transport hazards. This is becoming now-a-days the most risky situation all over the world. This environmental situation is causing death to thousands of people everyday, whether there is any adaptation level in the environmental situation or not. There is mostly over stimulation in this situation. There should be optimum level of stimulation to environment behaviour relationship. Human behaviour takes place in the environmental situation according to the level of the stimulation in the environmental factors. The optimum or the most desirable level of stimulation alone can establish environment-behaviour relationship. Wohlwill classifies this into three categories. All the environmental stimulations fall under these categories. They are (1) sensory stimulations, (2) social stimulations and (3) movement stimulations.

The sensory stimulation too low or too high is harmful. Neither too low nor too high can arouse response or behaviour in an organism. A sensory stimulation whether it is seeing, or hearing, or tasting or smelling or touching; will not arouse response in an organism if it is too low or too high. A very dim light will not help us to see anything in the dark room. Similarly a very bright light will be glaring and we could not see anything in the front. For example, two cars on the road during night time with bright head lights cause confusion. It may lead to only accident. In the same way in the sensory experience of hearing, too feeble or too high sound does not make any sense. Therefore, any sensory stimulation should be optimum if response is to be aroused or if environment behaviour relationship should exist. Therefore, these stimulations on the basis of the level of stimulation seem to function on three dimensions such as (1) intensity (2) diversity, and (3) patterning. These determine the characteristics of optimum level.

The dimension of intensity refers to the level of stimulation. What amount of pressure is given on the environmental factor?, for example the intensity of light, intensity of sound and so on. The intensity may be low or high, but what desired is the optimum for behaviour. In the case of social stimulation also there will be disturbance if there were too few people in a situation which needs more people. So also too many people in a situation, which need few people are causing disturbance. Social stimulation with too few people and too many people depends upon the situation. There are instances in which one would be happy with the stimulation of very few people. Similarly there are instances in which one would be happy with the social stimulation of too many people, for example a sports event or a football match. The stimulation of movement is not always optimum. The roadway sometimes may be like a barren desert. No transports or people moving there. It is too lonely and the person who goes on the road may not be happy or find interest in moving alone. But sometimes there may be too many people on the road and too many vehicles causing too much of noises and the road is tensely packed and very difficult to move about. Therefore, the movement stimulations also cause in either way disturbing behaviour.

According to Wohlwill (1974) diversity is one of the dimensions of environmental situation. Diversity in one surroundings produce pleasant feeling. But if there is no diversity it may cause disinterestedness for us. If there is nothing to see or buy or eat the surroundings becomes dull. The other extreme is too much of diversity in many towns and cities. Diversity refers to variety of elements in various structures. It is mainly environmental novelty. But much of diversity causes strain to eyes. The rows of bright neon lights and other decorative illuminations, the various types of structures — all are more than enough and have effects on behaviour. The man-made scenes and attractive designs have degrees of pleasant feelings.

The patterning is the third dimension of environmental stimulation. It is with reference to perception of structure. Anything can be designed with constant and diffused intensity of light or sound, with a single tone of the same volume cause disliking. Modern environment mostly are built-in-environment and have so much diversifications and complexities that it is very difficult to have clear perception. Sometimes we could not make out any idea. They also cause stress in the perceptual process. The pattern of roads and streets and the same type of designs and buildings and the numbering system to the streets and houses cause strain and confusion in predicting the place and the house number. A design of moderate pattern with sequence in the name and number of the streets and house may be comfortable and all persons may have easy access to reach the place and find out what they wanted. Therefore, the environmental stimulation should be of optimum level for environment-behaviour relationships.

We have stated that too little and too much of environmental stimulations cause strain and there cannot be optimum environment - behaviour relationship. But Wohlwill states that there are environmental situations in which inadequate environmental stimulation becomes the optimum level for behaviour. For example, Tibetans on the north of Himalayas live in an altitude so high that there is oxygen level below the normal level. People from plains who go there find it difficult to live there. But the Tibetans live there very comfortably. They adapt to the low level of oxygen concentration which is not possible for others.

What is optimum level for comfortable living or for behaviour of one group is not optimum level for other groups. The city group could tolerate very much for the tense crowds and also not so much tolerant in isolating, whereas the rural people are on the other way. They could tolerate isolation but intolerant of tense crowds. But the experience one gains for the environmental stimulation may learn to adapt to extreme environmental stimulations. A city born and bred person who migrated to some typical rural area and lived there for few years probably might have acquired greater amount of tolerance for too small crowds and for isolation than the persons who have never been to rural areas. Here what was optimum level of stimulation for the city bred persons has become extreme. There is a shift from the optimum level of stimulation. Wohlwill calls this process as adaptation. The continued exposure to extreme environmental stimulations makes one to adapt himself to the new situation. Adaptation level varies from time to time. For example in India in the months of May and June there will be extreme heat and persons who could not withstand the heat die of sun-strokes. Every year 200 to 250 people die of sun-strokes. Their level of adapting to extreme heat is poor and experience is inadequate. In the months of

December and January there will be extreme cold. But very few die of extreme cold. They adapt to the cold climate by wearing woollen clothes. Adaptation levels also vary from person to person to heat and cold. Similarly adaptation levels vary to different environmental stimulations.

Man in the extreme environmental stimulations finds it a difficult for a normal way of life. But if he is exposed to such situations continuously he learns to adapt himself. Wohlwill calls this process ***adaptation*** while Sonnenfeld (1966) calls it as ***adjustment***. According to Wohlwill adaptation is changing the individual's response to the stimulus. Sonnenfeld states that adjustment is changing the stimulus concerned. In extreme cold temperature we slowly make ourselves to suit to the atmosphere by putting on thick woollen suit and we do not shiver from the cold. The environmental condition is there but our response to it is changed. This is adaptation. But in adjustment we try to change the stimulus. In the extreme cold, the room is very chill and we use some heater and produce heat wave which would increase the room temperature. In hot climatic condition we use air-conditioner to reduce the heat and the room temperature is brought down. This is adjustment.

In the ancient times adaptation was the only option and man must learn to adapt himself to environmental situations. Since there is chance for escape, adaptation was the process for survival. Maladaptation caused extermination in the ancient societies. But in the modern society it has become easy for man to adjust himself to the environmental situations. Various technological developments have taken place to change the situation to suit to human requirements. A huge building in which thousands of people live and in big industries where thousands of employees work are air-conditioned and the temperature is brought down to optimum level in tropical countries and cold countries the temperature is raised to the optimum level in such places. In view of this explanation the adaptation level theory becomes out of place in modern times. Thus the environment-behaviour relationships are established. Various types of environmental conditions are having effects on human behaviour, whether the environmental conditions are natural physical types or man-made or built in environments. Environmental psychologists have studied various aspects of environment-behaviour relationships. But the reverse effects also are there. That is the effects of behaviour on environment. Sonnenfeld's (1966) adjustment theory is outcome of studying the effects of behaviour on the environments. The field that studies this aspect is called ***ecological psychology***. It is sometimes stated as social ecological psychology.

The ecological psychology studies are two-way reactions, the effects of environment on behaviour and also the effects of behaviour on the environment. We have discussed how man changes the extreme environmental conditions for optimum level to suit to his requirements. Reger Barker (1968) was the champion of ecological approach for many years. He and his colleagues in the University of Kansas made many contributions. Barker based his ecological view points on the studies he made on the behaviour of children in their natural environment. This was the basis for his ecological approach to know the kind of enivornment-behaviour relationships. Barker's ecological psychology is the study of environment-behaviour interdependencies. It focuses its attention on the behaviour setting as the unit of study. According to Barker the behaviour setting influences large number of people. In a built in environment or inside the room or hall many behaviour settings are possible.

Behaviour setting refers to the type of function that is going on in the hall. There can be more individuals. Barker states this, as extra individuals, which means a group of people. The group may be small group like the company conference, or it may be a big group like the audience of the football match or it may be a classroom. The functional purpose of the structure may be anything like a classroom, or a dance performance, or a concert, or a meeting to honour a person or a indoor basket ball match, or a prayer or a factory and so on. In a behaviour setting there will be always a group of people or extra individuals. The behaviour of extra-individuals in a built-in environment is different from the individuals of a natural environment. For example, the behaviour of the extra-individuals in a desert or in a forest, on a mountain or in the sea or in the sky may be completely different from the built-in environment, for example, a room or hall.

According to Barker the behaviour setting consists of two aspects such as ***standing patterns of behaviour*** and the ***physical milieu***. The standing patterns of behaviour is the collective behaviour of the group in a built-in environment. It is not referring to the individual behaviour. Because the type of behaviour of the standing pattern is not placing on the individual. The behaviour is designed for the group. For example, a classroom, a prayer hall, a company meeting, a musical performance, an indoor badminton match and so on need a group of people inside the room. The function for example in a classroom is mainly lecturing, then listening, taking down notes, writing on the board and so on. The other aspect of the physical milieu consists of the room where the behaviour setting is going on, the furnitures, books and notebooks, the black board, chalk piece, maps and so on. When the functions are over the individuals or the group of people go away. But the physical milieu or the materials remain there. Thus the standing patterns of behaviour is different from the physical milieu.

Bell *et al.* (1978) state that Barker's approach is mainly field investigation method and it cannot be satisfactorily tested in a controlled laboratory situation. It aims at the world real behaviour. Barker's method is very broad in its scope and certain specific cases cannot be tested and predicted. It deals with group behaviour as it is so designed. It is not dealing with individual behaviour. Therefore, it has certain advantages and disadvantages.

EXPERIMENT

Problem

Assess the attitude of an individual toward environmental pollution problems and compare it with the attitude of a small group.

Method

Participant

Find out a person who can answer the statement of the attitude scale and collect the personal information such as his age, sex, qualifications, home town, whether he/she is a smoker or not, whether he/she is from industrial area or agricultural area. Whether his/her home town is having modern sanitary facilities and the type of house he/she is living and so on.

Material

Prof. Rajamanickam's Environmental Pollution Attitude Scale, Manual, Test Sheet and the Scoring Key.

Description of the Material

The environment pollution attitude scale was designed and developed by Prof. M. Rajamanickam (1995) for measuring the attitude toward environment. It also deals with environmental effect on behaviour. This scale consists of 30 statements, each one expressing one idea regarding environmental effect. Under each statement there are five response categories, like, strongly agree, agree, unable to decide, disagree and strongly disagree. There are both positive (pro-pollution) statements and negative (anti-pollution) statements toward environmental effect on behaviour. The positive statements are referring to favourable attitude toward the environmental pollution. This means that environment is polluted. The negative statements are referring to unfavourable attitude toward the environmental pollution. This means that environment is not polluted. The respondent has to give his/her opinion on the idea expressed in the statement. If he/she strongly agrees with the idea he/she has to underline the response strongly agree, if he simply agrees with it has to underline agree. If he/she is not able to decide, then he/she has to underline the response unable to decide. If he/she disagrees with the idea he/she has to underline the response disagree and if he/she strongly disagrees with the idea he/she has to underline the response, strongly disagree. The strongly agree response of the positive (pro-pollution) statement is given 5 marks, agree 4, unable to decide 3, disagree 2 and strongly disagree one. The reverse order of the marks is for negative (anti-pollution) statements. That is, the strongly agree response is given 1 mark, agree 2, unable to decide 3, disagree 4 and strongly disagree response is given 5 marks. The maximum possible score for all the 30 statements is 150 and the minimum possible score is 30. The individual score is the total score for all the 30 statements. The scale was validated and its reliability is 0.76 and its intrinsic validity is 0.87.

Procedure

The experimenter should check first of all whether the test materials are correct, the manual, test sheet and the key. Having satisfied with the test materials, the experimenter should make the participant to sit comfortably in a well illuminated room in front of him/her at a distance and give him/her the following instructions.

> "This is a psychological test to assess your attitude towards environment. It is not an intelligence test nor personality test. Read the instruction at the top of the test sheet (showing the instructions in the test sheet) carefully and carry out the instructions. There are 30 statements. In each statement one idea is expressed regarding the

environment or its effect on human behaviour. The environments are both the natural and built-in environments. Under each statement five responses such as strongly agree, agree, unable to decide, disagree and strongly disagree are given left to right. You may agree with certain ideas and disagree with certain ideas. Whatever feeling you have, express it through one of the responses. If you strongly agrees with the idea in a statement underline the response, 'strongly agree'. If you agree with the response agree underline 'agree'. If you are not able to decide, then underline 'unable to decide'. In case if you disagree with the ides underline the response 'disagrees' and if you strongly disagree with underline 'strongly disagree'. You must carry out this procedure to all the 30 statements. As soon as you answered all the statements like this return the test sheet to me. The experimenter hands over the test sheet to the participant. 'Now, ready go'. "

When the participant started answering the items, the experimenter should watch how much time the participant takes to complete it. The experimenter should collect the test sheet from the participant as soon as he/she completed it and obtain the introspective report from his/her and allow him/her to go away.

Method of Scoring

The scoring should be made according to the procedure with the help of the key. The strongly agree response of the positive (pro-pollution) statement should be given 5 marks, agree 4, unable to decide 3, disagree 2, and strongly disagree one. The strongly agree of the negative (anti-pollution) statement should be given one mark, agree 2, unable to decide 3, disagree 4 and strongly disagree 5. The individual score is obtained by adding the marks for all 30 statements. The lowest possible score is 30 and the highest possible score is 150. The high score indicates the favourable attitude toward environmental pollution, i.e., there is environmental pollution. The low score indicates unfavourable attitude toward the environmental pollution, *i.e.* there is no environmental pollution. A score falling in between 30 and 60 is unfavourable attitude toward environmental pollution, a score falling in between 61 and 90 neutral attitude and a score falling in between 91 and 150 is favourable attitude toward environmental pollution.

Results

The responses of the participant should be scored according to the procedure and the total score should be obtained and presented in Table 20.1. The positive responses, the negative responses and neutral responses also may be presented separately in Table 20.1.

Table 20.1 : Showing the individual score on the attitude scale toward environment

S. No.	Type of Attitude	No. of Items	Score
1.	Favourable attitude in environmental effect on behaviour (accept pollution)		
2.	Unfavourable attitude to environmental effect on behaviour (not accepting pollution)		
3.	Neutral attitude to environmental effect on behaviour		
	Total		

State the individual's total score. Is it high or low? State whether it is favourable toward the environmental pollution or unfavourable? State his/her score favourable toward environment, unfavourable score and neutral score. State in which category the score is high.

Collect the group data from the other experimenters and present them in Table 20.2. Work out the mean score for the total and for other three categories of attitudes.

Table 20.2 : Showing the Group Data on the Environmental Pollution Attitude Scale

S. No.	Name	Urban or Rural	Industry or Agricultural	Total Score	Favourable Attitude Score	Unfavourable Attitude Score	Neutral Attitude Score
1.	S.V.						
2.	R.M.						
3.	M.K.						
4.	T.V.						
5.	K.R.						
6.	P.K.						
7.	M.R.						
8.	B.V.						
9.	V.S.						
10.	S.P.						
	Total						
	Mean						

State whose total score is the highest and whose score is the lowest in the attitude toward environmental pollution. State the mean score and how many in group scored above the group mean and how many below the group mean. State the highest favourable attitude score and the lowest score. Who are those obtained such scores. State the highest unfavourable scores and lowest unfavourable scores. Who are those obtained these scores. Similarly state the highest neutral score and the lowest score. Who are those obtained these scores. The neutral score is an ambiguous score. Therefore, this score should be as low as possible.

Discussion

Discuss the individual's attitude toward environmental pollution and its effect on behaviour citing the score in various categories. State whether he/she is having more of favourable attitude toward environmental pollution or more of unfavourable attitude to pollution. On the whole; is he/she favourable toward the environmental effect on behaviour and pollution or unfavourable? State the reason if any of such attitude. Do you find any of the personal factors having any influence in your participant's attitude? If so state and discuss the factors.

Has your participant scored above the group mean or below the group mean? If so where he/she stands on the favourable attitude toward environmental pollution or unfavourable to pollution. Discuss the individuals' scores taking into account their scores in all the categories and as a whole. State who are those favourable towards the environmental effect on behaviour and pollution and who are those unfavourable to the effect and pollution. Discuss in what context they are favourable and unfavourable. Do you find that any of their personal factors or social or environmental backgrounds are having any effect in their attitude.

Compare the attitudes of the urban participants with that of the rural. Who are having more favourable attitude toward the environmental effect on behaviour and pollution. Do you find whether the rural and urban backgrounds have any effect upon the respondent's attitude? Discuss.

Compare the attitude scores of the subjects who come from industrial areas with that of the subjects from agricultural areas. Do you find any difference between these two subjects in their attitudes toward pollution. Discuss it, citing the scores. Is there any difference between man and woman in their attitude? If so discuss it citing evidences. Compare your results with the results of some earlier studies in this area and state whether your result are in concurrent with those results.

Since the scale is dealing with both natural environment and built-in environment the subjects might have preferred either the natural environment or the built-in environment. Do you find whether your participant is more favourably inclined toward natural environment or built-in environment. State the reasons, if any.

Conclusions

1. State whether your participant is favourably disposed of or unfavourably disposed of toward the environmental pollution.
2. In the group how many subjects are favourable toward the environmental pollution and how many are unfavourable to environmental pollution.
3. State whether there is any urban-rural difference in their attitude toward environmental pollution.
4. State whether there is any difference between the subjects form industrial areas and agricultural areas in their attitude toward environmental effects and pollution.
5. State whether there is any difference in their attitudes due to sex variation toward environmental effect and pollution.

APPENDIX - 1

ENVIRONMENTAL POLLUTION ATTITUDE SCALE

DIRECTION FOR ANSWERING THE ITEMS

There are 30 statements below regarding the environmental pollutions. In each statement one idea is expressed about the environmental pollution. You may agree with some ideas and disagree with some. Whatever you feel, please indicate it by underlining *any one of the five responses,* given under each statement (*e.g.* Strongly agree; Agree; Unable to decide; Disagree; Strongly disagree). Please use the answer sheet for answering the items.

Example

1. People in our country throw all sorts of rubbish and wastages everywhere indiscriminately and pollute the environment without realizing the values of public health.
   <u>Strongly agree</u>; Agree; Unable to decide; Disagree; Strongly disagree.

2. The man-made physical environments, *i.e.*, the cities and towns provide all the requirements without any difficulties to the people for happy living.
   Strongly agree; Agree; Unable to decide; Disagree; <u>Strongly disagree</u>.

Meaning for Key Words

Polluted (Pollution) : It means that something is made unclean, impure, and contaminated with poisonous and harmful substances or mixed with garbage, wastages and dusts causing very bad smells.

Man-made environment : All those environmental components created by man, such as the buildings constructed for living or for office, the Schools, Colleges, Universities, Hospitals, Courts, Prisons, Play grounds, Museums, Auditoriums, Stadiums, Railway stations, Aerodromes, Bridges, Roads, Canals, Sea ports and so on.

STATEMENTS

1. The urban life is becoming very miserable in these days due to continued exposure to the tension ridden physical environments, like the loud noises, traffic congestions, air pollution by smokes, gases and dusts, unclean and contaminated water supply and so forth.
   Strongly agree; Agree; Unable to decide; Disagree; Strongly disagree.

2. We are living in a clean and unpolluted airy urban environmental conditions and we are maintaining healthy life.
   Strongly agree; Agree; Unable to decide; Disagree; Strongly disagree.

3. The urban environments, *i.e.*, the cities and towns in our country now-a-days are full of stinking garbage collections, smokes and dusts which make the whole areas badly polluted, causing health hazards to the people who live there.
   Strongly agree; Agree; Unable to decide; Disagree; Strongly disagree.

4. The Central and the State Governments spend much money on the basic amenities to the people and provide clean and protective water and other facilities and keep cities and towns free from pollutions.
   Strongly agree; Agree; Unable to decide; Disagree; Strongly disagree.

5. Pure water is an important component for human life, but the water resources in our nation are not carefully protected from various types of pollutions and contaminations by individuals and the industries. By using such contaminated water for drinking and for other purposes people develop all sorts of diseases.
   Strongly agree; Agree; Unable to decide; Disagree; Strongly disagree.

6. Our emotion of anger is more often aroused when we see the fresh river water is badly polluted by draining down all sorts of wastages and toxic materials into the river from leather and other chemical industries.
   Strongly agree; Agree; Unable to decide; Disagree; Strongly disagree.

7. There are many man-made environments, clean and unpolluted, and provide happiness to the people by drawing them from various groups into the stadium, auditorium and museum and so forth.
   Strongly agree; Agree; Unable to decide; Disagree; Strongly disagree.

8. The constant power failures, and sometimes prolonged, in our homes and in work places disrupt the very normal life or the people and cause heavy loss to the industries.
   Strongly agree; Agree; Unable to decide; Disagree; Strongly disagree.

9. There are many built-in unpolluted environmental conditions such as nurseries, clean hostels, libraries, museums and all public housing projects promoting good social interactions and happiness to the people.
   Strongly agree; Agree; Unable to decide; Disagree; Strongly disagree.

10. The beautiful natural environments such as the green forests, perches, greenfields, ponds and pools are mercilessly destroyed and polluted by man on the name of modernization and provide us with foul air, dusts, wastages and toxic materials in place of them which are injurious to our health.
    Strongly agree; Agree; Unable to decide; Disagree; Strongly disagree.

11. The cities and towns that looked green with bushy trees have become barren sterile and in place of them, we see only smokes and dusts, breathe gas polluted air, in addition we hear horrible noises.

Strongly agree; Agree; Unable to decide; Disagree; Strongly disagree.

12. The environmental conditions like the well designed buildings with beautiful looks, more comforts, natural illumination, fresh air ventilation and noise free provide happiness without any pollution.
Strongly agree; Agree; Unable to decide; Disagree; Strongly disagree.

13. The city and the town environments have become very much tensely populated, very difficult to move about, everywhere the transport gas smokes, road dusts, poor sewage and drainage systems, dirty water supply and pollutions of all types cause tension, stress and irritation in our life.
Strongly agree; Agree; Unable to decide; Disagree; Strongly disagree.

14. The man-made city and town environments are giving us very pleasant feelings with their exciting and beautiful shopping centres, good roads, magnificent and huge buildings, the zoos and museums well laid play grounds without any dust or pollution.
Strongly agree; Agree; Unable to decide; Disagree; Strongly disagree.

15. Many people who sought the city life at their financial sacrifice became highly disgusted as they were subjected to continuous exposure to various types of unhygienic and unhealthy way of life by the dirty lanes and roads and the poor oozing sewage systems polluting the whole area with very bad smells.
Strongly agree; Agree; Unable to decide; Disagree; Strongly disagree.

16. People could lead a happy life only in a man-made environment, where they find all their requirements without any difficulties. The modernized city and town provide everything they require.
Strongly agree; Agree; Unable to decide; Disagree; Strongly disagree.

17. Children's behaviour is very much restricted in urban environmental condition from the security point of view, whereas in rural natural environmental condition children have free and safe movement.
Strongly agree; Agree; Unable to decide; Disagree; Strongly disagree.

18. Human life can be healthy and secured only in man-made environments like the cities and town, which are clean with dust free roads and also supplying protected and unpolluted water and clean and unadulterated food commodities.
Strongly agree; Agree; Unable to decide; Disagree; Strongly disagree.

19. The Municipality and Corporation administrations in cities do not clear the garbage collections on the roads and in lanes regularly and attend to the oozing of the sewage systems, which all resulting in pollution of the whole area and produce disease causing germs.
Strongly agree; Agree; Unable to decide; Disagree; Strongly disagree.

20. I have not experienced any means of environmental pollutions like the contaminated water or air or splashy roads in urban or rural areas as some speak about.
    Strongly agree; Agree; Unable to decide; Disagree; Strongly disagree.

21. Many people in our country are still not aware of the consequences of the environmental pollutions which resulted in poisoning of food and water, causing kidney and brain damages, anemia, mental retardation and death.
    Strongly agree; Agree; Unable to decide; Disagree; Strongly disagree.

22. Man's scientific attitude toward the environmental conditions has brought out many healthy changes which are seen in public health systems and hygienic way of life of the people.
    Strongly agree; Agree; Unable to decide; Disagree; Strongly disagree

23. Human beings today are free from *en masse* killings by epidemic and other fast spreading fatal diseases because of the clean and hygienic modern environmental conditions.
    Strongly agree; Agree; Unable to decide; Disagree; Strongly disagree.

24. People who pollute the living environments by throwing away all sorts of rubbish, waste products, garbage, spoiled eggs, decomposed vegetables, bird's feathers, rotten mutton, bones, cotton pieces and all kinds of waste household things should be severely punished.
    Strongly agree; Agree; Unable to decide; Disagree; Strongly disagree.

25. The modern living environmental conditions everywhere, as far as I know are clean and not polluted by any means.
    Strongly agree; Agree; Unable to decide; Disagree; Strongly disagree.

26. The rural people who migrate to big cities should be prevented, if we want to keep the cities clean and unpolluted. Because when they go there, construct dirty slums everywhere and live unhygienic way of life, pollute the living environment which become the cause of epidemic diseases and also big problem for the city administration.
    Strongly agree; Agree; Unable to decide; Disagree; Strongly disagree.

27. I have not experienced so far any of the bad effects caused by the environmental pollutions.
    Strongly agree; Agree; Unable to decide; Disagree; Strongly disagree.

28. The scientific and technological developments have made vast changes in our living environmental conditions for a healthy way of life and have not polluted the living environments in any way.
    Strongly agree; Agree; Unable to decide; Disagree; Strongly disagree.

29. We are hale and healthy and are living happily in the existing environment and never experienced any environmental pollution or impure, unclean and contaminated substances causing any disease.
    Strongly agree; Agree; Unable to decide; Disagree; Strongly disagree.

30. The rapidly developed modern industries and the enormous amount of transports have become death traps and making life very miserable by causing death to thousands of people everyday by accidents in badly designed man-made environmental situations.
    Strongly agree; Agree; Unable to decide; Disagree; Strongly disagree.

Write down here comments and remarks, if any :

APPENDIX - II

ANSWER SHEET

ENVIRONMENTAL POLLUTION ATTITUDE SCALE

Please provide the following personal information before you start answering the items of this questionnaire.

1. Your age :
2. Gender :
3. Qualifications :
4. Religion :
5. Married or single :
6. Number of members in the family :
7. Your occupation or your Parent's occupation :
8. Approximate monthly income to your family :
9. Name of your Home Town :
10. Approximate population of your Home Town.
11. Whether your home town is an industrial or agricultural or metropolitan area :
12. Are the homes in the area where you live get regular clean water supply from the Corporation or Municipality?
13. Are the homes in your area get uninterrupted supply of electricity?
14. Do the Public Health Department people clear the garbage everyday in your area?
15. Do the sewage and the drainage systems work without any problems?
16. Are the petty shopkeepers and vegetable vendors encroaching into your street or road and pollute the area?
17. Are there many transport vehicles moving fast spreading dusts and smokes in your area?
18. Are there industries throwing dusts and smokes in your area?

ANSWER SHEET

ENVIRONMENTAL POLLUTION ATTITUDE SCALE

Please write down your answer in this answer sheet by underlining any one of the five responses with which you agree for each statement.

Core	State-ment	Responses				
	1	Strongly Agree	Agree	Unable to decide	Disagree	Strongly Disagree
	2	Strongly Agree	Agree	Unable to decide	Disagree	Strongly Disagree
	3	Strongly Agree	Agree	Unable to decide	Disagree	Strongly Disagree
	4	Strongly Agree	Agree	Unable to decide	Disagree	Strongly Disagree
	5	Strongly Agree	Agree	Unable to decide	Disagree	Strongly Disagree
	6	Strongly Agree	Agree	Unable to decide	Disagree	Strongly Disagree
	7	Strongly Agree	Agree	Unable to decide	Disagree	Strongly Disagree
	8	Strongly Agree	Agree	Unable to decide	Disagree	Strongly Disagree
	9	Strongly Agree	Agree	Unable to decide	Disagree	Strongly Disagree
	10	Strongly Agree	Agree	Unable to decide	Disagree	Strongly Disagree
	11	Strongly Agree	Agree	Unable to decide	Disagree	Strongly Disagree
	12	Strongly Agree	Agree	Unable to decide	Disagree	Strongly Disagree
	13	Strongly Agree	Agree	Unable to decide	Disagree	Strongly Disagree
	14	Strongly Agree	Agree	Unable to decide	Disagree	Strongly Disagree
	15	Strongly Agree	Agree	Unable to decide	Disagree	Strongly Disagree
	16	Strongly Agree	Agree	Unable to decide	Disagree	Strongly Disagree
	17	Strongly Agree	Agree	Unable to decide	Disagree	Strongly Disagree
	18	Strongly Agree	Agree	Unable to decide	Disagree	Strongly Disagree
	19	Strongly Agree	Agree	Unable to decide	Disagree	Strongly Disagree
	20	Strongly Agree	Agree	Unable to decide	Disagree	Strongly Disagree
	21	Strongly Agree	Agree	Unable to decide	Disagree	Strongly Disagree
	22	Strongly Agree	Agree	Unable to decide	Disagree	Strongly Disagree
	23	Strongly Agree	Agree	Unable to decide	Disagree	Strongly Disagree
	24	Strongly Agree	Agree	Unable to decide	Disagree	Strongly Disagree
	25	Strongly Agree	Agree	Unable to decide	Disagree	Strongly Disagree
	26	Strongly Agree	Agree	Unable to decide	Disagree	Strongly Disagree
	27	Strongly Agree	Agree	Unable to decide	Disagree	Strongly Disagree
	28	Strongly Agree	Agree	Unable to decide	Disagree	Strongly Disagree
	29	Strongly Agree	Agree	Unable to decide	Disagree	Strongly Disagree
	30	Strongly Agree	Agree	Unable to decide	Disagree	Strongly Disagree

APPENDIX - III

ENVIRONMENTAL POLLUTION ATTITUDE SCALE

KEY FOR SCORING THE RESPONSES

Statement	Weightages for the Responses				
	Strongly Agree	Agree	Unable to decide	Disagree	Strongly Disagree
1. Favourable	5	4	3	2	1
2. Unfavourable	1	2	3	4	5
3. Favourable	5	4	3	2	1
4. Unfavourable	1	2	3	4	5
5. Favourable	5	4	3	2	1
6. Favourable	5	4	3	2	1
7. Unfavourable	1	2	3	4	5
8. Favourable	5	4	3	2	1
9. Unfavourable	1	2	3	4	5
10. Favourable	5	4	3	2	1
11. Favourable	5	4	3	2	1
12. Unfavourable	1	2	3	4	5
13. Favourable	5	4	3	2	1
14. Unfavourable	1	2	3	4	5
15. Favourable	5	4	3	2	1
16. Unfavourable	1	2	3	4	5
17. Favourable	5	4	3	2	1
18. Unfavourable	1	2	3	4	5
19. Favourable	5	4	3	2	1
20. Unfavourable	1	2	3	4	5
21. Favourable	5	4	3	2	1
22. Unfavourable	1	2	3	4	5
23. Unfavourable	1	2	3	4	5
24. Favourable	5	4	3	2	1
25. Unfavourable	1	2	3	4	5
26. Favourable	5	4	3	2	1
27. Unfavourable	1	2	3	4	5
28. Unfavourable	1	2	3	4	5
29. Unfavourable	1	2	3	4	5
30. Favourable	5	4	3	2	1

EXPERIMENT 60 : MEASUREMENT OF ATTITUDE TOWARD NOISE

INTRODUCTION

Noise is one of the environmental components that humans often face either for advantage or for disadvantage. The effect of noise is more disadvantageous than advantageous. Because noise is exerting influence upon the behaviour of the people. This influence is not facilitating human behaviour, instead it is having harmful effect upon the sensory apparatus of human beings.

Noise is a sound, especially, one that is loud or disturbing. Noise may be clamorous or may be loud shouting. Noise is having varying levels of intensity. Man can tolerate certain level of noise in the environmental situation, so long as it is not disturbing his sensory apparatus of hearing. Man is living in the midst of various types of noises in his environmental condition. Of course, man is responding to environment stimulations either by seeing the visual stimulations or by hearing the auditory stimulations, or by touching the cutaneous sensations or by smelling the olfactory sensations and so on. In the case of noise man's response depends upon the optimum stimulation. Neither loud noise nor feeble sound cannot arouse suitable response. The stimulation in an environmental situation may be excessive stimulation or inadequate stimulation.

The Anatomy of Auditory Mechanism of Sound

The sounds or noises, whether they are wanted or unwanted are judged and evaluated by our ear and our brain structure. We perceive sounds or noises with our ears. Since noise is a physical component, it is also important that we must know something more about the structure of auditory system. We study sound components in physics. The principle that lies in the study of sound is that any solid body or matter when struck or hit will vibrate. Or any striking on a metal plate will produce vibrations. These vibrations of the solid body or metal are consistently, periodically and regularly made and they may be also called sound waves and these sound waves have frequency and amplitude. These sound waves causing pressure in the air and make the air molecules to move fastly and hit the eardrum. The air molecules together when pressed the eardrum make positive pressure and when they are pulled apart they make negative pressure.

Thus the air molecules carry the sound vibrations to the eardrum with both positive and negative pressures. In sound waves there are peaks and valleys. The positive pressures are peaks and the negative pressures are valleys. These pressures are occurring regularly and periodically. As a result of this, sound waves reach the eardrum periodically. The eardrum is a membrane and when the sound waves attack it or give pressure on it the membrane vibrates. This is how the membrane reacts to the pressures of the sound waves.

The human ear contains three parts, the outer ear, which is consisted of the Pinna (auricle), the external auditory meatus (or the canal leading from outside to the eardrum) and the eardrum or the tympanic membrane. Therefore, the outer ear consists of these three parts called the (1) Pinna or auricle, (2) the external auditory meatus, and (3) the eardrum. The eardrum is the inner end of the external auditory meatus. The second part is called the middle ear which contains three small bony parts. These are called ossicles. The first bony part is called malleus or the (1) hammer. This is connected to the eardrum and this is vibrated by the eardrum and the vibrations are passed on to the second part called (2) anvil or incus which in turn sends the vibration to the third part (3) the stirrup or stapes. These three bony parts are connected by tissues and function almost as a single unit. Thus when the eardrum or tymponic membrane is vibrated, all the three parts of the middle ear are also vibrated or made to move at the same time. The outer ear and the middle ear are filled with air.

The inner ear also consisted of three parts, called (1) the cochlea, (2) the vestibule, and (3) the three semicircular canals. All these three parts would look like a coiled moving snail. The cochlea is a fluid-filled structure of the inner ear. This is very important part in human sense of hearing. The coiled structure cochlea is very sensitive to a great range of sound frequencies. Cochlea is divided into three compartments or chambers. They are called (1) scala vestibule, (2) scala tympani, and (3) scala media or cochlear duct. The scala vestibule is the upper chamber, starts at the oval window. The oval window is situated at end of the middle ear. Just below the oval window, there is a round window. The upper chamber (scala vestibule) communicates with the scala tympani, which is the lower chamber through a small opening called *helicotrema* at the apex of the cochlea. The upper chamber (scala vestibule) and the lower chamber (scala tympani) are filled with a fluid called *perilymph.* The third chamber scala media is filled with a fluid called *endolymph.* This is independent and does not communicate with the other two chambers. The function of these parts of the cochlea contribute mainly to our sense of hearing.

How the Sound is Experienced

The sound waves from outside world attack the eardrum or tympanic membrane and make it to vibrate. The vibrations from the eardrum are immediately transmitted to the ossicles, the three bony parts in the middle ear. The three parts of the middle ear are the hammer (melleus), the anvil (incus) and the stapes (stirrup) function as a single unit. First the hammer (melleus) is made to move, this makes the other part anvil (incus) to vibrate and finally the anvil makes the stapes (stirrup) to vibrate. This stirrup is connected with the oval window of the inner ear and makes to vibrate the oval window. The middle ear and the inner ear are separated by rigid walls except at the place, *i.e.,* at the oval window and at the round window. The three parts of the middle ear form a bridge between the eardrum and the oval window. It is a membrane covering the opening of the fluid filled inner ear. When the oval window is pushed we can at once find an important clue as to how the ear discriminates the pitch. The oval window moves back and forth, because of the movement of the fluid and every time when the oval window is pushed in, then the round window also functions by bulging outward. This means the fluid in the oval window and round window of the inner ear vibrates with the eardrum. The basilar membranes separating the three chambers of the inner ear are flexible. A wave of compression is transmitted by the eardrum does not go all the way to the end of the inner ear. It manages to take a shorter way through the basilar membranes.

When the fluid in the oval window vibrates the basilar membranes also vibrate and move back and forth. The vibrations of the basilar membranes are responsible for stimulating the nerve endings. The fluid filled cochlea is divided into three compartments by two flexible membrane. One is the basilar membrane having connection with the nerve endings. These nerve endings transmit the impulses or vibrations from ear to the brain. The vibrations of the fluid in the cochlea of the inner ear cause different parts of the membrane to vibrate. The basilar membrane consists of many hair cells and any part of it may be stimulated depending upon the pitch of the sound. The vibrations from the basilar membrane stimulate the nerve endings of the hair cells. There are about 25,000 hair cells put together to form the auditory nerves and send their impulses to the temporal lobe of the brain. These stimulations are the impulses which finally go to the brain and provide the basis for auditory experience.

Studies on the Effects of Noise

The study of noise is a difficult task. The problem of noise is not easily studied like all other psychological phenomena. A few of the studies tried are, comparing one noise or sound with another which is a standard noise or sound. However, results of some studies are available on the effects of noise. It was observed that constant exposure to noises which were above the normal amplitude (dB) often caused damage to the human sensory apparatus of hearing. It was experienced that noise interferes also with feeling aspects. Since noise when it is very loud cause annoyance or irritation when people were forced to be in that situation they get frustrated. Such situations leave people with a feeling of hatred for loud noises.

There are events to which the modern youths are constantly exposed to loud noises on the name of modern music. Particularly, there is increased tendency to loud rock music or pop music. The rock-and-roll or rock-'n' roll is a type of pop music originated in the 1950s. It is a blend of rhythm-and-blues, that is the blend of country and western music. It is generally based upon the twelve-bar blues, the second and fourth beats in each bar being heavily accented and dancing performed to such music with exaggerated bodily movements stressing the beat. Blues meant something functioning as singing. It is the depression of the feeling of depression or deep unhappiness. Blues are a type of folk songs originated among the blacks and the Americans at the beginning of the twentieth century. In this a basic twelve bar chorus is employed. Lobo and Oliphant (1968) have found that the youth groups who play this type of rock music, sing and dance in clubs are exposed to loud noises to the decibel (dB) 110 to 120 sounds for one hour to two hours of non-stop period. Persons who participate in these often are driven to serious loss of hearing. Those who participate in high sounding pop music and dance often put into severe stress. This later on results in hypertension, increase of heart rate, sore secretion of adrenal glands and then they become restless. They even begin to hate normal life. Some even become abnormal in their behaviour.

There are also several other physical disorders, like, cardiovascular disorders, allergic reactions, intestinal disorders. Cohen (1973) and Janson (1973) have observed all these in persons who were constantly exposed to louder noises. The effects of noise not only cause physical disorders mentioned above, but also cause certain mental diseases, symptoms of such diseases also were observed in those people.

There are difference of opinions among the scholars who have studied about the effect of noises. Some are of the view that noise interferes with work performance and some disagree with this view. They are of the view that noise increases the efficiency in performance. Broadbent (1954) in one of his studies has verified this phenomenon. He found in his studies that the loud noise raised suddenly of unpredictable types cause much disturbance to the individuals who are performing some task. This type of unpredictable sudden noise often directs that attention to something else and made the person to commit more errors in the work situation. But, Corcoran's (1962) study reveals an opposite result. According to him the noise really improves the individual's performance in the work. But Corcoran has not stated whether it is predictable noise or unpredictable noise.

In the communication area it was observed that noise interference cause lots of problems in the communication processes in the business and industrial organizations. For example, while sending message through telephone any interference of noise may cause lot of confusion and the message may sometimes be distracted and the meaning of the massage get confused. This type of noise interference may occur in telecommunications, and auditory signals. Persons working in louder noisy situations often develop severe stress as they could not avoid the impairing noises. Because of the stress and tension people lack social facility. They get irritated and even become aggressive and do not have any interaction with others. Appleyard and Lintell (1972) have studied this aspect of the effects of noise.

Further, the effects of noise are also having influence upon one's physical health, mental health, temperament, social behaviour and so on. More persons are admitted in mental hospitals, according to the observations made by Meecham and Smith (1977). There were also increased heart strokes and death cases due to severe stress effects of chronic abnormal levels of noise. In Dellinger's (1979) study, cases of these types were observed.

Therefore, the effects of high levels of noise cause increasing rates of physical illness and also psychological illness to those who are constantly exposed to such situations.

Attitude toward Noise

Whether one accepts that noise is having harmful effect or not, it depends upon the person's attitude toward it a great deal. How far his attitude is related to noise. We tolerate many noises or sounds in our everyday life in various places like the work place, and outside but not at our homes. The meaning of noise and attitude toward it of many people varies. The noises caused in pop music is acceptable in the clubs, but not in other places. The concept of movie songs and dances are thoroughly changed to nothing but noise in these days. The pop music or rock-n-roll music and dance lovers seek noise in high pitch, even exceeding the legal limit. Studies reveal that these people suffer from partial or complete physical impairment of deafness and psychological illness like stress and tension. Sound as an environmental component is not a pleasant variable. Therefore, it is important to know what people have understood about noise or sound and what type of attitude they have toward noise. Hence an attempt is made to measure the attitude toward noise among the students and among the general public.

EXPERIMENT

Problem

Assess the attitude of a person toward noise and compare his/her attitude with a small group of persons. Test the hypothesis that noise is an unwanted sound.

Method

Participant

Choose a person who has a knowledge about this experiment and collect the personal information such as age, sex, qualifications, place of his residence, the mode of coming to the college/university and going home, his/her personal interests, the type of music he/she likes, socio-economic status, whether has sound sense of hearing etc.

Material

Prof. Rajamanickam's Noise Pollution Attitude Scale, Manual, Test Sheet and Scoring Key.

Description of the Scale

The noise pollution attitude scale was designed and developed by Prof. M. Rajamanickam in 1995 for the purpose of measuring attitude toward noise. The scale consisted of 30 statements each expressing one idea about noise or sound. It is a five point scale designed on the pattern of Likert Scale technique. There are five response categories such as strongly agree, agree, unable to decide, disagree and strongly disagree. There are both positive and negative statements. Points are assigned to each response category. For the pro-noise (positive) statements, the strongly agree response will receive 5 points, agree 4, unable to decide 3, disagree 2 and strongly disagree 1. The positive statements are referring to the approving of the presence of noise. For them noise is harmful. For the anti-noise (negative) statements the points are assigned in the reverse way. The strongly agree response is given 1 point, agree 2, unable to decide 3, disagree 4, and strongly disagree 5. The negative statements are referring to the disapproval of the noise. That is, noise is not existing. The participant has to read each statement and express his/her opinion about the idea expressed in the statement by underlining any one of the responses with which he/she agrees. The individual score is the total points to all the 30 statements. The lowest score in this scale is 30 and the highest is 150. The scale has significant reliability (0.82) and also validity (.90). It can be used as the individual test and also as a group test.

Procedure

The experimenter, first of all should check-up the test materials whether he/she has the manual, test sheet and the scoring key. The experimenter should make the participant to sit comfortably in a well illuminated room in front of him at a distance. Then the following instructions should be given.

> "This is a test to assess your attitude toward noise. It is not an intelligence test nor a personality test. There are 30 statements, each one expressing one idea about noise. Read each statement carefully and the idea may be agreeable to disagreeable to you. Whatever you feel, express it through the response categories. There are five responses under each statement. If you strongly agree with the idea underline strongly agree response; if you simply agree with the idea underline agree response; if you are not able to decide it, underline unable to decide; if you simply disagree with the idea underline disagree; if you strongly disagree with the idea underline strongly disagree. Carry out the same procedure to all the 30 statements. As soon as you complete answering in this manner you hand over it to me. Two examples are given at the top of the test sheet. You can answer everything in similar manner".

Now the experimenter can hand over the test sheet to the participant and can give the order. "Now, ready go". As soon as the participant start answering, the experimenter can also watch, how much time the participant takes to answer all 30 statements. As and when the participant has completed the test the experimenter should collect it from him/her and get the introspective report from him/her and may be allowed to go away.

Method of Scoring

The experimenter should scrutinize whether the participant has answered all the items according to the instructions given. The responses may be scored according to the procedure given in the key. In the case of pro-noise statements the strongly agree response gets a score of 5, the agree response 4, unable to decide 3, disagree response 2 and the strongly disagree response 1. In the case of anti-noise statements the strongly agree response gets the score of 1, the agree response 2, unable to decide 3, disagree response 4, and strongly disagree response 5. The individual score is the total score for all the 30 statements. The high score indicates the existence of noise, not favourable toward noise and noise is harmful. The low score indicates favourable attitude toward noise, that is, for them noise is not harmful. The lowest possible score in this scale is 30 and the highest possible score is 150. The individual's score between 30 and 60 is favourable to noise, between 61 and 90 neutral and between 91 and 150 is unfavourable.

Results

The individual's total scores for all the 30 statements should be presented in Table 20.3. The scores may be also classified into three categories on the basis of favourable and unfavourable and neutral responses given to the statements. In how many statements the participant is favourable towards noise, in how many statements he/she is unfavourable towards noise and in how many he/she is neutral towards the noise. The scores may be given in Table 20.3.

Table 20.3 : Showing the scores of the individual on the Noise Attitude Scale

S. No.	Type Attitude	No. of Items	Score
1.	Favourable attitude to Noise		
2.	Unfavourable attitude to Noise		
3.	Neutral attitude to Noise		
	Total		

State, on the whole from the total score whether the individual is having favourable attitude to noise or unfavourable attitude. State in how many items he is favourable and in how many items he is unfavourable and in how many neutral. Explain in what type of ideas he/she is favourable and in what type of ideas he/she is unfavourable and neutral. State whether scores of the individual is in confirmation with the hypothesis or against it. State whether the individual's scores are in terms of his native atmospheric condition. Is he/she from tropical area or from hilly area or from plains, or industrial areas?

Collect the group data from other experimenters and present them in Table 20.4. Work out the mean for the type of responses and also for the whole score. Compare the scores of your participant with the group scores. What is highest score and who has scored it? What is the lowest score and who has scored it?

Table 20.4 : Showing the group data on the Noise Attitude Scale

S. No.	Name	Total Score	Favourable Attitude Score	Unfavourable Attitude Score	Neutral Attitude Score
1.	P.V.				
2.	S.N.				
3.	K.R.				
4.	M.S.				
5.	P.K.				
6.	B.V.				
7.	T.S.				
8.	V.S.				
9.	R.M.				
10.	C.R.				
	Mean				

Discussion

The definition of noise is that it is an unwanted sound. Any sound that is causing disturbance and damage to our senses is called noise. What is the attitude your participant has developed toward noise? Has he/she favourable attitude toward noise or unfavourable attitude? The hypothesis of this study is that the participants should have unfavourable attitude toward noise. This means that the participant's scores should be high. Those who score high on the noise attitude scale above 91 are having unfavourable attitude toward noise, for them noise is harmful and those who score below 60 are having favourable attitude toward noise, for them noise is not harmful.

State who are those unfavourable toward noise and who are those favourable toward noise. Discuss in what context they are unfavourable toward noise and favourable toward noise. State who is the most unfavourable toward noise and who is the most favourable toward noise. Analyze their personal backgrounds and state what caused them to develop such attitudes toward noise.

Is there any person auditorily impaired? What is his score? Is it high or low? Discuss his score in relation to other personal factors. Is there any difference in the favourable and unfavourable attitudes on the basis of sex, age, religion or other factors. If so discuss in detail. Has any of the subjects stated in the introspective report anything about their experience with noise? If so, what he/she has stated. Discuss it. Is there anybody in the group who is interested in Karnatic music or in rock music? What is his score in every category and also in the whole scale. Is he/she favourable to noise or unfavourable to noise. Discuss whether the whole group's attitude is in confirmation of the hypothesis that noise is an unwanted sound. Compare your results with the results of earlier studies.

Conclusions

1. State whether your participant's attitude is favourable to noise or unfavourable to noise. State the reason, if any.
2. Whether the group as whole is having favourable attitude toward noise or unfavourable attitude.
3. Among the group members who is more favourable to noise and who is most unfavourable to noise.
4. Is there sex difference in the favourable and unfavourable attitude toward noise.
5. What is attitude of a hearing impaired person if any in group.

APPENDIX - I

NOISE ATTITUDE SCALE

DIRECTION FOR ANSWERING THE ITEMS

There are 30 statements given below, each one expressing one idea about Noise. Read each statement carefully. You may agree with some ideas and disagree with some ideas. Whatever opinion you have about the idea of noise, please indicate the level of your agreement or disagreement by underlining *any one of the five responses* given under each statement (*i.e.* Strongly agree; Agree; Unable to decide; Disagree; Strongly disagree). Please use the Answer Sheet for underlining your answer.

Example

1. Louder noises rupture our eardrum and destroy the other parts of the ear.
   <u>Strongly agree</u>; Agree; Unable to decide; Disagree; Strongly disagree.
2. Noises are produced for some purposes and we like them as a source of communication.
   Strongly agree; Agree; Unable to decide; Disagree; <u>Strongly disagree</u>.

Meaning for Key Words

1. *Noise :* Noise refers to the loud sound, generally disturbing or discardant or harsh connected with human feeling of irritation and having impairing effect in humans.
2. *Hypertension :* Abnormally high blood pressure.
3. *Cardiovascular disorder :* A kind of disorder relating to heart and the blood vessels.
4. *Infentile mortality :* A condition causing death on a large scale in very young children.
5. *Expected mothers :* Mothers who are going to give birth very soon or pregnant mothers.
6. *Nausea :* A feeling of sickness at the stomack with an impulse for vomitting.

STATEMENTS

1. The loud noises cause much problems to human beings in man-made byproducts of civilization and environment.
   Strongly agree; Agree; Unable to decide; Disagree; Strongly disagree.

2. Noise is not harmful, it is the indication of life and communication.
   Strongly agree; Agree; Unable to decide; Disagree; Strongly disagree.

3. Physically and rapidly produced noises give pressure on the molecules of the eardrum and damage it.
   Strongly agree; Agree; Unable to decide; Disagree; Strongly disagree.

4. Noises do not damage the parts of our ear as some people speak about.
Strongly agree; Agree; Unable to decide; Disagree; Strongly disagree.

5. Exposed to noises repeatedly for eight or more hours, damage physically the sense of hearing.
Strongly agree; Agree; Unable to decide; Disagree; Strongly disagree.

6. Noises are not always causing disturbances.
Strongly agree; Agree; Unable to decide; Disagree; Strongly disagree.

7. Irregular and unpredictable noises cause annoyance and irritation.
Strongly agree; Agree; Unable to decide; Disagree; Strongly disagree.

8. Loss of hearing in humans is due to several causes and not due to louder noises.
Strongly agree; Agree; Unable to decide; Disagree; Strongly disagree.

9. The unwanted noises over which we do not have any control, cause more severe tension and irritation.
Strongly agree; Agree; Unable to decide; Disagree; Strongly disagree.

10. It is the noise that makes human life lively and pleasant.
Strongly agree; Agree; Unable to decide; Disagree; Strongly disagree.

11. The louder rock-music is becoming the source of noise and the college students and the youths who are enamoured of it are getting their sense of hearing damaged very early.
Strongly agree; Agree; Unable to decide; Disagree; Strongly disagree.

12. Noises are not unwanted sounds, we like noises.
Strongly agree; Agree; Unable to decide; Disagree; Strongly disagree.

13. Man is put into severe stressful condition when exposed to high level of noises for longer duration.
Strongly agree; Agree; Unable to decide; Disagree; Strongly disagree.

14. Noises are not having any bad effect on human beings.
Strongly agree; Agree; Unable to decide; Disagree; Strongly disagree.

15. Men, who are working in heavy noisy conditions, often subjected to severe stress followed by hypertension and ulcers.
Strongly agree; Agree; Unable to decide; Disagree; Strongly disagree.

16. Noises and sounds are natural events and we cannot avoid them.
Strongly agree; Agree; Unable to decide; Disagree; Strongly disagree.

17. Medical researchers state that high concentration of noises cause higher diastolic and systolic blood pressure and increased cases of cardiovascular disorders.
Strongly agree; Agree; Unable to decide; Disagree; Strongly disagree.

18. All the noises are very much useful for us and we must tolerate them.
Strongly agree; Agree; Unable to decide; Disagree; Strongly disagree.

19. Surveys on noise report that frequent exposure to choring noises is strongly associated with acute and chronic illness and sleeplessness.
Strongly agree; Agree; Unable to decide; Disagree; Strongly disagree.

20. Noises and sounds are used as medium of verbal communicative process and we should not avoid them.
Strongly agree; Agree; Unable to decide; Disagree; Strongly disagree.

21. Medical researchers reveal that there is close association between aircraft noises and infantile mortality and also abortion in expected mothers, which are most dangerous effects on humans.
Strongly agree; Agree; Unable to decide; Disagree; Strongly disagree.

22. The musical notes we enjoy are born only out of noises and sounds and, therefore, noises are pleasant to hear.
Strongly agree; Agree; Unable to decide; Disagree; Strongly disagree.

23. Prolonged exposure to heavy noises often cause severe physical and mental stress which is one of the causative factors of mental illness.
Strongly agree; Agree; Unable to decide; Disagree; Strongly disagree.

24. We will be missing many things in this world if we avoid noises and sounds.
Strongly agree; Agree; Unable to decide; Disagree; Strongly disagree.

25. Persons who are exposed to high intensity of noise in their work places often suffer from headache, nausea, instability, sexual impotence and change of moods.
Strongly agree; Agree; Unable to decide; Disagree; Strongly disagree.

26. We must try to train ourselves to adopt to noises instead of avoiding them.
Strongly agree; Agree; Unable to decide; Disagree; Strongly disagree.

27. The unwanted loud choring noises directly contribute to mental tension and disgust in life.
Strongly agree; Agree; Unable to decide; Disagree; Strongly disagree.

28. Noises and sounds are necessary evils, we cannot reject them.
Strongly agree; Agree; Unable to decide; Disagree; Strongly disagree.

29. The consequences of noise are such that workers from noisy metal industries retire with deaf ears.
Strongly agree; Agree; Unable to decide; Disagree; Strongly disagree.

30. Noise is not having any bad effect in the performance of some tasks, instead it improves the performance.
Strongly agree; Agree; Unable to decide; Disagree; Strongly disagree.

Write down here comments and remarks, if any :

APPENDIX - II

ANSWER SHEET

NOISE ATTITUDE SCALE

This is a questionnaire in the form of statements relating to the environmental problem of noise. In each statement one idea about the problem of noise is expressed, either favourable or unfavourable. You may agree with some idea and disagree with some. Whatever you feel, please indicate your agreement or disagreement according to the direction given in the test booklet supplied to you. This is not an intelligence test nor a personality test and there is no right or wrong answers. This is only to assess your opinion about the problem of noise. Your answers and other personal information given by you are used only for the academic and research purposes and would be treated strictly confidential and will not be revealed. Therefore, you should be quite free and frank in answering this questionnaire. Please provide the following informations accurately.

1. Age :
2. Gender :
3. Religion :
4. Qualifications :
5. Married or Single :
6. Your or your parent's occupation :
7. What type of work performed in the work place?
8. Approximate monthly income of your family :
9. Approximate population of your Home Town :
10. Is your Home Town an agricultural or Industrial or Metropolitan area?
11. Are there any industries producing noise in your living area?
12. What are the noises you hate mostly? Name them :
13. How do you feel when there are loud thunder sounds?
14. How do you feel when the loud chord music is made on through the loudspeakers in the streets?
15. Are you a member of any music group? If so, state the role you are playing there :
16. Have you ever experienced any horrible disturbing noise? If so, state it in detail:

ANSWER SHEET

Please write down your answer in this ANSWER SHEET by underlining any one of the five responses with which you agree with the idea expressed in the statement.

Core	State-ment	Responses				
	1	Strongly Agree	Agree	Unable to decide	Disagree	Strongly Disagree
	2	Strongly Agree	Agree	Unable to decide	Disagree	Strongly Disagree
	3	Strongly Agree	Agree	Unable to decide	Disagree	Strongly Disagree
	4	Strongly Agree	Agree	Unable to decide	Disagree	Strongly Disagree
	5	Strongly Agree	Agree	Unable to decide	Disagree	Strongly Disagree
	6	Strongly Agree	Agree	Unable to decide	Disagree	Strongly Disagree
	7	Strongly Agree	Agree	Unable to decide	Disagree	Strongly Disagree
	8	Strongly Agree	Agree	Unable to decide	Disagree	Strongly Disagree
	9	Strongly Agree	Agree	Unable to decide	Disagree	Strongly Disagree
	10	Strongly Agree	Agree	Unable to decide	Disagree	Strongly Disagree
	11	Strongly Agree	Agree	Unable to decide	Disagree	Strongly Disagree
	12	Strongly Agree	Agree	Unable to decide	Disagree	Strongly Disagree
	13	Strongly Agree	Agree	Unable to decide	Disagree	Strongly Disagree
	14	Strongly Agree	Agree	Unable to decide	Disagree	Strongly Disagree
	15	Strongly Agree	Agree	Unable to decide	Disagree	Strongly Disagree
	16	Strongly Agree	Agree	Unable to decide	Disagree	Strongly Disagree
	17	Strongly Agree	Agree	Unable to decide	Disagree	Strongly Disagree
	18	Strongly Agree	Agree	Unable to decide	Disagree	Strongly Disagree
	19	Strongly Agree	Agree	Unable to decide	Disagree	Strongly Disagree
	20	Strongly Agree	Agree	Unable to decide	Disagree	Strongly Disagree
	21	Strongly Agree	Agree	Unable to decide	Disagree	Strongly Disagree
	22	Strongly Agree	Agree	Unable to decide	Disagree	Strongly Disagree
	23	Strongly Agree	Agree	Unable to decide	Disagree	Strongly Disagree
	24	Strongly Agree	Agree	Unable to decide	Disagree	Strongly Disagree
	25	Strongly Agree	Agree	Unable to decide	Disagree	Strongly Disagree
	26	Strongly Agree	Agree	Unable to decide	Disagree	Strongly Disagree
	27	Strongly Agree	Agree	Unable to decide	Disagree	Strongly Disagree
	28	Strongly Agree	Agree	Unable to decide	Disagree	Strongly Disagree
	29	Strongly Agree	Agree	Unable to decide	Disagree	Strongly Disagree
	30	Strongly Agree	Agree	Unable to decide	Disagree	Strongly Disagree

APPENDIX - III

NOISE ATTITUDE SCALE

KEY FOR SCORING THE RESPONSES

S. No.	Statement	Weightages for the Responses				
		Strongly Agree	Agree	Unable to Decide	Disagree	Strongly Disagree
1.	Favourable	5	4	3	2	1
2.	Unfavourable	1	2	3	4	5
3.	Favourable	5	4	3	2	1
4.	Unfavourable	1	2	3	4	5
5.	Favourable	5	4	3	2	1
6.	Unfavourable	1	2	3	4	5
7.	favourable	5	4	3	2	1
8.	Unfavourable	1	2	3	4	5
9.	favourable	5	4	3	2	1
10.	Unfaourable	1	2	3	4	5
11.	Favourable	5	4	3	2	1
12.	Unfavourable	1	2	3	4	5
13.	Favourable	5	4	3	2	1
14.	Unfavourable	1	2	3	4	5
15.	Favourable	5	4	3	2	1
16.	Unfavourable	1	2	3	4	5
17.	Favourable	5	4	3	2	1
18.	Unfavourable	1	2	3	4	5
19.	Favourable	5	4	3	2	1
20.	Unfavourable	1	2	3	4	5
21.	Favourable	5	4	3	2	1
22.	Unfavourable	1	2	3	4	5
23.	favourable	5	4	3	2	1
24.	UnFavourable	1	2	3	4	5
25.	favourable	5	4	3	2	1
26.	UnFavourable	1	2	3	4	5
27.	favourable	5	4	3	2	1
28.	Unfavourable	1	2	3	4	5
29.	favourable	5	4	3	2	1
30.	UnFavourable	1	2	3	4	5

REFERENCES
(Volume – II)

CHAPTER 11

Creativity and its Measurement

Arieti, S. *Creativity: The Magic Synthesis* New York: Basic Books, 1976

Barron, F. The psychology of imagination. *Scientific American,* 1958,*199*,150-156.

Bruner, J.S. Goodman, J.J. and Austin, G. A. *A Study of Thinking.* New York: John Wiley, 1956.

Bruner, J.S. *Knowing: Essays for the Left Hand.* Cambridge : Harvard University Press, 1963.

Bruner, J.S. The act of discovery. *Harvard Educational Review*, 1961, *31*, 21-32.

Bruner, J.S. The course of cognitive growth. *American Psychologist*, 1964, *19*,1-15.

Bruner, J.S. The growth of mind. *American Psychologist*, 1965, *20*, 1007-1017.

Cowell, H. The process of musical creation. *Amer. J. Psychologist*, 1926, *37*, 233-236.

Cropley, A. Originality, intelligence and personality. Unpublished Ph.D. dissertation. University of Alberta, Edmonton, Alberta, Canada, 1969.

Dorothy, S. and Singer, J.L. *Partners in Play,* New York: Harper & Row, 1977.

Edwards, A. J., and Scannell, D. P. *Educational Psychology.* Bombay: Allied Publishers Private Limited, 1975.

Gagne, R.M. *The Conditions of Learning* (1st ed) New York: Holt, Rinehart and Winston, 1965.

Gallapher, J.J. *Productive Thinking of Gifted Children in the Classroom.* Paper read at the American Psychological Association meeting, Chicago, September 1960.

Getzel, J.W. and Jackson, P.W. *Creativity and Intelligence.* New York: John Wiley, 1962.

Guilford, J.P. Creativity. *American Psychologist,* 1950, *5*, 444-454.

Guilford, J.P. *Factors that Aid and Hinder Creativity.* Teacher's College Record, 1962, *63*, 380-392.

Guilford, J.P. Three faces of intellect. *American Psychologist*, 1957,*14*, 469-479.

Guilford, J.P. Traits of creativity. In H.H. Anderson (Ed.) *Creativity and its Cultivation.* New York : Harper, 1959.

Hadamard, J. *The Psychology of Invention in the Mathematical Field*, Princeton University Press, 1945.

Haimowitz, N.D. and Haimowitz, M.L. (Eds.) *Human Development : Selected Readings.* New York : Crowell, 1960.

Helson, H. Adaptation level theory. In S. Koch (Ed.) *Psychology: A Study of Science.* Vol. 1. New York : McGraw-Hill, 1959.

Hollingworth, L.S. *Children Above 180 I.Q. Standford Binet,* Yonkers, World Book, 1942.

Kellogg, R. and O'Dell, S. *The Psychology of Children's Art.* New York : Random House, 1967.

Kimble, G.A.; Garmezy, N. and Zigler, E. *Principles of Psychology.* New Delhi : Wiley Eastern Limited, 1985.

Lefrancois, G.R. *Psychology for Teaching* (Third Edition) Belmont, California : Wadsworth Publishing Company, Inc., 1979.

Lowenfeld, V. *Creative and Mental Growth,* (Third Edition) New York : Macmillan, 1951.

Mackinnon, D.W. The Nature and Nurture of Creative Talent. *American Psychologist,* 1962, *17,* 484-495.

Madnick, S.A. The associative basis of the creative process. *Psychological Review,* 1962, 69, 220-282.

Maslow, A.H. *Motivation and Personality* (Second Edition) New York : Harper & Row Publishers, 1970.

Meer, B. and Stein, M.L. Measurement of intelligence and creativity. *Journal of Psychology,* 1955, *60,* 117-126.

Morse, W.C. and Wingo, G.M. *Psychology and Teaching* (Third Edition) Bombay : D.B. Taraporevala Sons & Co. Private Ltd., 1970.

Murphy, G. *An Introduction for Psychology.* New Delhi : Oxford and IBH Publishing Co., 1964.

Newall, A.; Shan, J.C. and Simon, H.A. The Processes of Creative Thinking. In Howard E. Gruber, Glenn Terrell and Michael Wertheimer, Eds. *Contemporary Approaches to Creative Thinking.* New York : Atherton Press, 1962.

Osborn, A.F. *Applied Imagination: Principles and Procedures of Creative Thinking.* New York : Scribner's 1953.

Paramesh, C.R. *Creativity and Personality.* Madras : Janata Book House, 1972.

Patrick, C. *Creative Thought in Poets. Archaeology Psychology 1935 No. 178,* pp. 53-54.

Peel, Paul, Jr. Time effects on the creative writing of sixth grade children. Unpublished M.Ed. thesis. University of Alberta, Edmonton, Alberta, Canada, 1968.

Pribram, K.N. *The New Neurology : Memory, Novelty, Thought and Choice in EEG and Behaviours.* Ed. C.N. Glaser. New York : Basic Book, 1963.

Rossam, J.J. *The Psychology of Inventor,* Washington : Inventors Publishing Co., 1931.

Skinner, B.F. *The Technology of Teaching.* Englewood Cliffs, New Jersey : Preintice Hall, Inc., 1968.

Spaulding, R.L. What teacher attributes bring out the best in gifted children? *Gifted Child Quarterly,* 1963, *7,* 150-156.

Spearman, C. *Creative Mind,* New York: Appleton, 1931.

Stern, W. *Psychology of Early Childhood* (trans. by A. Barwell) New York : Holt, 1924.

Taylor, C.W. and Holland, J.W. Development and application of tests of creativity, *Review of Educational Research,* 1964, *33,* 91-102.

Thomson, G. *Marxism and Poetry.* New York : International Publishers, 1946.

Thorndike, R.L. *The Measurement of Creativity.* Teacher's College Record, 1963, *54,* 422-424.

Torrance, E.P. Factors effecting creative thinking in children. Paper read at the *American Psychological Association Meeting.* Chicago, September 1960.

Torrance, E.P. *Guiding Creative Talent.* Englewood Cliffs, N.J. Prentice Hall, 1962.

Torrance, E.P. *Rewarding Creative Behaviour.* Englewood Cliffs. N.J. Prentice Hall, 1965.

Torrance, E.P. *Torrance Test of Creative Thinking. Norms Technical Manual,* Princeton, N.J. Personal Press, 1966.

Wallach, M.A. and Kogan, N. A new look at the creativity intelligence dimension. *Journal of Personality,* 1965, *33,* 348-369.

Wallach, M.A. and Kogan, N. *Modes of Thinking in Young Children : A Study of the Creativity Intelligence Distinction.* New York: Holt, Rinehart and Winston, 1965 (a).

Winston, S. *Days with Bernard Shaw.* New York: Vanguard Press, 1949.

Yamamoto, Kaaru. *Experimental Scoring Manual for Minnesota Tests of Creative Thinking and Writing.* Kent, Ohio : Bureau of Educational Research, Kint State University, May 1964.

References for the Experiment 30 : Study of Mental Imagery

Boring, E.G. *History of Experimental Psychology.* New York: Appleton, 1969.

Ellson, D.G. Critical conditions influencing sensory conditioning. *Journal of Experimental. Psychology,* 1942, *31*, 353-378.

Galton, F. *Inquiries into the Human Faculty,* London : Macmillan, 1883.

Garrett, H.E. *General Psychology.* (2nd ed.), New Delhi. Eurasia Publishing House Pvt. Ltd., 1975.

Haber, R.N. Eidetic Images. *Scientific American,* 1969, *200*, 36-55.

Marks, L.E. On coloured-hearing Synasthesia : Cross Model translations of sensory dimensions. *Psychological Bulletin,* 1975, *82,* 303-327.

Meenes, M. and Morton, M.A. Characteristics of the eidetic phenomenon. *J. Geno. Psychology,* 1936, *14,* 374-391.

Morris, C.G. *Psychology: An Introduction.* Englewood Cliff, New Jersey: Prentice-Hall, Inc. 1973.

Moskowitz, M.J. and Orgel, A.R. *General Psychology.* Boston: Houghton Mifflin Company, 1969.

Murphy, G. *An Introduction to Psychology.* New York: Harper & Row Publishers, 1951.

Reik, T. *Listening with the Third Ear.* New York: Farrer, 1949.

Warner, H. *Comparative Psychology of Mental Development (Revised).* New York: International University Press, 1957.

Woodworth, R.S. and Marquis, D.G. *Psychology.* London: Methuen & Co. Ltd. 1952.

CHAPTER 12

The Measurement of Human Aptitudes and Special Abilities

Farina, A.J. Jr. Development of taxonomy of human performance: A review of descriptive schemes for human task behaviour. Maryland: *American Institute for Research. Technical Report – 2* (R-69-8) January 1969.

Fleishman, E.A. Performance assessment based on an empirically derived task taxonomy, *Human Factors,* 1967, *9,* 349-366.

Friedman, G. and Detter, H.M. Factor analysis of airman classification battery and selected air force and civilian tests. *Research Bulletin,* 1954. AFPTRC-TR -54 -75.

Guilford, J.P. *The Nature of Human Intelligence.* New York: McGraw-Hill, 1967.

Guilford, J.P. Three faces of intellect. *American Psychologist,* 1959, *16,* 469-479.

McCormick, E.J. and Tiffin, J. *Industrial Psychology* (6th Edition) New Delhi: Prentice-Hall of India Private Limited, 1979.

Murphy, G. *An Introduction to Psychology.* Calcutta: Oxford & IBH Publishing Co. (Indian Ed.), 1964.

Spearman, C.E. *The Nature of Intelligence.* London: Macmillan, 1927.

The Ortho-Rater is manufactured by Bausch & Lamb Optical Company, Rochester, New York: The Sight - Screener by American Optical Company, Meadville, Pennsylvania, the Vision Tester by the Titmus Optical Co., Petersburg, Virginia.

Thrustone, L.L. *Primary Mental Abilities, Psychometric Monographs,* No. 1, Chicago: University of Chicago Press, 1935.

Tyler L.E. *The Psychology of Individual Differences.* New York: Meredith Publishing Company, 1965.

Vernon, P.E. *The Structure of Human Abilities.* London : Methuen & Co., 1950.

Woodworth, R.S. and Marquis, D.G. *Psychology.* London : Methuen & Co., Ltd., 1952.

References for the Experiment – 31: Measurement of Basic Abilities

Bennett, G.K., Seashore, H.G. and Wesman, A.G. *Manual for Differential Aptitude Test* (Third Edition) New York : The Psychological Corporation, 1959.

Bingham , W.V. *Aptitude and Aptitude Testing.* New York: Harper & Brothers, 1937.

Thorndike, R.L. and Hagen, E. *Measurement and Evaluation in Psychology and Education* (Third Edition) New York: Wiley Eastern Private Limited, 1970.

Thurstone, L.L. *Primary Mental Abilities,* Chicago: Psychometric Laboratory, University of Chicago: Report No. 30, 1948.

Woodworth, R.S. and Marquis, D.G. *Psychology.* London: Methuen and Co., Ltd., 1952.

References for the Experiment – 32 : Manual Dexterity Test

Adiseshiah, W.T.V. and Parry. J.B. *Experimental Psychology.* Bombay: Allied Publishers Private Limited, 1977.

Blum, M.L. *Industrial Psychology and its Social Foundations.* (Revised Ed.). New York: Harper and Row Publishers, 1956.

Gilmer, H. *Industrial Psychology.* New York: McGraw-Hill Book Co., 1961.

Thurstone, L.L. Primary Mental Abilities: Psychometric Monographs, 1938, No. 1.

Tiffin, J. and McCarmick, E.J. *Industrial Psychology.* Englewood Cliffs: Prentice-Hall, Inc., 1974.

References for the Experiment – 33 : Finger Dexterity Test

Fleishman, E.A. Performance assessment based on empirically derived task taxonomy. *Human Factors,* 1967, *9*, 349-366.

Friedman, G., and Detter, H.M. Factor analysis of airman classification battery and selected air force and civilian tests. *Research Bulletin,* AFPTRC – TR – 1954-75.

Theologus, G.C. Romashko, T. and Fleishman, E. A. Development of a taxonomy of human performance: A feasibility study of ability dimensions for classifying tasks, Silver Spring, Merryland: American Institute for Research, *Technical Report 85* (R 70-1), January, 1970.

References for the Experiment – 34 : Assessing Hand-Tool Dexterity

Bennett, G.K., and Fear, R.A. Mechanical Comprehension and Dexterity. *Personnel Journal,* May, 1943.

McCormick, E.J. and Tiffin, J. *Industrial Psychology.* New Delhi : Prentice-Hall of India Private Limited, 1984.

References for the Experiment – 35 : Assessing Hand - Steadiness of a Person.

Woodworth, R.S. and Schlosberg, H. *Experimental Psychology.* London: Methuen & Company, 1955.

References for the Experiment – 36 : Study of Accuracy of Psychomotor Skills in a given Task

Cronbach, L.J. *Essentials of Psychological Testing.* New York: Harper and Row, 1966.

Drews, D.W. Development and validation of synthetic dexterity tests based on elemental motion analysis. *Journal of Applied Psychology,* 1961, *45,* 179-185.

Fleishman, E. The description and prediction of perceptual motor skills learning in R. Glaser (Ed). *Training Research and Education.* Pittsburg: University of Pittsburg Press, 1962.

Humphreys, L.G.; Buxton, C.E.; and Taylor, H.R. Steadiness and rifle markmanship, *Journal of Applied Psychology,* 1939, *20,* 680-688.

McCormick, E.J. *Human Factors in Engineering.* Englewood Cliffs; N.J: Prentice-Hall, Inc., 1974.

Vernon, P.E. *The Structure of Human Abilities,* London: Methue & Co., 1950.

Wolins, L. and Mackinney, A.C. Validity Information exchange, No. 15-04, *Personal Psychology,* 1961, *15,* 227-229.

References for the Experiment – 37 : Assessing Mechanical Skills

Bennett, G.K. *Test of Mechanical Comprehension Form AA Manual,* New York: Psychological Corporation, 1940.

Firedman, G. and Ivens, F.C. Factor analysis of the airman classification battery AC-IB, the uses General Aptitude Tests, and Battery, experimental visualization and spatial tests, and psychomotor tests, *Research Bulletin,* AFRTRC-TR-1954-67.

Friedman, G. and Detter, H.M. Factor analysis of airman classification battery and selected air force and civilian tests, *Research Bulletin,* AFPTRC-TR. 1954-75.

O'Connor, J. *Born That Way.* Baltimore: Williams and Wilking, 1928.

O'Connor, J. Measure of Mechanical Aptitude. *Personnel Journal,* 1927, *6,* 15-24.

Reynolds, H.J. and Spencer, G.M. Validity information exchange No. 14-04, *Personnel Psychology*, 1961, *14,* 456-458.

Tylore, L.E. *The Psychology of Individual Differences.* New York: Meredith Publishing Company, 1965.

CHAPTER 13

Motivation and Achievement

Atkinson, J.W. *An Introduction to Motivation.* N.J: Van Nostrand, 1964.

Bernard, L.L. *Instinct: A Study in Social Psychology.* New York: Holt, Rinehart and Winston, 1924

Candland, D.K. *Psychology: The Experimental Approach.* New York: McGraw-Hill Book Company, 1968.

Clark, K.B. Some factors influencing the remembering of prose material. *Archaeology Psychology,* 1940, No. 253.

Cofer, C.N. and Appley, M.H. *Motivation: Theory and Research.* New York: Wiley, 1964.

Davids, A., and Sidmon, J. A Pilot study – Impulsivity, Time Orientation and Delayed Gratification in Future Scientists and in Underachieving High School Students. *Exceptional Children,* 1962, *29,* 170-174.

Duffy, E. *Activation and Behaviour.* New York: Wiley, 1962.

Duffy, E. Emotion: An example of the need for reorientation in Psychology. *Psychology Review,* 1934, *41,* 184-198.

Duffy, E. Leeper's motivational theory of emotion. *Psychology Review,* 1948, *55,* 324-328.

Duffy, E. The psychological significance of the concept of arousal or activation. *Psychology Review,* 1957. *64,* 265-275.

Garrett, H.E. *General Psychology.* New Delhi: Eurasia Publishing House (Pvt.) Ltd., 1975.

Guthrie, E.R. *The Psychology of Learning (*Rev. Ed.) New York: Harper & Row, 1952.

Hall, J.W. Elements of General Psychology. In Morse, W.C. and Wingo, G.M. *Psychology and Teaching.* Bombay.

Hilgard, E.R. *Introduction to Psychology,* New York: Harcourt, Brace and World, 1962.

Hull, C.L. *A Behaviour System: An Introduction to Behaviour Theory Concerning the Individual Organism.* New Haven Conn: Yale, 1952.

Hurlock, E.B. An evaluation of certain Incentives used in School Work. *Journal of Education Psychology,* 1925, *16,* 145-159.

Kimble, G.A. Garmezy, N. and Zigler, E. *Principles of Psychology.* New Delhi: Willey Eastern Limited, 1984.

Larner, E. and Murphy, L.B. (Eds.). Methods for the study of personality in young children. Monography. *Social Research Child Development,* 1941, *6,* No. 4.

Leeper, R. A motivational theory of emotion to replace emotion as disorganized response. *Psychology Review,* 1948, *55,* 5-21.

Lindsley, D.B. Psychophysiology of motivation. *In Nebraska Symposium on Motivation.* (ed.) M.R. Jones. Lincoln University of Nebraska Press, 1957.

Lorenz, K. *King Solomon's Ring.* London: Methuen, 1952.

MaClalland, D.C. Some Social Consequences of Achievement Motivation, In M.R. Jones (ed.) *Nebraska Symposium on Motivation.* Lincoln: University of Nebraska Press, 1955, 41-65.

MaClalland, D.C.; Atkinson, J.W.; Clark, R.V.; and Lowerll, E.L. *The Achievement Motive.* New York: Appleton - Century-Crofts, 1953.

Mandler, G., and Sorason, S.B. A study of Anxiety and Learning. *Journal of Abnormal and Social Psychology,* 1952, *46,* 166-173.

Maslow, A.H. *Motivation and Personality* (2nd edition). New York: Harper and Row Publishers, 1970.

McDougall, W. *An Introduction to Social Psychology.* London: Methuen, 1908.

Moskowitz, M.J. and Orgal, A.R. *General Psychology.* Boston: Houghton Mifflin Company, 1969.

Murphy, G. *An Introduction to Psychology.* New Delhi: Oxford & IBH Publishing Company, 1964.

Murray, H.A. *Explorations in Personality.* London : Oxford University Press, 1938.

Olds, J. Physiological Mechanisms of Reward, in M.R. Jones (ed.) *Nebraska Symposium on Motivation.* Lincoln: University of Nebraska Press, 1955, 73-139.

Richter, C.P. Animal behaviour and internal drives. *Quarter Rev. Biol.,* 1927, *2,* 307-343.

Sealaman, V. The influence of attitude upon the remembering of pictorial material. *Arch. Psychology,* 1940. No. 258.

Skinner, B.F. *The Behaviour of Organisations: An Experimental Analysis.* New York: Appleton-Century-Crofts, 1938.

Taylor, J.A. and Spence, K.W. The relationship of Anxiety Level to Performance and Serial Learning. *Journal of Experimental Psychology.* 1952, *44,* 61-64.

Thorpe, W.H. *Learning and Instinct in Animals* (2nd ed.) London: Methuen & Co., 1963.

Tinbergen, N. *The Study of Instinct.* London: Oxford University Press, 1951.

Tolman, E.C. A psychological mode. In T. Parsons and E. Shils (Eds.) *Towards a General Theory of Action.* Cambridge, Mass Harvard, 1951.

Tolman, E.C. *Purposive Behaviour in Animals and Man;* New York: Appleton-Century-Crofts, 1932.

Wardan, C.J. and others, *Animal Motivation.* New York: Columbia University Press, 1931.

Watson, D.L. *The Study of Human Nature.* Yellow Spring, O: Antioch Press, 1953.

Web, W.B. "A motivation theory of emotion. "*Psychology Review,* 1949, *55,* 329-335.

Wenger, M.A. An attempt to appraise individual differences level of muscular tension, *J. Psychology,* 1943, *32,* 213-229.

White, R.W. Motivation reconsidered the concept of competence, *Psychological Review,* 1959, *66,* 297-333.

Woodworth, R.S. and Marquis, D.G. *Psychology.* London: Methuen & Co. Ltd. 1952.

Reference for the Experiment – 38 : Assessing Level of Aspiration and Achievement

Child I.V. and Whiting, J.W.M. Determinants of Level of Aspiration: Evidence from Everyday Life. *Journal of Abnormal and Social Psychology,* 1949, *44,* 303-314.

Church, R.M. Millward, R.B. and Miller, P. Prediction of Success in a Competitive Reaction Time Situation. *Journal of Abnormal and Social Psychology,* 1963, *67,* 234-240.

Coleman, J.C. *Abnormal Psychology and Modern Life.* Bombay: D.B. Taraporevala Sons & Co. Private Ltd. 1974.

Davids, A. and Sidman, J. Level of Aspiration and Motor Performance in Prospective Scientists and in Underachieving Secondary School Boys, *Psychonomic Science,* 1965, *2,* 371-372.

Diggory, J.C. and Morlock, H.C. Jr. Level of Aspiration or Probability of Success, *Journal of Abnormal and Social Psychology,* 1964, *69,* 282-289.

English, H.B. *Dynamics of Child Development.* New York : Holt, Rinehart and Winston, 1962.

Heider, F. *The Psychology of Interpersonal Relations.* New York : John Wiley, 1958.

Hoppe, F. and Gould, R. An experimental analysis of level of aspiration, *Genet. Psychology* 1939, *21*, 3-115.

Medow, H. and Zander, A. Aspirations for the Group chosen by Central and Peripheral Members, *Journal of Personality and Social Psychology,* 1965, *1*, 224-228.

Moulton, R.W. Effects of Success and Failure on Level of Aspiration, as related to Achievement Motives. *Journal of Personality and Social Psychology,* 1965, *1,* 399-406.

Murphy, G. *An Introduction to Psychology.* New York: Harper & Row Publishers, 1951.

Rosenfeld, H. M. and Zander, A. The Influence of Teachers on Aspiration of Students, *Journal of Educational Psychology,* 1961, *51*, 1-11.

Rosenfeld, H.M. Social choice conceived as a Level of Aspiration, *Journal of Abnormal and Social Psychology,* 1964, *68,* 491-499.

Rotter, J.B. Generalized expectancies of internal *versus* external control reinforcement. *Psychological Monograph.* 1966, *80,* No. 1.

Rotter, J.B. *Social Learnings and Clinical Psychology.* Englewood Cliffs, NJ: Prentice-Hall, 1954.

Schroder, H.M. and Rotter, J.B. Rigidity as Learned Behaviour, *Journal of Experimental Psychology,* 1952, *43,* 141-150.

Veblen, T. *The Theory of the Leisure Class.* New York: Macmillan, 1899.

Weigand G. Good Aspiration and Academic Success. *Personnel and Guidance Journal.* 1953, *31,* 458-461.

References for the Experiment – 39 : Study of Goal Aspiration and Achievement

Coleman, J.C. and Brown, W.E. Jr. *Abnormal Psychology and Modern Life.* Bombay : Taraporevala Sons & Co. Private Ltd. 1974.

Davids, A. and Sidman, J. Level of Aspiration and Motor Performance in Prospective Scientists and in Underachieving Secondary School Boys. *Psychonomics Science.* 1965, *2,* 371-372.

Dembo, T. Der arger als dynamisches problem. *Psychologische Forschung,* 1931, *15,* 1-44.

Festinger, L. A Theoretical Interpretation of Shifts in Level of Aspiration. *Psychological Review,* 1942, *49,* 235-250.

Frank, J.D. Individual Differences in Certain Aspects of the Level of Aspiration. *American Journal of Psychology,* 1935, *47,* 119-128.

Gould, R. An experimental analysis of level of aspiration. *Gene. Psychology Monograph,* 1939, 21, 3-115.

Hilgard, E.R. *Introduction to Psychology* 4th edition New York: Harcourt, Brace and World, 1962.

Hoppe, F. Untensuchunger zur Handlungs-und affeckt – Psychologie, IX. Erflog and Musserflog (Investigations in the Psychology of Action and Emotion, IX. Success and Failure). *Psychologische Forschung,* 1930, *14,* 1-63.

Jucknat, M. Performance, Level of Aspiration and Self-consciousness. *Psychologische Forschung.* 1937, *22,* 89-179.

Lewin, K. Ramara Dembo, Leon Festinger and Pauline S. Sears. Level of Aspiration. In J. Mcv. Hunt (Ed.) *Personality and Behaviour Disorders.* Vol. 1. New York: Ronald Press, 1944.

McCoy, C.F. Some Ego Factors Associated with Academic Success and Failure of Educable Mentally Retarded Pupils. *Exceptional Children.* 1963, *30,* 80-84.

Medow, H. and Zander, A. Aspirations for the Group chosen by the Central and Peripheral Members. *Journal of Personality and Social Psychology.* 1965, *1,* 224-228.

Monte, C.F. *Beneath the Mask: An Introduction to Theories of Personality.* New York: Praeger Publishers, 1977.

Moulton, R.W. Effects of Success and Failure on level of Aspiration as Related to Achievement motives. *Journal of Personality and Social Psychology,* 1965, *1,* 395-406.

Rosenfeld, H.M. and Zander, A. The Influence of Teachers on Aspirations of Students. *Journal of Educational Psychology,* 1961, *52,* 1-11.

Rosenfeld, H.W. Social Choice conceived as a Level of Aspiration, *Journal of Abnormal and Social Psychology,* 1964, *68,* 491-499.

Weigand, G. Goal Aspiration and Academic Success, *Personnel and Guidance Journal,* 1953, *31,* 458-461.

Worrell, L. Level of Aspiration and Academic Success, *Journal of Educational Psychology,* 1959, *50,* 47-54.

References for the Experiment – 40 : Assessing the Strength of Various Motives

Hilgard, E.R. *Introduction to Psychology* (3rd ed.) New York : Harcourt, Brace and World 1962.

Maslow, A.H. *Motivation and Personality* (2nd ed.). New York : Harper & Row. Publishers, 1970.

Thorndike, E.L. *Human Nature and the Social Order.* New York: Macmillan, 1940.

Troland, L.T. *The Fundamentals of Human Motivation.* New York : Van Nostrand, 1928.

Woodworth, R.S. and Marquis, D.G. *Psychology.* London: Methuen & Co. Ltd. 1952.

References for the Experiment – 41 : Assessing the Influence of Knowledge of Results in Learning

Angell, G.W. The effect of Immediate Knowledge of Quiz Results of Final Examination Scores in Freshman Chemistry, *Journal of Educational Research,* 1949, *42*, 391-394. Garverial.

Blaker, K.E. The Effects of Delayed Knowledge of Results on Learning and Retention for Difffering Age Levels of School Age Children, Unpublished Ph.D. dissertation, Southern Illinois University, 1965.

Brackbill, Y.W.; Wenger, J.E.; and Wilson, D. Feedback Delay and the Teaching Machine, *Psychology in the Schools,* 1964, *1*, 148156.

Bruner, J.S. Some Theorems on Instruction Illustrated with Reference to Mathematics. In Hilgard (ed.). *Theories of Learning and Instructions,* 63rd year book, National Society for the Study of Education. Part I. Chicago: University of Chicago Press, 1964.

Candland, D.K. *Psychology: The Experimental Approach.* New York: McGraw-Hill Book Company, 1968.

Edwards, A.J. and Scannell, D.P. *Educational Psychology.* Bombay: Allied Publishers Private Limited, 1975.

English, R.A. The Effect of Immediate and Delayed Feeddback on Retention of Subject Matter, *Unpublished Ph.D. Dissertation.* University of Arizona, 1965.

Garrett H.E. *General Psychology.* New Delhi: Eurasia Publishing House (Pvt.) Ltd., 1975.

Jadd, C.H. Practice without knowledge of Results, Psychological Review, *Monograph Supplement* No. *29*, 1905, 185-198.

Krumboltz, J.D. and Weisman, R.G. The Effect of Intermittent Confirmation on Programmed Instruction. *Journal of Educational Psychology,* 1962, *53*, 250-253.

McGeoch, J.A. and Irion, A.L. *The Psychology of Human Learning,* New York: McKay, 1952.

Peterson, L.R. and Brewer, C.L. Confirmation, Correction and Contiguity. *Journal of Verbal Learning and Verbal Behaviour.* 1963, *1,* 365-371.

Sassenrath, J.M. and Garverial, C.M. Effects of Differential Feedback from Examination on Retention and Transfer. *Journal of Educational Psychology*, 1965, *56,* 259-263.

Seagoe, M.V. *A Teacher's Guide to the Learning Process.* Dubuaue, Iowa: William C. Brown, 1956.

Spitzer, H.F. Studies in Retention. *Journal of Educational Psychology,* 1939, *30,* 641-656.

Stolurow, L.M. Teaching by machine. Washington: V.S. Department of Health, *Education and Welfare,* 1961, 104-117.

Thorndike, E.L. *The Fundamentals of Learning,* New York: Teachers College, 1932.

Wallace, J. Concept of Dominance, Type of Feedback and Intensity of Feedback as Related to Concept Attainment, *Journal of Educational Psychology,* 1964, *55,* 159-166.

CHAPTER 14

Study of Fatigue

Brown, J.M. Berrien, F.K. and Russell, D.L. *Applied Psychology,* New Delhi : Amerind Publishing Co., Pvt. Ltd. 1966.

Candland, D.K. *Psychology: The Experimental Approach,* New York: McGraw Hill Book Company, 1968.

Crawley, S.L. An experimental investigation of recovery from work. *Archaeology Psychology,* No. *85*, 1926.

Geldard, F.A. *The Human Senses.* New York: Wiley, 1953.

Harrell, T.W. *Industrial Psychology.* New Delhi : Oxford and IBH, 1956.

Rawdon-Smith, A.F. Auditory fatigue, *Brit. J. Psychology,* 1934, *25,* 77-85.

Rawdon-Smith, A.F. Experimental deafness. *Brit. J. Psychology* 1936, *26,* 233-244.

Vernon, H.M. Prevention of accidents. *British J. Indust. Med.,* 1945, *2,* 3.

Wegel, R.L. and Lane, C.E. The auditory masking of one pure tone by another and its probably relation to the dynamic of the inner ear, *Phys. Rev.,* 1924, *23,* 266-283.

Wood, A.G. *A. Quantitative Account of the Course of Auditory Fatigue,* Master's thesis, University of Virginia, 1930.

Wyatt, S. *et al*. Fatigue and boredom in repetitive work. London : *Industrial Health Research Board,* Report No. 77, 1938.

References for the Experiment – 42 : Assessing Muscular Fatigue in a Continuous Physical Work

Crawford, A. Fatigue and driving. *Ergonomics,* 1961, *4,* 143-154.

Harrell, T.W. *Industrial Psychology.* New Delhi : Oxford and IBH, 1956.

Heimstra, N.W. The effects and stress fatigue on performance in a stimulated driving situation, *Ergonomics,* 1970, *13(2)* 209-218.

Safford, R.R. and Rockwell, T.H. *Performance Decrement in Twenty Four Hours Driving.* Presented at the 45th Annual meeting of the Highway Research Board, Washington; DC, January 1966.

Tiffin, J. *Industrial Psychology.* Englewood Cliffs. N.J: Prentice-Hall, 1952.

Vernon, H.M. Prevention of accidents. *British J. Ind. Med.* 1945, *2,* 3.

Vernon, H.M. *The Speed of Adaptation of Output to Altered Hours of Work,* London: Industrial Fatigue Research Board, No. 6, 1920.

Viteles, M.S. *Industrial Psychology.* New Yrok : Norton, 1932.

References for the Experiment – 43 : Assessing the Effects of Continuous Mental Work

Griffith, J.W. Kerr; W.A. Mayo; T.B. Jr.; and Topal, J.R. Changes in subjective fatigue and readiness for work during the eight-hour shift, *Journal of Applied Psychology,* 1950, *34,* 163-166.

Harrell, T.W. *Industrial Psychology.* New Delhi: Oxford and IBH, 1956.

Kossoris, M.D. and Kohler, R.F. *et al.* Hours of work and output. U.S. Department of Labour. *Bureau of Labour Statistics* Bulletin, No. 917, 1947.

Nelson, T.M. and Bartley, S.H. The pattern of personal response arising during the office work day, *Occupational Psychology,* 1968. *(1)* 77-83.

Tiffin, J. and McCormic, E.J. *Industrial Psychology* (2nd Edition) London : Allen and Unwin, 1965.

Viteles, M.S. *Motivation and Morale in Industry,* New York : Norton, 1953.

Viteles, M.S. *Psychology in Industry,* New York: Norton, 1935.

References for the Experiment – 44 : Testing Endurance – Muscular Fatigue (Tapping Test)

Brouha, L. Fatigue – *Measuring and Reducing It.* Advance Management 1954, *19,* 9-19.

Brown, J.A.C. *The Social Psychology of Industry.* Baltimore, Md : Penguin Books, 1954.

Brown, J.M.; Berrian, F.K.; Russill, D.L.; and Wells, W.D. *Applied Psychology.* New Delhi : Amerind Publishing Co. Pvt. Ltd.; 1966.

Crawford, A. Fatigue and driving. *Ergonomics.* 1961, *4,* 143-154.

Dashill, J.I. *An Experimental Manual in Psychology.* Boston: Hughton Mifflin Company, 1931.

Hebb, D.O. *The Organization of Behaviour,* New York : Wiley, 1949.

Lawson, R.B. *et al. Experiments in Psychology.* New York: Oxford University Press, 1975.

Lewin, K.K. *A. Dynamic Theory of Personality.* New York : McGraw Hill, 1935.

McClelland, D.C.; Atkinson, J.W.; Clark, R.A.; and Lowell, E.L. *The Achievement Motive.* New York : Appleton-Century-Crofts, 1953.

Milton Metfassal – *Student's Guide for Demonstration of Psychological Experiments.* New York: McGraw Hill Book Company Inc. 1932.

CHAPTER 15

Personality Types and Traits

Cattell, R.B. *Description and Measurement of Personality.* Yunkers World Book Co., 1946.

Cattell, R.B. The sixteen personality factors and basic personality structure. A reply to Eysenck. *Journal of Behavioural Science.* 1972, *1,* 169-187.

Claridge, G.S. *Personality and Arousal,* Oxford : Pergamon, 1967.

Easton, D.M. *Mechanisms of Body Function,* New Delhi : Prentice Hall of India Private Limited, 1978.

Eysenck, H.J. and Eysenck, S.B.C. *Psychoticism – A Dimension of Personality.* London : Oxford University Press, 1975.

Eysenck, H.J. and Eysenck, S.B.G. *Reference Manual for Eysenck's Personality Questionnaire-R.* London : University of London Press Ltd., 1980.

Eysenck, H.J. *Dimension of Personality.* London: Routledge & Kegan Paul, 1947.

Eysenck, H.J. *Psychology is About People.* New York: The Library Press, 1972.

Eysenck, H.J. *The Biological Basis of Personality.* Springfield, Ill. : Charles C. Thomas, 1967.

Eysenck, H.J. *The Dynamics of Anxiety and Hysteria: An Experimental Application of Modern Learning Theory to Psychiatry.* London : Routledge & Kegan Paul, 1957.

Eysenck, H.J. *The Scientific Study of Personality.* London : Routledge & Kegan Paul, 1952.

Eysenck, H.J. *Uses and Abuses of Psychology,* Baltimore; Penguin (Pelicon Ed.), 1953.

Guilford, J.P. Unitary traits of personality and factor theory, *American Journal of Psychology,* 1936, *48,* 673-680.

Hathaway, S.R. and McKinley, J.C. *Manual for the Minnesota Multiphasic Personality Inventory.* New York: Psychological Corporation, 1951.

Heidbreder, E. Measuring introversion and extroversion. *Journal of Abnormal and Social Psychology,* 1926, *21,* 120-134.

Jung, C.G. *Psychological Types.* New York: Harcourt Brace, 1923.

Jung, C.G. Two Essays on Analytical Psychology. *In Vol. 7 of the Collected Works of C.G. Jung,* 1917. Princeton, N.J. Princeton University Press, 1953.

Kant, E. *Critique of Pure Reason,* 1788.

Leukel, F. *Introduction to Physiological Psychology.* Saint Louis : The C.V. Mosley Company, 1968.

Moruzzi, G. *Brain Mechanism and Consciousness.* E.D. Adrian, F. Bremer, H.H. Jesper and J.F. Dalafaesnaye (Eds) Springfield Ill. : Thomas, 1954.

Pavlov, I.P. *Conditioned Reflexes: An Investigation into the Physiological Activity of the Cortex.* C.V. Anrap (Trans.) New York: Dever, 1927.

Pavlov, I.P. *Lectures on Conditioned Reflexes.* Vol. 1. W. Horsley Gantt (Trans.) New York: International Publishers, 1928.

Prince, M. *Clinical and Experimental Studies in Personality.* Cambridge : Mass. Sci-Art, 1929.

Teplov, B.V. The Historical Development of Pavlov's Theory of Typological Differences in the Dog. In J.A. Gray (Ed. and Trans.) *Pavlov's Typology.* New York: Macmillan, 1964.

Thompson, R.F. *Foundations of Physiological Psychology.* New York: Harper & Row, 1967.

Woodworth, R.S. and Marquis, D.G. *Psychology.* London : Methuen & Co. Ltd. 1952.

Wundt, W. *Outlines of Psychology* (Translated by C.H. Judd) Leipzig : Engelmann, 1897.

References for the Experiment – 45 : Assessing Dimensions of Personality

Eysenck, H.J. and G.D. Wilson. *The Experimental Study of Freudian Theories.* London: *Methuen,* 1973.

Eysenck, H.J. and S.B.G. Eysenck, *Personality Structure and Measurement.* San Diego, Calif: Robert R. Knaff, 1969.

Eysenck, H.J. and S.B.G. Eysenck. *Reference Manual for Eysenck's Personality Questionnaire-R.* London : University of London Press Ltd., 1980.

Eysenck, H.J. and S.B.G. Eysenck. *The Manual of the Eysenck Personality Questionnaire.* San Diego, Calif: Educational and Industrial Testing Service, 1975 b.

Eysenck, H.J. and sybil B. Eysenck. *The Eysenck Personality Inventory.* San Diego, Calif: Educational and Industrial Testing Service, 1968.

Eysenck, H.J. *Dimensions of Personality.* London : Routledge & Kegan Paul, 1947.

Eysenck, H.J. Dynamics of *Anxiety and Hysteria : An Experimental Application of Modern Learning Theory of Psychiatry.* London : Routledge & Kegan Paul, 1957.

Eysenck, H.J. *Eysenck Personality Questionnaire.* San Diego, Califf. Educational and Industrial Testing Service, 1975 a.

Eysenck, H.J. T*he Maudley Personality Inventory.* Sen Diego, Calif; Educational and Industrial Testing Service, 1959.

Garrett, H.E. *General Psychology.* New Delhi: Eurasia Publishing House (Pvt.) Ltd., 1961.

Kimble, G.A.; Garmezy, N. and Zigler, E. *Principles of Psychology* (6th ed.) New Delhi: Wiley Eastern Limited, 1984.

Woodworth, R.S. and Marquis, D.G. *Psychology.* London : Methuen & Co. Ltd. 1957.

References for the Experiment – 46 : Assessing Introversion-Extroversion

Cronbach, L.J. The Two Disciplines of Scientific Psychology. *American Psychologist,* 1957, *12,* 671-684.

Eysenck, H.J. and Sybil B.G. Eysenck. *Personality Structure and Measurement.* Sen Diego, Calif : Robert R. Knapp, 1969.

Eysenck, H.J. and Sybil, B.G. Eysenck. *Eysenck Personality Questionnaire.* San Diego, Calif : Educational and Industrial Testing Service, 1975 (a).

Eysenck, H.J. and Sybil, B.G. Eysenck. *The Manual of the Eysenck Personality Questionnaire,* San Diego, Calif : Educational and Industrial Testing Service, 1975 (b).

Eysenck, H.J. *Dimensions of Personality.* London : Routledge & Kegan Paul, 1947.

Eysenck, H.J. *The Maudley Personality Inventory.* San Diago, Calif. Educational Industrial Testing Service, 1959.

Eysenck, H.J. *The Scientific Study of Personality.* London : Routledge & Kegan Paul, 1952.

Eysenck, H.J. *Uses and Abuses of Psychology.* Baltimore : Penguin (Pelican) Ed. 1953.

Eysenck. H.J. *Sense and Nonsense of Psychology.* Baltimore : Penguin, 1957.

Heidbreder, E. Measuring introversion and extroversion. *Journal of Abnormal and Social Psychology,* 1926, *21,* 120-134.

Jung, C.G. *Psychological Types,* New York : Harcourt Brace, 1923.

Jung, C.G. The archetypes and the collective unconscious. Vol. 9. *The Collected Works of C.G. Jung.* Princeton, N.J : Princeton University Press, 1959.

Monte, C.F. Beneath the Mask : *An Introduction to Theories of Personality.* New York : Praeger Publishers, 1977.

Neymann-Kohlstedt *Diagnostic Test Manual.*

References for the Experiment – 47 : Assessing Personality Traits

Adlar, A. *The Fundamental Views of Individual Psychology.* In H.L. Ansbacher and R. R. Ansbacher (Eds.). 1955, New York : Harper, 1956.

Allport, G.W. *Personality : A Psychological Interpretation.* New York : Henry Holt, 1957.

Bernreuter, R.G. *Manual for the Personality Inventory. Palo Alto, California :* Consulting Psychologists Press, 1938.*

Cattell, R.B. *Description and Measurement of Personality,* Yonkers, World Book Co., 1946.

Eysenck, H.J. Dimensions of Personality. London : Routledge and Kegan Paul, 1947.

Freud, S. *The Ego and the Id.* In Vol. XIX of Standard Editors, 1923. London : Hogarth Press, 1961.

Guilford, J.P. *Personality*, New York: McGraw Hill, 1959.

Jung, C. *Psychological Types,* Vol. 6 of the Collected Works of C.G. Jung, 1923, Princeton, N.J. Princeton University 1971.

Kraeplin, E. *Lectures of Clinical Psychiatry* (2nd Ed.) London: Bailere, Tindell and Cox, 1905 (as cited by R.D. Laing, 1959).

Murray, J.A. *Exploration of Personality.* New York: Oxford University Press, 1938.

Thurstone, L.L. *Thurstone Temperament Schedule.* Chicago: Science Research Association, INC, 1950.

(*Note :* This material is available with Consulting Psychologists Press, Palo Alto, California, U.S.A.)*

References for the Experiment – 48 : Assessing Temperamental Traits

Birns, B.; Berten, S.; and Bridger, W.H. Individual differences in Temperamental Characteristics of Infant. *Transactions of the New York Academy of Sciences.* 1969, pp. 1071-1082.

Brazelton, T.B. Infants of Mothers: *Differences in Development.* New York : Dell, 1969.

Brofenbrenner, V. Toward an Experimental Ecology Human Development. *American Psychologists.* 1977, pp. 513-531.

Hall, C.S. Temperament : A Survey of Animal Studies. *Psychological Bulletin.* 1941, *38,* pp. 909-943.

Jung. C.G. *Psychological Types.* New York : Harcourt, Brace, 1923.

Sameroff, A.J. Concepts of humanity in primary prevention. In G. Albee and J. Rolf (Ed), *Primary Prevention of Psychopathology.* Burlington : 1977. Vol. 1.

Sameroff, A.J., and Zax, M. Neonatal characteristics of offsprings of schisophrenic and neurotically depressed mothers. *Journal of Neurosis and Mental Diseases.* 1973, *157*, pp. 191-199.

Scarr, S. and Kidd, K. Developmental behaviour genetics in M.M. Haith and J.J. Compos (Eds.), *Mussen Hand Book of Child Psychology.* New York : Wiley, 1983.

Slater, E., and Cowie, V. *Genetic of Mental Disorders.* London : Oxford University Press, 1971.

Stern, C.G.; Cald Well, B.M.; Hersher, L.; Lipton, E.L.; and Richmand, J.B. A Factor Analytic Study of the Modern Infant dyad. *Child Development.* 1969, *40,* pp. 163-181.

Thomas A. and Chess, S. *Temperament and Development.* New York : Brunner – Mazel, 1977.

Thomas, A., Chess, S., and Birch, H.G. *Temperament and Behaviour Disorders in Children.* New York : New York University Press, 1968.

Thomas, A., Chess, S., and Birch, H.G. The Origin of Personality. *Scientific American.* August 1970, pp. 102-109.

Thurstone, L.L. *Thurstone Temperament Schedule.* Chicago : Science Research Associates, Inc., 1950.

References for the Experiment – 49 : Assessing Psychosomatic Disorders

Alexander, F. Fundamental Concepts of Psychosomatic Research, Psychogenesis, Conversion, Specificity, *Psychosoms. Med:* 1943, *5,* 205-210.

Coleman, J.C. and Brown, W.E. *Abnormal Psychology and Modern Life.* Bombay: D.B. Taraporevala Sons & Co. Private Ltd. 1974.

Dunbar, F. *Psychosomatic Diagnosis.* New York: Harper, 1943.

Fenichel, O. Problems of Psychoanalytic technique, parts, *Psychoanal Quart.* 1939, *8,* 1-4.

Glover, E.; Fenichel, O.; Strachey, J.; Bergler, E.; Nunberg, E.; and Bibring, E.; Symposium on the theory of the therapeutic results of psychoanalysis 1937. *Int. J. Psychoanal.*

Halliday, J.L. Concepts of a psychosomatic affection. *Lancet.* 1943, 6, 205-210.

Lazarus, R.S. *Patterns of Adjustment* (3rd ed.) Tokyo: McGraw-Hill Kogakusha Ltd., 1976.

Rajamanickam, M. and Samraj, P., Neuropsychiatric and psychosomatic Symptoms in athletes and non-athletes. *Indian Journal of Community Guidance Service,* 1984, 1 & 2, 55-63.

Rajamanickkam, M and Mercy, R., Arunmozhi, Adjustment Problems in relation to psychosomatic disorders among the adolescent girls. *Journal of Community Guidance and Research,* 1992, *9*, 2, 151-179.

Reiser, M.F. Emotional stress and the precipitation of Congestive heart failure. *Psychosom. Med.,*1953, *15,* 38-60.

Stalker, E. The psychosomatic diseases: A hypothetical formulation. *J. Ment. Sci.* 1950, *6,* 355-368.

Wider, A.; Wolf, H.C.; Broadman, K.; Mittleman, B.; and Wechsler, D. *Cornell Medical Index Health Questionnaire,* New York: The Psychological Corporation, 1949.

Wolff, H.G. *Life – Situations Emotions and Bodily Disease Feelings and Emotions,* Remert, New York: McGraw-Hill, 1950.

CHAPTER 16

Counselling and Guidance : An Introduction

Bernard, H.W. *Mental Hygiene and Classroom Teachers (2nd Ed.)* New York: McGraw-Hill, 1961.

Carter, M. *Guidance and Counselling in British Schools* (Eds) Lytton, H and Craft, M. London: Arnold Ltd., 1969.

Fullmer, D. W. and Bernard, H.W. *Counselling: Content and Process.* New Delhi: Thomson Press, 1972.

Garrett, H.E. *General Psychology.* New York: American Book Company, 1961.

Gary, R. *Promise at Dawn.* New York : Harper, 1961.

Houghes, P.M. *Guidance and Counselling in Schools.* Oxford Pergamon Press, 1971.

Jones, A.J. Steffler, B. and Stewart, N.R. *Principles of Guidance.* New Delhi: Tata-McGraw-Hill Publishing Company Limited, 1979.

MacKinnon, D. What makes a person creative: *Saturday Review.* 1962, *45,* 15-17.

Maslow, A.H. *Motivation and Personality (2nd Ed)* New York: Harper and Row Publishers, 1970.

McCormick, E.J. and Tiffin, D. *Industrial Psychology, 7th Ed.* New Delhi: Prentice Hall of India Private Limited, 1984.

McMurry, R.N. Recruitment, dependency and morale in the Banking industry, *Admin. Sci. Quarterly,* 1958, 3, 87-117.

Narayana Rao, S. *Counselling Psychology.* New Delhi: Tata McGraw-Hill Publishing Company Limited, 1981.

Rogers, C.R. Client-Centred Therapy. Boston: Houghton Mufflin, 1939.

Super, D.E. and Bohn, M.J. *Occupational Psychology.* London : Tavistock Publication, 1971.

Super, D.E. *The Psychology of Careers.* New York: Harper and Row Publishers, 1957.

Zaran, F.R. and Riccio, A.C. Organization and Administration of Guidance Service. Chicago: Rand McNally, 1962.

Zudick, L. Implementing Guidance in the Elementary School. Illinois, F.E. Peacock Publishers, Inc., 1971.

References for the Experiment – 50 : Assessing the Problems in Various Areas for Counselling

Berdie, B.F. and Layton, W.L. *The Minnesota Counselling Inventory,* New York: American Psychological Corporation, 1957.

Kranzler, G.D., *et al.* Counselling with Elementary School Children: An Experimental Study. *Personnel and Guidance Journal.* 1966, 44, 9, 944-49.

Rothney, J.W.M. *Guidance Practice and Results.* New York: Harper Brothers, 1958.

Wall, W.D. *Adolescents in Schools and Society.* N.F.E.R., 1968.

Williamson, E.G. and Bordin, E.S. Evaluating Counselling by means of controlled group experiment. *School and Society,* 1940, *52*, 434-440.

Williamson, E.G. *Counselling Adolescents.* New York: McGraw-Hill Book Company, 1950.

References for the Experiment – 51 : Understanding Vocational Interest

McCormick, E.J. and Tiffin, D. *Industrial Psychology,* (7th Ed.) New Delhi : Prentice – Hall of India Private Limited, 1984.

Super, D.E., and Bohn, M.J. *Occupational Psychology.* London: Tavistock Publication, 1971.

Thurstone, L.L. *Thurstone Interest Schedule–Manual.* New York : The Psychological Corporation, 1941.

References for the Experiment – 52 : Study of Occupational Interests

Blum, M.L. and Balinsky, B. *Counselling and Psychology.* Bombay: Asia Publishing House, 1961.

Brainard, P.P., and Brainard, R.T. *Brainard Occupational Preference Inventory.* New York : The Psychological Corporation, 1945.

CHAPTER 17

Social Behaviour

Adorno, T.N. Frenkel–Brunswik, E., Levinson, D., and Sanford, N. *The Authoritarian Personality.* New York : Harper and Brothers, 1950.

Allport, F.H. *The Nature of Institutions.* Social Forces, 1927, *6*, 167-179.

Anderson, T.W.; Erwin, N.; Flynn D.; Lewin; L.; and Erwin, J. Effects of short-term crowding on aggression in captive groups of pigtail monkeys. *Aggressive Behaviour,* 1977, *3,* 33-46.

Baren, R.A. and Byrne, D. *Social Psychology.* (5th ed.) New Delhi: Prentice-Hall of India Private Limited, 1988.

Bogardus, E.S. Measuring social distance. *Journal of Applied Sociology,* 1925, *9,* 299-308.

Calhoun, J.B. Population density and social pathology. *Scientific American.* 1962, 206, *2,* 139-148.

Calhoun, J.B. Population. In A. Allison (Ed.) *Population Control.* Baltimore: Penguin Books, 1970, pp. 110-130.

Candland, D.K. *Psychology : The Experimental Approach.* New York : McGraw-Hill Book Company, 1968.

Cartwright, D. and Zander, A (Eds.) *Group Dynamics : Research and Theory.* New York: Harper and Row, 1960.

Festinger, L., and Thibant, J. Interpersonal communication in small groups. *Journal of Abnormal and Social Psychology.* 1951, *46,* 92-99.

Fiedler, F.E.; Meuwese, W.A.T.; and Oonk, S. An exploratory study of group creativity in laboratory task. *Acta Psychology,* 1961, *18,* 100-119.

Fromm, E. *Escape from Freedom.* New York. Rinehart and Company, 1941.

Goldman, M.A. Comparison of individual and group performance for varying combinations of initial ability. *Journal of Personality and Social Psychology.* 1965, *1,* 210-216.

Haythorn, W.W.; Couch, A.; Haefner, D.; Langham, P.; and Carter, A. The effects of varying combinations of authoritarian and equalitarian leaders and followers. *Journal of Abnormal and Social Psychology,* 1956, *53,* 210-219.

Heslin, R. Predicting group task effectiveness from member characteristics. *Psychology Bull.* 1964, *62,* 248-256.

Krech, D and Crutchfield, R.S. *Theory and Problems of Social Psychology.* New York: McGraw-Hill Book Company, Inc., 1948.

Lewin, K., and Lippitt, R. An experimental approach to the study of autocracy and democracy: A preliminary note. *Sociometry,* 1938. *1,* 292-300.

Lippitt, R. An experimental study of the effect of democratic and authoritarian group atmospheres. *University Iowa. Study. Child Welfare,* 1940, *16,* 43-195.

Lippitt, R. and White, R.K. An experimental study of leadership and group life. In T.M. Newcomb and E.L. Hartley (Eds.) *Readings in Social Psychology.* New York: Holt, 1947.

Mann, R.D. A review of the relationships between personality and performance in small groups. *Psychological Bulletin* 1959, *56,* 241-270.

Massey, A. and Vanderbergh, J.G. Puberty delay by a urinary cure from female house mice in feral populations. *Science.* 1980, *209,* 821-822.

Moreno, J.L. *Who Shall Survive?* Washington: Nerv. Ment. Dis. Publ. Co., 1934.

Newcomb, T.M. *The Acquaintance Process.* New York: Holt, 1961.

Paulus, P.B. *Crowding.* In P.B. Paulus (Ed.) Psychology of group influence, Hilsdale, N.J. : Erlbaum, 1980.

Rohde, K.J. Theoretical and experimental analysis of leadership ability. *Psycho. Report.* 1958, *4,* 243-278.

Sanford, F.H. The follower's role in leadership phenomena. In G.E. Swanson, T.M. Newcomb and E.L. Hartley (Eds.) *Readings in Social Psychology.* (Rev. ed.) New York: Holt, 1952.

Seaman, J.M.; Michel, J.B.; and Dillchay, R.C. Membership in Orthodox Christian Groups, adjustment and dogmatism. *The Sociological Quarterly,* 1971, *12,* 252-59.

Spitzer, S.P. Consensual states and communicative behaviour. *Sociometry,* 1964, *27,* 510-515.

Thalen, H.A. Group dynamics in institution: Principles of least group size. *School Review.* 1949, *57,* 139-148.

Torronse, E.P. The behaviour of small groups under the stress of conditions of survival. *Amer. Sociol. Review,* 1954, *19,* 751-755.

Triandis, H.G.; Hall, E.R.; and Ewin, R.B. Member homogeneity and dyadic creativity. *Human Relations.* 1965, *18,* 33-56.

Willis, F.N. Initial speaking distance as a function of the speaker's relationship. *Psychonom. Science* 1966, *5,* 221-222.

Zlutrich, S., and Altman, S. Crowding and human behaviour. In J. Wohlwill and D. Carson (eds.) *Environment and the Social Sciences: Perspective and Applications,* Washington, D.C: American Psychological Association, 1972.

References for the Experiment – 53 : Measurement of Social Distance

Bird, C. *Social Psychology,* New York : Appleton Century Co., 1940.

Bogardus, E.S. *A Social Distance Scale. Sociology and Social Research,* 1933, *17,* 265-271.

Bogardus, E.S. Measurement of social distance. *Journal of Applied Sociology*. 1925, *9*, 299-308.

Brown, R. *Social Psychology*. New York: Free Press, 1965.

Krech, D. and Crutchfield, R.S. *Theory and Problems of Social Psychology*. New York: McGraw-Hill Book Co., 1948.

Murphy, G., Murphy, L.B., and Newcomb, T.M. *Experimental Social Psychology*. New York: Harper & Brothers, 1947.

Sherif, M. *The Psychology of Social Norms*. New York: Harper & Brothers, 1936.

References for the Experiment – 54 : Sociometry

Baron, R.A. and Byrne, D. *Social Psychology* (5th ed.) New Delhi : Prentice-Hall of India Private Limited, 1988.

Bogardus, E.S. Social Distance and its Practical Implications. *Sociology and Social Research,* Vol. *XVII,* 1933, 265-271.

Festinger, L.; Schachter, S.; and Bach, K. *Social Pressures in Informal Groups : A Study of Human Factors in Housing:* New York : Harper and Row, 1950.

Goode, W.J. and Hatt, P.K. *Methods in Social Research.* New York: McGraw-Hill Book Company, Inc. 1952.

Homans, G.C. *The Human Group.* New York: Harcourt Brace, 1950.

Jennings, H.H. *Sociometry in Group Relation.* Washington DC : American Council on Education, 1948.

Junnings, H.H. *Leadership and Isolation;* New York: Longmans, 1943.

Lingren, H.C. *An Introduction to Social Psychology,* 2nd ed. New Delhi; Wiley Easter Private Ltd., 1974.

Moreno, J.L. *Who Shall Survive?* 2nd ed. Beacon, New York; Beacon House, 1953.

Newcomb, T.M. Stabilities underlying changes in interpersonal attraction, *J. Abnorm. and Soc. Psychology,* 1963, *66,* 326-386, *55,* 161-163.

Walster, E. and Festinger, L. The effectiveness of overheard persuasive communications, *Journal of Abnor nal and Social Psychology,* 1962, *69,* 395-402.

Whyte, W.H. Jr. *The Organisation Man.* New York: Simon & Schuster, 1956.

CHAPTER 18

Attitudes, Related Traits and Measurement

Adorno, T.W. Frenkel-Brunswick, E.; Levinson, D.H. and Sanford, R.N. *The Authoritarian Personality.* New York: Harper and Row, 1950.

Allport, F.H. Social Psychology. Boston: Houghton Mifflin, 1924.

Allport, G.W. *"Attitudes" Handbook of Social Psychology,* Edited by Carl Murchison. Worcester, Massachusetts: Clark University Press, 1935.

Allyn, J. and Festinger, L. The effectiveness of unanticipated persuasive communications. *Journal of Abnormal and Social Psychology,* 1961, *62,* 35-40.

Argyle, M. *Religious Behaviour.* London : Routledge and Kegan Paul, 1958.

Bain, A *Mental Sciences.* New York: D. Appleton & Co., 1868.

Baron, R.A. and Byrne, D. *Social Psychology* (5th ed.) New Delhi. Prentice-Hall of India Private Limited, 1988.

Baumeister, R.F. and Tice, D.M. Role of self-presentation and choice in cognitive dissonance under forced compliance: Necessary or sufficient causes? *Journal of Personality and Social Psychology,* 1984, *46,* 5-13.

Bern, D.J. and McConnell, H.K. Testing the self-perception explanation of dissonance phenomena : On the salience of premanipulation attitudes. *Journal of Personality and Social Psychology,* 1970, *14,* 23-31.

Bird, C. *Social Psychology.* New York : D. Appleton-Century, 1940.

Breckler, S.J. Empirical validation of effect, behaviour and cognition as distinct components of attitude. *Journal of Personality and Social Psychology,* 1984, *47,* 1191-1205.

Cherrington, B.M. Methods of education in international attitudes. *Teach. Coll. Cont. Educl.,* 1934, No. 595.

Christie, R. Eysenck's treatment of personality of Communists. *Psychology Bull.* 1956, *53,* 439-450.

Cooper, J. and Fazio, R.H. A new look at dissonance theory. In L. Berkowitz (Ed.) *Advances in Experimental Social Psychology.* Vol. *17.* New York: Academic Press, 1984.

Croyle, R. and Cooper, J. Dissonance arousal : Physiological evidence. *Journal of Personality and Social Psychology,* 1983, *45,* 782-791.

Dexter, E.S. Pesonality traits related to conservatism and radicalism. *Character and Personality.* 1939, *7,* 230-237.

Feshback, S. and Singer, R. The effects of personal and shared threat upon social prejudice. *Journal of Abnormal and Social Psychology,* 1957, *54,* 411-416.

Festinger, L. *A Theory of Cognitive Dissonance.* Evanston. Ill : Row Peterson, 1957.

Garrett, H.E. *General Psychology* (2nd ed.) New Delhi : Eurasia Publishing House (Pvt.) Ltd., 1975.

Geothals, G.R.; Cooper, J.; and Naficy, A. Role of foreseen, forseeable and unforeseeable behavioural consequences in the arousal of cognitive dissonance. *Journal of Personality and Social Psychology,* 1979, *37,* 1179-1185.

Guttman, L. The basis of scalogram analysis. In Stauffer *et al.*, eds., *Measurement and Prediction.* Princeton : Princeton University Press, 1950.

Hoviland, C.I.; Lumsdaine, A.; and Sheffield, F. Experiments on mass communication. *Studies in Social Psychology in World War II,* 226. Princeton, NJ: Princeton University Press, 1948.

Hovland, C.I. and Weiss, W. The influence of source credibility on communication effectiveness. *Public Opinion Quarterly,* 1952, *15,* 635-650.

James W. *The Principles of Psychology.* 2 Vols., New York: Holt, 1890.

Janis, I.L. Personality correlates of susceptibility to persuasion. *Journal of Personality,* 1954, *22,* 504-518.

Jones, V. Attitudes of college students and their change: A 37-year study. *Genet. Psychol. Monogr.,* 1970, *81,* 3-81.

Kiesler, C.A. and Kiesler, S.B. *Conformity: Reading,* M.A. : Additison–Wesley, 1969.

Krech, D. and Crutchfield , R.S. *Theory and Problems of Social Psychology.* New York: McGraw-Hill Book Company, Inc., 1948.

Lapiere, R.T. and Farnsworth, P.R. *Social Psychology.* New York: McGraw-Hill Book Company, 1949.

Leventhal, H.; Singer, R.; and Jones, S. The effects of fear and specificity of recommendations upon attitudes and behaviour, *Journal of Personality and Social Psychology,* 1965, *2,* 20-29.

Likert, R. A technique for the measurement of attitudes. *Arch. Psychology,* (N.W.) 1932, 28, No. 194.

Linder, D.E.; Cooper, J.; and Jones E.E. Decision freedom as a determinant of the role of incentive magnitude in attitude change. *Journal of Personality and Social Psychology,* 1967, *6,* 245-254.

Lindgren, H.C. *An Introduction to Social Psychology* (2nd ed.) New Delhi : Wiley Eastern Private Limited, 1974.

MacIver, C.M. and Page, C.H. *Society.* London: Macmillan, 1959.

Macrone, I.D. *Race Attitude in South Africa.* London : Oxford University Publication, 1937.

McDougall, W. *An Introduction the Social Psychology* (23rd Ed.) London : Methuen and Co., 1936.

Miller, N.; Maruyama, G.; Bealer, R.J.; and Valone, K. Speed of speech and persuasion. *Journal of Personality and Social Psychology,* 1976, *34,* 615-624.

Morgan, J.B. *Keeping a Sound Mind.* New York: The Macmillan Company, 1934.

Murphy, G. *An Introduction to Psychology,* New Delhi : Oxford & IBH Publishing Co, 1964.

Murphy, G.; Murphy, L.B.; and Newcomb, J.M. *Experimental Social Psychology,* New York : Harper Brothers, 1937.

Petty, R.E. and Cacioppo, J.T. The elaboration likelihood model of persuasion. In Berkowitz (ed.) *Advances in Experimental Social Psychology.* Vol. 19. New York: Academic Press, 1985.

Petty, R.E.; Ostrom, T.M.; and Brock, T.C. *Cognitive Responses in Persuasion.* Hillsdale, N.J. : Erlbaum, 1931.

Rokeach, M. *The Open and Closed Mind : Investigations into the nature of belief systems and personality systems.* New York : Basic Books, 1960.

Ruch, F.L. *Psychology and Life* (5th ed.) Chicago : Scott Foreman and Company, 1958.

Sargent, S.S. and Williamson, R.C. Social Psychology (2nd Ed.), New York: The Ronald Press, 1958.

Sherif, M. *An Outline of Social Psychology,* New York : Harper, 1948.

Sherif, M. and Cantril, H. *The Psychology of Ego Involvements.* New York: John Wiley and Sons, INC, 1949.

Sumner, F.B. A statistical study of belief. *Psychological Review.* 1898, *5,* 616-631.

Thomas, W.I. and Znaniaki, F. T*he Polish Peasant in Europe and America.* Boston : Badger, 1918.

Thurstone, L.L. and Chave, E.J. *The Measurement of Attitude,* Chicago, University of Chicago Press, 1929.

Vetter, B.M. and Green, M. Personality and group factors in the making of atheists. *The Journal of Abnormal and Social Psychology,* 1932, *27,* 179-194.

Walster, E. and Festinger, L. The effectiveness of "overheard" Persuasive communications. *Journal of Abnormal and Social Psychology,* 1962, *65,* 395-402.

Young, K. *Social Attitudes,* New York: Holt, 1931.

References for Experiment – 55 : Measurement of Attitudes (Religious)

Allport, G.W. "Attitudes", *Hand Book of Social Psychology* (Ed.) Carl Murchison Worcester Massachusetts : Clark University Press, 1935.

Argyle, M. *Religious Behaviour.* London : Routledge and Kegan Paul, 1958.

Briddle, W.E. *Integration of Religion and Psychiatry.* New York : The Macmillan Company, 1956.

Clark. W.H. *The Psychology of Religion.* New York : The Macmillan Company, 1958.

Flugel, J.C. *Man. Morals and Society.* London : Duckworth, 1945.

Freud, S. *Basic Writings of Sigmund Freud.* Translated and Edited by A.A. Brill. New York: Modern Library, INC., 1921.

Garrison, K.C. *Psychology of Adolescence* (5th ed.) Englewood Cliffs, New Jersey: Prentice-Hall, INC., 1959.

Hadfield, J.A. *Psychology and Mental Health.* London : George Allen and Unwin Ltd., 1956.

Henric, I. *Facts and Theories of Psychoanalysis.* New York : Knopf, 1957.

Jones, V. Attitudes of College Students and their Changes : a 37-year study. *Genet. Psychol. Monogr,* 1970, *81,* 3080, 93-94, 96.

Jung, C.G. Instinct and Unconscious. *The British Journal of Psychology,* 1932, *X.I.* 15-23.

Jung, C.G. *Psychology of Religion,* London : Oxford University Press, 1939.

Katz, D. and Allport, F.R. *Students Attitudes :* A Report of the Syracuse University Reaction Study, 1931.

Kuhlen, R.G. and Arnold, N. Age differences in religious beliefs and problems during adolescence, *Journal of Genetic Psychology,* 1944, *6,* 291-300.

Leuba, J.H. *A Psychological Study of Religion,* New York : The Macmillan Company, 1912.

McDougall, W. *Social Psychology,* London : Methuen & Co. Ltd. 1948.

Newcomb, T.M. and Sevhla, G. Inter-family relationship in attitude, *Sociometry*, 1937, *1,* 180-205.

Pratt, J.B. *Religious Consciousness,* New York: The Macmillan Company, 1934.

Rajamanickam, M. *A Psychological Study of Religious and Related Attitudes of the Student and Professional Groups in South India.* Annamalainager : Annamalai University, 1976.

Rajamanickam, M. *Handbook of Religious Attitude Scale.* Madras : Institute of Community Guidance and Research, 1989.

Rajamanickam, M. Study of Student Attitudes towards Religion, the Spiritual and the Supernatural. *The Journal of Social Psychology,* 1962, *7,* 105-111.

Ruch, F.L. *Psychology of Life.* (5th ed.) Chicago : Scott Foreman & Company, 1958.

Sappenfield, B.R. The attitudes and attitude estimates of Catholic, Protestant and Jewish students. *The Journal of Social Psychology,* 1942, *15,* 173-197.

Shuttleworth, F.K. The influence of early religious home training on College sophomore men. *Religious Education,* 1927, *2,* 57-60.

Summer, F.B. A Statistical study of belief *Psychological Review.* 1898, *5,* 615-631.

Thouless, R.H. *An Introduction to Psychology of Religion.* Cambridge : Cambridge University Press, 1956.

Thrustone, L.L. and Chave E.J. *The Measurement of Attitude.* Chicago : University of Chicago Press, 1929.

Thrustone, L.L. *Multiple Factor Analysis.* Illinois ; The University of Chicago Press, 1947.

Trotter, W. *Instinct of Herd in Peace and War.* London : 1916.

Vetter, B.W. and Green, M. Personality and Group factors in the making of atheists. *The Journal of Abnormal and Social Psychology.* 1932, *27,* 179-194.

Weiser, G.A.A. *Early Civilization*. New York : Knopf, 1922.

Wilson, B.R. Social aspects of religious sects : A study of some contemporary sects in Great Britain. *Unpublished Ph.D thesis.* London University.

References of Experiment – 56 : Assessing Reactionary and Radical Tendencies

Allport, G.W. Attitudes. In C. Murchison (Ed.) *A Hand Book of Social Psychology.* Worcester, Mass : Clark University Press, 1935.

Bogardus E.S. Social Distance and its Practical Implications. *Sociology and Social Research,* Vol *XVII,* 1933, 267-271.

Krech, D. and Crutchfield, R.S. *Theory and Problems of Social Psychology,* New York : McGraw-Hill Book Co., 1940.

Likert, R.A. Technique for measurement attitudes. *Arch. Psychology.* 1932, No. 140.

Murphy, G. and Likert, R.S. *Public Opinion and the Individual* New York : Harper, 1938.

Newcomb, T.M. and Jvehla, G. Intra-family relationships in attitudes. *Sociometry,* 1938, *1,* 180-205.

Rajamanickam, M. *Reactionism-Redicalism Scale.* Madras : Institute of Community Guidance and Research, 1989.

Sherief, M. and Candril, H. T*he Psychology of Ego Involvement,* New York : John Wiley & Sons, Inc., 1947.

Thorndike, R.L. and Hagen, E. *Measurement and Education in Psychology and Education* (Third Edition), New York; Wiley Eastern Private Ltd., 1970.

Thrustone, L.L. and Chave, E.J. T*he Measurement of Attitudes.* Chicago : University of Chicago Press, 1929.

References for Experiment – 57 : Assessing Authoritarian Personality

Adorno, T.W., Frenkal-Brunswick, Levinson, D. and Sanford, N. The *Authoritarian Personality*, New York, Harper & Brothers 1950.

Brown, R. *Social Psychology,* New York : Free Press, 1965.

Chistie, R. and Jahoda, M. Studies in the scope and Method of *"The Authoritarian Personality"*. New York : Free Press, 1954.

Fromm, E. *Escape from Freedom.* New York : Rinehart & Company, 1941.

Harvey London and Exner J.E. Jr. (Ed.) : *Dimension of Personality.* New York : John Wiley & Sons, 1978.

Hoffer, E. *The True Believer.* New York : Harper and Row, 1951.

Kirscht, J.P. and Dillchay, R.C. *Dimensions of Authoritarianism : A Review of Research and Theory.* Laxington : University of Kentricky Press, 1967.

Lawson, R.B. Goldstein, S.G. and Musty, R.E. *Principles and Methods of Psychology,* New York. Oxford University Press, 1975.

Rokeach, M. T*he Open and Closed Mind.* New York : Basic Books, 1960.

Shils, E.A. *Authoritarianism : Right and Left.* In Christie and M. Jahoda (Eds.) *Studies in the Scope and Method of Authoritarian Personality,* Glenco, Ill : Free Press, 1954.

Titus, H.E. and Hollander, E.P. The California F. Scale in psychological research, 1950-1955. *Psychological Bulletin,* 1957, *54,* 45-64.

CHAPTER 19

Developmental Psychology and References for the Experiment – 58 : Assessing of the Level of Motor Activity in the Development Process

Anishworth, M.D.S. and Bell, S.M. Attachment exploration and separation : Illustrated by the behaviour of one year old in a strange situation. *Child Development,* 1970, *41,* 49-67.

Anisworth, M.D.S. and Bell, S.M. Mother-infant interaction and the development of competence. In K. Connolly and J. Bruner (Eds.) *The Growth of Competence.* New York : Academic Press, 1973.

Ausubel, D.P. Cognitive structure and the facilitation of meaningful verbal learning. *Journal of Teacher Education,* 1963, *14,* 217-221.

Ausubel, D.P. Use of advance organizers in the learning and retention of meaningful material. *Journal of Educational Psychology.* 1960, *51,* 267-272.

Babson, S.; Kangas, J.; Young. N.; and Bramhall, J. Growth and development of twins of dissimilar size at birth. *Pediatrics,* 1964, *33,* 327-333.

Bandura A.; Grusec, J.E.; and Menlove, F.L. Observational Learning as a function of symbolization and incentive set. *Child Development,* 1966, *37,* 499-506.

Barry, H.; Child I.L.; and Bacon, M.K. Relation of child training to subsistence economy. *American Anthropologist,* 1959, *62,* 51-63.

Bernard, L.L. Instinct : *A Study in Social Psychology*, New York : Rinehart and Winston, 1924.

Bourne, L.E.; Dominowski, R.L.; and Leftus, E.F. *Cognitive Processes.* Englewood Cliffs, N.J. Perentice-Hall, 1979.

Bowlby, J. Attachment and love. *Vol. I., Attachment.* New York : Basic Books, 1969.

Brackbill, Y. Extinction of the smiling response in infants as a function of reinforcement schedule. *Child Development,* 1958, *29,* 115-124.

Bridges, K.M.B. Emotional development in early infancy. *Child Development,* 1932, *3,* 324-341.

Brown, R. and Bellugi, V. Three processes in the child's acquisition of syntax. *Harvard Educational Review,* 1964, *34*, 133-151.

Brown, R. The development of questions in child speech. *Journal of Verbal Learning and Behaviour,* 1968, *7,* 279-290.

Bruner, J.S. and Olver, R. Development of equivalence transformation in children. *Mongr. Soc. Res. Child Development,* 1963, *28,* No. *2*, 125-141.

Bruner, J.S.; Goodman, J.; and Austin, G. *A Study of Thinking.* New York: Wiley, 1966.

Bruner, J.S. The Course of Cognitive growth. *Amer. Pscyhologist,* 1964, *19,* 1-15.

Caputo, D.V., and Mandall, W. Consequences of low birth weight, *Development Psychology,* 1970, *3,* 363-383.

Caven, Ruth S. Burgess, E.W. Havghurst, R.J. and H. Goldhammer, *Personal* Adjustment in Old Age. Chicago : *Science Research,* 1949.

Churchill, J.A. The relationship between intelligence and birth weight in twins, *Neurology*, 1965, *15*, 341-347.

Collins English Dictionary. London : William Collins Sons & Co. Ltd., 1979.

Darwin, C.L. (1859) *The Expression of the Emotions in Man and Animals.* Chicago : Chicago University Press, 1965.

De Hirsh, K.; Jansky, J.; and Langford, W.S. Comparisons between Prematurely and maturely born children at three age level. *American Journal of Orthopsychiatry,* 1966, *36,* 616-628.

Drillien, C.M. The incidence of mental and physical handicaps in school-age children of very low birth weight. *Pediatrics,* 1967, *17,* 452-464.

Dubos, Rene. *So Human An Animal*, New York : Scribner, 1968.

Duncker, K. On problem solving. Translated by L.S. Lees from the 1935 Original. *Psychology Monogr.* 1945, *58,* No., 270.

Ericson, M.C. Child rearing and social status. *American Journal of Sociology,* 1946, *52,* 190-192.

Erikson, E.H. *Childhood and Society,* New York : 1956.

Evans, W.F. *Anatomy and Physiology.* Englewood Cliffs : Prentice-Hall, 1971.

Fantz, R.L.; Ardy, J.M.; and Udelf, M.S. Maturation of pattern vision in infants during the first six months. *Jl. Comp. Physical. Psychology.* 1962, *55,* 907-917.

Flakner, F. *Human Development.* Philadelphia of Saunders. 1966.

Flavell, J.B. *The Development Psychology of Jean Piaget.* New York : Van Nontrand, 1963.

Freedman, D.G. Smiling in blind infants and the issue of innate *vs*. acquired. *Journal of Child Psychology and Psychiatry,* 1964, *5,* 171-184.

Freedman, D.G.; Loring, C.B.; and Martin, R.M. Emotional behaviour and personality development. In Y. Brackbill (Ed.). *Infancy and Early Childhood.* New York ; Free Press, 1967.

Freud, S. *An Outline of Psychoanalysis.* New York : Norton, 1949.

Gagne, R.M. *The Conditions of Learning.* Rev. Ed. New York : Holt, Rinehart and Winston, 1970.

Gesell, A. and Thompson, H. Learning and growth in identical twins, *Genetic Psychology Monograph,* 1929, *6,* 1-124.

Gesell, A.; Helverson, H.M.; Thompson, H.; Ilg, F.L.; Castener, B.M.; and Amer, L.B. *The First Four Years of Life.* New York : Harper and Row, 1940.

Glaser, R., and Resnick, L.B. Instructional Psychology. *Ann. Rev. Psychology.*, 1972, *23,* 207-276.

Glueck, S. and Glueck, E. *Unravelling Juvenile Deliquency.* New York : Commonwealth Fund. 1950.

Gorman, J.J.; Cogan, D.G.; and Gellis, S.S. An apparatus for guarding the visual acuity of infants on the basis of optokinetic mystagmus. *Pediatrics,* 1957, *19,* 1088-1092.

Govetos, L.A. Relationships and age differences in growth measures and motor skills. *Child Development,* 1959, *30,* 333-340.

Gutteridge, M.V. A study of motor achievements of young children. *Arch. Psychology,* 1939, No. 44 : 178.

Harlow, H.F. and Harlow, M.K. Learning to think. *Scientific American* 1949, *181,* 36-19.

Heb, D.O. *The Organization of Behaviour : A Neuropsychological Theory.* New York : Wiby, 1949.

Hetherington, E.M. The effects of father absence on personality development in adolescent daughters. *Developmental Psychology,* 1972, *7,* 313-326.

Hilgard E.R. and Bower, G.H. *Theories of Learning.* (4th ed). Englewood Cliffs, N.J. : Prentice-Hall, 1975.

Horlock, E.B. Developmental Psychology. (3rd Ed.) New York : McGraw-Hill, 1968.

Hunt, E.B. and Hovland, C.I. *Programming a Model of Human Concept Formation.* Proceedings of the Western Joint Computer Conference. New York : IRE, 145-156.

Hunt, E.S.; Marin, J.; and Stone. P. *Experiments in Induction.* New York : Academic Press, 1966.

Ingle, D.J. Endocrine function and personality. *Psychology Review.* 1935, *42,* 466-479.

Inhelder, B., and Piaget, J. *The Growth of Logical Thinking from Childhood to Adolescence,* New York : Basic Books, 1958.

Jackson, R.L., and Kelly, H.C. Growth charts for use in pediatric practice. *Journal of Pediatrics,* 1945, *27,* 213-299.

Jarsild, A.T. and Holmes, F.B. *Children's Fears.* New York : Bureau of Publications, Teacher's College, Columbia University, 1935.

Keeney, T.J.; Cannizzo, S.R.; and Flavell, J.H. Spontaneous and induced verbal rehearsal in a recall task. *Child Development,* 1967, *38,* 953-966.

Kimble, G.A.; Garmezy, N.; and Zigler, E. *Principles of Psychology* (6th ed.) New Delhi : Wiley Eastern Limited, 1985.

Knobloch, H. and Pasamanick, B. Seasonal variations in the births of the mentally dificient. *American Journal of Public Health,* 1958, *48,* 151-157.

Krogman, W.M. The physical growth of child : An appraisal of studies, 1950-1955. *Monograph. Soc. Res. Child Development,* 1956, *20,* 1.

Lamb, M.E. Father-infant interaction in the first year of life. *Child Development,* 1972, *48,* 167-181.

Landreth, C. *Early Childhood : Behaviour and Learning.* New York : Kinoft, 1967.

Lansing, A.I. (Ed.) *Problems of Aging.* 3rd Ed. Baltimore : Williams Wilkins, 1952.

Lavin, H. and Turgeon, V.F. The influence of the mother's presence on children's doll-play aggression, *Journal of Abnormal and Social Psychology,* 1957, *55,* 304-308.

Lawrance, D.H. and Festinger, L. *Deterrents and Reinforcement : The Psychology Insufficient Reward.* Stanford : Stanford University Press, 1962.

Lefranceis, G.R. *Psychology for Teaching.* Belmont, California : Wadsworth Publishing Company, Inc., 1979.

Luria, A.R. *The role of speech in the regulation of normal and abnormal behaviour.* New York : Liveright, 1961.

McDougall, W. *An Introduction to Social Psychology.* London : Methuen, 1908.

McGraw, M.G. *Growth : A Study of Johnny and Jimmy.* New York : Appleton-Century, 1935.

McGraw, Myrtle, B. *The Neuromuscular Maturation of the Human Infant.* New York : Columbia University Press, 1943.

Merrifield, P.R. Analysis of Concepts from the point of view of the structure of intellect. In H.J. Klausmeir and C.W. Harris (Eds) *Analysis of Conceptual Learning,* New York : Academic Press, 1966.

Morgan, J.J.B. *Keeping a Sound Mind.* New York : *Macmillan*, 1934.

Murphy, G. *An Introduction to Psychology.* New Delhi. Oxford and IBH Publishing Co., 1964.

Neisser, U. *Cognitive Psychology.* Englewood Cliffs, N.J. : Prentice-Hall, 1967.

Newberry, H. Studies in foetal behaviour, II : The measurement of three types of foetal activity. *Journal of Comparative Psychology.* 1941, *32,* 521-530.

Peterson, L.R. and Peterson, N.J. Short-term retention of individual verbal items. *Journal of Experimental Psychology,* 1959, *58,* 193-198.

Piaget J. Autobiography in History of Psychology. In *Autobiography,* Vol. IV ed. E.G. Boring *et al.* Worcester, Mass : Clark University Press, 1952.

Piaget, J. and Inhelder, B. *The Psychology of the Child* . New York : Basic Books, 1969.

Piaget, J. Inhelder, B., and Szeminska, A. *The Child's Conception of Geometry.* London : Routledge and Kegan Paul, 1960.

Piaget, J. *Judgement and Reasoning in the Child.* New York : Harcourt, Brace and World 1928.

Piaget, J. *Play, Dreams and Imitation in Childhood,* New York : Norton, 1951.

Piaget, J. *The Child's Conceptions of Physical Causality.* London : Kegan Paul, 1930.

Piaget, J. *The Construction of Reality in the Child.* New York : Basic Books, 1954.

Piaget, J. *The Language and Thought of the Child,* New York, Harcourt, Brace and World, 1926.

Piaget, J. *The Moral Judgement of the Child.* London : Kegan Paul, 1932.

Piaget, J. *The Origins of Intelligence.* Trans, by Margaret Cook, New York : International Universities, 1952.

Piaget, J. The stages of the intellectual development of the child. *Bulletin of the Menninger School of Psychiatry,* March, 1961.

Prechtl, H.F.R. The mother-child interaction in babies with minimal brain damage. In B.M. Foss, ed., *Determinants of Infants Behaviour.* New York : Wiley, 1963, Vol. II.

Ray, M.K., and Destidar, D.G. *Cobol Programing* (2nd Ed.). New Delhi. Tata-MaGraw-Hill Publishing Company Limited, 1933.

Rutter, M. Parent-Child separation : Psychological effects on the children, *Journal of Child Psychology and Psychiatry,* 1971, *260*, 520-522.

Shultz, T. and Zigler, E. Emotional concomitants of visual mastery in infants : The effects of stimulus movement on smiling and vocalizing. *Journal of Experimental Child Psychology.* 1970, *10*, 309-402.

Sigel, I.E. The attainment of concept. In M.L. Hoffman and Lois W. Hoffman (eds.), *Review of Child Development Research.* New York : Russell Sage Foundation, 1964.

Skinner, B.F. The *Technology of Teaching.* Englewood Cliffs, New Jersey : Prentice-Hall, Inc., 1968.

Skinner, B.F. *Verbal Behaviour.* New York : Appleton-Century-Crofts, 1987.

Sokolov, A.N. Studies on the problems of the speech meachanisms in thinking in Vol. I of *Psychological Science in the U.S.S.R.* Ed. B.G. Anan'yev *et al.* Moscow : Scientific Council of the Institute of Psychology, Academy of Pedagogical Sciences, U.S.S.R. 1959, 669-701.

Stern, W. *General Psychology from Personalistic Standpoint.* Trans : by H.D. Sporel. New York : Macmillan, 1938.

Stevenson, H.W. Developmental Psychology. In D. Sills (ed.) *International Encyclopedia of Social Sciences.* New York : Macmillan, 1968.

Stott, L.H. *The Psychology and Human Development.* Delhi : The Macmillan Company of India Limited, 1974.

Stuart, H.C. and Prugh, D.G. (Eds). *The Healthy Child : His Physical, Psychological and Social Development.* Cambridge, Mass : Harvard University Press, 1960.

Tanner, J.J. Physical growth. In P.H. Mussen, Ed., Carmichael's *Manual of Child Psychology* (3rd ed.) New York : Wiley, 1970, Vol. I.

Todd, T.W. *Atlas of Skeletal Maturation.* St. Louis : Modby, 1937.

Tolman, E.D. *Collected Papers in Psychology.* Berkeley : University of California Press, 1951.

Vigotsky, L.S. *Thought and Language.* Translated from the Russian and edited by E. Mamsman and G. Vankan, Cambridge, Mass : MIT Press, 1962.

Walsh, F.B. *Clinical Neuro-Opthalmology.* (2nd ed.) Baltimore : Williams and Wilkins, 1957.

Watson, E.N. and Lowery, G.H. *Growth and Development of Children.* Chicago : Year Book Publishers, 1954.

Watson, J.B. and Morgan, J.J.B. Emotional reactions and psychological experimentation. *American Journal of Psychology.* 1917, *28*, 163-179.

Watson, J.B. *Psychology from the Standpoint of the Behaviourist.* Philadelphia : Lippincott, 1924.

Werner, H. *Comparative Psychology of Mental Development.* Chicago : Follet, 1948.

Wetzel, N.C. physical fitness in terms of physique : Development and basal metabolism. *Jl. Amer. Med. Ass.,* 1941, *116* : 1187-1195.

Wiener, G. Scholastic achievement at age 12-13 of prematurely born infants. *Journal of Special Education,* 1968, *2*, 237-250.

Wiener, G.; Rider, R.V.; Oppel, W.E.; Fisher, L.K.; and Harper, P.A. Correlates of low birth weight : Psychological status of 6-7 years of age. *Pediatrics,* 1965, *35,* 434-444.

Willerman, L., and Churchill, J.A. Intelligence and Birth Weight in Identical Twins, *Child Development,* 1967, *38,* 623-629.

CHAPTER 20

Environment and Behaviour

Altman, I. *The Environment and Social Behaviour.* Monterey, Calif : Brooks Cole Publishing Company, 1975.

Ashton, N.L.; Shaw, M.E.; and Worsham, A.P. Affective reactions to Interpersonal distances by friends and strangers, *Bulletin of Psychonomic Society,* 1980, *15,* 306-308.

Bell, P.A.; Fisher, J.D.; and Loomis, R.J. *Environmental Psychology,* Philadelphia : W.D. Saunders Company, 1978.

Berkowitz, L. The contagion of violence, In W.J. Arnold and N.M. Page (Eds). *Nebraska Symoosium on Motivation,* Vol. *18.* Lincoln : University of Nebraska Press, 1970.

Blot, W.J.; Brinton, L.A.; Fraumeni, J.F., Jr.; and Stone, B.J. Cancer mortality in U.S. Countries with Petroleum Industries. *Science* 1975, *198,* 51-53.

Brickman, L.; Jeger, A.; Gabriale, T.; Mchaughlin, G.; Berger, M; and Sunday, E. Dormitory density and helping behaviour. *Environment and Behaviour,* 1973, *5,* 465-490.

Broadbent, D. *Decision and Stress.* New York : Academic Press, 1971.

Broadbent, D. *Perception and Communication.* Oxford : Pergamon, 1958.

Cohan, S. Environmental load and the allocation of attention. In a Baun and S. Valins (Eds.). *Advances in Environmental Research,* Hillsdale, N.J. Lawrence Erlbaum, 1977.

Cunnigham, M.R. Weather, mood and helping behaviour : Quasi experiments with the sunshine Samaritan. *Journal of Personality and Social Psychology,* 1979, *37,* 1947-1956.

Fisher, J.D., Situation-specific variables – a determinants of perceived environmental aesthetic quality and perceived crowdness. *Journal of Research Personality,* 1974, *8,* 177-188.

Garland, H. and Pearce, J.N. Neurological complications of carbon monoxide, poisoning. *Quarterly Journal of Medicine,* 1967, *36*, 445-455.

Glass, D.C and Singer, J.E. *Urban Stress.* New York : Academic Press, 1972.

Green, R.G. and O'-Neal, E.C. Activation of one elicited aggression by general arousal. *Journal of Personality and Social Psychology,* 1969, *11,* 287-292.

Griffitt, W. Environmental effects on interpersonal effective behaviour : Ambient effective temperature and attraction. *Journal of Personality and Social Psychology,* 1970, *15,* 240-244.

Hall, E.T. *The Hidden Dimension,* New York : Doubleday, 1966.

Harridge, C.F., Aircraft noise and mental health. *Journal of Psychosomatic Research,* 1974, *18,* 239-243.

Hart, R.H. The Concept of APS : Air Pollution Syndromes. *Journal of the South Carolina Medical Association,* 1970, *66,* 71-73.

Hebb, D.O. *Textbook of Psychology,* (3rd Ed.). Philadelphia, W.B. Saunders, 1972.

Heimstra, N.W. and McFarling, L.R. *Environmental Psychology,* Montercy, California : Brooks Cole Publishing Company, 1974.

Heimstra, N.W., and McDonald, A.L. Psychology and Contemporary problems. Montercy, California : Brooks Cole Publishing Company, 1973.

Helson, H. *Adaptation Level Theory.* New York : Harper & Row, 1964.

Holohan, C.J. Environmental Psychology. In M.R. Rosenzweig and L.W. Porter (Eds). *Annual Review of Psychology,* Vol. 27. 381-407. Palo Alto, C.A. : Annual Reviews, Inc., 1986.

Hurt, H. The hottest place in the whole of U.S.A. *Texas Monthly* 1975, *3,* 50th.

Jansan, G. Non-auditory effects of noise : Physiological and Psychological reactions in man. In W.D. Ward (Ed.). *Proceedings of the International Congress on Noise as a Public Health Problem.* Washington, D.C. : U.S. Government Printing Office, 1973.

Kahneman, D. *Attention and Effort.* Englewood Cliffs, N.J. : Prentice-Hall, 1973.

Kantowitz, B.H. Environmental Psychology : Behaviour in the physical World. In R.A. Baron, Byrane, D., and B.H. Kantowitz, (Eds.). *Psychology Understanding Behaviour.* Philadelphia : W.B. Saunders. 1977.

Konecni, V.J.; Libuser, L.; Morton, H.; and Ebbesen, E.B. Effects of a violation of personal space on escape and helping responses. *Journal of Experimental Social Psychology,* 1975, *11,* 288-299.

Kryter, K.D. The effects of noise on man, New York : Academic Press, 1970,

LaVerne, A.A. Non-specific Air Pollution Syndrome (NAPS) : Preliminary Report, *Behavioural Neuropsychiatry,* 1970, *2,* 19-21.

Lazarus, R.S. *Patterns of Adjustment* (3rd Ed.) New York : McGraw-Hill, 1961.

Mendelson, R., and Orcut, G. An empirical analysis of air pollution dose-response curves. *Journal of Environmental Economics and Management.* 1979, *6,* 85-106.

Milgram, S. The experience of living in cities. *Science,* 1970, *167.*

Page, R.R. *Environmental Influences on Pro-social Behaviour : The Effects of Temperature.* Paper presented at the Midwestern Psychological Association Meeting, Chicago, 1978.

Parr, A.E. Psychological aspects of Urbanology. *Journal of Social Issues,* 1966, *22,* 39-45.

Proshansky, H.M. Environmental psychology and the real world. *American Psychologists,* 1976, *31,* 303-310.

Rose, E.F. and Rose, M. Carbon monoxide : A challenge to the Physician, *Clinical Medicine,* 1971, *78,* 12-18.

Schulte, J.H. Effects of mild carbon monoxide intoxication. *Archives of Environmental Health.* 1963, *7,* 524-530.

Webb, E.J., Campbell, D.T., Schwart, R.D., and Sachrest, L. *Unobtrusive Measures : Non-reactive Research in the Social Science,* Chicago : Rand McNelly, 1966.

Wohlwill, J.F. Human adaptation to levels of environmental stimulation. *Human Ecology.* 1974, *2,* 127-147.

Woodworth, R.S. *Psychology,* London : Methuen & Co., 1952.

Worchel, S. and Brown, E.H. The role of plausibility in influencing environmental attributions. *Journal of Experimental Social Psychology,* 1984, *20,* 86-96.

Worchel, S., and Teddlie, C. The experience of crowding : A two factor theory. *Journal of Personality and Social Psychology,* 1973, *74,* 30-40.

References for the Experiment – 59 : Measurement of Attitude toward Environmental Pollution

Barker, R.G. *Ecological Psychology : Concepts and Methods for Studying the Environment of Human Behaviour.* Standford : Standford University Press, 1968.

Bell, P.A.; Fisher, J.D.; and Loomis, R.J. *Environmental Psychology.* Philadelphia : W.B. Saunders Company, 1978.

Helson, H. *Adaptation Level Theory.* New York : Harper & Row, 1964.

Rajamanickam, M. *Environmental Pollution Attitude Scale.* Chennai : Institute of Community Guidance and Research, 1995.

Sonnenfeld, J. Variable values in space and landscape : An inquiry into the nature of environmental necessity. *Journal of Social Issues,* 1966, *22,* 71-82.

Wohlwill, J.F. Human adaptation to level of environmental stimulation. *Human Ecology,* 1974, *2,* 127-147.

References for the Experiment – 60 : Measurement of Attitude toward Noise

Appleyard, D., and Lintell, M. The Environmental quality of city streets : The residents' view point. *Journal of American Institute of Planners,* 1972, *38,* 84-101.

Broadbent, D.E. Some effects of noise on visual performance. *Quarterly Journal of Experimental Psychology,* 1954, *6,* 105.

Cohan, A. Industrial noise and medical absence and accident record data on exposed workers. In W.D. Ward (Ed.) *Proceedings of the International Congress on Noise as a Public Health Problem.* Washington, D.C. U.S. Government Printing Office, 1973.

Corcoran, D.W.J. Noise and loss of sleep. *Quarterly Journal of Experimental Psychology,* 1962, *14,* 178-182.

Dellinger, R.W. Jet Roar : Health Problems take off near airports. *Human Behaviour,* 1979, *8,* 50-51.

Jansen, G. Non-auditory effects of noise : Physiological and psychological reactions in man. In W.D. Ward (Ed.) *Proceedings of the International Congress on Noise as a Public Health Problem.* Washington, D.C. U.S. Government Printing Office, 1973.

Lebo, C.P. and Oliphant, K.P. Music as a source of acoustical trauma. *Laryngoscope.* 1968, *78,* 1211-1218.

Meecham, W.C., and Smith, H.G. Noise drives you crazy-jets and Mental hospitals, *British Journal of Audiology,* Psychology today, 1977.

Rajamanickam, M. Noise Pollution Attitude Scale, Chennai : Institute of Community Guidance and Research, 1995.

INDEX